FACILITY
PIPING SYSTEMS
HANDBOOK

FACILITY PIPING SYSTEMS HANDBOOK

Michael Frankel, CIPE, CPD
President, Utility Systems Consultants

Second Edition

McGraw-HILL

New York   Chicago   San Francisco   Lisbon   London   Madrid
Mexico City   Milan   New Delhi   San Juan   Seoul
Singapore   Sydney   Toronto

Cataloging-in-Publication Data is on file with the Library of Congress

McGraw-Hill

A Division of The McGraw-Hill Companies

1 2 3 4 5 6 7 8 9 0   DOC/DOC   0 7 6 5 4 3 2 1

ISBN 0-07-135877-3

The sponsoring editor for this book was Larry S. Hager and the production supervisor was Pamela A. Pelton. It was set in Times Roman by Pro-Image Corporation.

Printed and bound by R. R. Donnelley & Sons Company.

McGraw-Hill books are available at special quantity discounts to use as premiums and sales promotions, or for use in corporate training programs. For more information, please write to the Director of Special Sales, McGraw-Hill Professional, Two Penn Plaza, New York, NY 10121-2298. Or contact your local bookstore.

This book is printed on acid-free paper.

*For all the architects,
engineers, and designers
who make concepts
into solutions*

CONTENTS

Chapter 3   Solid-Liquid Separation and Interceptors   3.1

Chapter 4   Water Treatment and Purification   4.1

Chapter 7 Turf Irrigation Systems 7.1

Chapter 8 Cryogenic Storage Systems 8.1

Chapter 9 Plumbing Systems 9.1

Chapter 10   Special Waste Drainage Systems   10.1

Chapter 11 Facility Steam and Condensate Systems   **11.1**

pter 12 Liquid Fuel Storage and Dispensing Systems   **12.1**

Chapter 13   Fuel Gas Systems

Chapter 16　Animal Facility Piping Systems　　　　　　　　**16.1**

Chapter 17　Life Safety Systems　　　　　　　　**17.1**

PREFACE TO THE SECOND EDITION

The acceptance of the first edition of the *Facility Piping Systems Handbook* by engineers throughout the world has been very gratifying. This second edition expands on the previous work by providing additional information to the chapters on individual private sanitary sewage disposal systems, installation of natural gas piping on a site, and fixture and equipment mounting heights. More specific information regarding acids and acid drainage is also provided. Codes have been updated for these topics.

Additional chapters have been written on drinking water, heat exchangers, and measurement equipment and methods.

Where applicable, metric units have been included along with the Inch Pound numbers to make their conversion easier. A fourth appendix has been added to facilitate the conversion of Inch Pound (IP) to Metric (SI) units.

I would like to thank the National Fire Protection Association for allowing me to use the diagrams that appear throughout this book. I also thank the NFPA for the material reproduced with their permission. It is important to note that this material does not represent the official and complete position of the NFPA on the particular subject to which the material refers. The reader should consult the specific code or standard concerning the subject to which the material refers to fully understand the NFPA's position on that subject.

Michael Frankel

FACILITY
PIPING SYSTEMS
HANDBOOK

CHAPTER 1
CODES AND STANDARDS

This chapter describes the codes and standards used and referenced most often that affect the materials, design, and installation of the service and utility systems described in this handbook.

GENERAL

Codes relating to piping provide specific design criteria such as allowable materials, working stresses, seismic loads, thermal expansion, and other imposed internal or external loads as well as fabrication, installation, and testing for many aspects of a total piping system. Code compliance is mandated by various federal, state, and local agencies that have jurisdiction and enforcement authority. Each code has precisely defined limitations on its jurisdiction. Familiarity with these limitations can be obtained only after a thorough reading of the code.

These codes often refer to standards prepared by nationally recognized organizations. The term *nationally recognized* is defined as a group or organization composed of a nationwide membership representative of its members' views. To achieve nationally recognized status, an association must have been in existence for a reasonable period of time, be active in research and other issues relating to its area of interest, and be generally regarded by its peers to be scientifically accurate.

Standards provide specific design criteria and rules for specific components or classes of components such as valves, joints, and fittings. Dimensional standards provide control for components to assure that components supplied by different manufacturers are physically interchangeable. Pressure integrity standards provide performance criteria so that components supplied by different manufacturers will function and be service rated (pressure and temperature) in a similar manner. Standards compliance is usually required by construction or building codes or purchaser specifications.

In any piping system design, if different code requirements are discovered, the most stringent requirements must be followed.

The applicability of various codes and standards must be ascertained before the start of a project, because submission of plans is often required for approval prior to construction and installation of the piping systems. This requires a code search and consultation with the various authorities having jurisdiction.

Fire insurance carriers are another consideration in the area of standards. They very often have more restrictive requirements than the building and construction codes that are normally applicable to every project, particularly in the area of water supply storage and distribution for fire protection purposes, which may be combined with the domestic water system.

AMERICAN GAS ASSOCIATION (AGA)

The AGA advances the safe, economical, and dependable transport of gas to the public. In conjunction with the NFPA, it publishes NFPA-54, the Fuel Gas Code.

AMERICAN NATIONAL STANDARDS INSTITUTE (ANSI)

ANSI serves as the national coordinating institution for voluntary standardization and related activities in the United States. Through ANSI, organizations concerned with such activities may cooperate in establishing, improving, and approving standards and certification that such activities remain dynamically responsive to national needs and prevent duplication of work. ANSI's goals are to further the voluntary standards movement as a means of advancing national economy and benefiting public health, safety, and welfare; to facilitate domestic and international trade; to assure that the interests of the public, including consumers, labor, industry, and government, have appropriate protection, participation, and representation in standardization and certification; to provide the means for determining the need for new standards and certification programs; to assure activities by existing organizations competent to resolve the need; to establish, promulgate, and administer procedures and criteria for recognition and approval of standards as American National Standards and to encourage existing organizations and committees to prepare and submit such standards for approval by the institute; to cooperate with departments and agencies of the federal, state, and local governments in achieving optimum use of ANSI in regulation and procurement; and to serve as a clearinghouse for information on standards and standardization, certification, and related activities in the United States and abroad.

AMERICAN PETROLEUM INSTITUTE (API)

This organization affords a means to cooperate with the government in all matters of national concern relating to American petroleum products; to foster foreign and domestic trade in American petroleum products; to promote in general the interests of all branches of the petroleum industry; to promote the mutual improvement of its members and the study of the arts and sciences connected with the petroleum industry.

AMERICAN SOCIETY OF HEATING, REFRIGERATING, AND AIR CONDITIONING ENGINEERS (ASHRAE)

The purpose of ASHRAE is to advance the art and science of heating, ventilation, and air conditioning and allied arts and sciences, as well as related human factors for the benefit of the general public. To fulfill its role, the society recognizes the effect of its technology on environmental and natural resources.

AMERICAN SOCIETY OF PLUMBING ENGINEERS (ASPE)

The purpose of this organization is to develop and disseminate technical information in the field of engineered plumbing systems and to provide a forum for exchange of this information with other technical and construction code organizations.

AMERICAN SOCIETY OF TESTING AND MATERIALS (ASTM)

ASTM's objectives are to develop full consensus standards for the characteristics and performance of various materials, products, standards, and services; to develop and publish information designed to promote the understanding and advancement of technology; and to ensure the quality and safety of products and services.

ASTM has developed standards that consist of 67 volumes divided into 16 sections. Each volume is published annually to incorporate new standards and revisions.

AMERICAN SOCIETY OF MECHANICAL ENGINEERS (ASME)

ASME promotes the arts and sciences connected with engineering and mechanical construction for scientific purposes.

The ASME publishes the two principal codes used in facility systems: the Boiler and Pressure Vessel Code and ASME B31 series, Code for Pressure Piping. The Code for Pressure Piping has the following published sections:

B31.1. Power Piping

B31.3. Chemical Plant and Petroleum Refinery Piping

B31.4. Liquid Transpiration Systems for Hydrocarbons, Liquid and Petroleum Gas, Anhydrous Ammonia, and Alcohols

B31.5. Refrigeration Piping

B31.8. Gas Transmission and Distribution Piping

B31.9. Building Services Piping

B31.11. Slurry Transportation Piping

AMERICAN SOCIETY OF SANITARY ENGINEERS (ASSE)

ASSE promotes the welfare, health, and safety of the public through better sanitary engineering principles.

AMERICAN WATER WORKS ASSOCIATION (AWWA)

The purpose of the AWWA is to promote public health, safety, and welfare by improving the quality and increasing the quantity of water delivered to the public and to further an understanding of the problems involved by:

1. Advancing the knowledge, design, construction, operation, water treatment, and management of water utilities
2. Developing standards for procedures, equipment, and materials used by public water supply systems
3. Advancing the knowledge of problems involved in the development of resources, production, and distribution of safe and adequate water supplies
4. Educating the public on the problems of water supply and promoting a spirit of cooperation between consumers and suppliers in solving problems
5. Conducting research to determine the causes of the problems of providing a safe and adequate water supply and proposing solutions in an effort to improve the quality and quantity of the water supply

AMERICAN WELDING SOCIETY (AWS)

The purpose of the AWS is to encourage in the broadest sense the advancement of welding, to encourage and conduct research in sciences related to welding, and to engage and assist in the development of sound practices for the application of welding and related processes.

COMPRESSED GAS ASSOCIATION (CGA)

The CGA is a trade organization that writes and publishes guides, in the form of pamphlets, that include all aspects of compressed gas storage, distribution, and purity.

CURRENT GOOD MANUFACTURING PRACTICE (cGMP)

cGMP is a regulation established by 21 CFR 210 and 211 in the Federal Food, Drug, and Cosmetic Act. It establishes minimum good practices for methods, controls, facilities, manufacturing, processing, and packing of items to assure they meet the requirements covered by the act. Enforcement is provided by the FDA.

cGMP has no strict guidelines and is completely subject to reviewers of the FDA. It is intended to detail what must be done rather than how. The manufacturers must establish the durability and safety to the public of all products, and prove they meet the purity and efficacy characteristics that they represent to possess by application to the FDA.

CODE OF FEDERAL REGULATIONS (CFR)

The Code of Federal Regulations is the collection of the general and permanent rules and regulations originally published in the Federal Register by agencies of the federal government. There are 50 separate titles that are revised once a year. The most up-to-date information will be found in the Federal Register on a daily basis until revision is made to the CFR.

ENVIRONMENTAL PROTECTION AGENCY (EPA)

Created in 1970, the EPA is the government organization responsible for the prevention of pollution to the environment. The EPA creates national pollution standards and criteria, creates compliance and enforcement plans, and performs research and development for identifying pollution-related risks. Criminal enforcement is also within the jurisdiction of the EPA.

FOOD AND DRUG ADMINISTRATION (FDA)

The FDA is a government agency originally created by the Federal Food, Drug, and Cosmetic Act and charged with the responsibility to see that all drugs are safe, effective, and properly labeled. The regulation implementing its authority is 21 CFR 211.

ISO 9000

ISO is the English translation of the International Organization for Standardization based in Geneva, Switzerland. ISO consists of national standards organizations from approximately 100 countries throughout the world. The United States is represented by ANSI. Internationally recognized and accepted standards are required to establish a minimum level of consistency and standard of quality (quality assurance) for any product to be sold internationally. These are called ISO 9000 standards. Conformance with these standards is assured by audit, inspection, and review from third-party organizations, called registrars, which receive their accreditation through individual countries' accreditation bodies, approved by the ISO.

ISO standards are voluntary between members and are not a legal requirement. They assure that a manufacturer has a quality assurance system in place and that the procedures are written, documented, and observed by all employees. The ISO 9000 series consists of five quality standards:

1. *ISO 9000, ANSI/ASQC Q90* defines the terms and presents principal quality management and quality assurance practices used in the ISO 9000 series of standards and establishes guidelines for their selection and use. This standard is applicable to all industries.

2. *ISO 9001, ANSI/ASQC Q91* establishes models for quality assurance in the design, development, production, manufacture, installation, and service sectors of an organization. This standard, which is the most comprehensive of the three external quality assurance standards, is applicable to organizations that develop and produce their own products. This also applies to construction and engineering services.

3. *ISO 9002, ANSI/ASQC Q92* establishes models for quality assurance in production and installation. This standard is applicable to service industries and manufacturers that produce designs and specifications for other organizations.

4. *ISO 9003, ANSI/ASQC Q93* establishes models for quality assurance during final inspections and testing. This standard is applicable to testing laboratories, small shops, divisions within a firm, and equipment distributors that inspect and test supplied products.

5. *ISO 9004, ANSI/ASQC Q94* establishes internal organization management guidelines for design and implementation of quality systems; it is applicable to all industries.

OCCUPATIONAL SAFETY AND HEALTH ADMINISTRATION (OSHA)

The purpose of OSHA, a division of the Department of Labor, is to establish regulations that control and promote safety in the workplace. These regulations primarily concern the manufacturing, construction, transportation, and agricultural industries. OSHA also determines permissible exposure limits for chemicals and establishes norms for safety and monitoring procedures where workers are exposed to hazardous and toxic chemicals. These regulations require that all chemicals and hazardous materials be labeled and defined by material safety data sheets (MSDS).

MODEL REGIONAL BUILDING CODES

There are six regional building codes that along with their associated plumbing, mechanical, and fire protection codes have found general acceptance over various large areas of the country. They are:

1. *Building Officials Code Authority (BOCA)*. Plumbing Code: BOCA National Plumbing Code

2. *International Association of Plumbing and Mechanical Officials (IAPMO)*. Plumbing Code: Uniform Plumbing Code

3. *National Association of Plumbing-Heating-Cooling Contractors (PHCC)*. Plumbing Code: National Standard Plumbing Code

4. *Southern Building Code Congress International (SBCCI)*. Plumbing Code: National Standard Plumbing Code

5. *Council of American Building Officials (CABO)*. One and Two Family Dwelling Code

6. *International Conference of Building Officials (ICBO)*. Plumbing Code: Uniform Plumbing Code

MANUFACTURERS STANDARDIZATION OF THE VALVE AND FITTINGS INDUSTRY (MSS)

The MSS is a technical industry association organized for the development and improvement of industrial, national, and international codes and standards for valves, valve actuators, pipe fittings, flanges, pipe hangers, and seals. Society membership is composed of companies involved in the manufacture of these products. This society is recognized as the technical counterpart of the Valve Manufacturers Association and the American Pipe and Fittings Association, two nationally recognized trade associations.

Development of standards is a major part of its activities. The MSS provides technical assistance to other standards writing bodies in need of the expertise provided by its members. Many standards developed by MSS have been adopted as national standards, referenced by many codes.

NATIONAL ELECTRICAL MANUFACTURERS ASSOCIATION (NEMA)

NEMA is a nonprofit trade organization that establishes standards for motors, motor dimensions, and enclosures and sets minimum performance standards for many electrical devices.

NATIONAL FIRE PROTECTION ASSOCIATION (NFPA)

The NFPA (also abbreviated NFiPA to avoid conflict with the National Fluid Power Association) is a scientific and educational organization concerned with the causes, prevention, and control of destructive fire. Its purpose is to facilitate and encourage information exchange and to enhance the standards development process by providing the broadest possible forum for the consideration of proposed fire safety standards.

The NFPA is the principal source of consensus fire protection standards and codes. These codes and standards are written by voluntary technical committees and have been recognized by their adaptation and reference by statutory and regulatory law at all levels of government. More than 250 separate standards and codes have been published and are codified in the volumes of the *National Fire Codes.*

NATIONAL INSTITUTES OF HEALTH (NIH)

The NIH, a division of the Public Health Service, is a government agency responsible for biomedical research and science. It is one of eight agencies, comprising 24 separate institutes, centers, and divisions each devoted to a separate disease or disease group. Its mission is to discover new knowledge, conduct and fund research, and train research personnel.

NATIONAL OCEANIC AND ATMOSPHERIC ADMINISTRATION (NOAA)

Established as part of the Department of Commerce, NOAA monitors and predicts the state of the solid earth, the oceans and their living resources, the atmosphere, and the space environment of the earth and assesses the socioeconomic impact of natural and technological changes on the environment.

NOAA publishes various types of scientific and technical information. Of primary importance to facilities are technical memoranda, which report on research and technology results, and the atlas, which presents analyzed data generally in the form of maps showing distribution of rainfall.

NUCLEAR REGULATORY COMMISSION (NRC)

Created in 1975, the NRC is the government agency responsible for protecting the public health and safety relating to the use and disposal of nuclear material. The NRC regulates all industrial, commercial, and institutional uses of nuclear material including power plants. Services include the establishment of standards and regulations for the use and disposal of nuclear material, licenses for the use of nuclear material, and inspection of users to assure compliance with the applicable rules and regulations and the terms of individual license agreements. The NRC also provides services to states when a request is made and regulatory criteria and regulations are in place.

NATIONAL SANITATION FOUNDATION, INTERNATIONAL (NSF)

The NSF is an independent, not-for-profit organization of scientists, engineers, and educators. It is a neutral agency serving government, industry, and consumers in achieving solutions to problems relating to public health and the environment. Services include development of consensus standards, voluntary product testing, and certification of products in conformance with NSF standards.

In general, compliance with NSF standards is required for any material or component that is intended to process or prepare food, clean food-processing equipment, or carry potable water.

UNDERWRITERS LABORATORIES (UL)

Underwriters Laboratories is an independent, public service corporation, originally founded in 1968 by the life insurance industry.

Its purpose is the promotion of public safety through scientific investigation, study, experiments, and tests to determine the relative safety of various materials, devices, products, equipment, constructions, methods, and systems.

UL develops specifications and standards for materials, products, and equipment affecting the safety of the public. It tests items to conform with nationally recognized standards, and approves such items if acceptable.

3-A STANDARDS

3-A Standards have been prepared by three organizations with input from the Public Health Service:

1. International Association of Milk, Food, and Environmental Sanitarians
2. The Milk Industry Foundation
3. Dairy and Food Industries Supply Association

In an abbreviated form, the standard mandates the following:

1. Material of construction shall be 18-8 stainless steel with a carbon content of not more than 12 percent, or of a material of equal corrosion resistance.
2. Thickness or gauge of the material shall be sufficient for the purpose intended.
3. Surfaces in contact with product shall have a number 4 finish or smoother. This is the equivalent of an 80 to 150 grit finish, or an Ra value of 20 to 25 microinches (μin).
4. No threads shall contact product.
5. Square corners shall be avoided.
6. Piping shall be sloped to drain properly.
7. Design shall permit interchangeable parts.

Many pharmaceutical applications, particularly for pure water systems, are required by cGMP to exceed these requirements. This requires that metal surfaces be polished to a 150 to 240 grit (32 to 18 μin) finish to eliminate bacterial growth.

CHAPTER 2
PIPING

This chapter will describe the materials used most often for piping networks and systems described in this handbook. This chapter includes information on pipes, fittings, valves, jointing methods, hangers, and supports.

PIPE AND COMPONENT SELECTION CRITERIA

CODE ACCEPTANCE

Plumbing Systems

When selecting a piping material for any component of plumbing-related systems, the applicable plumbing code is of primary importance. Here, the allowable piping materials will be listed as well as any restrictions for their use. In addition, the code will also stipulate various accepted standards that govern the manufacture, tolerances, and installation of all components.

When renovating an existing facility or where circumstances require a unique design, it may be necessary to request a deviation from the accepted list of materials in order to match existing piping or obtain special design characteristics. The authorities reviewing such requests require enough information to determine that the intent of the applicable code provisions is followed in terms of safety and suitability of materials for the purpose intended.

Applicable plumbing and model building codes are discussed in Chap. 9, "Plumbing Systems."

Other Code Requirements

Codes required for systems other than plumbing are generally the following:

1. ANSI B31.1-9 series
2. FDA requirements, such as cGMP
3. Sanitary 3-A Standards for the food and pharmaceutical industries

PRESSURE AND TEMPERATURE RATINGS

The pressure rating for a wide range pipe and other components is provided by the ANSI Pressure Classification System, which publishes qualitative performance standards. The design pressure is the maximum sustained (steady-state) pressure that a piping system must contain without exceeding its code-defined stress limits.

The design temperature of a component must be equal to or greater than the maximum sustained temperature that will be experienced during any normal or abnormal mode of operation. Normal operating temperature is the temperature maintained by the system while it is operating at steady-state, full-load, nontransient operation. It is the temperature at which the allowable stress is calculated.

COST

The total installed cost of the network includes the piping material cost, assembly of the joint selected, handling due to the weight of the material, physical damage allowance, and the support system required.

CORROSION RESISTANCE

Corrosion is the thinning of a pipe wall (usually) caused by a chemical reaction from a corroding fluid or agent and is limited almost exclusively to metal products. Corrosion resistance is the ability of a pipe to resist the internal corrosive effects of the fluid flowing through it, as well as the external corrosive forces on the pipe, for example, soils (if underground) or surrounding atmospheric conditions (when installed above ground). Corrosion can be reduced or eliminated with the application of suitable coatings, linings, and cathodic protection, depending on the nature of the problem. Since there is no universal corrosion-resistant material, the selection of a specific grade or class of pipe material with specific alloys to resist expected corrosion resulting from specific fluids within a pipe is a matter of tradeoffs unique to each project.

Corrosion is separated into two basic types: general and localized. General corrosion describes the potential dissolution of pipe over its entire exposed surface. Localized corrosion affects only a small area of the pipe surface.

General Corrosion

This is a breakdown of the pipe material at a uniform rate over its entire surface by direct chemical attack. It is caused by the loss of the protective passive film that forms on the surface of the pipe coupled with a chemical reaction occurring between the pipe material and the chemical in the fluid.

Galvanic Corrosion. This type of corrosion occurs in a liquid medium (called an electrolyte) when a more active metal (anode) and a less active metal (cathode) come in contact with one another and form an electrode potential. When this occurs, the more active (noble) metal will tend to dissolve in the electrolyte and go into solution. This is shown in the galvanic series of metals (Table 2.1).

Intergranular Corrosion. This type of corrosion occurs in the pipe wall when material in the grain boundary of some alloys is less resistant to the corroding agent than the grains themselves, and the bond between the grains is destroyed.

Erosion Corrosion. This is caused by a wearing away of the pipe wall, usually as a result of excessive fluid velocity or impingement by suspended solids.

Localized Corrosion

This takes place on small areas of the surface, usually at high rates, and takes various forms:

TABLE 2.1 Galvanic Series of Metals and Alloys

Corroded end (anodic, or least noble)
Magnesium alloys
Zinc
Beryllium
Aluminum alloys
Cadmium
Mild steel, wrought iron
Cast iron, flake or ductile
Low-alloy high-strength steel
Nickel-resist, types 1 & 2
Naval bronze (CA464), yellow bronze (CA268), aluminum bronze (CA687), red bronze (CA230), admiralty bronze (CA443), manganese bronze
Tin
Copper (CA102, 110), silicon bronze (CA655)
Lead-tin solder
Tin bronze (G & M)
Stainless steel, 12–14% chromium (AISI Types 410, 416)
Nickel silver (CA732, 735, 745, 752, 764, 770, 794)
90/10 Copper-nickel (CA706)
80/20 Copper-nickel (CA710)
Stainless steel, 16–18% chromium (AISI Type 430)
Lead
70/30 Copper-nickel (CA715)
Nickel-aluminum bronze
Inconel alloy 600
Silver braze alloys
Nickel 200
Silver
Stainless steel, 18 chromium, 8 nickel (AISI Types 302, 304, 321, 347)
Monel alloys 400, K-500
Stainless steel, 18 chromium, 12 nickel-molybdenum (AISI Types 316, 317)
Carpenter 20 stainless steel, Incoloy alloy 825
Titanium, Hastelloy alloys C & C276, Inconel alloy 625
Graphite, graphitized cast iron
Protected end (cathodic, or most noble)

Stress-Corrosion Cracking. This type of corrosion is the physical deterioration and cracking of the pipe wall caused by a combination of high operating temperature, tensile stress on the pipe, and chemicals in the fluid stream.

Pitting. This is characterized by deep penetration of the metal at small areas of the surface, concentrating in small cells, without affecting the entire surface.

Crevice Attack Corrosion. This occurs at junctions between surfaces (often called crud traps) where a crack exists that allows an accumulation of a corroding agent.

Corrosion failure occurs if any pipe material is reduced to a minimum thickness mandated by code.

Manufacturers of pipes and fittings publish corrosion-resistance tables concerning chemical resistance and compatibility with fluids carried by the piping they produce. These relate to temperature, concentration, and specific chemicals. These tables should be used for selection of a specific pipe to transport any specific chemical.

PHYSICAL STRENGTH

Physical strength is the capability of any pipe to resist the damage that might occur either during the construction phase or after the pipe is placed in service.

FIRE RESISTANCE

Fire resistance is the capability of a piping system to simply remain intact and not fall during a fire or, in some cases, to also retain the ability to carry water. Where this is a factor, pipes, joints, and supports strong enough for this purpose should be selected.

AVAILABILITY

Any pipe is considered available when it can be supplied at a competitive price to the area where the project is under construction. In addition, obtaining the mechanics and special tools needed to assemble the piping system must also be possible.

METALLIC PIPE AND PIPING MATERIALS

ALUMINUM (Al)

Aluminum pipe is manufactured in various wall thicknesses similar to copper and in sizes ranging from ⅛ to 12 in. Sizes above 3 in are not readily available. Aluminum is manufactured in various alloys; the most commonly used for facilities conforms to ASTM B-210. Joints are made using brazing or welding with special aluminum alloy filler metals.

Aluminum tubing is light in weight and generally used for specialty services, such as cryogenics (where ductility and strength are necessary) and for carrying compressed specialty gases (because of its corrosion resistance). It also resists many specialty chemicals and is resistant to atmospheric corrosion. It is not suitable for acids, mercury, and strong alkalis. It has a high rate of expansion.

For a list of various alloys, refer to Table 2.2.

BRASS (BR)

Brass is an alloy of copper and zinc. The proportion varies from *85%* copper to 67% copper. Pipe with a high copper content is known as red brass, and that with a lower content is known as yellow brass. When used for drainage systems, it is obtained plain end. Joints for this pipe can be either threaded, flanged, brazed, or soldered. Brass pipe for utility piping systems shall conform to ASTM B 43: Red Brass Tube, Seamless.

Brass is generally used for local branch drainage lines, where this alloy will resist specific corrosive drainage effluent and, in larger sizes, for potable and other water supply lines and to match existing work for alterations. Its advantages and disadvantages are the same as for copper.

Brass castings for pipe fittings and components of plumbing fixtures are not made with the same alloy as pipe and often contain lead. Pipe is lead free and brass fittings with lead are no longer permitted by code to be used for potable water.

TABLE 2.2 Aluminum Alloys

Alloy number	UNS number series	Major alloying elements
1XXX	A91NNN	Aluminum with 99.00% minimum purity
2XXX	A92NNN	Copper
3XXX	A93NNN	Manganese
4XXX	A94NNN	Silicon
5XXX	A95NNN	Magnesium
6XXX	A96NNN	Magnesium plus silicon
7XXX	A97NNN	Zinc
8XXX	A98NNN	Miscellaneous

CAST IRON SOIL PIPE (CI)

Technically known as gray cast iron, this pipe is a ferrous material alloyed with carbon in the form of free graphite flakes, silicon, and other impurities. It is available in three classifications: service (standard) weight, extra heavy, and hubless. The pipe is commonly lined internally with cement or coal tar enamel, and coated externally with a variety of materials to reduce corrosion by soils.

Joints require two types of pipe ends: hub and spigot or hubless. The hub and spigot ends can be joined either by caulking or by the use of an elastomeric compression gasket. Hubless pipe is joined by an external compression coupling.

Cast iron soil pipe should conform to the following standards:

1. ASTM A 74: Hub and Spigot Cast Iron Soil Pipe and Fittings
2. CISPI 301: Hubless Cast Iron Soil Pipe
3. CISPI 310: Hubless Cast Iron Fittings for Soil Pipe

Cast iron is well suited for sanitary effluent and can be used in any part of a gravity drainage and vent system. Advantages include an ability to withstand moderate external pressure (such as direct burial in soil), good fire resistance, good flow characteristics, and good corrosion resistance in most natural soils. Piping in use for over 100 years has been documented. Disadvantages are that the pipe is brittle and subject to breakage when roughly handled, it is subject to corrosive attack by aggressive soils and highly septic effluent, it is heavy, and it has a high initial material cost.

Cast iron pipe is manufactured with both inside and outside coated for corrosion resistance. A PE wrapping is often used to eliminate external corrosion of cast iron pipe buried underground on a site.

ACID-RESISTANT CAST IRON (AR)

Commonly referred to as high silicon iron pipe, acid-resistant CI is an alloy of gray cast iron containing between 14.25 and 15% silicon, and small amounts of manganese, sulfur, and carbon. It is available only in extra heavy pattern, with the same dimensions as CI piping.

Joints require two types of pipe ends: hub and spigot or hubless. The hub and spigot ends can be joined by caulking. Rubless pipe is joined by a compression coupling.

Acid-resistant cast iron pipe shall conform to ASTM A 861 and ASTM A 518.

This specialty piping material is used for drainage of various corrosive liquids, and since it is stronger than glass, is recommended for exposed or underground applications where there is a possibility of physical damage.

CARBON STEEL (ST)

Steel is a very broad category of piping because of the large number of alloys that have been produced. It is divided into two broad categories according to the method of manufacture: mill pipe and fabricated pipe. Mill pipe is produced to meet finished pipe specifications. Fabricated pipe is manufactured from steel plate with spiral or straight welded seams.

Steel pipe is manufactured by either the seamless (extruded) or welded method, and is available either plain (black) or galvanized (zinc plated inside, outside, or both). Wall thickness is expressed as "schedule," and ranges from schedule 5 (lightest) to schedule 160. The wall thickness varies with the size of the pipe. The larger the schedule number, the thicker the pipe wall for a specific pipe size.

Steel pipe, depending upon type, can be obtained with threaded ends for screwed fittings, plain ends, and flanged and beveled ends for welding.

There is an extremely large number of steel pipe alloys available. The selection depends on the intended service. The steel pipe alloys most commonly used for service and utility systems conform to ASTM A 53: Steel Pipe, Welded or Seamless, Black or Galvanized and ASTM A-106: Steel Pipe, Welded or Seamless, Black or Galvanized.

Steel pipe is generally used for pressure piping. Its advantages are long laying lengths, high internal and external strength, and the availability of varying pipe thickness to meet almost any design pressure. It has good flow characteristics and fire resistance and is low in initial cost. The most serious disadvantage is low corrosion resistance. This requires internal and external protection, with galvanizing the most commonly used method.

COPPER

Copper tubing is seamless, made from almost pure copper (99.9 percent), and is available in hard (annealed) and soft (drawn) form. It is manufactured in sizes ranging from $\frac{1}{8}$ to 12 in, but sizes over 6 in are not generally available. All tubing is manufactured with plain ends only. Joints are made by soldering, and brazing, and with flared and flanged fittings.

The six types of copper tubing used most often are:

1. ASTM B-88 is the grade used most often for potable water, and also for compressed gases and vacuum systems where high purity is not a factor. It is seamless, available in nominal pipe sizes from $\frac{3}{8}$ to 12 in, in hard and soft temper, and in three wall thickness grades—K (thickest wall), L, and M (thinnest wall). If patented flare fittings are used, the pipe must be obtained without outside diameter embossing, which would interfere with the sealing of the pipe wall against the side of the fitting.

2. ASTM B-819 is similar to B-88, except that it is available only in grades K and L and, in addition, the pipe is factory cleaned, capped, and specially marked. This pipe is required to be used for medical gas systems in health-care facilities. It should also be considered as the primary copper pipe for gases in laboratories as well.

3. ASTM B-75 is seamless, available as either hard or drawn, and in nominal sizes from $\frac{1}{8}$ to 2 in O.D. The smaller sizes are often referred to as capillary tubing. This is the grade most often used for very small diameter pipe [$\frac{1}{4}$ in O.D. (6 mm) or less] connecting instruments to the distribution piping. This pipe is commonly joined by patented flare joints, which require temper 060 to seal correctly.

4. ASTM B-280, type ACR (air conditioning and refrigeration), is available cleaned and capped for field refrigeration piping and could also be considered for laboratory use. It is available only in hard temper, and its size is the actual O.D. of the pipe. This requires that all non-ACR fittings used in the system, which are manufactured in nominal pipe size, be dimensionally compatible with the ACR pipe, which is manufactured in actual pipe size.

5. ASTM B-306, copper drainage tube, is known as DWV (drainage, waste, and vent). This designation applies to copper tube used for nonpressure drainage systems. This tube has the thinnest wall of any copper product. The preferred jointing method is soldering, which is adequate in strength and the least costly. Primary use is in residential construction as indirect waste lines and in larger projects for local branch lines where human waste is not discharged. Advantages are its light weight, ease of assembly, and smooth interior. Disadvantages include corrosive attack by ordinary sewage, poor fire resistance, and the need for dielectric connections to eliminate galvanic corrosion where this material is connected to any iron piping.

6. ASTM B-837 Type G is seamless tube, available in either hard (drawn) or soft (annealed) temper, and in nominal sizes from ⅜ to 1⅛-in O.D. This tubing is identified by the O.D. rather than nominal size. This grade is manufactured specifically for natural gas and LP fuel gas systems. Fittings shall be similar to those used with type ACR copper. The pipe is joined by either brazing or flare type joints. Brazed joints are required for system pressures above 14 in WC where installed in inaccessible locations. Brazing alloys shall contain less than 0.05 percent phosphorus.

DUCTILE IRON PIPE (DI)

Ductile iron is a cast ferrous material alloyed with free nodular or spheroidal graphite in lieu of the flakes that are present in cast iron. This is achieved by the addition of a magnesium inoculant. It is used either as a gravity sewer or pressure pipe. Sizes available range from 3 to 54 in. There are eight pressure ratings—class 50 (125 psi) to class 56 (350 psi), and gravity sewer pipe. A cement or bituminous lining can be provided to resist internal corrosion. This pipe can be assembled with mechanical, gasketed, or flanged joints. Ductile iron pipe shall conform to ASTM A 518 and ASTM A 861.

The advantages of ductile iron are the same as those for CI pipe, except that it is far stronger in terms of allowable pressure ratings and external load-bearing capacity. It is also not as brittle, allowing rougher handling. It has a higher initial cost than CI.

LEAD (LD)

Lead pipe is made from 99.7% pig lead with various alloys available for special applications. Joints for this pipe are either wiped, burned, or mechanically flanged. Lead pipe shall conform to WW-P-325a: Lead Pipe, Bends and Traps.

Uses of lead pipe include existing connections to floor-mounted water closets, radioactive wastes, and special laboratory corrosive wastes. It has very limited use in modern drainage systems.

STAINLESS STEEL (SS)

The term *stainless steel* encompasses a wide variety of alloys containing 11 to 30% chromium (Cr), 0 to 35% nickel (Ni), and 0 to 6% molybdenum (Mo) in various combinations as well as small amounts of other elements such as titanium, manganese, niobium, and nitrogen. It is widely used in the chemical, pharmaceutical, and food processing industries.

Pipe is available in sizes ranging from ⅛ to 48 in, and is manufactured in plain end, prepared end for welding, and flanged. Joints can be welded, threaded, or flanged. Wall thickness is expressed as a "schedule," and ranges from schedule 5 (lightest) to schedule 160. The wall thickness varies with the size of the pipe. The larger the schedule number, the thicker the pipe wall for a specific pipe size. Stainless steel pipe is also available as tubing with wall thickness designated as a decimal.

The composition of a stainless steel alloy determines its metallurgical structure or grade, and therefore its properties. There are five groups of stainless steel based on metallurgical structure: ferritic, austenitic, superaustenitic, martensitic, and duplex.

1. Ferritic stainless steels contain 12 to 30% (more typically 16 to 18%) Cr, 0 to 4% Ni, and 0 to 4% Mo with a low carbon content. This material is magnetic and has good ductility and cold formability but is not hardenable by heat treatment. This class is generally less vulnerable to chloride-induced stress corrosion cracking. Its primary use is in transport of strong oxidizing fluids (such as nitric acid) in process environments, machinery, and kitchen equipment. This class is exemplified by type ASTM grade 430.

2. Austenitic stainless steels contain 17 to 27% Cr, 8 to 35% Ni, and 0 to 6% Mo. This material is typically nonmagnetic and readily weldable, and has good ductility (even at cryogenic temperatures) and cold formability but is not hardenable by heat treatment. This class is more generally corrosion resistant, but is generally vulnerable to chloride-induced stress corrosion cracking. Regular carbon grades are susceptible to corrosion around welded joints due to migration of Cr away from the weld site. These problems are overcome by using a low carbon grade, indicated by an L suffix, that reduces carbon to below 0.035 percent. Grades within this class are the most commonly used stainless steels. This class is exemplified by type ASTM grades 304 and 316 (304L and 316L). A superaustenitic grade is also available with superior resistance to chloride pitting.

3. Superaustenitic stainless steel alloys were created to better withstand corrosion in a more severe environment than conventional stainless steel. They are alloys of Ni, Cr, Mo, copper, and iron typified by UNS (Unified Numbering System) alloys N08020, N08024, and N08026.

4. Martensitic stainless steels contain 11 to 18% Cr, 0 to 6% Ni, 0 to 2% Mo, and 0.1 to 1% C. This class is magnetic, oxidation resistant, and hardenable by heat treatment. Little used in piping applications, its primary uses are in cutlery, turbine blades, and high-temperature parts. This class is exemplified by type ASTM grade 410.

5. Duplex stainless steel is characterized by a microstructure containing both ferritic and austenitic types with different grades, having a mixture of 40 to 60% of each for various alloys. Its advantages are good resistance to chloride-induced stress corrosion cracking and high mechanical strength properties along with good ductility and impact strength. Disadvantages include corrosion of pipe by reducing acids and weld site corrosion by oxidizing acids.

Stainless steel is available in various wall thicknesses. Pipe is commonly available from schedule 5 to 80, and tubing from 0.028 to 0.188.

Stainless Steel Finishes

For stainless steel piping used in pharmaceutical, food-processing, chemical, and electronics applications, the interior and exterior of the piping are often required to be finished as required by FDA, EPA, USDA, or other applicable codes. Finishing the exterior makes it easier to keep clean. Finishing the interior will prevent the adherence of any solids, increase corrosion protection, and shorten pipe interior cleaning procedures. Finishing can be abrasive, electropolished, or both.

Abrasive finishes are mechanically produced by polishing and wearing away of the surface. This is often specified by a particular size or "grit" of the abrasive used, such as a 220 grit. This signifies the size of the abrasive passing through a specific size mesh. The larger the number, the finer the finish. Electropolishing is an electrochemical process using an electrical current to deposit metal from an anode to a cathode. Electropolishing, which is the opposite of electroplating, removes surface metal from microscopic high points faster than from low points. The metal to be polished is the sacrificial anode. Surface ions of iron are removed leaving a chromium-rich surface resulting in a smooth, corrosion-resistant pipe interior. Another method used to specify standard sheet and pipe exterior finishes uses numbers 1 to 8. An explanation of sheet finishes is given in Table 2.3. Table 2.4 gives the grit equivalent of microinch measurement.

Finishes are often indicated as Ra (arithmetic mean roughness average, or roughness average) and Rq, the equivalent of RMS (root mean square). Both of these are measured in microinches and denote the smoothness of the surface. The smaller the number, the finer and finish. The Ra reading is approximately 87.5 percent of the Rq (RMS) reading. Other methods of expressing smoothness are centerline average (CLA) and arithmetic average (AA).

Passivation

When the interior surface of stainless steel piping is required to have a very low rate of corrosion, it must be made passive. This is accomplished by the formation of a thin surface film that acts as a barrier to corrosion. The surface film (or passive film) is made thicker by exposing it to oxidizing and chelant solutions or electropolishing. Oxidizing solutions frequently used are nitric acid, ammonium persulfate, hydrogen peroxide, and citric acid. Chelants are nontoxic organic acids or their viable salts and nontoxic synergizing agents. ASTM A 380 suggests other passivation chemical combinations that have proven successful.

TABLE 2.3 Table of Standard Sheet Finishes

	Number	Description
Unpolished (as rolled) finishes	1	A rough, dull finish that results from hot rolling to the specified thickness, followed by annealing and descaling.
	2D	A dull finish that results from cold rolling, followed by annealing and descaling, and perhaps a final light roll pass through unpolished rolls. This finish is used where appearance is of no concern.
	2B	A bright cold-rolled finish resulting from the same process as that for number 2D finish, except that the annealed and descaled sheet receives a final light roll pass through polished rolls. This is the general-purpose, cold-rolled finish that can be used as is or as a preliminary step to polishing.
Polished finishes	3	An intermediate polished surface obtained by finishing with a 100-grit abrasive. This finish is generally used where a semifinished polish surface is required and usually receives additional polishing during fabrication.
	4	A polished surface obtained by finishing with a 120- to 150-grit abrasive, following initial grinding with coarser abrasives. This is a general-purpose bright finish with a visible "grain" that prevents mirror-like reflection.
	7	A highly reflective finish obtained by buffing finely ground surfaces, but not to the extent of removing the scratch pattern.
	8	A reflective surface obtained by polishing with successively finer abrasives and buffing extensively until all scratch patterns from preliminary grinding are removed.

Source: Data from *Finishes for stainless steel*, Publication SS201-683-14M-EB, American Iron and Steel Institute.

TABLE 2.4 Grit to Microinch Conversion

Grit	Microinch (μin)
120	45–48
150	30–33
180	20–23
240	15–18
320	9–11

CORRUGATED STEEL PIPE

Corrugated steel pipe is available from 6 in (125 mm) to 96 in (2.66 m). It is fabricated from flat steel that has been rolled into various shapes and impressed with grooves around the circumference of the pipe, generally described as circular arcs connected by tangents. Corrugations are measured by pitch (the dimension from crest to crest at right angles to the corrugations) and depth.

Longitudinal pipe seams are riveted, welded, bolted or have helical lock seams. Joints are generally steel bands with a gasket under the sleeve and tightened by bolts inserted in an integral angle. Fittings are usually made from straight piping and shop fabricated into the desired shape.

Corrugated steel pipe shall conform to different AASHTO standards depending on the actual pipe material specification.

OTHER METALLIC PIPING MATERIALS

The chemical compositions of various metal piping materials selected for special corrosion-resistant characteristics are given in Table 2.5.

TABLE 2.5 Alloy Composition of Miscellaneous Metallic Piping

Material	C, %	Cr, %	Cu, %	Fe, %	Mn, %	Mo, %	Ni, %	Si, %	Other elements, %
Admiralty metal	—	—	70	—	—	—	—	—	Zn, 29; Sn, 1
Aluminum 2S	—	—	—	—	—	—	—	—	Al, 99.5+
Aluminum 3S	—	—	0.1	0.5	1.25	—	—	0.5	Al, 97.7
Brass, admiralty	—	—	71	—	—	—	—	—	An, 28; Sn, 1
Brass, aluminum	—	—	76	—	—	—	—	—	Al, 2; Zn, 22
Bronze, silicon	—	—	96	—	—	—	—	3	
Carpenter 20 alloy	—	20	4	44	—	3	29	—	
Copper, arsenical	—	—	99.9+	—	—	—	—	—	As, 0.04
Croloy 2¼ alloy	0.15	2.25	—	96.1	—	1.0	—	0.5	
Cupro-nickel, 70-30	—	—	69.3	0.75	—	—	30	—	
Cupro-nickel, 90-10	—	—	88.8	1.25	—	—	10	—	
Durimet 20 alloy	0.07	20	3.5	45.7	—	1.75	29	—	
Duriron	0.85	—	—	84.7	—	—	—	14.5	
Hastelloy B alloy	0.1	—	—	5	—	28	62	—	
Hastelloy C alloy	0.1	16	—	5	—	16	55	—	W, 4
Incoloy alloy 800	0.04	21	0.3	45.3	1	—	32	0.4	
Inconel alloy 600	0.04	15.8	0.1	7	0.20	—	76	20	
Iron, cast	3.4	—	—	94.3	0.5	—	—	1.8	
Lead	—	—	—	—	—	—	—	—	Pb, 99.95
Monel alloy 400	0.12	—	31.5	1.4	1	—	66	0.15	
Monel alloy K-500	0.15	—	29.5	1.0	0.60	—	65	0.15	
Nickel 200	0.06	—	0.05	0.15	0.25	—	99.4	—	
Nickel silver, 18%	—	—	65	—	—	—	18	—	Zn, 17
NI-Hard alloy	3.4	1.5	—	89.5	0.5	—	4.5	0.6	
NI-Resist, Type I	2.8	2.5	6.5	69.6	1.3	—	15.5	1.8	
NI-Resist, Type II	2.8	2.5	—	71.9	1.0	—	20.0	1.8	
Steel, carbon-molybdenum	0.15	—	—	99.4	—	0.5	—	—	
Steel, low-carbon, electric-resistance-welded, ASTM A587-68	0.1	—	—	99.5	0.4	—	—	—	P, 0.01; S, 0.02; Al, 0.05
Steel, mild (SAE 1020)	0.2	—	—	99.1	0.45	—	—	0.25	
Steel, Ni-Cr (SAE 3140)	0.4	0.65	—	96.7	0.8	—	1.25	0.25	
Tantalum	—	—	—	—	—	—	—	—	Ta, 99.9+
Titanium (Ti-50A)	0.08 max.	—	—	0.20 max.	—	—	—	—	
Worthite alloy	0.07	20	1.75	46.7	—	3	25	3.5	N, 0.05 max.; H, 0.015 max.

PLASTIC PIPE AND PIPING MATERIALS

This section will discuss various commercially available plastic piping materials and their properties.

Plastics, because of their unique chemical resistance and other characteristics, are widely used as raw material for both piping and elastomeric sealing and gasketing compounds. It is no longer justifiable to think of plastic pipe as merely a cheap substitute for other piping materials. Plastic has become the material of choice for piping systems used to convey various liquids, chemicals, pharmaceuticals, liquid fuels, and fuel gases, and those used for underground sewer water.

Plastic pipe is available in a great variety of compositions. When used for plumbing systems, the restrictions imposed by the applicable code will be the single most important determining factor in the use and selection of any plastic pipe material.

Plastic pipe is manufactured in two types: thermoset (TS) and thermoplastic (TP). Thermoset piping is permanently rigid; examples are epoxy and phenolics. Thermoplastic material will soften when subject to any degree of heat and reharden upon removal of the heat. This will affect the strength of the pipe. Therefore, extreme care must be used when selecting the material type and support system for the material.

In general, the advantages of plastic pipe are excellent resistance to a very wide range of sanitary and chemical effluent and aggressive soils, long laying lengths, good flow characteristics, and economical initial system costs. Disadvantages are poor structural stability requiring close support, susceptibility of some materials to changes resulting from exposure to ultraviolet rays or sunlight, poor fire resistance, lowering of pressure ratings with elevated temperature, and production of toxic gases released by some materials when burning.

KEY PROPERTIES OF PLASTIC PIPE

The advantages of plastic pipe include:

1. Resistance to a very wide range of sanitary and chemical effluents
2. Resistance to aggressive soils
3. Availability in long lengths
4. Light weight
5. Low resistance to fluid flow
6. Generally low initial cost

Disadvantages include:

1. Poor structural stability requiring additional support
2. Susceptibility of some types of plastics to physical changes resulting from exposure to sunlight
3. Generally low resistance to solvents
4. Poor fire resistance

5. Lowered pressure ratings at elevated temperature
6. Production of toxic smoke and gases, which are released upon combustion of some types of plastic pipe

DESCRIPTION AND CLASSIFICATION

Plastic pipe is as descriptive a phrase as metallic pipe. The properties of various plastic materials are obtained from the basic chemical composition of the polymer resin, additives, and the manufacturing process itself. In order to better understand the material called "plastic," definitions of the basic terms and ingredients used by the plastic piping industry are necessary. Please understand that these are simplified definitions.

Plastic is a material whose essential ingredient is an organic substance of large molecular weight which at some stage in its manufacture can be shaped by flow and becomes solid in its finished state.

A *polymer* is a material consisting of molecules with a high molecular weight. A *monomer* is a chemical compound capable of reacting to form a polymer. *Polymerization* is a chemical reaction in which molecules of a monomer are linked together to form a polymer. When two or more monomers are used, the process is called *copolymerization.*

The following are common additives used in the manufacture of plastic piping: flame retardants, plasticizers to increase flexibility and workability, antioxidants to retard degradation from contact with air, stabilizers to retard degradation at higher temperatures, lubricants to aid in the extrusion process, pigment or dies to color the final product and protect against ultraviolet light, fillers to modify strength or lower cost, and modifiers to produce a special property response.

Another type of plastic is an *elastomer.* Used mostly for gaskets, an elastomer is a material that is capable of being repeatedly stretched to at least twice its original length at room temperature and which will return to its approximate original length upon release.

The following is a partial list of plastic pipe and elastomer materials available from all sources. The names in parentheses are trade names patented by various manufacturers. Only those piping materials that are commonly available will be discussed. Elastomers, indicated as (E), are listed only for reference since they are outside the scope of this chapter:

ABS = acrylonitrile butadiene styrene; also (Buna-N) (E)
BR = butadiene (E)
CAB = cellulose acetate butyrate (Celcon)
CIIR = chlorinated isobutene isoprene (E)
CPE = chlorinated polyethylene (E)
CPVC = chlorinated polyvinyl chloride
CR = chloroprene rubber (Neoprene) (E)
CSP = chlorine sulphonyl polyethylene (Hypalon) (E)
ECTFE = ethylenechlorotrifluoroethylene
EP = epoxide, epoxy
EPDM = ethylene propylene-diene monomer (E)

EPM = ethylene propylene terpolymer (E)
FEP = fluorinated ethylene propylene
FPM = fluorine rubber (Viton) (E)
HDPE = high-density polyethylene
IIR = isobutene isoprene (butyl) rubber (E)
IR = polyisoprene (E)
PA = polyamide
PAEK = polyaryl etherketone
PB = polybutylene
PC = polycarbonate
PCTFE = polychlorotrifluoroethylene (Halar)
PE = polyethylene
PEX = cross linked polyethylene
PF = phenol-formaldehyde
PFA = perflouralkoxy
PP = polypropylene
PPS = polyphenylene sulfide
PTFE = polytetrafluoroethylene (Teflon)
PEEK = polyether etherketone
PFA = perfluoroalkoxy
PS = polysulfone
PVC = polyvinyl chloride
PVDC = polyvinylidene chloride
PVDF = polyvinylidene fluoride
SBR = styrene butadiene (E)

Plastic materials used for piping are divided into two basic groups, thermoplastic and thermosetting. Thermoplastics soften upon the application of heat and reharden upon cooling. This permits pipe to be extruded or molded into shapes. The most common piping materials are thermoplastic. Thermosetting plastics form permanent shapes only when cured by the application of heat or the use of a curing chemical. Once shaped, they cannot be reformed.

There are subclassifications of pipe based on the material used for the pipe itself. The two most common are polyolefins and fluoroplastics. Polyolefins, which are plastics formed by the polymerization of certain straight chain hydrocarbons, include ethylene, propylene, and butylene. Piping includes PP, PE, and PB. Fluoroplastics are polymers containing one or more atoms of fluorine. Piping includes PTFE, PVDF, CTFE, ETFE, PFA, and FEP.

PLASTIC PIPE STANDARDS AND NOMENCLATURE

A variety of standards and nomenclature is used to designate pressures and standard dimensions used for the procurement and identification of plastic pipe. Some are used to match existing metallic pipe specifications and others are unique to the plastic pipe industry. The following is an explanation of the terms used in various standards:

SDR. The standard dimensional ratio is the most commonly accepted measure for providing a pipe wall thickness category and constant mechanical properties for many plastic pipe materials. Used for solid homogeneous pipe, the SDR is found by dividing the average outside diameter of a pipe by the wall thickness. This designation has resulted in a series of preferred industry standard numbers that are constant for all sizes of pipe. It is possible for a pipe to have different SDRs depending on whether the I.D. or O.D. is the controlling factor.

DR. The dimensional ratio is often incorrectly used interchangeably with SDR. The DR is found in the same manner as above and means the same thing, but is used when the product does not have the preferred SDR number established by other prevailing standards. Pipe manufactured to pressure ratings for AWWA C-900 series standards uses this designation.

O.D. controlled. This designation is used when the outside diameter of the pipe is the controlling factor in the selection of the pipe.

I.D. controlled. This designation is used when the inside diameter of the pipe is the controlling factor in the selection of the pipe.

P.R. Pressure rated is used when the pressure rating is the controlling factor in the selection of the pipe rather than the dimensions of the pipe itself.

PS. Pipe stiffness is used only for sewer pipe. This designation is in PSI. The higher number has a thicker pipe wall.

Schedule. This designation is used to match the standard dimensions for metallic pipe sizes. The pressure rating of the pipe varies with pipe size. Some standards use iron pipe size (IPS) instead of schedule to keep the wall thickness consistent with iron pipe.

PSM. This is an arbitrary designation for products having certain dimensional characteristics unique to a very specific product.

The AWWA has several proprietary dimensional standards that are used to specify plastic pipe used only for pressurized potable and fire water main distribution and transmission systems. The composition of the plastic piping material is referenced to ASTM standards. Since plastic pipe connects to or replaces cast iron and ductile iron pipe, these standards are O.D. controlled for use with O-ring gasketed joints only and are dimensionally compatible with these joints. These standards are:

AWWA C-900 = 4 to 12 in PVC
AWWA C-901 = ½ to 3 in PE
AWWA C-902 = ½ to 3 in PB
AWWA C-903 = Deleted
AWWA C-904 = Fittings for C-900 pipe
AWWA C-905 = 14 to 36 in PVC
AWWA C-906 = Larger diameter PE pipe

Trade laws allow import of Canadian pipe materials into the United States. The Canadian Standards Association (CSA) has standards of their own, but many of them have not been completely coordinated with the United States standards for similar products. At this time, using CSA standards as reference for plastic products is not recommended.

There are three designations used in plastic drainage pipe standards: DWV, sewer, and drain. All standards are O.D. controlled and are non-pressure-rated. The only differences between identical materials with different designations are dimensions. Different materials with the same designations have the same dimensions.

For the most part, the codes are very specific about acceptable materials for use in plumbing systems, such as potable water and drainage, where approval of the authorities is required. Regarding uses such as draining chemicals from laboratories or industrial work, the codes are vague. When the chemical waste will be treated inside the project boundary, usually the materials (and design) used for the waste system do not fall under the jurisdiction of the plumbing code. In these cases, the engineer has the most latitude in the selection of materials used for drainage piping.

ELEMENTS FOR SELECTION

Due to the differences in manufacture, grade, and chemical composition of the pipe, test data must be obtained from the local supplier or manufacturer. Properties of similar materials from different manufacturers are often not the same. Very often, a range of values for properties such as tensile strength, maximum operating temperature, and hardness is given.

Elevated Temperature Considerations

Service temperatures in plastic piping systems depend on the type of plastic used. A maximum service temperature is generally fixed for thermoplastics, and identifies the upper limit to which the pipe may be heated without damage. When heated above this temperature, the pipe will soften and deform. Upon cooling, it will harden to the deformed shape and dimensions.

Long-Term Hydrostatic Strength

The design pressure for plastic pipe is based on long-term hydrostatic strength, which is determined by finding the estimated circumferential stress that, when applied continuously, will produce failure of the pipe after 100,000 h at a specified temperature. In addition, a service factor is included in the design calculations. This factor takes into account certain variables together with a degree of safety appropriate to the installation. The service factor is usually selected by the design engineer, and referenced to a service design life of about 50 years. This design method does not include the fittings, joints, or cyclic effects such as water hammer.

Most pressure ratings for thermoplastic pipes are calculated assuming a water environment. As the temperature rises, the pipe becomes more ductile and loses strength, and therefore the rating of thermoplastic pipe must be decreased to allow for safe operation. These factors are different for each pipe material.

Fatigue Behavior

When surges and water hammer are likely to be encountered, additional allowance should be made or protective devices installed in the piping system to reduce the pressure.

Aging and Long-Term Degradation

Aging is the change in physical and chemical properties during storage or use, and is generally dependent on temperature. These changes can occur naturally through normal atmospheric or building temperature fluctuations, or can be developed artificially due to elevated temperatures of the fluid in the pipe. The origin of these changes within the pipe are in the molecular or crystallographic structure of the pipe. Plastics and elastomers experience chemical changes due to the influence of light, heat, oxygen, humidity, and radiation, all of which cause breaks in their molecular chains. One criterion for determining the onset of aging is the measurement of thermal stability (oxygen induction time) using differential scanning calorimetry.

Ultraviolet Radiation (UVR)

UVR is a known source of degradation to plastic pipe. The effects can be reduced by adding pigments or covering the installed pipe with a jacket.

Flammability

During fire conditions, the degradation of plastics is greatly accelerated. In the early stages of a fire, most plastics melt and lose their structural shape and strength. As heat is added at a rising rate, plastics undergo a series of typical changes, which include chemical decomposition often releasing toxic chemicals. This decomposition occurs at a lower temperature than ignition. By the time ignition occurs or is possible, a relatively long period of chemical emission has elapsed.

When thermoplastic pipe burns, it releases smoke and toxic gases, provides heat that increases the intensity of a fire, and may provide a path for flame to spread along its length. In addition, open holes may develop at wall or ceiling penetrations which could provide a route for the passage of gases between rooms.

All organic materials are flammable, but this is particularly true of polyolefins. It is well proven that many polymers are as a result of their chemical composition difficult to ignite. Polymers can also be made much more difficult to ignite by the addition of flame retardants.

Acoustic Transmission

Because of its light weight, thermoplastic piping does little to reduce airborne sound. An appropriate thickness of insulation must be used to reduce noise.

Thermal Expansion

The amount of movement resulting from thermal effects is relatively high, thus requiring special attention to installation. As a general rule, runs in excess of 20 ft should be checked for the necessity of expansion offsets.

Corrosion Resistance

Corrosion occurs in two ways, as chemical and stress corrosion. There are two general types of chemical attack on plastic piping. The first is called solavation, which is the solubility, or absorption, of chemicals into the piping material from the fluids inside the pipe. This causes swelling and softening. The second type of attack occurs where the polymer or base resin molecules are somehow changed by a chemical agent, and the original properties of the plastic pipe cannot be restored upon removal of that chemical. Stress (or strain) corrosion weakens the pipe due to constant and repetitive movement and/or pressure surges.

The chemical resistance of the various types of plastics varies greatly not only among different types of plastic, but among different grades of the same type of plastic. Achieving full resistance is a function of the resistance of the compound used to make the pipe and the processing of the plastic.

The factors that determine the suitability and service life of any specific plastic pipe are

1. The specific chemicals and their concentrations
2. The jointing method
3. Dimensions of the pipe and fittings
4. Pressure inside the pipe
5. Ambient temperature and temperature of the fluid
6. Period of contact
7. Stress concentrations in the pipe and fittings

Abrasion

If a material such as sand, gravel, or slurry is transported in the piping system, or frequent cleaning with mechanical equipment is anticipated, resistance to abrasion should be investigated. Mechanical cleaning equipment manufacturers have data available from tests on various piping materials. Pipe manufacturers are also a source of information regarding various effluents. Additives can be used to increase the abrasion resistance of any pipe.

Biological Resistance

Very few types of plastic pipe can be degraded and/or deteriorated by the action of micro- or macroorganisms. For the most part, plastic pipe shows negligible or no susceptibility to bacterial attack. Refer to manufacturers for specific data.

Electrical Properties

Because plastic pipe is nonconductive, electrostatic charging of a pipeline is possible if dry, electrically nonconductive material is transported. All pipe materials with a specific resistance of 106 ohms per centimeter (Ω/cm) are considered nonconductive. Plastic pipe is generally not recommended to carry ignitable mixtures or electrically nonconductive dry substances due to potential electrostatic charging and possible damage to the dry material inside the pipe.

Static electrical charges can be prevented by providing the pipe with a conductive coating of metallic powder or lagging the pipe with metallic foil that has been grounded.

Permeability

Permeation is a process where fluids pass either into or out of a piping system through the walls of a pipe. Permeation can occur through the walls of a susceptible plastic pipe, through gaskets or other jointing material, and through defects or inappropriately or incorrectly sealed pipe.

Organic matter that migrates from soil through the plastic pipe is called permeate, and the process is called permeation. Until additional scientific work is completed on permeation through plastic pipe, it is not recommended that plastic pipe be used to carry potable water in areas of contaminated soil.

Leaching

Leaching is a process where substances sometimes called extractables are released from the walls of the pipe material into the fluid, but not through the pipe walls. The most common extractables are inorganic chemicals and volatile organic compounds (VOCs).

Tests have shown that the rate of leaching from plastics in contact with high purity water usually decreases with time. The time it takes for any specific plastic to reach a steady state after being subject to immersion in the fluid (elution) in dynamic systems is a function of the water temperature and velocity. Experience has shown that leach-out in the first five days is considered a burn-out period. The release of VOCs from various plastic pipes in contact with high purity water at 74°F is shown in Fig. 2.1. Figure 2.2 shows calculated lengths of thermoplastics that will increase total organic compound (TQC) level of high purity water by 1 ppt when used as a transfer medium (assuming 4-in pipe and a water velocity of 6 ft/s).

Creep

When a load is continuously applied on a plastic material, it creates an instantaneous initial deformation that further increases at a decreasing rate. This further deformation is called creep. If the load is removed at any time, there is partial immediate recovery followed by a gradual creep recovery. If, however, the plastic is deformed (strained) to a given value that is maintained, the initial load (stress) created by the deformation slowly decreases at a decreasing rate. This is known as the stress relaxation response. The ratio of the actual values of stress to strain for

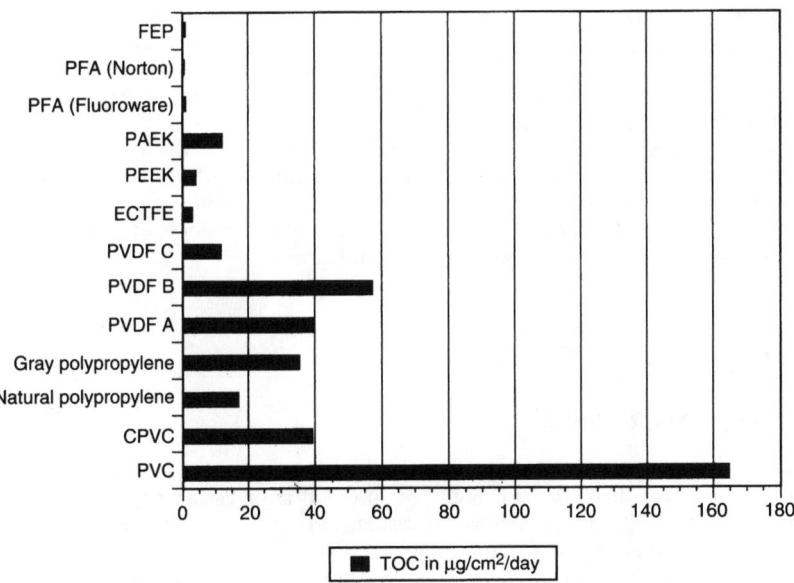

FIGURE 2.1   Release of VOCs from plastic pipe.

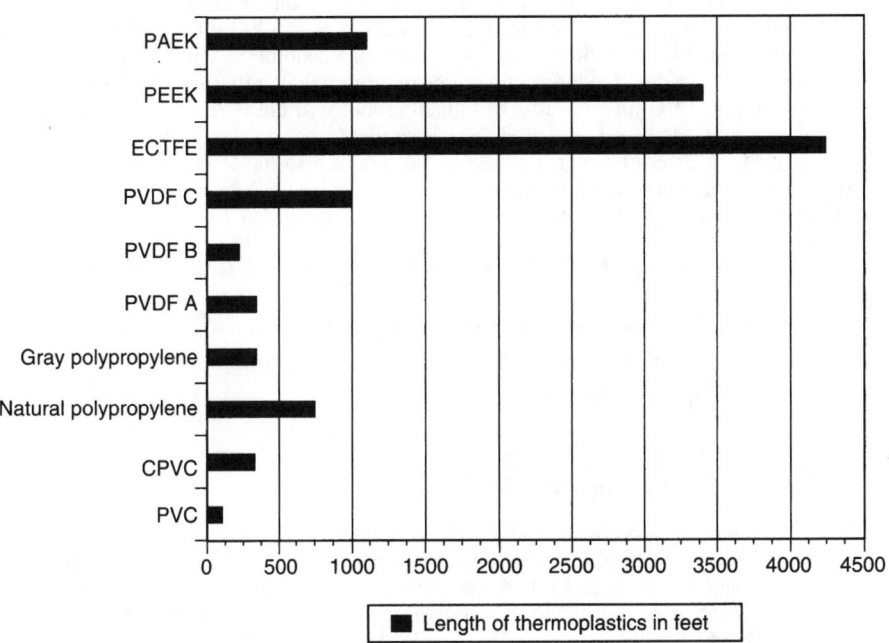

FIGURE 2.2   Leaching from 4-in plastic piping.

a specific time under continuous stressing or straining is commonly referred to as the effective creep modulus, or the effective stress-relaxation modulus. This modulus is significantly influenced by time. Most pipe manufacturers are willing to provide values of effective moduli for specific materials and loading conditions. Experience has shown that all plastic pipe will creep.

The properties of plastic pipe are influenced by time of loading, temperature, and environment. Therefore, standard data sheet values for mechanical properties may not be satisfactory for design purposes. The stress-strain responses of plastic reflect its viscoelastic nature. The viscous, or fluidlike, component tends to dampen or slow down the response between stress and strain.

PLASTIC PIPE MATERIALS

Polyvinyl Chloride (PVC)

PVC is the strongest and the most widely used plastic pipe. It has very poor resistance to solvents. It is used for both pressure and nonpressure applications and is approved by the NSF for potable water systems. Principal uses are for potable water, other fluids, and all types of drainage applications within chemical and temperature limits. A generally accepted upper service temperature limit is 150°F.

Pipe is made only from compounds containing no plasticizers and a minimum amount of other materials. Piping is referred to as rigid in the United States and unplasticized in Europe. Neither designation is required today if ASTM classifications are used. Pipe is available in schedule 40 and 80 and in diameters up to 20 in (500 mm). Fittings for water piping are available in diameters up to 6 in (150 mm). Drainage fittings are made in diameters up to 8 in (200 mm). Schedule 80 pipe can be joined by threading, solvent cement, or elastomeric seals. Schedule 40 pipe cannot be threaded. Underground water-main pipe generally uses push-on gaskets for jointing. PVC pipe is self-extinguishing and will not burn. However, when subject to conditions of a fire, a toxic gas is emitted.

Pipe used for exterior water mains may be required to have special approval if used to supply fire protection water.

PVC pipe should conform to the following ASTM standards:

1. D-2241, D-1785, and D-272: interior water pipe pressure. Fittings shall conform to D-2665.
2. Pipe used for exterior water mains shall conform to AWWA standards.
3. D-2665, D-3034, F-679, and F-769: drainage pipe and fittings.
4. CS 272: PVC plastic DWV pipe and fittings.

Chlorinated Polyvinyl Chloride (CPVC)

CPVC is a chemical modification of PVC, with an extra chlorine atom in its structure that extends its service temperature limitations to about 200°F, 50°F more than PVC pipe. Available sizes and other properties are very similar.

CPVC pipe should conform to the following ASTM standards: ASTM D-2846, F-441, and F-442 for piping and F-437, F-438, and F-439 for fittings.

Polypropylene (PP)

PP is a polyolefin available in two types, type 1, homopolymer, and type 2, copolymer. It has superior resistance to sulfur-bearing compounds and is capable of withstanding a wide range of both corrosive and sanitary effluents. It is the most resistant to organic solvents of all the common plastic pipe materials, and is only slightly less rigid than PVC. Its principal use is for laboratory drainage purposes. It is superior to PE at higher temperatures. When manufactured with no pigment, it is widely used for pure water system distribution.

PP is available in schedule 40 and 80 and in diameters up to 12 in (300 mm). Joining methods include solvent-welded socket joints, heat-fused socket-welded joints, heat-fused butt-welded joints, threaded joints, or proprietary mechanically threaded joints. Only schedule 80 can be threaded.

PP pipe does not have any ASTM standards for dimensions or wall thickness of pipe or fittings. Each manufacturer provides exact dimensions. Some joints refer to D2657 for heat-fused socket joints and ASTM 2.1 for threaded joints.

Polyethylene (PE)

PE is probably the best known polyolefin. It is widely used because of its superior toughness, ductility, flexibility, and ability to dampen water shock. The upper service temperature limit is around 140°F. It is used for underground fuel gas distribution piping and site irrigation systems and it has NSF approval for potable water piping. It is also used for all types of gravity and pressure drainage applications within chemical and temperature limits where flexibility is an advantage, particularly in foundation drains. It has the least mechanical strength of all the common plastic pipe materials, and is only moderately resistant to solvents.

There are four types of PE manufactured, rated by the density of the natural resin used, types 1, 2, 3, and 4. Types 3 and 4 have the highest density and have been given the abbreviation HDPE. The most commonly used pipe is composed of types 3 and 4. It is available in schedule 80 in diameters up to 6 in (150 mm). It is joined by socket and butt heat-fusion and insert fittings for water irrigation systems.

PE pipe should conform to the following ASTM standards:

1. D-2513: underground fuel gas pipe, D-2683 and D-2513: fittings
2. D-2239, D-2447, and CS-255: pressurized water piping
3. D-2609: fittings
4. D-3350: PE pipe material

Acrylonitrile Butadiene Styrene (ABS)

ABS is used for interior and exterior sanitary and industrial drainage and vent systems and pressurized liquid lines such as those used for salt water, irrigation, crude oil, and pumped waste. NSF has approved this material for potable water systems. It is slightly more rigid than PVC and is the least resistant to solvents of all the popular plastic materials. The generally accepted upper service temperature limit is 175°F. ABS is available in schedule 40 and 80, and in diameters up to 12 in (300 mm). Drainage fittings are available only in diameters up to 6 in (150 mm). Joints are made by solvent cement, threaded, or compression joints. Only schedule 80 can be threaded.

ABS pipe should conform to the following ASTM standards:

1. D-2751 and F-628: sewer pipe and fittings
2. D-2661 and D-275 1: pressurized water pipe
3. ASTM D-2661: fittings

Polybutylene (PB)

PB is a polyolefin having slightly less stiffness than regular PE, but with a higher strength than HDPE. It has good abrasion resistance, is resistant to creep, and retains its strength at elevated temperatures. This material is resistant to soaps, most acids and bases, and polar solvents at lower temperatures. It is attacked by some aromatic hydrocarbons and chlorinated solvents. PB pipe is used mainly for water and, to a lesser degree, for pressurized liquids, gases, slurry lines, and chemical waste lines. Its maximum temperature is around 200°E

Pressure piping is available in coils in sizes up to 2 in (50 mm) or in straight lengths in sizes up to 18 in (450 mm). Joining methods include butt and socket heat-fusion, insert fittings with hose clamps, flared joints, and compression fittings with sleeve inserts.

PB pipe should conform to the following ASTM standards:

1. D-2666: coiled tubing, F-845: fittings
2. D-2662 and D-3000: pipe and fittings
3. F-809: large diameter pipe and fittings

Fluoroplastics

Fluoroplastics are a broad class of thermoplastics that have some or all of their hydrogen replaced by fluorine. Their useful temperature range is from −100 to 300°F.

Polytetrafluoroethylene (PTFE) piping is a fully fluorinated fluorocarbon polymer. This family of piping material is known as "Teflon," and includes PFA and FEP. It is resistant to fats and oils, fuels, aromatic compounds, amines, ketones, and esters. It is attacked by sodium, potassium, and fluorine gas. It is used for standard and high purity pressurized water distribution systems, drainage systems for many chemical effluents, and liquid chemical systems. Joining methods include butt and socket heat fusion and threaded and triclamp fittings for ultrapure water systems. The piping is pressure rated and manufactured in sizes up to 4 in, and tubing up to 1 in. The initial cost of this class of pipe is among the highest of all the plastic pipe materials.

Polyvinylidene fluoride (PVDF) piping is a thermoplastic partially fluorinated fluorocarbon polymer that is resistant to most inorganic acids and bases, aliphatic and aromatic hydrocarbons, halogens (except fluorine), alcohol, and halogenated solvents. It is attacked by alkalis, ketones, and some amines, The PVDF family of piping includes ETFE, CTFE, ECTFE, and Kynar. It is manufactured in several grades for specific uses. It is used for standard and high-purity pressurized water distribution systems, drainage systems for many chemical effluents, and liquid chemical systems. Joining methods include butt and socket heat fusion and threaded and triclamp fittings for ultrapure water systems. The piping is pressure rated and manufactured in sizes up to 10 in. The initial cost of this pipe is among the highest of all the plastic pipe materials.

Fluoroplastic pipe does not have any dimensional standards except those established by individual manufacturers. The piping shall be composed of materials and tested by the manufacturer as specified by the ASTM. The pressure rating is referenced to the schedule rating. Heat-fused socket joints shall be referenced to ASTM D-2657.

Reinforced Thermosetting Resin Pipe (RTHP)

RTHP is a class of composite pipe that consists of a reinforcement either imbedded in, or surrounded by, cured thermosetting resin. The most common reinforcement is fiberglass. This type of pipe is known as fiberglass reinforced pipe (FRP). Reinforcement can be composed of any mineral fiber. RTHP is also available in a variety of resins, liners, and wall construction. It has a high strength-to-weight ratio. The generally accepted maximum temperature rating is 250°F.

Pipe is available in sizes from 1 to 48 in and is used for pressure and gravity drainage purposes. Because of its construction, the piping is capable of withstanding much higher temperatures than most plastic pipe, and is much stronger both physically and mechanically while being resistant to a wide variety of chemicals. Joints are bell and spigot secured with adhesive, threaded and secured with adhesive, butt and wrapped, and mechanical joints. Adhesives selected depend on the specific service intended and type of pipe.

The resin is the binder that holds the composite structure together. It supplies the source of temperature and chemical resistance. There are four resin types: epoxy, polyester, vinyl ester, and furans. Furans are difficult to work with and are rarely used. Epoxy and vinyl ester are the most widely used. Epoxy resin is stronger than vinyl ester. The curing agent (or catalyst) also has an effect on the chemical resistance of the pipe. Aromatic amine-cured epoxy has better chemical resistance than polyamide-cured epoxy.

The various combinations of materials are resistant to a great variety of chemicals and suitable for many services. Consult manufacturers' information to select the appropriate pipe material combination.

RTHP pipe should conform to the following ASTM standards:

1. D-3262: sewer pipe, D-3840 and D-4160: fittings
2. D-3754 and D-3517: pressure pipe
3. D-2996: filament wound pipe
4. D-299: centrifugally cast pipe

OTHER PIPING MATERIALS

ASBESTOS CEMENT PIPE (ACP)

Asbestos cement pipe is manufactured by mixing portland cement (with or without silica) and asbestos fiber. This pipe is widely used for pressurized systems such as potable water mains and noncorrosive, nonpressurized sewer systems. Piping is available in sizes ranging from 4 to 42 in (100 to 1000 mm), and in pressure ratings ranging from nonpressure (gravity) pipe to 100-, 150-, and 200-psi pressure-rated pipe and fittings. Gravity pipe is made in five strength classes, each denoting the three edge-bearing strengths regardless of size. These strength classes are 1500, 2400, 3300, 4000, and 5000 lb per linear foot.

Joints are made using compression gasket joints or asbestos-cement/rubber-gasket joints.

Asbestos cement pipe shall conform to the following standards:

1. AWWA C-400, 401, 402, and 403: pressure pipe
2. ASTM C-428: gravity pipe

GLASS (GL)

Glass pipe is made from a low expansion borosilicate glass with a low alkali content. It is used for laboratory gravity waste service and is available in sizes up to 6 in. The pipe is joined by an elastomeric compression gasket secured by a sleeve. Glass pipe conforms to the following standards: ASTM C 599 and Federal Specification DD-G-541B.

This pipe is primarily used for drainage of various corrosive liquids. It is considered very brittle, and should be used where it has some measure of protection, such as in pipe spaces or behind laboratory furniture.

VITRIFIED CLAY (VC)

Vitrified clay pipe is made from selected clay and shale mixed together with water, formed into pipe, and fired (baked) at 2000°F. At this temperature the material fuses together (vitrifies) to form a homogeneous mass.

VC pipe is suitable for gravity underground service only. Sizes range from 3 to 54 in. The pipe is made in two grades—standard and extra strength. It is available only in bell and spigot ends, with compression gaskets used for joint assembly. A new development is a pipe with the joints flush with the pipe exterior, thus gaining superior strength to resist jacking and external damage as the pipe is pushed through the soil. This pipe has not yet received approval from all authorities and its use has been limited.

Vitrified clay pipe shall conform to the following standards:

1. ASTM C 4
2. ASTM C 700
3. ASTM C 1208: Vitrified clay pipe and joints for use in jacking, sliplining, and tunnels

Its primary advantage is the ability to accept a broad range of effluent with a diverse chemical and acid content, especially where the generation of hydrogen sulfide gas could result in the formation of sulfuric acid in the piping system. It has excellent resistance to scour and corrosion from soil. The disadvantages are that the pipe is very brittle and has a high initial cost due to the intensive labor involved in the installation.

CONCRETE PIPE

Concrete pipe is manufactured from either nonreinforced or reinforced concrete. Concrete pipe is suitable for underground (clear water) gravity drainage systems and pressurized water service, typically for nonpotable cooling or process water from surface sources. Sizes for concrete pipe range from 4 to 36 in. Reinforced concrete pipe is available in 12- to 144-in sizes. Concrete pipe is manufactured in two strengths, standard and extra strength. Reinforced pipe is manufactured in five pressure ratings, classes I-V, increasing in strength from low to high. Concrete pipe is not manufactured to nationally accepted dimensional standards. Refer to individual manufacturers.

Joints are tongue and groove or bell and spigot. Tongue and groove joints are sealed with rubber gaskets and are generally used for gravity systems. Bell and spigot joints use compression gaskets and are generally used for pressurized systems. There is no standard for fitting dimensions. When dimensions are required, they must be obtained from the proposed manufacturer.

Concrete pipe shall conform to the following standards:

1. ASTM C 76: Reinforced Concrete Culvert, Storm Drain and Sewer Pipe
2. ASTM C 14: Concrete Culvert, Storm Drain and Sewer Pipe
3. ASTM C 443: Joints for Circular Concrete Sewer and Culvert Pipe Using Rubber Gaskets
4. ASTM C 361: Reinforced Concrete Low Head Concrete Pressure Pipe
5. AWWA C 301: Prestressed Concrete Cylinder Pipe and Fittings

FITTINGS

Fittings are used to connect pipes to one another and to change the direction of straight runs of pipe. An alternative to fittings is bending the pipe to allow changes in direction.

CODE CONSIDERATIONS

Codes require that any change in direction of plumbing piping in a drainage system be made with fittings. The fittings shall have long radius bends so as not to allow solids any place to accumulate and to provide good hydraulic flow characteristics. These are known as drainage pattern fittings, and they are required to be used for drainage systems. Threaded drainage fittings shall be of the recessed type. Drainage pattern fittings are not required to be used for vent piping. Pressure systems can use any fitting type.

FITTING TYPES

A small number of fittings used either singly or in combination with one another are the most commonly used.

1. Bends or sweeps are used to change the direction of a pipe. These fittings come in various angles. A ¼ bend is a 90° fitting, and comes either as a short or long sweep (or radius). A ⅛ bend is a 45° angle and a ¹⁄₁₆ bend is a 22.5° angle.
2. The wye is used to connect a branch pipe into a straight run of piping at a 45° angle. These fittings are available with either all similar sizes or various combinations of pipe sizes in any connection.
3. The tee fitting is used to connect a branch pipe into a straight run of piping at a right angle, where the flow characteristic of liquid is not a factor. These fittings are available with either all similar sizes or various combinations of pipe sizes in any combination.
4. The elbow is a 90° change of direction with a very short radius. They are available with either all similar sizes or combinations of pipe sizes.

Refer to Fig. 2.3 for an illustration of these four basic types of fittings. In addition to those discussed here, there is a large variety of special combination drainage fittings manufactured for use either where space considerations are of primary importance or where combination fittings will result in considerable savings in time and labor.

FITTING MATERIALS

Cast Iron

Fittings of this material can be used with CI or steel pipe. They are available with screwed, hub and spigot, no hub, or flanged ends. These fittings should conform to the following standards:

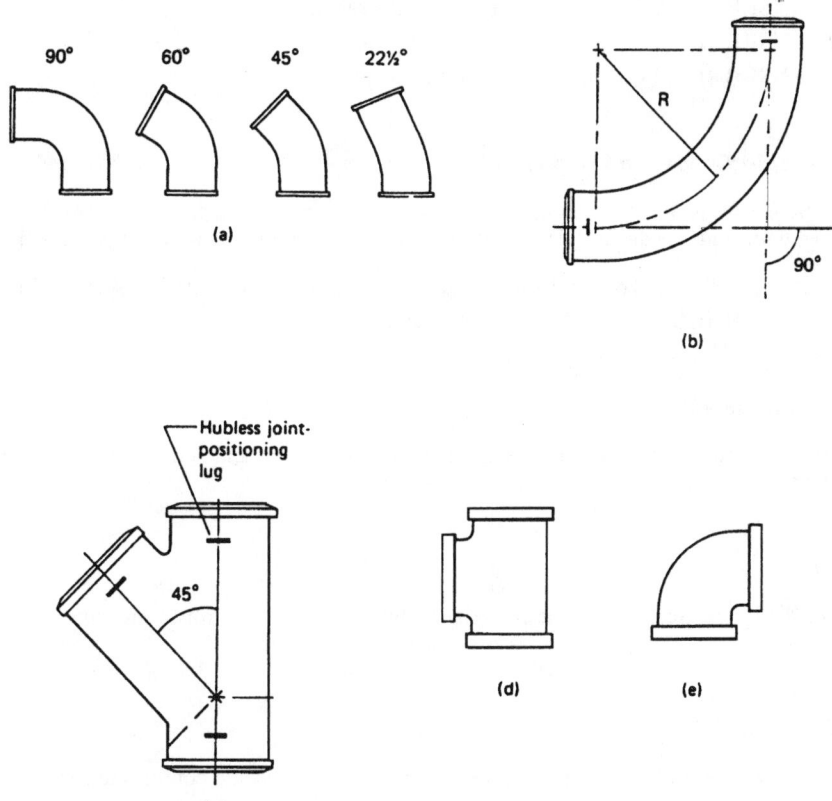

FIGURE 2.3 Typical fitting shapes. (*a*) 90°, 60°, 45°, and 22½° bends; (*b*) 90° sweep; (*c*) 45° sanitary wye; (*d*) standard tee; (*e*) elbow.

1. ANSI B 16.12: cast iron drainage fittings, threaded

2. ASTM A 74: cast-iron drainage fittings, hub and spigot

3. CISPI 310: cast-iron soil pipe and fittings, hubless

Malleable Iron

Fittings of this material can be used with ST or CI pipe. They are available with screwed ends. These fittings should conform to ANSI B 16.3: Malleable Iron Screwed Fittings.

Cast Copper Alloy

Fittings of this material can be used with any copper or brass pipe. They are available with solder ends only. These fittings should conform to the following standards:

1. ANSI B 16.18: Cast Copper Solder Joint Pressure Fittings
2. ANSI B 16.23: Solder Joint Drainage Fittings
3. ANSI B 16.26: Cast Copper Alloy Flared Fittings

Wrought Copper and Copper Alloy

Fittings of this material can be used with any copper or brass pipe. They are available with solder ends only. These fittings should conform to the following standards:

1. ANSI B 16.22: Wrought Copper and Copper Alloy Solder Joint Pressure Fittings
2. ANSI B 16.29: Solder Joint DWV Drainage Fittings

Acid-Resistant CI

Acid-resistant fittings are of the same material as the pipe and conform to the same standards.

Glass

Glass fittings are of the same material as the pipe and conform to the same standards.

Stainless Steel

Stainless steel fittings are of the same material as the pipe and conform to the same standards.

Bronze

Cast bronze fittings shall conform to ANSI/ASME B 16.15 and are available in class 150 and class 250 threaded ends. Flanges conforming to ANSI B 16.24 are available in sizes 2½ in (65 mm) and larger.

Dielectric

Dielectric fittings are actually adapters used to connect any pipe containing copper with any pipe containing iron to prevent the galvanic action between these dissimilar metals from causing corrosion failure. The fitting itself is constructed so that insulation prevents either connecting pipe from coming into direct contact with the other.

Plastic Fittings

Plastic fittings can only be used with the individual plastic pipe materials from which they are made. Nationally recognized standards are not available for many plastic fitting materials and should be obtained from individual manufacturers. Those that are available shall conform to the following standards:

1. ASTM D-2665: PVC Fittings
2. ASTM D-2852: Styrene Fittings
3. ASTM D-2661: ABS Fittings
4. ASTM D-2609: PE Fittings

Unions

Unions are fittings used when it is necessary to easily separate piping, usually for the purpose of removing a piece of equipment (such as a pump) without having to take apart a joint. A union is a screwed fitting. These fittings should conform to the following standards:

1. ANSI B 16.39: Unions, Steel or Malleable Iron
2. ANSI B 16.41: Unions, Brass or Bronze

JOINTS

A joint is required each time it is necessary for piping to be connected either to itself, a fitting, or a piece of equipment. It must be able to withstand the greatest possible pressure exerted upon it by either the design or test pressure of the individual system. Piping systems use both pressure and nonpressure piping. The same joints may be utilized for similar piping on any system.

Most plumbing codes refer to recognized standards that govern the methods and materials used to fabricate each type of joint. The selection of the jointing methods will be determined by the type of pipe used, the type of fittings available for the pipe, the highest pressure expected in the system, and the possible need for disassembly.

METALLIC PIPE JOINTS

Caulked

The caulked joint is used only for piping with hub and spigot ends. After the spigot end is placed inside the hub, a rope of oakum or hemp is packed into the annular space around the spigot end, until it reaches 1 in from the top. (For AR pipe, the packing is hydrous magnesium aluminum silicate reinforced with fiberglass.) Then, a 1-in thickness of molten lead is poured into the annular space on top of the rope. The lead is then pounded further into the joint by a caulking iron until it is at a level $\frac{1}{8}$ in below the rim of the hub. In use, the hemp or oakum swells when it absorbs water and further increases the joint's ability to resist leaking.

There are no widely recognized standards that govern the fabrication of caulked joints. Requirements controlling installation are often provided in plumbing codes.

This rigid, nonpressure joint is suitable for all above-ground and underground drainage installations. Because they are very labor intensive, gasketed and coupling joints have replaced caulked joints for most CI joint applications. Refer to Fig. 2.4 for an illustration of a caulked joint.

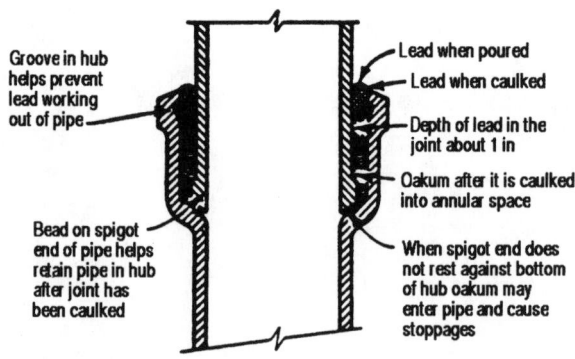

FIGURE 2.4   Caulked joint.

Compression Couplings

Often referred to as a sealing sleeve, this joint is used to connect hubless CI, AR, and GL piping. The same sleeve assembly is used for CI pipe with either plain ends (cut) or factory ends. A separate assembly is required for the two different end types of GL pipe.

Compression couplings depend on the friction between the sleeve and the pipe exterior for sealing and resistance to being pulled apart. The coupling assembly consists of an inner elastomeric gasket and an outer metallic sleeve with an integral bolt used for tightening. The two ends of the pipe to be joined are butted together, the entire sleeve assembly is placed over both ends, and the bolt is securely tightened according to the torque requirements established by the manufacturer.

Standards governing the fabrication of this joint are: ASTM C 564 and CISPI 310 for cast iron pipe.

The compression coupling is a rigid, nonpressure joint preferred for aboveground installations because of its ease of assembly and strength. Underground, the metallic sleeve often fails after years of service due to corrosion of the bolts by surrounding soil or fill. Dresser couplings are accepted for use in pressurized fuel gas systems.

Refer to Fig. 2.5 for an illustration of a compression coupling.

Screwed

A screwed joint can be used for any plain end pipe that has the necessary wall strength to have threads cut into it. A screwed joint requires that threads be cut on the outside of a pipe (male threads) and the inside of a fitting (female threads). The threads used for pipes are known as American Tapered Pipe Threads (APT), and may be referred to as internal (APTI) or external (APTE). The joint is assembled by placing the male thread inside the female thread and turning the fitting or pipe until the joint is tight. The male thread is sealed prior to assembly by pipe joint compound or Teflon tape in order to prevent water from seeping past the threads. It is important to clear the pipe of any burrs or chips resulting from the cutting process.

The standard governing the fabrication of this joint for service and utility systems is ANSI B 2.1: American Standard Tapered Pipe Threads.

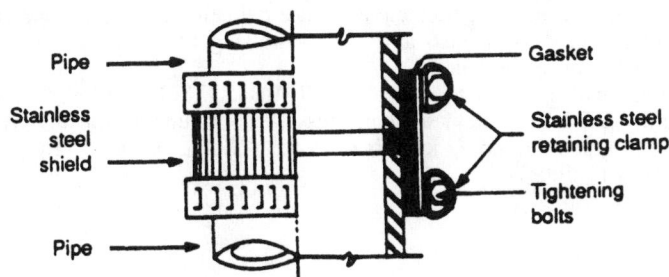

FIGURE 2.5   Compression coupling joint.

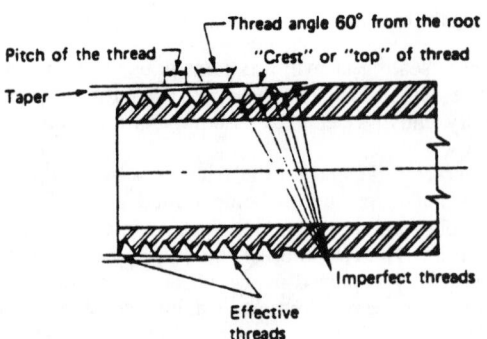

FIGURE 2.6   Tapered thread screwed joint.

This type of joint is inexpensive and easy to fabricate. It is generally restricted to piping 3 in or smaller, because of the great effort required to turn larger sized pipe when making up the joint.

Refer to Fig. 2.6 for an illustration of a tapered thread screwed joint.

Soldered

Also known as a "sweat joint," soldering is a process used to join metallic pipe below its melting point using an alloy that produces a metal solvent action between pipe and solder at a relatively low temperature.

Solder is an alloy that melts at a temperature of 840°F or less. Nonpotable water lines can use a solder of 50-50 tin-lead. The Safe Drinking Water Act banned the use of lead in solder and flux, requiring the use of lead-free solder. A commonly used lead-free filler alloy is tin and antimony (95% tin, 5% antimony) and other filler metals, often with proprietary alloys. Solder comes in the form of a flexible roll approximately ⅛ in in diameter. The joint is fabricated by placing flux on the clean male pipe and inserting it fully into a clean socket end on a fitting. The assembly is heated enough to melt the solder. The solder is then applied completely around the perimeter. Capillary action draws the solder throughout the entire joint. When the solder cools, it adheres to the walls of both the pipe and fitting, creating a leakproof joint.

Flux is a material with a consistency of paste, spread over the entire pipe end and used to remove an oxide film and eliminate oxidation during the heat generated by the jointing process. It also aids in the drawing of solder into the joint by producing a "wetting" action.

Soldering shall conform to standard ASTM-B8Z8. Solder alloy shall conform to standard ASTM B 32. It is a rigid, pressure joint suitable for any type of installation for which the piping itself is acceptable. Refer to Fig. 2.7 for an illustration of a solder joint.

Brazed

Also known as a "silver solder," brazed joints are used to join plain end copper or copper alloy pipe with solder end fittings. They are also used to join aluminum

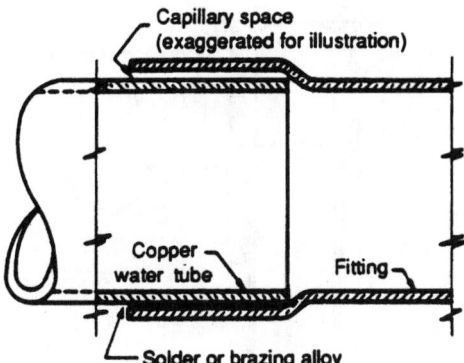

FIGURE 2.7   Soldered and brazed joint.

piping with special filler metals. The brazed joint is far stronger than the soldered joint and produces joints in copper tube stronger than the pipe itself.

Similar to a solder joint, brazing alloy is available in a variety of compositions that melt at a temperature of 850°F or more. It is commonly used for medical and specialty gas applications and other systems that require a high working pressure when using copper tube or copper alloy piping.

Brazing alloys are available in two classes. Those containing 30 to 60% silver are in the BAG class, those with copper alloy and phosphorus fall in the BCuP class. Flux is also required for some alloy metals but not all. When joints are made up for medical gas and other pipelines that require a high degree of cleanliness, the use of flux is prohibited. This requires the use of appropriate BCuP alloys and wrought copper fittings containing phosphorus normally provided in the flux.

Brazing alloys shall conform to ANSI/AWS A 5.8. Refer to Fig. 2.7 for an illustration of a brazed joint.

Flared

The flared joint, illustrated in Fig. 2.8, is a rigid, pressure joint used only with soft (annealed) piping. It is made by first placing a loose, threaded coupling nut on one end of the pipe, then cold-forming that end with a mandrel that enlarges the pipe end to fit a mating end on a threaded coupling shank. The screwed coupling nut and shank are then turned in opposite directions drawing the pipe together to form a leakproof seal.

Compression Gasket

Sometimes erroneously referred to as a mechanical type of joint, compression gaskets are used to join CI soil pipe, DI and FRP sewer and pressure pipe, and VC pipe. Each pipe requires a specific type of end suitable for the individual joint type.

Although the shape of the gasket differs according the application, the fabrication is the same. The gasket is a ring or sleeve of elastomeric material of the required shape to fit the bell end of the specific pipe. First, the gasket is inserted

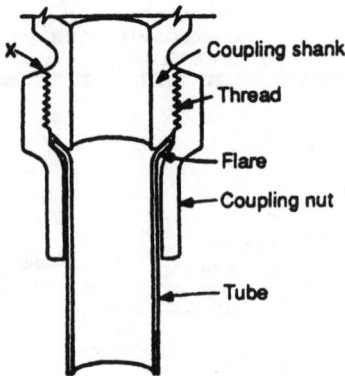

FIGURE 2.8 Flared joint.

into the bell end of the fitting. An approved lubricant is spread on the male pipe which is then inserted into the fitting by use of a mechanical assist. The barrel of the pipe compresses the gasket forming a leakproof joint.

Standards governing the fabrication of this joint are:

1. ASTM C 564: metallic pipe
2. ASTM D 3212: plastic pipe
3. CISPI HSN: gaskets for hubless cast iron soil pipe

This flexible joint allows some deflection of the pipe' during installation. For soil pipe, it is not considered a pressure fitting, but when used on DI pipe, it is acceptable for a pressurized system. It is well suited for both above-ground and underground installations, and has replaced caulked joints for underground piping because of its ease of fabrication and the inert nature of the gasket that prevents corrosive attack. Refer to Fig. 2.9 for an illustration of a gasketed joint.

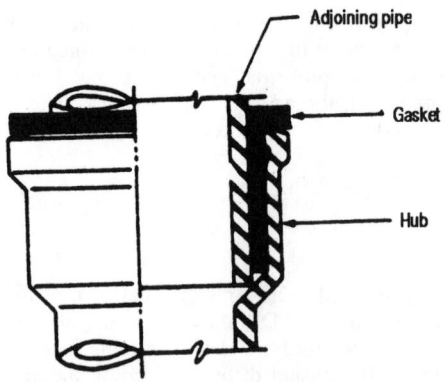

FIGURE 2.9 Compression gasket joint.

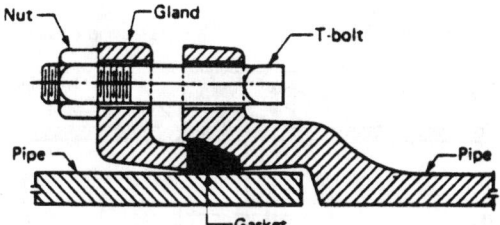

FIGURE 2.10   Typical mechanical joint.

Mechanical Joint

The mechanical joint is used for DI pressure or drainage pipe. There are many different kinds of joints. All mechanical joints, no matter what shape, are similar in principle. Refer to Fig. 2.10 for an illustration of a typical mechanical joint.

Grooved

There are two pipe-end preparations acceptable for a grooved joint: roll grooving (shoulder) and cut grooving. Roll grooves are used when the pipe is too thin to have a groove cut into it.

The coupling assembly consists of an inner elastomeric gasket and an outer split metallic sleeve with an integral bolt used for tightening. The outer sleeve has extensions at each end that fit into grooves cut or formed around the perimeter near the ends of the pipes to be joined. The two ends of the pipe are butted together. The coupling assembly is then placed over both the ends and the extensions are mated with the grooves in the pipe. The bolt is tightened to the torque requirements established by the manufacturer to form a watertight joint.

Standards governing the fabrication of this joint are:

1. AWWA C 606: couplings
2. ASTM D 735: gaskets
3. ASTM D 183: bolts

This rigid, pressure joint is well suited for both pressure and nonpressure lines. The cut grooving method is stronger but cannot be used on some pipe judged to have thin walls. Refer to Fig. 2.11 for an illustration of a grooved joint.

Wiped

The wiped joint is used only to connect lead pipe with itself or any other adapter for lead pipe used in drainage systems. The joint is assembled by first placing the ends of the lead pipe and other pipe together. Molten lead is then poured onto the joint and spread evenly, or wiped, around the pipe with a hand-held pad until the minimum thickness is reached. There shall be a minimum covered surface on each

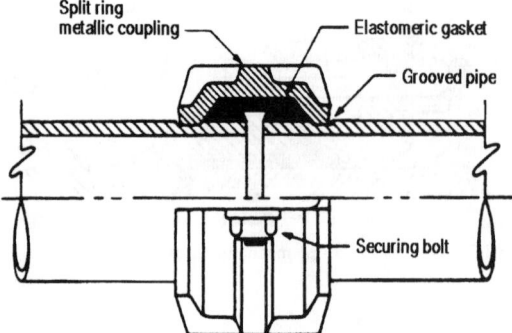

FIGURE 2.11   Grooved joint.

side of the joint of not less than ¾ in, and the added lead shall be a minimum of ⅜-in thick at the thickest part.

There are no recognized standards that govern the fabrication of the wiped joint. Requirements are usually provided in plumbing codes.

This is a rigid, nonpressure joint. Since lead pipe is rarely encountered in modern drainage systems, there are few experienced mechanics capable of making up this joint.

Burned

The burned joint is used only to connect lead pipe to itself. It is made by flaring one end of a pipe larger than the outside diameter of the other pipe to be joined. The two pipes are then brought together with the ends overlapping. Heat is applied evenly around the perimeter, melting the overlapping edges and fusing them together.

There are no recognized standards that govern the fabrication of the burned joint. This is a rigid, nonpressure joint. Since lead pipe is rarely encountered in modern drainage systems, there are few experienced mechanics capable of making up this joint.

WELDING

In the welding process, both pipe walls, at the joint, are brought to the melting point and fused together with the addition of filler metal to allow for correct wall thickness and strength. The necessary amount of heat for welding is produced by either a high temperature flame or an electric arc formed between the welding electrode and the pipe. Refer to Fig. 2.12 for illustrations of welded joints.

Because oxidation causes poor weld quality, an inert gas displacing the air, or a coated electrode that creates such a gas when heated, is used to keep air away from the molten metal. In order to properly weld pipe, the pipe ends must be specially prepared according to pipe thickness, composition, and welding method. The proper weld end preparation is critical to proper welding and must be diagrammed or described in the specifications.

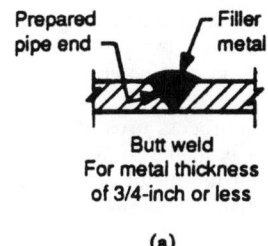

Butt weld
For metal thickness
of 3/4-inch or less

(a)

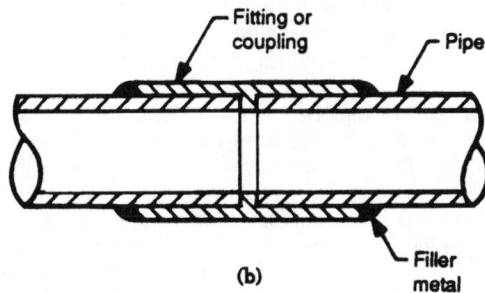

(b)

FIGURE 2.12   Typical welded joints. (*a*) Butt weld; (*b*)
socket weld.

Types of Welding

Shielded Metal Arc Welding (SMAW).   This is the standard electric arc welding.
This process uses a coated rigid metal stick electrode of the proper composition.
The stick electrode is an anode (+) and the pipe is the cathode (ground). The
electrode is used to produce the arc and is also consumed as filler metal in the
process of welding the joint. This electrode is coated with a substance that, when
heated by the electric arc, creates a shielding gas and, in addition, makes a flux for
a protective slag over the cooling weld metal of the joint. The coated electrode
forms a flux on top of the weld that must be removed before the next pass can be
made.

   This process is widely used because it is the least expensive, requires the least
clearance between obstructions, and is the easiest to manipulate the actual welding
tool in the field.

Gas Metal Arc Welding (GMAW).   This process, also known as metallic inert gas
(MIG) welding, uses a metal electrode wire of the proper composition to produce
the arc. The electrode wire also becomes the filler metal in the process of welding
the joint. Inert gas, usually argon, is piped to the weld area from a remote tank and
is delivered to the weld site through the electrode holder. This process is the next
least expensive method and produces a better quality weld faster than SMAW.

Gas Tungsten Arc Welding (GTAW).   Also known as tungsten inert gas (TIG)
welding, the electrode is long-lasting tungsten which is not consumed during the
welding process. The inert gas, usually argon, is piped to the weld area from a
remote tank through the electrode holder. Filler metal is added to the joint either

automatically through the electrode holder or separately by hand. This process is the most costly and slowest method, but produces the finest weld quality of all. This is used most commonly as an automated shop-welding process.

Oxy-fuel Torch Welding. This process uses oxygen in combination with another fuel gas (usually acetylene, but propane, MAPP, and natural gas are also used) to produce the heat needed to melt the metal of the pipe and filler metal. The filler metal is added to the joint by hand. This process is mostly used for welding pipe in the field.

Orbital Welding. Orbital welding is a welding process used to join stainless steel tubing, mainly in the pharmaceutical and food-processing industries. It is a TIG process that uses a welding rig that automatically welds the pipe and produces welds with little or no obstruction to the pipe interior.

Welded Joint Types. There are two types of joints used for welding, butt joint and socket joint. Butt welding is the term used to describe two pipes placed end to end and joined with no overlapping. Socket welding describes one pipe being placed on top of the other when only one end of the exposed pipe is actually welded to the bottom pipe. This is like a coupling, but with the joint on the outside of the pipe. In general, only pipe 2 in or less (called small bore pipe) can be socket welded.

In butt welding, the pipe ends must be separated by a fixed distance. This can be done either with short pieces of wire or a ring called an insert placed between the two ends of the pipe before the weld is started. Because incorrectly placed wire may extend into the pipe after the weld is completed, the insert should be the only method specified for welding drainage pipe. This insert melts and becomes part of the joint, and so is called a consumable insert.

Since metal that is too thick does not make a proper weld, several thin welds on a single pipe may have to be made. Each weld is called a pass. Multiple passes are usually required.

Welded joints can be made in the shop or in the field. Shop welding is the least expensive and highest quality because there is more control over the entire welding process by the use of automated welding machines. As many pieces of pipe are assembled that can fit on a truck (usually about 40 ft long and about 16 ft wide) or that will fit into the building site. The whole assembled pipe is called a spool. Doing as much work in the shop as possible will limit the number of field welds.

In general, the following welding processes are recommended for the specific types of work: TIG welding, with a consumable insert for the first pass of shop-welded joints done by machine, because it gives the smoothest interior; MIG welding, to weld the rest of the joint when done in the shop by machine; electric arc welding, the most common method used in the field to join piping, with consumable inserts used when a smooth interior is desired.

Welder Qualification

Specifications for and approval of the entire welding process are necessary for both shop welding and field welding. It is also necessary to qualify welding personnel to ensure that they have sufficient training and knowledge to produce a weld of the required quality. Qualifications of welding personnel are difficult to assess. High-temperature, high-pressure pipe is covered by ASME codes that specify the selection of successive welding type passes, filler metal composition, joint preparation, movement and handling of the pipe, tack welding and clamping, welding currents, metal deposit rates, and weld inspection. None of these code requirements apply to welded nonpressure drainage pipe. If the engineer does not have the knowledge to specify the minimum requirements for welders and the welding process, it could be left up to the contractor to determine the correct specifications for the project and recommend them to the engineer for approval. When this is done, the contractor establishes minimum criteria that will qualify any individual for welding on this particular project. It is now up to the contractor to test a welder's ability to make sound welds under the actual working conditions and using the same equipment expected to be used on the job and certify that person as being qualified. These criteria should be reviewed by the engineer for acceptability. It is common practice to use an outside, knowledgeable third party for this review process and to establish welding criteria.

Weld Testing

Defects in welded piping must be found and corrected. Defects occur because the weld does not actually create a monolithic piece of pipe. The flaws are cracks or voids in the joint. Nondestructive testing (NDT) methods are:

1. *Visual inspection of the weld.* This is the least expensive and the least informative. It is the most widely used, and coupled with hydrostatic tests of the completed system, will find most defects.

2. *Dye penetration.* This requires the use of two liquids. The first, which is a super-wetting agent, is applied to the weld, allowed to find its way into any defects, and then wiped off. The second fluid is a developer, which when applied, changes the color of the first liquid, creating a red mark where any defect occurs. This is used to check flaws that may not be visible to the eye.

3. *Magnetic testing.* First, an electrical current is passed through a pipe to induce a magnetic field in the pipe. Colored metal chips are then applied to a pipe joint. Since the magnetic field is strongest where the voids and defects occur, an accumulation of them indicates a flaw. Since stainless steel is nonmagnetic, this method is not used for lines made of this material.

4. *Ultrasonic testing.* This method uses an ultrasound generator and a screen to display reflected images from the weld. The most difficult part of this technique is to find well-trained and qualified operators. This is a costly procedure and not used for drainage lines.

5. *X-ray testing.* This technique uses an x-ray picture of the weld to disclose defects. This is the best of the various types but the most costly. Due to its cost, it is not used for drainage lines. This is the most commonly used when a record of the weld is required, but this is rarely required for drainage piping.

FLANGED

Sometimes referred to as a bolted joint, a flange is a perpendicular projection of a pipe. This projection is sufficiently long to allow holes drilled in the projection to secure to another mating surface. Often, a gasket is required to assure proper seating

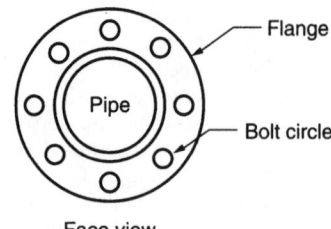

Face view

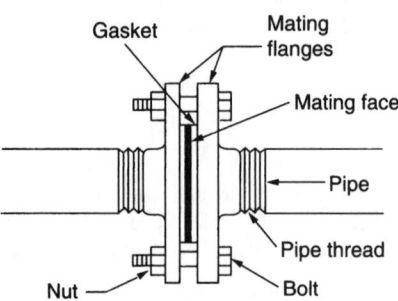

FIGURE 2.13   Flanged joint.

of the mating surfaces. Bolts through each projection secure the pipe ends together. A typical flanged joint is illustrated in Fig. 2.13.

Piping can be manufactured with flanges cast or forged to the pipe end or attached to the pipe in a separate operation. Attachment can be made by welding, threading, or brazing, depending on the pipe material. Flange dimensions shall comply with ANSI/ASME B 16.5.

Flanged joints are available in a large variety of mating surfaces (plain, serrated, grooved, seal welded, or ground and lapped for metal-to-metal contact) and in either flat-face or raised-face configurations. Gasket materials must be capable of resisting temperature, pressure, and fluid in the pipe. Bolting material is also available in various alloys and sizes.

HINGED CLAMP

The hinged clamp fitting is made by using pipe ends having small flanges that, when aligned, are secured with a hinged joint placed over both flanges and tightened to form a watertight joint. It is easily disassembled. A gasket can be used if desired for specific service requirements. This joint can be used for both pressure and nonpressure systems. There are several proprietary types of similar fittings using the same principle. A typical hinged clamp joint is illustrated in Fig. 2.14.

This type of fitting is generally used on stainless steel tubing in the food and pharmaceutical industries where a high degree of cleanliness is required; it conforms to Sanitary 3-A standards. This joint leaves no crud traps and permits piping to be easily disassembled for cleaning.

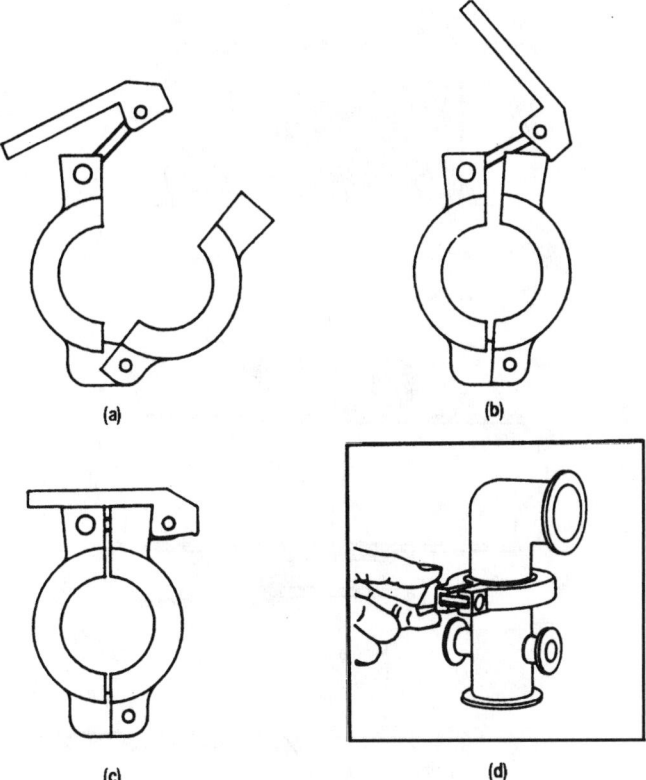

FIGURE 2.14   Hinged clamp joint. (*a*) Clamp open; (*b*) part closed; (*c*) closed; (*d*) completed joint.

Mechanically Pressed Joint

This is a proprietary joint called "Pressfit" and is used to connect schedule 5 or lighter carbon steel and stainless steel, plain end, small bore pipe, ½ in (DN15) to 2 in (DN50) in size.

Pressed joints depend on the friction between the housing and the pipe, along with an indent to resist being pulled apart. The addition of a compressed O-ring seal provides a pressure seal. Refer to Fig. 2.14.1(a) for an illustration of a pressed joint.

There are no codes or standards governing the fabrication of this joint.

Compression Sleeve Coupling

The compression sleeve coupling is a rigid, pressure-type joint often used to connect plain end fuel gas, air, oil and water piping. This is a proprietary joint also known as "Dresser" couplings, after the manufacturer of the joint.

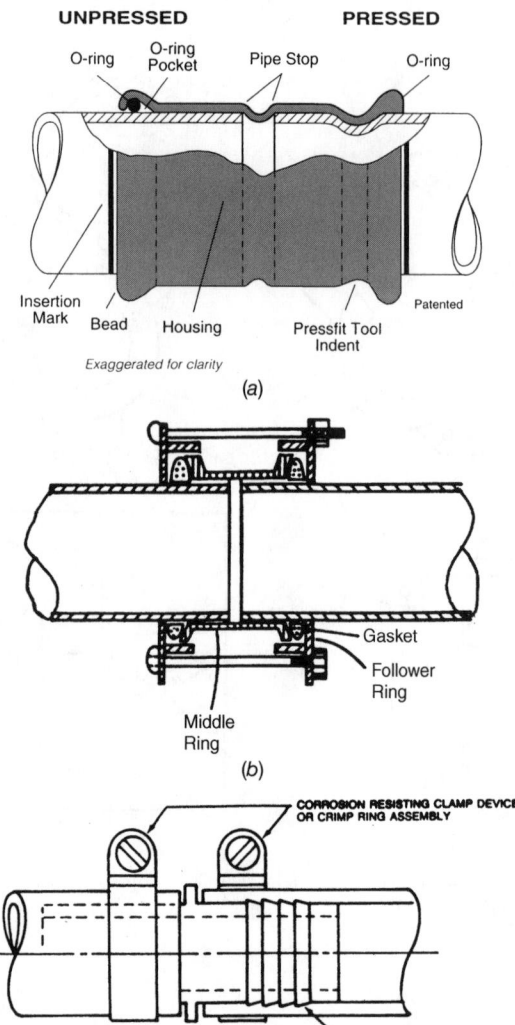

UNPRESSED **PRESSED**

O-ring O-ring Pocket Pipe Stop O-ring

Patented

Insertion Mark Bead Housing Pressfit Tool Indent

Exaggerated for clarity

(a)

Gasket

Follower Ring

Middle Ring

(b)

CORROSION RESISTING CLAMP DEVICE OR CRIMP RING ASSEMBLY

INSERT FITTING

NOTE: SOME MANUFACTURERS SUGGEST TWO CLAMPS ON EACH SIDE OF THE FITTING FOR ADDITIONAL JOINT STRENGTH.

(c)

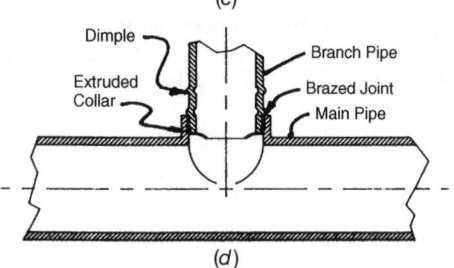

Dimple Branch Pipe

Extruded Collar Brazed Joint Main Pipe

(d)

FIGURE 2.14.1 (*a*) mechanically pressed joint; (*b*) compression sleeve coupling; (*c*) plastic insert joint; (*d*) mechanically formed joint.

This coupling uses a metallic middle ring with a gasket placed at each end. The gaskets are compressed by means of a separate metallic gland placed over the gaskets and bolted to each other across the joint.

This is a proprietary joint that has no standards governing fabrication. Refer to Fig. 2.14.1(b) for an illustration of a compression sleeve coupling

Plastic Insert Joint

The plastic insert joint is a rigid, pressure-type joint used to connect plain end, flexible plastic tubing such as PEX for potable water service.

The joint is fabricated by placing an insert into both ends of the pipe to be connected, bringing the ends together and securing the insert to each pipe end by means of a clamping device or crimp assembly. For added strength, two clamping devices on each end are often used. Refer to Fig. 2.14.1(c) for an illustration of a plastic insert joint.

Mechanically Formed Fittings

The mechanically formed joint is a rigid, pressure-type joint commonly used to provide a branch connection in copper water piping. This joint shall only be brazed. This is a proprietary joint also known as "T-Drill", after the manufacturer of the joint.

These joints are formed using a proprietary tool that first drills a pilot hole in the main pipe and upon withdrawal extrudes a collar. The branch pipe to be inserted shall be dimpled to form a depth stop preventing the branch from being inserted too far into the main pipe collar and restricting flow.

This is a proprietary joint that has no standards governing fabrication. Refer to Fig. 2.14.1(d) for an illustration of a mechanically formed joint.

ADAPTERS

Adapters are required when joining pipes with different dimensions or different joint types. Most plumbing codes require the use of approved adapters when joining two different pipe materials or piping with dissimilar joint ends.

JOINTS FOR PLASTIC PIPING

Plastic piping can be joined by a variety of methods. All of the joints except the solvent cement type are similar to those used for metallic pipes.

1. Solvent cement

2. Heat fusion
   a. Buttjoints
   b. Socket joints

3. Flanged
4. Insert-type mechanical fittings
5. Threaded
6. Threaded-type mechanical fittings
7. Grooved joints
8. Dresser type joints
9. Mechanical stab-type coupling
10. Elastomeric compression couplings
11. Adapter fittings from plastic pipe to other piping
12. FRP piping
    a. Tapered adhesive joint
    b. Butt and strap joint

Solvent Welding

Also known as solvent cementing, this process can be used only with specific types of plastic pipe and fittings with plain and socket ends. Specific solvents and cements can be used only with specific plastic pipe types.

The cement used to fabricate the joint reacts chemically with the pipe and fitting and dissolves the material it contacts. It is spread on the male section, which is then inserted in the female portion of the joint. The dissolved portions in contact with each other flow together and, when dry, are fused into a single mass, producing a leakproof joint.

There are different standards governing the solvent cement depending on the pipe used. They are:

1. No standard: PVDF
2. ASTM D 2235: ABS
3. No standard: PP
4. ASTM D 2855: PVC
5. ASTM D 2846 or ASTM F 493: CPVC

This process produces a rigid pressure joint that is suitable for any type of installation for which the piping itself is acceptable. Refer to Fig. 2.15 for an illustration of a solvent welded joint.

Heat Fusion

This joint can be used only with specific types of plain end thermoplastic plastic pipe and fittings manufactured for this purpose. The socket joint is fabricated by first placing a fitted wire with multiple loops around the outside of the plain end of the pipe. The wire will conduct electricity and has two leads, called pigtails, about 2 ft long. The pipe and wire are then placed into the socket of a fitting. These leads are connected to a carefully controlled source of electricity provided by the pipe manufacturer for this purpose. When the electricity is turned on, the wires inside the joint are heated, causing those portions of the pipe and the fittings

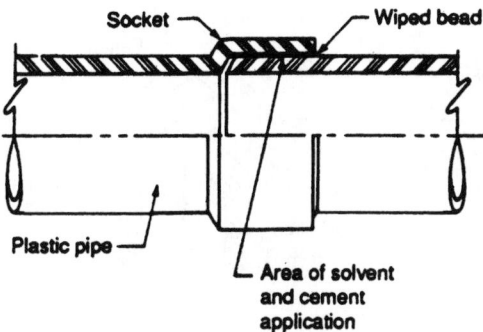

FIGURE 2.15   Solvent cement joint.

contacting each other to melt and fuse together. After the electricity is turned off, the plastic hardens, creating a watertight joint. The embedded wires become part of the joint. The leads are cut off to complete the process. The fused butt joint is made by separately heating the pipe ends to the melting point and then bringing them together. When the joint hardens a leakproof seal is formed.

There are no widely recognized standards that govern the fabrication of this joint. The manufacturer's instructions must be carefully followed.

This rigid pressure joint is suitable for all above-ground and underground installations. Refer to Fig. 2.16 for illustrations of heat-fused joints.

JOINTS FOR FRP PIPING

Tapered Adhesive Joint

This type of joint, illustrated in Fig. 2.17, is made by machine tapering the outside of the male pipe and inserting it into a matching bell-shaped female end. An adhesive spread on the male section secures the joint. A coupling with tapers at each side is used to join two lengths of tapered male pipe ends. A straight coupling (used with straight pipe ends) with no taper is also available.

Butt and Strap Joint

This type of joint, illustrated in Fig. 2.18, is made by squaring and deburring each end of the pipes to be joined. The ends are aligned and then the outside is wrapped with layers of glass mat and cloth saturated with resin. The width of the wrapping depends on the pressure rating of the pipe.

Elastomeric Compression Joint

The joint shall conform to ASTM D-3212.

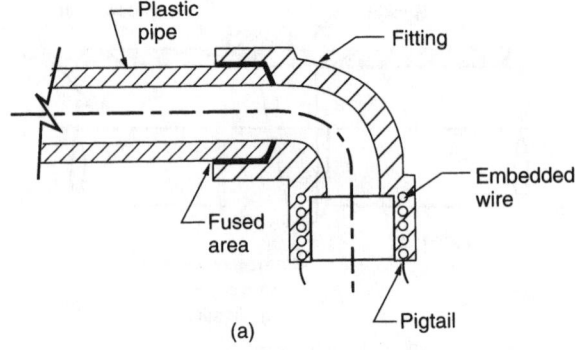

(a)

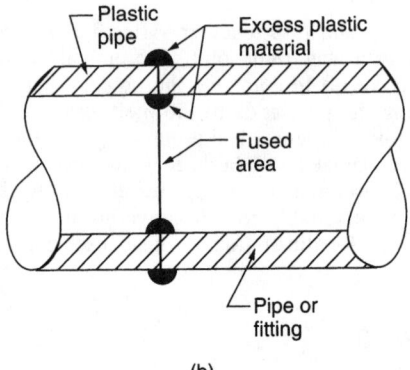

(b)

FIGURE 2.16   Heat-fused joints. (*a*) Socket-fused joint; (*b*) butt-fused joint.

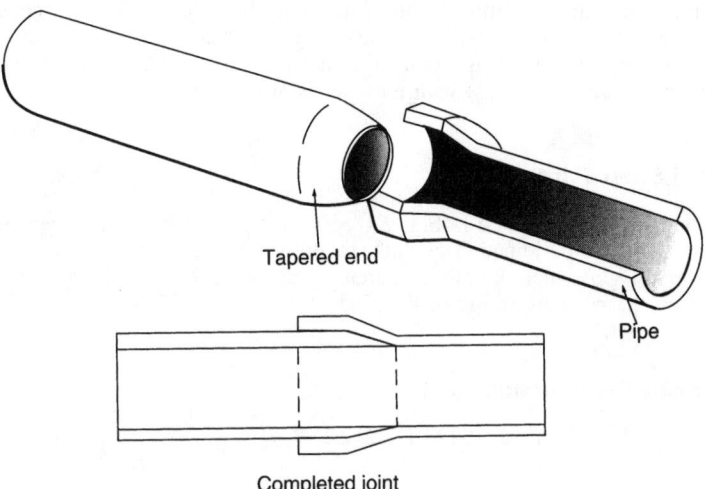

FIGURE 2.17   Tapered adhesive joint.

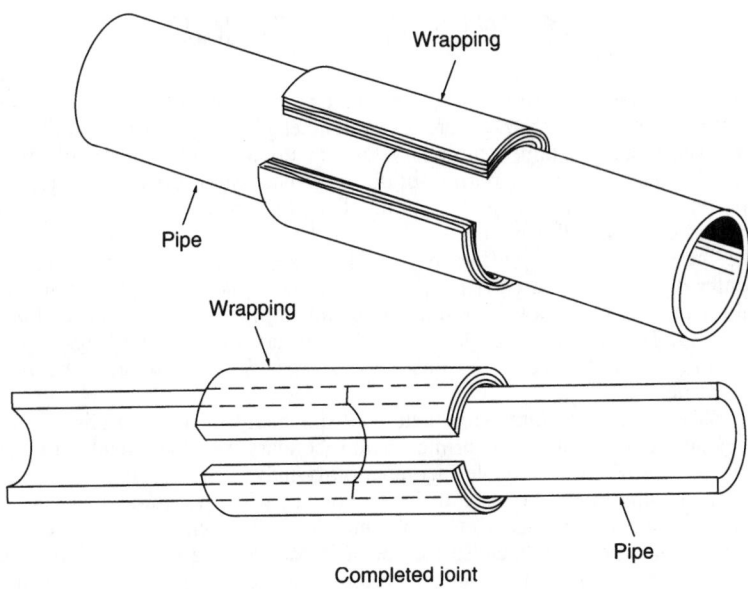

FIGURE 2.18 Butt and strap joint.

DOUBLE-WALL PIPING

Installing an outer pipe around an inner pipe has been found to prevent the release of hazardous liquids being transported in the inner pipe. This system as a whole is called a secondary contained piping system, or double-walled pipe. A major use for these systems is for transporting liquid fuel. The use of double-wall piping for this purpose appears in Chap. 12, Liquid Fuel Systems. There are no generally recognized codes for dimensions.

The interior pipe transporting the liquid is called the carrier or primary pipe. The outer pipe around the carrier pipe is called the containment pipe or secondary containment. A detail of a typical double-contained pipe is schematically illustrated in Fig. 2.19. The two pipes are kept apart by spacers, often called frogs or spiders by manufacturers. Figure 2.20 illustrates a cross section through the double-contained pipe.

Double-wall systems are manufactured from many different piping materials. There is no requirement for the primary and secondary pipe to be made of the same material except where a possible incompatibility may exist. Because the secondary containment pipe does not have to be in contact with the fluid, it is very cost-effective to have the secondary containment pipe made from a different, less costly material. This is possible because the selected outside pipe will not have to be in constant contact with the fluid and, therefore, may be acceptable for only limited contact at a lower temperature and pressure.

The major problem in the design of double-wall systems is thermal expansion and contraction of the primary and secondary pipes. If transporting hot liquids, the primary and secondary pipes will expand at different rates, even if they are made of the same materials, since the secondary pipe is at a lower temperature. Compensation methods for expansion and contraction include expansion loops between restraints with oversize containment, changes of direction with oversize containment

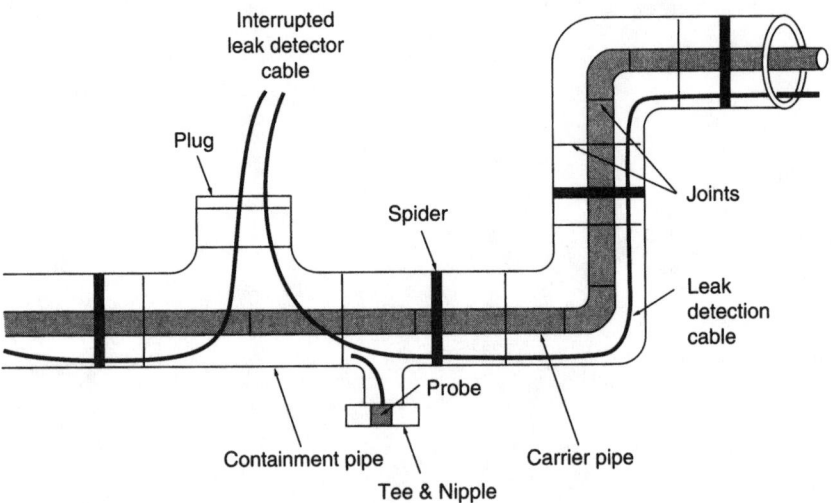

FIGURE 2.19 Typical double-contained piping.

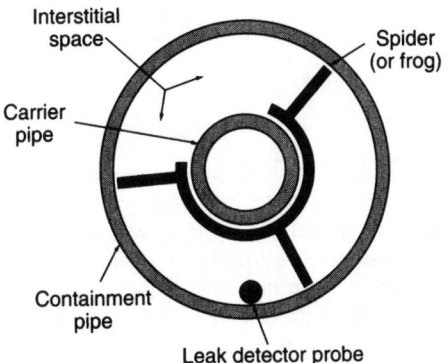

Interstitial space

Spider (or frog)

Carrier pipe

Containment pipe

Leak detector probe

FIGURE 2.20   Section through double-contained pipe.

elbows, expansion offsets between restraints, expansion joints, and proprietary fittings to keep the containment pipe in alignment if the temperature difference is not too large. Since each manufacturer uses different materials and jointing methods, expansion compensation methods for each specific system must be obtained from specific manufacturers.

With the potential for polluting the environment if any product leaks from the piping system, a leak detection system is mandatory to detect leaks from the primary pipe. For facilities, the two methods used most often are (1) an electronic resistance or capacitance cable with sensing panel and (2) capped tees with a probe installed so that any product leaking into the containment will spill into the tees and be detected by a immersion probe. In the first method, a cable is installed throughout the pipe run and is located at the bottom of the pipe. If the cable cannot be correctly routed through changes of direction, it must be interrupted (illustrated in Fig. 2.19). The cable has the ability to detect moisture anywhere along its length and the sensing panel will indicate where the leak is located along the cable. In the second method, a probe in the tee is connected by cable to a panel that will show which tee probe detected the liquid. Experience has shown that the cable is difficult to install and is prone to false annunciation due to condensation in the pipe. Because of this, the second method (the probes installed in a tee) is preferred as of this writing. The tees are generally spaced from 20 to 50 ft apart depending on economics and, often, on distances established by client preference.

VALVES

GENERAL

Valve functions can be defined as ON/OFF service, throttling service (flow control), prevention of reverse flow (or back flow), pressure control, regulation and pressure relief. Valves can be classified as either linear (gate valve) or rotary (ball valve) based on the action of the closure member. They are also classified by the shape of their closure member such as gate, globe, butterfly, ball, plug, diaphragm, pinch, and check.

Their primary function, however, is to control the flow of liquids and gases, including plain water, corrosive fluids, steam, toxic gases, or any number of fluids with widely varying characteristics. Valves must also be able to withstand the pressure and temperature variations of the systems in which they are used. Some valves on combined water service mains, and those handling flammable material, may be required to be fire safe or approved for fire protection use.

Pressure regulating valves for water service are discussed in Chap. 9, Plumbing Systems, in the section entitled Water. Pressure regulating valves are discussed in their respective chapters. Pressure relief valves are outside the scope of this handbook.

CODES AND STANDARDS

The following standards apply to valve construction:

1. AWWA C 500: gate valves for water and sewage systems
2. AWWA C 504: rubber seated ball valves
3. MSS SP 67: butterfly valves
4. MSS SP 80: bronze gate, globe, angle, and check valves

VALVE COMPONENTS

The following are the primary components of a valve.

1. A valve body is the housing for all the internal working components of a valve and it contains the method of joining the valve to the piping system.
2. The closure element, known as the disk or plug, is a valve component that, when moved, opens or closes to allow the passage of fluid through the valve. The mating surface of the disk bears against the seat.
3. The actuator is a movable component that, when operated, causes the closure element to open or close.
4. The stem is a movable component that connects the actuator to the closure element.

5. The bonnet is a valve component that provides a leakproof closure for the body through which the stem passes and is sealed.

6. The seat is a component that provides a surface capable of sealing against the flow of fluids in a valve when contacted by a mating surface on the disk. The seat is attached to the valve body.

7. The stuffing box is the interior area of the valve between the stem and the bonnet that contains the packing.

8. Packing is the material that seals the stem from leaking to the outside of the valve. The packing is contained by the packing nut on the bonnet.

9. The backseat is a seat in the bonnet used in the fully open position to seal the valve stem against leakage into the packing. A bushing on the stem provides the mating surface. Backseating is useful if the packing begins to leak and it provides a means to prevent the stem from being ejected from the valve. Backseating is not provided on all valves.

10. The stroke of a closure member is the distance the member must travel from the fully opened to the fully closed position.

VALVE BODY MATERIALS

Valves are manufactured in both metallic and nonmetallic materials. Nonmetallic materials consist primarily of thermoplastics.

Metallic

Valve bodies are manufactured of the following materials:

1. Bronze valves are usually limited to the smaller sizes and are used in water service up to 450°F. Different alloys are available for higher temperature and pressure applications.

2. Cast iron is a commonly used material for water and steam up to 450°F and is generally limited to smaller size castings. High tensile strength iron may be used in large sizes.

3. Malleable iron is characterized by pressure tightness and resistance to stress and shock.

4. Ductile iron casting has high tensile strength, good ductility, and good corrosion resistance.

5. Steel is available in a wide variety of alloys that are recommended for high temperature and pressure applications and conditions that may be too severe for iron and bronze bodies.

6. Stainless steel is available in a wide variety of alloys and is often used for pure water and other services requiring noncorrosive materials.

Thermoplastics

Thermoplastics are rapidly gaining popularity in many utility systems. They have proven very successful for carrying corrosive fluids where conventional valves are not suitable or special alloy metals are very costly. It is estimated that they are suitable for use in 85 to 90 percent of all utility services.

There are many different types of thermoplastic materials capable of carrying most chemicals. All standard valve types are available. The valve materials often have additives different from those used for manufacturing piping. Generally available valve types and their sizes are given in Table 2.6. In general, plastic valves are limited to a maximum temperature of 250°F and pressure of 150 psig (1035 kPa). General properties of plastic material used for valves are given in Table 2.7.

VALVE ACTUATORS

Valve Categories

There are three operating methods for valve actuators: multiturn (used for gate, globe, and diaphragm valves), quarter-turn (used for plug, ball, and butterfly valves), and linear (used for gate, diaphragm, and globe valves). The valves can either operate manually or be power actuated. There is no generally recognized code or standard for valve operators outside of those concerned with specific, high-risk industries, for example, nuclear work.

Manual Operation

Manually operated valves are usually used when the valve is easily accessible, does not require automatic operation or is operated infrequently. Multiturn valves use hand-wheels, gears, or levers. The most common is the handwheel. If the operating effort is too high for a handwheel, a gear box could be installed. A less common gear arrangement is the impact gearbox. This consists of a free-wheeling handwheel for part of its rotation that then imparts a hammer blow to the stem to break loose

TABLE 2.6 Thermoplastic Materials and Valve Types

Valve design	Materials	Size range, in
Ball (union design)	PVC, CPVC, PP, PVDF	$\frac{1}{4}$–4
Ball (compact design)	PVC, CPVC	$\frac{1}{2}$–3
Ball, multiport	PVC, CPVC, PP, PVDF	$\frac{1}{2}$–3
Diaphragm	PVC, CPVC, PP, PVDF	$\frac{1}{2}$–10
Butterfly	PVC, CPVC, PP, PVDF	$1\frac{1}{2}$–24
Globe	PVC, CPVC, PP	$\frac{1}{2}$–4
Gate	PVC	$1\frac{1}{2}$–14
Ball check	PVC, CPVC, PP, PVDF	1–4
Swing check	PVC, PP, PVDF	$\frac{3}{4}$–8
Labcock	PVC	$\frac{1}{4}$
Foot	PVC, CPVC, PVDF	$\frac{1}{2}$–4
Pressure relief	PVC, CPVC, PP	$\frac{1}{2}$–4
Solenoid	PVC, CPVC, PP	$\frac{1}{2}$–1

TABLE 2.7 General Properties of Thermoplastic Materials Used for Valves

Properties	Unit	PVC	CPVC	PP	PPG	PVDF
Specific gravity	—	1.43	1.45	0.92	1.03	1.76
Tensile strength	psi	7,000–7,800	8,500–9,200	4,300–5,000	9,950–11,000	7,000–7,800
Elongation	%	60–120	20–40	150–200	3–5	30–50
Tensile modulus	10^3 psi	455–485	485–510	130–170	12,800–14,220	170–200
Flexural strength	psi	14,220–15,560	14,220–17,060	7,820–9,240		13,510–14,930
Flexural modulus	10^3 psi	384–412	426–455	213–228	540–570	227–256
Compressive strength	psi	12,800–14,220	14,220–15,560	8,530–9,950		12,800–14,220
Compression modulus	10^3 psi	242–256	256–284	120–156		14–199
Poisson's ratio	—	0.37	0.35	0.44		0.28
Hardness (Rockwell R)	degree	115	118	95	107	110
Impact strength (Izod) w/V-notch	kg-cm/cm²	4–5	4–5	4–7	6–8	10–12
Resistance to heat (continuous)	°F	140	194	194	221	248
Deflection temperature (at 66 psi)	°F	165	250	230	302	302
Thermal conductivity	kcal/mh°C	0.13	0.12	0.08		0.11
Dielectric strength	kV/m	0.90	0.90	1.02	1.18	1.18
Volume resistivity	ohm-in	2.0×10^{15}	2.3×10^{15}	1.9×10^{15}	0.4×10^{15}	1.95×10^{15}
Dielectric constant						
10 Hz		2.8–3.0				
60 Hz		3.15	2.93	2.42		9.8
103 Hz		3.14	2.92	2.41		9.5
106 Hz		2.85	2.69	2.41	2.2	7.5
Dissipation factor						
60 Hz	10^{-2}	1.18	1.09			0.05
10^3 Hz	10^{-2}	1.91	1.10	0.044		0.048
10^6 Hz	10^{-2}	1.72	0.92	0.063	0.02	0.160
Water absorption 24 h, ⅛ in thickness	%	0.07	0.15	0.01	0.02	0.03

Note: For thermal expansion, see Table 2.15.

a stuck gate or globe valve. For the torque necessary to open a gate valve against an unbalanced force, refer to Fig. 2.21. This is generally referred to as "breakaway torque."

Quarter-turn and linear valves use a lever.

Power Actuators

Power actuators are used where valves are remotely located, frequent operation is required, or automatic operation is necessary due to system considerations.

Depending on the type of valve, the activator will be required to deliver output for rotary or linear motion. This may be for ON/OFF (fully opened to fully closed) service, or stroke functions such as those required to keep a valve partially open, or a combination of both. The power source must be capable of exceeding the torque requirements needed by the actuator by an adequate safety factor. In the case of throttling, a detailed analysis may be required. The worst case is in providing the breakaway torque. Valve-operating torque is never constant, varying with closure member position. Typically, the peak torque is required at breakaway, reduces to about 30 percent of breakaway in the half-open position and increases to about 90 percent of breakaway at closure.

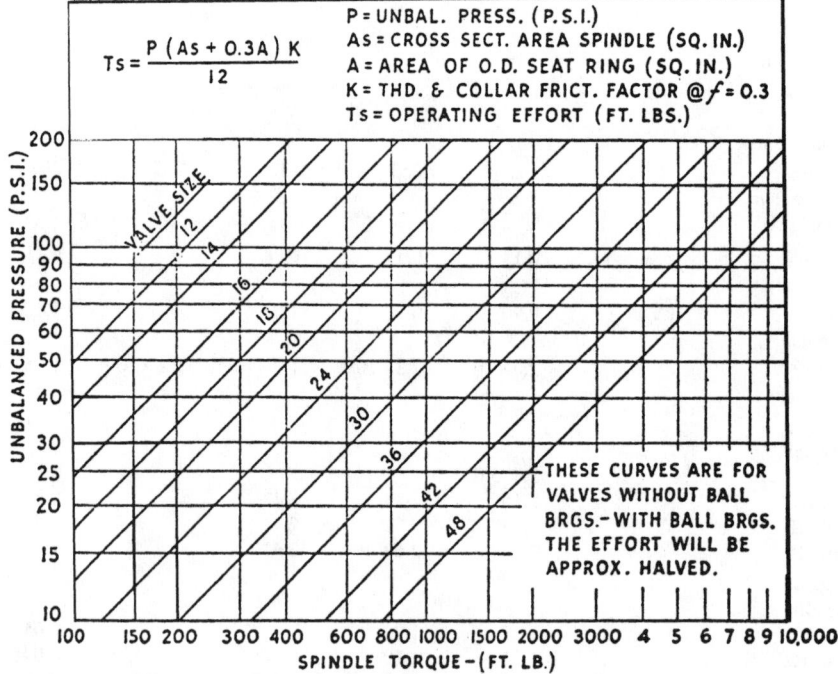

FIGURE 2.21   Force required to open a gate valve against unbalanced pressure. (*Data compiled by The Chapman Valve Manufacturing Company.*)

Actuators are classified by their source of power: electric, pneumatic, or hydraulic.

Electric Actuators.   These are the most commonly used power source. They can be solenoid or motor operated. They usually have the least total installed cost because the source of electricity is generally available throughout a facility and the wiring and control instrumentation is relatively simple. Solenoid operation is generally limited to smaller lines, ON/OFF service, and systems in which the water hammer produced by quick closing can be easily attenuated because of the systems' low pressure and velocity requirements. Motor-operated actuators tend to be bulky and slow, particularly when large gear reduction is used to increase torque. Their advantage is that the torque output is constant throughout their stroke, and their response is linear. For critical systems, there must be an emergency power supply.

The speed at which a valve closure member is operated is important. The figure generally used for a gate valve closure member is 12 in (300 mm) per min and for a globe valve closure member, 4 in (100 mm) per min. Higher rates are available, but exceeding the maximum specified speed will damage the seat and disk. Gate and globe valves are torque-seating valves when closed. In the open direction, a limit switch is often provided to protect the seat against backseat overtightening. Quarter-turn valves are position-limited open and closed because seating is based on position, not force.

Electric motors do not stop instantaneously, but coast to a stop. The use of a solenoid brake to prevent the motor from overtightening the closure member should be ascertained from the manufacturer of the valve on which the actuator is installed.

Unless an emergency power source is available, electric motors should not be used where cycling to a fully opened or fully closed position is a requirement in the event of a power failure. Motor operators should be limited to moderate cycling functions. They are not recommended for service where severe cycling is necessary. Generally, 2-in (50-mm) valves do not have sufficient strength for a standard electric motor actuator. Electric motor actuators have been extensively tested in seismic and harsh ambient environments.

Pneumatic Actuators.   Depending on the system selected, these air-driven devices generally operate in a range of between 30 and 120 psig (210 and 830 kPa), with 30 to 60 psig (210 to 415 kPa) being the most common. The compressed air supply should be a dedicated one, conforming to ISA requirements, and preferably supplied from a control air compressor assembly. An alternative, but less desirable means, would be to obtain the air supply from the facility system, using additional compressed air-conditioning filters, dryers, and PRVs as required.

Pneumatic actuators are well suited for frequent operation and fast response times. There are two types of actuators, piston and diaphragm. The piston actuator is generally used for ON/OFF operation. The piston stroke can be long, making it suitable for large valves. The diaphragm actuator is appropriate for modulating service but because the travel is short, the valve size on which it can be used is limited.

The fail-safe mode is usually accomplished by using either an internal spring or a secondary accumulator tank to provide the necessary power to cycle to an opened or closed position. The internal spring may cause the assembly to be flexible, which may be a problem for seismic installation. The accumulator tank is externally mounted, often on a nearby wall or column.

Pneumatic actuators are large in size and require frequent maintenance because of air leakage over time (particularly piston types) that also makes response time longer. There is a limitation on maximum valve differential pressure.

Hydraulic Actuators. These devices produce torque by using an electric pump to push fluid to a piston. They are capable of providing fast actuation and are suitable for modulating service. They allow operation on large valves with high pressure differentials and are appropriate for frequent cycling. They have no fail-safe mode unless emergency electrical power is available. The stroke is easily adjustable in service.

Typically, these actuators operate slowly. Their cost is between the electric and pneumatic systems.

OPERATING AND TORQUE CONSIDERATIONS

The following factors require consideration in the selection of an actuator:

1. *Valve type and size.* The valve-operating torque results from the inherent size and characteristics of the valve itself and the type of seat. The amount of torque necessary to overcome static imbalance must be obtained from the manufacturer.

2. *Pressure drop.* The operating torque increases with an increase in pressure drop across the valve. A valve operating at full-rated pressure will require significantly more operating torque than one operating at a low-pressure drop. Depending on the source of pressure, it is probable that the pressure differential will vary throughout the valve's entire stroke. This condition is important if the actuator torque output must be carefully matched with that of the valve.

3. *Service-operating conditions.* Will the valve be required to be only opened or closed or will it also be required for throttling flow? Actuators for ON/OFF service will be selected only on breakaway torque. For quarter-turn valves requiring throttling, calculating the torque is more complicated because additional torque is required to counterbalance the momentum of the flowing fluid. Unbalanced forces generate "hydrodynamic torque." The actuator torque output must be well above the operating torque to achieve smooth operation.

4. *Seat material.* Most valves have a metal closure member sealing on a soft seat made of elastomers. Metal-seated valves may require as much as 50 percent more seat material as needed for soft seat valves.

5. *Fluid being transported.* Since air and gas do not provide any lubrication, their operating torque requirements add to the frictional forces. Water and other media may provide excellent lubrication. Liquids carrying solids clog clearances between stem and bearings. The fluid may also corrode internal parts, so that in time the torque valve may rise considerably, up to twice that when new. An adequate safety factor should be considered to assure reliable and continued operation.

6. *Bidirectional seating.* If operating conditions require the reversal of flow, additional torque may be required for seating.

7. *Fire safety.* The valve may require secondary metal-to-metal seating if the primary seat is destroyed by fire. This will require more operating torque.

8. *Fail-safe operation.* With the automatic fail-safe operation, the energy necessary to close or open the valve requires a larger size actuator than one without a failsafe requirement.

9. *Temperature of fluid.* Torque requirements are lowest at room temperature. High temperature and cryogenic bearings require higher operating torque. Fluid temperatures above 300°F may require a special operating and mounting assembly, often a stem extension. Ambient temperatures must also be considered, for example, actuators located outdoors require special consideration.

10. *Cycling rate.* Pneumatic and hydraulic actuators cycling in excess of 30 cycles per h are considered to have high operating rates. The same is true for electric actuators cycling in excess of 10 percent of their duty cycle (operating for 1 cycle and resting for a time equivalent equal to 9 cycles). An extended duty motor should be obtained for this condition.

11. *Cycle speed.* Fast cycle speeds of less than one-half standard cycle times require special consideration. The sudden physical shock associated with fast operating speed combined with fast cycling rates can damage valve and actuator parts. Pneumatic actuators may need quick exhaust valves, special solenoids, and larger actuators. Higher speeds are accomplished using different gearing devices, which may increase torque output, or an electronic speed control, which will not affect torque output.

12. *Stem orientation.* Orientation of the valve stem in a position other than vertical will require mounting in a manner that may cause stem seal leakage or galling due to side thrusts induced by an overhung load on the actuator. The use of heavy-duty couplings and mounting brackets will minimize these problems.

FIRESAFE VALVES

By nature of their service, some valves require a firesafe designation. There is no single generally recognized definition of firesafe or a code that can be used to determine suitability or acceptance. A simplified definition is that a valve must not melt in a fire or leak after a fire and that the seat must close adequately.

The standard used most often for the CPI is the API 607 rating. For water fire-service lines, FM is the most conservative, although a listing with UL may be acceptable depending on the specific insurance carrier used. Specific companies often have ratings that must be used when projects are designed for them.

Firesafe valves require testing to meet minimum recommended performance standards when operating in a firesafe environment. These recommendations are:

1. *Minimum internal leakage.* A valve must offer acceptable seating prior to and after exposure to high temperatures without depending on supplementary pressure from spring-loaded or other devices and without depending on a critical seal.

2. *Minimal external leakage.* The valve body design should minimize external leakage by using fire-resistant stem seals and avoiding large gasketed body joints.

3. *Continued operability.* A valve must be operable despite fire damage. The body and actuator must resist warpage and damage from high temperatures.

VALVE RATINGS

There are a number of designations used to indicate the pressure ratings of valves. Valves are pressure rated by their ability to withstand pressure within a range of temperatures. Standard pressure ratings have been established to match ANSI ratings of flanges and fittings and are designated by class, conforming to ANSI B 16.34 ratings. Two types of designation are WSP and WOG. WSP, or working steam pressure, rates the ability to handle steam at the specified working pressure. WOG, or water, oil, and gas, rates the ability to handle cold water, oil, and gases at the assigned working pressure. When the two ratings are given, WSP is called the primary rating. When only one rating is given, the valve is not generally used for the service not mentioned. The rating 150 lb refers to the working pressure in psig for which the valve is rated. If a valve is primarily used for water service, a common designation is WWP, or water working pressure. This designation rates the ability to handle cold water. The valve class designates the working pressure of a valve. A class 300 rating indicates a valve with a working pressure of 300 psig.

Cold temperatures mean ambient temperatures from 32 to 90°F For high temperatures, the valve pressure shall be derated. For high pressures, the temperature rating shall be derated. The temperature limitation on most metallic valves is generally based on the capabilities of the seat and interior trim materials.

PRESSURE LOSS THROUGH VALVES

In general, valves used for utility piping are rarely selected based on pressure drop through the valve but rather for their suitability in service. Calculations are not needed since established equivalent lengths of pipe for each type of valve are sufficiently accurate for determination of the approximate pressure drop through the valve. Refer to Fig. 9.1 to determine the equivalent length of pipe for common valves.

There may be occasions where precise determination of the pressure drop through any valve would be desired, such as in cases where pressure drop must be kept to a minimum or the exact determination of the pressure drop is necessary. This is done by using the standard measure of valve flow, the coefficient C_v. This coefficient is the flow in gallons per minute that will pass through a valve in the wide-open position with a pressure drop of 1 psi (6.9 kPa). This coefficient is determined by the valve manufacturer using actual flow tests. With the C_v known, the pressure differential can be found using the following formula:

$$\Delta P = \frac{S\ F^2}{C_v^2} \tag{2.1}$$

where ΔP = pressure difference, psig
S = specific gravity of fluid (the value for water is 1)
F = flow rate in gpm
C_v = valve flow coefficient (obtained from valve manufacturer)

The C_v is calculated by dividing the flow in GPM by the square root of the pressure difference across the valve.

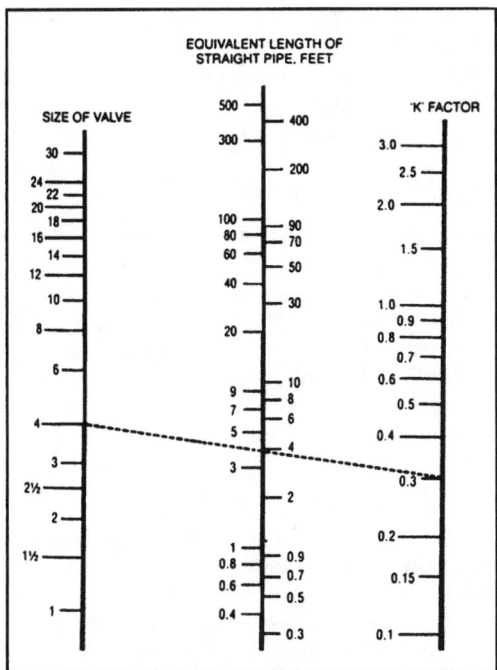

FIGURE 2.22   Equivalent run determination from K factor. (*Courtesy Stockham.*)

Another designation used by some manufacturers is the K factor, which will allow determination of the pressure drop through a valve in equivalent length of a straight run of pipe. The K factor is obtained from the valve manufacturer and is determined by actual tests. Figure 2.22 is a chart to determine the equivalent length from the K factor of a valve. For example, the dashed line shows the resistance of a 4-in valve to be equivalent to approximately 3.8 ft of 4-in standard steel pipe.

VALVE SELECTION CONSIDERATIONS

The following general items must be considered in selecting valves.

1. *Temperature.* The valve bodies, trim, and operating parts must be capable of withstanding the highest temperature expected during sustained normal and transient operating conditions.
2. *Pressure.* The valve must be rated for the highest transient pressure that might be expected.
3. *Shutoff.* The degree of allowable shutoff must be known. For utility piping, some minor leakage should be allowed and would prove extremely costly to eliminate. Bubbletight valves are those that exhibit no visible leakage through the elastomeric seat of the valve for the duration of a test as defined by MSS SP 82.

4. *Valve operation.* It must be determined whether the valve be used only for ON /OFF use or for throttling.

5. *Pressure drop.* Allowable pressure drop must be established and the size (equal to or less than that figure) selected.

6. *Corrosion resistance.* This is affected by the nature, concentration, and temperature of the fluid.

7. *Velocity.* The velocity of the fluid through the valve must be considered.

8. *Firesafe.* It must be known if this is a requirement. (Refer to previous discussion.)

9. *Hazardous material.* When the fluid being transported is considered hazardous or lethal, valves must be specifically designed to handle these materials. Redundant stem packing and leak detection ports are typical design features. ANSI B 31.3, category M, defines this category of fluids.

GATE VALVES

Gate valves, illustrated in Fig. 2.23, use a wedge-shaped disk or gate as the closure member operating perpendicular to the flow; it is raised to open and lowered to close the valve. As the disk closes, it fits tightly against the seat surfaces in the valve body. A gate valve is used fully opened or closed only. It should not be used for throttling service (partly open), as the gate will vibrate and quickly become damaged and subject to wire drawing caused by the velocity of the liquid flowing past the disk.

Primary Gate Valve Components

There are four main features that dictate gate valve design: the disk, stem, bonnet/ body connection, and body. The body materials and end connections have been previously discussed.

Often, iron body valves use a combination of materials to provide corrosion-resistant bearings for stems and other wear points such as seating surfaces. These valves are called iron body bronze mounted (IBBM).

Disk Design. There are three types of disk constructions: solid wedge, split wedge, and flexible wedge (illustrated in Fig. 2.24).

Solid Wedge. Solid wedge disks are most prevalent because of their simple and usually less expensive design.

Split Wedge. Split wedge disks, also called double disks, have somewhat better sealing characteristics than solid disks because the two disk halves are forced outward against the body seats by a spreader after the disk has been fully lowered into its seating position. When the valve is opened, pressure on the disk is relieved before it is raised, eliminating the friction and scoring of body seats and disk.

Flexible Wedge. Flexible wedge disks are solid only at the center and are flexible at the outer edge and seating surface. This design enables the disk face to overcome the tendency to stick in high-temperature service where wide swings in temperature occur. This type of disk is generally found only in steel valves.

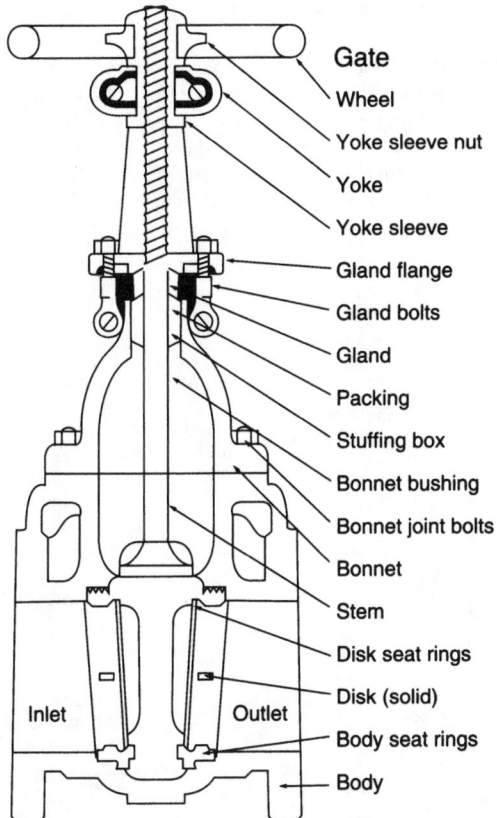

Gate

Wheel

Yoke sleeve nut

Yoke

Yoke sleeve

Gland flange

Gland bolts

Gland

Packing

Stuffing box

Bonnet bushing

Bonnet joint bolts

Bonnet

Stem

Disk seat rings

Disk (solid)

Body seat rings

Body

Inlet   Outlet

FIGURE 2.23   Gate valve. (*Courtesy Stockham.*)

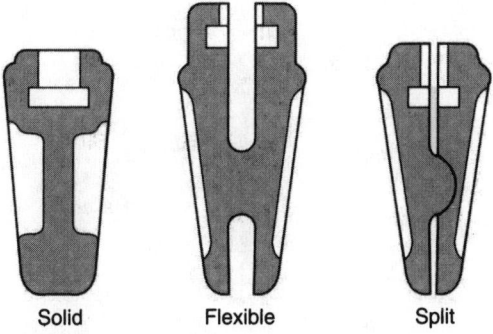

Solid   Flexible   Split

FIGURE 2.24   Gate valve disk construction.

Stem Construction.   There are also five basic types of stem construction, shown in Fig. 2.25.

Rising Stem/Outside Screw and Rising Stem/Outside Screw and Yoke.   These two types of stem construction keep stem threads outside the valve and are recommended where high temperatures, corrosives, and solids in the line might damage stem threads inside the valve. When the hand wheel (nonrising) is turned, the stem rises as the yoke bushing engages the stem threads. The external threads enable easy lubrication; however, care must be taken to protect the exposed stem threads from damage.

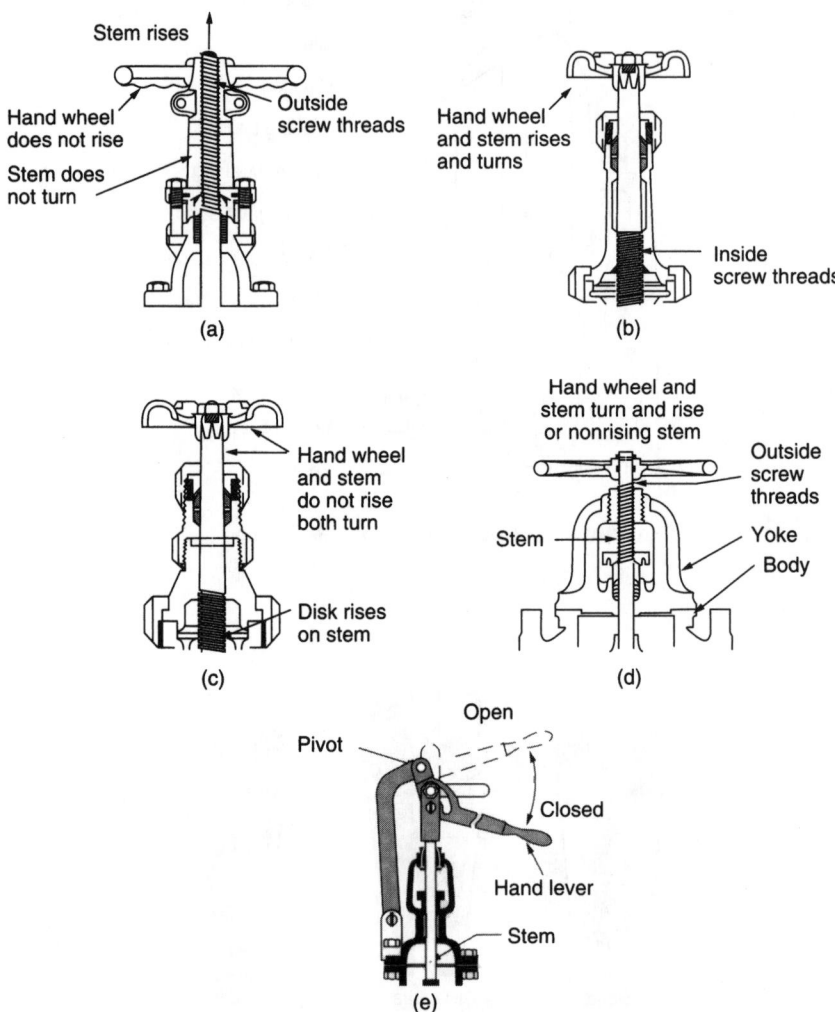

FIGURE 2.25   Basic stem construction. (*a*) Rising stem, outside screw; (*b*) rising stem, inside screw; (*c*) nonrising stem, inside screw; (*d*) rising stem, outside screw and yoke; (*e*) sliding stem.

Rising Stem/Inside Screw. Rising stem/inside screw is the most common stem design in bronze gate valves. Because the hand wheel and stem both rise, adequate clearance must be provided for operation. The stem and hand wheel positions indicate the position of the disk inside the valve. In the open position, the backseat helps protect the stem threads, but care must be taken to protect the stem externally.

Nonrising Stem/Inside Screw. Nonrising stem/inside screw design has the chief advantage of requiring minimum head room for operation. Since the stem does not travel vertically, packing wear is reduced. Heat, corrosion, erosion, and solids may damage the stem threads inside the valve and cause excessive wear. Also, it is impossible to determine the disk position since the hand wheel and stem do not rise.

Sliding Stem. The operation of the stem is linear, straight up and down. A lever takes the place of a hand wheel and there are no threads on the stem. Available on gate and globe valves, this type of stem is useful where quick closing or opening of a valve is desired.

Bonnet Construction. The basic types of bonnet construction include union, screwed, and welded designs (shown in Fig. 2.26). Union bonnets are preferred for rugged service. Screwed bonnets are the least expensive design and should be used for lower pressures only. Welded bonnet construction provides the most leak-free body-to-bonnet joint. The disadvantage of the welded bonnet is that it provides no access to the trim parts if repairs are needed.

GLOBE VALVES

Globe valves are so named due to the globular shape of the valve body. Globe valves are used where throttling and/or frequent operation is desired. Each uses the same method of closure-a round disk or tapered plug-type disk that seats against a round opening (port). This design deliberately restricts flow, so globes should not be used where full, unobstructed flow is required. There are three basic types of globe valve: the standard globe valve (Fig. 2.27), the angle globe valve (Fig. 2.28), and the needle valve (Fig. 2.29).

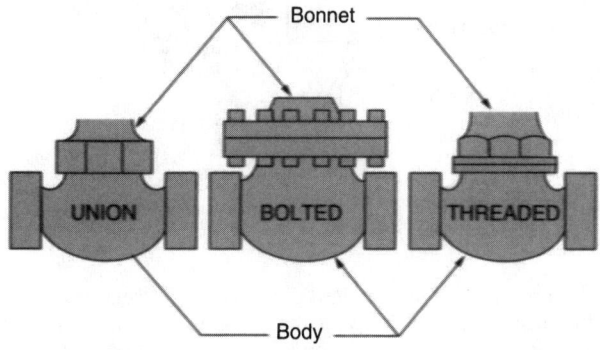

FIGURE 2.26 Typical bonnet construction.

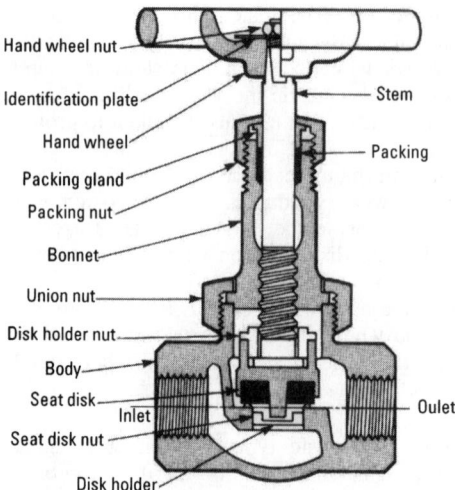

FIGURE 2.27   Standard globe valve.

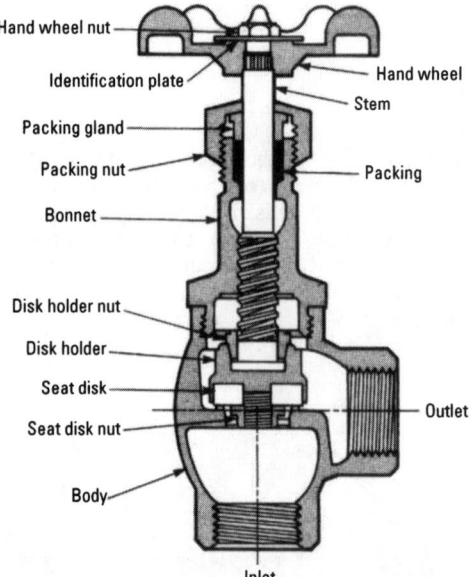

FIGURE 2.28   Typical angle globe valve.

Angle valves are identical to standard globe valves in seat design and operation. The basic difference is that the body of the angle valve acts as a 90° elbow, eliminating the need for a fitting at that point in the system. Angle valves also have less resistance to flow than the combination of globe valves and the fittings they replace. Needle valves are generally small in size and are intended to provide

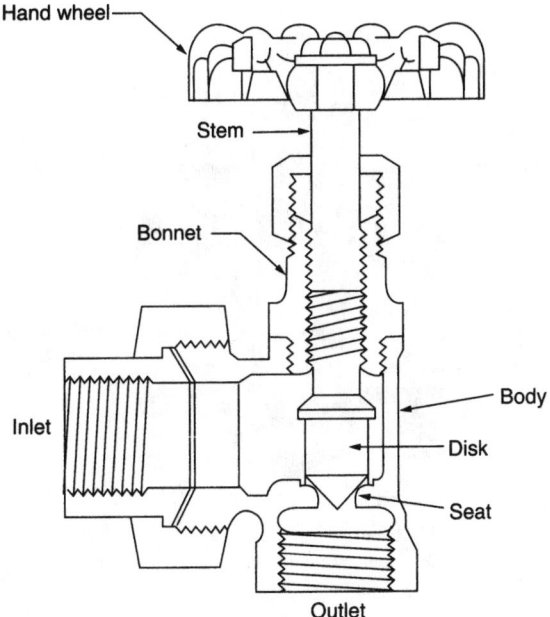

FIGURE 2.29 Typical needle valve.

precise flow control. Many turns of the handle are required to adjust flow in order to achieve precise control.

A globe valve should be installed with the flow entering under the disk. The end of the valve through which you can see the disk seat when the valve is closed is the end where the flow should enter. This is necessary for good throttling control of flow and best shutoff conditions. Globe valves should also be installed with the disk closed to prevent seat damage during installation.

Most globe valve leakage is due to foreign matter settling on the area between the disk and seat. When this occurs, it can often be corrected by opening the valve slightly and then closing it.

PLUG VALVES

A plug valve, shown in Fig. 2.30, is a quarter-turn valve that uses a tapered cylindrical plug that fits a body seat of corresponding shape. When the port in the plug is aligned with the body opening, flow is permitted in a way similar to a ball valve. A one-quarter (90°) turn operates the valve from opened to closed and vice versa.

Plug valves fall into two basic categories, lubricated and nonlubricated. A lubricated plug valve is designed with grooves in the surface of the plug. The grooves are connected to a lubricant channel in the stem. When the grooves are filled with lubricant (also called sealant), a tight seal develops between the plug and valve body. Lubricant is usually applied with a hand pump, providing a hydraulic jacking force and lifting the plug slightly for easier turning. When properly lubricated, this valve gives tight shutoff and is easily operated. Proper lubrication requires addi-

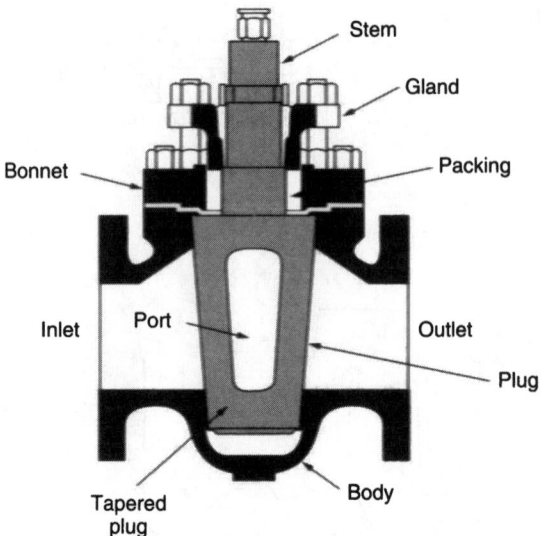

FIGURE 2.30   Typical plug valve.

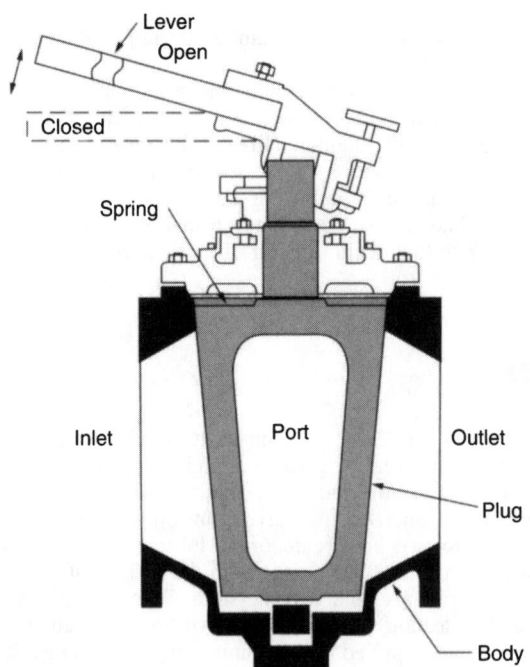

FIGURE 2.31   Typical lift-type plug valve.

tional maintenance. This valve is often used in natural gas service. A disadvantage is that the lubrication may enter the product stream and is not recommended where purity is a primary concern.

Nonlubricating plug valves have two basic designs, lift type and sleeved. In the lift type (Fig. 2.31), the plug is mechanically lifted while being turned to disengage it from the seating surface, thereby reducing seating force. The sleeve type generally has a fluorocarbon sleeve (retained in the body) that surrounds the plug, giving a continuous seal.

There are three port sizes: 100, 70, and 40 percent of inlet pipe size opening, as shown in Fig. 2.32. The size of the port determines the physical size of the valve, with the larger port having the largest valve size. The 70 percent port is normally supplied.

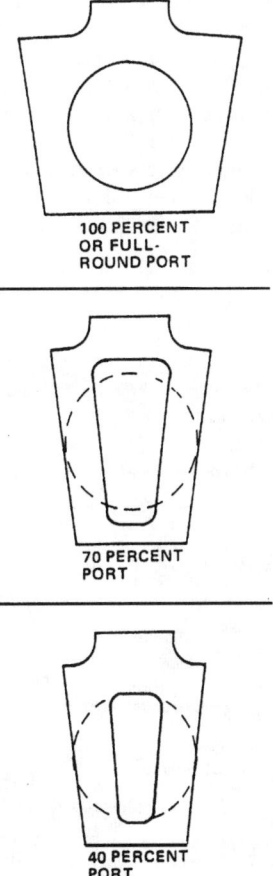

FIGURE 2.32 Plug valve port sizes.

BALL VALVES

A ball valve utilizes a ball with a hole drilled through it as the opening/closing device. It is a quarter-turn valve. The ball seals by fitting tightly against resilient seat rings on either side. Flow is straight through, and pressure loss depends on the size of the opening in the ball (port).

Ball valves are available in one-, two-, or three-piece body types (Fig. 2.33). The one-piece body, also called "end entry," is machined from solid bar stock material or is a one-piece casting. The ball is inserted into one end for assembly, and a body insert that acts as a seat ring is threaded in against the ball. The two-piece body, also an end-entry design, is the same as a one-piece valve except the body insert is larger and acts as an end bushing. The three-piece body consists of a center body section containing the ball that fits between two body end pieces. The entire assembly is held together by two or more body bolts. This design allows the valve to be repaired without disassembling surrounding piping. This type is recommended for utility services.

There are three port sizes, standard, reduced, and full port. Standard port is generally one pipe size smaller than the valve size, reduced port is up to two sizes smaller, and full port has the same opening as the connecting pipe.

Ball valves are generally selected for ON/OFF service. They are easily adapted to power actuation and are generally less expensive than equivalent sizes of gate and globe valves. With the development of high temperature and superior grade elastomeric seating material, tight seating problems have been overcome.

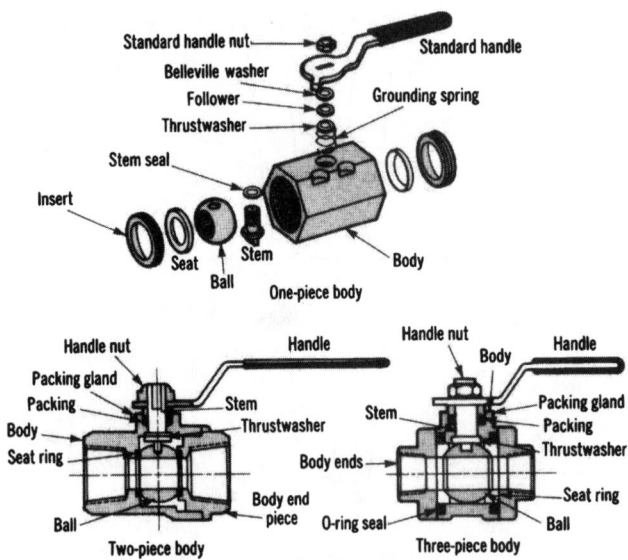

FIGURE 2.33   Typical ball valves.

BUTTERFLY VALVES

A butterfly valve has a wafer-shaped body with a thin rotating disk as the closing device. Like the ball valve, the butterfly operates with a one-quarter turn from fully opened to fully closed. The disk is always in the flow path, but since it is relatively thin, it offers little restriction to the flow. When the valve is closed, the disk edge fits tightly against a ring-shaped liner (seat).

These valves generally have one-piece bodies that fit sandwich-style between two pipe flanges. The two most common body types are wafer body and lug body, illustrated in Fig. 2.34. The wafer body is placed between pipe flanges, and the flange bolts surround the valve body. The lug body has protruding lugs that provide bolt holes matching those in the flanges.

Another design has an extended body for connections to grooved end piping. In this valve, the sealing member is the disk itself, which is fully encapsulated with a resilient material selected for the service conditions at hand.

Butterfly valves have continued to grow in popularity, generally at the expense of gate valves, because they are lightweight, easy to install, low in cost, easy to actuate, and easy to insulate and also because they feature one-quarter turn operation, tight shutoff (due to resilient seal), and a variety of construction materials.

DIAPHRAGM VALVES

A diaphragm valve uses a rubber, plastic or elastomer diaphragm to seal the stem. The diaphragm not only seals the stem but forms the closure element.

There are two styles of diaphragm valves, one having a body with a weir and the other having a straight-through body. On the weir type, shown in Fig. 2.35, the stem is connected to a finger arrangement, which in turn presses the diaphragm down onto a weir. This creates an extremely tight seal that will seal even on some solids. This valve has been used extensively in radwaste services and maintenance

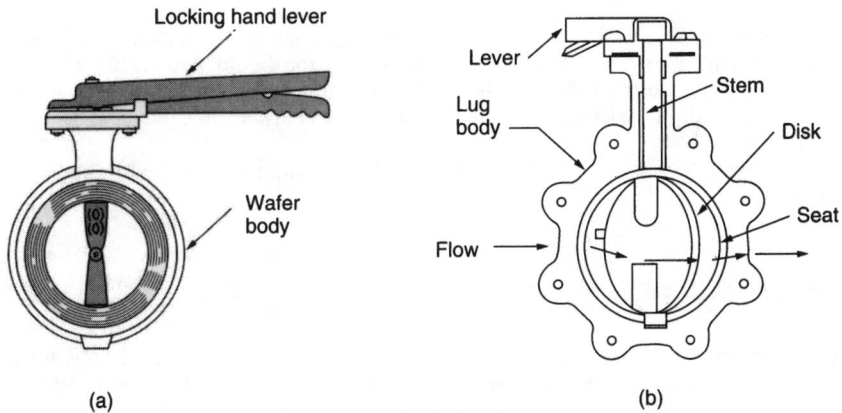

(a)   (b)

FIGURE 2.34 Butterfly valves. (*a*) Wafer body; (*b*) lug body.

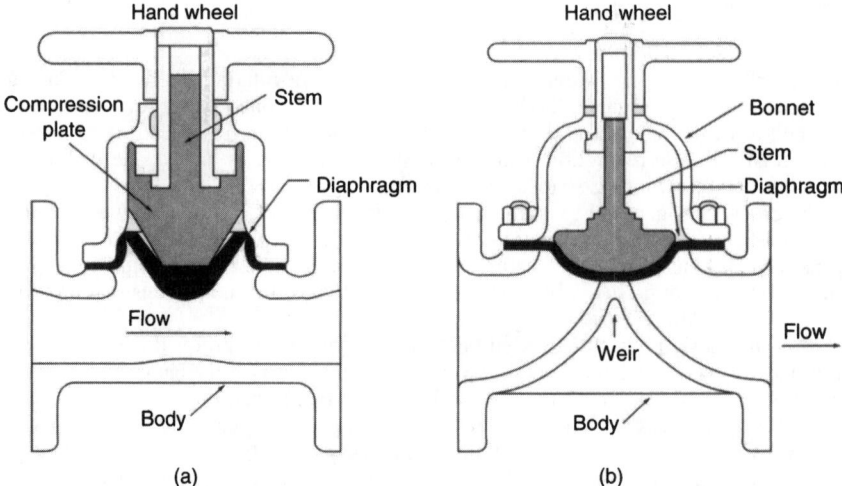

FIGURE 2.35   Diaphragm valves. (*a*) Straight-through valve in open position; (*b*) weir-type valve in closed position.

is extremely simple. On the straight-through type (Fig. 2.34), the diaphragm presses onto the bottom of the valve body for seating.

Since the diaphragm is not metallic and forms the closure, the valve is severely limited in pressure and temperature. A wide variety of diaphragm materials are available for use with different fluids. To enable draining of horizontal pipelines, the weir valve must be mounted 15° from the horizontal plane (because of the weir). This complicates installation, especially with air actuators.

PINCH VALVES

A pinch valve (Fig. 2.36) uses a round elastomeric sleeve connected to the valve body from inlet to outlet that completely isolates the liquid passing through the valve from all internal valve components. Closure is made by a movable closure element outside the sleeve that pinches the sleeve between the element and the valve body.

This type of valve is used for slurry and other liquids with highly corrosive properties.

CHECK VALVES

Check valves (Fig. 2.37) automatically check or prevent the reversal of flow. Basic types are the swing check, lift check, ball check, and wafer check designs. Another designation used for sanitary waste systems is a backwater valve. The swing check has a hinged disk, sometimes called a flapper, that swings on a hinge pin. When flow reverses, the pressure pushes the disk against a seat. The flapper may have a

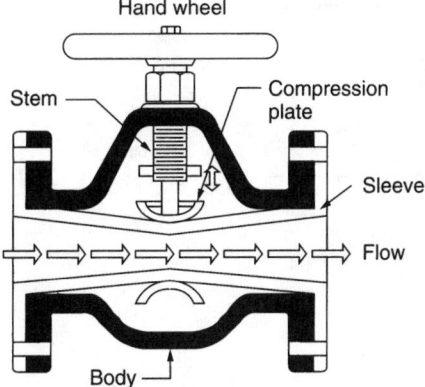

FIGURE 2.36 Pinch valve.

composition disk, rubber or Teflon, rather than metal when tight closure is required. Swing checks offer little resistance to flow.

The lift check has a guided disk that is raised from the seat by upward flow pressure. Reversal of flow pushes the disks down against the seat, stopping back flow. Lift checks have considerable resistance to flow, similar to that of a globe valve. They are well suited for high-pressure service.

Another common check is a wafer design which fits between flanges in the same fashion as a butterfly valve. Wafer checks come in two types: a dual flapper that is hinged on a center post and a single flapper that is similar to the standard swing check. They are generally used in larger size piping (4 in and larger) because they are much lighter and less expensive than traditional flanged end swing check valves.

A demand check value is of two-piece construction, with one piece having a spring-loaded closure similar to the air values found on automobile tires. The second piece, when inserted into the first, opens the valve, allowing free passage of air. The demand check valve is used for connecting gauges, allowing removal without permitting air to escape from the pipe.

MISCELLANEOUS VALVE TYPES

Various other types of valves are often used in utility systems. They can be either independently installed to operate as self-contained units or controlled electronically from a panel, system signal, or other remote source.

Pump Control Valve

This type of valve is used on pumped systems to control or eliminate surges caused by pump start and stop. It operates by using a spring-loaded closure member that opens or closes slowly to restrict the initial flow of water when a pump starts and stops.

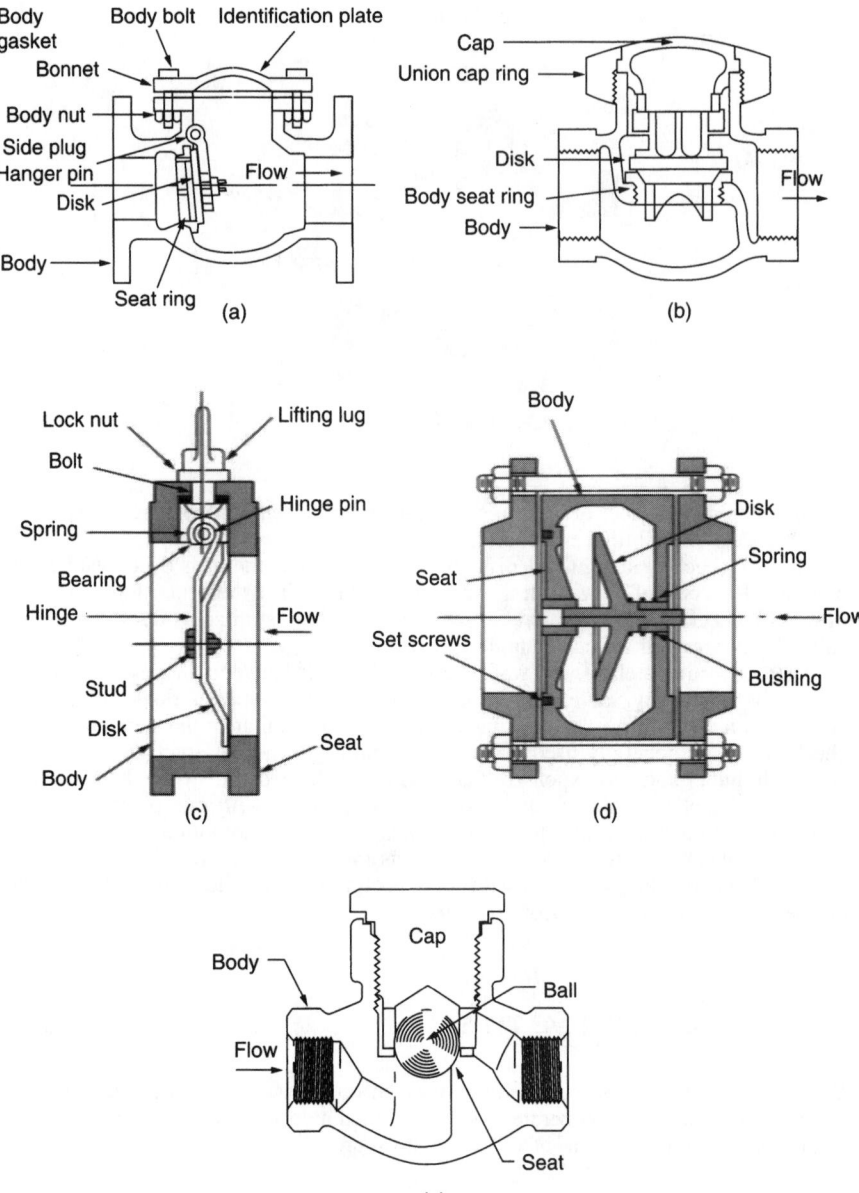

FIGURE 2.37 Check valves. (*a*) Swing check; (*b*) lift check; (*c*) single flapper wafer check; (*d*) double flapper wafer check; (*e*) hall check.

Flow Control Valve

This valve operates by using a calibrated orifice or venturi tube to control the flow of liquid to a predetermined set point regardless of fluctuating line pressure.

Pressure Control Valve

Similar to the flow control valve, this valve limits the pressure of a flowing liquid to a predetermined set point regardless of fluctuating flow rate.

Level Control Valve

This valve accurately controls the level of liquid in a tank or vessel. An altitude valve uses a controlling device to maintain the level, and a float valve uses a movable float on an arm (similar to that in a water closet) to stop the flow at a predetermined level.

Conduit Valve

A conduit valve (Fig. 2.38) is used where an unobstructed opening through the valve is required, such as when pigs are used to clean the pipeline.

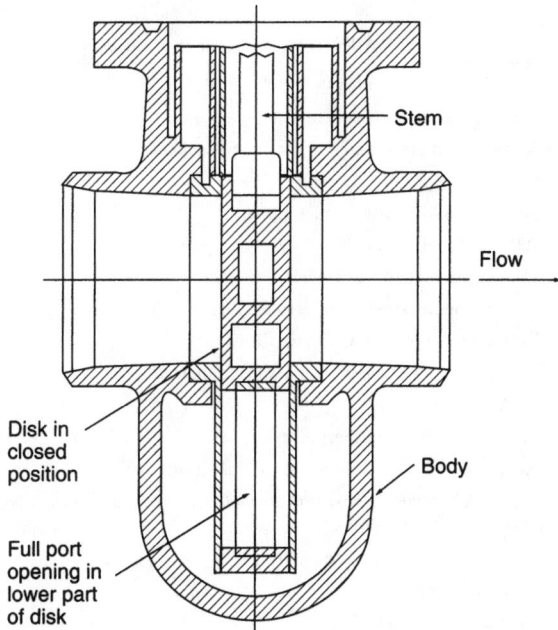

FIGURE 2.38   Lower portion of conduit valve.

METALLIC ALLOY DESIGNATIONS

The Unified Numbering System (UNS) is intended to correlate the many metal alloy numbering systems currently administered by North American technical societies, trade associations, individual users, and ASTM. Prior to the introduction of the UNS by the ASTM, a confusing array of names and numbers for metal and alloy designations had been developed independently over a 60-year period. Proprietary alloys were designated by trade names and some were registered trademarks. Often, different companies manufactured the same alloy under different trade names. Another factor is that the old numbering system does not allow the use of computer databases requiring uniform designations. Although it originated in the United States, the UNS is gaining national and worldwide acceptance. The book, *Metals and Alloys in the Unified Numbering System,* ASTM DS56B, published by ASTM, is now in its fifth edition.

The UNS is a simple alphanumeric system consisting of a letter prefix followed by five digits. This requires only six spaces to identify any metal or alloy. The letter prefix is usually suggestive of the family of metals identified. The system establishes 17 categories or families of metals. The categories are given in Table 2.8.

As an example, stainless steel ASTM type 304 is now S30400 and type 316 is S31600. If 304L is required, it is called S30403.

TABLE 2.8 UNS Metal Family Designations

The first letter (followed by five digits)	Alloy category (assigned to date)
Axxxxx	Aluminum and its alloys
Cxxxxx	Copper and its alloys
Exxxxx	Rare-earth metals, and similar metals and alloys
Fxxxxx	Cast irons
Gxxxxx	AISI and SAE carbon and alloy steels
Hxxxxx	AISI and SAE H-steels
Jxxxxx	Cast steels (except tool steels)
Kxxxxx	Miscellaneous steels and ferrous alloys
Lxxxxx	Low-melting metals and their alloys
Mxxxxx	Miscellaneous nonferrous metals and their alloys
Nxxxxx	Nickel and its alloys
Pxxxxx	Precious metals and their alloys
Rxxxxx	Reactive and refractory metals and their alloys
Sxxxxx	Heat- and corrosion-resistant steels (including stainless), valve steels and iron-based "superalloys"
Txxxxx	Tool steels (wrought and cast)
Wxxxxx	Welding filler metals
Zxxxxx	Zinc and its alloys

The UNS designation is not a specification and it does not establish a requirement for form, condition, properties, or quality. These are all established by individual specification literature such as ASTM alloy standards and those obtained from manufacturers.

There is no UNS for plastic pipe or related materials.

METRIC PIPE SIZES

The United States is the only major industrial country where inch sizes are commonly used. This is commonly referred to as the inch/pound, or IP system. All U.S. government projects are required to be in metric sizes, and many individual states are also requiring project documents to use metric units. Metric units are referred to as SI units, or International System of Units. Table 2.9 is compiled from several sources and presents the different conversions of pipe IP size to SI metric units. DN refers to nominal dimension, which is a soft, rounded standard obtained by multiplying the nominal pipe size by 25. Refer to Appendix D for additional conversion factors.

TABLE 2.9   Equivalent Metric (SI) Pipe Sizes

Nominal pipe size (NPS), in IP	ASHRAE std. wt. size, mm	AWWA pipe size, mm	NFPA pipe size, mm	ASTM copper tube size, mm	Nominal pipe size DN
$\frac{1}{8}$	—	—	—	6	6
$\frac{3}{16}$	—	—	—	8	8
$\frac{1}{4}$	8	—	—	10	10
$\frac{3}{8}$	10	—	—	12	12
$\frac{1}{2}$	15	12.7 & 13	12	15	15
$\frac{5}{8}$	—	—	—	18	18
$\frac{3}{4}$	20	—	—	22	20
1	25	25	25 & 25.4	28	25
$1\frac{1}{4}$	32	—	33	35	32
$1\frac{1}{2}$	40	45	38 & 38.1	42	40
2	50	50 & 50.8	51	54	50
$2\frac{1}{2}$	65	63 & 63.5	63.5 & 64	67	65
3	80	75	76 & 80	79	80
$3\frac{1}{2}$	—	—	89	—	90
4	100	100	102	105	100
$4\frac{1}{2}$	—	114.3			115
5	—	—	127	130	125
6	150	150	152	156	150
8	200	200	203	206	200
10	250	250	—	257	250
12	300	300	305	308	300
14	—	350	—		350
16	—	400	—		400
18	—	—	—		450
20	—	500	—		500
24	—	600	—		600
28					700
30					750
32					800
36					900
40					1000
44					1100
48					1200
52					1300
56					1400
60					1500

HANGERS AND SUPPORTS

This section will provide a general overview of basic methods and devices required to properly support the piping network within a facility. The utility systems that are the subject of this handbook generally are low-pressure and low-temperature networks, whose support requirements are not considered complex.

GENERAL

The entire piping network must be attached to the building structure in a manner that will ensure adequate support under all static and dynamic operating conditions. These conditions include:

1. Adequate connection to both the structure and the pipe
2. An allowance for weight of the filled pipe
3. Slope
4. Expansion of the pipe run
5. Seismic forces
6. Pipeline temperature

CODES AND STANDARDS

The following standards are often referred to regarding the selection and design of pipe supports:

1. ANSI B 31: series.
2. MSS SP-58: material and design of pipe supports.
3. MSS SP-69: selection and application of pipe supports.
4. MSS SP-89: fabrication and installation of pipe supports.
5. NFPA-13: sprinkler systems. This standard has wide application to pipe supports other than those for sprinkler systems.

SYSTEM COMPONENTS

The pipe support is an assembly of components including a device or method used as a direct attachment to the structure, a means of securing the pipe, and a connecting member extending from the structure attachment to the device used to secure the pipe. Other devices include pipe restraints or anchors and pipe guides.

Attachment to Structure

Various methods and devices are used to attach the support to the structure. Typical attachments include:

1. An insert (Fig. 2.39a) is installed at the time the slab is poured. An anchor or expansion bolt is installed after the slab is poured.
2. A beam clamp (Fig. 2.39b) provides attachment to exposed structural members. Beam clamps can also be welded to the beam.
3. Brackets (Fig. 2.39c) attach to walls.

The manufacturer of each type of attachment will have specifications for the maximum loading permitted for each type of attachment.

Hanger Rods

The hanger rod is usually threaded and connects the attachment to the hanger that is threaded to receive the rod. The diameter of the rod is selected by the amount of weight it will support, which is determined by the area of the rod at the root of the thread. The safe weight capable of being supported by different rod diameters is shown in Table 2.10. Table 2.11 conforms to UL and FM requirements for minimum acceptable rod diameters for various sizes supporting individual pipes. A rod diameter less than ⅜ in (12 mm) is not recommended to be used as support for any pipe inside a facility.

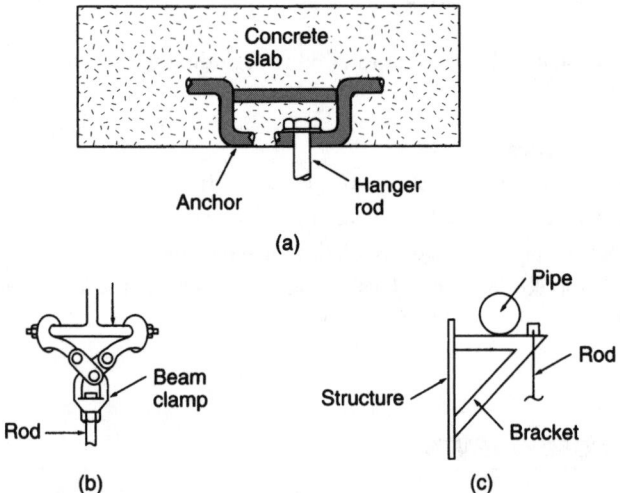

FIGURE 2.39 Typical structure attachments. (a) Insert; (b) beam clamp; (c) bracket.

TABLE 2.10 Load Rating of Threaded Rods

Nominal rod diameter, in	Root area of thread, in^2	Maximum safe load at rod temperature of 650°F, lb
$\frac{1}{4}$	0.027	240
$\frac{5}{16}$	0.046	410
$\frac{3}{8}$	0.068	610
$\frac{1}{2}$	0.126	1,130
$\frac{5}{8}$	0.202	1,810
$\frac{3}{4}$	0.302	2,710
$\frac{7}{8}$	0.419	3,770
1	0.552	4,960
$1\frac{1}{8}$	0.693	6,230
$1\frac{1}{4}$	0.889	8,000
$1\frac{3}{8}$	1.053	9,470
$1\frac{1}{2}$	1.293	11,630
$1\frac{5}{8}$	1.515	13,630
$1\frac{3}{4}$	1.744	15,690
$1\frac{7}{8}$	2.048	18,430
2	2.292	20,690
$2\frac{1}{4}$	3.021	27,200
$2\frac{1}{2}$	3.716	33,500
$2\frac{3}{4}$	4.619	41,600
3	5.621	50,600
$3\frac{1}{4}$	6.720	60,500
$3\frac{1}{2}$	7.918	71,260

TABLE 2.11 Recommended Rod Size for Individual Pipes

Pipe size, in	Rod size, in
2 and smaller	$\frac{3}{8}$
$2\frac{1}{2}$ to $3\frac{1}{2}$	$\frac{1}{2}$
4 and 5	$\frac{5}{8}$
6	$\frac{3}{4}$
8 to 12	$\frac{7}{8}$
14 and 16	1
18	$1\frac{1}{8}$
20	$1\frac{1}{4}$
24	$1\frac{1}{2}$

Hangers

A hanger is the device used to secure the pipe to the hanger rod. It must not distort, cut, or abrade any pipe while allowing free movement. There is a wide variety to choose from, including:

1. Pipe clamps (Fig. 2.40a) support pipes passing through openings in floors.

2. Saddles (Fig. 2.40b) support pipe from floors.

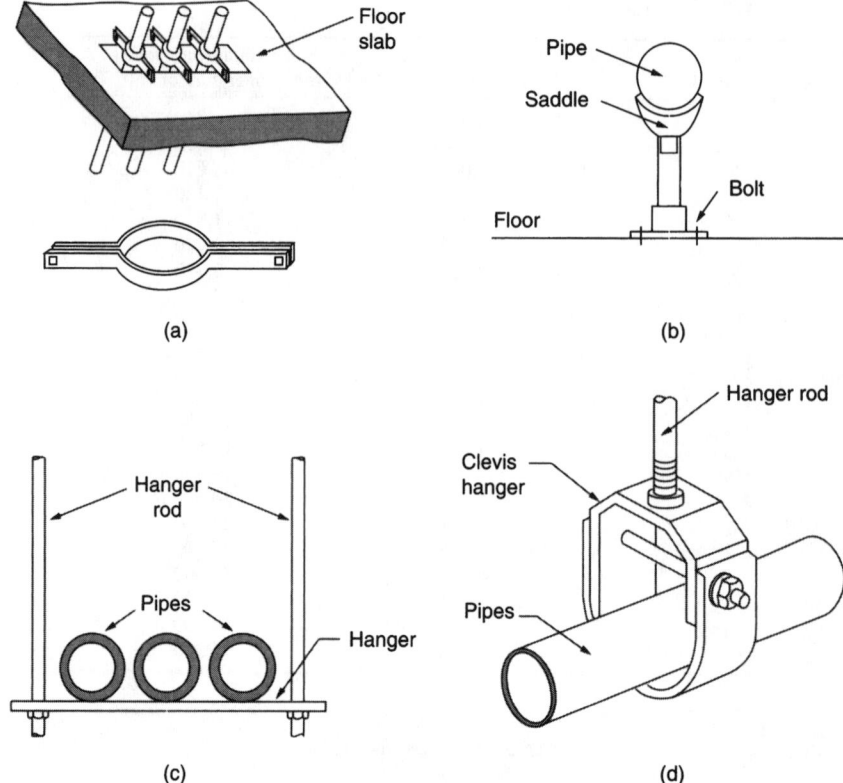

FIGURE 2.40   Typical pipe hanger types. (*a*) Pipe clamp; (*b*) adjustable saddle; (*c*) trapeze hanger; (*d*) adjustable clevis hanger.

3. Trapeze hangers (Fig. 2.40*c*) support multiple pipes.

4. A clevis hanger (Fig. 2.40*d*) is an adjustable hanger for a single pipe.

When pipes are insulated, precautions must be taken in order not to distort softer fiberglass insulation or create a break in the vapor barrier, if one is provided. Refer to Fig. 2.41 for an illustration of insulation over a split ring hanger and Fig. 2.42 for an illustration of protective inserts used with a clevis hanger.

The manufacturer of each type of hanger will have specifications for the maximum loading permitted for each type.

Hanger Spacing

The horizontal spacing of hangers depends on the size of the pipe and the requirement to prevent sagging based on the weight of the pipe filled with water, weight

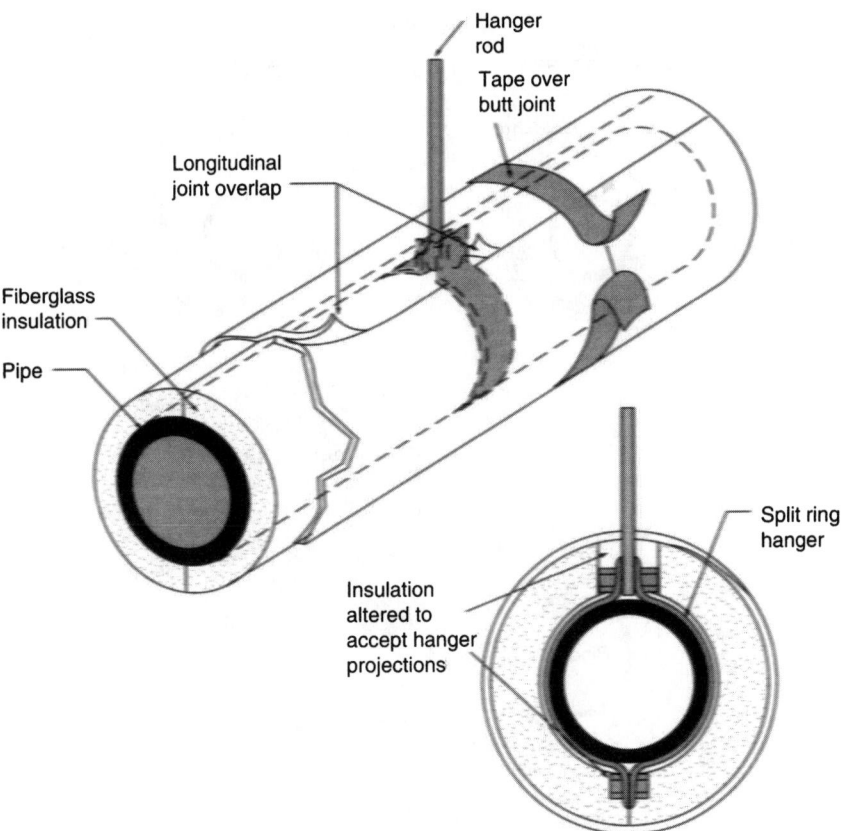

Hanger
rod

Tape over
butt joint

Longitudinal
joint overlap

Fiberglass
insulation

Pipe

Split ring
hanger

Insulation
altered to
accept hanger
projections

FIGURE 2.41 Insulation over a split ring hanger. (*Courtesy Mica.*)

of insulation, and other heavy valves and fittings on the pipe run. For pipe where
no other requirements exist, the recommended maximum spacing is as follows:

1. For steel piping 1 in and smaller, 8 ft O.C.
2. Steel pipe 1¼ in and larger, 12 ft OC.
3. Copper tubing 1¼ in and smaller, 6 ft O.C.
4. Copper tubing 1½ in and larger, 10 ft O.C.
5. SS tubing, same as copper pipe
6. Cast iron pipe, every 5 ft and at each fitting. For pipe length of 10 ft, hangers
   shall be equal to pipe length.
7. Glass pipe, 10 ft O.C. and extra hangers at places where three joints exist in a
   10-ft section.
8. Plastic pipe-all sizes and types, continuously supported to prevent sagging by
   means of an angle iron or 18-gauge angle, supported every 10 ft

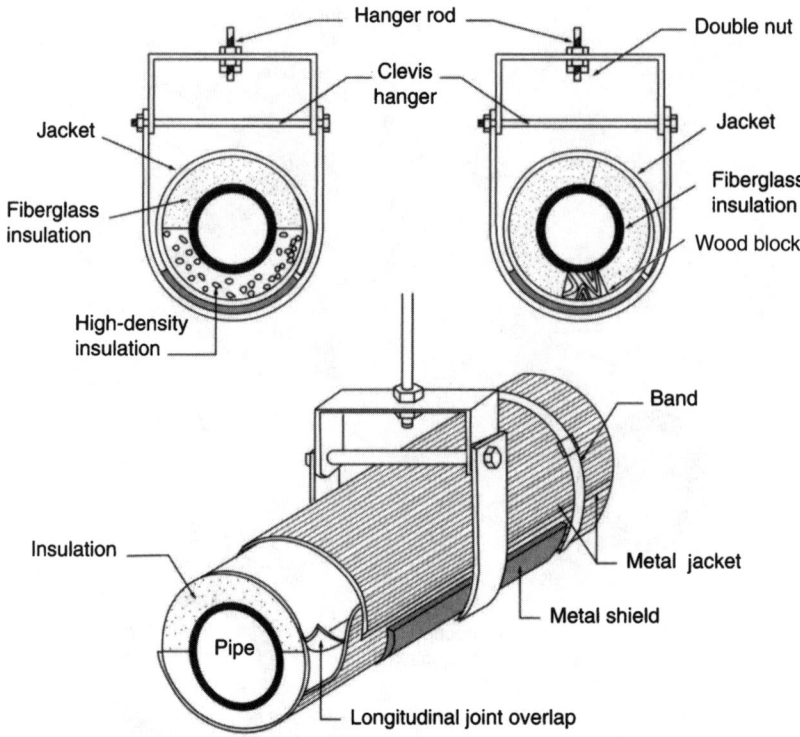

FIGURE 2.42   Clevis hanger with protective inserts. (*Courtesy Mica.*)

Vertical piping shall be supported as follows:

1. Cast iron pipe with caulked joints, each floor or 20 ft O.C.
2. Cast iron pipe with no-hub joints, each floor or 10 ft O.C.
3. Copper and SS tubing, 10 ft O.C.
4. Glass pipe 2 in and smaller, alternate floors or 20 ft O.C.
5. Glass pipe 2½ in and larger, each floor or 10 ft O.C.
6. All other piping, alternate floors or 25 ft O.C.
7. A base support or hanger support at the offset of all piping from vertical to horizontal.

Calculating Weight of Pipe and Contents

For systems with limited complexity that comply with conventional spacing, the hanger assembly shall be selected to support the total weight of the piping. The procedure used to calculate the total weight is as follows:

1. Calculate the weight of the pipe using the following formula:

$$W = F \times 10.68 \times T \times (\text{O.D.} - T) \qquad (2.2)$$

where W = weight of pipe, lb/ft
  F = relative weight factor, see Table 2.12
  T = wall thickness of pipe
  O.D. = outside diameter of pipe, inches

Note: Add a safety factor of 5 percent for tolerance in pipe manufacturing.

2. Determine the weight of the water contained in the pipe. Refer to Table 2.13 for the water weight based on the size of the pipe. The quantity and weight of

TABLE 2.12 Relative Weight Factor for Metal Pipe

Pipe	Weight factor*
Aluminum	0.35
Brass	1.12
Cast iron	0.91
Copper	1.14
Stainless steel	1.0
Carbon steel	1.0
Wrought iron	0.98

*Average plastic pipe weighs one-fifth as much as carbon steel pipe.

TABLE 2.13 Weight of Steel Pipe and Contained Water

IPS, in	Weight per foot, lb	Length in feet containing 1 ft³ of water	Gallons in 1 linear ft
1/4	0.42		0.005
3/8	0.57	754	0.0099
1/2	0.85	473	0.016
3/4	1.13	270	0.027
1	1.67	166	0.05
1 1/4	2.27	96	0.07
1 1/2	2.71	70	0.1
2	3.65	42	0.17
2 1/2	5.8	30	0.24
3	7.5	20	0.38
4	10.8	11	0.66
5	14.6	7	1.03
6	19.0	5	1.5
8	25.5	3	2.6
10	40.5	1.8	4.1
12	53.5	1.2	5.9

TABLE 2.14 Relative Weight Factor for Insulation

Nominal pipe size*	Nominal insulation thickness*										
	1"	1½"	2"	2½"	3"	3½"	4"	4½"	5"	5½"	6"
1	0.055	0.10	0.16	0.23	0.31	0.40					
1¼	0.048	0.12	0.15	0.22	0.30	0.39					
1½	0.063	0.11	0.21	0.29	0.38	0.48					
2	0.080	0.14	0.21	0.29	0.37	0.47	0.59				
2½	0.091	0.19	0.27	0.36	0.46	0.58	0.70	0.83			
3	0.10	0.17	0.25	0.34	0.44	0.56	0.68	0.81			
3½	0.14	0.23	0.31	0.41	0.54	0.66	0.78	—	0.97		
4	0.13	0.21	0.30	0.39	0.51	0.63	0.77	0.96	1.10		
5	0.15	0.24	0.34	0.45	0.58	0.71	0.88	1.04	1.20		
6	0.17	0.27	0.38	0.51	0.64	0.83	0.97	1.13	1.34		
8	—	0.34	0.47	0.66	0.80	0.97	1.17	1.36	1.56	1.75	
10	—	0.43	0.59	0.75	0.93	1.12	1.32	1.54	1.76	1.99	
12	—	0.50	0.68	0.88	1.07	1.28	1.52	1.74	1.99	2.24	2.50
14	—	0.51	0.70	0.90	1.11	1.34	1.57	1.81	2.07	2.34	2.62
16	—	0.57	0.78	1.01	1.24	1.49	1.74	2.01	2.29	2.58	2.88
18	—	0.64	0.87	1.12	1.37	1.64	1.92	2.21	2.51	2.82	3.14
20	—	0.70	0.96	1.23	1.50	1.79	2.09	2.40	2.73	3.06	3.40
24	—	0.83	1.13	1.44	1.77	2.10	2.44	2.80	3.16	3.54	3.92

*For pipe sizes and insulation thicknesses not shown in above chart, use Eq. (2.4).

water shown for steel pipe should be used for all pipe. They are well within the accuracy and safety factors expected for selection of hangers.

3. Calculate the weight of insulation on the pipe. For standard insulation and thickness, use the following formula:

$$W = F \times D \qquad (2.3)$$

where W = weight of insulation, lb/ft^3
       F = relative weight factor for insulation, see Table 2.14
       D = density of insulation, lb/ft^3

TABLE 2.15 Thermal Expansion of Piping Materials

Piping Material	Coefficient of Thermal Expansion in/in/°F per ASTM 0696
Aluminum	0.000098
Brass-red	0.000009
Brass-yellow	0.000001
Copper	0.00001
Cast iron	0.0000056
Carbon steel	0.000005
Ductile iron	0.0000067
Stainless steel	0.0000115
Borosilicate (glass)	0.0000018
ABS	0.00005
CAB	0.00008
CPVC	0.000035
FRP-epoxy	0.000012
FRP-polyester	0.000017
FRP-vinyl ester	0.00001
FEP	0.000005
HDPE	0.00011
PAEK	0.000023
PB	0.00015
PE	0.00008
PEEK	0.000026
PEX	0.00093
PFA	0.000066
PP	0.000065
PPS	0.00003
PVC	0.00004
PS	0.000031
PTFE	0.000038
PVDF	0.000096
Saran	0.000038
Styrene	0.00006

When insulation thickness is not standard or the pipe size is outside the values in Table 2.11, use the following formula:

$$W = D \times 0.0218 \times T \times (\text{O.D.} \times T) \qquad (2.4)$$

where W = weight of insulation, lb/ft^3
      D = density of insulation, in lb/ft^3
      T = wall thickness of insulation, inches
  O.D. = outside diameter, inches

Add all three of the above figures together to find the total weight for any run of pipe that must be supported. Then, determine the number of hangers required so that the weight on any single hanger is less than the smallest weight allowed for any component of the specific hanger assembly chosen.

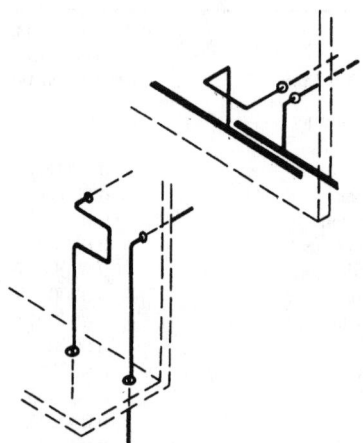

FIGURE 2.43 Methods of providing pipeline flexibility.

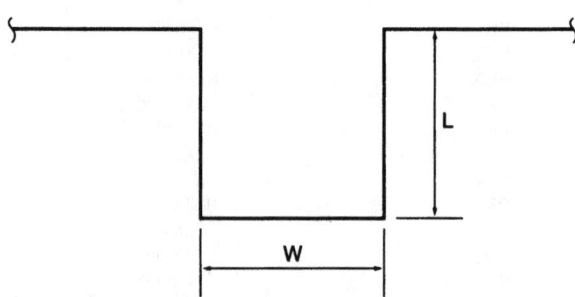

FIGURE 2.44 Expansion loop.

Pipe Expansion

All piping materials will change dimensions as they expand and contract due to temperature changes above and below the temperature at which they were installed. These dimensional changes result in a change in the length of run of the pipe. If the piping network does not have the necessary flexibility to compensate for the expansion and contraction, the resultant stress could cause damage to connected equipment or cause the pipe to leak or fail completely. Thermal expansion of piping materials is given in Table 2.15.

The thermal expansion for plastic pipe is not linear, but increases with increasing temperature. In the range of temperature encountered in general utility piping, this is not a consideration.

Expansion Loops

The stress and forces resulting from thermal expansion must be compensated for by providing flexibility in the piping network. Flexibility is generally provided by changes in direction of the pipe and with offsets. These are generally enough to allow adequate compensation for normal expansion. Branch piping connections to provide flexibility are illustrated in Fig. 2.43. Where their installation is not possible, a mechanical expansion joint must be provided.

For long runs of straight pipe or when there is doubt as to adequate flexibility, the method used most often to provide required flexibility is the expansion loop (Fig. 2.44). In the figure, there is no specific dimension for W; however, this dimension is generally the same as L, but could be as little as $\frac{1}{2}L$ when necessary for installation space requirements. To calculate the length required to absorb movement without damage, the following formula is used:

$$L = 1.44 \sqrt{D \times \Delta C} \qquad (2.5)$$

where L = pipe leg length, ft
  D = nominal outside diameter of pipe, in
  ΔC = change of dimension of pipe run, in (This figure is calculated by multiplying the pipe run by the rate of thermal expansion found in Table 2.12.)

Seismic Considerations

All utility piping must be installed in conformance with code requirements for the earthquake zone where the facility is located. A map of the United States showing the earthquake zones is shown in Fig. 2.45.

The model and local codes provide guidance and criteria for the design of piping supports. In general, sway bracing is provided to prevent excessive pipe movement. The design of pipe supports to resist earthquake loads is outside the scope of this handbook.

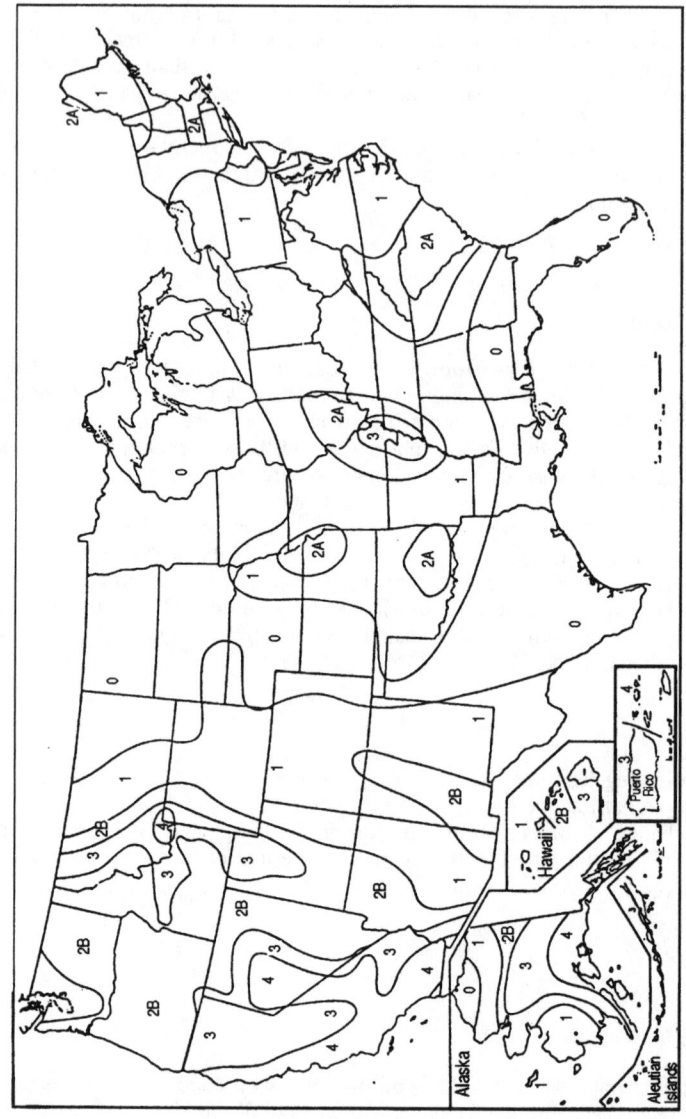

FIGURE 2.45 Earthquake zone for the United States. Zone 0: no earthquake damage areas. Zone 1: minor damage areas. Zone 2A: minor to moderate damage areas. Zone 2B: moderate to major damage areas. Zone 3: major damage areas. Zone 4: areas near earthquake faults.

REFERENCES

Air Conditioning, Heating and Ventilating Magazine data sheet, "Formulas for Sizing Control Valves," April 1959.

Bhasin, V. C., "Actuator Selection," *Chemical Engineering Magazine,* pp.140–145, November 1990.

Coleman, D. C., and R. W. Evans, "Fundamentals of Passivation and Passivity in the Pharmaceutical Industry," *Pharmaceutical Engineering Magazine,* March/April 1993.

Delstar Electropolish, "Electropolishing."

Dillon, C. P., et al, "Stainless Steels for Bioprocessing," *BioPharm Magazine,* April–July 1992.

Fagan, D., "Piping Materials and Jointing Methods for Plumbing Systems," *Heating, Piping and Air Conditioning Magazine,* October 1993.

Frankel, M. "Selecting Plastic Pipe," *Plumbing Engineer Magazine,* pp.26–31, April 1993.

Harris, C. M. (ed.), *Handbook of Utilities and Services for Buildings,* McGraw-Hill, New York, 1990.

International Copper Association, "Water Tube from Copper and Other Materials," April 1990.

Henon, B., "Welding of WDI and WFI Piping Systems for a Bioprocess Application," *Pharmaceutical Engineering Magazine,* pp.18–24, December 1993.

Issacs, M., and R. B. Setterlund, "A Guide to Pipe Jointing Methods."

Nayar, M. (ed.), *Piping Handbook,* 6th ed., McGraw-Hill, New York, 1992.

Pollock, W. I., "Baffled by How Alloys are Labeled?," *Chemical Engineering Magazine,* pp. 169–174, October 1992.

Reid, R. N., "Double-Walled Pipe and Plumbing Systems," *Heating, Piping and Air Conditioning Magazine,* pp. 53–56, April 1993.

Reschenberg, A., "Taking the Mystery Out of Specifying Sanitary Stainless Steel Finishes," *Powder and Bulk Engineering Magazine,* pp. 21–32, April 1993.

Robinson, T., "Compare Nonmetal and Metal Valves," *Chemical Process Magazine,* pp. 25–30, December 1993.

Sinha, D., "Choose the Right Thermoplastic for High Purity Water," *Chemical Process Magazine,* pp. 84–87, October 1991.

Ulanski, W., *Valve and Actuator Technology,* McGraw-Hill, New York, 1991.

CHAPTER 3
SOLID-LIQUID SEPARATION AND INTERCEPTORS

This chapter will describe the methods used to separate suspended solids (particulates) and liquids from a feedwater stream. Separation is characterized by the recovery of all of the processed water and having the flow of the feedwater, in general, pass through the filter perpendicular to the filter bed or medium. General selection criteria will also be discussed.

FILTER CLASSIFICATION AND TESTING

There are many ways to classify filters:

1. *Filtration types.* Depth, surface, and screen are general filter types.
2. *Driving force.* Flow through the filter can be induced by pumps (pressure), centrifugal force, or gravity.
3. *Function.* The goal of the filtration process is either retention of the dry solid when the filter cake is of value or disposal of the filter cake when process liquid is of value.
4. *Operating cycle.* The cycle of operation can be batch mode or continuous.
5. *Nature of the solid.* The accumulation of solids within a filter matrix can be either deformable (compressible) or rigid (incompressible).

The classification of filters is not exclusive and the distinction between them is arbitrary. Here the characterization of filters will be based on the type of filtration, the characteristic generally used in utility and service water filtration systems.

Interceptors, strainers, and filters are all devices used to reduce (or remove) and retain suspended solids. Other separation processes, such as sedimentation and centrifugation, that are used to treat large quantities of water or for dewatering, are outside the scope of this book. Design and selection criteria for specific contaminant removal are provided where appropriate in various other chapters discussing individual systems.

The distinction between filtration and water purification is arbitrary. Methods such as membrane and membrane exchange filtration that removes ions, allows preferential passage of specific substances, and does not conform to the previous definition of filtration are considered water purification methods and are discussed in Chap. 4.

GENERAL

Feedwater, raw water, and source water are various ways of referring to a solution whose components are intended to be separated. *Filtration* is the process used for separation and retention of suspended and colloidal particles by mechanical capture and adsorption from fluids by passage through a porous medium. Mechanical capture physically prevents a contaminant particle from passing through a barrier with openings (pores). Adsorption is the attraction to and adhering of a particle to the surface of the filter medium. Adsorption can occur even if the pore is larger than the particle. This attraction is due to a variety of surface chemical forces between the particle and filter medium.

The mechanical properties of the particles suspended in the water stream must also be considered. At one extreme are solid, undeformable particles such as sand or quartz, and at the other extreme are gelatinous or deformable materials such as synthetic colloids and bacteria. Because they can deform, they are more likely than hard particles of the same size to pass through a filter.

FILTER CATEGORIES AND DEFINITIONS

Screen, surface, and depth filtration are the three broad categories of the filtering process.

A screen filter is best thought of as a single, thin layer of a material that has a symmetrical arrangement of openings or passages called pores. These pores trap all particles larger than the pore size on the surface of the filter. This process is called sieving, or size exclusion, and is the classic filtration method. Sieving can also be referred to as screening or straining. Screen filtration is essentially absolute because any particle larger than the pore size cannot pass through. Another mechanical capture mechanism, called bridging, occurs as particles captured by direct interception form a particle mat, or bridge, across the filter medium. By partially blocking the filter pores, this bridge or filter cake may produce a smaller filter pore structure that will aid in particle capture. Examples of screen filters are woven metal, nylon, and dacron mesh. Cast polymeric membranes are used where the smallest size pores are required for submicronic and macromolecular separations.

A surface filter is thicker than the screen filter and constructed from thick or multiple layers of filter media, often glass or polymeric fibers. When the water passes through a surface filter, particles larger than the spaces within the fiber matrix are retained, primarily on the surface. Smaller particles are trapped within the matrix, giving this type of filter the properties of both a screen and depth filter.

A depth filter relies on the density and thickness of the layers to mechanically trap the particles, and it will retain relatively large quantities of them. Depth filtration occurs on the surface and throughout all or part of the filter medium as the water passes through a complex network of flow channels. The particles are retained by random adsorption and mechanical entrapment. Depth filters can be of two types, granular and preformed. Preformed depth filters are composed of fibrous or sintered materials that have a random pore structure. Granular depth filters have either a graded or consistent density of granular media and typically are long in length. Graded granular filters have layers of media that become progressively denser through the matrix as water flows through them. Constant density granular filters' have the same size filter media or openings throughout the matrix.

A filter that is hydrophilic is one that has an affinity for water; it can be wetted with almost any liquid. A hydrophobic filter is one that cannot be wetted by an aqueous solution. Some filter materials may leach substances into the fluid as it is processed, thereby affecting its purity. Such substances, called extractables, can be minimized by preflushing. There is a test for plastics conforming to USP class VI that is used to ensure that there will be no adverse reaction of body fluids to extractables from filter housing or media materials.

The molecular weight of any compound is measured in daltons. Some filter media measure passage through the filter by molecular weight for separation of one compound from another.

FILTER RATINGS

Filters and strainers are rated in several ways. Absolute and nominal ratings are based on the size particle the filter is expected to capture and retain. Particles are measured in micrometers (microns), which is 1/1,000,000 m (1/25,000 in) and abbreviated μm. This rating is a single number called the micrometer (micron) rating. The micron rating of a filter or strainer can be absolute or nominal. These ratings are often misunderstood and this is an area of confusion in the filtration industry. Another method is called the beta rating, which is based on actual particle counts of different particle sizes of both the influent and effluent liquid stream. The beta rating is considered the most accurate rating measurement of a filter. Refer to Table 3.1 for the relationship between beta value and percent removal efficiency.

Efficiency is a measure of particle removal. It indicates what percent of particles above a certain size will be retained. For absolute rated filters, the rated pore size indicates 100 percent removal and is based on the log reduction values associated with bacterial retention testing. Because the pore size of some filters is not well defined, it is not possible to assign those filters an absolute rating. Instead they are given a nominal pore rating, which indicates the particle size above which a predictable percentage of particulates will be retained. As an example, a nominally rated 1.0-μm depth filter will remove 90 to 95 percent of all particles 1.0 μm or larger. For a surface filter of the same rating, the efficiency would be 99.99 percent.

An absolute micron rating indicates the smallest size particle that the filter will capture; no particles of that diameter or larger will pass through the filter. The absolute rating generally depends on sieving, since the capture of particles by adsorption is never assured. Since absolute ratings are generally unrealistic for most services, nominal ratings are the most common method used to rate filters. One exception is in pharmaceutical service, where absolute ratings are required to assure that all particulates of a certain size are removed.

Nominal ratings allow the filter rating to consider particles retained by adsorption. The nominal rating has no generally accepted definition in the industry, and there are no industry standards. As defined by ANSI, the nominal rating is an arbitrary micrometer value indicated by the filter manufacturer. Due to its lack of reproducibility, this rating is depreciated. The ambiguity of this rating method makes it difficult to achieve reliable and consistent results. Many manufacturers use different methods to rate their filters, for example, expressing the results gravimetrically, which does not represent the particle size and number in the effluent stream. Some have specific test conditions that do not represent the actual conditions for which the filters will be used. These test conditions may use fine or coarse particles such as AC test dust, latex beads, carbon fines, or bacteria. The nominal rating can be used as a guideline, provided that the micron rating includes the percent removal efficiency rating of that micron size.

TABLE 3.1 Relationship Between
Beta Value and Removal Efficiency

Beta ratio	Removal efficiency %
1	0
2	50
10	90
100	99
1,000	99.9
5,000	99.98
10,000	99.99

Void volume of preformed fibrous media is the ratio of pore area to the fiber diameter of the filter media. If all other factors are equal, the medium with the greatest void volume will have the longest life and lowest initial clean pressure drop per unit thickness. Factors such as strength, compressibility of the fiber material under pressure (which reduces void volume), cost, and compatibility of the media with the water contaminants being removed should all be considered when selecting a filter for a specific application.

MEMBRANE FILTER TESTING

An important feature of a membrane filtration system is its ability to be tested before and after filtration runs. Testing can detect a damaged membrane, ineffective seals, or a system leak that may result in passage of contaminants that the filter is designed to trap. These tests are commonly called integrity tests. Testing before and after a run will ensure that the entire system is intact, thereby validating the process. Prior to testing, cleaning to remove large-scale contamination (and sterilizing the filter and apparatus if necessary to ensure elimination of microbial contamination) is required.

The type of test selected is dependent on the specific filter chosen. However, if the previous history of a specific filter is not available, the only accurate method of testing the filter is to place it in service and run an on-site fouling and compatibility test.

Air Permeability Test

An air permeability test is normally used to test wound cartridges. It is a simple, nondestructive test that correlates well with filter performance and it is considered more revealing than micron rating.

Bubble Point Test

Membrane filters have discrete, uniform passages from one side of the membrane to the other which, in effect, are fine uniform capillaries. The bubble point test is based on the fact that a liquid is held in these capillary-like structures by surface tension and the minimum pressure required to force this liquid out of the capillary space is a measure of the capillary diameter. The pressure required is inversely proportional to the largest pore size. After the filter is wetted, air pressure upstream of the filter is increased.

There are two widely used variations of the bubble point test. The first is the visual test. For this variation, the downstream side is watched for the appearance of bubbles, which indicate that the air is passing through the capillaries. The pressure that produces a steady, continuous stream of bubbles is the bubble point pressure. The second variation is the monitored method, where a pressure drop will occur as the gas begins to flow through the filter.

It is not necessary to determine the exact pressure of a given filter to prove its integrity. If the pressure exceeds the minimum point determined by the manufacturer of the filter, its integrity is assured. The bubble point test is also used to test the integrity of the filter cartridge.

Diffusion Test

In a high volume system where a large volume of water must be displaced before bubbles can be detected, a diffusion test should be performed instead of the bubble point test. This test is based on the fact that in a wetted membrane filter under pressure, air flows through the water-filled pores of the filter at a differential pressure below the bubble point pressure by a diffusion process following Fick's law. In small filters, the flow of air is very slow. But in a large filter it is significant and can be measured to perform a sensitive filter integrity test. In a wetted filter, a constant air pressure is applied at approximately 80 percent of the bubble point pressure established for that particular filter.

There are two widely used variations of the diffusion test. The first is the forward flow test, which relies on direct measurement of the diffusive gas flow rate. This flow rate is measured either by instruments placed in the gas flow upstream of the filter or by calculating the volume of airflow according to the rate of flow of displaced water downstream of the filter. The second method, called the pressure decay method, calculates the loss of diffusion gas pressure from a known volume of gas over a period of time.

Water Breakthrough Test

Similar to the bubble test except that water is used instead of air. Water pressure is increased on the upstream side of the filter, and the pressure that results in a steady stream of water downstream of the filter is recorded. The breakthrough pressure must be correlated to empirical data on contaminant retention from the manufacturer in order to be a valid test.

Water Intrusion Test

Also called the water pressure integrity test, this is often used for hydrophobic filters (which resist wetting by water). This test requires that the filter be wetted by an alcohol/water mixture. Water pressure is applied upstream of the filter and the pressure decay is measured. Care must be taken to test with water from the same source because of variations in surface tension. When used as a vent filter, the membrane must be dried before being placed back in service.

STRAINERS

GENERAL

A strainer is a closed vessel with a cleanable screen or mesh generally used to remove and retain foreign particles larger than 45 μm (325 mesh) from flowing liquids. If a particle is visible to the naked eye, a strainer should be chosen to remove it from the liquid stream. If the device retains particles finer than 45 μm, it is generally considered a filter. The relationship between mesh and opening size is given in Table 3.2. The difference between filters and strainers is one of semantics; a strainer could be considered a coarse filter.

TABLE 3.2 Relationship Between Mesh and Opening Size

Mesh size	Particle diameter, μm
Strainer	
4	5,205
8	2,487
10	1,923
14	1,307
18	1,000
20	840
25	710
30	590
35	500
40	420
45	350
50	297
60	250
70	210
80	177
100	149
120	125
140	105
170	88
200	74
230	62
270	53
Filter	
325	44
400	37
550	25
800	15
1,250	10

TYPES OF STRAINERS

Types of strainers include Y and basket (or bucket) types. Basket strainers are available in self-cleaning models. All types are available with a large variety of jointing methods for insertion into a pipeline, including soldered types available for copper piping. They are available as off-the-shelf models capable of meeting any reasonable need. Manufacturers can construct units for special conditions.

The Y-type strainer, illustrated in Fig. 3.1, is compact in design and is considered for use where space is at a premium. Frequent cleaning is often required. Its construction makes it a good choice for high pressure applications and for gases such as steam, natural gas, and compressed air where pressures are higher and amount of dirt present is low. It has a smaller dirt-holding capacity than similar sizes of basket strainers. It is installed in a pipeline with its strainer element in the down position, and can be positioned either horizontally or vertically. Very often, a valved pipe will be put in the removable end of the strainer so that the accumulated debris can be easily blown out while keeping the line in service.

The basket strainer, illustrated in Fig. 3.2, gets its name from the upright, perforated basket used to trap particles. It is installed upright in the strainer body, and the top of the strainer must be removed for cleaning. Because of its large size, it has the ability to store large quantities of dirt and so has a lower pressure loss than a similar sized Y strainer.

MATERIALS OF CONSTRUCTION

Strainer Bodies

The materials used most commonly for strainer bodies are cast iron, bronze, carbon steel, stainless steel, and plastic.

Because of its low initial cost, cast iron is the most popular strainer body. It is used in systems where the pressure and temperature of the water are not high and the system is not subject to high thermal or mechanical shock. Cast iron is mostly

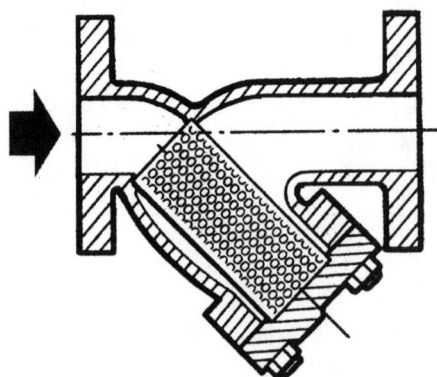

FIGURE 3.1 Y-type strainer.

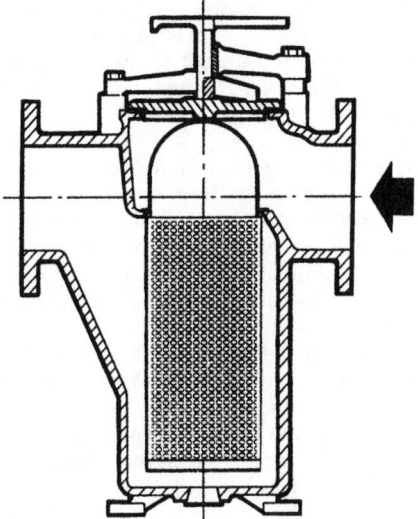

FIGURE 3.2 Basket-type strainer.

used for larger size potable water lines and many nonpotable water systems in addition to a variety of other product and process uses.

A bronze body is preferred for brackish, saline, and seawater service. It is also often used for potable water services in smaller sizes. Its cost is about double that of cast iron.

Carbon steel bodies are used where moderately high temperature and pressure conditions are encountered, and where resistance to high thermal and mechanical shock is required. Carbon steel components are the materials of choice where fire hazards exist, such as in the petroleum and petrochemical industries. Their cost is equal to bronze. For high pressures and temperatures over 1000°F, chrome-moly steel is usually specified for bodies.

Stainless steel is the preferred body, basket, and screen material for the pharmaceutical, food-processing, and chemical industries because of its resistance to corrosion and contamination and ease of cleaning. Stainless steel costs about four times as much as cast iron.

Plastic strainer bodies are available in all of the materials used for pipelines. Baskets and screens of all metallic materials are also available.

Basket and Screen Construction and Materials

The actual collection and retaining of the debris, dirt, and other particles in all strainers is done by the basket or screen that is placed inside the body of the strainer. The size of the openings through a screen is referred to as mesh, and the size of the openings through a basket is referred to as perforations.

The term *mesh* describes a screen that uses a woven wire cloth manufactured from the material chosen for the intended service. The most common material for all applications is stainless steel. Mesh screens are generally available in standard sizes from 20 to 200 mesh, with a variable wire size used depending on the mesh. The mesh size does not indicate the particle size retention, since the size of the mesh opening is determined by the diameter and number of wires per inch. For example, a 100 mesh means 100 vertical and 100 horizontal strands of wire per inch. For critical applications, a screen should be selected on particle retention capability, not mesh size.

Standard perforated screens are manufactured from a light or heavy gauge sheet metal available in standard sizes generally ranging in diameter from $\frac{1}{32}$ to $\frac{1}{4}$ in. The baskets should be assembled by either welding or brazing. The strongest are of welded construction. Solder is much weaker and more easily broken. If a basket strainer is required for fine straining, it is common practice to add a wire mesh liner inside the perforated bucket since the screen alone is too weak to provide the necessary mechanical strength.

The material to be used depends both on the intended water service and the body that the strainer is installed into. The most commonly used basket materials are brass and stainless steel. Cast iron strainer bodies commonly use baskets of brass and stainless steel, depending on the intended service. Bronze bodies usually require Monel metal baskets because of the severe service required. Stainless steel baskets are used with stainless steel bodies.

The baskets are made from sheet metal with a wide variety of diameter perforations. A Y strainer is generally furnished with $\frac{1}{16}$-in perforations in sizes up to 4 in, $\frac{1}{8}$-in perforations in larger sizes for liquid service, and $\frac{3}{64}$-in perforations for steam service. If finer straining is required, a wire mesh screen fitted inside a basket is used. In this case, generally accepted practice limits the perforations in the basket to 50 percent of the wall area in order not to lose strength. In addition, this combination usually provides the best ratio of maximum flow rate with adequate strength. The mesh and basket should be an integral unit, with the mesh fastened to the basket both at the top and the bottom to prevent any debris from bypassing the unit.

By generally accepted practice, the open area perforation ratio should be about 4:1 to avoid excessive pressure drop through the unit. A smaller ratio will require frequent cleaning. Additional strainer basket area can be obtained by using a pleated basket. If finer filtration is required after the strainer has been in service, a mesh liner can be added inside the basket. If the size particle to be removed is known, the perforations should be slightly smaller.

Fluid streams may contain iron or steel particles that are small enough to pass through the finest screens. If this is a problem, a strong magnet capable of lifting several times its own weight should be suspended in the basket. The magnet should be installed so that all the water passes over it. This magnet should be encased in an inert material to prevent corrosion.

Baskets, especially when full, are not capable of withstanding the same pressure as the body. A particular phenomenon called runaway buildup is possible, in which the dirt builds up and plugs the mesh or perforations, thereby reducing the free area. The pressure in the strainer increases slowly at first, but faster and faster over time. The water velocity and pressure inside the basket escalate quickly, which causes the resultant flow to stop or be reduced to a trickle. This full-line pressure can burst the basket.

Covers

A cover is provided in order to remove and clean the basket. The most common type is bolted; the bolts must be loosened and removed to provide access to the basket. This type of cover is the strongest and should be used for high pressure applications. However, its removal is time consuming. Another type is the clamping yoke, in which threaded, tee-shaped handles are used to secure the cover to the body. Often, the cover is attached to the body with a hinge mechanism, making it very easy to remove. This type of cover is more expensive than the bolted type.

Another type of strainer is the automatic type, which does not require manual cleaning. A rotating, circular screen is used as the basket. The water inlet is directed to the inside of the basket. A rotating backwash inlet inside the basket uses the differential pressure between the atmosphere and line pressure to produce a localized reverse flow across only a portion of the basket, thereby allowing continuous cleaning. This type of strainer is appropriate for large consumers of water such as raw water inlets from rivers and lakes used for cooling and process. Automatic strainers are available to 60-in (150-mm) size.

DESIGN CONSIDERATIONS AND SELECTION CRITERIA

When selecting a strainer, the four main considerations are its physical size, friction loss through the unit, price, and ease of cleaning. The viscosity and specific gravity of the fluid, and the degree of perforation of the basket (or size mesh) all influence the pressure drop. In many cases, the size of the unit is a consideration if it is to fit into an existing space. Generally accepted practice limits the pressure drop to a maximum of 2 psi (13.8 kPa). Another general rule is to have a minimum 4:1 open area ratio of the perforations in the basket. In order to reduce the friction loss, a strainer one or more sizes larger than the pipeline into which it is installed should be selected. In some cases, the basket may become large and unwieldy when filled with debris. If this is a possibility, several smaller units in parallel should be considered. Typical friction loss through a Y-type strainer is given in Fig. 3.3, and for a basket strainer in Fig. 3.4.

Another consideration is the length of time required between cleanings. Past experience with the specific application (if available) should give a good idea of the size and amount of debris expected. Examination of the total suspended solids present in the water analysis allows calculation of the amount of debris to be expected over a period of time. This, along with any specific preference, should provide good guidelines for the proposed size of the unit to be selected.

If continuous operation of the units is critical, duplex strainers should be used. This allows one to be cleaned while permitting full use of the system. Another method is to install two strainers in line, the first with larger openings to trap large particles and the second with a finer mesh or smaller openings for small particles. This would prevent the loading up and frequent cleaning of the strainer in a one-strainer system by spreading out the cleaning load to two strainers.

There are a number of methods used to divert flow between strainers for cleaning. The most common is a multiport plug valve. For large pipe, the plug valve may get too large. For larger strainers, sliding gate valves that have synchronized discs operated by hand wheels to divert flow are often used. Another method is to use synchronized butterfly valves. The actual method selected depends on the available friction loss, available room, and cost.

Basket strainers are usually large units with a high initial cost. In general, Y strainers have a lower initial cost for the same size unit, smaller dirt-holding capacity, and larger friction loss because of their smaller bodies. It is good practice

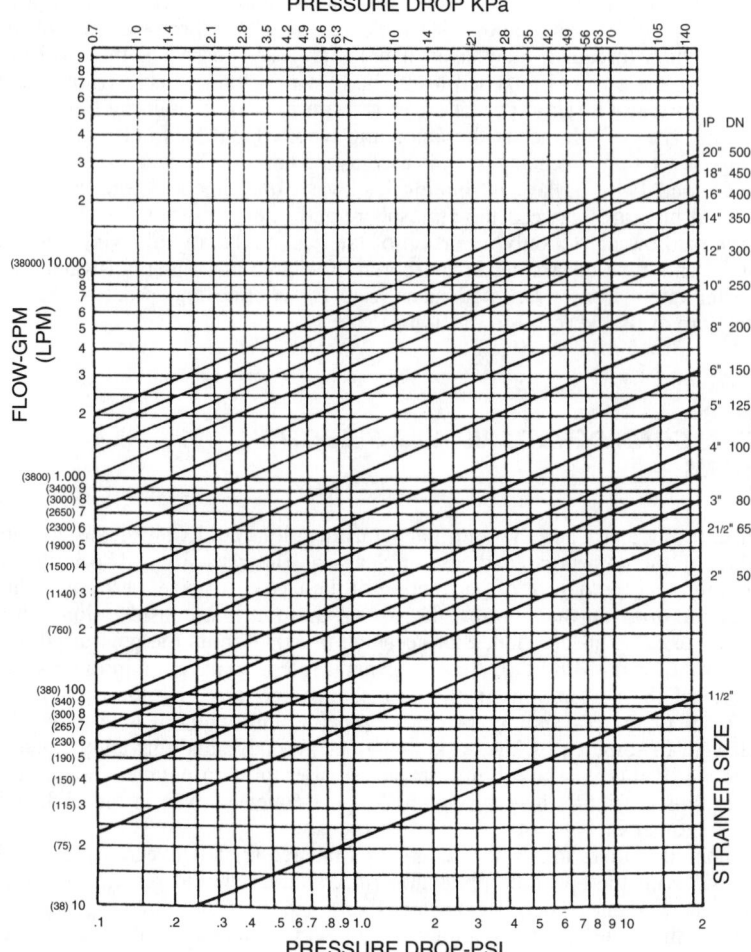

FIGURE 3.3 Pressure drop through Y-type stainers.

to allow higher pressure safety margins due to the possibility that the strainers will be placed in a position to receive water hammer shocks and water slugs. There must be enough room around the unit(s) for the necessary frequent cleaning and laydown of the basket.

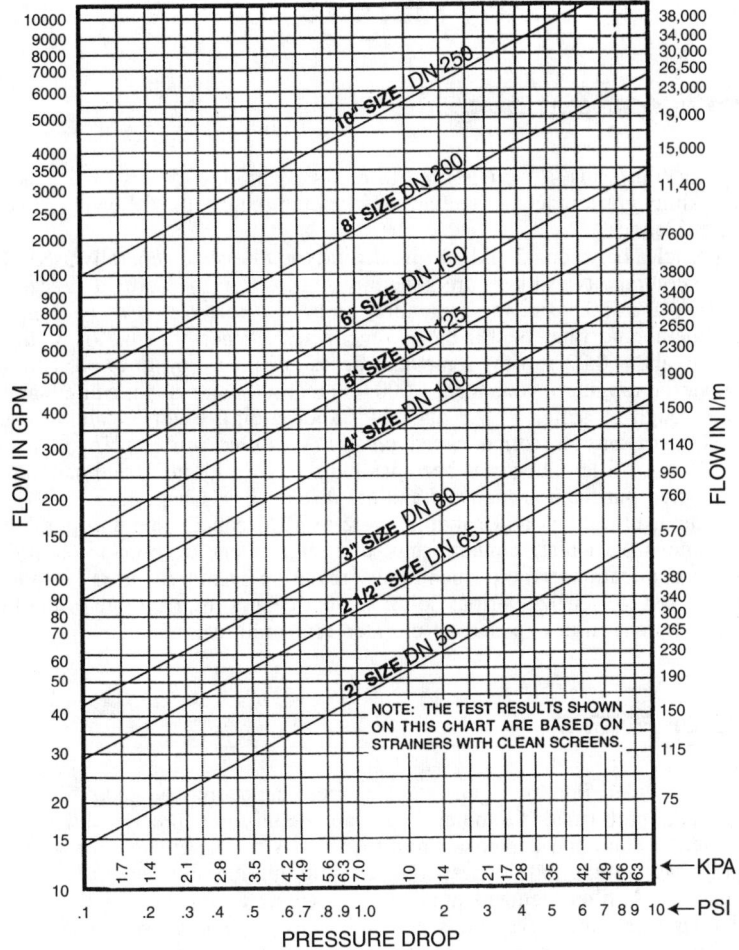

FIGURE 3.4 Presure drop through basket strainers (⅛ perforations).

FILTERS

FILTER CLASSIFICATIONS

In very general terms, a filter is used to remove particles 45 μm (325 mesh) and smaller. Although in practice there are overlapping ranges, the following represents general guidelines.

For particles between 45 and 10 μm, depth-type filters are generally used. Below 10 μm, membrane-type filters are generally selected. In the particulate size range between 10 and 0.02 μm, the separation process is referred to as *microfiltration* and the membrane filter is still the most effective. When the size of particulates falls below about 0.02 μm the removal process is called *ultrafiltration*. Ultrafilters retain material ranging in size from 1,000,000 to 1000 daltons while allowing water to pass through (1 dalton = $\frac{1}{12}$ mass of carbon atom). Filters retaining material below 1000 daltons are often called *nanofilters*. Often, different manufacturers' nomenclature of filter categories becomes blurred. Ultrafiltration and nanofiltration are considered purification methods and are discussed in Chap. 4.

In general, ultrafilters and nanofilters are used to concentrate and purify fluids and to remove particulate contaminants, and microfilters are used to clarify a solution for applications where quantitative retention is not required. In addition, filters are used to sterilize various solutions that cannot use heat due to the loss of biological activity after exposure to elevated temperatures.

TYPES OF FILTERS

Filters are divided into two general categories, depending on their filter media, granular and preformed. Granular filters are depth-type filters using individual grains such as sand and charcoal. Preformed filters can be either screen-, surface-, or depth-type, ranging in thickness from a single thin membrane element to a thick filter mat. Often, filter elements are contained within a housing called a cartridge.

Granular filters are larger units generally used to remove suspended particles larger than 10 μm. Examples of granular filter media are single or multimedia sand and activated charcoal contained in a vessel, tank, or column and septum filters.

Cartridge filters are relatively small, and generally used to clarify a previously filtered stream of water containing suspended particles smaller than 10 μm. When the media is plugged, the cartridge is replaced. Another type of cartridge filter is the capsule filter, in which the filter media is contained in a sealed housing. When plugged, the entire capsule is discarded. Commonly used materials for cartridge filters are paper, cloth, polymetric fibers, and various combinations of these.

DEEP BED GRANULAR (SAND) FILTRATION

Deep bed sand filters consist of a tank containing either silica or garnet sand of constant size (grade), or layers of a multimedia type consisting of a variety of graded material such as anthracite, silica sand, garnet sand, and quartz. This type of filter is most often used as a prefilter to remove larger-sized suspended solids in order to extend the duty cycle of finer filters downstream. It has a relatively large retention capacity of solids and removes particles 10 μm and larger. During normal operation, the raw water to be treated enters at one end (or the top) of the unit, the suspended solids adhere to the media, and the clear water collects at the other end or on the bottom.

Sand filters are either gravity or pressure type. If the tank is atmospheric and water flows through the unit with no assistance from pumps, it is a gravity filter. If the filter is in line and uses the pressure of the water supply to force its way through the filter, it is a pressure type. The pressurized filter is the most commonly used because of its smaller size and higher flow rate. A typical pressure sand filter is illustrated in Fig. 3.5.

Types of Granular Filter Media

The filter media type, particle size, and specific gravity are primarily selected for particle retention capability and ease of restratification of the media in the filter tank after backwash. A typical multimedia filter arrangement has a top bed of anthracite, a middle layer of sand, and a bottom layer of garnet. An additional layer of gravel may sometimes be used as an underlayer on the bottom of the unit to support the media bed. Anthracite usually has 1.1-mm grain size with a specific gravity of not less than 1.4. Typical depth is usually between 8 and 12 in. Silica sand has an effective size of between 0.35 and 0.50 mm with a specific gravity of 2.6. The sand layer is usually between 8 and 12 in in depth. Garnet or ilmenite typically has 0.2-mm grain size and 4.2 specific gravity. Ilmenite is commonly substituted for garnet of the same grain size, has a specific gravity of 4.5, and has

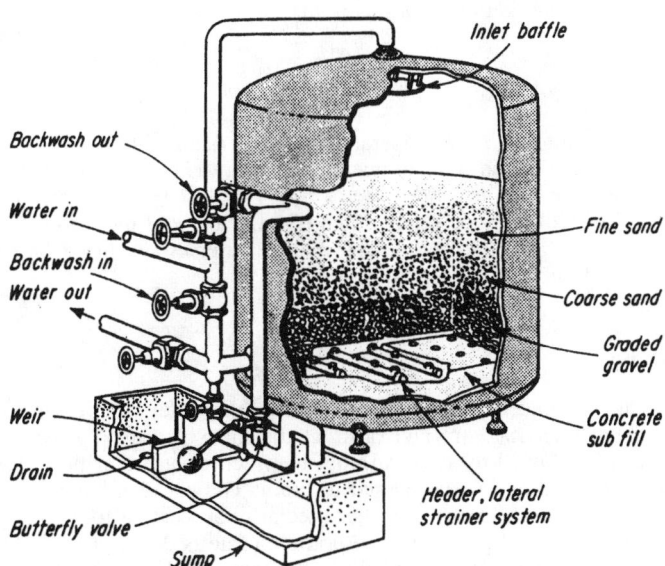

FIGURE 3.5   Typical pressure sand filter.

a typical depth of about 4 in. The gravel depth can range from 6 to 24 in, but usually falls between 10 and 18 in.

Gravity Filters

Roughing-type filters of the gravity type are used for initial treatment (pretreatment) of surface water, and are made to handle large flows economically and remove large particles. These types of filters are most often rectangular, concrete tanks. Conventional gravity filters usually operate on pretreated water and are housed in steel tanks. Gravity filters are usually rated at 2 to 4 gallons per minute per square foot (gpm/ft^2) of cross-sectional bed area. Aluminum or ferric sulfate (which is composed of multivalent ions) could be added as a coagulant to the feedwater in order to neutralize the surface charge of the colloids, thus making their removal easier. The resulting flocculated aggregates removed by the filter are discharged to drain. This type of filter is used to treat large volumes of water for process use and is outside the scope of this handbook.

Pressure Filters

Pressure filters are the most common types used for general utility and service systems. They usually operate under normal water pressure, but may require pumps to overcome excessive friction loss through the unit if available water pressure is too low. Commonly called high rate or rapid sand filters, different designs are available that offer flow rates higher than those available for gravity designs. Three different kinds of granular filters are used: single medium filters with sand media; dual media filters, consisting of a top layer of anthracite and a bottom layer of sand; and multimedia filters, consisting of a three-layer bed of anthracite on top, sand in the center, and garnet on the bottom.

Granular Filter Flow Rates

A single medium filter is usually operated at a flow rate of 3 to 4 gpm/ft^2 of cross-section area. A dual media filter is operated at a flow rate of about 6 gpm/ft^2. A multimedia filter has a typical flow rate of 6 to 15 gpm/ft^2. Another consideration is the face velocity of the feedwater, which is the velocity of the feedwater through the surface layer area of the vessel. The manufacturer establishes a face velocity, based on actual tests, that cannot be exceeded.

Backwashing

Eventually, the suspended particles removed from the water will accumulate in the top layer of the filter medium and obstruct the flow of water through the filter. These solids are removed by backwashing, in which clean water flows backwards through the filter bed and is discharged to drain. The backwash flow and volume should be capable of expanding the filter media by 50 percent to permit complete dislodging of the trapped particles. The optimum rate is determined by the manufacturer of the equipment and is usually in the range of 10 to 15 gpm/ft^2 of filter area for multimedia filters. For single and dual medium filters the rate is lower than that of multimedia filters. A backwash pump provides the pressure needed if the water pressure is not enough, and the typical length of backwash time is between 5 and 15 mm. Where surface water is being filtered and to speed the cleaning process, compressed air mixed with the backwash water may be used to effectively remove algae and other organisms. This is called air scouring. The need to backwash is indicated when, generally the pressure drop falls to between 7 and 11 psi (49 to 77 kPa) above the clean pressure value.

During the backwash cycle, the action of the water in the single medium type filter distributes the coarsest sand at the bottom and the finest at the top. This distribution will result in the accumulation of a majority of particulates removed from the raw water in the fine sand layer on the top where the raw water enters the filter. This quickly reduces its effectiveness, thereby requiring frequent backwash.

Granular Filter Selection Criteria

The advantage of the granular filter is its low initial cost. The advantage of the multimedia filter is the consistent distribution of the various sized media after backwashing, thereby increasing the time between backwash cycles. Tests have shown that the multimedia filter is more effective in removing particulates than the single medium sand filter. It is often cost effective to prefilter a liquid containing large amounts of particulates with a high solid capacity sand filter before putting it through a cartridge filter.

Sand filters can be obtained with either vertical or horizontal tanks depending on the headroom available. The horizontal tank has a larger filter bed area, which requires a higher backflow rate. Compartmentalization of the vessel may offset this requirement somewhat. The key to operation for this type of filter is the arrangement of the filter media, best arranged from coarse to fine in the direction of water flow. In addition, activated carbon added to the filter bed will remove excessive odor and bad taste, although this is usually done in a separate filter.

Problems have resulted from failure of the interior lining of the filter due to abrasion of the lining by the sand during backwash. This abrasion process creates iron corrosion products that can pass through the filter and interfere with downstream purification processes.

It is important to operate these units at the face velocities recommended by the manufacturer to prevent the forcing of the impurities through the filter. This phenomenon is called breakthrough.

Precoat (Septum) Filters

Precoat filters, often called diatomaceous earth filters, are depth-type filters used to clarify water when the solids concentration is low. One of the most frequent uses is to clarify water for swimming pools. A *septum* is a thin, porous membrane with relatively large pores that has little resistance to the flow of water. The filter itself consists of a tank containing one or more vertical filter leaves or plates (septa) onto which a thin coating of filter aid is evenly deposited to form a filter surface, or precoat. The precoat could be pearlite or diatomaceous earth. The most commonly used material is diatomaceous earth. As the feedwater flows through the filter, it passes through the filter precoat first, which traps the particulates. Although pressurized filters are available, most precoat filters use a pump to create a vacuum to pull the water through the filter, with the inlet water level controlled by a float valve. When the filter coat becomes plugged, it must be replaced. This is determined by excessive pressure loss through the filter. The filter is cleaned by backwashing, and the filter aid along with trapped solids are discharged to sewer. Care must be used in redepositing the new filter aid to ensure uniformity. Generally accepted filter aid coverage is 0.1 lb/ft^2 of filter septum surface. The diatomaceous earth media usually have a dry weight of 8 to 10 lb/ft^3 and range from 5 to 64 μm in size.

ACTIVATED CARBON FILTERS

Activated carbon is a depth-type granular filter that depends on adsorption to separate contaminants from the feedwater. It is discussed in this chapter because it falls in the granular filter category and conforms to the definition for filtration. It is not specifically used for particulate removal but rather for reduction of free chlorine, removal of TOCs, soluble organics, and trihalomethanes. Chlorine removal is necessary to protect some RO membranes and ion-exchange resins from attack.

Activated carbon media are carbon granules that have been processed from raw high-carbon materials, such as lignite and bituminous coal, wood, peat, and coconut husks. It is available in powdered and granular forms. Another rarely used method is to have powdered carbon added to a process stream and filtered out downstream. This method will not be discussed.

The most common type of activated carbon for water treatment is the granular form often called activated charcoal, which will be the only one discussed here. Activated carbon is manufactured by grinding the raw material into uniform sizes, adding a binder if necessary, and reducing (burning) the mixture in the presence of steam. This creates skeletal granules with a very large network of micro- and macropores, thereby becoming activated. Bituminous-based coal has the highest bed density and therefore a larger number of pores available for attraction of contaminants. In addition, the bituminous product is more amenable to reactivation. Because of these attributes, this is the material most often used. Granular carbon is usually 8 to 30 mesh, with an effective size of 0.9 mm and a uniformity coefficient of 1.8.

The granules are placed within a housing and the feedwater to be treated is passed through the granules. The organic removal capacity of the media depends on the diffusion rate of the organic molecules through the pores of the media, the surface area of the media, the pore size, and the source method used to manufacture the carbon. The characteristics of organic impurities also play a role in removal. The impurities' molecular weight, polarity, pH, temperature, and concentration are important factors in the rate of adsorption. This filter is much more effective in stripping compounds of low molecular weight because those of high molecular weight tend to be poorly adsorbed.

One problem with activated carbon filters is the growth of bacteria due to trapped organics, mostly because of the removal of chlorine. This can be controlled by frequent cleaning of the filter with steam, hot water (80°C or higher), or dilute caustic soda backwash.

The activated carbon media gradually lose their adsorptive capacity and have to be periodically replaced or reactivated. Small amounts of media cannot be economically reactivated. Coal-based media can be activated, but all other types are usually replaced. Reactivation is accomplished by heating the media to a temperature of

1600 to 1800°F. Expect a 10 to 20 percent loss of media during reactivation. After treatment, the carbon will lose some of its smaller adsorption pores, reducing its capacity to retain trace level contaminants. If this is an important factor, the media must be replaced with virgin material rather than reactivated. The indication for replacement is an excessive pressure drop through the unit found through periodic testing of the water quality.

A generally accepted conservative method used to select the size of the filter is to use a figure of 5 gpm/ft^3 of media. Use manufacturers' specifications to select the housing size, amount, and type of filter media for specific applications and flow rates.

Activated carbon filters used for pretreatment should allow for a face velocity of about 2 to 4 gpm/ft^2 of cross-sectional area. A minimum depth of 24 in is recommended for most feedwater streams, but the actual depth should be selected to achieve the recommended contact time with the flowing water. These two figures should be adjusted depending on the quantity and type of contaminants. Tests have shown that 1 g of activated carbon is capable of removing 1 g of residual chlorine. Because of anticipated breakthrough of organic material, the media should be replaced approximately every 6 months.

CARTRIDGE DEPTH FILTERS

Cartridge depth filters consist of comparatively thick, replaceable, preformed filter media that are placed inside a housing. The particle size retained by a depth filter is not precisely defined because of the random nature of the fiber matrix. The advantage of the depth filter is that it has superior particle load because the filter is capable of holding particles throughout its entire matrix.

Filter Media

Depth filter media vary and consist of wound fiber (such as cotton, polypropylene, or rayon) or resin-bonded laminates (such as cellulose, acrylic rayon, or fiberglass). Filter cartridges are of the wound or pleated type. The wound fiber type is used most often.

Wound filters have the same effect as a stack of woven cloth, with the filter wound around a round mandrel instead of flat. The characteristics that affect filter operation are the type of fiber, fiber diameter, and cartridge-winding techniques such as yarn tension, winding pattern, and spacing.

Fibers are available in two types, staple and monofilament. Staple fibers such as cotton and wool are spun into a thread or yarn, which is then wound onto the mandrel. Staple fibers have a fuzzy surface, or nap, that provides a high surface area for adsorption in addition to a long, random path for particle interception.

Monofilament fibers, such as those of polypropylene and other plastics, are not spun but are manufactured in any desired diameter. These fibers must be texturized in some fashion to make them suitable for filter media. They can also be produced in smaller diameters and spun into "yarn."

Pleated filters are generally made from resin-bonded materials, with the pore size determined by the type and size of the fiber and the binding method used.

Pleated cartridges usually depend more on sieving than do the wound type, and can better remove particles of a specific size.

Specific Materials

Polysulfone.   Polysulfone is a membrane that is basically hydrophilic and has excellent flow rates, low extractability, broad chemical resistance, high mechanical strength, and heat resistance, permitting a variety of sterilization methods.

Nylon.   Nylon is a hydrophilic membrane with generally very high flow rates, high tensile strength, low extractables, and limited chemical resistance.

PTFE.   PTFE is a naturally hydrophobic membrane often laminated to a polypropylene support for added strength. It has excellent chemical and heat resistance.

Acrylic Copolymer.   Acrylic copolymer is a membrane that is basically hydrophilic and has excellent flow rates, low extractables, and a low differential pressure.

Polypropylene.   Polypropylene is naturally hydrophobic and is available as a membrane that is chemically inert, has a broad pH stability and high flow rate, and is considered very durable.

Glass.   Glass fibers are usually manufactured from borosilicate glass. Thicker fibers are spun and thinner fibers are made into mats with the addition of a binder. They typically have low differential pressures, good wet strength, and high dirt-holding capacity.

Housings

Housing selection for filter cartridges is based on the proposed application. Although housings most often contain a single cartridge, units that hold multiple filters are available. General industrial requirements differ greatly from the sanitary requirements of the pharmaceutical and food service industries. Manufacturers provide filters and housings of standard diameter and length.

There are no industry standards for general industrial housings. The material can range from plastic to stainless steel. For safety, pressurized housings for compressed gas and high-pressure liquid applications must be ASME code stamped.

Sanitary requirements require conformance to 3-A Sanitary standards, such as an interior surface finish of a minimum 150-grit polish, the capability of being completely disassembled and cleaned, welds ground smooth, and flush sanitary connections to piping.

Housings come in three basic styles: tee, in-line, and L-shaped. The tee style offers easy filter changing, and it is easier to fit into an existing pipeline. It usually has the highest pressure drop of the three styles. The in-line style offers the lowest pressure drop of the three styles. The L-shaped style is used most often for multiple cartridge installations in industrial applications.

Seals

Seals are used to prevent the feedwater from bypassing the filter medium. They must be reliable enough to withstand repeated changing of the cartridge. The most common sealing system uses a double open-ended cartridge with a top compression spring on the cartridge. This provides a knife-edge seal on a seat at the top of the cartridge. Single open-ended cartridges, most often used in single cartridge housings, use a piston type O-ring seal contained within the cartridge itself. The material of the seal must be compatible with the type of fluid being processed, the fluid temperature, and the proposed use.

Filter Selection Criteria

There is a wide variety of filter options available to accomplish separation; and many of these options have overlapping effectiveness. Choosing the appropriate method will often involve tradeoffs between performance and value. The following factors should be taken into consideration when selecting a filter based on the intended service and the quality of the feedwater.

1. The first step to decide the suitability of any filter is to obtain an accurate analysis of the feedwater and to determine the size and quantity of the particulates to be removed.

2. It should be decided whether absolute or nominal filter ratings will be used. For filtering service, most filters are selected by using nominal ratings, with the intended use of the water dictating the various degrees of removal.

3. If the particles are deformable, the wound depth filter is more efficient in removing them. They tend to be extruded through a thin, solid, pleated filter. For hard particles, either wound or pleated can be used.

The selection of a pleated or wound filter should be based on the type and quantity of solids to be removed. The pleated filter has about 16 times the effective filtration area of a similar wound filter and so is much more effective in removing smaller particulates captured on the surface. For larger sized particulates or for a large amount of dirt to be held in the filter, the depth-type filter is more effective. Pleated filters are usually more costly than wound filters.

Depth filters have high dirt-holding capacity and are used most often as prefilters for clarification of water prior to treatment by water purification equipment. They are less costly than other types of filters.

4. The removal efficiency of the filter must be known. This value is typically specified by the end user and often based on past experience. Since the micrometer rating alone is not a fair representation of filter efficiency, the degree of removal must also be specified. A rating of 90 percent efficiency at 5 μm means that 90 percent of the particles of that size will be retained on the filter medium.

5. The flow rate must be determined. This is also called the flow density, which relates flow rate to unit of filter area. As the flow rate increases, the contaminant capacity and service life of the filter decrease. Manufacturers have established flow rates for filters. Housings containing multiple cartridges can be selected to lower the flow rate of individual cartridges.

6. The pressure drop across the filter is another consideration. The smaller the pore diameter, the higher the pressure drop across the filter. If the ratio of the initial clean pressure drop through the filter to the total available pressure is high, unacceptably low flow will quickly result even when the particulate-holding capacity is not fully realized. Pressure requirements across the filter from new to plugged condition must also be considered, with the filter selected based on the highest allowable pressure drop. Accepted practice is to have a maximum of 80 percent filter plugging. If the pressure drop is too high or if the amount of plugging is not acceptable, the size of the filter can be increased or multiple filters used. Housings that hold multiple filters are used to reduce the pressure drop when available water pressure is low.

7. If the filter is intended to be an initial filter in the system, the beta rating should be obtained from the manufacturer to determine its removal efficiency and performance characteristics. Ratings for a minimum of five or more particle sizes should be provided. If the filter is used in an intermediate location, the particle sizes will be reduced and the nominal rating of specific particulate sizes will allow for efficient selection.

8. The maximum temperature that the filter media can withstand will determine if they can be used for service where they must be sterilized and if the media are acceptable for sterilization using high-temperature water.

9. The maximum differential pressure is the difference in pressure between the inlet and outlet water pressure beyond which the filter will fail structurally, and is established by the manufacturer.

10. Effective filtration area is the measure of the usable area of the filter media independent of the material or type.

11. The toxicity and pyrogenicity of the filter materials must be compatible with the intended use of the product water.

12. Compatibility with any chemicals intended to be used for sterilization that will come in contact with the filter must be considered.

13. Extractables from the filter media must be reduced to acceptable limits to prevent problems with any downstream process. The method used most often to control and reduce extractables is to preflush the filters prior to use.

INTERCEPTORS

GENERAL

An interceptor is broadly defined as a device that separates and/or retains a specific substance from a liquid effluent stream without impairing the ability of the remaining effluent to be discharged into a drainage system. It is generally used where the effluent is intended to be discharged into a public sewer system, although it is not limited to that use. The interceptors discussed here are intended to be used inside a building.

There are a number of substances that have the potential to create safety, health, or mechanical problems within a piping network if they are allowed to be discharged directly into a drainage system. For other substances, such as precious metals, recovery is desirable for economic reasons. It is therefore necessary for them to be intercepted, retained, and (possibly) neutralized before this occurs. This is accomplished by interceptors that are designed to protect against specific hazards.

CODE CONSIDERATIONS

Plumbing codes usually have a requirement that any liquid waste containing grease, flammable materials, sand, or any other substance that, in the opinion of the local authorities, is harmful to the building drainage system, the public sewer system, or the sewage treatment process be prevented from being discharged. If there is any question as to the need for a separator, the local authorities should be consulted. Approval of specific separators may also be required.

DESIGN CONSIDERATIONS

The size and shape of the interceptor must be based on the nature of the waste (either lighter or heavier than water), the amount of material to be retained, rate and volume of total discharge into the interceptor, and the corrosion potential of the effluent (acid, caustic, or abrasive).

STRUCTURAL CONSIDERATIONS

In many cases, the interceptor may be installed either on top of or under the slab, with the cover or grate extending up to the finished floor above. Many interceptors are very heavy, and when filled with water they require special support. If the entire cover extends through the slab, additional strengthening of the slab may be required because of the size of the required penetration.

LOCATION

The interceptor must be located so that it is easily accessible for cleaning, servicing, and maintenance. The use of ladders or the need to clear away stored articles in order to remove the intercepted material is a violation of many codes. In addition, if it is easy to clean out the interceptor, it will get done regularly.

Most interceptors require manual removal of trapped substances, and so they must be located in areas where ongoing access will not interfere with normal operations. This must be balanced with the need to locate these interceptors as close as possible to the source of the substances that are being removed.

SPECIFIC SUBSTANCES

Food Related Grease Traps and Interceptors

Grease is most commonly discharged from establishments where food is prepared and/or consumed, such as restaurants, butcher shops, supermarkets and specialty stores. The hazard created is mechanical, because the grease that enters a drainage pipe suspended in hot water will harden as the water cools and accumulate to form a blockage. To be effective, enough retention time through the unit must be allowed in order for grease to float to the top. For smaller units, a flow control device on the inlet is required to reduce the flow to that required to meet unit design requirements. For larger units, the length is usually sufficient to avoid a flow control device.

For purposes of this handbook, a *grease trap* shall be a unit capable of retaining up to approximately 100 lbs. (45 kg) grease holding capacity and is intended to be placed inside a facility adjacent to the fixture or equipment discharging grease. A *grease interceptor* shall be a larger unit retaining more than 100 lbs. (45 kg) intended to be installed outside the facility on the site.

Grease Traps. Grease traps are divided into four categories based on the method used to remove grease from the unit. First is the manual type, in which the unit's cover must be taken off in order to remove the grease by hand. This is the type used most often for smaller installations. Second is the semi-automatic type, in which the grease stored in the top of the unit is discharged through a special valved connection into a separate container. This is done by running hot water through the unit for the purpose of removing the stored grease. Third is an automatic unit that continuously removes grease from the effluent. This type is only used for very large amounts of grease in large projects. A fourth type uses enzymes added to the grease trap to break up the grease inside the unit and allow it to directly enter the sanitary piping system.

The grease trap can be placed either under the floor or above the floor depending on space conditions available and method of grease removal preferred. The semi-automatic removal requires an above-the-floor installation, usually under a sink or in a storage room.

The most common material used for a grease trap is cast iron, although steel can be used for less server service. Refer to Fig. 3.6 for an illustration of a typical grease trap. Figure 3.7 illustrates typical grease trap installations.

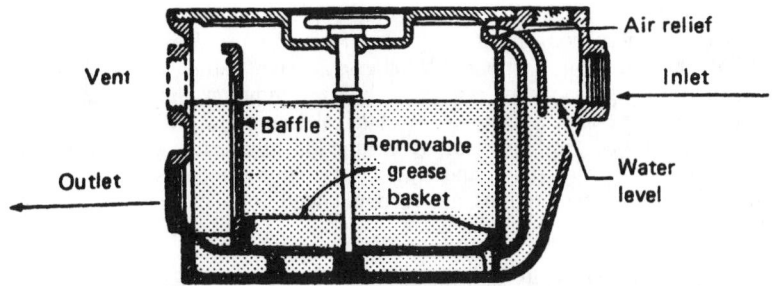

FIGURE 3.6 Grease trap.

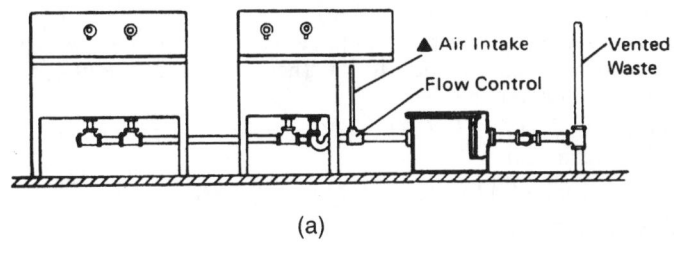

(a)

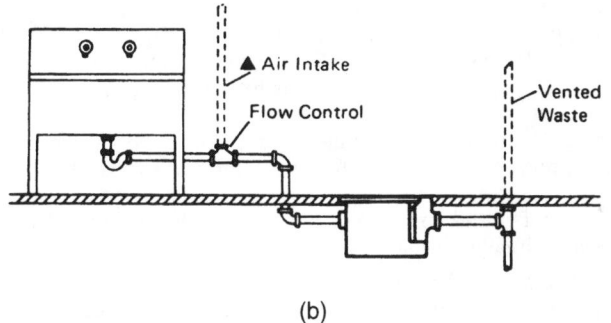

(b)

FIGURE 3.7 Typical grease trap installations. (*a*) On floor; (*b*) under floor.

Grease Interceptors.   Field testing has proven that solidified grease blockages are common in many public sewer systems. Very often, grease traps installed adjacent to fixtures and equipment do not adequately separate grease from the waste water due to inadequate or no maintenance. It must be emphasized that grease traps adhering to applicable certification standards and properly cleaned remove grease very effectively. Reports indicate they are rarely emptied and maintenance is generally poor. As a result, animal and vegetable fats, oils and grease (AVFOG) flow through the grease trap into the public sewer system causing stoppages. This situation has led many local jurisdictions to require the installation of a larger grease interceptor outside the building on the site. This allows inspection by public authorities to be more convenient, and the larger size provides adequate AVFOG separation and retention. Compared to other extensive testing, rating and certification standards for grease traps originated by the Plumbing and Drainage Institute (PDI) and the American Society of Mechanical Engineers (ASTM), grease interceptors have not had the benefit of wide attention to research concerned with establishing design, configuration and effluent discharge standards.

The following considerations are recommended for the selection of a grease interceptor:

1. The top of concrete units shall meet AASHTO A-20 guidelines where there is a possibility that trucks could pass over the unit. If the unit is in a non-traffic area, the unit shall meet minimum top loading standards.

2. An inlet diverter is necessary to increase the retention time and avoid short circuiting.

3. A rectangular interceptor is believed to be the optimum shape. An approximate ratio of depth to width shall be 1 to 1.5. There should be at least 4 inches above the water level for venting.

4. An allowance of 6 to 12 in (152 to 300 mm) shall be provided on the bottom for sludge accumulation.

5. A minimum depth of 42 in (1070 mm) is suggested.

6. A sample port should be provided to allow ease of sample taking.

Interceptor Sizing Guidelines.   The most important consideration is conformance with any code requirements or standards of the authorities having jurisdiction. When these codes or standards are found, they must be followed.

The following guidelines have been adapted from various published standards and codes and are intended to be used only when there are no other applicable requirements.

For establishments other than restaurants, the maximum flow rate method was selected because it relates the flow rate for fixtures to the size of the interceptor.

1. Determine the number and size trap of fixtures and the size of a dishwasher (if any) in the establishment discharging into the interceptor.

2. Add the dishwasher and only the single largest sink together. If there is no dishwasher, add the two largest trap size gpm requirement together. This is the maximum probable flowrate into the interceptor. This figure is based on the probability that no more than two fixtures could discharge at the exact same time.

3. Multiply the flowrate by 30 to calculate the minimum pounds of grease required to be retained. (This figure has a proven past history of success). Pick a standard size interceptor with a capacity equal to or larger than the calculated size.

4. A minimum size shall be chosen on a sliding scale based on the size of the establishment, as follows:

  a. For small shops, such as a pizza parlor or other similar establishments—200 gallon (760 l) capacity;

  b. For medium-size shops and those with higher FOG potential, such as meat markets, etc.—500 gallon (1900 l) capacity;

c. For large establishments, such as regular supermarkets—1,000 gallon (3800 l) capacity;

d. For very large supermarkets and other similar establishments—1500 gallon (5700 l) capacity.

5. A determination of the maximum possible flow rate shall be based on the following table:

Drain outlet or fixture trap size		Drainage fixture unit valve	Peak flow per minute	
in	DN		GPM	DN
1½	40	3	22.5	85
2	50	4	30	115
2½	65	5	35	133
3	80	6	45	170
4	100	8	60	228

Dishwasher tank capacity	GPM	L
Up to 30-gallon (115 l) water tank capacity	15	(57)
Up to 50-gallon (190 l) water tank capacity	25	(95)
Up to 100-gallon (380 l) water tank capacity	40	(152)

For restaurants, the size of the interceptor should be based on the number of seats and the number of meals served, as indicated in eq. 3.1. This formula is based on information taken from the manual written by the EPA. This method was chose because it appears to have the most realistic sizing criteria of all methods examined.

$$D \times GL \times \frac{HR}{2} \times LF \tag{3.1}$$

where D = number of seats in dining room
  GL = 5 gallons (20 l) of waste per meal served
  HR = number of hours restaurant is open
  LF = loading factor*
  1.25 where located on interstate freeway
  1.00 for other freeways and recreational areas
  0.8 for main highway
  0.5 other highways (most often used regardless of location)
  *The figure should be sufficient for most applications.

The recommended configuration of the interceptor shall be based on the following general criteria. These recommendations should be adjusted to suit specific job conditions and the selection of a specific interceptor.

1. 50 percent of the wetted height of the entire interceptor (both compartments) shall be allowed for the storage of grease.

2. 6 to 12 in on the bottom shall be allowed for the accumulation of settled solids. The smaller figure is usually applied to interceptors smaller than 1,000 gallons.

3. The invert of the inlet pipe shall be 6 to 12 in (152 to 300 mm) off the bottom, clear of the settlement zone. An inlet baffle shall be provided, such as a tee

facing sideways or other acceptable method, to direct flow to the side of the interceptor and avoid short circuiting.

4. There should be 4 in (100 mm) freeboard above the top of the outlet pipe as a vent space.

5. Two compartments should be used. A baffle should be installed to divide the compartment into approximate ⅔ (inlet retention) and ⅓ (outlet) sections.

6. A 30-in (760 mm) manhole shall be provided directly over the baffle. A better alternative would be to have a manhole over the center of each compartment.

7. A rectangular interceptor shall be consider a standard shape.

Refer to Fig. 3.8 for an illustration of a typical grease interceptor.

Flammable and Volatile Liquids

Federal, state, and local regulations have established standards for the discharge of volatile liquids, particularly oil, into storm water and sanitary sewage discharges. These standards vary, and the responsible enforcement and code authorities must be consulted to determine the level of removal required.

The most common flammable liquid is oil. The most common sources of oil are automobile-related facilities such as parking garages, service stations, and car washes. Industrial facilities also create oil waste; for example, volatile liquids such as kerosene, gasoline, naphtha, and trisodium phosphate are released from dry-cleaning establishments and other industrial processes. Laboratories also may release various solvents.

The hazard created is that of either safety (e.g., vapors can create an explosive condition or oil can float on water and then be set on fire) or medical (e.g., inhaled vapors can be dangerous to health or chemicals ingested by humans, fishes, and wildlife can be toxic). The common characteristic of all of these substances is that they are lighter than water. Their removal is similar to that of oil.

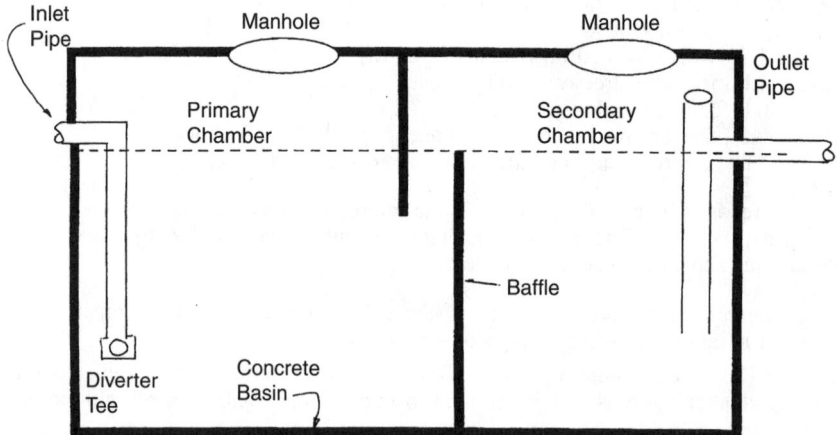

FIGURE 3.8   Typical grease interceptor.

Oil in Water.   Oil in water exists in several forms:

1. Free oil.

2. Mechanically dispersed oil is made up of fine droplets ranging in size from micrometers to fractions of a millimeter. It is stable due to electrical charges and other forces, but not because of the presence of surface active agents.

3. Chemically stabilized emulsions are made up of fine droplets that are stable because of surface active agents.

4. Dissolved and dispersed oil is either actually dissolved or suspended in such a small size (typically 5 μm or less) that ordinary filtration is not possible.

5. Oil-wet solids are particulates in which oil adheres to their surface.

Methods of Separation and Treatment.   Oil spills and leaks are best treated in their most concentrated state, which is at their source or as close as is reasonable. The primary methods used to separate and remove free oil and oil-wet solids are floatation and centrifugation. Secondary treatment, such as chemical treatment/coalescence or filtration, is then used to break up oil-water emulsions and remove dispersed oil. Finally, tertiary treatment, such as ultrafiltration, biological treatment, or carbon adsorption, will remove the oil to required levels prior to discharge. This section will discuss the general principles of the primary and secondary separation methods and devices only.

The American Petroleum Institute (API) has established criteria for the large-scale removal of globules larger than 150 μm. In abbreviated form, they are:

1. The horizontal velocity through the separator may be up to 15 times the rise velocity of the slowest rising globule, up to a maximum of 3 ft/s.

2. The depth of flow in the separator shall be between 3 and 8 ft.

3. The width of the separator shall be between 6 and 20 ft.

4. The depth-to-width ratio shall be between 0.3 and 0.5.

5. An oil-retention baffle should be located no less than 12 in downstream from a skimming device.

Gravity Separators.   Gravity separation is the primary separation method. It is based on the specific gravity difference between immiscible oil globules and water. Since all of these liquids are lighter than an equal volume of water, gravity separators operate on the principle of floatation. As water and oil flow through the unit, the oil floats to the top and is trapped inside a series of internal baffles. Since the oil remains liquid, it is easily drawn off.

Floatation Devices.   For service on a larger scale, the floatation of oil and oil-wet solids to the top of the floatation chamber can be increased by the attachment of small bubbles of air to the surface of the slow-rising oil globules. This is done by adding compressed air to the bottom of the floatation chamber in a special manner that will create small bubbles which will mix with, and attach themselves to, the oil globules.

Centrifugal Separators.   For service on an even larger scale, the centrifugal separator is used. This device operates on the principle of inducing the combined oil and water mixture to flow around a circular separation chamber. The lighter oil globules will collect around a central vortex, which contains the oil removal mechanism, and the clear water will collect at the outer radial portion of the separation chamber. Methods have evolved that can produce effluent water with only 50 to 70 parts per million (ppm) of oil; proprietary devices exist that can lower oil content to 10 ppm.

Filtration.   Using chemical methods first to break oil-water emulsions and then using depth-type filters to remove the destabilized mixture have proven effective in removal of oil globules in sizes between 1 and 50 μm. The velocity and flow rate of the mixture must be carefully controlled to allow optimum effectiveness of the system.

Smaller Systems.   Oil separators for small flows usually take the form of a single unit consisting of a drain grating into which the effluent flows and inside which the oil remains to be drawn off manually. Another type of unit uses an overflow arrangement which sends the trapped oil to a remote oil storage tank.

Because there is the possibility that the vapor given off by the flammable liquid could also ignite, it is important to provide a separator vent that terminates in the open air at an approved location above the highest part of the structure. Some codes require that a flame arrester be installed on the vent.

The most common material used for an oil interceptor is cast iron, although steel can be used for less severe service. Gratings must have the strength for the type of vehicle expected.

Refer to Fig. 3.9 for an illustration of a typical small oil interceptor. Figure 3.10 shows the installation of a typical oil interceptor with gravity oil draw-off for garage floor drains.

Sand

Whenever a potential source of solids discharges into the drainage system, a sand interceptor, or trap, should be provided. A mechanical hazard could be created, since the sand could create a blockage in the piping system.

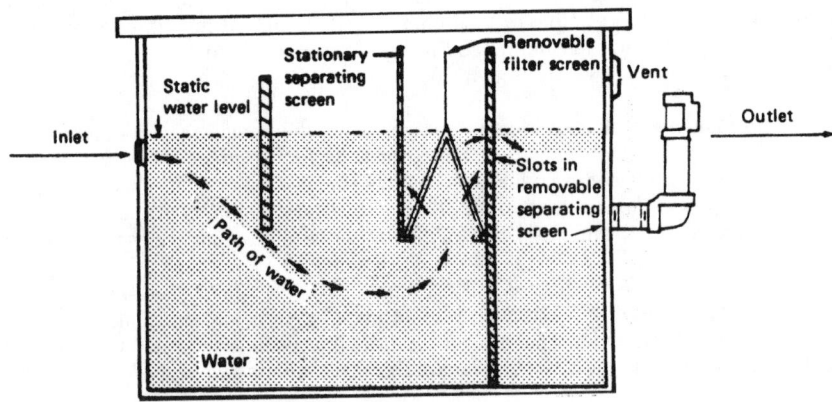

FIGURE 3.9 Typical oil interceptor. (*Courtesy of Rockford Co.*)

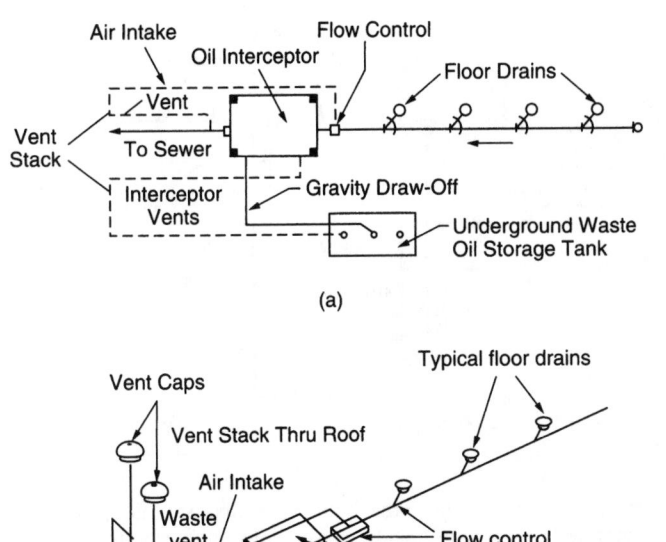

(a)

(b)

FIGURE 3.10   Typical gravity draw-off installation. (*a*) Plan; (*b*) isometric.

Since the solids are heavier than water, these traps operate on the principle of settlement, permitting the solids to accumulate at the bottom as the effluent flows through the device. The outlet from the sand trap should be located so that the accumulated material is prevented from being discharged. The solids must be removed from the trap by hand. Sufficient space must be available around the trap to make this easy to accomplish.

Sand traps are commonly constructed of masonry, but prefabricated units made of cast iron or steel are also used. Refer to Fig. 3.11 for an illustration of a typical sand interceptor.

Precious Metals

The most common source of gold, silver, or platinum is from jewelry establishments. The small amount of metal discharged would not be detrimental to any drainage system, but it should be recovered because of its value.

A solids interceptor for this type of service is a small, in-line unit, using either a fine wire mesh screen or stainless steel wool as a filter inserted inside a small housing. This housing is installed instead of a trap on a fixture.

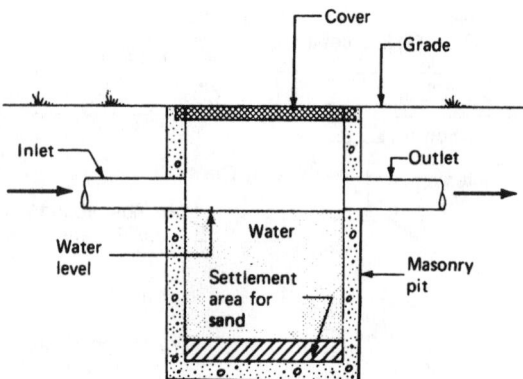

FIGURE 3.11 Sand interceptor.

To recover the precious metals, the housing is disassembled, and the filter is removed and emptied at a remote location. The housing can be made of brass, cast iron, steel, plastic, or any material permitted by local code.

Hair

Hair is discharged from barber shops and beauty parlors, and has the potential to accumulate at some minor obstruction that would not ordinarily cause trouble. The hair will accumulate and eventually cause a blockage.

An interceptor for this type of service is a small, in-line unit, using a perforated basket strainer inserted inside a small housing. This housing is installed instead of a trap on a fixture. To remove the accumulated hair, the housing is disassembled, and the strainer is removed and emptied at a remote location. The housing can be made of brass, cast iron, steel, plastic, or any material permitted by local code. The strainer is usually made of stainless steel. Refer to Fig. 3.12 for an illustration of a typical hair trap.

Acid Neutralizers

Whenever the possibility exists that acid of any kind may be discharged into the drainage system, an acid neutralizer must be provided. As a general rule, many authorities permit waste with a pH of 4 or above to enter the drainage system, where it will be further diluted with other effluent. Acid will attack ordinary piping material and cause it to fail prematurely.

One method used for small, isolated, and intermittent discharges is to percolate the acid through a tank containing limestone chips. These chips range from 1 to 3 in (25 to 75 mm). They are placed in a tank until it is filled to approximately 50 percent of its volume. There is a baffle arrangement to ensure that the effluent is in continuous contact with the chips. Generally used contact periods of between 10 and 15 min are sufficient to neutralize the acid, with shorter times used based on the individual manufacturer's recommendations. The actual required contact time

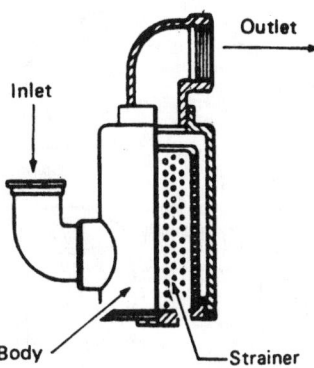

FIGURE 3.12 Typical hair trap.

for proper neutralization should be based on specific interceptors, effluent dilution, and pH values. Neutralization is accomplished by chemical reaction of the acid with the chips. There is no residue. However, the chips must be replenished periodically, depending on the amount of acid that is treated. General figures require 100 lb (45 kg) of chips to treat 98 lb (44 kg) of sulfuric acid and 73 lb (33 kg) of hydrochloric acid.

Unit sizes vary, ranging from small units that can be placed under individual sinks to tanks suitable for large facilities. They must be located in areas where their covers can be readily removed to add new chips. The larger tanks are heavy and need additional structural support. They can be installed either above the floor or below the slab, with the cover extending up to the finished floor above.

Acid-neutralizing basins are made in a variety of materials, depending on the amount and type of acid expected to be encountered. However, for all but the smallest laboratories and pharmaceutical facilities, the neutralizing basin is no longer considered acceptable. The acid must be treated in a manner that will ensure neutralization of the acid to a pH required by the FDA, EPA, and local authorities. This requires the use of a tank or chamber that introduces varying amounts of

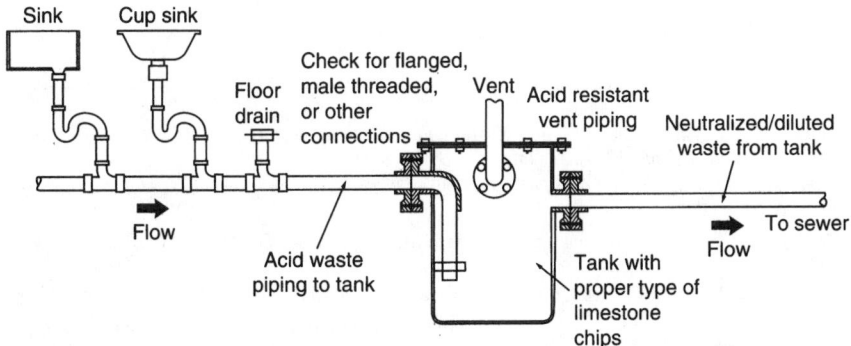

FIGURE 3.13 Small acid-neutralizing basin.

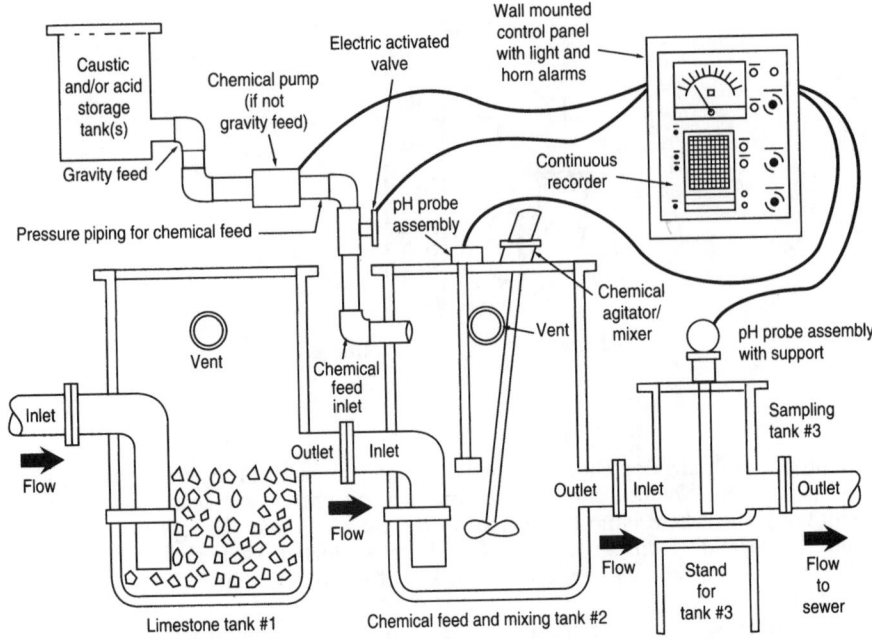

Note: Variations of this setup are available, including one large tank with three compartments instead of three separate tanks.

FIGURE 3.14 Acid-neutralizing system installation.

caustic neutralizing agent to mix with a variable flow rate of acid. A method of sampling must be provided and a record of the pH of the effluent from the facility and after neutralization should be kept.

Refer to Fig. 3.13 for an illustration of a typical acid-neutralizing basin. Figure 3.14 illustrates the installation of an acid-neutralizing system.

REFERENCES

Batty, B., "Solving Boiler/Cooling System Water Problems," *Heating/Cooling/Air Conditioning Magazine,* pp. 125–129, January 1991.

Cheremisinoff, Paul N., "Oil/Water Separation," *The National Environmental Journal,* pp. 32–36, May/June 1993.

Departments of Army and Air Force Technical Manual TM 5-813-3/AFM 88-10, *Water Treatment,* 1985.

Frankel, M., National Precast Concrete Association, Grease Interceptor Design Seminar, 2000.

Frankel, M., "The Problems with Exterior Grease Interceptors," *Plumbing Engineer Magazine,* pp. 36–40, April 2001.

Hartung, R., and D. Marturana, "Water Treatment," *Chemical Engineering Magazine,* pp. 98–105, January 1992.

Hayward Industrial Products, Inc., "Basket Strainers."

Kemmer, Frank N. (ed.), *The NALCO Water Handbook,* 2nd ed., McGraw-Hill, New York, 1988.

Kononov, Anthony, "Optimal Sterile Filter Processing for Pharmaceutical Manufacturing," *Pharmaceutical Engineering Magazine,* pp. 30–34, May/June 1994.

Lewis, Sidney A., "Take the Stress out of Strainer Selection," *Chemical Engineering Magazine,* pp. 110–115, June 1991.

Nayar, M. L. (ed.), *Piping Handbook,* 6th ed., McGraw-Hill, New York, 1992.

Olsen, Kevin R., "How to Remove Fouling Solids from Cooling Water," *Chemical Engineering Magazine,* pp. 165–168, May 1992.

Parekh, Bipin S., "Get Your Process Water to Come Clean," *Chemical Engineering Magazine,* pp. 71–85, January 1991.

Uberoi, T., "Effectively Select In-line Filters," *Chemical Engineering Progress Magazine,* pp. 75–80, March 1992.

Water Conditioning Manual, Dow Chemical Co., 1985.

CHAPTER 4
WATER TREATMENT AND PURIFICATION

This chapter describes water treatment methods used to partially remove and replace various impurities from a feedwater stream, water-conditioning additives used to neutralize impurities and inhibit corrosion, and pure water systems used to remove all impurities from the feedwater to a level at or below limits desired by the end user.

Information is presented on basic water chemistry, impurities found in water, water analysis, and impurity measurement. Also discussed are general selection criteria for the removal or reduction of specific impurities from water. Separate sections cover water-conditioning systems for boiler feedwater, water-conditioning systems for cooling water, and the generation of purified water.

For purposes of discussion only, the term *water treatment* is intended to mean only the removal and/or replacement of undesired impurities in feedwater with more desirable impurities. The resultant water is not pure. *Water conditioning* is intended to mean the addition of chemicals to water for the purposes of inhibiting corrosion and neutralizing undesirable impurities. *Pure water systems* is intended to mean the production of water pure enough for pharmaceutical and laboratory applications. The explanations and definitions given are simplified ones.

CODES AND STANDARDS

There are many codes and standards that apply to various systems. Among them are:

1. Potable water treatment shall comply with the 1986 Safe Drinking Water Act and Amendments.
2. Potable water treatment shall comply with EPA technologies describing contaminant removal.
3. Pure water treatment shall comply with the following, depending on the purity of the water desired:

  a. CAP and ASTM reagent-grade water
  b. USP standards for water purity
  c. AAMI standards
  d. NCCLS standards
  e. SEMI and ASTM electronics-grade water standards (outside the scope of this handbook)

BASIC WATER CHEMISTRY

Water to be treated is known variously as raw water, feedwater, or source water. Water that has been treated is known as treated water, product water, or solute. Impurities that dissolve in water are called electrolytes. The dissolved impurities dissociate (separate) to form negatively and positively charged atoms called ions. The positively charged atoms are called cations because they migrate to the cathode electrode; the negatively charged atoms are called anions because they migrate to the anode electrode. The ions in solution act almost independently. For example, magnesium sulfate dissociates to form positive magnesium ions and negative sulfate ions. Ions are often referred to as salts in reverse osmosis water production.

When a compound dissolves in water, the molecules of the compound separate, disperse among the molecules of water, and then are held in suspension. When the solubility limit is reached, those same compounds become suspended solids because there is no chemical reaction. Dissolved materials cannot be removed by filtration.

Some types of liquids cannot be dissolved. They break down into extremely small particles and then disperse into water, but do not dissolve. These liquids are called immiscible liquids, and the resulting small particles are known as colloids. Colloidal material is composed of suspended particles at the upper end of the size range for ions and molecules, from 0.001 to 1.0 μm. Other liquids, such as oil and grease, cannot separate into smaller particles and become dispersed, but simply remain in suspension. If they are lighter than water, they float on top. If heavier, they sink below the surface.

All acid compounds in water chemistry consist of hydrogen combined with an acid radical. Since the acid radical moves around as a unit, it is convenient to view an acid radical as an integral anion unit. When a metal radical and an acid radical combine, they form a class of chemicals called salts. When a metal cation and a hydroxide anion combine, a base results.

The valence of any element is the measure of its chemical combining power compared to that of a hydrogen atom, which has the assigned value of 1. An element with a valence of +2 can replace two hydrogen atoms in a compound; with a valence of −2, it can react with two hydrogen atoms. When atoms combine to form compounds, a cation atom must combine with an anion atom and, in addition, the valences of the two basic elements must be equal in order to form the compound.

Most of the basic chemical reactions in water treatment consist of rearranging cation and anion atoms using their valences. As can be seen in Table 4.1, hydrogen has a valence of one and sulfate has a valence of two. When combined to form sulfuric acid, two hydrogen atoms are required to form the combination, resulting in the formula H_2SO_4. The chief exception is the case where carbonates and bicarbonates are destroyed by heating or aeration, giving off carbon dioxide.

The term *equivalent weight* is the weight in pounds of any element that could combine with 1 pound of hydrogen. Since the valence of an element is proportional to its combining power, the equivalent weight is based on its valence. This is illustrated in Table 4.1.

TABLE 4.1 Important Elements, Acid Radicals, and Acids in Water Chemistry

Element	Symbol	Atomic weight	Valence	Equivalent weight
Aluminum	Al	27.0	3	9.0
Barium	Ba	137.4	2	68.70
Calcium	Ca	40.1	2	20.05
Carbon	C	12.0	Variable	—
Chlorine	Cl	35.46	Variable	35.46
Fluorine	F	19.0	1	19.0
Iron (ferrous)	Fe″	55.8	2	27.9
Iron (ferric)	Fe‴	55.8	3	18.6
Hydrogen	H	1.0	1	1.0
Magnesium	Mg	24.3	2	12.15
Nitrogen	N	14.0	Variable	—
Potassium	K	39.1	1	39.1
Oxygen	O	16.0	2	8.00
Phosphorus	P	31.02	Variable	—
Sodium	Na	23.0	1	23.0
Sulfur	S	32.0	Variable	—
Silicon	Si	28.06	4	7.01

Acid radicals	Formula	Molecular weight	Valence	Equivalent weight
Bicarbonate	HCO_3	61.0	1	61.0
Carbonate	CO_3	60.0	2	30.0
Chloride	Cl	35.46	1	35.46
Nitrate	NO_3	62.0	1	62.0
Hydroxide	OH	17.0	1	17.0
Phosphate	PO_4	95.0	3	31.66
Sulfite	SO_3	80.0	2	40.0
Sulfate	SO_4	96.06	2	48.03

Acid	Formula	Molecular weight	Equivalent weight
Carbonic acid	H_2CO_3	62.0	31.0
Hydrochloric acid	HCl	36.46	36.46
Phosphoric acid	H_3PO_4	98.0	32.67
Sulfuric acid	H_2SO_4	98.1	49.05
Sulfurous acid	H_2SO_3	82.1	41.05

Miscellaneous compounds	Formula	Molecular weight	Equivalent weight
Aluminum hydroxide	$Al(OH)_3$	78.0	26.0
Calcium bicarbonate	$Ca(HCO_3)_2$	162	81.0
Calcium carbonate	$CaCO_3$	100	50.0
Calcium sulfate	$CaSO_4$	136	68.0
Carbon dioxide	CO_2	44.0	22.0
Ferric hydroxide	$Fe(OH)_3$	107	35.6
Magnesium carbonate	$MgCO_3$	84.3	42.1
Magnesium hydroxide	$Mg(OH)_2$	58.3	29.1
Magnesium sulfate	$MgSO_4$	120	60.1
Sodium sulfate	Na_2SO_4	142	71.0

WATER IMPURITIES

Natural or source water is never pure. Water picks up impurities as it comes into contact with the ground or, when percolated through the earth, mineral formations. It also contains dissolved gases and dust picked up by rain, snow, and hail or by surface water in contact with the air above the water level. Water is classified as surface water when it comes from sources such as lakes and rivers and as ground water when it comes from streams, wells, or other aquifers originating underground.

SUSPENDED MATTER (PARTICULATES)

Turbidity

Turbidity is a general term used to describe any form of insoluble matter, or suspended solids, suspended in water. Color is often used to describe turbidity and may be used when referring to water containing decaying vegetation. However, turbidity is used most often when referring to mineral particulates, such as silt, that are usually the most plentiful substances in the water. Other commonly occurring impurities are liquids, such as oil, and the residue caused by decaying vegetation. Coarse particles that settle rapidly when water is standing are referred to as sediment; fine particles that remain in suspension are called silt.

Microorganisms

Microorganisms are bacteria and viruses. They are living forms of particulate matter. Their unusual physiology allows them to grow and multiply in water containing only trace levels of nutrients. The presence of these nutrients in untreated water is an indicator of the presence of microorganisms (if the temperature is favorable for their growth). Although microorganisms are suspended solids, the treatment required for their removal or neutralization puts them in a separate category.

Pyrogens cause fever, and pathogenic organisms (such as leageonella) cause diseases of any kind. Endotoxins, which are fragments derived from the cell walls of gram-negative bacteria, are considered the most important and widely occurring group of pyrogens. Other organic growths include algae (a primitive form of plant life), fungi (plants that lack the chlorophyll required for photosynthesis), and bacteria that exhibit both plant and animal characteristics. Bacteria are further subdivided into slime bacteria, which secrete slime; iron bacteria, which thrive on iron; sulfate-reducing bacteria, which live by consuming sulfate and converting it to hydrogen sulfide gas; and nitrifying bacteria, which use ammonia and whose byproduct results in the formation of nitric acid.

Several methods of measurement are used, including viable count assays, direct count epifluorescent microscopy, scanning electron microscopy, and biochemical techniques. The most common means of measuring bacterial contamination is the viable count method, in which the water being measured is passed through a sterile nutrient medium, and the number of colonies appearing on the medium is counted after a period of time is allowed for growth. These are called colony forming units or CFUs. Endotoxins are measured in endotoxin units per milliliter (EU/mL). A popular form of measurement for endotoxins is the limulus amoebocyte lysate

(LAL) test, in which a blood extract of the horseshoe crab becomes turbid in the presence of bacterial endotoxins. This detection technique uses optical density (turbidity level) measured over a period of time.

Other Organisms

This form of impurity is also applied to larger living things such as clams, mussels, and their larvae. They tend to clog water inlets of salt- and fresh-water and may also find their way into the piping system of a facility. See the section on nonpotable water systems for further discussion.

DISSOLVED MINERALS AND ORGANIC SUBSTANCES

Organic Substances

Dissolved organic substances typically found in water include both manmade and natural substances. Manmade impurities include herbicides, pesticides, trihalomethanes, surfactants, and detergents. Naturally occurring impurities include lignins, tannins, humic and fulvic acid, and other biodecomposition products.

Alkalinity

All natural water contains some alkalinity. Alkalinity is a measure of the quantity of dissolved earth minerals in water, and reflects the water's ability to neutralize acids. It is mainly the sum of the carbonate, bicarbonate, and hydroxide ions in water, with borate, phosphate, and silicate ions partially contributing to the total. It is reported as the ppm equivalent of calcium carbonate. Alkalinity is regarded as the most important characteristic in determining the scale-forming tendency of water.

Alkalinity is measured using two end-point indicators. The phenolphthalein alkalinity, or P alkalinity, measures the strong alkali in the solution. The methyl orange alkalinity, or M alkalinity, measures all of the alkalinity present in the solution. M alkalinity is often called total alkalinity because it also includes P alkalinity. Alkalinity is not a measure of pH.

Iron

The most common form of iron is ferrous bicarbonate. It is also considered a form of hardness of water. Iron causes problems with many ion exchange resins.

Magnesium

The most common forms are magnesium carbonate, magnesium bicarbonate, and magnesium chloride. These impurities tend to deposit scale on surfaces they come in contact with.

Silica

The three most common kinds of silica are soluble, colloidal, and particulate. Soluble silica is often referred to as reactive silica, and colloidal silica is sometimes called nonreactive or polymeric. The most common form in solution is silicon oxide; in suspension it is found as a fine colloid. These impurities tend to deposit a scale on surfaces they come in contact with and form a gelatinous mass on RO membranes.

Sodium and Potassium

Both elements form similar salts, with the three most common being sodium or potassium chloride, sodium or potassium carbonate, and sodium or potassium bicarbonate.

Chlorides and Sulfates

The most common forms are dissolved salts of sodium, potassium, calcium, and magnesium. These impurities tend to deposit a scale on surfaces they come in contact with.

Hardness

Hardness is the total amount of calcium, magnesium, iron, and other metallic elements that contribute to the "hard" feel of water. Carbonate, sulfate, and chloride salts of these elements are responsible for most of the scaling deposited on pipe walls and boilers. Generally accepted practice limits the term *hardness* to include only calcium and magnesium. Often, water is characterized in general terms by the amount of hardness, expressed in mg/L as $CaCO_3$, as follows:

Soft	0 to 75 mg/L as $CaCO_3$
Moderate	76 to 150 mg/L as $CaCO_3$
Hard	151 to 300 mg/L as $CaCO_3$
Very hard	Over 300 mg/L as $CaCO_3$

Trace Elements

Trace elements are present in very small quantities and are only considered problems if the amount is above an accepted level for the purpose for which the water is to be used. Examples are lead, cadmium, copper, barium, silver, lithium, zinc, chromium, mercury, arsenic, and selenium.

DISSOLVED GASES

The most common dissolved gases in natural raw water are oxygen, carbon dioxide, nitrogen, and hydrogen sulfide. In addition, water obtained from a potable water supply usually has chlorine and fluorides present, added for public health. Of increasing concern is the presence of radon gas in many water supplies obtained from wells.

Oxygen is the basic factor in the corrosion process and must be present for the corrosion of metals. Removal or reduction reduces the corrosiveness of the water.

For chlorine, no pretreatment is usually necessary for feedwater containing less than 1 ppm. When more than 1 ppm of chlorine is present, an activated carbon filter is recommended.

WATER ANALYSIS AND IMPURITY MEASUREMENT

GENERAL

Analyzing a water sample is the process of finding the quantity of various impurities present. In order to accomplish this, the quantities must be presented in a logical and understandable manner to allow for easy and practical interpretation. It is of utmost importance that the initial analysis of incoming water be accurate and contain a worst-case scenario and that the desired output quality be established prior to the selection of any treatment system.

The most accurate analysis of a water sample is done by laboratories specializing in this type of work. Sterile containers must be used and several samples must be taken over a period of time to ensure that peak readings and average values will be obtained. There are also many field tests that are not as accurate as laboratory tests but give an accuracy acceptable to the user.

The results of the analysis can be expressed in many ways. A common method is to report the concentration of ions in solution by the weight of an element or compound per liter of water, expressed as milligrams per liter of water (mg/L). Another method is expressed in parts per million (ppm), which can be expressed either by the weight of an impurity compared to the weight of water, abbreviated w/w (weight to weight), or by the volume of the impurity compared to the volume of water, abbreviated v/v. Other units are also used, such as grains per gallon (gpg) and equivalents per million (epm). mg/L differs from ppm in expressing a proportion in weight per volume. This finds specific use in analysis of saline waters. Where the specific gravity of a liquid is around 1, mg/L and ppm are equal. Grams per gallon (gpg) is a term used in the discussion of ion-exchange equipment capabilities, where 1 gpg = 17.1 ppm.

As previously explained, compounds break down into ions when dissolved. Although chemists can measure the amounts of each ion present in a sample, it is not practical to find the total amount of each compound that actually went into solution. In practice, the actual method of analysis measures ions only. Using the ionic form in measurement when reporting impurities makes it easier to interpret the results.

To further simplify reporting, it is desirable to reduce all ions present in solution to a common denominator. The common denominator is calcium carbonate. This is accomplished by comparing the equivalent weight of all ions present and expressing them as the ppm anion and cation equivalent of calcium carbonate. The main reason for this is that the molecular weight of calcium carbonate is 100 and its equivalent weight is 40. This method of expression is a widely accepted, but not universal, standard for reporting water analysis. Table 4.2 presents the conversion factors used for major impurities. Figure 4.1 illustrates a typical water analysis report indicating impurities in ppm, equivalents useful in calculating reacting chemicals, and a comparison of positive and negative ions.

pH

When alkalines (bases) are mixed in water, hydroxyl ions result. In a mixture of acid and water, hydrogen ions result. pH is a measurement of the hydrogen-ion

TABLE 4.2 Converting ppm of Impurities to ppm of Calcium Carbonate

Cations	Ionic ppm multiplier	Anions	Ionic ppm multiplier
Hydrogen	50.00	Hydroxide	2.94
Ammonium	2.78	Chloride	1.41
Sodium	2.18	Bicarbonate	0.82
Potassium	1.28	Nitrate	0.81
Magnesium	4.10	Bisulfate	0.52
Calcium	2.50	Carbonate	1.67
Ferrous	1.79	Sulfate	1.04
Ferric	2.69		
Cupric	1.57		
Zinc	1.53	Other	
Aluminum	5.55	Carbon dioxide	2.27
Chromic	2.89	Silica	1.67

WATER ANALYSIS REPORT

Sample No. __605__   Collected _____

For __ABC Co.__   Analyzed _____

                          Reported _____

Ion			epm	ppm as CaCO₃
Cations	Calcium as Ca	__62__ ppm	3.10	155
	Magnesium as Mg	__31__ ppm	2.54	127
	Sodium and potassium as Na	__38__ ppm	1.64	83
	Total cations		7.28	365
Anions	Bicarbonate as HCO₃	__250__ ppm	4.10	205
	Carbonate as CO₃	__0__ ppm	0	0
	Hydroxide as OH	__0__ ppm	0	0
	Chloride as Cl	__11__ ppm	0.31	15
	Sulfate as SO₄	__138__ ppm	2.87	145
	Nitrate as NO₃	_____ ppm		
	Total anions		7.28	365

Silica as SiO_2 __5__ ppm   Total hardness __282__ ppm CaCO₃

Iron as Fe_2O_3 __1.2__ ppm   Methyl orange
                                            alkalinity __205__ ppm CaCO₃
Total dissolved solids   __536__ ppm

Suspended solids (weight) __5__ ppm   Phenolphthalein
                                            alkalinity __0__ ppm CaCO₃
Chloroform-extractable matter _____ ppm

Turbidity (after shaking) __5__ ppm   pH __7.7__   Color _____

Carbon dioxide as CO_2 __10__ ppm   Sp conductance _____ μmhos

FIGURE 4.1 Typical water analysis report.

concentration of a solution. Since the balance of hydroxyl (cation) and hydrogen ions (anion) must be constant, changes in one ion concentration produces a corresponding change in the other. The pH value is calculated from the logarithmic reciprocal of the hydrogen-ion concentration in water. The pH scale ranges from 0 to 14, with 0 being acid and 14 being alkaline. 7.0 is neutral. A change of one unit represents a tenfold increase (or decrease) in strength. pH is not a measure of alkalinity.

Specific Resistance

Specific resistance is a measure of the amount of electrolytes in water. It measures the ability of 1 cm^3 of a sample solution at a given temperature to resist the flow of an electrical current. It is based on the activity of the compounds, i.e., ionized salts, dissolved in water and varies with the temperature of the water. It is the most practical method of measuring impurities from a given sample. Resistance is given in ohms; pure water has a resistance of 18.3 megaohms (MΩ). Resistivity conversions are given in Table 4.3.

TABLE 4.3 Resistivity and Conductivity Conversion

Grains/gal* as CaCO$_3$	ppm as CaCO$_3$	ppm NaCl	Conductivity, μmho/cm	Resistivity, MΩ/cm
99.3	1700	2000	3860	0.00026
74.5	1275	1500	2930	0.00034
49.6	850	1000	1990	0.00050
24.8	425	500	1020	0.00099
9.93	170	200	415	0.0024
7.45	127.5	150	315	0.0032
4.96	85.0	100	210	0.0048
2.48	42.5	50	105	0.0095
0.992	17.0	20	42.7	0.023
0.742	12.7	15	32.1	0.031
0.496	8.5	10	21.4	0.047
0.248	4.25	5.0	10.8	0.093
0.099	1.70	2.0	4.35	0.23
0.074	1.27	1.5	3.28	0.30
0.048	0.85	1.00	2.21	0.45
0.025	0.42	0.50	1.13	0.88
0.0099	0.17	0.20	0.49	2.05
0.0076	0.13	0.15	0.38	2.65
0.0050	0.085	0.10	0.27	3.70
0.0025	0.042	0.05	0.16	6.15
0.00099	0.017	0.02	0.098	10.2
0.00070	0.012	0.015	0.087	11.5
0.00047	0.008	0.010	0.076	13.1
0.00023	0.004	0.005	0.066	15.2
0.00012	0.002	0.002	0.059	16.9
0.00006	0.001	0.001	0.057	17.6
none	none	none 0.055	18.3†	

*Grains per gal = 17.1 ppm (CaCO$_3$).
†Theoretical maximum.

Specific Conductance

Specific conductance measures the ability of 1 cm^3 of a sample solution at a given temperature to conduct an electrical current. It is the reciprocal of the resistance, in ohms. Since it is the opposite of resistance, it is given the name mho, which is ohm spelled backwards. The actual conductance is so small it is measured in micromhos (μmho, one millionth of a mho). For example, at 70°F (19°C) demineralized water with ½ ppm dissolved salt has a conductance of 1 μmho. Pure water has a conductance of 0.036 μmho. Conductivity conversions are given in Table 4.3. Specific conductance in actual practice is normally measured by probes suspended in the stream of water.

Total Suspended Solids

This figure is the sum of all of the suspended material found in a water sample and is commonly measured in either parts per million (ppm, w/w), or milligrams per liter (mg/L). For all practical purposes, these two forms of measurement are equal to each other (1 ppm 1 mg/L).

Turbidity in water is classified by the size of the particulates in micrometers (1/1000 of an inch) and tested by a light interference method, known as nephelometric. This test compares the color of the water sample to a standard color scale, which indicates the total suspended solids based on this comparison. The most common reporting method is the nephelometric turbidity unit (NTU). The higher the number, the more turbid the water.

The NTU measures the color of a beam of light passed through a water sample. A common standard for potable water is the Standard Method for the Examination of Water and Wastewater by the American Public Health Service, which uses formazin as the standard for producing a known volume of turbidity. The standard color scale to which it is compared is derived from the platinum cobalt unit (PCU). Other methods used less frequently are the comparator tube determination using formazin, called the formazin turbidity unit (FTU), and the original test, the Jackson turbidity unit (JTU), named for the man who developed the standard candle used to compare the color of the candlelight through the sample to a color standard. The most accurate method of measuring solids is the gravimetric method, in which a known quantity of water is evaporated and the resulting solids weighed.

The most effective devices for removing turbidity are filters and strainers. The equipment chosen to accomplish this task depends for the most part on the size and type of solids to be retained. Other factors include materials composing the device, the nature of the raw water, flow rate requirements, particle removal target, initial and operating costs, and maintenance requirements. Filters and strainers are discussed in Chap. 3.

Total Dissolved Solids (TDS)

Often referred to as dissolved inorganics and mineral salts, TDS is the sum of all the dissolved minerals including chlorides, sulfates, and carbonates. Dissolved solids contribute to scale deposit and corrosion of piping and equipment. When dissolved in water, mineral salts form positively charged ions, mostly sodium and calcium, and negatively charged ions, mostly chlorides and sulfates.

Total Organic Carbon (TOC)

TOC is a measurement of the carbon level in water and is widely used to determine the level of contamination of water by organic compounds. These compounds contribute to corrosion, cause problems in manufacturing, and usually indicate the presence of endotoxins in water for pharmaceutical use.

The measurement is generally complicated and dependent on the expected level. For high levels, the organic compound is first converted to carbon dioxide which is then measured by infrared absorption. Gas stripping is required to remove other forms of carbon ions from dissolved mineral compounds. For PPB levels, photolytic oxidation is used and the resultant carbon dioxide measured.

Silt Density Index (SDI)

The SDI is a measure of the fouling potential of a feedwater source. Since colloids and other solids can be any size in the submicrometer range, there is no direct method to measure their concentration in feedwater. The SDI is found by passing the feedwater through a 0.45-μm rated Millipore filter at 30 psi (207 kPa), and then applying the following formula:

$$\text{SDI} = \frac{1 - \dfrac{t_1}{t_2} \times 100}{T} \qquad (4.1)$$

where t_1 = initial time, in seconds, needed to collect 500-mL sample of water through fresh 0.45-μm filter, 47 mm in diameter
t_2 = time in seconds to filter and collect second 500-mL sample after exposing same filter as above for 15 min to flow of feedwater
T = total test time, in minutes, typically 15 min; for high SDI, T may be less

To obtain an accurate test, the filter should not have become more than approximately 74 percent plugged by the end of the elapsed time. If this figure is exceeded, the test should be repeated using a shorter overall elapsed time. A Millipore filter is the only membrane currently approved by the ASTM for determining the SDI. The higher the SDI, the greater potential for fouling.

Many manufacturers of RO cartridges recommend allowable SDI figures for feedwater. For hollow-fiber modules the maximum SDI is 3, and for spiral-wound modules the maximum SDI is 4. For continuous deionization, an SDI of 4 or less is recommended. In practice, when water has an SDI greater than 4, a 4-μm depth prefilter is recommended. In addition to the 4-μm filter, an additional 1-μm filter is recommended downstream. The use of a 4-μm filter in the feedwater stream is always recommended as a precaution against fouling regardless of the potential SDI.

DEPOSITS AND CORROSION

The contaminants previously discussed will cause piping system fouling by depositing materials on the walls of the pipe, thereby reducing the efficiency of the system and reducing the thickness of the pipe wall by corrosion, which will eventually cause failure of the piping system. Following is a brief discussion of these categories and treatment methods as they apply to most systems.

DEPOSITS

Scale and Sludge

Scale is a solid deposit on the walls of a pipe resulting from precipitation of dissolved solids in the fluid stream. Scale reduces heat transfer and interferes with the flow of water by increasing the friction of the fluid against the walls of the pipe. Boiler scale consists of calcium, magnesium, iron, and silica minerals. It is prevented by pretreating water before it enters the boiler to remove much of the scale-forming ingredients or by adding effective chemicals to feedwater to adjust pH, prevent corrosion, and prevent deposits from occurring.

Sludge is a sticky and adherent deposit in the feedwater, which results from the settling out of suspended matter from several sources. One source is an excess of iron, generally iron oxide (rust) and iron carbonate (a corrosion product). Other sources are mud, dirt, and clay that tend to collect and adhere in areas of low circulation. Sludge is prevented by filtering the incoming feedwater and adding chemical dispersants to keep the solids in suspension. They are removed by blowdown. Mud, dirt, and clay are rarely encountered except where the feedwater comes from surface sources.

Condenser scale deposits consist of calcium carbonate, calcium sulfate, or silica minerals when their concentrations exceed their solubility or their pH exceeds saturation, causing these minerals to come out of solution. This is prevented by controlling the pH, diluting the circulating water to prevent concentration, and adding chemicals to inhibit and prevent scale formation.

Biological Fouling

Microbiological fouling is caused by the growth of bacteria, algae, fungi, and other organisms. Their growth is helped by a favorable water temperature, favorable pH, and the presence of oxygen and food. Slime is the buildup of microbes and their waste products and also dust and other suspended matter.

Microbial control consists of sterilization, disinfection, and sanitation. Sterilization is defined as the lethal disruption of all bacteria, molds, and yeasts and the elimination of biofilm and spores. Numerically, it is a 12 log reduction in bacteria. Disinfection is a 6 log reduction of microbials. Sanitation is generally considered to be the killing of the vegetative organisms and it minimizes the presence of bacteria and endotoxins. Numerically, it is a 3 log reduction in bacteria.

There are different methods of controlling biofouling. The method selected depends on the intended use of the treated water and the proposed materials of the system components. Chemicals, ultraviolet radiation, heat, filtering, and ozone are the most commonly used.

CORROSION

Corrosion is the loss and eventual failure of metals and alloys from the electrochemical reaction between water and the pipe material. It is separated into two basic types: general and localized. General corrosion describes the potential dissolution of pipe over its entire exposed surface. Localized corrosion affects only a small area of the pipe surface.

General Corrosion

This is a breakdown of the pipe material at a uniform rate over its entire surface by direct chemical attack. It is caused by the loss of the protective passive film that forms on the surface of the pipe coupled with a chemical reaction occurring between the pipe material and the chemical in the fluid.

Galvanic Corrosion. This type of corrosion occurs in a liquid medium (called an electrolyte) when a more active metal (anode) and a less active metal (cathode) come in contact with one another and form an electrode potential. When this occurs, the more active (noble) metal will tend to dissolve in the electrolyte and go into solution.

Intergranular Corrosion. This occurs in the pipe wall when material in the grain boundary of some alloys is less resistant to the corroding agent than the grains themselves, and the bonds between the grains are destroyed.

Erosion Corrosion. This is caused by a wearing away of the pipe wall, usually as a result of excessive fluid velocity or constant wearing away by solids in the water striking the walls of the pipe.

Localized Corrosion

This takes place on small areas of the surface, usually at high rates, and takes various forms:

1. Stress-corrosion cracking is the physical deterioration and cracking of the pipe wall caused by a combination of high operating temperature, tensile stress on the pipe and chemicals in the fluid stream.
2. Pitting is characterized by deep penetration of the metal at small areas of the surface, concentrating in small cells, without affecting the entire surface.
3. Crevice-attack corrosion occurs at junctions between surfaces (often called crud traps) where a crack exists that allows an accumulation of a corroding agent.

Conventional corrosion treatment of feedwater for boilers and cooling water systems consists of pH control and chemical corrosion inhibitors. Dissolved gases are removed by deaeration.

PREDICTING SCALE FORMATION AND CORROSION TENDENCIES

GENERAL

A common and costly water-caused problem is the formation and deposit of mineral scale. Although scale deposits may contain a complex mixture of mineral salts, the primary constituent is calcium carbonate.

Most salts are more soluble in hot water than cold water. Calcium and magnesium salts, on the other hand, dissolve more readily in cold water than hot. As a result, they tend to deposit on surfaces when there is a rise in temperature. The following are the primary factors which affect this tendency:

1. Alkalinity
2. Hardness (calcium)
3. pH
4. Total amount of dissolved solids
5. Temperature

pH

The pH value reflects the ratio of bicarbonate to carbonate alkalinity. A higher pH value indicates a greater carbonate content of the water, and the increased tendency of calcium and magnesium carbonates to precipitate out of solution. Also, the higher the temperature, the greater the tendency of dissolved solids to precipitate out of solution because of their property of inverse solubility.

LANGELIER SATURATION INDEX (LSI)

In the 1930s, W. F. Langelier studied the primary factors that effected the tendency of water to form deposits of mineral scale on heat transfer equipment. As a result of this work, the Langelier index was created. The index is based on numerical values given to the factors that affect deposits. This index is best known as the Langelier saturation index (LSI).

The LSI is actually a calcium carbonate saturation index. It is based on the assumption that water with a scaling tendency will tend to deposit a corrosion-inhibiting film of calcium carbonate and thus be less corrosive. A water with non-scaling tendency will tend to dissolve protective films and thus be more corrosive. The interpretation of the LSI is based on numerical values given in Table 4.4. The LSI is calculated as follows:

$$LSI = pH - pHs \qquad (4.2)$$

TABLE 4.4 Prediction of Water Tendencies by the Langelier Index

Langelier saturation index	Tendency of water
2.0	Scale-forming and for practical purposes noncorrosive
0.5	Slightly corrosive and scale-forming
0.0	Balanced, but pitting corrosion possible
−0.5	Slightly corrosive and nonscale-forming
−2.0	Serious corrosion

where LSI = Langelier saturation index
  pH = value obtained from testing the water in question
  pHs = value calculated for the pH of saturation for the calcium carbonate present in the water in question

The most accurate method for calculating pHs is to use the following formula:

$$pHs = (9.3 + A + B) - (C + D) \tag{4.3}$$

The numerical values of A, B, C, and D for substitution into Eq. (4.3) are found in Table 4.5. A more empirical method to find pHs is to use the chart in Fig. 4.2.

TABLE 4.5 Numerical Values for Substitution in Eq. (4.3) to Find the pHs of Saturation for Water

Total Solids	A	Ca as $CaCO_3$	C	M. Alkalinity	D
50–330	0.1	10–11	0.6	10–11	1.0
400–1000	0.2	12–13	0.7	12–13	1.1
		14–17	0.8	14–17	1.2
Temp, °F	B	18–22	0.9	18–22	1.3
32–34	2.6	23–27	1.0	23–27	1.4
36–42	2.5	28–34	1.1	28–35	1.5
44–48	2.4	35–43	1.2	36–44	1.6
50–56	2.3	44–55	1.3	45–55	1.7
58–62	2.2	56–69	1.4	56–69	1.8
64–70	2.1	70–87	1.5	70–88	1.9
72–80	2.0	88–110	1.6	89–110	2.0
82–88	1.9	111–138	1.7	111–139	2.1
90–98	1.8	139–174	1.8	140–176	2.2
100–110	1.7	175–220	1.9	177–220	2.3
112–122	1.6	230–270	2.0	230–270	2.4
124–132	1.5	280–340	2.1	280–340	2.5
134–146	1.4	350–430	2.2	350–440	2.6
148–160	1.3	440–550	2.3	450–550	2.7
162–178	1.2	560–690	2.4	560–690	2.8
178–194	1.1	700–870	2.5	700–880	2.9
194–210	1.0	880–1000	2.6	890–1000	3.0

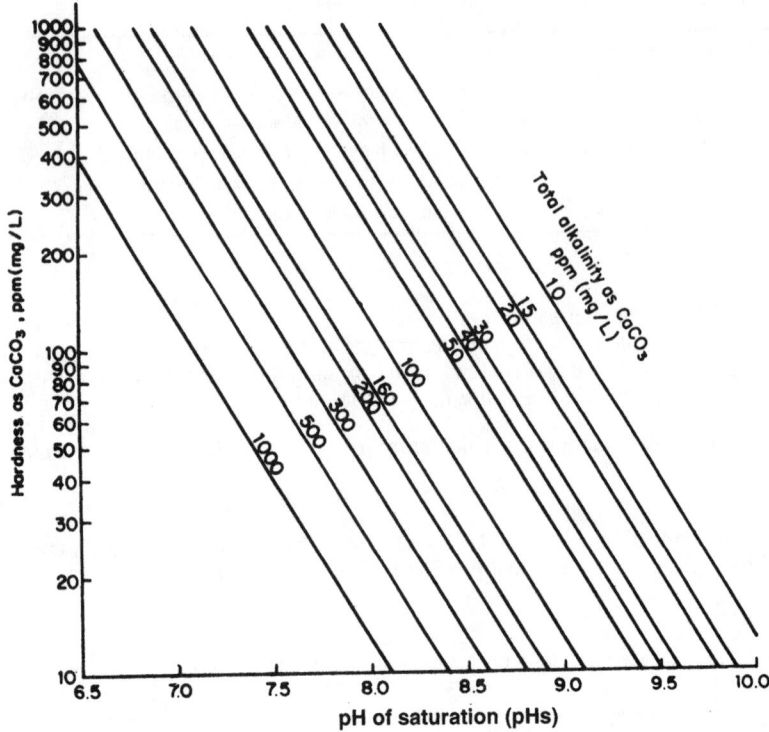

FIGURE 4.2 pH of saturation for water.

RYZNAR STABILITY INDEX (RI)

Also referred to as simply the stability index (SI), the RI is an empirical method used to predict scale-forming tendencies of water. The RI is calculated from the following formula using the same definitions used for the LSI:

$$RI = 2 \text{ pHs} - pH \tag{4.4}$$

The RI is always positive. When it falls below 6.0, scale formation is possible and becomes more probable with decreasing RI. For an interpretation of the RI, refer to Table 4.6.

AGGRESSIVENESS INDEX (AI)

Developed by the EPA, the aggressiveness index is used rarely as a guideline to find the corrosive tendency of potable water. AI is calculated as follows:

TABLE 4.6 Prediction of Water Tendencies by
the Ryznar Index

Ryznar index	Tendency of water
4.0–5.0	Heavy scale
5.0–6.0	Light scale
6.0–7.0	Little scale or corrosion
7.0–7.5	Significant corrosion
7.5–9.0	Heavy corrosion
9.0 and higher	Intolerable corrosion

$$\text{AI} = \text{pH} + \log_{10} (\text{alkalinity} \times \text{hardness}) \qquad (4.5)$$

Values less than 10 indicate aggressive water, values between 10 and 12 indicate moderately aggressive water, and values more than 12 indicate nonaggressive water.

TECHNOLOGY AND COMPONENT DESCRIPTION

AERATION

Aeration is a gas transfer process in which water is brought into contact with air for the purpose of transferring volatile substances to or from the raw water. It is used most often to remove undesirable gases such as carbon dioxide, hydrogen sulfide, and methane. Aeration, by introducing oxygen, is also used to remove iron and manganese and to lower the amount of volatile organic compounds (VOCs) from ground water. The following are criteria for its use:

1. Reduction of carbon dioxide by 90 percent is obtained by near saturation with oxygen. Carbon dioxide dissolved in ground water will consume lime in the lime-soda softening process without any accompanying softening. Generally accepted practice indicates that aeration is not economical for carbon dioxide concentrations in water of less than 10 mg/L. The cost of the lime saved should be compared to the cost of purchasing and operating the proposed aerator.

2. Aeration will partially remove VOCs from raw water by oxidation, making them insoluble. They would then be coagulated and removed from the water.

3. Aeration alone could be used for the removal of hydrogen sulfide in concentrations of 2 mg/L or less. Above this level, it could be used in conjunction with chlorination, which oxidizes hydrogen sulfide.

4. Iron and manganese can be removed by aeration if this cannot be done by other methods. These metals are oxidized to form insoluble hydroxides which precipitate out at proper pH levels. They can then be removed by settlement or filtered out of the water stream. Each ppm of dissolved oxygen will oxidize about 7 ppm of iron or manganese.

Several types of aerators are commonly used: waterfall, diffusion (or bubble), and mechanical. The most commonly used system in utility water treatment is the waterfall.

Waterfall aerators are made in several types: cascade, spray nozzle, and multiple tray. They operate by having the raw water enter the unit from the top and fall by gravity to the bottom, using various methods to evenly distribute and disperse the water throughout the unit. The most common waterfall aerator is the naturally ventilated, multiple tray type. This unit consists of a series of trays, one above the other, with perforated, slot, or mesh bottoms. The trays are filled with 2 to 6 in (40 to 140 mm) of a medium such as coke, stone, or ceramic balls to improve water distribution and gas transfer as well as providing catalytic oxidation in the medium. The vertical openings between trays range from 12 to 30 in, depending on the number of trays required. Water loading on the trays is usually in the range of 10 to 20 gpm/ft^2. Efficiency can be improved by the use of enclosures and forced air blowers to provide counterflow ventilation.

The simplest type of aerator is the diffusion type which bubbles compressed air up through the water tank. The large volume of air required limits the diffusion aerator to smaller flows of water. Air requirements vary from 0.1 to 0.2 SCFM per gallon of water aerated. A detention time is necessary, ranging from 10 to 30 mm. Advantages of this type of aerator are the freedom from cold weather operating problems, very low head loss, and the possibility of using this process for chemical mixing.

The mechanical aerator consists of an open impeller operating on the water surface of a tank. It is not as efficient as either of the two systems previously described, and so, longer retention times are necessary. This unit is also free from cold weather problems and can be used for chemical mixing.

CLARIFICATION

Clarification is a process to reduce or remove turbidity, silt, and sediment present in the raw water supply. The water can be treated with chemicals or filtered, depending on the amount of impurities present and the volume of water to be treated. If treated with chemicals, time is required for the solids to settle out of suspension. The chemical treatment process is usually reserved for large volumes of water.

Suspended and colloidal particles will normally be in the range of 0.1 to 10 μm in size. They stay apart because negatively charged ionized matter is absorbed on their surfaces and thus they repel each other. A measure of the charge that surrounds the colloid is known as the "zeta potential." These negative charges are reduced by the use of positively charged chemicals called coagulants. Coagulants, through a chemical reaction, reduce the zeta potential and allow the colloids to cluster to form a larger, jellylike mass called floc. In a process called flocculation, colloidal particles, bacteria, and organic matter are mechanically brought together into larger and heavier particles. These particles are now too heavy to remain suspended, and so, they settle out of the water to the bottom. This process is called sedimentation.

Clarification of large volumes of water is done in a basin. First, the coagulant(s) must be introduced with a strong action in order to completely disperse and mix the chemicals with the incoming water. Often, some of the sludge produced by flocculation is mixed with the coagulant. After coagulation is accomplished, the mixture is then gently mixed in order to allow the now larger particles to settle rapidly to the bottom of the tank. This settling process is called sedimentation. The particles and chemicals that settle out are called sludge, which must be removed from the tank and disposed of. Most clarifier designs do both processes in the same operating unit.

The chemical coagulant used most frequently is aluminum sulfate, called alum or filter alum. Other coagulants used often are sodium aluminate and potash alum. Iron coagulants include ferric sulfate, ferris sulfate, and ferric chloride. Organic polyelectrolytes, which are high molecular weight polymers, are also employed in low dosages to increase the effectiveness of treatment.

It is usually a matter of trial and error to find the correct amount of alum. This is because the amount depends on the size and quantity of suspended solids of the raw water, retention time prior to flocculation, water temperature, and the amount of mixing. All of these conditions change from day to day.

As with almost every process, there is more than one method appropriate to achieve a desired goal when the demand for water is not very large. One alternative flocculation method is to produce a finer floc which can be removed by an appropriate filter system. By eliminating a large settling basin, a high quality water can be produced more quickly with smaller equipment.

DEAERATION

Dissolved gases in the water supply can produce corrosion and pitting. They must be removed prior to using the water for most purposes. These gases are oxygen, carbon dioxide, hydrogen sulfide, and ammonia.

Although carbon dioxide and hydrogen sulfide can be removed by aeration, in many cases it is easier and less expensive to remove them by deaeration. This process operates on the principle of raising the water temperature to the saturation point for the existing pressure. There are two types of deaerators, steam and vacuum. When heated water is needed, for example, for boilers, the steam type is preferred. When cold water is required, the vacuum type is used.

Steam deaerators break up water into a spray or thin film, then sweep the steam across and through it to force out the dissolved gases. Using this method, oxygen can be reduced to near the limit of detectability. Designs fall broadly into spray, tray, and combination units. Typical deaerators have a heating and deaeration section and a storage section for hot, deaerated water. Often, a separate tank is provided to hold an additional 10-mm supply of deaerated water. The direction of steam may be crossflow, downflow, or countercurrent. The majority of the steam condenses in the first section of the unit. The remaining mixture of noncondensable gases is discharged to the atmosphere through a vent condenser.

Vacuum deaerators use a steam jet or mechanical vacuum pumps to develop the required vacuum used to draw off the unwanted gases. The vessel has a packing material inside, and the inlet water is introduced to the top of the unit and passed down through this packing. Deaerated water is stored at the bottom of this vessel. The steam or vapor vacuum located at the top of the unit discharges the unwanted gases. The vacuum unit is far less efficient than the steam one, and is most often used in demineralizer systems to reduce the chemical operating cost and the demineralizer size, and to protect anion exchange resins from possible oxidation damage.

DECARBONATION

Decarbonation is rarely used but should be considered if the bicarbonate level in the feedwater is in the range of 14 to 20 mg/L or higher. Decarbonation is usually accomplished by the use of filtered air counterflowing through the water stream and stripping out the carbon dioxide.

DEALKALIZING

Dealkalizing is a process that reduces the alkalinity of feedwater. This can be done either with or without the use of acid regenerent. Without the acid, the regenerent used is salt, and the process is often called salt splitting. The salt-splitting process exchanges all bicarbonate, sulfate, and nitrate anions for chloride anions. For best results, it is recommended that deionized water be used as feedwater. Where hard, alkaline, low sodium water is available, the use of a weak acid resin should be considered.

The entire regeneration cycle is similar to that previously described for a water softener. It is not uncommon to use the same salt and regenerent piping to accommodate both the softener and dealkalizer. Some caustic soda may be added (1 part caustic soda to 9 parts salt) to reduce leakage of alkalinity and carbon dioxide. Since this can cause hardness leakage from the dealkalizer, a filter downstream of the processed water is necessary.

A weak acid resin can also be used. The process transfers the alkaline salts of calcium and magnesium to the weak acid resin. This process should include degasification if required by the product water. The weak acid process operates at a very high utilization factor, near the theoretically required amount. Hydrochloric acid is preferred for regeneration rather than sulfuric acid. This process is very sensitive to the flow rate, temperature, and contaminant level of the feedwater. These changing conditions must be considered by the manufacturer in the design of the process. The entire regeneration cycle is similar to that of cation ion exchange column.

DISTILLATION

In its basic form, distillation is the boiling of feedwater, condensing the steam produced from the feedwater, and collecting the condensate which yields a water product that is theoretically free of nonvolatile impurities. There are three methods currently used to produce distilled water: single-stage, vapor compression, and multieffect distillation.

Single-Stage Distillation

This is the most simple still. Feedwater enters the still, then is evaporated and condensed in a single stage. Cooling water is required to condense the steam produced. This type of still will produce water of approximately 1 MΩ/cm, with higher purity possible with optional equipment that removes dissolved gaseous impurities. This still has a small footprint, is not labor intensive, and will tolerate feedwater with a high level of impurities.

Vapor Compression Distillation

Vapor compression, sometimes called thermocompression distillation, is a method of evaporation in which a liquid is boiled inside a bank of tubes. The vapor generated then passes through a mist eliminator that removes any water droplets. The pure vapor is withdrawn by a compressor in which the energy imparted results in a compressed steam with increased pressure and temperature. The high-energy-compressed steam is discharged into an evaporator. At this point, the steam gives up most of its energy (latent heat) to the water inside the tubes. More vapor is generated and the process is repeated. The condensate (distilled water) is withdrawn by the distillate pump and is discharged through a two-stream heat exchanger.

The excess feedwater that did not evaporate is also pumped through an exchanger. Both the distillate and the blowdown are cooled, and the feedwater is preheated prior to entering the evaporator. These exchangers minimize the energy consumption of the system and eliminate the need for additional cooling water. The system operates continuously once it is started. Additional makeup heat, usually supplied by steam, is required for continuous operation. Vapor compression is generally considered more economical for large quantities of water and does not require a high quality feedwater for proper operation. The vapor compression still is moderate both in initial and operating costs.

Refer to Fig. 4.3 for a schematic diagram of a vapor compression distillation unit.

Multieffect Distillation

Multieffect distillation units use the principle of staged evaporation and condensation to generate distilled water. Each stage is called an effect. Distilled water is produced in each effect by condensing the steam generated by the evaporation of high purity feedwater in the previous stage. The initial driving force for the evaporation is power steam applied to the shell side of the first effect vessel. The multieffect still has the highest initial cost and lowest operating cost, and it requires the highest quality feed-water of all the stills.

The feedwater enters the vessel and its pressure is boosted by the feed pump. The feedwater flows through a coil in the condenser, which allows it to pick up heat from the condensing steam. This preheated feedwater flows through the feed control valve and into the tube side of the first effect. The first effect level controller senses the feedwater level and signals the feed control valve to maintain the desired level. Power steam is introduced into the unit and flows through the steam control valve and into the shell side of the first effect.

Temperature sensors sense the temperature on the tube side of the first effect and signal the steam control valve to maintain the required temperature. This steam condenses on the outside of the tubes of the first effect, giving up its latent heat of vaporization to the feedwater inside of the tubes, causing the water to boil and generate vapor.

The pure steam generated in the first effect is introduced into the shell side of the second effect. The pure steam condenses, producing distilled water, while giving up its latent heat to the high purity feedwater inside the second effect tubes, which in turn causes the feedwater to boil and generate vapor. Each effect operates at a lower pressure than the previous effect in order to provide the temperature difference that allows the transfer of heat. The pure steam generated in the tube side of the first effect by the condensing power steam passes through the mist eliminator to remove any entrained water droplets. Feedwater from the first effect passes through an orifice and into the tube side of the second effect. The first effect pure steam enters the shell side of the second effect and is condensed on the outside of the tubes.

The condensate (distilled water) passes through an orifice and enters the shell side of the third effect. Feedwater in the second effect passes through an orifice and into the tube side of the third effect.

After passing through the mist eliminator, the last effect pure steam enters the condenser and condenses on the outside of the condenser coils. This distilled water from the last effect and the distilled water from the previous effects is cooled by the cooling water of the condenser. The distilled water exits the condenser and enters the distillate pump, where it is pumped through the distillate control valve and through the storage-dump valve. The condenser level controller senses the distillate level and signals the control valve to maintain the desired level.

Noncondensable gases in the condenser are vented to the atmosphere. The condenser temperature is maintained at a predetermined level by the cooling water flow. The unit is protected by pressure relief valves along with high and low level alarms.

Refer to Fig. 4.4 for a schematic diagram of a multieffect distillation unit.

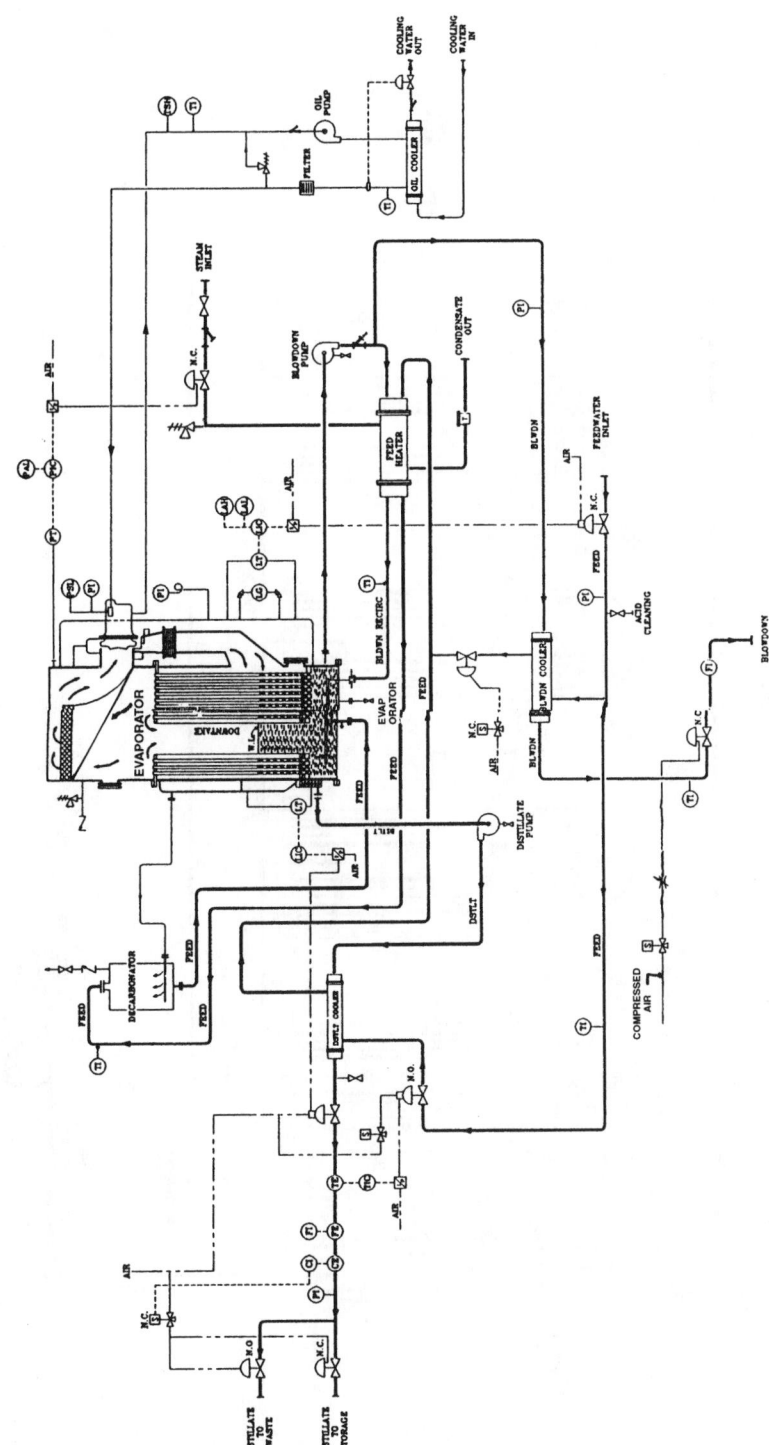

FIGURE 4.3 Detail of vapor compression still.

4.25

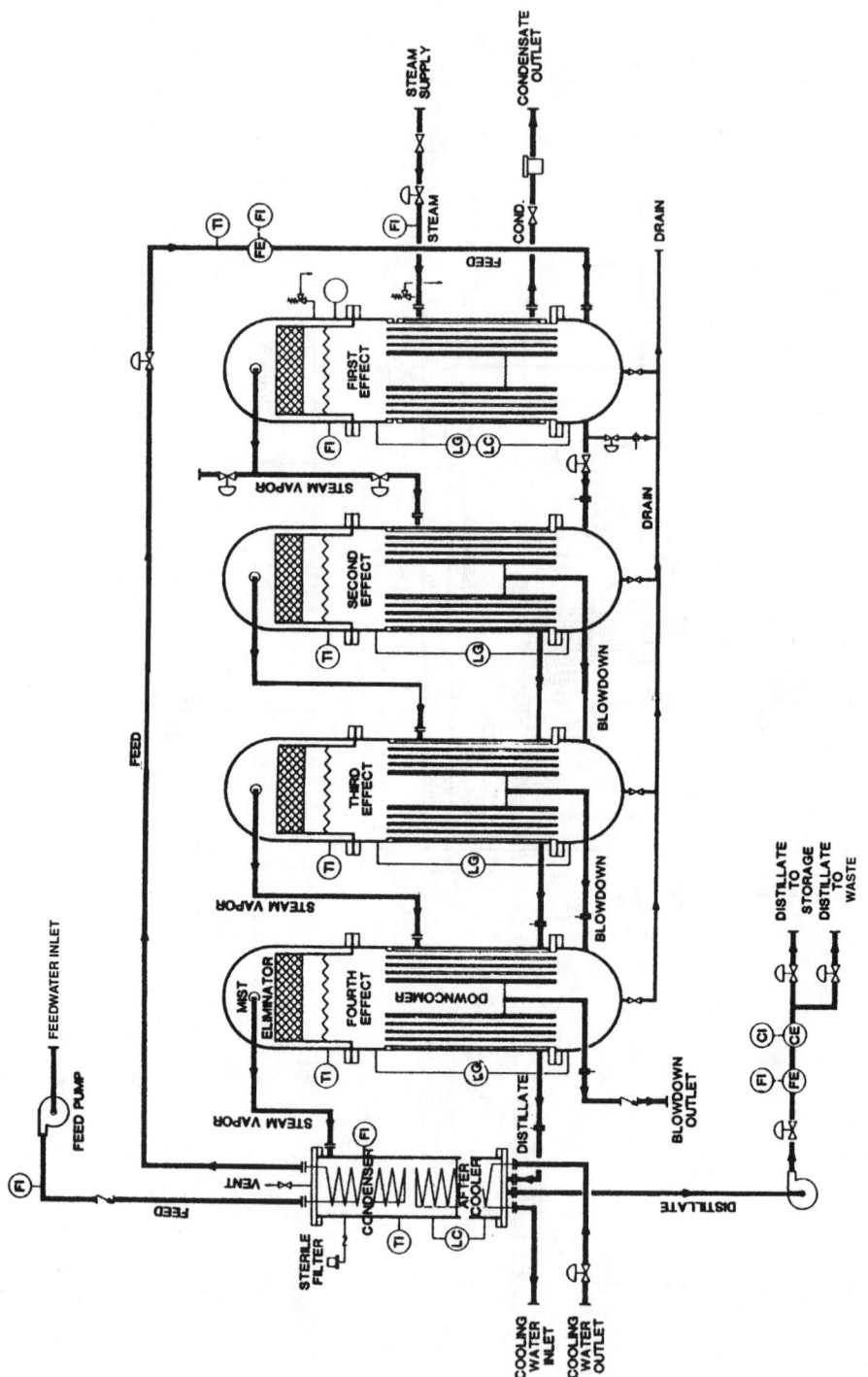

FIGURE 4.4 Detail of multieffect still.

4.26

FILTRATION

Deep Bed Sand Filtration

Deep bed filters are designed to remove coarse suspended particulates larger than 10 μm in size. They are pressure-type filters that use either multigraded sand or multimedia in the filter. Particulate removal on the order of 98 percent should be expected. For additional discussion, refer to Sand Filters in Chap. 3.

Sand-only filters for laboratory water systems should generally operate at a face velocity of about 4 gpm/ft^2 of cross-sectional bed area. Multimedia filters operate at about 6 gpm/ft^2 of cross-sectional bed area. Since these values are general in nature, it is important to operate these units at the velocities recommended by the individual manufacturers.

The multimedia filter achieves a more uniform distribution of filter media throughout the bed and is considered a more effective type of filter. A typical multimedia filter for laboratory use consists of a top layer of anthracite having a 1.1-mm grain size and 1.5 specific gravity, a middle layer of sand having a 0.5-mm diameter grain size and a specific gravity of 26, and a bottom layer of garnet having a 0.2-mm grain size and 4.2 specific gravity. Normal operational flow rate is from 6 to 15 gpm/ft^2 of bed area.

Backwashing is required to clean the filter, with the effluent discharged to the sanitary drainage system. A backwash flow rate of 10 to 15 gpm/ft^2 is generally required for effective cleaning.

Activated Carbon Filtration

Activated carbon is used to remove dissolved, nonionic organics such as residual chlorine disinfectants, trihalomethanes, and chlorimines, as well as a major portion of naturally occurring dissolved organic material, from municipal water supplies. The non-ionic organics tend to coat ion-exchange resins and all types of membranes. For additional discussion, refer to Activated Carbon Filters in Chap. 3.

System designers are reluctant to use the activated carbon filter in the generation of PW because of the development of significant levels of bacteria in the unit itself. This can be controlled by periodically sanitizing with pure steam or hot water with a temperature greater than 80°C. The need for sanitizing can be determined only by testing the water. Because of this sanitizing, the interior of the filter housing should be lined or coated. When using potable water as feedwater, stainless steel housings should be avoided because of the possible chloride stress corrosion and chloride pitting resulting from the chlorine in the feedwater. A typical detail of a packed bed activated carbon unit is illustrated in Fig. 4.5.

ION EXCHANGE AND REMOVAL

Ion exchange is the basic process where specific ions in a feedwater stream are transferred onto an exchange medium called resin, and exchanged for different ions of equal charge. When the ion exchange process is used to treat water only for removal of hardness, it is generally known as water softening. When the ion exchange process is used to treat water for the removal of ions to produce pure water, it is often referred to as deionization or demineralization.

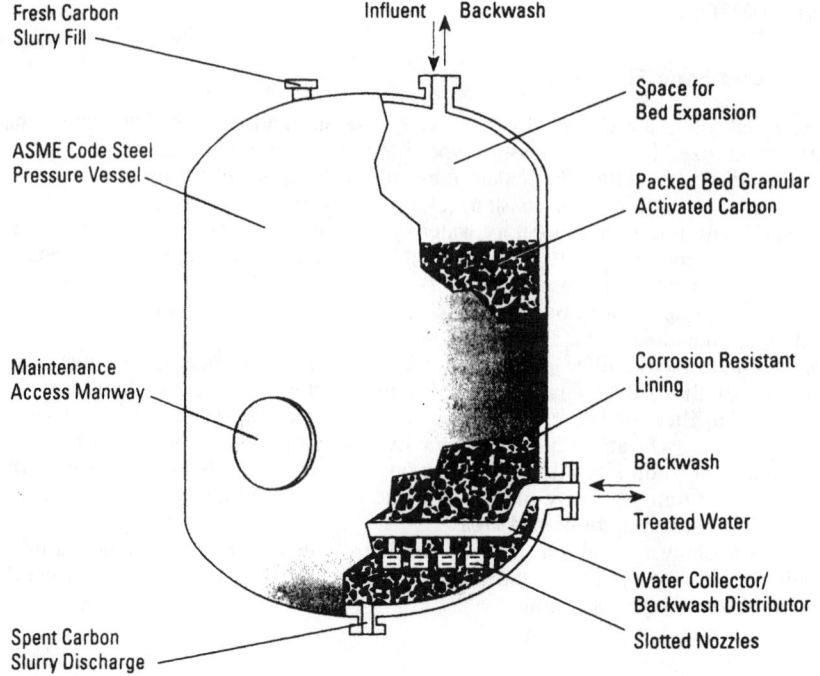

FIGURE 4.5 Schematic detail of large-scale granular activated carbon filter.

The deionization/demineralization process uses different types of resin to exchange first anions and then cations, which will result in the removal of all ions from feedwater when carried to completion. When all of the ionic components devolved in water are removed by ion exchange, the water is said to be deionized or demineralized. The ion exchange process is also used to remove dissolved inorganics. Water softening only exchanges some types of ions for others less detrimental for the intended end use of the water. Table 4.7 lists the common anions and cations typically found in water. Ion exchange will not remove significant amounts of organics, bacteria, pyrogens, or turbidity.

TABLE 4.7 Typical Cations and Anions Found in Water

Cations	Anions
Calcium	Carbonates
Magnesium	Bicarbonates
Sodium	Sulfates
Potassium	Chlorides
Iron	Nitrates
Manganese	Silica

Regenerable Ion Exchange

Regenerable ion exchange is a batch process where ions in raw water are transferred onto a resin medium in exchange for other ions bonded to that medium as the raw water percolates through them. This is accomplished by having the ions in the raw water adsorbed onto a bed of exchange resins and replaced with an equivalent amount of another ion of the same charge. This action continues until the medium has reached its exchange capacity, where it is no longer capable of exchanging ions. Water softening and deionization are the two most common ion exchange processes.

There are two general types of deionizers, working and polishing. The working type is used for the initial removal of the bulk of ions from feedwater or as an ion exchange process alone (such as hardness removal) if the purification is a single process. A polishing type is used to purify feedwater after an initial run through a working ion exchange system.

Resins

Resin exchange media include natural inorganic aluminum silicates (sometimes called zeolites or greensands), bentonite clay, synthetic gelatinous resins, and synthetic organic resins. Most processes use the synthetic resins. Resins are graded by purity and consistency of bead size.

Resin is manufactured in the form of a large number of spherical beads, typically about 0.4-mm diameter. These beads have weakly bonded ions present on their surface that are used for the exchange process. Because the process must exchange ions of the same charge, ion exchange resins are composed of either anion or cation exchange resins. Manufacturers are constantly making new resins for different ion removal purposes.

Traditional deionization exchanges cations with H^+ ions (acids) and anions with OH^- ions (bases). Although not 100 percent effective, the end result of these two exchange processes combines the ions remaining in the feedwater to create deionized or demineralized water as the end product.

There are a large number of ion exchange resins available. Each of them has been formulated to obtain optimum performance for different impurities. The affinity for specific ions in solution is termed the *selectivity coefficient.* The number of charges (valence) available on a particular ionic medium is a major factor in the selection of specific resins to remove the desired impurities, and is based on an analysis of raw water. These resins are contained in a vessel often referred to as a column. The actual resin bed could be supported by a mat of graded gravel, screen-wrapped pipe, or perforated plates, which also act to evenly distribute feedwater over the entire resin bed. The size of the resin beads in the vessel also creates an effective depth filter. This filtering action leads to fouling and unpredictable operating runs because of an accumulation of particulates.

Anion resins could be either strong or weak bases. A common anion resin is divinyl benzene, a gelatinous bead. Anion resin type 1 premium has a very close tolerance of bead size. Anion resin type 1 regular is generally used for maximum silica reduction. Resin type 2 is used most often unless type 1 is specifically requested. There is a difference in cost and capacity between the two resins. In general the higher cost of the type 1 resin is considered acceptable in order to obtain a more efficient and longer-lasting resin. Weak base exchangers are not effective in the removal of carbon dioxide or silica. They remove strong acids more by adsorption than by ion exchange. The end result is the same, and the efficiency of weak base regeneration for acid salt removal is far superior to that of the strong base material doing the same job. Thus weak base units are superior when the feedwater is high in sulfates and chlorides.

The two cation exchange resins used most often are strong or weak acids. Strong cation resins remove all cations regardless of the anion with which they are associated. These resins have a moderate exchange capacity and require a strong acid regenerant such as hydrochloric or sulfuric acid.

The deionization process can be arranged as either a two step (dual bed) or single step (mixed bed) process. In the dual bed process, one vessel contains the anion exchange resins and the second vessel, the cation exchange resin. In the mixed bed unit, a single vessel contains a mixture of both resins. The dual bed arrangement produces water that is less pure than that produced by a mixed bed, but the dual bed has a greater removal capacity. A typical mixed bed contains 40 percent cation resins and 60 percent anion resins. Dual beds are easier to regenerate. It is not uncommon to have a dual bed exchanger, often referred to as a working exchanger, installed in front of a mixed bed to remove the bulk of the impurities, and then have the mixed bed, often called a polishing exchanger, further purify the water to the desired high level. A typical single-bed ion exchange unit is illustrated in Fig. 4.6. A typical dual-bed ion exchange unit is illustrated in Fig. 4.7. A typical mixed-bed ion exchange unit is illustrated in Fig. 4.8. The piping and valve arrangement for different manufacturers may be different.

Regeneration Cycle

The ion exchange process is reversible. As the water continues to pass through the ion exchange resin beds, the number of ions on the resin beads available for

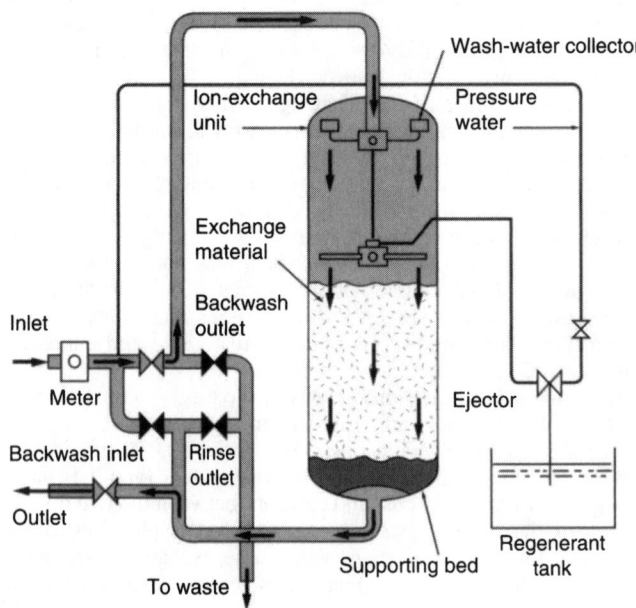

FIGURE 4.6 Typical single bed ion exchanger.

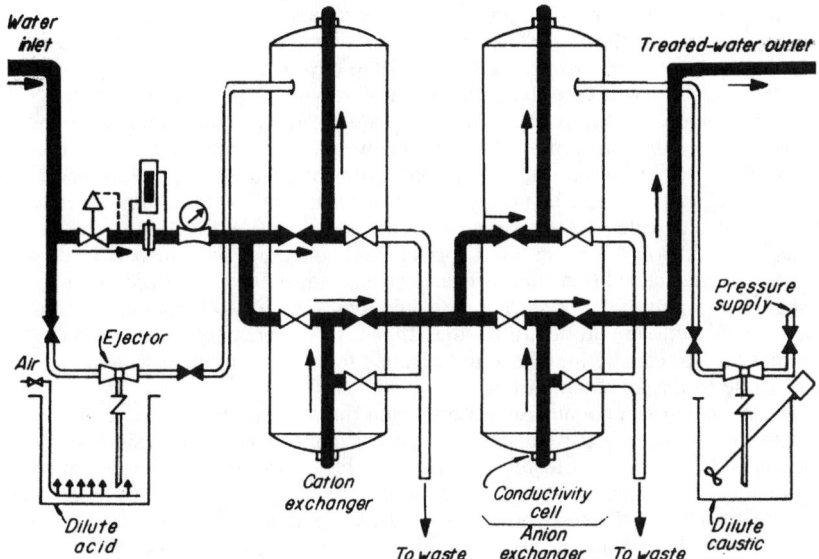

FIGURE 4.7 Typical dual bed ion exchanger.

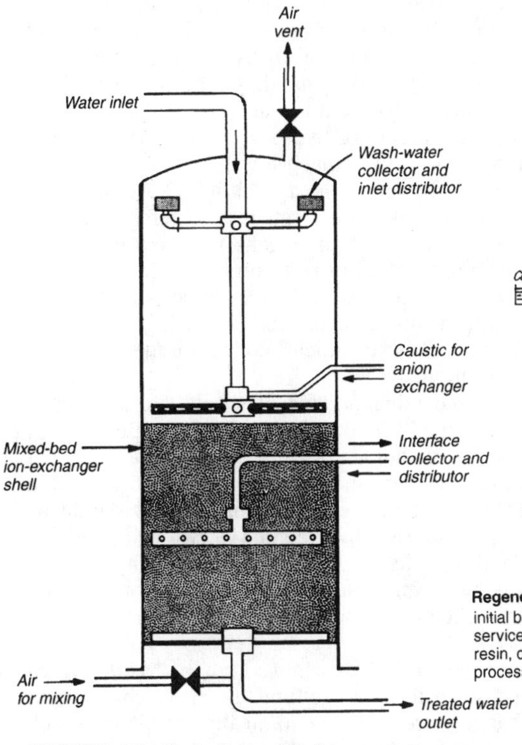

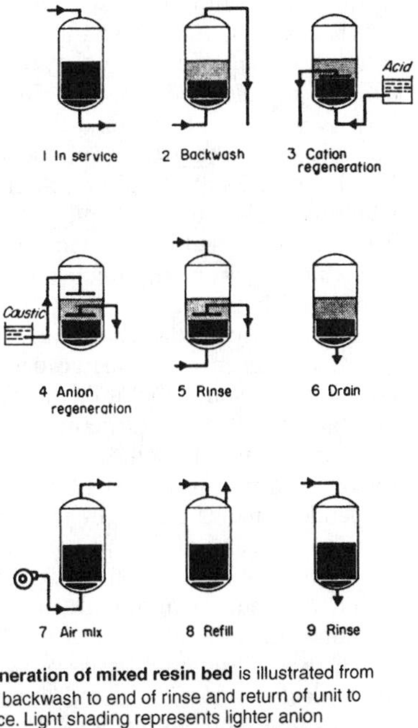

Regeneration of mixed resin bed is illustrated from initial backwash to end of rinse and return of unit to service. Light shading represents lighter anion resin, dark tint is heavier cation beads. Entire process can take anywhere from two to four hours

FIGURE 4.8 Typical mixed bed ion exchanger.

exchange declines with time and gradually becomes exhausted. This starts first at the water entry to the vessel and progresses down the bed. When the resin has reached the limit of exchange, the bed is said to have reached its exchange capacity. It is then necessary to take the column out of service to be regenerated.

Regeneration, which is the reverse of deionization, is the term used for the replacement of the ions removed by the feedwater. Regeneration generally consists of three steps: (1) backwashing, (2) application of regenerating solution, and (3) rinsing. Regeneration can be performed either cocurrently (in the same direction as the flow of feedwater) or countercurrently (in the opposite direction of the flow of feedwater). All of the water used for regeneration must be routed to a drain of adequate size. In addition, the acid and caustic must be neutralized prior to discharge into a public sewer system. It is common practice to combine the acid and caustic waste streams to neutralize the effluent to the greatest extent possible. Additional acid or caustic may have to be added to the final effluent to produce a pH acceptable to the local authorities.

Backwashing is a countercurrent operation that accomplishes two purposes. The first is to remove any particulates that have accumulated in the resin bed and on the beads. The second is to regrade the resin beads so that new beads are on the top of the bed, which is where the heaviest duty from the beads is required. This is done by having the resin bed expanded by the reverse flow of water from its normal packed condition during use. Individual manufacturers have established the required flow rate of backwash that should be maintained. Too high a flow rate will blow resin out of the tank and into the drain. The flow rate of water should be high enough to scrub the beads together to increase the cleaning action. Higher than recommended flow rates will only waste water and provide no additional benefits.

The two chemicals used to regenerate cation resin beds are either a 93 percent solution of sulfuric acid or a 30 to 32 percent solution of hydrochloric acid, also called mureatic acid. As they flow through the columns, they replace the retained cations with hydrogen ions from the acid. Hydrochloric acid is used most often because it has the greatest efficiency; the amount of hydrochloric acid used is one-fourth the amount of sulfuric acid. Sulfuric acid is much lower in cost and is used when there is a large quantity of resin to be regenerated, making its use practical. The chemical used most often for regenerating anion resins is a 40 percent mixture of sodium hydroxide, which replaces the retained anion ions with hydroxyl ions. For mixed bed units, the resins must be separated prior to regeneration.

The quality of the chemicals used for regeneration has an important effect on the maintenance of exchange capacity. Although chemically pure ingredients are not required, some contaminants found in these chemicals collect on the resins and eventually will cause difficulty in operation.

Technical grade acids, free of oils and other organic materials, are acceptable for regeneration of cation resins. They should be 66 degrees Baumé (°Bé), free of suspended matter and light in color. They should mix freely with water and not form any precipitate. Acid-containing inhibitors should not be used. Sulfuric acid is usually the most economical for large-scale use. Hydrochloric acid should be technical grade, a minimum of 30 percent HCl by weight (18°Bé) and shall not contain excessive amounts of iron and organic materials. HCl obtained by the salt-aid or the hydrogen-chlorine process has been found satisfactory. HCl obtained by the hydrolysis of chlorinated organic chemicals should be avoided, particularly if used to treat potable water.

Anion exchange resins are regenerated with 76 percent sodium hydroxide, which must be low in iron, chlorides, and silica to avoid fouling the strong base anion exchangers. Weak base anion exchangers are most economically regenerated with technical grade flake sodium hydroxide. Strong base exchangers are best regenerated using nylon or rayon grade sodium hydroxide, also 76 percent. If purchased in a 40 percent solution, use the grade previously indicated. All caustics must have a maximum of 2 ppm chlorates.

The flush cycle is the shortest. It is a cocurrent process whose purpose is to flush away any remaining residue of the regeneration liquids and to repack the bed in preparation for the new run.

The entire regeneration cycle typically takes about 1 h. If the process requires continuous operation, a duplicate set of equipment is installed so that one is in use while the other is being regenerated.

To estimate the frequency of regeneration, first, find from the manufacturer the exchange capacity in grains of the selected resin bed. Next, from the analysis of the raw water find the average TDS level and convert this figure into grains per gallon. Dividing the flow rate in gallons into the grain capacity of the resin bed will give the time it takes to saturate the resin bed before regeneration is required.

Service Deionization

Service deionization is not another form of deionization but rather, a different type of equipment arrangement. In the regenerable type, the DI equipment is permanent and the regeneration is done on site by operations or maintenance personnel, who must handle and store the chemicals used for regeneration.

The service DI system uses individual cartridges or tanks for the anion, cation, and mixed beds. When the individual cartridges are exhausted, the supplier replaces them with recharged units on site and the exhausted cartridges are removed to be regenerated at the supplier's premises. This arrangement considerably reduces the initial cost of the equipment, eliminates the need to store chemicals, and frees the operations or maintenance personnel from the job of regenerating the units. In addition, it saves water that does not have to be used for backwash. However, the operating costs are higher than for the permanent bed type.

Continuous Deionization

Continuous deionization (CDI), also known as electrodeionization (EDI), is a continuous water purification process that uses direct current, an alternating arrangement of cation- and anion-permeable ion exchange membranes that form parallel flow compartments (concentrating compartments) on either side of an additional flow compartment containing a thin layer of mixed bed ion exchange resin (diluting compartment). The components, called a cell pair, are installed in a plate and frame device where the flow compartments form various flow paths for the product and wastewater. This arrangement is illustrated schematically in Fig. 4.9.

The feedwater flows through the diluting compartment. When a DC field is applied across the pair of membranes, ions move from the diluting stream, through the ion beads and the membrane into the concentrate stream, thereby producing separation. With an alternating arrangement of cell pairs, the cations and anions are trapped in the concentrate stream where they are routed to drain. The resin bed serves as a highly conductive medium through which the ions flow. The various flow streams are hydraulically independent allowing a high volume of high purity water (product) and a low volume of concentrate (waste).

The resin-filled diluting compartment (cell) creates a low level resistance path for ions. At the outlet of the diluting cell, under the proper combination of flow, temperature, water conductivity, and voltage, the ion exchange resins will regenerate

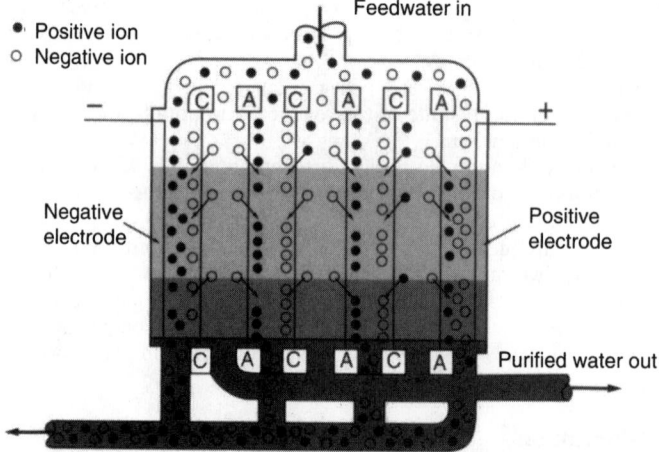

FIGURE 4.9 Schematic operation of a continuous deionization unit. In a continuous deionization cell, electric field pulls positive ions in feedwater through cation-permeable membranes (C), negative ions through anion-permeable membranes (A). Blockage by next membrane confines all ions to one set of channels, purified water to other set. Result: one purified stream, one waste stream.

automatically without the use of added chemicals. This process is continuous, and results in a steady supply of high purity water from the diluting compartment.

CDI is very sensitive to feedwater impurities. Experience has shown that very few natural or potable feedwater supplies can meet the required feedwater specifications without softening. Typical feedwater system requirements are the same as those for potable water. Individual manufacturers may have other specific requirements. Because of these requirements, most processes that purify water for pharmaceutical applications use CDI for polishing purposes after RO. Another disadvantage of this process is that the membrane used is incompatible with standard chemical sanitizing agents and sensitive to the frequency of sanitizing.

Water Softening

Water softening is a process that reduces or removes the dissolved impurities that cause hardness in water. This is commonly done by either of two methods: adding lime-soda ash to the raw water (for very large volumes) or passing the raw water through an ion exchange process.

Lime-Soda Addition. The lime-soda ash method uses either hydrated lime or quicklime along with soda ash. When added to water, these chemicals react with the dissolved calcium and magnesium carbonate to form insoluble compounds. These compounds precipitate out of solution and are passed through a filter to remove them. They are then discharged to drain. This process is usually carried out during the clarification process rather than as a separate process and is reserved for large volumes of water.

Ion Exchange. The ion exchange process is a cation exchange used to remove insoluble and scale-forming iron, calcium, magnesium carbonate, and other multivalent cations, which are the primary causes of hardness, and replace them with sodium ions, which do not contribute to hardness. The removal of these impurities prevents the buildup of insoluble scale precipates on the piping and reverse osmosis membrane. Ion exchange is accomplished by passing the water through a bed of granular sodium cation exchange resin. This process is commonly called sodium cycle ion exchange.

The resin bed typically occupies about two-thirds of the tank. The other one-third is needed for expansion of the resin bed during backwash. A generally accepted range of 0.4 to 3 gpm/ft^3 of resin is used to determine the volume of resin and cycle time of the unit.

Microbial growth inside the unit is a concern in softening systems used for pharmaceutical and some laboratory purposes. The water softener is regenerated with a brine solution that does not destroy bacteria. The liquid brine solution storage and regeneration equipment also allow microbial growth in storage tanks that are exposed to the atmosphere. An alternative is to use a dry storage system, which generates a salt solution from water mixed with salt pellets only when necessary for regeneration. The dry storage system controls microbial growth better than wet systems, but constant maintenance is required to monitor the brine tank. The quality of the salt in all systems should be checked periodically to ensure that there are no added substances present.

The regeneration cycle is similar to that previously discussed. The difference is that salt is used to regenerate the resin bed. Industrial water softeners use rock salt for economy. Rock salt, because of its high mineral content, requires a special tank called a dissolver to dissolve the rock salt in water prior to use. The water softener is similar to the schematic single bed ion exchanger illustrated in Fig. 4.6.

Ion Exchange System Design Considerations

One of the major decisions that must be made when selecting an ion exchange system is how much leakage to allow. Leakage is the presence of unwanted ions in the final treated water. The amount of leakage is a function of the completeness of the resin regeneration. For water softening, generally accepted leakage amounts range from 0.1 to 1 ppm. Since total regeneration of the resin bed is impractical, most water softeners operate at one-half to two-thirds of their ultimate capacity. There is sodium leakage from cation exchangers and silica leakage from anion exchangers. Normally, mixed bed units have negligible leakage.

In general, adequate purification of the water stream for high purity applications is not possible in one pass through the system, so a polisher is necessary. A mixed-bed ion exchange system has an initial cost about 74 percent less than a dual bed system when used as a polisher. A single pass RO system is about equal to a dual-bed ion exchange system.

Usually, if the water demand for a facility is less than about 40 gpm, the greatest benefit will be derived from simpler and less costly equipment at the expense of higher operating costs. For this water demand, it is usual to have a mixed bed unit without a degasifier, which is not required. For systems with a requirement of 200 gpm or more, most of the systems installed have multiple bed units and a degasifier.

Manufacturers must be consulted for the specific system, resin selection, and the required equipment.

Where applicable, the use of weakly acidic and weakly basic resins minimizes chemical costs and reduces losses to waste because of high regenerative capacity.

There are some problems, such as microbial growth, associated with water softeners. Sanitization is usually accomplished during regeneration. Iron buildup in the unit could pass through to downstream purification equipment unless operating personnel constantly monitor the water quality.

MEMBRANE FILTRATION AND SEPARATION

Membrane filtration and separation is a general term for a water purification process that removes contaminants from feedwater by means of a thin, porous barrier called a membrane. When used as a filter, a membrane is capable of removing impurities of a much smaller size than other types of filters. Filters of this nature are often called ultrafilters and nanofilters.

A semipermeable membrane limits the passage of selected atoms and/or molecules in a specific manner. Membrane filtration and separation, when used to produce pure water, is characterized by having the feedwater flow parallel to the membrane (often called tangential flow). Not all of the feedwater is recovered. Many of the membranes used are also available as both depth filters and, in single thickness, disk filters. There are two general categories of membrane filtration: reverse osmosis using a semipermeable membrane and filtration using ultrafiltration and nanofiltration membranes.

Reverse Osmosis

Reverse osmosis (RO) is a broad-based water purifying process involving osmosis and ionic repulsion. Osmosis is the spontaneous passage of nonvolatile solute molecules (impurities such as sodium chloride) in a solvent (such as water) through a semipermeable membrane. This membrane is called semipermeable because it allows the solvent to diffuse, or pass through, but is impervious to the solute.

In the natural osmosis process, when two solutions of different concentrations are separated by a semipermeable membrane, water molecules from the less concentrated solution will spontaneously pass through the membrane to dilute the more concentrated solution. This continues until a rough equilibrium is achieved. The driving force is a difference of pressure, called the osmotic pressure or concentration gradient, that exists across the membrane and is based on the degree of concentration of contaminants. The pressure in the stronger solution is lower than that in the weaker solution. This pressure is what drives the flow of solvent. The flow or flux will continue until the osmotic pressure is equalized, which then results in a higher pressure on the concentrated solution side that is equal to the osmotic pressure.

Reverse osmosis is the flow of solvent in the direction opposite to that of natural osmosis. If enough pressure is applied to the more concentrated solution, which in these discussions is water, pure water is diffused through the membrane leaving behind the bulk of the contaminants. These contaminants are continuously flushed to drain, thereby removing them from the system. The purified water is called permeate and the contaminant-containing water is called reject or the reject stream. In some cases, the reject stream is referred to as salt. The performance characteristics of the selected membrane determine how large a system is required. The flow rate is measured in membrane flux, which is a measurement of the flow rate of permeate that will pass through a given area of the membrane at a specific temperature and pressure. The ratio of purified water flow to the feedwater flow is called recovery. Most applications require a minimum 40 percent recovery rate to be considered practical. Rejection characteristics are expressed as a percent of the specific impurities retained and depend on ionic charge and size.

Membrane Module Configurations.   There are four types of membrane module configurations used for RO applications: hollow fiber, spiral wound (SWRO), tubular (TRO), and plate and frame. Hollow fiber and spiral wound are the most common configurations. In each design, maximum turbulence is necessary to avoid concentration polarization.

Hollow-Fiber Reverse Osmosis.   The hollow-fiber configuration, illustrated in Fig. 4.10, consists of a perforated tube manufactured from ceramic, carbon, or porous plastic with inside diameters ranging from ⅛ to 1 in (8 to 25 mm). They require rigid support when mounted inside the pressure vessel. Feedwater can be introduced either into the center or outside depending on the manufacturer of the RO module. Fouling resistance is low.

Spiral Wound Reverse Osmosis (SWRO).   This configuration, illustrated in Fig. 4.11, typically achieves a large surface area per unit volume. In this design, a flat membrane is formed around a fabric spacer closed on three sides with the open side terminating in a perforated produce water tube. The unit is then placed in a pressure vessel. Feedwater permeates through the membrane and flows radially inside the enclosure toward the product tube.

Tubular Reverse Osmosis (TRO).   This configuration, illustrated in Fig. 4.12, consists of a perforated tube manufactured from ceramic, carbon, or porous plastic with larger inside diameters than the hollow fiber configuration. The membrane is installed on the inside of the tube. A number of tubes are installed inside a pressure vessel. Feedwater enters the tube and permeates through the membrane to be collected on the outside. The feedwater channels are much more open than the SWRO and less subject to fouling.

Plate and Frame.   This configuration, illustrated in Fig. 4.13, consists of a membrane that is fixed to a grooved plastic or metal plate with several plates stacked together in a frame that includes feedwater and drain ports. As the feedwater flows across the membrane surfaces, the purified water penetrates the membrane and flows along the frame for collection. The retentate continues to flow and could be recirculated or directed to drain.

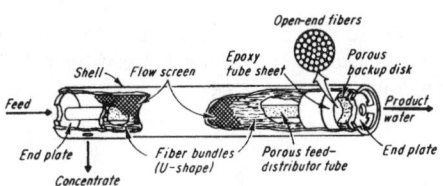

FIGURE 4.10   Hollow fiber reverse osmosis configuration.

Spiral-wound

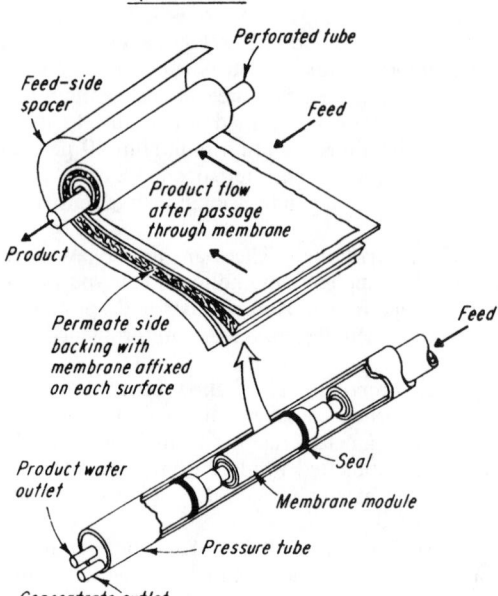

FIGURE 4.11   Spiral wound reverse osmosis configuration.

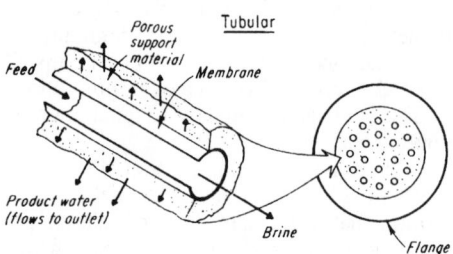

FIGURE 4.12   Tubular reverse osmosis configuration.

This configuration is used mostly for filtration and rarely for RO systems. Packing density is low and resistance to fouling is very high. It is used for small to medium volumes, generally less than 20 gpm.

Membrane Selection

System performance is determined by considering the following factors that influence the capacity of the individual membranes selected:

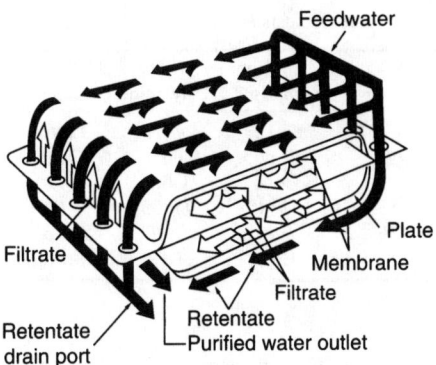

FIGURE 4.13 Plate and frame reverse osmosis configuration.

1. Operating pH
2. Chlorine tolerance
3. Temperature of the feedwater
4. Feedwater quality, usually measured as SDI
5. Types of impurities and previous feedwater treatment
6. Membrane flux
7. Number of operating hours
8. Resistance to biodegradation and ability to be sanitized
9. Rejection characteristics. Typical RO systems remove the following contaminants to the following levels:

   Inorganic ions   93–99%
   Dissolved organics (>300 MW)   >99%
   Particulates   >99%
   Microorganisms   >99%

The selection of a system configuration must consider the following:

1. Maximum recovery
2. Fouling properties and resistance
3. Production rate per unit volume

There are only a small number of polymers that have the necessary characteristics to function as semipermeable membranes:

1. Thin film composite of various polymer materials
2. Polyamide
3. Cellulose acetate
4. Cellulose triacetate
5. Polysulfone

TABLE 4.8 Comparison of Reverse Osmosis Polymers

	Thin film composite	Polyamide	Cellulose acetate	Cellulose triacetate	Polysulfone
pH Stability	2–12	4–11	2–8	4–7.5	3–11
Chlorine tolerance	Fair-poor	Poor	Good	Fair-good	Good
Biological resistance	Good	Good	Poor	Fair-good	Good
Temperature limit for stability, °F (°C)	122 (45)	95 (35)	95 (35)	86 (30)	95 (35)
Typical rejection ionic species, %	>90	>90	90	90	90
Flux	High	Low	Low-medium	Low-medium	High

Typical characteristics and a comparison of these membranes are given in Table 4.8.

Filtration

Ultrafiltration and nanofiltration describe membranes that are categorized by their pore size. Ultrafiltration membrane pore sizes range from 0.001 to 0.02 μm. Nanofiltration membranes have pore sizes that will allow the passage of solids to 10,000 daltons. The 10,000-dalton cutoff is recommended for the complete removal of pyrogens. Typical recovery rates for ultrafilters range between 95 and 98 percent, with the remainder flushed to drain.

A membrane is manufactured by bonding the membrane onto a porous, supporting substrate and then configuring it into elements. These filters are generally used for pretreatment in the removal of colloids, bacteria, pyrogens, particulates, and high molecular weight organics. Spiral wound and hollow fiber are the two configurations used most often.

MICROBIAL CONTROL

Chemicals

The most common disinfection method is the addition of oxidizing or nonoxidizing chemicals. Chemicals could be either biocides, which are substances that kill microbes, or biostats, which prevent the further growth of microbes. Commonly used chemicals are chlorine, chlorine compounds, hydrogen peroxide, and acid compounds.

In order to be effective, the chemical must have a minimum contact time in the water. In addition, a residual amount of the chemical must be present to maintain disinfection.

Chemicals add impurities to the water and are not generally suitable for a pure water environment. They disinfect potable and process water and equipment by injection directly into the fluid stream by means of a metering pump. They must be removed from feedwater used for purification. Chlorine may produce trihalomethanes.

Ultraviolet Radiation

Ultraviolet (UV) radiation is an in-line process. UV light is generated using mercury vapor lamps. The UV spectrum is divided into three wavelengths: UVA (315–400 nm), UVB (280–315 nm), and UVC (less than 280 nm). Only UVB and UVC wavelengths can economically produce the intensity and energy output necessary for the intended germicidal treatment. Federal standard 209E and aseptic guidelines issued by the FDA provide some guidance for the use and application of UV irradiation.

The 254-nm wavelength operates in the germicidal region, sterilizing by destroying bacteria, mold, viruses, and other microorganisms. This wavelength is preferred for pure water systems and will significantly reduce the multiplication of organisms.

The 185-nm wavelength operates in the ozone-forming region, where it has the ability to break down organic molecules to carbon dioxide by the photooxidation process. It slowly breaks the bonds in organic molecules by direct radiation, and also oxidizes organisms by the formation of hydroxyl radicals. The UV spectrum is illustrated in Fig. 4.14.

A flow rate of approximately 2 ft/s is an industry standard for effective sanitization of purified water. Flow rate through the UV device should be reduced compared to that of the circulation loop to extend the necessary contact time. The recommended location is ahead of deionization equipment.

Problems include generation of ions that lower the resistivity of water and the possible leaching of silica from the quartz sleeve of the UV device. Glass, plastic, rubber, and similar materials exposed to UV radiation will eventually crack, etch, discolor, and flake. Tests have shown that only half the energy used by a new bulb

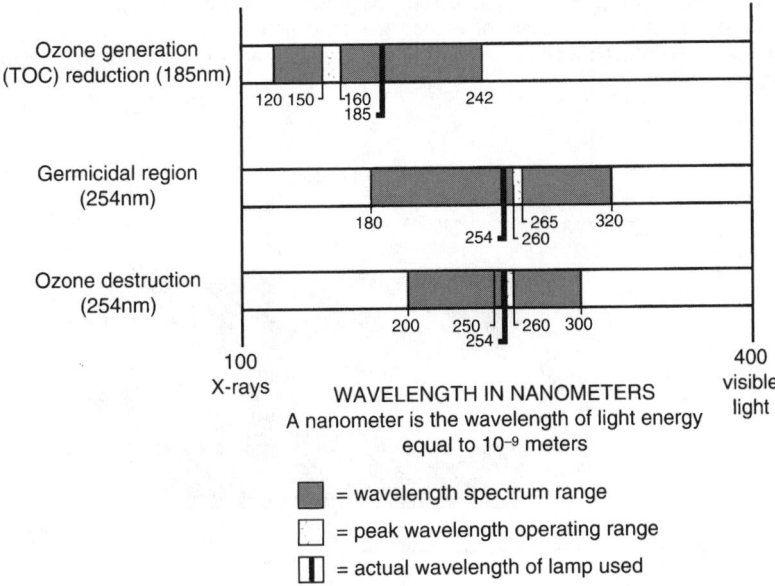

FIGURE 4.14 UV wavelength spectrum.

is actually transmitted to the water and that in time an additional 25 percent of the output will be lost compared to a new bulb.

Filtering

A filter will remove organisms from the fluid stream. Generally accepted practice is to use a 0.2-μm absolute filter for the removal of bacteria. However, some authorities question the effectiveness of this practice. Recommended current practice is to use a membrane filter with an absolute rating (cutoff) of 10,000 daltons. Cartridge filtration is the most common method.

Heat

Heating to 80°C (175°F) will effectively sanitize water under pressure. The heating can be accomplished by using steam, electricity, or other heat exchangers. It is common practice to circulate purified water at this temperature, and use heat exchangers to lower the water temperature at each point of use if necessary.

Ozone

Ozone (O_3) is an oxidizing gas generated from gaseous oxygen or catalytically from water. The most popular method to produce ozone is by a corona-discharge generator converting the oxygen in air to ozone. The air is passed between two electrodes where an electrostatic discharge across the gap converts oxygen to ozone. A schematic drawing of a corona-discharge generator is illustrated in Fig. 4.15.

The ozone system consists of a feed gas treatment unit, an ozone generator, a water-ozone contact mechanism, and a destruction unit to eliminate any residual ozone.

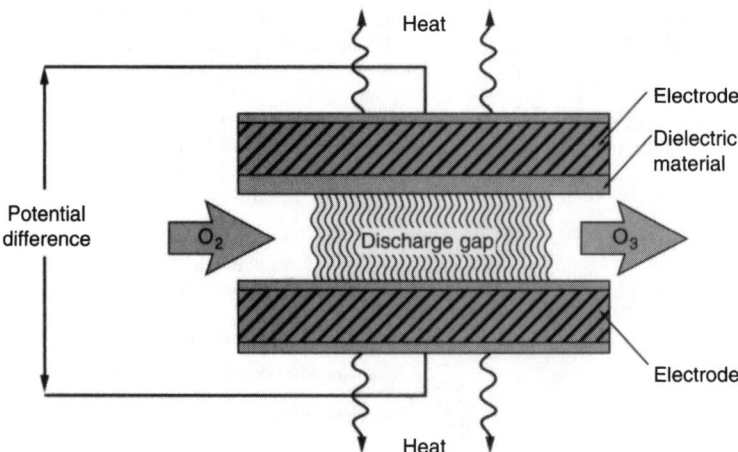

FIGURE 4.15 Principle of corona-discharge ozone generator.

1. *Feed gas treatment.* The gas reaching the generator must have all particles larger than 0.3 μm and 95 percent of particles larger than 0.1 μm removed. In addition, aerosols, moisture, and hydrocarbons must be removed, as required by the manufacturer. This purification is usually supplied as part of a package.

2. *Ozone generator.* The three basic types of generators are the Lowther plate unit, Otto plate unit, and tube units. They differ only in the manner they are cooled. Ozone generators use large amounts of electrical power, generally between 15 and 26 kWh per kilogram of O_3.

3. *Water-ozone contact mechanism.* Ozone and water at a pH of 7.5 or less are mixed in direct contact with one another by the use of static or mechanical mixers, injectors, or columns that optimize the dissolution of the gas.

4. *Destruction unit.* Depending on the generator, ozone concentrations can vary from 100 to 3000 ppm. Because high concentrations are harmful to humans and metals, the ozone should not be allowed to escape to the atmosphere without being treated to a level below 0.1 ppm. Destruction can be accomplished by catalytic, thermal, or activated carbon. Thermal units operate at a temperature of 300°C and generally require a 3 to 5 min contact time to be effective.

The mechanism for ozone oxidation is through the generation of hydroxyl radicals. The gas is directly injected into the water stream. Problems with this method include incomplete oxidation of organic compounds and slow operation. Action is enhanced with the application of UV radiation at 185 nm.

WATER TREATMENT

This section will describe the various methods and equipment used to treat water for various purposes.

UTILITY WATER TREATMENT

Water from wells, rivers, lakes, and streams is commonly used for cooling and washing. Clarifying and treating this water in order to meet the purity requirements of the proposed end use requires good monitoring and quick reactions to raw and treated water fluctuations. If the water is to be recirculated, the treatment methods are more stringent. If the water is to be discharged into the environment, local codes must be followed regarding allowable chemical amounts present in the wastewater in order to avoid the need for waste treatment.

Initial Filtering

If the supply is from surface water, a coarse or fine screen is usually placed at the intake to keep out fish and other large debris. Coarse screens are usually ½-in diameter bars with a clear opening of 1 to 3 in. Fine screens could have openings of approximately ⅜-in square, with the water velocity through the screen limited to about 2 ft/s. If the quantity of water is small enough, basket strainers can be used. In climates where freezing may occur, the inlet should be placed far enough below the low water level to prevent freezing.

Clarification

After initial filtering, clarification is required to obtain water that meets the standards for proposed use. The selection of the clarifier is based on the volume of water to be treated and the final quality desired. If the volume of water is small and the raw water is not very turbid, filters may be used.

Biological Control

In order to control microorganism fouling of the system, they must be destroyed, if possible, or at least inactivated (to keep them from reproducing) and then removed from the water stream. This is usually accomplished by chlorination, filtration, UV radiation, ozone generation, and special adsorbents.

Chlorination is the least costly and most common method. The action of chlorine requires a contact time and the establishment of a residual chlorine amount. A range of 0.5 to 1.0 ppm is generally acceptable for typical water.

Water Softening

Water softening should be considered to reduce the hardness in the utility water system when used for recirculated cooling purposes.

BOILER FEEDWATER CONDITIONING

In a boiler, energy in the form of heat is transferred from a fuel source to water to create steam. When subject to elevated temperature and pressure, the composition of the raw boiler water supply undergoes radical changes.

The effects of dissolved gases are magnified. The dissolved minerals may deposit a scale on the transfer surface, which will effect the heat transfer process. The result will be a reduction in the heat transfer rate, reduction in the flow rate, and increased damage from corrosion. A wide range of treatment methods is available depending primarily on the impurities found in raw water, operating pressure of the boiler, and makeup rate of water. These methods include mechanical treatment; addition of chemicals to prevent deposits, inhibit corrosion, or neutralize impurities detrimental to the proposed end use of the water; and the use of dispersants to keep particulates in suspension.

The treatment starts with an accurate feedwater analysis. This is compared to final treatment objectives established by the user and the boiler manufacturer. The degree of removal is determined by the difference in the two analyses. Recommended standards for boiler feedwater and steam quality are given in Table 4.9.

A boiler water treatment program must do the following:

1. Reduce or remove hardness to control scale by either mechanical (external) or chemical (internal) treatment
2. Maintain proper levels of alkalinity to assure that proper chemical reactions can occur
3. Control dissolved oxygen and carbon dioxide through deaerating and adding an oxygen scavenger
4. Maintain proper levels of conditioners so that the suspended solids remain in suspension and can be easily eliminated through blowdown
5. Optimize boiler blowdown

Mechanical removal of hardness is accomplished most often by water softening using an ion exchange process, where the insoluble calcium and magnesium ions are replaced with highly soluble sodium ions. A strong acid resin in sodium form, often referred to as zeolite, is commonly used. Real zeolite is in reality a naturally occurring mineral that is no longer widely used because of its high cost. Artificial resins are more efficient and cost effective for most uses.

The adjustment of pH is accomplished by the injection of dilute sulfuric or hydrochloric acid, sodium hydroxide, or sodium carbonate. This requires close monitoring.

Feedwater oxygen and carbon dioxide are normally removed with deaerators before the feedwater enters the boiler. In addition to mechanical deaeration, it is recommended that a chemical oxygen scavenger be added to the water to quickly eliminate any remaining traces of oxygen. Many scavengers are in use but the most

TABLE 4.9   Recommended Boiler Feedwater Limits and Steam Purity

Drum pressure, psig	Range TDS* boiler water, ppm max.	Range total alkalinity† boiler water	Suspended solids boiler water, ppm max.	Range TDS‡ steam, ppm (max. expected value)
		Drum-type boilers*		
0–300	700–3500	140–700	15	0.2–1.0
301–450	600–3000	120–600	10	0.2–1.0
451–600	500–2500	100–500	8	0.2–1.0
601–750	200–1000	40–200	3	0.1–0.5
751–900	150–700	30–150	2	0.1–0.5
901–1000	125–625	25–125	1	0.1–0.5
1001–1800	100		1	0.1
1801–2350	50		n/a	0.1
2351–2600	25		n/a	0.05
2601–2900	15		n/a	0.05
		Once-through boilers		
1400 and above	0.05	n/a	n/a	0.05

*Actual values within the range reflect the total dissolved solids (TDS) in the feedwater. Higher values are for high solids, lower values are for low solids in the feedwater.

†Actual values within the range are directly proportional to the actual value of TDS of boiler water. Higher values are for the high solids, lower values are for low solids in the boiler water.

‡These values are exclusive of silica.

n/a = not available.

Source: American Boiler Manufacturers Association.

popular scavengers are sulfites (up to 1000 psig) and compounds of hydrazine. Hydrazine is being replaced by diethylhydroxylamine (DEHA), carbohydrazide, hydroquinone, and isoascorbic acid.

Chemicals can be added to the boiler feedwater for two purposes. One is to keep the hardness salts in solution so they will not cause scaling. The second is to precipitate them out of solution to allow the particulates to be disposed of during blowdown. There are many effective materials, generally blends of component chemicals used to accomplish this. The most common are polymers, phosphates, and carbonates that are designed to optimize the precipitation of calcium, magnesium, and iron. Modern treatment has made hardness deposits much less common, but iron deposits more common because of the return of condensate to the boiler.

Boiler blowdown is the bleeding off of some water from the boiler. When steam is generated from feedwater, impurities are left behind and will accumulate in time. When the level of impurities becomes too high, a portion of the water is discharged to drain. This is called blowdown. Blowdown could be intermittent or continuous.

COOLING WATER CONDITIONING

This section will discuss the basic principles used for controlling scale, corrosion, and biological fouling of water used in evaporative cooling towers and condensation systems.

The basic reason for treatment of cooling water is to keep any dissolved solids from depositing onto any piping or equipment. This is accomplished by not allowing the dissolved solids to reach the saturation point. This is usually expressed as the cycle of concentration, which compares the amount of dissolved solids in the feedwater with the amount of dissolved solids of the circulating water. As an example, if the feedwater has a TDS of 100 ppm and the circulating water has a TDS of 300 ppm, a cycle of concentration of 3 has been reached. The cycle of concentration is reduced by bleed-off (blowdown) from the system.

Scale

The basic treatment for scale in cooling water systems is to add inhibitors to keep the scale from depositing itself on the walls of the pipe. The inhibitors are similar to those used for boiler water treatment, with the addition of surfactants that change the surface characteristics of the pipe to prevent deposition and aid in removal if deposits occur.

Corrosion

The corrosion potential of the water is indicated by the Langelier or Ryznar indexes. Corrosion treatments consist of pH control and chemical inhibitors. These are recommended by the manufacturers of the chemicals and the equipment to ensure compatibility. The pH of the circulating water is usually controlled to a point near 7.5 to decrease the scale-forming tendency. The addition of a dilute acid is used most often. A corrosion inhibitor often adds compounds that increase the tendency of scale to occur; therefore, a balance must be reached.

Biological Fouling

Microbial control is achieved with the addition of biocides, either oxidizing or nonoxidizing depending on their chemistry and killing action. Ozone is also commonly used.

Biocide treatment is intended to shock microbes initially with a heavy dose and then allow the concentration of the compounds to fall to a level of 25 percent of the initial dose. This is a generally accepted lower limit below which the biocide is not considered effective. Often, because of several different strains of microbes present, more than one biocide may be required. A contact time is required that must be approved by the chemical manufacturer.

These chemicals are added by means of a chemical feed pump discharging directly into the piping system. The levels must be closely monitored with alarms established by performance indicators based on operating experience.

POTABLE WATER TREATMENT

Water used for human consumption or intended to be part of food products must be treated to comply with the Safe Drinking Water Act (SDWA), local regulations, and also the Surface Water Treatment Rule, which is a part of the SDWA. In addition, compliance with regulation 10CFR141 is required. Substances that affect the quality of potable water are classified in four major categories:

1. *Physical.* Physical characteristics related to the appearance of the water include color, turbidity, taste, and odor. Physical quality is corrected mostly through the use of various types of filters.

2. *Chemical.* The chemical characteristics of water are related to dissolved minerals (mostly hardness), gases, and organics. The chemical quality of water is adjusted by the use of water softeners, ion exchange, RO units, and activated charcoal units to remove organic impurities.

3. *Biological.* This characteristic is concerned with microorganisms that will affect the health of the consumer. Biological treatment requires the use of biocides

TABLE 4.10 Water Treatment Technology for Small Potable Water Systems

Technology	Advantages	Disadvantages
	Filtration	
Slow sand	Operational simplicity and reliability, low cost, ability to achieve greater than 99.9 percent *Giardia* cyst removal	Not suitable for water with high turbidity, requires large land areas
Diatomaceous earth (septum filter)	Compact size, simplicity of operation, excellent cyst and turbidity removal	Most suitable for raw water with low bacterial counts and low turbidity (<10 ntu), requires coagulant and filter aids for effective virus removal, potential difficulty in maintaining complete and uniform thickness of diatomaceous earth on filter septum
Reverse osmosis membranes	Extremely compact, automated	Little information available to establish design criteria or operating parameters, most suitable for raw water with turbidity <1 ntu, usually must be preceded by high levels of pretreatment, easily clogged with colloids and algae, short filter runs, concerns about membrane failure, complex repairs of automated controls, high percent of water lost in backflushing
Rapid sand/ direct filtration package plants	Compact, treats a wide range of water quality parameters and variable levels	Chemical pretreatment complex, time-consuming; cost

(Continued)

TABLE 4.10 Water Treatment Technology for Small Potable Water Systems *(Continued)*

Technology	Advantages	Disadvantages
	Disinfection	
Chlorine	Very effective, has a proven history of protection against waterborne disease, widely used, variety of possible application points, inexpensive, appropriate as both primary and secondary disinfectant	Potential for harmful halogenated by-products under certain conditions
Ozone	Very effective, no THMs formed	Relatively high cost, more complex operation because it must be generated on-site, requires a secondary disinfectant, other by-products
Ultraviolet radiation	Very effective for viruses and bacteria, readily available, no known harmful residuals, simple operation and maintenance for high-quality waters	Inappropriate for surface water, requires a secondary disinfectant
	Organic contaminant removal	
Granular activated carbon	Effective for a broad spectrum of organics	Spent carbon disposal
Packed-tower aeration	Effective for volatile compounds	Potential for air emissions issues
Diffused aeration	Effective for volatile compounds and radionuclides	Clogging, air emissions, variable removal efficiencies
Advanced oxidation	Very effective	By-products
Reverse osmosis	Broad spectrum removal	Variable removal efficiencies, wastewater disposal
	Inorganic contaminant removal	
Reverse osmosis	Highly effective	Expensive waste removal
Ion exchange	Highly effective	Expensive waste removal
Activated alumina	Highly effective	Expensive waste removal
GAC	Highly effective	Expensive waste removal

and biostats to eliminate and reduce the number of microbes present in the water and to create a residual amount of the chemical to maintain the required level of action required by code.

4. *Radiological.* This is concerned primarily with radon in areas where the water may have come in contact with radioactive substances. Retention and aeration will lower the radon count to acceptable limits in approximately 8 h.

A synopsis of general treatment methods for small potable water systems that shows general advantages and disadvantages is given in Table 4.10.

WATER PURIFICATION

This section will discuss pure water used for laboratory and pharmaceutical purposes. The various systems will be broadly defined and general guidelines for their production, storage, and distribution will be provided.

Ultrapure water used in production of food products and electronics industries is outside the scope of this handbook. For information on guidelines for water purity in the electronics industry, refer to the Semiconductor Equipment Manufacturers Institute, Mountain View, California.

The total water treatment system consists of three interrelated phases: pretreatment, purification, and distribution (including posttreatment). Purification methods include distillation, ionization, membrane filtration, and other approved processes. WFI water can only be produced by distillation or membrane filtration. Processes needing ultrapure water used for specific applications often use pure water as feedwater and they further purify it to meet those specific requirements at the point of use.

CODES AND STANDARDS

The required quality of purified water depends on the application. Various codes have specifically defined water quality for use in various industries. Among them are:

1. 21 CFR 210 or cGMP for drugs
2. 21 CFR 211 or cGMP for finished pharmaceuticals
3. USP/NF official water nomographs
4. Federal Food, Drug, and Cosmetic Act

LABORATORY SYSTEMS

For laboratory work, all applications do not require the same quality of water. The American Society for Testing and Materials (ASTM), the College of American Pathologists (CAP), the National Committee for Clinical Laboratory Standards (NCCLS), and the Association for the Advancement of Medical Instrumentation (AAMI) have all developed standards for water used in laboratories depending on their intended use. These standards are summarized in Tables 4.11 (ASTM, CAP), 4.12 (NCCLS), and 4.13 (AAMI). ASTM electronics grade water standard is given in Table 4.14 for reference only. There are two types of pure water categories in the NCCLS specifications:

Type I, called reagent grade water, is used for analysis of trace matter and other critical applications.

Type II, called analytical grade water, is suitable for all but the most critical procedures.

TABLE 4.11   ASTM Reagent-Grade Water Specifications

	Type I	Type II
Resistivity (Megohm-cm compensated to 25°C)	18.0	1.0
TOC (ppb)	100	50
Sodium (ppb)	1	5
Chlorides (ppb)	1	5
Silica (ppb)	3	3

TABLE 4.12   NCCLS/CAP Reagent Water Specifications

	Type I	Type II
Resistivity (Megohm-cm @ 25°C)	10	1.0
Silicate (mg/L SiO$_2$)	0.05	0.1
TOC (ppb)	Activated carbon (not required by CAP) or distillation or RO	Not specified
Particulates	0.22 micron filter	Not specified
Bacteria (cfu/mL)	10	1000

Additional purification may be required for selected clinical laboratory procedures such as:
1. Preparation of water with minimal pyrogen levels for cell culture
2. Preparation of bacteria-free water for direct fluorescent detection of bacteria as in Legionella direct fluorescent antibody testing or direct fluorescent stains of mycobacteria
3. Preparation of water with minimal organic content for HPLC.

TABLE 4.13   AAMI/ANSI Water Quality Standards

Contaminant	Suggested maximum level, mg/L
Calcium	2 (0.1 meq/L)
Magnesium	4 (0.3 meq/L)
Sodium	70 (3 meq/L)
Potassium	8 (0.2 meq/L)
Fluoride	0.2
Chlorine	0.5
Chloramines	0.1
Nitrate (N)	2
Sulfate	100
Copper, barium, zinc	0.1 each
Arsenic, lead, silver	0.005 each
Chromium	0.014
Cadmium	0.001
Selenium	0.09
Aluminum	0.01
Mercury	0.0002
Bacteria	200 (cfu/mL)

Source: Association for the Advancement of Medical Instrumentation (AAMI) "Hemodialysis Systems Standard," March 1990. Adopted by American National Standards Institute (ANSI), 1992.

TABLE 4.14 ASTM Electronics-Grade Water Standard*

Assay	Grade			
	E-I	E-II	E-III	E-IV
Resistivity, minimum,	>18.0*	17.5†	12	0.5
$M\Omega \cdot cm$ at 25°C	>17.0*	>16.0†		
SiO_2 (total), max, $\mu g/L$	5	10	50	1000
Particle count per milliliter	1	3	10	100
Particle size limit, μm	0.10	0.5	1.0	10
Viable bacteria, max	1/1000 mL	10/1000 mL	10/1 mL	100/1 mL
Copper, max, $\mu g/L$	1	1	2	500
Zinc, max, $\mu g/L$	0.5	1	5	500
Nickel, max, $\mu g/L$	0.1	1	2	500
Sodium, max, $\mu g/L$	0.5	1	5	1000
Potassium, max, $\mu g/L$	2	2	5	500
Chloride, max, $\mu g/L$	1	1	10	1000
Nitrate, max, $\mu g/L$	1	1	5	500
Phosphate, max, $\mu g/L$	1	1	5	500
Sulfate, max, $\mu g/L$	1	1	5	500
Total organic carbon, max, $\mu g/L$	25	50	300	1000
Endotoxins	0.03EU‡	0.25EU‡	N/A§	N/A§

*Above 18 $M\Omega \cdot cm$ 95% of the time, not less than 17.
†17.5 or greater 90% of the time, not less than 16.
‡EU = Endotoxin unit.
§N/A = not applicable.
Source: ASTM Standard D5127-90.

PHARMACEUTICAL WATER

The pharmaceutical industry in most countries is regulated. In the United Sates, the industry is regulated by the Food and Drug administration (FDA), which takes guidance from several sources, including the Pharmaceutical Research and Pharmaceutical Manufacturers of America (PhRMA), the United States Pharmacopoeia (USP) and updates from the Pharmacopoeia Forum (PF), published by the U.S. Pharmacopeial Convention. Each country has its own governing and guiding agencies.

Information in this handbook refers to standards of the United States. The standards include USP 24/NF 19, January 1, 2000, as given in Tables 4.15 and 4.16.

PURIFIED WATER TYPES

1. *Compendial water.* This is a general term that includes all types of purified water and water for injection intended to be used in any final pharmaceutical drug dosage form.

2. *Purified water* (PW). The quality of the feedwater shall meet drinking water standards. The final product shall contain no added substances. The USP re-

TABLE 4.15 Standards for Purified Water and Water for Injection

Parameter	Standard
TOC	550 ppb Carbon
Conductivity	1.3 μS/cm @ 25°C
	0.1 μS/cm @ 20°C
Bacteria	PW: 100 cfu/mL
(guidelines <1231>)	WFI: 10 cfu/100 mL
Bacterial endotoxins	WFI: 0.25 USP EU/mL

Notes: μS/cm = microSiemens per centimeter
cfu/mL = colony forming units per milliliter
EU/mL = endotoxin units per milliliter

TABLE 4.16 Standards for Packaged PW, WFI, and Sterile PW

Parameter	Standard
pH	5.0 to 7.0
Chloride (mg/L)	0.5
Sulfate (mg/L)	1.0
Ammonia (mg/L)	0.03
Calcium (mg/L)	1.0
Carbon dioxide (mg/L)	5.0
Oxidizable substances	pass USP Permanganate test
Bacteria	
*Purified Water (cfu/mL)	100
*WFI (cfu/100 mL)	10
Sterial PW	pass the USP Sterility test <71>
Endotoxin WFI (EU/mL)	0.25

Note: *Limits are guidelines in General Chapter <1231>
Source: U.S. Pharmacopeial Convention Inc., Reference 8.

quirement with added substances has always been concerned with additions to the final product and not to the feedwater. It is commonly interpreted by USP that substances may be added to the feedwater provided that they are removed to an acceptable level during the final treatment process. PW is used as the feedwater for preparation of compendial water.

3. *Sterile purified water.* This type, including some Sterile Water for Inhalation, is made using purified water as a raw water source, sterilized and suitably packaged. It contains no antimicrobial agent.

4. *Bacteriostatic purified water.* PW sterilized and suitably packaged. It contains no antimicrobial agent.

5. *Water for injection* (WFI). This type is made using PW as a raw water source that must be further purified by distillation or RO. Bacteria and endotoxins must be reduced to the required level.

6. *Sterile bacteriostatic water for injection.* WFI sterilized and suitably packaged. It contains no antimicrobial agent.

7. *Sterile water for inhalation.* WFI sterilized and suitably packaged. It contains no antimicrobial agent except when used in humidifiers or similar devices subject to contamination, or other added substances.

8. *Sterile water for irrigation.* WFI sterilized and suitably packaged. It contains no antimicrobial agent.

PHARMACEUTICAL WATER TREATMENT PROCESS

Pharmaceutical water treatment takes raw water that meets drinking (potable) water standards and removes sufficient contaminants from that water in a treatment plant to meet the standards for the various types of compendial waters. It takes a number of treatment steps in order to process raw water into PW and other subsequent pure water types.

Each individual treatment plant will have a different configuration. A typical configuration is illustrated in Fig. 4.16. The intent of Fig. 4.16 is to show a process flow and include most of the equipment and general arrangement found in a typical plant. It is not intended to indicate that this is the best or only manner to arrange the process.

All pharmaceutical water shall start with potable water as a raw water source. Potable water usually contains a residual of some oxidizing biocidal agent to control disease causing microorganisms. Typically either chlorine or chlorimine is the oxidizing agent, and it is used in sufficient quantities to achieve a free chlorine residual necessary to achieve it purpose. There are no requirements regarding the amount

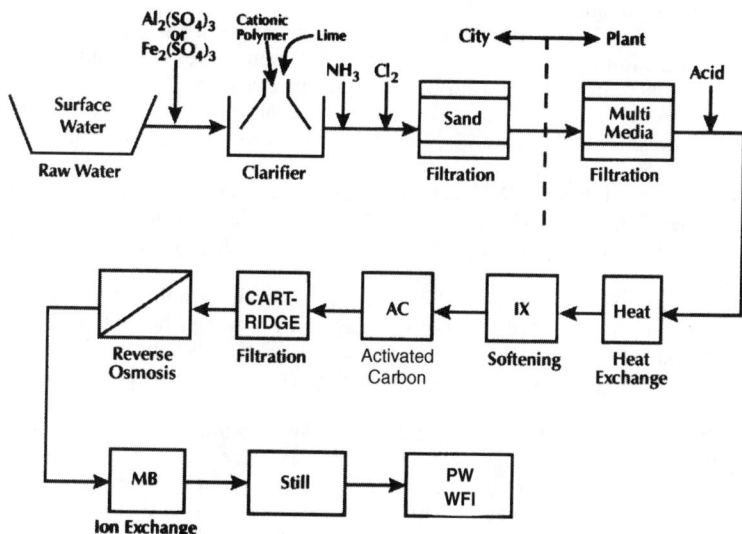

FIGURE 4.16 Pharmaceutical water treatment plant configuration.

of TOC or endotoxins present. The TOC is subject to seasonal variances, with the lowest level usually occurring in the wintertime.

Multimedia filtration. One of the first steps commonly found in the treatment process is multimedia filtration. Its purpose is to remove the bulk of suspended contaminants larger than 30 microns.

Acid injection. Some treatment plants must add acid injection somewhere in the system. The purpose of the acid is primary for scale control It is shown in Fig. 4.16 before the heat exchanger because calcium carbonate is more likely to scale out when water is hot. If there is little hardness (calcium, and magnesium ions) or alkalinity (hydroxide, bicarbonate and carbonate ions) in the feedwater, acid may be eliminated. Another common reason to use acid is to minimize damage to cellulose acetate membranes if it is a downstream RO process component. These membranes are easily damaged at any pH level, but at a pH level of 5.5 to 6, the damage is minimized.

Heat exchange. Feedwater is heated to lower the pumping costs for an RO unit because the warmer the water, the less pressure is required to pump water through an RO membrane. Another reason is that warmer water accelerates the rate of diffusion and chemical reactions. Generally speaking, for every 18°F (10°C) rise in temperature, the speed of chemical reactions doubles.

Softening. A water softener controls scale by removing the hard, scale forming cations like calcium and magnesium, and exchanging (replacing) them with non-scale forming sodium ions.

Activated carbon units. Often called activated carbon beds or activated carbon filters, these filters remove chlorine or chlorimine compounds from the feedwater to protect both RO membranes and DI resins from the oxidizing action of the chlorine and chlorimine compounds. A second, less common reason is to remove certain organic compounds. Organic compounds are molecules that always contain carbon, usually contain hydrogen and frequently contain other atoms. They are called organic because prior to human intervention all organic compounds came from living organisms.

Cartridge filtration. This filter is installed upstream of the RO units as an additional protection against suspended solids. RO membranes will foul if sufficient suspended solids are not removed by the multi-media filter.

Reverse osmosis. RO is a membrane that removes a bulk of the suspended solids and contaminants, which is, typically, 98 to 99+ percent of ionic contaminants and 95 to 99+ percent of dissolved TOC. A single pass through the RO will typically not meet PW or WFI standards. A double pass system, where the processed water from one RO unit is processed again through another unit, may be allowable. Distillation units are more commonly used to provide PW and WFI quality water.

Ion exchange. Following an RO unit, it is generally required to further reduce the contaminant level. This is done typically using a mixed bed deionization unit. This cold be combined with electrodeionization (EDI), also known as continuous deionization (CDI) to further polish RO water. The resultant water is not pyrogen free.

Distillation. In order to achieve PW and WFI quality, most treatment plants use distillation (stills) as a final step. A still heats the feedwater to the boiling point, and the resulting steam condensate typically meets required standards. There are various types of stills that require different feedwater quality; therefore, alternative upstream requirements other than that shown in Fig. 4.16 may be necessary. Distillation produces water that is pyrogen free.

FEEDWATER

Feedwater quality, which is source dependent, is the first parameter to be identified in the design of a pure water system. The source strongly influences the pretreatment options and may dictate the purification methods. There is a wide difference in types and concentrations of various impurities possible, depending on whether the feedwater is obtained from a public utility or privately, such as from a well or other surface or groundwater source. Surface waters are usually high in particulates, colloids, and organics. Underground waters tend to have low levels of particulates, colloids, and organics, and relatively high levels of calcium and magnesium hardness and alkalinity (bicarbonates). Source water from agricultural areas often has high levels of nitrates, phosphates, and organic pesticides. Water from public utilities has residual chlorine, fluorides, and chlorimines as well as iron oxides and other pipe-related impurities. It is important to establish or obtain historical water analysis data from different times and seasons of the year in order to properly design a system with enough flexibility to obtain the required purity under worst case conditions.

It is an FDA requirement that the feedwater for USP Purified Water and WFI systems meet the EPA guidelines for potable water. Of particular concern is the microbial level. The water should be virtually free of coliform, which is a pathogenic marker organism. The feedwater must also meet the 400 cfu/ml, the maximum value specified in USP XXII. Another requirement is that it shall contain no added substances. This is an issue that appears to be interpreted differently by individuals and organizations involved in the design of PW systems. There are systems presently in use that appear from an overall perspective to violate this position. Items such as the chlorination of raw water supplies, acid addition to adjust pH, and the use of ozone to control microbial growth are subject to interpretation. All of the concerns should be resolved during the application phase for FDA approval and validation of the system.

PURIFICATION SYSTEM DESIGN

General

Specific methods of purification are capable of removing various types of impurities better than others. None can be depended on to remove all the impurities necessary to achieve the purity level required for USP purified water. It is accepted practice to use a combination of technologies, each of which is designed to remove a specific type of impurity.

The methods used to produce this water are dependent on the feedwater supplying the facility. In most instances, that feedwater has a high level of some classes of contaminants. It is much more economical to pretreat that water to remove the bulk of large quantities and/or the more concentrated of those impurities, and then use the purification equipment to "polish" the water to the desired purity level. In some cases, the removal of individual impurities is necessary to allow the use of specific types of purification equipment.

Pretreatment

General. Pretreatment is considered for two reasons: to prevent potential damage to the membrane selected and to increase membrane filtration quality. The need for pretreatment is determined by an analysis of the raw water supply. The decision is generally governed by the cost efficiency of the pretreatment method, specifically whether the cost of purchasing and installing the pretreatment equipment will reduce the initial cost of the main treatment equipment and lower the operating cost of the system as a whole enough to justify its installation.

Water Temperature. Membrane productivity (flux) is usually rated with feedwater at 77°F, and is inversely proportional to the feedwater viscosity. When the water temperature is lower, additional membrane area is required. The flux increases with increased feedwater temperature. Heating the feedwater lowers the viscosity. The water can be heated with a separate water heater or with a blending valve using domestic hot water mixed with feedwater to provide the necessary temperature.

pH Adjustment. The selected membrane will have an optimum pH operating range. When using drinking water standards, this problem is rarely encountered. In order to achieve it, a dilute acid is injected into the feedwater if necessary.

Filtration. Filters are used to remove suspended solids originating from any source. If continuous production is required, a duplex arrangement should be installed so that the filters can be backwashed or replaced with no interruption in service.

Sand and Multimedia Filters. A common initial method for gross particulate removal from source water is a pressure multimedia sand filter if suspended solids are greater than 0.2 ppm. If lower, a 5- to 10-μm cartridge filter is often used. Experience has found that cartridge replacement is not economical as compared to the cost of backwashing.

Cartridge Filters. Used to remove lesser amounts of particulates prior to the central purification equipment, an in-line 5-μm cartridge depth filter should be installed to eliminate any particulates that would clog or interfere with operation of the central purification equipment. This filter is recommended if the SDI is less than 4, and is generally required if the level is more than 4. If the level is more than 4, an additional 1-μm cartridge filter is recommended downstream of the main filter.

Carbon Filtration. Following the removal of gross particulates, a granulated carbon filter is provided to remove residual disinfectants (chlorine, chlorimine, etc.), dissolved organics (oils, pesticides, surfactants, etc.), and suspended organics (humic and fulvic acids, etc.).

Flow rates through this filter are usually in the range of 1.0 to 4.0 gpm/ft^2 of filter area depending on the quantity of organics and chlorimines of the entering water. The higher the quality, the higher the flow rate. Also the higher the quality of water, the slower the flow rate.

The problem with carbon filtration is its tendency to harbor microbial growth due to the removal of chlorine. Frequent sanitizing is necessary, usually with potassium permanganate, sodium hydroxide, or steam. If steam is selected, the tank should be constructed of type 316L stainless steel.

Water Softening. If the hardness is high, it is necessary to provide a water softener to reduce the calcium and magnesium present to the level required by the membrane selected. Water softening is recommended if the iron content exceeds 0.4 ppm.

The softener is another device that harbors microbial growth. Sanitizing with potassium permanganate is generally used.

Biological and TOC Reduction. UV units and ozone generators are generally used to remove microorganisms and to remove TOC prior to the feedwater reaching the RO units. Materials used in the system must be compatible with them. There is a reluctance to use any chemical to remove microorganisms due to FDA restrictions against adding chemicals to the feedwater. However, the addition of chlorine to disinfect the feedwater is commonly used because it is cost effective and can be easily removed by GAC filters.

Biocide Removal. A far greater source of feedwater for most facilities is potable water. The water contains a residual amount of chlorine necessary to comply with code for drinking water quality. To remove this residual chlorine, a granulated activated charcoal filter is provided. An organic trap may also be required if organics are very high.

CENTRAL PURIFICATION EQUIPMENT

The basic methods used to produce high purity water are distillation, ion exchange, and reverse osmosis. The method used will depend on the purity desired and limitations on initial or operating cost.

Storage

Storage of water will reduce the size of the purification equipment. Storage tanks are constructed from FRP, PP, and SS. The bottom of the tank is dished or conical to aid in complete drainage. It is an established fact that high purity water degrades in storage. The four major sources of this degradation are:

1. Water will extract contamination from any container.
2. Bacteria will grow and secrete waste products.
3. Organics from solvents and shedding of clothing can diffuse through the air and dissolve in the water.
4. Laboratory personnel secrete urea in perspiration and respiration that can cause the formation of ammonia in stored laboratory water.

The tank for USP water should be airtight and equipped with a nonshedding, 0.2-μm hydrophobic filter for venting. For WFI systems, the tank should be type 304L or 316L stainless steel, pressure rated to a minimum of 35 psig pressure, 30 in Hg vacuum, 180 grit, and electropolished. It is common practice to have a nitrogen gas blanket in the vapor space above the water in the storage tank to reduce the possibility of airborne contamination. The tank should have a jacket to maintain a temperature of 80°C and be insulated and provided with a rupture disk.

The discharge from the tank should be from the center bottom of the tank to allow complete circulation of the stored water. The water return should be at the top of the tank and be equipped with a spray ball or spray ring. This minimizes microbial growth by continuously washing the upper areas of the tank.

Sterilization. It is common practice to provide an in-line UV sterilization device to reduce microorganisms that may be present in the water. Another common method is to heat purified water to a temperature of 80°C (177°F) to prevent microorganism growth, and circulate it to maintain the sterile condition. Another possibility is installation of a filter that will remove any organic particulates.

Piping Distribution Network

The piping material for USP water should be fabricated from virgin PP, PVDF, or SS. Plastic pipe should be butt-joint heat-fused and SS pipe should be orbital welded. Piping material for WFI water should be SS, 180 grit and electropolished. Fittings must have extended ends for orbital welding. All couplings must be sanitary triclamp. Insulation must be nonchloride bearing and designed for a temperature of 80°C. Velocity in the system should be approximately 6 to 9 ft/s for supply from the storage tank to the system and 3 to 6 ft/s in the return leg. Th4s is accomplished by the installation of a back-pressure regulator. These velocities are recommended to scour the pipe interior and prevent the formation of biofilm.

The pipe should be sloped at a ⅛-in pitch to allow complete drainage of the network. Dead legs of more than 6 pipe diameters are not permitted by cGMP requirements. To prevent future plastic pipe sagging from interfering with drainage, continuous support is recommended. Using PP as a baseline, PVDF is twice the cost and SS four times the cost installed.

Valves must be consistent with the piping materials. Prior to the RO or DI units, full bore ball valves are recommended. In the purified loop, SS diaphragm valves with an EPDM or Teflon diaphragm and a backing ring should be used. Sample valves (usually needle type) should be provided in areas of the system to allow samples to be taken at strategic points.

Pumps for purified water should be of sanitary design using a double mechanical seal with product water as the lubricant for the seals. WFI pumps should be type 316L SS, 180 grit and electropolished. A casing drain should be provided.

System Design Considerations

USP/WFI water is used in the pharmaceutical manufacturing process and often becomes part of the product. Because of this, all aspects of the purification system and distribution network are subject to inspection and validation by the FDA, which has the responsibility to determine if the quality of water used is adequate.

The FDA has specific guidelines for selection of stills and RO equipment used for production of WFI. In contrast, PW systems can utilize different types of purification equipment, since the microbial and chemical quality can vary depending on the proposed use of the water.

The purpose of any guidelines or standards is to verify that all pertinent purity requirements of the equipment and distribution system conform to current good manufacturing practice (cGMP), are capable of consistently producing water of the required quality, and are capable of delivering water that meets the acceptance criteria for water that comes in contact with product. This investigation also includes verification that the purification equipment selected be capable of producing water of the required purity, that the quality of installation for the distribution network produces a piping network capable of delivering water of the required quality to all outlets, and that the continuing quality of system operation, maintenance, and ongoing testing will consistently provide water of the desired purity.

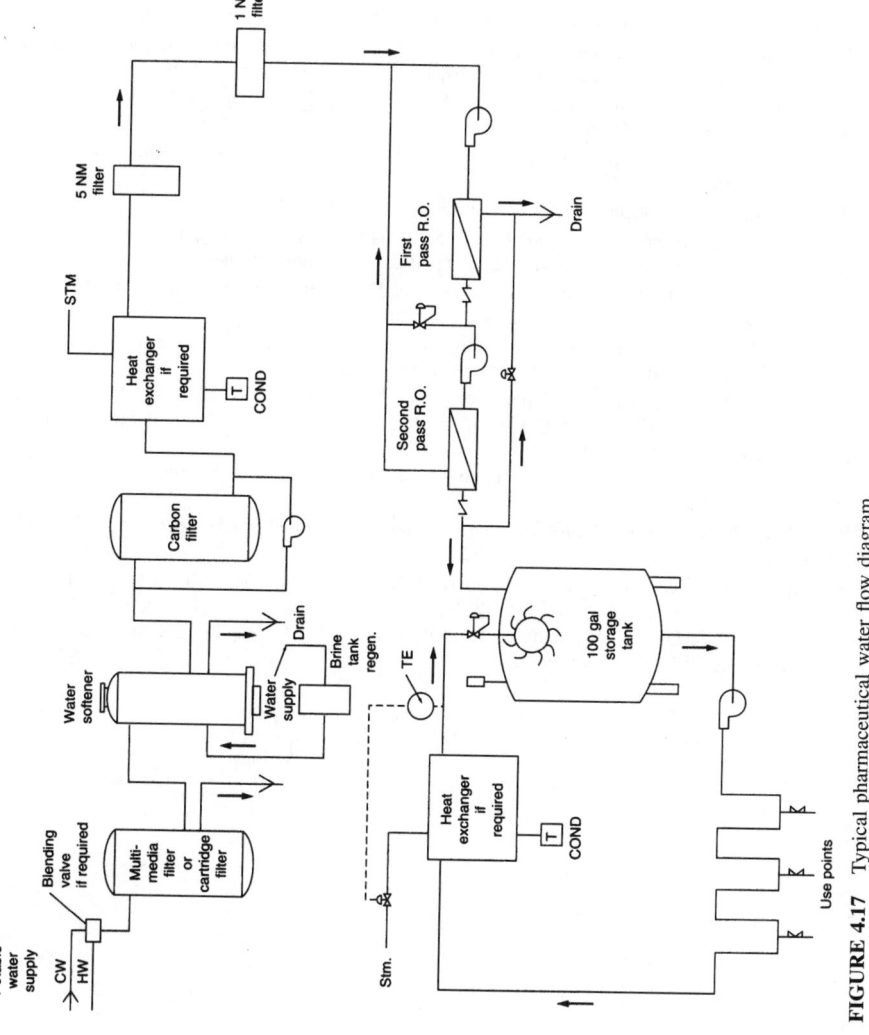

FIGURE 4.17 Typical pharmaceutical water flow diagram.

One typical pharmaceutical water purification flow diagram is shown in Fig. 4.17.

REFERENCES

Blake, R. T., *Water Treatment for HVAC and Potable Water Systems,* McGraw-Hill, New York.

Brown, J., N. Jayawardena, and Y. Zelmanovich, "Water Systems for Pharmaceutical Facilities," *Pharmaceutical Engineering Magazine,* pp. 14–23, July/August 1991.

Cartwright, Peter S., "Reverse Osmosis and Nanofiltration System Design," *Plumbing Engineer Magazine,* pp. 45–49, March 1994.

Collentro, W. V., "Pharmaceutical Water," Parts 1 and 2, *Ultrapure Water Magazine,* November/December 1992.

Dow Chemical Corporation, "Water Conditioner Manual," 1988.

Gorry, M., P. Amin, and D. W. Richardson, Sr., "Take the Guesswork out of Demineralizer Design," *Chemical Engineering Magazine,* pp. 112–116, March 1994.

Janoschek, R., and G. C. du Moulin, "Ultraviolet Disinfection in Biotechnology: Myth vs Practice," *BioPharm Magazine,* pp. 24–27, January/February 1994.

Meyrick, C. E., "Practical Design of a High Purity Water System," *Pharmaceutical Engineering Magazine,* pp. 20–27, September/October 1989.

Nalco Chemical Company, *The Nalco Water Handbook,* 2nd ed., McGraw Hill, New York, 1988. Nussbaum, O. J., "Treating Cooling Water," *Heating/Piping/Air Conditioning Magazine,* pp. 63–67, February 1992.

Parekh, B. S., "Get Your Process Water to Come Clean," *Chemical Engineering Magazine,* pp. 71–85, January 1991.

Sendelbach, M. G., "Boiler Water Treatment," *Chemical Engineering Magazine,* pp. 127–132, August 1988.

Stenzel, M. H., "Remove Organics by Activated Carbon Adsorption," *Chemical Engineering Progress Magazine,* pp. 36–43, April 1993.

Yeh, K. L. and S. H. Lin, "Looking to Treat Wastewater-Try Ozone," *Chemical Engineering Magazine,* pp. 113–116, May 1993.

CHAPTER 5
HEAT TRANSFER, INSULATION, AND FREEZE PROTECTION

This chapter will discuss the basic fundamentals of heat transfer, thermal insulation, freeze protection, and heat tracing.

CODES AND STANDARDS

Insulation is often regulated by code requirements. The individual sections of any code must be carefully read in order to determine how flame-spreading or smoke-developing characteristics restrict the use of particular materials in specific areas of a building.

Typically, codes define the class of any building erected into one of four types: I, II, III, or IV. Type I is fireproof, Type II is noncombustible and either protected or nonprotected, and Types III and IV are combustible. The allowable flame-spread rating for any construction material is based on each class of construction.

The specific areas within a building also have restrictions pertaining to the allowable smoke-developed rating. These areas are:

1. Concealed spaces such as chases and shafts, not serving the HVAC system as air supplies or returns.

2. The same spaces as above, but used for air supplies or returns.

3. Rooms and spaces that may have a higher allowable smoke-developed or flame-spread rating, such as mechanical equipment rooms. Here the interior finish and the insulation fire code requirements are relaxed because of the nature of the space.

The building class as a whole usually determines the flame spread, and the specific area within the building usually determines the smoke developed. It is important to determine the most cost-effective insulation based on the code interpretations.

FUNDAMENTALS OF HEAT TRANSFER

BASICS

Heat is a type of energy that is produced by the movement of molecules. The greater the movement, the greater the heat. All molecular motion stops at absolute zero.

Temperature measurement finds only the level (or intensity) of the heat energy, not the amount. There are four methods used to express temperature:

1. Degrees Fahrenheit (F) shows the temperature expressed in English units. The scale is determined by dividing the actual temperature difference between the ice point and steam point of water into 180 divisions. The ice point is 32°F and the steam point is 212°F. Absolute zero is −459.7°F.

2. Degrees Celsius (C) shows the temperature expressed in metric units. The scale is determined by dividing the actual temperature difference between the ice point and steam point of water into 100 divisions. The ice point is 0°C and the steam point is 100°C. Absolute zero is −273.2°C.

3. Degrees Rankine (R) shows an absolute temperature scale starting at absolute zero, using the same division units as the Fahrenheit scale. The ice point of water is 491.7°R, the steam point is 671.7°R, and absolute zero is 0°R.

4. Degrees Kelvin (K) shows an absolute temperature scale starting at absolute zero and uses the same division units as the Celsius scale. The ice point of water is 273.2 K, the steam point is 373.2 K, and absolute zero is 0 K.

The Rankine and Kelvin scales were created for laboratory use and are rarely used in facility design engineering calculations.

The quantity of heat is measured either in Btu, which is an abbreviation for British thermal units, or kcal, which means kilocalorie. A less frequently used form of measurement is the watt-hour.

One Btu is the amount of heat needed to raise 1 lb of water 1°F. One kcal is the heat needed to raise 1 kg of water 1°C.

There are two kinds of heat flow, transient and steady state. For transient heat flow, the temperature varies with time. For general engineering applications, the transient method is too complex. This book will use the steady state method of calculating the heat loss through insulation. The results of these calculations are well within the range of accuracy sufficient for engineering purposes. Steady state heat flow is further divided into series and parallel types. In order to simplify the calculations, the following conditions will be assumed:

1. The series concept of heat flow will be used. As the heat flows through continuous layers, the total resistance is the sum of all of the individual layers.

2. Heat in equals heat out. There will be no accumulation of heat in the system.

3. All factors, including temperature, have reached equilibrium.

4. The temperature and heat flow of all components remain constant with time.

5. The same heat flow exists through any plane in the system.

Heat transfer, or the movement of heat energy, always flows from a higher temperature to a lower temperature. It can occur in any one of three ways: convection, radiation, or conduction.

Convection is a large-scale movement of liquids or gases. It cannot occur in solids. Density differences between hot and cold fluids produce a natural gravity movement. When a fluid is heated, it becomes less dense. The lighter fluid will move upwards in the absence of forced circulation. Heat is transferred faster when forced circulation is created by means of a fan or pump. When a liquid or gas moves from one place to another, it takes heat energy with it. Convection is always accompanied by conduction in the region where warmer and colder fluids interact.

Radiant heat is similar to radio waves, but has a shorter wavelength. They are capable of traveling through air, some solids and liquids, and vacuums. All substances at a temperature above absolute zero will radiate heat. Radiant heat increases with temperature and is proportional to the fourth power of the absolute temperature of the radiating body.

Solids transfer heat by conduction. Whenever the molecules move about, their constant vibration or oscillation causes a collision between them and other adjacent molecules. The heat does not remain constant because faster-moving molecules strike the slower-moving ones with a resulting decrease in the speed of the faster molecule. This will eventually equalize the temperature throughout the solid unless heat is added or removed.

There are several methods used to calculate the flow of heat through material, or the material's resistance to such flow.

The ability of a specific solid to conduct heat is called thermal conductivity, and is measured in Btu/h. This is referred to as the k factor. The standard that is used to find k determines the amount of heat that will flow in 1 h through material measuring 1 in thick and 1 ft^2 in area and having a temperature difference of 1°F (0.5°C) between the faces of the material being measured. The total heat flow in Btu/ft^2/h through any material is calculated by the following formula:

$$k = \frac{KI \times \Delta T \times A}{t} \tag{5.1}$$

where k = heat flow in Btu/h/ft^2/°F
  KI = k factor of the insulation material
  ΔT = temperature difference, °F
  A = area in square feet, at insulation exterior
  t = thickness of insulation, in

As the k factor increases, so does the flow of heat. Conductivity (conductance) measures heat flow through a standard arbitrary thickness sample of a specific material. Conductance (C) is measured in Btu/h/ft^2/°F, calculated with the following formula:

$$C = \frac{\text{thermal conductivity}}{\text{Thickness (in)}} \tag{5.2}$$

The resistance to the flow of heat through any individual solid is called thermal resistance R, and is measured in Btu/h/ft^2/°F. It is the reciprocal of the conductance value.

$$R = \frac{1}{\text{Conductance}} \tag{5.3}$$

As the resistance increases, the heat flow is reduced.

Total thermal resistance, RT, of a system measures heat flow in several materials layered together in series. It is found by adding together all the individual resistances R to obtain the total resistance of the entire system.

Thermal transmittance U is the rate of heat flow through several materials layered together, and is measured in Btu/h/ft^2/°F. It is the reciprocal of the total thermal resistance, and is calculated from the following formula:

$$U = \frac{1}{\text{Total thermal resistance}} \tag{5.4}$$

For insulation, conduction is the primary method in the transmission of heat. True thermal conductivity takes place only in homogeneous materials. In the range of materials used for thermal insulation, a homogeneous material is defined as a substance whose thermal conductivity does not change within the range of thickness normally used. For the most part, building materials such as brick and lumber are considered to be homogeneous. However, most thermal insulation is porous and actually composed of solid material surrounding small pockets of air. Therefore, conduction is not the sole means of heat transfer. In addition, a surface film of air, liquid, or even solid matter is almost always present around both the insulation and the pipe on which it is installed. This impedes the flow of heat. For this reason, the term *apparent thermal conductivity* is scientifically correct when referring to the k values used. With this explanation in mind, please note that the word *apparent* will be omitted in future discussions.

The term *heat flow* is defined as the total heat gain or loss from an entire system or component of that system, and is measured in Btu/h. The term *heat flux* is used to measure the heat gain or loss from only 1 ft^2 of a system or component. Heat flux is measured in Btu/h · ft^2 and is the product of the temperature differential and the conductance C.

Water Vapor Migration

Thermal insulation is fully effective only when completely dry. If water in either the form of vapor or liquid is present in or on insulation, it will have a serious effect both physically and thermodynamically. The loss of insulating capacity is well documented. If enough water is allowed to accumulate, it will cause rotting or other possible corrosive effects in most types of insulation.

Water vapor present in air has a measurable vapor pressure that is a function of both temperature and relative humidity. The lower the relative humidity and/or temperature, the lower the vapor pressure. Applying this fact specifically to an insulated body, the movement of water vapor is proportional to the difference in temperature between the ambient air and the wall of the pipe or vessel.

When the temperature of the insulated pipe or vessel is above ambient temperature, vapor pressure is higher at the pipe wall than on the outside surface of the insulation. This means that a vapor pressure differential exists between the pipe wall and the surrounding air, driving the water vapor away from the inner surface and towards the outside.

However, when the temperature of the pipe is below ambient, the opposite is true. The pressure is lower at the pipe wall than on the outside surface of the insulation, and the direction of water vapor flow is reversed. It is then possible for the vapor to permeate the insulation, allowing water to be absorbed and retained by the insulating material, and thereby reducing its effectiveness. It is also possible for water to actually condense as a liquid on the pipe wall. Certain types of insulation can become saturated over a period of time. If the insulation material does not absorb water, the air cells may become saturated. Another possibility is that the water may start corroding the pipe exterior.

The water vapor transmission rate (WVTR) is a measure of water vapor diffusion into or through any total insulation system. This flow of water vapor, called permeance, is measured in U.S. perms, or perm. A *perm* is the weight of water, in grains, that is transmitted through a 1-in (25 mm) thickness of the material in question in 1 h/ft^2, having a pressure difference between faces of 1 in of mercury. There are 7000 grains in 1 lb.

In order to restrict the flow of water vapor from the warm to the cold side of the object being insulated, a vapor barrier is installed on the warm side of the insulation. Since there is no perfect vapor barrier, the insulation materials used should also have some resistance to moisture in addition to having a good thermal resistance. The ideal material would absorb little or no moisture, would allow quick elimination of any that did enter, and would not be affected by moisture during the time it was present. A generally accepted figure of 0.30 perm is the maximum rate of an effective vapor barrier.

In addition to the primary purpose of retarding the flow of heat and water vapor, there are other important secondary factors that must be considered when selecting the type of insulation or covering for a particular application. They are:

1. Smoke and fire requirements
2. Space limitations
3. Personal protection
4. Protection from physical damage
5. Acoustical properties
6. Health and safety requirements
7. Dimensional stability
8. Corrosion resistance
9. Hygiene
10. Installed cost

If a fire were to start inside a building, it is possible that any part of the structure including its contents can contribute to the fire either by supporting combustion or generating smoke if noncombustible. In order to control the amount of combustible material inside buildings, code limitations for flame spread and smoke developed have been established for the components used in the interior of fireproof and noncombustible buildings. Material ratings have been established as follows.

The so-called tunnel test is used to obtain fire-related data for different kinds of insulation. The same test has been given different names by each agency conducting it. They are ASTM E-84, NFPA 255, and UL 723. The particular material is tested for flame spread, fuel contributed, and smoke developed. The materials tested are compared to red oak flooring (rated at 100) and asbestos-cement board (rated at zero) in all three categories. Caution should be used when trying to establish these values separately for the insulation, jacket, and some adhesives rather than as a complete system. Tests have shown that elements of an insulation system that have the same values when tested separately may have quite different ratings when tested as a composite unit.

The generally accepted maximum value for flame spread is 25, and the value for both fuel contributed and smoke developed is 50. These values are used when the building is of fireproof construction, which is the predominant type of building. For other kinds of construction, these values may be different.

Space limitations can be a factor in areas where insulated pipe is to be installed. Different types of insulation must be evaluated in order to obtain the best compromise between thickness and the k value.

Insulation will protect personnel from being scalded by touching a bare pipe. The insulation and jacket should be selected such that the surface temperature of its exterior will be no higher than 110°F (43°C).

Protection of the insulating material from physical damage may be a consideration when insulated piping is installed in an area of a building where there is storage of material that is regularly moved or where maintenance operations are carried out next to the pipe on a regular basis. This protection can be obtained by using a strong jacket or other insulating material that will not deform after repeated blows. Another consideration might be the strength of the insulation needed during the time it is being stored, transported, and installed in a possibly adverse climate or location.

The properties that make insulation effective will also attenuate sound made by the flow of contents through the pipe. When a pipe is installed where sound transmission could be a problem, for example, as a theater, adding extra thickness of insulation or special jacketing could reduce the sound to an acceptable level.

The potential health and safety hazards of insulation and accessories fall into two categories: (1) those related to storage, handling, and installation and (2) those that occur after installation. Correct procedures can reduce or eliminate most or all of the problems in the first category. However, exposure to materials such as asbestos could cause extremely serious health hazards.

The rate at which any particular insulation system will expand or contract has a definite effect on efficiency. The possible difference in expansion between the pipe and the insulation may eventually produce voids or gaps after repeated flexing. An insulation with a high k factor that might fail because of a large difference in the rates of expansion between it and the pipe will prove less economical over time than another insulation system with a lower k factor.

When an insulated pipe is to be installed in a corrosive atmosphere, the insulation system, particularly the jacket, must be capable of resisting whatever substance is causing that corrosion.

Insulation that is used for the food-processing, chemical, cosmetics, and pharmaceutical industries (or in other similar operations), must have the ability to withstand cleaning or sterilizing by a wide variety of methods. The important properties are:

1. Smooth finish
2. Resistance to fungus, bacteria, or mildew growth
3. Resistance to washing by detergents, steam, and chemicals
4. No chipping, cracking, or peeling
5. Nontoxicity and fire resistance

TYPES OF INSULATION

The following general designations are generic names of the materials. The individual manufacturers have different trade names for each of them. For each of the separate types, various properties will be compared, with the following properties common to all.

1. They must have been tested for fire-related values by the ASTM, NFPA, and UL as previously discussed, and meet the minimum standard for a flame spread of 25 and smoke-developed rating of 50 or less except where otherwise noted.
2. The temperature at which the k and R figures have been calculated is 75°F (24°C).

Fiberglass

Fiberglass insulation (ASTM C 547) is fibrous glass, made either plain or with a heat-resistant binder in order for the fiberglass to hold its shape. Typical values for material with a density of 3 to 5 lb/ft^3 (48 to 80 kg/m^3) are $k = 0.22$ to 0.26 and $R = 3.8$ to 4.5.

Fiberglass is the most popular insulation, and it comes in many forms. Felted glass fiber without any binder is available in rolls. Made with a thermosetting resin binder, it comes in several different stiffnesses. In the form most commonly used for pipe, it is molded and shaped into semicircular sections. The binder is the critical factor for the ultimate temperature for which it can be used.

Fiberglass by itself is not strong enough to stay permanently on a pipe without falling off in layers. Since fiberglass is porous, there is no way to seal it to prevent water vapor from flowing freely from the air to the pipe and then condensing on the pipe and saturating the fiberglass. In addition, there is no way to finish fiberglass that would be considered pleasant to look at. For these reasons, a covering, or jacket, must be added to protect it from physical damage, allow it to be firmly and permanently attached to the pipe, and prevent the penetration of water vapor.

Fiberglass is recommended for temperatures ranging from 35 to 800°F (1.5 to 422°C). It is available in thicknesses from ½ to 5 in (DN 15 to 125) for ½- to 33-in (DN 15 to 750) piping, with manufacturers providing various thicknesses for certain size pipes. For insulating tanks, fiberglass is also available in boards 48 to 96 in (1200 to 2400 mm) long, 12 to 24 in (300 to 2400 mm) wide, and up to 4 in (100 mm) thick. A high temperature, flexible blanket can be used with temperatures up to 1000°F (530°C).

Cellular Glass

Cellular glass insulation (ASTM C 552) is pure glass with closed cell air spaces. This material has a flame spread of 5 and smoke developed of 0. It also has a 0 perm rating. The typical k value is 0.38 and the R value is 2.6. A jacket is necessary for abrasion resistance; the type used depends upon the expected severity of service.

This extremely rigid and strong insulation is available for pipe sizes up to 36 in (900 mm) and 1 to 4 in (25 to 100 mm) thick. Form-fitting covers are used for any standard component. Flat blocks come in sizes 12 × 18 in (300 × 450 mm), 1½ to 5 in (40 to 125 mm) thick, and in sizes 18 × 24 in (450 × 450 mm). Factory fabricated shapes can be made to fit specific requirements. Recommended applications include temperatures ranging from −450 to +450°F (−265 to +230°C), with limitations based on the type of adhesive used with the material. Cellular glass is used where an extremely strong and impermeable material is required. It is also impervious to common acids and corrosive environments, and must be cut with a saw. It is available either plain or with a variety of factory-applied jackets.

Expanded Plastic Foam

Elastomeric plastic insulation (ASTM C 534) is an expanded foam, closed cell material, made from nitrile rubber and polyvinyl chloride resin. The typical k value is 0.27 and the R value is 3.6. This material has a perm rating of only 0.17, and does not require a jacket except for appearance; it can also be painted. The flame ratings of 50 are valid for all thicknesses. For material ½ in (15 mm) thick and less, a smoke-developed rating of 100 has been established; up to 1 in (25 mm), the rating is close to 150. Because of the high rating, building codes do not allow it to be used in all types of construction. A recent development has enabled manufacturers to reduce the smoke-developed rating down to 50 or below.

Commonly called rubber, this flexible insulation is available for pipe sizes up to 5 in ips (DN 125), and ½- and ¾-in (15 and 20 mm) thicknesses. In sheets, it is available in sizes up to 36 × 48 in (900 × 1200 mm) with 24 × 48 being the most common. Sheet thickness ranges from ⅛ to 1½ in. It is also available in 48 in (1200 mm) wide rolls, with thicknesses of ½, ¾ and 1 in (15, 20 or 25 mm). Yet another product is a 2-in-wide roll, 1/8 in thick, with self-sealing adhesive. Recommended applications for pipes include temperatures from 35 to 220°F (1.5 to 103°C), and for sheets up to 180°F (81°C), due to the adhesive required to apply it to a tank. It is used in pipe spaces and boiler and mechanical equipment rooms, where code requirements may be relaxed and the ease of application could make it more cost effective.

Foamed Plastic

Foamed plastic insulation is a continuously molded, rigid product made from foaming plastic resin, which results in a closed cell material. Typical insulation materials are polyurethane (ASTM C 591), polystyrene (ASTM C 578), and polyethylene. A factory-applied jacket is usually provided. The typical k value is 0.15; R value is 6.7.

Due to the possibly wide variations in the composition of the materials that fall into this category of insulation, the fire rating varies between manufacturers. Although the materials are combustible, they can be made self-extinguishing. Foamed plastic is recommended for low temperatures including cryogenic, and for moderate temperatures, generally up to a maximum of about 220°F (103°C).

Calcium Silicate

Calcium silicate (ASTM C 533) insulation is a rigid material compounded from silica, asbestos-free reinforcing fibers, and lime. At 500°F (260°C), it has a k value of 0.5 and an R value of 2.0. A field-applied jacket is required. This insulation is commonly referred to as "calsil."

Mineral Fiber

Mineral fiber (ASTM C 553) insulation is a rigid material composed of rock and slag made into fibers bound together with a heat-resistant inorganic binder. The typical k value is 0.28, and the R value is 4.9. This material is very well suited for high temperature work.

Insulating Cement

Insulating cement is produced from fibrous and/or granular insulation and cement, then mixed with water to form a plastic mass. Typical k values range between 0.65 and 0.95, depending upon the composition of the cement. They can be of either the hydraulic setting or the air drying type. This material is best suited for irregular surfaces or as a finish for other insulation applications. It can also be used in situations where space is at a premium and some kind of insulation is required. Installation costs are very high.

JACKETS

In order to function more efficiently and extend service life, most insulation must be protected from damage and degradation by the application of an effective cover, or jacket material. A *jacket* is defined as any material, except cements and paints, that can be used to cover or protect insulation installed on a pipe or vessel. The choice of jacketing will depend upon its use, which can be divided into seven general functional categories:

1. Weather barriers are used to prevent the entry of liquid water into insulation and also the entry of chemicals that would affect the inside or outside of the insulation. Materials include plastic, aluminum, and stainless steel as well as weather barrier mastics.

2. Vapor barriers are used to reduce the entry of water vapor into the surface of the insulation. In order to be effective, the vapor barrier must be completely sealed at every opening. A vapor barrier is used on cold surfaces primarily for eliminating the possibility of entrapped water vapor condensing on the pipe.

3. Mechanical abuse-resistant coverings are used to protect the underlying insulation from mechanical damage due to abuse or accidental contact by personnel or equipment. The compressive strength of the insulation used should be considered when selecting a jacket. Metal products are most commonly used.

4. Corrosion- and fire-resistant coverings are used as part of a complete hazard resistance system. Almost any type of jacket or mastic increases the fire rating. The most successful corrosion jackets are plastic or stainless steel depending upon the nature of the spill, leak, or atmosphere expected. Some mastics are also useful.

5. The visual appearance of some jackets over piping in exposed areas is an important feature in the selection of various coatings, finishes, cements, and covers. Since this consideration must often be approved by an architect or client, he or she should be consulted before final selection.

6. Jackets capable of being disinfected are used to present a smooth surface that will resist fungal and bacterial growth. They must withstand cleaning with powerful detergents coupled with steam and high pressure water. This requires a jacket with high mechanical strength.

7. Plain jackets are used on hot services and in other cases when a jacket is desired for ease of installation and appearance.

Jackets come in various forms and types, and can be divided into three general categories: rigid (plastic, aluminum, or stainless steel), membrane (glass cloth, coated papers, treated papers, and papers laminated with foils and/or cloth), and mastic. Jackets can be specified separately or factory applied. Separate jackets are used for special situations when a factory applied jacket is not available or possible, for example, jackets made of aluminum or plastic sheets. The factory applied jacket is by far the most common and is available in three types: the so-called all-purpose jacket, which has a vapor barrier, a plain jacket, and a weatherproof jacket.

Each manufacturer has a different combination of materials that are laminated to each other to provide flexibility, strength, and fire resistance. Kraft paper that has been coated or treated with chemicals is the most common base. The next layer is usually fiberglass cloth, used for strength, and the third layer is usually an aluminum foil. All three layers are permanently bonded together with a special adhesive to give the desired strength and water vapor retardation characteristics.

All-Service Jacket (ASJ)

The all-purpose, or all-service, jacket has a vapor barrier. The complete jacket is a lamination of kraft paper, fiberglass cloth (skrim) and either aluminum foil or metalized film. This is commonly referred to as an FSK jacket, for foil, skrim, and kraft.

The kraft paper is a bleached 30-lb (13.5 kg) basis weight material, which means 30 lb (13.5 kg) for each 30,000 ft² (2790 m²) area (or one ream). There is also a 45-lb (20.2 kg) basis weight paper available if a heavier paper is desired.

The fiberglass scrim is used for strength and reinforcement of the paper. The standard weave is 5×5, which means five lines per inch. Other weaves are available, ranging from 1×1 to 10×10, and also a 10×20. Also available is a bias weave, which adds diagonal threads in a third direction. The closer the weave, the stronger the jacket.

The foil used is aluminum, ranging in thickness from 0.35 to 1.0 mil. The standard thickness is 0.50 mil. Metalized film is also available. Although thinner than foil, it retains its shape better under impact. One manufacturer described its product as a white, metalized polypropylene film with a perm rating of 0.02.

The composition of the adhesive and the actual methods used to bind the components together are proprietary. It is the adhesive that imparts the fire resistant rating to the entire jacket system. After a layer of adhesive is applied to the kraft paper, the scrim is added and the adhesive forced through the weave. Finally, the foil or metalized film is put on next. The three layers are then laminated together to form the complete jacket system.

Lagging

Lagging is the process of insulating a pipe or vessel and then covering the insulation with a cloth jacket. The jacket is primarily used to improve the surface appearance of any insulation, offering very little in the way of protection. Lagging materials are available in a full spectrum of colors and may eliminate the need for painting. This cloth can be canvas or fiberglass, for example, and is secured to the insulation with lagging adhesive and/or sizing. Also available is a combination system that serves as both an adhesive and protective coat.

Aluminum Jackets

Aluminum jackets (ASTM B-209) are available in a corrugated or smooth shape and in thicknesses ranging from 0.010 to 0.024 in, with 0.016 in being the most commonly used. Also available are different tempers or hardness. These range from H 14 (half hard) to H 19 (full hard), with H 14 being the most common. Aluminum jackets can be secured by one of three different methods: banded by straps on 9-in centers, by a patented seam in an S or Z configuration, or by sheet metal screws. The ends are overlapped 2 in and secured with straps or screws (or nothing for the interlocking type). Since they are usually applied over insulation, a variety of vapor barrier materials can be factory applied to the aluminum jackets, which may be necessary if the insulation has any ingredient that causes galvanic or corrosive attack on the aluminum, or if an additional vapor barrier is thought to be necessary. Fittings are fabricated from roll material in the shop.

There are four different alloys of aluminum commonly used for jacketing material: 1100, 3003, 3105, and 5005. Although there are differences among them, it is not usually necessary to specify which alloy is to be used. The properties of all types are so closely matched that the service or performance of the material is not affected by different choices. It is common practice for the fabricator of the jacket to interchange any of the four types depending upon availability and price. By specifying the ASTM code alone, the engineer is allowing the contractor to use any of the types (since they are all acceptable), and avoids the possibility of a delay caused by waiting for the particular alloy specified and the extra cost involved.

One alloy, 1100, is mostly used for fittings because it is the most malleable of the four. If the jacketing is used on a pipe that may expand and contract often because of system operation, corrugated aluminum jackets should be used. These jackets easily expand and contract.

Aluminum jackets have the following advantages:

1. Easy application in any weather
2. Easy formation into different shapes
3. Good resistance to abuse
4. Ready availability

Aluminum jackets have the following disadvantages:

1. Low resistance to pH ranging from 7 to 11
2. Low fire rating
3. Low emittance value

4. High initial cost

5. Low resistance to strong cleaning chemicals

Stainless Steel Jackets

Stainless steel jackets (ASTM A-240) are available in either flat or corrugated forms and in standard thicknesses of 0.010, 0.016, and 0.019. They are secured in the same manner as aluminum jackets. A factory applied moisture barrier can also be added.

The most commonly available alloys are types 301, 302, 303, 304, 305, and 316; 304 is the most popular. It is best to consult with the manufacturer for the criteria that will help determine which alloy would be best for any particular application. Several types of finishes are available, from polished to dull. Stainless steel jackets have the following advantages:

1. Excellent fire rating

2. High resistance to mechanical abuse

3. Excellent corrosion and weather resistance

4. Easy application in any weather

5. Excellent hygienic characteristics

Stainless steel jackets have the following disadvantages:

1. High initial cost

2. Corrosion cracking where chlorine or fluorine exists

3. Low emittance value

4. Long lead time

There are often strict union regulations requiring that stainless steel jackets over 0.20 in thick be installed by sheet metal workers. Jackets 0.20 in thick or less can be installed by the insulation contractor. The insulation contractor is more knowledgeable about this kind of work, so when job conditions permit, it is usually more cost effective to specify the thinner thickness to ensure that the work will be done by the insulation contractor.

Wire Mesh

Wire mesh is a little-known jacket material. It's mainly used when a strong, flexible, abrasion-resistant covering that must be easily removed is needed. It is available in widths from 1 to 43 in (25 to 1075 mm), with 12, 18, 24, and 30 in (300, 450, 600, and 750 mm) used most often. Common wire diameter of the mesh is either 0.008 or 0.011 in. The thicker wire is used where greater strength is needed or heavy use expected. The openness of the weave is expressed in density, which gives the number of openings per inch. Densities of 48 to 130 are used, with 60 being the most common. Material of the mesh can be Monel, Inconel, or stainless steel. It is attached with lacing hooks or sewn with stainless steel wire. In addition, it must be secured with either tie wires or metal straps.

Plastic Jackets

Plastic jackets are manufactured in a great variety of materials, including PVC, ABS, PVF, PVA, and acrylics. Thickness ranges from 3 to 35 mils. The manufacturers should be consulted to determine the criteria necessary to select the best material and thickness for any particular application. Plastic jackets have the following advantages:

1. Lowest cost of any solid jacket
2. Best resistance to chemical corrosion
3. Excellent resistance to bacterial and fungal growth

Plastic jackets have the following disadvantages:

1. Poor fire rating
2. Low impact resistance
3. Softening at high temperatures
4. Vulnerability to infrared and ultraviolet rays and ozone
5. Cold weather embrittlement

COATINGS, ADHESIVES, AND SEALANTS

There are a large number of products available. Rather than listing them here, the design criteria necessary to the selection of the proper material will be discussed.

Manufacturers specify where to use specific products, but regulations from government and, in some cases, private agencies may dictate the choice of product that will be used. Some of the considerations are the following:

1. Flammability in the wet or dry state
2. Type of system—solvent or water based (depending on whether the material will attack the substrate to which it is applied)
3. Recommended dry film thickness
4. Temperature conditions required for application
5. Limitations on toxicity levels while being applied and drying
6. Method of application—brush, trowel, spray, or palm
7. Resistance to chemical and mechanical factors, such as abrasion, temperature, impact, and expansion

The choice of material is also governed by the size, shape, and location of the surface to be protected. Large irregular shapes may require a thicker material, which will cling to surfaces offering no other means of application.

Adhesives are used to permanently bond the insulation either to itself or to the surface on which it is applied. Each different type of insulation requires its own special type of adhesive. This information should be obtained directly from the insulation or adhesive manufacturer, who will recommend the adhesive best suited for the material and service conditions established by the engineer.

Weather barrier coatings are used to seal an insulation system from the elements, thereby protecting the underlying insulation from damage. In cold service the use of a breather final coat over a vapor barrier will prevent liquid water from penetrating the coating while allowing water vapor to pass through. When used on a hot service, a breather coat will allow the escape of the minimal amount of water vapor trapped inside. When additional strength or protection is required, glass cloth membranes as well as metal mesh should be used to reinforce the weather barrier application.

Vapor barrier coatings are used on the outer layer of cold service insulation to inhibit the passage of water vapor. These coatings are formulated to be used alone as the top coat, or with factory or field-applied jackets made of FSK, cloth, or other membranes as an adhesive or sealer.

Lagging adhesives are used to apply cloth lagging both to itself and to the insulation surface. In some cases, the surface may be rough or irregular, and a sizing may be necessary to seal the substrate in order for the lagging to adhere properly or for a finish to be correctly applied. The correct adhesive can fill the rough surfaces and gaps to provide a smooth and decorative finish. Coatings for application over almost any material used for insulation can be obtained. Acoustical mastics are available for use as sound barrier coatings, and can be used alone or with acoustical jacket material.

INSULATION MATERIAL AND THICKNESS SELECTION

The general criteria needed to make a choice among various insulation materials are as follows:

1. The reason insulation is needed
2. Service temperature expected
3. Code requirements
4. The location where insulation will be installed
5. Accessibility for the insulated pipe
6. Installed cost of the complete insulating system

Reasons for Using Insulation

There are four reasons to use insulation:

1. Condensation prevention
2. Reduction of heat loss
3. Personnel protection
4. Noise reduction

Condensation Prevention.   Insulation applied to pipes carrying storm water, city water, chilled water, and drinking water is done to prevent condensation. On pipes containing chilled and drinking water condensation prevention is the secondary consideration after heat gain prevention. Each of the various insulating manufacturers has prepared charts giving the necessary thickness of the insulation in question to prevent condensation. The design temperature of the fluids is: storm water, 35°F (2°C); domestic water, 60°F (15°C); and chilled water, 50°F (10°C). The design ambient temperature is 90°F (32°C), the relative humidity, 90 percent. A vapor barrier jacket is required on fiberglass, mineral wool, and calsil. It is not necessary over rubber or cellular glass. The total system perm rating must be no more than 0.30.

For insulation thickness to prevent condensation, refer to Tables 5.1 and 5.2 for service temperatures of 50°F (10°C) and 34°F (2°C), respectively. Find the pipe size, relative humidity, ambient temperature of the space where the pipe is located, and lowest service temperature of the pipe in the table. Read the recommended minimum thickness of insulation at the intersection of the pipe size and humidity columns.

Reduction of Heat Loss. For hot systems, the most important consideration is the reduction of heat loss. As a result of this reduction, operating cost will be lowered, due to savings in fuel and increase in process efficiency. Capital costs may also be reduced.

The economic evaluation of a particular insulation system includes either the selection of the optimal thickness of insulation for a particular service or the comparison between two or more different insulation systems to find which will return the most savings in conserved energy over a specific period of time. For a given set of economic variables, there is only one solution. The reason for this is that

TABLE 5.1 Insulation Thickness to Prevent Condensation, 50°F Service Temperature and 70°F Ambient Temperature*

Nom. Pipe		Relative humidity, %																
		20			50			70			80			90				
DN	size, in	TNK†	HG‡	ST§	THK	HG	ST	THK	HG	ST	THK	HG	ST	THK	HG	ST
15	0.50				0.5	2	66	0.5	2	66	0.5	2	66	1.0	2	68
20	0.75				0.5	2	67	0.5	2	67	0.5	2	67	0.5	2	67
25	1.00				0.5	3	66	0.5	3	66	0.5	3	66	1.0	2	68
32	1.25				0.5	3	66	0.5	3	66	0.5	3	66	1.0	3	67
40	1.50				0.5	4	65	0.5	4	65	0.5	4	65	1.0	3	67
50	2.00				0.5	5	66	0.5	5	66	0.5	5	66	1.0	3	67
65	2.50				0.5	5	65	0.5	5	65	0.5	5	65	1.0	4	67
75	3.00				0.5	7	65	0.5	7	65	0.5	7	65	1.0	4	67
90	3.50	Condensation			0.5	8	65	0.5	8	65	0.5	8	65	1.0	4	68
100	4.00	control not			0.5	8	65	0.5	8	65	0.5	8	65	1.0	5	67
125	5.00	required for this			0.5	10	65	0.5	10	65	0.5	10	65	1.0	6	67
150	6.00	condition			0.5	12	65	0.5	12	65	0.5	12	65	1.0	7	67
200	8.00				1.0	9	67	1.0	9	67	1.0	9	67	1.0	9	67
250	10.00				1.0	11	67	1.0	11	67	1.0	11	67	1.0	11	67
300	12.00				1.0	12	67	1.0	12	67	1.0	12	67	1.0	12	67

*25 mm = 1 in.
†THK—Insulation thickness, inches.
‡HG—Heat gain/lineal foot (pipe) 28 ft (flat), BTU
§ST—Surface temperature, °F

TABLE 5.2 Insulation Thickness to Prevent Condensation, 34°F Service Temperature and 70°F Ambient Temperature*

Nom. Pipe		Relative humidity, %																
		20			50			70			80			90				
DN	size, in	TNK†	HG‡	ST§	THK	HG	ST	THK	HG	ST	THK	HG	ST	THK	HG	ST		
15	0.50				0.5	4	64	0.5	4	64	0.5	4	64	1.5	2	68		
20	0.75				0.5	4	64	0.5	4	64	0.5	4	64	1.5	3	67		
25	1.00				0.5	6	63	0.5	6	63	1.0	4	66	1.5	3	67		
32	1.25				0.5	6	63	0.5	6	63	1.0	5	65	1.5	3	67		
40	1.50				0.5	8	62	0.5	8	62	1.0	5	66	1.5	4	67		
50	2.00				0.5	8	63	0.5	6	63	1.0	6	66	1.5	4	67		
65	2.50				0.5	10	63	0.5	10	63	1.0	6	66	1.5	5	67		
75	3.00				0.5	12	62	0.5	12	62	1.0	8	65	1.5	6	67		
90	3.50	Condensation			0.5	14	61	0.5	14	61	1.0	7	66	1.5	6	67		
100	4.00	control not			0.5	15	62	0.5	15	62	1.0	9	65	1.5	7	67		
125	5.00	required for this			0.5	16	63	0.5	16	63	1.0	11	65	2.0	7	67		
150	6.00	condition			0.5	22	61	0.5	22	61	1.0	13	65	2.0	8	67		
200	8.00				1.0	16	65	1.0	16	65	1.0	16	65	2.0	10	67		
250	10.00				1.0	20	65	1.0	20	65	1.0	20	65	2.0	11	67		
300	12.00				1.0	22	65	1.0	22	65	1.0	22	65	2.0	13	67		

*25 mm = 1 in.
†THK—Insulation thickness, inches.
‡HG—Heat gain/lineal foot (pipe) 28 ft (flat), BTU
§ST—Surface temperature, °F

increasing thickness beyond the optimal thickness does not give increased return on investment.

A high design service temperature of piping could eliminate some insulating material from consideration. The manufacturer's technical literature will indicate the highest temperature recommended for a particular insulating system. The adhesive is usually the weakest link in the insulating system chain.

To find the amount of heat lost through piping installed outdoors covered with various insulation types and thicknesses, refer to Table 5.3. Heat loss tables for various types and thicknesses of insulation installed indoors are available from insulation manufacturers. Although the base figures are for fiberglass, insulation factors are provided for other types of insulation. Multiply the figures in the table by the factor for other insulation types to obtain the figure for the new insulation. Also included is heat loss in Btu/in of insulation. Use the average thickness obtained from Table 5.6.

Personnel Protection. If system economy is not a consideration for any particular piping system, a hot pipe must be insulated to bring the surface temperature of the insulation to 120°F (48°C) or below, the point that would burn a person's skin. Refer to Table 5.4 for the thickness of fiberglass insulation necessary to achieve a surface temperature of 110°F (43°C) with a conservative ambient temperature of 80°F (26°C). Enter the table with the service temperature and size of the pipe. Read the required thickness and resulting surface temperature for the thickness selected at the intersection of the appropriate columns. Use the 250°F column for lower service piping temperatures.

TABLE 5.3 Heat Loss from Piping*

Insulation type	Insulation factor	Heat loss per inch thickness, based on K factor @ 50°F mean temp (Btu/hr · °F · ft^2)
Glass fiber (ASTM C547)	1.00	0.25
Calcium silicate (ASTM C533)	1.50	0.375
Cellular glass (ASTM C552)	1.60	0.40
Rigid cellular urethane (ASTM C591)	0.66	0.165
Foamed elastomer (ASTM C534)	1.16	0.29
Mineral fiber blanket (ASTM C553)	1.20	0.30
Expanded perlite (ASTM C610)	1.50	0.375

Insula-tion thick-ness, in	ΔT, °F	½	¾	1	1¼	1½	2	2½	3	4	6	8	10	12
		¾	1	1¼	1½	2								
0.5	10	0.5	0.6	0.7	0.8	0.9	1.1	1.3	1.5	1.8	2.6	3.3	4.1	4.8
	50	2.5	2.9	3.5	4.1	4.8	5.5	6.5	7.7	9.6	13.5	17.2	21.1	24.8
	100	5.2	6.1	7.2	8.6	9.9	11.5	13.5	15.9	19.9	28.1	35.8	43.8	51.6
	150	8.1	9.5	11.2	13.4	15.5	17.9	21.0	24.8	30.9	43.8	55.7	68.2	80.2
	200	11.2	13.1	15.5	18.5	21.4	24.7	29.0	34.3	42.7	60.4	76.9	94.1	110.7
	250	14.6	17.1	20.2	24.1	27.9	32.2	37.8	44.7	55.7	78.8	100.3	122.6	144.2
1.0	10	0.3	0.4	0.4	0.5	0.6	0.6	0.7	0.8	1.0	1.4	1.8	2.2	2.6
	50	1.6	1.9	2.2	2.5	2.9	3.2	3.7	4.4	5.4	7.4	9.4	11.4	13.4
	100	3.4	3.9	4.5	5.2	5.9	6.8	7.8	9.1	11.2	15.5	19.5	23.8	27.8
	150	5.3	6.1	7.0	8.2	9.3	10.5	12.2	14.2	17.4	24.1	30.4	37.0	43.3
	200	7.4	8.4	9.7	11.3	12.8	14.6	16.8	19.6	24.0	33.4	42.0	51.2	59.9
	250	9.6	11.0	12.6	14.8	16.7	19.0	22.0	25.6	31.4	43.6	54.9	66.9	78.2
1.5	10	0.3	0.3	0.3	0.4	0.4	0.5	0.5	0.6	0.8	1.0	1.3	1.6	1.8
	50	1.3	1.5	1.7	1.9	2.2	2.4	2.8	3.2	3.9	5.3	6.6	8.0	9.3
	100	2.7	3.1	3.5	4.0	4.5	5.1	5.8	6.7	8.1	11.1	13.8	16.7	19.5
	150	4.3	4.8	5.5	6.3	7.1	7.9	9.1	10.4	12.6	17.2	21.5	26.0	30.3
	200	5.9	6.7	7.6	8.7	9.8	11.0	12.5	14.5	17.5	23.8	29.7	36.0	41.9
	250	7.8	8.7	9.9	11.4	12.8	14.4	16.4	18.9	22.8	31.1	38.9	47.1	54.8
2.0	10	0.2	0.2	0.3	0.3	0.4	0.4	0.4	0.5	0.6	0.8	1.0	1.2	1.4
	50	1.1	1.3	1.4	1.6	1.8	2.0	2.3	2.6	3.1	4.2	5.2	6.3	7.3
	100	2.4	2.7	3.0	3.4	3.8	4.2	4.8	5.5	6.5	8.8	10.9	13.1	15.2
	150	3.7	4.2	4.7	5.3	5.9	6.6	7.5	8.5	10.2	13.7	17.0	20.4	23.6
	200	5.2	5.8	6.5	7.4	8.2	9.1	10.3	11.8	14.1	19.0	23.5	28.2	32.7
	250	6.8	7.5	8.5	9.6	10.7	11.9	13.5	15.4	18.5	24.8	30.7	36.9	42.7
2.5	10	0.2	0.2	0.2	0.3	0.3	0.3	0.4	0.4	0.5	0.7	0.8	1.0	1.2
	50	1.0	1.1	1.3	1.4	1.6	1.8	2.0	2.3	2.7	3.6	4.4	5.2	6.0
	100	2.2	2.4	2.7	3.0	3.3	3.7	4.1	4.7	5.6	7.4	9.1	10.9	12.6
	150	3.4	3.7	4.2	4.7	5.2	5.8	6.5	7.3	8.7	11.5	14.2	17.0	19.6
	200	4.7	5.2	5.8	6.5	7.2	8.0	9.0	10.2	12.1	16.0	19.6	23.5	27.1
	250	6.1	6.8	7.5	8.5	9.4	10.4	11.7	13.3	15.8	20.9	25.7	30.7	35.4
3.0	10	0.2	0.2	0.2	0.3	0.3	0.3	0.3	0.4	0.5	0.6	0.7	0.9	1.0
	50	1.0	1.1	1.2	1.3	1.4	1.6	1.8	2.0	2.4	3.1	3.8	4.5	5.2
	100	2.0	2.2	2.4	2.7	3.0	3.3	3.7	4.2	4.9	6.5	7.9	9.4	10.8
	150	3.1	3.4	3.8	4.3	4.7	5.2	5.8	6.5	7.7	10.1	12.3	14.7	16.8
	200	4.3	4.8	5.3	5.9	6.5	7.2	8.0	9.0	10.7	14.0	17.0	20.3	23.3
	250	5.7	6.2	6.9	7.7	8.5	9.4	10.5	11.8	13.9	18.3	22.3	26.5	30.5

*Pipe heat loss (Q_B) is shown in watts per foot. Heat loss calculation is based on metal pipes insulated with glass fiber (ASTM C547) and located outdoors in a 20 mph wind. A 10% safety factor has been included. (Note: Watts/ft = Btu/ft × 0.293.)

Source: RAYCHEM.

TABLE 5.4 Insulation Thickness for Personnel Protection

120° Maximum Surface Temperature, 80° Ambient

Nom. pipe size, in	250 HL†				350 HL				450 HL				550 HL			
	TH*	LF‡	SF§	ST¶	TH	LF	SF	ST	LF	SF	ST	TH	TH	LF	SF	ST
0.50	0.5	25	51	109	1.0	30	40	104	48	64	118	1.0	1.5	55	52	113
0.75	0.5	25	41	104	0.5	42	68	120	45	43	107	1.5	1.5	64	61	118
1.00	0.5	34	55	112	1.0	37	40	105	60	66	120	1.0	1.5	69	58	117
1.25	0.5	37	49	109	1.0	47	51	112	55	42	107	1.5	1.5	77	59	118
1.50	0.5	46	61	117	1.0	48	46	109	62	47	110	1.5	2.0	70	40	106
2.00	0.5	50	55	114	1.5	56	47	110	70	48	111	1.5	2.0	84	48	112
2.50	0.5	59	56	115	1.0	45	26	97	72	41	107	1.5	1.5	102	59	119
3.00	0.5	75	64	120	1.0	76	52	114	93	53	115	1.5	2.0	110	55	117
3.50	1.0	43	25	96	1.0	71	41	107	93	46	111	1.5	2.0	112	49	113
4.00	0.5	89	61	119	1.0	90	52	114	112	56	117	1.5	2.0	131	58	119
5.00	1.0	67	33	102	1.0	110	55	117	134	59	120	1.5	2.5	131	46	112
6.00	1.0	79	35	103	1.0	130	57	119	124	44	110	2.0	2.5	150	48	114
8.00	1.0	95	33	103	1.0	157	55	118	153	45	112	2.0	2.5	177	48	114
10.00	1.0	121	36	105	1.5	136	37	106	179	45	112	2.0	2.5	215	51	117
12.00	1.0	129	32	103	1.0	212	54	118	207	46	113	2.0	2.5	248	52	118

*TH—Thickness of insulation, inches.

†HL—Heat loss, Btu/h.

‡LF—Heat loss per lineal foot of pipe, Btu/h.

§SF—Heat loss per square foot of outside insulation surface, Btu/h.

¶ST—Surface temperature of insulation, °F.

Noise Reduction.   Some types of insulation and/or special jackets can reduce the noise that may be generated or transmitted by pipes, depending on the degree of reduction desired. Manufacturers have compiled data on the reduction factors of various insulation and jackets, which will aid in the selection of materials for that purpose.

Expected Temperature

The expected temperature of the pipe is an important factor in the selection of insulation systems. For hot pipes, the highest expected temperature will affect the choice of thickness and type of insulation and the adhesives used to adhere the insulation or jackets to the pipe and itself. For cold pipes, the coldest expected temperature will affect the selection of the insulation and vapor barrier required to stop condensation. The expansion of the pipe and insulation should also be compared to see if the difference is excessive.

Location of Insulation

Where the insulated pipe or equipment is located may dictate the type of insulation, jacket, or covering and the method used to attach it to the pipe. Factors such as outdoor installation, high humidity, and low ambient temperature will affect the selection of system components.

Accessibility

There may be a need to remove portions of the insulation to allow maintenance, inspection, or repair to a section of the piping system on a regular basis. Fiberglass and rubber are the easiest to remove, simply by cutting the cover and insulation with a knife and removing them. Putting them back is just as simple as installing them originally, and no harm is done to the surrounding insulation and covering on either side. If this is to be done often or on a regular basis, another popular method is to use a blanket insulation and wrap it with a wire mesh that is secured with straps. This last method is limited to hot lines that are not subject to any physical abuse whatsoever.

CALCULATION OF INSULATION THICKNESS

The following examples will illustrate the use of previously presented formulas of heat transfer fundamentals involving practical solutions to selecting the actual thickness of the insulation intended to be used for any specific project.

In all probability, there are charts and tables available to determine some of the heat loss figures to be calculated. But, there are bound to be situations that will not allow the use of standard conditions or standard figures. Also, most charts do not go into enough detail or provide criteria for the specific conditions that may be present in the design of a specific project. It is for these circumstances that the formulas will be used. Before proceeding with actual calculations, the following paragraphs will describe several typical factors and basic formulas from which the design criteria are derived.

First is the surface film factor (SFR). A film of air exists on the surface of any solid in direct contact with air. This surface film has a definite resistance to the passage of heat. Table 5.5 gives the surface resistance (film factor) for a variety of conditions. In order to simplify the figures, the pipe is assumed to be at the same temperature as its contents.

When calculating the heat flux for piping (as compared to flat surfaces), the fact that the inner surface of the insulation has a different area than its outer surface must be taken into account. Since the measurement is the amount of heat loss per square foot of exposed surface, a means must be found to determine the actual area that should be used. This is done by using an equivalent insulation thickness (ET), which is equal to the logarithmic mean of the inner and outer surfaces (see Table 5.6). Enter this table with the pipe size and thickness of insulation intended to be used.

Also included in Table 5.6 is the actual square foot area (A) along the outside of the selected insulation per linear foot of pipe length. This will be necessary to find the actual heat loss for the installation. Simply multiply the calculated heat loss by A to obtain the heat loss per foot of pipe.

The variables below will be used in the following series of formulas:

HF = heat flux in Btu/h/ft^2
T_1 = temperature of pipe
T_2 = ambient temperature
ET = equivalent thickness of insulation (Table 5.6)
AT = actual thickness of insulation
ST = surface temperature at insulation exterior
k = insulation k factor (manufacturer's rating)
SFR = surface film factor, or resistance (Table 5.5)
HL = heat loss per linear foot of pipe
A = area of insulation exterior, ft^2 (Table 5.6)

TABLE 5.5   Surface Film Factor

Condition	Resistance R
Still air (0 mph)	
Heat flow up	0.61
Heat flow down	0.92
Heat flow horizontal	0.68
Moving air	
7.5 mph (12 km/hr) (summer)	0.25
15.0 mph (24 km/hr) (winter)	0.17
Round pipe	0.65

Note: Surface resistances decrease as air velocities increase. All values are taken from *ASHRAE Handbook of Fundamentals*. Some of these examples consider only the insulation surface-to-ambient film factor. It is assumed that the inside surface area is at the same temperature as the contents (air, gas, or liquid), such as in a duct, pipe, or tank. Generally, the inside air film factor is used only for cases involving occupied spaces, 60 to 90°F (15 to 32°C).

Courtesy: Owens/Corning.

TABLE 5.6 Equivalent Thickness

NPS	$\frac{1}{2}$		1		$1\frac{1}{2}$		2		$2\frac{1}{2}$		3	
	L_1	A	L_1	A	L_1	A	L_1	A	L_1	A	L_1	A
$\frac{1}{2}$	0.76	0.49	1.77	0.75	3.12	1.05	4.46	1.31	—	—	—	—
$\frac{3}{4}$	0.75	0.56	1.45	0.75	2.68	1.05	3.90	1.31	—	—	—	—
1	0.71	0.62	1.72	0.92	2.78	1.18	4.02	1.46	—	—	—	—
$1\frac{1}{4}$	0.63	0.70	1.31	0.92	2.76	1.31	3.36	1.46	—	—	—	—
$1\frac{1}{2}$	0.60	0.75	1.49	1.05	2.42	1.31	4.13	1.73	—	—	—	—
2	0.67	0.92	1.43	1.18	2.36	1.46	3.39	1.73	4.43	1.99	—	—
$2\frac{1}{2}$	0.66	1.05	1.38	1.31	2.75	1.73	3.71	1.99	4.73	2.26	—	—
3	0.57	1.18	1.29	1.46	2.11	1.73	2.96	1.99	3.88	2.26	4.86	2.52
$3\frac{1}{2}$	0.92	1.46	1.67	1.73	2.46	1.99	3.31	2.26	4.22	2.52	5.31	2.81
4	0.59	1.46	1.28	1.73	2.01	1.99	2.80	2.26	3.65	2.52	4.68	2.81
$4\frac{1}{2}$	0.94	1.74	1.61	1.99	2.35	2.26	3.15	2.52	4.11	2.81	5.02	3.08
5	0.58	1.74	1.20	1.99	1.89	2.26	2.64	2.52	3.54	2.81	4.40	3.08
6	0.54	2.00	1.13	2.26	1.79	2.52	2.60	2.81	3.36	3.08	4.17	3.34
7	—	—	1.11	2.52	1.84	2.81	2.54	3.08	3.27	3.34	4.25	3.67
8	—	—	1.18	2.81	1.81	3.08	2.49	3.34	3.39	3.67	4.15	3.93
9	—	—	1.17	3.08	1.79	3.34	2.62	3.67	3.32	3.93	4.06	4.19
10	—	—	1.09	3.34	1.85	3.67	2.50	3.93	3.18	4.19	3.90	4.45
12	—	—	1.22	3.93	1.82	4.19	2.45	4.45	3.10	4.71	3.79	4.97
14	—	—	1.07	4.19	1.65	4.45	2.26	4.71	2.90	4.97	3.57	5.24
16	—	—	1.06	4.71	1.63	4.97	2.23	5.24	2.86	5.50	3.50	5.76
18	—	—	1.05	5.24	1.62	5.50	2.21	5.76	2.82	6.02	3.45	6.28
20	—	—	1.05	5.76	1.61	6.02	2.19	6.28	2.79	6.54	3.41	6.81
24	—	—	1.04	6.81	1.59	7.07	2.16	7.33	2.74	7.59	3.35	7.85

NPS = nominal pipe size

L_1 = equivalent thickness in inches

$$L_1 = r_2 \ln\left(\frac{r_2}{r_1}\right)$$

where r_1 = inner radius of insulation, inches

r_2 = outer radius of insulation, inches

ln = log to the base e (natural log)

A = square feet of pipe insulation exterior per lineal foot of pipe

The formula used to find heat flux from a round surface is:

$$\text{HF} = \frac{T_1 - T_2}{(\text{ET}/k) \times \text{SFR}} \qquad (5.5)$$

The following examples will demonstrate the use of the formulas. The first example will involve the following conditions:

Temperature of pipe contents = 200°F
Ambient temperature = 80°F
Pipe size = 8 in
Insulation = fiberglass, 2 in thick

Substituting in Eq. (5.5)

$$HF = 200 - \frac{80}{(2.49/0.23) + 0.68} = \frac{120}{9.2 + 0.68} = \frac{120}{9.88}$$

$$= 12.1 \text{ Btu/h} \cdot \text{ft}^2$$

To find the actual heat loss per linear foot, use the formula:

$$HL = HF \times A \tag{5.6}$$

$$= 12.1 \times 3.34 = 40.4 \text{ Btu/h} \cdot \text{ft}$$

To find the heat loss from the entire system, find the total length of run and multiply that figure by HL. In order to do an accurate calculation, the different lengths of pipe run in various temperature-controlled areas will each have to be calculated separately.

To find the surface temperature of the insulation exterior, use the formula:

$$ST = T_2 + (HF \times SFR) \tag{5.7}$$

$$= 80 + (12.1 \times 0.68) = 80 + 8.2 = 88.2°F$$

In order to calculate the heat loss from a tank, the only substitution in the above formula should be AT instead of ET, and there is no need to find HL. To find the heat loss from the whole tank, multiply HF by the area of the tank.

When calculating the heat gain for a cold line, the same formula as for a hot pipe is used, except that the result is the amount of heat gained by the pipe from the ambient air.

General Design Considerations

The following paragraphs describe the general selection criteria for different materials under average situations.

1. *Domestic cold water or chilled water service.* Fiberglass with ASJ should be secured with staples. The reasons are that fiberglass is the least expensive insulation to install and the staples can be applied in any weather and regardless of the dust conditions on the job. Fittings should be covered with plastic-fitting covers. The staples must be coated with a vapor barrier mastic. If the working conditions are reasonably dust-free, self-sticking lap joints are a cost-saving alternative.

2. *Domestic hot water, hot water return, and other services under 240°F (114°C).* The same materials as for cold water service (see Number 1 above) are used.

3. *Hot tanks under 240°F (114°C).* Stiff fiberglass board covered with a breather coat of mastic or fiberglass cloth should be used. Insulating cement can be used if additional protection from abuse is needed.

4. *Cold (not cryogenic) tanks.* Rubber insulation with no jacket can be used if the area where it will be installed is reasonably clean. A jacket is not required and the surface can be painted. If there is a problem with dust or dirt, the material of choice would be fiberglass boards with a coating of cut back mastic as a vapor barrier.

5. *Outdoor service.* Rubber insulation with an ASJ should be used in order to resist moisture penetration into the insulation. The final cover should be aluminum secured with bands or a lock seam. Field experience has shown that screws often fall out if subject to vibration.

6. *Sanitary exterior.* The best choice is a stainless steel jacket if the cleaning is to be severe. If not, a plastic jacket is more cost effective.

7. *Hazardous environment.* Where there is a danger that the contents of the pipe or chemicals in the surrounding area could possibly penetrate the jacket and cause a potentially dangerous situation, cellular glass insulation should be used because it is nonadsorbent. A jacket capable of resisting the chemical hazard, usually plastic, must be used over the glass insulation.

8. *Special fire resistance.* Again, cellular glass is the best material because it is completely noncombustible. A jacket of stainless steel provides the best protection.

9. *Very hot piping (450°F (230°C) or higher).* Calsil or mineral wool is the material of choice because of its superior qualities at higher temperatures. A plain jacket or a breather coating is a good covering. Insulation thickness should be determined to ensure a low surface temperature, if the piping is accessible to people.

10. *Cryogenic piping.* Four-inch (100 mm) thick polypropylene should be used.

11. *Steam and condensate piping.* Calsil or mineral wool should be used.

There are four specific problem areas that have been found to cause the most failures. First is where workers can walk on insulated pipe after it has been installed, for example, in boiler rooms and MERs. This will ruin any insulation in short order. Observation of seams no longer bound together is good evidence that this has occurred. Second is a failure to properly and completely seal a vapor barrier, particularly around valves and fittings. Water vapor will enter at these points and saturate the insulation very quickly. On occasion, the water will run along the pipe and drip far from the actual fault. Third is the failure to properly support the insulated pipe on a hanger. In order to keep the insulation from becoming compressed (causing tearing and a loss of insulating value), the weight of the pipe must be distributed by using a metal shield (if the pipe is small), a length of rigid insulation between the hanger and the insulated pipe, or, if the pipe and contents are large and heavy, a block of wood placed on the bare pipe (under the vapor barrier) to support the weight of the pipe at the hanger. Fourth is the installation of adhesives and self-sticking jacket seals under extremely dusty conditions, causing dust to be deposited on the adhesive before it is installed permanently.

FREEZE PROTECTION

FREEZING OF WATER

Much time, effort, and money must be spent to restore service when pipes and mains freeze. In addition, many times these efforts must be made under very difficult working conditions and with great urgency. Because of this, it is preferable to design systems that will not freeze.

Studies at the U.S. Army Cold Region Research and Engineering Laboratory (CRREL) have shown that the dynamics of the freezing process of water in pipes are much more complicated than originally thought, and that this process is different for static and flowing water. It was found that when water starts to freeze, flow in a piping system can become blocked much earlier than previously believed. It was also established that the actual mechanics of the freezing process are less predictable in terms of both time and heat loss than was previously believed.

Previously, it was believed that when ice formed inside a pipe or vessel, it started on the pipe wall and grew uniformly inward until the entire pipe was blocked. It was also thought that the only apparent difference between the freezing that occurred in static and flowing water was a difference in the rate due to the heat created by the flowing water.

THE MECHANICS OF THE FREEZING PROCESS

Static Water

When the temperature outside a pipe falls below freezing, before any ice can form, the water inside must be cooled below the freezing point. This is called supercooling, and in the CRREL studies the temperature fell as low as 29°F (-1.6°C) before ice started to form. The initial ice formation is called *nucleation*. As the process continues, the ice takes the shape of thin feathery crystals interspersed with water, similar in shape to a Christmas tree. This new formation is called *dendritic* ice. As the dendritic crystals rapidly grow larger, they release latent heat of fusion due to the change of state of the water. Since the surrounding soil or air cannot absorb all this heat, the temperature of the water is then raised back to the 32°F (0°C) level. This brings the dendritic phase to an end. Only after all the heat is absorbed by the surrounding medium does the annular growth of ice actually start. The dendrils become more and more dense and are gradually incorporated into the growing annulus. Eventually, the annulus increases in size, becomes solid, and occupies the entire cross section of pipe. The pipe and its contents will continue to cool until the ambient temperature is reached. The freezing of still water is illustrated in Fig. 5.1.

Flowing Water

Prior to the CRREL studies, it was believed that flowing water began to freeze in a pipe with the formation of annular ice along the inside of the pipe perimeter, assuming a tapered cross section and having the smaller end at the downstream

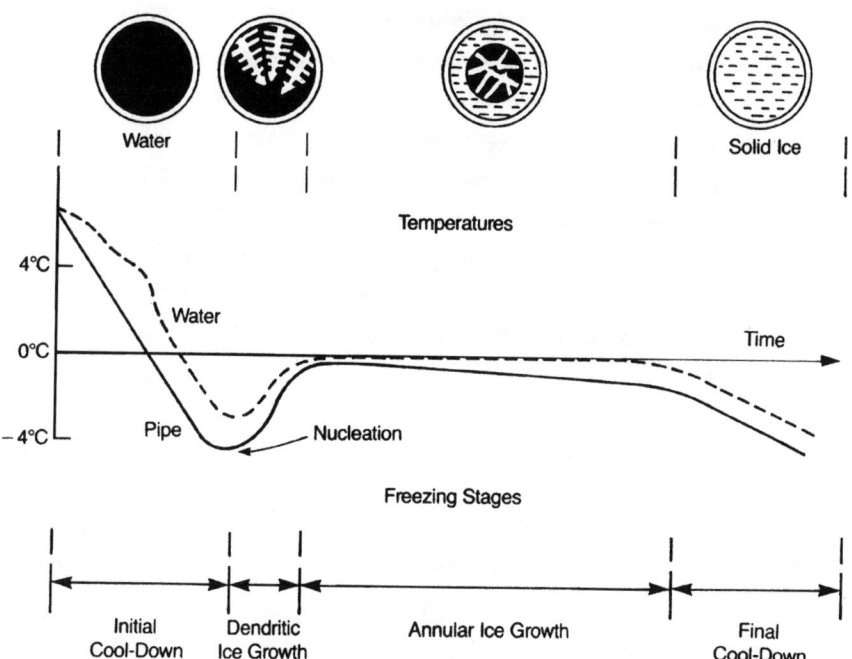

FIGURE 5.1 Freezing of still water.

side. Blockage was thought to occur when the smaller cross section thickened and eventually closed, filling the pipe at that point. As the ambient temperature dropped, the remainder of the water in the pipe froze until the entire pipe became filled.

In actuality, it was found that as the ambient temperature fell, an ice-water mixture was produced in the shape of rippled ice surfaces that moved along with the flow of water. This mixture was found to be in the shape of a taper in cross section with the smaller end pointed downstream. At regular intervals there was sudden enlargement. A further lowering of the ambient temperature did not lead to a thickening of the ice but rather, a progressively closer spacing of the narrow ice bands. The friction factor of the flowing water was found to be greatly affected by this type of ice formation. Since the friction head loss of pressure consists primarily of wall drag and the nozzle losses that occur at each band, the hydraulics become very unstable. Any further decrease in the discharge rate of the system or an increase in the friction loss due to more ice will quickly create a condition where the total head requirements for flow are greater than the source can provide. It is at this point that all flow would stop and the water would start to freeze immediately. The freezing of flowing water is illustrated in Fig. 5.2.

PIPE DAMAGE DUE TO FREEZING

Water falls into the category of substances that have a temperature of maximum density. Above that temperature it has a positive coefficient of expansion, which

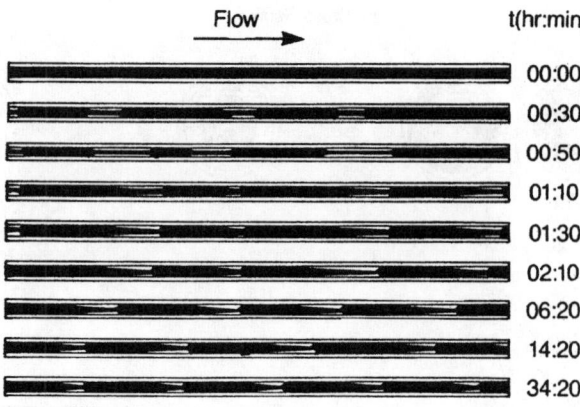

FIGURE 5.2   Freezing of flowing water.

means that water will expand upon heating. However, if water is cooled below that temperature, it has a positive coefficient of expansion since it will also expand when cooled. The relationship between temperature and volume is shown in Table 5.7. Because of this expansion, water (or ice) will exert considerable pressure when confined within any enclosed space. The relationship between temperature and pressure is shown in Table 5.8.

A mathematical analysis of the actual stress produced by the freezing of water in an enclosed space has shown that the circumferential stress exceeds the longitudinal stress in all cases. What this means in practical terms is that any pipe can be expected to fail along its length in the absence of a weak joint or a flaw in the piping material. In tests conducted at the National Bureau of Standards, it was observed that in all cases of failure due to freezing, liquid water was released from the pipe at the point of failure.

Applying these facts to a piping system, differences in heat transfer, caused by variations present in all piping systems, will result in the freezing of pipe in some locations but not others along the same run. This will cause a rupture in one or more of the unfrozen sections of the piping network that are the last to freeze.

TABLE 5.7   Relationship between Water Temperature and Volume

Temperature		Specific volume, mL/g	
°C	°F	Water	Ice
−10	14	1.002069	
−5	23	1.000825	
0	32	1.0001324	1.0908
3	37.4	1.0000078	
4	39.2	1.0000000	
5	41	1.0000081	
10	50	1.0002720	

Source: Data of J. F. Mohler, at 1 atm pressure.

TABLE 5.8 Relationship between Freezing Point and Pressure for Water

Freezing point		Pressure	
°C	°F	atm	psi
0	32	1	0
− 5	23	590	8660
−10	14	1090	16000
−22	−7.6	2047	30000

THE FREEZING OF WATER IN ATMOSPHERIC VESSELS

General

The most common method used to store water is in an on ground or elevated water storage tank made of either wood or steel. If enough heat cannot be added by means of new water flowing into the tank to keep the water from freezing, it must be added by some other means. In addition to the tank itself, the riser from grade up to the bottom of an elevated tank must be protected also. It is generally thought that adequate heating of any tank is almost as important as proper structural design.

The Mechanics of Ice Formation

Water in an open tank or vessel loses heat on all of the sides exposed to the atmosphere. For a tank on grade, the loss of heat into the surrounding soil is not as great due to the insulating factor of that soil. However, the greatest amount of heat is lost through the surface of the water. As the water cools, an internal circulation takes place because the surface layer of water cools faster than the deeper water. This causes the surface layer to become denser, resulting in that water migrating to the bottom of the tank and displacing the warmer water to the top. This circulation will continue to take place until the temperature of most of the water reaches about 32°F. When this occurs, the water is at its maximum density. At this point, an inversion takes place. The coldest water, at a lesser density, now rises to the surface. Continued cooling of the surface water further reduces its density, and the temperature rapidly falls to the point of freezing.

Ice generally appears first at the sides of the tank, then quickly forms a continuous layer over the surface. This same progression is true for similar installations such as pools and lagoons. A formula for the determination of ice formation on the surface of an open vessel has been developed. It considers the insulating effect that the forming layer of ice will have on the transfer of heat from the surface of the water to the air. One assumption is that the time is constant. The following formula calculates the thickness of ice formation in any period of time. The formula is:

$$X = \frac{25\ l}{32 - T} \qquad (5.8)$$

where X = time, h
  I = ice thickness, in (A generally accepted figure of $\frac{1}{16}$ in is required to stop operation of float valves, etc.)
  T = design air temperature, °F

A design air temperature value based on the mean of the high and low reading is considered adequate for design of tanks storing potable water. For tanks used to store fire protection water only, it would be advisable to use the lowest one-day mean temperature. The lowest one-day mean temperature can be found in Fig. 5.3.

Selecting the Heating System Type

Selecting the most economical type of heating system depends on the tank height, amount of heat required, and the availability and cost of any particular fuel or heating medium. The basic methods or devices are:

1. Direct discharge of steam into the water
2. Steam coils inside the tank
3. Hot water coils inside the tank
4. Electric immersion heating elements inside the tank

The direct discharge of steam into a tank is the method used most often for nonpotable water where steam is available. A steam supply line of adequate size is piped directly into the tank. This line ends in a tee placed about one-third of the height of the tank from the bottom. A condensate return line may be required, but it does not penetrate the tank.

When there is a potable water supply in combination with fire reserve water in a tank, the possibility of cross-contamination from an outside source must be minimized to the greatest extent possible. As a result, the following three methods should be considered.

The use of steam in coils at the bottom of a tank is mostly limited to tanks that have a flat bottom, are not elevated to a great height, and do not need too much heat transferred. Based on past experience, this method is not considered very reliable because a large number of problems have been reported.

Gravity circulation of hot water requires a heat exchanger or a hot water generator to be placed in close proximity to the tank, usually under the tank in the valve pit if provided or in a separate building adjacent to the tank. Cold water is taken from the discharge pipe and run through the heater. Since the now heated water is lighter than the colder supply, a natural circulation is obtained. Long runs of heater piping are not practical due to the heat loss from the piping and the expense of insulation needed to keep the heat from being lost. In addition, the hot water generator takes space in what might be a small valve pit under the tank.

The use of electric immersion heaters is practical only when the cost of the electricity is low. An advantage is the lowest initial cost of the heating methods discussed when used for potable water.

Calculating Heat Loss from Elevated Tanks

A method to determine the amount of heat lost from water tanks developed by the National Fire Protection Association (NFPA) has become the standard for design,

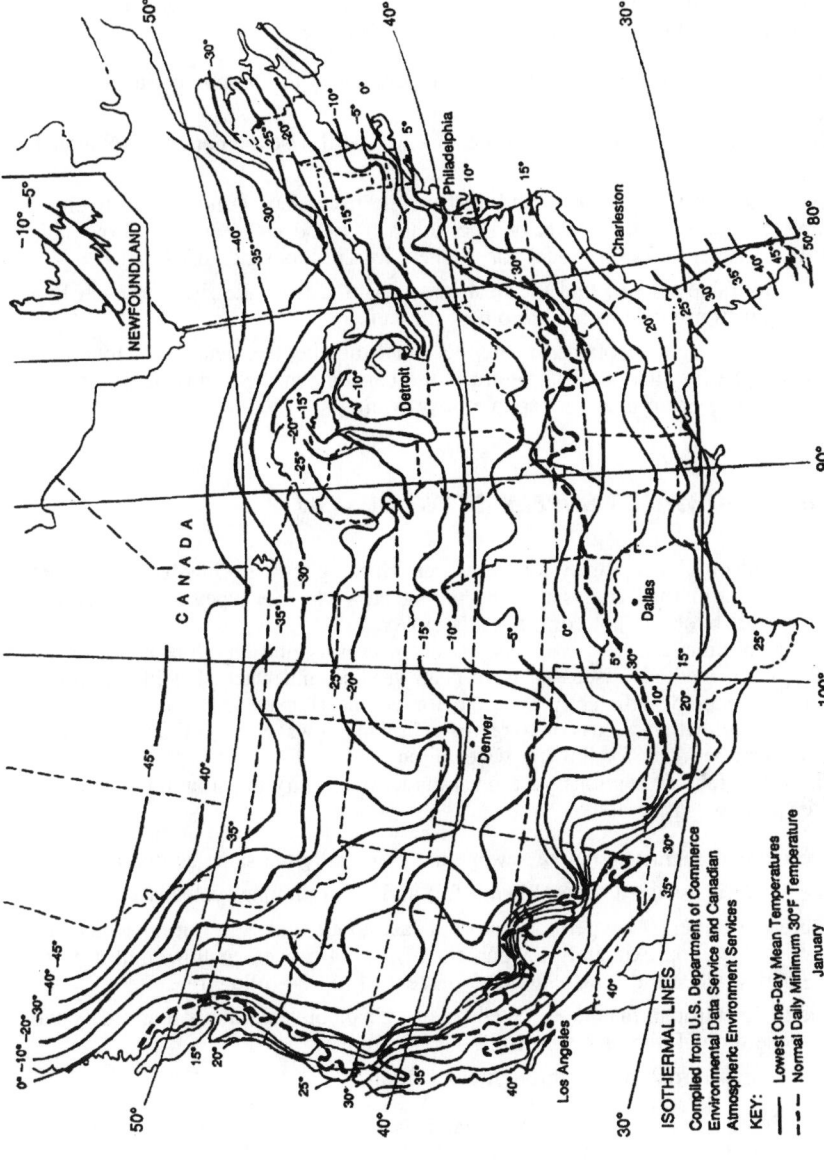

FIGURE 5.3 Lowest one-day mean temperature. (*NFPA*.)

NEWFOUNDLAND

CANADA

Detroit

Philadelphia

Charleston

Denver

Dallas

Los Angeles

ISOTHERMAL LINES

Compiled from U.S. Department of Commerce
Environmental Data Service and Canadian
Atmospheric Environment Services

KEY:

——— Lowest One-Day Mean Temperatures
— · — · Normal Daily Minimum 30°F Temperature
January

5.29

and is very conservative in its approach. The basic consideration is not to allow the temperature of the water to fall below 42°F. This is a safe temperature above freezing that makes allowance for tolerances in various equipment and for variations in weather design data.

The procedures required to calculate the amount of heat necessary to replace the heat that was lost to the atmosphere are as follows:

1. Determine the lowest one-day mean temperature at the proposed site. Refer to Fig. 5.3 for this information.
2. Determine the number of gallons the tank will hold. Use the nominal, not actual, capacity.
3. Using the lowest temperature and the capacity from steps 1 and 2, refer to either Table 5.9 (for elevated wood or steel tanks or wood tanks on grade) or Table 5.10 (for steel tanks on grade) for the heat loss in thousands of Btu/h.
4. Refer to Table 5.11 to find the heat loss in Btu/h from the riser for each foot of length from grade to the bottom of the tank.
5. Add the two figures obtained in steps 3 and 4 together to calculate the total heat loss for the installation. This figure is the capacity in thousands of Btu/h that the heating system must be capable of providing.

SEWER AND WATER SUPPLY PIPING DESIGN

Designing a complex piping system to prevent freezing is a very involved procedure that requires a thermal analysis of the piping network. The types of systems discussed in this book are not considered complex.

In current practice, precise analysis is not necessary and therefore is not made. The empirical methods discussed have been verified under actual field conditions and are considered sufficiently accurate for engineering design calculations. Although the following discussions are for a flowing sewage system, the same calculations can be made for flowing water piping.

The following assumptions have been made to simplify the solution of the heat transfer process. These are

1. The thermal characteristics of sewage are essentially the same as that of water.
2. The sewer is assumed to be flowing full and at a minimum velocity of 2 ft/s.
3. There are critical flow periods that will last about 6 to 8 h, during which the flow will be approximately 25 percent of the 24-h average flow, and a 1- to 2-h period when the flow is about 5 percent of the 24-h flow rate.
4. The losses through manholes, catch basins, cleanouts, and the like are the same as those through the pipe run itself.
5. Heat is transferred mainly through conduction.

A basic equation the steady state heat flow by conduction is used to calculate the heat loss from sewers. It is based on the Fourier law, and is stated:

$$Q = \frac{K}{X} \times A \times (T_1 - T_2) \tag{5.9}$$

TABLE 5.9 Heat Loss from Typical Elevated Wood or Steel Tanks or Wood Tank on Grade

Thousands of British thermal units lost per hour when the temperature of the coldest water is 42°F (4°C)

°F	°C	Wooden tanks—capacities, in thousands of gallons								
		10	15	20	25	30	40	50	75	100
35	2	8	10	11	13	14	19	21	28	33
30	−1	11	14	16	19	21	27	31	40	49
25	−4	15	20	21	25	28	36	42	54	65
20	−7	19	25	27	32	35	46	54	69	83
15	−10	24	31	34	39	44	57	66	85	102
10	−13	28	36	40	46	51	68	78	100	121
5	−15	33	43	47	54	60	78	92	117	142
0	−18	38	49	53	62	69	90	106	135	164
−5	−20	43	56	61	71	79	103	120	154	187
−10	−23	49	63	69	80	89	116	136	174	211
−15	−26	54	71	77	89	100	130	153	195	236
−20	−29	61	79	86	99	111	145	169	217	262
−25	−33	68	87	95	110	123	160	188	240	291
−30	−34	74	96	104	121	135	176	206	264	319
−35	−37	18	105	115	133	148	193	226	289	350
−40	−40	88	114	125	144	162	210	246	317	382
−50	−45	104	135	147	170	190	246	290	372	450
−60	−51	122	157	171	197	222	266	307	407	490

°F	°C	Steel tanks—capacities, in thousands of gallons							
		30	40	50	75	100	150	200	250
35	2	43	51	59	77	92	120	145	168
30	−1	62	72	83	110	132	171	207	242
25	−4	82	96	111	146	175	228	275	323
20	−7	103	120	139	183	220	287	346	405
15	−10	145	146	169	222	267	267	347	419
10	−13	147	172	200	263	316	411	496	582
5	−15	171	200	233	306	367	478	577	676
0	−18	197	231	268	352	423	551	664	779
−5	−20	224	262	304	400	480	626	755	884
−10	−23	253	296	344	452	543	707	853	1000
−15	−26	283	331	384	506	607	790	954	1118
−20	−29	314	368	427	562	674	878	1059	1241
−25	−33	348	407	473	622	747	972	1173	1375
−30	−34	382	447	519	683	820	1068	1288	1510
−35	−37	419	490	569	749	900	1171	1413	1656
−40	−40	456	534	620	816	979	1275	1538	1803
−50	−45	538	629	731	962	1154	1503	1814	2126
−60	−51	624	730	848	1116	1340	1745	2105	2467

TABLE 5.10 Heat Loss from Typical Steel Tank on Grade

°F	°C	Tank capacities—in thousands of gallons							
		100	150	200	300	400	500	750	1000
35	2	85	114	135	175	206	238	312	380
30	−1	121	162	193	248	294	340	445	543
25	−4	161	216	257	330	393	453	594	722
20	−7	202	271	323	414	493	568	745	907
15	−10	245	329	391	502	537	689	904	1099
10	−13	290	389	463	595	707	816	1071	1302
5	−15	337	452	539	691	822	949	1244	1514
0	−18	388	521	620	796	947	1093	1434	1744
−5	−20	441	592	705	905	1076	1241	1628	1981
−10	−23	498	669	797	1023	1216	1403	1841	2239
−15	−26	557	748	891	1143	1360	1569	2058	2503
−20	−29	619	830	989	1270	1510	1742	2286	2781
−25	−33	685	920	1096	1406	1673	1930	2532	3080
−30	−34	752	1010	1203	1545	1837	2119	2781	3383
−35	−37	825	1108	1320	1694	2015	2325	3050	3710
−40	−40	898	1206	1437	1844	2193	2531	3320	4039
−50	−45	1059	1422	1694	2175	2586	2984	3915	4762
−60	−51	1229	1651	1966	2524	3002	3463	4544	5528

TABLE 5.11 Heat Loss in Riser of Elevated Storage Tanks

Btu loss per hour through 4-ft-diameter (3.3 m) riser per foot-length

Atmospheric temperature, °F	°C	Btu/h
35	2	69
30	−1	192
25	−4	340
20	−7	506
15	−10	692
10	−13	893
5	−15	1092
0	−18	1309
−5	−20	1536
−10	−23	1771
−15	−26	2020
−20	−29	2291
−25	−33	2568

where Q = heat loss from pipe, Btu/h · ft
K = mean thermal conductivity, Btu/ft² · h · °F
T_1 = temperature at outside of pipe, °F
T_2 = ambient temperature at a distance from pipe that would not be affected by the heat of the pipe. (This would be either the ground temperature for buried pipe or the design temperature of the air.)

A = area, in square feet, normal to the direction of heat flow of a 1-ft long section of pipe. (Where there is insulation, the cylinder of insulation should be used to calculate the arithmetic mean of the area between the pipe wall and the outside of the insulation. Refer to Table 5.6.)

X = distance, in inches, from the outside of the pipe wall where T_1 is measured to where T_2 is taken.

Equation (5.9) has been used to create Fig. 5.4, which is a nomogram that shows the temperature drop of flowing water in a pipeline. Before Fig. 5.4 can be entered, the following values must be known:

V = velocity of flow, ft/s

H = heat-transfer coefficient (By field experience, a value of 6 has been found to be an average value.)

S = length of pipe run, ft

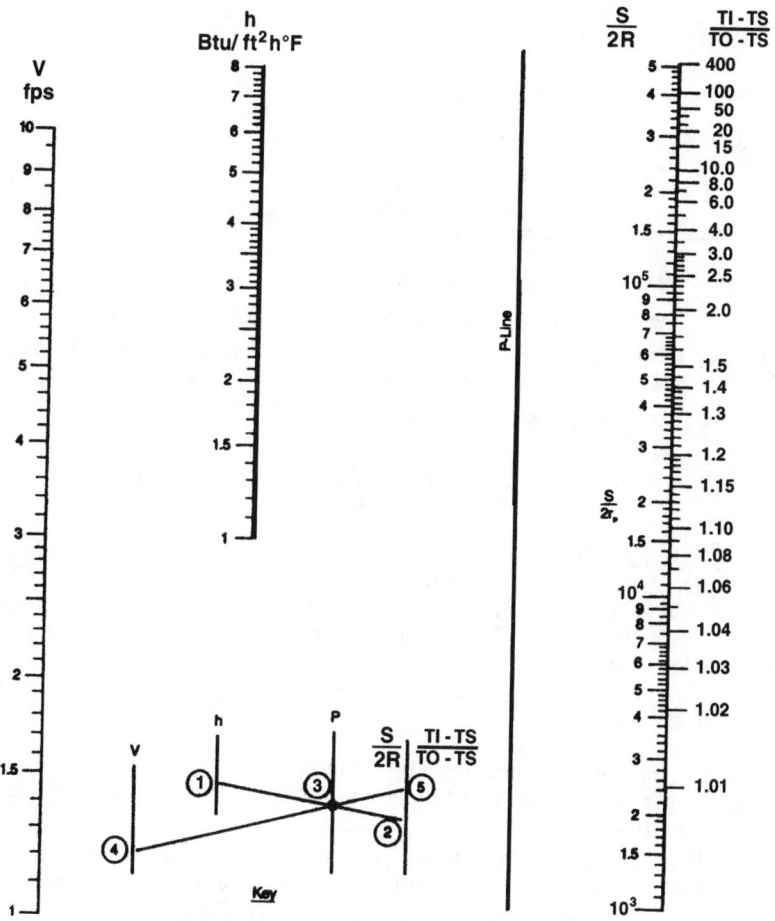

FIGURE 5.4 Temperature drop of flowing water in a pipeline.

R = radius of pipe, ft
TI = inlet water temperature, °F
TO = outlet water temperature, °F (It is recommended that a minimum temperature of 35°F be used to provide a safety factor to allow for variations in the thermal properties of soil along the run, intermittent flow in the pipe, and temperature fluctuations of both the air and ground.)
TS = ambient temperature of either air or soil around pipe (Soil temperature can be approximated by adding 5°F to the lowest monthly mean temperature.)

The following example illustrates the use of the nomograph in Fig. 5.4 to find the temperature drop in flowing water in a sewer, given the following conditions:

$$\text{Pipe diameter} = 12 \text{ in}$$

$$\text{Inlet water temperature} = 40°F$$

$$\text{Proposed velocity} = 2 \text{ ft/s}$$

$$\text{Length of run} = 2.2 \text{ mi}$$

$$\text{Estimated ground temperature} = 25°F \ (T_2)$$

Step 1.   Find

$$\frac{S}{2R} = \frac{2.2 \times 5280}{2 \times 0.5} = 1.16 \times 10^4$$

Step 2. $H = 6.0$ (from previous discussion).

Step 3.   Enter Fig. 5.4 and connect points 1 and 2 (6.0 and 1.16×10^4).

Step 4.   Using pivot point 3 as the anchor, connect points 3 and 4. On the other side of the line, read a value of 1.77. We will call this value Z.

Step 5.   Use the following formula to find TO.

$$Z = \frac{\text{TI} - \text{TS}}{\text{TS} - \text{TO}} \qquad 1.77 = \frac{40 - 25}{\text{TO} - 25} \qquad \text{TO} = 33.5°F \tag{5.10}$$

The recommended minimum temperature of water flowing in a sewer pipe is 35°F, which allows a small safety factor. The calculated temperature of 33.5°F is therefore considered unsafe. Using Fig. 5.4, and a temperature of 35°F as a guide, the velocity can be raised to 3 ft/s, or perhaps the inlet temperature raised to about 43°F by adding water.

To prevent the freezing of water in a sewer line, the following methods have proven successful:

1. Providing insulation around the pipe to limit heat loss. The addition of insulation will not prevent the freezing of the water in time, but will considerably delay it from occurring.

2. Providing an enclosure around the pipe in such a manner that the enclosure does not touch the pipe walls. This will provide additional insulation to the system.

3. Heat tracing the pipe with electric cable, steam, or hot water.

4. Providing sufficient velocity to the liquid so that it will reach its terminus without freezing.

5. Adding warm water at the origin of the sewer, or at several places along the run (if possible), to counteract the heat loss of the water to the surrounding soil.

FROST CLOSURE OF VENTS

During cold weather, exposed vents on a roof may become wholly or partially blocked by frost on the inside portion of the exposed pipe. This is due to the fact that in cold weather, a current of warm, moist air rises through the plumbing piping when there is little or no flow through the system. This upward flow of air is caused by the temperature difference between the air outside the building and the air inside the pipe. Since the air inside the building is warmer than the free air, the inside air is lighter, causing the upward current. This is the so-called chimney effect. When this warm, moist air reaches the chilled surface of the exposed pipe, moisture condenses on the colder surface of the pipe in the form of droplets. If the correct conditions exist, these droplets will freeze. If this continues long enough, the pipe will become blocked.

There are actually two phenomena that can occur. The more common is the formation of ice in the form of an annular ring in the pipe interior. Another kind of blockage takes the form of a frost cap on top of the pipe. This problem of frost closure was the subject of a study at the National Bureau of Standards in 1922. The conclusions resulting from that study are still valid today.

The best method to prevent frost closure is to increase the size of the vent pipe just below the roof level, allowing the warm air to bypass the side of the exposed pipe as much as possible. It was shown that a 4-in (DN 100) pipe would not close up except under the most adverse conditions. One of these conditions was observed in the vicinity of Niagara Falls, New York, where the spray from the falls solidified on the pipes regardless of their size. The solution was to run either an electric heat trace line or circulate hot water around the perimeter of the pipe to keep it warm. For more unusual conditions, it was found that keeping the vent pipe as low to the roof as possible was acceptable. Blocking by snow at that low height never occurred, because the snow was porous enough to pass air and had a tendency to melt very quickly.

DEPTH OF FREEZE IN SOILS

In order to prevent the freezing of water in underground piping systems, the pipes must be buried far enough below grade so that the soil used as backfill provides enough insulation from the air temperature to prevent the freezing process from starting. This depth is called the frost depth, which is the level to which the 32°F isotherm will penetrate.

There are various methods used to determine the frost depth in a particular area: the local authorities such as the building department or fire marshal for information and advice based on past experience; recommendations of fire insurance carriers; or the use of maps such as the one shown in Fig. 5.5.

The following discussions will show how to calculate the frost depth and how to obtain the required criteria in order to perform the calculations with an acceptable degree of accuracy.

DERIVATION OF THE BASIC FORMULA FOR FROST DEPTH

The U.S. Army Cold Region Research and Engineering Laboratory (CRREL) has developed a practical method for calculating the frost depth of soil. The method used to calculate frost penetration is based on heat transfer principles involving a phase change (from water to ice) of pore water, or water held in the voids of soil. The actual freezing process under these conditions is very complex and does not lend itself to a fixed mathematical solution. In order to reduce the complexities

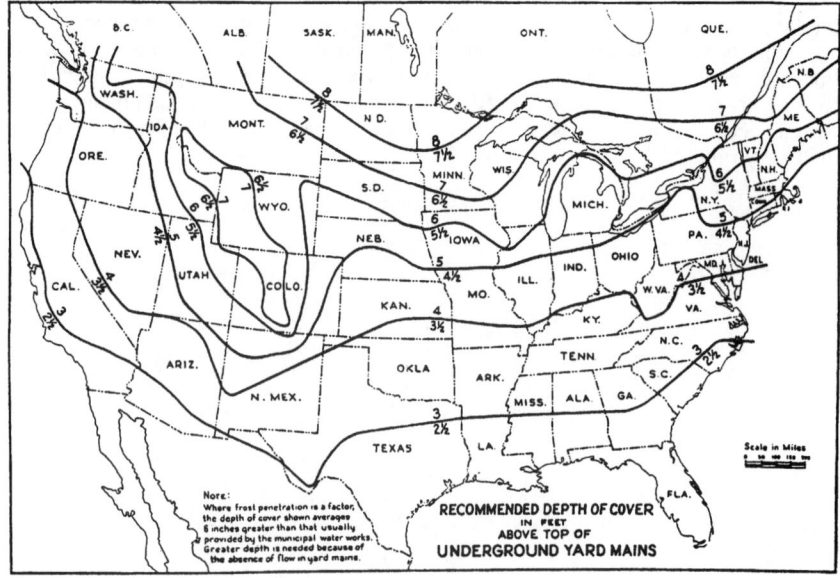

FIGURE 5.5   Approximate frost depth in feet (*NFPA*).

involved for the exact determination of each value that follows, several assumptions have been made. These assumptions have the effect of simplifying the solution, but also introduce some slight error. All of the following derivations have been verified under actual field conditions and found to be well within the accuracy necessary for engineering design purposes. These assumptions are:

1. The soil is considered homogeneous throughout.
2. All pore water is converted to ice at 32°F.
3. Average thermal properties are applicable.

The depth to which the 32°F isobar will penetrate can be calculated by using the modified Berrigan equation, which is:

$$X = C \sqrt{\frac{48K \cdot NF}{L}} \qquad (5.11)$$

where X = penetration of frost into soil, ft
C = coefficient, dimensionless (This is a function of the freezing index, the mean winter temperature at the site, and the thermal properties of the soil.)
N = conversion factor, air index to surface index
AF = air freezing index
SF = surface freezing index
K = thermal conductivity of the soil, Btu/ft²/h/°F
F = air freezing index, degree days Fahrenheit/year
L = volumetric latent heat of fusion, Btu/ft³

Each of the elements of the formula will now be discussed and defined.

1. *Coefficient C.* This is a general coefficient that considers the overall effect of the temperature change in the soil mass around a buried pipe. It is a function of the site freezing index, the mean annual site air temperature, and the thermal properties of the soil. To find this coefficient, two factors must be determined. They are the thermal ratio and the fusion parameter.

The thermal ratio TR is calculated from the following formula:

$$TR = \frac{ST}{AD} \qquad (5.12)$$

where TR = thermal ratio
ST = mean annual site air temperature, °F, minus 32 (The mean annual site air temperature can be obtained from the NOAA publication "Comparative Climatic Data." It is available only in tabular form and much too large to be reproduced in this book.)
AD = average surface temperature differential [This is found by dividing the surface freezing index (SF) by the number of days in the freezing season.]

The surface freezing index is derived from the formula:

$$SF = N \times AF \qquad (5.13)$$

The number of days in the freezing season, which is defined as the estimated number of days the temperature will fall to 32°F or below, can be found in Fig. 5.9.

The fusion parameter FP is calculated from the following formula:

$$FP = AD \times \frac{\text{volumetric heat capacity}}{\text{latent heat of fusion in soil}} \qquad (5.14)$$

AD is calculated from Eq. (5.13). Volumetric heat capacity is found in Fig. 5.6. The latent heat of fusion is found in Fig. 5.7. Table 6.8 gives the soil weight and average value for moisture content.

2. *Air ground conversion factor N.* This factor is necessary in order to approximate the temperature of the ground when only the air temperature is known. The combined effects of radiative, convective, and conductive heat exchange at the ground air junction have been considered in the determination of the actual value. Refer to Table 5.12 for values that have been established by experimentation under freezing conditions.

3. *Air freezing index AF.* The penetration of freezing temperatures into the soil is partially dependent upon the duration and magnitude of the temperature difference between the air and the ground. The lower the air temperature and the longer the freezing persists provide a cumulative increase in the penetration of frost. The

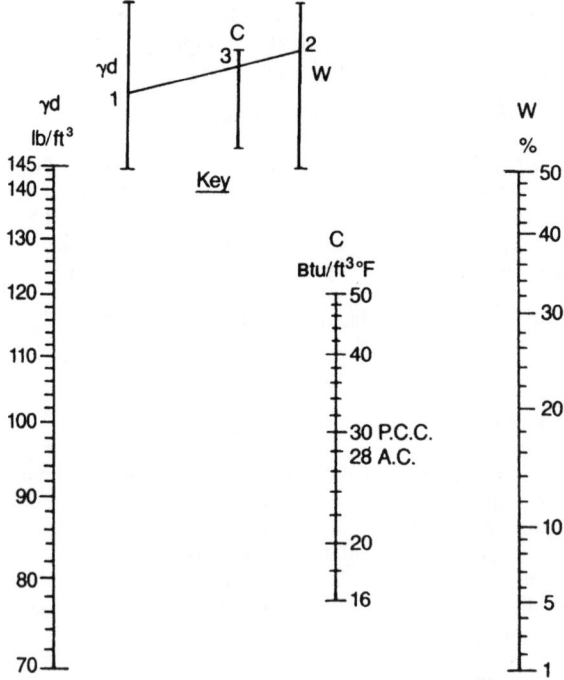

Note: Specific heat of soil solids assumed to be 0.17 Btu/lb.°F

FIGURE 5.6   Volumetric heat capacity for soils.

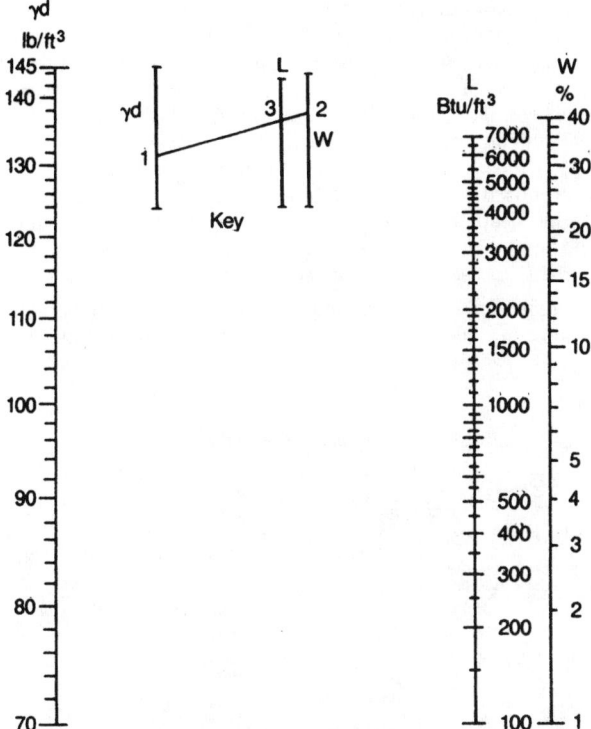

FIGURE 5.7   Latent heat of fusion for soils.

TABLE 5.12   Air–Ground Conversion Factor

Surface type	N Factor
Snow	1.0
Pavement free of snow and ice	0.9
Sand and gravel	0.9
Turf	0.5

method used to express the freezing temperature-duration measurement is in degree days Fahrenheit. In order to find the value, the high and low temperatures for any 24-h period are averaged. Then 32 is subtracted from that number to obtain the freezing index for that day. If it is found to be 10°F below freezing for one day, this amounts to 10 degree days. If the number calculated comes to 1°F below freezing, and if this continues for 10 days, the figure would also be 10 degree days. The freezing index should not be confused with the normal degree-day measurement used by HVAC engineers for heating calculations.

Four separate methods have been established to determine the freezing index, each based on the number of years that weather information has been available. The first uses the single coldest year in a 30-yr period. The second uses an average of the three coldest years in a 30-yr period. The third method calculates the average of the five coldest years in a 30-yr period. The fourth method, called the design freezing index, uses the single coldest year in a 10-yr period of time. The relationship between all three of these methods appears in Fig. 5.8. Isobaric maps of the North American continent show the mean freezing index, illustrated in Fig. 5.9, and the design freezing index, illustrated in Fig. 5.10. With all of this information available, what is the "correct" value? The best possible information is data obtained at the proposed site or the records of the nearest U.S. National Weather Service first order weather station. If obtaining this information is not practical, use Fig. 5.9 to find the mean index, then refer to Fig. 5.8 to obtain a value. Generally accepted practice is to use the coldest index possible.

4. *Thermal conductivity K.* Thermal conductivity is a measure of heat flow through a substance, given as Btu/ft of material per unit of time per degree tem-

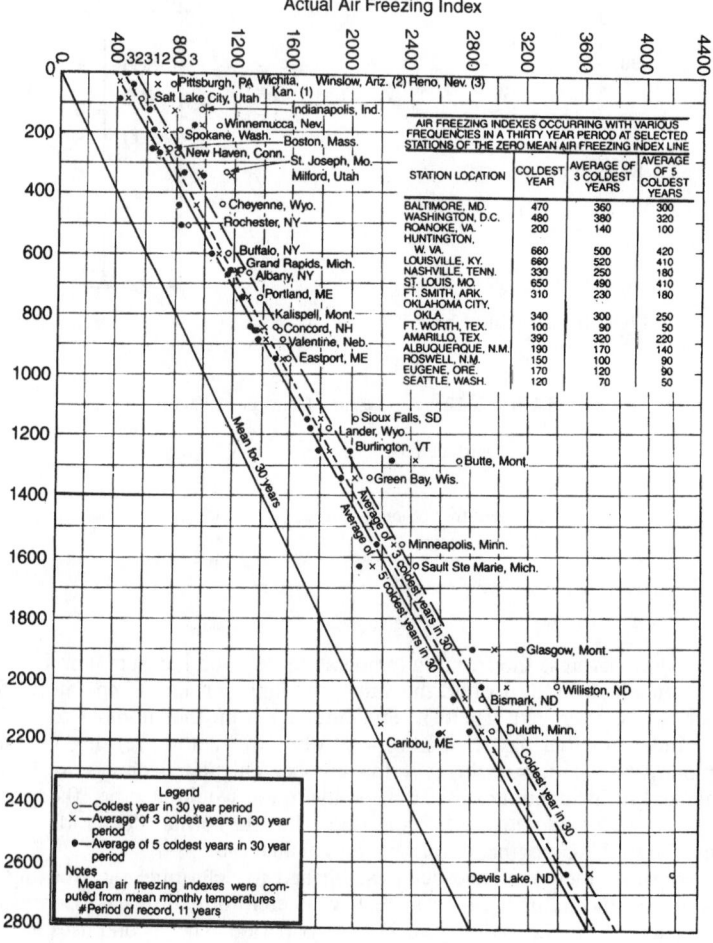

FIGURE 5.8 Relationship between mean and other freezing indexes.

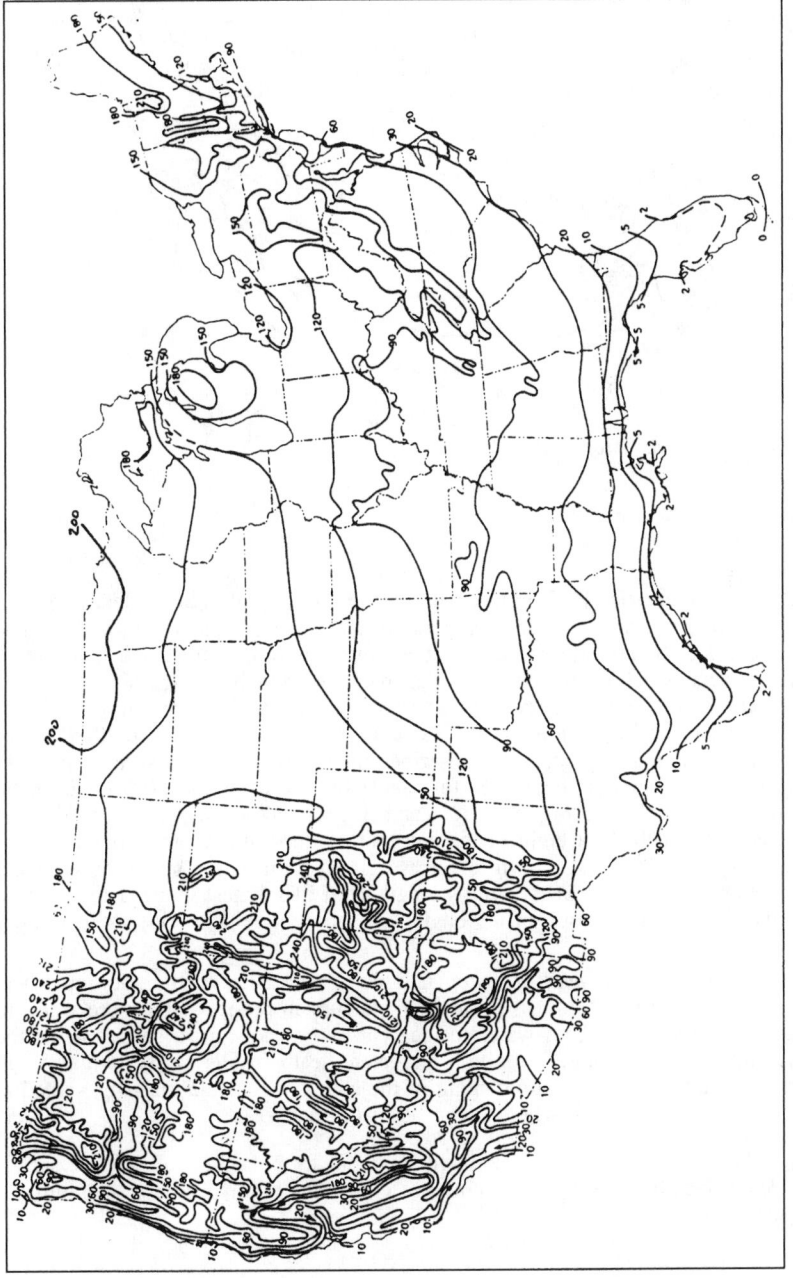

FIGURE 5.9 Mean freezing index. Mean annual number of days with minimum temperature 32°F and below.

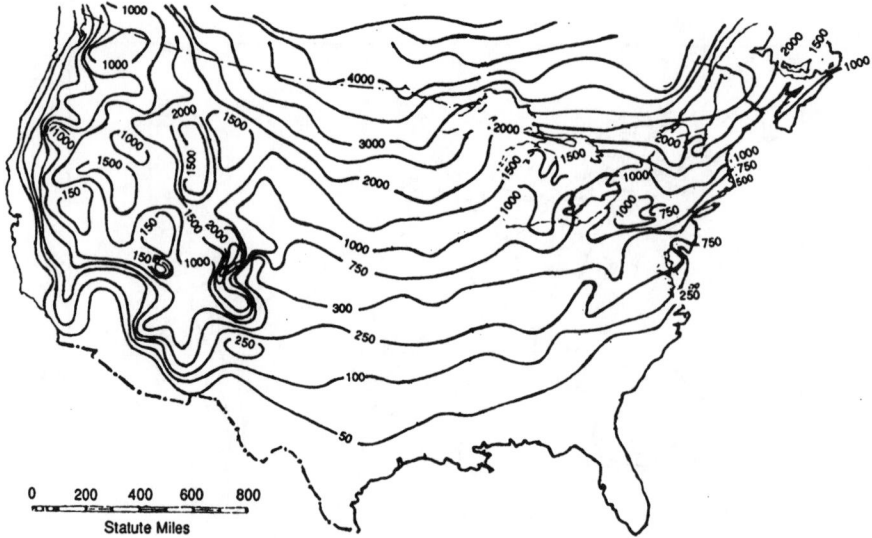

FIGURE 5.10   Design freezing index.

perature difference (in °F). The thermal conductivity of soil is dependent upon soil density, actual moisture content, temperature and state of pore water, and particle shape. Field tests have determined that the average values presented here are of sufficient accuracy for engineering design purposes. For this discussion only, soil groups are divided into three general categories, sand, silt with clay, and peat. Table 6.8 lists the soil weights used in the calculations following. Table 6.20 separates soils into unified soil groups and describes their composition. Average moisture content of these soils is about 15% for sand, 30% for clay, and 80% for peat. To find the value of K, use Fig. 5.11 for sand, Fig. 5.12 for clay, and Fig. 5.13 for loam, entering with the weight of the soil and the moisture content. If the project under design has building construction associated with it, the structural engineer will have the soil tested for bearing strength. Part of this test is the determination of moisture content. A copy of the engineer's report should provide the necessary information.

5. *Volumetric latent heat of fusion L.* This is a measure of the additional heat that must be absorbed by the surrounding medium in order for the pore water to change its phase state from water to ice. The more moisture a soil contains the higher the L value. To find L, multiply the latent heat of water (144) by the weight of the soil and by the moisture content of the soil. This calculation is presented in graphic form in Fig. 5.14, where L can be read directly.

To illustrate the use of the modified Berrigan formula, the following example will determine the frost depth for a project in Green Bay, Wisconsin, with the following conditions:

Mean annual temperature   43.7°F (from NOAA)
Length of freezing season   200 days (obtained from Fig. 5.10)

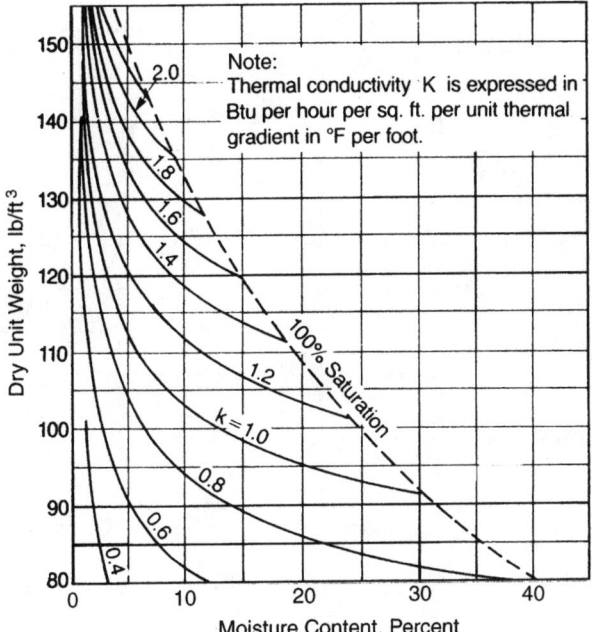

FIGURE 5.11 Average thermal conductivity for sandy soils.

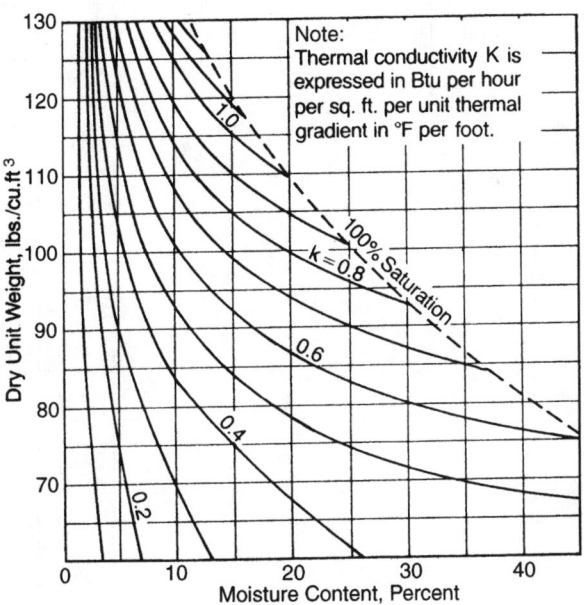

FIGURE 5.12 Average thermal conductivity for silt and clay soils.

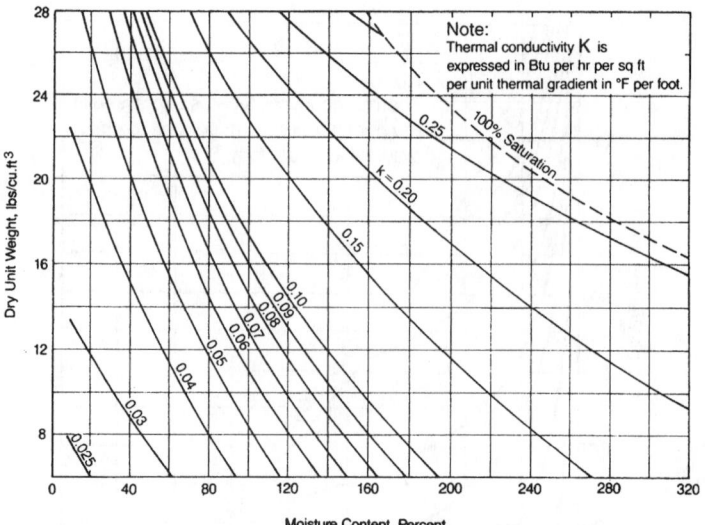

FIGURE 5.13 Average thermal conductivity for peat.

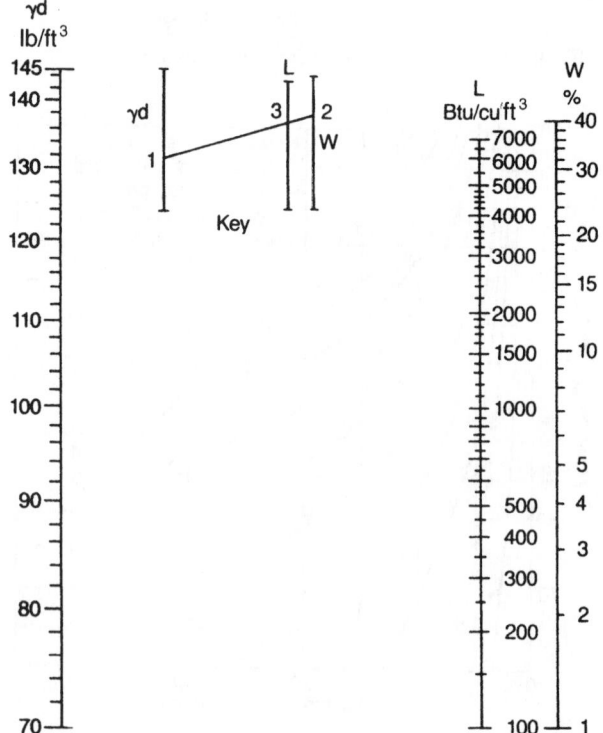

FIGURE 5.14 Volumetric latent heat of fusion.

Soil properties	Sandy soil, snow cover
Weight	100 lb/ft³
Moisture content	15%
Latent heat of fusion	2160 Btu/ft³ (obtained from Fig. 5.14)
Thermal conductivity	1.01 Btu/h/ft²/°F (obtained from Fig. 5.13)
Surface freezing index	N 3 AF = 2150 degree days. (AF is obtained from Fig. 5.9, coldest reading; N is obtained from Table 5.12, snow cover; therefore, 2150 × 1.0 = 2150 surface freezing index.)

To obtain the coefficient C refer to Fig. 5.15. Enter with the fusion parameter and thermal ratio. To find the fusion parameter [FP in Eq. (5.14)], first calculate AD:

$$AD = \frac{2150}{200} = 10.7$$

Therefore $$FP = 10.7 \times \frac{28.2}{2160} = 10.7 \times 0.013 = 0.14$$

To find the thermal ratio [TR in Eq. (5.12)] calculate the following:

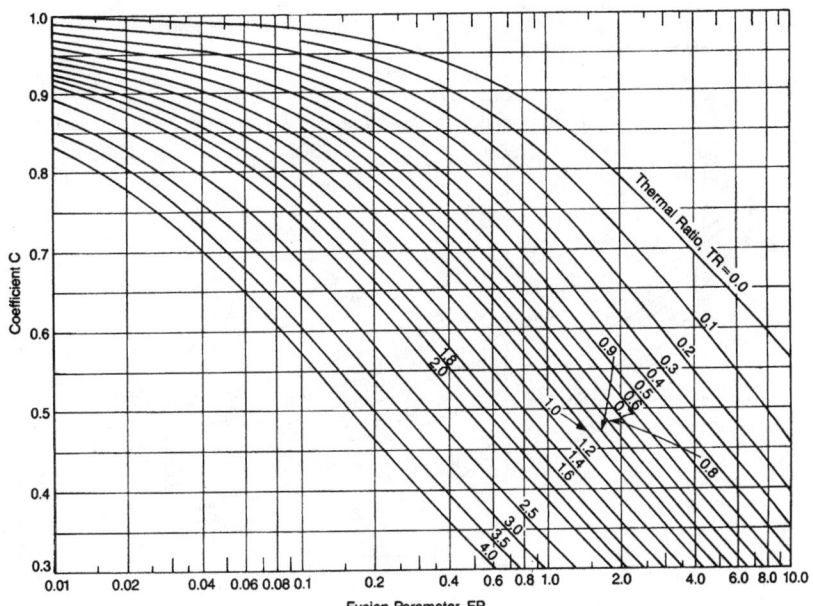

FIGURE 5.15 Coefficient in the modified Berrigan formula.

$$TR = \frac{ST}{AD} = \frac{(43.7 - 32)}{10.7} = \frac{11.7}{10.7} = 1.09$$

It is now possible to enter Fig. 5.13 with 0.14 and 1.09. Read $C = 0.8$. This gives us all the information needed to solve the modified Berrigan formula, Eq. (5.11),

$$X = 0.8 \sqrt{\frac{48 \times 1.01 \times 2150}{2160}} = 0.8 \sqrt{\frac{104.232}{2160}}$$

$$= 0.8 \times \sqrt{48.25} = 0.8 \times 6.94 = 5.55 \text{ ft}$$

Estimated depth of frost penetration is 5.55 ft.

Several nomographs have been prepared based on the modified Berrigan formula that are easier to use than the original formula. The first is shown in Fig. 5.16, giving the frost depth for various conditions as a direct reading.

In some cases, a pipe may have been buried under a road and a nonsusceptible fill used. This means that the fill will not be affected by frost, and so the condition known as frost heave will not occur. Use Fig. 5.17 to find the depth of fill that will prevent frost penetration below the fill. For a direct reading of the actual frost penetration under pavements, refer to Figs. 5.18, 5.19, and 5.20. Use the combination of soil weight and moisture content applicable to the project under design.

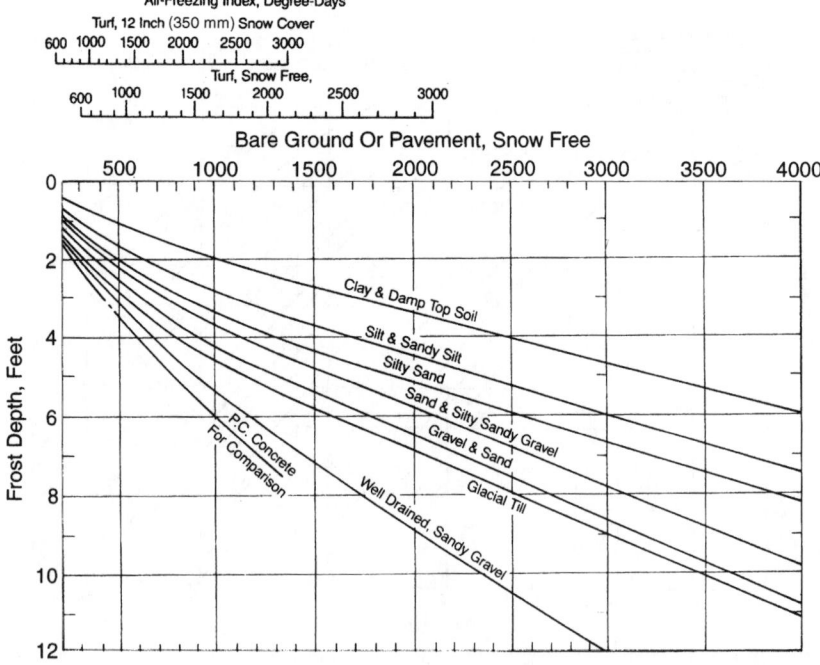

FIGURE 5.16 Nomograph to determine frost depth using modified Berrigan formula.

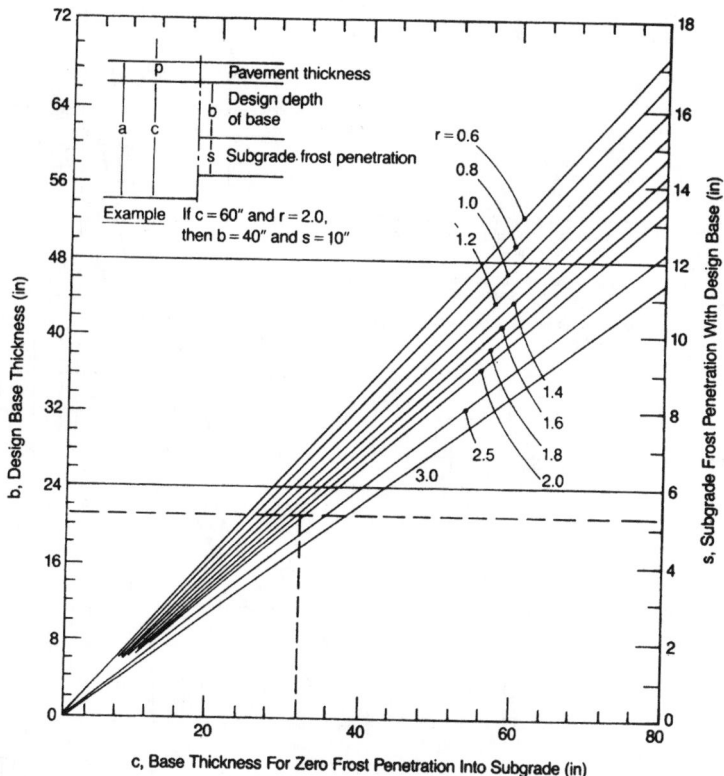

FIGURE 5.17   Frost penetration in a nonsusceptible base.

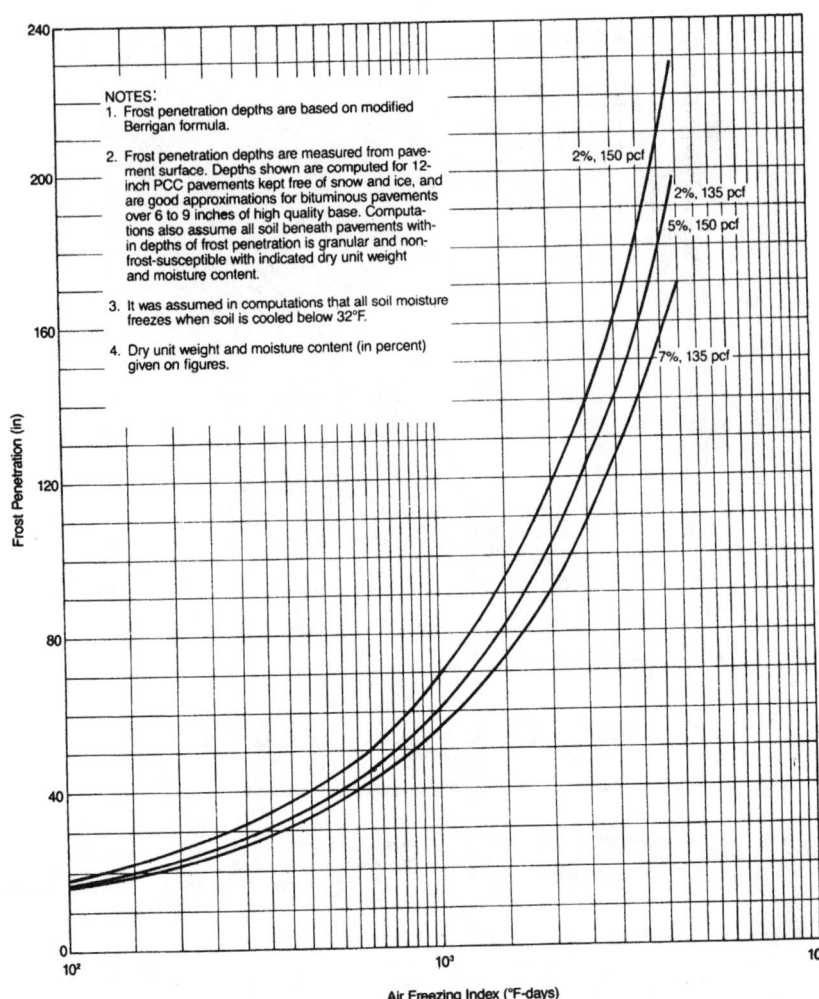

FIGURE 5.18 Frost penetration beneath pavement: 135 to 150 PCF base material.

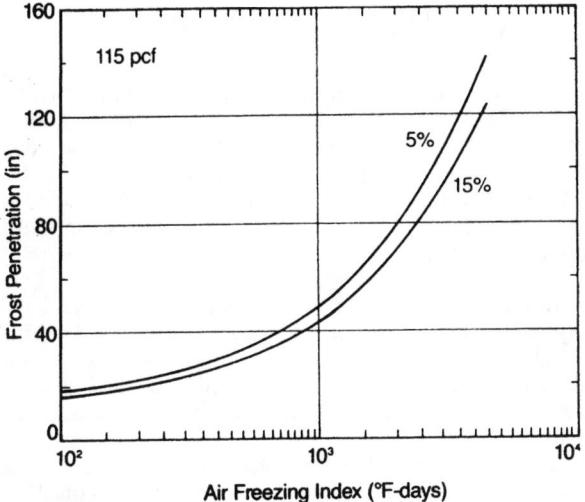

FIGURE 5.19   Frost penetration beneath pavement: 115 PCF base material.

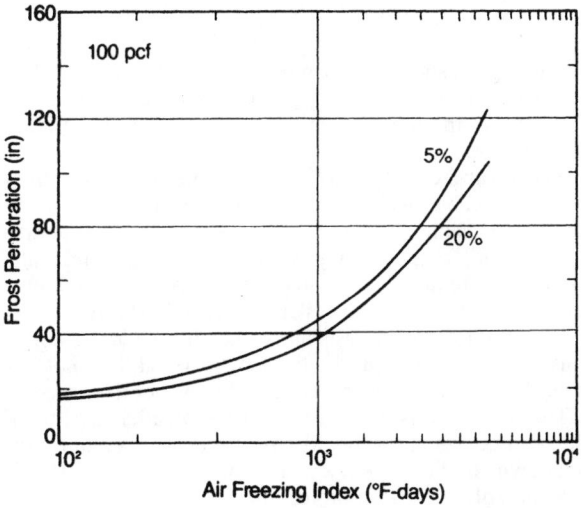

FIGURE 5.20   Frost penetration beneath pavement: 100 PCF base material.

HEAT TRACING

Heat tracing is the continuous or intermittent application of heat to a pipe or vessel in order to replace the heat lost to ambient air. Heat tracing is used for preventing freezing, for thawing, for maintaining the temperature of pipe contents, and for facilitating product transport in a pipe (for example, by increasing the viscosity of Number 6 fuel oil). Heat tracing used for domestic hot water temperature maintenance is discussed in Chap. 9, Plumbing Systems. This section will be concerned with freeze protection. Other aspects of this subject are outside the scope of this book.

There are two broad classes of heat tracing methods, electric and fluid. Electric heat tracing systems convert electric energy into heat. Fluid methods generally utilize water or steam at an elevated temperature to transfer heat from one pipe to another. Fluid heating media are usually contained in a small tube or pipe directly attached to the pipe being protected. Of the two fluids, steam is the more prevalent method of heat tracing.

Steam is more costly to install and maintain than electrical resistance. Periodic steam leaks and failed steam traps waste energy and require constant maintenance. In addition, a steam tracer produces 2 to 10 times more heat than is actually required for most applications. Choosing the most appropriate and cost-effective option for any specific application involves many factors. If both systems are available, Table 5.13 can be used as a checklist of the most important information needed to decide which option to choose. In general, if steam is not present in sufficient quantities to provide the necessary heat, it is almost always more practical and far less costly to use electric heat tracing.

ELECTRIC HEAT TRACING

General

This type of system consists of a heat-producing conductor (or cable), a controller to sense the temperature of the air or pipeline, and a relay to turn on the current. As the electricity flows through the cable, the resistance of the conductors causes the wire to become hot.

There are three general categories of heating cable. The first is mineral insulated, series circuit-resistant cable, which has a constant heat output for its full length. It has a metal sheath for cable protection. It is used when close temperature control is necessary, the liquid (or pipe) may reach a very high temperature, protection from physical abuse is required, the cable may be placed in a wet or moist environment, or close supervision of the system is needed. The second type is constant wattage, parallel circuit-constant resistant cable. It should be considered for use when long runs of cable are required, the length of cable is not known, or close monitoring is necessary. Care must be exercised so as not to exceed the temperature rating in use. The third type is the self-regulating, parallel circuit-variable resistant cable. It could be used when the liquid has a wide temperature range, the length of cable is unknown, or liquid temperatures will not exceed 185°F. All three of these cables are suitable for freeze protection.

In order to design and specify an efficient and cost-effective system, the following information should be obtained or calculated:

TABLE 5.13 Checklist for Steam and Electric Heat Tracing

	Electric	Steam
Preexisting project constraints		
Heat tracing recommended in company engineering practices (Yes/No)		
Other		
Project design phase criteria		
Total feet of traced pipe		
Cost of electricity, $/kWh, and steam energy ($/1000 lb)		
Estimated annual maintenance cost per foot of tracing, $/ft		
Number of heat-tracing circuits required		
Needed accuracy of temperature control, degrees		
Temperature control cost per circuit for needed accuracy, $		
Cost of monitoring one circuit in distributed control system, $		
Design time per circuit, h		
Design tools available for engineers (Good, Fair, Poor, None)		
Capital cost of steam capacity, condensate return, etc., $		
Capital cost for required electrical power, $		
Other		
Project installation criteria		
Labor cost to install tracing and accessories, $/ft		
Training time per plant laborer, h		
Labor and overhead costs, $/h		
Total installed heat-tracing costs, $/ft		
Other		
Project operation criteria		
Annual maintenance required, h/ft		
Annual cost for replacement parts, $/ft		
Annual energy cost, $/ft		
Total annual maintenance cost, $		
Total annual operating cost, $		
Other		

1. Minimum temperature of the fluid to be maintained
2. Ambient temperature of the area where installed
3. Wind velocity of the area where installed, if outdoors
4. Pipe or vessel size and material
5. Tolerance of components allowed
6. Type and thickness of insulation over heating cable
7. Whether monitoring of operation will be required
8. Whether an alarm or enunciating of alarms will be required in the area in which the cable is installed
9. Area where the cable is to be installed
10. Type of temperature control, if any
11. Whether a dedicated or emergency power supply will be needed

These factors are discussed in detail in the following paragraphs:

1. The minimum temperature to which water should be allowed to fall is 40°F. This is a safe temperature, which provides a safety factor to allow for variables over which there is no control, such as new record low temperatures, higher than expected wind velocities, uneven heat distribution of the cable selected, and the types of control devices used.

2. The ambient temperature in the area where the heating cable is to be installed can be obtained from Fig. 5.3. This is the mean low temperature, not the absolute low.

3. A wind velocity of 20 mph (9 mls) (12 km/hr) has been selected as an average value. For indoor installations, the requirements for heating can be reduced by 10 percent. Conversely, if a velocity of 40 mph is used, the value should be increased by 10 percent.

4. Pipe or vessel size speaks for itself. The material of the container affects the transmission of heat from the cable to the liquid. The charts provided here are based on metal. When plastics are used, the heat produced must be increased by 30 percent. This value is acceptable for pipes up to 6 in and most plastic tanks. Be careful to check the surface temperature of the cable and make sure the plastic material is capable of withstanding that temperature without harm.

5. Tolerance of various control elements, such as sensing devices (used to detect the temperature of liquids) and control thermostats, is important if the liquid temperature must be closely maintained.

6. The type and thickness of the insulation is an important factor in the heat loss from the pipe and cable. The charts provided are based on fiberglass. If another type of insulation is used, refer to Table 5.6 for the insulation factor.

7. Special monitoring of the operation of the heating cable system will affect the selection of the heating cable type. If close control is required, the use of an ammeter and an adjustable current sensing monitor (set to between 3 and 20 percent of base current) will be required. If a self-regulating system is used, there will be no base current since the system current varies. Therefore, the only kind of monitoring that could be used would be the continuity type, which detects a break in the conductor.

8. If an alarm is desired, the type (for example, a bell, a light, or both) and the location must be chosen.

9. The area in which the cable will be installed will affect the choice of cable. The conditions of the area, such as whether the cable will be immersed or merely become wet, whether there is very high humidity or not, whether the location is hazardous or subject to explosions are very important when specifying the system.

10. The type of temperature control desired will affect the specific device used to turn the system on. There are two possible modes of sensing: the air temperature or the surface temperature of the pipe. Although there may be some special case in which the air temperature is most important, sensing the actual temperature of the liquid in the pipe is usually much more significant. There are two types of sensors, capillary-bulb and electronic. Both of these types can be used for either application, but the electronic device is selected mostly for surface measurement. The bulb has a tolerance of 2 percent full scale. The electronic type is very accurate, on the order of hundredths of a degree.

Sensing bulbs generally have a tolerance of about 4°F, and a repeatability error of about 1°F. The longer the distance from the bulb to the actual controller, the greater the possibility of error. Therefore, it is not good practice to use these types of sensors if the distance is longer than 10 ft.

The electronic, or solid-state, sensor is extremely accurate, far beyond the normal requirements of a freeze protection system, but has the advantage of maintaining the programmed accuracy for a distance of longer than a mile, if necessary.

11. Very often, the line being heat traced is a critical one and must be assured of continuous operation. The options available are an emergency power supply or two separate supplies each connected to a different circuit.

System Design Procedure

The following example will describe the design procedure for an electric heat tracing system to provide freeze protection for an exterior potable water pipe. The installation will be in the city of Chicago. The pipe will be installed on an exposed roof, run for 100 ft, be 3 in in size, be made of copper, and have 1-in mineral fiber insulation. There is one valve and 10 supports.

1. Determine the true heat loss from the pipe and insulation. Referring to Fig. 5.3, the low temperature on the map is −15°F (−5°C). The water temperature is to be maintained at 40°F. This is a difference, or ΔT, of 55°F (−26°C). Referring to Table 5.3, using the insulation type and thickness, pipe size, and temperature difference of 50°F (30°C) (closest to 55), read 4.4 watts per foot (W/ft) of pipe. Because of the additional 5°, add another 0.6 W to make the value 5.0 W/ft. Since cellular glass is being used and not fiberglass (which the chart is based on), refer to the insulation correction factor in Table 5.3. For mineral fiber, read 1.20. Therefore, to find the true heat loss, multiply 5.0 × 1.20 to calculate the true loss of heat from the pipe and insulation together of 6.0 W/ft of pipe. This is the minimum amount of heat the cable must be capable of producing.

2. Select the cable type. With the true heat loss calculated, the type of environment where it will be installed, and the maximum temperature that the cable might attain, consult a manufacturer's catalog and select a cable type.

3. If the proposed cable is not capable of providing the necessary heat, for example, the cable selected is capable of producing only 4.0 Btu/ft, there are three options available:

A. Select a thicker or different insulation.

B. Install two cables to give the required heat.

C. Spiral the heating cable around the pipe.

The spiral option will be selected. To design for the spiral condition, the following steps are required:

(1) Determine the required length of cable, since some cables have a maximum permitted length.

(2) Starting with a run of 100 ft, add the additional cable required because of spiraling, valves, and supports. To find the spiraling factor, divide the required heat by the available cable heat, $6 \div 4 = 1.5$. Thus, for a cable capable of 4 W/ft if installed straight to now give 6 W/ft, more than 1 ft of cable must be placed on 1 ft of pipe. This is done by spiraling the cable around the outside of the pipe. The figure just calculated (1.5) is the actual number of feet of cable per foot of pipe that will give the required amount of heat. Therefore, we must add 50 feet for spiraling.

(3) For installation purposes, use Table 5.14 to find the spiral pitch factor, which is the actual installed distance between cable spirals. Enter with the calcu-

TABLE 5.14 Spiral Pitch Factors P

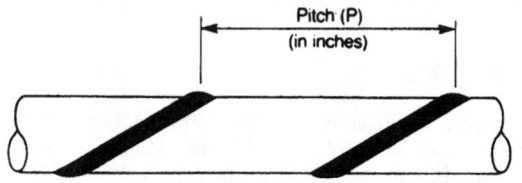

DN	Nominal pipe, in	Feet of pipe tracing per foot of pipe						
		1.1	1.2	1.3	1.4	1.5	1.6	1.7
13	0.50	5.8	4.0	3.2	2.7	2.4	2.1	1.9
20	0.75	7.2	5.0	4.0	3.4	3.0	2.6	2.4
25	1.00	9.0	6.2	5.0	4.2	3.7	3.3	3.0
32	1.25	11.4	7.9	6.3	5.3	4.7	4.2	3.8
40	1.50	13.0	9.0	7.2	6.1	5.3	4.8	4.3
50	2.00	16.3	11.2	9.0	7.6	6.7	6.0	5.4
65	2.50	19.7	13.6	10.9	9.2	8.1	7.2	6.6
75	3.00	24.0	16.6	13.2	11.2	9.8	8.8	8.0
90	3.50	27.4	18.9	15.1	12.8	11.2	10.1	9.1
100	4.00	30.8	21.3	17.0	14.4	12.6	11.3	10.3
125	5.00	38.1	26.3	21.0	17.8	15.6	14.0	12.7
150	6.00	45.4	31.4	25.1	21.2	18.6	16.7	15.1
200	8.00	59.1	40.8	32.6	27.7	24.2	21.7	19.7
250	10.00	73.7	50.9	40.7	34.5	30.2	27.0	24.6
300	12.00	87.4	60.4	48.2	40.9	35.8	32.1	29.1

lated spiraling factor (1.5) and the pipe size (3 in), read 9.8 at the intersection. That is the separation, in inches, that the cable should use when wound around the pipe.

4. If spiraling is not necessary, refer to Table 5.15 for the additional valve losses, in Btu. Multiply the factor found by the Btu heat loss from the pipe.

5. To find the additional length of cable needed for valves, fittings, and supports, refer to Table 5.16. Entering with the pipe size, find 3 ft for the valve and three times the diameter of the pipe for each support. Ten supports multiplied by 9 in equals 90 in, or about 8 ft. Adding all of the above together:

Basic run of pipe	100 ft (33 m)
Spiraling	50 ft (17 m)
Valve	3 ft (1 m)
Supports	8 ft (2.0 m)
Total	161 ft (53 m) of cable required

Heat Tracing for Indoor Tanks

The information presented previously applies to tanks as well as piping. The pitch most commonly used for winding the cable around a tank is two revolutions per foot length of the tank. The design procedure is as follows:

1. Find the area of the tank in square feet. For a square or rectangular tank, multiply the length, width, and height dimensions of the tank. For a round tank, refer to Table 5.17.

2. Determine the difference in temperature between the tank wall and the ambient air (ΔT).

3. Select the thickness and type of insulation.

4. Refer to Fig. 5.21 to find the heat loss in watts per square foot from the tank. If the insulation is other than fiberglass, use the correction factor in Table 5.6 to find the actual heat loss. If two lengths of cable are used for each square foot of the tank area, remember to divide the heat requirements in half to determine the watts per foot of the cable itself.

5. Calculate the amount of heat necessary to replace that lost and select an electric cable capable of providing that amount.

TABLE 5.15   Valve Heat Loss Factor

Valve type	Heat loss factor
Gate	4.3
Butterfly	2.3
Ball	2.6
Globe	3.9

Courtesy: Raychem.

TABLE 5.16 Cable Footage Allowance for Valves and Fittings

	Nominal pipe size, in								
	0.5	0.75	1.0	1.5	2.0	2.5	3.0	4.0	6.0
Screwed or welded	0.5	0.75	1.0	1.5	2.0	2.5	3.0	4.0	7.0
Flanged valves*	1.0	1.5	2.0	2.5	2.5	3.0	3.5	5.0	8.0
Butterfly	0	0	1.0	1.5	2.0	2.5	2.5	3.0	3.5
Flanges	Use two times the nominal pipe diameter								
Pipe supports	Use three times the nominal pipe diameter								

*Valves include: gate/globe wedge plug.

Note: If cables are spiraled on pipe, use the amount of cable shown times the spiral factor. Example: 2-in gate valve—(screwed)—2.0 ft. Spiral factor 1.1 ft/ft of pipe 2 × 1.1 = 2.2 ft.

Courtesy: Smith Gates Corp.

Safety Factors

Various safety factors have been included as part of the design criteria. The water maintenance temperature of 40°F is used, which gives a safety factor of 8°F. Table 5.17 has a built-in safety factor of 10 percent. Additional safety factors are not necessary or economical.

STEAM TRACING

For several reasons, steam is the most practical and economical method to provide heat to piping systems. Being a gaseous vapor, steam is easy to distribute—it requires no pumping, and because it is under pressure, it can be piped to remote locations in lines of relatively small diameter.

A steam tracing system consists of a pipe of small diameter (carrying the steam) attached to the outside of the pipe being protected, a connection to an adequate steam supply, a pressure-reducing valve assembly from the steam supply if required, a control valve or device to turn the steam on and off or modulate the amount of steam at predetermined set points, and a steam condensate return and disposal system (if desired). A simplified detail of a typical steam tracing system is illustrated in Fig. 5.22.

Another type of steam heating for a pipeline is a steam jacket, in which the steam for heating is introduced into the space between the pipe to be protected and an outer jacket placed around the pipe for this specific purpose. This is a very costly method, used generally for precise product temperature maintenance, and is outside the scope of this book.

General

It is generally accepted practice to install tracing lines on horizontal runs symmetrically at the bottom of the pipe being protected. Typical placement of tracer pipes

	Nominal pipe size, in								
	8.0	10.0	12.0	14.0	16.0	18.0	20.0	24.0	30.0
Dn	200	250	300	350	400	450	500	600	750
Screwed or welded	9.5	12.5	15.0	18.0	21.5	25.6	28.6	34.0	40.0
Flanged valves*	11.0	14.0	16.5	19.5	23.0	27.0	30.0	36.0	42.0
Butterfly	4.0	4.0	5.0	5.5	6.0	6.5	7.0	8.0	10.0

is illustrated in Fig. 5.23. The tracer should be placed at the center, but if this is not practical because of supports, it should be installed as close as possible to the center line of the pipe. Some engineers, however, consider it satisfactory to install multiple tracer pipes equally around the circumference of the pipe. Vertical pipe should have the tracer line(s) placed as illustrated in Fig. 5.24. It is recommended that a condensate trap be provided at the base of the vertical riser. Table 5.18 gives the recommended number of tracer lines and their size that should be used to protect various size pipes.

Another method of tracing a pipe is spiral tracing, where the tracer is wound around the pipe. This is used when a single tracer is desired but cannot supply enough heat if run straight.

Steam Tracing Pipe Size and Materials

The most popular size for tracer pipe is ½ in (DN 15). Smaller ⅜-in (DN 6) pipe is more easily plugged by sediment or debris in the system and is generally not recommended. However, since the amount of heat available from ½-in (DN 15) pipe is generally much more than required, and if the steam is clean, ⅜-in (DN 6) pipe should be considered. Larger sizes are more costly and generally not necessary unless a very large amount of heat is required for a specific purpose. The main advantage of using a single size pipe is that a facility can stock pipes, fittings, and valves in one standard size.

The most common materials used for tracing service lines are copper, stainless steel, and carbon steel. Copper tube is the most popular because of its heat transfer characteristics. Copper type L, ASTM B-88, is used often for piping and ASTM B-75 for tubing. Stainless steel tubing should be selected if there is the possibility of a corrosive environment. Carbon steel, ASTM A-53, schedule 40, is often a client preference based on existing standard pipe specifications, but is not generally recommended for tracer lines because of its tendency to corrode easily and produce rust that flakes off inside the tracer pipe and produces stoppages. In addition, the carbon steel pipe is subject to corrosion, which will cause the pipe to fail much more quickly than copper or stainless steel. Carbon steel is frequently used for steam and condensate headers.

Fittings for copper and stainless steel are usually the compression type, similar to Swagelok. Screwed or welded joints are used to install steel piping. Bends in copper pipe are made with a tube bender, rather than fittings, with care taken not

TABLE 5.17 Square Foot Surface Area of Round Tanks

Two end areas, ft²													
6.3	9.8	14.1	19.2	25.1	31.8	39.3	47.5	56.5	66.3	77.0	88	101	113
Tank height, *H*, ft					Tank diameter, *D*, ft								
2	2.5	3	3.5	4	4.5	5	5.5	6	6.5	7	7.5	8	8.5
3.1	3.9	4.7	5.5	6.3	7.1	7.8	8.5	9.4	10	10.9	12	12.6	13
6.3	7.8	9.4	11	12.6	14.1	15.7	17	18.8	20	22	24	25.1	27
12.6	15.7	18.8	27	25.1	28	32	35	37.7	41	44	47	50.2	53
18.9	23.5	28.3	33	37.7	42	47	52	56.5	61	66	71	75.4	80
25.1	31.4	37.7	44	50.2	57	63	69	75.4	82	88	94	101	107
31.4	39.3	47.1	55	62.8	71	79	86	94.2	102	110	118	126	133
37.7	47.1	56.5	66	75.7	85	95	104	113	123	132	141	151	160
44.0	54.9	66.9	77	87.9	99	110	121	132	143	154	165	176	187
50.2	62.8	75.4	88	101	113	126	138	151	163	176	188	201	213
56.5	70.6	84.8	99	113	127	142	155	170	184	198	212	226	240
62.8	78.5	94.2	110	126	141	158	173	188	204	220	236	251	267
75.4	94.2	113	132	151	170	189	207	226	245	264	283	301	320
87.9	110	132	154	176	198	221	242	264	286	308	330	352	374
101	126	151	176	201	226	252	276	301	327	352	377	402	427
113	141	170	198	226	254	284	311	339	367	396	424	452	480
126	157	188	220	251	283	315	354	377	408	440	471	502	534
188	236	287	330	377	424	473	518	565	612	659	707	754	800
251	314	377	440	502	565	630	691	754	816	879	942	1005	1068
314	393	471	550	628	707	788	864	942	1021	1099	1178	1256	1335

Tank height, *H*, ft values (left column): 0.5, 1, 2, 3, 4, 5, 6, 7, 8, 9, 10, 12, 14, 16, 18, 20, 30, 40, 50

Two end areas, ft²													
126	142	157	226	308	402	509	628	981	1413	1923	2512	3179	3925
Tank height, *H*, ft					Tank diameter, *D*, ft								
9	9.5	10	12	14	16	18	20	25	30	35	40	45	50
14	15	15.7	18.8	21.9	25.1	28.3	31.4	39.3	47.1	55	62.8	70.6	78.5
28	30	31.4	37.7	43.9	50.2	56.5	62.8	78.5	94.2	110	126	141	157
57	60	62.8	75.5	87.9	101	113	126	157	188	220	251	283	314
85	90	94.2	113	132	151	170	188	236	283	330	377	424	471
114	119	126	151	176	201	226	251	314	377	440	502	565	628
141	149	157	188	220	251	283	314	393	471	550	628	707	785
170	179	188	226	264	301	339	377	471	565	659	753	848	942
198	209	220	264	308	352	396	440	550	659	769	879	989	1099
226	239	251	301	352	402	452	502	628	754	879	1005	1130	1256
254	268	283	339	396	452	509	565	707	848	989	1130	1272	1413
283	298	314	377	440	502	565	628	785	942	1099	1256	1413	1570
339	358	377	452	515	603	678	754	942	1130	1319	1507	1697	1884
396	418	440	528	615	703	791	879	1099	1319	1539	1758	1978	2198
452	477	502	603	703	804	904	1005	1256	1507	1758	2010	2261	2512
509	537	565	678	791	904	1017	1130	1413	1696	1978	2261	2543	2826
565	597	628	754	879	1005	1130	1256	1570	1884	2198	2512	2826	3140
848	895	942	1130	1319	1507	1695	1884	2355	2826	3297	3768	4239	4710
1130	1193	1256	1507	1758	2010	2260	2512	3140	3768	4396	5024	5652	6280
1413	1492	1570	1884	2198	2512	2825	3140	3925	4710	5495	6280	7065	7850

Tank height, *H*, ft values (left column): 0.5, 1, 2, 3, 4, 5, 6, 7, 8, 9, 10, 12, 14, 16, 18, 20, 30, 40, 50

Source: *Power Trace.*

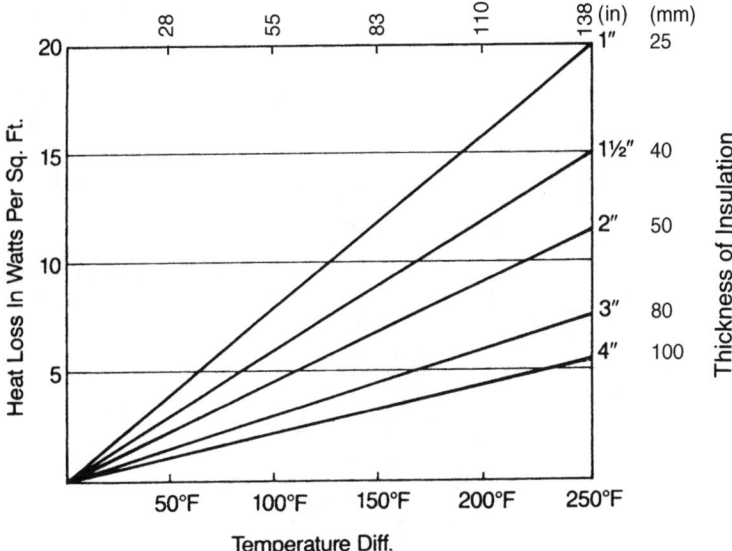

FIGURE 5.21 Heat loss through insulated vessel walls.

to crimp the tracer pipe. Unions are commonly provided in tracer lines at flanges and other connections. The purpose of the joints at these points is to allow replacement of the pipe in the future and to allow for expansion. Often, a union is installed to allow for changes, testing, and blowout, eliminating the need to make up another fitting whenever these actions are required.

Valves include gate valves, IBBM for main supplies, and bronze for tracer lines, with screwed or flanged ends.

The tracer pipe is commonly attached to the main line with thin galvanized or stainless steel bands about ½ in (15 mm) wide, between 18 to 20 gauge. The spacing between bands depends on the size main line, with ⅜-in (DN 12) tracers secured 12 to 18 in apart, ½-in (15 mm) tracers secured 18 to 24 in apart and ¾-in tracers secured 24 to 36 in apart. Fittings should have three bands to assure close contact between the tracer and the pipe. Where it is not desirable to use bands, such as at valve bodies, soft annealed 18-gauge stainless steel wire is an acceptable alternative.

Application of Insulation

Insulation must cover both the pipe being protected and the tracer line. It is important that the air space is kept clear. This can be achieved in several ways.

The first method is to wrap both the pipe and tracer with aluminum foil or a thin galvanized steel sheet attached by wire. The insulation is applied over this wrapping. Small mesh galvanized netting can also be used instead of the foil. The second method is to use insulation one or two sizes larger than the pipe being protected. This method is the least costly but the insulation can be easily crushed. The third method is to use special preformed insulation designed to cover both the pipe and tracer.

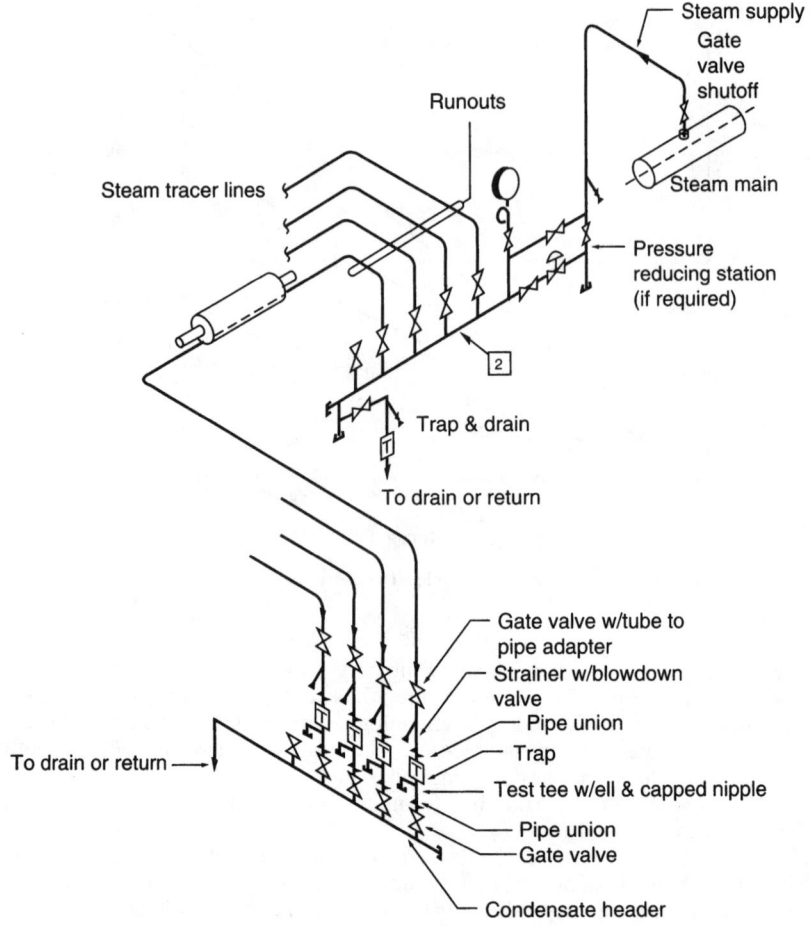

FIGURE 5.22 Simplified detail of a typical steam heat tracing system.

The insulation must be covered or finished in a manner that will protect it from mechanical damage and the elements, if installed outdoors.

Steam Pressure

The choice of steam pressure used for tracing must be consistent with the temperature to be maintained in the pipe to be protected. Low pressure steam of less than 100 psi (680 kPa) is used unless the temperature must be kept very high. Normally, saturated steam is used for tracing. Table 5.19 gives the temperature of saturated steam as a function of steam pressure. If the only steam pressure available is higher than that needed for tracing, a steam pressure-reducing station should be provided. In general, most steam tracing systems use a pressure of 50 psi (340 kPa) or less.

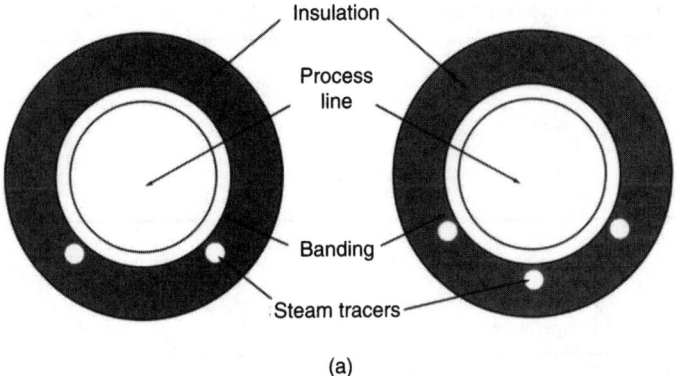

(a)

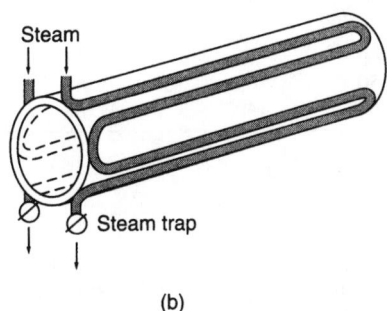

(b)

FIGURE 5.23   Typical installation of steam tracing lines for horizontal piping.
(*a*) Single and multiple tracing; (*b*) multiple tracing.

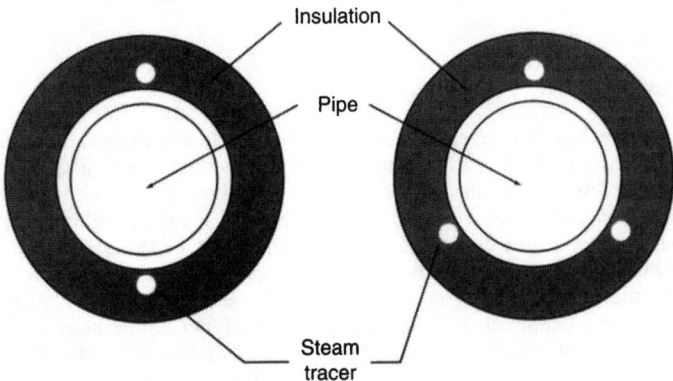

FIGURE 5.24   Typical installation of steam tracing lines for vertical piping.

TABLE 5.18   Recommended Number of Tracer Lines Based on Pipe Size

Pipe size, in	DN	Number of tracer lines	Nominal pipe size, tracer, in	DN
1–3	25–80	1	3/8	12
4–6	100–150	1	1/2	15
8–20	200–500	2	1/2	15
24–30	600–750	3	1/2	15

TABLE 5.19   Relationship of Steam Pressure to Temperature

Pressure, psig	kPa	Steam temperature, °F	°C
15	105	250	120
30	210	274	133
50	350	298	146
75	525	320	160
100	700	338	168
150	1050	366	183
200	1400	388	196

Steam Supply

Steam for tracing is generally obtained from the facility's steam service. Normally, saturated steam is used for tracing purposes. If multiple tracing lines are required, it is standard practice to connect to the main steam supply once and use a tracer supply manifold, or header, to feed all the individual tracers. The connections to the header should be made from the top. This steam supply to the header should have a manual shutoff valve, and each of the individual tracers should also have its own isolation valve. Refer to Table 5.20 for a general guide to the number of 1/2-in tracer lines that can be supplied from a header with a maximum steam pressure of 50 psi (350 kPa). A steam trap assembly should be provided on the header to remove condensate. A typical steam supply manifold is illustrated in Fig. 5.25.

Condensate Return

The condensate can be either returned for reuse or, when a condensate return is not practical, disposed of (to drain). Exterior tracer lines commonly discharge condensate into a dry well placed into the ground. Condensate is pure water and thus causes no damage to the environment. When located inside a building, condensate is often discharged into a floor drain. If the temperature of the condensate is above 140°F (60°C), mixing with cold water may be necessary for the drain to be routed into the sanitary drainage system. Refer to Table 9.30 for mixture proportions. If discharged into a chemical or industrial system, a higher temperature is acceptable up to the temperature limit of the piping system and jointing method into which it is discharging. For a detail of nonreturned condensate, refer to Fig. 5.26.

A problem potentially occurs in the tracer line after the control valve closes. Because the steam in the tracer line will condense to water at a much lower specific

TABLE 5.20 Recommended Number of Tracer Lines Based on Class of Service

		Number of ½-in (DN 15) tracers		
		Type A	Type B	Type C
Product line size, in	DN	Gneral frost protection or where solidification may occur at temps below 75° F (55°C)	Where solidification may occur at temps between 75–150°F (55–64°C)	Where solidification may occur at temps between 150–300°F (64–147°C)
1	25	1	1	1
1½	40	1	1	2
2	50	1	1	2
3	80	1	1	3
4	100	1	2	3
6	150	2	2	3
8	200	2	2	3
10–12	250–300	2	3	6
14–16	350–400	2	3	8
18–20	450–500	2	3	10

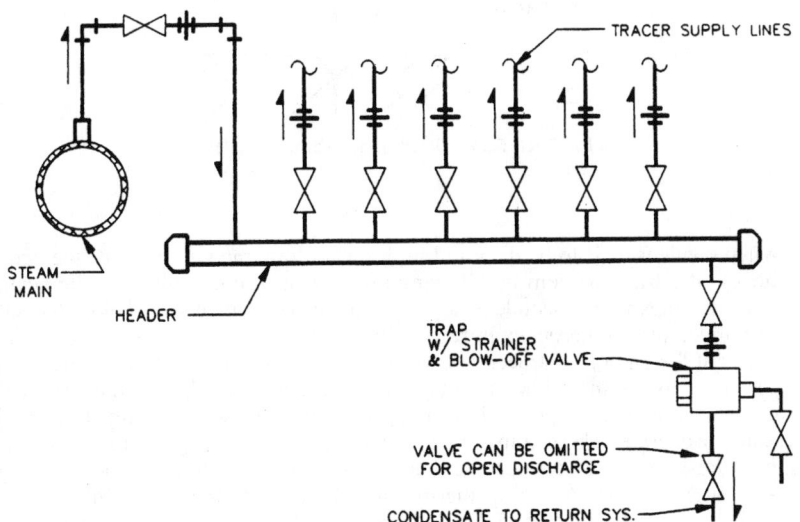

FIGURE 5.25 Typical detail of steam supply manifold.

volume (0.017 ft^3/lb for 50 psi condensate) than saturated steam (6.8 ft^3/lb), a vacuum will be created in the tracer line. To eliminate this vacuum, a vacuum breaker should be installed after the control valve to allow air to enter the tracer pipe to replace the volume formerly filled with steam.

There are conditions where branch lines are distant from the steam header or are installed in cold climates. In these cases, it is recommended that the steam supply line have a trap immediately before the temperature control. Each separate tracer should be provided with a steam trap and a strainer with a blowdown valve should be installed before the steam trap at the end of each tracing run.

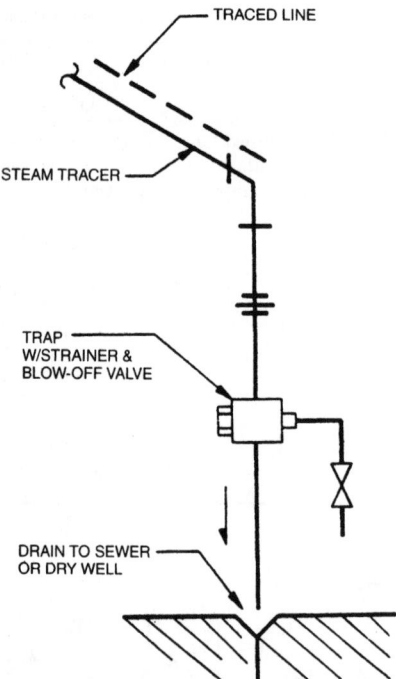

FIGURE 5.26 Detail of nonreturned condensate.

When condensate is to be disposed of, a subcooling trap will enhance the energy efficiency of a tracer system by allowing sensible heat to contribute to the heating duty. These traps release condensate only after it has cooled well below the saturation temperature. Where supply temperature control valves are used, the condensate traps at the end of the tracer run will be in the closed position when the tracing duty is satisfied and the steam is shut off. Since the pressure driving the condensate through the trap will be zero, it is important that the selected trap not require pressure to operate. These traps are classified as free-draining, and are known as temperature-sensing or thermostatic steam traps. The trap should be installed in a free-draining position. A typical thermostatic condensate trap installed at a tracer line end is illustrated in Fig. 5.27. A typical condensate header is shown in Fig. 5.28.

Temperature Control

For most systems, the simplest way to control the temperature is to use an adjustable steam pressure reducing assembly on the steam supply to the tracer line. The pressure can be adjusted based on operating experience to produce the required temperature. This method allows only approximate temperature control and so is used when the line to be protected has a fairly constant flow and the heat makeup is

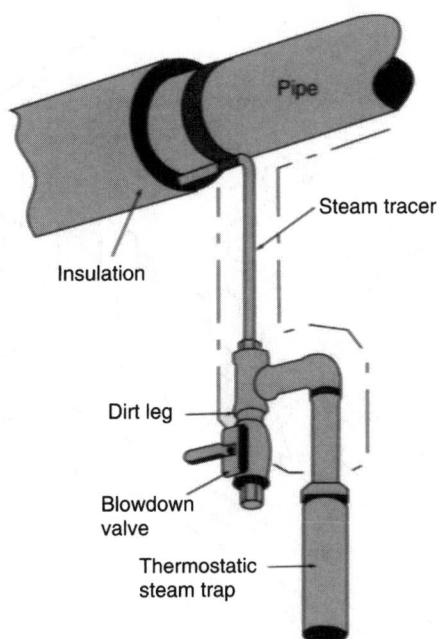

FIGURE 5.27   Detail of thermostatic steam trap.

constant. Operating experience has shown that reliable temperature control of steam is not practical below 250°F (121°C). Above this temperature, control to within 50°F (10°C) is reasonably achievable, but expensive.

When closer control is necessary, an automatic, direct-acting temperature control valve often provides an economical solution. One advantage of this type of valve is that it does not require either electricity or compressed air to operate. The valve operation is controlled by an attached sensor that can be arranged to sense the appropriate relevant temperature—ambient air, surface wall of the protected pipe, or the fluid stream to be protected. Modulating temperature control valves can reduce shocks to the system by opening and closing slowly, thus reducing thermal shock, water hammer, and abrupt temperature changes. An additional advantage is that the automatic feature relieves operating personnel from having to manually turn on and off steam supply lines at various points throughout the facility.

Other methods used less often include electric, pneumatic, and electric-pneumatic controlling devices that also require positioners, set point controllers, temperature sensors, and power supplies. These methods of controlling the steam supply to the tracer line are more costly, but generally more accurate, and their use depends on the specifics of the application.

Freeze protection requirements for piping within a facility should be controlled by a temperature-sensing device attached directly to the pipe being protected. When the pipe temperature falls below a set point, the steam is turned on, and when it rises above a set point, the steam is turned off. This sensor should be located as far away from the tracing line as possible.

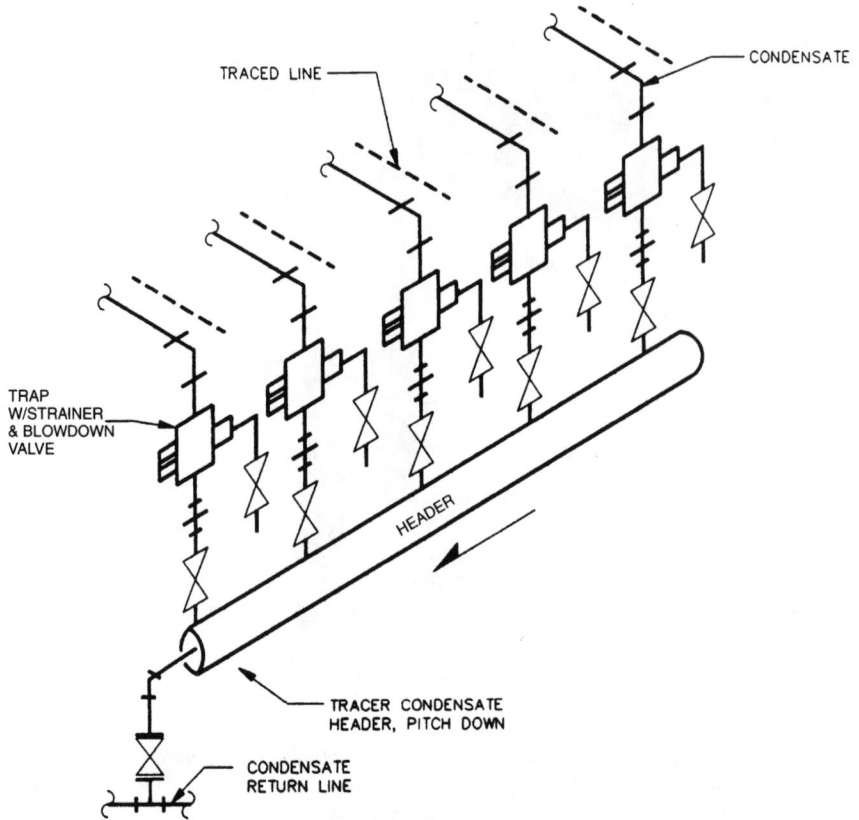

FIGURE 5.28   Detail of condensate return manifold.

System Design Procedure

1. Determine the number and size(s) of the pipe(s) to be protected. Use Table 5.18 or 5.20 as a guide to select the number and pipe size of the tracer lines. Freeze protection is considered type A, noncritical, in Table 5.20. Establish the total number of tracer lines required, the steam supply location and pressure, and the method of condensate removal or return. The steam supply must be taken from a source that is continuously available.

2. From the steam supply location, determine whether a pressure-reducing station is required. Locate the manifold for a multiple tracer system. Based on the pressure available, select a subheader size from Table 5.21 (30 to 50 psig [210 to 350 Kpa) steam) or Table 5.22 (100 to 150 psig [700 to 1050 kPa] steam). With the proposed tracer line size and the steam pressure available, select the subheader size to supply the selected number of tracer lines based on the steam pressure. Accepted practice limits subheader size to approximately 50 ft (15 m).

3. To find the longest allowable heat tracing pipe run, use Table 5.23, entering the table with the available steam pressure and the line size. Another factor in the

TABLE 5.21 Subheader Size Using 30 to 50 psig (210–350 kPa) Steam Pressure

Maximum number of tracers ³/₈-in (12 mm) O.D.	Maximum number of tracers ¹/₂-in (15 mm) O.D.	Subheader pipe size, in	DN
1–4	1–2	³/₄	20
5–9	3–5	1	25
10–22	6–16	1¹/₂	40

TABLE 5.22 Subheader Size Using 100 to 150 psig (700–1000 kPa) Steam Pressure

Maximum number of tracers ³/₈-in (12 mm) O.D.	Maximum number of tracers ¹/₂-in (15 mm) O.D.	Subheader pipe size, in	DN
1–5	1–3	³/₄	20
7–12	4–7	1	25
13–30	8–24	1¹/₂	40

TABLE 5.23 Heat Tracer Length
Tracing design: ¹/₂-in O.D. tubing—parallel to pipe maximum tracing run per trap (ft)[1]

Line size, in	DN	Steam pressure, kPag					
		10 70	50 350	100 700	150 1050	200 1400	250 1750
Single tracer							
1	25	230	270	290	360	430	450
1¹/₂	40	190	230	250	310	360	390
2	50	170	210	220	270	320	350
3	80	140	170	180	240	260	280
4	100	130	150	160	210	220	230
6	150	100	120	130	160	170	200
Double tracer							
8	200	50	60	70	80	90	100
10*	250	100	120	130	160	170	200
12*	300	100	120	130	160	170	200
14*	350	50	60	70	80	90	100

1 ft × 0.305 = meters
*Use ³/₄-in O.D. tubing for tracer.

design of the tracer line is the sharp vertical rise and drop of the pipe (a gradual pitched line is not considered). It is recommended that friction loss from the combined sharp vertical rise and drop not exceed 45 percent of the steam gauge pressure, and that any one vertical dimension not exceed 20 ft (6.2 m). If it is not possible to stay within these requirements, a second tracer line should be used.

4. To size the condensate header pipe, use Table 5.24. Enter the table with the number of tracer line traps (one for each tracer line) and the steam pressure. Ac-

TABLE 5.24   Condensate Header Size

Maximum number of ⅜-in and ½-in (12 and 15 DN) traps

30–50 psi (210–350 kPa) STM	100–150 psi (700 to 1000 kPa) STM	Condensate header pipe size, in	DN
1–3	1–2	¾	20
4	3	1	25
5–11	4–8	1½	40
12–22	9–15	2	50
23–50	16–40	3	80

cepted practice limits length of header run to approximately 60 ft (18 m), and header pressure to 30 percent of the lowest inlet trap pressure.

5. If the installation requires maximum heat transfer between the tracer and the pipe, a heat-conducting paste can be used to fill any gaps along the length of the tracer. Care must be taken to adequately clean both pipe surfaces before applying the paste.

6. If the installation has pipe or product that may be sensitive to the high temperature resulting from direct contact with the tracer, a strip of insulating material, such as mineral wool or fiberglass, could be installed between the tracer and the pipe.

7. To calculate the heat lost in Btu through common insulation materials, refer to Table 5.25.

Expansion

The tracer line will expand at a different rate than the pipe being protected if they are made from different materials. This difference must be provided for. When flanged and screwed pipe is being protected, it is common practice to provide an expansion loop at each joint. For long runs of straight welded pipe, expansion loops are provided every 50 ft, usually as loops in the tracing line. Expansion loops are not required on vertical piping where spiral tracing is used.

TABLE 5.25   Heat Loss in Btus Through Common Insulation Materials

	100°F 37°C	200°F 92°C	300°F 150°C	400°F 200°C	500°F 260°C	600°F 310°C
Fiberglass	0.26	0.30	0.34			
Polyurethane	0.16	0.16	0.16			
Calcium silicate	0.33	0.37	0.41	0.46	0.57	0.60
Cellular glass	0.39	0.47	0.55	0.64	0.74	0.85

Note: These are representative values per inch thickness for one square foot of area. Exact values should be confirmed by the insulation manufacturer.

Special Installation Considerations

Large pumps, instruments, and other irregularly shaped pieces of equipment must also be protected. These pieces of equipment generally require that the tracer be either directly in contact with the object being protected or, in cases where overheating cannot be tolerated, isolated from that equipment by spacers between the tracer and equipment or by tape wound around the tracer to reduce the surface temperature of the tracing pipe. Special design details for tracing methods must be custom made for specific pieces of equipment.

Because it is common to find dirt and debris in the tracer line, it is important that it be blown out with steam before being placed in service.

ACKNOWLEDGMENTS

Isseks Brothers Tank Corporation, Mr. Scott Hochhauser
Raychem, Mr. Fred Storey

REFERENCES

Alter, Amos, "Sewage and Sewage Disposal in Cold Regions," CRREL, U.S. Army Corps of Engineers.

Care, Kevin L., "The Freezing and Blocking of Water Pipes," CRREL.

Depth of Freeze and Thaw in Soils, U.S. Army Technical Manual, TM 5-852-6.

"Design of Fluid Systems Hook-ups," Spirix Sarco Inc., Allentown, PA.

Dill, Richard S., "The Freezing of Pipes and Vessels," National Bureau of Standards.

Fire Protection Handbook, 16th Ed., National Fire Protection Association.

Frankel, M., "Cold Weather Design of Plumbing Systems," parts 1 and 2, *Plumbing Engineer,* November 1987–February 1988.

Frankel, M., "Thermal Insulation for Plumbing Systems," parts 1 and 2, *Plumbing Engineer,* July–September 1988.

Kenny, Thomas M., "Steam Tracing: Do It Right," *Chemical Engineering Progress,* pp. 40–45, August 1992.

Sanger, F. J., "Foundations for Structures in Cold Weather," CRREL.

CHAPTER 6
SITE UTILITY SYSTEMS

This chapter will describe various pressurized and gravity flow utility and service systems that are found outside of buildings on the site. These also include services that extend from buildings to points of connection or disposal. A common connection point for site work is considered to be 5 ft from the building wall.

Site utilities discussed here include subsurface drainage, storm water drainage and retention methods, sanitary drainage, water supply, and industrial and laboratory site drainage systems. Also included are sections that provide information and fundamentals of hydrology and design of buried piping.

BASIC GEOLOGY AND HYDROLOGY

Because of the close relationship of geology and hydrology to site utility work, it is useful to have an elementary knowledge of the hydrological cycle and underlying geological formations.

HYDROLOGIC CYCLE

The continuous circulation of water through surface water, the atmosphere, and the land is called the *hydrologic cycle.* Inflow to the hydrologic system arrives as precipitation (rain or snowmelt) on the surface. Three things may then happen to this water. It may be pulled into the soil surface by capillary action and evaporated back into the atmosphere, be absorbed by plant roots and evaporated back into the atmosphere by transpiration through leaves and roots, or infiltrate down through the soil until it reaches the groundwater table. The hydrologic cycle is schematically illustrated in Fig. 6.1.

From surface sources, including rainfall, streams, rivers, and lakes, water infiltrates down into the soil. Basic geological formations are illustrated in Fig. 6.2. By infiltration, capillary action, and percolation, some portion of that water eventually reaches the groundwater table. Water in the soil between the surface of the ground and the water table is called subsurface water. The slow movement of subsurface water through the ground places this water in direct contact with soluble minerals that make up Earth's crust. The water, therefore, may have a wide variation in its chemical character even within small geographic regions.

In some areas, water may accumulate in a local zone of saturation above an impervious stratum and be prevented from reaching the level of the water table. This is called *perched water,* and its free surface is called the *perched water table.*

The term *groundwater* is classically defined by geologists as water found below the water table or in a geological formation that is fully saturated. *Subsurface water* is defined as near surface water that infiltrates the soil but is not absorbed into the

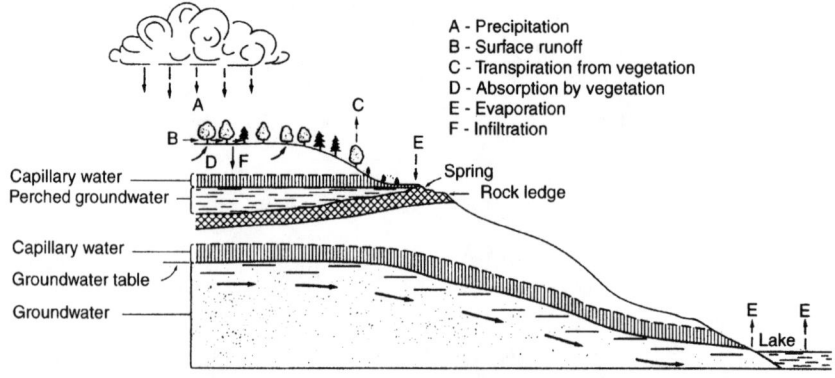

FIGURE 6.1 The basic hydrologic cycle. (*Courtesy of U.S. Army Corps of Engineers.*)

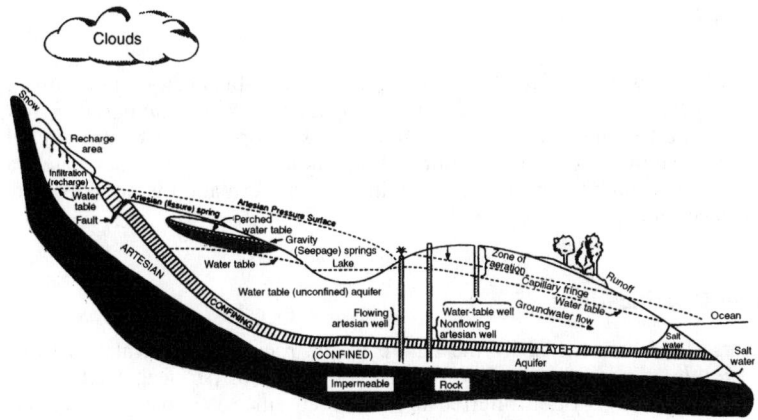

FIGURE 6.2 Basic geological formations.

ground at a lower level. It is primarily subsurface water that requires draining from excavations.

Groundwater and subsurface water are part of the hydrologic cycle and have two important applications. The first is using groundwater to obtain a supply of domestic water. The second is the necessity of collecting and removing subsurface water when required.

The *lithosphere* is the geological term for Earth's crust. All earth materials are known collectively as rock. *Consolidated* rock is the common hard rock such as granite or limestone, which will be called simply rock in this book. It is classified by origin into three main groups: igneous, sedimentary, and metamorphic. Unconsolidated rock is made up of small particles created from rock and composed of soils such as gravel, clay, and sand. It is classified into groups described in Table 6.20.

Water falling on the ground either collects on or near the surface or infiltrates downward into the subsurface. Subsurface water exists in pores and occurs in two distinct zones. The upper zone is called the *unsaturated zone,* with pores that contain both air and water. It is also called the zone of aeration. In the lower one, called the *saturated zone,* all the pore spaces are completely filled with water.

The unsaturated zone is divided into different layers. The porous, upper part of the unsaturated zone is called the *soil water belt.* This belt, composed mostly of soil, is a layer that extends down from the ground surface at least to the bottom of major plant roots, and often, much lower. This layer is where water is held in suspension by capillary action against the force of gravity. Below the soil water belt is an area called the vadose zone, which is the lower part of the unsaturated zone. This is the area where water under hydrostatic pressure that cannot be held in the belt by gravity moves downward into the saturated zone.

The saturated zone is the region where all the pores are completely filled with water. The top of the saturated zone is called the *water table.* Water found in the saturated zone is called groundwater and is the source of drinking water obtained from wells. Groundwater is replenished, or recharged, from precipitation that infiltrates down into the saturated zone. Groundwater is discharged through wells.

AQUIFERS

An *aquifer* is a water-bearing formation or stratum capable of storing or transmitting water in sufficient quantities to permit development for a specific purpose. To qualify as an aquifer, the formation must have pores or open spaces large enough for water to move in reasonable quantities. Where pores are so small that water cannot easily move through them, for example, in clay, the formation is not considered an aquifer even though it is saturated.

Aquifer Classifications

Aquifers can be classified by the nature of the rock that the water is stored in. *Unconsolidated* aquifers consist of a high percentage of permeable granular material. This material is often referred to as alluvial since it is usually sediment deposited by running water. This type of aquifer produces the greatest amount of water and is recharged directly from precipitation or streams.

The *semiconsolidated* aquifer is similar to the unconsolidated type except for weak bonding materials, usually calcium carbonate and iron oxide, that fill a portion of the pore spaces. These materials tend to be geologically older than unconsolidated materials and are usually found at a greater depth.

Consolidated aquifers consist of different types of solid rock that have some porosity, with the available water depending on the number of fractures present in the layer.

Another possible classification of aquifers is based on the intended use of the water. These classifications, or similar ones, are determined by local and state agencies. The general classifications are

1. Class 1: groundwater of special ecological significance
2. Class 2: groundwater for potable water supply
3. Class 3: groundwater for uses other than potable water

Aquifer Categories

Aquifers are often categorized by the conditions in which water is stored and whether a free water surface (or water table) exists under atmospheric conditions. *Unconfined* aquifers occur in the unconsolidated layer and have water and air filling the pores, and the top of the layer is at atmospheric conditions. The top of the layer is the water table, which fluctuates with the amount of water present.

When an aquifer is found between impermeable layers, the water is confined in the same way as it would be inside a pipe. Because of the confining strata, the aquifer is under pressure greater than atmospheric. This aquifer is called an *aquiclude,* but is better known by the more popular terms *confined* aquifer or *artesian* aquifer. Hydrostatic pressure within an artesian aquifer is sometimes high enough to cause water in a well to rise above the surface level of the ground above it.

When a confining bed is located beneath an aquifer, it prevents water from continuing its downward movement. The limited amount of water that accumulates above this confining bed is known as a *perched aquifer.*

SUBSURFACE DRAINAGE

This section discusses the drainage aspects of subsurface water entering excavations and the drainage requirements resulting from intrusion of subsurface water into footing drains.

When water in any form interferes with the construction of a project or could potentially cause damage to any structure or installation placed in the ground, it becomes a problem rather than a resource and must be removed. Problems occur when the presence of water could cause a structure to float or when groundwater could cause soil-bearing resistance to be lost.

Subsurface water is removed by placing a drainage system below the level required to be kept dry. In many cases, this drainage system will remove excess water by gravity. Two of the most common systems make use of trenches filled with pervious backfill and drain pipes. The drain pipes may have holes in them or may be installed with open joints to allow water to enter, and are pitched to provide a flow path. A special layer of pervious backfill is placed above the trench bottom or drain pipes to allow subsurface water to flow easily through them. This pervious backfill is often called *filter material.*

There are three general methods of draining subsurface water, categorized according to their purpose. The first is *subbase drainage,* which is generally used for draining road subbases. This system consists of shallow drain pipes laid near and parallel to the edge or end of the pavement to be drained. The second is called *subgrade drainage,* which is generally used when it is necessary to lower a high water table from around and inside the project area. It consists of either trenches filled with filter material or pipes placed to collect and route the water away. The third is called an *interceptor drainage* system, which consists of pipes or trenches placed in a manner that will intercept the groundwater before it gets to the project site.

When none of the gravity drainage methods are possible, a pumped system is required. The most common method for draining excavations of a project under construction uses well points, which serve the same purpose as intakes for a well. A *well point* is a reinforced, pointed metal cylinder with holes in it for water to enter. The points are driven into the ground below the level of the excavation and connected to the surface with a discharge pipe. The top of the pipes are connected by a header to the suction side of a pump, which collects the groundwater and discharges it to an approved location. The groundwater level is brought below the bottom of the excavation, keeping it dry. A typical well-point system is illustrated in Fig. 6.3.

INFLOW INTO EXCAVATIONS

The development of predictive methods for the inflow of water into excavations or from foundation drains has not kept up with the solutions to other problems in groundwater hydrology. It is safe to say that soil exploration must be done before a structure is built on any site. Analysis of the borings will establish if the site has groundwater that must be removed. In addition, after the excavation has been started, the amount of water drained from well points (if the site is a wet one) is the most accurate indication of the quantity of water that must be drained after the

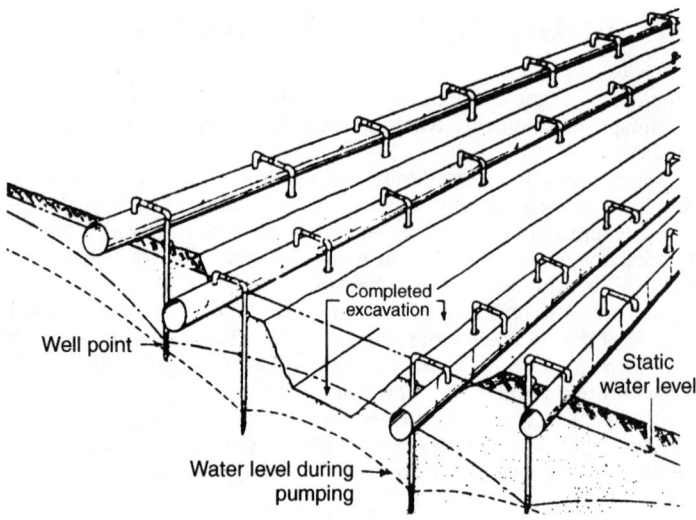

FIGURE 6.3 Typical well-point system.

foundation is in place. In addition to this information, the engineer must research the prior weather conditions to determine if there has been a dry spell or drought that might have resulted in a smaller amount of groundwater present. Allowance must also be made for additional water that could result from storms creating a higher groundwater level.

In the absence of such information, the following is a method that is useful in estimating the probable water quantity discharged from a subsurface drainage pipe. This method was established by the U.S. Army Corps of Engineers. The formula is:

$$Q = \frac{KHC}{60} \tag{6.1}$$

where K = horizontal permeability of soil, ft/min (This is a measure of how fast the water will flow through any type of soil. Refer to Table 6.1 and Fig. 6.4 depending on whether the material is backfill or undisturbed soil.)

H = difference in elevation between the center of the pipe and the ground surface, L distance from the drain (An example is illustrated in Fig. 6.5.)

C = shape factor dependent upon L and H. L is the distance to the edge of the excavation around a building. (Use $L = 50$ when K is greater than 10^3 ft/min. Refer to Fig. 6.6 to find the shape factor.)

Q = discharge quantity in cubic feet of water per second (cfs) for each foot of subsurface drainage pipe

TABLE 6.1 Permeability of Backfill

The following tabulation may be used for preliminary estimates of average coefficients of permeability for remolded samples of sand and gravel bases:

Coefficient of permeability for remolded samples,* fpm	
Percent by weight passing 200-mesh sieve	Coefficient of permeability for remolded samples,* fpm
3	10^1
5	10^2
10	10^3
15	10^4
25	10^5

*The coefficient of permeability of crushed rock and slag, each without many fines, is generally greater than one foot per minute.

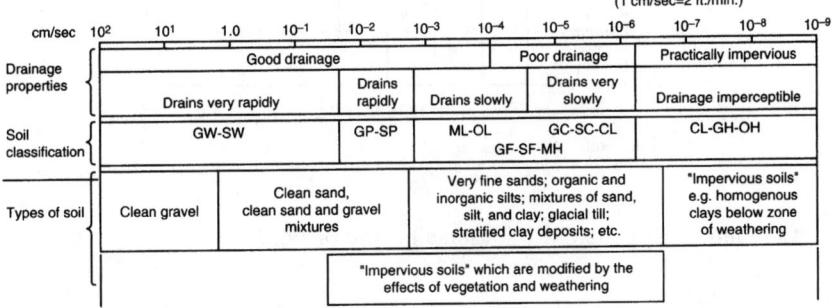

FIGURE 6.4 Permeability of soils.

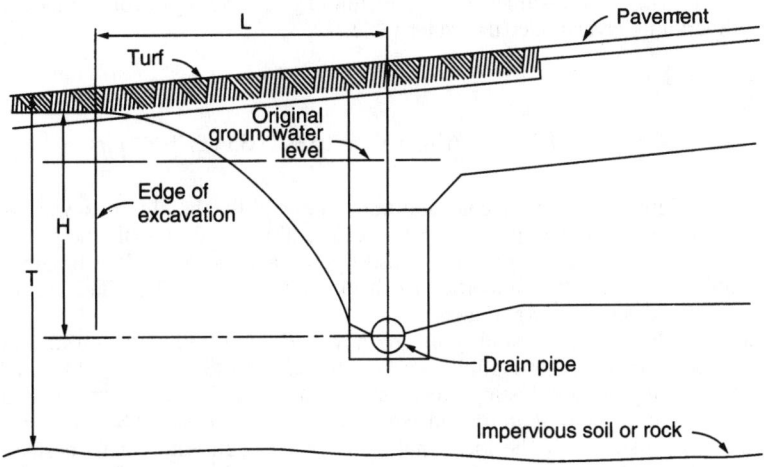

FIGURE 6.5 Example diagram.

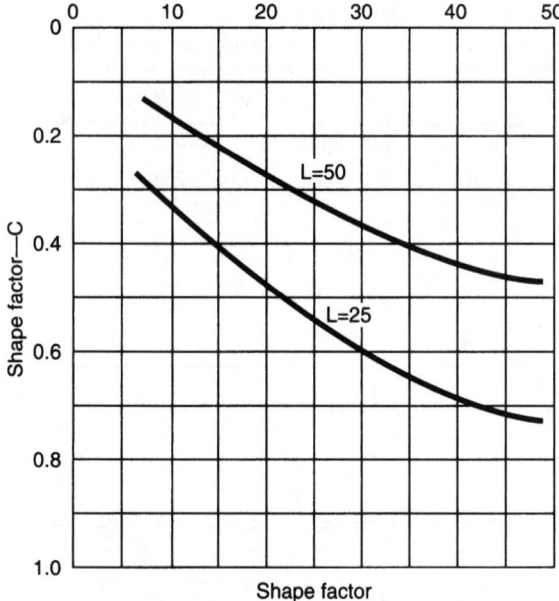

FIGURE 6.6   Shape factor.

The following example will illustrate the use of Eq. (6.1):

$K = 10^{-4}$ fpm

$H = 6$ ft

$C = 0.53$ (from Fig. 6.5)

$L = 25$ ft (This is the dimension by examining the borings from grade to the impervious layer under the water table.)

Therefore:

$$Q = \frac{1}{60} \times 0.0001 \times 6 \times 0.53 = 0.0000053 \text{ cfs/ft of pipe}$$

To this figure, add 25 percent additional flow if the height of a wet period occurred in the immediate past and 50 percent additional flow if the height of the dry season is occurring at present. These figures are the author's estimate for a reasonable allowance. It is important to check with local building officials for confirmation based on their experience.

There is always the possibility of groundwater being polluted. Before groundwater is discharged into the public sewer, samples of the water should be taken. Acceptable sampling and testing procedures must be used. The key terms used are sampling for *priority pollutants* and *contaminants of concern*. After the tests are run, a full assessment by the local, state, and federal authorities must be made to determine both the extent of any contamination and the ultimate method of disposal.

Because of sand present in discharge water coming from subsurface drainage piping, a sand interceptor is usually required by plumbing codes. The sand interceptor should be placed in the drainage line before it discharges into a building sump or ejector pit for disposal. This will avoid the necessity of installing the sand interceptor at the public sewer, where discharge of sand is prohibited. There are several commercially available sand interceptors, or one can be built on site.

The drainage lines should be sized using standard flowcharts, such as nomographs using the Manning, Kutter, or Hazen-Williams formula, with a roughness coefficient corresponding to the type of pipe used in the system and the pitch of the drainage pipe.

SURFACE DRAINAGE

This section is intended as a guide for development of a storm water drainage system for smaller uncomplicated sites of less than 2 square miles. Procedures follow established engineering principles and allow a simplified, conservative method of design and piping layout for small watersheds.

Included is rain water removal from parking lots, roadways, and undeveloped areas of the site. The selection and location of inlets, catch basins, manholes, and piping along with storm sewer design must be accomplished. The design of storm water removal, for example, from building roofs, is discussed in Chap. 9, in the section entitled Interior Storm Water Drainage.

When systems are located on larger sites or complications are encountered, it may be necessary to request the help of a consultant specializing in this type of work to ensure that the design is adequate.

PRELIMINARY INVESTIGATIONS

The following investigations must be concluded before designing the proposed system:

1. Determination of the site storm water discharge location or method of disposal, such as public sewers, waterways, surface absorption, recharge basins, or dry-wells.

2. If streams or rivers are to be used as a final outfall, local, state, or federal agencies must be contacted regarding permitted flow, outfall structures, and required permits. Such agencies may be the U.S. Army Corp of Engineers, Environmental Protection Agency (EPA), or local sewer department. EPA permits are required if such a waterway is considered navigable.

3. If a public sewer is to be the final outfall, public agencies must be contacted regarding permitted flow, connection details, plan submittal, and required approvals.

4. High and low water levels of all streams and rivers must be known, including past history of floods.

5. If site storm water will be discharged directly into smaller streams, it must be determined whether a possible flood condition downstream could be created.

6. Provision should be made for any future development of surrounding site areas of building additions.

Layout of System

At this point, it is assumed a survey plan and a preliminary site plan with contours are available for a general layout of the system. Runouts from buildings should now be located, outfall of the storm sewer selected, and work started on a trial location of storm water inlets. A trial layout of the site piping should be developed at this time and coordinated with water mains, electrical or telephone duct banks, sanitary sewers, gas mains, and other piping to ensure proper clearance and allow for proper and reasonable slope for the storm sewer lines. Obstacles to be avoided, such as trees, underground structures, and rock, should be located. Recommendations for final grading could be made to the landscape architect to possibly reduce the cost of storm water sewers by advantageous placement of inlets to minimize piping runs.

In general, the following should be considered when locating and designing inlets:

1. Where streets or roads intersect, inlets must be located upstream of traffic flow.

2. Where a series of inlets are located in a road with continuous slope, each intermediate inlet should be designed for approximately 75 to 90 percent of design flow, passing 10 to 25 percent on to the next downstream inlet. If the slope is shallow and flow is not great, 100 percent capacity should be considered. More than one inlet should be provided at the bottom to accept the additional flow.

3. Inlet capacity should be limited to approximately 5 cfs (0.15 m^3/s) of water.

4. All site low points must be provided with an inlet if off-site gravity flow is not possible. Combination inlets are used for streets or roadways. Parking lot inlets should be flat grates only, even when located at a curb.

5. Distance between drainage inlets should be a maximum of about 300 ft.

6. In a roadway with a sag vertical curve, more than one inlet should be considered if a large quantity of runoff is anticipated. Additional inlets should be located on either side of the low point inlet to minimize flooding and sediment buildup resulting from large flows.

7. Manholes should be used at all changes of direction and slope of pipe and at multiple pipe connections. Drain structures also may serve the same purpose. Distances between manholes should be 250 to 600 ft (75 to 188 m) depending on sewer size.

8. Steep slopes should have inlets located closer together than normally required to limit inflow.

9. A hooded catch outlet should be used as a final structure before connection to a combined sewer to prevent the passage of unwanted gases.

10. The efficiency of inlets in roadways depends on gutter flow, water velocity, roadway slope, and inlet depression, if any. In many cases, freedom from clogging or interference with traffic may take precedence over hydraulic considerations.

In general, the following should be considered:

- The road crown should be as steep as possible.
- Where traffic will not travel close to curb and clogging is not a problem, use a depressed gutter inlet. Use a depressed curb or combination-type inlet where clogging may occur.
- Where traffic moves close to the curb, use an undepressed inlet and gutter grate with longitudinal bars only.

STORM WATER INLET SELECTION

Drainage Structures

A drainage structure is an assembly of elements outside of piping built or installed below ground. Inlet structures are intended to collect surface storm water and route the effluent into a piping system.

While the piping system is being designed, after the storm water drainage inlet structures have been located, the type and size of inlets and grates must be selected. In most cases, the authority having jurisdiction has design standards available for such purposes. If no standards are available, the following can be used as a guide. Storm water drainage inlet structures are classified according to their functions as follows:

1. *Drainage inlets* (DIs) are structures that admit storm water into the piping system. They are generally located in areas that are reasonably free from sediment or debris such as paved, lightly vegetated, or unimproved areas. The bottom of a DI is level with the invert of the outlet pipe. A shallow DI is illustrated in Fig. 6.7 and a deep DI is illustrated in Fig. 6.8.

2. *Catch basins* (CBs) are similar to drainage inlets except that there is a space below the inlet and outlet pipes for retention of debris or sediment. A CB is usually located in paved areas where debris can be easily washed in. Experience has indicated that inadequate maintenance negates their benefits, and catch basins are not generally used unless good maintenance can be assured. A typical CB is illustrated in Fig. 6.9.

3. *Manholes* (MHs) do not allow surface rainfall to enter but are provided for ease of pipe connections for cleaning purposes. Drop manholes provide for pipe connections where a difference of more than two feet in elevation exists between inlet and outlet pipe. A precast MH is illustrated in Fig. 6.10 and a drop connection into a MH is illustrated in Fig. 6.11.

Gratings

Gratings allow water to enter the drainage structure, and some gratings will prevent the entrance of debris into the piping system. Gratings are classified as plain, curb, or combination.

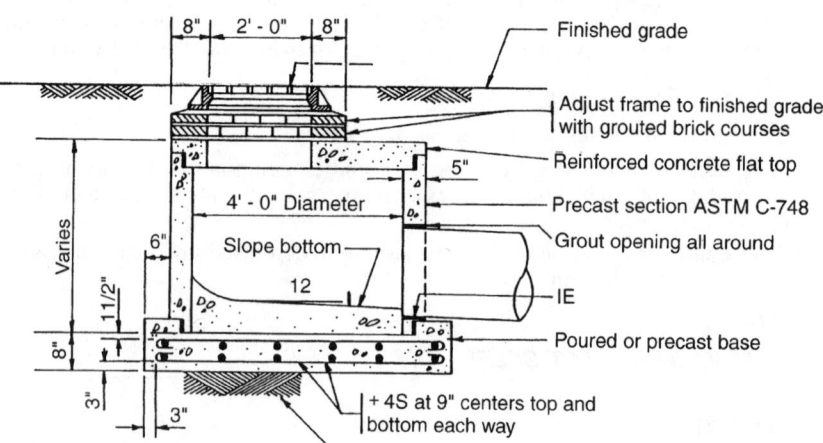

FIGURE 6.7 Typical shallow drainage inlet.

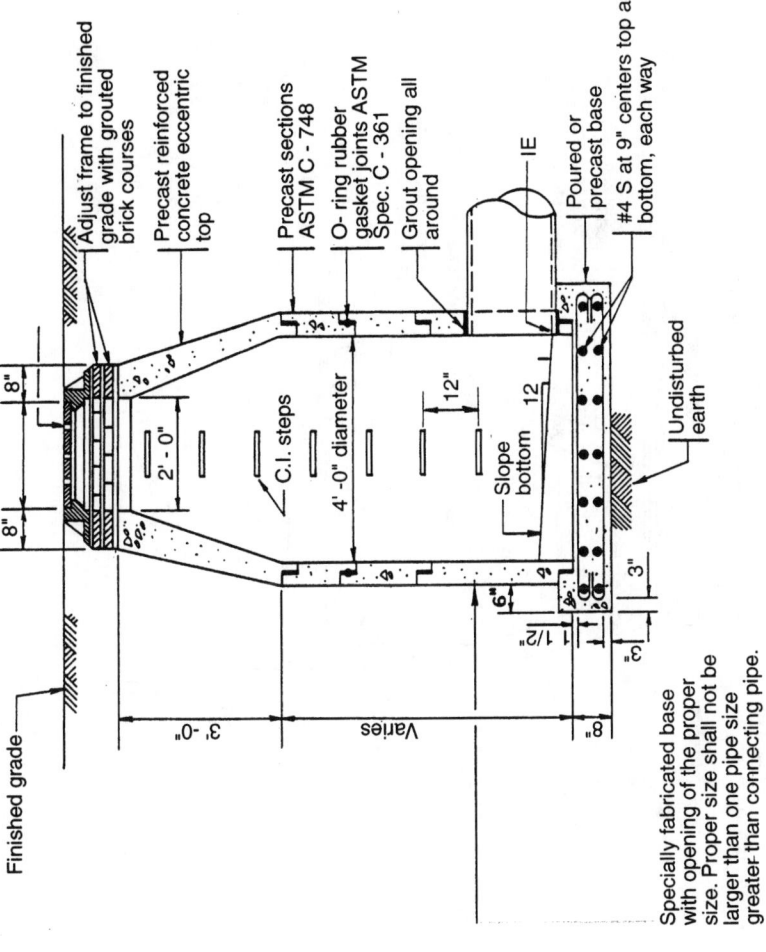

Finished grade

Adjust frame to finished grade with grouted brick courses

Precast reinforced concrete eccentric top

Precast sections ASTM C - 748

O- ring rubber gasket joints ASTM Spec. C - 361

Grout opening all around

IE

Poured or precast base

#4 S at 9" centers top and bottom, each way

Undisturbed earth

8"

8"

2' - 0"

C.I. steps

4' - 0" diameter

12"

Slope bottom

12

3' - 0"

Varies

6"

1 1/2"

3"

3"

8"

Specially fabricated base with opening of the proper size. Proper size shall not be larger than one pipe size greater than connecting pipe.

FIGURE 6.8 Typical deep drainage inlet.

6.13

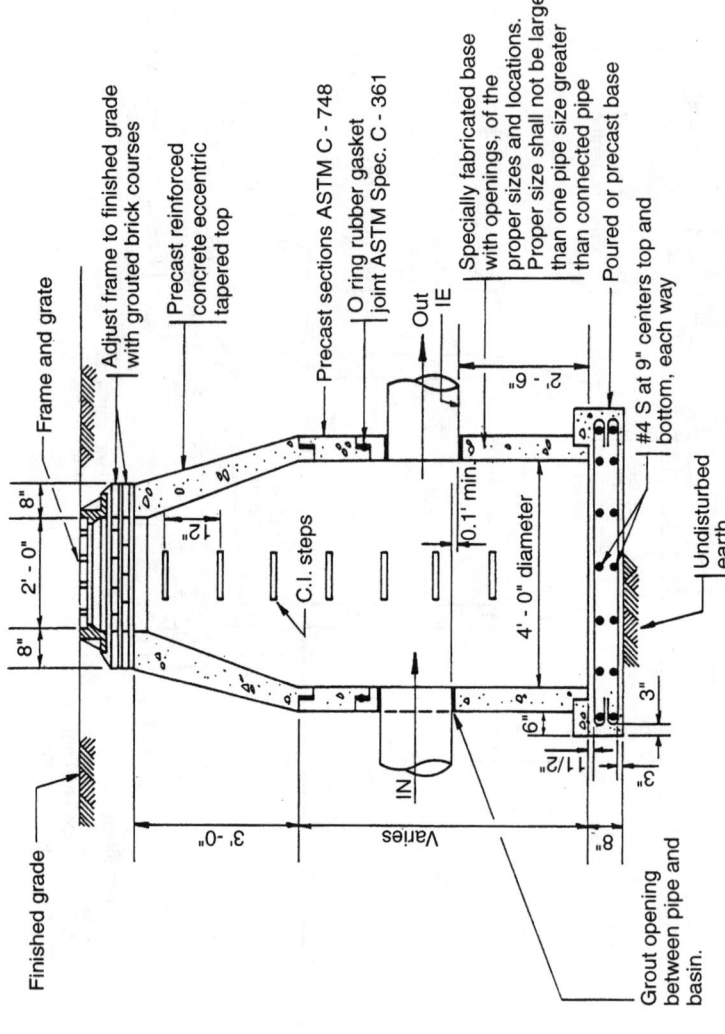

Finished grade

Frame and grate

Adjust frame to finished grade
with grouted brick courses

Precast reinforced
concrete eccentric
tapered top

Precast sections ASTM C - 748

O ring rubber gasket
joint ASTM Spec. C - 361

Specially fabricated base
with openings, of the
proper sizes and locations.
Proper size shall not be larger
than one pipe size greater
than connected pipe

Poured or precast base

#4 S at 9" centers top and
bottom, each way

Undisturbed
earth

Grout opening
between pipe and
basin.

C.I. steps

Out

IE

IN

8"

2' - 0"

8"

12"

0.1' min.

4' - 0" diameter

2' - 6"

3"

6"

1 1/2"

3"

3' - 0"

Varies

8"

FIGURE 6.9 Typical catch basin.

6.14

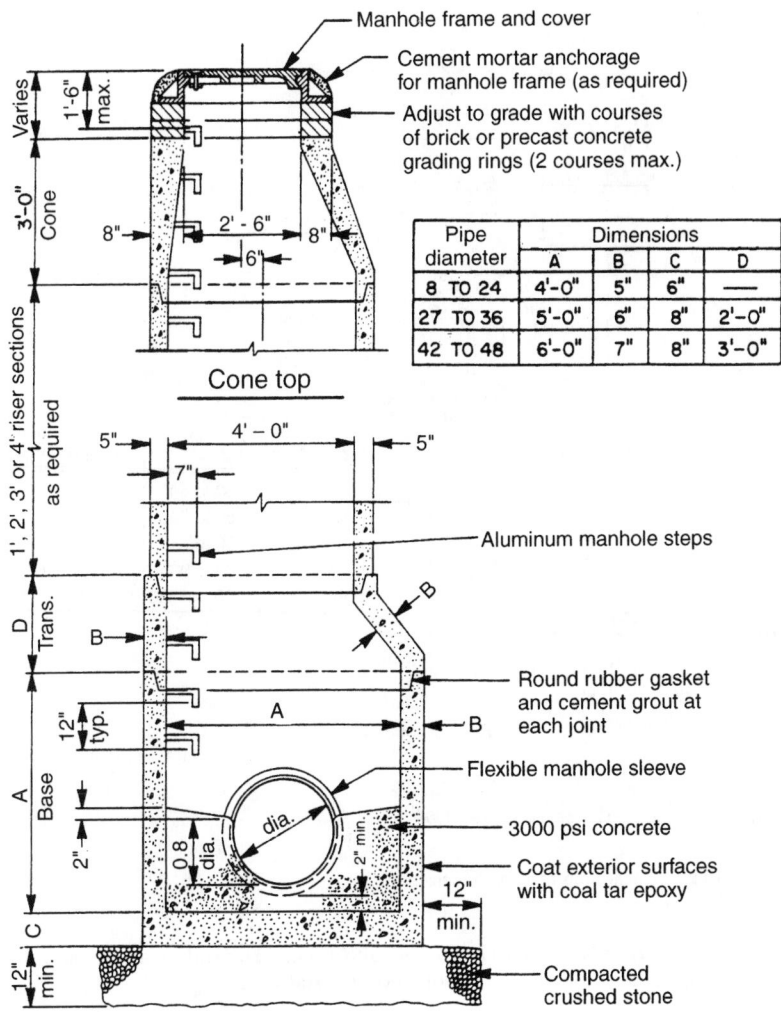

FIGURE 6.10   Typical precast manhole.

Pipe diameter	Dimensions			
	A	B	C	D
8 TO 24	4'-0"	5"	6"	—
27 TO 36	5'-0"	6"	8"	2'-0"
42 TO 48	6'-0"	7"	8"	3'-0"

- A plain grate (or gutter grate) is flat and allows water to enter from all unobstructed sides. Bars prevent the entrance of debris.
- A curb grate has a single vertical opening in the side of a curb. Water inters only from the front. Debris can enter freely.
- A combination grate is a plain grate and curb grate acting as a single unit with one placed in front of the other.
- Multiple inlets are two closely placed inlets acting together.
- A round grating cannot fall into a drainage structure.
- Flat grates are assumed to be flush with grade and flooded.

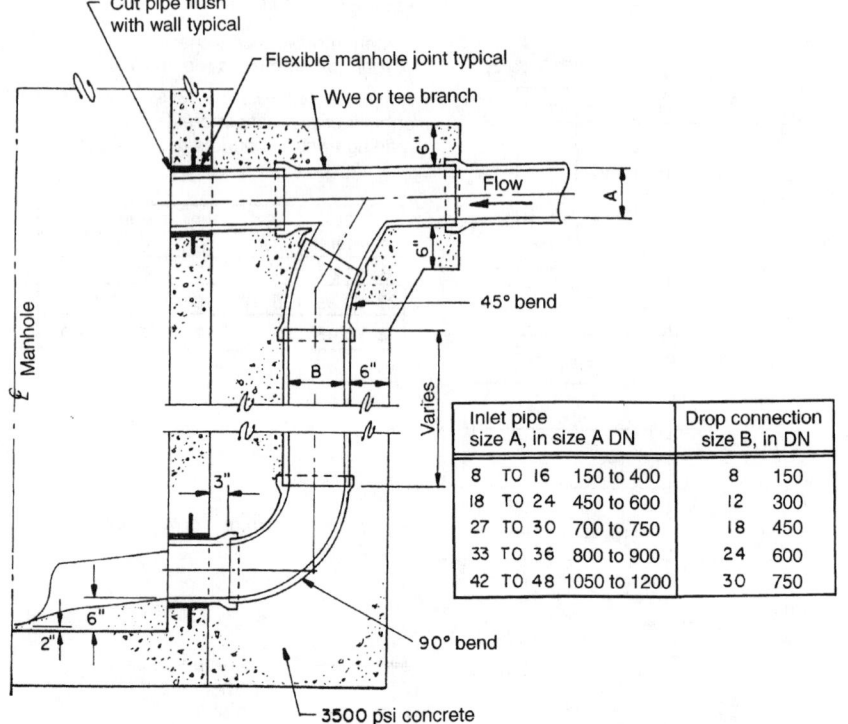

FIGURE 6.11   Typical drop manhole connection.

Grates Installed in Flat Areas (1 Percent or Less).   The following empirical formulas have been developed for the selection of gratings:

Plain Grate

$$SF \times Q = C L H^{2/3} \qquad (6.2)$$

where Q = quantity of water entering grate, cfs
       C = constant of 3.0
       L = total perimeter of grate, ft
       H = depth of water, above level of road, ft
       SF = safety factor

Discussion:

1. H is usually 2 to 3 in, but never more than 6 in.
2. A safety factor is required to allow for blockage by debris. In paved areas, use 1.25. Other areas use 1.5 or 2.0, according to conditions of possible debris accumulation. 2.0 is not unusual.

3. Where one side of a grate is against a curb, only three sides are used to compute the perimeter.

4. Figure 6.12 depicts a graphical solution of inlet capacity for a flooded plain grate, based on Eq. (6.2).

5. Use a manufacturer's catalog to select a standard size grate with the calculated perimeter as a minimum starting point.

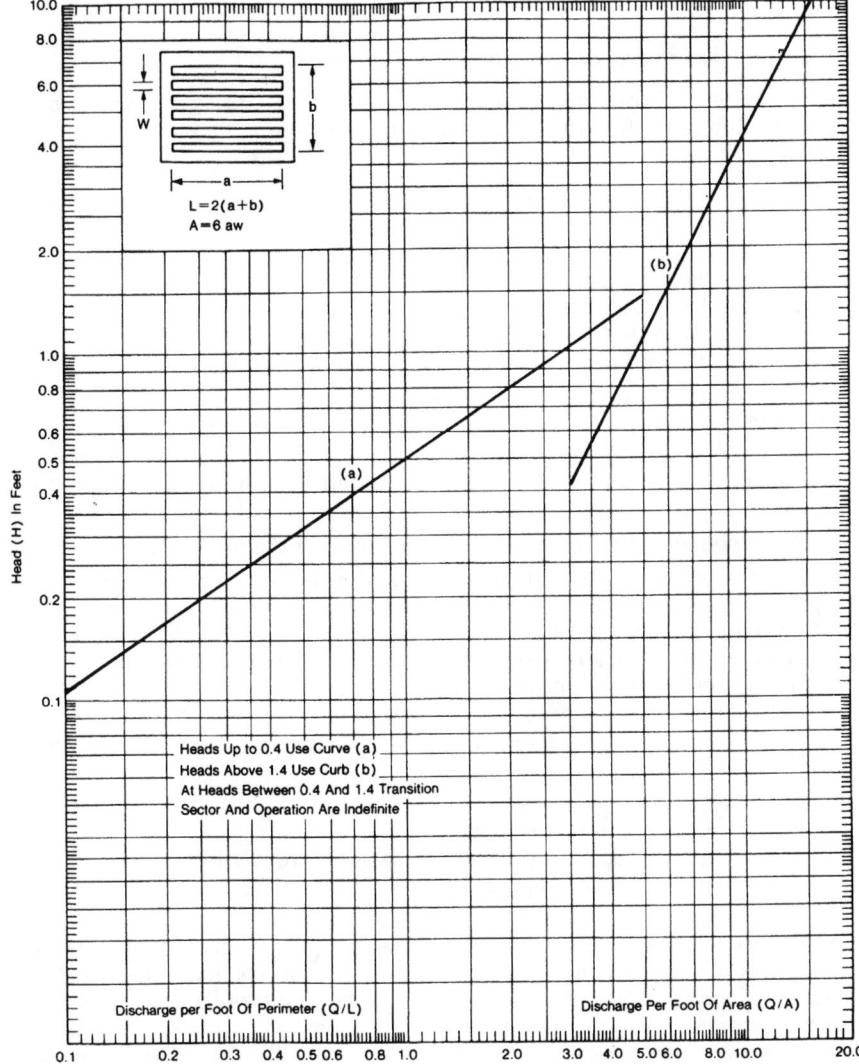

FIGURE 6.12 Graphical solution for a plain flooded grate. (*Courtesy Bureau of Public Roads, Baltimore County.*)

Curb Grate. This type of inlet is rarely used alone unless conditions do not permit any other type and, in addition, favorable inlet depression and contour is possible. Figure 6.13 can be used to determine flooded curb grating size.

Combination Grate. Experiments have shown that combination grates do not admit substantially more water than a flat grate; therefore, Eq. (6.2), for the plain grate, should be used to determine the size. An additional 10 percent for paved areas and up to 75 percent for other areas should be added as safety factors plus an additional 25 to 75 percent as a debris factor. This type of grate is preferred where provision to collect all storm water is required. The curb grate acts as a safety valve when a flat grate is clogged with debris.

Grates Installed in Sloped Areas (Greater Than 1%)

Plain Grate. The size of a flat grate located in a sloping roadway can be determined from the trial length derived from the following formula.

$$L = \frac{V}{2} \sqrt{Y + D} \qquad (6.3)$$

where L = minimum length of opening, ft
   V = velocity of water in gutter, fps
   Y = depth of water in curb, ft
   D = thickness of grate, ft (from a manufacturer's catalog)

Discussion:

1. A minimum length of 3 ft (1 m) and a minimum width of 2 ft (1.3 m) are recommended.

2. Net opening of the grate should be 50 percent of width or greater.

3. The nomograph in Fig. 6.14 should be used to determine water velocity in a gutter or channel.

Curb Grate. This type of inlet is very inefficient, and a long length is usually required to collect the entire design flow. The capacity for complete interception is determined from the following formula.

$$Q = 0.7L(A + Y)^{1.5} \qquad (6.4)$$

where Q = capacity of grate, cfs
   L = length of clear opening, ft
   A = depth of depression at inlet, ft (below roadway level)
   Y = depth of water in gutter, ft (above roadway level)

Combination Grate. The capacity of combination inlets has been determined by a series of complex equations. Since actual experiments have proven that the capacity of combination inlets is not much greater than that of flat inlets, it is much easier to design the grate as if it were flat. Since some allowance can be made for the combination grate, the size of the flat grate portion can be reduced if the calculated size is larger than that of the standard size produced by a manufacturer.

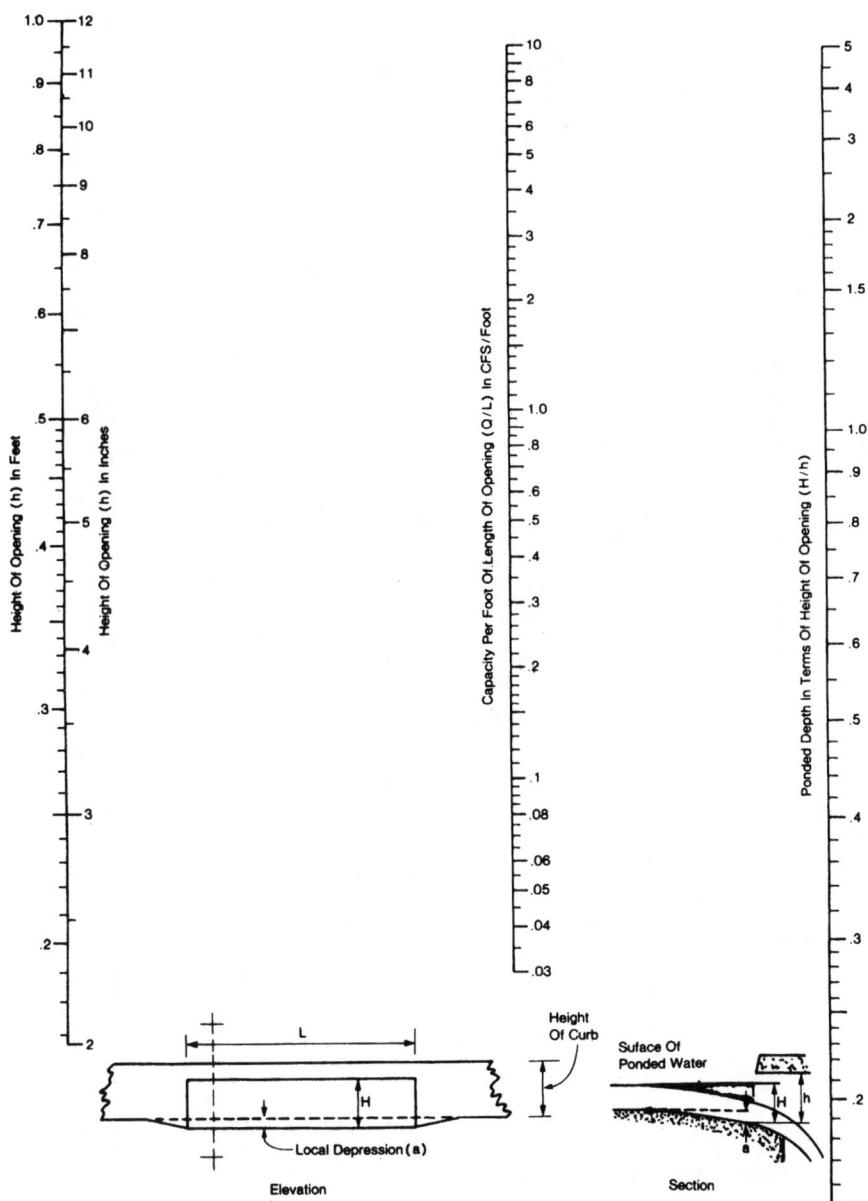

FIGURE 6.13 Graphical solution for a curb grate.

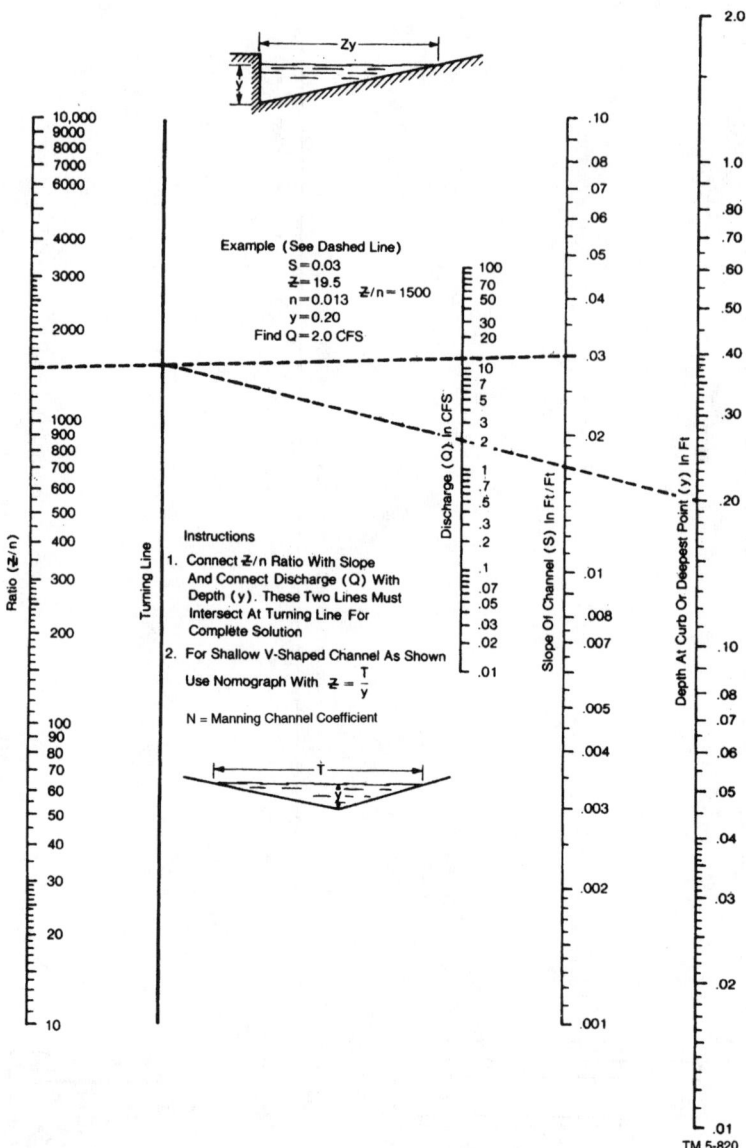

FIGURE 6.14 Water velocity in a gutter.

SYSTEM DESIGN CRITERIA

General

The *Rational Method* will be used to determine theoretical water inflow into a storm sewer system. It has been widely used since its introduction slightly before the turn of the century and has a proven record of acceptability. It is an empirical formula well suited for the design of small watersheds. The formula is

$$Q = A \times I \times R \tag{6.5}$$

where Q = quantity of storm water runoff, cfs
 A = area to be drained, acres
 I = imperviousness factor of surface comprising drainage area, dimensionless
 R = rate of rainfall, in/h

The following discussion will provide an understanding of the basic values that will be used in the Rational Method formula.

Design Storm

The Rational Method reduces an inexact set of conditions into an exact formula. Such variables as rainfall rate, overland water flow, and the amount of storm water that actually reaches the DI can never be exactly determined. Because of this unpredictability, it is important to realize that some judgment is required in applying the calculated values and given information in the design of a storm water system.

The storm water drainage system is designed to remove the maximum amount of expected runoff as quickly as it falls to avoid ponding or flooding. The ability to calculate the flow rate is complicated by the inability to accurately predict many of the factors affecting the actual amount of runoff resulting from any given storm. In order to calculate the maximum estimated runoff, an artificial "design storm" must be created. This simulation can be used to predict runoff volume accurately enough to provide a basis for the piping network design.

The design storm is based on actual rainfall records, and has been presented in a convenient form by the National Oceanic and Atmospheric Administration/National Weather Service (NOAA). Design storms are available as either intensity-duration-frequency curves or as charts and formulas appearing in several technical memoranda covering different areas of the United States. The intensity-duration-frequency charts are in a very convenient, easy-to-use format and are considered accurate for the small watersheds that are the subject of this handbook. Charts for various cities throughout the United States are presented in Fig. 6.15. However, because of the longer period of data collection, the charts have been superseded by technical memorandum *NWS Hydro 35,* which must be considered the latest information available. This handbook presents the intensity-duration-frequency charts only because the values obtained from them fall well within the acceptable accuracy for calculations for small watersheds and because the charts are easy to use.

Imperviousness Factor

The imperviousness factor allows for the loss of rainwater as it flows over the ground from a remote point of the DI tributary area until it enters the DI. The causes for this loss include infiltration of water into the soil, ponding, and water remaining on vegetation. The average figures for imperviousness of various surfaces are found in Table 6.2.

Inspection, if possible, is the best way to determine the nature of an existing surface. In places where the exact nature of future development is uncertain, the figure selected should represent the minimum expected loss, to allow the largest quantity of storm water to reach the DI. Where any area consists of different types of surfaces and/or soil combinations, a weighted overall value may be assigned for ease of calculation.

Rate of Rainfall (Rainfall Intensity)

The rate, or intensity of rainfall, is obtained from the rainfall intensity-duration-frequency curves, shown in Fig. 6.15. The intensity is measured in in/h. In order

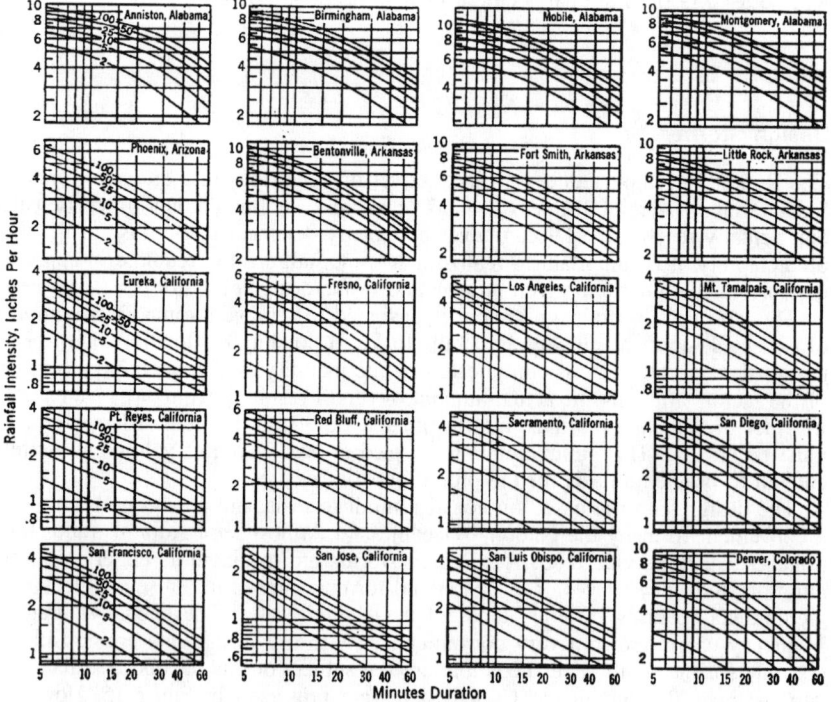

FIGURE 6.15 Rainfall intensity-duration-frequency charts.
The rainfall intensity-duration-frequency curves are abstracted from Technical Paper No. 25 of the Weather Bureau. Included are data for 200 selected stations in the United States and Puerto Rico. The data are substantially in the form found in the Technical Paper, except that the curves are cut off at the 60-minute duration line, enabling data to be presented in six pages.

Rainfall data in this form are intended for use in designing the modest storm drainage system associated with buildings and with industrial plants and their surrounding parking and lawn areas.

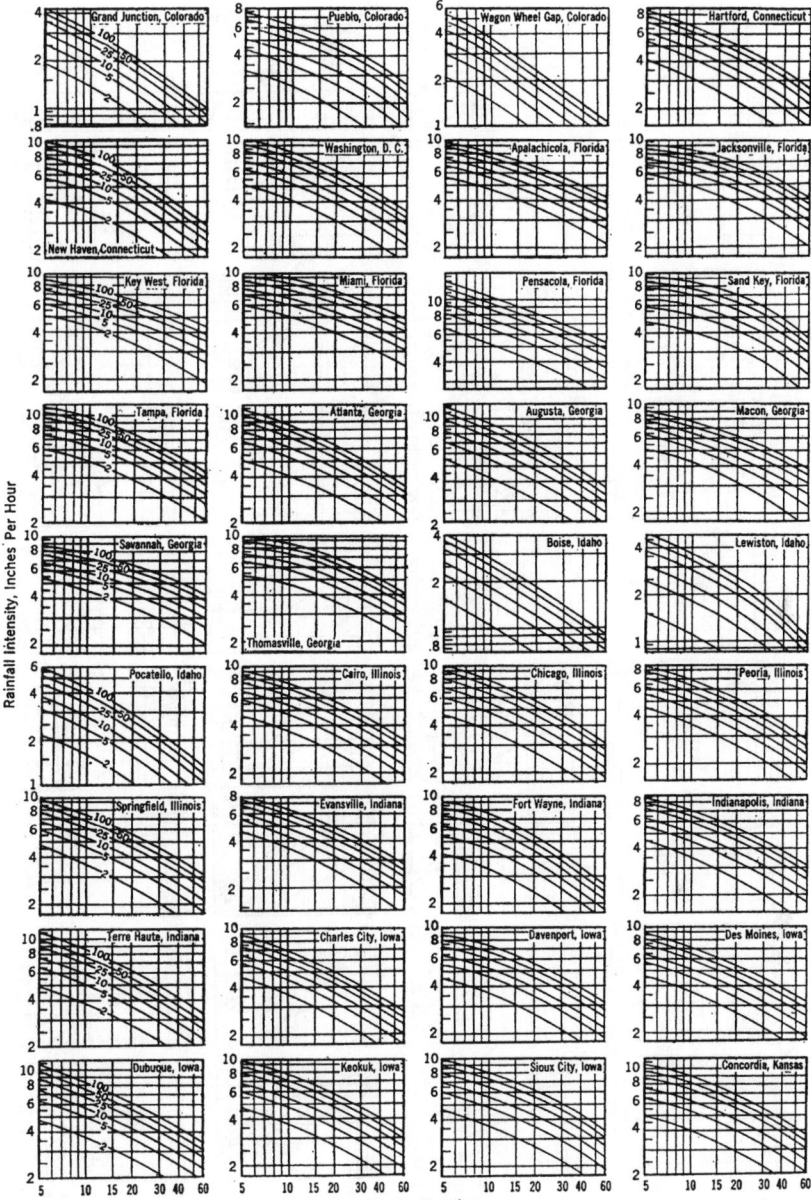

FIGURE 6.15   (*Continued*)

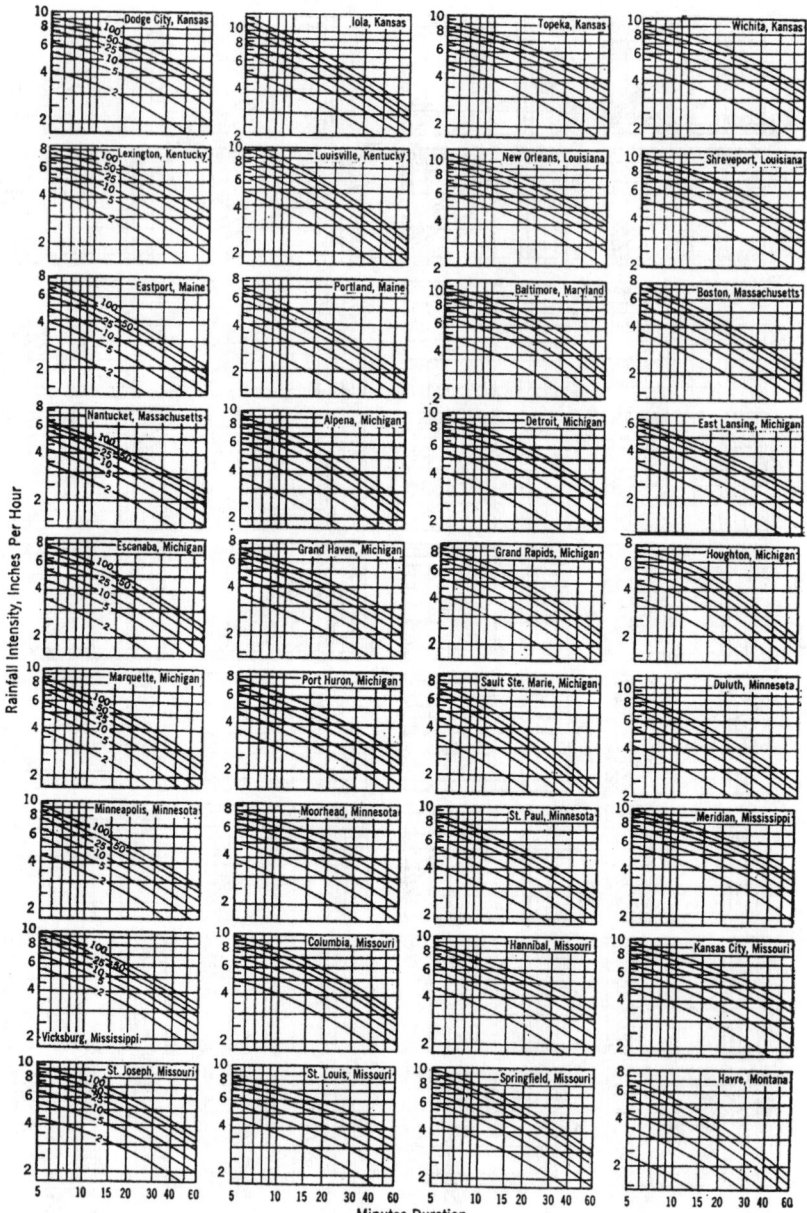

FIGURE 6.15 (*Continued*)

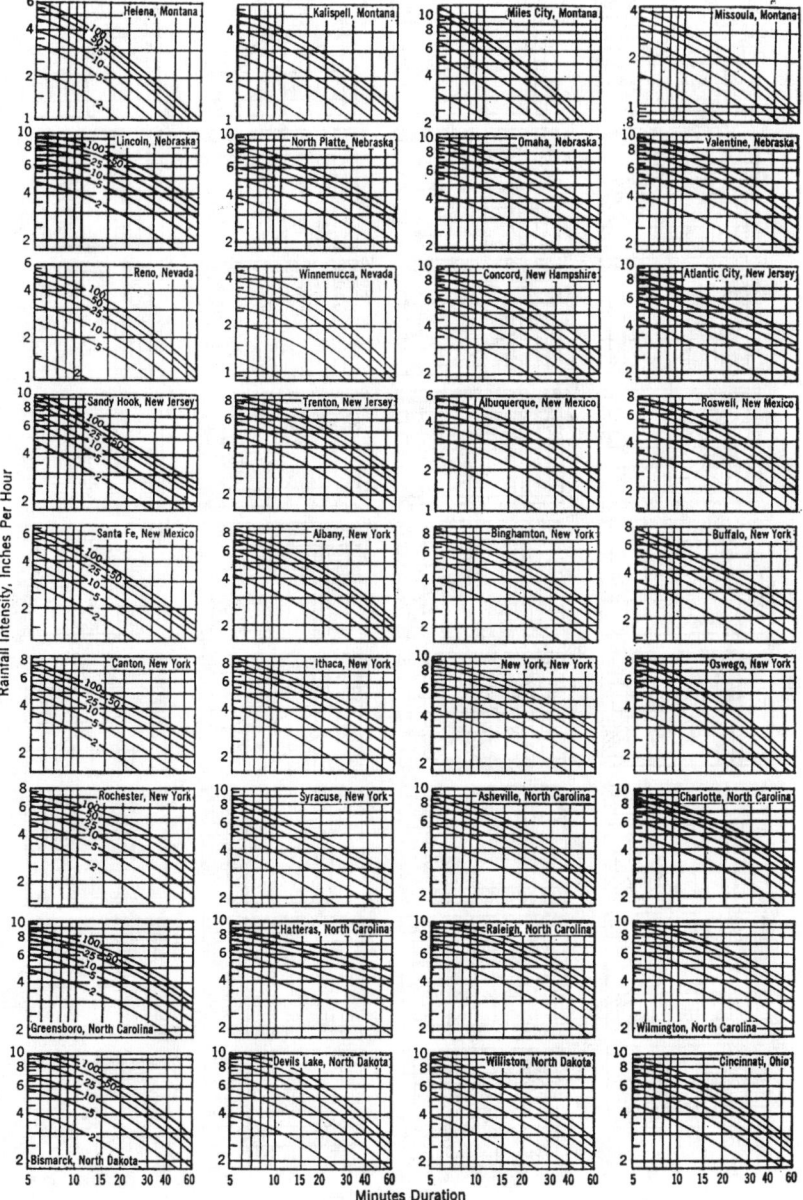

FIGURE 6.15 (*Continued*)

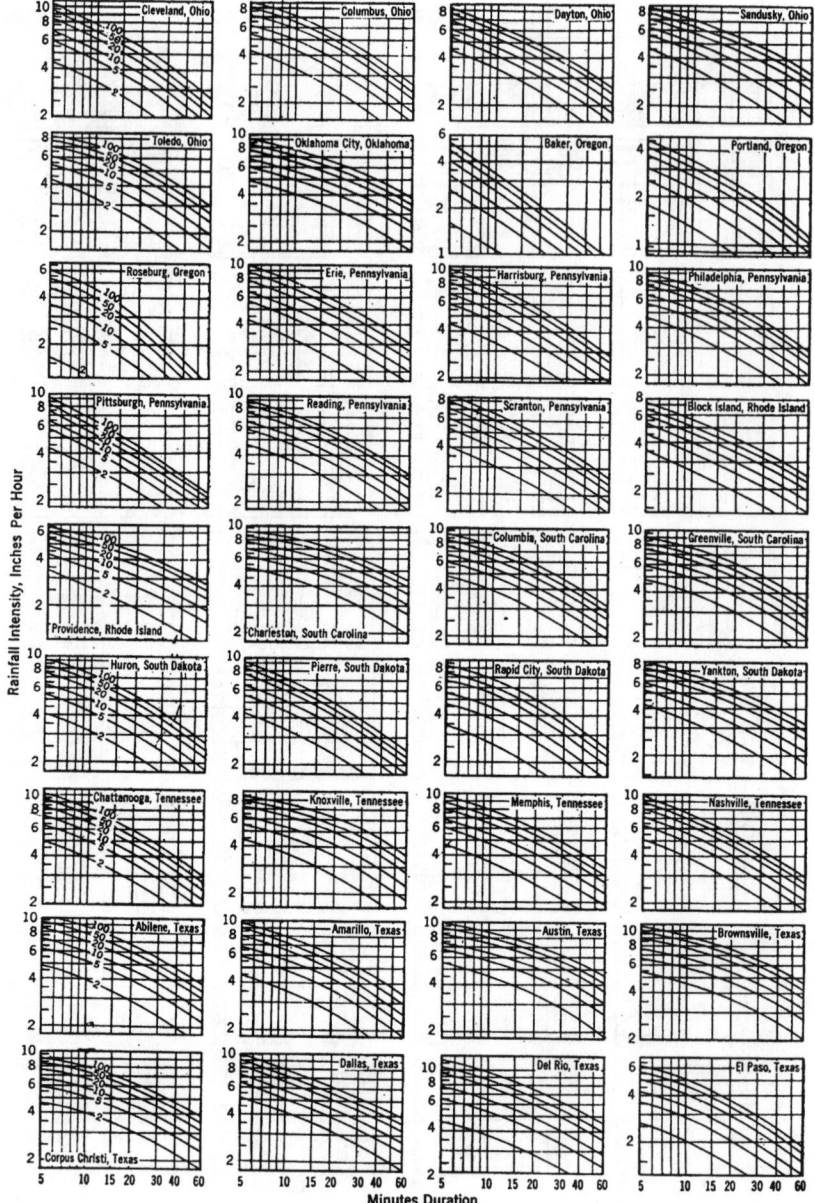

FIGURE 6.15   (*Continued*)

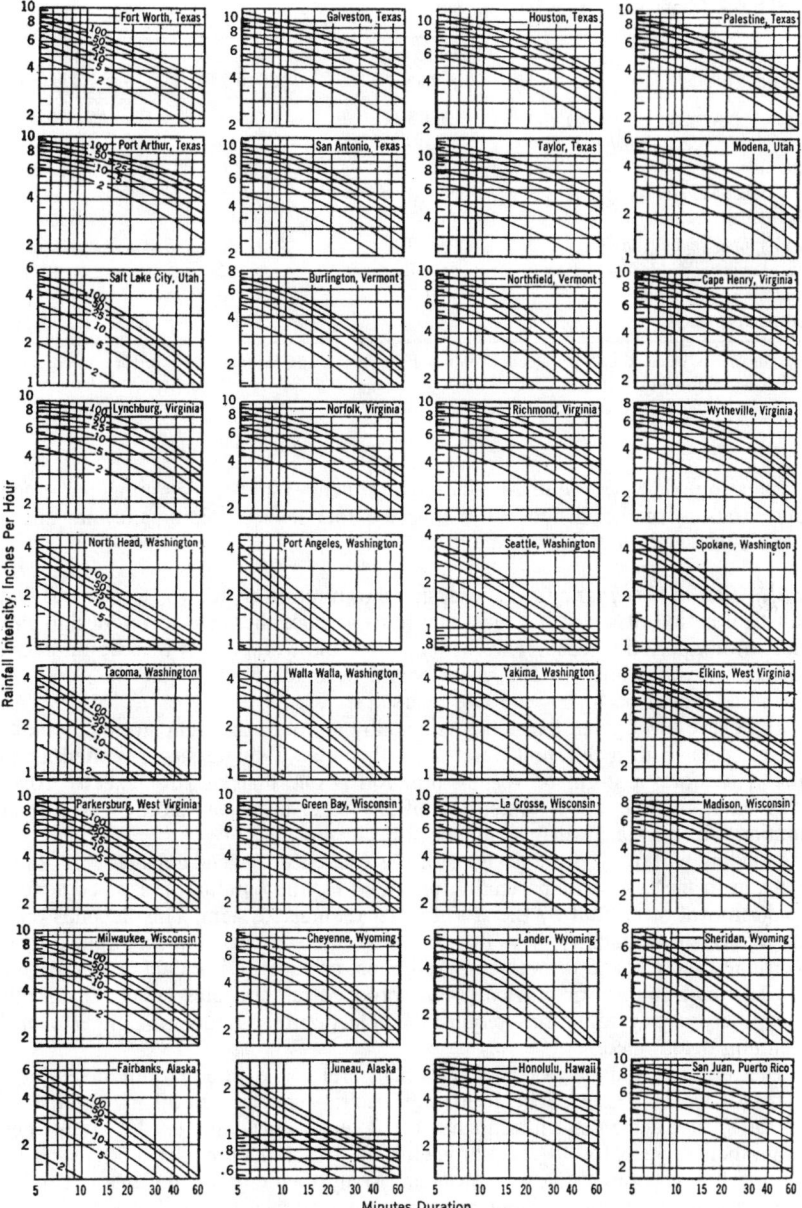

FIGURE 6.15 (*Continued*)

TABLE 6.2 Imperviousness Factor for Surfaces

Surface	Flat slope, less than 2%	Average slope, 2.0 to 7.0%	Steep slope, 7.1% or more
Roofs & pavement (all types)*	0.95	0.95	0.95
Clay-sparse vegetation	0.40	0.55	0.70
Clay-lawn	0.15	0.20	0.30
Clay-dense vegetation	0.10	0.15	0.20
Clay-dense woods	0.07	0.12	0.17
Sand-sparse vegetation	0.20	0.30	0.40
Sand-lawn	0.07	0.12	0.17
Sand-dense vegetation	0.05	0.10	0.15
Sand-dense woods	0.03	0.08	0.13

*For ease of calculation, the factor for roofs and pavement could become 1.0 if excessive runoff will not result in a significant overdesign.

Source: Baltimore County Design Standards.

to enter Fig. 6.15, the return period (or frequency) and the duration (or time of concentration) of the design storm must be found. Determining the intensity requires consideration of many factors.

Return Period (Frequency).   The design storm return period (or frequency) is the statistical period of years that must elapse to produce the most severe design storm once in that period of time. The theory is that the worst storm that would ever be expected to occur once in 100 years would be much more severe than the worst storm that would occur once in two years.

The curve for a given frequency is actually a plot of different storms of varying lengths, each with a different duration, rather than a single storm plotted against time. Therefore, a storm of 10-min duration is one that lasts for exactly 10 min and then stops. It is not an extension of any other plot. Any storm with a longer duration will have a lower instantaneous peak flow.

Flooding will result if the design storm is exceeded. Special consideration should be given to the degree of protection provided for the building and its contents by rapid removal of rainfall by the storm water drainage system. This depends on the importance of the facility, how flooding may affect access, the importance of uninterrupted service, and the value of equipment or material installed or stored. Since a severe thunderstorm or hurricane may produce rainfall rates greater than anticipated, the property value may require the selection of a frequency longer than the minimum determined.

The return period can be based on client preference, code requirements, or the degree of safety desired. The following criteria may be helpful as a guide. For average sites, a design storm frequency of 10 years is generally used. Certain clients design their projects for 40 years of useful life, which may require a 50-year design storm if flooding might cause a problem. Many municipal authorities use a 2- or 5-year design storm for city sewers. Some use a 10-year storm. The U.S. Corps of Engineers normally uses a 10-year storm, but for military installations a 25-year storm is used.

Duration (Time of Concentration).   The duration (or time of concentration) is measured in minutes. It is found by calculating the overland flow time and the time in pipe of a theoretical drop of water from the most remote point on the site to any design point on a branch or main drainage line.

It is important because the shorter the time, the more intense the rate of rain. This heavy rainfall, multiplied out to 1 h, would be much greater than the rainfall expected for a 1-h period. As an example, for New York City, the heaviest total amount of rain on record for a 5-min period is 0.75 in. Multiplied out to determine the 1-h rate using the 5-min reading gives an hourly figure of 9 in of rain per hour. Yet the largest amount of rainfall every recorded over a 1-h period is 3.11 in. It can thus be seen that the rate of rainfall measured during a short period is much more intense than the rate measured over a longer period of time.

The underlying principle is that the rainfall will stop at the exact moment the entire design area is contributing to the flow in this particular section of the sewer. Therefore, calculations must establish the shortest amount of time a drop of water takes to run from the farthest point of the site to any design point of the sewer. This is necessary so that the shortest time of concentration is obtained and the largest hourly rainfall rate is used for design purposes.

Overland Flow Time (Inlet Time). The overland flow time, often called the inlet time, is the number of minutes it takes a drop of water to travel from the farthest point of the area contributing flow to a DI until it spills into the inlet. The impedance to overland flow of water by a grass surface is greater than that of an asphalt surface. For equal distances and slope, it takes longer for a drop of water to enter a DI from the farthest point of a lawn than from a paved parking lot. Slope of the land is also a factor, since for any given surface, the steeper the slope, the faster water will flow.

Figure 6.16 offers a solution to determine overland flow time when slope of grade and type of surface are known. During a rainstorm, the flow of water from surfaces covered with pavement is in the form of a very shallow sheet of water covering the surface of the ground. This phenomenon is known as *sheet flow.* Finding the velocity of sheet flow in feet per second and multiplying that figure by the actual distance in feet gives the time in seconds. This should be converted to minutes to calculate the overland flow time.

Time in Pipe. The time in pipe is the number of minutes the theoretical drop of water takes to flow from the DI to the design point. This requires that the pipe be sized and the velocity of the storm water known.

Finding the velocity in feet per second and multiplying that figure by the actual pipe length in feet gives the time in pipe, in seconds. This should be converted to minutes to calculate the time in pipe.

Final Calculation. Adding the overland flow time and the time in pipe will give the time of concentration. Experience has shown that for most projects, use of a short duration will result in excessive runoff volume and excessive pipe sizes. It is recommended that a 10-min duration be used as the minimum for typical design conditions.

SYSTEM DESIGN PROCEDURE

1. The drainage structures should be located, and final coordination completed, with site planner. Select the type of drainage structure, either DI or CB.

2. The return period for the design storm should be chosen.

3. A site plan showing the location of the drainage structures and contour lines should be used to determine the exact surface area and type of surface contributing to flow into each DI. A chart or tabulation in any convenient form should be prepared indicating the type of surface and the area in square feet of each type of

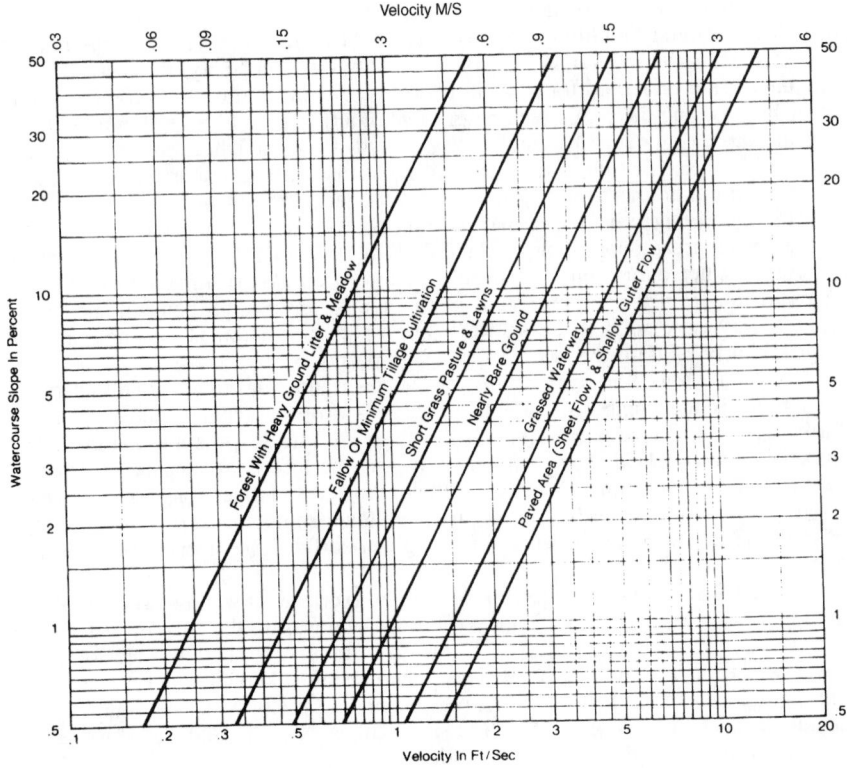

FIGURE 6.16 Overland flow time.

surface draining into each DI. Room should be left for additional information that will be added later. For each DI, the distance from the farthest point on the watershed that could contribute water to the DI should be measured. Also, the slope of the land into each DI should be calculated.

4. The imperviousness factor of the various surfaces in the watershed should be determined, keeping in mind the average slope of the area and the type of surface or surface combinations. A weighted factor for areas having more than one surface type should be used and recorded on the chart for reference.

5. Using Fig. 6.16, the overland flow time into the furthest DI should be determined using the slope of the ground and the type of surface. The minimum flow time should be 10 min. Rarely will a smaller time be economically justifiable.

6. Utilizing the rainfall intensity-duration-frequency curves (Fig. 6.15) for the city closest to the project location, the rainfall intensity should be selected using the return period and the time of concentration previously chosen. For each station, six curves are given. Although specifically so designated in the first column only, each set corresponds to storm frequencies (or return periods) of 2, 5, 10, 25, 50, and 100 years. For the normal industrial site, the 10-year curve should be used. The more conservative the design, the larger the design period that will be chosen.

7. From data in the form just prepared, and using the rational formula, the total inflow to each individual DI should be calculated. A weighted factor should be used for areas having more than one surface type, and this information placed in the form for reference. The individual pipe from each DI will be sized from this information.

8. With the total inflow in cfs to individual DIs now known, the grate type and size can be selected. The grate type should be entered on the form for reference. DI locations should be adjusted for flow requirement if necessary. Most manufacturers have developed inlet flow charts for specific grates and their catalogs will give the grate size that can accept the calculated flow.

9. The piping system from the DI to the point of disposal can now be laid out. MHs can be located and the pipe material selected.

10. Building roof drain runouts shall be connected to the storm water drainage lines and their drainage areas noted.

11. The critical inlet is the one that produces the maximum combination of inlet time plus flow time in sewer to the very first connection with any other branch. The critical inlet should be chosen now. The inlet at the furthest end of the drainage system should be selected as the starting point. This point will be the longest in time, not necessarily in distance. This may require some trial calculations of several DIs at various far ends in order to determine which drainage inlet actually is critical. Distance will not be the only criterion. An area composed of asphalt, which has a fast inlet time, might be a much greater distance from an inlet than an area consisting of grass, which has a longer inlet time. The slope and length of sewer pipe must also be considered.

12. With the selection of the critical inlet, the individual sewer pipe line can now be sized from the critical inlet up to the first point of intersection with any other contributing source of storm water. The layout of the sewer system will establish the slope of the pipe, and the form will provide the flow rate of storm water in cfs. The pipe material will have been selected and value of n for the pipe for use in the Manning formula will be known. Refer to Table 6.3 to select the n value. Entering Fig. 6.17 with the pipe material, connecting a straight line from the flow rate to the slope of piping will now establish the pipe size and water velocity. The necessary adjustment must be made, as indicated by the conversion factor if a different n is used from that of the chart. The desired figure is now multiplied by

TABLE 6.3 n Value of Pipe Used in the Manning Formula

Pipe material	Range of n values	Generally accepted value*
Asbestos-cement	0.011–0.015	0.013
Corrugated metal pipe	0.022–0.026	0.024
Cast iron	0.011–0.015	0.013
Concrete pipe	0.011–0.015	0.013
Ductile iron (cement lined)	0.011–0.015	0.013
Plastic pipe, all kinds	0.010–0.015	0.011
Steel pipe	0.012–0.020	0.015
Vitrified clay	0.011–0.015	0.013

*Values will vary based on condition of pipe.

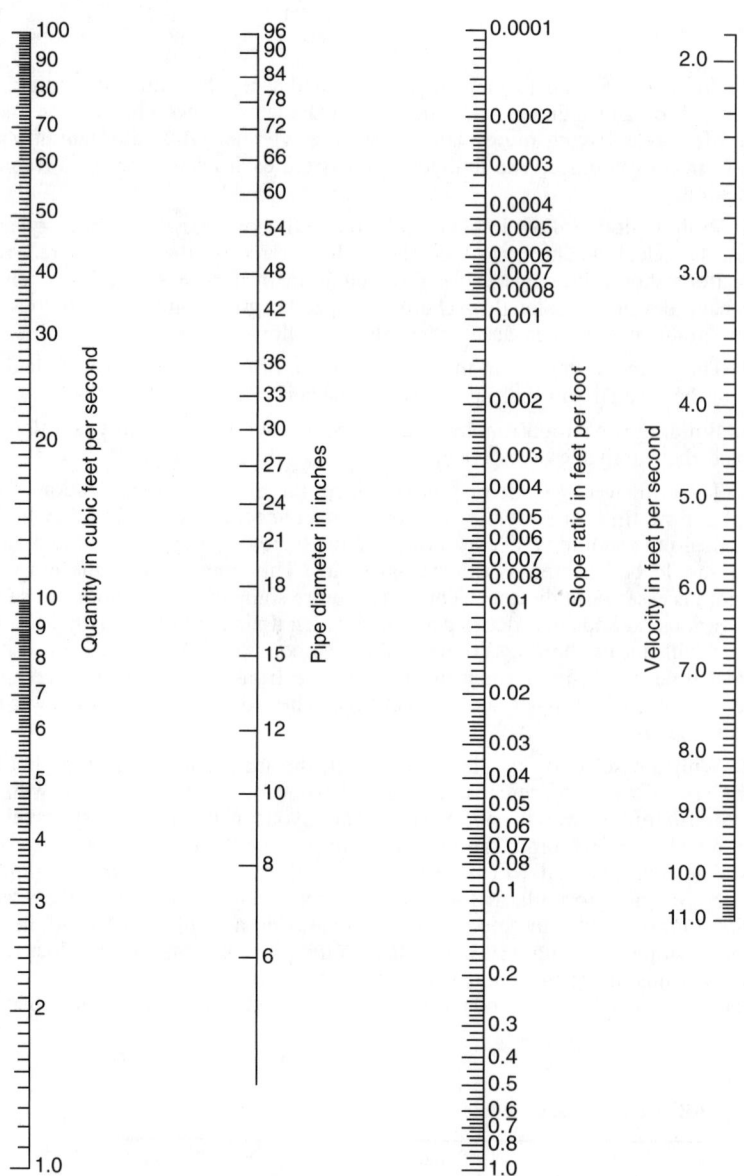

Value of n	0.008	0.010	0.011	0.012	0.013	0.015	0.019	0.021	0.024
Conversion factor for discharge and velocity	1.62	1.30	1.18	1.08	1.00	0.87	0.68	0.62	0.54

FIGURE 6.17 Pipe flow chart. Diagram for solution of Manning formula for circular pipes flowing full ($n = 0.013$).

the conversion factor found under the actual n. Generally accepted practice assigns a value of 0.013 to most pipe materials.

13. When the first junction with another contributing source has been reached, the entire area of all contributing sources up to this first junction should be totalled. The new time of concentration should be calculated from the most remote inlet and the time in pipe to the design point of the storm sewer. Entering the intensity-duration-frequency chart to find the rainfall rate will now give all necessary information required for insertion into the Rational Method formula [Eq. (6.5)] to find total cfs. The resultant flow into the pipe is used to size that particular segment downstream from the design point. This is continued in succession to the end of the system, requiring a longer time of concentration each time a new design point is reached.

14. As progress is made along the route of the sewer, all branches to the main sewer line should be sized individually as their connection is reached. The branch is sized using the time of concentration required for that branch only. To size the main, all the areas contributing to the flow at any particular junction should be added up, using the longer time of concentration determined from the initial overland flow time from the critical inlet added to the flow time of the water in the pipe.

SYSTEM DESIGN CONSIDERATIONS

1. The velocity in the piping system should not be less than 2 fps (0.6 m/s), to allow for the movement of sediment and any other solids in the pipe.

2. The maximum velocity should be limited to approximately 30 fps to prevent erosion of the pipe interior. If large quantities of sand are present in the water, the scouring action will increase and so the velocity should be reduced for this condition.

3. To limit velocity, the slope of the piping system should be decreased rather than using oversized pipe. There are other methods available, also. If the slope of the ground is steep, a drop manhole should be used where the difference in elevation between inlet and outlet inverts is more than 2 ft (0.6 m).

4. A manhole should be located at every change in pipe size, slope, or direction. In many cases, a drainage structure can be used for this purpose if such placement is practical.

5. The minimum size pipe should be 12 in (DN 300) to reduce the possibility of stoppage by debris. A smaller size for runout from a building roof storm water system is generally acceptable.

6. When a change in pipe size occurs, the pipe should be installed crown to crown to eliminate a surcharge of the upstream portion of the pipe.

7. Factors such as snowmelt may add additional unexpected quantities of water. Snowmelt does not usually add a significant amount of inflow, but it may if the depth of snow is large. However, increasing one pipe size may lead to a problem by reducing velocity too much. In that case, it is advisable to increase the slope of the sewer line. As a general rule, one should be a little generous with pipe sizes. The slight increase in cost will be more than made up by the safety factor of additional capacity provided.

8. Exfiltration occurs when water leaks out of a pipe through bad joints or cracks into the surrounding ground. When a pipe passes under a road or railroad, a joint should be chosen that will permit very little or no leakage to avoid washing away the subgrade.

9. Piping should be installed roughly parallel to the slope of the ground. Standard percentages of slopes if possible should be used to permit contractors to easily install the pipe.

STORM WATER RETENTION METHODS

There are circumstances where it is not possible or desirable to remove storm water from a site as quickly as it collects. Some built up urban areas generally have sewer systems originally designed for low projected land development and population density, which may result in the design outflow from the site exceeding the permitted flow into the sewer or the environment. Therefore, many cities have adapted a storm water management program to limit the outflow of storm water from developed sites to a flow rate equal to that before the development was started. Similarly, if a small stream is used as an outfall, it may not be possible to drain a large paved area that was formerly woodland into this stream without flooding the downstream portion.

In these circumstances, some method of temporarily storing the storm water must be provided. The methods used most often are to retain the storm water on the roof, on the site, or both.

ROOF RETENTION

Refer to Chap. 9, Plumbing Systems, section entitled, Limited Discharge Roof Drainage Systems, for the method used to retain water on the roof.

SITE RETENTION METHODS

Storage of water on roofs, if allowed, provides only a small portion of the retention generally required for an entire site drainage system. Therefore, additional provisions must be made. For smaller sites where space is limited, this is best accomplished by temporarily retaining the excess storm water in a retention basin either at or below ground. Another popular method is to greatly oversize the discharge piping, in effect, using the pipe to store the required volume. On-ground retention is called *ponding*.

The least costly method is the oversized pipe. Substituting a 72-in (DN 1800) or larger pipe for the designed size is a possibility if the ground is deep enough to accommodate it. If not, a retention basin should be considered.

Several factors should be considered in the selection of a retention basin. First is cost. An owner may want a pond in an out-of-the-way corner of the site rather than the more expensive retention basin. Second is available depth. The depth of an underground basin may be limited by the invert of the sewer to which it would be connected, or the depth of the stream or river which would provide the final outfall. In many cases, an underground basin may be the only solution in the development of an urban site, possibly with a pump to reach a higher sewer line.

The design of a retention basin begins with the concept that the total volume of storage required will be the difference between storm water inflow and allowable outflow for a given period of time. This requires the calculation of three variables:

1. Outflow from basin, which varies as a function of time
2. Inflow to basin, which varies as a function of time
3. Storage, which is the difference between 1 and 2 above for a specific time period

The information for retention basin design has been extracted from Technical Release Number 55, U.S. Department of Agriculture, Soil Conservation Service. The methodology differs from that of the Rational Method for storm water design. For example, total volume of rainwater produced by a given storm during a particular time is used rather than rate of rainfall in inches per hour.

When original agricultural and wooded areas are replaced with new, less impervious surfaces such as buildings and roads, both the volume and the peak rate of discharge will increase. The additional peak rate of discharge for the storm period is the amount of water to be stored on site while being released at a lower predetermined flow. Following are brief explanations of the factors to be considered in obtaining a reliable estimate of the peak rate of discharge and total runoff volume:

1. *Storm distributions.* There are two storm distributions that are used for different climatic regions: Type I and Type II. Type I storms are located in maritime climates and are found only in limited areas west of the Rocky Mountains. Type II storms are found in regions where high rates of runoff are usually generated from summer thunderstorms or tropical storms.

2. *Hydrologic soil group.* There are groups of soils that have the same runoff potential under similar storm conditions. Nationwide, over 8000 specific soils have been classified into four soil groups as follows:

Group A: Low runoff potential soils having a high infiltration rate even when thoroughly wet and consisting chiefly of deep, well- to excessively drained sand or gravel

Group B: Soils having a moderate infiltration rate when wet and consisting chiefly of moderately deep to deep, moderately well- to well-drained soil with moderately fine to moderately coarse texture

Group C: Soils having a slow infiltration rate when wet and consisting of soil with a layer that impedes downward movement of water, or soil with moderately fine to fine texture

Group D: High runoff potential soils having a very slow infiltration rate when wet and consisting chiefly of clay soil with a high swelling potential, soil with a high permanent water table, soil with a clay pan or clay layer at or near the surface, and shallow soil over nearly impervious material

Caution is advised when determining the soil group for the final developed conditions, as machine compaction or removal of upper soil layers can change the group drastically. Advice of the Soil Conservation Service is recommended when determining the soil group.

3. *Average watershed slope.* The steeper the average slope of the watershed, the greater the peak discharge will be.

4. *Land use.* The most common effect of urbanization is reduced infiltration, resulting in increased runoff volume, decreased overland travel time, and higher peak rates of flow. The volume of runoff is governed primarily by infiltration characteristics, and is related to soil type, type of vegetative cover, impervious surfaces, and surface retention. Travel time is governed by slope, flow length, and surface roughness. The peak discharge is based on the relationship of these factors.

5. *Runoff curve number (CN).* This is a dimensionless number selected for use in calculations that takes into account the previous land use information. The CN allows information to be conveyed in easy tabular or graphical form.

6. *Amount of rainfall.* The criteria for the selection of rainfall quantity used to determine the design parameters for the storage requirements are different than those used to design the piping system. Total rainfall is used rather than the instantaneous rate. In fact, the total rainfall for an entire 24-h period is used because of a phenomenon called *abstraction,* which is the sum of interception, depression storage, and infiltration. *Interception* is rain caught by foliage, leaves, and twigs that evaporates before it reaches the ground. *Depression storage* is rain caught in low points on grade and thus not available as runoff. *Infiltration* is water absorbed into the ground. Investigations have shown that for areas having a CN of 60 to 65, 2 in (50 mm) of rain must fall before runoff even starts. For this reason, a 24-h period is chosen. A heavy concentration of rain falling later in a given storm event will produce a greater peak discharge than at the beginning of the same storm.

The 24-h time period is also used to determine total volume of runoff for the same reasons. As will be seen, modifications and assumptions must be made to properly estimate runoff from the rainfall amount chosen.

The first step in retention basin design is to calculate the peak rate of discharge from the existing site and then from the site as it will be developed. To do this, the CNs of both the original and developed site must be found. When there is only one type of land use and one soil group, this is easy. Otherwise, a weighted CN must be found. This is done by determining the CN for each type of soil condition found and the number of acres for each CN, and multiplying these two figures. The sum of all the products is divided by the total area to find the weighted CN.

EXAMPLE A site consists of 100 acres with two types of land use. One type is 70 acres of pasture land in good condition with soil group C, and the other type is 30 acres of grain field with soil group B. Find the weighted CN for the site.

solution Refer to Table 6.4 to find the runoff CN. The pasture has a CN of 88, the grain field, a CN of 71.

$$
\begin{array}{r}
70 \text{ acres} \times 88 = 6160 \\
\underline{+30 \text{ acres} \times 71 = +2130} \\
100 \qquad\qquad 8290
\end{array}
$$

$$\frac{8290}{100} = 82.9 \text{ (rounded to 83)}$$

Therefore, 83 is weighted CN for the entire site.

A similar calculation is made to find the CN for the developed site.

1. After the CN has been determined for both the undeveloped site and the developed site, the rate of rainfall and return period should be selected. Local rainfall standards must be used. However, if no standards exist, the maps in Figs. 6.18 (10-year, 24-h rainfall), 6.19 (25-year, 24-h rainfall), 6.20 (50-year, 24-h rainfall), and 6.21 (100-year, 24-h rainfall) should be used. Interpolation between isobars is necessary for intermediate values. A recommended average storm would be a 50-year, 24-h rainfall, Fig. 6.20.

2. Next, an adjustment to the 24-h rainfall figure must be made to allow for abstraction and the other considerations that were made when the CN was determined. Table 6.5 reduces the amount of runoff depth according to the various factors previously discussed.

TABLE 6.4 Runoff Curve Number

Land use description		Hydrologic soil group			
		A	B	C	D
Cultivated land*:					
without conservation treatment		72	81	88	91
with conservation treatment		62	71	78	81
Pasture or range land:					
poor condition		68	79	86	89
good condition		39	61	74	80
Meadow: good condition		30	58	71	78
Wood or forest land:					
thin stand, poor cover, no mulch		45	66	77	83
good cover†		25	55	70	77
Open spaces, lawns, parks, golf courses, cemeteries, etc.					
good condition: grass cover on 75% or more of the area		39	61	74	80
fair condition: grass cover on 50% to 75% of the area		49	69	79	84
Commercial and business areas (85% impervious)		89	92	94	95
Industrial districts (75% impervious)		81	88	91	93
Residential‡:					
Average lot size	Average % impervious§				
⅛ acre or less	65	77	85	90	92
¼ acre	38	61	75	83	87
⅓ acre	30	57	72	81	86
½ acre	25	54	70	80	85
1 acre	20	51	68	79	84
Paved parking lots, roofs, driveways, etc.¶		98	98	98	98
Streets and roads:					
paved with curbs and storm sewers¶		98	98	98	98
gravel		76	85	89	91
dirt		72	82	87	89

*For a more detailed description of agricultural land use curve numbers refer to *National Engineering Handbook,* Section 4, Hydrology, Chapter 9, Aug. 1972.

†Good cover is protected from grazing and litter and brush covered soil.

‡Curve numbers are computed assuming the runoff from the house and driveway is directed toward the street with a minimum of roof water directed to lawns where additional infiltration could occur.

§The remaining pervious areas (lawn) are considered to be in good pasture condition for these curve numbers.

¶In some warmer climates of the country a curve number of 95 may be used.

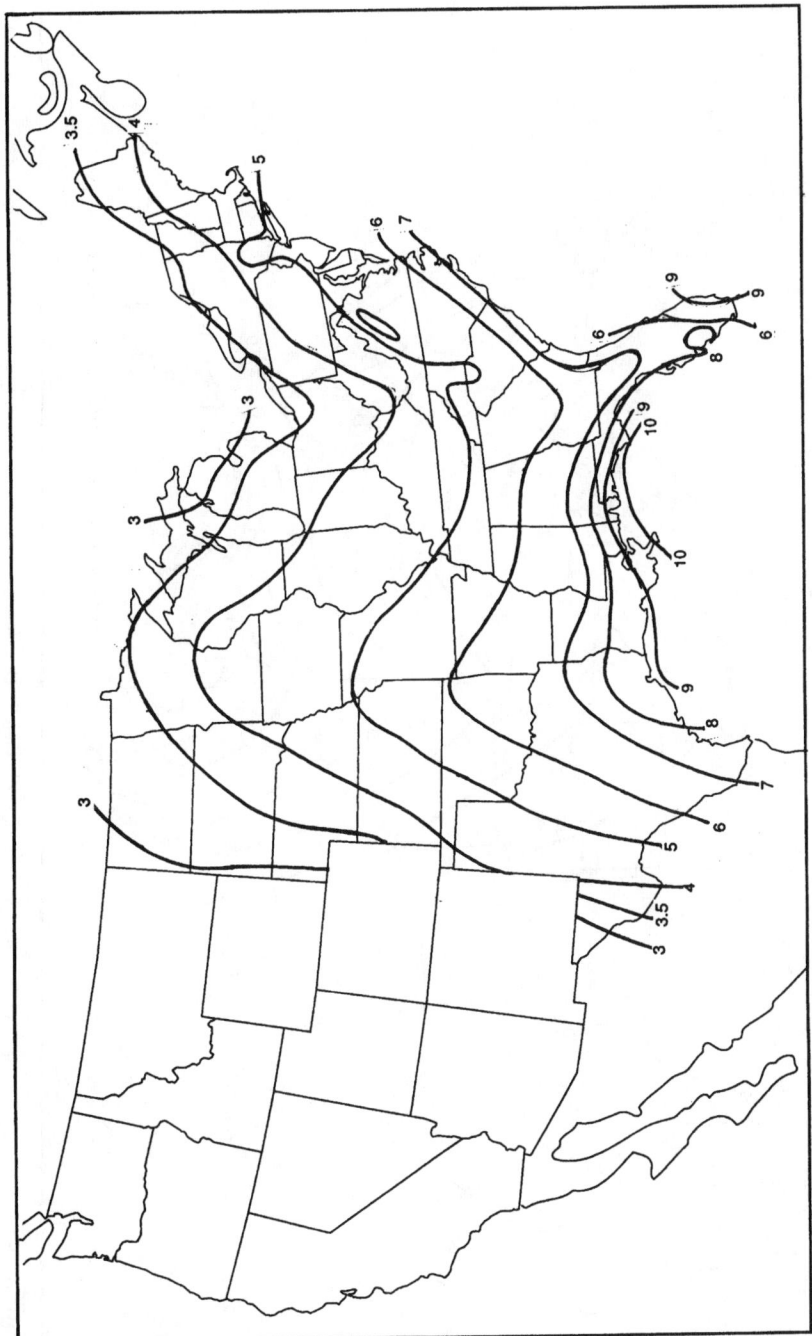

FIGURE 6.18 10-year, 24-hour rainfall. (*Adapted from U.S. Weather Bureau map.*)

6.39

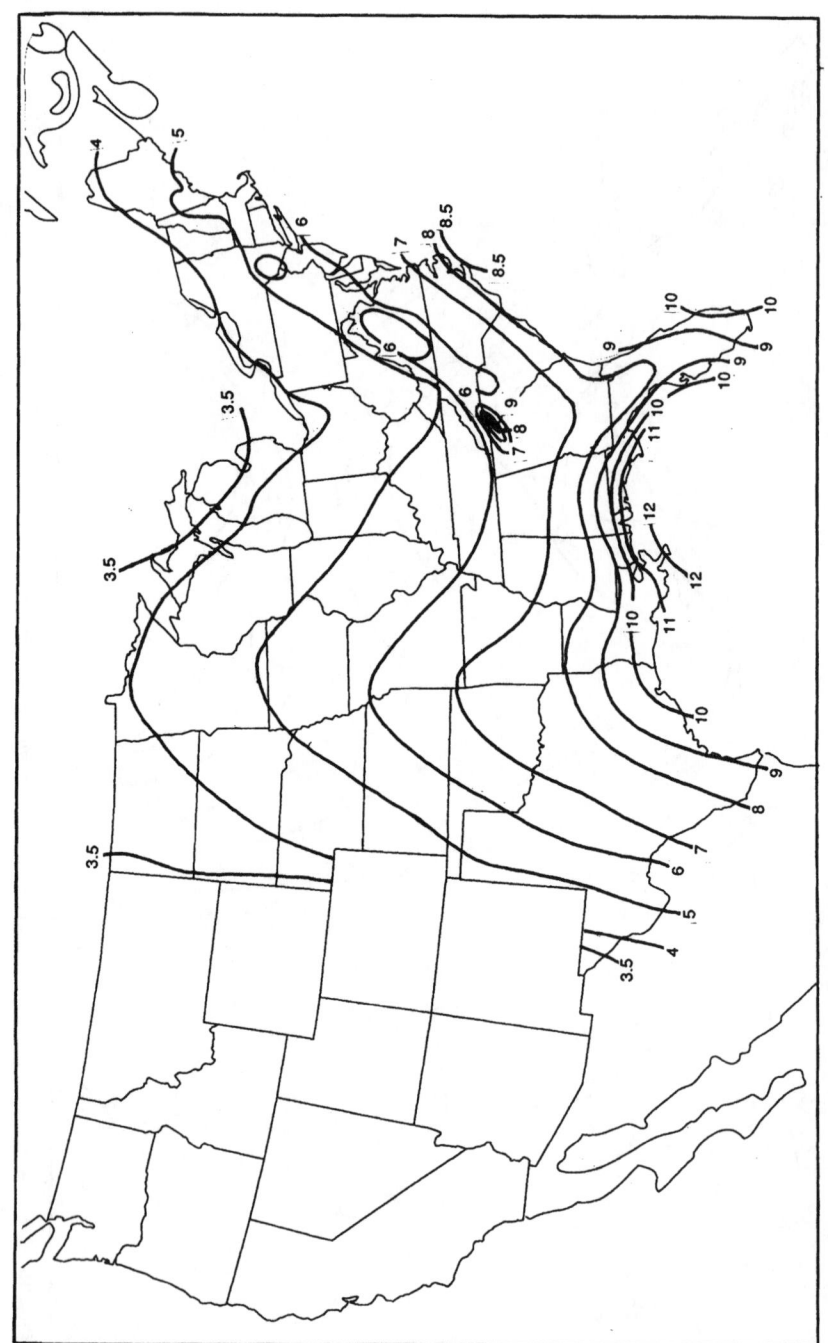

FIGURE 6.19 25-year, 24-hour rainfall. (*Adapted from U.S. Weather Bureau map.*)

6.40

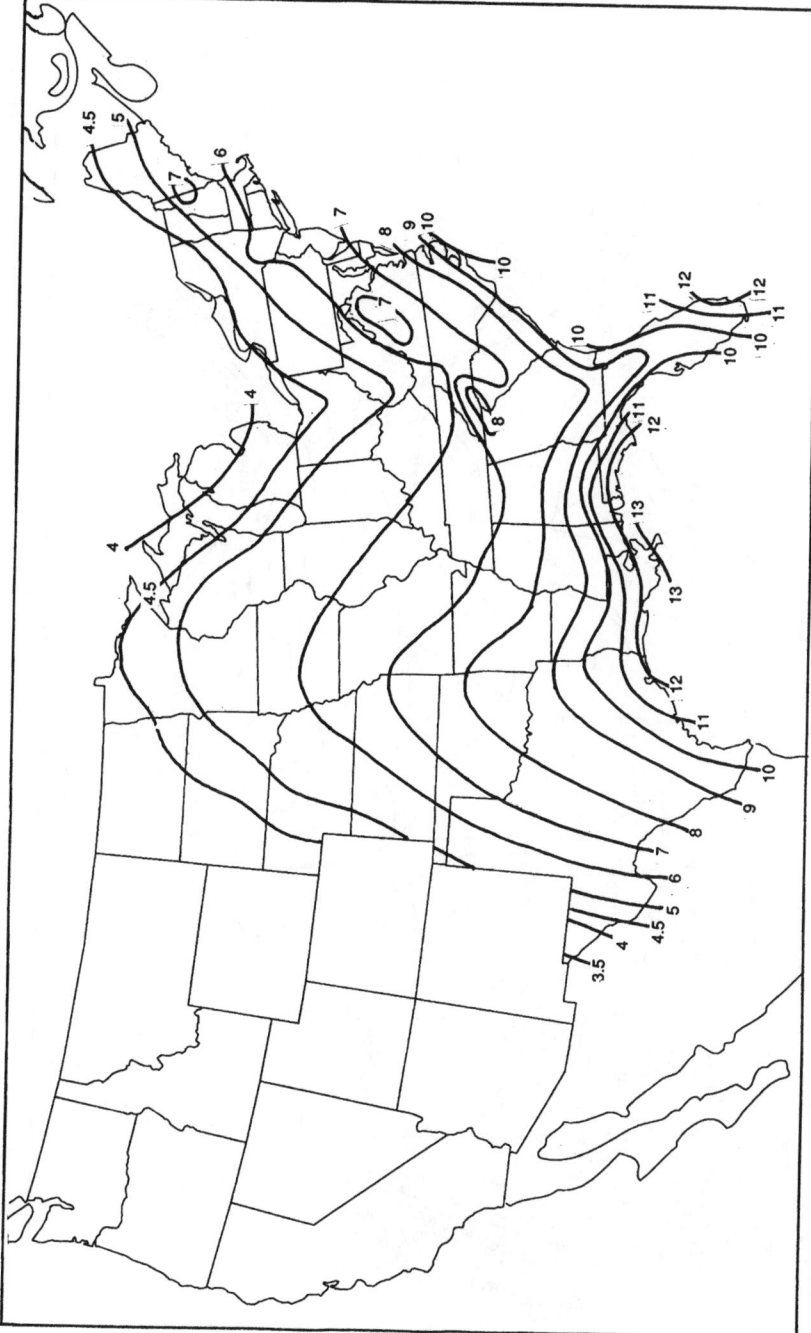

FIGURE 6.20 50-year, 24-hour rainfall. *(Adapted from U.S. Weather Bureau map.)*

6.41

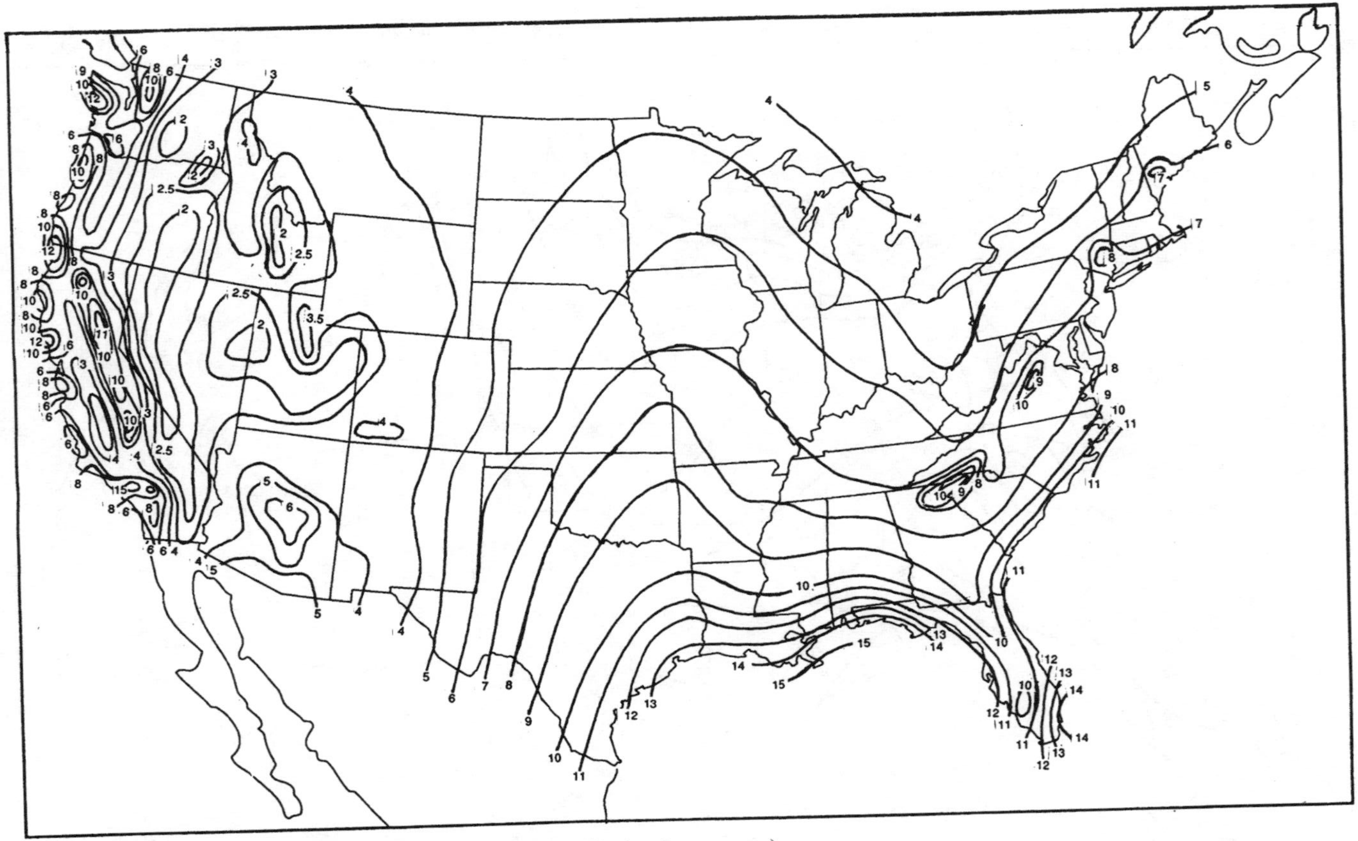

FIGURE 6.21 100-year, 24-hour rainfall. (*Adapted from U.S. Weather Bureau map.*)

6.42

TABLE 6.5 Reduction of Runoff Depth

Rainfall, in	Curve number (CN)*								
	60	65	70	75	80	85	90	95	98
1.0	0	0	0	0.03	0.08	0.17	0.32	0.56	0.79
1.2	0	0	0.03	0.07	0.15	0.28	0.46	0.74	0.99
1.4	0	0.02	0.06	0.13	0.24	0.39	0.61	0.92	1.18
1.6	0.01	0.05	0.11	0.20	0.34	0.52	0.76	1.11	1.38
1.8	0.03	0.09	0.17	0.29	0.44	0.65	0.93	1.29	1.58
2.0	0.06	0.14	0.24	0.38	0.56	0.80	1.09	1.48	1.77
2.5	0.17	0.30	0.46	0.65	0.89	1.18	1.53	1.96	2.27
3.0	0.33	0.51	0.72	0.96	1.25	1.59	1.98	2.45	2.78
4.0	0.76	1.03	1.33	1.67	2.04	2.46	2.92	3.43	3.77
5.0	1.30	1.65	2.04	2.45	2.89	3.37	3.88	4.42	4.76
6.0	1.92	2.35	2.80	3.28	3.78	4.31	4.85	5.41	5.76
7.0	2.60	3.10	3.62	4.15	4.69	5.26	5.82	6.41	6.76
8.0	3.33	3.90	4.47	5.04	5.62	6.22	6.81	7.40	7.76
9.0	4.10	4.72	5.34	5.95	6.57	7.19	7.79	8.40	8.76
10.0	4.90	5.57	6.23	6.88	7.52	8.16	8.78	9.40	9.76
11.0	5.72	6.44	7.13	7.82	8.48	9.14	9.77	10.39	10.76
12.0	6.56	7.32	8.05	8.76	9.45	10.12	10.76	11.39	11.76

*To obtain runoff depths for CNs and other rainfall amounts not shown in this table, use an arithmetic interpolation.
† 1 in = 24.5 mm.

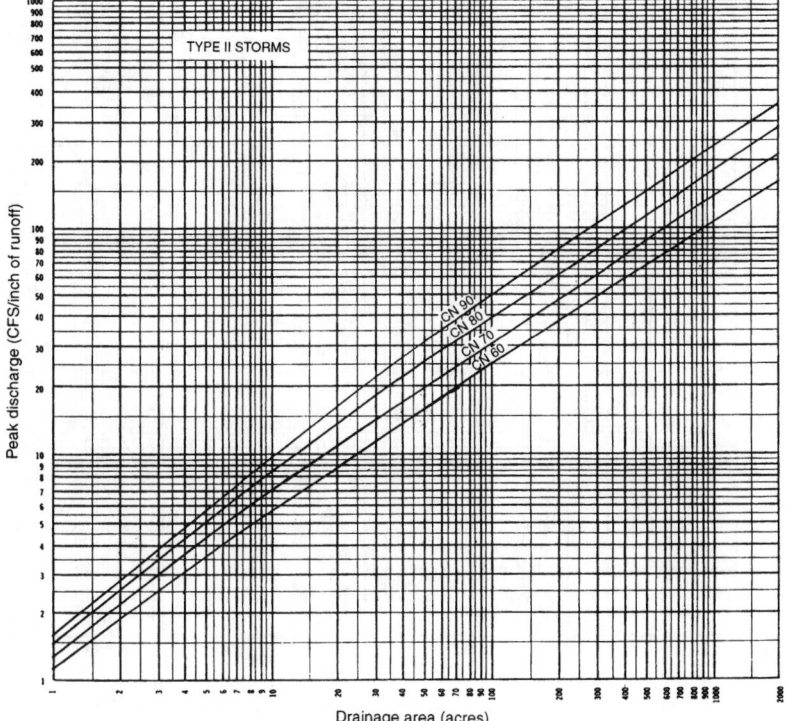

FIGURE 6.22 Adjustment to peak rate of discharge, flat slope, up to 2 percent.

6.43

3. Another item required for the determination of peak rate of discharge for the site is the adjustment to rainfall based on the average slope of the watershed. This is the average slope of grade to the storm water inlets, not of the piping system. Figures 6.22 (up to 2 percent slope), 6.23 (up to 7 percent slope), and 6.24 (up to 50 percent slope) give the discharge in cfs per inch of adjusted rainfall on the site, according to the average slope for Type II storms. Table 6.6 gives adjustment factors for intermediate slopes other than those found in Figs. 6.22 to 6.24. Type I storm tables are not available.

With the CN calculated, it is now possible to calculate the peak rate of discharge for the site. For example, a 300-acre site is to be developed in the northern part of New Jersey. The present weighted CN has been determined to be 75. The average slope of the watershed is found to be 4 percent. A 50-year, 24-h storm has been selected.

1. The actual rainfall must be found. From Fig. 6.20, for a 50-year, 24-h storm, 6 in of rain is read.

2. The actual rainfall must be adjusted. From Table 6.5, for present CN of 75 and 6 in of rain, an adjusted figure of 3.28 in of rain is used. This number is known as VR.

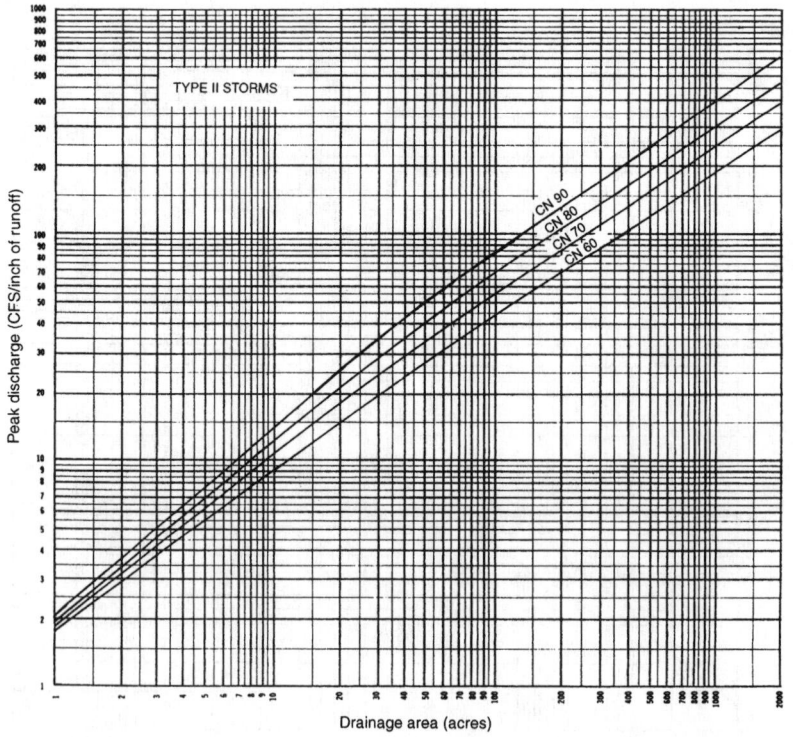

FIGURE 6.23   Adjustment to peak rate of discharge, moderate slope, up to 7 percent.

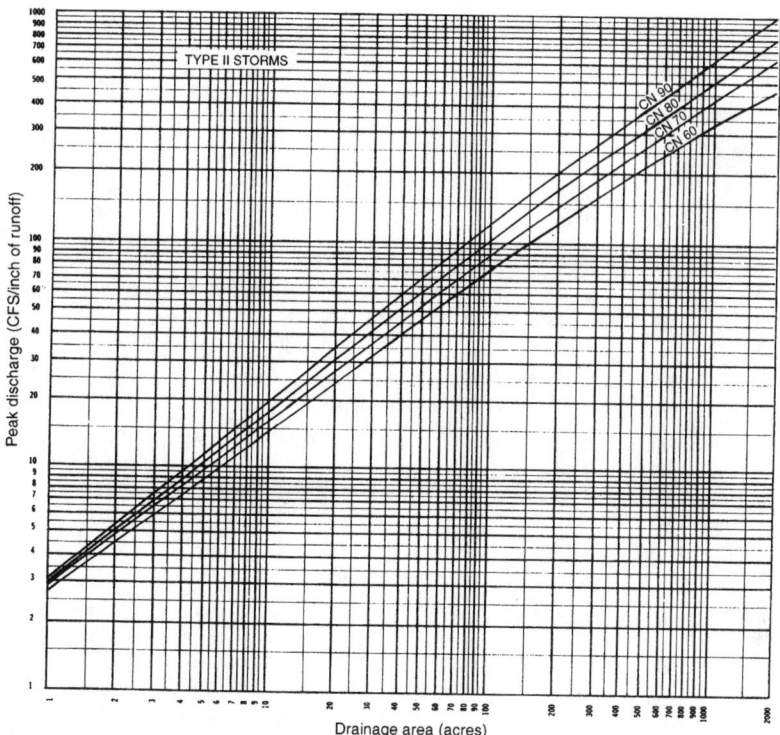

FIGURE 6.24 Adjustment to peak rate of discharge, steep slope, up to 50 percent.

3. The actual peak discharge must be determined based on original slope of the site. From Fig. 6.23 (for a 4 percent slope) enter with a CN of 75 and a watershed area of 300 acres. Read 125 cfs per in of rainfall. Therefore, $125 \times 3.28 = 399$ cfs peak discharge for the undeveloped 300-acre site. (Table 6.6 should be used for intermediate slope values, if required.)

To establish the peak discharges for the proposed development, two additional adjustments must be made to the rainfall amount. The first adjustment will increase the peak runoff due to the amount of impervious area that will replace the undeveloped land. The second will increase the peak runoff due to the faster overland flow rate that will result from a change in surface characteristics. This results in a shorter time of concentration. Continuing with the site as before, the next part of the procedure follows.

1. The future condition weighted CN must be calculated. It is found to be 80.

2. Assuming the same slope, the actual peak discharge must be found for future condition. Using Fig. 6.23, this time entering with a CN of 80 and 300 acres, 140 cfs per inch of rainfall is read. The future base discharge is $140 \times 3.28 = 459$ cfs peak discharge.

TABLE 6.6   Adjustment to Peak Rate of Discharge, Intermediate Slopes

Slope (percent)	10 acres	20 acres	50 acres	100 acres	200 acres	500 acres	1000s acres	2000 acres
				Flat slopes				
0.1	0.49	0.47	0.44	0.43	0.42	0.41	0.41	0.40
0.2	0.61	0.59	0.56	0.55	0.54	0.53	0.53	0.52
0.3	0.69	0.67	0.65	0.64	0.63	0.62	0.62	0.61
0.4	0.76	0.74	0.72	0.71	0.70	0.69	0.69	0.69
0.5	0.82	0.80	0.78	0.77	0.77	0.76	0.76	0.76
0.7	0.90	0.89	0.88	0.87	0.87	0.87	0.87	0.87
1.0	1.00	1.00	1.00	1.00	1.00	1.00	1.00	1.00
1.5	1.13	1.14	1.14	1.15	1.16	1.17	1.17	1.17
2.0	1.21	1.24	1.26	1.28	1.29	1.30	1.31	1.31
				Moderate slopes				
3	0.93	0.92	0.91	0.90	0.90	0.90	0.89	0.89
4	1.00	1.00	1.00	1.00	1.00	1.00	1.00	1.00
5	1.04	1.05	1.07	1.08	1.08	1.08	1.09	1.09
6	1.07	1.10	1.12	1.14	1.15	1.16	1.17	1.17
7	1.09	1.13	1.18	1.21	1.22	1.23	1.23	1.24
				Steep slopes				
8	0.92	0.88	0.84	0.81	0.80	0.78	0.78	0.77
9	0.94	0.90	0.86	0.84	0.83	0.82	0.81	0.81
10	0.96	0.92	0.88	0.87	0.86	0.85	0.84	0.84
11	0.96	0.94	0.91	0.90	0.89	0.88	0.87	0.87
12	0.97	0.95	0.93	0.92	0.91	0.90	0.90	0.90
13	0.97	0.97	0.95	0.94	0.94	0.93	0.93	0.92
14	0.98	0.98	0.97	0.96	0.96	0.96	0.95	0.95
15	0.99	0.99	0.99	0.98	0.98	0.98	0.98	0.98
16	1.00	1.00	1.00	1.00	1.00	1.00	1.00	1.00
20	1.03	1.04	1.05	1.06	1.07	1.08	1.09	1.10
25	1.06	1.08	1.12	1.14	1.15	1.16	1.17	1.19
30	1.09	1.11	1.14	1.17	1.20	1.22	1.23	1.24
40	1.12	1.16	1.20	1.24	1.29	1.31	1.33	1.35
50	1.17	1.21	1.25	1.29	1.34	1.37	1.40	1.43

3. The number of additional acres that will be covered with an impervious cover must now be determined and expressed as a percent compared to the existing condition. This includes areas that will be developed by adding buildings, roads, parking lots, and the like. Assuming that 25 percent will be covered, Fig. 6.25 gives the adjustment factor based on percent of imperviousness. Entering with the CN of 80 and percentage of new impervious area of 25, 1.16 is read.

4. The length of run along the slope of the new development that is modified, and which will affect the overland flow time, must also be determined and expressed as a percent of the total length. Assuming that 50 percent of the original hydraulic length is modified, Fig. 6.26 gives the adjustment factor based on hydraulic length modified. Entering with a CN of 80 and 50 percent length, 1.31 is read.

5. The future condition peak discharge can now be found by multiplying the base peak discharge by both the impervious area factor and the hydraulic length modified factors. Thus,

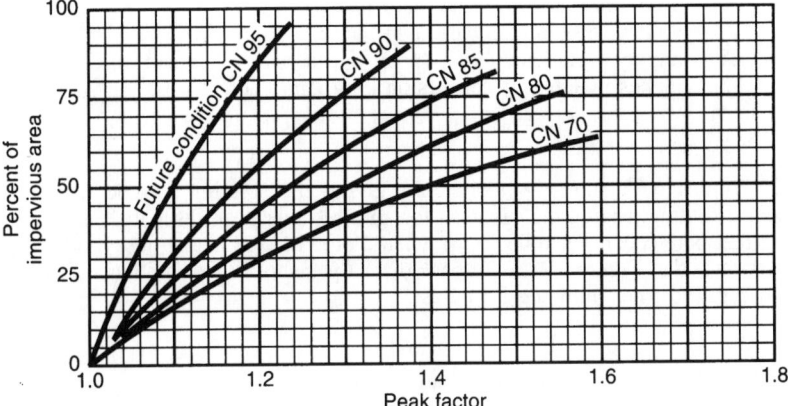

FIGURE 6.25 Factor for adjusting future condition runoff CN based on percent of imperviousness.

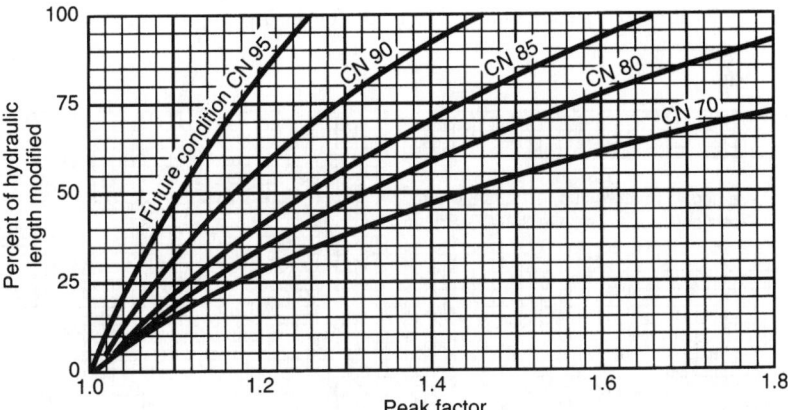

FIGURE 6.26 Factor for adjusting future condition runoff CN based on hydraulic length modified.

$$459 \times 1.16 \times 1.31 = 697.5 \text{ (rounded to 698 cfs)}$$

The effect of this proposed development is to increase the peak discharge from 399 cfs to 698 cfs.

Now that the basic watershed parameters have been established, the final design of the storage basin can proceed. This method is not as accurate as the computer program developed by the Soil Conservation Service for this purpose, but Eq. (6.6) will provide a conservative approach based upon average storage effects on peak discharges.

$$V_B = \frac{V_S \times A}{12} \tag{6.6}$$

where V_B = volume of storage basin, acre ft
$\quad\quad\quad V_S$ = volume of storage, watershed in
$\quad\quad\quad A$ = area of watershed, acres (note: 640 acres = 1 square mile)
$\quad\quad\quad 12$ = inches in 1 ft (for conversion of VS into acre ft)

Discussion:

1. Figures 6.27 and 6.28 are graphs giving the solution for the single-stage structure routing. Figure 6.27 is used for an allowable peak discharge of up to 300 cubic feet per second per mile (csm) when a pipe is used as the outlet, or 150 csm for a weir outlet. Figure 6.28 is for peak discharges above those amounts. To convert cfs to csm, use the following formulas:

$$\frac{cfs \times 640}{\text{watershed area in acres}} = csm \tag{6.7}$$

$$\frac{cfs}{\text{area in miles}^2} = csm \tag{6.8}$$

2. To use Fig. 6.27, two parameters must be known before entering the graph. One is the allowable peak discharge from the basin in csm, and the second is the

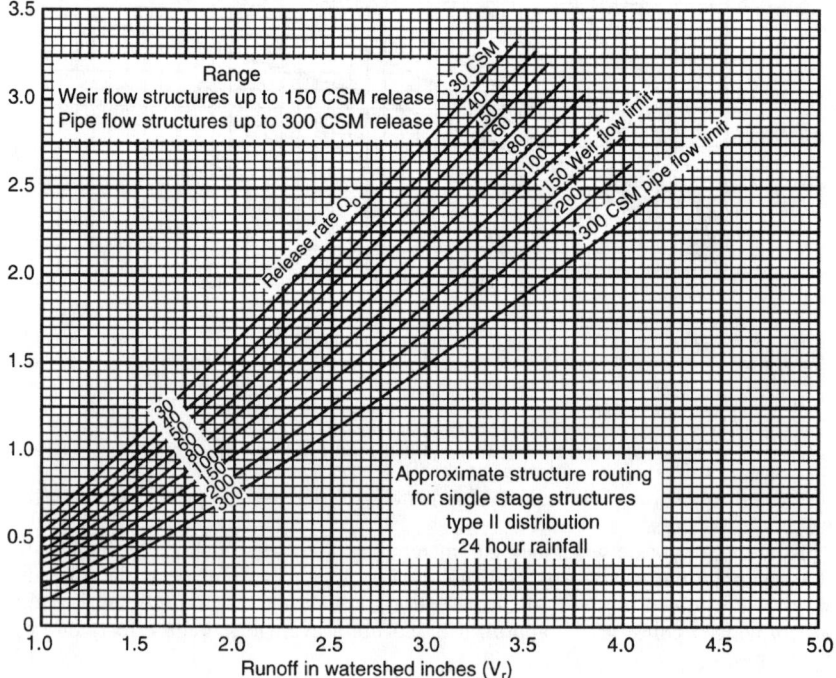

FIGURE 6.27 Single-stage structure routing, up to 300 csm.

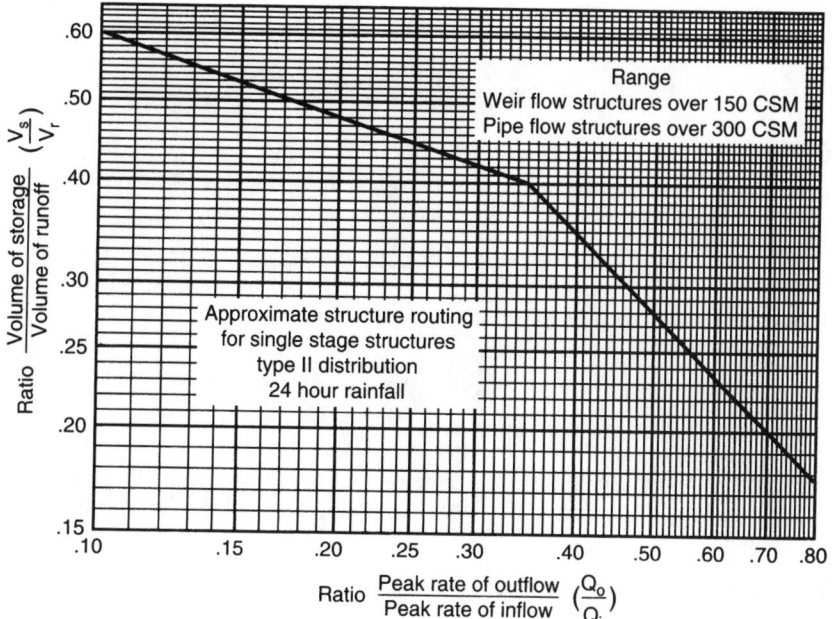

FIGURE 6.28   Single-stage structure routing, over 300 csm.

adjusted rainfall figure (VR) from Table 6.5. Entering the figure with each of these numbers, V_S is found for use in Eq. (6.6).

3. To use Fig. 6.28, additional steps are required due to different parameters used to construct the graph. The peak rate of outflow (Q_O) and peak rate of inflow (Q_I) are known from previous calculations. The allowable peak discharge from the basin in csm must also be calculated, but only to use the appropriate figure. Also, the adjusted rainfall figure (VR) from Table 6.5 must be found. Entering Fig. 6.28 with the ratio of Q_O over Q_I, reading up to where it intersects the heavy line, and then left, the actual ratio of storage volume over runoff volume is found. With the ratio now known, the ratio just found is multiplied by VR (adjusted rainfall figure from Table 6.5) to find V_S used in Eq. (6.6). The following examples show how Figs. 6.27 and 6.28 are used.

EXAMPLE 1   Continuing with the 300-acre site, assume the maximum permissible discharge from the site equal to that established for the undeveloped site. This is 399 cfs. A pipe will be used for discharge.

1. Convert cfm to csm (using Eq. 6.7):

$$\frac{399 \times 640}{300} = \frac{255,360}{300} = 851.2 \text{ (rounded to 852 csm)}$$

Figure 6.28 must be used since the result is greater than 300 csm.

2. Find ratio:

$$\frac{Q_I}{Q_O} = \frac{399}{698} = 0.507 \text{ (rounded to 0.51)}$$

3. Determine VR: From above, 3.28 in of rain is used.

4. Find V_S: First enter Fig. 6.28 with $Q_O/Q_I = 0.51$. Read 0.27 ratio on the left side of chart. Therefore:

$$V_S = 3.28 \times 0.27 = 0.88$$

5. Find basin volume [Eq. (6.6)]:

$$V_B = \frac{0.88 \times 100}{12} = \frac{88}{12} = 7.3 \text{ acre ft}$$

EXAMPLE 2 A 0.4 square mile watershed with an adjusted rainfall of 3.28 in is required to limit discharge to 103 cfm through a piped outlet. What basin volume is required?

1. Find csm [use Eq.(6.8)]

$$\frac{103}{0.4} = 258 \text{ csm}$$

Therefore use Fig. 6.27 since this is less than 300 csm.

2. Find V_S. Entering Fig. 6.27 with V_R of 3.28, find V_S of 1.55.

3. Find V_B. Use Eq. (6.6).

$$V_B = \frac{V_S \times A}{12} = \frac{1.55 \times 640 \times 0.4}{12} = 33.1 \text{ acre ft}$$

To convert from acre feet to cubic feet, multiply acre feet by 43,560 and divide the product by 9.

The following design criteria must be established to select the size of the retention basin and the size of the outflow pipe:

1. Allowable cfs to outflow pipe

2. Height from the top of the basin to centerline of outlet pipe

3. Location of the retention basin

4. Completed site drainage system

5. Trial size of outlet pipe

6. Trial slope of outlet pipe

First, the retention basin must be located on the site plan. The size restrictions must be determined, if any, and then a trial depth found. Next, assuming a trial size pipe, one-half of the pipe diameter is subtracted from the basin depth. Also, at this time, the rate of discharge from the site to the outfall must have been definitely established. With the above information, the size discharge pipe can be calculated from the following orifice formula. Note that the allowable discharge will only occur when the design head is reached.

$$A = \frac{Q}{C \times 2GH} \tag{6.9}$$

where A = area of outlet pipe, ft² (refer to Table 6.7)
 Q = cfs discharge allowable
 C = orifice coefficient (use 0.60, an average value)
 G = 32.2, acceleration due to gravity, 32.2 ft/sec²
 H = design head from surface of water to center of outlet pipe

To convert pipe size from square feet to diameter in feet the following formula is used:

$$D = \frac{4A}{\sqrt{\pi}} \tag{6.10}$$

where D = interior diameter of pipe, ft (Table 6.7)
 A = interior area of pipe, ft² (Table 6.7)
 π = 3.14

Refer to Table 6.7 to select design properties of pipe. If an exact cfm discharge to the outfall is required, using a standard size pipe, the depth of the basin can be adjusted to achieve the desired result. If the height of the basin is fixed, an orifice plate of the exact calculated size can be cut and installed in the outlet pipe. If a weir outlet is to be used, the outflow is determined by the shape of the weir and height of water over the weir bottom. Standard references should be consulted for discharge through such weir openings.

TABLE 6.7 Design Properties for Pipe

DN	Nominal size, in	Inside area, ft²	Nominal I.D., ft	Cast iron	Ductile iron	Plain concrete	Reinforced concrete	Clay	Corrugated steel (⅛-in corr.)	Composite
							Outside diameter, ft			
250	10	0.545	0.84	0.89	0.93	1.04		1.04	0.92	0.94
300	12	0.785	1.00	1.06	1.10	1.29	1.33	1.25	1.08	1.13
350	14	1.18	1.18	1.32	1.28					1.33
375	15	1.227	1.25			1.56	1.62	1.54	1.33	1.44
400	16	1.53	1.33		1.46					1.52
450	18	1.767	1.50		1.62	1.87	1.92	1.86	1.58	1.68
500	20	2.26	1.70		1.82					1.87
525	21	2.405	1.75			2.20	2.20	2.14	1.83	1.98
600	24	3.142	2.00		2.12	2.62	2.62	2.39	2.08	2.27
700	27	3.976	2.25			2.92	2.92	2.77	2.33	2.55
750	30	4.909	2.50		2.66	3.20	3.20	2.08	2.58	2.84
800	33	5.940	2.75			3.50	3.50	3.31	2.83	3.14
900	36	7.069	3.00		3.19	3.79	3.79	3.66	3.08	3.39

DESIGNING BURIED PIPING

This section describes the criteria and methods used in the design and installation of underground piping on a site. The loads placed on the pipe, the method of calculating these loads, and determination of the required supporting strength of the pipe will be discussed.

GENERAL

All buried pipe is subject to stresses imposed upon it by the nature of burial. Such factors as bedding methods, type of backfill material, type and shape of trench, loading from live loads, and storage of material over the trench will each contribute to the total load transmitted to the pipe and the ability of the pipe to support such loads. The pipe selected, combined with the method of installation, must have the strength to withstand the entire load placed upon it without crushing, cracking, or deforming.

CODES AND STANDARDS

1. ASTM and ANSI standards for pipe and installation
2. AWWA standards for pipe, installation, and disinfection
3. NFPA-20 standard for installation of private water mains

PIPE AND INSTALLATION CLASSIFICATIONS

Pipe Installation

Pipe installation is divided into three general classifications:

1. *Pipe in a tunnel.* This is a classification for pipe forced through the earth by a ram or other means that leaves the surrounding earth undisturbed.

2. *Pipe in an embankment condition.* This is a classification for pipe installed on the surface of the ground (positive projection) or in a trench (negative projection) with a layer of earth placed above the original ground surface.

3. *Pipe in a trench or ditch.* This is a classification for pipe installed in an excavation below grade, with the trench backfilled approximately to the original ground line. Trench loading will be the only condition considered here, since it is the most common means of installation.

Pipe Classifications

Piping is divided into two classifications, rigid and flexible. Rigid pipe material is concrete, composite, metal, or clay. Rigid pipe fails when it breaks under a three-edge-bearing test. Concrete pipe fails when the three-edge-bearing test produces a crack 0.01 in wide and 2 in long. Flexible pipe is plastic, composite, or corrugated steel. Flexible piping is considered to fail when loads produce a vertical deflection of 5 percent or more of pipe diameter.

LOADS ACTING ON A PIPE

Definitions

The *earth load* is the weight of all earth backfill over the pipe. The *superimposed load* is produced by either a moving object (live load) passing over the pipe or a uniformly distributed (dead) load placed at ground level over the pipe (in addition to the earth load). A uniformly distributed load and a live load acting concurrently on a pipe will not be considered.

Earth Loads

General. The earth load is the vertical force earth transmits to a pipe buried in a trench. The pipe is subject to a very complex relationship among many factors. The type of fill and the depth and width of a trench must all be considered. Original research at Iowa State College under the direction of Professor Anson Marston produced a formula for determination of earth loads. This, and additional research, resulted in Eqs. 6.11 and 6.12.

Earth Load Calculation. The earth load is obtained by selecting the lower result from either Eq. (6.11) or (6.12). Actual tests have proven that the loads for a positive projection condition should be used when it is the lower of the two even if the pipe is placed in a trench.

$$L_e = C_p\, W\, D \text{ (positive projection condition)} \tag{6.11}$$

$$L_e = C_d\, W\, B \text{ (ditch condition)} \tag{6.12}$$

where L_e = earth load on the pipe (lb/ft of pipe length)
  C_d = coefficient for load calculation of ditch condition
  C_p = coefficient for load calculation of positive projection condition
  W = weight of backfill, lb/ft^3
  B = width of trench at top of pipe, ft
  D = outside diameter of pipe, ft

Discussion:

1. To determine C_d or C_p, refer to Figs. 6.29 and 6.30 or 6.31,

where H = height of backfill over top of pipe, ft
  D = outside diameter of pipe, ft
  B = width of trench at top of pipe, ft

Curves A, B, C, D, and E represent the following soils in Fig. 6.30.

A = sand and sandy loam

B = sand and gravel

C = saturated topsoil

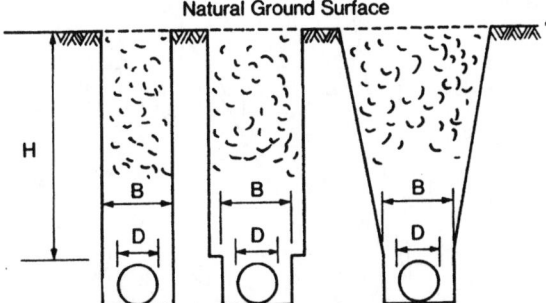

FIGURE 6.29 Installation conditions for earth load calculations.

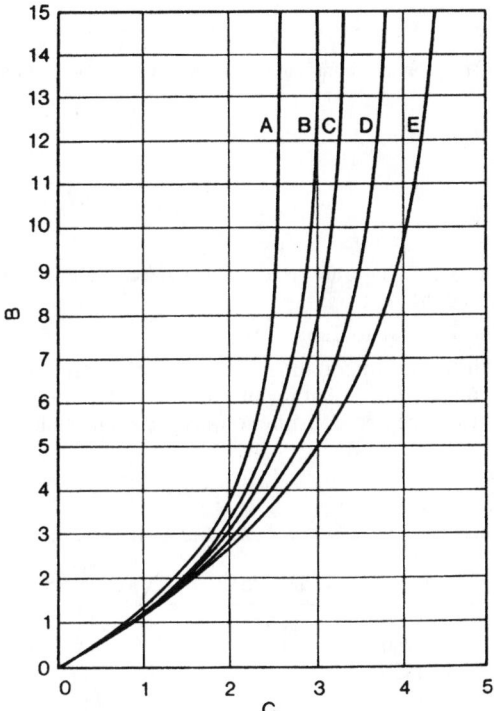

FIGURE 6.30 Calculation coefficients for ditch condition. (*Courtesy: ANSI 21.1.*)

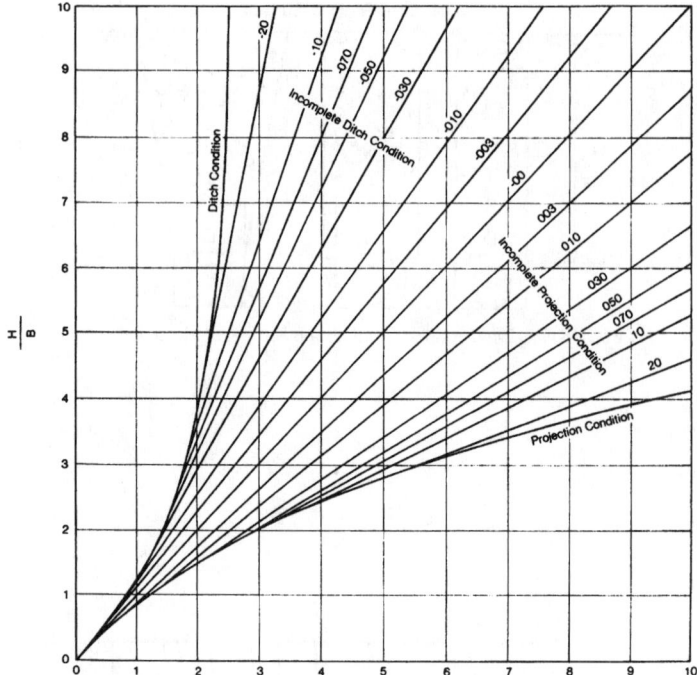

FIGURE 6.31 Coefficient for positive projection condition. (*Courtesy: ANSI 21.1.*)

D = average clay

E = saturated clay

Curve 0.70 is an average value used for standard calculations in Fig. 6.31 (incomplete projection condition).

2. To determine *W*, refer to Table 6.8.

3. Trench width has an important influence on the soil load transmitted to the pipe. Any trench must provide sufficient working space for the installation with tolerance allowed for accepted construction practices and field conditions. Experi-

TABLE 6.8 Weight of Soil

Type of soil	Kg/m³	Weight, lb/ft³	Correction factor	Average moisture content, %
Sand and sandy loam	1600	100	0.83	15
Sand and gravel mix	1760	110	0.90	20
Saturated topsoil	1840	115	0.95	25
Average clay	1920	120	1.00	30
Wet clay, peat	2080	130	1.10	100

ments have proven that only the trench width at the top of the pipe need be considered in calculating the earth load. If the trench is widened above the top of pipe, it does not contribute any additional load to the buried piping. If sheeting is used, the width of the trench is computed to the inside of the sheeting if it is to be removed. It is recommended that the sheeting remain, if possible.

At any given depth, for a particular size pipe and type of soil, a trench width is reached beyond which there is no longer any additional load added to the buried pipe. This is called *transition width,* and is the widest dimension that need be considered to compute the earth load. Table 6.9 gives the transition width for various combinations of pipe size and depth of burial.

It is common practice to allow 1 ft (0.3 m) on either side of a pipe in a trench for working space. Figure 6.32 provides direct determination of earth load for this condition. If soil weight is other than 120 lb/ft³ (1920 kg/m³), proportionally decrease or increase the resulting weight according to the correction factor in Table 6.8.

SUPERIMPOSED LOADS

Live Loads

Whenever pipe is buried in the ground there is always the possibility of heavy equipment or trucks riding over the buried pipe during construction. When pipe is placed under a road or railroad, this is a normal occurrence. This load must be added to the earth load to determine the total force exerted on the buried pipe. The general contractor should be consulted to determine if piping will be buried to its full depth during construction or if additional cover will be added later.

A live load consists of two separate types of loads: the actual weight of any vehicle passing over the pipe and the impact load that places additional stress on the pipe due to the fact the load is moving.

Live Load from Trucks. The type of road surface over the pipe may have a reducing effect on the amount of live load reaching the pipe. A flexible surface such as light-duty asphalt will have little or no reducing effect. A concrete road or heavy-duty asphalt will greatly reduce the load intensity. Railroad tracks are constructed over a standard bed consisting of rock ballast and ties, which gives a uniform distribution of load to the piping.

For calculating the live load transmitted to a pipe under a road, the following formulas are used.

Unpaved or light-duty pavement:

$$L_t = CRPF \qquad (6.13)$$

Rigid or heavy-duty pavement:

$$L_t = CBPF \qquad (6.14)$$

where L_t = truck superload, lb/ft of pipe
C = surface load factor
R = reduction factor
P = wheel load

TABLE 6.9 Transition Width for Trenches

Cover over top of pipe*	Nominal pipe size, in												
	10	12	14	15	16	18	20	21	24	27	30	33	36
2'0"	1'9"	1'11"	2'1"	2'2"	2'3"	2'5"	2'8"	2'9"	3'1"	3'5"	3'7"	3'10"	4'3"
2'6"	1'10"	2'0"	2'3"	2'4"	2'5"	2'8"	2'10"	2'11"	3'3"	3'6"	3'10"	4'1"	4'4"
3'0"	1'11"	2'3"	2'5"	2'6"	2'7"	2'10"	3'0"	3'2"	3'5"	3'8"	3'11"	4'3"	4'7"
4'0"	2'0"	2'5"	2'8"	2'10"	2'11"	3'1"	3'4"	3'6"	3'9"	4'1"	4'5"	4'6"	4'11"
5'0"	2'2"	2'7"	2'11"	3'1"	3'2"	3'5"	3'8"	3'10"	4'1"	4'6"	4'8"	5'1"	5'3"
6'0"	2'5"	2'9"	3'1"	3'2"	3'4"	3'8"	3'11"	4'1"	4'5"	4'8"	5'0"	5'5"	5'8"
7'0"	2'5"	2'10"	3'2"	3'4"	3'6"	3'10"	4'1"	4'4"	4'8"	5'1"	5'5"	5'8"	6'0"
8'0"	2'6"	2'11"	3'3"	3'6"	3'7"	4'0"	4'3"	4'6"	4'10"	5'3"	5'7"	6'1"	6'2"
9'0"	2'8"	3'0"	3'5"	3'7"	3'9"	4'1"	4'5"	4'7"	5'0"	5'6"	5'10"	6'3"	6'8"
10'0"	2'9"	3'1"	3'6"	3'8"	3'11"	4'3"	4'6"	4'9"	5'2"	5'7"	6'1"	6'5"	6'10"
12'0"	3'0"	3'4"	3'7"	3'10"	4'0"	4'5"	4'10"	5'0"	5'5"	6'0"	6'5"	6'10"	7'4"
14'0"	3'1"	3'5"	3'11"	4'0"	4'2"	4'7"	5'1"	5'3"	5'8"	6'3"	6'9"	7'2"	7'8"
16'0"	3'2"	3'8"	4'0"	4'3"	4'5"	4'9"	5'2"	5'5"	5'11"	6'6"	7'0"	7'5"	7'11"
18'0"	3'3"	3'10"	4'3"	4'5"	4'7"	5'0"	5'4"	5'6"	6'1"	6'8"	7'3"	7'9"	8'2"
20'0"	3'5"	4'1"	4'4"	4'6"	4'9"	5'2"	5'7"	5'8"	6'2"	6'10"	7'5"	7'11"	8'6"

*1 ft = 0.3 m.

6.58

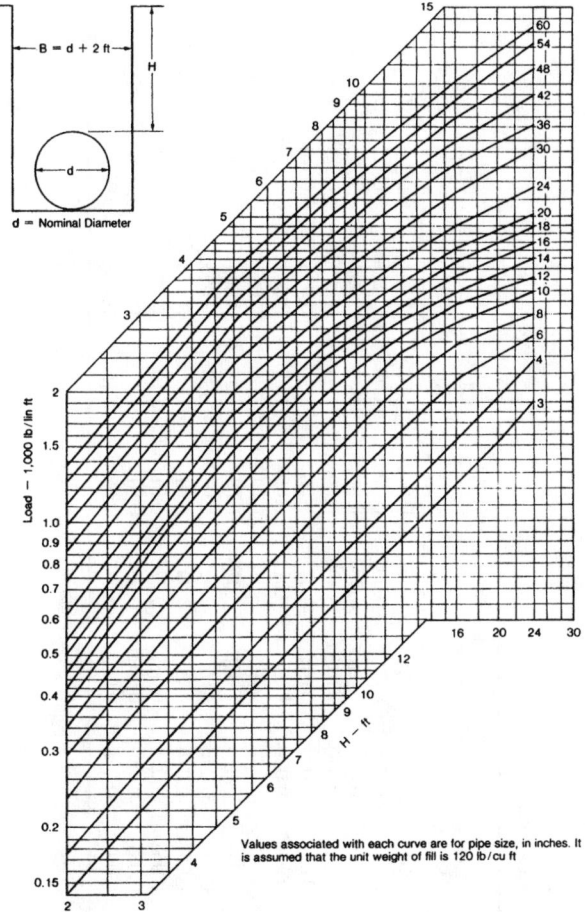

FIGURE 6.32 Earth load on pipe in trench. (*Courtesy: ANSI 21.1.*)

F = impact factor
B = outside diameter of pipe, ft

Discussion:

1. The surface load factor is the weight transmitted to the pipe from a truck on a road passing over the pipe. Tables have been prepared for both one truck and two trucks passing over the pipe simultaneously. The proper chart should be consulted based on road construction and number of trucks. Refer to Table 6.10 (one truck on flexible pavement, Table 6.11 (two trucks on flexible pavement), or Table 6.12 (one or two trucks on rigid pavement).

2. The reduction factor allows for the fact that adjacent portions of pipe that are not directly under the load assist in carrying some portion of the live load. Refer to Table 6.13.

TABLE 6.10 Surface Load Factor—One Truck on Flexible Pavement*

DN	Pipe size, in	Depth of cover, ft†												
		2	2½	3	3½	4	5	6	8	10	12	16	20	24
80	3	0.028	0.020	0.014	0.011	0.009	0.006	0.004	0.002	0.0015	0.001	0.0006	0.0004	0.0002
100	4	0.034	0.024	0.017	0.013	0.011	0.007	0.005	0.003	0.002	0.0015	0.0008	0.0005	0.0003
150	6	0.048	0.034	0.025	0.020	0.015	0.010	0.007	0.004	0.003	0.002	0.001	0.0007	0.0004
200	8	0.062	0.044	0.033	0.026	0.020	0.013	0.009	0.006	0.0035	0.0025	0.0013	0.0008	0.0005
250	10	0.074	0.054	0.040	0.031	0.025	0.016	0.012	0.007	0.004	0.003	0.0016	0.001	0.0006
300	12	0.087	0.063	0.048	0.036	0.030	0.019	0.014	0.008	0.005	0.0035	0.002	0.0012	0.0007
350	14	0.099	0.072	0.055	0.042	0.034	0.022	0.016	0.010	0.006	0.004	0.0025	0.0015	0.0008
400	16	0.110	0.082	0.061	0.047	0.038	0.025	0.018	0.011	0.007	0.005	0.003	0.0017	0.001
450	18	0.122	0.090	0.068	0.052	0.042	0.028	0.020	0.012	0.008	0.0055	0.0035	0.002	0.0012
500	20	0.132	0.098	0.075	0.058	0.046	0.031	0.022	0.013	0.009	0.006	0.004	0.0025	0.0015
600	24	0.150	0.113	0.087	0.068	0.054	0.037	0.026	0.015	0.010	0.007	0.0045	0.003	0.0017
750	30	0.171	0.132	0.102	0.081	0.065	0.045	0.031	0.019	0.012	0.009	0.005	0.0035	0.002
900	36	0.188	0.148	0.117	0.093	0.076	0.052	0.037	0.022	0.015	0.010	0.006	0.004	0.0025

*These factors are for a single concentrated wheel load centered over an effective pipe length of 3 ft.
†1 ft = 0.3 m.
Source: ANSI 21.1.

TABLE 6.11 Surface Load Factor—Two Truck on Flexible Pavement*

DN	Pipe size, in	Depth of cover, ft†												
		2	2½	3	3½	4	5	6	8	10	12	16	20	24
80	3	0.019	0.012	0.008	0.006	0.005	0.004	0.0035	0.003	0.0025	0.002	0.001	0.007	0.0005
100	4	0.032	0.022	0.016	0.012	0.009	0.006	0.005	0.004	0.0035	0.003	0.002	0.0013	0.0010
150	6	0.058	0.042	0.032	0.024	0.020	0.014	0.010	0.007	0.006	0.005	0.003	0.0019	0.0015
200	8	0.076	0.058	0.044	0.036	0.030	0.022	0.017	0.011	0.009	0.007	0.004	0.0027	0.0021
250	10	0.092	0.072	0.056	0.046	0.039	0.028	0.021	0.014	0.011	0.008	0.005	0.0033	0.0026
300	12	0.108	0.086	0.070	0.056	0.047	0.034	0.027	0.018	0.012	0.009	0.006	0.0039	0.0030
350	14	0.122	0.098	0.078	0.065	0.055	0.040	0.031	0.020	0.014	0.010	0.007	0.0046	0.0036
400	16	0.136	0.110	0.090	0.074	0.062	0.046	0.036	0.024	0.016	0.012	0.009	0.0060	0.0046
450	18	0.149	0.122	0.101	0.084	0.070	0.052	0.041	0.027	0.019	0.014	0.010	0.0066	0.0051
500	20	0.162	0.136	0.115	0.096	0.080	0.060	0.048	0.032	0.022	0.016	0.012	0.0079	0.0061
600	24	0.185	0.152	0.126	0.106	0.091	0.067	0.053	0.036	0.026	0.119	0.013	0.0086	0.0067
750	30	0.212	0.176	0.146	0.124	0.107	0.080	0.064	0.044	0.032	0.024	0.016	0.0106	0.0081
900	36	0.235	0.202	0.169	0.146	0.127	0.095	0.075	0.052	0.038	0.028	0.019	0.0125	0.0097

*These factors are for two trucks with 6-ft rear wheel spacing passing with inside rear wheels 3-ft apart. Effective pipe length is 3 ft coinciding with the distance between the adjacent inside wheels.
† 1 ft = 0.3 m.

TABLE 6.12 Surface Load Factor—One or Two Trucks on Rigid Pavement*

Depth of cover, ft‡	Pavement thickness, in†							
	One truck				Two passing trucks			
	4	6	8	10	4	6	8	10
2	0.0244	0.0149	0.0101	0.0076	0.0410	0.0263	0.0186	0.0142
2½	0.0213	0.0139	0.0097	0.0072	0.0364	0.0246	0.0177	0.0136
3	0.0186	0.0126	0.0090	0.0070	0.0333	0.0228	0.1067	0.0129
3½	0.0164	0.0114	0.0085	0.0066	0.0290	0.0206	0.0156	0.0122
4	0.0144	0.0102	0.0079	0.0061	0.0262	0.0187	0.0146	0.0117
5	0.0114	0.0084	0.0066	0.0054	0.0210	0.0156	0.0123	0.0102
6	0.0093	0.0071	0.0057	0.0047	0.0170	0.0133	0.0107	0.0088
8	0.0065	0.0052	0.0043	0.0036	0.0114	0.0097	0.0081	0.0069
10	0.0046	0.0039	0.0033	0.0029	0.0080	0.0070	0.0062	0.0055
12	0.0034	0.0030	0.0026	0.0023	0.0059	0.0054	0.0049	0.0045
16	0.0022	0.0019	0.0017	0.0016	0.0034	0.0032	0.0030	0.0028
20	0.0013	0.0011	0.0010	0.0009	0.0024	0.0023	0.0022	0.0021
24	0.0008	0.0007	0.0006	0.0005	0.0015	0.0014	0.0013	0.0012

*These factors were computed by the methods explained in "Vertical Pressure on Culverts under Wheel Loads on Concrete Pavement Slabs," Portland Cement Assn., Chicago, IL in Bulletin ST65.
† 1 in = 25 mm.
‡ 1 ft = 0.3 m.
Source: ANSI 21.1.

TABLE 6.13 Reduction Factor

Pipe size, in	Depth of cover, ft*			
	2½–3½	4–7	8–10	>10
3–12	1.00	1.00	1.00	1.00
14	0.92	1.00	1.00	1.00
16	0.88	0.95	1.00	1.00
18	0.85	0.90	1.00	1.00
20	0.83	0.90	0.95	1.00
24–30	0.81	0.85	0.95	1.00
36–60	0.80	0.85	0.90	1.00

*1 ft = 0.3 m.
Source: ANSI 21.1.

3. Wheel load is the actual weight of the truck and its cargo on a wheel passing over the pipe. Refer to Table 6.14. The H-20 loading is most commonly used.

4. Impact factor is a dynamic load caused by a moving object. Refer to Table 6.15 to determine the factor based on depth of burial.

5. Table 6.16 gives the outside diameter of pipe, based on pipe size and material.

TABLE 6.14 Truck Wheel Load

AASHTO truck	Gross weight, tons	Wheel load (P), lb*
H-10	10	8,000
H-15	15	12,000
H-20	20	16,000

*1 lb = 2.2 kg.

TABLE 6.15 Truck Impact Factor

Depth of cover*	Impact factor (F)
Up to 1 ft-0 in	1.30
1 ft-1 in to 2 ft-0 in	1.20
2 ft-0 in to 2 ft-11 in	1.10
3 ft-0 in or more	1.00

*1 ft = 0.3m.

TABLE 6.16 Outside Diameter of Pipe

Nominal size, in	Inside area, ft²	Nominal I.D., ft	Cast iron	Ductile iron	Plain concrete	Reinforced concrete	Clay	Corrugated steel, ⅛ in	Composite
10	0.545	0.84	0.89	0.93	1.04	—	1.04	0.92	0.94
12	0.785	1.00	1.06	1.10	1.29	1.33	1.25	1.08	1.13
14	1.18	1.18		1.28	—	—	—	—	1.33
15	1.227	1.25	1.32		1.56	1.62	1.54	1.33	1.44
16	1.53	1.33	—	1.46	—	—	—	—	1.52
18	1.767	1.50	—	1.62	1.87	1.92	1.86	1.58	1.68
20	2.26	1.70	—	1.82	—	—	—	—	1.87
21	2.405	1.75	—		2.20	2.20	2.14	1.83	1.98
24	3.142	2.00	—	2.12	2.62	2.62	2.39	2.08	2.27
27	3.976	2.25	—		2.92	2.92	2.77	2.33	2.55
30	4.909	2.50	—	2.66	3.20	3.20	3.08	2.58	2.84
33	5.940	2.75	—		3.50	3.50	3.31	2.83	3.14
36	7.069	3.00	—	3.19	3.79	3.79	3.66	3.08	3.39

Live Load from Trains. For calculating the live load transmitted to a pipe under a railroad, refer to Table 6.17. Recommendations of the American Railway Engineering Association are used. This figure should be added to the weight of the track structure. The following criteria are used:

1. Copper E72 railroad design load is assumed.

2. The weight of track structure and ballast is assumed to be 200 lb/ft

3. An impact factor of 1.4 is used, decreasing to 1.0 with 10 ft of cover. The impact factor has been included in the calculations used to prepare Table 6.17.

TABLE 6.17 Railroad Loads on Circular Pipe, Pounds per Linear Foot*

Pipe size, in	Height of fill H above top of pipe, 10 ft																
	1	2	3	4	5	6	7	8	9	10	12	14	16	18	20	25	30
12	3270	3059	2750	2431	2109	1808	1550	1330	1138	982	788	646	528	442	380	259	185
15	3986	3729	3352	2964	2571	2204	1890	1622	1388	1198	960	788	644	540	463	315	226
18	4702	4400	3955	3497	3033	2599	2229	1913	1637	1413	1133	930	759	636	546	372	266
21	5416	5067	4555	4027	3493	2904	2568	2204	1886	1627	1305	1071	874	733	629	428	307
24	6132	5738	5158	4560	3955	3390	2908	2495	2135	1842	1478	1212	990	830	712	485	348
27	6849	6408	5760	5093	4417	3786	3247	2786	2384	2058	1650	1354	1106	927	796	542	388
30	7562	7075	6360	5623	4877	4180	3586	3077	2633	2272	1822	1495	1221	1024	879	598	428
33	8279	7746	6963	6156	5339	4576	3925	3368	2882	2487	1995	1637	1336	1120	962	655	469
36	8995	8416	7565	6689	5801	4972	4265	3660	3132	2702	2167	1778	1452	1217	1045	711	510
42	10420	9754	8768	7752	6724	5763	4943	4242	3630	3132	2512	2061	1683	1411	1211	824	591

*Cooper E72 design loading consisting of four 72,000-lb axles spaced 5 ft c/c. Locomotive load assumed uniformly distributed over an area 8 ft × 20 ft. Weight of track structure assumed to be 200 pounds per linear foot. Impact included. Height of fill measured from top of pipe to bottom of ties. Interpolate for intermediate pipe sizes and/or fill heights.

Source: Concrete Pipe Design Manual.

Loads from Other Equipment. For calculating the live load transmitted to a pipe from any other heavy equipment (crane, bulldozer, or others) the total weight of the vehicle must be found. The weight on each of its tracks or wheels is calculated by dividing the number of wheels or tracks into the vehicle weight. With the track or wheel weight and the pipe depth of bury known, refer to Table 6.18 for the percent of load transmitted to the pipe. The track or wheel weight is then multiplied by the percent load transmitted to calculate the actual load on the pipe. Regardless of how slowly the vehicle is capable of moving, an impact factor must be used. The calculated actual load on the pipe is then multiplied by the impact factor obtained from Table 6.15. The result of this calculation is the total actual live load on the buried pipe.

Uniformly Distributed Load

When an additional load, such as fill or lumber, is stored at grade on top of the buried pipe, this is called a uniformly distributed load. This load is added to the earth load over the pipe. The formula for calculating the uniform load is:

$$L_s = CBW \qquad (6.15)$$

where L_s = superimposed load on pipe, lb/ft of pipe
C = coefficient, dimensionless
B = width of trench at top of pipe, ft
W = weight of superimposed load, lb/ft^2

Discussion:

1. The additional load L_s is limited in dimension.
2. To find C, refer to Table 6.19 to determine the coefficient for vertical load on pipe installed in trench.
   Table 6.19 is entered with the following values:

TABLE 6.18 Percentage of Wheel Loads Transmitted to Pipes*

Tabulated figures show percentage of wheel load applied to one linear foot of pipe.

Depth of backfill over top of pipe, ft	Pipe size, in													
	6	8	10	12	15	18	21	24	27	30	33	36	39	42
	Outside diameter of pipe, ft (approximate)													
	0.64	0.81	1.0	1.2	1.5	1.8	2.1	2.4	2.7	3.0	3.3	3.5	3.9	4.2
1	12.8	15.0	17.3	20.0	22.6	24.8	26.4	27.2	28.0	28.6	29.0	29.4	29.8	29.9
2	5.7	7.0	8.3	9.6	11.5	13.2	15.0	15.6	16.8	17.8	18.7	19.5	20.0	20.5
3	2.9	3.6	4.3	5.2	6.4	7.5	8.6	9.3	10.2	11.1	11.8	12.5	12.9	13.5
4	1.7	2.1	2.5	3.1	3.9	4.6	5.3	5.8	6.5	7.2	7.9	8.5	8.8	9.2
5	1.2	1.4	1.7	2.1	2.6	3.1	3.6	3.9	4.4	4.9	5.3	5.8	6.1	6.4
6	0.8	1.0	1.1	1.4	1.8	2.1	2.5	2.8	3.1	3.5	3.8	4.2	4.3	4.4
7	0.5	0.7	0.8	1.0	1.3	1.6	1.9	2.1	2.3	2.6	2.9	3.2	3.3	3.5
8	0.4	0.5	0.6	0.8	1.0	1.2	1.4	1.6	1.8	2.0	2.2	2.3	2.5	2.6

*These figures make no allowance for impact.
Source: *Clay Pipe Engineering Manual.*

TABLE 6.19 Vertical Load Coefficients

D/H	L/2H													
	0.1	0.2	0.3	0.4	0.5	0.6	0.7	0.8	0.9	1.0	1.2	1.5	2.0	5.0
0.1	0.019	0.037	0.053	0.067	0.079	0.089	0.097	0.103	0.108	0.112	0.117	0.121	0.124	0.128
0.2	0.037	0.072	0.103	0.131	0.155	0.174	0.189	0.202	0.211	0.219	0.229	0.238	0.244	0.248
0.3	0.053	0.103	0.149	0.190	0.224	0.252	0.274	0.292	0.306	0.318	0.333	0.345	0.355	0.360
0.4	0.067	0.131	0.190	0.241	0.284	0.320	0.349	0.373	0.391	0.405	0.425	0.440	0.454	0.460
0.5	0.079	0.155	0.224	0.284	0.336	0.379	0.414	0.441	0.463	0.481	0.505	0.525	0.540	0.548
0.6	0.089	0.174	0.252	0.320	0.379	0.428	0.467	0.499	0.524	0.544	0.572	0.596	0.613	0.624
0.7	0.097	0.189	0.274	0.349	0.414	0.467	0.511	0.546	0.584	0.597	0.628	0.650	0.674	0.688
0.8	0.103	0.202	0.292	0.373	0.441	0.499	0.546	0.584	0.615	0.639	0.674	0.703	0.725	0.740
0.9	0.108	0.211	0.306	0.391	0.463	0.524	0.574	0.615	0.647	0.673	0.711	0.742	0.766	0.784
1.0	0.112	0.219	0.318	0.405	0.481	0.544	0.597	0.639	0.673	0.701	0.740	0.774	0.800	0.816
1.2	0.117	0.229	0.333	0.425	0.505	0.572	0.628	0.674	0.711	0.740	0.783	0.820	0.849	0.868
1.5	0.121	0.238	0.345	0.440	0.525	0.596	0.650	0.703	0.742	0.774	0.820	0.861	0.894	0.916
2.0	0.124	0.244	0.355	0.454	0.540	0.613	0.674	0.725	0.766	0.800	0.849	0.894	0.930	0.956

Source: *Clay Pipe Engineering Manual.*

a. *D* is either length or width of material at grade, in ft.

b. *L* is either length or width of material at grade, in ft.

c. *H* is the height of backfill over top of pipe in ft, prior to placement of new load.

BEDDING

General

Bedding is the contact between pipe and earth and is used when pipe is installed in a trench, prior to the major backfilling operation. The type of bedding used has an important influence on the total load any pipe can support.

Bedding Methods

Rigid Pipe. The four most popular bedding methods are shown in Fig. 6.33. The load factors are indicated in each detail. The most common are the Class B methods. The least desirable is Class D, which is not recommended. In all cases, the bell holes are dug out prior to placement of pipe.

Flexible Pipe. The recommended bedding method for flexible pipe is shown in Fig. 6.34.

Pipe Bedding Material Classification

ASTM D 2321 presents a method of classifying soils and aggregates that are used as bedding and backfill around pipes. Refer to Table 6.20 for the classification of embedment and backfill material.

Selection of Bedding and Backfill Material

General recommendations for the selection and installation of soils for bedding and backfill are given in Table 6.21.

Rigid Pipe

The load factor for a bedding condition is used to determine the actual supporting strength of a pipe. As will be seen, the load factor increases the total load a pipe can support. The laboratory-calculated pipe strength multiplied by a load factor will give the *field supporting strength* of a pipe. These load factors have been determined experimentally for the various bedding methods described and are indicated in Fig. 6.33 in each bedding diagram. However, they do not contain a safety factor. The result of the calculations will determine whether the field supporting strength of the pipe is great enough to resist the imposed load.

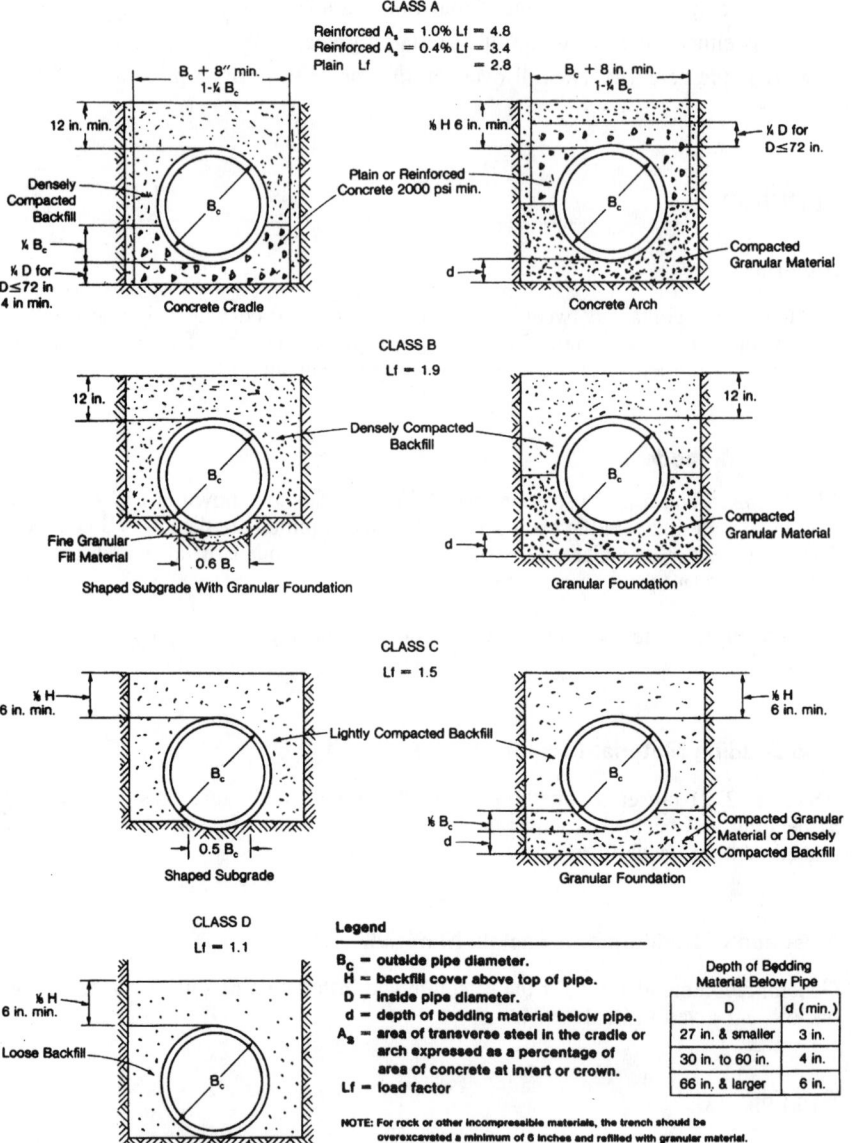

FIGURE 6.33 Rigid pipe bedding methods. (*Courtesy: ASPE Data Book.*) Note: 1 in = 25 mm.

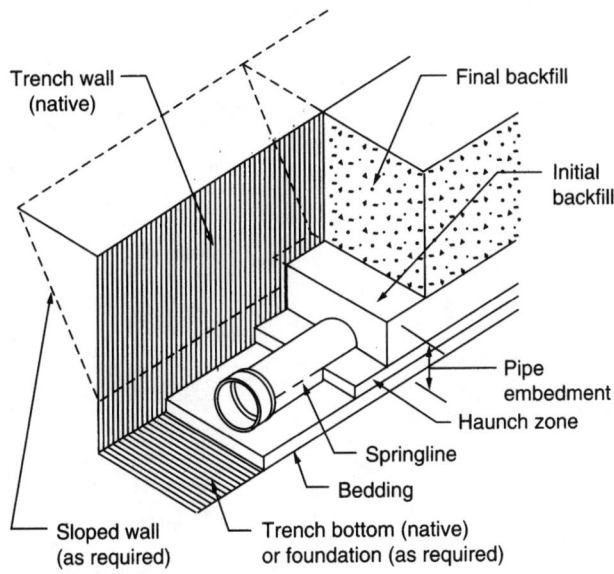

FIGURE 6.34 Flexible pipe bedding.

Flexible Pipe

Pipe Deformation (Deflection)

1. *Radial deflection.* Commonly called *ovalling,* radial deflection is a change in pipe diameter caused by an exterior load on a portion of the pipe. This will occur in buried pipe from a load on the top portion of the pipe if the support of the surrounding soil is not enough to resist the force.

2. *Longitudinal deformation.* Often referred to as *sagging,* longitudinal deformation is a change along the length of a pipe that causes the pipe to bend.

3. *Deformation at structures.* Not limited to flexible pipe, this type of deformation is caused by uneven settlement at wall penetrations and manholes.

The structural behavior of flexible pipe is not the same as for rigid pipe. Flexible pipe does not have the structural strength and stiffness to resist traffic and backfill loads without support from surrounding embedment. Being flexible, the pipe deforms (or deflects) under load. This generally takes the form of radial elongation. This elongation structurally stresses the surrounding embedment material. The resistance to this stress is required for long-term support for the pipe. In order to achieve long-term resistance, proper placement of acceptable material around the pipe is required.

STRENGTH OF PIPE MATERIALS

Rigid Pipe

All rigid pipe has been tested for strength in the laboratory by means of a three-edge-bearing test. In order to simplify selection of some piping, a method was

TABLE 6.20 Classification of Soils for Embedment and Backfill

Class	Type	Soil group symbol D 2487	Description	1½ in (40 mm)	No. 4 (4.75 mm)	No. 200 (0.075 mm)	LL	PI	Uniformity C_u	Curvature C_c
IA	Manufactured aggregates: open-graded, clean	None	Angular, crushed stone or rock, crushed gravel, broken coral, crushed slag, cinders or shells; large void content, contain little or no fines.	100	≤10	<5	Nonplastic			
IB	Manufactured, processed aggregates; dense-graded, clean	None	Angular, crushed stone (or other Class 1A materials) and stone/sand mixtures with gradations selected to minimize migration of adjacent soils; contain little or no fines (see X1.8).	100	≤50	<5	Nonplastic			
II	Coarse-grained soils, clean	GW	Well-graded gravels and gravel-sand mixtures; little or no fines.	100	<50 of "coarse fraction"	<5	Nonplastic		>4	1 to 3
		GP	Poorly-graded gravels and gravel-sand mixtures; little or no fines.						<4	<1 or >3
		SW	Well-graded sands and gravelly sands; little or no fines.		>50 of "coarse fraction"				>6	1 to 3
		SP	Poorly graded sands and gravelly sands; little or no fines.						<6	<1 or >3
	Coarse-grained soils, borderline clean to w/fines	e.g. GW-GC, SP-SM.	Sands and gravels which are borderline between clean and with fines.	100	Varies	5 to 12	Nonplastic		Same as for GW, GP, SW and SP	

	Category	Symbol	Typical names			12 to 50		Plasticity criteria
III	Coarse-grained soils with fines	GM	Silty gravels, gravel-sand-silt mixtures.	100	<50 of "coarse fraction"	12 to 50		<4 or <A line
		GC	Clayey gravels, gravel-sand-clay mixtures.					<7 and >A line
		SM	Silty sands, sand-silt mixtures.		>50 of "coarse fraction"			>4 or <A line
		SC	Clayey sands, sand-clay mixtures.					>7 and >A line
IVA*	Fine-grained soils (inorganic)	ML	Inorganic silts and very fine sands, rock flour, silty or clayey fine sands, silts with slight plasticity.	100	100	>50	<50	<4 or <A line
		CL	Inorganic clays of low to medium plasticity, gravely clays, sandy clays, silty clays, lean clays.					>7 and >A line
IVB	Fine-grained soils (inorganic)	MH	Inorganic silts, micaceous or diatomaceous fine sandy or silty soils, elastic silts.	100	100	>50	>50	<A line
		CH	Inorganic clays of high plasticity, fat clays.					>A line
V	Organic soils	OL	Organic silts and organic silty clays of low plasticity.	100	100	>50	<50	<4 or <A line
		OH	Organic clays of medium to high plasticity, organic silts.				>50	<A line
	Highly organic	PT	Peat and other high organic soils.					

*Includes Test Method D 2487 borderline classifications and dual symbols depending on plasticity index and liquid limits.

NOTE—"Coarse fraction" as used in this table is defined as material retained on a No. 200 sieve.

Source: Copyright ASTM. Reprinted with permission.

TABLE 6.21 Recommendations for Installation and Use of Soils and Aggregates for Foundation, Backfill, and Embedment

Use	Soil class (see Table 6.20)*					
	Class IA	Class IB	Class II	Class III	Class IV-A	

Use	Class IA	Class IB	Class II	Class III	Class IV-A
General recommendations and restrictions	Do not use where conditions may cause migration of fines from adjacent soil and loss of pipe support. Suitable for use as a drainage blanket and underdrain in rock cuts where adjacent material is suitably graded.	Process materials as required to obtain gradation which will minimize migration of adjacent materials. Suitable for use as drainage blanket and underdrain.	Where hydraulic gradient exists check gradation to minimize migration. "Clean" groups suitable for use as drainage blanket and underdrain.	Do not use where water conditions in trench may cause instability.	Obtain geotechnical evaluation of proposed material. May not be suitable under high earth fills, surface-applied wheel loads, and under heavy vibratory compactors and tampers. Do not use where water conditions in trench may cause instability.
Foundation	Suitable as foundation and for replacing over-excavated and unstable trench bottom as restricted above. Install and compact in 6-in maximum layers.	Suitable as foundation and for replacing over-excavated and unstable trench bottom. Install and compact in 6-in maximum layers.	Suitable as a foundation and for replacing over-excavated and unstable trench bottom as restricted above. Install and compact in 6-in maximum layers.	Suitable as foundation and for replacing over-excavated trench bottom as restricted above. Do not use in thicknesses greater than 12 in total. Install and compact in 6-in maximum layers.	Suitable only in undisturbed condition and where trench is dry. Remove all loose material and provide firm, uniform trench bottom before bedding is placed.
Bedding	Suitable as restricted above. Install in 6-in maximum layers. Level final grade by hand. Minimum depth 4 in (6 in for rock cuts).	Install and compact in 6-in maximum layers. Level final grade by hand. Minimum depth 4 in (6 in for rock cuts).	Suitable as restricted above. Install and compact in 6-in maximum layers. Level final grade by hand. Minimum depth 4 in (6 in for rock cuts).	Suitable only in dry trench conditions. Install and compact in 6-in maximum layers. Level final grade by hand. Minimum depth 4 in (6 in for rock cuts).	Suitable only in dry trench conditions and when optimum placement and compaction control is maintained. Install and compact in 6-in maximum layers. Level final grade by hand. Minimum depth 4 in (6 in for rock cuts).

Haunching	Suitable as restricted above. Install in 6-in maximum layers. Work in around pipe by hand to provide uniform support.	Install and compact in 6-in maximum layers. Work in around pipe by hand to provide uniform support.	Suitable as restricted above. Install and compact in 6-in maximum layers. Work in around pipe by hand to provide uniform support.	Suitable as restricted above. Install and compact in 6-in maximum layers. Work in around pipe by hand to provide uniform support.	Suitable only in dry trench conditions and when optimum placement and compaction control is maintained. Install and compact in 6-in maximum layers. Work in around pipe by hand to provide uniform support.
Initial backfill	Suitable as restricted above. Install to a minimum of 6 in above pipe crown.	Install and compact to a minimum of 6 in above pipe crown.	Suitable as restricted above. Install and compact to a minimum of 6 in above pipe crown.	Suitable as restricted above. Install and compact to a minimum of 6 in above pipe crown.	Suitable as restricted above. Install and compact to a minimum of 6 in above pipe crown.
Embedment compaction†	Place and work by hand to ensure all excavated voids and haunch areas are filled. For high densities use vibratory compactors.	Minimum density 85% Std. Proctor.‡ Use hand tampers or vibratory compactors.	Minimum density 85% Std. Proctor.‡ Use hand tampers or vibratory compactors.	Minimum density 90% Std. Proctor.‡ Use hand tampers or vibratory compactors. Maintain moisture content near optimum to minimize compactive effort.	Minimum density 95% Std. Proctor.‡ Use hand tampers or impact tampers. Maintain moisture content near optimum to minimize compactive effort.
Final backfill	Compact as required by the engineer.	Compact as required by the engineer.	Compact as required by the engineer.	Compact as required by the engineer.	Suitable as restricted above. Compact as required by the engineer.

*Class IV-B (MH-CH) and Class V (OL, OH, PT) materials are unsuitable as embedment. They may be used as final backfill as permitted by the engineer.

†When using mechanical compactors avoid contact with pipe. When compacting over pipe crown maintain a minimum of 6-in cover when using small mechanical compactors. When using larger compactors maintain minimum clearances as required by the engineer.

‡The minimum densities given in the table are intended as the compaction requirements for obtaining satisfactory embedment stiffness in most installation conditions.

Source: Copyright ASTM. Reprinted with permission.

developed to eliminate any difference due to pipe size. Therefore, a D load has been developed by dividing the actual laboratory strength by the pipe size in feet. In cases where different sizes of pipe have different strengths, this is not possible. Table 6.22 gives the laboratory design loads for various pipe material.

Flexible Pipe

Flexible pipe is plastic or corrugated steel pipe. The failure of this type of piping is considered to occur when the total calculated load on the pipe produces a radial deflection of 5 percent or more of the pipe diameter. Refer to Table 6.23 for the maximum allowable deflection of various size pipes.

TOTAL LOAD ON BURIED PIPE

To calculate the actual load used to select any rigid pipe, the following formulas can be used:

For pipe using D load:

$$L_D \times F = \frac{L_T}{D} \times \text{SF} \tag{6.16}$$

For pipe using ordinary load:

$$L_A \times F = L_T \times \text{SF} \tag{6.17}$$

where L_D = maximum allowable D load on pipe, lb/ft (Table 6.22)
L_A = maximum allowable load on pipe, lb/ft (Table 6.22)
L_T = total calculated combined load (earth, superimposed, railroad), lb/ft
D = outside diameter of pipe, ft (Table 6.16)
F = load factor (Fig. 6.33)
SF = safety factor

Discussion:

1. L_D is the allowable D load for pipe.

2. L_A is the allowable load for all other types of rigid pipe.

3. L_T is the actual total combined load on pipe after determining the earth load, superimposed load, and/or railroad load.

4. F is the load factor determined from the pipe bedding method (Fig. 6.33).

5. SF is the safety factor. A practice normally followed is to allow for the possibility of all extreme conditions occurring simultaneously, thereby producing maximum loading on the buried pipe. Therefore, a safety factor is added to the calculated loads on the buried pipe. This factor ranges between 1.25 and 1.50. How closely and competently the excavation and backfill specification is both written and supervised will determine the number selected. When using reinforced concrete pipe, with the 0.01 crack as failure criterion, no safety factor is required. This is the only exception. An SF of 1.5 is to be used unless close supervision of the job will be maintained and a very descriptive and tight specification is written. If such

TABLE 6.22 Pipe Design Loads, lb

DN	Pipe size, in	Cast iron	Clay pipe, regular strength	Clay pipe, extra strength	Concrete pipe, standard strength	Concrete pipe, extra strength	Composite pipe, strength class	Reinforced concrete pipe		
										D loads
								Strength class	0.01 crack	Ultimate
250	10	2265	1600	2400	1400	2000	1500	I	800	1200
300	12	2231	1800	2600	1500	2250	2400	II	1000	1500
375	15	2302	2000	2900	1750	2750	3300	III	1350	2000
450	18		2200	3300	2000	3300	4000	IV	2500	3000
525	21		2400	3850	2200	3850	5000	V	3000	3750
600	25		2800	4700	2400	4000				
700	27		2800	4700						
750	30		3300	5000						
900	36		4000	6000						

TABLE 6.23 Maximum Allowable Deflection for Pipe

DN	Size, in	Maximum deflection, in
300	12	0.60
375	15	0.75
450	18	0.90
525	21	1.05
600	24	1.20
700	27	1.35
750	30	1.50
800	33	1.65
900	36	1.80

is the case, a 1.25 SF can be used. In addition, a 1.25 SF should be used for cast iron pipe in all conditions.

6. The allowable (or field supporting strength) must be greater than the actual load so that the pipe will not fail under design conditions.

A formula has been developed to calculate the deflection under the total calculated load transmitted to the buried pipe. It is called the Iowa formula.

$$\text{DEF} = \frac{\text{LF} \times K \times L_T \times R^3}{E \times I + 0.061 \times S \times R^3} \tag{6.18}$$

where DEF = deflection of pipe, in
      LF = deflection lag factor
      K = bedding constant
      L_T = total load transmitted to pipe, lb/in
      R = mean radius of pipe, in
      E = modules of elasticity
      I = moment of inertia
      S = modulus soil reaction, psi

Discussion:

1. LF is the lag factor, which allows for continued deflection of pipe after the total load has been developed. It is an empirical number, with a recommended value of 1.25 for good backfill (85 percent density), or 1.50 for excellent backfill (95 percent density).

2. K is a factor depending on the width of bedding. An average value of 0.1 is recommended.

3. L_T is the total load (earth, superimposed, railroad) in lb/in of pipe length.

4. R is the radius of pipe measured from the center of the corrugation. For this formula, one-half of the nominal size should be used.

5. To determine I, refer to Table 6.24 for the correct value of corrugated pipe, based on the gauge of steel selected. For plastic pipe, individual manufacturers should be consulted because of the great diversity of materials available.

6. E is 30 3 10^6 for all pipe sizes.

TABLE 6.24 Moment of Inertia for Corrugated Steel Pipe

Corrugation pitch & depth	16 gauge	14 gauge	12 gauge
$2 \times \frac{1}{2}$	0.00194	0.00246	0.00354
$2\frac{2}{3} \times \frac{1}{2}$	0.00189	0.00239	0.00343

7. S is soil reaction modulus in psi. S is dependent upon the degree of compaction of backfill around the pipe. This number has not been fully correlated for backfilled soil. Estimates by the Bureau of Public Roads are 700 for good backfill at 85 percent compaction and 1400 for excellent backfill at 95 percent compaction. The value established for good or excellent backfill should be used in Eq. (6.18).

In certain cases, other structural failures for corrugated steel pipe may have to be considered during design. Seam, buckling, handling, and installation strengths and conduit wall compression strength may have to be determined. Refer to the *Handbook of Steel Drainage and Highway Construction Products,* published by the American Iron and Steel Institute, for a description of the method of calculating all the above design parameters, if they are required.

SEWER CLEANING AND REPAIR

CLEANING METHODS

There are many options available to clean sewer piping, manholes, and drainage inlets. The following paragraphs will describe the methods and applications associated with each.

Cleaning Ball

The cleaning ball method is limited to light cleaning of sewer pipe, for example, new piping recently installed or pipe in which almost no accumulation of debris is present. This method uses a ball with serrated ridges around its circumference, which is slightly smaller than the pipe to be cleaned. The ball is placed in the pipe and water is added behind the ball. The water pushes forward and forces its way through the serrations. The water pressure on the ball causes the ball to rotate and dislodge any light dirt. The water forcing its way around the ball flushes the loosened debris away.

Hydraulic Jet Rodder

Application of the hydraulic jet rodder (HJR) is limited to pipes filled to no greater than 30 percent of their depth with dirt or debris. This method uses a bullet-shaped head, with water nozzles both in front and in the rear, attached to a flexible hose. The head is inserted into a sewer through a manhole and uses the water jet to force its way into the pipe. This action drives some debris in front of it, but also leaves much debris at its rear, behind the head. The head is then withdrawn back to the manhole where it was inserted, using the rear-facing water jets to force the loosened dirt and debris to a point where they can be easily removed from the manhole. Depending upon the particular machine used, water pressure up to 4000 psi can be developed. Care must be exercised to avoid damage to the pipe, which is a limiting factor to the amount of debris which can be loosened. In addition, a circular rotating blade, operated hydraulically, can be attached to the head for cleaning grease, with the rear-facing jets of water removing it back to the manhole. It is anticipated that the jet rodder will become the most popular method of pipe cleaning.

Bucket Machine

Application of the bucket machine is limited to piping filled to between 30 and 99 percent with debris. An open bucket capable of remote closing and attached to a cable is inserted into the sewer pipe from one manhole with its open end into the pipe. The cable must be pulled from two ends, usually from one manhole to another. The cable, with bucket attached, is pulled forward into the sewer. When filled with debris, the bucket end closes and the other end of the cable pulls the bucket backwards to the manhole, where it is emptied. This process is repeated until the line is clear. If a sewer line is completely blocked, preventing a cable from being stretched from one point to another, this cleaning method cannot be used.

Excavation and Disassembly

When a pipe is completely blocked, and standing water is observed in the manhole, there may be no other alternative than to excavate around the pipe at a joint, disassemble the joint, and remove the accumulated debris from a trench by hand. This method also requires a portable pump.

Manhole or Drainage Inlet Cleaning

This is accomplished with a "clam digger," a small-diameter, articulated finger-type machine that can reach down into a manhole or inlet for debris removal.

Vapor Rooting

Vapor rooting is used only to chemically remove roots inside a sewer and retard their future growth outside the sewer for several years. Its application is limited to sewers that rarely run full. This method uses a foam dispenser that is run through the sewer pipe, producing a chemical foam having the consistency of thick soap-suds. Since roots do not occur in a pipe full of water, the foam adheres to the roots and pipe above the water line, killing the roots and retarding future growth for approximately three years. This chemical does not remove the roots immediately, but is very effective in about three to six months. If the roots must be removed immediately, the use of an HJR is necessary. In addition to being used alone, this chemical can be added to the grout used to repair cracks.

REPAIR OF CRACKS AND JOINT SEPARATIONS IN PIPING

In many cases, cracks have appeared in pipes and joints have separated due to uneven settlement of the earth surrounding the sewer pipe. When this occurs, the pipe will leak, washing away the supporting earth fill and also allowing roots to enter the pipe. Repair of these deficiencies is of the highest priority.

The most common method used for the repair of piping up to 24 in diameter is called *chemical grouting* (CG). This method uses a polymer adhesive with the viscosity of water that hardens into a solid in a period of time determined by the mixture of ingredients. It is applied from inside a pipe by a device called a *packer,* which is, in essence, a 2½-ft length of pipe, slightly smaller than the pipe being repaired, with an inflatable donut at either end. The packer must be positioned correctly by the use of a video camera. Once in position, the donuts inflate creating a seal, and the adhesive is injected under pressure between them. The adhesive is forced out of the pipe into the surrounding soil, where it hardens, forming a tough elastic and waterproof barrier to inhibit further leakage. If the rooting chemical is added to this adhesive, any roots growing toward the hardened soil will die. This is considered a permanent repair. If any adhesive hardens on the inside of the pipe, it must be cleaned, generally with an HJR with a rotating blade.

REPAIR OF MANHOLES

Manholes are generally repaired using a waterproof epoxy cement or other hydraulic coating applied to all areas of the manhole until a smooth finish is achieved. This is very labor intensive.

PREVENTIVE MAINTENANCE

After the piping network is initially cleaned and repaired, the following scheduled preventive maintenance program is recommended:

1. After the initial vapor rooting procedures, a second application is recommended after one year and again every three years. This should be applied only where there are trees. The price would be the same as for VR cleaning.

2. A regularly scheduled cleaning using the HJR once every two years is recommended.

3. A video inspection of the piping system once every five years is recommended to see if any problems have developed in the piping network.

SANITARY GRAVITY SEWERS

This section will discuss gravity sewers collecting domestic sanitary waste and acceptable industrial discharge. These sewers carry this combined waste from small areas or buildings from the property line or building wall to a connection with a public sanitary sewer. Sanitary sewers are called house sewers for private residences and service laterals for larger facilities. On large sites, long runs of private sewers receiving discharge from various buildings route the sewage into the public sewer for disposal.

If a public sewer is available, it is the least expensive method of disposing sanitary waste. In urban areas, a sewer could be considered available if it is within 500 ft of a property line. This distance varies based on local requirements. It may be necessary for a house sewer to travel a considerable distance, often in a public street.

If a public sewer is not available, the sanitary effluent must be treated by a septic tank or a sewage treatment system to the extent required by local authorities. The treated effluent can then be discharged to the environment.

Plumbing codes govern the size and installation of house sewers and building sewers, or building laterals. Any discharge into the public sewer other than sanitary effluent must be treated so that the quality of the effluent is within guidelines established by the local authorities. These guidelines vary based on the municipal sewage measurements. It could combine with the sanitary house sewer prior to discharge into the public sewer to save installation costs.

The following information should be obtained before designing a sewer:

1. The size, location, and depth of the existing public sewer, and whether the public sewer is a sanitary or combined sewer

2. The location and size of spurs and manholes in the public sewer

3. The allowable method of connection to the public sewer

   • Do the local authorities require a manhole? Will the local authority build the manhole? If so, who will bear the cost?
   • If a pipe is installed in a public street, are there special installation or other requirements?

4. Topographical map along the proposed route of the sewer

5. Minimum depth of bury if mandated by the local authorities

SEWER COMPONENTS

The components of the sewer system consist of appurtenances, or structures, such as manholes, cleanouts to grade for smaller lines, and the piping network.

MANHOLES

A manhole is an underground structure that facilitates pipe connections, access into the sewer for observation and maintenance, and keeps hydraulic interference from connections and changes in direction to a minimum. A typical manhole is illustrated in Fig. 6.10.

There are several different types of manholes, each used for a different purpose. A shallow manhole (Fig. 6.35) is used when the depth from the invert of the sewer to grade is less then approximately 3 ft (1 m). A drop manhole (Fig. 6.11) should be used when the invert of the incoming pipe is more than 2 ft (60 cm) higher than the discharge pipe. This is to avoid the inflow spilling onto the channel of the manhole because it would quickly wear the channel away. Therefore, a drop pipe is provided to bring the effluent into the manhole to lessen the hydraulic impact. Two pipe entries into the manhole are provided in case the drop pipe has a stoppage. When the peak flow fills the pipe to more than three-quarters full, it is common practice to make the drop pipe one size larger than the incoming sewer line to minimize stoppages.

Manholes consist of a frame and cover, structure body, steps, pipe connections, sewage channel, and bottom slab.

Frame and Cover

Frames and covers are available in round and square configurations and in a wide variety of sizes. The advantage of a square cover is the large space for entry into the manhole. The advantage of the round cover is that it cannot fall into the structure. The frame and cover materials is normally cast iron, but ductile iron and aluminum are also commonly available. They are manufactured in light- and heavy-duty, based on the load that could be safely placed on them.

It is common practice to install the top of the cover 1 in above the road surface or grade to keep infiltration of storm water between the frame and cover to a minimum. If distance is greater than 3 in, use concrete rings or brick firmly bedded in mortar to raise the frame. The casting is set to final grade elevation by means of a mortar bed. Care should be taken to select frames and covers that fit well in order to prevent rattling in traffic, provide reasonably tight closure, and resist unauthorized entry.

The size of the cover is determined by the intended use. For workers to enter, the minimum size is 2 ft (60 cm), which is considered small. The common size is 2 ft, 6 in (75 cm). It is common practice to cast into the cover the word "sanitary" or "storm" (or other) as required for each utility to easily identify the system.

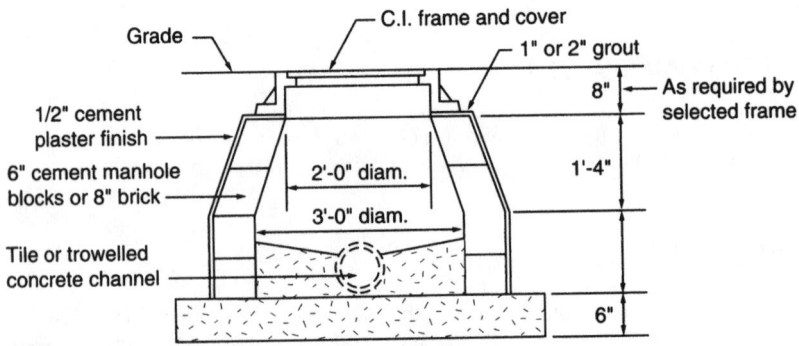

FIGURE 6.35   Detail of shallow manhole.

Manhole Construction

Most manholes are circular, with the inside dimensions large enough for a worker to perform necessary functions without difficulty. The minimum recommended diameter of the major section is generally 3 ft (1 m), with a 4-ft (1.25 m) inside diameter used most often. It is common practice to use eccentric conical sections at the top to give a straight vertical side for the steps. If the cover is 2 ft, 6 in (75 cm), the clear inside dimension of the top section should be a minimum of 3 ft (1 m).

Manholes can be constructed of brick, precast concrete sections, poured (cast-in-place) concrete, and concrete block. The most common method of construction is with precast sections called risers. The typical precast manhole consists of a base, riser sections, and a conical top section ready for installation. The sections fit together with tongue-and-groove joints with gaskets used in the joints to make them watertight. Gaskets should be installed in accordance with manufacturers' recommendations regarding lubricants, cements, and other special installation requirements. Risers are available in a large variety of sizes and shapes.

Poured-in-place or precast reinforced concrete structures must use concrete with a compressive strength of 4000 psi after 28 days in accordance with ASTM C-478. Manhole steps should be installed into forms. A concrete base should be provided under the structure. Rebars must conform to ASTM A-615, grade 40, and wire fabric, to ASTM A-185.

Brick and block manhole walls should be constructed 8 in (20 mm) thick for up to 8 ft of depth, and 12 in (30 mm) for greater depths. The outside of the manholes should be parged (coated) with at least one coat of cement mortar at least one-half in thick for protection against deterioration. Two coats are often used. For additional protection in wet soils, a coating of coal tar epoxy is often applied over the parging after it has dried for 30 days.

Steps

Manhole steps are generally constructed of cast iron or aluminum. Common step diameters are three-quarters in (19 mm) or 1 in (25 mm). Width is 16 in (40 cm) if it is desired to have two feet on one rung; 12 in (30 cm) rungs are the most common size. Steps are spaced 12 in (30 cm) to 16 in (50 mm) apart, with 12 in the most common. Good practice is to have a distance of approximately 6½ in (16 mm) from the wall to the inside of the rung.

The Base and Channel

Structures are built on concrete bases or footings. For precast manholes a recessed center is provided to receive the male tongue and groove joints of a riser section. The invert channel is formed either by shaping poured concrete after the drainage lines have been installed or by a cast iron pipe that has the top cut away after installation.

A channel is the lowest internal part of the manhole and should be a smooth continuation of the pipe. It provides a U-shaped open flow path for the liquid. Its height should be at least one-half the pipe size for 12 in (30 cm) pipe and three-quarters the height for 15 in (38 cm) and larger.

SIZING THE SANITARY SEWER

The sanitary sewer pipe size is selected after calculating the peak flow rate, determining the slope of the pipe, and selecting the sewer pipe material.

Peak Flow Rate Determination

Four factors must be considered when calculating the peak flow:

1. Peak sanitary discharge and flow from fixtures
2. Peak nonsanitary discharge, i.e., drainage and estimated leakage flow from all process, utility, and manufacturing sources, as well as blowdown and similar sources
3. Allowance for future
4. Infiltration for long runs of piping

Peak Sanitary Discharge. Peak sanitary discharge is calculated using fixture units in accordance with local plumbing codes. A method of converting sanitary fixture units into gpm is presented in Fig. 6.36. Private sewers serving multiple buildings are also sized in the same manner until the fixture unit count becomes too large. After the code fixture unit value is exceeded, the drainage flow rate figure is then calculated by using the facility water fixture unit flow rate demand.

Peak Nonsanitary Discharge. The process, utility, and manufacturing discharge can only be calculated by a study of the entire facility and a determination of the total gpm expected to be discharged from all sources. It is doubtful that all discharges will occur at the same time, therefore some diversity might be used to reduce the peak flow. If there is any doubt, use the highest figure.

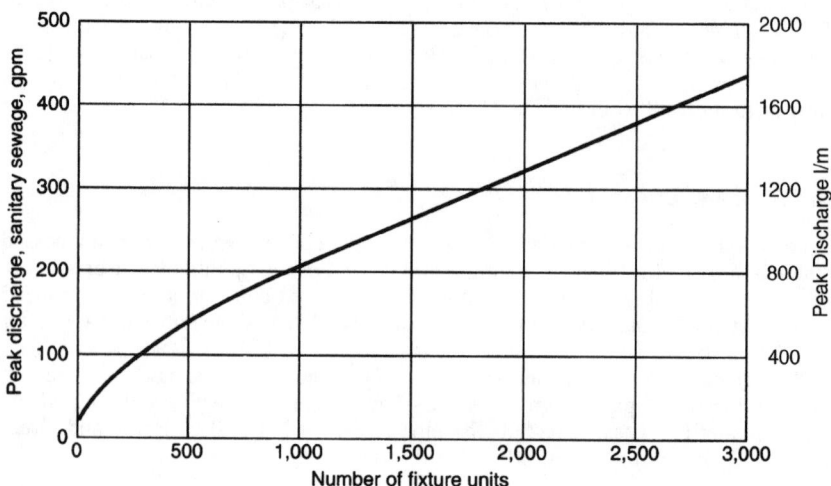

FIGURE 6.36   Fixture unit conversion to gpm.

It is often necessary to prepare a preliminary estimate for pipe size before the exact information is available. Table 6.25 provides the gallons per day for various types of facilities based on population or other easily found criteria.

To estimate a size, find the total gallons per day for the facility. Divide that figure by the number of hours the facility is in operation to give the average hourly flow. Double the hourly flow rate and divide by 60 to find the average peak flow rate in gpm. This is only an average figure. To determine pipe size, allow an additional 10 percent safety factor for peak flow rate.

Allowance for the Future.   Allowance for the future should be obtained and added to the peak discharge.

Infiltration.   Infiltration is the amount of groundwater or storm water entering the sewer piping network from faulty joints in underground piping, leakage around manholes and manhole covers, or cracked pipe. Another major contributor to infiltration is poorly installed laterals. Infiltration should be considered only for a very long run from the building to the sewer or in a large network serving multiple facilities on a single site. Existing sewers are frequently found to be quite leaky.

The allowance for infiltration is based on the sewer system when it is reaching the end of its useful life, not when it is new. The types of joints and the pipe material used have an effect on the figure selected. Common figures used for private sewer systems are 500 gallons per day per mile for smaller sewers and 1000 gal/day/mi for larger sewer lines. The figure selected for infiltration does not relate to the infiltration allowance used to test for sewer acceptance.

For a single facility with a short run to a public sewer, no allowance should be made. If the run is more than about 1500 ft (450 m), some allowance should be made. For lengths of run shorter than 1500 ft (450 m), some consideration should be made depending on the local authorities, stability of ground conditions, and the pipe-jointing methods selected.

Slope of the Piping.   The pipe slope is generally determined either by the topography where the sewer will be installed or the elevation of the outfall where the sewer line will discharge. Often, it is a combination of both. Keeping a fairly uniform depth of bury will establish the general slope, with adjustment made to keep the slope uniform in different sections of the run. If grade is flat, the slope should be based on the optimum velocity, if possible. If the outfall elevation is the controlling factor, there is little choice in the slope. If an adequate slope to provide the necessary velocity can't be maintained, it will be necessary to use a lift station and pump the effluent. The slope of the line shall be steep enough to provide a minimum scouring velocity of 2 fps.

Selecting the Pipe Material.   The pipe material selected is based on many factors. Since no single material will meet all conditions, the material selected should be based on the most important characteristics. Some factors to consider are:

1. Flow characteristics (friction coefficient)
2. Life expectancy
3. Resistance to scour
4. Resistance to effluent and surrounding soil
5. Ease of handling and installation
6. Physical strength to resist loading

TABLE 6.25　Quantities of Sewage Flow

Type of establishment	Gallons per person per day (unless otherwise noted)	
Miscellaneous facilities	Typical	Range
Apartments—multiple family (per resident)	60	
Apartments—multiple family (per apartment)	110	79–132
Bathhouses and swimming pools	10	
Camps		
Campground with central comfort stations	35	
With flush toilets, no showers	25	
Construction camps (semipermanent)	50	
Day camps (no meals served)	15	11–18
Resort camps (night and day) with limited plumbing	50	
Luxury camps	100	
Labor camp	45	37–53
Cottages and small dwellings with seasonal occupancy	50	
Country clubs (per resident member)	100	80–105
Country clubs (nonresident)	20	16–26
Dwellings:		
Boarding houses	50	
Additional for nonresident boarders	10	
Luxury residences and estates	150	
Multiple family dwellings (apartments)	60	
Rooming houses	40	
Single family dwellings	75	
Factories (gallons per person, per shift, sanitary only)	25	11–30
Institutions other than hospitals (per bed space)	125	106–160
Laundries, self-service (gallons per wash, i.e., per customer)	50	
Mobile home parks (per space)	250	
Motels (per bed space)	50	
Picnic parks (toilet wastes only) (per picnicker)	5	
Picnic parks with bathhouses, showers, and flush toilets	10	
Restaurants (toilet and kitchen wastes per patron)	10	
Restaurants (kitchen wastes per meal served)	3	
Restaurants additional for bars and cocktail lounges	2	
Schools:		
Boarding	100	
Day, without gym, cafeteria, or showers	15	
Day, with gym, cafeteria, and showers	25	
Day, with cafeteria, but without gym or showers	20	
Service stations (per vehicle served)	10	
Swimming pools and bathhouses with toilet and shower	14	11–16
Theaters:		
Movies (per auditorium seat)	5	
Drive-in (per car space)	5	
Per seat	4	3–5
Travel trailer parks without water and sewer hook-ups (per space)	50	
Travel trailer parks with individual water and sewer hook-ups (per space)	140	130–160
Workers:		
Construction	50	
Day, at schools and offices (per shift)	15	

(Continued)

TABLE 6.25 Quantities of Sewage Flow (*Continued*)

Source	Unit	Wastewater flow, gallons per day per unit	
		Range	Typical
Commercial sources			
Airport (per passenger)	4	3–5	
Automobile service station	Vehicle services	7.9–13.2	10.6
	Employee	9.2–15.8	13.2
Bar & cocktail lounge	Customer	1.3–5.3	2.1
	Employee	10.6–15.8	13.2
Bowling Alley (per alley)	20	16–26	
Bowling Alley (per sq. ft.)	0.256 GPD		
Hotel	Guest	39.6–58.0	50.1
	Employee	7.9–13.2	10.6
Industrial building (excluding industry and cafeteria)	Employee	7.9–17.2	14.5
Laundry (self-service)	Machine	475.0–686	580.0
	Wash	47.5–52.8	50.1
Motel	Person	23.8–39.6	31.7
Motel with kitchen	Person	50.2–58.1	52.8
Office	Employee	7.9–17.2	14.5
Restaurant	Meal	2.1–4.0	2.6
Restaurant (per seat)	40	32–48	
Rooming house	Resident	23.8–50.1	39.6
Service area, roadway			
Counter seat	350	265–420	
Table seat	175	150–210	
Store, department	Toilet room	423.0–634	528.0
	Employee	7.9–13.2	10.6
	Sq. ft	0.22 GPD	
Shopping center	Parking space	0.5–2.1	1.1
	Employee	7.9–13.2	10.6
	Sq. ft	0.160 GPD	
Institutional and recreational sources			
Hospital, medical	Bed	132.0–351.0	250.0
	Employee	5.3–15.9	10.6
Hospital, mental	Bed	79.3–172.0	106.0
	Employee	5.3–15.9	10.6
Prison	Inmate	79.3–159.0	119.0
	Employee	5.3–15.9	10.6
Rest home	Resident	52.8–119.0	92.5
	Employee	5.3–15.9	10.6
School, day:			
With cafeteria, gym, showers	Student	15.9–30.4	21.1
With cafeteria only	Student	10.6–21.1	15.9
Without cafeteria, gym, showers	Student	5.3–17.2	10.6
School, boarding	Student	52.8–106.0	74.0
Restort	Person	52.8–74.0	58.1
Barber shop	Chair		55
Beauty parlor	Chair		250

TABLE 6.25 Quantities of Sewage Flow (*Continued*)

Source	Unit	Wastewater flow, gallons per day per unit	
		Range	Typical
Institutional and recreational sources			
Cabin, resort	Person	34.3–50.2	42.3
Cafeteria	Customer	1.1–2.6	1.6
	Employee	7.9–13.2	10.6
Campground (developed)	Person	21.1–39.6	31.7
Cocktail lounge	Seat	13.2–26.4	19.8
Coffee shop	Customer	4.0–7.9	5.3
	Employee	7.9–13.2	10.6
Country club	Member present	66.0–132.0	106.0
	Employee	10.6–15.9	13.2
Dining hall	Meals served	4.0–13.2	7.9
Dormitory, bunkhouse	Person	19.8–46.2	39.6
Hotel, resort	Person	39.6–63.4	52.8
Public park with toilets	Visitor	5–11	9
Store resort	Customer	1.3–5.3	2.6
	Employee	7.9–13.2	10.6
Swimming pool	Customer	5.3–13.2	10.6
	Employee	7.9–13.2	10.6
Theater	Seat	2.6–4.0	2.6
Visitor center	Visitor	4.0–7.9	5.3

7. Type and flexibility of joints
8. Requirements of the local authorities
9. Cost

Laterals and service connection runs for sanitary discharge are often made of extra-heavy cast iron with compression gasket joints. Another popular material is plastic, usually PVC with butt-fused joints.

Sizing the Sewer Pipe

The pipe size is selected after determining the pipe material, laying out the run of the sewer to determine the slope, and calculating the peak flow rate.

Sanitary effluent has the same characteristics as storm water. Sewers are sized using the Kutter or Manning formula. Both will yield the same results. The easiest method, rather than using the actual equation, is to find readily available prepared charts. Refer to Fig. 6.17, which solves for the Manning equation. Another set of charts (also based on the Manning formula) providing a direct reading of the size based on gpm flow rate, pitch, and depth is given in Table 6.26.

TABLE 6.26 Drainage Pipe Sizing Tables

Discharge of circular sewer N = .013

Pipe size, in	Grade: Inch per foot	Grade: %	½ Full: Discharge gpm	½ Full: Discharge cfs	½ Full: Velocity ft/s	⅔ Full*: Discharge gpm	⅔ Full*: Discharge cfs	⅔ Full*: Velocity ft/s	Full: Discharge gpm	Full: Discharge cfs	Full: Velocity ft/s
2	⅛	1.0	7	0.02	1.3	10	0.02	1.5	13	0.03	1.3
	¼	2.1	9	0.02	1.8	14	0.03	2.0	18	0.04	1.8
	½	4.2	14	0.03	2.9	22	0.05	3.2	28	0.06	2.9
	1	8.3	20	0.05	4.0	32	0.07	4.5	40	0.09	4.0
3	⅛	1.0	18	0.04	1.7	29	0.06	1.9	36	0.08	1.7
	¼	2.1	26	0.06	2.4	41	0.09	2.7	51	0.11	2.4
	½	4.2	40	0.09	3.7	64	0.14	4.1	80	0.18	3.7
	1	8.3	57	0.13	5.3	90	0.20	5.9	114	0.25	5.3
4	⅛	1.0	39	0.09	2.0	61	0.13	2.2	77	0.17	2.0
	¼	2.1	55	0.13	2.8	87	0.20	3.1	110	0.25	2.8
	½	4.2	87	0.20	4.5	138	0.31	5.0	174	0.39	4.5
	1	8.3	123	0.28	6.3	194	0.44	7.1	245	0.55	6.3
6		0.5	79	0.18	1.8	124	0.28	2.0	157	0.35	1.8
	⅛	1.0	110	0.25	2.5	174	0.39	2.8	220	0.49	2.5
		1.5	135	0.30	3.1	213	0.47	3.5	269	0.60	3.1
	¼	2.0	157	0.35	3.6	248	0.55	4.0	314	0.70	3.6
		2.5	175	0.39	4.0	277	0.62	4.5	350	0.78	4.0
	⅜	3.0	193	0.43	4.4	305	0.68	4.9	386	0.86	4.4
		3.5	207	0.46	4.7	327	0.73	5.3	413	0.92	4.7
	½	4.0	225	0.50	5.0	355	0.79	5.6	449	1.00	5.0
	⅝	5.0	247	0.55	5.6	391	0.87	6.3	494	1.10	5.6
	¾	6.0	270	0.60	6.1	426	0.95	6.8	539	1.20	6.1
	⅞	7.0	292	0.65	6.6	461	1.03	7.4	583	1.30	6.6

See last page of table for footnotes.

(*Continued*)

TABLE 6.26 Drainage Pipe Sizing Tables (*Continued*)
Discharge of circular sewer N = .013

Pipe size, in	Grade		½ Full			⅔ Full*			Full		
	Inch per foot	%	Discharge gpm	cfs	Velocity ft/s	Discharge gpm	cfs	Velocity ft/s	Discharge gpm	cfs	Velocity ft/s
8		0.2	108	0.24	1.6	170	0.38	1.8	215	0.48	1.6
		0.4	153	0.34	2.0	241	0.54	2.2	305	0.68	2.0
		0.6	191	0.43	2.4	302	0.67	2.7	382	0.85	2.4
		0.8	236	0.53	2.9	372	0.83	3.2	471	1.05	2.9
	⅛	1.0	247	0.55	3.2	391	0.87	3.6	494	1.10	3.2
		1.5	303	0.68	3.8	479	1.07	4.3	606	1.35	3.8
	¼	2.0	348	0.78	4.5	550	1.22	5.0	696	1.55	4.5
		2.5	392	0.88	4.9	621	1.38	5.5	785	1.75	4.9
	⅜	3.0	427	0.95	5.4	674	1.50	6.0	853	1.90	5.4
		3.5	449	1.00	5.8	710	1.58	6.5	893	2.00	5.8
	½	4.0	494	1.10	6.2	780	1.74	6.9	987	2.20	6.2
		4.5	516	1.15	6.6	816	1.82	7.4	1032	2.30	6.2
10		0.2	211	0.47	1.7	334	0.74	1.9	422	0.94	1.7
		0.4	303	0.68	2.4	479	1.1	2.7	606	1.35	2.4
		0.6	359	0.80	2.9	568	1.3	3.2	718	1.60	2.9
		0.8	438	0.98	3.5	692	1.5	3.9	875	1.95	3.5
	⅛	1.0	472	1.05	3.8	745	1.7	4.3	943	2.10	3.8
		1.5	561	1.25	4.5	887	2.0	5.0	1122	2.50	4.5
	¼	2.0	651	1.45	5.3	1029	2.3	5.9	1302	2.90	5.3
		2.5	741	1.65	5.7	1170	2.6	6.4	1481	3.30	5.7
	⅜	3.0	808	1.80	6.4	1277	2.8	7.2	1616	3.60	6.4
		3.5	853	1.90	6.8	1348	3.0	7.6	1706	3.80	6.8

Section 12:

12		0.2	337	0.8	1.9	533	1.2	2.1	674	1.5	1.9
		0.4	472	1.1	2.7	745	1.7	3.0	943	2.1	2.7
		0.6	584	1.3	3.3	922	2.1	3.7	1167	2.6	3.3
		0.8	718	1.6	4.1	1135	2.5	4.6	1436	3.2	4.1
	1/8	1.0	763	1.7	4.3	1206	2.7	4.8	1526	3.4	4.3
		1.2	831	1.8	4.7	1313	2.9	5.3	1661	3.7	4.7
		1.4	898	2.0	5.0	1409	3.2	5.6	1795	4.0	5.0
		1.6	965	2.2	5.3	1525	3.4	5.9	1930	4.3	5.3
	1/4	1.8	1010	2.3	5.7	1596	3.6	6.4	2020	4.5	5.7
		2.0	1077	2.4	6.0	1702	3.8	6.7	2154	4.8	6.0
		2.2	1122	2.5	6.2	1773	4.0	6.9	2244	5.0	6.2
		2.4	1167	2.6	6.6	1844	4.1	7.4	2334	5.2	6.6

Section 14:

14		0.1	382	0.8	1.6	603	1.3	1.7	763	1.7	1.6
		0.2	516	1.2	2.2	816	1.8	2.5	1032	2.3	2.2
		0.3	651	1.5	2.7	1029	2.3	3.0	1302	2.9	2.7
		0.4	763	1.7	3.1	1206	2.7	3.5	1526	3.4	3.1
	1/16	0.5	853	1.9	3.5	1348	3.0	3.9	1706	3.8	3.5
		0.6	920	2.1	3.8	1454	3.2	4.3	1840	4.1	3.8
		0.7	1010	2.3	4.1	1596	3.6	4.6	2020	4.5	4.1
		0.8	1077	2.4	4.4	1702	3.8	4.9	2154	4.8	4.4
		0.9	1145	2.6	4.6	1816	4.0	5.2	2299	5.1	4.6
	1/8	1.0	1212	2.7	4.8	1915	4.3	5.4	2424	5.4	4.8
		1.1	1257	2.8	5.1	1986	4.4	5.7	2513	5.6	5.1
		1.2	1324	3.0	5.3	2092	4.7	5.9	2648	5.9	5.3
		1.3	1369	3.1	5.5	2163	4.8	6.2	2738	6.1	5.5
		1.4	1436	3.2	5.8	2269	5.1	6.5	2872	6.4	5.8
		1.5	1481	3.3	5.9	2340	5.2	6.6	2962	6.6	5.9
		1.6	1526	3.4	6.0	2412	5.4	6.7	3052	6.8	6.0
		1.7	1571	3.5	6.3	2483	5.5	7.1	3142	7.0	6.3

(Continued)

TABLE 6.26 Drainage Pipe Sizing Tables (*Continued*)

Discharge of circular sewer N = .013

Pipe size, in	Grade Inch per foot	Grade %	½ Full Discharge gpm	½ Full Discharge cfs	½ Full Velocity ft/s	⅔ Full* Discharge gpm	⅔ Full* Discharge cfs	⅔ Full* Velocity ft/s	Full Discharge gpm	Full Discharge cfs	Full Velocity ft/s
15		0.1	439	1.0	1.6	694	1.6	1.8	878	2.0	1.6
		0.2	628	1.4	2.3	993	2.2	2.6	1257	2.8	2.3
		0.3	763	1.7	2.7	1206	2.7	3.0	1526	3.4	2.7
		0.4	898	2.0	3.2	1419	3.2	3.6	1795	4.0	3.2
		0.5	1010	2.3	3.6	1596	3.6	4.0	2020	4.5	3.6
		0.6	1100	2.5	3.9	1738	3.9	4.4	2199	4.9	3.9
		0.7	1190	2.7	4.3	1880	4.2	4.8	2379	5.3	4.3
		0.8	1279	2.9	4.6	2021	4.5	5.2	2558	5.7	4.6
		0.9	1347	3.0	4.8	2128	4.7	5.4	2693	6.0	4.8
	⅛	1.0	1437	3.2	5.2	2270	5.1	5.8	2873	6.4	5.2
		1.1	1481	3.3	5.3	2340	5.2	5.9	2962	6.6	5.3
		1.2	1549	3.5	5.5	2447	5.5	6.2	3097	6.9	5.5
		1.3	1616	3.6	5.8	2553	5.7	6.5	3231	7.2	5.8
		1.4	1683	3.8	5.9	2660	5.9	6.6	3366	7.5	5.9
		1.5	1751	3.9	6.2	2766	6.2	6.9	3501	7.8	6.2
		1.6	1796	4.0	6.4	2837	6.3	7.2	3591	8.0	6.4
		1.7	1863	4.2	6.6	2943	6.6	7.4	3725	8.3	6.6

16

0.1	539	1.2	1.7	851	1.9	1.9	1077	2.4	1.7
0.2	786	1.8	2.4	1242	2.8	2.7	1571	3.5	2.4
0.3	967	2.2	2.9	1528	3.4	3.2	1933	4.3	2.9
0.4	1122	2.5	3.4	1773	4.0	3.8	2244	5.0	3.4
0.5	1257	2.8	3.8	1986	4.4	4.3	2513	5.6	3.8
0.6	1392	3.1	4.2	2199	4.9	4.7	2783	6.2	4.2
0.7	1504	3.4	4.5	2376	5.3	5.0	3007	6.7	4.5
0.8	1594	3.6	4.8	2518	5.6	5.4	3187	7.1	4.8
0.9	1683	3.8	5.1	2660	5.9	5.7	3366	7.5	5.1
1.0	1796	4.0	5.4	2837	6.3	6.0	3591	8.0	5.4
1.1	1885	4.2	5.6	2980	6.6	6.3	3770	8.4	5.6
1.2	1975	4.4	5.8	3121	7.0	6.5	3950	8.8	5.8
1.3	2020	4.5	6.1	3192	7.1	6.8	4040	9.0	6.1
1.4	2110	4.7	6.3	3333	7.4	7.1	4219	9.4	6.3
1.5	2177	4.9	6.6	3440	7.7	7.4	4354	9.7	6.6

$\frac{1}{8}$

18

0.1	719	1.6	1.8	1136	2.5	2.0	1437	3.2	1.8
0.2	1010	2.3	2.6	1596	3.6	2.9	2020	4.5	2.6
0.3	1257	2.8	3.3	1986	4.4	3.7	2514	5.6	3.3
0.4	1459	3.3	3.7	2306	5.1	4.1	2918	6.5	3.7
0.5	1616	3.6	4.1	2554	5.7	4.6	3232	7.2	4.1
0.6	1796	4.0	4.5	2837	6.3	5.0	3591	8.0	4.5
0.7	1953	4.4	4.8	3085	6.9	5.4	3905	8.7	4.8
0.8	2043	4.6	5.2	3228	7.2	5.8	4085	9.1	5.2
0.9	2155	4.8	5.5	3405	7.6	6.2	4309	9.6	5.5
1.0	2244	5.0	5.7	3546	7.9	6.4	4488	10.0	5.7
1.1	2357	5.3	6.2	3724	8.3	6.9	4713	10.5	6.2
1.2	2469	5.5	6.8	3901	8.7	7.6	4937	11.0	6.8
1.3	2581	5.8	7.3	4078	9.1	8.2	5162	11.5	7.3
1.4	2693	6.0	7.8	4255	9.5	8.7	5386	12.0	7.8

$\frac{1}{8}$

(Continued)

TABLE 6.26 Drainage Pipe Sizing Tables *(Continued)*

Discharge of circular sewer N = .013

Grade			½ Full			⅔ Full*			Full		
Pipe size, in	Inch per foot	%	Discharge gpm	cfs	Velocity ft/s	Discharge gpm	cfs	Velocity ft/s	Discharge gpm	cfs	Velocity ft/s
20		0.1	965	2.2	2.0	1525	3.4	2.2	1930	4.3	2.0
		0.2	1392	3.1	2.8	2199	4.9	3.1	2783	6.2	2.8
		0.3	1706	3.8	3.5	2695	6.0	3.9	3411	7.6	3.5
		0.4	1975	4.4	4.0	3121	7.0	4.5	3950	8.8	4.0
		0.5	2200	4.9	4.5	3476	7.7	5.0	4400	9.8	4.5
		0.6	2401	5.4	4.8	3794	8.5	5.4	4802	10.7	4.8
		0.7	2648	5.9	5.3	4184	9.3	5.9	5296	11.8	5.3
		0.8	2805	6.3	5.6	4432	9.9	6.3	5610	12.5	5.6
		0.9	2918	6.5	6.0	4610	10.3	6.7	5835	13.0	6.0
	⅛	1.0	3150	7.0	6.3	4977	11.1	7.1	6300	14.0	6.3
		1.1	3254	7.3	6.5	5142	11.5	7.3	6508	14.5	6.5
		1.2	3366	7.5	6.8	5319	11.9	7.6	6732	15.0	6.8
21		0.1	1100	2.5	2.1	1738	3.9	2.4	2200	4.9	2.1
		0.2	1571	3.5	2.9	2483	5.5	3.2	3142	7.0	2.9
		0.3	1930	4.3	3.6	3050	6.8	4.0	3860	8.6	3.6
		0.4	2200	4.9	4.1	3476	7.7	4.6	4400	9.8	4.1
		0.5	2491	5.6	4.6	3936	8.8	5.2	4982	11.1	4.6
		0.6	2693	6.0	5.0	4257	9.5	5.6	5386	12.0	5.0
		0.7	2918	6.5	5.5	4610	10.3	6.2	5835	13.0	5.5
		0.8	3150	7.0	5.8	4977	11.1	6.5	6300	14.0	5.8
		0.9	3299	7.4	6.2	5212	11.6	6.9	6597	14.7	6.2
	⅛	1.0	3479	7.8	6.5	5497	12.2	7.3	6957	15.5	6.5
		1.1	3658	8.2	6.7	5780	12.9	7.5	7316	16.3	6.7
24		0.05	1122	2.5	1.6	1773	4.0	1.8	2244	5.0	1.6
		0.1	1616	3.6	2.3	2554	5.7	2.6	3232	7.2	2.3
		0.2	2244	5.0	3.2	3546	7.9	3.6	4488	10.0	3.2
		0.3	2805	6.3	3.9	4432	9.9	4.4	5610	12.5	3.9

0.5	3591	8.0	5.1	5673	12.6	5.7	7181	16.0	5.1
0.6	3927	8.8	5.5	6205	13.8	6.2	7854	17.5	5.5
0.7	4264	9.5	6.0	6738	15.0	6.7	8528	19.0	6.0
0.8	4488	10.0	6.4	7092	15.8	7.2	8976	20.0	6.4
27									
0.05	1482	3.3	1.7	2341	5.2	1.9	2963	6.6	1.7
0.1	2110	4.7	2.5	3333	7.4	2.8	4219	9.4	2.5
0.2	3130	6.8	3.5	4787	10.7	3.9	6059	13.5	3.5
0.3	3703	8.3	4.3	5851	13.0	4.8	7406	16.5	4.3
0.4	4309	9.6	4.8	6808	15.2	5.4	8617	19.2	4.8
0.5	4713	10.5	5.5	7446	16.6	6.2	9425	21.0	5.5
0.6	5162	11.5	6.0	8156	18.2	6.7	10323	23.0	6.0
0.7	5610	12.5	6.5	8864	19.8	7.3	11220	25.0	6.5
30									
0.05	2020	4.5	1.9	3192	7.1	2.1	4040	9.0	1.9
0.1	2805	6.3	2.7	4432	9.9	3.0	5610	12.5	2.7
0.2	4040	9.0	3.8	6383	14.2	4.3	8079	18.0	3.8
0.3	4938	11.0	4.6	7802	17.4	5.2	9876	22.0	4.6
0.4	5610	12.5	5.3	8864	19.8	5.9	11220	25.0	5.3
0.5	6508	14.5	6.0	10282	22.9	6.7	13016	29.0	6.0
0.6	7181	16.0	6.5	11346	25.3	7.3	14362	32.0	6.5
33									
0.05	2581	5.8	2.0	4078	9.1	2.2	5162	11.5	2.0
0.1	3703	8.3	2.8	5851	13.0	3.1	7406	16.5	2.8
0.2	5162	11.5	4.0	8156	18.2	4.5	10323	23.0	4.0
0.3	6512	14.5	4.9	10289	22.9	5.5	13023	29.0	4.9
0.4	7406	16.5	5.6	11701	26.1	6.3	14811	33.0	5.6
0.5	8303	18.5	6.4	13119	29.2	7.2	16606	37.0	6.4
0.6	9201	20.5	7.0	14537	32.4	7.8	18401	41.0	7.0
36									
0.05	3254	7.3	2.1	5142	11.5	2.4	6508	14.5	2.1
0.1	4713	10.5	3.0	7446	16.6	3.4	9425	21.0	3.0
0.2	6733	15.0	4.3	10638	23.7	4.8	13465	30.0	4.3
0.3	8303	18.5	5.2	13119	29.2	5.8	16606	37.0	5.2
0.4	9425	21.0	6.0	14892	33.2	6.7	18850	42.0	6.0
0.5	10772	24.0	6.8	17019	37.9	7.6	21543	48.0	6.8

*The depth of flow is equal to two-thirds the pipe diameter.

Source: After Chezy.

SANITARY SEWER DESIGN CONSIDERATIONS

Minimum and Maximum Effluent Velocity

The minimum velocity of the effluent should be sufficient to prevent solids from being deposited on the bottom of the pipe. This is called a *scouring velocity.* Tests have established this minimum velocity at 2 fps or 0.62 meter per second (mps) when the pipe is flowing full. Accepted practice has a recommended minimum of 2.5 fps (0.77 mps). Since pipes do not run full most of the time, it may be necessary to flush the sewer in order to remove accumulated sediment from time to time.

The maximum velocity for clear water in hard-surfaced pipe is quite high. Tests have shown that storm water velocities in excess of 40 fps (12 mps) have been found harmless to concrete channels. In practice, sanitary sewers that have continuously high velocities and where grit is expected to be a problem should limit the highest velocity to 10 fps (3 mps). For ordinary sewers without grit and only occasional periods flowing full, a maximum velocity of 20 fps (7 mps) would be considered acceptable.

For large facilities, if the house sewer is sized for the peak flow rate, the velocity during peak flow rate should be adequate to flush the pipe clean of any deposits left during periods of lower flow, for example, during the night.

Tests have shown that pipe size is not a factor in determining the velocity, provided that the pipe is large enough to prevent surcharging.

Connections to the Public Sewer

Connecting the house lateral or the private sewer to the public sewer is generally governed by the local sewer authority.

Most public sewers in public streets have built-in connections, called *spurs,* installed during construction. The location, size, and invert of spurs is often available on utility survey maps. If the size of the lateral is small enough to use a spur, it should be used.

Local authorities usually require that the connection to public sewers be made using manholes where spurs are not available. If the public sewer is very large compared to the house sewer, a special method of connection is required to pierce the main sewer and install the new pipe. One such method is illustrated in Fig. 6.37.

Depth of Bury

There are no hard and fast rules to determine the depth at which sanitary sewers should be buried. Often, local authorities have established this depth. When no guidance exists, a reasonable starting point would be 3 ft of cover, with a depth of 2 ft in areas where the pipe will not be disturbed. Consideration should be given to the depth necessary to resist possible pipe breakage by traffic or other vehicles passing over the pipe. When crossing other utilities, the sewer should be placed below them whenever possible. Since it is good practice to allow building laterals to have a ¼-in pitch when connecting to a sewer, the sewer should be deep enough to accommodate this pitch.

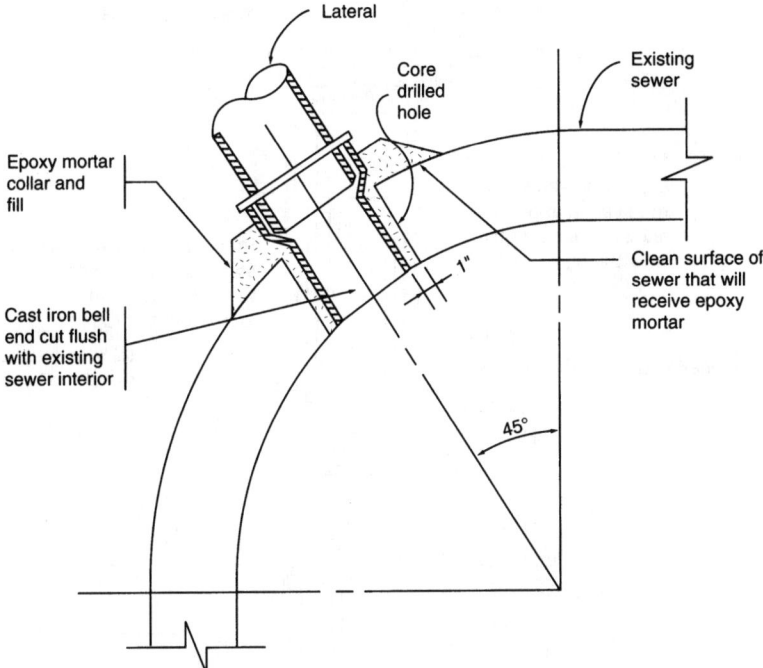

FIGURE 6.37 Method of connection to a large sewer.

Design Depth of Flow

Sewers serving facilities should be designed to carry the peak design flow rate, including allowance for the future, flowing between one-half to two-thirds full. This practice allows a safety factor for higher than expected peak flow rates. Another reason for this practice is that the air space above the flowing effluent allows for ventilation of the piping network.

Spacing of Manholes

When manholes are installed in a public street, local authorities usually mandate the distance between manholes based on sewer size. When sewers are installed on private property, the following guidelines are used for placement and spacing of manholes for sewer pipes up to 24 in (60 cm) in diameter. A maximum spacing of between 150 and 400 ft (80 and 125 m) along straight runs of pipe. Generally accepted practice has manholes spaced between 200 and 250 ft apart, with the longer distance for the larger pipe size. Sewers larger than 24 in usually have manholes 500 to 600 ft (155 to 190 m) apart. Smaller building laterals should have cleanouts brought up to grade 50 to 100 ft apart rather than manholes. Cleanouts to grade are illustrated in Fig. 6.38.

Manholes should be placed at:

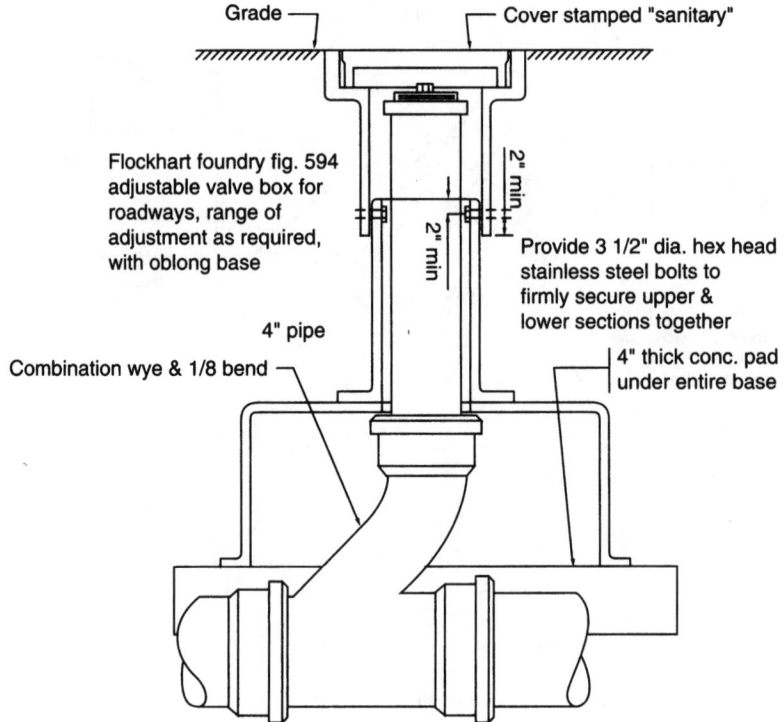

FIGURE 6.38 Cleanout to grade.

1. All changes in direction 45° or greater
2. Changes of sewer alignment
3. Changes in the size of the sewer
4. Changes of grade
5. Intersections with other sewers
6. The end of the sewer system

Clearances

Clearance is the distance between the sewer pipe and any other pipe, measured from exterior to exterior. Clearances are usually mandated either by code requirements or local authorities. There are different dimensions required horizontally and vertically.

The most stringent requirements are between potable water pipes and the sewer. Generally accepted practice is to have a minimum 10-ft (3-m) separation between water and sewer lines horizontally, with the sewer line below the water line. If that horizontal distance cannot be maintained, the sewer line should be considerably lower than the parallel water main, at the often recommended minimum distance of 4 to 6 ft (1.25 to 1.85 m) below the water line. When crossing, the sewer should be a minimum of 1 ft (0.30 m) below the water main. If this is not possible, the sanitary pipe shall be encased in concrete for a distance of 1 ft (0.30 m) past the water main. Consult local authorities for minimum distance required.

For other utilities, a 6-in clearance is adequate and encasement is not necessary.

SEWAGE LIFT STATIONS

When due to topography or low elevation the discharge of sanitary effluent from a facility is lower than the public sewer intended for disposal, pumping up to the sewer point is required. A sewage lift station, or sewage pumping station, is the system used to accomplish this.

There are three general categories of sewage pumping stations: municipal systems that are designed to serve a specific drainage area and are part of the public sanitary sewer system; industrial types that serve a single facility or site with multiple buildings; and residential types that serve a group of individual or multiple dwellings. This section will discuss methods used to pump sewage discharged from industrial/commercial types of facilities.

The difference in terminology between a sewage ejector and a lift station is one of scope. In general, the ejector system is used to pump discharge from a portion of a building up to the main house sewer for disposal, while the lift station is used to pump discharge from an entire building or site to a public sewer for disposal. The components of sewage ejector systems and sewage lift station systems are very similar to the ejector systems discussed in Chap. 9, Plumbing Systems.

CODES AND STANDARDS

There are requirements in regional model plumbing codes that apply to sewage pumping installations. The specific plumbing code used for the area where the facility is built must be followed.

SYSTEM COMPONENTS

The components of a sewage lift station are sewage pumps, basin, discharge pipe (routed to the public sewer or point of disposal), controls, and alarms. The sewage pump discharge line is often referred to as a force main.

Sewage Pumps

The types of pumps used in sewage lift stations are similar to the ejector systems discussed in Chap. 9, Plumbing Systems.

The pump selected most often is the submersible type, because of its low initial cost, wide range of capacities, and tolerance of many starts. Another advantage is that a smaller basin can be used because additional height below grade is not required to house the motor of a conventional vertical, submerged ejector pump. A typical submersible pump assembly is illustrated in Fig. 6.39.

Pump casings, impellers, and other components are available in a wide variety of materials to resist chemical corrosion, abrasion, and suspended solid size.

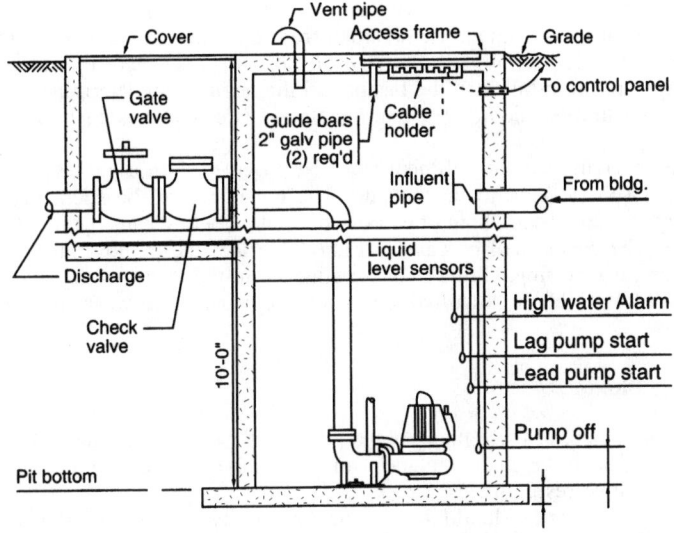

FIGURE 6.39   Detail of typical exterior submersible pump.

Storage Basin

Basins, often referred to as sumps or wet wells, are typically manufactured from cast iron, fiberglass, poured-in-place concrete, and steel.

Large basin capacities require that the design of the basin include principles important to proper pump functioning. These principles are:

1. The flow of effluent from the liquid entrance should be directed to the pump inlet in a manner that will reduce swirl and hydraulic loss.

2. Water depth shall be the minimum established by the manufacturer to avoid surface vortices.

3. Excessive turbulence should be avoided. A small amount of turbulence shall be limited to that required to prevent stagnation of stored water.

4. To prevent sedimentation and accumulation of solids, the floor of the basin should be sloped to the pumps and the floor joints with walls shall have a fillet.

5. Where the inflow into the sump is at a high elevation, the falling water entrains air as it spills onto the surface of water in the sump. This entrained air should be allowed to rise to the surface before reaching the pump. This is accomplished by providing a baffle wall in the sump and/or a dissipation arrangement at the inflow. If the basin is large, there should be a long distance from the point of liquid entry to the pumps to permit air to dissipate.

6. Pumps should be installed as close to the outside walls as practical to avoid stagnant areas that allow solids to accumulate.

7. A small sump should be installed in the basin to permit total drainage for maintenance.

A majority of systems consist of relatively small units. For this type of installation, a standard basin should be obtained from the pump manufacturer. The basin
sizes correspond to the number and size of the pumps selected; the inlet and discharge pipe are preinstalled in the basin; and the depth is predetermined to contain
the desired quantity of liquid. The complete basin is installed as a single, integrated
unit.

Basins receiving sanitary effluent shall be provided with a gasketed cover that
has an atmospheric vent to the outside. The location shall be such that an odor
emanating from the basin would be easily dissipated before causing any discomfort
to people. The cover shall be easy to remove so that the pumps can be raised to
the surface for servicing. Guide rails are attached to the inside of the basin for this
purpose. A hose bibb or lawn hydrant should be provided to wash off the pump.

Discharge Piping

The discharge pipe, or force main, is the pipe from the pump discharge to the point
of disposal. The force main includes valves, clean-outs, air and vacuum release
valves, and thrust restraints.

The piping material should be selected primarily based on soil and effluent
corrosion resistance and pressure rating. Other considerations are strength of pipe
material and what materials are allowed by the local authorities. When there is a
choice among several different materials, the total cost of installation will be the
deciding factor.

Commonly used materials include carbon steel, ductile iron, PVC, CPVC, FRP,
and PE.

COMPONENT DESIGN AND SELECTION

General

The design of the complete sewage lift station as a whole is an iterative one, where
selection of each component is somewhat dependent on other components for size
and capacity.

Pump Selection

Pump Flow Rate.   For a facility sewage pump, it is accepted practice to have a
pump capacity equal to the highest instantaneous flow rate expected from that
facility. The reasoning is that in the event the basin is incapable of being used, a
single pump must be capable of discharging all possible effluent from all sources.
It is also critical that there be at least a duplex set of pumps, each with the same
capacity, in order to ensure that the facility will be kept in operation if one pump
is out of service. If there is the possibility of a wide range of instantaneous flow
rates, a three-pump system, with each pump sized at 75 percent of maximum flow,
should be considered. One pump would be on standby.

The maximum discharge is calculated for two conditions. The first is for the
maximum possible inflow. This is done by adding the plumbing fixture load in gpm
obtained from Hunter's WFU curve to the discharges from other sources to arrive
at a maximum instantaneous flow rate. If the facility is mostly discharging effluent

from plumbing fixtures, add 10 percent to that figure as a safety factor. The second condition is based on a reasonable number of starts per hour.

Required Head.   The required system head is calculated by adding the static height (in feet) from the bottom of the basin to the point of discharge (or the highest point of the force main run) to the friction loss (in feet of head) of the liquid flowing through the pipe based on the equivalent length of run. The friction loss is obtained from standard engineering charts for the material selected, using water as the liquid for sanitary effluent. For more viscous liquids, appropriate charts shall be used.

Calculation of Basin Capacity

The capacity of the basin is based on a single primary pump discharge flow rate and the desired number of starts per hour. Recommended practice is to have an average of six or seven starts per hour up to a maximum of 12 starts per hour, and a minimum running time of 1 min. Within these parameters, and with consultation of the pump manufacturer, the storage capacity can be selected. Cost will also be a factor if the basin is large. Since the exact inflow into the basin in any time period will never be known, a minimum running time shall be selected based on the number of gallons stored, and if additional flow into the basin occurs, the running time of the pump is extended, which is an added benefit.

The actual size may be limited by space conditions at the location where the basin is located. Additional depth may be used to obtain the desired volume if the length or width cannot be adjusted. To size the basin, refer to Table 9.8 for capacities in gallons per foot of various basin sizes.

Force Main Pipe Material Selection and Sizing

The discharge pipe, commonly called a force main, is sized to convey the effluent economically from the pump to its ultimate point of discharge. The pipe material selected is based on flow rate, friction loss of the fluid in the pipe, chemical resistance to the effluent, and fluid velocity. A minimum size of 4 in is highly recommended to allow solids to pass easily through the pipe with little chance of producing stoppages.

Selection of Pipe Material.   The selection of pipe material depends on the type of effluent expected and the strength of the pipe, based on its installed condition. For sanitary effluent, pressure-rated PVC or PE is often selected. Soil contamination may not permit plastic to be used. When the force main is to be routed in public streets, the local authorities may have pipe material and specific installation requirements. Where strength of pipe material is a factor (for example, burial close to the surface of a public road), ductile iron should be considered. It is common practice to coat the exterior of the pipe or protect it with a film of PE in corrosive or contaminated soils. Several methods of installation are described in ANSI A-21.3/AWWA C-105. A typical installation of a PE protective wrap is shown in Fig. 6.40.

Flow Rate.   The flow rate used to size the force main is based on two conditions. The first (which is the normal condition) has one pump of a multiple pump system running. The second (emergency condition) has all the pumps running at the same

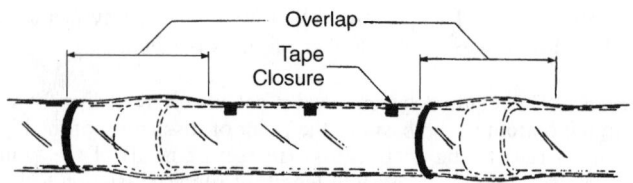

FIGURE 6.40   Typical polyethylene pipe wrap.

time. In the second case, the flow rate is the sum of the flow rates of all the pumps together.

Selecting the pipe size is an iterative procedure done when selecting the pump. A comparison of the pump system head curve is made to see "what if" when the pipe size is changed. The total friction loss is selected to give the optimum choice between the cost of the pipe and the cost of the additional horsepower required to pump the fluid at the higher head due to the smaller pipe size.

Friction Loss.   The friction loss shall be found by using the maximum flow possible, that is, with the maximum number of pumps running. Use friction loss charts similar to Fig. 9.28 for water piping, based on the pipe material selected, flow rate, and its size.

Velocity.   The velocity in the pipe should be within a range of between a minimum of 2 fps and the generally accepted maximum of about 10 fps. The low flow velocity is based on the flow rate when one pump is running, and the highest velocity occurs when all pumps are running together. Since the flow rate for all pumps running is considered a rare occasion, a slightly higher velocity should be considered acceptable if the possibility of the development of excessive water hammer does not occur. One recommendation is to use a slow-closing check valve to limit the generation of excessive pressures every time the pump(s) stop.

Pipe Sizing.   The force main pipe size is based on flow rate, velocity, and friction loss. The criteria used to size the pump are capacity and total discharge head. After deciding on the pump capacity, the only criterion that can be adjusted is the pipe friction loss. A balance must be made between a large pipe size with low friction loss and a small pipe size with large friction loss. A small pipe size may also require a high velocity, which should be avoided. With different pipe sizes, at some point the friction loss will change the horsepower of the pump. The optimum economical point is reached when a larger pipe size will no longer reduce the horsepower of the pump(s). It is also possible that for a long pipe run, a smaller pipe size will justify a larger pump size. The force main size depends upon whichever condition is present.

SYSTEM DESIGN CONSIDERATIONS

Lift Station Location

The basin should not be located at a point on the site that is subject to flooding unless it is the only possible location. Borings of the area should be made to determine if the area has high groundwater or other underground obstacles that would interfere with the installation of the basin. The basin shall be installed with its top several inches above grade so that storm water will not be directed toward the basin.

Floatation

Any structure extending below the groundwater table, or extending into a soil that may become saturated, must be checked to ensure that the buoyant force will not cause the basin to float. To determine the buoyant force, calculate the upward and downward forces. If the sum of the upward force exceeds the downward force, the structure will float.

The downward forces are the sum of the equipment and basin weight, the weight of the concrete base slab under the basin, if present, and the dry weight of the soil on top of the basin, if any. The dry weight is calculated by determining the wet weight minus the weight of the water. For typical soils, refer to Table 6.8 to obtain the weight per cubic foot or meter, and subtract the weight of a cubic foot (or meter) of water, which is 62.4 lb/ft^3. The effective weight of concrete is 87 lb/ft^3 in water (150 lb/ft^3 of concrete minus 63 lb/ft^3 for water). For displaced saturated soil, the figure used is 50 lb/ft^3, with the assumed weight of soil of 100 lb/ft^3.

The buoyant force is the volume of displaced weight of groundwater measured to the highest level possible or the weight of the displaced saturated soil.

The above calculations do not consider additional restraining forces such as skin friction between the basin and the soil and the shear between overburden and adjoining soil, which may have to be considered for large basins. Common practice regards these figures as optional safety factors for smaller basins.

If it is determined that the basin will float, the most common method to offset floating is to use a concrete slab under the basin to provide the necessary added weight. The basin must be anchored to the slab. It is common practice to use a 25 percent safety factor when calculating the slab weight.

Control Considerations

The basic operating levels are decided in a manner similar to those of an ejector system, discussed in Chap. 9. If mercury switches are used, experience dictates that no less than 6 in (150 mm) between levels should be used.

Vacuum and Pressure Venting

It is good practice to provide an air pressure vent at the high point of the piping run. If the high point of the piping system occurs prior to the point of discharge, it is desirable to place a combination pressure/vacuum device in the pipe to allow accumulated air at the high point to be eliminated, and since the remainder of the run is downhill, a vacuum may be produced by the flowing liquid that will have to be broken to allow free flow of the liquid. Since there is a probability that air relief devices will discharge some water, the device is usually put in a large valve box with a gravel bottom to adsorb the liquid. Some authorities require that air release devices be installed in manholes. Ease of maintenance is the most important consideration. A detail of a typical device in a manhole is shown in Fig. 6.41, and a

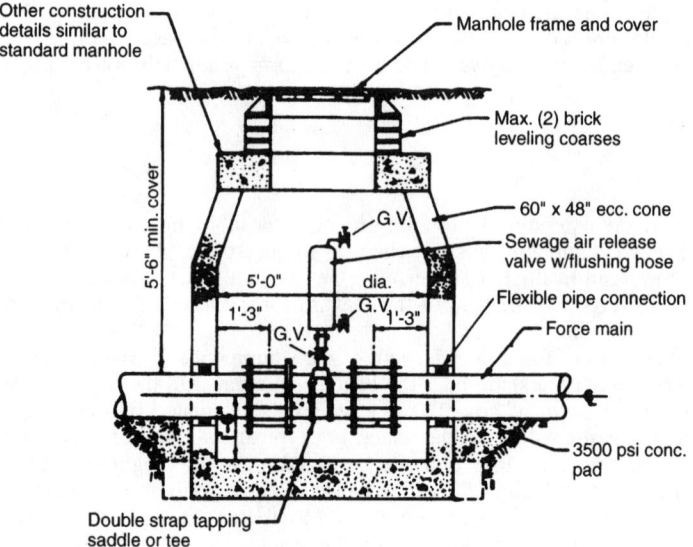

Section B-B

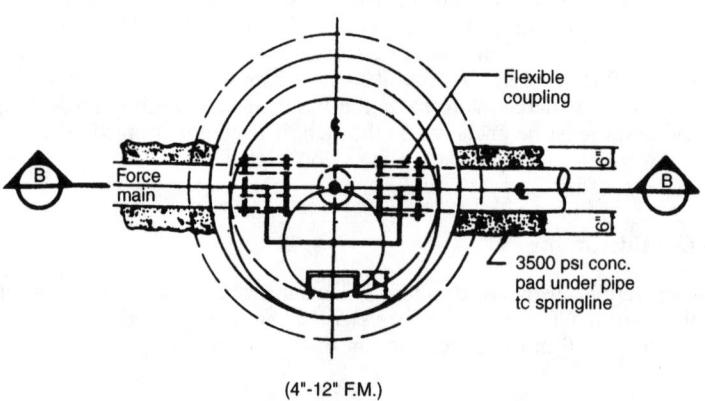

(4"-12" F.M.)

FIGURE 6.41 Air and vacuum release in manhole.

recommended air release at the final discharge into a public sewer manhole is illustrated in Fig. 6.42.

STORM WATER DISPOSAL

Storm Water Discharge Permit

Historically, the EPA National Pollutant Discharge Elimination System (NPDES) has focused primarily on the discharge of industrial process wastewater from mu-

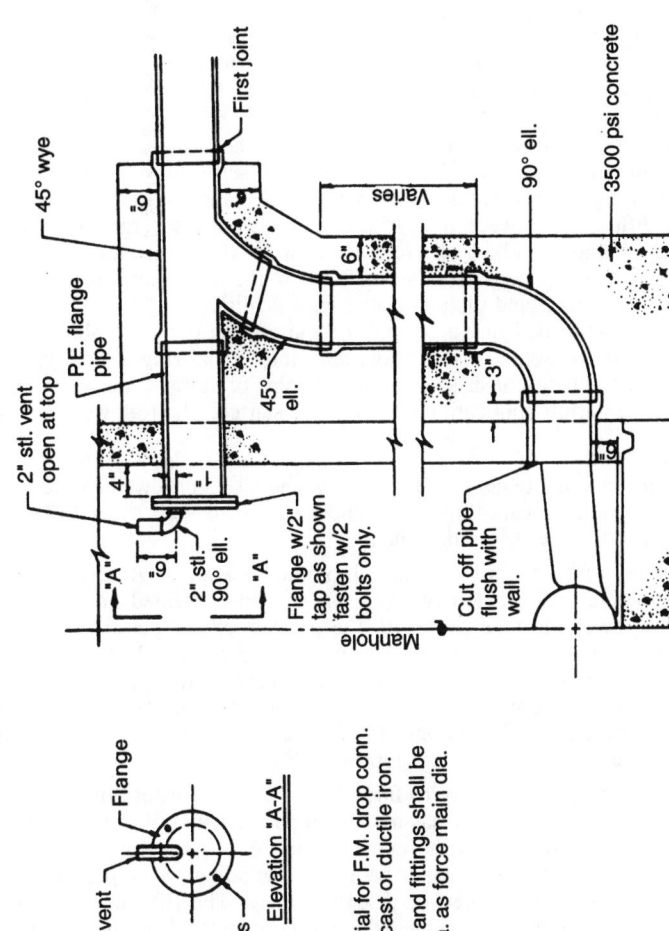

Notes

1. All material for F.M. drop conn. shall be cast or ductile iron.

2. All pipes and fittings shall be same dia. as force main dia.

FIGURE 6.42 Air release valve at manhole.

6.107

nicipal wastewater treatment plants. Between 1978 and 1983, the EPA funded additional studies to measure the pollutants in storm water runoff.

The results of these studies concluded that a considerable amount of pollutants were carried to nearby lakes, streams, and the ocean. In 1990, the EPA issued storm water regulations that apply to both municipal and industrial storm water discharge. These regulations define who must apply for and obtain a NPDES permit for storm water discharge. Additional information can be obtained from the *Guidance Manual for the Preparation of NPDES Applications for Storm Water Discharge Associated with Industrial Activity,* published by the U.S. EPA Office of Water Enforcement and Permits.

Industrial facilities conforming to a specific "industrial activity" list and that discharge either directly into U.S. waters or into separate municipal storm water systems must apply for permits. Table 6.27 presents the list of industrial activities. There is a difference between industrial and municipal discharges. Only the industrial type discharges will be discussed. Storm water discharges into combined sewers do not require any permits. Also included in the list of industrial activities are construction sites and landfills associated with facilities.

Once it is established that a permit is required, the permit application must be prepared, usually by some person or organization specializing in this type of work. The field work (which consists of taking samples of storm water discharge) associated with the permit application must also be started. The following activities are necessary:

1. All survey and site plans of the facility, and also storm water sewer site plans, must be reviewed. A walk-through should be done to verify that the system has been installed the way it was designed.

2. A survey of the facility is required to note the amount, if any, of dry weather flow. In addition, any previously unknown or undocumented connections to the storm water sewer must be found, to the greatest extent possible. If any illegal connections to this system are found, they must be corrected. Another possibility is to permit each of the illegal connections separately, thereby making them "legal." Experience with this approach has proven very time consuming and expensive. One of the requirements is to fill out an EPA form certifying that all outfalls have been tested for non-storm-water and nonapproved discharges.

3. It is an absolute necessity to find all of the storm water outfalls since experience has shown that the EPA has a very broad definition of outfalls. One way to identify outfalls is to find all the conveyances. A *conveyance* is any channel or passage that conducts or carries storm water, including any pipe, ditch, channel, tunnel, conduit, well, container, or discrete fissure. The EPA may determine that some outfalls are identical, thereby reducing the number that has to be monitored.

4. Any kind of flow from any conveyance, including that from grassy areas, are included. However, common sense must be applied. If a sample can't be collected from a particular area, then it probably would not be considered an outfall.

5. Samples taken must report actual quantitative data resulting from any storm water event. There are two types of samples required. The first type is called a *first flush grab sample* and must be taken during the first 30 min of a storm event. The second type is a *flow-weighted composite sample* for the entire event, and requires that a sample be taken during the event based on a predetermined volume of water flowing past a meter. A summary of pollutants to be analyzed are listed in Table 6.28.

TABLE 6.27   Industrial Activity for Storm Water Runoff

1. Facilities subject to storm water effluent limitations guidelines, new source performance standards, or toxic pollutant effluent standards under 40 CFR, subchapter N.

2. Facilities classified as:

SIC 24	Lumber and wood products (except 2434)	SIC 32	Stone, clay, and glass products (except 323)
SIC 26	Paper and allied products (except 265 and 267)	SIC 33	Primary metal industries
SIC 28	Chemicals and allied products (except 283)	SIC 3441	Fabricated structural metal
SIC 29	Petroleum and coal products	SIC 373	Ship and boat building and repairing
SIC 311	Leather tanning and finishing		

3. Facilities classified as SIC 10 through 14 including active or inactive mining operations and oil and gas exploration, production, processing, or treatment operations, or transmission facilities that discharge storm water contaminated by contact with or that have come into contact with, any overburden, raw material, intermediate products, finished products, by-products or waste products located on the site of such operations.

SIC 10	Metal mining	SIC 13	Oil and gas extraction
SIC 12	Coal mining	SIC 14	Nonmetallic minerals, except fuels

4. Hazardous waste treatment, storage, or disposal facilities, including those that are operating under interim status or a permit under subtitle C of RCRA.

5. Landfills, land application sites, and open dumps that receive or have received any industrial wastes including those that are subject to regulation under subtitle D or RCRA.

6. Facilities involved in the recycling of materials, including metal scrapyards, battery reclaimers, salvage yards, and automobile junkyards, including but not limited to those classified as:

SIC 5015	Motor vehicle parts, used
SIC 5093	Scrap and waste materials

7. Steam electric power generating facilities, including coal-handling sites.

8. Those portions of transportation facilities that are involved in vehicle maintenance, equipment cleaning operations, or airport deicing operations, or which are otherwise identified as industrial activities.

SIC 40	Railroad transportation	SIC 44	Water transportation
SIC 41	Local and interurban passenger transit	SIC 45	Transportation by air
SIC 42	Trucking and warehousing (except 4221-25)	SIC 5171	Petroleum bulk stations and terminals
SIC 43	U.S. Postal Service		

9. Wastewater treatment works with a design flow of 1 million gallons per day or more or required to have an approved pretreatment program under 40 CFR part 403.

10. Construction activity including clearing, grading, and excavation activities except operations that result in the disturbance of less than five acres of total land area which are not part of a larger common plan of development or sale.

(Continued)

TABLE 6.27   Industrial Activity for Storm Water Runoff   (*Continued*)

11. Facilities where material handling equipment or activities, raw materials, intermediate products, final products, waste materials, by-products, or industrial machinery are exposed to storm water.

SIC 20	Food and kindred products	SIC 31	Leather and leather products (except 311)
SIC 21	Tobacco products	SIC 323	Products of purchased glass
SIC 22	Textile mill products	SIC 34	Fabricated metal products (except 3441)
SIC 23	Apparel and other textile products	SIC 35	Industrial machinery and equipment
SIC 2434	Wood kitchen cabinets	SIC 36	Electronic and other electric equipment
SIC 25	Furniture and fixtures	SIC 37	Transportation equipment (except 373)
SIC 265	Paperboard containers and boxes	SIC 38	Instruments and related products
SIC 267	Misc. converted paper products	SIC 39	Miscellaneous manufacturing industries
SIC 27	Printing and publishing	SIC 4221	Farm product warehousing and storage
SIC 283	Drugs	SIC 4222	Refrigerated warehousing and storage
SIC 285	Paints and allied products	SIC 4225	General warehousing and storage
SIC 30	Rubber and misc. plastics products		

TABLE 6.28   Summary of Pollutants to Be Analyzed

	Industrial	
Pollutant	First flush grab sample	Flow-weighted composite sample
Oil and grease	√	
pH	√	
Biological oxygen demand (BOD)	√	√
Chemical oxygen demand (COD)	√	√
Total suspended solids (TSS)	√	√
Total phosphorous	√	√
Nitrate and nitrite nitrogen	√	√
Total kjeldahl nitrogen	√	√
Any pollutant in the facility's effluent guideline	√	√
Any pollutant in the facility's NPDES permit	√	√
Any pollutant in EPA Form 2F Tables believed to be present	√	√

6. A system for monitoring storm water runoff shall measure rainfall and flow rates from outfalls and take a first flush grab sample and flow-weighted composite sample.

7. The flow meter measures open channel flow in a pipe or channel. There are a variety of methods used to measure flow rates in open channels, including ultrasonics, submerged pressure transducers, and bubblers. Weirs and flumes can also be used. When it is not practical to install any of the above, the Manning formula can be used. This is done by relating the channel shape, slope, roughness, and level to the flow rate. A meter can also be connected to a rain gauge. This combination would report on the date, duration, and amount of rainfall from each event. The flow rate and the known pipe size give the total flow from the event. Another function of the flow meter could be to automatically activate or signal a flow rate sampler when a flow-weighted sample must be taken.

8. A sampler is used to collect water samples from a flow stream. It consists of a pump, glass or plastic sample containers, and a controller. Samples can be taken at specific time intervals or after a set volume of water has passed the monitoring point. Another requirement may be to take a sample one time only (first flush) and keep it in a separate container. Regulations require that samples be taken from storm events ranging from ½ to 1½ times the average storm event.

In many cases, it is far less expensive and more desirable to train plant personnel to collect samples, rather than to rent and maintain automatic samplers and rain gauges. In order to do this, the following precautions must be observed:

1. Additional personnel must be available for the first flush grab sample. This is the most hectic time.

2. Collections should be made in pairs. Avoid collecting samples during thunderstorms and at night.

3. Extension rods should be used to take samples from manholes and ditches.

4. Holding times should be limited. Biological oxygen demand (BOD) tests must be made within 48 h and pH within 6 h. Preservatives in bottles are time limited.

POTABLE WATER SUPPLY

This section will discuss the source of potable water supply to a facility. This could be water obtained from a public utility, wells, or an approved surface water source. Included will be the distribution pipe extended from the source to a point just outside the building wall. Very often, the source will supply both the facility potable water and fire protection water. Potable water systems inside buildings are discussed in Chap. 9, Plumbing Systems.

Surface water sources are rarely used to supply potable water. This source of water is used most often for various process and production applications, cooling, and for fire protection. The use of water for these purposes is outside the scope of this book.

WATER SUPPLIED FROM A PUBLIC UTILITY

General

Public utilities are the most reliable sources of water supply. The utility company is required to provide water of sufficient purity to the facility, which relieves the facility of the requirements of monitoring and treating the water for impurities, thereby eliminating ongoing maintenance and technical problems. This service has a price, which must be balanced against the cost of electrical power, chemicals, and maintenance of supplying water from alternate sources.

When a single service is used to provide potable and fire protection water, the supply is considered a combined service and requires conformance to additional code requirements. It is recommended that separate supplies be provided for fire protection and potable water if possible. Some reasons for the separation are:

1. High pressures used for fire fighting may be detrimental to domestic piping and equipment. A pressure-reducing valve may be required on the domestic service.
2. Process use during an emergency may reduce the flow available for fire-fighting purposes.
3. System pressure may vary widely and be difficult to control.
4. Constant flow of water will increase corrosion of the combined main, in contrast to the relatively stagnant water in a dedicated fire main where much of the oxygen-causing corrosion is dissipated.

Codes and Standards

1. NFPA-24 for private water mains for fire service (if required)
2. AWWA codes for pipe materials, installation, and testing requirements
3. ASTM codes for pipe materials, installation, and testing requirements
4. Local plumbing and fire department requirements

5. Fire insurance carrier requirements

6. Authorities responsible for backflow prevention

System Components

In addition to the piping and valves, the system as a whole consists of backflow preventers, water meters, pipe joint restraints, fire hydrants, and air and vacuum release valves.

Connection to Utility Company Mains

Connection to a public main can be made either with the water shut off in the main or under pressure. Connections made with the water shut off are usually reserved for large building water services connected to large utility mains, but are also used if the main is too small to accept a new service line that is considered too large for a pressurized connection. A corporation cock that screws directly into a tapped hole made in the main is used for smaller connections. Very often, the utility company leaves branch connections in the main when the main is constructed. Their presence and location can be verified with the utility company.

Connection to a pressurized metallic main is done with proprietary methods using a tapping machine. For larger mains, the hole is made by the tapping machine through a tapping valve and sleeve that mates with the main and retains water after the drilling bit and pipe coupon are removed. The valve has an outlet flange for a future connection. For a detail of a typical tapping arrangement under pressure, refer to Fig. 6.43. For recommended minimum excavation dimensions, refer to Fig. 6.44.

For branch connections 2 in (50 mm) or smaller, a service clamp with a threaded outlet for connection to a corporation cock is used. The corporation cock is a fitting that has a male thread on one end which screws into the outlet on the clamp assembly, an integral shutoff, and on the other end a connection to match the future building service pipe material.

The best water main arrangement is in the form of a loop around the entire site within section valves. This will permit the flow of water in two directions, allowing service to most of the site if there is a break in any portion of the site main.

Valves

Valves in common use are gate and butterfly valves approved for underground service by the local authorities. Valves can be directly buried or installed in a valve pit. Installation of a valve in a pit permits easy access for repair and inspection, but is far more costly than direct burial.

For valves directly buried, the operator can be provided with a nut instead of a handle or be extended above grade on a post called a *post indicator valve* (PIV), which indicates whether the valve is open or closed. Approval of the valve type and manufacturer by the insurance carrier or other authorities may be required.

The operating nut is made accessible through a valve box extending up to the surface. Operation is accomplished by use of a T handle wrench. Valves directly

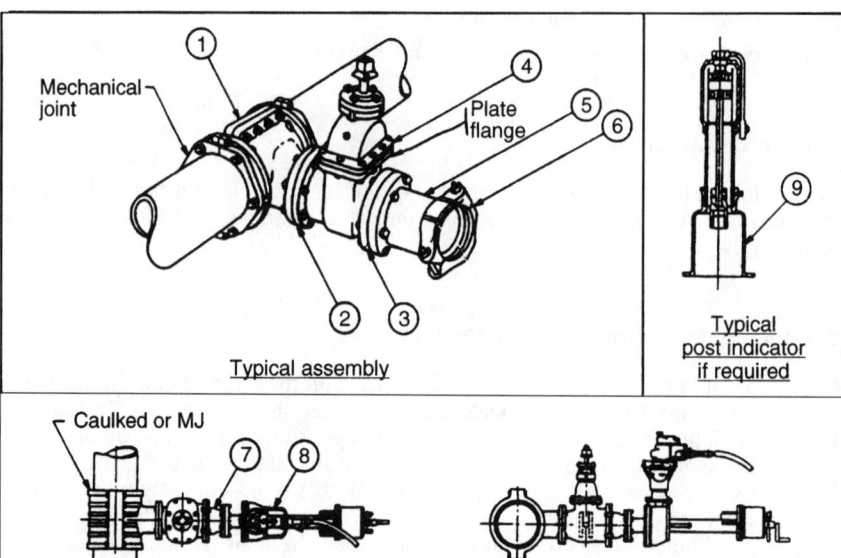

Mechanical joint

Plate flange

Typical assembly

Typical post indicator if required

Caulked or MJ

Tapping equipment assembly

1. Tapping sleeve, caulked type is usually available in line sizes 3" thru 48". Mechanical joint type is usually available in sizes 4" thru 12". Full line size outlets are not recommended. Tapping saddles may be used in place of tapping sleeves.

2. ANSI 125 lb flange inlet. Test tapping sleeve prior to installing tapping valve.

3. ANSI 125 lb outlet

4. Tapping valve. ANSI 125 lb flange inlet and flange outlet in sizes 2" thru 12". Conventional full-port gate valves may be used as tapping valves if hole is cut one size smaller than the valve size and stem threads are protected with a valve box.

5. Pipe–ANSI 125 lb flange one end, split coupling groove other end.

6. Split coupling in accordance with underground valve and piping specifications.

7. Adapter piece for attaching tapping equipment. Specify flange adapter (MJ is normally supplied).

8. Tapping machine.

9. Adjustable post indicator mounted on plate flange provided with tapping valve. Specify with valve and include depth of bury, centerline of valve to grade.

FIGURE 6.43 Pressure tapping of water main.

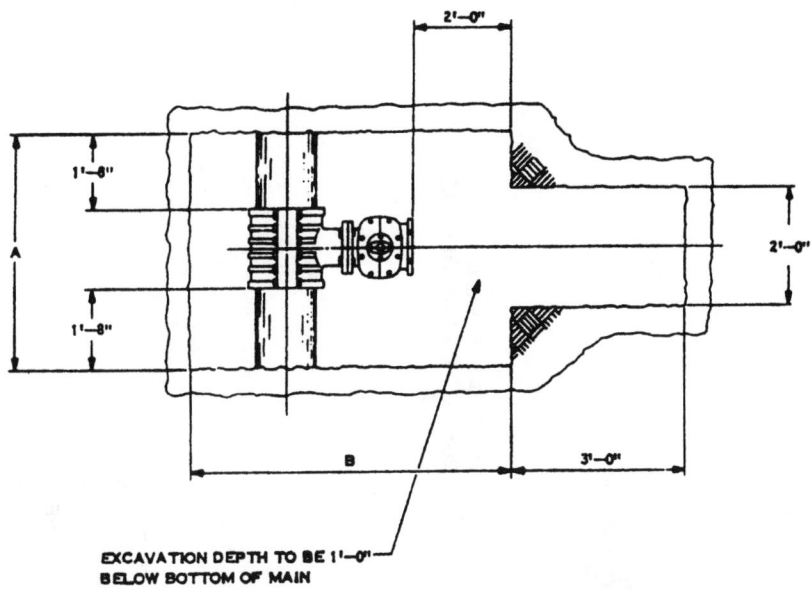

MINIMUM EXCAVATION DIMENSIONS

HEADER SIZE	A	B
3" THRU 8"	5' – 0"	5' – 6"
10" THRU 12"	5' – 6"	6' – 0"
14" THRU 24"	6' – 0"	7' – 0"

FIGURE 6.44 Pressure water main tapping. Excavation dimensions.

buried must have nonrising stems. The PIV valve operator extends above grade on a post that makes its location obvious, is easily accessible, and has an indicator to visually show if the valve is open or closed. A typical valve and valve box are illustrated in Fig. 6.45 and a typical PIV is illustrated in Fig. 6.46. If the PIV is located near a road, it is common practice to protect the valve with guard posts. Refer to Fig. 6.47 for a detail of typical guard posts. In general, the use of curb boxes instead of PIVs on a combined building service is discouraged for the following reasons:

1. They are difficult to locate and supervise, since they are often covered by dirt, snow, ice, or paving materials.
2. Constant care is required to prevent dirt and stones from entering the hole and preventing valve operation.
3. Delays may be encountered if the operating T wrench is misplaced or stored at a remote location.

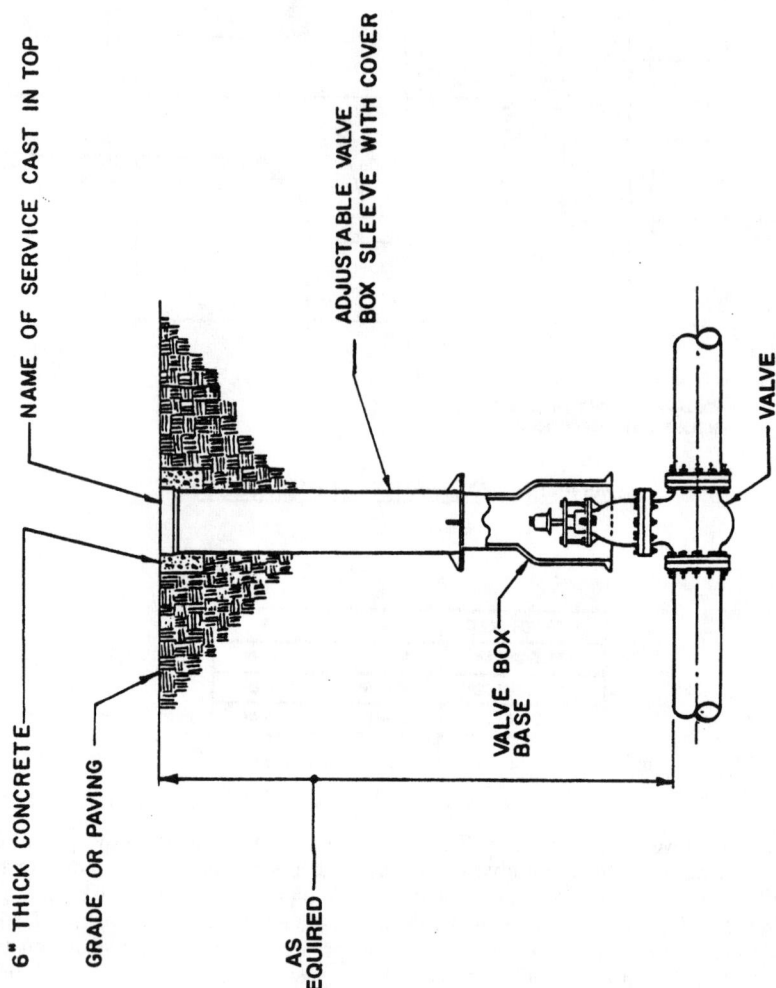

6" THICK CONCRETE

NAME OF SERVICE CAST IN TOP

GRADE OR PAVING

ADJUSTABLE VALVE
BOX SLEEVE WITH COVER

AS
REQUIRED

VALVE BOX
BASE

VALVE

FIGURE 6.45 Valve box.

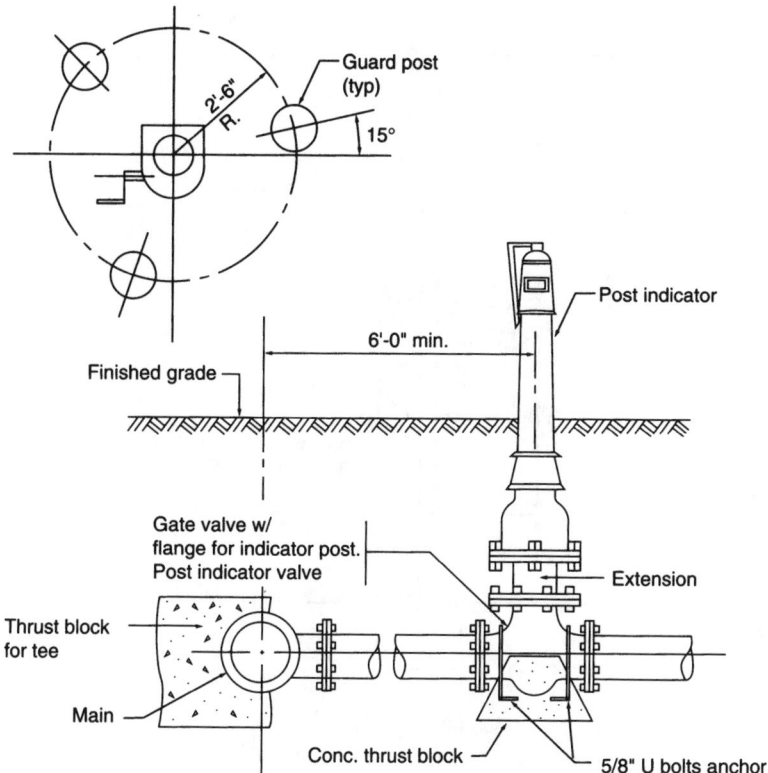

FIGURE 6.46 Detail of post indicator valve.

If the connection is for service to a site consisting of several buildings, the utility company requires that a shutoff or isolation valve be provided at the point of connection. Based on utility company requirements, a valve pit may be necessary or the valve could be directly buried.

On the site where water mains divide, it is good practice to provide an isolation valve, or section valves on all branch connections. For combined services, a post indicator valve is required.

Building Service Shutoff. All building water service branches must have a means of shutting off the water to the building from the site without the necessity of personnel having to gain access to the inside of a building. When the building service main is a combined service, a PIV located 50 ft (15.5 m) from the building wall is recommended. If this distance is not possible, the PIV shall be located as far away as practical. In urban locations, a shutoff valve is usually installed in a public street or adjacent sidewalk. This shutoff valve is commonly called a curb valve and is installed using a curb box, as shown in Fig. 6.45.

In addition to the shutoff valve outside the building, code requires that another shutoff valve inside the building must be provided as close to the water service entrance as space permits.

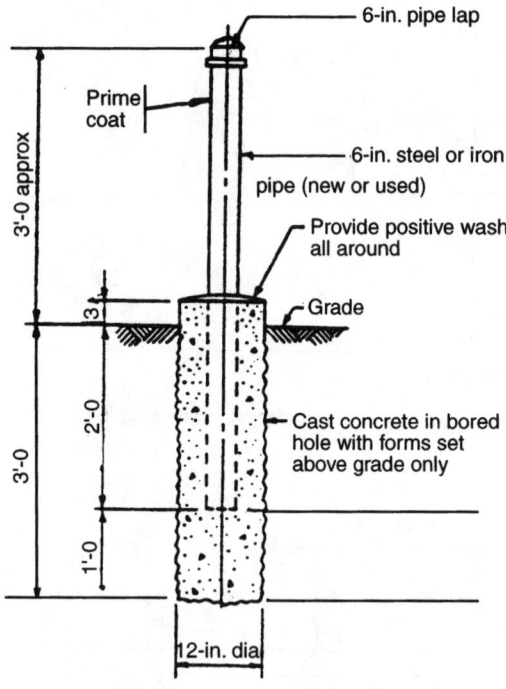

6-in. pipe lap

Prime coat

6-in. steel or iron pipe (new or used)

Provide positive wash all around

Grade

3'-0 approx

3

2'-0

1'-0

3'-0

Cast concrete in bored hole with forms set above grade only

12-in. dia

Capped hollow pipe

FIGURE 6.47   Detail of typical guard post.

Backflow Prevention

Because of concerns about contamination, the installation of backflow preventers on the facility service line is generally required by local authorities if they determine that a potential for contamination is present. External backflow prevention installed on a building or site service is different in concept from the internal backflow prevention required on or at equipment or devices inside buildings. The reasoning is that any possible contamination of public water mains that could originate from the building or facility must be prevented. This is achieved by isolating the facility service connection from the public main and containing any contamination within the facility service piping. Isolation and containment are accomplished by the installation of an approved device on the facility service pipe. The only acceptable backflow preventers (BFPs) are the reduced pressure zone (RPZ), the double check valve (DCV), and the air gap. The type of device used is based on the potential hazard. The types and categories of hazards vary widely and are known by various names in different localities.

There are two primary causes of contamination. The first is the fire department. If the service main is combined, the fire department may use a nearby source of water instead of the public water supply. A pumper may also have its own water supply for use immediately upon arrival to fight a fire until other apparatus arrives to make a connection to the public water supply. When the apparatus pumps contaminated water into the facility service, the water under high pressure must be stopped from entering the public mains. The other cause of contamination is the building itself, because if the public main breaks, it creates a vacuum which draws contaminated water from the facility into the public main.

For industrial facilities using a storage tank, an air gap must be provided in the water supply into the tank. This will provide the best possible protection of the public water supply.

Determining the Degree of Hazard. The choice of protective device is based on an evaluation of the degree of potential hazards. The following should be considered:

1. Use, toxicity, nature, and availability of contaminants
2. Availability of a supplemental supply of water
3. Fire-fighting system evaluation

Hazard Classifications. The facility rating or classification should be made only after consulting with the proper authorities. The categories mentioned cannot list every circumstance or facility type. Judgment must be used in the final selection. The following ratings are often used based on the use, toxicity, nature, and availability of contaminants:

1. *High (severe) hazard.* Any facility that uses chemicals considered toxic or has the potential for discharge of toxic waste is considered a high hazard. Typical facilities are hospitals, chemical processors and manufacturers, pharmaceutical processors and manufacturers, laboratories, food processors and manufacturers, industrial manufacturers and processors, and water and sewage treatment plants.

2. *Medium (moderate) hazard.* Commercial buildings and establishments, fire protection storage tanks and mains with no additives, and facilities that discharge water at higher than normal temperatures are medium hazards. The fire protection system will have only stagnant water present in the pipe.

3. *Nonhazardous (minor) hazard.* Private homes and commercial establishments without complex plumbing or fire protection systems are minor hazards.

The availability of a supplemental or auxiliary supply of water for fire-fighting purposes could create a situation that affects the hazard classification. If the facility is located close to a river or lake, the fire department could draw from this source to provide additional water during a fire and then this water of indeterminate toxicity could be pumped back through the combined facility service into the public main (if not protected). Also, well water of unsatisfactory purity that is used for any purpose could find its way back into the public water supply.

An evaluation of the fire protection system shall be made separately because of the number of variables involved. Factors such as whether the system is wet or dry, the presence of storage tanks and pumps, whether there is a direct connection to the public mains and interconnection with other water supplies must be studied in order to determine the degree of hazard. It is accepted practice to provide any water storage tank with an air gap between the pipe supplying the source of water and the highest possible level of water inside the tank. A gravity overflow two sizes larger than the supply pipe will provide sufficient protection from tank flooding.

Selection of the Appropriate BFP Device. Any facility classified as hazardous must be protected with either a reduced pressure zone (RPZ) BFP assembly or an approved air gap on the facility supply prior to connection to the public water main. For medium hazard, the selection of a double check valve (DCV) assembly is usually acceptable. For nonhazardous facilities, internal BFPs are usually sufficient, with no requirement for external BFP in the facility service line.

The above selections are generalities and must be approved and accepted by all responsible code officials and facility authorities.

Location and Installation of a BFP Device.   For a site service main, the external BFP must be located at or very close to the property line of the facility. The installation requirements are obtained from the responsible authorities, which may be the local or state health department, the water utility, or the local building department. BFPs inside buildings must be located downstream of the meter or as close to the incoming service entrance as possible.

The installation of the BFP device depends on its type. RPZ BFPs must be installed above grade or above the level of the water main inside the building. This requirement is to ensure that the water released as a result of backflow will be freely discharged from the device with no restriction. The wastewater must flow away from the device into a drain and not form a pool where the device is located. The drain must discharge by gravity and be large enough to accept the full flow expected from the device. The drain cannot be flooded and must be screened to prevent the entrance of dirt and vermin. For most installations, this requires that the device be installed in an above-ground structure. The enclosure must be protected from freezing. A schematic illustration of an exterior RPZ BFP in a pit is shown in Fig. 6.48. A schematic illustration of an exterior RPZ BFP above grade is shown in Fig. 6.49. A schematic illustration of an interior RPZ BFP is shown in Fig. 6.50.

The DCV does not discharge water and can be installed in a pit below grade. It must be protected against freezing and access must be provided for operation of the test cocks. If the service main is installed below the frost line, this is considered acceptable freeze protection. A schematic illustration of an exterior DCV BFP in a pit is shown in Fig. 6.51.

The BFP must be installed in such a manner as to allow easy access for testing and maintenance.

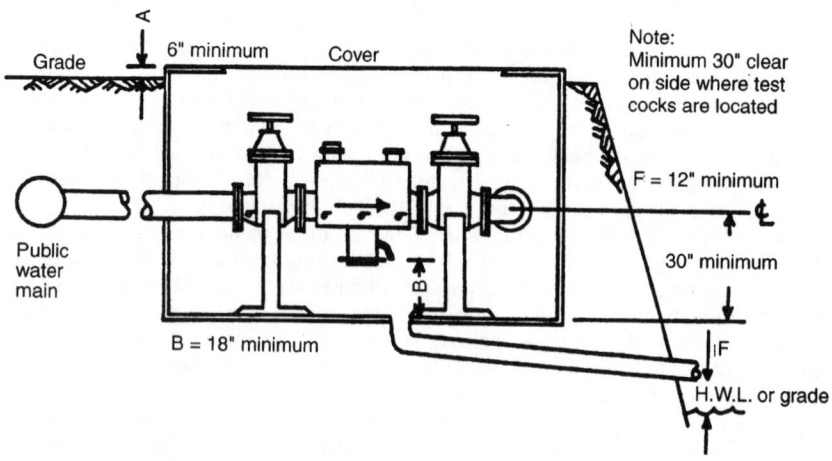

FIGURE 6.48   Detail of RPZ BFP in a pit.

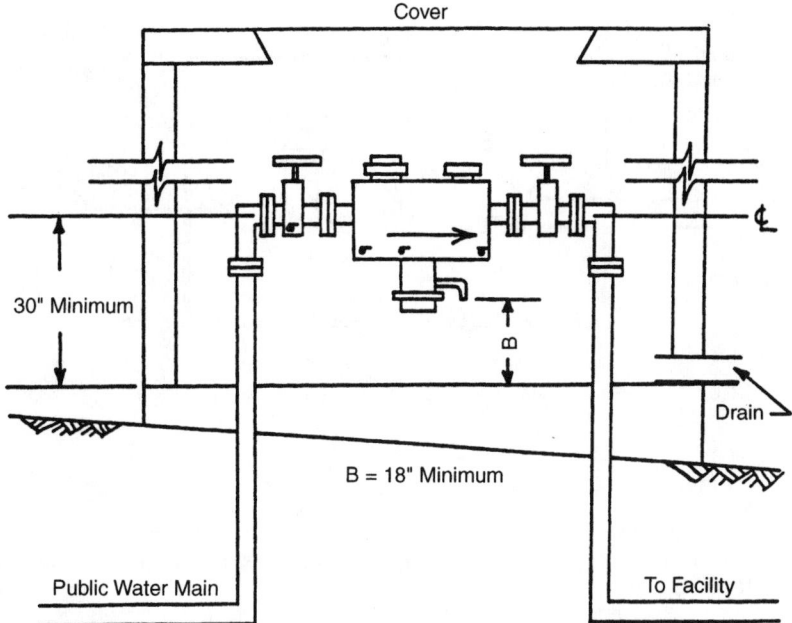

FIGURE 6.49 Detail of RPZ BFP above grade.

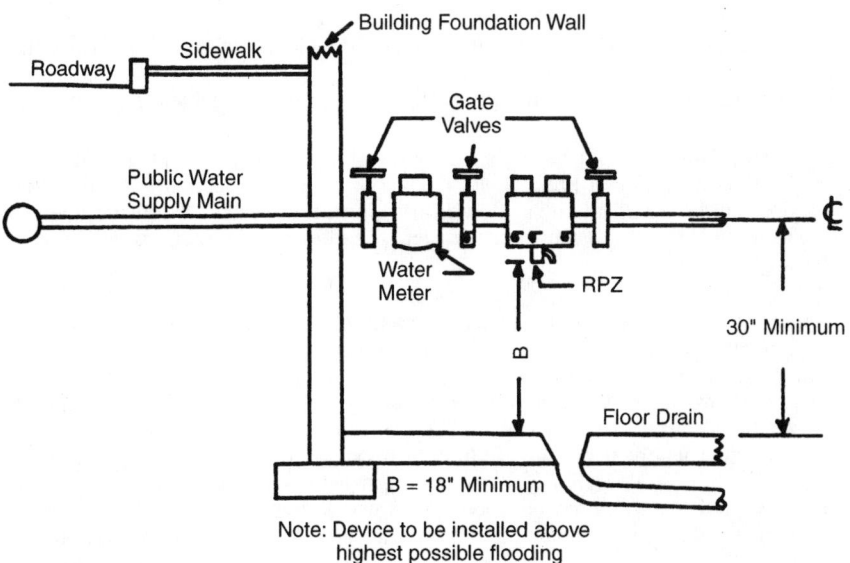

FIGURE 6.50 Interior detail of RPZ BFP.

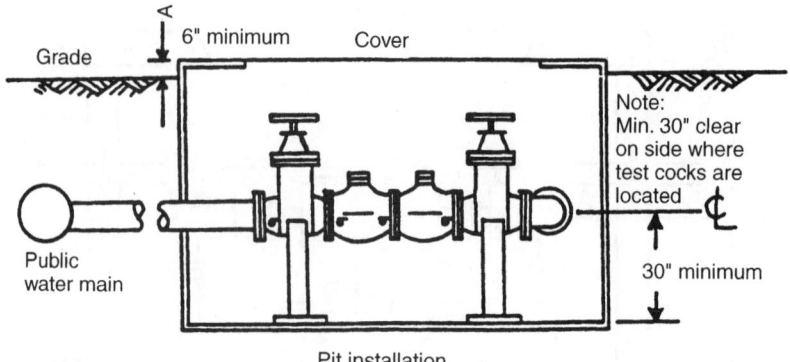

Pit installation

FIGURE 6.51 Detail of double check valve BFP in pit.

Water Meters

Since water obtained from a public utility must be paid for, a water meter is required to record the amount used. When there is no cost, meters are often installed by many facilities simply to record the amount of water used. Meters are selected by their function, accuracy, and pressure loss through the meter assembly. Registers on the meter are available to record usage in cubic feet or gallons (cubic meters or liters); remote reading and strip recording are usually available as options.

Meters can be installed at the property line, adjacent to a building, or inside a building. They can be installed aboveground or underground in a pit. The meter installation, called a meter assembly, usually includes valves, test tees, and a bypass around the meter to allow maintenance or replacement with no disruption to facility operation. The assembly arrangement is regulated by local authorities and the utility company.

Separate water meters are designed for fire protection or domestic water uses.

Fire Protection. Meters used for fire protection purposes are suitable only for large flows and minimal pressure loss. They are not very accurate. A "detector" function can be added to signal flow in a dedicated fire main. A check valve can also be an integral part of a fire water meter. The most popular meter is the detector-check type.

Domestic. The meters for domestic use should be very accurate and selected based on accuracy at designed maximum flow and pressure loss through the meter. Where wide extremes of flow are expected, a compound-type meter is used. This meter consists of two separate meters: one to accurately register low flow and another to register higher flow.

Where reasonably steady low to high flows are expected, turbine or disk meters are likely choices. Propeller-type meters are usually selected for large pipe sizes and very high, steady flows, or where low flows do not occur.

Smaller meters can be obtained to protect against frost with a breakaway bottom. If frozen, this is the only component that could have to be replaced. Special meters are required for hot water above 140°F (60°C).

Friction loss figures through meters at various flows are available from meter manufacturers.

Pipe Restraints

Mechanical joints and push-on joints seal when the elastomeric gasket becomes compressed in the space between the spigot end of the pipe and the bell. These joints are leaktight when installed on a straight line. When a fitting introduces a change of direction, the flowing water exerts a pressure on that fitting. (Pressures are listed in Table 6.29.) The joint is not capable of resisting this pressure, therefore a suitable means must be provided to restrain the joint from coming apart. The most common methods are integral joint restraint, clamps and tie rods, and thrust blocks.

Plastic piping with butt-fused joints has the same resistance as piping and does not require additional restraint. Split couplings with grooved pipe are also capable of resisting this pressure without restraint, if couplings designed for this purpose are used. This jointing method is generally recommended for new underground pipe installation.

Clamps and Tie Rods. Resistance is provided by clamps attached to the pipe at each side of the joint and rods connecting the clamps across each joint. Standard charts giving the required restraint for different size pipes, depth of bury, number of joints to be restrained, and different fittings are available.

Integral Joint Restraint. This type of joint uses integrally cast glands for mechanical joint pipe and internally locked, grooved, and keyed joints for push-on joints.

TABLE 6.29 Pressure Exerted at Joints by Flowing Water

Total pounds resultant thrust at fittings at 100 psi water pressure

Nom. pipe dia., in	Dead end	90° bend	45° bend	22½° bend	11¼° bend
4	1,810	2,559	1,385	706	355
6	3,739	5,288	2,862	1,459	733
8	6,433	9,097	4,923	2,510	1,261
10	9,677	13,685	7,406	3,776	1,897
12	13,685	19,353	10,474	5,340	2,683
14	18,385	26,001	14,072	7,174	3,604
16	23,779	33,628	18,199	9,278	4,661
18	29,865	42,235	22,858	11,653	5,855
20	36,644	51,822	28,046	14,298	7,183
24	52,279	73,934	40,013	20,398	10,249
30	80,425	113,738	61,554	31,380	15,766
36	115,209	162,931	88,177	44,952	22,585
42	155,528	219,950	119,036	60,684	30,489
48	202,683	286,637	155,127	79,083	39,733
54	256,072	362,140	195,989	99,914	50,199

Note: To determine thrust at pressures other than 100 psi, multiply the thrust obtained in the table by the ratio of the pressure to 100. For example, the thrust on a 12-in pipe with a 90° bend at 125 psi is 19,353×125/100 = 24,191 lb.

Thrust Blocks.   Thrust blocks consist of cast-in-place concrete blocks, size and weight of which depend on the pipe size, water pressure (including surge loads), and safe soil-bearing load. Recommended locations for thrust blocks are illustrated in Fig. 6.52. These blocks must rest on undisturbed soil in order to resist the force exerted. Average safe soil-bearing loads are given in Table 6.30. These loads are general and must be verified for specific projects.

To determine the size of a thrust block, the thrust developed at the joint must be divided by the soil resistance. This will give the required area to be provided by the thrust block. General dimensions of thrust blocks for different fittings are given in Fig. 6.53.

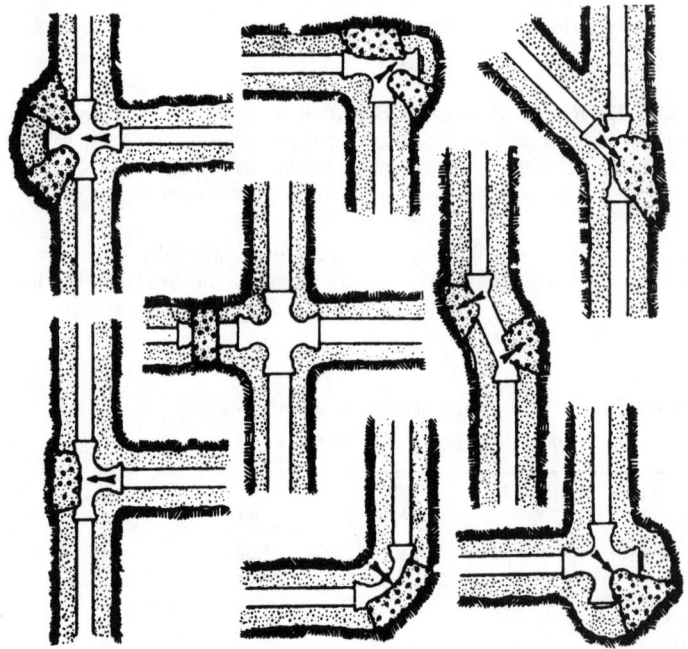

FIGURE 6.52   Locations for thrust blocks.

TABLE 6.30   Soil-Bearing Loads

Soil	Bearing load, lb/ft^2
Muck	0
Soft clay	1000
Silt	1500
Sandy silt	3000
Sand	4000
Sandy clay	6000
Hard clay	9000

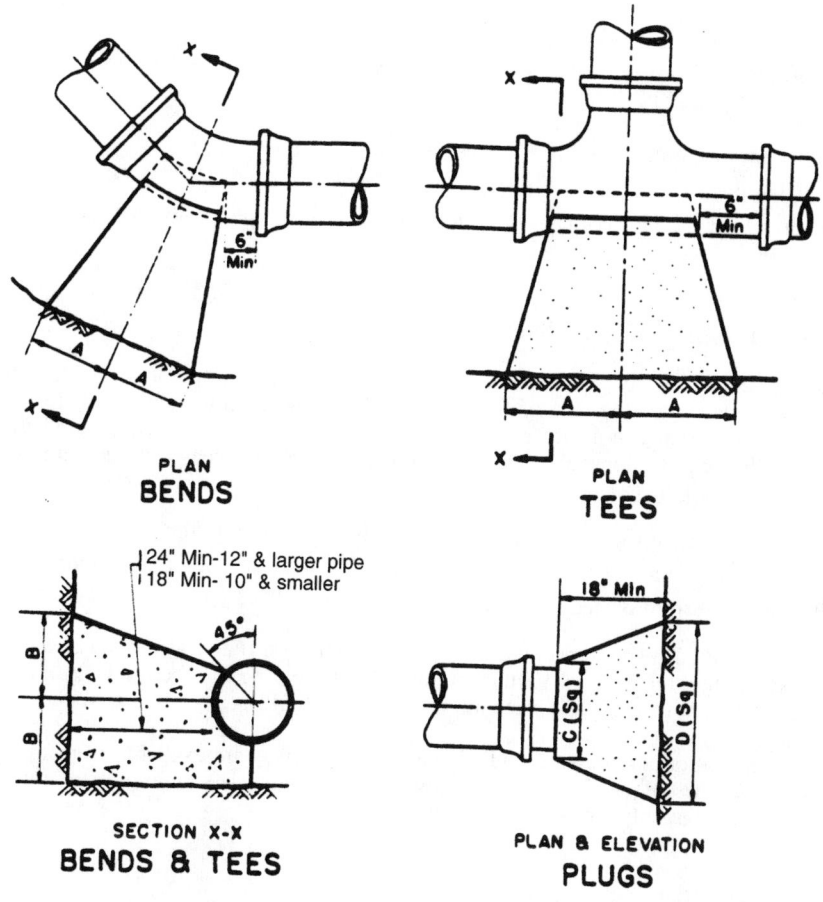

TYPE	SIZE	1/4 BENDS		1/8 BENDS		1/16 BENDS		TEES		PLUGS	
		A	B	A	B	A	B	A	B	C	D
TYPE I **4000 PSF SOIL**	6"	8"	10"	6"	8"	3"	8"	8"	8"	10"	15"
	8"	12"	12"	8"	10"	5"	9"	9"	12"	12"	20"
	10"	16"	14"	10"	12"	6"	10"	11"	14"	14"	25"
	12"	19"	16"	12"	14"	8"	11"	14"	16"	16"	30"
	14"	23"	18"	14"	16"	10"	12"	16"	18"	18"	34"
	16"	26"	20"	16"	18"	11"	13"	18"	20"	20"	38"
TYPE II **2000 PSF SOIL**	6"	16"	10"	9"	10"	6"	8"	10"	12"	10"	21"
	8"	22"	13"	12"	13"	8"	10"	13"	16"	12"	29"
	10"	26"	17"	14"	17"	10"	13"	16"	20"	14"	36"
	12"	29"	21"	16"	21"	11"	16"	18"	24"	16"	41"
	14"	35"	24"	19"	24"	12"	20"	22"	27"	18"	48"
	16"	38"	27"	21"	27"	12"	24"	24"	30"	20"	54"

Note: Based on 100 p.s.i. static pressure plus AWWA water hammer
All bearing surfaces to be carried to undisturbed ground

FIGURE 6.53 Typical thrust block dimensions.

In general, thrust blocks are not recommended for the following reasons:

1. The blocks are comparatively large.
2. Often, undisturbed soil is not present.
3. High cost.
4. Space for the blocks may not be available due to the proximity of adjacent site utilities.

Air and Vacuum Relief Valve

When water is run in pipes, air pockets often develop at high points in the system. Air pockets reduce the area of the pipe thus lowering the flow rate and increasing the pressure loss. The device used to eliminate the air is called an *air release valve.* The following locations may require the installation of air release valves, but rarely at every location:

1. High points of the pipeline, in particular the hot water system
2. At aboveground locations, where the pipe rises up and then returns underground
3. Upstream from orifices, reducers, or other similar obstructions

A vacuum can be created when a pipeline is emptied. If a service line is intended to be emptied regularly, a vacuum relief valve should be installed. Also, for long runs of thin-walled pipe, a vacuum relief valve should be installed to avoid pipe collapse.

Combination air pressure and vacuum release valves are available from manufacturers. These devices are selected based on main pipe size, main pressure rating, main pipe material, and wall thickness.

Fire Hydrants

Fire hydrants are found only on combined services. A fire hydrant is a type of aboveground valve that provides a dedicated means for fire department apparatus to connect directly to a piped source of water.

There are two categories of fire hydrants, dry barrel to prevent freezing and wet barrel. The dry barrel has a device that allows the water in the barrel to drain down to the frost line after the valve that controls water to the outlets has been shut off. The dry type is recommended for all areas that have any remote possibility of freezing. Experience has shown that freezing only once in 25 years is sufficient reason to install dry-type hydrants. For a detail of a typical dry-type hydrant, refer to Fig. 6.54. Shutoff valves for the hydrant are optional. Check local code and client preference for installation requirements.

The sizes of connections available on hydrant barrels vary. All connections have a threaded end onto which fire hose is connected. These threads must be compatible with the local fire department and the facility on whose property the hydrant has been installed. The threaded ends are protected by a cap attached to the hydrant by a chain. A single internal valve, terminating in an operating nut on top of the hydrant, controls all of the connections. A special wrench is used to turn the nut. Special hydrants can be obtained with individual valves on each outlet. The most common sizes are nominal 2½ in (60 mm), 4 in (100 mm), and 6 in (150 mm). The two larger sizes are called pumper connections. The most common arrangement

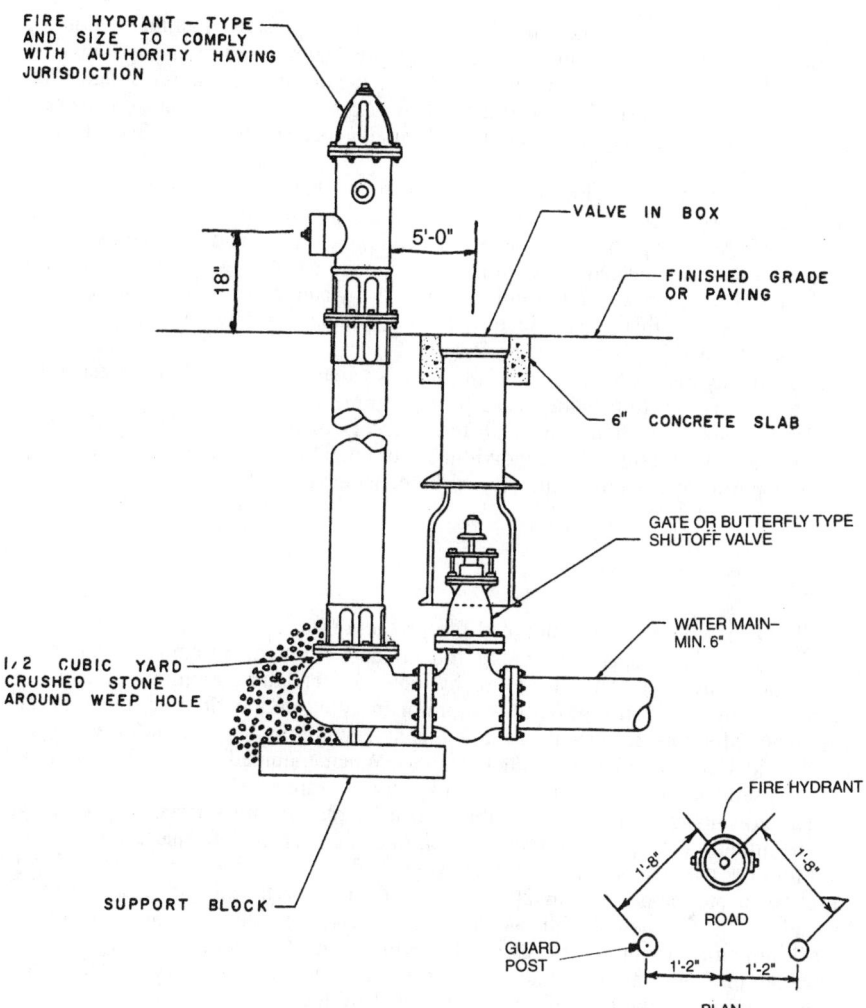

FIRE HYDRANT — TYPE
AND SIZE TO COMPLY
WITH AUTHORITY HAVING
JURISDICTION

VALVE IN BOX

5'-0"

18"

FINISHED GRADE
OR PAVING

6" CONCRETE SLAB

GATE OR BUTTERFLY TYPE
SHUTOFF VALVE

WATER MAIN–
MIN. 6"

1/2 CUBIC YARD
CRUSHED STONE
AROUND WEEP HOLE

FIRE HYDRANT

1'-8"   1'-8"

ROAD

GUARD
POST

1'-2"   1'-2"

SUPPORT BLOCK

PLAN

FIGURE 6.54 Typical fire hydrant.

consists of two 2½-in (60-mm) connections, mounted at a 90° angle, facing the road. When a larger size pumper connection is added, the pumper connection faces the road and the other, smaller connections, are arranged at 90° angles on either side. Generally accepted practice is not to provide a pumper connection unless specifically required by local code or client preference. Pumper connections are used when a cross-connection is needed between two water mains. Fire hydrants shall be rated for a minimum of 300 psig (2000 kPa).

Locating and Installing Hydrants. Hydrants are typically located 250 to 300 ft apart (78 to 90 m). For large sites, it is recommended that two hydrants be installed 200 ft (65 m) from any point where a significant hazard exists. They should be located about 40 ft (13 m) from a building wall under normal conditions, with a longer distance required if the building stores or uses explosives, flammable liquids, chemical products, petroleum products, or high piled storage. Excessive radiant heat and the danger from falling building walls are factors in considering longer distances. Hydrants are usually placed 2 to 10 ft (0.6 to 3 m) from center of hydrant to the curb of a road. Hydrant outlets are usually located from 1 ft, 3 in, to 2 ft (37 to 60 cm) above grade.

Hydrants are generally located adjacent to roads that will allow fire department apparatus to approach near enough to the hydrant to allow easy hookup. If this is not possible, a clear path over a lawn, for example, from a road shall be available.

A valve is often located between the water main and the hydrant to allow servicing of the hydrant without shutting down the water main. Opinion is divided about the need for this valve, since hydrant servicing is rarely necessary. The installation of the valve is a client preference.

Since the hydrant is located adjacent to roads where trucks travel, it is a generally accepted precaution to provide guard posts to protect the hydrant.

Fire hydrants are often painted different colors or marked to identify the flow capacity and/or pressure of the individual hydrant. This is regarded as a great help to fire department personnel and should be considered.

Monitors

A *monitor* is a permanently mounted fire protection nozzle assembly, connected to a water main and capable of being rotated and elevated. The nozzle can be a standalone unit or attached to a fire hydrant. A standalone monitor for a warm climate is illustrated in Fig. 6.55. A monitor installed on a hydrant is illustrated in Fig. 6.56. Monitors are usually installed as safeguards for significant hazards such as chemical and petroleum product storage. When mounted on towers, monitors can be used for vapor suppression from specific hazards.

The advantage of a monitor is that it can be placed into service very quickly, can be aimed in the proper direction, and, once set, can be left unattended.

Monitors are available in 500 to 2000 gpm (3150 to 12,600 lpm) models. They shall be located about 150 to 200 ft (45 to 62 m) apart, with all protected areas capable of being reached with two streams. A general requirement of 80 psig (550 kPa) minimum pressure is necessary for proper operation, which usually requires a fire pump. Consideration shall be given not to place the monitors too close to a hazard because of the potential heat generated by fire.

Pipe Material Selection

The pipe material selected depends first on whether the system is combined or potable only. Other factors are availability, pressure rating, water quality, soil conditions, depth of bury, pipe strength requirements, and approvals of various local authorities.

For potable water piping 2 in (50 mm) and smaller, the most commonly used metallic piping is soft temper K copper tubing and copper fittings. Plastic piping includes PP and HDPE plastic pipe and fittings. For sizes 2½ in and larger, the most commonly used metallic piping is cement-lined ductile iron pipe and fittings. Plastic piping includes FRP and PVC pipe and fittings.

For combined systems, the two piping materials used most often are ductile iron and PVC, approved for use as fire protection mains.

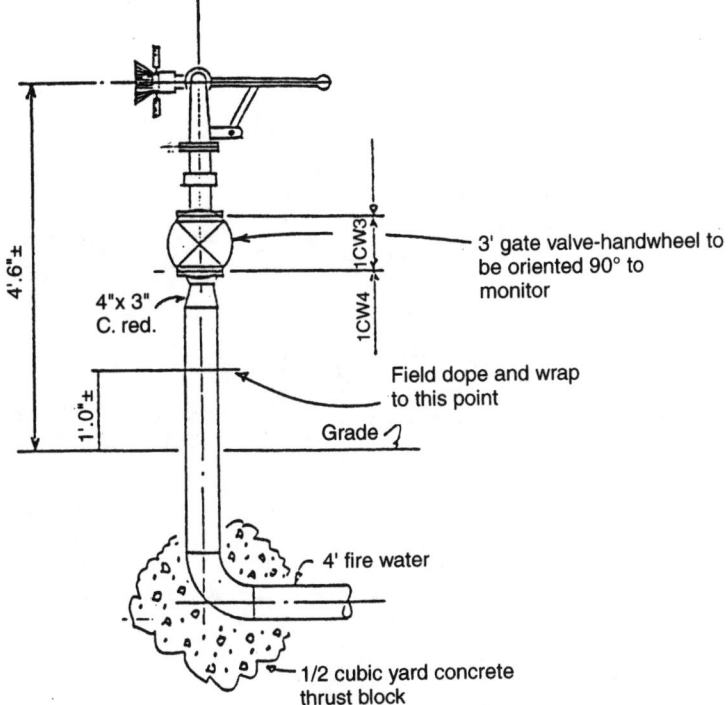

FIGURE 6.55 Typical standalone monitor for warm climate.

System Design Procedure

Preliminary Information Required

1. The location, size, material, and depth of the public water main available to the project under design must be known.

2. The water pressure must be determined for the specific elevation closest to the point where the proposed service will be connected. This requires that a flow test be conducted in order to be accurate. This test will give the static pressure (with no water flowing), the residual pressure (the water pressure observed with a flow rate), and the time of day when the test is conducted. Several flow rates and residual pressures should be recorded to allow plotting on a flow chart.

3. Determine the authority responsible for installation and approval of (and the need for) any BFP devices for the specific facility.

4. Obtain from the utility company or local authorities the type of water meter required, the location of the meter, and the extent of work provided by the utility company versus the amount of work required to be done by a contractor. Usually, the utility company has standard details for meter assembly installations and the names of specific devices required to be part of the meter assembly and their arrangement.

5. The frost depth must be found for the area as well as the minimum cover over the proposed water line.

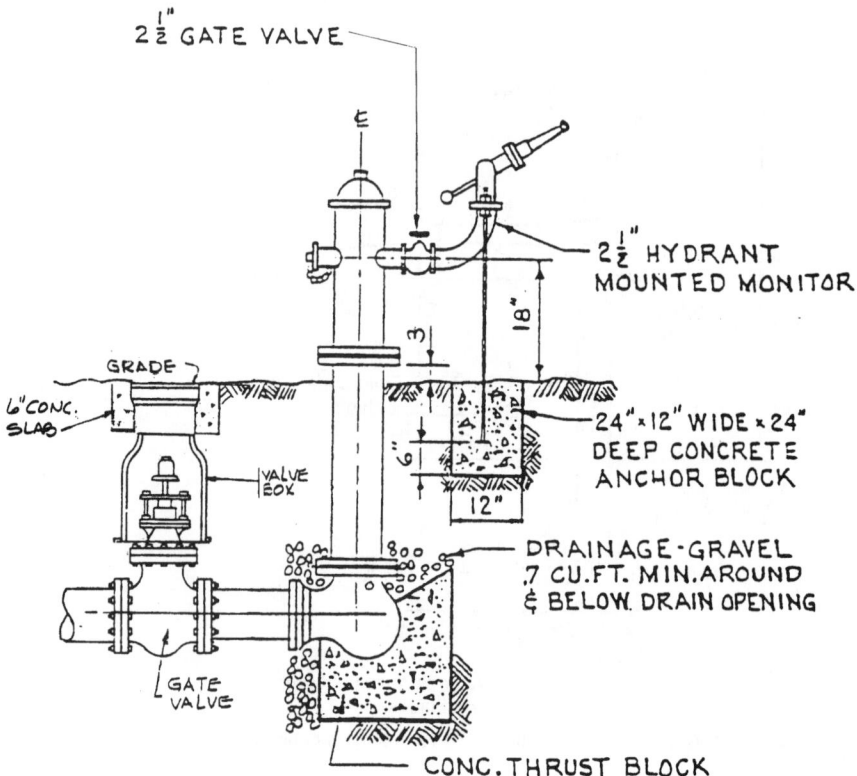

FIGURE 6.56 Typical hydrant-attached monitor.

6. Other special items or criteria that will affect the building service installation or add to the friction loss of the water service must be determined.

7. The maximum flow rate for the project must be calculated.

Calculation of the Pipe Size

1. Find the residual water pressure and the elevation at which the pressure was taken.

2. Calculate the maximum flow rate for the domestic water service.

3. Locate and lay out the service main from the building wall to the point of connection with the utility service or site main. Locate all devices (curb valve, BFP if required, meter, strainer, building valves, etc.).

4. Calculate the static pressure loss based on the elevation of the water main and the elevation of the service entrance into the building.

5. Select the pipe material.

6. Add up all the losses in the service main:

- Losses from valves, meters, BFP devices
- Pressure loss (or gain) from the difference in elevation (from step 4)
- Friction loss in service main to building entrance

Add an allowance of between 5 to 10 psig loss for future pressure drop in the public main. The pressure used shall reflect the possibility of other projects being built in the area. Size the pipe based on maximum flow rate and the above criteria, using the friction loss as a variable along with the velocity of the pipe size selected.

7. Determine if the available pressure is high enough for the selected building water distribution system.

WATER SUPPLIED FROM WELLS

A well is a vertical hole or shaft, excavated or bored through the earth for the purpose of bringing water to the surface from an aquifer. A major consideration is that construction of a well must protect the overall quality of the groundwater in the aquifer. Wells are often classified as deep or shallow. A shallow well is considered to be about 50 to 100 ft (15 to 30 m) deep. Deep wells are generally considered a superior source of water because the water is less susceptible to contamination and the depth of the aquifer usually fluctuates less than water in a shallow well.

Codes and Standards

1. The Safe Drinking Water Act
2. Local requirements

Water Well System Components

The total well system consists of the well pump and the well structure, which is made up of the borehole and casing, intake section, grout, and the distribution pipe that brings the water to the surface.

Well Pumps. The well pump is the means used to bring water in the aquifer to the surface. Three types of well pumps are generally used: vertical turbine, jet pumps, and submersible pumps.

Vertical turbine pumps are centrifugal pumps with the motor mounted on the surface over the borehole and the impeller suspended in the aquifer. The impeller is enclosed in a bowl and connected to the motor by a long shaft, positioned by bearings in the discharge pipe. This type of pump is well suited for larger flow rates, deep wells, and high discharge heads. A typical vertical pump installation is illustrated in Fig. 6.57.

Jet pumps are centrifugal pumps that use the flow of water through a special fitting to create a partial vacuum at the bottom of the well, which draws additional amounts of water into the discharge pipe. This is a two-pipe system—one pipe supplies water down into the well and the second discharges the water to the surface for use. The disadvantages of jet pumps are that since a portion of the discharge

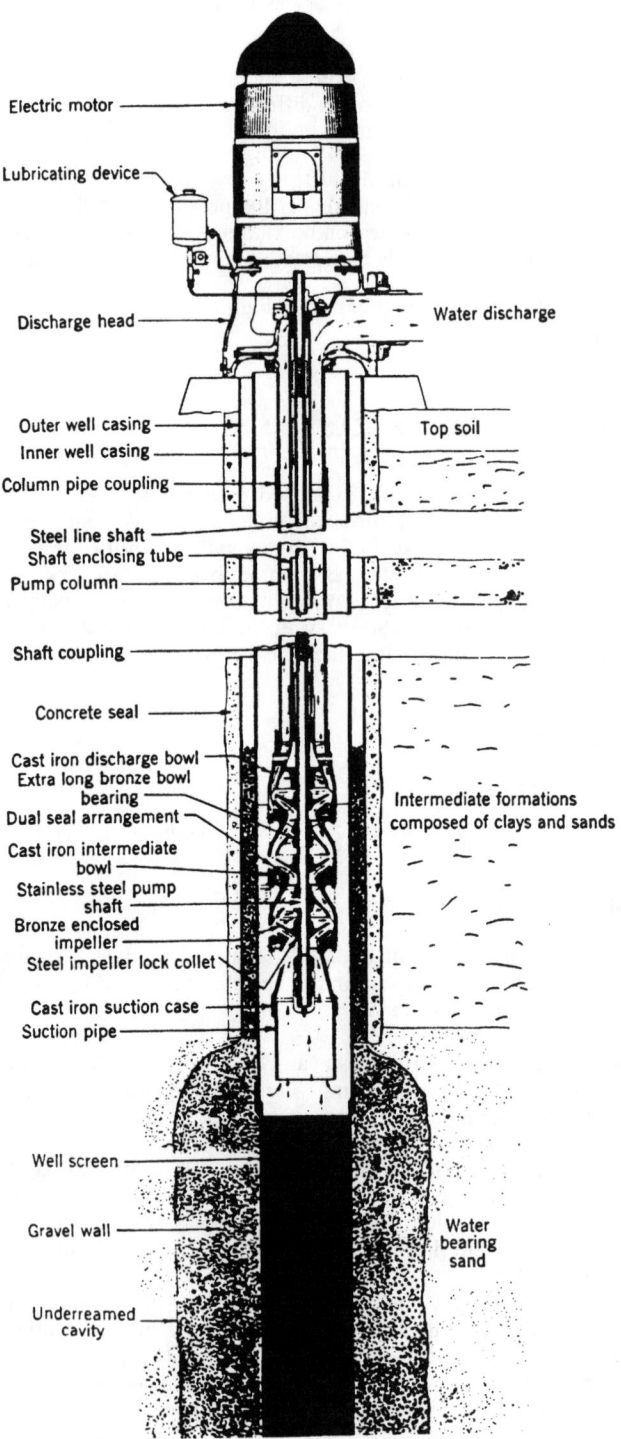

Electric motor

Lubricating device

Discharge head — Water discharge

Outer well casing — Top soil
Inner well casing
Column pipe coupling

Steel line shaft
Shaft enclosing tube
Pump column

Shaft coupling

Concrete seal

Cast iron discharge bowl
Extra long bronze bowl
bearing — Intermediate formations
Dual seal arrangement composed of clays and sands
Cast iron intermediate
bowl
Stainless steel pump
shaft
Bronze enclosed
impeller
Steel impeller lock collet

Cast iron suction case
Suction pipe

Well screen

Gravel wall — Water
bearing
sand

Underreamed
cavity

FIGURE 6.57 Typical vertical turbine installation.

6.132

must be pumped back down the well, the jet pump is generally used for smaller flow rates. In addition, a larger casing is usually required. The advantage is that it is located at the surface for ease of maintenance. The pump could be located over the well or in a remote location. A simplified, remote deep well jet pump installation is illustrated in Fig. 6.58.

The *submersible pump* system (Fig. 6.59) is a small diameter, totally self-contained centrifugal impeller in a housing that is close-coupled to an electric motor. It is placed inside the casing near the bottom of the well. The pump is supported by the discharge pipe and available in a wide range of flow rates and pressures. The electrical supply is attached to the discharge pipe. The submersible pump must be installed lower than the expected lowest drawdown level of the well. Advantages are avoidance of aboveground installation enclosures, installation in wells that have

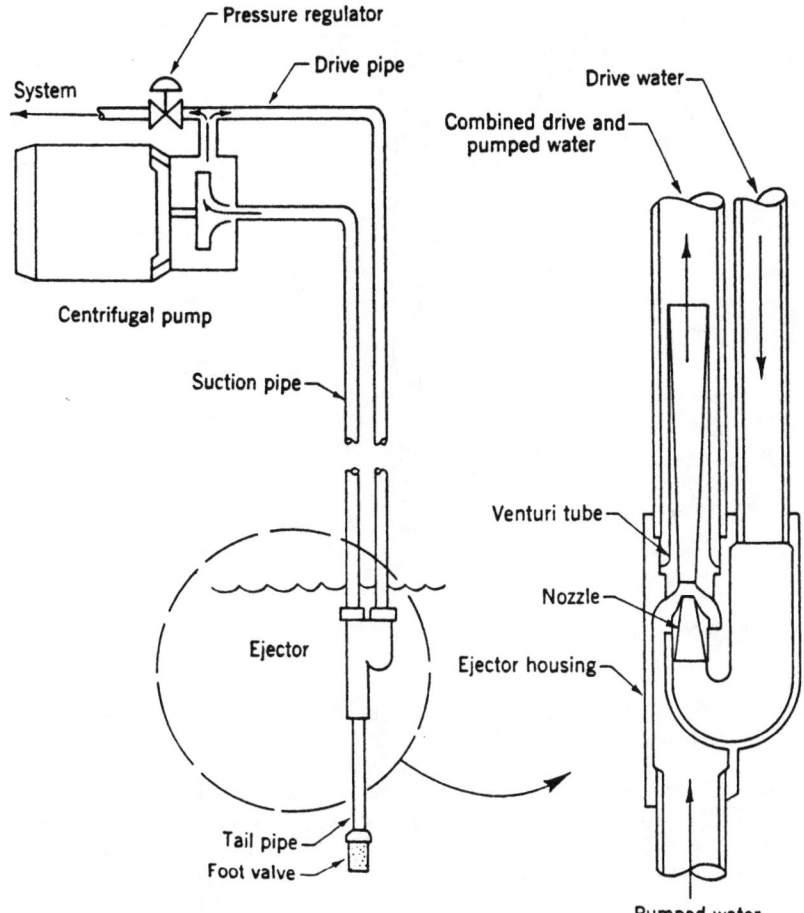

FIGURE 6.58 Simplified deep jet pump installation.

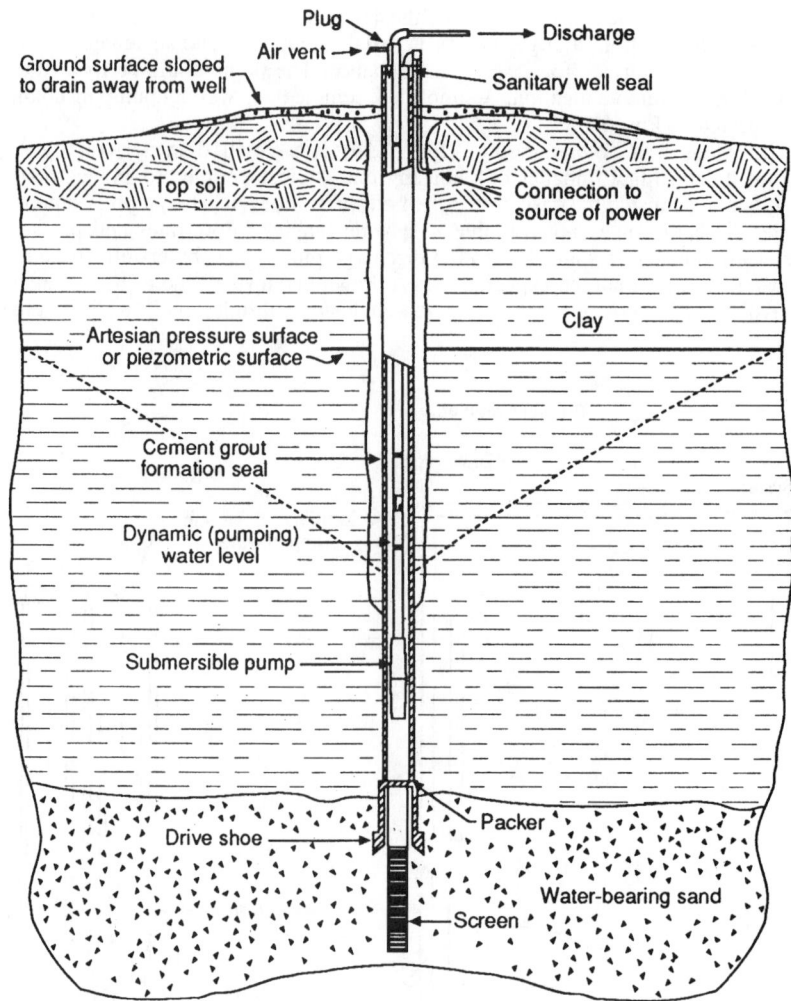

FIGURE 6.59   Typical submersible pump installation.

crooked casings, little noise, and ease of maintenance. An exploded view of a submersible pump is illustrated in Fig. 6.60.

Casing.   The casing is a thin-walled cylinder placed in the borehole of the well, usually as it is drilled. It acts as a lining and extends from ground surface to the bottom of the well where the intake screen is located. It helps to withstand shifts in the earth and prevent cave-ins. It also helps to eliminate seepage of contaminants into the well from undesirable aquifers. Casing material is available in steel, fiberglass, or plastic.

Casing is subject to physical and chemical forces. The physical forces are tension, column loading, and collapse pressures. The chemical forces are those from

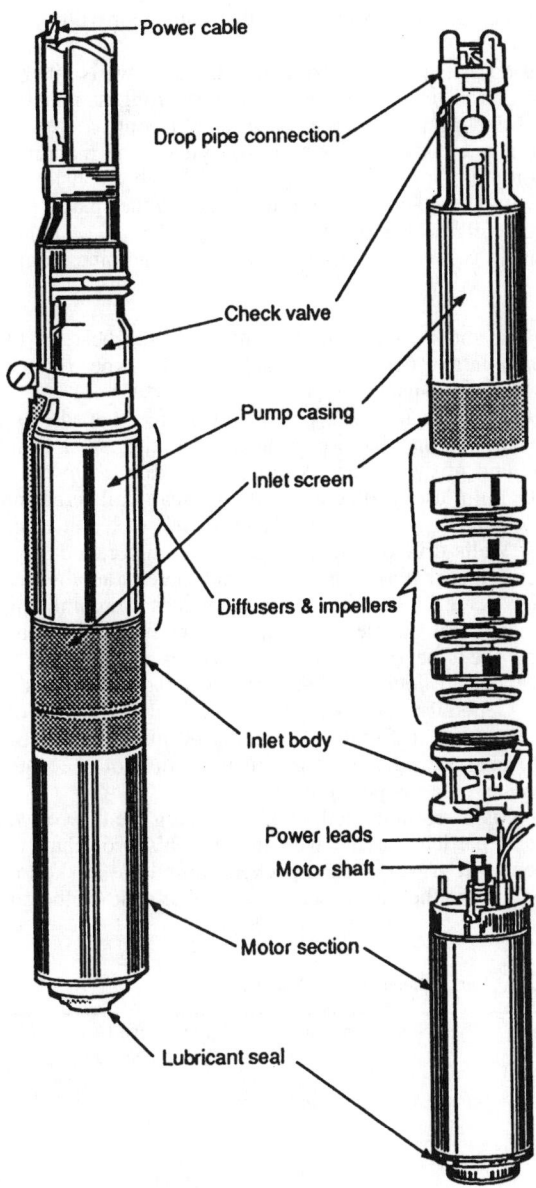

Power cable

Drop pipe connection

Check valve

Pump casing

Inlet screen

Diffusers & impellers

Inlet body

Power leads

Motor shaft

Motor section

Lubricant seal

FIGURE 6.60 Exploded view of submersible pump.

contaminants in the soil and groundwater that cause corrosion and degradation of the casing.

Plastics are used primarily in short, small diameter wells, fiberglass is primarily used for highly corrosive waters, and steel, the strongest material, is used most often. Cable tool drilling requires the use of steel casing.

The casing diameter is generally based on yield. The recommended diameters of wells based on yield are listed in Table 6.31. Another consideration in selecting the casing size is to keep the maximum flow through the smallest part of the casing at or below a generally accepted velocity of 3 fps (1 mps). Another common practice is to use casing two sizes larger than the discharge pipe required by the pump selected.

Well Intakes. The intake is regarded by many as the heart of the well. Two of the most common intake constructions are the tube type, often called naturally developed, and the gravel-packed type.

The *well screen,* or intake, is a perforated device that allows water to move freely from the aquifer into the well while preventing sand from entering. It is connected to the end of the casing. Without the screen, many wells would not be capable of producing a useful flow of water. Typical well screens are slotted, louvered, and wire-wrapped. The wire-wrapped screen is well suited for large diameter, high yield wells. Wells drilled in consolidated rock are called *naturally developed* because the fractured rock acts as an intake and there is an absence of sand. There is little need for a screen intake in this case. Intakes are commonly manufactured from plastic, bronze, steel, stainless steel, and Monel metal. The length of the screen is determined by the type of aquifer (confined or unconfined), diameter of the casing, and the expected flow rate. For large flow rates, lengths of 30 to 40 ft are not uncommon. Generally accepted criteria limit the velocity of water passing through the screen to about 2 to 4 in/s (50 to 100 mm/s). It is recommended that the intake be located at the lowest practical portion of the aquifer so that the drawdown can be as great as possible.

Gravel packing is a common method of enhancing the flow of water by reducing the intake of sand particles at the well intake. This procedure replaces material around the screen with gravel or sand packed around the intake to maximize yield and prevent clogging of the intake openings by fine sand in the aquifer.

TABLE 6.31 Recommended Well Diameters

Anticipated well yield, gpm	Nominal pump bowl size, in	Optimum well casing size, in	Smallest well casing size, in
Less than 100	4	6 I.D.	5 I.D.
75 to 175	5	8 I.D.	6 I.D.
150 to 400	6	10 I.D.	8 I.D.
350 to 650	8	12 I.D.	10 I.D.
600 to 900	10	14 O.D.	12 I.D.
850 to 1300	12	16 O.D.	14 O.D.
1200 to 1800	14	20 O.D.	16 O.D.
1600 to 3000	16	24 O.D.	20 O.D.

Grouting.   Grout is a material used to fill the annular space between the casing and the borehole and also to seal the well at the surface. This material must be able to be pumped far underground and capable of hardening sufficiently to form a seal. The material used most often is a slurry of plain concrete or a mixture of concrete and up to 8% bentonite. Other additives are available that resist water intrusion, add strength, and resist corrosion. The concrete/bentonite mixture has less shrinkage than plain concrete. Grouting and sealing are done for the following reasons:

1. To prevent seepage of polluted surface water down into the well along the outside of the casing
2. To seal out water of unsuitable quality from the strata above the aquifer
3. To stabilize and secure the casing in the drilled hole
4. To form a protective barrier around the casing, thereby increasing its life and protecting it against corrosion

It is not necessary to place grout down to the lowest level of the well, but only to the lowest level where it is necessary to keep out pollutants of concern.

Well Drilling

There are two primary methods used to drill a water well, cable tool drilling and rotary drilling. The choice of method depends on the expected borehole depth, access to the site, purity of the water required, and, often, the driller's preference.

Other methods, such as digging wells by hand, using water jets (jetted wells), and using an earth auger turned by hand or light machines (bored wells) are rarely used to obtain water for any facility, and are outside the scope of this book.

Cable Tool Drilling.   The cable tool rig is often referred to as the "yo-yo" method. This system is based on the simple lifting and dropping action of the tool string in the borehole. Gravity, the weight of the drill bit, and the rotation of the drill string break up and cut through the underground formations. The *tool string* consists of the rope and connecting socket that connects the string to the surface engine, drilling jars (optional), the drill stem which serves as a guide to the string, and the drill bit. Water is added to produce a slurry mixture of the cuttings at the borehole bottom. All of this is connected at the surface to a derrick that lifts and drops the entire string.

Cuttings from the borehole must be removed periodically. This is done by means of a *bailer,* which is a 10- to 20-ft (3- to 6-m) section of pipe with a check valve that holds the cuttings along with the slurry. The tool string is raised clear of the hole and the bailer is lowered down for this purpose.

Although not used very often, the advantage of the cable tool method is the low cost of the entire rig as a whole, simplicity and sturdiness of the design and its ability to move over rough terrain. It is also able to collect qualitative data on the water-bearing characteristics of various strata as the casing is being inserted. Disadvantages are relatively slow drilling speed and the need for a casing to be inserted when drilling through unconsolidated formations.

Rotary Drilling. The rotary drilling method is used to drill approximately 80 percent of all water wells. The *rotary drilling* process bores a hole by means of a rapidly rotating drill bit that has a constant downward pressure exerted on it. The rotary drilling system consists of the rig, a pump to circulate or create the drilling fluid, the drill stem, and a drill bit. The rig is made up of a derrick and hoist, a motor, and a removable, rotating table placed directly over the borehole through which the drill stem passes. The drill stem is composed of the drill collar which connects the bit to the pipe; the drill pipe, which connects the bit to the surface; and the kelly, which is a fitting at the top of the pipe that imparts rotary motion to the stem from the table.

The use of drilling fluids is required in rotary drilling. Drilling liquid or air performs the following four primary functions:

1. Removes cuttings produced by the bit.
2. Transports cuttings up to the surface. The fluids can be interrupted for information about the geophysical features of the borehole.
3. Maintains borehole stability.
4. Cools the drill bit.

Rotary drilling is separated into two primary categories based on the method used to drill the borehole, liquid (or hydraulic) and air rotary drilling.

Hydraulic Rotary. This method can be used to drill in most formations. It functions by using a rotary drill along with the pumping of the drilling fluid which is generally drilling mud, referred to in the field simply as mud. This could be any type of mixture. The filter cake that forms around the length of the borehole wall from the mud flow assists in retaining soft formations and provides a protective layer that resists corrosive effects of the drilling mud.

Air Rotary. This method is mostly used to drill in consolidated formations. Often, mud or water is also used in the unconsolidated strata before bedrock is encountered. The same type of tools are used as for drilling. The difference is that compressed air, usually around 50 psig (340 kPa), is used to depths of about 250 ft (80 m).

Water Well Design and Construction

The design of a well includes criteria for present and future flow rates, the purpose of the well, and its expected useful life.

The drilling or construction of a well requires prior planning. Criteria such as probable depth of water-bearing formations and required permits and approvals of various agencies must be obtained. It must be understood that as the drilling of the well progresses, plans and methods may have to be altered or modified based on actual formations and drilling conditions encountered.

There are several issues that should be considered in the design of a well:

1. Minimum desired output (yield) of the well and the conditions under which the flow is desired.
2. Proposed diameter of the well casing.
3. Out-of-plumbness of the borehole. A common maximum figure is 1 well diameter per 100 ft (33 m).
4. Out-of-straightness. Generally accepted requirements should be limited to a dimension that would permit passage of a 33-ft (10-m) long blank, ½ in (13 mm)

less in diameter than the casing inside diameter. Straightness is more important than being plumb.

5. Screen characteristics, including size and type of openings, length, and net open area.
6. Casing material, type, and weight.
7. Cleaning and disinfection.
8. Development and testing.

During the drilling of the well, constant analysis of the drilling fluid to find an aquifer should be made as the bit passes through each formation. There are no set standards to recognize water-bearing strata. The experience of the well driller is important in this respect. Often, a pilot well drilled for this purpose is highly recommended.

When the borehole is complete, the casing shall be installed (if not previously done), the water level in the well accurately measured, gravel packing done, the intake connected, and the well grouted. The last step in the construction process is disinfection, where a strong chlorine solution is introduced into the well. It is also recommended that intermediate sterilization be performed during the drilling operation.

Well Development and Testing

Every type of drilling causes some kind of disturbance to the aquifer by clogging the pores of the formation where the aquifer exists. After the borehole is completed, the casing is inserted, the intake is attached and the gravel packing installed (if desired). The next step is to have it "developed." Development is the process that removes finer material from the natural formation around the intake, enlarging it and leaving only larger gravel and stones around the screen. The larger pores allow the water in the aquifer to flow into the intake at maximum capacity. Development is the last stage of construction and is regarded as much an art as a science. The experience of the driller is an important factor. After development, the well is capable of providing water.

Development. There is no single, effective method but rather several available, each with its own effectiveness. Most of the methods discussed are used in combination with others. The most common methods include:

1. *Overpumping.* This is a process in which the capacity of the well is greatly exceeded by pumping. All backflow preventing devices are in place. This process cleans out the fine materials and begins to collapse the aquifer strata material around the well. A separate pump is usually selected because one impeller would be excessively worn by the additional sand brought into the pump by the large volume of water.

2. *Rawhiding.* This is a variation of overpumping except that all backflow-preventing devices are removed. The pump is started and stopped, and when the water in the discharge pipe falls to the bottom of the borehole, it loosens the aquifer material adjacent to the well screen. Starting and stopping the pump are done many times.

3. *Surging.* This is a process that uses a plungerlike tool moved up and down inside the well and intake. A tight-fitting insert called a *surge block* creates an alternate vacuum and pressure that force water in and out of the intake.

4. *Air lifting.* This is a process in which an air rotary drill string is placed at the bottom of the well without the bit. Compressed air is pumped down the well allowing air pressure to loosen the material around the bottom of the borehole. This method often results in too great a flow out of the well. This method has many drawbacks and is used primarily in large production wells at the end of the development process.

5. *Air surging.* This method is a variation of air lifting that uses two types of air rotary drills, one within the other. The inner one carries air and the outer one carries water to the surface. A valve is alternately opened and closed after compressed air is allowed to enter the well, resulting in a surge of water that provides the desired results.

6. *High velocity jetting.* Generally regarded as the most effective method of development, it uses multiple nozzles to spray water at 150 to 300 fps (50 to 100 mps) through the intake while it is raised and lowered. Often, the pump will be turned on to take some water to the surface.

Testing for Yield. Following development of the well, the next procedure is testing the well by measuring the flow rate of pumping and the level of water at the various pumping stages. Testing will determine the quantity of water, or yield, that can be produced in a given period of time and the acceptable drawdown level. Testing will provide a basis to estimate the water supply quantity available from the well and to determine the type of pump to be used.

Water Level Measurement. The static water level is the level to which the water reaches in a well under atmospheric conditions when the well is not being pumped. The dynamic water level is the elevation to which water falls in the well during pumping at a given flow rate. The distance between the static level and the dynamic level is *drawdown.* This level will change depending on the pumping flow rate. When water is pumped from a well, the level of water adjacent to the intake creates an inverted cone of depression of the water level in the aquifer (see Fig. 6.61).

To determine drawdown, it is important that this level be checked often over a relatively long period of time at various flow rates. This test shall be recorded in a log. No calculation of capacity should be accepted unless it has been made for at least several hours at hourly intervals. For practical purposes, the pumping level is considered established when three test levels are the same for a minimum of 3 h. Often, a 24-h test is conducted. A common method used to describe the yield is to express the discharge capacity in relation to a distance of drawdown. This relationship is called the specific capacity of the well and is expressed as gpm per foot (m^3/m) of drawdown.

Yield. The character of the aquifer and the construction of the well each will affect the yield. In determining yield, the result that pumping has on the aquifer shall also be determined. The following are definitions of different types of yield:

1. *Safe yield* is the quantity of water that can be withdrawn annually without the ultimate depletion of the aquifer.

2. *Permissive sustained yield* is the maximum rate at which water can be withdrawn from an aquifer on a continuing basis without developing undesirable results.

3. *Maximum sustained yield* is the maximum rate at which water can be withdrawn from the aquifer on a continuing basis.

The yield can be effectively increased by lengthening the intake and changing the type or number of openings in the intake screen. Another method is to make

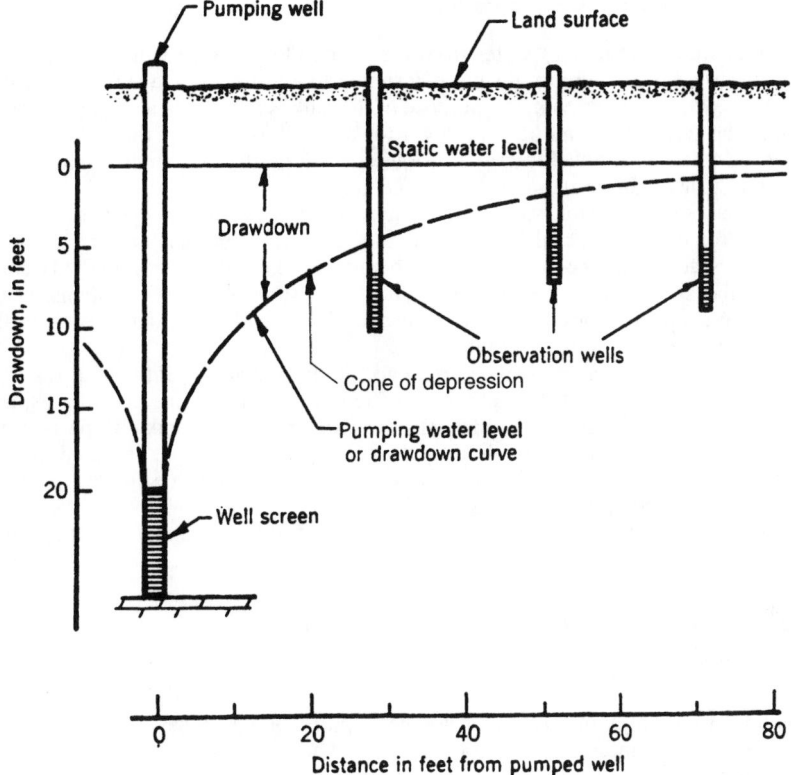

FIGURE 6.61 Cone of depression.

the diameter of the borehole larger, but studies have shown that doubling the diameter only increases the yield by 10 percent. In most cases, this is not an economical solution.

Water Pumps

The selection of the pump depends on many factors, including the following:

1. Flow rate and total head
2. Well diameter
3. Depth of the well
4. Straightness of the well
5. Presence of sand in the water
6. Duration of the pumping cycle

Well System Design Considerations

The total well system takes water from the well and distributes it in a manner that is useful to a facility, almost always as a supply to a water storage tank from which the water will be withdrawn. The purpose of the tank is to provide a storage capacity so that the well pump can be shut down for a period of time. If the tank is large, it can be filled by the well pump at a lower flow rate than the high instantaneous flow rate that might be required by the facility.

For very small systems, similar in capacity to private residence systems, the well water is routed into a small hydropneumatic tank. This tank has an air space that is compressed by the incoming water. When the volume of water in the tank compresses the air to a high enough level, the pump stops. For large systems, water should be supplied to ground-level or elevated water tanks in order to prevent the well pumps from running continuously.

The pump head, or total dynamic head, is calculated by adding the distance between the impeller and the surface, the friction loss in the pipe and fittings, and the height from 1 ft above the inlet into the elevated storage tank (if one is used) or the amount of pressure required in the hydropneumatic tank. If the system is used to directly supply a specific piece of equipment, the value for the pressure required by the equipment is used.

The discharge pipe material is the same as used for normal plumbing site piping.

WATER SUPPLIED FROM SURFACE WATER SOURCES

It is very rare that a surface water source is pure enough to serve as a source of potable water for individual facilities. If one is chosen, it must be approved by the local authorities.

PROJECT WATER SUPPLY

This section will describe the methods and criteria necessary to obtain the size and flow rate for potable water and fire protection water supply.

DOMESTIC WATER SERVICE

The first requirement is to find the water purveyor or utility company. The next requirement is to calculate the preliminary maximum instantaneous service requirements, in GPM. This is done from a preliminary (or final, if available) fixture count for the project under design and adding 10% to allow for estimated additional usage, such as cooling tower fill, boiler fill or other miscellaneous use of water in addition to the domestic use. The water supply fixture unit count can be obtained from Table 9.1. For specialized facilities, add a figure obtained from the owner for additional equipment flow rate.

Once this is calculated, a letter should be written to obtain the following information for your files. This cannot be overemphasized. The following should be requested:

1. A site plan from the utility company showing the water main
2. Other information

Other information shall be requested in a typical water utility letter, as follows:

To: Water Utility Company
Re: Project name
Lot and block number
Address
Dear sir or madam:
    We are the engineers responsible for the design of the mechanical work for this proposed new [addition to an existing] project.
    We are enclosing three site plans, which give a detailed location of the proposed structure and also our desired location for the point of entry into the building. At this time our preliminary water flow rate estimate for this project is a maximum instantaneous demand of _____ GPM domestic and _____ GPM fire protection flow, for a total of _____ GPM.
    Based on the above data, we request the following information.

A. What are the size and location(s) of any existing water mains adjacent to the project site that can be utilized for both domestic and fire protection water service. If at present no such service exists, what is the expected date of completion for such services? Please mark up one copy of the enclosed site plan with this information.

B. What is the depth of bury of the water mains based on the datum taken from the enclosed site plans?

C. What is the static and residual pressure in these mains? If there is any cost involved for obtaining this information, or if you do not normally conduct such tests, please advise us. If weather conditions do not permit such tests at this time, please advise us when such tests would be conducted.

D. Please advise us about the requirements of, and any specific required locations for, both domestic and fire protection meter assembly installations. Is a separate meter required for both domestic and fire water service? Please include dimensions for all such meter assembly installations.

E. Are there any requirements or preferred piping materials and jointing methods for this service?

F. Please provide us with a breakdown of the work provided by you, the utility company, and all work required to be performed by the plumbing contractor relating to the domestic and fire protection service from the main into the building and the installation of the water meter assembly. Who is responsible for connecting to the public water main? If it is our contractor, please provide us with any rules and regulations regarding the installation.

G. Please provide us with any rules and regulations regarding the requirements for and installation of backflow preventing devices. If you are not responsible for such installations, please provide us with the name and address of the agency or department having jurisdiction in this matter. If there are none, please so state.

H. Based on your experience, would you please advise us of the minimum depth of bury acceptable for water mains in your jurisdiction?

I. At the present time, we expect construction to start on or about _____.

End of letter

After the utility company has answered this letter, they will have provided most of the information requested. Based on this information, the following major design items shall be determined.

1. Based on the available pressure, a decision regarding the method used to increase water pressure if required for the building will be decided. The space requirements for the necessary equipment shall be determined. The location inside the building shall be selected.

2. We have a fixture count and have calculated the maximum water supply flow rate.

3. The run into the building has been determined and the meter assembly has been selected and located.

A discussion follows for a typical domestic water service and all of the possible installed devices in detail. First will be a description of the various required information, installed components and their design parameters. Figure 6.62 is a typical sheet used to arrange and calculate all necessary data to determine the service water pressure.

1. *Difference in height* from the centerline of the main to the centerline of the service inside the building where it connects to the pressure increasing device or at the point where the distribution network begins for a street pressure system.

WATER SERVICE CALCULATION SHEET

SYSTEM: DOMESTIC WATER

SHEET 1

PREPARED BY: CHECKED BY: DATE:

PROJECT:

STATIC: _____ PSI

RESIDUAL: _____ PSI BASED ON _____ GPM FOR BLDG AT ELEV. _____

ACTUAL STREET PRESSURE _____ PSI [FT] (1)

SERVICE LOSSES

ST. PR. GPM_____ _____

PUMP GPM _____ _____

HVAC GPM _____ _____

MISC. CONTINUOUS GPM _____ _____

TOTAL GPM _____ _____

SIZE OF SERVICE TO BLDG _____" VEL.° _____ FPS, MATERIAL _____

FRICT. LOSS / 100' _____ RUN INTO BLDG_____FT. TOTAL FRICTION LOSS_____FT.HD.

ALLOWANCE FOR FUTURE DROP IN ST. PR. _____ PSI '_____ FT. HD

DIFFERENCE IN ELEVATION_____FT.

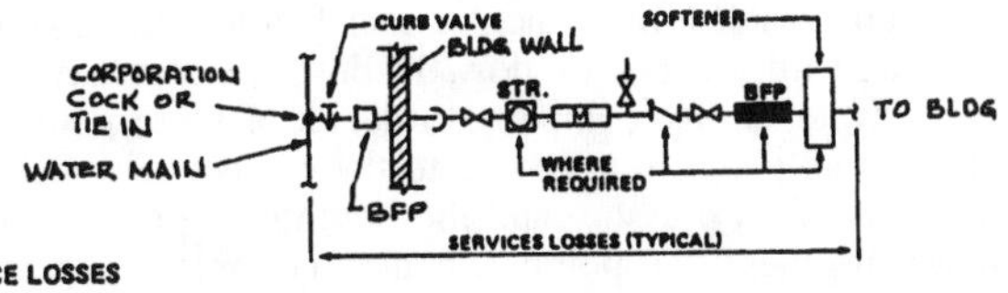

SERVICE LOSSES

QUANTITY	SIZE"	ITEM		
		TIE IN @ _____ PSI = _____ FT.		
		VALVES @ _____ = _____ FT.		
		MISC FTGS @ _____ = _____ FT.		

STRAINER LOSS @ _____ PSI x 2.31 = _____ FT.

METER   LOSS @ _____ PSI x 2.31 = _____ FT.

BFP   LOSS @ _____ PSI x 2.31 = _____ FT.

SOFTENER LOSS @ _____ PSI x 2.31 = _____ FT.

TOTAL SERVICE LOSSES [FT.] (2)

AVAILABLE PRESSURE
ACTUAL PRESSURE - SERVICE LOSS [FT.] (3)
        (1)   (2)

FIGURE 6.62   Water service calculation sheet.

2. *Water pressure, both static and residual.* A hydrant flow test is the most often used method to obtain this information. It may be possible to obtain a map of the piping system of the area where the project is located in order to chose the hydrants where the flow test will be conducted.

This test consists of choosing two fire hydrants closest to the site where the project is located and taking three pressure readings. The first reading is the pressure from a pressure gauge connected to one hydrant with no water flowing. This is called the static pressure because there is no flow. The second reading uses both hydrants. The hydrant with the pressure gauge remains unchanged. The second hydrant is now opened and a velocity pressure reading is taken using a pitot tube held directly in the water stream. A pitot tube is illustrated in Fig. 6.63, and the method of taking the reading is illustrated in Fig. 6.64. At the same time, another reading is taken from the first hydrant while the second hydrant is flowing water. The actual flow rate must be known for this test to be meaningful. This is calculated by converting the velocity pressure into GPM by referring to Fig. 6.65, which is a relative discharge curve using the diameter of the nozzle from which the water is flowing and the pressure reading from the pitot tube held in the water stream.

The time of the year and time of day the test is taken are considerations that will affect the flow test data. There are seasonal variations, mainly in the summer, where the flow is generally regarded as greater than other times of the year in generally residential areas. Another factor is the time of day, which also often has large differences of flow, leading to lower residual pressures. Consult with the water utility company to decide if these items are important in deterring the actual pressure available to the project.

The static and residual pressure and the flow rate represent only two pieces of information. With these two points it is now possible to determine the pressure available at any flow rate. This is important since the flow rate for the project under design is almost certainly less than the flow rate at which the hydrant flow test was conducted. The method used is to plot these two points on hydraulic graph paper, illustrated in Fig. 6.66. The vertical axis has the pressure, in psi, and the horizontal axis has the flow rate, any one scale of

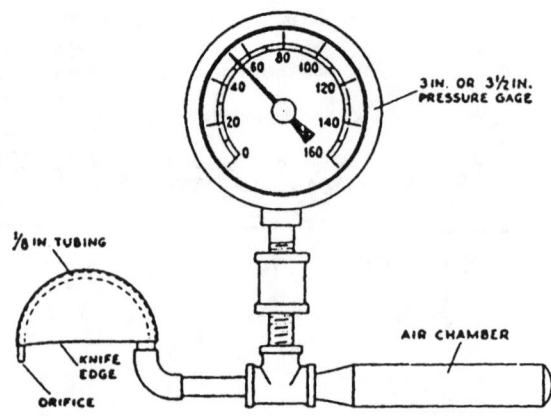

FIGURE 6.63 Pitot tube with gage and air chamber.

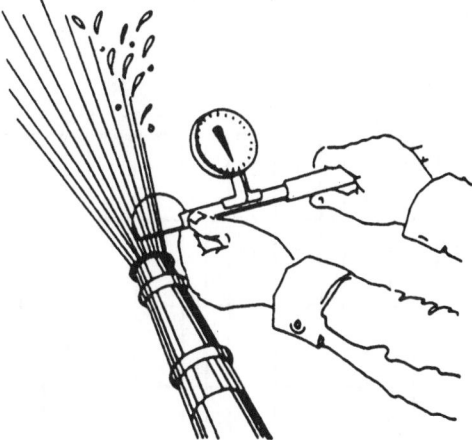

FIGURE 6.64 Taking nozzle pressure with a Pitot tube.

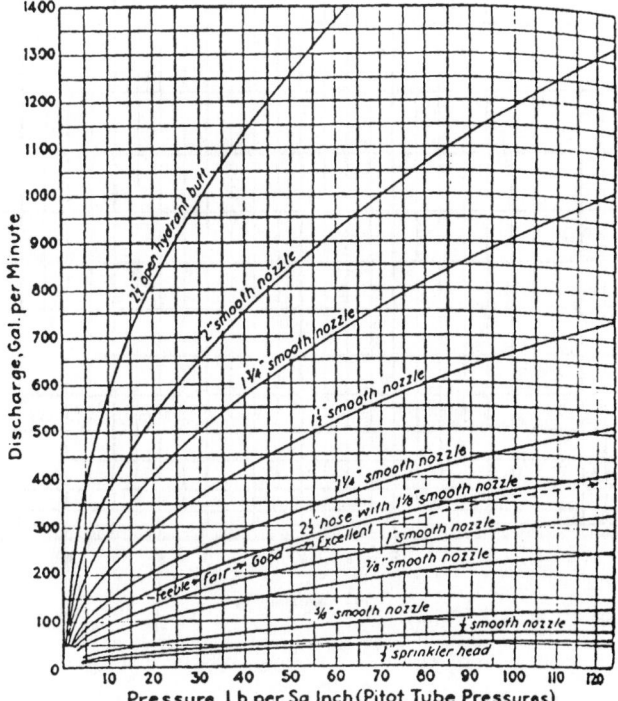

FIGURE 6.65 Relative discharge curves. (Chemical Engineers' Handbook, McGraw-Hill Book Co.)

WATER SUPPLY GRAPH NO. N 1.85

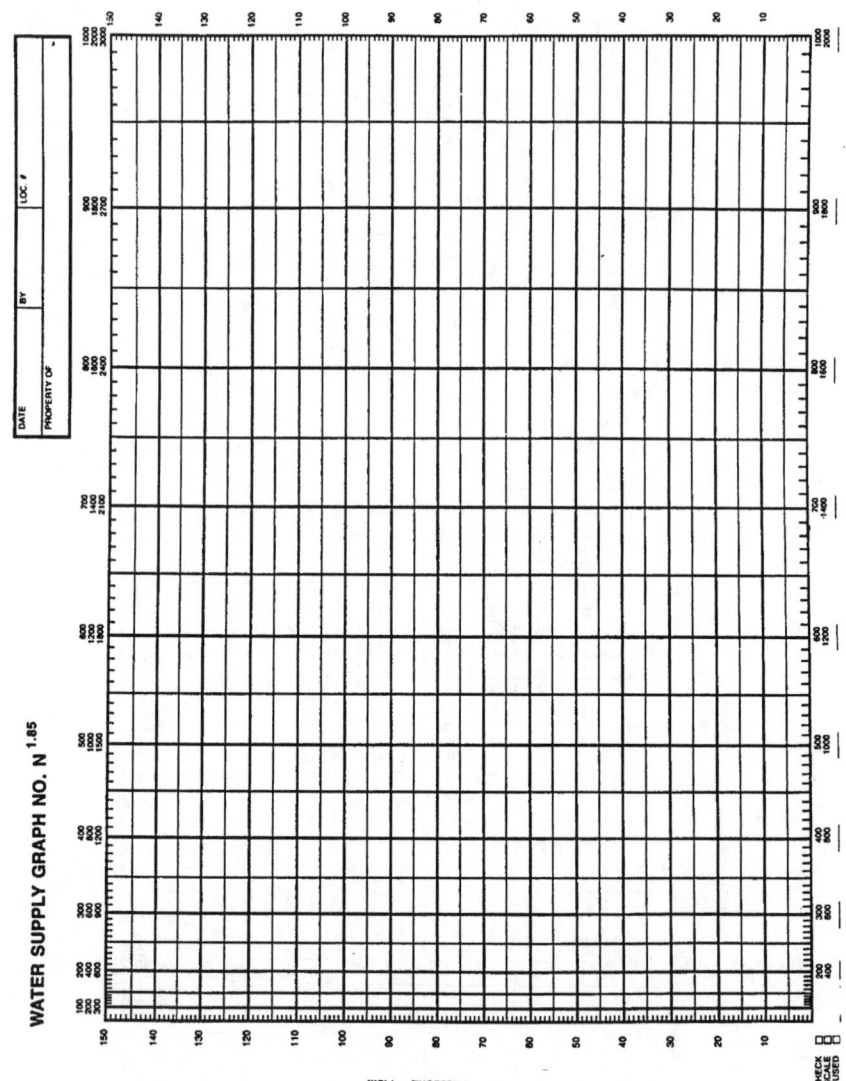

PRESSURE – P.S.I.

FIGURE 6.66 Water supply graph No. N1.85.

6.148

which can be selected for the flow rate established. By connecting the two points, any other point on the line will give the pressure using the flow rate desired.

3. *Pressure loss through ordinary taps* into the water main is given in Table 6.32. Pressure lost through larger wet tap is best obtained from the manufacturer.

4. *Pressure lost through valves and fittings* shall be obtained from Fig. 9.1.

5. *The total friction loss in the piping run* from the main to the last item in the water service assembly can be obtained from standard texts, such as Camerons Hydraulic Data.

6. *The difference in elevation* is the number of feet, either positive or negative, from the centerline of the water main to the centerline of the service to the last item in the water service assembly.

7. *Backflow preventers* (BFP). The water utility company is responsible for protecting the public water supply from any possibility of backflow contamination containing contaminants and pollutants. The most common reasons for the utility company to require a BFP is a close supply of unpure water (like from a stream) that could be used by a fire department to supply water that will be pumped into a building to fight a fire, and a potential source of contaminated water from a facility that could find its way back into a public main if there is a break in the main. For these reasons, a reduced pressure zone BFP will be required near the property line at the connection to the public main. Refer to Chapter 9, Potable Water Systems, for a further discussion of BFPs.

At typical pressure loss through a RPZ BFP is 10 psi. A typical pressure

TABLE 6.32 Loss of Pressure through Taps and Tees in Pounds per Square Inch (psi)

Gallons per minute	Size of trap or tee (inches)						
	⅝	¾	1	1¼	1½	2	3
10	1.35	0.64	0.18	0.08			
20	5.38	2.54	0.77	0.31	0.14		
30	12.1	5.72	1.62	0.69	0.33	0.10	
40		10.2	3.07	1.23	0.58	0.18	
50		15.9	4.49	1.92	0.91	0.28	
60			6.46	2.76	1.31	0.40	
70			8.79	3.76	1.78	0.55	0.10
80			11.5	4.90	2.32	0.72	0.13
90			14.5	6.21	2.94	0.91	0.16
100			17.94	7.67	3.63	1.12	0.21
120			25.8	11.0	5.23	1.61	0.30
140			35.2	15.0	7.12	2.20	0.41
150				17.2	8.16	2.52	0.47
160				19.6	9.30	2.92	0.54
180				24.8	11.8	3.62	0.68
200				30.7	14.5	4.48	0.84
225				38.8	18.4	5.6	1.06
250				47.9	22.7	7.00	1.31
275					27.4	7.70	1.59
300					32.6	10.1	1.88

For SI: 1 inch = 25.4 mm, 1 psi = 6.895 kPa, 1 gpm = 3.785 L/m.

loss through a DCV BFP is 5 psi. Since there are many types of each BFP, check with the manufacturer regarding the exact amount of discharge and pressure loss. It varies with size. Testes at the independent Foundation for Cross-Connection Control and Hydraulic Research at the University of Southern California have established that various manufacturers do not represent their BFPs' pressure loss correctly. It would be appropriate for the design engineer to request the flow curves produced at the Foundation for the most accurate method of comparing various devices.

8. *Strainer losses* are given in Figs. 3.3 for "Y" strainers and 3.4 for basket strainers. Strainers are rarely used except for water with known problems.

9. *Meter losses* for typical meters shall conform to AWWA standards. Refer to manufacturers for exact information.

10. Water softener and other treatment systems shall be obtained from the manufacturer.

FIRE PROTECTION WATER SERVICE

The fire protection water supply to a building starts from connection from a source of water to a point above ground inside a building. This source is usually a public water main but could also include other sources. This building service includes water storage tanks, backflow preventers, meters, valves and other devices that may be required based on the nature of the water service, insurance company requirements and local regulations. This section is not intended to cover private fire service man. For private service mains, refer to NFPA-24 and Factory Mutual Loss Prevention Data Sheet 3-10.

Facility Water Supply

If the water supply to a facility is a combined service, that is both domestic and fire service, the main shall be considered a fire service until the domestic connection to the building service is made. If the source is from a well, an aquifer performance analysis and investigation of the history of adjacent wells shall be made. For many facilities, multiple water supplies from two separate public mains is a very desirable feature. This is to allow water to be supplied to the facility from either of two directions. If these connections are made to a single main, a sectionalizing valve installed in the source main somewhere between the connections will be necessary or this will not be possible. Another desirable feature would be a fire hydrant installed on the building supply adjacent to a hydrant on the public main. This would allow a fire department pumper to increase the volume and pressure to the building if necessary.

Contamination of the public water supply is a prime concern of the water company that supplies water to the project. The determination of the acceptable device for prevention of the fire protection water supply will be made by either the health department, the water purveyor or the plumbing sub-code official. The two most often used are the double check valve assembly or a reduced pressure zone back flow preventer. These devices are discussed in the domestic water supply section. The mains shall be buried below the frost line.

In many cases, it is not possible to supply the required water supply for fire fighting purposes from the public supply. For these situations, the use of a water storage tank is called for. These tanks can be either elevated or installed on the ground. The design of these tanks is outside the scope of this handbook. Connections from the water supply to a storage tank shall terminate 1 ft, 0 in over the overflow level. The overflow shall be two pipe sizes larger than the water supply pipe size. Where necessary, the tanks shall be protected against freezing.

Ancillary Devices

Fire Hydrants. Fire hydrants directly connected to the building supply shall be installed adjacent to roadways that will allow easy access by the fire department. If reasonable, they shall be located on all sides of the building being protected. A desired separation from the building wall is 50 ft, 0 in to provide some protection from building wall collapses. Recommended separation between hydrants would be 200 ft, 0 in apart at a building location. Hydrants located near a road shall be protected by guard posts. It is good practice to provide a shut-off valve on the branch line to a hydrant to allow east repair without having to shut down the main. For some industrial applications, fire hydrants with an attachment called a monitor shall be placed adjacent to specific hazards.

Guard Posts. Guard posts are necessary to protect any device that is installed above grade near roads.

Post Indicator Valve. A post indicator valve is used to shut off the supply of water. It is used as a section valve on water mains and also on building services. Since it is critical for emergency personnel to be certain that any control valve is open or closed, an indicator is positioned above grade with a window or some other method to allow easy observation of the valve position.

Joint Restraint. When run in a straight line, properly specified joints will provide a leak-proof installation. When there is any change in direction of the pipe, a great pressure is exerted by the flowing water on the fitting. The joints may not provide resistance to the force exerted by the pressure of the water against the fitting. In order to prevent joint failure due to the pressure exerted, most pipe joints must be restrained in some manner.

The preferred method of restraining joints is by means of the split ring couplings for grooved pipe. The use of properly selected joints will not require any additional method of restraint. If steel piping is used, welded joints do not require restraints.

Mechanical joints and push-on joints seal by a rubber type gasket compressed in a space between the spigot end of the pipe and the bell. These joints will not resist the pressure of the water on the joint. The method of restraint is with external clamps and rods. Friction clamps are bolted around the pipe on both sides of the joint or joints. These clamps engage steel tension rods, sometimes called tie rods, across the joint preventing it from separating. This will combine the resistance of these joints to prevent separation.

Another method is the use of integrally cast glands for mechanical joint pipe with internally laced grooved and keyed push on joints. This type of restraint is usually recommended only for repair of existing systems.

The most common but generally the least desirable method is to use a block of concrete contacting the fitting and poured against undisturbed soil. The size of the block varies with the water pressure (the higher the pressure, the larger the block), pipe size (the larger the pipe the larger the block), and bearing pressure of the soil (the less the soil bearing pressure the larger the block). The main problems are that on many project sites there is often no undisturbed soil against which to base the thrust block and the size of the concrete block prevents piping from being placed adjacent to the run of pipe being protected.

Sizing the Fire Protection Water Service

All of the information required for design has previously been obtained for the domestic water system. The fire marshal, the fire protection sub-code official and the insurance company shall be contacted for their installation requirements.

What constitutes an adequate water supply has generated much discussion over time. There are many factors that, when added together, will provide an adequate water supply. The water supply should be capable of supplying the largest demand, which is usually the largest sprinkler system and expected hose flow under reasonably adverse conditions. Other factors, such as building occupancy, yard storage, external structures that must be protected, exposure protection and a catastrophic hose demand will all add to the possible maximum flow rate.

Demand Flow Rate. The following figures are presented for discussion purposes only. They are not to be used for actual design, which should only be made on a specific project basis.

The flow requirements for the sprinkler system is the hydraulically calculated flow rate.

The base hose streams depend on the occupancy hazard. These are:

1. Light and ordinary hazard—500 gpm (1900 L/m)
2. Extra hazard, group 1—750 gpm (2900 L/m)
3. Extra hazard, group 2—1000 gpm (3800 L/m)
4. High piled storage—as required by insurance company

The hose demand shall be increased by 25% for the following conditions:

1. Combustible construction
2. Possible delay in response by public fire department
3. Minimum protection less than recommended by insurance company requirements
4. Limited access to remote interior sections

Additional requirements for monitors: allow 500 gpm for each monitor.

Storage Tank Capacity. If the gravity tank is the sole source of water, the tank shall be capable of being filled in 8 hours. Evaluation of the total capacity should consider the following storage capacity:

1. Light and ordinary hazard occupancies, group 1—2 hours
2. Ordinary hazard occupancies, group 2 & 3—3 hours
3. Extra hazard occupancies—4 hours

GROUNDWATER MONITORING WELLS

Monitoring wells are required to detect contamination in groundwater. Because the integrity of samples depends in large part on well construction, the drilling methods and materials used in the construction must have no influence on the quality of groundwater samples brought to the surface.

CODES AND STANDARDS

1. ASTM D 5092
2. ASTM, *Standard Guide for the Development of Groundwater Monitoring Wells in Granular Aquifers*

DRILLING METHODS

The most critical aspect of the drilling method selected is that it must not have any effect on groundwater quality near the borehole. The drilling methods must not use liquids. Cable tool, dual tube, and reverse circulation air rotary methods are acceptable. A typical monitoring well is illustrated in Fig. 6.67.

CASING

The material selected for casing material must be structurally strong, easy to handle, and durable enough to resist the long-term subsurface environment. Durability depends on groundwater and soil chemistry.

The casing material must be capable of resisting any degradation, adsorption, and reaction to the possible contamination products being measured. If it is for water only, it should not contribute any chemicals of its own.

PVC, PE, and other plastic materials are structurally strong and resistant to normal groundwater conditions, but should not be used when solvents might be encountered. Fluoropolymer materials such as Teflon are chemically inert, but their high cost and low tensile strength make them poor choices. PVC is used most often because solvents are rarely encountered. Stainless steel is also considered acceptable for many applications.

GROUT

The most common materials are straight (neat) cement or bentonite. A surface seal of cement must be placed so that surface runoff is prevented from running down the well.

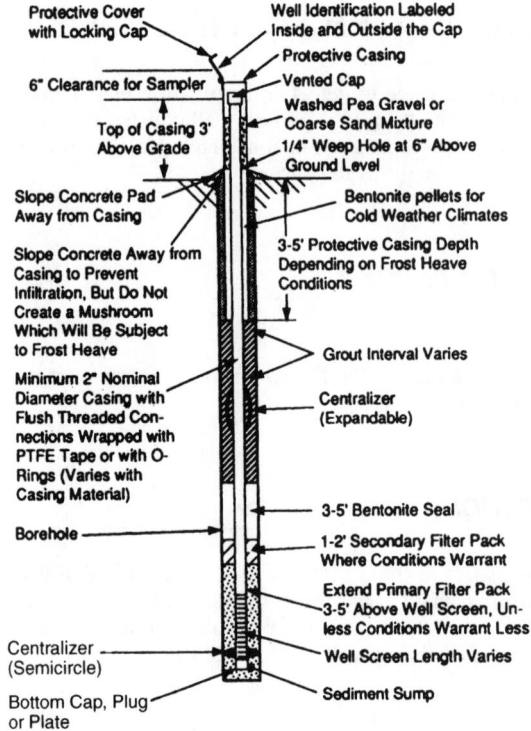

FIGURE 6.67 Typical monitoring well components.

INTAKES

If the intake screen and filter pack allow sediment into the well, increased sampling time and cost will result due to the additional filtering required. Factory-slotted casings and continuous slot wire wound intakes are preferred. For formations less than 50 ft, screen lengths are recommended. For larger formations, proportionally longer lengths can be used. Filter packs generally consist of medium sand to fine gravel with a thickness of 2 to 3 in surrounding the intake. They should extend about 3 ft above the intake to allow for settlement.

WELL DEVELOPMENT

The most effective and popular development technique is surging, although hydraulic jetting and overpumping can be used. Well development is not effective in clay formations or in wells that straddle the water table, such as those drilled to monitor petroleum hydrocarbons.

FINISHING THE MONITOR WELL

A surface seal of cement or concrete may be required by local codes to extend some distance beyond the borehole. Two types of well completion, sometimes called finishes, are common: the aboveground and flush-to-ground completions. In the aboveground type a protective casing is installed around the well casing. It should be either painted or manufactured from a corrosion-resistant material. It is important to drill a drainage hole in the protective casing to drain any water that might accumulate and possibly freeze in colder climates. If no enclosure is installed around the monitor well, bumper guards are recommended if the well is in an area well traveled by vehicles.

For a flush-to-ground completion, a small protective box the size of a small meter is set into the casing with the top flush with grade. This type is typically used in urban areas, parking lots, and streets where an aboveground completion is impractical. It must be watertight.

REFERENCES

5 to 60 Minute Precipitation Frequency for 11 Western States, NOAA Technical Memorandum, National Oceanic & Atmospheric Administration.

5 to 60 Minute Precipitation Frequency for Eastern and Central U.S., NOAA Technical Memorandum, NWS Hydro-35, National Oceanic & Atmospheric Administration, 1977.

ANSI A21.1 (AWWA C101-67), Thickness Design of Cast Iron Pipe, 1967.

Cast Iron Soil Pipe and Fittings, National Engineering Manual, Creative Printing, 1972.

Concrete Pipe Design Manual, 1st Ed., American Concrete Pipe Association, 1970.

Corbitt, R. A., Ed., *Standard Handbook of Environmental Engineering,* McGraw-Hill, New York, 1989.

Data Book, Vol. 1, American Society of Plumbing Engineers, 1975/76.

Design and Construction of Sanitary and Storm Sewers, American Society of Civil Engineers, 1970.

Design and Construction of Water Wells, National Water Well Association, Van Nostrand Reinhold, New York, 1988.

Design Manual, Johns-Manville Transit Pipe.

Dillon, Norman, "Storm Drainage Design Made Adaptable for Computers," *Building Systems Design,* October/November 1973.

Drainage for Areas Other Than Airfields, Dept. of the Army TM 5-820-4.

Fannon, William W., "Rainfall Intensity-Duration-Frequency Curves," *Air Conditioning, Heating, and Ventilating,* June/July/August 1961.

Fannon, William W., "Storm Sewer Design," *Air Conditioning, Heating, and Ventilating,* January 1961.

Handbook of Cast Iron and Ductile Iron Pipe, 5th Ed., Cast Iron Pipe Research Association, 1978.

Handbook of Steel Drainage and Highway Construction Products, American Iron and Steel Institute, 1967.

Manas, Vincent T., *National Plumbing Code Handbook,* McGraw-Hill, New York, 1957.

Nachbar, J., "Storm Drainage Problems in Industrial Plant Design," *Air Conditioning, Heating, and Ventilating,* June 1957.

Rainfall Frequency Atlas of the United States, Technical Paper 40, U.S. Dept. of Commerce, Weather Bureau, May 1961.

Rainfall Intensity-Duration-Frequency Curves, Technical Paper No. 25, U.S. Dept. of Commerce, Weather Bureau, December 1955.

Sewer Manual for Corrugated Steel Pipe, National Corrugated Steel Pipe Association.

Storm Water Management Pond Design Manual, Maryland Association of Soil Conservation Districts, 1976.

Surface Drainage Facilities for Airfields and Heliports, Dept. of the Army, TM 5-820-1.

Urban Hydrology for Small Watersheds, Technical Release No. 55, U.S. Dept. of Agriculture, Soil Conservation Service, January 1975.

Zoeller, John, "Design of Industrial/Commercial Sewage Pumping Stations," *Plumbing Engineer,* November/December 1993.

CHAPTER 7
TURF IRRIGATION SYSTEMS

This chapter will provide the information necessary to design an economical irrigation system for delivering the water required to keep plants healthy and to ensure that this water will be applied evenly to all cultivated areas of the site. Generally accepted means of estimating the amount of water needed, selection of equipment (heads and piping), head-spacing criteria (to produce an even distribution of water), and design criteria for sizing the piping system, choosing timers, and sectioning will be described.

GENERAL

The Department of Agriculture has estimated that grass is the largest irrigated crop in the country. Much of the time, the water used for irrigating this crop is the most expensive water available. Watering on a regular basis is necessary for the continued beauty and health of the grass in these areas, not to mention protecting the investment that the owner has made in the cultivated areas.

In order to find the amount of water that is considered adequate for irrigation purposes, the expected rainfall, soil type, plant type, and irrigation requirements for the site must be known. The total overall quantity of water required to keep plants healthy is based on the difference between the natural rainfall and the minimum amount of water necessary for a healthy lawn. The design must consider the worst condition, which is during times of drought when no rain has fallen. This is the basic criterion used to determine the irrigation rate.

The design of cultivated lawn and turf areas requires the planning and advice of landscape architects. The assistance of a landscape architect will be required in order to answer questions regarding plant types and to confirm the criteria that will be developed for each specific project. In some cases, the design of an extensive and diverse site may be beyond the ability of any design professional. It may then be necessary for an irrigation specialist to be called in to assist in the design.

SYSTEM COMPONENTS

A complete irrigation system is composed of a network of piping, valves, sprinkler heads, electrical controls, timers, and wiring. If the water supply is connected to a potable water source, a water meter and backflow preventer will also be required.

CODES AND STANDARDS

There are no mandated standards concerning the design of turf irrigation systems. If connected to the potable water system, the installed system must conform to plumbing and building codes and other ordinances of the locality.

GENERAL DESIGN CONSIDERATIONS

Examine the local codes for any restrictions that would prohibit the use of potable water for irrigation purposes or ordinances that require the application of water be done only during certain times of the day. Inquire about requirements for the use and installation of meters and backflow preventers. Restrictions may also exist that limit the amount of water available for irrigation. If there is an ideal time to irrigate, it is in the early morning when there is generally less wind and there is a minimum loss of water to evaporation before the water droplets reach the ground. In tests conducted at the University of Wyoming, there was a 15 percent increase in efficiency in the amount of water reaching the ground between midday and morning or evening. This percentage decreases in the Northeast.

The location, pressure, and availability of the water must be determined. If pressure is not adequate, a pump may be required. Separate metering of the water may be required.

The quality of the water source is an important consideration. If the project is to use potable water, there is no question as to its adequacy. If, however, a private well, river, or other nonpotable source is used, the water should be tested and the results discussed with the landscape architect. Using gray water in arid parts of the country has proven successful. The reason for this is that some plants cannot tolerate certain minerals or salts, so either the plants or source have to be changed.

It is impossible to work without an accurate site plan, complete with the proposed location and designation of various types of plants and the topographical elevations indicated to develop the slope of the land. Unlike a cash crop, the color of grass is the determining factor for the care given to a lawn. The green color is controlled by a combination of nutrients and water. If the grass is overirrigated, the excess water will carry away nitrogen, slowing growth and causing the grass to turn yellow. In addition, grass roots use oxygen from the air along with water. Excess water robs the roots of oxygen, restricting their growth. If less than the ideal amount of water is put on a lawn, the grass will grow more slowly, lengthening the time between mowings.

THE SOIL

Soil is composed of particles of sand, silt, and clay called *separates,* and organic matter. The percentage of separates (also called *factions*) determines the texture of a soil. With the addition of organic matter, the separates and organic matter form a complex called the *soil structure.* The spaces between the separates are called *pores.*

Water Absorption

When water first moves into average soil, it forms a thin film around the soil particles and is held in place by adhesion. As more water is added, the film enlarges to fill the pores and becomes heavier. Gravity pulls the water deeper into the ground. The speed at which the water travels deeper into the soil structure varies with the texture and structure of the soil, and is called the *percolation* (or *infiltration*) rate. At the point of balance between the cohesiveness of the water (keeping it adhered to the soil particles) and gravity (pulling the water down), the soil is holding the maximum amount of water possible.

Water Loss

Water is lost from the soil in two ways. The main loss of water is from evaporation. Water in the soil moves by capillary action to the surface and is evaporated. Water is also absorbed by plant roots. The water travels upward through the plant and is lost to the air through transpiration. Evaporation and transpiration are combined into a single factor called *evapotranspiration* or ET. If water is not replaced in the soil, a point will be reached where the adhesion of water to the soil particles is stronger than the plants' ability to capture it. This lack of water stresses the plants causing them to wilt. Refer to Fig. 7.1 for a graphical diagram of the evapotranspiration cycle.

Irrigation Requirements

Water must be applied to the soil in a greater quantity than the ET rate, or the reservoir of water necessary to avoid constant irrigation will not be met. Over an extended period of time, the difference between the maximum water capable of being held in the soil and the water needed by plants to avoid the stress point is the actual amount of water required for irrigation. A primary goal is to determine the instantaneous rate, while also considering the total amount of water used for irrigation purposes so that water will not be wasted.

System Efficiencies

More water than the actual amount absorbed into the ground must be discharged by the sprinkler system due to the following system distribution inefficiencies:

1. The uniformity of water applied throughout the entire individual sprinkler head pattern is not even. The difference ratio can be as large as 9 to 1. A coefficient of uniformity is used to calculate this loss, and is dependent upon the pressure at a head, symmetry and speed of rotation if applicable, and the symmetry of the discharge orifice of the head.

2. Water is lost through evaporation into the air from the time it is discharged from the head until it lands on the ground. This loss is greater during the daytime than at night.

3. Loss of water through wind drift must be allowed for.

FIGURE 7.1   The evapotranspiration cycle. The sun (1) and wind (2) evaporate water from both the soil and the plant. Water is absorbed by the plant roots (3), is passed upward via its tissues (4), and evaporates through stomata in leaves (5). This transpiration process cools the leaves and aids the absorption of nutrients. The two systems form evapotranspiration (ET). (*Courtesy: The Toro Company.*)

4. Systems are rarely shut off at the exact instant that the soil demand has been satisfied.

5. Water should not be applied faster than the infiltration rate of the soil (the rate at which the soil can accept the water), otherwise water will be lost through runoff on the surface of the ground. The infiltration rate and the slope of the land determine the maximum infiltration rate.

6. The difference in water pressure between the first head in any branch and the last head causes the head with less pressure to discharge less water. Good practice calls for a maximum difference of 10 percent.

Sprinkler Head Selection

The sprinkler head is the most visible and critical of the components making up the irrigation system. Deciding which sprinkler head to use depends upon the following criteria:

1. Size and shape of the area to be irrigated
2. Type of plants to be irrigated
3. Water pressure and flow available for system
4. Required rate of head discharge
5. Compatibility of sprinkler type with others in area
6. Obstructions preventing proper distribution of water
7. Client preferences

There are three general types of sprinkler heads used to apply water for lawn irrigation purposes:

Spray Heads. As the name indicates, spray heads are stationary and discharge water continuously over a fixed area in fanlike sheets in the form of either fine droplets or small individual jets. Surface spray heads were the first type to find widespread use. They are fixed in height and are rarely used today. By far the most popular are the "pop-up" spray heads. They are retractable and are mounted below the top of the grass at ground level. When water pressure is applied, the head rises (or pops up) above the top of the grass and starts to spray in a fanlike pattern. When the pressure stops, the heads retract back to grade level and do not provide any obstruction to lawn mowers or people. These heads are available to cover a full circle or any part of an arc.

Rotating Heads. Rotating heads discharge water in a single, large stream, similar to that of a garden hose. The heads rotate to throw water evenly over the distribution area, and the arc can be adjusted. Impact heads have a separate piece of freely attached metal that gets in the way of the stream (or impacts the stream), causing water to deflect randomly. These heads are available to cover a full circle or any part of an arc. Rotating heads can be the pop-up type, exposed stationary, or removable from the piping system with a quick disconnect coupling.

Bubbler and Drip Heads. Bubbler heads are similar to spray heads. They discharge a fan of very small diameter jets of water in smaller amounts and have a smaller pattern. Drip heads are actually lengths of hose with holes that allow water to drip out over their entire length. The amount of water is controlled by the diameter of the pipe, diameter of the holes, and the water pressure.

SYSTEM DESIGN

To minimize the amount of time spent in designing a system, there is a progression that should be followed. The one assumption made is that there is a reasonably unlimited supply of water available for the irrigation system. The progression is an iterative one and should generally follow this order.

1. Determine the source, location, pressure, and quality of the water supply.
2. Obtain the local code and permit requirements for the system.

3. Obtain the regulatory requirements for mandated components of the system, such as meters, backflow preventers, and pipe materials. Locate these components inside or outside the building as required by the specific design or code requirements of the project. Select a tentative location for the control panel and valve boxes for control of different piping zones.

4. Using the complete site plan, select and lay out the irrigation heads and the piping system. Divide the entire site into zones as necessary. Identify the walkways, patios, and so on, that will not need to get water. Identify where the meter and BFP (if required) will be located.

5. Calculate the water flow requirements for the system as a whole and the various sections in particular, select the piping material, and size the piping system.

6. Select desired automatic features of the irrigation scheduling system and schedule the various sections.

7. Make adjustments to the pipe sizes based on the actual calculated flow, head requirements, and actual friction losses in the system and components.

Determination of Available Water Pressure

The available water pressure for the system is found by using standard flow tests and other information that give static and residual flow rates. First, select the source of water. Then determine the residual pressure at the source (accepted practice is to reduce the static pressure by about 15 psi if the residual pressure is not available). From the code, list the mandated components (such as additional meters, backflow preventers, and the like) and find the preliminary pressure losses through them. Add the pressure losses in the piping system and other components from the main to the connection point of the irrigation system. Subtract from the residual pressure the losses just calculated. The result is the pressure in the service line at the connection point for the irrigation system. The above method is an iterative one, with substitutions made to various items as the detailed design progresses.

Water Supply Requirements

There is no quick and easy method available to calculate the exact amount of water needed for irrigation purposes with any degree of accuracy. Specialists use a plant list along with a soil chart based on climatic conditions to establish an accurate ET. More general information will be discussed here. If a project presents a particularly complex problem, a specialist in irrigation should be consulted. For design purposes, a general figure for the necessary amount of water can be used to design a simple system with acceptable accuracy.

For lawn irrigation purposes, it is easiest to find the amount of water to be applied if it is calculated on a weekly basis to discharge enough water to replace slightly more than that lost to ET.

A device called a U.S. Class A Pan Evaporator can be used to establish the ET. This device is a pan of a certain size, which is partially filled with water, set out on the site for a period of time, and the amount of water evaporated from the bucket is measured. Experimentation has determined that the rate of evaporation from a pan evaporator is equal to 80 percent of the actual ET rate. The weekly amount of water to be applied can be derived.

There have been studies made to determine the ET rate for general areas of the United States. One of them divides the country into "hardiness zones," which give the estimated or potential ET, called PET, throughout all areas. Generally, the further south, the greater the water requirements. In addition to the hardiness zone concept, another table has been developed based on the average temperature and humidity of an area. The climate PET values are listed in Table 7.1, which gives the PET rate for one day. Multiply the figure found in Table 7.1 by 7 to calculate the weekly PET rate. The criteria in Table 7.1 correlate with known criteria successfully used for the weekly rate of irrigation. In order to enter Table 7.1 to calculate the actual water flow in the system, the local temperature and humidity conditions must be known. Check with the appropriate sources to find the temperature and humidity design conditions for the project site as it applies to this table.

Another factor to be considered is that a well designed and engineered irrigation system is approximately 80 percent efficient, taking into account all of the inefficiencies previously discussed. This requires that an additional 20 percent be added to the PET to establish the actual amount of water to be discharged from all the heads in a circuit.

Separation into Design Areas

Start with the site plan showing the location of the different plant types and topography, and use the following general guidelines for separation of the site into various zones (or circuits). Consider this a preliminary effort at this point.

1. Separate the areas containing different plant types.
2. Separate any smaller or odd-shaped areas that may cause difficulty due to irregular spacing of the heads.
3. Try to separate the shaded areas from the sunny portions of the site, if practical.
4. Separate any area subject to higher wind velocities than other parts of the site.
5. Exclude all areas that are not to receive water, such as walkways, sidewalks, etc.

TABLE 7.1 Climate PET Table

Climate*	Inches daily	Gallons per ft^2
Cool humid	0.10–0.15	0.062–0.09
Cool dry	0.15–0.20	0.09–0.125
Warm humid	0.15–0.20	0.09–0.125
Warm dry	0.20–0.25	0.125–0.156
Hot humid	0.20–0.30	0.125–0.187
Hot dry	0.30–0.45	0.187–0.218
	↑	
	Worst case	

*Cool is under 70°F as an average midsummer high. Warm is between 70° and 90°F as midsummer highs. Hot equals over 90°F. Humid is over 50% as average midsummer relative humidity (dry = under 50%).

Source: Rainbird, Inc.

6. Locate zone control valves in underground enclosures that will provide easy access for the wiring to the solenoid operators and for maintenance.

7. Provide drain valves on the low points of all zones or piping laterals that will allow the entire piping system to be drained. This is particularly important in freezing climates where the system must be drained each winter.

One point to be remembered about selecting the different areas or zones is that the actual amount of water applied to any zone changes, depending on decisions by the building management, and will eventually be based on actual operating experience over a period of time. The zones will be scheduled to be watered on a regular basis at first, and the results will be observed. After some time, the condition of the lawn will show if any trouble spots exist. At that time, adjustments of the scheduling control to different zones will be made. The engineer must give the operator enough flexibility to adjust an individual zone without requiring major alterations to the schedule of adjacent areas.

Sprinkler Head Selection and Layout

Heads do not distribute water evenly over their entire design area. This is because the total quantity of water is spread out in ever-increasing concentric areas. Since the difference in area from the closest to the farthest point for a head with a 50-ft diameter spray is about 9 to 1, the same amount of water is spread over 9 times the area. In order to even out the distribution, heads should be spaced so that adjacent sprinklers overlap each other and the end of the spray radius of one head hits the adjacent head's spray. There is still enough adjustment permitted so that an additional 10 percent separation can be used to even out the area covered by the actual spacing requirements.

Sprinkler heads can be located in either a square or triangular pattern. The triangular method results in wider spacing and is usually better for irregular area boundaries.

If there is a wind condition and the prevailing direction is known, shorten the spacing perpendicular to the wind direction. Refer to Table 7.2 for adjustments to spacing recommended for estimated wind velocities.

1. In each zone, select the different types of sprinkler heads suitable for each plant type, water application rate, geometrical shape of water distribution, and distance of water throw. Use the manufacturers' specification guides to obtain this information.

2. Try not to mix different heads on the same circuit. This is not because the different heads will not work with each other, but they may require adjustment at

TABLE 7.2   Adjustment in Head Spacing for Wind Conditions

For wind velocities of:	Use maximum spacings of:
0 to 3 mph	60% of diameter
4 to 7 mph	55% of diameter
8 to 12 mph	50% of diameter

Source: Rainbird, Inc.

different times in the future. If necessary, it is acceptable to put a few heads of a different type on a circuit as long as the application rate and pressure requirements are the same.

3. Keep the discharge of a different arc head relative to that of a full circle head. For example, the 180° head should have half the discharge of a full head; a 90° head, one-quarter of the discharge of a full head; and so forth.

4. Select a head that operates in the pressure range calculated for the individual zone and is capable of discharging the estimated flow at that pressure.

5. Where heads that deliver a very small amount of water are selected, it is a good idea to provide a strainer in the supply line to prevent the heads from clogging.

Scheduling the System

The piping system as a whole is sized to provide the maximum instantaneous amount of water discharged by the selected heads at the calculated pressure. Scheduling is meant to provide the minimum overall amount of water necessary for a healthy lawn over a period of time. It is quite common to overirrigate a cultivated area. Overwatering is potentially more dangerous to plants than underwatering.

Most irrigation systems are automatically controlled by the use of preassembled and programmable timing devices. They consist of electrically operated time clocks with multiple programmable features and self-diagnostic capability. The system should include the possible connection of moisture sensors to avoid watering in the rain or when there is enough moisture in the soil. The controller is usually connected to a 120 V power supply through a plug. The purpose of the controllers is to turn on and off the zone control valves at predetermined intervals. Locate these controllers in an accessible location at eye level, allowing for easy changes to the timing programs. An electrical engineer will usually be responsible for wiring the controller to the valves and sensors throughout the site.

The most important aspect of the scheduling process is the determination of exactly what the minimum amount of water is. This is accomplished either through experience or with the use of moisture sensors. Other important considerations are that the watering times are consistent with good practice (such as when wind velocity and evaporation losses are at a minimum), the watering is done within the required "water window" of the time frame required by the specific project, the application rate does not exceed the absorption of water into the soil, and the total water applied falls within the range of water capable of being held in the 4- to 6-in soil root zone, which is the depth of grass roots.

Moisture sensors called tensionometers measure the adhesion of water to the soil particles. Since the force required by the roots to take the water from the soil particles is known, the amount of water available for the plant roots can be determined. These sensors, which are placed in the ground and wired to the control panel, will automatically open zone valves to start the water supply. If the soil water tension is too great, more water must be added to the soil. At this time, there are different products available, and a difference of opinion exists between experts as to the number of sensors required to provide adequate information for scheduling the system. Obtain sensor spacing and location criteria from the individual sensor manufacturer. It has been found that when using moisture sensors, more frequent and shorter watering periods work better.

Another way to determine the overall watering requirement is to compare the average rainfall data for the area where the project is located with the PET. The difference is the amount the irrigation system must provide. The Toro Company has a tabular compilation for the United States and Canada (form number 490.1358). If such a document is not available, the National Oceanographic and Atmospheric Administration has the rainfall data for specific areas, and the Soil Conservation Service can provide the PET rate. Shrubs and areas with plants other than grass should use 50 to 75 percent of the grass figure for the amount of water, depending upon the plants.

Many of the figures given are in inches of precipitation or water application. To convert gpm into inches of precipitation, use the following formula:

$$\text{inches per hour} = \frac{96.3 \times \text{gpm}}{\text{ft}^2 \text{ area of head}} \qquad (7.1)$$

The Piping System

After the heads have been selected, the spacing of the heads established, and the zones (and zone valves) defined, the next step is to select the piping materials and calculate the water supply requirements for the individual zones in order to size the piping system.

Plastic pipe, because of its cost, ease of installation, and imperviousness, is the most popular pipe material. Polyvinyl chloride (PVC) is generally used for mains under constant pressure and polypropylene or polyethylene for the branches beyond the zone control valves.

The following general rules for sizing the piping system are suggested:

1. In any one zone, keep the friction loss difference to 10 percent between the last head and the first head of the zone. For the system as a whole, use a figure of 20 percent from the last head to the supply source. This is done by selecting pipe sizes that give the required pressure losses.

2. Keep the application rate below the rate where runoff would occur. Table 7.3 lists the maximum precipitation rate based on slope.

3. Use pipe sizing to adjust friction loss between the various branches to meet the 10 percent allowable variation in pressure.

TABLE 7.3 Maximum Precipitation Based on Slope

Soil texture	0 to 5% slope	5 to 8% slope	8 to 12% slope	12%+ slope
Coarse sandy soils	2.00	2.00	1.50	1.00
Coarse sandy soils over compact subsoils	1.75	1.25	1.00	0.75
Light sandy loams uniform	1.75	1.25	1.00	0.75
Light sandy loams over compact subsoils	1.25	1.00	0.75	0.50
Uniform silt loams	1.00	0.80	0.60	0.40
Silt loams over compact subsoil	0.60	0.50	0.40	0.30
Heavy clay or clay loam	0.20	0.15	0.12	0.10

TABLE 7.4   Checklist for Irrigation Designers

Water source/location

☐ Meter and size
     or
☐ Pump and size

Point of connection

☐ Size of pipe
☐ Type of pipe
☐ Minimum and maximum pressure available
☐ Who is responsible for this connection

Backflow prevention assembly

☐ Property sized (low friction loss)
☐ Meets local code

Master valve

☐ Manual
☐ Automatic
☐ Drain or blowout point

Total system pressure requirement

☐ Sprinkler operating pressure
     +
☐ Water meter pressure loss
     +
☐ Service line loss
     +
☐ Control valves loss
     +
☐ Backflow preventer loss
     ±
☐ Elevation changes
     +
☐ Mainline loss and fittings
     +
☐ Lateral line loss and fittings (most critical zone)
     +
☐ Pump: filter, check valve; regulator: lift requirements

Mainlines and laterals

☐ Type of material
☐ Depth of bury

Sleeving locations (wire and pipe)

☐ Size
☐ Depth of bury

Control valves

☐ Location
☐ Size/flow rate
☐ Controller station noted
☐ Isolation valve location and size
☐ Quick couplers location and size
☐ Valve box designation

Control system

☐ Controller and location
☐ Electrical power and who is responsible for hookup
☐ Operations manual with owner documents

Legend and other supporting information

☐ North arrow
☐ Prevailing wind indicator
☐ Scale
☐ Water pressure static and design
☐ Precipitation rates by zone or area
☐ Water schedule and programming calculation
☐ Soil type considerations
☐ Utilities noted/critical areas
☐ Name of firm and designer on all sheets

Detail sheet

☐ Rotor detail/swing joint
☐ Pop-up sprays/swing joint
☐ Stationary sprays/bubbler detail
☐ Valve/quick coupler/valve box detail
☐ Point of connection detail
☐ Controller mounting detail
☐ Trenching detail
☐ Backflow detail
☐ Wire requirements detail
☐ Minimum warranty requirements

Source: Used with permission of Eastern Irrigation Consultants.

Table 7.4 can be used as a checklist for any additional basic information that will aid in the preparation of the drawings and the irrigation system as a whole.

ACKNOWLEDGMENTS

The Toro Company, Irrigation Division, Riverside, CA

Rainbird Sprinkler Manufacturing Corp., Glendora, CA

Mr. Richard VanKlein, National Irrigation Engineer, U.S. Soil Conservation Service

Mr. Brenden E. Lynch and Mr. Brian E. Vinchesi, Eastern Irrigation Consultants Inc., Pepperell, MA

CHAPTER 8
CRYOGENIC STORAGE SYSTEMS

Cryogenic gas, as used in this handbook, is defined as any gas in a liquid state at or below −20°F. This chapter describes the bulk storage of cryogenic liquids used for laboratory and light industrial purposes and the piping of cryogenic liquids from storage tanks. There are two applications of cryogenic storage. The first is for a facility that uses a large volume of gas and needs storage on site for practical and economical reasons. The second is for cold liquids required for research, cooling, and other purposes.

Cryogenic storage systems for gases in health care facilities and distribution of gases created from cryogenic storage are discussed in Chap. 14, Compressed Gas Systems.

SYSTEM COMPONENTS

The major components of a cryogenic storage system include the bulk storage tank containing gas in liquid form, a vaporizer (if a gas is desired), and the piping network conveying either gas or liquid to the point of use. The vaporizer is directly connected to the storage tank and is used to convert the liquid gas into its gaseous state (see Fig. 8.1).

CODES AND STANDARDS

The following are codes used for the design and fabrication of cryogenic systems:

1. Underwriters Laboratories, UL-644
2. ASME Code for Unfired Pressure Vessels
3. NFPA 50, 50A, and 50B
4. NFPA 99 Health Care Facilities (often used as a standard for laboratory piping)
5. ASTM G-88
6. ASME/ASTM B3 1.3 Code for Pressure Piping

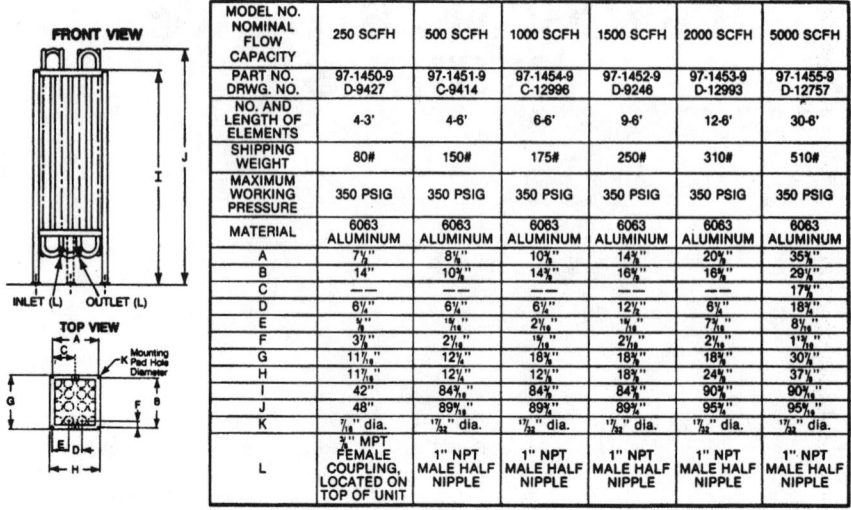

MODEL NO. NOMINAL FLOW CAPACITY	250 SCFH	500 SCFH	1000 SCFH	1500 SCFH	2000 SCFH	5000 SCFH
PART NO. DRWG. NO.	97-1450-9 D-9427	97-1451-9 C-9414	97-1454-9 C-12996	97-1452-9 D-9246	97-1453-9 D-12993	97-1455-9 D-12757
NO. AND LENGTH OF ELEMENTS	4-3'	4-6'	6-6'	9-6'	12-6'	30-6'
SHIPPING WEIGHT	80#	150#	175#	250#	310#	510#
MAXIMUM WORKING PRESSURE	350 PSIG	350 PSIG	350 PSIG	350 PSIG	350 PSIG	350 PSIG
MATERIAL	6063 ALUMINUM	6063 ALUMINUM	6063 ALUMINUM	6063 ALUMINUM	6063 ALUMINUM	6063 ALUMINUM
A	$7\frac{1}{2}$"	$8\frac{1}{8}$"	$10\frac{3}{8}$"	$14\frac{3}{8}$"	$20\frac{3}{8}$"	$35\frac{3}{8}$"
B	14"	$10\frac{3}{8}$"	$14\frac{3}{8}$"	$16\frac{3}{8}$"	$16\frac{3}{8}$"	$29\frac{1}{4}$"
C	--	--	--	--	--	$17\frac{3}{8}$"
D	$6\frac{1}{4}$"	$6\frac{1}{4}$"	$6\frac{1}{4}$"	$12\frac{1}{2}$"	$6\frac{1}{2}$"	$18\frac{1}{4}$"
E	$\frac{3}{4}$"	$\frac{15}{16}$"	$2\frac{1}{16}$"	$\frac{15}{16}$"	$7\frac{3}{16}$"	$8\frac{1}{4}$"
F	$3\frac{7}{8}$"	$2\frac{1}{4}$"	$2\frac{7}{8}$"	$2\frac{1}{4}$"	$2\frac{1}{4}$"	$1\frac{15}{16}$"
G	$11\frac{7}{16}$"	$12\frac{1}{4}$"	$18\frac{3}{8}$"	$18\frac{3}{8}$"	$18\frac{3}{8}$"	$30\frac{7}{8}$"
H	$11\frac{7}{16}$"	$12\frac{1}{4}$"	$12\frac{1}{4}$"	$18\frac{3}{8}$"	$24\frac{7}{8}$"	$37\frac{1}{4}$"
I	42"	$84\frac{5}{8}$"	$84\frac{5}{8}$"	$84\frac{5}{8}$"	$90\frac{5}{8}$"	$90\frac{5}{8}$"
J	48"	$89\frac{1}{16}$"	$89\frac{1}{16}$"	$89\frac{1}{16}$"	$95\frac{1}{4}$"	$95\frac{1}{4}$"
K	$\frac{7}{8}$" dia.	$\frac{17}{32}$" dia.	$\frac{17}{32}$" dia.	$\frac{17}{32}$" dia.	$\frac{17}{32}$" dia.	$\frac{17}{32}$" dia.
L	$\frac{3}{4}$" MPT FEMALE COUPLING, LOCATED ON TOP OF UNIT	1" NPT MALE HALF NIPPLE	1" NPT MALE HALF NIPPLE	1" NPT MALE HALF NIPPLE	1" NPT MALE HALF NIPPLE	1" NPT MALE HALF NIPPLE

FIGURE 8.1   Ambient air vaporizers.

BULK STORAGE

Gas is often stored as a cryogenic liquid when the anticipated volume of gas usage is large enough to make cryogenic storage economical and practical. The reason for this is volume. The cold liquid occupies considerably less volume than a comparable quantity of compressed gas. The gases most commonly stored as cryogenics are nitrogen, argon, and oxygen. Also available but less common are carbon dioxide, hydrogen, and helium. The storage tanks can be generally categorized as either large bulk tanks or smaller dewers.

Large Bulk Tanks

Refer to Table 8.1 for standard large tank sizes. These tanks are highly insulated, generally with Pearlite or other proprietary high efficiency insulation in a vacuum, and can be installed either horizontally or vertically. Vertical installation is most common because a vertical tank occupies less site area than a horizontal one. Another reason is that the vertical tank presents less wetted area for the liquid to vaporize, and it is desirable to keep the stored liquid in that state as long as possible. The dimensions of the tanks vary slightly if installed horizontally or vertically, but the greatest difference is only about 6 in in length for the largest tanks. The horizontal tank is slightly smaller and lighter. All capacities are given in gallons of water. Manufacturers have similar tanks models, with some standard sizes having a capacity as low as 300 gallons. All tanks are ASME rated as unfired pressure vessels.

When locating the tanks on a facility site, there must be enough room for the delivery truck to approach close to the tank. Access to the tank must be easy because the operator must vent gas from the hose connection of the truck to the storage tank by means of the manual vent before filling can start. Two connections

TABLE 8.1   Typical Bulk Cryogenic Storage Tank Dimensions

Nominal capacity, gal	Diameter	Height	Working pressure, psig	Nominal tare weight, lb	Normal evaporation rate, %/day LOx
315	4'	8'1"	250	2,600	.90
525	4'	11'1"	250	3,600	.55
900	5'	11'7"	250	5,500	.40
1,500	5'6"	15'0"	250	9,100	.35
1,500	5'6"	15'8"	150	10,800	1.5
1,500	5'6"	6'6"	150	11,000	1.5
3,000	8'	16'7"	50	14,900	.17
3,000	8'	16'7"	250	20,360	.50
3,000	8'	16'4"	250	17,340	.17
6,000	8'	9'0"	250	34,500	.30
6,000	8'	27'0"	50	19,900	.15
9,000	9'6"	30'9"	250	53,500	.26
9,000	9'6"	30'9"	250	51,300	.10
11,000	9'6"	35'7"	75	34,900	.10
11,000	9'6"	35'7"	250	65,900	.25
11,000	9'6"	35'7"	250	60,000	.10
13,000	10'	36'7"	72	41,000	.10
13,000	10'	36'7"	250	68,300	.10
13,000	10'	36'7"	250	74,100	.25

Source: Minnesota Valley Engineering, Inc.

to the storage tank are desirable, one to the top of the tank (in the vapor space) and one to the bottom (in the liquid). This allows the operator to adjust the pressure in the tank during filling. Filling from the bottom will compress the vapor on top, increasing pressure. Filling from the vapor space on top will lower the pressure because some of the vapor will condense and turn back to liquid thereby reducing the volume of vapor. If filling to a set level is required, a level gauge must be installed on the tank.

There will be leakage of gaseous product from the cryogenic tank if there is no withdrawal of liquid product for a period of time. Each tank is insulated and not intended to have a high internal pressure. The rising of pressure inside the tank resulting from internal vaporization of product will raise the pressure higher than the allowable tank working pressure and will have to be vented to atmosphere from the relief vent. For the actual losses from specific tanks and gas products, consult the manufacturer of the tank or supplier of the gas. For comparison, refer to Table 8.1 for vaporization of oxygen in terms of percent loss of gases.

Dewers

Dewers are smaller insulated tanks used to store smaller quantities of cryogenic gases in individual laboratories or outdoors if required by space conditions. They can be manifolded together for larger storage capacities, if desired. Standard size dewers are illustrated in Fig. 8.2.

SIZING THE LARGE BULK TANK

The amount of liquid to be stored is based on the anticipated volume of gas to be used between deliveries. The delivery schedule represents a compromise on the length of time between deliveries preferred by the supplier and those preferred by the client. The period of time most often suggested between deliveries ranges between once or twice a month. Proceed with the sizing as follows:

1. Determine the proposed usage of each gas per day, shift, or week as closely as possible. This is best done based on past experience. If the installation is new or past information is not available, calculate the expected usage based on the total number of outlets and connected equipment, the quantity of gas used by each, and amount of time each day they are expected to be used.

2. Contact the intended supplier (or interview several if one supplier is not being used at the present time) to obtain the intended delivery schedule and price. Agree on a tank size based on keeping the tank as small as possible and yet having a reasonable supply between deliveries. A minimum of two weeks between supply is a good starting point. Once a month is also common.

3. Calculate the actual usable capacity of the storage tank(s) based on the proposed usage per day multiplied by the number of days between deliveries. Use Table 8.2 to convert gallons of liquid to cubic feet of gas.

4. Add 25 percent to the actual usable capacity found in step 3. This figure will allow 15 percent for the additional empty volume used as vaporization space above the high allowable liquid level when full, and 10 percent additional volume actually occupied by the liquid gas present but not intended to be used, since this is a reserve capacity. This 10 percent should allow a two-day reserve supply of liquid in the tank when the actual low level alarm point is reached in order to allow enough time for the supplier to send more product. An absolute low-level point, which would trigger an emergency call to the supplier, should be 5 percent of capacity. In summary, select a total storage tank volume based on actual usage plus 25 percent.

5. When filling a tank, consider the fact that the density of cryogenic gases varies with pressure. The higher the storage pressure the more liquid will be in the tank. For nitrogen, refer to Fig. 8.3 for the density-pressure relationship.

6. Install vertical tanks on a concrete pad. The strength of the concrete should be a minimum of 3000 psi. The size of the pad should be a minimum of 6 in larger than the diameter of the tank, and if there is additional equipment (like a vaporizer), the pad should be enlarged accordingly. Recommendations of the manufacturers indicate that the minimum soil-bearing strength of the soil should be 2000 psi. In addition, the concrete should be poured over a 6-in layer of crushed stone or gravel. The thickness of the pad depends on the capacity of the tank and the pad should be reinforced. The reinforcement should be wire mesh for small pads and rebar each way (top and bottom) for large pads. Since the tanks may be required to be tested with water some time, use water weight to determine the tank foundation requirements. To estimate the total weight to be supported, find the tare (empty)

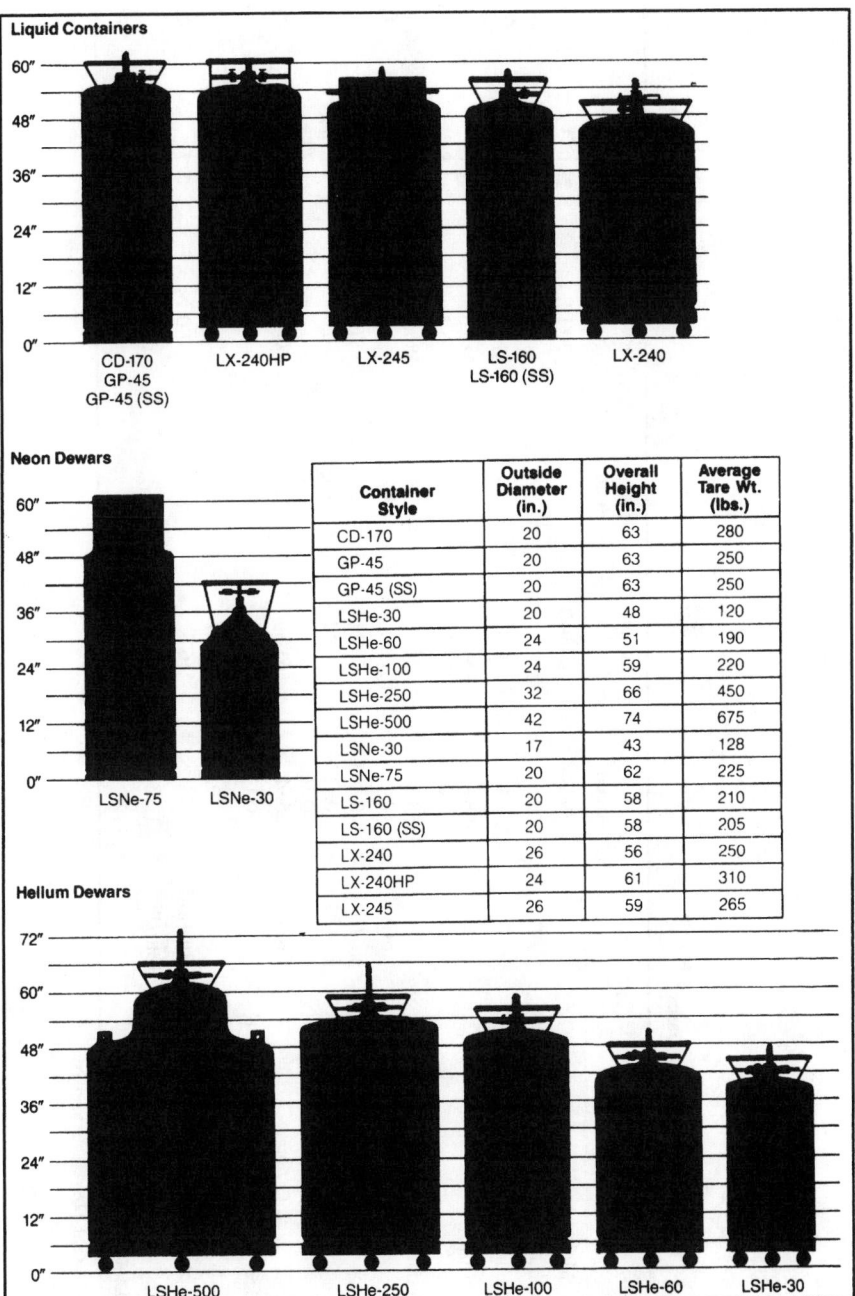

Container Style	Outside Diameter (in.)	Overall Height (in.)	Average Tare Wt. (lbs.)
CD-170	20	63	280
GP-45	20	63	250
GP-45 (SS)	20	63	250
LSHe-30	20	48	120
LSHe-60	24	51	190
LSHe-100	24	59	220
LSHe-250	32	66	450
LSHe-500	42	74	675
LSNe-30	17	43	128
LSNe-75	20	62	225
LS-160	20	58	210
LS-160 (SS)	20	58	205
LX-240	26	56	250
LX-240HP	24	61	310
LX-245	26	59	265

FIGURE 8.2   Cryogenic containers.

TABLE 8.2 Physical Properties of Gases

Gas		Argon	Carbon dioxide	Helium	Neon	Nitrogen	Oxygen
Atomic or Molecular Weight		39.948	44.010	4.0026	20.183	28.013	31.999
Normal Boiling Point (nbp)	°F	−302.6	−109.3	−452.1	−410.9	−320.4	−297.3
Freezing Point	°F	−308.9	−109.3	—	−415.5	−346.0	−361.8
Triple Point (tp)	°F	−308.9	−69.8	—	−415.5	−346.0	−361.8
	psia	10.0	75.2	—	6.3	1.8	0.022
Critical Point	°F	−188.1	87.9	−450.3	−380.0	−232.5	−181.1
	psia	710.0	1071.0	33.2	395.0	493.0	737.0
Density	Gas, NTP lbs/ft³	0.1034	0.1144	0.01034	0.05215	0.07245	0.08281
	Gas, STP lbs/ft³	0.1114	0.1234	0.01114	0.05618	0.07805	0.08921
	Vapor, nbp lbs/ft³	0.363	—	1.04	0.596	0.287	0.279
	Liquid, nbp lbs/ft³	86.98	63.36	7.798	75.35	50.46	71.27
Specific Heat, Cp, Gas, NTP Btu/lb, °F		0.125	0.20	1.25	0.246	0.247	0.220
Specific Heat Ratio, Cp/Cv, Gas, NTP		1.67	1.31	1.66	1.66	1.41	1.40
Heat of Vaporization, nbp	Btu/lb	70.2	246.6	8.72	37.0	85.7	91.7
Heat of Fusion, tp	Btu/lb	12.7	85.6	—	7.1	11.1	6.0
1 Gallon Liquid to cu. Ft Gas, NTP		112	92	100	95	93	115

NTP = Normal Temperature and Pressure

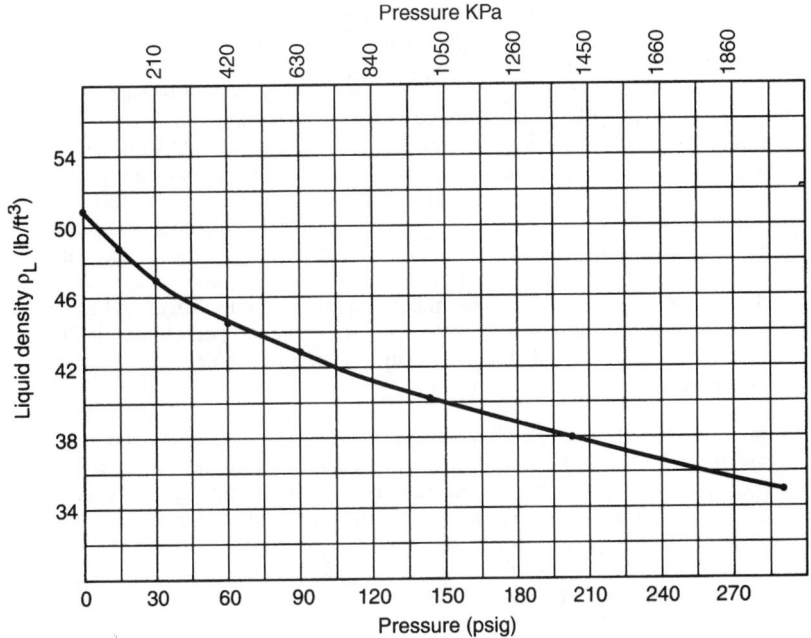

FIGURE 8.3 Density-pressure relationship for nitrogen.

weight of the tank and use the capacity as if filled with water. For a quick estimate, multiply the capacity in gallons by 15.

Tank pad thickness should conform to these minimums:

Capacity, gallons	Thickness	
	inches	mm
Less than 900	12	300
1500 to 3000	15	375
6000 to 9000	18	450
11,000	21	525
13,000 and larger	24	300

Large horizontal tanks usually have piers located 8 ft from each end of the tank. Smaller tanks have supports located one-quarter of the distance from each end. It is then necessary to design foundations based on the weight of the tank and its contents spread out between the two piers. The foundations for the piers shall extend below the frost level.

Vertical tanks are attached to the foundations by bolts connecting legs on the tanks to the concrete pad. The bolts are supplied by the manufacturer and installed by the contractor supplying the pad. The specifications should require the contractor to install the tanks based strictly on manufacturers' requirements and under their supervision.

7. Cryogenic storage tanks are actually two tanks, one inside the other. The annulus is filled with insulation and a vacuum in order to provide a high degree of insulation. Storage tanks used only for gas supply are normally constructed for a working pressure range of between 150 and 250 psig, with the pressure relief valve set to 10 percent over the working pressure. If the low temperature were not maintained, the liquid gas introduced into the tank during filling would vaporize instantly and the pressure inside the tank would quickly rise above the working pressure. If a filled tank is not used for a period of time, usually about two days, approximately 0.5 percent of the contents of the tank per day will be lost through the relief valve in order to maintain the set pressure. This is called the normal evaporation rate. For the approximate normal vaporization rate of oxygen from storage tanks, refer to Table 8.1. Other gases are approximately equal except hydrogen, where experience has shown that the evaporation rate is approximately 3 percent per day because of the ability of the atoms of the gas to slip through the molecules of the metallic tanks and piping.

Another option for storage tank system installation is to allow the supplier to size, design, and install the tank based on a performance specification. It is then the responsibility of the supplier to accurately size the storage system based on the provided usage criteria. In many cases, it is also possible for the installation of the storage tank and equipment to be paid for by the supplier and the cost paid out over 7 to 10 years along with the cost of the gases.

The information required by suppliers is as follows:

1. *Location of the facility (to determine outside air conditions).* If site temperature ranges are available, provide this also.

2. *Peak quantity of gas to be used in cfm and cfh (Lpm or Lph).* Indicate peak usage per day and week if possible. If this quantity is not obtainable, then state how the gas will be used and the total number of outlets or stations in the facility.

3. *Constant or intermittent use of gas.* If only a portion of the use is constant, the use is considered constant. This information is used to determine if a vaporizer is necessary, and to size it based on usage.

4. *Required pressure range of the gas.*

5. *The proposed location of the tank on the site.* Consider easy road access to the location by the supply truck.

6. *Flammable or reactive gas.* If the gas is either one, then separate the tank and any other material storage as required by NFPA and local codes.

SIZING THE VAPORIZER

Bulk gases are stored as liquid, and must be converted to gas prior to being used. If a large volume is used by the facility, neither the storage tank nor the pipeline gain sufficient heat to convert liquid into enough gas to satisfy demand. To convert the necessary volume of liquid into gas, a device called a vaporizer is required.

The most popular vaporizer has no moving parts and uses ambient air to warm the cryogenic liquid as it passes through a long length of finned tubing. The vaporizer is installed close to the storage tank, usually on the same pad. Each manufacturer has standard size units that are selected based on the SCFH, type of cryogenic liquid to be vaporized, and the lowest expected outdoor temperature. For typical sizes of atmospheric vaporizers, refer to Fig. 8.1. If a large volume is expected, a vaporizer using an additional source of heat obtained from steam, electricity, or fuel gas may be required.

A common material for the vaporizer is aluminum. If high purity is required, stainless steel is used. For preliminary sizing based on a temperature of 70°F and 70 percent relative humidity, an approximate figure is that each 8-ft length of 8-in diameter aluminum tube with 8 fins per foot will vaporize 500 cfh.

Manufacturers have proprietary methods for sizing vaporizers. Two factors are necessary for sizing: the lowest mean ambient air temperature for 72 h and the flow rate of the gas. The flow rate is given in cubic feet of gas per hour. For example, if there is a requirement for 100 ft³ of gas in only 10 mm, the flow rate of vaporization must be 600 cfh. Aluminum vaporizers are generally rated at 400 psig, with Monel, Inconel, and stainless steel inserts inside the fin tube available for higher pressures.

The area under the vaporizer is constantly wet. This is due to the ice melting and condensing on the fins and dripping to the ground when the ambient temperature becomes warm. Because of this, the immediate area must have good slope drainage or be provided with a drainage inlet.

PRESSURE FROM LIQUID GASES

The vapor pressure inside of a dewer or tank depends on the outside ambient temperature. Figure 8.4 is the vapor pressure curve for nitrogen and 8.5 for carbon dioxide.

NOTE: TYPICAL SERVICE CONDITIONS 87 PSIG & – 320°F

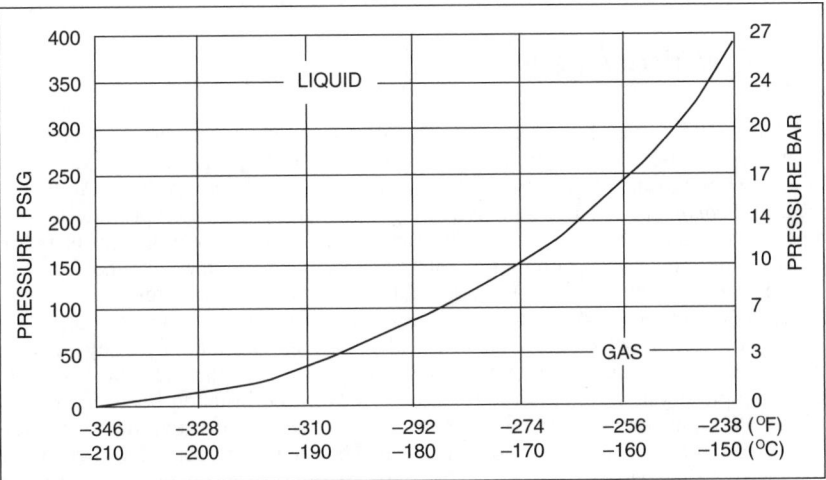

FIGURE 8.4   Vapor pressure curve for nitrogen.

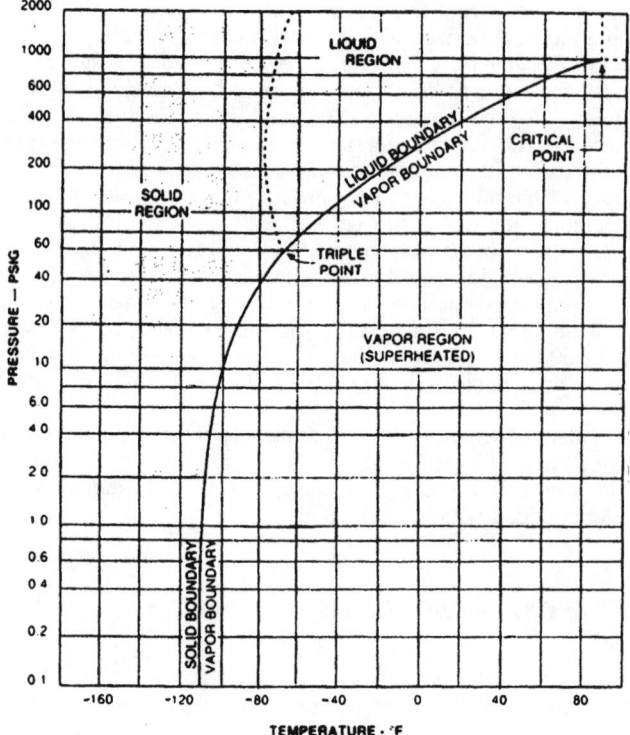

FIGURE 8.5   The pressures and temperatures in all CO_2-storage vesssels and transfer piping must be closely monitored, to keep the substance from changing phase. Source: Airco Gases.

PRESSURE RELIEF VALVES

Pressure relief valves for overpressure protection must be provided in all cases where some gases or cryogenic liquids can become trapped between valves. In oxygen service (and other gases supporting combustion), a condition known as *adiabatic compression* can occur. When a gas at high pressure escapes into a lower pressure pipeline and encounters an obstruction, the temperature of the gas is greatly raised. This may cause autoignition of the metal used for the pipeline (see Fig. 8.6). For noncombustible or nonflammable gases, the rise in pressure of liquid turned to gas by adsorbing heat could exceed the design pressure of the pipe.

Pressure relief valves must be installed on a thermal standoff. They must be sized to vent the required amount of gas needed to lower the pressure inside the tank or piping system. A pressure release valve must be installed between any two valves. When the pressure relief valve goes off, it does not reseat at the set pressure but instead at a pressure about 20 psig lower. This may cause excessive loss of gas if allowed to happen regularly. Another type of relief used on tanks is a *rupture disk*. This is a nonresettable device set to break at a predetermined pressure and is considered an emergency vent that automatically operates when all other pressure

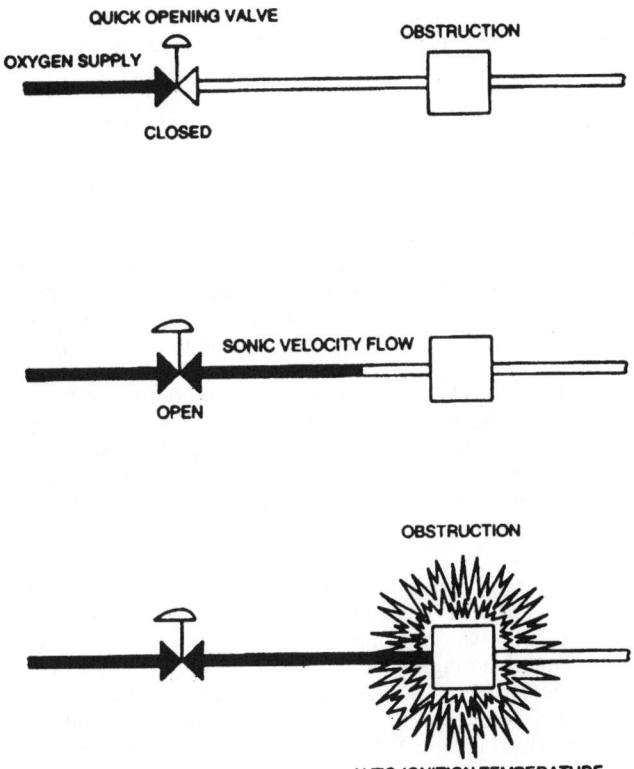

FIGURE 8.6 Adiabatic compression.

relief devices have failed to relieve pressure. All contents of the tank will be lost if the rupture disk is used.

PIPE MATERIALS AND INSULATION

General

Liquid gas piped directly from storage is often needed for laboratory and industrial purposes. It often requires a relatively long run of pipe. When a cryogenic liquid is piped from the storage tank to its point of use, a heavily insulated pipe must be used to avoid having product lost through flashing into a gas due to the high temperature of the pipe and absorbing heat from the ambient air. Most flashing occurs during the initial cooling of the pipe by the cryogenic liquid after a period of nonuse.

For cryogenic work, major consideration must be given to the material of the piping system and the insulation used to keep the heat loss to a minimum. It is impossible to separate the piping material from insulation because, in most cases, the insulation is an integral part of the pipe itself or is factory applied.

Selection of Pipe and Insulation

The criteria for selecting pipe and insulation depend on the extent of the pipe run and the use of the piping.

When only gas is to be produced from cryogenic liquids, no insulation is used so that the product can absorb as much heat as possible. The type of piping on the tank is chosen by the manufacturer. For runs of piping carrying cryogenic liquid, the piping system is selected based on the amount of product allowed to be lost by flashing and adsorption of heat and whether the piping will be in continuous or intermittent use.

There are two general classes of insulated piping: double-wall piping with the annulus vacuum insulated and single-wall piping with the pipe exterior thermally insulated. Vacuum-insulated piping is double walled. It could be either a static vacuum system, where the annulus vacuum is factory assembled during manufacture, or a dynamic vacuum system, where a vacuum is applied to the annulus by continuous vacuum pumping.

Double-Wall Piping with Annulus under Static Vacuum.   This double-wall system is usually manufactured from schedule 40, 304 SS, with nonmetallic spacers used to keep the piping apart. The pipe is manufactured in sections with the vacuum applied to the annulus during manufacture and sealed. Pipe joints are assembled using a mechanical "bayonet" joint with an O ring to connect the sections, which consists of a long male section being inserted into a female section and physically connected with a bolted flange. This provides a long heat leakage path.

Double-Wall Piping with Annulus under Dynamic Vacuum.   This double-wall system is often manufactured from corrugated copper tube with nonmetallic spacers used to keep the piping apart. Pipe joints are assembled using a bayonet joint as previously discussed or brazed joints using a silver brazing alloy. A connection to the outer double-wall pipe provides the inlet to a vacuum pump which is used to produce the vacuum-insulating pressure.

Single-Wall Pipe with Exterior Thermal Insulation.   Typical pipe could be made of copper water tube (hard drawn type K or L) or stainless steel (type 304 or 316 SS preinsulated with PE having a density of 2 lb/ft^3), either 1½, 2 or 3 in thick. Fittings are usually wrought copper. The pipe is jacketed with PVC, .06 in thick. The pipes and fittings are preinsulated, with the ends bare to allow jointing. Additional insulation is installed after the pipe is tested. Copper pipe is jointed by using a silver brazing alloy. SS pipe is usually welded.

PIPE SIZING METHODS

The cryogenic lines are sized on flow rate and head loss. Because the specific gravity of the liquid is almost, but always less, than that of water, the same charts can be used for velocity and friction loss with the assurance of a safety factor. As

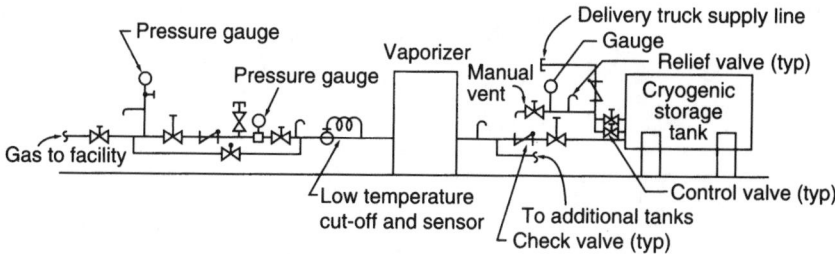

FIGURE 8.7 Detail of large-scale storage tank.

an example, the specific gravity of liquid nitrogen is 0.81. The pressure in the storage tank can be adjusted to produce the pressure necessary to overcome the static lift and friction loss through the lines.

A combination of vapor and liquid in the same line is called *two-phase flow.* This flow is less than that of liquid alone. For laboratories that use small amounts of cryogenic liquid for freezer cooling, or to fill dewers, two-phase flow is not important because only a small amount of liquid discharge, from the line is desired after the gas is eliminated. It is important for facilities that use an open discharge port to fill dewers, that the room or area where this is done be very well ventilated to avoid the accumulation of gas that invariably is discharged along with the liquid.

SYSTEM DESIGN CONSIDERATIONS

For a detail of a typical large-scale storage tank, refer to Fig. 8.7.

REFERENCES

Compressed Air and Gas Handbook, 4th Ed., Compressed Air and Gas Institute.
Nayar, M. L., Ed., *Piping Handbook* 6th Ed., McGraw-Hill, New York, 1992.
"Oxygen Systems," Technical Bulletin No. 5, Swagelok Company.
"Vertical Storage Systems," Trinity Industries Inc.

CHAPTER 9
PLUMBING SYSTEMS

This chapter describes various interior plumbing systems and provides the system design criteria necessary for accurate and cost-effective sizing of equipment and piping for each of the systems. For a continuation of some of these systems outside of a building on the site, refer to Chap. 6, Site Utility Systems, for potable water supply, sanitary drainage, and storm water disposal.

GENERAL

Plumbing systems directly affect the health and safety of the public, and thus are distinguished from other piping systems by the following general requirements:

1. The design, materials, and installation of the systems are directly regulated by a plumbing code.
2. System design must be approved by an authorized code official charged with the responsibility of ensuring plumbing code compliance.
3. A permit for installation of the systems must be obtained from the authority having jurisdiction.
4. The systems shall be installed by an individual duly licensed by the authority having jurisdiction for determining the competence of an individual to obtain a plumbing installation license. This may not be required in some jurisdictions.
5. The installed systems must be inspected and approved by an authorized code official charged with the responsibility of code enforcement.

The basic plumbing systems are

1. Sanitary drainage
2. Sanitary vent
3. Storm water drainage
4. Potable water
5. Fuel gas (Refer to Chap. 13.)

CODES AND STANDARDS

Plumbing codes establish minimum acceptable standards for the design and installation of the various plumbing systems and for the components that compose them. There are six regional building codes that, along with their associated plumbing codes, have found general acceptance over various large areas of the country. They are

1. Building Officials Code Authority (BOCA), plumbing code: BOCA National Plumbing Code
2. International Association of Plumbing and Mechanical Officials (IAPMO), plumbing code: Uniform Plumbing Code
3. National Association of Plumbing-Heating-Cooling Contractors (PHCC), plumbing code: National Standard Plumbing Code
4. Southern Building Code Congress International (SBCCI), plumbing code: National Standard Plumbing Code
5. Council of American Building Officials (CABO), plumbing code: One- and Two-Family Dwelling Code
6. International Conference of Building Officials (ICBO), plumbing code: Uniform Plumbing Code
7. International Plumbing Code (IPC)
8. National Plumbing Code (A40)

Some states and large cities have adopted codes other than these building codes. In addition, some local authorities using specific regional building codes have adopted plumbing codes other than the one usually associated with that building code. Because of this nonstandardization, the plumbing code required to be used for each specific project must be obtained from a responsible code official.

The information pertaining to those systems included in the approved local plumbing code is the primary criteria for accepted methods and sizing. The tables and charts appearing in this chapter are used only to illustrate and augment discussions of sizing procedures and design methods, and should not be used for actual design purposes.

There are many nationally recognized standards that establish dimensions, manufacturing methods, material composition, tests, and numerous other details specific to individual components of the plumbing system. A partial list of organizations originating such standards adapted by various plumbing codes are

- American National Standards Institute (ANSI)
- American Society of Mechanical Engineers (ASME)
- American Society of Sanitary Engineers (ASSE)
- American Society of Testing and Materials (ASTM)
- American Water Works Association (AWWA)
- American Welding Society (AWS)
- Cast Iron Soil Pipe Institute (CISPI)
- National Fire Protection Association (NFPA)
- National Sanitation Foundation (NSF)

- Plumbing and Drainage Institute (PDI)
- Underwriters Laboratories (UL)

FIXTURE UNITS

The *fixture unit* (FU) is an arbitrary, comparative, and dimensionless value assigned to a specific plumbing fixture, device, or piece of equipment. FU values represent the probable flow the fixture will discharge into a drainage system or use (demand) from a potable water supply system, compared to other fixtures.

The use of fixture units for plumbing systems was expanded upon by the late Roy B. Hunter of the former Bureau of Standards. If the original criteria were used, the results would be oversized water and drainage systems, since new fixtures have been developed and changes in the patterns of water use and conservation have evolved. Since the development of the Hunter Method, long-term data and modern statistical methods and analyses have resulted in revised figures, which are used for this book.

Since sanitary discharge and water demand FUs are different, the designation DFU for *drainage fixture units* and WFU for potable *water fixture units* will be used to differentiate between them.

Table 9.1 lists average drainage and vent DFUs, hot and cold water WFUs, and branch size information for typical fixtures.

PLUMBING FIXTURES

A plumbing fixture is any approved receptacle specifically designed to receive human and other waterborne waste and discharge that waste directly into the sanitary drainage system, usually with the addition of water. Ideal fixture materials should be nonabsorbent, nonporous, nonoxidizing, smooth, and easily cleaned.

Plumbing codes usually mandate the number and type of fixtures that must be provided for specific building usage, based on the proposed population. Provisions for the handicapped have been made an integral part of code requirements, mandating the number and layout of and barrier-free access to those fixtures.

Potable water discharged from specific plumbing fixtures may be restricted to a maximum flow rate mandated by water conservation requirements. Refer to specific code provisions for these restrictions.

EQUIVALENT LENGTH OF PIPING

When calculating the pressure loss through a pressurized piping system, one of the factors to be considered is the equivalent length of pipe. This is the actual pipe run plus an additional length, expressed as a number of feet of straight pipe that would have the same friction loss as that occurring through various fittings, valves, and so on. Figure 9.1 gives the straight run of pipe for both water and gas systems equal to various valve types and fittings for different pipe sizes. A popular and

TABLE 9.1 Typical Plumbing Fixture Schedule*

Fixture type	DFU	Trap	Vent	WFU	Cold	Hot	Flow, gpm†
		Drainage			Water		
		Size			Size		
Automatic clothes washer	3	2	$1\frac{1}{2}$	2	$\frac{1}{2}$	$\frac{1}{2}$	5
Bathroom group (WC, LAV, SH/BT) FV	8			8			
Bathroom group (WC, LAV, SH/BT) tank	6			6			
Bathtub (BT), with or without SH	2	$1\frac{1}{2}$	$1\frac{1}{2}$	2	$\frac{1}{2}$	$\frac{1}{2}$	5
Bidet	1	$1\frac{1}{4}$	$1\frac{1}{4}$	1	$\frac{1}{2}$		2
Clinic sink	6	3	$1\frac{1}{2}$	2	$\frac{1}{2}$	$\frac{1}{2}$	3
Dishwasher, domestic	2	$1\frac{1}{2}$	$1\frac{1}{2}$	1	$\frac{1}{2}$	$\frac{1}{2}$	3
Dental lavatory, cuspidor and unit	1	$1\frac{1}{2}$	$1\frac{1}{2}$	1	$\frac{1}{2}$	$\frac{1}{2}$	1
Drinking fountain	$\frac{1}{2}$	$1\frac{1}{4}$	$1\frac{1}{4}$	$\frac{1}{2}$	$\frac{1}{2}$		$1\frac{1}{2}$
Floor drain	5	3	$1\frac{1}{2}$				
Kit. sink & tray, with food grinder	4	2	$1\frac{1}{2}$	2	$\frac{1}{2}$	$\frac{1}{2}$	3
Kit. sink & tray, single $1\frac{1}{2}$ trap	2	$1\frac{1}{2}$	$1\frac{1}{2}$	2	$\frac{1}{2}$	$\frac{1}{2}$	3
Kit. sink & tray, multiple $1\frac{1}{2}$ traps	3	$1\frac{1}{2}$	$1\frac{1}{2}$	2	$\frac{1}{2}$	$\frac{1}{2}$	3
Lavatory, private	1	$1\frac{1}{4}$	$1\frac{1}{4}$	1	$\frac{3}{8}$	$\frac{3}{8}$	2
Lavatory, public	2	$1\frac{1}{4}$	$1\frac{1}{4}$	2	$\frac{3}{8}$	$\frac{3}{8}$	2
Laundry tray, 1 or 2 compartments	2	$1\frac{1}{2}$	$1\frac{1}{2}$	2	$\frac{1}{2}$	$\frac{1}{2}$	5
Shower (SH) per head or stall	2	2	$1\frac{1}{2}$	2	$\frac{1}{2}$	$\frac{1}{2}$	3
Service sink (SS), trap standard	3	3	$1\frac{1}{2}$	3	$\frac{3}{4}$	$\frac{3}{4}$	4
Service sink, P trap	2	$1\frac{1}{2}$	$1\frac{1}{2}$	2	$\frac{1}{2}$	$\frac{1}{2}$	4
Sink, pot & scullery	2	$1\frac{1}{2}$	2	2	$\frac{1}{2}$	$\frac{1}{2}$	$4\frac{1}{2}$
Sink, bar	$1\frac{1}{2}$	$1\frac{1}{2}$	$1\frac{1}{2}$	1	$\frac{1}{2}$	$\frac{1}{2}$	2
Sink, flushing rim	6	3	$1\frac{1}{2}$	5	1		15–30
Sink, surgeon's	3	2	$1\frac{1}{2}$	2	$\frac{1}{2}$	$\frac{1}{2}$	$2\frac{1}{2}$
Sink, wash fountain, per faucet	2	$1\frac{1}{2}$	$1\frac{1}{2}$	2	$\frac{1}{2}$	$\frac{1}{2}$	$2\frac{1}{2}$
Urinal pedestal, blowout	6	3	$1\frac{1}{2}$	10	1		15–40
Urinal washout	4	2	$1\frac{1}{2}$	5	$\frac{3}{4}$		10–20
Water closet private flush valve	6	3	$1\frac{1}{2}$	10	1		15–40
Water closet private tank type	4	3	$1\frac{1}{2}$	5	$\frac{1}{2}$		3–5
Water closet private pressure tank	4	3	$1\frac{1}{2}$	4	$\frac{1}{2}$		3–5
Water closet public flush valve	6	3	$1\frac{1}{2}$	10	1		15–40
Water closet public tank type	4	3	$1\frac{1}{2}$	5	$\frac{1}{2}$		3–5
Water closet public pressure tank	4	3	$1\frac{1}{2}$	4	$\frac{1}{2}$		3–5
Fixture not listed	1	$1\frac{1}{4}$	$1\frac{1}{4}$				
Fixture not listed	2	$1\frac{1}{2}$	$1\frac{1}{2}$				
Fixture not listed	3	2	$1\frac{1}{2}$				
Fixture not listed	5	3	$1\frac{1}{2}$				
Fixture not listed	6	4	2				
Hose bibb or sill cock, public				5	$\frac{3}{4}$		5
Hose bibb or sill cock, private				3	$\frac{1}{2}$		3
Water supply not listed				1	$\frac{3}{8}$		
Water supply not listed				2	$\frac{1}{2}$		
Water supply not listed				3	$\frac{3}{4}$		
Water supply not listed				10	1		

*Refer to Table 2.9 for conversion of NPS to DN pipe sizes.
†1 GPM = 3.8 I/M or 0.63 I/S

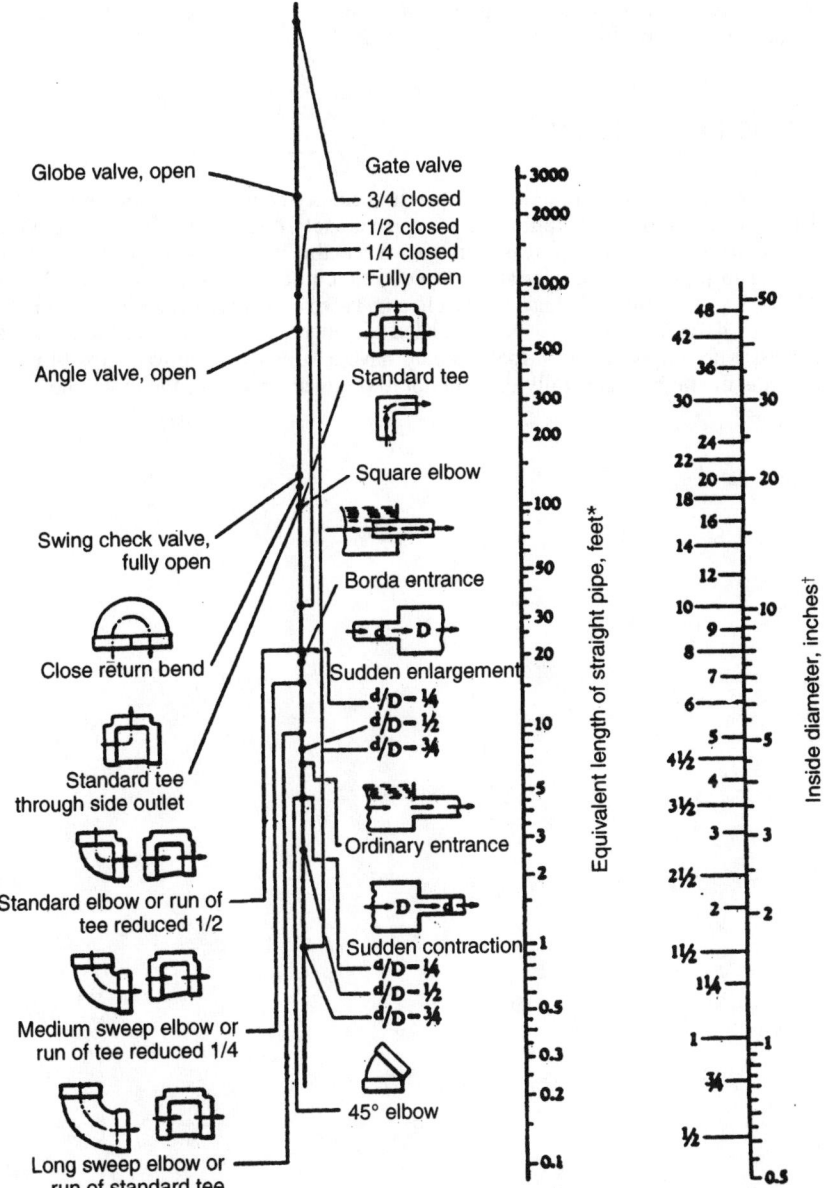

FIGURE 9.1   Equivalent lengths of pipe for valves and fittings. Note: For sudden enlargements or sudden contractions, use the smaller diameter on the nominal pipe size scale.
   *1 ft = 0.3 m.
   †For conversion of NPS to DN pipe size, refer to Table 2.9.

conservative method of quickly finding the equivalent run is to add 50 percent to the actual measured pipe run.

VELOCITY OF WATER

It is generally accepted practice to limit the velocity of water in commercial establishments, such as office buildings, to approximately 6 to 8 ft per second (fps) (1.8 to 2.4 m/s). The primary reason is to reduce the noise produced by the water flowing in pipes. For theaters and other quiet facilities, 2 to 4 fps (0.6 to 1.2 m/s) should be maintained. Where quick closing valves are installed, a maximum velocity of 4 fps (1.2 m/s) is recommended to reduce water hammer. Industrial facilities, where noise is not a factor, could have a velocity of up to 12 fps (4 m/s), which is the highest generally thought not to produce erosion of the piping network.

SANITARY DRAINAGE

SYSTEM DESCRIPTION

The sanitary drainage system conveys waterborne effluent discharged from plumbing fixtures and other equipment to an approved point of disposal. The sanitary system receives all liquid waste except storm water or unacceptably treated process or chemical drainage. This effluent must be treated prior to liquid discharge into the environment. Treatment can be accomplished by either a public utility or treatment on site by a private plant that must receive approval from authorities having jurisdiction.

MAJOR SYSTEM COMPONENTS

Cleanouts

Codes mandate that cleanouts generally be provided at the base of stacks before the pipe changes direction from vertical to horizontal, at changes in horizontal pipe direction greater than 45° and along horizontal runs of pipe every 50 ft (15 m). Typical cleanouts are illustrated in Fig. 9.2*a* and cleanout components in Fig. 9.2*b*.

Floor Drains

A floor drain is a receptacle used to remove liquid effluent from building interior floor areas and other similar locations. A typical floor drain is illustrated in Fig.

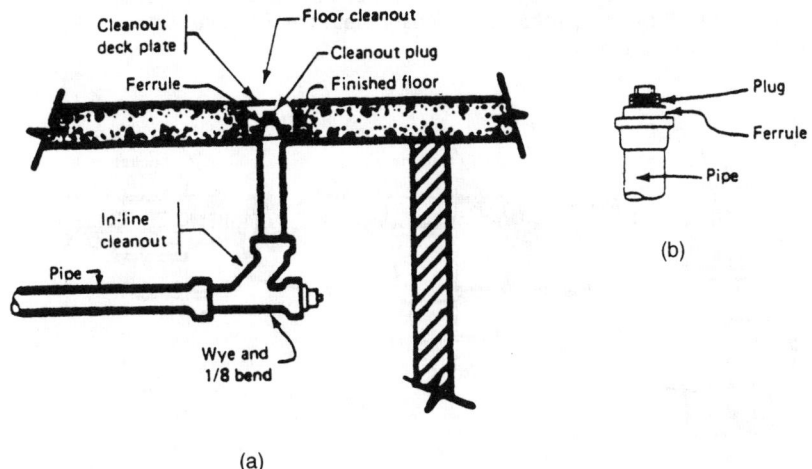

(a)

FIGURE 9.2 (*a*) Typical cleanouts; (*b*) cleanout components. (*Courtesy of Jay R. Smith Co.*)

9.3. It provides a receptacle for spills, washdown, and effluent to be collected and routed directly into the sanitary drainage piping system. Code provisions do not specify where a drain should be located. However, most codes regulate the minimum seal requirements for drain traps, the minimum open area of grates and strainers, and the mandatory inclusion of certain individual components (such as removable secondary strainers or sediment buckets) for drains in some locations. A standard commonly cited in the selection of floor drains is ANSI A112.21-1, Floor Drains. Drains consist of the following components:

1. Drain body
2. Grates located at the top of a drain permit liquid effluent to enter the drain body, while excluding larger solids and foreign matter. Grates are classified as follows:
   a. Light-duty: foot traffic only
   b. Medium-duty: live wheel loads up to 2000 lb (4400 kg)
   c. Heavy-duty: live wheel loads up to 5000 lb (11,000 kg)
   d. Extra-heavy-duty: live wheel loads of more than 5000 lb (11,000 kg)
3. A secondary strainer may be installed below the grate in a drain that does not have a sediment trap.
4. A sediment trap (or bucket) is a removable device inside the drain body that may be installed to trap and retain small solids that pass through the grate.
5. A flashing ring or clamp is a device used to secure flashing directly to the body of the drain.

Interceptors

Plumbing codes require that any substance harmful to the building drainage system, the public sewer, or the municipal sewage treatment process be prevented from being discharged into the public sewer system. Such materials are grease, flammable liquids, sand, and other substances designated by the local authorities. Another reason to provide an interceptor is to recover any precious material or valuable substance that may be lost through discharge into the drainage system. Interceptors are discussed in Chap. 3.

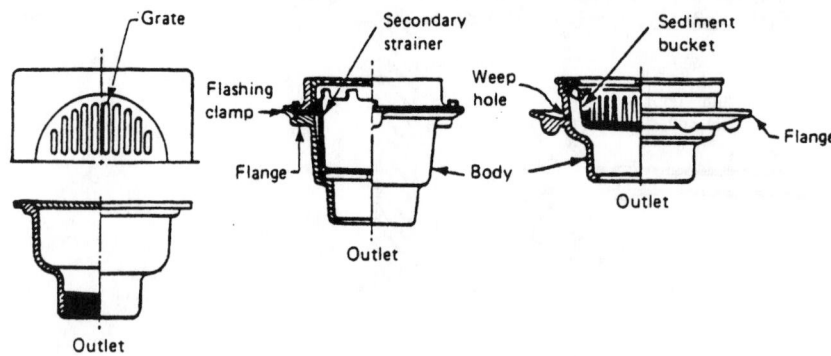

FIGURE 9.3 Typical floor drains.

Traps

A fixture trap (Fig. 9.4) is a U-shaped section of pipe of the necessary depth to retain sufficient liquid as required by code. It prevents odors that originate in the drainage system from being discharged from the fixture and provides a seal to stop vermin passage. All direct connections into the sanitary drainage system are required to have traps.

In general, traps must (1) be self-cleaning, (2) provide a liquid seal of at least 2 in with larger seals where required, (3) conform to local code requirements regarding minimum size, (4) provide an accessible cleanout, and (5) be capable of rapidly draining a fixture. All traps must be vented in some manner, except for specific conditions waived by local code requirements or authorities.

Traps that are prohibited by code include traps requiring moving parts to maintain the seal, full S-type traps (Fig. 9.5), crown vented traps (Fig. 9.6), and drum traps (Fig. 9.7). Drum traps may be permitted by some codes for special use sinks, such as in laboratories. The branch drainage line extending from the trap to the vent is called the trap arm (Fig. 9.8). The maximum length of the trap arm is shown in Table 9.2.

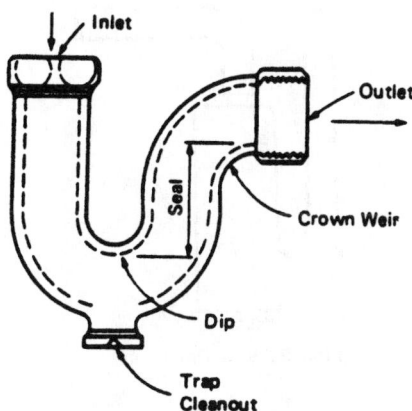

FIGURE 9.4 Typical fixture trap.

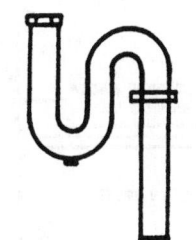

FIGURE 9.5 Full S trap.

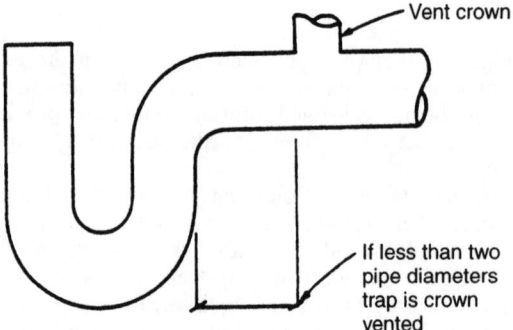

FIGURE 9.6 Crown vented trap.

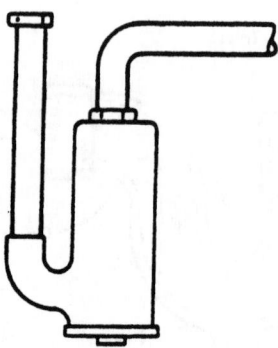

FIGURE 9.7 Drum trap.

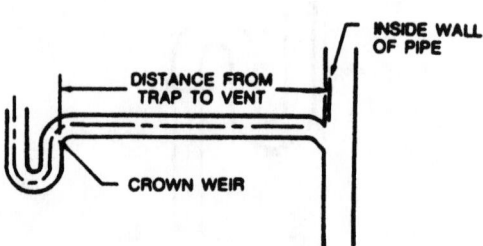

FIGURE 9.8 Detail of trap arm.

TABLE 9.2 Maximum Length of Trap Arm

MM	Diameter of trap arm, in	Distance of trap to vent, ft	M
32	1¼	3½	1.4
40	1½	5	1.5
50	2	8	2.4
75	3	10	3
100	4	12	3.6

SANITARY SYSTEM DESIGN

The design of the gravity drainage piping system is strictly regulated by the applicable plumbing code. All codes include charts similar to those presented here, which permit the design engineer to properly size all horizontal and vertical pipes based on the accumulated fixture unit discharge and slope of the pipe.

The pitch of the drainage system must provide sufficient velocity to produce a scouring action that will convey all solids along with the liquid stream. The recommended minimum velocity for ordinary sewage is 2 fps (0.6 m/s) to prevent the settlement of solids out of the effluent stream. When grease is suspended in the effluent, the velocity should be at least 4 fps (1.2 m/s).

Accepted practice for low-rise buildings with relatively uniform discharge is to size horizontal drainage lines to flow half full under average design conditions. High-rise buildings produce higher velocities and turbulence in building drains that can fill portions of the piping system as much as two-thirds full, with completely full pipes expected for short distances at stack runouts.

The following is a simplified method for sizing the drainage piping system:

1. Establish the location, size, and invert of the point of ultimate disposal of sanitary effluent. Determine if sump or ejector systems will be required and locate them.

2. Locate and lay out branch lines, stacks, and the house sewer.

3. Start with the individual device or fixture at the farthest and most remote point of the system or branch for which the code specifies two drainage values. The first is the drainage fixture unit (DFU) value that will be used to size the drainage piping system. For this value, refer to Table 9.1. The second is a minimum size of the trap, which is the minimum individual branch pipe size. Refer to Table 9.1 for typical values. If a fixture or device is not listed, use either the unlisted value based on the trap size or ask the local code official for the accepted value.

4. The size of the horizontal drainage line is determined by both the pitch of the drainage line and the accumulated total number of DFUs discharging into it. Refer to Table 9.3 for sizes of branch lines. When the drainage line is considered a main building drain or building sewer, refer to Table 9.4 for the size of building drains and sewers.

5. To determine the size of a drainage line based on a given flow in gpm rather than DFUs, refer to Table 9.5. Use the appropriate pitch and velocity combinations necessary to select a size. When there is a combination flow of both DFUs and gpm into a horizontal line or stack, generally accepted practice is to assign 2 DFUs for each gpm in order to use the code charts to determine a size based on DFUs.

TABLE 9.3 Drainage Branches and Stacks

Maximum number of fixture units that may be connected to:

Diameter of pipe, in¶	Any horizontal fixture branch,* DFU	One stack of three branch intervals or less, DFU	Stacks with more than three branch intervals	
			Total for stack,‡ DFU	Total at one branch interval, DFU
1½	3	4	8	2
2	6	10	24	6
2½	12	20	42	9
3	20†	48†	72†	20†
4	160	240	500	90
5	360	540	1100	200
6	620	960	1900	350
8	1400	2200	3600	600
10	2500	3800	5600	1000
12	3900	6000	8400	1500
15	7000			

*Does not include branches of the building drain.

†Not more than 2 water closets or bathroom groups within each branch interval nor more than 6 water closets or bathroom groups on the stack.

‡Stacks shall be sized according to the total accumulated connected load at each story or branch interval and may be reduced in size as this load decreases to a minimum diameter of $\frac{1}{2}$ of the largest size required.

¶See Table 9.1.

TABLE 9.4 Building Drains and Sewers*

Maximum number of fixture units that may be connected to any portion of the building drain or the building sewer

Diameter of pipe, in†	Slope			
	$\frac{1}{16}$ in/ft .5 cm/m	$\frac{1}{8}$ in/ft 1 cm/m	$\frac{1}{4}$ in/ft 2 cm/m	$\frac{1}{2}$ in/ft 4 cm/m
2			21	26
2½			24	31
3			42‡	50‡
4		180	216	250
5		390	480	575
6		700	840	1,000
8	1,400	1,600	1,920	2,300
10	2,500	2,900	3,500	4,200
12	2,900	4,600	5,600	6,700
15	7,000	8,300	10,000	12,000

*On-site sewers that serve more than one building may be sized according to the current standards and specifications of the Administrative Authority for public sewers.

† See Table 9.1.

‡ Not over two water closets or two bathroom groups, except that in single family dwellings, not over three water closets or three bathroom groups may be installed.

TABLE 9.5 Drainage Discharge Rates of Sloping Pipes, gpm

*Flowing half-full**
Discharge rate and velocity‡

Actual inside diameter of pipe, in‡	Slope							
	$\frac{1}{16}$ in/ft .5 cm/m		$\frac{1}{2}$ in/ft 1 cm/m		$\frac{1}{4}$ in/ft 2 cm/m		$\frac{1}{2}$ in/ft 4 cm/m	
	Discharge, gpm¶	Velocity, fps§	Discharge, gpm	Velocity, fps	Discharge, gpm	Velocity, fps	Discharge, gpm	Velocity, fps
1¼							3.40	1.78
1⅜					3.13	1.34	4.44	1.90
1½					3.91	1.42	5.53	2.01
1⅝					4.81	1.50	6.80	2.12
2					8.42	1.72	11.9	2.43
2½			10.8	1.41	15.3	1.99	21.6	2.82
3			17.6	1.59	24.8	2.25	35.1	3.19
4	26.70	1.36	37.8	1.93	53.4	2.73	75.5	3.86
5	48.3	1.58	68.3	2.23	96.6	3.16	137.0	4.47
6	78.5	1.78	111.0	2.52	157.0	3.57	222.0	5.04
8	170.0	2.17	240.0	3.07	340.0	4.34	480.0	6.13
10	308.0	2.52	436.0	3.56	616.0	5.04	872.0	7.12
12	500.0	2.83	707.0	4.01	999.0	5.67	1413.0	8.02

* Half-full means filled to a depth equal to one-half of the inside diameter.
† Computed from the Manning formula of ½-full pipe, $n = 0.015$. For ¼ full: multiply discharge by 0.274; multiply velocity by 0.701. For ¾ full: multiply discharge by 1.82; multiply velocity by 1.13. For full: multiply discharge by 2.00; multiply velocity by 1.50. For smoother pipe: multiply discharge and velocity by 0.015 and divide by n value of smoother pipe.
‡ For conversion see Table 9.1. § 1 gpm = 3.8 L/m. § 1 fps = 0.3 m/s.

6. The size of a stack is governed by the total DFU discharge into it and its height. Refer to Table 9.3 using the applicable column and the total number of DFUs for the stack to find the stack size.

7. To size a stack based solely on gpm, refer to Table 9.6. Two generally accepted recommendations regarding the maximum proportion of cross-section area that may be occupied with water flowing down a stack are ¼ and 7⁄24, depending on

TABLE 9.6 Drainage Stack Capacity, gpm*

Pipe diameter, in†	¼ Full	7⁄24 Full
1¼	5	6.5
1½	8.1	10.5
2	17.5	22.6
2½	31.8	41
3	52.1	67.2
4	111	143
5	202	261
6	336	423
8	709	915

* 1 gpm = 3.8 L/m
† See Table 9.1.

the code used and the requirements of the local authority having jurisdiction. Separate columns are provided for each of these two values. Accepted practice is to use the ¼ full criteria, which closely matches the allowable flow from a horizontal pipe flowing full at ¼ in pitch (2 cm/m).

8. If a stack should offset more than 45° from the vertical, the horizontal offset portion of the stack must be sized as a house drain. If the offset size is larger than that portion of the stack higher than the offset, the larger size must be carried down from the offset to the lower level. The portion of the stack above the offset may remain unchanged.

9. The purpose in differentiating between branch intervals and the actual number of horizontal soil or waste branch lines entering the stack is prevention of stack overloading in a short distance. Many codes limit the number of DFUs allowed in a branch interval.

Flow conditions in the offset portion of a stack create severe turbulence. Because of the resulting pneumatic effects, all branch connections that normally would be made at the level of the offset should be carried down 10 pipe diameters of the stack below the level of the offset.

When ultra low flush (ULF) water closets are required by code, care should be taken not to place these fixtures at the end of a long run with shallow pitch and few other fixtures available to provide sufficient water to create the necessary scouring action within the drainage pipe. Field experience has shown that this condition regularly produces stoppages.

SUDS PRESSURE AREAS

Appliances and fixtures normally using detergents, such as kitchen sinks, bathtubs, showers, dishwashers, and clothes washers, can discharge a large quantity of detergents into the drainage system. When flowing through drainage piping, turbulence causes large amounts of suds to be generated. The suds accumulate in the lower portions of the drainage system, and can remain there for a considerable period of time. When additional liquids flow into these sections of the system, the suds are displaced and will follow the path of least resistance. Enough suds pressure can be built up to force the suds through a fixture trap. Suds pressure areas exist in the following parts of the drainage system, as illustrated in Fig. 9.9:

1. For an upper level stack offset serving fixtures on two or more floors above the offset, there are two suds areas. The first area (Fig. 9.9a) extends 40 pipe diameters of the stack upwards from the base of the offset. The second (Fig. 9.9b) extends 10 pipe diameters horizontally downstream from the point of change in direction.

2. For an upper level stack offset turning from horizontal back to vertical, there is one area (Fig. 9.9c) extending 40 pipe diameters of the stack upstream from the fitting changing direction from horizontal to vertical.

3. In the horizontal runout from a stack when the pipe changes direction horizontally with a fitting greater than 45°, there are two areas. The first (Fig. 9.9d) extends 40 pipe diameters of the horizontal pipe upstream from the change in direction. The second (Fig. 9.9e) is 10 pipe diameters downstream.

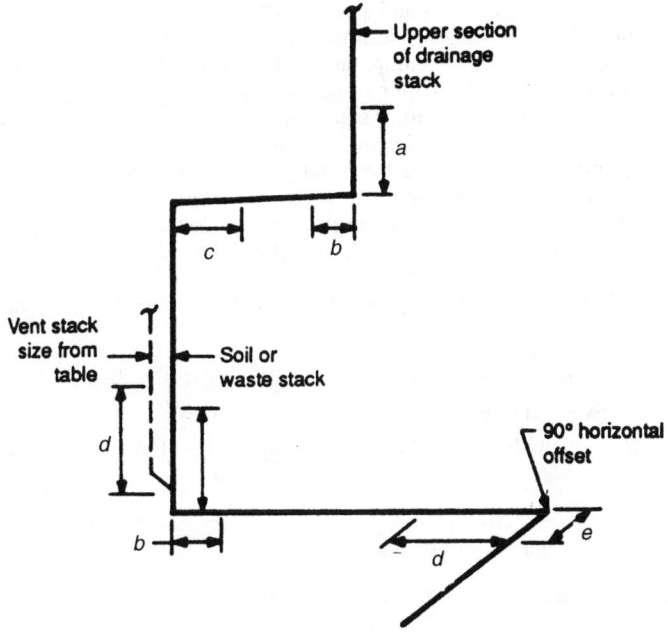

FIGURE 9.9 Suds pressure zones.

When suds pressure is anticipated, no pipe shall connect to any of the areas indicated as *a* to *e* in Fig. 9.9. Refer to Table 9.7 for actual distances based on pipe size.

SUMP PUMPS AND SEWAGE EJECTORS

When liquid waste must be removed from a level below the ultimate point of disposal, sump pumps or sewage ejectors are used.

TABLE 9.7 Length of Pipe for Suds Pressure Connection

Pipe size, in*	40 diameters, ft†	10 diameters, ft
1	5	1½
2	7	1½
2	8	2
3	10	2½
4	13	3½
5	17	4
6	20	5

* See Table 9.1.
† 1 ft = 0.3 m

The distinction between *sump* and *ejector* systems is mainly one of terminology, with the main difference being in the impeller. The impeller of the ejector pump is designed to pass solids. Sump pumps are primarily designed to transport turbid, nonsanitary waste water with smaller suspended solids. Ejector pumps are designed to transport sanitary waste with larger solids suspended in the effluent. Sump pumps designed to transport large quantities of nonsanitary water with larger suspended solids are sometimes referred to as *trash pumps*. The components of these systems are motors (drivers), impellers (pumps), basins (receivers), and controls.

Because pump failure could result in flooding the lower levels, it is highly recommended that two pumps be installed, each sized for full load and connected in parallel. This is referred to as duplex installation.

There are three systems of pumps generally used. These systems are categorized as follows:

1. Submersible pumps in a receiver (for a detail refer to Fig. 9.10)
2. Wet-pit receiver with vertical lift pumps (for a detail refer to Fig. 9.11)
3. Wet-pit receiver with cantilever, self-priming pumps (for a detail refer to Fig. 9.12)

Each system has specific advantages and disadvantages that must be evaluated based on the following major considerations:

1. Range of head and capacity for motor and impeller combinations
2. Initial cost
3. Floor space requirements
4. Materials available for pump and bearing construction
5. Ability to pump hot liquids

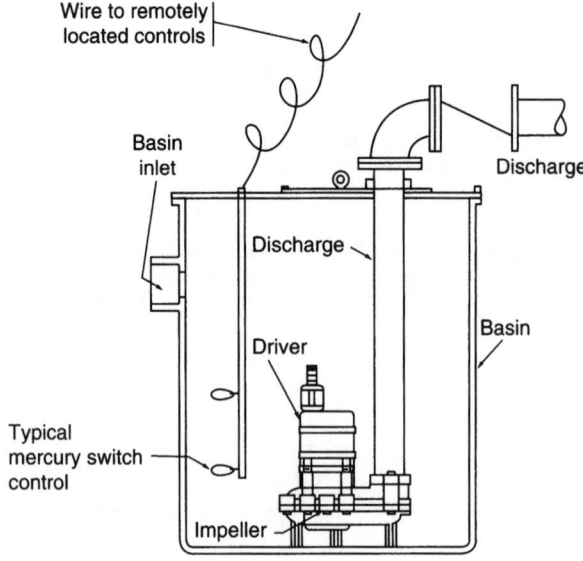

FIGURE 9.10   Submersible pump system.

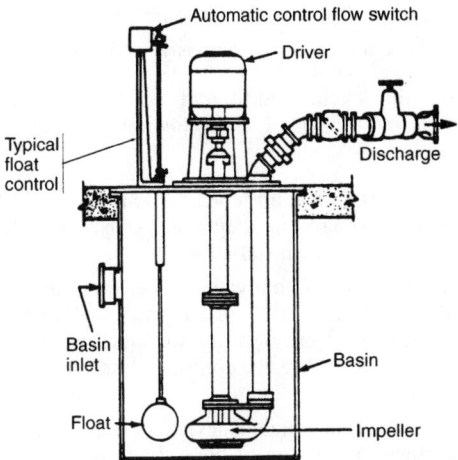

FIGURE 9.11   Vertical lift submerged pump system.

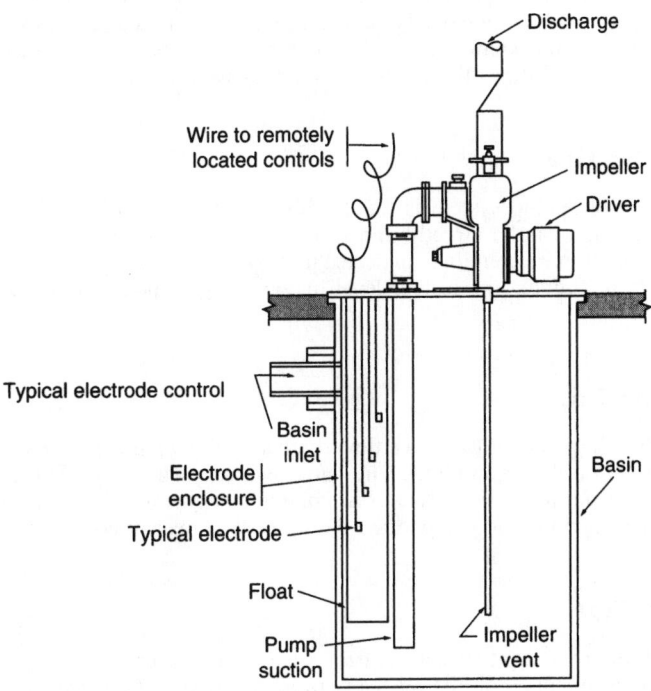

FIGURE 9.12   Cantilever type system.

6. Headroom required to remove pump and impeller

7. Ability to pump highly abrasive and corrosive effluent

Control of the operating cycle used to start and stop the pumps is based on the level of water in the basin. Alarms of the following types are also included in the control panel.

1. *Float.* This type uses a rod with an attached float on the surface of the effluent level, with a set point on the rod operating an exposed switch on the basin cover. Refer to Fig. 9.11 for a typical installation.

2. *Mercury switch.* This is a sealed, buoyant device containing mercury and integral contacts, attached at one end only and suspended at various levels inside the basin. As the effluent level rises and falls, so does the orientation of this device. The mercury then makes or breaks contact. Refer to Fig. 9.10 for a typical installation.

3. *Pneumatic.* This alarm is an inflated device suspended below the level of the effluent. The difference in pressure of the effluent on the inflated device is used to make or break remotely located contacts.

4. *Electrode type.* This alarm uses separate electrodes suspended in the effluent. As the level of effluent rises and falls, different sets of electrodes are immersed and send a signal to remotely located controls. Refer to Fig. 9.12 for a typical installation.

5. It is good practice to have one of these on a separate circuit. In addition, another alarm that is highly recommended is a *pump failure to start alarm light* for duplex installations. In many cases, if the first pump does not function, the second pump will automatically start and satisfy system requirements. Operating personnel will not know that the first pump has broken until the second pump breaks down and the entire pumping system does not function.

Submersible Type

The submersible pump system is totally submerged in the effluent within the receiver. The pump can be directly joined to the discharge pipe or indirectly joined by using a proprietary coupling connected to it by gravity, thus permitting the entire assembly to be brought up to floor level for servicing without personnel having to enter the basin.

Vertical Lift Type

The vertical lift system utilizes a vertical shaft centrifugal pump and separate driver, both mounted on the basin cover. The pump is immersed in the effluent and supported from under the cover by a special column that encloses the connecting shaft. The motor is independently supported on top of the cover directly over the pump.

Cantilever Type

The cantilever, or wet-pit system, utilizes a horizontal centrifugal pump and close-coupled driver mounted on the receiver cover. A suction pipe is extended (or cantilevered) from the pump down into the receiver, so that the effluent can be lifted up to the level of the impeller intake.

Pump and Receiver Design

Pump head is calculated by adding the static height from the bottom of the basin to a level 1 ft above the proposed highest point of discharge and the friction loss of effluent through the pump discharge piping system. The flow shall be calculated with both pumps running in a duplex installation.

The capacity of the pump is a function of the basin size. Generally accepted practice is to have a pump run from 1 to 5 min, with an optimum of six starts per hour. This is to avoid premature failure due to short cycling. If these conditions cannot be met, the least amount of starts should be used. This consideration shall be balanced against the cost of a basin sized to give fewer starts per hour.

To find the basin size, the space available for the basin must be considered, as well as the permissible depth. For duplex pumps, a diameter (or side) of 4 ft (1.2 m) is reasonable for most installations. Often, the basin is at the lowest area in the building, so depth is a factor. Refer to Table 9.8 for the capacity of various sized round and square basins.

To avoid overflow, the pump capacity should be capable of discharging the maximum possible inflow. This might result from a planned maintenance operation or the flow from the system itself. The flow from fixtures discharging into the basin is calculated from the gpm of the water supply (using WFUs from Tables 9.1 and 9.20 to convert to gpm), plus 10 percent. Known discharge might be condensate from HVAC or other equipment. A leakage rate from pumps of 2 to 5 gpm (8-20 l), depending on the size of the project, is reasonable.

To find the basin depth, use the following approximate dimensions as a guide, assuming duplex pumps and starting from the invert of the inlet pipe.

TABLE 9.8 Capacity of Sump and Ejector Basins, Gallons per Foot Depth

Circular		Square	
Diameter, ft	Gallons	Side, ft	Gallons
2	23.50	2	30.00
2½	36.72	2½	45.00
3	52.88	3	67.50
3½	71.91	3½	90.00
4	94.00	4	120.00
4½	110.32	4½	149.60
5	146.89	5	187.00
6	158.64	6	270.00
7	170.00	7	365.50
8	181.00		
9	193.00		
10	204.00		

1 ft = 0.3 m.
1 gal = 3.8 L.

1. From the invert of the inlet pipe, allow 6 in to the high water alarm.

2. From the high water alarm, allow 6 in to pump 2 start.

3. From pump 2 start, allow 6 in to pump 1 start.

4. Below pump 1 start, the dimension of the liquid capacity depends on a 1- to 5-min operating period of the selected pump. The lower level of the storage portion is pump(s) stop. Refer to Table 9.8 for storage capacity.

5. Allow 6 in from pump stop to the inlet of the pump.

6. Allow 1 ft to the basin bottom from the inlet of the pump. This dimension varies between various manufacturers.

Sewage and waste effluent has the same hydraulic characteristics as water; therefore, the pump discharge piping should be sized using the same criteria as for potable water. Ejector discharge lines should be a minimum of 3 in and sump discharge lines should be a minimum of 2 in (DN 50) to prevent stoppages. Pipe sizing should be made larger (if cost effective) to lessen the friction loss in the piping system, if this would result in a reduction of the motor size.

HOUSE SEWER DESIGN

The house sewer is that portion of the sanitary drainage system outside the building, generally extending from the building wall to the ultimate point of disposal. This point is either a public sewer, private treatment facility (such as a septic tank and field), or small sewage treatment plant discharging into the environment. A public sewer must be used if it is available. The definition of what is considered available varies greatly, depending on local codes and regulatory agencies.

There may be additional regulatory agencies, other than the plumbing code officials, that have jurisdiction over the design and installation of the building sewer. Consult with the local code officials to find the various agencies from which approval must be obtained, and their requirements for approval. Design criteria for house sewers are given in Chap. 6.

PRIVATE SEWAGE DISPOSAL SYSTEMS

INTRODUCTION

This sub-section will discuss private sewage disposal systems. They are required if a public sewer is not considered available either by circumstance or by the authorities having jurisdiction. A common definition of "available" could mean that a public sewer is located more than 500 ft (150 m) from a building wall.

CODES AND STANDARDS

1. Local and state codes and standards for private sewage disposal systems
2. International Private Sewage Disposal Code

SEPTIC SYSTEM

General

The collection and disposal of raw sanitary effluent where there are no sewers is by means of a private sewage disposal system. The most commonly used private method is a septic system consisting of a septic tank and a gravity flow soil absorption system illustrated in Fig. 9.13.

Most areas of the country have local regulations regarding the design and installation of the private sewage disposal system that must be followed. The examples given here are only general guidelines and illustrations of the criteria and methods to be used and, therefore, should not be used for actual project design purposes.

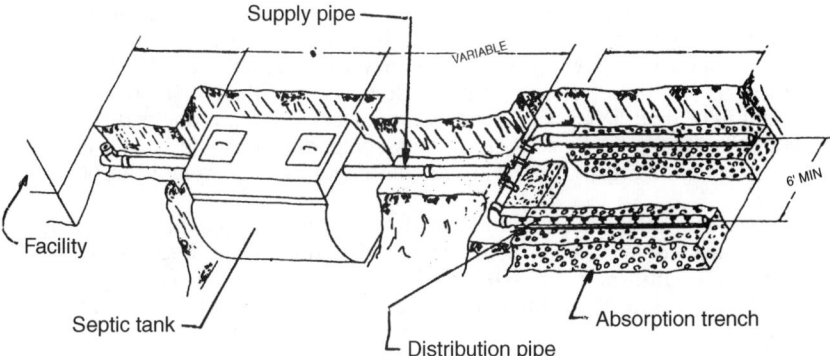

FIGURE 9.13 Typical septic system diagram.

System Description

The septic system receives raw sewage and waste from a facility or private dwelling, provides a method for decomposition of most solids into liquid and disposes of the liquid by absorption into the soil adjacent to the facility.

Where food-related areas are located within a facility, a grease trap is required to keep the food-related grease from entering the septic tank. This does not include kitchens in private residences, although separate disposal of grease from this source is encouraged.

Primary Collection and Treatment System

Septic Tanks.   A septic tank is a large storage tank that protects the soil absorption system by collecting raw, untreated sewage effluent, decomposing fecal and other organic matter and separating other solids from the effluent stream. If this were not accomplished, the soil absorption system would quickly clog up and fail. The discharge from a properly designed septic tank is mostly liquid. A septic tank functions as follows.

1. Heavier suspended solids and organic (fecal) matter settle to the bottom of the tank as sludge. This accumulated bottom sludge undergoes a bioseptic process where bacterial digestion changes most of the particles into either a gas or liquid.

2. Lighter particles float to the top to collect there as a surface mat. An interchange of particles takes place between the top mat and bottom sludge, allowing the digesting bacteria to continuously rise and settle, acting on both the top and bottom areas of the tank. Sludge is defined as an accumulation of solids at the bottom of the septic tank. Scum is a partially submerged mat of floating solids that forms at the surface of the fluid in the septic tank.

3. The remaining suspended organic particles, mostly colloidal in nature, are subjected to a biological decomposition process transforming them into liquid particles. This permits leaching into the soils without any clogging effect. The bacteria present in the liquid area of a variety called anaerobic because they thrive in the absence of free oxygen. This decomposition, or treatment, of the sewage under anaerobic conditions is termed septic. Hence, the name of the system.

A single compartment septic tank will provide acceptable performance. The available data indicate, however, that a two-compartment septic tank (with the first compartment equal to one-half to two-thirds of the total volume) will provide a better suspended-solids removal, and this may be especially valuable for the protection of the soil absorption system. Septic tanks with three or more equal compartments will give at least as good performances as single compartment septic tanks of the same total capacity. Each tank compartment should have a minimum plan dimension of 2 ft (0.6 m) with a liquid depth ranging from 30 to 60 in (0.8 to 1.5 m).

An access manhole should be provided to each tank compartment. Venting between the tank compartments should be provided to allow the free passage of gas. Inlet and outlet fittings in the septic tank should be proportioned (as for a single tank). The same allowance should be made for storage above the liquid line (as in a single tank).

A vented inlet tee or baffle should be provided to divert the incoming sewage downward. This device should penetrate at least 6 in (150 mm) below the liquid level of the septic tank, but in no case should the penetration be greater than that allowed for the outlet device.

It is important that the outlet piping arrangement penetrate just far enough below the liquid level of the septic tank to provide balance between the sludge and scum storage volume; otherwise, part of the capacity is lost. A vertical section of a properly operating septic tank would show it divided into three distinct layers: scum at the top, a middle zone free of solids (called clear space) and a bottom layer of sludge. Observations of the sludge accumulations in the field indicate that the outlet device should be extended to a distance below the surface equal to 40 percent of the liquid depth. The outlet device should be extended above the liquid line to approximately 1 in (25 mm) from the top of the septic tank. The space between the top of the septic tank and the baffle will allow any gas to pass through the septic tank into the house vent.

For private dwellings, the septic tank shall be sized on the number of bedrooms. In the absence of code requirements, use Table 9.9 as a guide. For other facilities, the starting point shall be a 750 gal (2850 L) tank to which an additional capacity is added. This additional capacity shall be as shown in Table 9.10. A typical septic tank is illustrated in Fig. 9.14.

Soil Absorption System. The soil absorption system receives the liquid discharge from the septic tank and, by means of supply piping, distributes the liquid into excavated trenches called absorption trenches. These trenches are designed and sized to adequately absorb the discharged liquid into the earth. The piping in the absorption trenches is called the distribution piping.

The successful operation of a subsurface waste disposal system requires a comprehensive site evaluation, good design criteria and careful installation. When evaluating a site the engineer must consider the following: lot size, soil, composition, slope, topography, surface water and the seasonal high-water table. Residential lost must be large enough to accommodate the projected area of seepage beds while maintaining minimum clearances from surface waters and wells, as mandated by local codes. In their absence, use Table 9.11 as a guide for residential properties.

For multiple dwellings, institutional, industrial and other types of facilities, it is common for codes to have separate formulas and absorption areas to allow for varying minimum daily flow for each type of facility. For a comprehensive listing of daily sewage flows from facilities other than private dwellings, refer to Table 6.25.

TABLE 9.9 Septic Tank Capacity for One- and Two-Family Dwellings

Number of bedrooms	Septic tank (gallons)	Liters
1	750	3000
2	750	3000
3	1,000	4000
4	1,200	4800
5	1,425	5700
6	1,650	6600
7	1,875	7500
8	2,100	8400

TABLE 9.10 Additional Capacity for Other Facilities

Building classification	Capacity (gallons)*
Apartment buildings (per bedroom—includes automatic clothes washer)	150
Assembly halls (per person—no kitchen)	2
Bars and cocktail lounges (per patron space)	9
Beauty salons (per station—includes customers)	140
Bowling centers (per lane)	125
Bowling centers with bar (per lane)	225
Campgrounds and camping resorts (per camp space)	100
Campground sanitary dump stations (per camp space) (omit camp spaces with sewer connection)	5
Camps, day use only—no meals served (per person)	15
Camps, day and night (per person)	40
Car washes (per car handwash)	50
Catch basins—garages, service stations, etc. (per basin)	100
Catch basins—truck washing (per truck)	100
Churches—no kitchen (per person)	3
Churches—with kitchen (per person)	7.5
Condominiums (per bedroom—includes automatic clothes washer)	150
Dance halls (per person)	3
Dining halls—kitchen and toilet waste—with dishwasher, food waste grinder or both (per meal served)	11
Dining halls—kitchen waste only (per meal served)	3
Drive-in restaurants—all paper service (per car space)	15
Drive-in restaurants—all paper service, inside seating (per seat)	15
Drive-in theaters (per car space)	5
Employees—in all buildings, per employee—total all shifts	20
Floor drains (per drain)	50
Hospitals (per bed space)	200
Hotels or motels and tourist rooming houses	100
Labor camps, central bathhouses (per employee)	30
Medical office buildings, clinics and dental offices	
Doctors, nurses, medical staff (per person)	75
Office personnel (per person)	20
Patients (per person)	10
Mobile home parks, homes with bathroom groups (per site)	300
Nursing and rest homes—without laundry (per bed space)	100
Outdoor sports facilities (toilet waste only—per person)	5
Parks, toilet wastes (per person—75 persons per acre)	5
Parks, with showers and toilet wastes (per person—75 persons per acre)	10
Restaurants—kitchen waste only—without dishwasher and food waste grinder (per seat)	9
Restaurants—toilet waste only (per seat)	21
Restraurants—kitchen and toilet wastes (per seating space)	30
Restaurants (24-hour)—kitchen and toilet wastes (per seating space)	60
Restaurants—dishwasher or food waste grinder or both (per seat)	3
Restaurants (24-hour)—dishwasher or food waste grinder (per seat)	6
Retail stores—customers	1.5
Schools (per classroom—25 pupils per classroom)	450
Schools with meals served (per classroom—25 pupils per classroom)	600
Schools with meals served and showers provided (per classroom)	750
Self-service laundries (toilet waste only, per machine)	50
Automatic clothes wasters (apartments, service buildings, etc.—per machine)	300
Service stations (per car)	10
Showers—public (per shower taken)	15
Swimming pool bathhouses (per person)	10

*1 gal = 3.785 L.

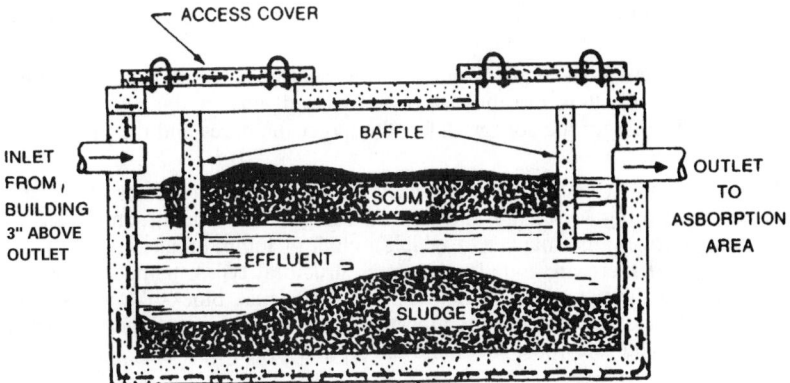

FIGURE 9.14   Typical septic tank.

TABLE 9.11   Distance from Objects to Soil Absorption Laterals

	Distance	
Object	feet	m
Cistern	50	15
Habitable building, below-grade foundation	25	7.6
Habitable building, slab-on-grade	15	4.5
Lake, high-water mark	50	15
Lot line	5	1.5
Reservoir	50	15
Roadway ditches	10	3.0
Spring	100	30
Streams or watercourse	50	15
Swimming pool	15	4.5
Uninhabited building	10	3
Water main	50	15
Water service	10	3
Water well	50	15

If gravity collection and/or distribution are not possible, pumped systems are used. When the area surrounding any facility is not suitable for excavated soil absorption methods, a mound system shall be provided.

Guide for Estimating Soil Absorption Potential.   A percolation test is the only known means for obtaining a quantitative appraisal of the soil absorption capacity. However, observation and evaluation of soil characteristics can provide very useful clues to the relative capacity of a soil to absorb a liquid. Most suitable and unsuitable solids can be identified without additional testing.

The following information is required in order to make a full evaluation of the site.

1. *Soil Maps.* Soil survey reports, soil categories and maps are published by the U.S. Department of Agriculture.

2. *Clues to Absorption Capacity.* This is obtained by a close visual inspection of the soil.

3. *Texture.* Texture is generally defined as the relative proportion of sand, slit and clay in the soil. It is the most commonly used clue to the water absorption capacity. The larger the soil particles, the larger the pores and the faster the rate of absorption.

4. *Structure.* Soil structure is characterized by the grouping together of the textual particles, forming secondary particles of a larger size. The structure can easily be recognized by the manner in which a clod, or lump, breaks apart.

In general there are four fundamental structural types, named according to the shape of the aggregate particles: platy, prism-like, block-like and spherical. A soil without structure is generally referred to as massive. Spherical structure tends to provide the most favorable absorption properties, and platy structure, the least.

5. *Color.* If the soil has a uniform reddish-brown to yellow oxidized color, this indicates that there has been a free movement of the air and water in and through the soil. Such a soil has some desirable absorption characteristics. At the other extreme, soils that have a dull-gray or mottled coloring, indicating a lack of oxidizing conditions or a very restricted movement of the air and the water, have poor absorption characteristics.

6. *Depth (or thickness) of Permeable Strat.* The quantity of water that may be absorbed by the soil is directly proportional to the thickness (or volume) of the absorbent stratum, when all other conditions are alike.

7. *Swelling Characteristics.* Soils that shrink appreciably when dry are soils that may given trouble in a distribution field when they are wet.

Procedure for Percolation Tests

Number and Location of Tests. Six or more tests should be made in separate test holes, spaced uniformly over the proposed absorption field site.

Type of Test Hole. Dig or bore a hole, with horizontal dimensions or diameter varying from 4 to 12 in (101.6 to 304.8 mm) and vertical sides to the depth of the proposed absorption trench. To save time, it is common practice to have the holes bored with a 4-in (101.6 mm) auger.

Preparation of Test Hole. Carefully scratch the bottom and the sides of the hole with a knife blade or sharp-pointed instrument in order to remove any smeared soil surfaces and to provide a natural soil interface into which water may percolate. Remove all loose material from the hole. Add 2 in (50 mm) of coarse sand or fine gravel to protect the bottom from scouring and sediment.

There are several types (procedures) of percolation tests that are acceptable to most local administrative authorities. However, in this sub-chapter the procedure for percolation tests was developed at the Robert A. Taft Co.

Saturation and Swelling of the Soil. It is important to distinguish between saturation and swelling. Saturation means that the void spaces between soil particles are full of water. This can be accomplished in a short period of time. Swelling is caused by intrusion of water into the individual soil particle. This is a slow process, especially in clay-type soil, and is the reason for requiring a prolonged soaking period.

In the conduct of the test, carefully fill the hole with clear water to a minimum depth of 12 in (304.8 mm) over the gravel. In most soils, it is necessary to refill the hole by supplying a surplus reservoir of water, possibly by means of an automatic siphon, to keep water in the hole for at least 4 hours and preferably overnight. Determine the percolation rate 24 hours after water is first added to the hole. This procedure is to insure that the soil is given ample opportunity to swell and to approach the condition it will be during the wettest season of the year. Thus, the test will give comparable results in the same soil, whether made in a dry or in a wet season. In sandy soils containing little or no clay, the swelling procedures are not essential and the test may be made as previously described, after the water from one filling if the hole has completely seeped away.

Percolation Rate Measurement. With the exception of sandy soils, the percolation rate measurements should be made on the day following the procedure described under saturation and swelling of the soil.

1. If the water remains in the test hole after the overnight swelling period, adjust and depth to approximately 6 in (152 mm) over the gravel. From a fixed reference point, measure the drop in the water level over a 30-minute period. This drop is used to calculate the percolation rate.

2. If no water remains in the hole after the overnight swelling period, add clear water in order to bring the depth of the water in the hole to approximately 6 in (152 mm) over the gravel. From a fixed reference point, measure the drop in the water level at approximately 30-minute intervals for 4 hours, refilling 6 in (152 mm) over the gravel as necessary. The drop that occurs during the final 30-minute period is used to calculate the percolation rate of the soil. The lowering of liquid level during the prior periods provide the information for a possible modification of the procedure in order to suit any local circumstances.

3. In sandy soils (or other types of soils in which the first 6 in (152 mm) of water seep away in less than 30 minutes after the overnight swelling period), the time interval between the measurements should be taken as 10 minutes and the test should be run one hour. The drop that occurs during the final 10 minutes is used to calculate the percolation rate.

Soil Absorption System Design

For areas where the percolation rates and the soil characteristics are good, the next step after making the percolation tests is to determine the required area of seepage trenches that will be satisfactory.

When a soil absorption system of seepage trenches is determined to be unusable, alternative types of systems should be considered by the engineer such as absorption trenches, seepage beds and seepage pits.

Absorption Trench Design Criteria. A leaching field is a number of absorption trenches called laterals, each containing distribution piping consisting of a length (or lengths) of 4-in (DN 100) agriculture drain tile of vitrified clay sewer pipe or perforated PVC pipe installed in absorption trenches. In areas having unusual soil or water characteristics, local experience should be reviewed before selecting the distribution piping materials. The individual laterals should not be over 100 ft (30 m) long, and the trench bottom and title distributing piping should be level. Use of more and shorter laterals is preferred because if something should happen to disturb one line, most of the field will still be serviceable. Many different types of engineering designs may be used in laying out the subsurface disposal fields. Typical layouts of absorption trenches are shown in Figs. 9.15 and 9.16, and a cross-section is illustrated in Fig. 9.17.

In considering the depth of the absorption field trenches, it is possible for the tile lines to freeze during a prolonged cold period. Freezing rarely occurs in a carefully constructed system kept in continuous operation. It is important during construction of the system to assure that the tile lines are surrounded by gravel.

The required absorption area of a trench is considered the bottom area only. These figures include a statistical allowance for absorption by the sidewall area of the trench. Trenches shall be designed for two conditions. One is for private dwellings and the second is for other types of facilities. Both are based on the results of the soil percolation test and the flow into the leaching field from bedrooms or people. The area requirements per bedroom shall be obtained from Table 9.13. The quantity of sewage from other sources shall be obtained from Table 6.25.

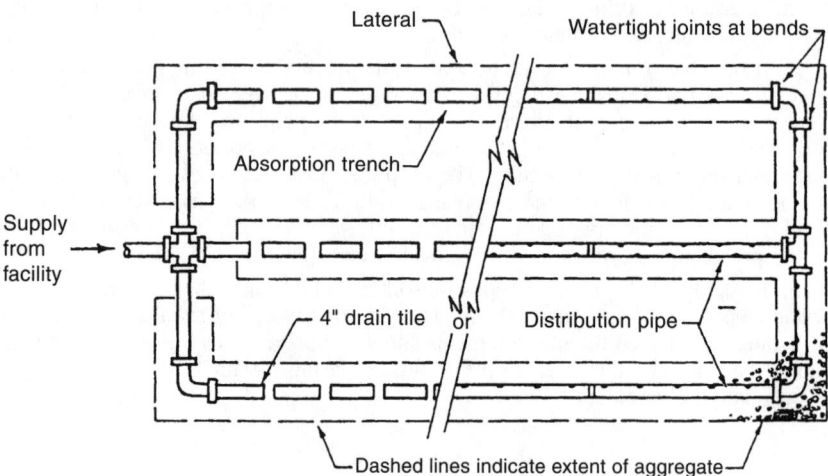

FIGURE 9.15 Distribution piping connected by a cross fititng.

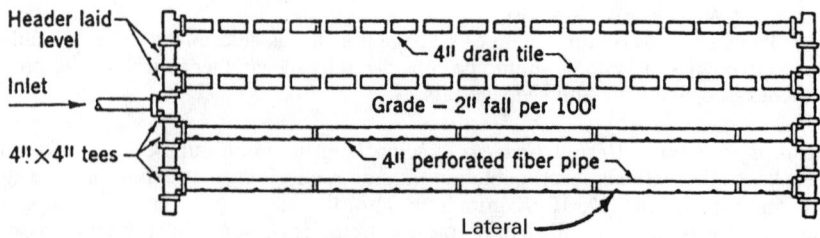

FIGURE 9.16 Distribution piping connected by a header.

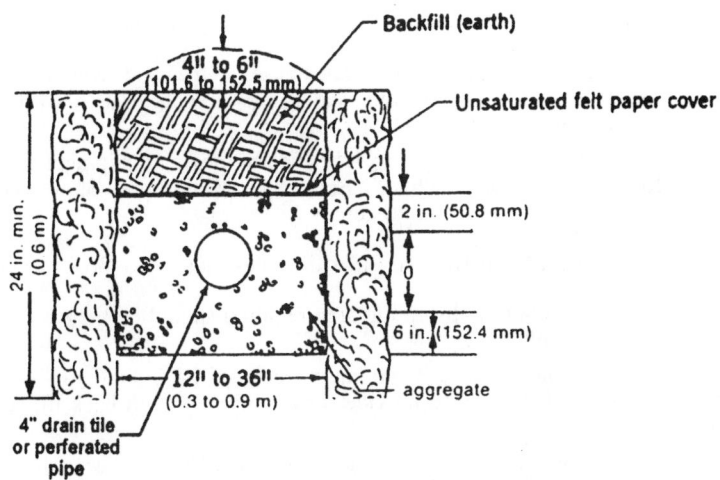

FIGURE 9.17 Cross section through typical absorption trench.

Trench Installation. The pipe trench shall be of sufficient width and depth and surrounded by clean, graded gravel. In order to provide the minimum required gravel depth and earth cover, the depth to the bottom of the absorption trenches should be at least 24 in (0.6 m). Additional depth may be needed for contour adjustment, extra aggregate under the tile or other design purposes. The aggregate may range in size form ¾ to 1.5 in (19.1 to 38.1 mm). Cinders, broken shell and similar materials are not recommended because they are usually too fine and may lead to premature clogging. The material should extend from at least 2 in (50.8 mm) above the top of the pipe to at least 6 in (152.4 mm) below the bottom of the pipe. If tile is used, the upper half of the joint openings should be covered. The top of the gravel should be covered with entreated building paper, a 2-in (50.8 mm) layer of hay or straw or a similar material. An impervious covering should not be used because it will interfere with the evapotranspiration at the surface. Although generally not figured in the engineering calculation is, evapotranspiration is often an important factor in the operation of horizontal absorption systems.

The top of a new absorption trench should be hand tamped and with 4 to 6 in (100 to 152 mm) of earth. Unless this is done, the top of the trench may settle to a point lower than the ground. This condition will cause the collection of storm water in the trench, and this can lead to the premature saturation of the absorption field and, possibly, to a complete washout of the trench. Machine tamping or hydraulic back filling of the trench should not be used. Caution shall be used to avoid root intrusion when constructing leaching fields near trees.

Seepage Beds. Absorption systems having trenches wider than 3 ft (0.9 m) are called seepage beds. Variations of design practices utilizing an increased width are being used in many areas with the approval of the local administration authority. The design of the trenches is based on an empirical relationship between the percolation test and the bottom area of the trenches.

Studies have demonstrated that a seepage bed is a satisfactory means for the disposal of the effluent in soils that are acceptable for soil absorption systems.

The use of a seepage bed results in the following advantages.

1. Wide seepage beds make more efficient use of the land available for the absorption system than a series of long and narrow trenches (with wasted land between these trenches).

2. Efficient use may be made by a variety of earth moving equipment employed at projects for other purposes, such as basement excavation and landscaping, resulting in savings on the cost of the system.

Design of the seepage bed shall consider the following:

1. The amount of the bottom absorption area required should be the same as that recommended in Table 9.12.

2. The seepage bed should have a minimum depth of 24 in (0.6 m) below the natural ground level in order to provide for a minimum earth back fill cover of 12 in (0.3 m).

3. The seepage bed should have a minimum depth of 12 in (0.3 m) of rockfall or packing material extending at least 2 in (50 mm) above and 6 in (150 mm) below the distribution pipe.

4. The bottom of the seepage bed and distribution tile (or perforated pipe) should be level.

5. The lines for distributing the effluent should be spaced less than 6 ft (1.8 m) apart and less than 3 ft (0.9 m) from the seepage bed sidewall.

6. When more than one seepage bed is used, there should be a minimum of 6 ft (1.8 m) of undisturbed earth between the adjacent seepage beds, and the seepage beds should be connected in series.

Seepage Pits. A seepage pit is a pit that receives the discharge of a septic tank and is commonly used either to supplement the subsurface disposal system or in lieu of a system where the conditions favor their operation. It is constructed with an open bottom and sides of perforated concrete, brick or other material that allows effluent to percolate out into the surrounding soil. It is sized using Table 9.12 to determine the required square ft (m²) area required and Table 9.13 to determine the size of the pit.

A cesspool is a pit that receives raw sewage from a facility that allows effluent to percolate out into the surrounding soil. It is constructed in a similar manner as a seepage pit. A cesspool is generally not permitted to be used.

Distribution Boxes

Distribution boxes are not recommended and are harmful to the systems for the following reasons:

1. Data indicates that, on level ground, equal distribution is not necessary if the system is designed so that an overloaded trench can drain back to the other trenches before a failure occurs.

2. On slopping ground, a method of distributions is needed to prevent an excessive build-up of head and the failure of any one trench before the capacity of the entire system is reached. It is doubtful that distribution boxes can provide an equal distribution.

TABLE 9.12 Allowable Area for Absorption Trenches

Percolation rate (time in min for water to fall 1 in. [25.4 mm])	Maximum rate of sewage application for absorption trenches,[a] seepage beds, and seepage pits[b], g/ft²/d (L/m²/d)[c]	Percolation rate (time in min for water to fall 1 in. [25.4 mm])	Maximum rate of sewage application for absorption trenches,[a] seepage beds, and seepage pits[b], g/ft²/d (L/m²/d)[c]
1 or less	5.0 (244.3)	10	1.6 (78.2)
2	3.5 (171.0)	15	1.3 (63.5)
3	2.9 (141.7)	30[d]	0.9 (44.0)
4	2.5 (122.2)	45[d]	0.8 (39.1)
5	2.2 (107.5)	60[d,e]	0.6 (29.3)

[a] Absorption area is figured as trench bottom area and includes a statistical allowance for vertical sidewall area.
[b] Absorption area for seepage pits is effective sidewall area.
[c] Not including effluents from septic tanks that receive wastes from garbage grinders and automatic washing machines.
[d] More than 30 is unsuitable for seepage pits.
[e] More than 60 is unsuitable for absorption systems.

9.31

TABLE 9.13 Square Feet Requirements for Seepage Pits

Inside diameter of chamber in feet plus 1 foot for wall thickness plus 1 foot for annual space	Depth in feet of perrmeable strata below inlet					
	3	4	5	6	7	8
7	47	88	110	132	154	176
8	75	101	126	151	176	201
9	85	113	142	170	198	226
10	94	126	157	188	220	251
11	104	138	173	208	242	277
13	123	163	204	245	286	327

For SI: 1 ft = 304.8 mm, 1 square ft = 0.0929 m².

Serial Distribution

Serial distribution is achieved by arranging the individual trenches of the absorption system so that each trench is forced to pond to the full depth of the gravel fill before the liquid flows into the succeeding trench. It minimizes the importance of variable absorption rates by forcing each trench to absorb the effluent until its complete capacity is utilized. The variability of the soils, even in the small area of an individual absorption field, raises a doubt about the desirability of uniform distribution.

Mound Systems

The mound system is used when there is not enough depth of naturally occurring soil over an impermeable formation or where high ground water exists. This usually requires the effluent to be pumped from the septic tank discharge to the soil absorption bed at a higher elevation.

A mound type soil absorption system is a last resort and should be used only when there is not enough naturally occurring soil to provide the required soil absorption bed area on a site. If the mound system can not be installed, the site will often be considered unbuildable. The following parameters should, in general, be considered unsuitable for a mound system:

1. Over an existing, failing soil absorption system
2. Where less than 2 ft (610 mm) of unsaturated soil is available over crevised or porous bedrock
3. Where less than 2 ft (610 mm) of unsaturated soil is available over high ground water
4. Over previously compacted soil
5. A percolation rate of greater than 120 in per min (4.7 mm/min)

Local codes must be used for all design and installation requirements.

Estimates of Sewage Quantities

Normally, disposal systems are designed prior to the actual construction of a facility. To estimate the quantity of sewage, the first consideration must be given to local code requirements. These requirements usually result in design flows considerably above average.

If there are no code requirements, the most reliable criteria would be readings from water meters at similar existing buildings. This can usually be obtained from the water utility company. It is recommended that a safety factor of 10% be added to the quantity of water obtained to account for any additional unknown requirements.

Where codes and actual usage data are not available, it is necessary to use other methods of estimating the amount of sewage to be discharged. Refer to Table 6.86 for typical waste water flows from commercial, institutional and recreational sources.

Design Considerations

Some authorities favor placing limits on the size and capacity of septic tank type disposal systems. Under such circumstances it may be desirable to provide separate systems for groups of limited numbers of apartments, mobile homes or other occupancies. Some favor limits approaching (not to exceed) 10 apartments per system. This has the advantage of limiting the number of units affected by temporary system malfunctioning as by overloading due to plumbing fixture malfunctioning (constantly running water closets, for instance).

COLLECTION AND TREATMENT ALTERNATIVES

The use of self contained sewage treatment plants, particularly the tertiary treatment types, has found wide acceptance for facilities. This method is outside the scope of this chapter but is discussed briefly under individual sewage treatment systems.

The alternatives to conventional primary and secondary treatment includes sand filtration and evapotranspiration. Sand filters have been used in sewage treatment for many years. More recently, the standard systems have been modified to recirculating sand filter systems and have proved that, if properly designed, installed and operated, they can produce effluents that will meet stringent effluent and stream quality national standards. Evapotranspiration, as a means of disposing of domestic wastes, has been researched at several locations and its use accepted by various local jurisdictions. However, this system is rarely used.

Alternatives to the typical gravity collection and distribution system that should be evaluated are small diameter gravity sewers, pressure sewers and vacuum sewers. If the septic tank effluents are collected rather than the raw wastes, small-diameter pipe (4-in nominal) may be used to reduce the cost of conventional gravity sewers. Pressure sewer systems generally consist of septic tanks at each facility, a small submersible pump and small diameter plastic mains. Grinder pumps may be used in place of the septic tank. Construction costs are reduced because the sewer main can follow the contour of the land just below the frost line.

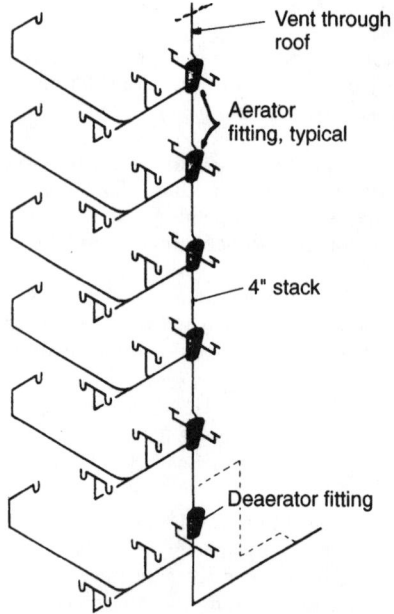

FIGURE 9.18 Sovent system.

SOVENT PLUMBING SYSTEM

The Sovent drainage system is a patented, combination drainage and vent system that uses a single pipe instead of the conventional two-pipe (vent and drainage) piping. The Sovent system is suited only for multistory buildings. Although widely installed in many countries throughout the world, it has found only limited acceptance in the United States. A typical Sovent stack is illustrated in Fig. 9.18. This system has the potential for substantial cost savings compared to a conventional two-pipe system.

The principal components of the Sovent system are standard pipes used for stacks and branch piping, aerator fittings used at the branch connection to a stack on each floor, and a deaerator fitting installed at the base of the stack before connecting to the house's drain.

The Sovent system shall conform to ANSI B-16.45 and CISMA Standard 177.

SANITARY VENT

SYSTEM DESCRIPTION

The sanitary vent system is a network of pipes terminating in the atmosphere and directly connected to the sanitary drainage piping system for the purpose of limiting air pressure fluctuations within the sanitary drainage piping to ±1 in of water column. There are two main reasons why the vent system is an integral and necessary adjunct to any drainage piping network:

1. It prevents the loss of fixture trap seals.
2. It permits the smooth flow of water in the drainage system.

Less important problems caused by excessive air pressure in the drainage system will also be prevented by vents:

1. Unsightly movement of water levels in water closet bowls
2. The possibility of sewer gases discharging through a fixture trap
3. Noise in the drainage system due to the "gurgling" of water

Most problems that do not result from blockages in the drainage system are caused by fluctuations in air pressure. These problems can be either eliminated or reduced to a level where they are no longer objectionable by designing a vent system that limits air pressure variations in the drainage piping network to a generally accepted figure of ±1 in of water column. This basic design criteria have been used to determine the vent sizing and allowable lengths that appear in modern codes.

NOMENCLATURE

Definitions are given in the glossary. Since it is possible that there are differences in terminology between some definitions used in this book and those appearing in local, regional, or national codes, it is necessary to check the codes to ensure a full understanding of all the components of the system.

SYSTEM DESIGN

General System Design Considerations

Differences in pressure within drainage piping are caused by the flow of water. When water is flowing under design conditions in a horizontal drain (approximately one-half full), the air above the liquid will be forced into movement by the friction between the flowing water and the air. In a stack, the water flows around the perimeter of the pipe leaving a central core of air except when overloaded.
The following principles govern the design of the vent system.

1. When design flow in the sanitary system is exceeded, the pipes are completely filled with water, forming a solid slug of water that compresses the air ahead of, and creates a vacuum behind, the slug.

2. The air moving in a vent pipe has friction losses similar to those of flowing water. Thus, the longer the pipe, the larger the diameter.

3. The amount of air displaced is proportional to the amount of water flowing in the drainage pipe. The flow is determined by using drainage fixture units (DFUs).

4. The size of a vent stack should be a minimum of one-half the size of a drainage stack. The size of a branch vent should be a minimum of one-half the size of the branch drainage line it serves.

5. The charts appearing in plumbing codes often contain a heading for soil or waste size that refers to "stack size." This heading should also be used for horizontal branch soil and waste stacks. Since the venting requirements for a stack are more severe than those for a horizontal drainage line, a small safety factor is required.

6. All fixture vents must rise above the flood level of the fixture served in order not to act as waste lines in the event the drain line becomes blocked.

Developed Length Measurement

The developed length of an individual or common vent is measured from its connection with the fixture trap arm to the connection with the branch vent or vent stack. The developed length of a branch vent is taken from the farthest connection with a waste branch from the point being sized. The developed length of a vent stack is taken from its connection with the soil or waste stack to its terminal above the roof.

Sizing Vent Stacks, Vent Branches, and Fixture Vents

Plumbing codes contain the information necessary to size the vent system. A typical vent-sizing chart is presented in Table 9.14.

In order to enter Table 9.14, there are three items that must be known: (1) the total DFU count of the soil or waste line associated with the vent being sized, (2) the developed length of the vent being sized, and (3) the size of the soil or waste branch or stack.

Having calculated these items, enter the table with the most severe condition of soil pipe size or DFUs. Then read horizontally across to the figure that meets or exceeds the calculated developed length. Read up to find the correct size of the vent. Use the following as a guide to sizing:

1. For vent stacks, use the total DFU load for the drainage stack and the full developed length of the vent to find the size. Vent stacks must be undiminished in size for their entire length.

2. For branch vents, use the longest developed length from the point where the size is being determined to the farthest connection to the waste line.

TABLE 9.14 Size and Length of Vents

Size of soil or waste stack, in†	Fixture units connected	Diameter of vent required, in*								
		1¼	1½	2	2½	3	4	5	6	8
		Maximum length of vent, ft‡								
1½	8	50	150							
2	12	30	75	200						
2	20	26	50	150						
2½	42		30	100	300					
3	10		30	100	100	600				
3	30			60	200	500				
3	60			50	80	400				
4	100			35	100	260	1000			
4	200			30	90	250	900			
4	500			20	70	180	700			
5	200				35	80	350	1000		
5	500				30	70	300	900		
5	1100				20	50	200	700		
6	350				25	50	200	400	1300	
6	620				15	30	125	300	1100	
6	960					24	100	250	1000	
6	1900					20	70	200	700	
8	600						50	150	500	1300
8	1400						40	100	400	1200
8	2200						30	80	350	1100
8	3600						25	60	250	800
10	1000							75	125	1000
10	2500							50	100	500
10	3800							30	80	350
10	5600							25	60	250

* See Table 2.9.
† See Table 9.1.
‡ 1 ft = 0.3 m.

3. For individual fixture vent size, refer to Table 9.1.

4. For building trap vents and fresh air inlets, the size should be half the size of the building drain, with a minimum size of 4 in (DN 100).

Vent Terminals

The vent pipe passing through the roof must remain open under all circumstances. The two conditions that would cause the exposed pipe to become blocked are frost closure and snow closure. Local codes and authorities should provide the minimum extension to avoid closure by accumulated snow on a roof.

In the absence of specific code requirements, the following could be used as a guide to locate vent extensions. The vent extension shall not be located under or within 10 ft of any window, door, or ventilating opening unless it is 2 ft (0.6 m) above the opening. If the terminal is through a building wall, it shall be located a minimum of 10 ft (3 m) from the property line, a minimum of 10 ft (3 m) above grade, and not under any overhang.

Relief Vents

Soil or waste stacks with no offsets in buildings having more than 10 branch intervals shall be provided with a relief vent at each 10th interval starting at the top floor. Offsets in the drainage stacks may also be required to have relief vents.

There are several acceptable configurations allowed by various codes. In general, the lower end of the relief vent shall connect to the soil or waste stack below the horizontal branch serving the floor required to have the relief vent. The upper end of the relief vent shall connect to the vent stack no less than 3 ft (1 m) above that same floor level. The size shall be equal to that of the vent stack to which it is connected or the drainage stack, whichever is smaller.

CIRCUIT AND LOOP VENTS

These venting schemes are intended to provide a more economical means of venting than using individual vents. They are allowed only for venting floor-mounted fixtures such as water closets, shower stalls, and floor drains and may not be acceptable in all code jurisdictions.

Circuit venting (Fig. 9.19) requires a uniformly sized drainage line with at least two, but no more than eight, fixtures connected in a battery arrangement. The circuit vent is connected on one end to the horizontal drain line between the two most

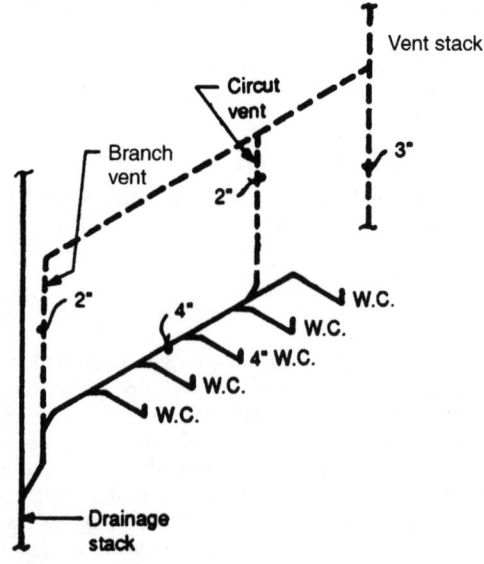

FIGURE 9.19   Detail of circuit vent.

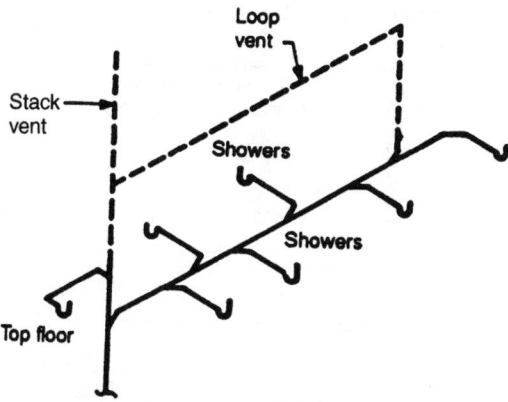

FIGURE 9.20 Detail of loop vent.

remote fixtures and is connected on the other end to the vent stack. In addition to the circuit vent, a relief vent is required to be connected to the horizontal drain line at the end of the battery, or every eight fixtures. The sizes of each shall be one-half the size of the horizontal drainage line or the full size of the vent stack, whichever is smaller.

Loop venting (Fig. 9.20) is the same as circuit venting except for the connection of the branch vent to the building system. This type of vent loops back to the stack vent instead of the vent stack.

WET VENTS

A wet vent is a combined vent/drain line that receives drainage from fixtures in addition to serving as a vent pipe. Wet vents are primarily used in residential projects, and permitted only by a limited number of codes. Wet vents must conform to the guidelines provided in each application code.

SUDS RELIEF VENTS

Suds pressure zones are illustrated in Fig. 9.9. If a drainage connection to these zones is made, a relief vent must be installed from the base of the suds pressure zone of the drainage stack to a nonpressure zone. The sizes of typical suds pressure relief vents are shown in Table 9.15.

SUMP AND EJECTOR VENTS

All codes require the venting of ejector pits, since they are gasketed and airtight. Many codes may also require the venting of sump pits. The vent pipe size is

TABLE 9.15   Suds Pressure Relief Vent Size

Waste size, in*	Relief vent size, in*
12	2
2½	2
3	2
4	2
5	3
6	4
8	5
	6

* See Table 2.9.

determined from the gpm discharge of the pump and the developed length from the pit to its connection with the building vent system or vent terminal. Table 9.16 is a typical guide to sizing such vents.

VENT HEADERS

When combining several vent stacks into a common header at the highest level to penetrate the roof only once, various codes require that the combined vent pipe be

TABLE 9.16   Size and Length* of Sump and Ejector Vents

Discharge capacity of sump pump, gpm[b]	Diameter of vent, inches[a]					
	1¼	1½	2	2½	3	4
	(Maximum equivalent length of vent, ft, given below)[c]					
10	N.L.†	N.L.	N.L.	N.L.	N.L.	N.L.
20	270	N.L.	N.L.	N.L.	N.L.	N.L.
40	72	160	N.L.	N.L.	N.L.	N.L.
60	31	75	270	N.L.	N.L.	N.L.
80	16	41	150	380	N.L.	N.L.
100	10‡	25	97	250	N.L.	N.L.
150	NP§	10‡	44	110	370	N.L.
200	NP	Np	20	60	210	N.L.
250	NP	NP	10	36	132	N.L.
300	NP	NP	10‡	22	88	380
400	NP	NP	NP	10‡	44	210
500	NP	NP	NP	NP	24	130

[a] See Table 2.9.
[b] 1 gpm = 3.8 L.
[c] 1 ft = 0.3 m.
　* Developed length plus an appropriate allowance for effects of entrance losses and friction due to fittings, changes in direction, and changes in diameter. Suggested allowances may be obtained from NBS Monograph 31 or other acceptable sources. An allowance of 50% of the developed length may be assumed if a more precise value is not available.
　† No limit; actual values greater than 500 ft.
　‡ Less than 10 ft.
　§ Not permitted.

TABLE 9.17 Cross-Section Area of Pipe

Pipe size, in*	Area, in²†
1½	1.767
2	3.1416
2½	4.908
3	7.068
4	12.566
5	19.635
6	28.274

*See Table 2.9.
†1 in² = 6.5 mm².

sized using the total number of DFUs of all the connected vents and the single longest developed length of all the vent stacks being combined. Other codes require that the combined vent stacks have a minimum cross-section area of all the separate vent pipes being combined. Table 9.17 lists the cross-section areas for pipes in square inches.

INTERIOR STORM WATER DRAINAGE

SYSTEM DESCRIPTION

This section describes the interior storm water drainage system that collects storm water runoff from building roofs and smaller ancillary areas exposed to the weather and conveys the runoff to an approved point of disposal. Ancillary areas of a building include areaways, walkways, canopies, and balconies.

Drainage from parking lots, roadways, and other larger areas of the site are discussed in Chap. 6, in the section entitled Surface Drainage.

GENERAL SYSTEM CRITERIA

General

The design of the storm water drainage system requires that the following information be obtained in order to establish design criteria.

1. Local climatic conditions
2. Local building and plumbing code restrictions
3. Building use
4. Roof pitch
5. Location, size, depth, type, and availability of the ultimate point of disposal
6. Total size of roof and other areas contributing to the storm water flow
7. Method of disposal or connection to public sewers. It is common practice for authorities concerned with storm water disposal to have a storm water management program to restrict the inflow of storm water to a predetermined value. This requires that storm water be retained somewhere on site. Refer to Chap. 6, section entitled Storm Water Retention Methods, for criteria.
8. Allowable methods of disposal and necessary permits required if public sewers are not available
9. Facility standards and preferences
10. For industrial sites, discharge of untreated storm water subject to contamination is prohibited. This could result from process equipment located on roofs. Refer to Chap. 6 for additional discussion.

ROOF DRAINAGE SYSTEMS

There are two types of systems used to remove storm water runoff, conventional and limited discharge. The conventional system removes runoff as quickly as it accumulates. The limited discharge system removes only a portion of the runoff, storing the remainder temporarily on the roof. Selection of the appropriate method depends on capacity and/or availability of disposal facilities and acceptance of the method by local codes and authorities.

Roof Drains

Roof drains allow rainwater to enter the drainage system after falling on a roof. Drains shall have a dome that extends above the roof level in order to allow water to enter the drain even if partially blocked by debris. It is good practice to install two drains in every area of a roof to allow drainage if one drain is completely blocked. Domes are manufactured in a variety of materials, with PP often selected. They shall have a free area 150 percent of the area of the drain discharge pipe. Domes on roof drains selected for limited discharge systems have a restriction that limits the amount of storm water entering the drain.

Drain bodies are available to match any roof construction and must be selected to specifically match that construction.

Overflows and Secondary Storm Drainage Systems

If roof water retention is not a design feature, a fail-safe method must be provided to remove excess runoff from a roof before the water level rises to a point where damage would result. *Scuppers* are commonly used to allow the excess runoff to discharge directly off the roof and down the side of the building. Codes often stipulate how high above the roof level scuppers must be installed. One disadvantage of scuppers is that wave action may allow water to spill out of the scuppers before the design depth is reached. The discharge from rectangular scoppers is given in Table 9.18.

Various codes mandate the use of a completely separate and independent overflow piping network complete with roof drains in addition to the regular storm water drainage system. This network could connect either to the regular system independently outside a building or directly to the point of disposal, depending on

TABLE 9.18   Discharge from Rectangular Scuppers-Gallons per Minute

Water head (inches)	Width of scupper—(inches)					
	6	12	18	24	30	36
0.5	6	13	19	25	32	38
1	17	35	53	71	89	107
1.5	31	64	97	130	163	196
2		98	149	200	251	302
2.5		136	207	278	349	420
3		177	271	364	458	551
3.5			339	457	575	693
4			412	556	700	844

NOTES:
   Table 9.18 is based on discharge over a rectangular weir with end contractions.
   Head is depth of water above bottom of scupper opening.
   Height of scupper opening should be 2 times the design head.
   Coordinate the allowable head of water with the structural design of the roof.

code requirements. Another type of overflow protection is the use of separate over-flow pipes or drains, in addition to the regular drains, connecting to the regular storm water drainage piping. Local codes and authorities often have preferences and should provide the preferred method for design. No additional roof area is required for a secondary drain.

Sidewall Area Calculation

Precipitation falling on vertical walls of structures located on roofs and other areas, such as penthouse walls and stair towers, will add to the runoff calculated for horizontal area only. It is necessary for these vertical areas to be included into the horizontal tributary area. To calculate the amount of sidewall area that shall be added to the horizontal roof segment, determine the square foot area of the single or two largest adjacent walls that would contribute runoff to any one roof drain. Then divide this area in half, because a smaller amount of rain falls on a vertical surface than on a horizontal surface. Add this vertical sidewall area to the horizontal tributary area for each roof drain to obtain the total tributary area.

Roof Drainage Design Procedures

Conventional Small Roof Drainage Systems. The "small" roof is one with a roof area less than the maximum square foot area appearing in the tables associated with the plumbing code used for the project. Larger roof areas are discussed later in this chapter. Design of a conventional roof drainage system consists of the following general steps:

1. Locate drains on roofs and ancillary areas throughout all areas of the project discharging into the system.

2. Determine the code-mandated overflow system requirements.

3. Route the storm water and overflow system piping and set pitch of pipe.

4. Size the piping network by first calculating the total tributary area for each individual drain located in each specific section of the roof and other areas. For horizontal pipe, the sizing procedure starts at the most remote part of the system, using the pipe pitch and total tributary area for each horizontal pipe section from design point to design point.

5. Storm water piping is sized using the total square feet of roof area at each design point and the pitch of the drainage pipe. Two different styles of tables are in general use by model codes. One style uses a single rainfall rate that is incorporated into the code charts along with the maximum allowable tributary area and various pitches of pipe. No further calculations are required. A second uses multiple rainfall rates with the maximum allowable tributary area and various pitches of pipe.

6. Enter the chart using a single rainfall rate with the pitch and total tributary area. For horizontal pipe, start from the most remote part of the system. Determine the pipe pitch and total tributary area for each horizontal pipe section from design point to design point. Select the figure at the intersection of the criteria used for that specific chart and choose a pipe size corresponding to a figure equal to or greater than the calculated tributary area. (Notes associated with different chart types allow conversion to another rainfall rate if different from that of the chart.)

TABLE 9.19   Size of Horizontal Storm Drains,* Single Rainfall Rate

Diameter of drain.		Maximum projected area for drains of various slopes†					
		⅛-in slope		¼-in slope		½-in slope	
in	DN	ft²	gpm‡	ft²	gpm	ft²	gpm
3	80	822	34	1,160	48	1,644	68
4	100	1,880	78	2,650	110	3,760	156
5	125	3,340	139	4,720	196	6,680	278
6	150	5,350	222	7,550	314	10,700	445
8	200	11,500	478	16,300	677	23,000	956
10	250	20,700	860	29,200	1,214	41,400	1,721
12	300	33,300	1,384	47,000	1,953	66,600	2,768
15	375	59,500	2,473	84,000	3,491	119,000	4,946

*Based upon a maximum rate of rainfall of 4 in/h. Where maximum rates are more or less than 4 inches per hour, the figures for drainage area shall be adjusted by multiplying by 4 and dividing by the local rate in inches per hour.
† 1 ft² = 0.093 m².
‡ 1 gpm = 3.8 L.

For vertical piping, use the tributary area discharging into the vertical leader, increasing the size as additional areas are added. Table 9.19 is a typical table used to size the horizontal drainage pipe based upon pitch of the pipe and tributary area. Table 9.20 is a typical table used to size the roof drain itself and the vertical pipe stack using the tributary area. Table 9.21 allows the sizing of gutters. These tables are reproduced from the Uniform Plumbing Code, which uses 4 in/h as the basis for the tables. Rates required by local authorities shall always take precedence over a regional or model code rainfall rate.

7. The second style of code has been developed for model codes using multiple rainfall rates. These charts are very similar to the regional ones except for the addition of multiple rainfall rates. A typical chart is shown in Table 9.22. Enter the chart using the selected rainfall rate, the pitch of the pipe and total tributary area. Select the figure at the intersection of the criteria used and choose a pipe size corresponding to a figure equal to or greater than the calculated tributary area. The rainfall rate selected is the rainfall rate used for the main site or a 10-year, 10-min storm.

8. In some areas of the country, the public sewer is a combined sanitary and storm water system. The sanitary and storm water systems can be combined either inside or outside the building wall. Good engineering practice has them leave the building separately and combined outside the building wall. The combined system is usually sized on the basis of DFUs. In order to size the combined system, the square feet of tributary area of the storm water system is converted to DFUs. To calculate the size of the combined building drainage network, use Table 9.23 as a guide for the conversion of square feet into DFUs. When a pump discharges into the storm water system, refer to Table 9.24 for the square ft equivalent for 1 gpm of pump discharge.

Conventional Large Roof Drainage Systems

1. When a very large building, such as a warehouse or factory, has a total square foot tributary area in excess of those appearing in the code sizing tables, another

TABLE 9.20 Size of Vertical Conductors and Leaders*

Diameter of leader or conductor,†		Maximum projected roof area	
DN	in	ft²	gpm
50	2	544	23
65	2½	987	41
80	3	1,610	67
100	4	3,460	144
125	5	6,280	261
150	6	10,200	424
200	8	22,000	913

*Based upon a maximum rate of rainfall of 4 inches per hour and on the hydraulic capacities of vertical circular pipes flowing between one-third and one-half full at terminal velocity, computed by the method of NBS Mono. 31. Where maximum rates are more or less than 4 inches per hour, the figures for drainage area shall be adjusted by multiplying by 4 and dividing by the local rate in inches per hour.

†The area of rectangular leaders shall be equivalent to that of the circular leader or conductor required. The ratio of width to depth of rectangular leaders shall not exceed 3 to 1.

TABLE 9.21 Size of Roof Gutters

Based on 4 in/h (100 mm/h) rainfall

Diameter of gutter,*		Maximum projected roof area for gutters, ft²			
IN	DN	1/16 in (.5 cm/m)	1/8 in (1.0 cm/m)	1/4 in (2.0 cm/m)	1/2 in (4.0 cm/m) & vertical
3	80	170	240	340	480
4	100	360	510	720	1,020
5	125	625	880	1,250	1,770
6	150	960	1,360	1,950	2,770
7	175	1,380	1,920	2,760	3,900
8	200	1,990	2,800	3,980	5,600
10	250	3,600	5,100	7,200	10,000

*Gutters other than semicircular may be used provided they have an equivalent cross-sectional area.
Source: Uniform Plumbing Code.

TABLE 9.22 Size of Horizontal and Vertical Storm Drains, Multiple Rainfall Rate

DN	Size of pipe, in; ⅛-in slope	Maximum rainfall, inches (mm) per hour				
		2 (51)	3 (76)	4 (101)	5 (127)	6 (152)
80	3	1,644	1,096	822	657	548
100	4	3,760	2,506	1,880	1,504	1,253
125	5	6,680	4,453	3,340	2,672	2,227
150	6	10,700	7,133	5,350	4,280	3,566
200	8	23,000	15,330	11,500	9,200	7,600
250	10	41,400	27,600	20,700	16,580	13,800
300	12	66,600	44,400	33,300	26,550	22,200
375	15	109,000	72,800	59,500	47,600	39,650

DN	Size of pipe, in; ¼-in slope	Maximum rainfall, inches (mm) per hour				
		2	3	4	5	6
80	3	2,320	1,546	1,160	928	773
100	4	5,300	3,533	2,650	2,120	1,766
125	5	9,440	6,293	4,720	3,776	3,146
150	6	15,100	10,066	7,550	6,040	5,033
200	8	32,600	21,733	16,300	13,040	10,866
250	10	58,400	38,950	29,200	23,350	19,450
300	12	94,000	62,600	47,000	37,600	31,350
375	15	168,000	112,000	84,000	67,250	56,000

DN	Size of pipe, in; ½-in slope	Maximum rainfall, inches (mm) per hour				
		2	3	4	5	6
80	3	3,288	2,295	1,644	1,310	1,096
100	4	7,520	5,010	3,760	3,010	2,500
125	5	13,360	8,900	6,680	5,320	4,450
150	6	21,400	13,700	10,700	8,580	7,140
200	8	46,000	30,650	23,000	18,400	15,320
250	10	82,800	55,200	41,400	33,150	27,600
300	12	113,200	88,800	66,600	53,200	44,400
375	15	238,000	158,800	119,000	95,300	79,250

Vertical leaders							
DN	Diameter of leader, in	Normal rate of rainfall, in (mm) per hour					
		2	3	4	5	6	8
		Roof area, ft²					
80	2	1440	960	720	576	480	260
100	2½	2000	1733	1300	1040	865	650
125	3	4400	2933	2200	1760	1470	1100
150	4	9200	6133	4000	3680	3070	2300
200	5	—	—	8650	6920	5765	4325
250	6	—	—	—	—	9000	6750

Source: *Uniform Plumbing Code.*
1 ft² = 0.093 m².

TABLE 9.23 DFU-Square Foot Equivalents

Drainage area, ft²*	Fixture unit equivalent
180	6
260	10
400	20
490	30
1,000	105
2,000	271
3,000	437
4,000	604
5,000	771
7,500	1,188
10,000	1,500
15,000	2,500
20,000	3,500
28,000	5,500
Each additional 3 ft²	1 fixture unit

*1 gpm = 19 ft².
1 ft² = 0.093 m².

TABLE 9.24 Square ft Equivalent Values for Continuous Flow into a Combined Storm Sewer

Where there is a continuous or semicontinuous discharge into the building storm drain or building storm sewer, as from a pump, ejector, air-conditioning plant, or similar device, each gallon per minute of discharge shall be computed as being equivalent to the square feet or roof area, based upon a rate of rainfall in inches per hour according to the following table:

Rate of Rainfall	Equivalent for one gpm roof area, sq ft
2	48
3	32
4	24
5	19
6	16
7	13
8	12

method must be used to calculate the volume of runoff and size the pipe. This is done by using the actual flow rate rather than square feet of tributary area.

2. The most widely accepted method for the calculation of runoff is the Rational Method, repeated here for convenience. Refer to Chap. 6 for further discussion.

$$Q = A \times I \times R \qquad (9.1)$$

where Q = quantity (flow rate) of storm water runoff, in cubic feet per second (cfs). 1 cfs = 448 gpm (1702 L/m).

A = tributary area in acres. 1 acre = 43,560 ft² (4051 m²).

I = imperviousness factor. Use a value of 1.0 for roofs.

R = rate (intensity) of rainfall, in inches per hour (in/h). When large roofs are encountered, the intensity requires judgment on the part of the design engineer. The following should be considered:

a. If the building is part of a larger site, the criteria shall be the same as those used for the site as a whole.

b. Generally accepted practice uses a 10-year return frequency for the design of roofs. Longer periods are not justified unless a greater degree of protection is desired.

c. When used for roof design, a 5-min duration is recommended. Some codes allow longer durations, producing less runoff. The decision depends on the degree of protection desired.

d. With the duration and frequency established, use Fig 6.15 to select an intensity.

e. Snowmelt should be considered only under very unusual circumstances. Snowmelt is greatest if a severe rainstorm occurs when there is snow on the roof. Tests have shown that a maximum figure of 0.2 in/h (0.2 mm/h) should be added to the intensity, with a figure of 0.1 in/h (0.1 mm/h) acceptable for most storms.

The design procedure is the same as that described for a conventional drainage system. Use Table 9.19 to find the pipe size up to the largest areas in the chart. When those are exceeded, use Eq. (9.1) to calculate the flow rate in gpm of the system at each design point. Determine the slope of the pipe. Use Fig. 6.17 to find the pipe size by connecting the slope and flow rate with a straight line and read the size on the chart where it intersects the pipe size line.

Limited Discharge Roof Drainage Systems.   There are two reasons to select a limited discharge roof drainage system: for economic reasons or where authorities mandated restricting discharge. The voluntary use of the limited discharge system for economic reasons will allow smaller piping throughout the entire network, resulting in a savings on the total installed cost of the piping.

The use of a mandated limited discharge system may be required where a storm water management program is in force. Areas may have overloaded sewers, allowing only a small portion of the runoff to be discharge into the public sewer. In suburban areas, if discharge into a small stream is proposed, the fact that new roofed areas produce a much greater runoff than former woodland must be considered, since the downstream portion of the stream could become flooded if the full amount of runoff discharged instantaneously into the stream. One of several options is the temporary storage of water on a roof. Other methods of storm water retention are discussed in Chap. 6.

In order to limit the discharge, roof drains are provided with factory preset grate openings that allow only a smaller, predetermined flow of storm water to enter the drain. The allowable amount to be discharged must be given to the manufacturer in order to be set correctly at the factory. The runoff not discharged must be temporarily stored on the roof.

A simplified design procedure for a limited discharge roof drainage system consists of the following general steps:

1. Determine if the code and local authorities will permit limited discharge roof drainage for the project.

2. Since the roof will have to be waterproof and need high curbs built around all openings, there is usually considerable opposition to the storage of water on the roof when based only on economic factors. Past experience has shown that an expensive roof covering is required and the possibility of leakage increases as the roof ages.

3. Establish the maximum allowable flow rate in gpm permitted to discharge into the ultimate point of disposal from all sources of water for the entire site. If the system is voluntary, some iterations will be necessary after the piping is run to find the optimum pipe size compared to the amount of storm water stored on the roof. If discharge is into a sewer, the authorities having jurisdiction will provide the allowable discharge amount. If discharge is into a waterway, the existing site conditions should be calculated. The difference between the existing and proposed discharge volume is the volume of water that should be stored on the roof. If drains in areas other than the roof are present, it is not generally possible to store water from those areas. The discharge from these drains will have to be separated from the total roof discharge to find the actual discharge allowed from the roof alone.

4. Decide on the length of time water will be allowed to remain on the roof. Generally accepted practice is to allow between 12 and 18 h, starting with as short a time as possible. Lengthen the time as necessary to obtain the calculated rate of discharge. Accepted practice is to use a maximum of 24 h.

5. Since the structural engineer must design the roof to support the additional load of stored water, a mutual agreement between the plumbing and structural engineers must be made to determine the depth. If the roof is flat, the generally accepted depth is 3 in. If the roof is pitched, add 3 in of water depth above the high point of the roof. Table 9.25 gives the weight for each inch of water.

6. Find the total amount of rainfall in inches that will fall on the roof for the time established in step 3. This is obtained from Fig. 9.21, which is a 10-year, 24-h rainfall or from Fig. 6.21, which is a 100-year, 24 h rainfall. The 24-h period is used because of its conservative design. A 10 percent reduction in the 24-h figure approximates an 18-h rainfall.

7. Divide the figure found in step 5 with the time found in step 3 (e.g., 8 in divided by 12 h equals 0.67 in/h). Then determine the gpm discharge. Use Table

TABLE 9.25 Weight of Rainfall

MM	Amount of rain, in	Weight of water, lb/ft^2*
150	6	31.21
125	5	26.01
100	4	20.81
75	3	15.61
50	2	10.40
25	1	5.20

*1 lb = 2.2 kg.

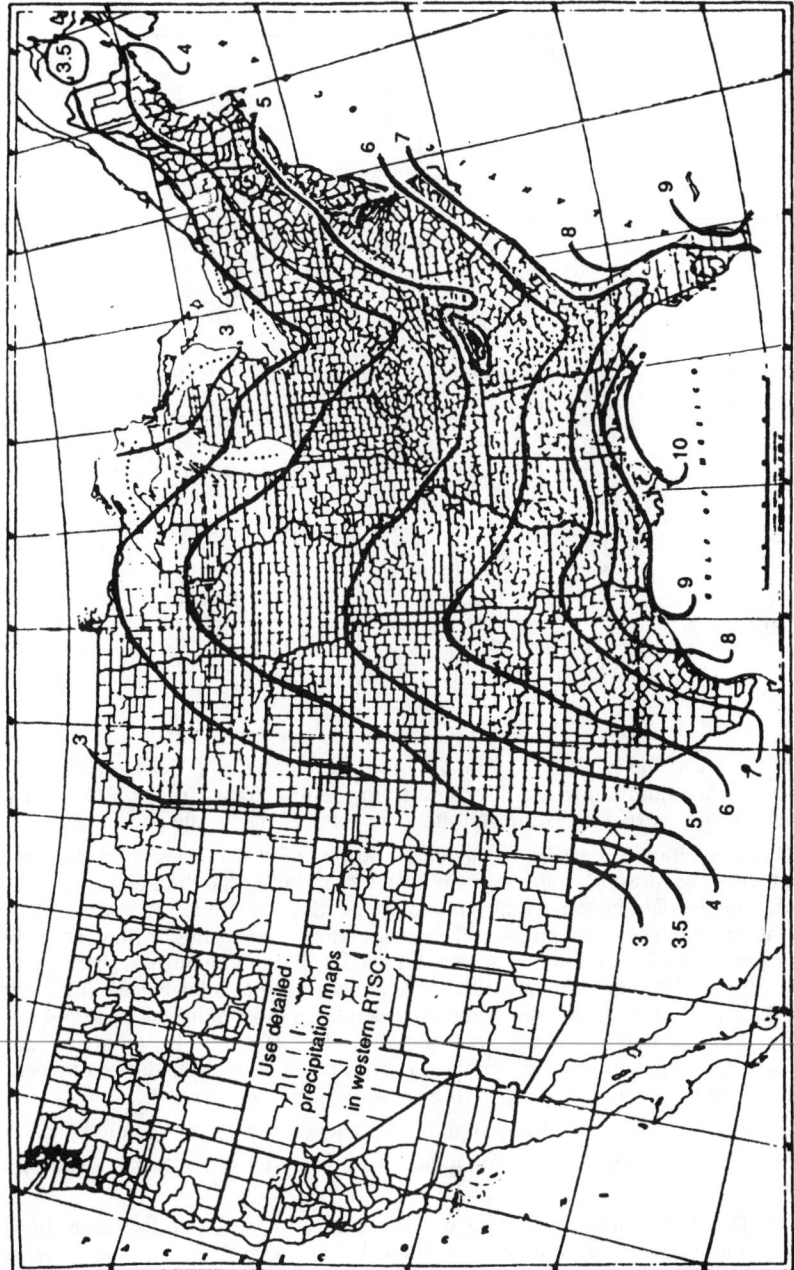

FIGURE 9.21   10-Year, 24-hour rainfall. (*Adapted from U.S. Weather Bureau map.*) In inches
(1 in = 25.4 mm).

TABLE 9.26 Rainfall Conversion Data

Rainfall, in/h[a]	gph per 1 ft²†	gpm per 1 ft²†
3.0	1.870	0.0312
2.9	1.808	0.0302
2.8	1.745	0.0291
2.7	1.683	0.0281
2.6	1.621	0.0270
2.5	1.558	0.0260
2.4	1.496	0.0250
2.3	1.434	0.0239
2.2	1.371	0.0229
2.1	1.309	0.0218
2.0	1.247	0.0208
1.9	1.184	0.0198
1.8	1.122	0.0187
1.7	1.060	0.0177
1.6	0.997	0.0166
1.5	0.935	0.0156
1.4	0.873	0.0146
1.3	0.810	0.0135
1.2	0.748	0.0125
1.1	0.686	0.0114
1.0	0.623	0.0104

*1 in = 25.4 mm.
†1 gal = 3.8 L.

9.26 to convert the rainfall from in/h to gpm (if appropriate). Compare this figure with the maximum allowable discharge from the roof found in step 2. Adjust the retention time as required to match the allowable discharge rate.

8. Consideration must be given for a heavy rainfall occurring after the design storm ends, while some water remains on the roof during the draindown time. This rainfall rate should be the actual 1-h duration, with the same return period selected for the project. This 1-h rainfall will deposit a number of inches of rain. The actual drainage rate will be the figure selected in step 2. By subtracting the drainage rate figure (in/h) from the 1-h rainfall figure, the actual total allowable rainfall depth will be found. This depth must not exceed the depth established in step 4. If it does, additional depth of storage is required, or the client must be willing to accept rainwater spilling out of the emergency overflow once during the return period selected, or the additional water will have to be stored on the site.

9. Locate drains on roof and find the total number of drains required.

10. Route the storm water piping and overflow systems. Set height of overflow drains.

11. Divide the number of drains into the amount of total roof discharge allowed to find the gpm flowing out of each drain. Using Table 9.27, size the individual drains, branches, and stacks using the gpm and pitch of the pipe. Limited discharge roof drains are set at the factory for the specified maximum flow rate established by the engineer.

12. Size the piping network based on accumulated gpm flow at each design point and the actual pitch of the pipe.

TABLE 9.27 Drainage Capacity of Piping

Flow capacity for storm water piping systems in gallons per minute, flowing full (n = .013)

Pipe diameter, in	DN	Roof drain and vertical downspouts	Horizontal storm drainage		
			Slope		
			cm/m in/ft	1.0 $\frac{1}{8}$	2.0 $\frac{1}{4}$	4.0 $\frac{1}{2}$
2	50	30	—	—	—
2½	65	54	—	—	—
3	80	92	36	51	80
4	100	192	77	110	174
6	150	563	220	315	449
8	200	1208	494	696	987
10	250	2600	943	1302	1800
12	300	6000	1526	2154	2800
15	375	—	2873	3500	4950

1 gpm = 3.8 L/m.

Conventional Roof Drainage Design Considerations

1. It is important that the local codes and authorities be consulted for the method used to connect to the public sewers.

2. It is good practice to place two roof drains in any area (except very small areas) to provide at least one open drain if the other becomes blocked by debris.

3. Limit the square footage of any individual roof drain to a maximum of 5000 ft² of tributary area.

Limited Discharge Roof Drainage Design Considerations

Drains should be no greater than 200 ft (165 m) apart and no greater than 50 ft (15 m) from the end of a tributary area.

Disposal Methods When There Are No Public Sewers

In the absence of public sewers, the most common methods of disposal are discharge into nearby watercourses, recharge basins, or drywells.

Discharge into watercourses presents unique problems. Usually, special permits must be obtained and special regulatory requirements followed. Often, oil or other interceptors may be required before the water can be discharged. Since impervious concrete and roof surfaces produce more runoff than the original ground, usually some method of retaining water will be required so the downstream portion of a small stream or creek will not flood during a severe storm. If the watercourse is navigable, permits will be required from the U.S. Army Corps of Engineers and other agencies of the government.

Recharge basins are depressed areas on a site that are normally dry. These basins are sized to collect all the estimated site runoff from a storm of a given duration, usually a 100-year, 24-h storm. Storm water runs into the basin at a high flow rate, is collected, absorbed into the ground, and evaporates. The size of the recharge basin is determined from a test of the soil's ability to absorb the calculated total volume of runoff plus a safety factor.

Drywells are round or square structures buried in the ground that have openings or porous sides and bottoms. During a rainstorm, the storm water fills the structure, escapes through the openings, and is absorbed into the ground. Local authorities that allow this means of disposal usually mandate the size and number of drywells permitted per square foot of roof area in their specific locality. A typical drywell is illustrated in Fig. 9.22. Other storm water retention methods are discussed in Chap. 6.

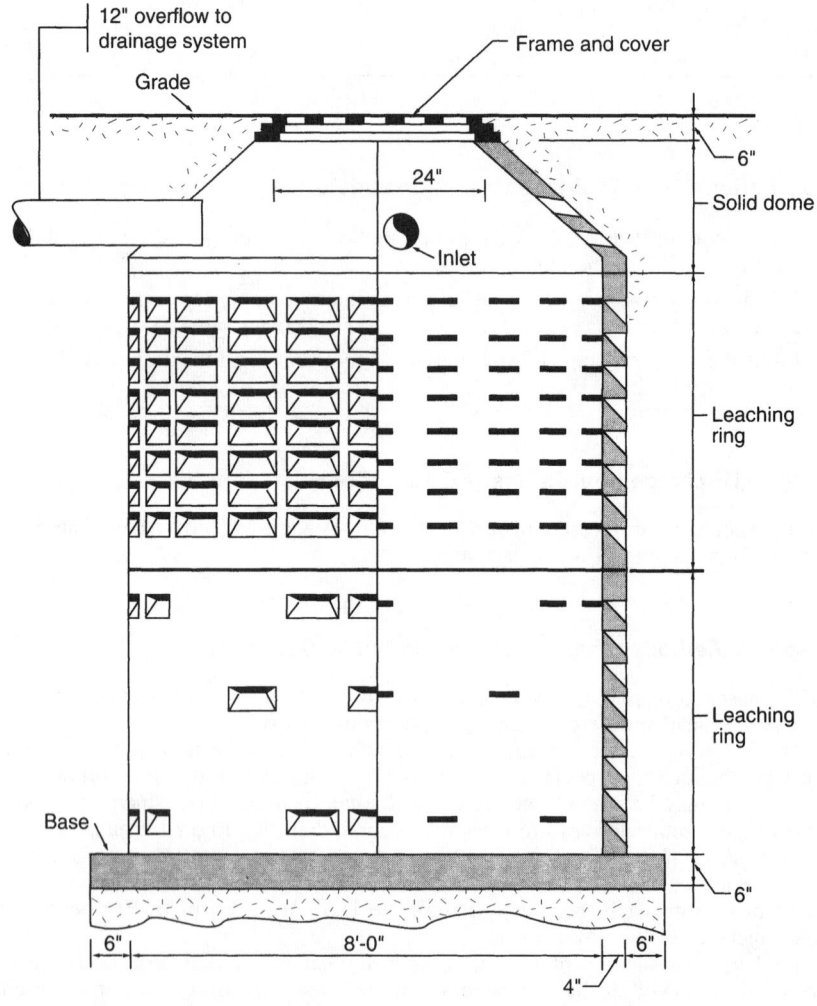

FIGURE 9.22 Detail of typical drywell.

POTABLE WATER SYSTEMS

SYSTEM DESCRIPTION

The potable water system provides hot and cold water that is suitable for human consumption and has adequate purity, pressure, temperature, and volume to satisfy all requirements for a specific project or purpose.

WATER TREATMENT

Raw water obtained from wells or other similar sources must be tested and treated, if necessary, to reduce or remove various impurities found in raw water. Specific treatment is based on the volume of water to be treated, type of impurities present and the degree to which various impurities are to be removed. Water treatment methods are discussed in Chap. 4.

WATER METERS

When water is provided by a utility company, meters are required for billing purposes. The utility company will usually require a specific location for the meter, such as inside the building or in a pit adjacent to the property or building, clearance around the meter needed for reading the meter, piping arrangement, and other regulations to discourage attempts to bypass the meter.

The water meter is usually a part of a water meter assembly that could include shut-off valves, strainers, a test tee for the local authorities to test the accuracy of the meter, and a meter bypass.

The selection of a meter type, if not mandated by the utility company or local authorities, is based on accuracy and pressure loss through the meter at the intended flow rates. Water meter types include disk meter, turbine meter, and compound meter.

Water Meter Types

The utility company may require the specific make, model, and type, depending on the anticipated usage. A remote-reading type of meter, capable of being read by electronic means at distances away from the meter, may also be required.

In the absence of any regulations, the following is a general guide for selection of a suitable potable water meter.

1. *Disk meter.* Normally ranging in size up to 2 in (DN 50), this type is used for smaller variable flows such as those occurring in residential and small commercial installations.

2. *Compound meter.* Normally ranging in size from 2 to 6 in (DN 50 to 150), this type is used where large volumes and wide variations in flow might be encountered. The initial cost of this meter is high, and it is the largest in physical size.

3. *Turbine meter.* Normally ranging in size from 2 to 10 in (DN 50 to ?), this type is used for variable flow in a narrower range of flow than that of a compound meter.

CONTAMINATION PREVENTION

Contaminants found in potable water are divided into three general classifications: severe, moderate, and minor. It is necessary to evaluate each facility as a whole and also at specific points of use to protect against potential hazards. Local health, building codes, and ordinances must be consulted for specific requirements and for the suitability of any device for a specific application.

There are five basic methods of preventing the contamination of a potable water system due to cross connections with, and backflow from, a potentially contaminated source:

1. *Air gap.* Suitable for severe hazards, the air gap is passive and is the only fail-safe method of preventing backflow. An air gap is defined as the clear distance between the end of a potable water supply pipe and the highest possible water level from a source of potential contamination. The minimum allowable air gap dimension is provided in local codes, and is usually either 2 in or 2 times the diameter of the supply pipe, whichever is larger. An air gap is illustrated in Fig. 9.23.

2. *Pressure type vacuum breaker.* Suitable for minor hazards, the pressure type vacuum breaker is a mechanical device designed to prevent backflow caused only by back pressure conditions. It is designed to operate under continuous pressure on both sides of the device. A typical pressure type vacuum breaker is illustrated in Fig. 9.24.

3. *Atmospheric type vacuum breaker.* Suitable for minor hazards, the atmospheric type vacuum breaker is a mechanical device designed to prevent backflow caused only by back siphonage conditions. It is designed to operate with pressure

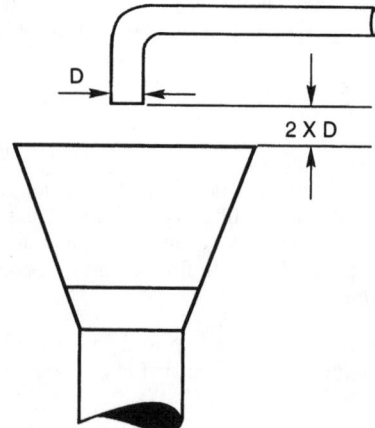

FIGURE 9.23 Detail of typical air gap.

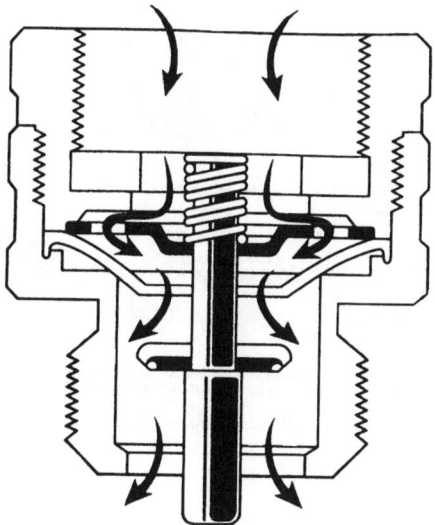

FIGURE 9.24   Pressure type vacuum breaker.

on only one side of the device. A typical atmospheric type vacuum breaker is illustrated in Fig. 9.25.

   4. *Double check valve assembly* (*DCVA*). Suitable for moderate hazards, a double check valve (Fig. 9.26) is a mechanical device consisting of two independently operating, soft seat, swing check valves. It prevents backflow from back siphonage and back pressure conditions. A variation adds a means of detection to this assembly.

   5. *Reduced pressure zone.* Suitable for severe hazards, a reduced pressure zone backflow preventer (Fig. 9.27) is a mechanical device consisting of two independently operating, soft seat, spring-loaded, check valves with the addition of an independent, differential pressure relief valve installed between the check valves,

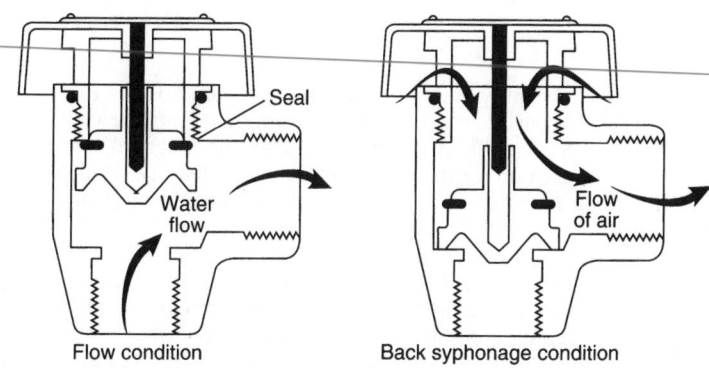

Flow condition   Back syphonage condition

FIGURE 9.25   Atmospheric type vacuum breaker.

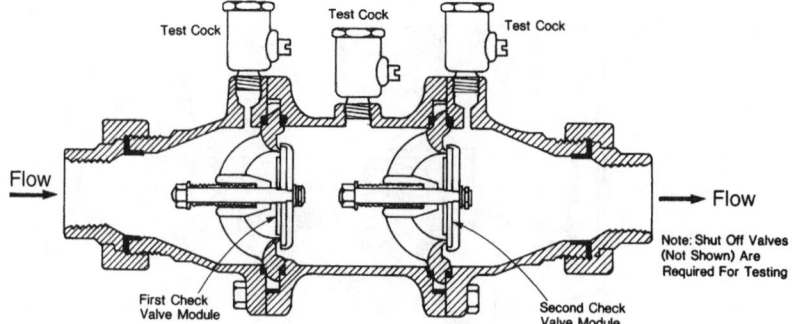

FIGURE 9.26 Double check valve backflow preventer.

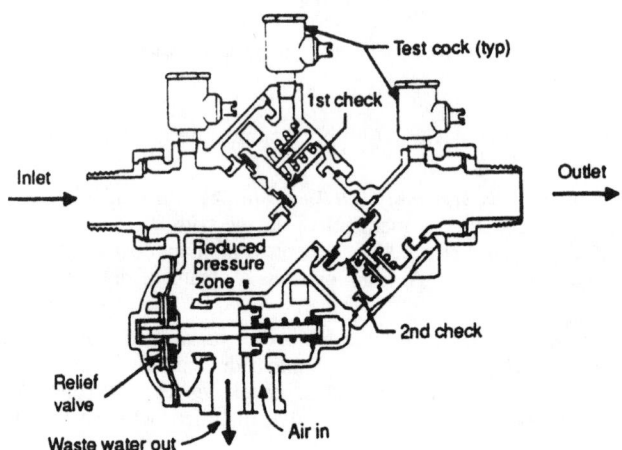

FIGURE 9.27 Reduced pressure zone backflow preventer.

in the region of the assembly called the reduced pressure zone (RPZ). The relief valve will open under backflow conditions, and discharge all the upstream water under pressure until the condition is corrected. With the potential for large flows, it is accepted practice to direct the discharge outside a facility. Typical discharge from an RPZ can be found in Fig. 9.28. Consult the manufacturer of the selected RPZ for actual values.

The supplier of public potable water is responsible for protecting the public distribution system from contamination by any nearby contaminated water source, such as a pond or stream, that may be used by a fire department to fight a fire at a project location. Depending on local regulations, it may be necessary to provide

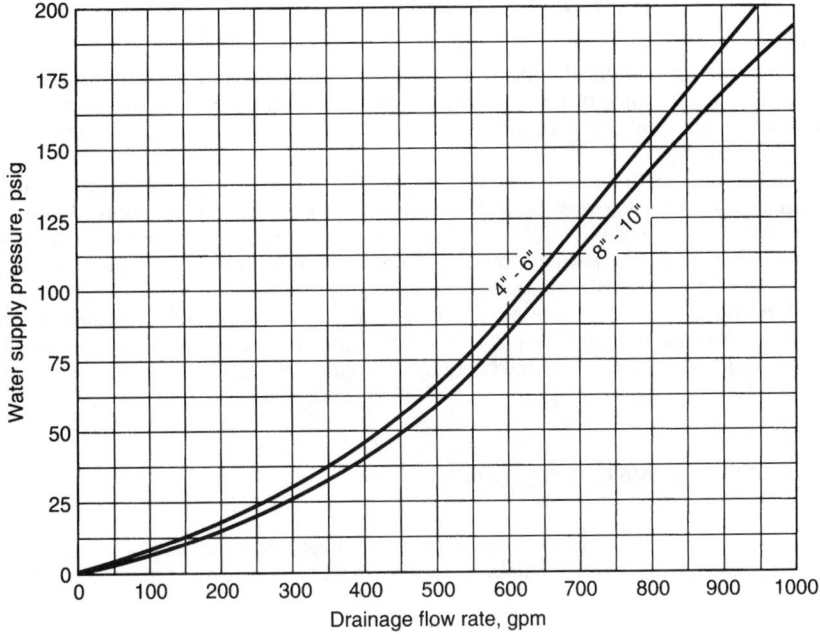

FIGURE 9.28   Discharge from a typical RPZ backflow preventer.

backflow protection on the site main before it connects to the public source of water.

Another aspect of contamination protection is the disinfection of the entire potable water system, including both the interior and site water supply. Code requires disinfecting interior water piping. A commonly referenced standard for site water mains is AWWA C651, Standard for Disinfecting Water Mains. Refer to Chap. 6 for additional discussion of backflow prevention for site water supply.

WATER VELOCITY

The velocity of water flowing through the piping system should be kept low enough to prevent objectionable noise, water hammer, and accelerated component wear due to erosion or nuisance splashing of water from fixtures. In order to avoid excessive noise, generally accepted practice for commercial buildings is to limit water velocity to between 6 and 8 fps (2 to 2.3 ms). For industrial projects, 10 fps (3 ms) is acceptable in work areas where the noise is not noticeable. For theaters and similar projects, 4 to 6 fps is desirable. If the water supply is controlled by a quick-closing valve, the velocity should be limited to approximately 4 fps (1.4 ms) to avoid water hammer.

WATER DISTRIBUTION SYSTEMS

The purpose of a water distribution system is to deliver both hot and cold water, in an acceptable range of pressure and temperature to fixtures and equipment in all parts of a building. All distribution systems can be divided into either upfeed or downfeed systems.

In the typical downfeed system (Fig. 9.29), water is supplied down through vertical pipes to the lowest point in the pressure zone. In a typical upfeed system (Fig. 9.30), the water is supplied upward through vertical pipes to the highest point in the pressure zone. If street pressure is not enough to provide sufficient pressure at the top floor, a booster pump system is required.

The choice of a system is based on several factors, such as the space requirements to run the hot and cold water distribution mains, origin of the source of supply, and economics. The downfeed system generally has smaller pipe sizes and is usually more economical than an upfeed system.

ESTIMATING WATER DEMAND

The calculation of maximum demand is approximate because it is not possible to predict exactly how many fixtures or various pieces of equipment may be in use at

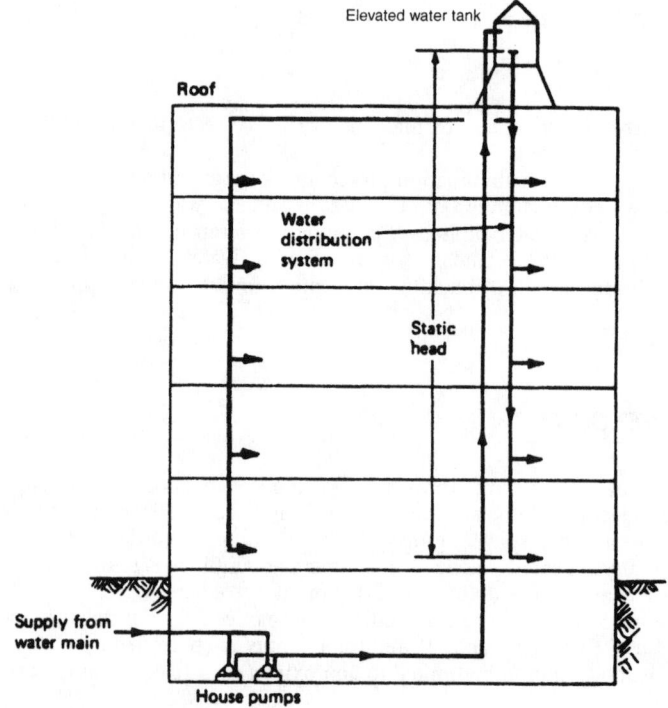

FIGURE 9.29 Simplified downfeed water supply system with simplified elevated water tank.

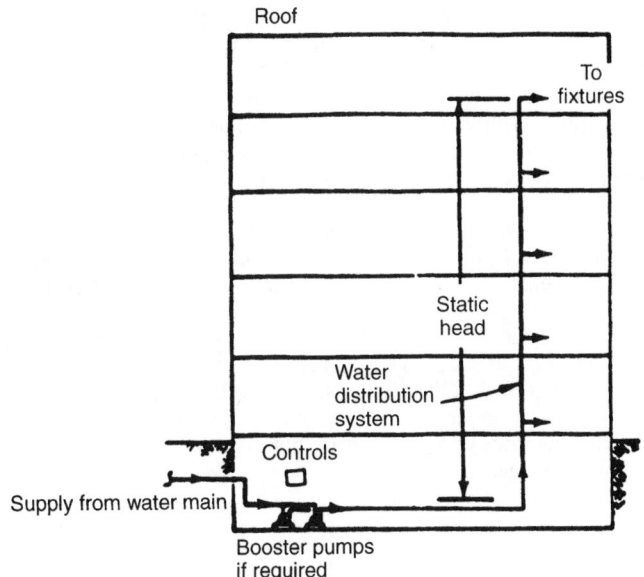

FIGURE 9.30 Simplified upfeed water supply system.

the same time. The estimate used for sizing should allow for the probability that the calculated flow will be exceeded occasionally, but not permit wasteful oversizing because the estimate will rarely be exceeded.

The method used to calculate the maximum probable demand is based on water fixture units (WFU) converted into gpm. The Hunter Method is used for this conversion. It has been long recognized that the Hunter Method, which is based on a statistical analysis of how often various fixtures are in use simultaneously, did not consider many types of modern facilities. In some cases, this method resulted in oversized piping and, in others, undersized piping. Studies have been undertaken to investigate the application of the Hunter curve for the purpose of providing a more accurate method of determining peak flow.

These studies have been completed using computers and a new methodology with statistical methods that were not available to Hunter. They have not been adapted by the model or regional codes, but are being considered. New building classifications have been proposed with the intention of revising or replacing the Hunter curve to provide more accuracy. They are:

1. *Residential.* The fixtures are limited to household and personal care. Examples are one- and two-family dwellings.

2. *Commercial.* Access to fixtures is limited to those people regularly occupying this type of building, including multiple dwellings, hotels, small business offices, and small professional buildings.

3. *Public access, need driven.* Access to centrally located fixtures is determined primarily by a biological need, with a social need also present. Examples are large office buildings, industrial facilities, shopping centers, and restaurants without banquet capabilities.

4. *Public access, event driven.* Access to centrally located fixtures is determined primarily by an event. Examples are schools, places of public assembly, stadiums, transportation terminals, and institutions.

The Hunter curve with variations is the method of choice in obtaining peak flow. In some cases, such as facilities with known periods of high use (such as in number 4 in the previous list), the flow shall be increased to allow for this intensive period of fixture usage.

Refer to Table 9.1 for WFU values for typical fixtures. To determine the estimated peak maximum flow rate in gpm for accumulated WFUs at any point in the system, use Table 9.21. This table is a variation of the original Hunter curve. Interpolate to find intermediate values.

Table 9.28 is divided into two sections, one for systems containing flush-valve-operated water closets in addition to other fixtures, and the other for flush-tank water closets in addition to other fixtures. Use the appropriate column for the specific branch or system under design. For the most accuracy, when calculating individual branch flow rates to fixtures using both hot and cold water, use 75 percent of the listed WFU value.

DESIGN OF THE WATER SUPPLY DISTRIBUTION SYSTEM

The water supply system must achieve the following basic objectives:

1. To deliver an adequate volume of water to the most hydraulically remote fixture during minimum pressure and maximum flow conditions

2. To provide adequate water pressure to the most hydraulically remote fixture during minimum pressure and maximum flow conditions

3. To prevent excessive water velocity during maximum flow conditions

The process of pipe sizing and component selection is an iterative one, requiring the design professional to assume initial values and recalculate if necessary using new values, if the initial assumptions prove wrong. Use the following simplified method as a guide to sizing. Additional criteria regarding system components are presented later in this chapter.

Pipe sizing is based on the maximum velocity of water in conjunction with the allowable friction loss in the piping system. The basic method of system design is first to establish values that are fixed, such as the minimum operating pressure of the farthest fixture and the difference in static height of that fixture from the pressure source. The pipe size would then be selected so that the remaining system pressure, in the form of friction loss of the water flowing through the pipe, would be used, but not to exceed recommended velocity figures. For a pumped system, the design engineer has the ability to increase the total dynamic head of the pump in order to provide additional pressure to the piping network if desired. The method for simplified design of a street pressure system is as follows:

1. Find the static and residual source water and the elevation at which the pressures are obtained. The residual pressure is the basis of the sizing procedure. This is discussed in Chap. 6, in the section entitled Potable Water Supply.

TABLE 9.28 Maximum Probable Flow, gpm (L/s)

Water supply fixture units	Tank-type water closets	Flushometer-type water closets	Water supply fixture units	Tank-type water closets	Flushometer-type water closets
1	1 (0.07)		120	25.9 (2.0)	76 (5.7)
2	3 (0.21)		125	26.5 (2.0)	76 (5.7)
3	5 (0.38)		130	27.1 (2.1)	77 (5.8)
4	6 (0.45)		135	27.7 (2.1)	78 (5.8)
5	7 (0.53)	27.2(2.2)	140	28.3 (2.1)	78.5 (5.8)
6	8 (0.60)	29.1(2.2)	145	29.0 (2.2)	79 (5.9)
7	9 (0.68)	30.8(2.4)	150	29.6 (2.2)	80 (6.0)
8	10 (0.70)	32.3(2.5)	160	30.8 (2.3)	81 (6.1)
9	11 (0.83)	33.7(2.5)	170	32.0 (2.4)	83 (6.2)
10	12.2(0.92)	35 (2.6)	180	33.3 (2.5)	84 (6.3)
12	12.4(0.94)	37.3(2.6)	190	34.5 (2.5)	85 (6.4)
14	12.7(0.96)	39.3(2.8)	200	35.7 (2.6)	86 (6.5)
16	12.9(0.98)	41.2(3.1)	220	38.1 (2.8)	88 (6.7)
18	13.2(1)	42.8(3.2)	240	40.5 (3.0)	90 (6.8)
20	13.4(1.01)	44.3(3.3)	260	43.0 (3.2)	92 (7.0)
22	13.7(1.02)	45.8(3.5)	280	45.4 (3.4)	94 (7.2)
24	13.9(1.03)	47.1(3.6)	300	47.7 (3.6)	96 (7.2)
26	14.2(1.07)	48.3(3.7)	400	59.6 (4.5)	102 (7.4)
28	14.4(1.09)	49.4(3.8)	500	71.2 (5.3)	108 (8.2)
30	14.7(1.1)	50.5(3.9)	600	82.6 (6.3)	113 (8.6)
35	15.3(1.1)	53.0(4.0)	700	93.7 (7.1)	117 (8.9)
40	15.9(1.2)	55.2(4.1)	800	105 (8.0)	120 (9.1)
45	16.6(1.3)	57.2(4.2)	900	115 (8.7)	123 (9.3)
50	17.2(1.3)	59.1(4.3)	1,000	126 (9.5)	126 (9.5)
55	17.8(1.4)	60.8(4.5)	1,500	175 (13.3)	175 (13.3)
60	18.4(1.4)	62.3(4.6)	2,000	220 (16.7)	220 (16.7)
65	190.(1.5)	63.8(4.7)	2,500	259 (19.7)	259 (19.7)
70	19.7(1.5)	65.2(4.9)	3,000	294 (22.3)	294 (22.3)
75	20.3(1.5)	66.4(5.0)	3,500	325 (24.7)	325 (24.7)
80	20.9(1.6)	67.7(5.1)	4,000	352 (26.7)	352 (26.7)
85	21.5(1.6)	68.8(5.2)	4,500	375 (28.5)	375 (28.5)
90	22.2(1.7)	69.9(5.3)	5,000	395 (30)	395 (30)
95	22.8(1.7)	71.0(5.3)	6,000	425 (32.3)	425 (32.3)
100	23.4(1.8)	72.0(5.4)	7,000	445 (34)	445 (34)
105	24.0(1.8)	73.0(5.5)	8,000	456 (34.6)	456 (34.6)
110	24.6(1.9)	73.9(5.6)	9,000	461 (35)	461 (35)
115	25.3(1.9)	74.8(5.7)	10,000	462 (35)	462 (35)

2. Determine by rough calculation if a water pressure booster or reducing systems are necessary. If pressure adjustment is required, select the appropriate system.

3. Locate main runs and route the water distribution system piping within the building.

4. Calculate pressure losses in the distribution system as follows:

    a. Estimate the maximum flow in the building water service. This is done by adding all WFUs and converting this number to gpm using Table 9.27.

    b. Calculate the loss of pressure in the building water service from the source into the building. Add (or subtract) the height difference between the source and height of main distribution piping inside the building, friction loss of the service line, meter, BFP, valves, and all equipment contributing to any loss of pressure. Allow an additional 5 to 10 psi (35 to 70 kPa) for future losses in water supply source pressure as a safety factor, if applicable.

   c. Find the height of the most hydraulically remote fixture from the height of main distribution piping.

   d. Find the pressure required to operate the most hydraulically remote fixture from Table 9.28.

5. Add the result of steps 4*b*, 4*c*, and 4*d* together and subtract from the figure obtained in step 1.

6. Calculate the total equivalent run of water piping to the farthest fixture.

7. Divide the pressure calculated in step 3 into the pipe run calculated in step 4 to find the friction loss allowable for the piping system. This figure shall be compatible with the following tables used to find pipe friction losses.

8. Using appropriate pipe friction loss charts or tables (such as Cameron's Hydraulic Data) or Fig. 9.31 (copper pipe), 9.32 (new steel pipe), or 9.33 (old steel pipe) and the estimated water demand, size the piping at each design point. It is helpful to make a specific project table using the pipe size that will provide the allowable friction loss, the maximum gpm for that loss, and the water velocity.

9. At each design point, do not exceed recommended water velocity.

For a pumped system, steps 1, 3, 4*a*, and 4*b* will determine the suction pressure at the inlet to the pump. Steps 4*c* and 4*d* will establish the fixed pressure requirements. The design engineer can then select a pump with enough pressure to allow a reasonable friction loss in the piping system.

ADJUSTING WATER PRESSURE

If the pressure in the water source is sufficient to supply the most hydraulically remote fixture in a building, a street pressure system is the most economical selection.

When the pressure is not adequate, it must be increased. Systems used to increase pressure include elevated water tanks, booster pumps, and hydropneumatic tank systems. If the pressure is excessive, it must be reduced to an acceptable level by a pressure-reducing valve.

Excessive Water Pressure

Water pressure in a water distribution system is considered to be excessive if it will damage, or create conditions that will damage, components of the water distribution system or create a nuisance, for example, water splashing out of a fixture during use.

There is no precise value of water pressure below which the pressure will never damage a water distribution system, and above which, will always damage the system. A widely accepted range is 70 to 80 psi (480 to 550 kPa) psi. Often, the maximum permissible water pressure is stipulated by code.

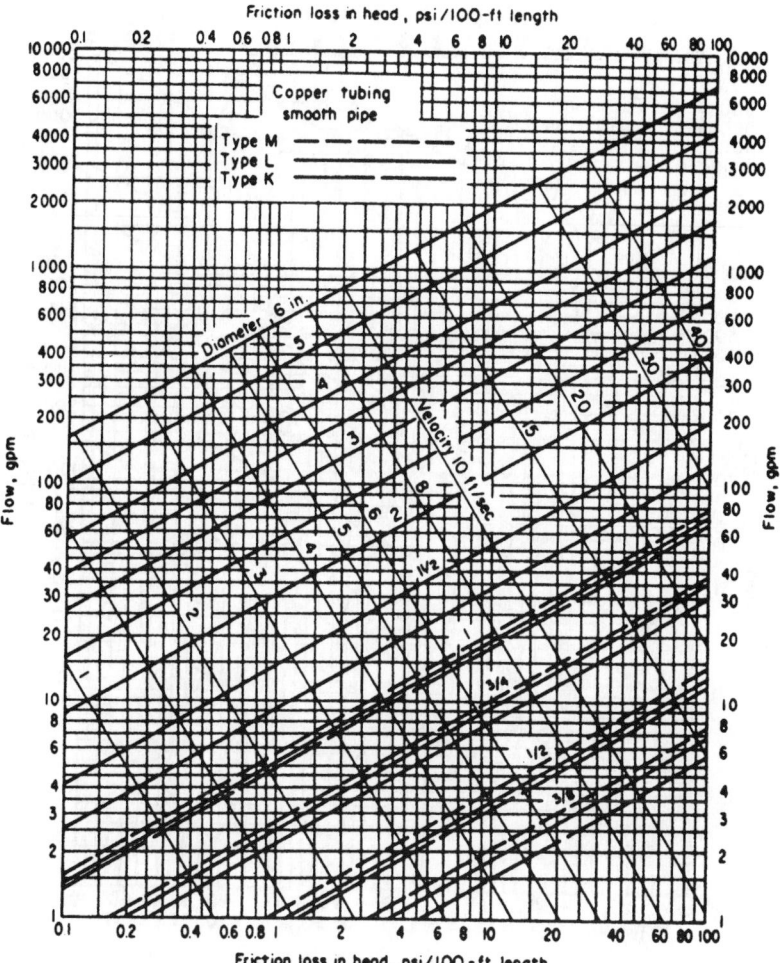

FIGURE 9.31 Flow of water in copper pipe.

A *pressure-regulating valve* (PRV) is a device used to lower and automatically maintain the water pressure within predetermined design parameters for both dynamic and static flow conditions. The PRV uses a closure device that opens and closes an orifice in response to fluctuations in outlet (regulated) pressure. The degree of closure depends upon the ability of a sensing mechanism to detect changes in water pressure at the outlet side of the valve. Pressure-regulating valves fall into two general categories: direct operated and pilot operated.

The *direct-operated* PRV has the closure member controller in direct contact with water pressure in the outlet (regulated) side of the valve. When the outlet pressure varies, the differing pressure causes the closure member to open or close by an amount necessary to achieve the desired outlet pressure. Direct-operated valves are lower in initial cost but provide less accuracy in regulating the outlet

FRICTION LOSS IN HEAD IN LBS. PER SQ. IN. PER 100 FT. LENGTH

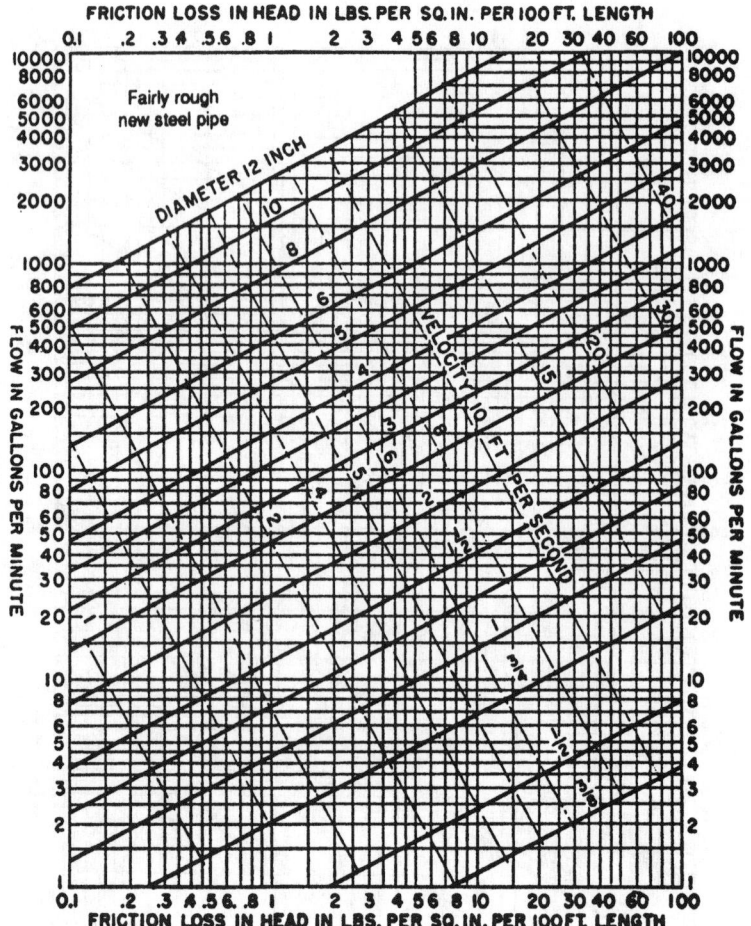

FIGURE 9.32   Flow of water in new steel pipe.

pressure. They produce a pressure reduction in proportion to the flow—the larger the flow, the less the pressure in the discharge line.

A *pilot-operated* valve is a combination of two pressure-regulating valves in a single housing. It consists of a primary valve (or pilot) that is in direct contact with water pressure in the outlet (regulated) side of the valve, and a main valve that contains the closure member. The pilot valve senses variations in the outlet pressure and magnifies closure member travel to achieve the desired outlet pressure.

Pilot-operated valves are higher in initial cost but provide a greater degree of accuracy over a wider range of pressure and flow conditions.

Selecting a Pressure-Regulating Valve.   Manufacturers offer different types of pressure-regulating valves. The different valves represent a compromise among price, capacity, accuracy, and speed of response. This information is provided by

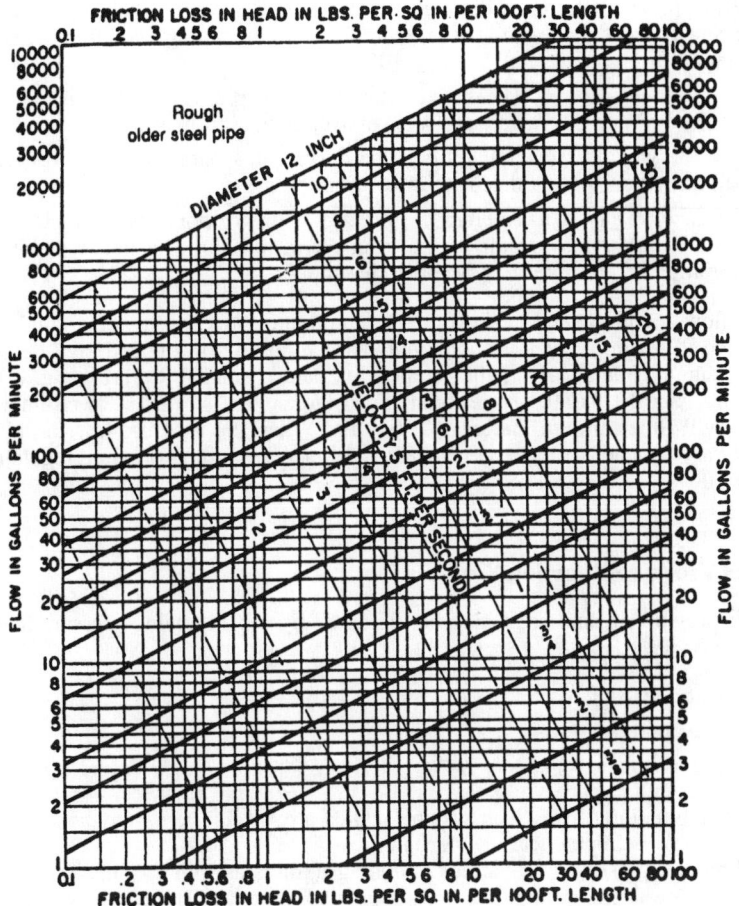

FIGURE 9.33   Flow of water in old steel pipe.

manufacturers for use in valve selection. The following considerations affect the selection of a PRV:

1. *Minimum flow rate.* The minimum rate of flow (other than zero) expected in the piping section under design.
2. *Maximum flow rate.* The maximum rate of flow expected in the piping section under design.
3. *Nature of flow.* Whether the flow rate is reasonably constant or intermittent.
4. *Location of the PRV.* The valve can be located at the beginning or end of a branch.
5. *Maximum inlet pressure.* The highest pressure expected at the inlet of the PRV.
6. *Outlet pressure.* The pressure that the PRV must maintain.

7. *Fall-off.* The difference between the design pressure at which the system has been set and the actual outlet pressure found in the piping—a difference usually limited to approximately 15 psi (105 kPa).

8. *Pressure differential.* The difference between the pressure in the inlet and outlet of the PRV. If this difference is excessive, cavitation may result.

9. *Accuracy of pressure regulation.* The degree of accuracy desired to be maintained within the regulated water distribution system.

10. *Speed of response to changes in pressure.* If this response is too rapid, a noise called *chatter* may result. If the response is too slow, an unacceptably wide variation of pressure may occur at the outlet.

If flow requirements are beyond the recommended capacity of a single PRV, multiple PRVs in parallel are commonly used to allow for low and high flows in a single supply branch. A smaller branch line and a small PRV are set at a lower pressure than the parallel larger branch and larger PRV so that a lower amount of water will flow through the small PRV first. When the pressure loss through the small valve is too great, the larger PRV will open, providing pressure regulation at higher flows.

Water Pressure-Boosting Systems

Water pressure in buildings may be increased by the following pressure-boosting systems:

1. Elevated water tank
2. Booster pump
3. Hydropneumatic tank

A combination of two such systems, called a hybrid system, may be used where a single type is impractical or uneconomical.

Minimum pressures for various plumbing fixtures as well as a maximum allowable pressure in the system are often stipulated in codes. These values of pressure must be used as a basis for system design unless more stringent values are required for the project under design. Generally accepted minimum values of operating pressure for various fixtures are given in Table 9.29. For special equipment not mentioned, consult the equipment manufacturer.

Elevated Water Tank. In an elevated water tank system, such as the simplified system shown in Fig. 9.29, water is pumped from the water main to an elevated water storage tank (commonly called a gravity tank or a house tank) located above the highest and most hydraulically remote point in the water supply system of the building. A schematic piping arrangement between the components of an elevated water tank system is illustrated in Fig. 9.34. The height of the tank provides additional static head, resulting in a higher pressure in the water distribution system.

The advantages of the elevated water tank system are:

1. It is less complex than either of the other two systems.
2. Fewer components are required to control and operate the system.
3. The efficiency is greater and operating costs are lower than for either of the other systems.

TABLE 9.29 Minimum Acceptable Operating Pressures for Various Fixtures

Fixture	Pressure, psi*
Basin faucet	8
Basin faucet, self-closing	12
Sink faucet, ⅜ in (0.95 cm)	10
Sink faucet, ½ in (1.3 cm)	5
Dishwasher	15–25
Bathtub faucet	5
Laundry tub cock, ¼ in (0.64 cm)	5
Shower	12
Water closet gravity, low consumption	20
Water closet flush valve	25
Urinal flush valve	25
Garden hose, 50 ft (15 m), and sill cock	30
Water closet, blowout type	30
Urinal, blowout type	25
Water closet, low-silhouette tank type	30
Water closet, pressure tank	35

*psi × 7 = kPa.

4. A smaller pump capacity is required than for either of the other two systems.
5. Pressure fluctuations in the system are small.
6. Maintenance requirements are minimal.

The disadvantages of the elevated water tank system compared with the other two systems are:

1. An exposed tank (or the enclosure around it) may be considered unsightly.
2. The building structure may require reinforcement to support the additional weight of the tank and water.
3. The water in the tank and the supply pipes from the tank are subject to freezing if the tank is exposed.
4. The water pressure on the highest floor(s) may be inadequate requiring an additional pumped system for the top several floors.
5. The possibility exists of a catastrophic tank failure flooding the roof with water.

The capacity of the house tank depends on the type of facility it will serve. Refer to Table 9.30 to find recommended minimum domestic storage volumes of gravity tanks for various building types except multiple dwellings. For multiple dwellings, refer to Fig. 9.35.

To use Table 9.30, first determine the total number of fixtures, regardless of type. Multiply the number of fixtures by the gpm per fixture. Then multiply the resulting figure by the tank size multiplier. This figure is for domestic storage only. To this, add constant uses of water, such as HVAC makeup, fire protection water storage, and any process requirements, if necessary. Select a standard tank size equal to or exceeding the storage required. Standard wood roof tank sizes are given in Table 9.31. If the building is occupied predominantly by women, add 15 percent to the number of gallons of water storage requirements.

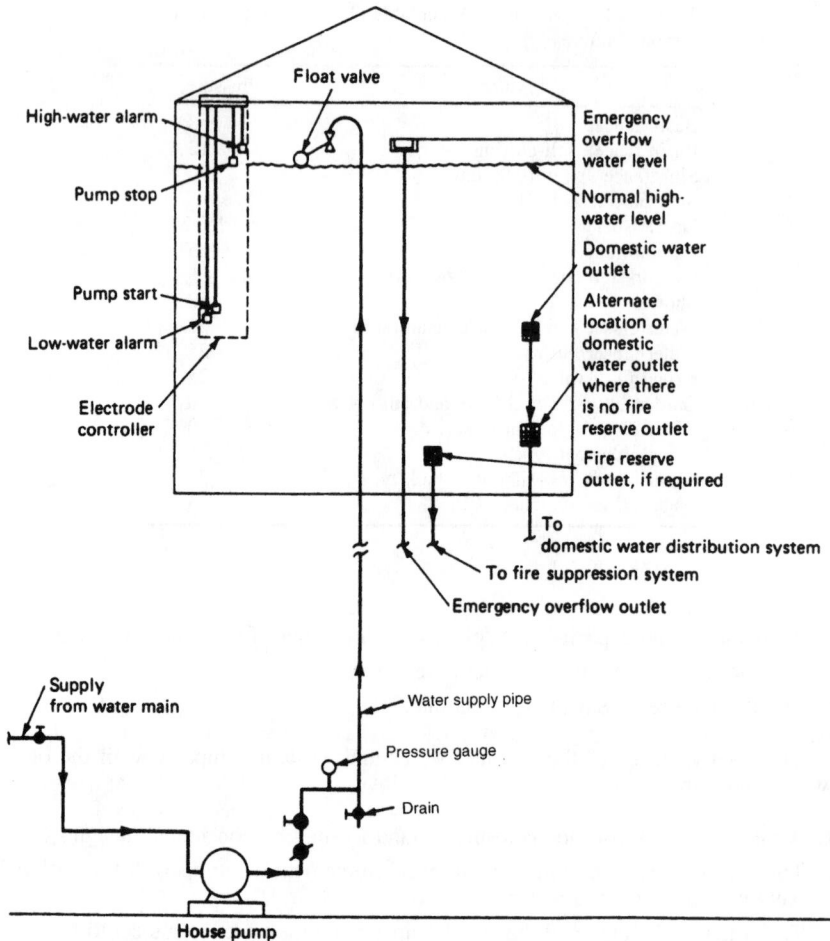

FIGURE 9.34   Piping arrangement of an elevated water tank.

For multiple dwellings, first find the number of apartments. Using Fig. 9.35, find the number of gallons storage per person. If the actual number of people is not available, an approximate number can be found by using two people per bedroom or four people per apartment, whichever is larger, and multiplying by that number. Then multiply the number of people by the number of gallons (liters) required per person to find the domestic storage capacity of the house tank. To this, add constant uses of water, such as HVAC makeup and fire protection water storage, if any. Then select a standard storage tank size equal to or larger than the required amount.

The capacity of the house pump(s) is determined by the quantity of water stored for domestic use. In general, a house pump should be capable of replacing the domestic reserve in about ½ to 2 h—with 1 h a generally accepted value. A duplex pump arrangement (i.e., two pumps in parallel), with each pump full size, should

TABLE 9.30   Size of Gravity Tanks

Number of fixtures	gpm per fixture	Tank size multiplier	Minimum domestic capacity, gal	Minimum pump capacity, gpm
		Hotels and clubs		
1–50	0.65	30	2000	25
51–100	0.55	30	2000	35
101–200	0.45	30	3000	60
201–400	0.35	25	3000	100
401–800	0.275	25	3000	150
801–1200	0.25	25	3000	225
1201–above	0.2	25	3000	300
		Hospitals		
1–50	1	30	2000	25
51–100	0.8	30	2000	55
101–200	0.6	25	3000	85
201–400	0.5	25	3000	125
401–above	0.4	25	3000	210
		Schools		
1–10	1.5	30	2000	10
11–25	1	30	2000	15
26–50	0.8	30	2000	30
51–100	0.6	30	2000	45
101–200	0.5	30	3000	65
201–above	0.4	30	3000	110
		Industrial buildings		
1–25	1.5	30	2000	25
26–50	1	30	2000	40
51–100	0.75	30	3000	60
101–150	0.7	25	3000	80
151–250	0.65	25	3000	110
251–above	0.6	25	3000	165
		Office buildings		
1–25	1.25	30	2000	25
26–50	0.9	30	2000	35
51–100	0.7	30	2000	50
101–150	0.65	25	3000	75
151–250	0.55	25	3000	100
251–500	0.45	25	3000	140
501–750	0.35	25	3000	230
751–1000	0.3	25	3000	270
1001 and above	0.275	25	3000	310

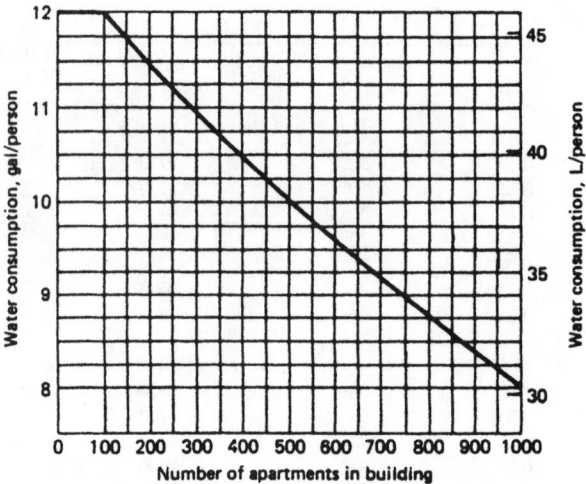

FIGURE 9.35 Estimated water house tank storage capacity, multiple dwellings.

TABLE 9.31 Size of Standard Wood House Tanks

Capacity, gal*	Dimension, ft†		Number of dunnage beams	Dunnage, in‡
	Diameter	Stave		
5,000	10 ×	10	5	4 × 6
5,000	9 ×	12	5	4 × 6
5,000	11 ×	8	6	4 × 6
6,000	11 ×	10	6	4 × 6
7,000	11½ ×	10	6	4 × 6
7,500	12 ×	10	6	4 × 6
7,500	11 ×	12	6	4 × 6
8,000	12½ ×	10	6	4 × 6
9,000	12 ×	12	6	4 × 6
10,000	13 ×	12	7	4 × 6
10,000	12 ×	14	6	4 × 6
10,000	11 ×	16	6	4 × 6
12,000	14 ×	12	7	4 × 6
12,000	13 ×	14	7	4 × 6
12,000	12 ×	16	6	4 × 6
15,000	14 ×	14	7	4 × 6
15,000	13½ ×	16	8	4 × 6
20,000	16 ×	16	8	6 × 6
25,000	17½ ×	16	9	6 × 6
30,000	18 ×	18	9	6 × 6
40,000	20 ×	20	10	6 × 6
50,000	22 ×	20	11	6 × 8

*1 gal = 3.8 L.
†1 ft = 0.3 m.
‡1 in = 25.4 mm.
Source: Isseks Bros. Co.

be provided in case one pump goes out of service. A schematic piping arrangement at the pump is shown in Fig. 9.34.

A common problem with the gravity tank system is the lack of adequate water pressure on the upper floors of the building unless the tank is elevated to an impractical height. Under these conditions, a hybrid system comprising the elevated tank plus a small booster pump system only for the top several floors can be used.

Booster Pump System.   In a booster pump system, illustrated schematically in Fig. 9.36, the booster pump(s) are supplied from a source with pressure that is both insufficient and variable. A control system senses variations in the distribution system and adjusts either the speed of the pump or a PRV to maintain a constant increased pressure of acceptable value. Two types of booster pump drives are available for adjusting the pressure and flow in the water distribution system, constant speed and constant pressure.

Constant-speed drive.   This type of drive should be considered where (1) water demand requirements are relatively constant, (2) a low-to-medium boost pressure is required, and (3) a low initial cost is important. It should also be considered where the frictional losses in the piping network are relatively minor.

Variable-speed drive.   This type of drive is recommended where (1) there are large fluctuations in the water main supply pressure to the pumps, (2) there is a requirement for a high pressure boost, or (3) a great variation is expected in the system water demand. Variable speed drives are also recommended where the friction losses in the water distribution system are relatively high. A variable-speed drive system is higher in initial cost than a constant-speed drive.

Number of Pumps.   A booster pump system that runs continuously should include pumps of various sizes, so that during periods of little use the smaller pump is used, resulting in a more economical operation. A hybrid system using a small hydropneumatic tank should be considered during off-peak hours to provide additional operating economy by allowing the pump system to be shut down during extended periods of little or no water demand.

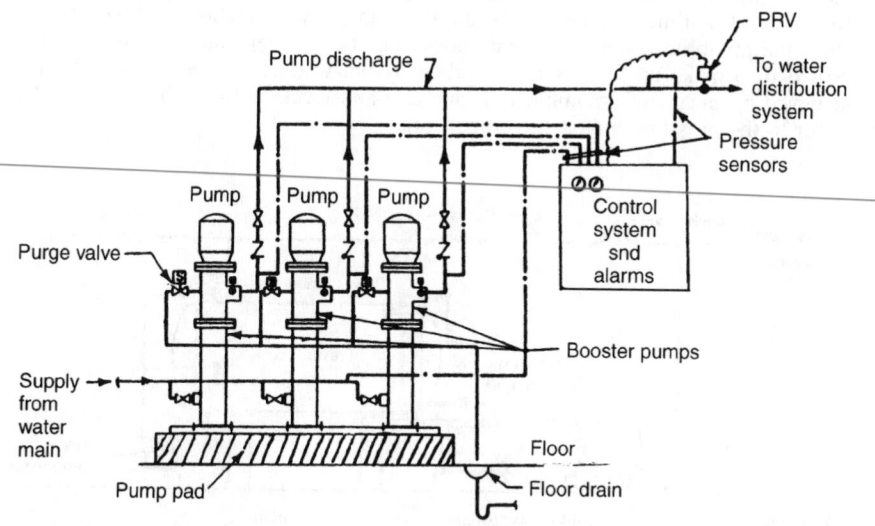

FIGURE 9.36   Schematic detail of booster pump system components.

A two-pump system is usually selected for both redundancy and economy. Each of the two pumps should be capable of supplying 60 percent of the total estimated peak water demand. If maintenance of a total supply of water is a prime consideration, then select three pumps, each of which could supply 50 percent of the total requirements. Then, full system capacity will be maintained even if one of the pumps is out of service.

Water usage studies have shown that in most buildings, the actual water demand is approximately 25 percent of estimated peak demand during three-quarters of a normal working day. If operating cost is a determining factor, the use of three pumps is desirable, sized to provide 50, 50, and 25 percent of the peak estimated demand flow. In this configuration, the smaller pump operates most of the time when the load is low. It also operates in conjunction with the larger pumps during periods of peak demand. A typical three-pump system is shown in Fig. 9.36.

Variable-Speed Control. The following control devices are commonly used to vary the speed of either the pump or the motor to maintain satisfactory pressure in the water distribution system:

1. *Electronic controller.* An electronic controller is a solid-state device that controls the current supplied to the motor, thereby varying its speed. This allows the motor to be connected directly to the pump.

2. *Fluid and magnetic couplings.* These variable-speed couplings are used to connect the motor to the pump. These devices use either a fluid (usually water or oil) or electrically induced magnetism to adjust the speed of the pump while the motor rotates at a constant speed.

Hydropneumatic Tank System. In a hydropneumatic tank system, such as that shown in the schematic diagram in Fig. 9.37, water is pumped from the water supply into a pressure tank for storage. The air in the tank is compressed by the water entering the tank. As the pressure in the tank is increased, the pressure in the water distribution system is also increased since it is also connected to the tank. The pressure tank may be located anywhere in the building. The water stored in the tank and the pressure in the tank are sufficient to allow the pumps to shut down for a period of time and yet satisfy the demand for water. This conserves energy since the pump need not be in continuous operation. A separate air compressor is provided to make up for any air absorbed by the water. Control of the system is achieved by means of a combination pressure and level control, which adds water or air to the pressure tank at the proper time.

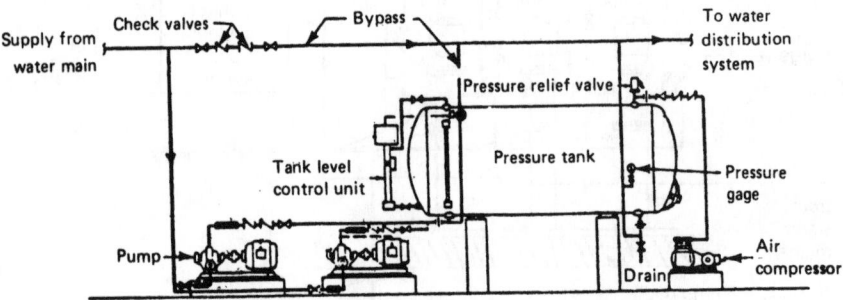

FIGURE 9.37 Schematic detail of hydropneumatic tank components.

Since the total amount of stored water is relatively small, a pressure tank may run out of water during periods of peak demand unless a very large tank is selected. Therefore, the fill pumps must be sized to supply the peak demands of the building at the lowest calculated system pressure. The piping arrangement of such a system must include a tank bypass arrangement to allow the fill pumps to supply the building directly if a sustained peak demand occurs and there is not enough water in the pressure tank to supply the peak demand for an extended period of time.

The size of the air compressor is determined by the size of the storage tank. Select an air compressor that can supply 2 cfm for a tank having a total capacity in excess of 4000 gal (16 000 L) or less.

Because of very high initial cost and the large amount of floor space required, a hydropneumatic system is no longer considered for an entire building supply. Booster pumps have replaced the hydropneumatic tank, although some of these systems still are operating in older buildings. Small hydropneumatic systems are often used as part of a hybrid system to allow booster pumps to be shut down occasionally.

Hybrid System. A hybrid system is a combination of two systems. The most common is a small hydropneumatic system used in conjunction with a booster pump for economy in facilities where little or no water is used for extended periods of time. Examples are office buildings or factories that are closed at night. Another hybrid system, consisting of a small booster pump assembly and a house tank, is used to provide a higher pressure to the upper floors where the height of the house tank is limited, such as for multiple dwellings.

PIPE SIZE SELECTION

Pipe sizing is accomplished using maximum allowable pressure loss, flow rate, and velocity at the design point. Two methods of determining these values are available. The first uses prepared tables for each pipe size, with the velocity and pressure loss for various flow rates. The second uses nomographs, where all the information is displayed on a single graph. The tables are more accurate but the nomographs are more convenient. The nomographs in Figs. 9.31, 9.32, and 9.33 are provided for smooth, fairly rough, and rough pipes. Special charts have been made for specific plastic types of pipe, but space limitations prevent reproduction in this handbook.

SERVICE HOT WATER SYSTEMS

SYSTEM DESCRIPTION

The service hot water system heats raw water from ambient temperature to a desired higher temperature and delivers the heated water to terminal points with a delay of less than 20 s. A booster hot water system heats primary service hot water to a higher temperature for a specific purpose.

CODES AND STANDARDS

In addition to the standard requirements of the applicable plumbing code, many other additional codes are used in the design and manufacture of specific equipment.

1. NFPA code for oil-burning equipment and gas-fired heaters
2. ASME code for pressure vessels, relief valves, etc
3. UL codes for electric and oil-burning heaters
4. ANSI Z-21 heaters and relief valves
5. ASHRAE code for energy conservation

WATER HEATING METHODS AND EQUIPMENT

Water heaters are categorized as fully instantaneous, semi-instantaneous, or storage type. They can be either directly or indirectly fired. Direct-fired heaters use a heat source, such as electricity, fuel gas, oil, and solar panels, to directly transfer heat in the heating vessel. Indirect heating uses the fuel to produce a heating medium, such as steam or hot water, that gives up its heat in a heat exchanger.

Fully instantaneous heaters consist of a unit that heats water as quickly as the demand flow rate requires. They have no storage and a high recovery rate. Advantages include little floor space required, low initial cost, and factory preassembly into a package ready to install. Disadvantages are difficult control of outlet water temperature and high instantaneous Btu requirements for the heating medium. For instantaneous heaters, an accurate approximation for steam can be calculated by multiplying the gpm requirements by 50 lb/h. This type of heater is almost always indirectly fired using either steam or hot water supplied from a central heating plant, steam utility system, or a boiler.

Semi-instantaneous heaters are similar to the fully instantaneous types except that they have a very small water storage capacity which permits better control of outlet water temperature. This type of heater can be either directly or indirectly fired, and is preferred over the fully instantaneous type. Refer to Fig. 9.38 for a detail of a typical semi-instantaneous water heater.

Storage heaters have a large storage capacity and low recovery rate. This system consists of either a combination storage tank and a direct or indirect immersion heater inside the tank or separate water heater and storage tank. This system should

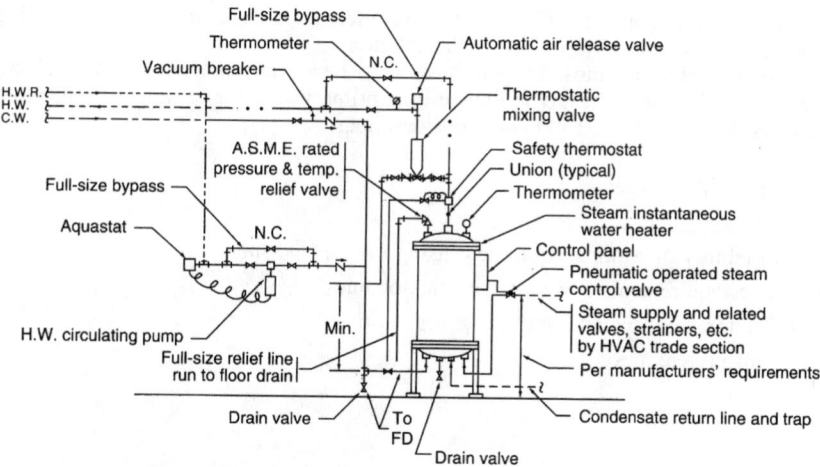

FIGURE 9.38   Detail of semi-instantaneous water heater.

be considered when high peak surge loads are encountered for short periods of time and when a limited source of heat energy exists. Disadvantages include a large amount of floor space required and high initial cost. Advantages include a low instantaneous heat energy demand rate. Refer to Fig. 9.39 for a detail of a typical storage heater.

Point-of-use heaters are usually direct fired and are used for isolated and remote locations where it is not economical to run piping from the central service hot water system. Popular units are electric in-line instantaneous heaters installed below an

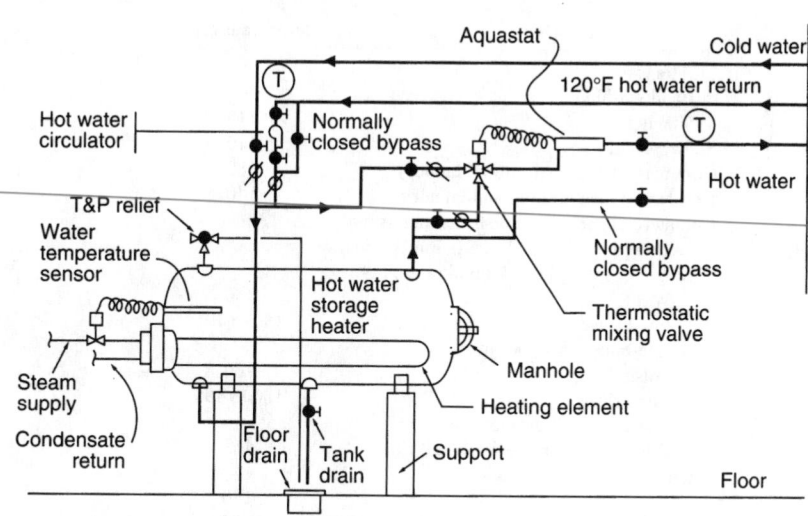

FIGURE 9.39   Detail of storage type water heater.

individual fixture or 5-gal (20-L) units installed on a shelf or in a hung ceiling adjacent to a service sink or similar installation.

The fuels that are most frequently used to heat water are electricity, fuel oil, fuel gas, and solar energy. The choice of a primary fuel for central water-heating systems depends on the following considerations:

1. Availability of fuel
2. Cost
3. Availability of heating equipment using the desired fuel
4. Space requirements and cost for vents or flues
5. Fuel storage space and applicable regulations
6. Client preferences

ACCEPTABLE HOT WATER TEMPERATURES

Table 9.32 lists the generally acceptable minimum temperatures for common purposes, various plumbing fixtures, and other pieces of equipment. Other codes and standards, particularly for hospitals and health care facilities, may further restrict the allowable temperature of the hot water supplied for various uses. These figures must be obtained from such codes and standards required by authorities.

TABLE 9.32 Minimum Hot Water Temperature for Plumbing Fixtures and Equipment

Use	Minimum temperature, °F
Lavatory:	
Hand washing	105
Shaving	115
Showers and tubs	110
Commercial and institutional laundry	180
Residential dishwashing and laundry	140
Commercial spray-type dishwashing as required by National Sanitation Foundation:	
Single or multiple tank hood or rack type:	
Wash	150
Final rinse	180 to 195
Single-tank conveyor type:	
Wash	160
Final rinse	180 to 195
Single-tank rack or door type:	
Single-temperature wash and rinse	165
Chemical sanitizing glasswasher:	
Wash	140
Rinse	75

WATER HEATER SIZING

Hot water usage and flow rates are characterized by intermittent periods of peak, sustained, and low to zero flow conditions. The pattern and usage requirements vary depending on building type and use, equipment, population, and time of day.

Each different type of water heater is sized using different criteria. Both types of instantaneous heaters are sized only on the maximum peak flow rate. Storage heaters are sized using the recovery rate in conjunction with the storage capacity to satisfy demand over a period of time ranging from 1 to 4 h. Determination of the recovery and storage criteria can be based on either population or the number of fixtures in the facility. Population is the more accurate criterion when available.

When designing a project for a client with a preference for specific heater types and sizing criteria, such as various city and state agencies, the federal government, and Department of Defense, the design engineer must follow the methods established by the client. When no criteria exist, sizing should be based on accepted practices, such as those published by the ASPE and ASHRAE. It must be emphasized that sizing water heaters is as much an art as a science, with many possible methods capable of specifying adequately sized equipment for facility needs.

Instantaneous and Semi-instantaneous Water Heater Sizing

Instantaneous and semi-instantaneous water heaters are sized by the fixture unit method, calculating the maximum hot water flow rate in gpm. The gpm for hot water alone is less than the value found for fixtures using both hot and cold water. The maximum fixture demand is found by using a specialized hot water WFU table and a modified Hunter curve converting hot water WFUs into gpm.

Table 9.33 assigns hot water demand in WFUs for various fixtures. Figures 9.40 and 9.41 convert the WFUs into gpm using a modified Hunter curve for hot water depending on the total WFUs calculated. To the maximum fixture demand, add other equipment, such as food service, laundries, and the like, used in the facility.

Storage-Type Water Heater Sizing

Storage-type heaters are sized using the total volume of hot water required for the estimated duration of maximum demand. This total volume can be provided by any combination of hot water storage capacity and water heater recovery rate. Storage and recovery values depend on the type of project. The major advantage of using a storage heater is that the instantaneous demand for steam (or other heating media) is considerably less than that for instantaneous heaters.

One disadvantage of a storage tank is that more cold water will enter the tank than can be heated to the desired temperature. An allowance called the cold water correction factor must be made to provide for the fact that the temperature of the water in the tank will be lowered when this happens. The generally accepted figure is that no more than approximately 30 percent cold water can enter the tank before the temperature of the water is lowered below acceptable limits. Conversely, only 70 percent of water in a storage tank is considered useful.

No single sizing method is considered suitable for all building types. Three criteria are commonly used: (1) number of fixtures, (2) population, and (3) flow rate.

TABLE 9.33 Hot Water Demand in Fixture Units at 140°F (60°C)

Type of fixture	Fixture units	Type of fixture	Fixture units
Hospital		Assoc. Bldg. YMCA	
Private lavatories	0.75	Private lavatory	0.75
Public lavatories	1.0	Public lavatory	1.0
Semiprivate lavatories	1.2	Private shower‡	1.5
Private shower‡	1.5	Tub and shower‡	1.7
Ward shower‡	2.5	Service sink	2.5
Semiprivate shower‡	1.5	Laundry tray	2.0
Private bath	1.5	Restaurant†	
Ward bath	2.0		
Sink—flushing rim	2.0	Private lavatory	0.7
Sink—scrub-up	1.5	Public lavatory	2.0
Sink—laboratory	1.5	Private shower‡	1.5
Sink—general purpose	1.0	Public shower‡	1.7
Bath—leg	6.0	Sink—kitchen	3.0
Bath—arm	4.0	Sink—pantry	2.5
Bath—sitz	3.0	Service sink	2.0
Bath—foot	3.0	Sink—pot (single)	2.5
Bath—emergency	2.0	Sink—pot (double)	3.5
Hydrotherapeutic showers:		Sink—pot (triple)	5.5
#1 shower head	8.0	Sink—vegetable	2.0
#2 spray	1.2	Sink—bar	2.5
Continuous flow bath		Washer—silver	2.0*
Continuous flow fill	2.0	Washer—glass	2.0*
Continuous flow operate	1.5	Washer—can	3.0
Hubbard	4.0	Coffee urn	1.2
Autopsy table	2.0	Baine marie	1.0
Autopsy sink and table	2.5	Pot and pan washer	2.0*
Club		Dish prerinse	2.5
		Prescraper	2.0
Private lavatory	0.75	Prescraper conveyor	2.5
Public lavatory	1.0	36″ half Bradley	1.0
Private shower‡	1.5	36″ full Bradley	1.5
Public shower‡	1.7	Typical dishwashers	
Tub and shower‡	1.5	(Use booster to heat from 140° to 180°F.)	
Service sink	2.5		
36″ half Bradley	1.0	Single tank stationary rack	
36″ full Bradley	1.5	16 × 16 rack	2.2
54″ half Bradley	1.5	18 × 18 rack	3.7
54″ full Bradley	2.0	20 × 20 rack	4.0
Gymnasium		Multiple tank conveyor type	
		Dishes—inclined	2.0
Private lavatory	0.75	Dishes—flat	2.5
Public lavatory	1.0	Single tank conveyor type	2.3
Private shower‡	1.5	Hotel-motel	
Public shower‡	3.0	Private lavatory	0.75
Sink—slop	1.5	Public lavatory	1.0
Basin—foot	1.2	Private shower‡	1.5
36″ half Bradley	1.0	Tub and shower‡	1.5
36″ full Bradley	1.5	Basin—barber	2.0
54″ half Bradley	1.5	Service sink	2.5
54″ full Bradley	2.0	Basin—beauty parlor	2.5

TABLE 9.33 Hot Water Demand in Fixture Units at 140°F (60°C) (*Continued*)

Type of fixture	Fixture units	Type of fixture	Fixture units
Office bldg.		Multiple dwelling	
Private lavatory	0.75	Private lavatory	0.75
Public lavatory	1.0	Public lavatory	1.0
Private shower‡	1.5	Private shower‡	1.5
Service sink	2.5	Public shower‡	1.5
Janitor drop	2.5	Tub and shower‡	1.5
36″ half Bradley	1.0	Sink—kitchen	0.75
36″ full Bradley	1.5	Sink—slop	1.5
Factory		Sink—pantry	1.5
		Domestic clothes washer	1.2
Private lavatory	0.75	Domestic dish washer	1.5
Public lavatory	1.0	Laundry tray	1.5
Private shower‡	1.5	Private—public school	
Public shower‡	3.0		
Service sink	2.5	Private lavatory	0.75
36″ half Bradley	1.0	Public lavatory	1.0
36″ half Bradley	1.5	Private shower‡	1.5
54″ half Bradley	1.5	Tub and shower‡	1.7
54″ half Bradley	2.0	Service sink	2.5
Correctional or mental institution		Janitor drop	1.5
		Domestic clothes washer	2.0
Private lavatory	0.7	Domestic dish washer	2.0
Public lavatory	1.0	Institution—home	
Private shower‡	1.5		
Public shower‡	3.0	Private lavatory	0.7
Tub and shower‡	1.5	Public lavatory	1.0
Sink—slop	2.0	Private shower‡	1.5
Janitor drop	2.0	Tub and shower‡	1.5
36″ half Bradley	1.0	Service sink	2.0
36″ full Bradley	1.5	Laundry tray	2.0
54″ half Bradley	1.5		
54″ full Bradley	2.0		

*These items require 180°F hot water. The consumption figures are based on supplying 140°F water with a booster heater used to obtain 180°F water.

†Add 20% to all figures when not used in combination with other building services from same heater.

‡The fixture units listed for shower heads are based on a flow rate of 3 gpm. These units should be corrected for other flow rates. Multiply the fixture units by Correction Factor "C" from the formula: C = G×.33. Where C = Correction Factor, G = gpm of shower head being used. Example: Shower head 4 gpm = C = 4×.33 or 1.32. From Fixture Capacity Table, Hotel-Motel (showers) which shows 1.5 fixture units, multiply 1.5×1.32 = 2.0 fixture units per shower head using 4 gpm.

Source: Patterson-Kelly.

Sizing Based on Number of Fixtures. Fixtures are often used as criteria for design of buildings such as health care facilities. The occupants of these facilities use the fixtures over an extended period of time, and the number of fixtures are generally proportional to the number of people. This sizing method calculates the total anticipated volume in gallons (liters) of hot water for the peak duration of use and a storage tank capacity. The following simplified procedure will find the volume of hot water required.

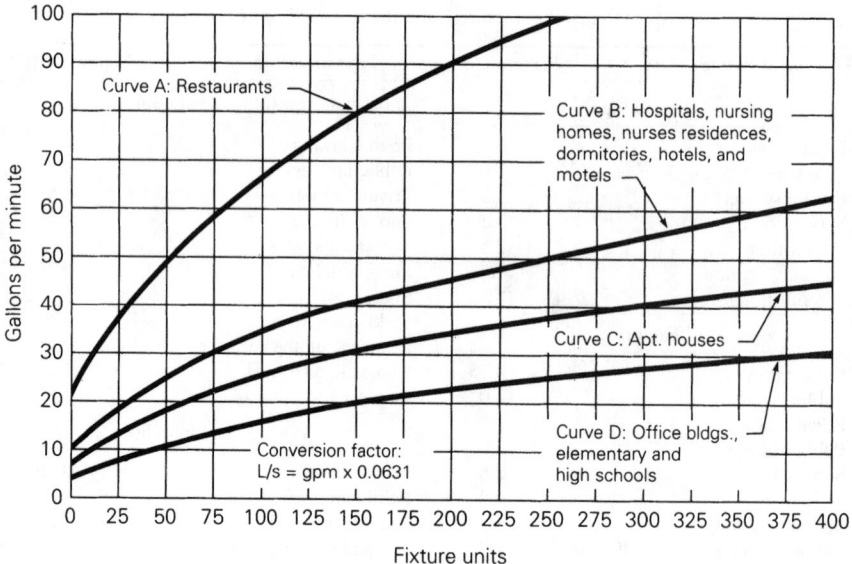

FIGURE 9.40 Modified Hunter curve for hot water, enlarged section.

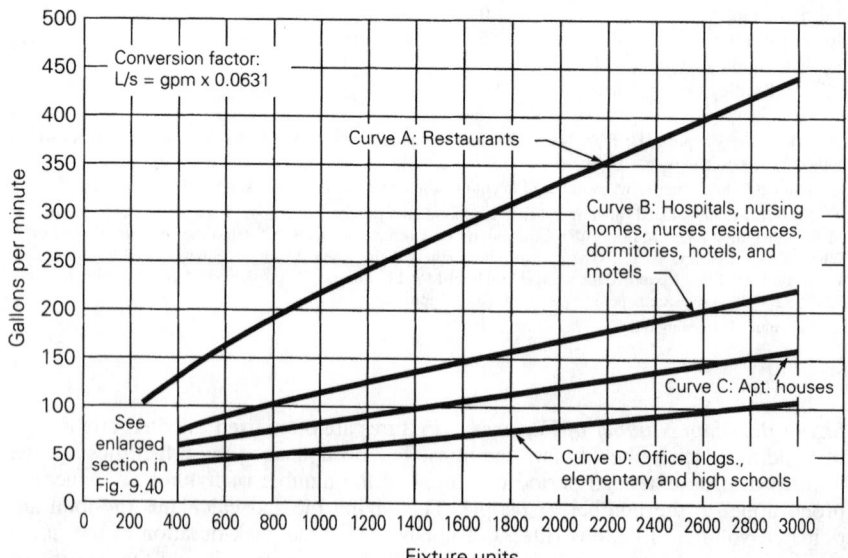

FIGURE 9.41 Modified Hunter curve for hot water.

1. Using Table 9.34, select the facility in the table that is closest to the actual facility. Count the total number of each type of fixture found at the facility and assign the gallons per hour value for each fixture found under the selected building type heading.

2. Multiply the value found for each separate fixture by the total number of individual fixtures.

3. Add the resultant figure from all individual fixture types together to get a total connected load in gallons (liters) of water per hour.

4. Find the actual hourly demand for the facility by multiplying the total connected load calculated in step 3 by the demand factor at the bottom of Table 9.34. The resulting figure is the actual volume of hot water that will be used by the facility during a 1-h period of time.

5. In conjunction with the total hourly demand, a recommended amount of hot water shall be stored. The volume of storage is a function of the actual hourly demand. The storage requirement is found by multiplying the total hourly demand (found in step 4) by the storage capacity factor found at the bottom of Table 9.34. The result is the usable volume of hot water required to be stored. To find the actual required capacity, add an additional 30 percent of the calculated volume of the usable capacity.

6. The recovery requirement and the storage requirement represent one hour's peak use of hot water. Select any combination of heating capacity (recovery) and storage capacity capable of meeting this peak hour. Consultation with the supplier of the heating medium would be helpful in determining the Btus available from the central plant. If the duration is longer than 1 h, proceed to the next step.

7. Refer to Table 9.35 to determine if the facility should have more than a 1-h peak usage duration. If so, multiply the actual hourly demand by the number of hours duration to find the total volume, in gallons (liters) of hot water required for the entire peak period. Select a recovery rate and storage tank capacity combination for that amount. As an example, if the actual hourly demand is found to be 1000 gal/h for a duration of 3 h, the total volume of hot water required will be 3000 gal. If the usable storage volume is 1000 gal, the recovery rate required for the 3-h period will be 2000 ÷ 3, or 670 gal/h. The total of 1000 gal storage + 2010 gal makeup (670 × 3) will equal slightly more than the 3000 gal required.

8. To determine the volume of water for any temperature other than 140°F, use the following formula:

$$CF = \frac{100 - IWT}{HDT - IWT} \div 0.6 \times \text{actual hourly demand} \qquad (9.2)$$

where CF = correction factor
IWT = inlet water temperature
HDT = heater water discharge temperature

Multiply the actual by the CF to find the adjusted hourly demand.

Sizing Based on Population. Population should be the criterion for design when the number of fixtures does not necessarily correlate to the number of people using the facility, for example, in residence-type projects, hotels, motels, schools, and nursing homes.

This method of sizing calculates the total anticipated volume in gallons (liters) of hot water, the recovery, and storage tank capacity for 1 h. Figures 9.42*a* to 9.42*h*

TABLE 9.34 Hot Water Demand in Gallons (Liters) per Hour at 140°F (60°C)

	Apartment house	Club	Gymnasium	Hospital	Hotel	Industrial plant	Office building	Private residence	School	YMCA
1. Basins, private lavatory	2 (7.6)	2 (7.6)	2 (7.6)	2 (7.6)	2 (7.6)	2 (7.6)	2 (7.6)	2 (7.6)	2 (7.6)	2 (7.6)
2. Basins, public lavatory	4 (15)	6 (23)	8 (30)	6 (23)	8 (30)	12 (45.5)	6 (23)	—	15 (57)	8 (30)
3. Bathtubs	20 (76)	20 (76)	30 (114)	20 (76)	20 (76)	—	—	20 (76)	—	30 (114)
4. Dishwashers*	15 (57)	50–150 (190–570)	—	50–150 (190–570)	50–200 (190–760)	20–100 (76–380)	—	15 (57)	20–100 (76–380)	20–100 (76–380)
5. Foot basins	3 (11)	3 (11)	12 (46)	3 (11)	3 (11)	12 (46)	—	3 (11)	3 (11)	12 (46)
6. Kitchen sink	10 (38)	20 (76)	—	20 (76)	30 (114)	20 (76)	—	10 (38)	20 (76)	20 (76)
7. Laundry, stationary tubs	20 (76)	28 (106)	—	28 (106)	28 (106)	—	20 (76)	20 (76)	—	28 (106)
8. Pantry sink	5 (19)	10 (38)	—	10 (38)	10 (38)	—	—	5 (19)	10 (38)	10 (38)
9. Showers	30 (114)	150 (568)	225 (850)	75 (284)	75 (284)	225 (850)	10 (38)	30 (114)	225 (850)	225 (850)
10. Service sink	20 (76)	20 (76)	—	20 (76)	30 (114)	20 (76)	30 (114)	15 (57)	20 (76)	20 (76)
11. Hydrotherapeutic showers				400 (1520)						
12. Hubbard baths				600 (2270)						
13. Leg baths				100 (380)						
14. Arm baths				35 (130)						
15. Sitz baths				30 (114)						
16. Continuous-flow baths				165 (625)						
17. Circular wash sinks				20 (76)		30 (114)	20 (76)		30 (114)	
18. Semicircular wash sinks				10 (38)		15 (57)	10 (38)		15 (57)	
19. DEMAND FACTOR	0.30	0.30	0.40	0.25	0.25	0.40	0.30	0.30	0.40	0.40
20. STORAGE CAPACITY FACTOR†	1.25	0.90	1.00	0.60	0.80	2.00	2.00	0.70	1.00	1.00

*Dishwasher requirements should be taken from this table or from manufacturers' data for the model to be used, if this is known.

†Ratio of storage tank capacity to probable maximum demand/h. Storage capacity may be reduced where an unlimited supply of steam is available from a central street steam system or large boiler plant.

Source: ASPE.

TABLE 9.35 Duration of Peak Use

Building type	Peak duration, hours
Motel	2
Hotel	3
Apartment house	3
Nursing home	3
Office	2
Food service	1 to 4 (varies with type)

provide infinite combinations of storage and recovery for use with this method. The chart ranges have been selected based on the studies that produced this method of design. Any requirements imposed by heating medium and space restrictions can be satisfied by selecting an acceptable combination.

1. Using Fig. 9.42, find the type of facility under design. Determine the population of the facility.

2. Entering the chart, select the combination of storage and recovery desired.

3. Using the chart curve, read the usable storage capacity and recovery figure per person in gal/h.

4. Calculate the total usable storage and recovery capacity that satisfies project requirements by multiplying the separate recovery and storage figures by the population.

5. Add an additional 30 percent to the calculated usable volume of storage to determine the total storage capacity to allow for the cold water correction factor.

6. Add a water temperature correction if necessary.

7. A system heat loss of 15 Btu/ft run of pipe shall be added to the gal/h recovery figure calculated to make up for the heat lost from the entire system. This is only necessary for the population method because no consideration for this heat loss was made in the hot water usage criteria.

Sizing Based on Flow Rate. Flow rate should be the criterion for design of specialized buildings, such as convention centers, prisons, sporting arenas, and gymnasiums, where predictable, concentrated peak usage periods occur. The following simplified procedure will find the volume of hot water required:

1. Determine the duration, in hours or minutes of peak use.

2. Find the typical flow rate of all fixtures contributing to flow during the peak duration from Table 9.1.

3. Establish the amount of time the fixtures are actually used during the peak duration.

4. Select a combination recovery rate and storage tank capacity that will supply the calculated amount of hot water for the peak duration.

Additional Sizing Criteria. Table 9.36 has been prepared from actual metered data from a limited number of the various facilities indicated. Because of the limited study it has not been generally accepted as a sizing guide but does allow a check

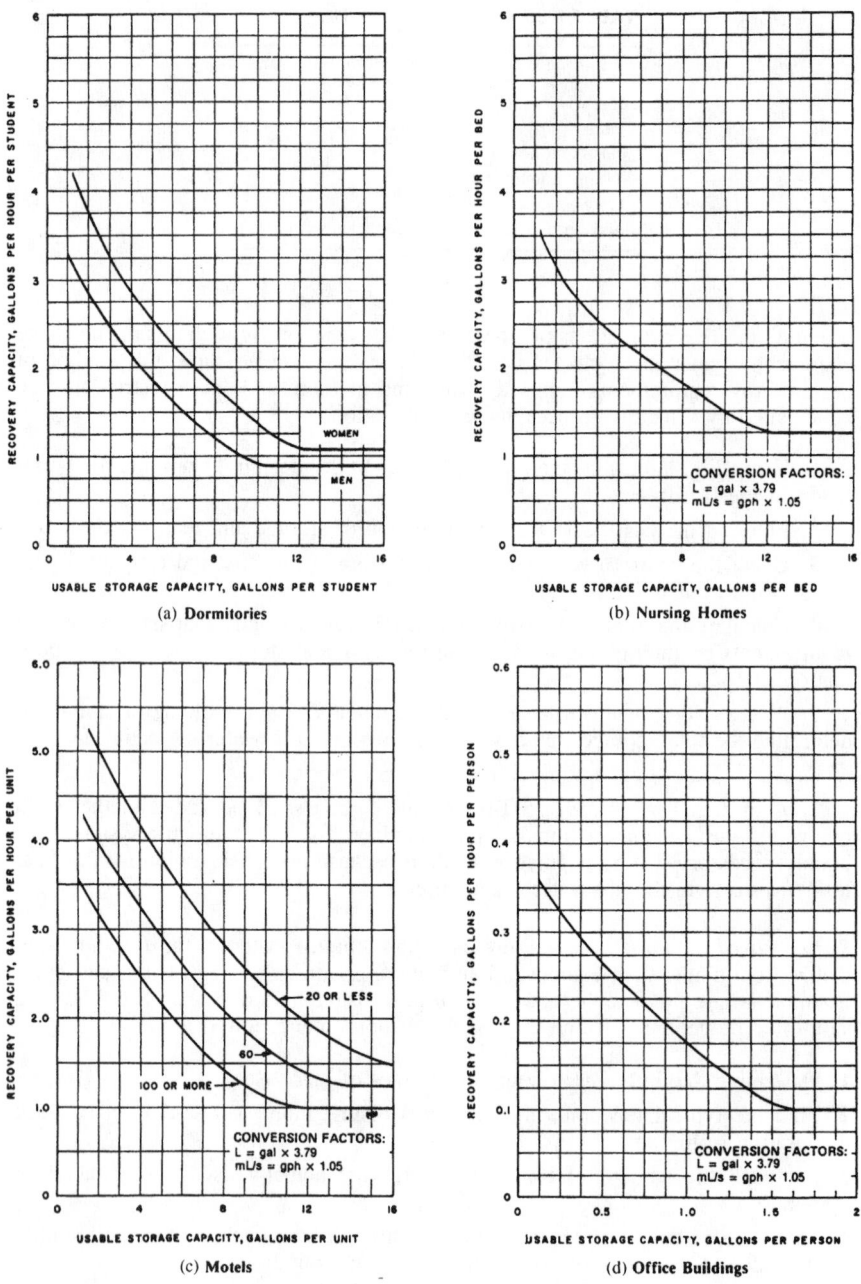

(a) **Dormitories**

(b) **Nursing Homes**

(c) **Motels**

(d) **Office Buildings**

FIGURE 9.42 Relationship between recovery and storage capacity.

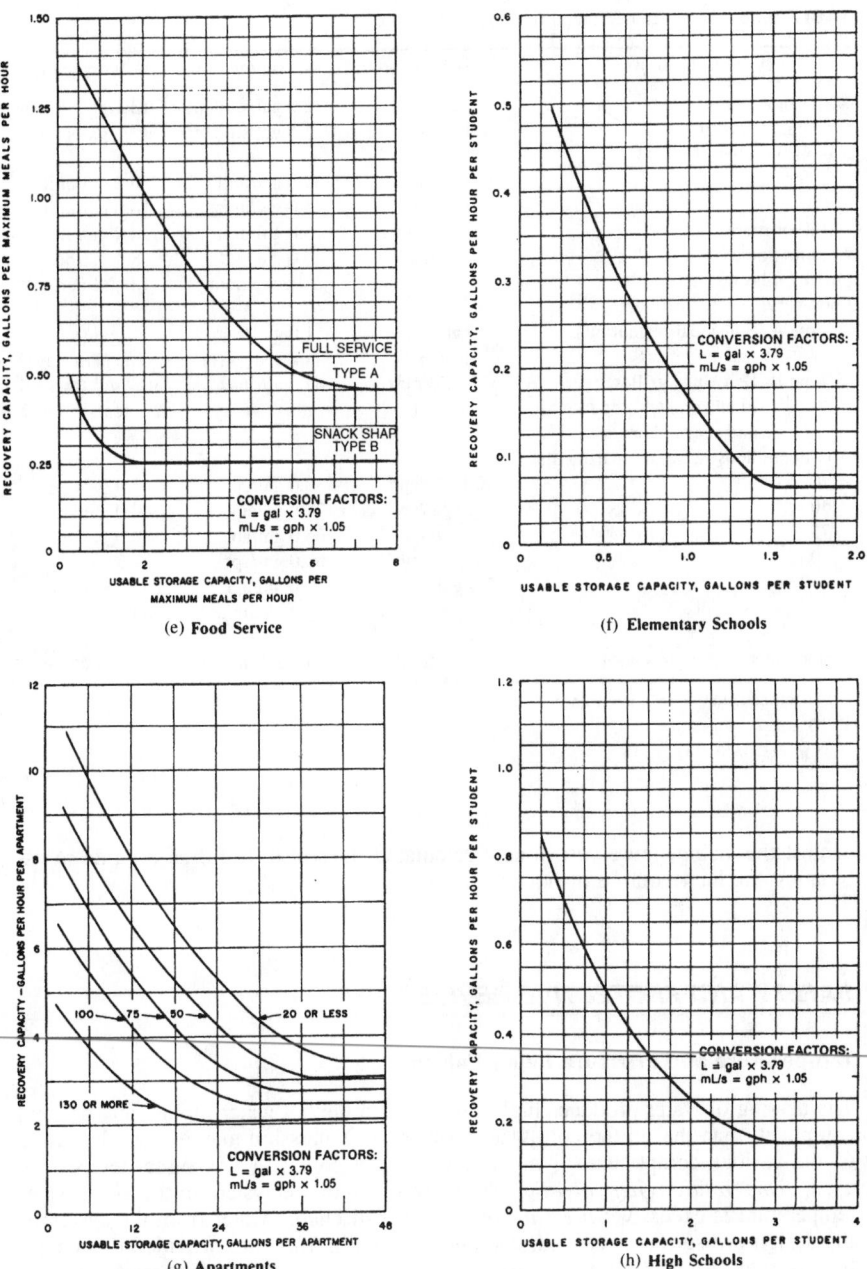

FIGURE 9.42 Relationship between recovery and storage capacity. (*Continued*)

TABLE 9.36   Hot Water Demand for Various Facilities

Type of building	Maximum hour	Maximum day	Average day
Men's dormitories	3.8 gal/student	22.0 gal/student	13.1 gal/student
Women's dormitories	5.0 gal/student	26.5 gal/student	12.3 gal/student
Motels: number of units*			
20 or less	6.0 gal/unit	35.0 gal/unit	20.0 gal/unit
60	5.0 gal/unit	25.0 gal/unit	14.0 gal/unit
100 or more	4.0 gal/unit	15.0 gal/unit	10.0 gal/unit
Nursing homes	4.5 gal/bed	30.0 gal/bed	18.4 gal/bed
Office buildings	0.4 gal/person	2.0 gal/person	1.0 gal/person
Food-service establishments:			
Type A: Full-meal restaurants	1.5 gal/(max	11.0 gal/(max	2.4 gal/(avg
and cafeterias	meals)/(h)	meals)/(day)	meals)/(day)†
Type B: Drive-ins, grilles,	0.7 gal/(max	6.0 gal/(max	0.7 gal/(avg
luncheonettes, sandwich,	meals)/(h)	meals)/(day)	meals)/(day)†
and snack shops			
Apartment houses: no. of apartments:			
20 or less	12.0 gal/apt	80.0 gal/apt	42.0 gal/apt
50	10.0 gal/apt	73.0 gal/apt	40.0 gal/apt
75	8.5 gal/apt	66.0 gal/apt	38.0 gal/apt
100	7.0 gal/apt	60.0 gal/apt	37.0 gal/apt
130 or more	5.0 gal/apt	50.0 gal/apt	35.0 gal/apt
Elementary schools	0.6 gal/student	1.5 gal/student	0.6 gal/student†
Junior and senior high schools	1.0 gal/student	3.6 gal/student	1.8 gal/student†

*Interpolate for intermediate values.
†Per day of operation.
Source: ASHRAE.

against the sizing figures previously calculated. It is also useful as a guide in estimating the total volume of hot water.

SAFETY AND PROTECTIVE DEVICES

Temperature and Pressure Relief Valves

The heating of water produces higher pressure in the hot water piping network and equipment than that in the cold water system. This pressure may exceed the rating of the various components if not relieved. To accomplish this three devices are used. *Temperature relief valves* discharge water from the system in the event water temperature is excessive; *pressure relief valves* discharge water from the system if the design pressure is exceeded; and *energy cutoff devices* are provided to stop the flow of fuel or a heating medium to the heater if the temperature or pressure value is exceeded anywhere in the system. It is common practice to combine the temperature and pressure relief into a single safety device.

The set temperature shall not exceed 210°F (99°C). The pressure shall be set at 5 percent lower than the pressure rating of the tank or heater. The relief valve discharge shall be routed to a safe place, visible to observe discharge, pitched down and the same size as the relief valve discharge. No obstruction shall be placed in the line.

Thermostatic Mixing Valves

It is common practice to provide a *thermostatic mixing valve* (TMV) on the hot water outlet of any hot water generator to prevent water that is too hot from being delivered to the facility. The TMV is a three-way mixing valve connected to the hot water supply, cold water supply, and the tempered water outlet supplying the facility. This device is generally considered mandatory on instantaneous heaters and strongly recommended on storage heaters.

Mixing valves are selected using the minimum flow rate (usually about 2 gpm), the maximum flow rate, and the pressure drop through the unit. If the range exceeds the capacity of a single unit, one small and one large TMV are installed in a parallel configuration. Use hot water DFUs (Table 9.33) and the modified Hunter curves (Figs. 9.40 and 9.41) to find the flow rate. To select a unit, consult manufacturers' catalogs, using the calculated figures.

Expansion Tanks

Water expands when it is heated. When a storage tank system is used and there is a no-flow condition, the excess heat in the heating medium will continue to heat the water in the storage tank. If the water is prevented from expanding, it will cause the temperature and pressure relief valve to open. An expansion tank will safely accept that additional water volume and is highly recommended.

An expansion tank is a pressure tank partially filled with air that is compressed by the addition of water. The volume of the tank is determined by giving to the manufacturer the following information:

1. Storage tank size
2. Maximum system temperature
3. Cold water inlet temperature
4. System pressure
5. Relief valve set point (allowable working pressure). If less than 125 psi, deduct 10 percent from 125.

Double-Wall Shell and Tube Heat Exchangers

A potential for contamination exists when the heating medium is at a higher pressure than the potable water being heated. Most codes now require the use of double-wall heat exchanger tubes with vented air space. Local codes must be checked for exact requirements.

SYSTEM DESIGN CONSIDERATIONS

There are heat losses in the hot water system that will create additional load for the heater.

1. The standby heat loss in the storage tank and piping network must be added to the recovery rate of the heater. It is strongly recommended that the storage tank be insulated.

2. Because the introduction of colder water into a storage tank will lower the water temperature, the actual usable quantity of the water in the storage tank is only about 70 percent of the total quantity of water stored. This reduced volume must be considered when determining storage capacity.

3. To find the exact mixture of hot and cold water necessary to produce an intermediate temperature, refer to Table 9.37 for a mixture based on percent and Table 9.38 for a mixture based on volume.

4. Special conditions, occupants, or processes might affect the usage parameters.

The temperature rise in water heaters is generally based on an incoming water temperature of 40°F (4.4°C).

HOT WATER TEMPERATURE MAINTENANCE

Generally accepted practice requires that hot water be delivered to a terminal point at utilization temperature within a reasonable period of time, usually 20 seconds. Some codes require that if the distance from a heater to the furthest fixture exceeds 50 ft (15 m) temperature maintenance is required.

Introduction

When the hot water supply system is stagnant and not used, the temperature of the water in the piping system will lose heat to the ambient air during periods of no draw. In addition, when sizing the water heater and the hot water supply piping system, the lack of a proper hot water temperature maintenance system can seriously impact the required heater and pipe size. Water temperature is maintained by one of two methods.

1. *Circulating water system.* In this method the water in the hot water supply piping system is continuously circulated through the water heater when the temperature of the water falls below a predetermined set point.

2. *Electric heating cable.* This method consists of the installation of a self-current-regulating electric heating cable on the hot water distribution piping that is set to start operation when the temperature of the water falls below a predetermined set point.

Systems Description

A circulating water system consists of a hot water circulating pump, a method to control this pump, an additional hot water recirculation pipe from the end of all branches or risers of the hot water supply piping and returning to the circulation pump, and an in-line flow balancing device on each return pipe branch to control the amount of water flowing through that specific branch (or riser) of the system. The methods utilized to create circulating systems are as varied as the imagination

TABLE 9.37 Mixing of Hot and Cold Water on a Percentage Basis*

Temperature of cold water, °F

210° hot water

Temperature of mixture, °F	40	50	60	70	80	90
200	94*	94	93	93	92	92
190	88	88	87	86	85	83
180	82	81	80	79	76	75
170	76	75	73	71	69	67
160	71	69	67	65	62	58
150	65	62	60	57	54	50
140	59	56	53	50	46	42
130	53	50	46	43	38	33
120	47	44	40	36	31	25
110	41	38	33	29	23	17
100	36	31	26	21	15	10

200° hot water

Temperature of mixture, °F	40	50	60	70	80	90
190	94	93	93	92	92	91
180	88	87	86	85	83	82
170	81	80	79	76	75	73
160	75	73	71	69	67	64
150	69	67	65	62	58	55
140	62	60	57	54	50	45
130	56	53	50	46	42	36
120	50	46	43	38	33	27
110	44	40	36	31	25	18
100	38	33	29	23	17	9

190° hot water

Temperature of mixture, °F	40	50	60	70	80	90
180	93	93	92	92	91	90
170	87	86	85	83	82	80
160	80	79	76	75	73	70
150	73	71	69	67	64	60
140	67	65	62	58	55	50
130	60	57	54	50	45	40
120	53	50	46	42	36	30
110	46	43	38	33	27	20
100	40	36	31	25	18	10

180° hot water

Temperature of mixture, °F	40	50	60	70	80	90
170	93	92	92	91	90	89
160	86	85	83	82	80	78
150	79	76	75	73	70	67
140	71	69	67	64	60	56
130	65	62	58	55	50	44
120	57	54	50	45	40	33
110	50	46	42	36	30	21
100	43	38	33	27	20	11

170° hot water

Temperature of mixture, °F	40	50	60	70	80	90
160	92	92	91	90	89	88
150	85	83	82	80	78	75
140	76	75	73	70	67	63
130	69	67	64	60	56	50
120	62	58	55	50	44	38
110	54	50	45	40	33	25
100	46	42	36	30	21	12

160° hot water

Temperature of mixture, °F	40	50	60	70	80	90
150	92	91	90	89	88	86
140	83	82	80	78	75	71
130	75	73	70	67	63	57
120	67	64	60	56	50	43
110	58	55	50	44	38	29
100	50	45	40	33	25	14

150° hot water

Temperature of mixture, °F	40	50	60	70	80	90
140	91	90	89	88	86	83
130	82	80	78	75	71	67
120	73	70	67	63	57	50
110	64	60	56	50	43	33
100	55	50	44	38	29	17

140° hot water

Temperature of mixture, °F	40	50	60	70	80	90
130	90	89	88	86	83	80
120	80	78	75	71	67	60
110	70	67	63	57	50	40
100	60	56	50	43	33	20

130° hot water

Temperature of mixture, °F	40	50	60	70	80	90
120	89	88	86	83	80	75
110	78	75	71	67	60	50
100	67	63	57	50	40	25

*Percentage of mixture which is hot water.

Source: *Air Conditioning, Heating, and Ventilating,* August 1958.

TABLE 9.38 Mixing of Hot and Cold Water on a Ratio of Volume to Volume

Temperature of cold water, °F

Temperature of mixture, °F	210° hot water						200° hot water						190° hot water					
	40	50	60	70	80	90	40	50	60	70	80	90	40	50	60	70	80	90
200	16.0*	15.0	14.0	13.0	12.0	11.0	—	—	—	—	—	—	—	—	—	—	—	—
190	7.5	7.0	6.5	6.0	5.5	5.0	15.0	14.0	13.0	12.0	11.0	10.0	—	—	—	—	—	—
180	4.7	4.3	4.0	3.7	3.3	3.0	7.0	6.5	6.0	5.5	5.0	4.5	14.0	13.0	12.0	11.0	10.0	9.0
170	3.3	3.0	2.8	2.5	2.3	2.0	4.3	4.0	3.7	3.3	3.0	2.7	6.5	6.0	5.5	5.0	4.5	4.0
160	2.4	2.2	2.0	1.8	1.6	1.4	3.0	2.8	2.5	2.3	2.0	1.8	4.0	3.7	3.3	3.0	2.7	2.3
150	1.8	1.7	1.5	1.3	1.2	1.0	2.2	2.0	1.8	1.6	1.4	1.2	2.8	2.5	2.3	2.0	1.8	1.5
140	1.4	1.3	1.1	1.0	0.9	0.7	1.7	1.5	1.3	1.2	1.0	0.8	2.0	1.8	1.6	1.4	1.2	1.0
130	1.1	1.0	0.9	0.8	0.6	0.5	1.3	1.1	1.0	0.9	0.7	0.6	1.5	1.3	1.2	1.0	0.8	0.7
120	0.9	0.8	0.7	0.6	0.4	0.3	1.0	0.9	0.8	0.6	0.5	0.4	1.1	1.0	0.9	0.7	0.6	0.4
110	0.7	0.6	0.5	0.4	0.3	0.2	0.8	0.7	0.6	0.4	0.3	0.2	0.9	0.8	0.6	0.5	0.4	0.3
100	0.6	0.5	0.4	0.3	0.2	0.1	0.6	0.5	0.4	0.3	0.2	0.1	0.7	0.6	0.4	0.3	0.2	0.1

Temperature of mixture, °F	180° hot water						170° hot water						160° hot water					
	40	50	60	70	80	90	40	50	60	70	80	90	40	50	60	70	80	90
170	13.0	12.0	11.0	10.0	9.0	8.0	—	—	—	—	—	—	—	—	—	—	—	—
160	6.0	5.5	5.0	4.5	4.0	3.5	12.0	11.0	10.0	9.0	8.0	7.0	—	—	—	—	—	—
150	3.7	3.3	3.0	2.7	2.3	2.0	5.5	5.0	4.5	4.0	3.5	3.0	11.0	10.0	9.0	8.0	7.0	6.0
140	2.5	2.3	2.0	1.8	1.5	1.3	3.3	3.0	2.7	2.3	2.0	1.7	5.0	4.5	4.0	3.5	3.0	2.5
130	1.8	1.6	1.4	1.2	1.0	0.8	2.3	2.0	1.8	1.5	1.3	1.0	3.0	2.7	2.3	2.0	1.7	1.3
120	1.3	1.2	1.0	0.8	0.7	0.5	1.6	1.4	1.2	1.0	0.8	0.6	2.0	1.8	1.5	1.3	1.0	0.8
110	1.0	0.9	0.7	0.6	0.4	0.3	1.2	1.0	0.8	0.7	0.5	0.3	1.4	1.2	1.0	0.8	0.6	0.4
100	0.8	0.6	0.5	0.4	0.3	0.1	0.9	0.7	0.6	0.4	0.3	0.1	1.0	0.8	0.7	0.5	0.3	0.2

Temperature of mixture, °F	150° hot water						140° hot water						130° hot water					
	40	50	60	70	80	90	40	50	60	70	80	90	40	50	60	70	80	90
140	10.0	9.0	8.0	7.0	6.0	5.0	—	—	—	—	—	—	—	—	—	—	—	—
130	4.5	4.0	3.5	3.0	2.5	2.0	9.0	8.0	7.0	6.0	5.0	4.0	—	—	—	—	—	—
120	2.7	2.3	2.0	1.7	1.3	1.0	4.0	3.5	3.0	2.5	2.0	1.5	8.0	7.0	6.0	5.0	4.0	3.0
110	1.8	1.5	1.3	1.0	0.8	0.5	2.3	2.0	1.7	1.3	1.0	0.7	3.5	3.0	2.5	2.0	1.5	1.0
100	1.2	1.0	0.8	0.6	0.4	0.2	1.5	1.3	1.0	0.8	0.5	0.3	2.0	1.7	1.3	1.0	0.7	0.3

*Number of gallons of hot water which, when mixed with 1 gal of cold water, produces a given mixture temperature.

of the engineer. Two basic circulating system diagrams are shown in Figs. 9.43 and 9.44.

The electric heating cable system consists of a self regulating cable installed directly onto the hot water distribution pipe. This cable is capable of gradually adjusting the amount of electrical current (and therefore the amount of heat produced) based on the temperature of the water in the pipe. When the temperature of the water is adequate, no electrical current will be required. No recirculation line is required.

Point of use water heaters are practical where there is a remote or very seldom used fixture that does not justify a long run of piping. They are cost effective because a hot water and hot water circulation line do not have to be installed. A point of use heater is illustrated in Fig. 9.45.

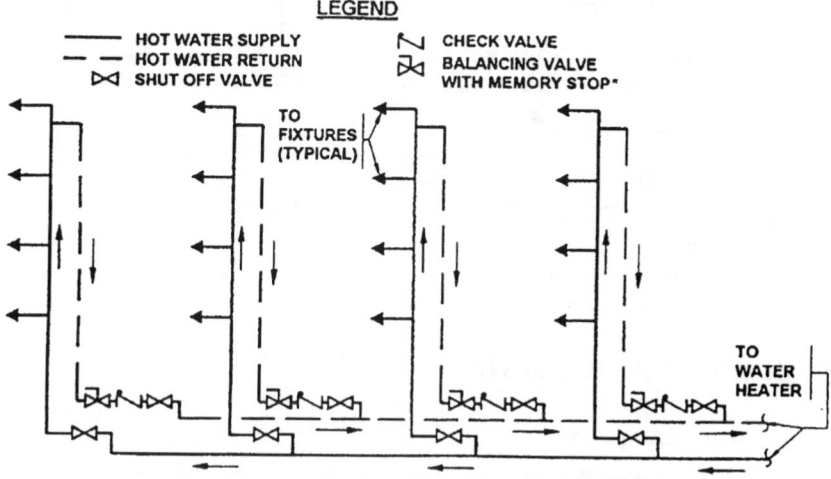

FIGURE 9.43   Upfeed hot water system with heater at bottom of system.

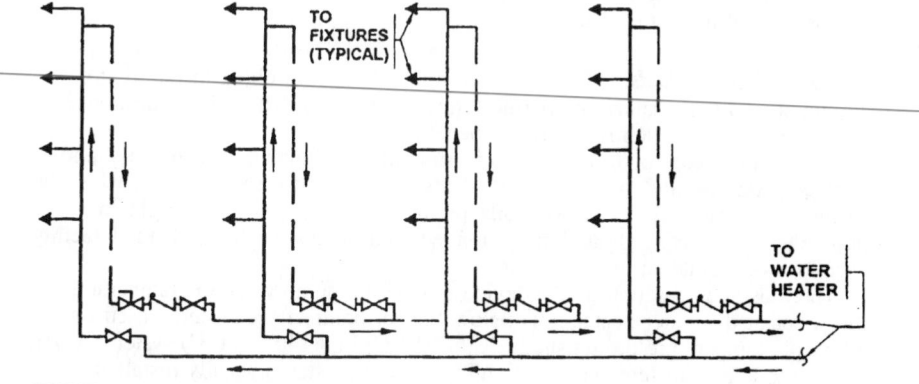

FIGURE 9.44   Upfeed hot water system with heater at bottom of system.

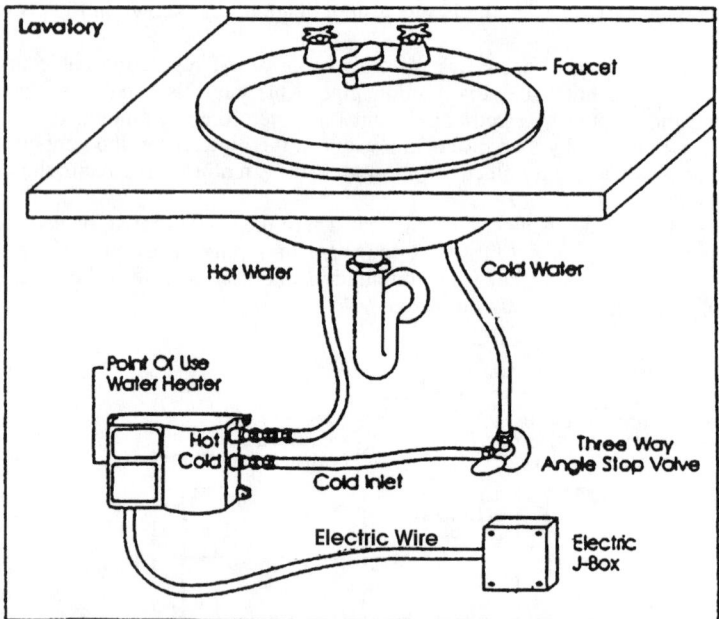

FIGURE 9.45   Point of use water heater.

CIRCULATING WATER SYSTEM

Component Description

Hot Water Circulating Pumps.   Hot water circulating pumps are centrifugal type, available either as in-line (for smaller systems) or base-mounted units for larger systems. In-line pumps are the most commonly used. Because of the corrosiveness of hot water, the pump internals should be bronze or brass. Conventional iron body pumps are not recommended.

Pump Control Methods.   There are two methods of controlling the starting of the circulating pump: manual or thermostatically operated. There is a difference of opinion regarding which is of greater benefit.

The manual control has the system running continuously when the pump is started by turning the power on or off. This system is of greatest benefit when the system is not used for relatively long periods of time, such as at night in office buildings, over weekends and for extended shut-downs, such as planned facility vacations or maintenance down-time.

Thermostatic operation uses a device to start the pump when the temperature of the water at the water heater falls below the set point of the system based on the allowable temperature drop, usually 130°F (52°C) in a 140°F (60°C) system, which equates to a 10°F differential (5.5°C ΔT). The thermostat is usually installed in the return piping using an oversized tee and nipple.

Flow Balancing Devices. It is extremely important that the circulated hot water system be balanced for its specified flows, including all of the various individual loops within the circulated system. If they are not, water will tend to short-circuit through the closest loops and create high velocities in the piping system, and this will result in people complaining about the long delays in getting hot water at the most remote loops. Because of the problems inherent in manually balancing hot water circulation systems, many professionals have incorporated factory preset flow control devices into their hot water systems. While the initial cost of the device is higher than a manual balancing valve, the present devices are normally less expensive when field labor cost to balance the entire hot water system is included. However, with the use of present flow control devices, the plumbing designer has to be far more accurate in selecting the capacity of the control device, as there is no field adjustment available. Therefore, if more or less hot water return flow is needed by the actual field installation, a new flow control device must be installed with the old flow control device being removed and junked. Following are the various flow control devices generally used.

1. *Fixed orifice.* These are devices that can be bought for the specific flow rates and simply inserted into the hot water return piping system. However, extreme care should be provided to be able to remove these devices and clean them out, as they may become clogged due to debris in the water. Therefore, it is recommended that a strainer with a blowdown valve be placed ahead of each of these devices or a strainer with a fine mesh screen be installed on the main water line coming into the building. Also, a shutoff valve should be installed ahead of and after these valves so that the entire loop does not have to be drained to service these devices.

2. *Factory preset automatic flow control valves.* The same admonition about strainers also applies to the installation of these devices.

3. *Ball valves.* These are valves that can be used to determine the flow rate by reading the pressure drop across the valve by some of the various devices that are available.

4. *Ball valves with memory stops.* These can be adjusted to the proper setting by use of temperature measuring devices inserted in the piping system or by the use of snap-on flow sensing devices that will indicate the flow rate in the pipe line.

Pump Control Methods. The starting of the circulating pump could be controlled either manually or by means of a thermostat set to start the pump when the temperature of the water in the pipe falls below the lower set point temperature of the system.

System Design Criteria

The following criteria shall be established for the design of a circulated water system:

1. Maximum length of uncirculated supply pipe
2. Amount of time allowed for proper temperature water to reach the terminal fixture
3. Allowable temperature trop of hot water supply to the terminal fixture

4. The flow rate to be circulated

5. The hot water circulation system pipe size

6. Velocity in the recirculation system piping

Maximum Length of Uncirculated Supply Pipe. Currently, due to not only energy conservation but to water shortages in numerous areas of the country, the wasting of water due to the long delay in obtaining hot water at the fixtures has become a critical aspect. Therefore, to reduce significantly the wasting of cooled hot water, the acceptable distance permitted for uncirculated, dead-end branches to plumbing fixtures must be kept to a minimum. These allowable distance for uncirculated, dead-end branches are a trade-off between the energy utilized in circulating the hot water (or keeping the water hot by other means) and the cost of the insulation versus the cost of the energy to heat the excess cold water makeup, the cost of wasted potable water, and extra sewer surcharge costs. There are some codes that set a maximum dead end hot water supply branch length to 50 ft (15 m) in length.

Acceptable Time Limits for Hot Water to Reach Terminal Fixtures. When designing the hot water system, the hot water must reach fixtures within acceptable time limits. For other than very infrequently used fixtures, such as industrial facilities or certain fixtures in office buildings, etc., a delay of up to ten seconds is normally considered acceptable for most residential or public fixtures in office building applications. A delay of 11 to 30 seconds is marginal but acceptable, and beyond 31 seconds is normally considered unacceptable and a significant wastage of water and energy. A maximum delay of 20 seconds is a generally accepted figure for most applications.

This normally means that the hot water circulated or heated by other means should be brought to within a maximum distance of 20 ft (7 m) from all the plumbing fixtures requiring hot water. The design professional may want to stay under this length limitation because the actual installation in the field may be slightly different than what the engineer designed, and there may be a longer delay caused by either routing of pipe or other problems. Furthermore, with today's low fixture flow rates mandated by national and local requirements, it takes considerably longer to obtain hot water from non-circulated hot water lines than it did in the past when fixtures had greater flow rates. In addition, when there is a long time delay in obtaining hot water at the fixture, a significant wastage of potable water results since the cooled hot water is wasted.

Allowable System Temperature Drop. The greater the design temperature drop across the system, the less water is required to be circulated, thereby saving on pumping costs. However, the 20°F (11°C) temperature drop previously used is now considered excessive. If the domestic hot water supply starts out at 140°F (60°C) with a 20°F (11°C) temperature drop across the supply system, the fixtures near the end of the circulating loop will only be provided with hot water supply of 120°F (52°C), which is considered too low a temperature. Furthermore, with a hot water supply delivery temperature of only 120°F (52°C) instead of 140°F (60°C), the fixtures will use a greater volume of hot water to get to the desired blended water temperature.

Therefore, the system temperature drop should be in the magnitude of 10°F (6°C). So, if the hot water system starts out at 140°F (60°C), 130°F (54°C) would be the coolest hot water supply temperature provided by the hot water supply system. This would still give good operation of the hot water system.

Circulating Water Flow Rate.   The flow rate of circulating water is a function of the allowable system temperature drop selected by the design professional and the actual heat loss through the hot water supply and recirculating piping. The most accurate method of calculating the heat loss through the piping system into a gpm equivalent is to relate it as a function of temperature drop. This relationship is derived from the following formula, and is tabulated in Table 9.39.

$$Q = [SH \times W] [60 \ (min/h)] [\Delta T] \qquad (9.3)$$

where Q = Quantity (flow rate) of recirculated water, in gpm
SH = specific heat of water, equal to 1
W = weight of water, 8.33 lbs/gal
ΔT = selected system temperature drop

There are two very general and arbitrary "rules of thumb" that are used for small, uncomplicated systems that experience has proven successful. One is to allow ½ gpm (2.0 L) for each 20–25 fixtures. Another figure, when larger systems with a number of risers or branches are present, is to allow ½ gpm (2.0 L) for each smaller riser or branch and 1 gpm (4.0 L) for each intermediate and larger riser or branch. It is highly recommended that scientific methods be used to calculate the flow rate rather than rules of thumb.

Hot Water Circulation System Pipe Size.   The selected size of the recirculation piping is an iterative one, depending on the flow rate and an acceptable friction loss in the piping network. The higher the friction loss, the more horsepower the circulating pump will require. As a general starting point, a single branch recirculation line shall be a minimum size of ½ in (DN 15). For multiple-branch systems when recirculating branches combine, the size should be one half of the hot water supply pipe. These sizes are starting point that could be adjusted as actual system design and the selection of the circulating pump progresses.

Velocity of Circulating Water.   For copper piping systems, it is important that the circulated hot water supply and hot water return piping be sized so that the water is moving at a low velocity. For copper piping, the recommended maximum rate is 8 ft per second (2.6 ms).

System and Component Sizing Procedure

The procedure for sizing the system components and entire piping network is as follows.

TABLE 9.39   Relationship of Temperature Drop to BTU/GPM

Temperature Drop	BTU/GPM Relationship
5°	2,500 B/h = 1 GPM
10°	5,000 B/h = 1 GPM
15°	7,500 B/h = 1 GPM
20°	10,000 B/h = 1 GPM

1. Layout the system and measure the equivalent pipe run for the entire hot water supply and return piping network.
2. Size the recirculation piping system. This assumes that the hot water supply piping has been previously sized.
3. Determine the system temperature drop.
4. Calculate the system heat loss.
5. Calculate the return system flow rate.
6. Calculate the friction loss in the entire hot water supply and return piping. Adjust the return system pipe size as required to reflect velocity considerations and pump head restrictions.
7. Size the circulating pump based on calculations from 5 and 6.

Layout the System.   Make a piping diagram of the hot water distribution system showing hot water piping and approximate lengths.

Size the Return System Piping.   Use the sizing method previously discussed under system design criteria to preliminarily size the return piping. This will give a starting point for the remaining calculations.

Select the System Temperature Drop.   The recommended drop should be 10°F (6°C). Use judgment when using another figure.

Calculate the System Heat Loss.   Tabulate the total system heat losses (heater and piping) in BTUs per hour. For piping, measure the length and size of pipe in the piping diagram; the total hourly heat loss in only the longest length of hot water supply and return system should be calculated. Enter Table 9.40 with the insulation thickness and pipe size for piping heat loss, and Table 9.41 for approximate tank heat loss. Add these figures together and arrive at a total system heat loss in BTU/h.

TABLE 9.40   BTU/h Heat Loss per ft for Tubing & Steel Pipe*

Pipe size (in)	Pipe size DN	Insulated steel pipe schedule 40	Insulated copper tube type L	Non-insulated steel pipe	Non-insulated copper tube
½″	15	10.0	8.2	4,0	2,3
¾″	20	11.5	10.3	5,0	3,0
1″	25	13.5	12.0	6,0	4,0
1-¼″	32	16.0	13.9	7,5	4,5
1-½″	40	18.6	15.8	8,5	5,5
2″	50	20.5	20.0	11,0	6,5
2-½″	65	25.0	22.6	12,0	8,0
3″	80	31.0	26.7	15,0	9,5
4″	100	35.0	33.8	19,0	12,0
5″	125	50.0		22,5	
6″	150	65.0		26,0	

*Based on 140°F hw and 70°F ambient air, ½ in fiberglass insulation. 1 ft = 0.3 m.

TABLE 9.41 Heat Loss from Various Size Tanks with Various Insulation Thicknesses

Insulation thickness (in)	Tank size (gal)	Approx. energy loss from tank at hot water temperature 140°F (BTU/h)*
1	50	468
1	100	736
2	250	759
3	500	759
3	1000	1273

Insulation thickness (mm)	Tank size (L)	Approx. energy loss from tank at hot water temperature 60°C (W)^a
25.4	200	137
25.4	400	216
50.8	1000	222
76.2	2000	373
76.2	4000	

*For unfired tanks, federal standards limit the loss to no more than 1.9 W/m² of tank surface.
Source: From Sheet Metal and Air Conditioning Contractors National Association (SMACNA) Table 2 data.

Calculate the System Flow Rate. The formula to find the system flow rate in gpm is:

$$\frac{\text{total system heat loss}}{\text{gpm relationship from Table 9.39}} = \text{system flow rate in gpm} \qquad (9.4)$$

As an example, if the total heat loss is 8,500 BTU/hr and the ΔT is 5°F.

$$\text{gpm} = \frac{8500}{2500} \text{ gpm} = 3.4$$

Once a hot water return required flow rate (in gpm) is established, then the size of the hot water return piping system should be adjusted based on piping flow rate, velocity and the available pump head obtained from a manufacturers catalog. Select the most cost-effective solution.

Size the Circulating Pump. The hot water return circulating pump is selected based on the required gpm flow rate previously calculated and system friction loss. The pump head is calculated using the friction loss only (the flow rate through the single longest equivalent length of pipe using standard references such as Cameron's Hydraulic Data or other proprietary calculators available from major manufacturers). This friction loss must include that occurring in the entire hot water piping system. The reason for this is that the hot water return system flow is only needed in order to keep the system up to temperature when there is no flow in the hot water supply piping. Therefore, when hot water is in use, there is usually

TABLE 9.42 Fixture and Equipment Mounting Heights

Component	Children	Standard	ADA complient	
LUC	13	14–15	17–19	FV of 26″ for PS K-6 29″ for adults
BIDET		15		
UR	18	24	17	To top of lip
UR FV		44		
LAV	2.4	36	33–36	
SK	24	36	33–36	
SS		26		
MOP SK		36		
SPOUT				
BT		14–17	14–17	
BT Handles		16	16	Above top of BT
WHORLPOOL		18–24		
SH HEAD	54–56	72–84		
SH HANDLES	45	48	36	
DF, EWC	24	40	34–36	To spout
ES HEAD		88	82–96	Bottom of head
EW		42	33	
ES		78	48–54	Bottom of pull
HB		24		
WH		26		
FAI		26		
FIRE EXT		60		Top of Extinguisher
SIAMESE		26		
HOSE CAB		74		Top of Cabinet
Hose Value		70		
Fire Dept Value		48		
Spir Test		84		

All dimensions are in inches.
Dimensions are from floor or grade to and of component.
Plumbing fixture dimensions are to the rim or top of fixture.

sufficient flow in the hot water supply piping to normally keep the system hot water supply piping up to temperature even without the flow in the hot water return piping.

The principal variance of this concept of ignoring the friction in the hot water supply piping is where the hot water return piping is connected to a relatively small hot water supply line. By relatively small is meant a line that is less than one pipe size larger than that to which the hot water return line is connected. The problem that occurs when the hot water supply line is too small for the hot water return line is that the hot water supply line will add friction head to the hot water circulating pump, and additionally the hot water circulating pump flow rate may deprive the plumbing fixtures on this hot water supply line from obtaining their required flow. Therefore, the hot water supply line supplying each hot water return piping connection point should be increased to eliminate this potential problem.

It is quite common for a plumbing designer to have a wrong preliminary assumption as to the size of the hot water return line to establish the initial heat loss figure. However, after establishing the required hot water return flow rate (gpm), it is determined that the hot water return line must be larger, the plumbing designer should simply recheck the calculations using the new data based on the correct pipe sizing and verify that all the rest of the calculations are correct.

In selecting the hot water circulating pump's head, be sure that only the restrictions placed on the circulating pump are calculated and do not include any static heads where none exist. For example, in Fig. 1 the hot water circulating pump only has to overcome the friction in the piping, but not the static head of pumping the water up to the fixtures. This is because in a closed system the entire system is in equilibrium; therefore, static head loss is offset by the static head gain when the hot water return piping drops down.

System Design Considerations

The following are some of the potential problems in hot water circulated systems that must be addressed by the plumbing system designer.

1. It is extremely important in circulated systems that shutoff valves be provided on every circulation branch in order to isolate the entire circulated loop, so that if individual fixtures need modification, that piping loop can be isolated out of the entire system and the entire hot water system does not have to be shut off and drained.

2. The means for ensuring hot water system balancing requires that the hot water return system be provided with either a separate balancing valve in addition to the shutoff valve or, if the balancing valve is also used as the shutoff valve, the balancing valve should have a memory stop. With a memory stop on the valve, the plumbers can return the system to its balanced position without having the whole piping system unbalanced when they leave. As previously indicated, unbalanced hot water systems create serious problems.

3. Provide a check valve on each hot water return line when it joins other hot water return lines. This is to ensure that a plumbing fixtures does not draw hot water return water instead of hot water supply water due to pressure drops in the various piping systems.

4. Keep the time delay in obtaining hot water at the fixtures to within the time parameters given previously to eliminate unhappy users and occupants of the hot water system and to prevent wasting of water by having it run for long periods of time.

5. For plumbing fixtures that are remote from the circulating system and/or plumbing fixtures that are infrequently used, the plumbing designer may wish to consider a point-of-use water heater near that fixture rather than extend the hot water system to circulate to a point near the remote fixtures.

6. When a central mixing valve is used to produce temperate hot water for a facility, the hot water return piping arrangement shown in Fig. 9.46 is recommended in order to eliminate fluctuations in temperature and pressure.

7. Do not have the return piping from a riser as the highest connection. If this is done, air binding will result.

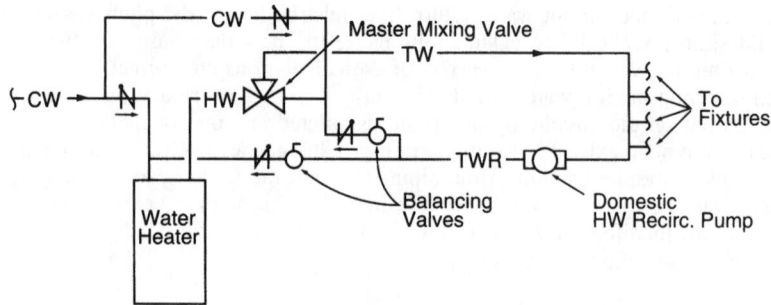

FIGURE 9.46 Schematic diagram of hot water return system.

ELECTRIC HEATING CABLE

General

Because of the problems of balancing circulated hot water systems and the loss of energy in the return piping, the use of electric heating cable has come into its own. In many installations, this concept provides the instant hot water required at all fixtures. This is because heat from the electric cable is put on the hot water lines right up to the fixture connection, so the hot water supply water is always kept up to temperature. The electric trace heating of domestic hot water systems is an outgrowth of snow melting and industrial trace heating utilized on specific piping lines to prevent freezing, etc., which have been used successfully for many years. The electric heat cable method can also be used in conjunction with a piped hot water circulated system with only dead-end branches to the plumbing fixtures provided with cable heat so the two systems are not mutually exclusive.

The electric heating system has the following advantages:

1. Allows a possible reduction in the heating of water to only the end-user-required temperature rather than higher to allow for drop in temperature of the water

2. Eliminates the hot water return piping and pump

3. Power is drawn only when the temperature of the water is low, and then only in proportion to the actual water temperature.

Component Description

The basic component of this system is an electric heating cable attached directly to the hot water piping system under the insulation. This cable is called self-regulating, which allows gradual and adjustable changes to the current in response to the gradual cooling of the water in the pipe. When the hot water supply temperature is adequate, there is no current flow. The adjustment is automatic and does not require a thermostat.

Other components of the heater cable system are power connections from the cable to the electrical power supply, splice and tee connections to the cable and end seals of the cable to protect the end of the cable from damage and prevent exposure to live electric wires.

Since it is not possible for an entire building to be heated on one electrical circuit, the system shall be divided into separate circuits, each controlled by a separate thermostat and having a separate circuit breaker.

System Design Criteria and Sizing

The following criteria shall be established for the design of a circulated water system.

1. Determine the extent and layout of the hot water circulating piping.
2. Select the system design temperature. This is usually 140°F (60°C).
3. Select the appropriate cable from the manufacturers catalog. This will also establish the current draw for the cable.
4. Determine the electrical requirements. Each different type of cable has a maximum allowable circuit length. Based on electrical requirements for the largest circuit breaker size (usually 30 amps), determine the number of circuits by dividing the maximum allowable circuit length into the total feet of cable. Maximum cable length should not be exceeded.
5. If the available voltage is not the same as the cable, a transformer must be provided.
6. Some larger size pipes require more than one cable. Refer to manufacturers recommendations.

System Design Considerations

1. In general, separate circuits should be provided for the following logical segments:
   a. Each main and long riser
   b. Each wing or building section
   c. Each pressure zone
2. Locate power connections near electrical panels to minimize the electrical costs.
3. Locate tees in accessible locations to make trouble-shooting and maintenance easy.

CHAPTER 10
SPECIAL WASTE DRAINAGE SYSTEMS

This chapter describes and discusses the collection and criteria necessary for the design of liquid effluent from various special waste drainage systems other than discharge from sanitary and storm water sources. Except for the neutralization of acid effluent, treatment methods are outside the scope of this chapter.

Each waste system discussed has unique properties that must be separately addressed. These systems are generally routed from fixtures and equipment into a local facility waste treatment system, with the treated effluent discharging directly into the public sanitary sewer system. Very often, untreated waste is stored on site and collected by approved waste removal contractors for disposal.

Unless specifically noted otherwise, all of the waste streams will be assumed to have the approximate flow characteristics of water. Pipe sizing criteria is based on this assumption.

CODES AND STANDARDS

There are two general jurisdictional bodies that regulate different aspects of special waste systems.

The first is the local and regional authorities that create and enforce plumbing and health codes. Included are the local authorities charged with review and approval of plumbing systems design. They are concerned primarily with regulating sizing and design of the piping systems within buildings. Their regulations generally mention special system piping material, treatment and system configuration. They do not mandate specific use of double-wall piping or leak detection. These requirements are mandated by federal, local and state agencies. Recommendations for piping material usually require that any pipe material be approved by the local authority and be capable of resisting degradation and the effects of the nature and concentration of any special waste. System design is concerned only with adequacy of the pipe to carry away the design flow. System configuration usually requires that traps be provided on fixtures and floor drains, and that the venting system conform to good plumbing design practice. The effect that the very general regulations have in terms of conformance with local and model plumbing codes is that all elements of the system design are left to the experience of the design engineer and are also to comply with the requirements of other regulating agencies.

The second jurisdictional body, and with far more stringent regulations, is the various agencies concerned with protecting workers, the public and the environment from the discharge of toxic substances. Included are federal, state and local authorities responsible for preventing discharge into the general environment, public sewers and public treatment systems of any substances considered harmful. Discharges can occur either lawfully or unlawfully, or as a result of spills and accidents. In order to prevent this discharge, it is common practice for these agencies to inspect facilities, mandate treatment systems, and require the use of specific piping systems such as double-wall piping and leak detection to advise of such leakage. The effect of some very specific regulations in terms of conformance with agency requirements is to rely on the experience of the design engineer to provide pipe materials and systems capable of complying with regulations and receiving approval from those agencies. Other regulations, such as cGMP and validation, are constantly being revised because of technological and design changes.

It is recommended that the services of an experienced environmental consultant, who is familiar with the latest applicable rules and regulations and their interpretations, be consulted for system compliance with the maze of regulations.

SYSTEM APPROVAL REQUIREMENTS

There are no applicable codes or standards relating directly to a chemical or acid drainage system. Any special drainage system effluent routed for treatment inside a facility or on the site does not require examination or approval by the local plumbing official. If any effluent is routed to the public sanitary sewer system for eventual treatment, maximum concentrations of any contaminant or pH levels will have been established by the local authorities having jurisdiction. The discharge then requires conformance with local regulations.

If the only required method of treatment is pH adjustment prior to discharging into the building sanitary drainage system or public sewer system, most authorities have requirements for acid drainage systems that must be followed. A generally accepted pH value of 4.0 is the lowest acceptable level for direct discharge into a public sewer system.

GENERAL MATERIAL, SIZING AND DESIGN CONSIDERATIONS

PIPE MATERIAL AND JOINT SELECTION CONSIDERATIONS

Selection of the most appropriate piping and jointing method is made by first establishing the composition and concentration of all chemicals that are to be expected and then establishing their expected flow rate and temperature range. All manufacturers of pipe have published chemical compatibility charts that give the effect various chemicals have on that particular pipe and recommendations for acceptance using these chemicals. For conditions that are not on these charts, direct contact with the manufacturer of the material giving the anticipated conditions will allow them to ask technical assistance from their organizations, which have the experience to provide answers.

When compatibility for various pipe systems is the same, total installed cost and possible ease of disassembly will be the deciding factors.

PIPE SIZING CONSIDERATIONS

System design is concerned only with adequacy of the pipe to carry away the design flow. System configuration usually requires that traps be provided on fixtures and floor drains, and that the venting system conform to good plumbing design practice, which limits the pressure inside the system. All elements of system design are left to the experience of the design engineer.

The various plumbing codes generally use only satisfactory performance as a sizing guide for special waste system pipe sizing, as compared to specific drainage requirements used for sizing sanitary drainage systems. A common exception is drainage and vent systems for laboratory fixtures, which may require sizing based on a fixture unit basis if mandated by the local code.

The reason for the lack of code requirements is that special drainage systems do not have a predictable or documented past usage history as do standard plumbing fixtures in sanitary drainage systems. Equipment drainage, spills, discharge from production facilities and discharge from fixtures within the facilities are not always planned. They occur mostly at random intervals dictated by cleaning, production and maintenance schedules and, often, accidents.

In addition, special drainage systems that are completely within the property of the facility do not fall under plumbing or other local code requirements for piping size or design. Because of these factors, the special drainage piping system is sized on the basis of "good engineering practice," which uses pipe slope, composition of the effluent and the expected flow rates rather than fixture units.

From each point in the system, the flow rate and pitch of the piping must be known in order to size the pipe. Determine the pipe size based on the following criteria.

1. Effluent has the characteristics of water.

2. The drainage system is sized on the basis of gravity drainage and maintaining a minimum scouring velocity of between 2 and 2.5 ft per second (fps) or 0.6 to 0.75 meters per second (ms) using the anticipated average flow rate and pitch of the pipe at each point of design.

3. Gravity drainage pipe size is based on flow rate, slope and velocity. Refer to Table 10.1, which is an abridged table for finding the pipe size based on flow rate, velocity and slope. Depending on the system, piping should be sized to flow between ½ to ¾ full to allow for unexpected larger discharges, future changes and accidents.

PH DEFINITION

Any dissolved impurity in water separates to form negative and positive charged atoms called ions. Negative ions are called cations because they migrate to the cathode, and positive ions are called anions because they migrate to the anode.

All acid compounds consist of hydrogen combined with an acid radical. In a mixture of acid and water, hydrogen ions result. pH is a measurement of the hydrogen ion concentration of a solution. Since the balance of hydroxyl (cation) and hydrogen (anion) ions must be constant, changes in one ion concentration produces a corresponding change in the other. The pH value is calculated from the logarithmic reciprocal of the hydrogen-ion concentration in water. The pH scale ranges from 0 to 14, with 0 being acid and 14 being alkaline. 7.0 is neutral. A change of one unit represents a tenfold increase (or decrease) in strength. pH is not a measure of alkalinity.

GENERAL SYSTEM DESIGN CONSIDERATIONS

It is good practice to separate each of the different systems inside the facility or building to a point outside the building so that the individual services can be isolated and allowed to be tested and sampled as may be required in the future by any local or national authority. Also, the system may, at some point in the future, require separate treatment because of a new substance that may be discharged. For this reason, the effluent should be routed outside the building separately and discharged into a manhole. This manhole could receive sanitary effluent, further diluting the treated chemicals or acid. A typical manhole capable of receiving acid is illustrated in Fig. 10.1 The elbow on the outlet would be eliminated if there were no sanitary inflow to the manhole.

One of the most constant aspects of the special drainage systems are future changes. In time, the processes will change, equipment will be more efficient, facilities will become larger and technology will be improved so that the effluent will be different from the time that the systems were originally designed. This change must be allowed for. It is common practice to size the drain one size larger than the design figures indicate, or, by not sizing the drain line to the exact point on the sizing chart where the figures indicate, especially where there is a probability of future expansion.

TABLE 10.1 Slopes of Cast Iron Soil Pipe Sanitary Sewers Required to Obtain Self-cleansing Velocities of 2.0 and 2.5 ft/sec. (Based on Mannings Formula with $N = .012$)

Pipe size (in.)	Velocity (ft./sec.)	¼ Full		½ Full		¾ Full		Full	
		Slope (ft./ft.)	Flow (gal./min.)	Slope (ft./ft.)	Flow (gal./min.)	Slope (ft./ft.)	Flow (gal./min.)	Slope (ft./ft.)	Flow (gal./min.)
2.0	2.0	0.0313	4.67	0.0186	9.34	0.0148	14.09	0.0186	18.76
	2.5	0.0489	5.84	0.0291	11.67	0.0231	17.62	0.0291	23.45
3.0	2.0	0.0178	10.11	0.0107	21.46	0.0085	32.23	0.0107	42.91
	2.5	0.0278	13.47	0.0167	26.82	0.0133	40.29	0.0167	53.64
4.0	2.0	0.0122	19.03	0.0073	38.06	0.0058	57.01	0.0073	76.04
	2.5	0.0191	23.79	0.0114	47.58	0.0091	71.26	0.0114	95.05
5.0	2.0	0.0090	29.89	0.0054	59.79	0.0043	89.59	0.0054	119.49
	2.5	0.0141	37.37	0.0085	74.74	0.0067	111.99	0.0085	149.36
6.0	2.0	0.0071	43.18	0.0042	86.36	0.0034	129.54	0.0042	172.72
	2.5	0.0111	53.98	0.0066	107.95	0.0053	161.93	0.0066	215.90
8.0	2.0	0.0048	77.20	0.0029	154.32	0.0023	231.52	0.0029	308.64
	2.5	0.0075	96.50	0.0045	192.90	0.0036	289.40	0.0045	385.79
10.0	2.0	0.0036	120.92	0.0021	241.85	0.0017	362.77	0.0021	483.69
	2.5	0.0056	151.15	0.0033	302.31	0.0026	453.66	0.0033	604.61
12.0	2.0	0.0028	174.52	0.0017	349.03	0.0013	523.55	0.0017	678.07
	2.5	0.0044	218.15	0.0026	436.29	0.0021	654.44	0.0026	872.58
15.0	2.0	0.0021	275.42	0.0012	550.84	0.0010	826.26	0.0012	1101.68
	2.5	0.0032	344.28	0.0019	688.55	0.0015	1032.83	0.0019	1377.10

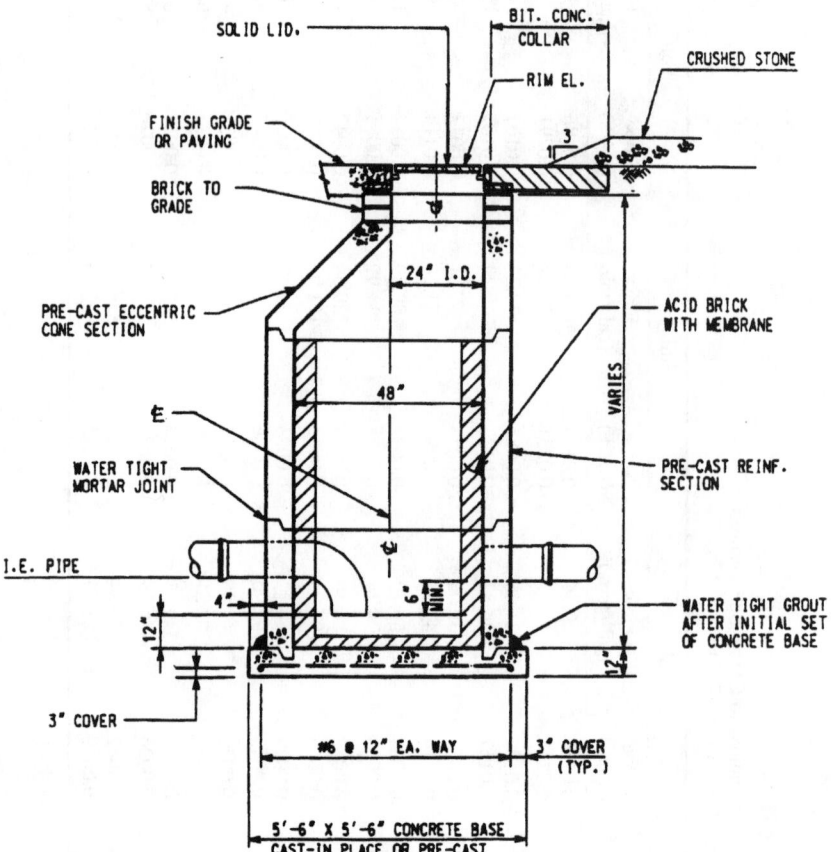

FIGURE 10.1 Typical acid manhole.

Selecting a pipe size slightly larger than required for the immediate flow rate or a material that is capable of resisting a greater selection of chemicals than necessary at the time of design should be considered. This, of course, must be verified with the client to assure that the extra cost is acceptable.

ACID WASTE AND VENT SYSTEMS

GENERAL

An acid waste drainage system collects and transports liquid wastes with a pH lower than 7.0 from laboratory fixtures, equipment and all areas of a facility for discharge into an appropriate treatment facility or the sanitary drainage system after local treatment. The acid vent system equalizes flow in the drainage system in the same manner as the sanitary drainage system. For purposes of this chapter, acid waste will be divided into two general categories: laboratory waste and industrial waste.

A laboratory could be considered as any room or area within a building where investigation, teaching, testing, experiments and research is conducted. Pharmaceutical facilities generally prepare, manufacture and package drugs of all kinds. Manufacturing is generally considered any facility where a product is the end result of having material or components packaged or assembled from parts obtained elsewhere or made within the facility. These facility definitions cover an extremely wide range of possible drainage categories discharging from various sources, each with different effluent characteristics and design requirements.

Laboratory waste consists primarily of diluted and concentrated mixtures of liquid chemical substances of mineral and organic origin, and water. Acids of many types are usually present. Laboratory waste is discharged from sinks, cup sinks, flume hoods, other similar fixtures and equipment. Discharge from floor drains, autoclaves, glass washers and condensed water from various sources are also included. All of these fixtures would require a separate vent, and should be considered a completely separate drainage system.

Acid waste drainage from industrial facilities will consist of accidental spills originating from tanks and piping and anticipated waste from equipment into drains. Very often, the drainage piping would have to carry many of the acids that are used as part of the process. Where spills are directed into holding tanks, the drainage piping, tanks, pumps and piping necessary to convey the effluent to treatment facilities will normally be part of the plumbing engineer's responsibility.

The most important considerations in the selection of piping, valves and tanks for acid are concentration and temperature of the acid. Acid wastewater from chemical and other facilities must be neutralized to a pH of 4.0 or higher prior to eventual disposal into the public sewer system.

Health and Safety Concerns

All grades and concentrations of acids can cause severe damage to eyes and tissues of the body. Contact with the skin will cause irritation and burns. Contact with the eyes could cause blindness. Inhaling the mist or vapors could cause lung irritation or burns. Ingestion will destroy tissue of the mouth, throat and stomach. Extreme care should be exercised in the handling and cleanup of all acids.

These concerns mandate that emergency drench equipment be provided immediately adjacent to all hazards and locations where spills and other accidents could occur. If several people are normally present at a hazardous location, multiple drench equipment should be provided. Where fumes may be given off, emergency breathing apparatus shall be provided. Refer to Chap. 17, Life Safety Systems, for a discussion of such emergency equipment.

For the laboratory environment, emergency showers shall be easily available to every room. Where rooms are adjacent, a single shower is acceptable. Floor drains for emergency drench equipment are not required but will prevent the surrounding floor from becoming wet and a hazard to helping individuals. Every room shall have an emergency eyewash inside the room usually mounted on a sink or free standing if sink mounting is not possible. In large rooms with more than one sink, it is good practice to install emergency eyewashes on each end.

Where the generation of acid vapor is possible, fog nozzles using water to suppress the vapor and foam systems to prevent vapor from rising should be considered to mitigate the effects of an accidental spill.

Common Types of Acids

Acids are widely used chemicals in the chemical processing industry. The most commonly used acids are as follows.

1. *Sulfuric acid* (H_2SO_4). Sulfuric acid, among the most used of the acids, is commercially available in many concentrations and as various percentages of oleum. Oleum is sulfuric acid containing sulfur trioxide dissolved in the acid, which are called "fuming" grades. Generally recommended piping materials for these acids at low temperatures (140°F [60°C] and lower) and up to 90% concentration could be PVC, CPVC, PP, PVDF, ETFE and HDPE plastic, glass, Alloy 20, Duriron and FRP piping with special resins. At 90% concentration and higher, carbon steel schedule 80 is often used. Stainless steel is generally unsuitable, except for oleum greater than 103% concentration. Vent lines should be of the same material as the drain line.

Valve types include ball, gate and diaphragm, with gate valves being the most commonly used. For low pressure and temperatures suitable for the specific plastic pipe, plastic is often used. For higher temperature and pressures, alloy 20 is preferred. In all cases, because of differences in manufacture, pipe vendors should be consulted as to the suitability of materials for specific acid piping service.

Centrifugal pumps constructed of SS alloy 320 with Teflon packing are in common use. Other manufacturers use FRP and plastic pumps. Also available are metallic pumps lined with plastic or glass. Temperature limits should be carefully checked for material suitability.

Spills of concentrated acids from tanks onto floors and on equipment should be washed off and flooded with water, which can then be routed to the acid drainage system for neutralization. Tanks that contain this spillage should be of a suitable plastic. Since water reacts rapidly with the acid and splatters, caution should be exercised. Heat and fumes are also given off. Breathing of the fumes will cause throat and lung injury. Where this situation is possible, suitable emergency breathing apparatus should be provided. An emergency shower should be provided in the immediate vicinity of acid storage and pipe routing.

Sulfuric acid is nonflammable, but it is highly reactive. Below a concentration of 75% it reacts with carbon steel and other metals to form hydrogen. It is particularly hazardous when in contact with carbides, chlorates, nitrates, fulminates, picrates and powdered metals. In higher concentrations it will ignite combustible materials such as oily rags and sawdust. Dry chemicals or carbon dioxide are the fire suppression methods of choice.

Oleum spills, because of the danger of fumes, should be contained by curbs and the liquid diverted away from the area of a spill to a containment area where the liquid will be neutralized. The resulting liquid should be absorbed with diatomiatus earth, expanded clay or other non-reactive material. This material will be carted away for suitable disposal.

2. *Phosphoric acid* (H_3PO_4). Phosphoric acid is available in concentrations of between 75 and 87%. Recommended pressure piping is SS type 316 Extra Low Carbon (FLC). Drainage and vent piping, valves and pumps are similar to those used for sulfuric acid. OSHA has limits for human exposure to this acid.

Spills, safety and health concerns are similar to those for sulfuric acid.

3. *Hydrochloric acid* (*HCL*). Hydrochloric acid, also known as Muriatic acid, is available in four strengths designated as degrees Baume (an equivalent notation of specific gravity).

Piping materials for drainage and vent piping, valves and pumps are similar to those used for sulfuric acid. Spills, safety and health concerns are similar to those for sulfuric acid, except that caustic soda should not be used because hydrochloric acid reacts with this chemical.

4. *Nitric acid* (HNO_3) Nitric acid is available in three grades, designated by percent of concentration by weight: 56–70, 70–84 and 97.5–100%.

Recommended pressure piping material for concentrations up to 95% is 304L SS. Above this concentration, aluminum piping is recommended. Pumps for concentrations up to 95% should be constructed of 304L SS. Above this concentration, titanium, aluminum type 3003 or silicon iron are commonly in use. Recommended materials for gate, ball, plug and globe valves are 347 SS or 304L SS. Drain lines should be glass.

Spills, safety and health concerns are similar to those for sulfuric acid, except that temperature and humidity has an effect on reaction of nitric acid on such metals as copper, brass and zinc. Nitric acid reacts violently with organic substances, occasionally causing explosions. Self-contained breathing apparatus is required when approaching spills because of the emission of nitrogen oxides, commonly called nitrous fumes, which are extremely hazardous.

5. *Hydrobromic acid* (*HBR*). Hydrobromic acid is commercially available in two concentrations: 70 and 99.95%.

Recommended pressure piping materials are glass and rubber lined steel pipe, PVC, PE and PTFE. Glass pipe could be used for drainage in addition to the pressure piping. Valve types are often ball and plug type with PVC, PE and PTFE lining. Rubber lined pinch valves are commonly used. Pumps are similar to those used for sulfuric acid, with the addition of Hasteloy B material.

Spills, safety and health concerns are similar to those of phosphoric acid, but, in addition, the vapors are much more hazardous. This acid reacts with metals and produces explosive hydrogen gas.

6. *Perchloric acid* ($HCLO_4$). Perchloric acid is available in a concentration of 69–72% strength and is the strongest of all the inorganic acids.

Recommended pressure piping materials are glass and PTFE. Drain lines could be glass or Duriron. Valve types are often ball and plug type manufactured from PTFE and Duriron. Pumps manufactured from PTFE are the most commonly used.

Spills, safety and health concerns are similar to those of phosphoric acid, except that when heated to 150°F (64°C), perchloric acid can cause objects not normally considered combustible, such as rubber gloves and human skin, to burst into flames.

SELECTION OF LABORATORY WASTE PIPING
AND JOINT MATERIAL

The majority of the effluent from an "average" laboratory consists of primarily water and acid. Chemicals used for experiments are usually confined to fume hoods if toxic to the staff. Obtain from the end user the extent and concentration of all the chemicals expected to be used in the laboratory.

The most cost effective piping above the floor from laboratory fixtures is generally fire retardant polypropylene, either with heat fused socket or proprietary "screwed mechanical" type joints. Other acceptable materials are glass with compression joints and high silicon cast iron with caulked or compression gasket joints. Although PVC and CPVC have the least initial cost, they also have a limited range of chemical compatibility, with PVC having a low temperature rating. PTFE is the most resistant to the widest variety of chemicals and has the highest temperature rating and highest cost. The piping should be continuously supported on angle irons to prevent future saging resulting from hot effluent.

Piping underground could also be polypropylene with heat fused socket joints or high silicon cast iron with compression gasket joints. Glass piping should be encased in a sleeve of polyethylene for protection.

Vent pipe shall be the same material as the drain pipe. The vent shall be carried up to above the roof level. Vent piping penetrating the roof shall not be glass. Use an adapter and provide any other acceptable acid resistant pipe material through the penetration.

SYSTEM DESIGN CONSIDERATIONS

The laboratory drainage system shall have the same general system design considerations as the sanitary drainage system, including placement of cleanouts. Each fixture shall be individually trapped and vented. Clean water, such as that discharged from air compressors and other condensate drains, could also spill into the laboratory drainage system when convenient. Because of possible stoppages that could flood all the pipe, the entire laboratory waste system shall be of the same acid resistant piping material.

Where the only waste discharge is from laboratory fixtures, the use of fixture unit schedules for pipe sizing is acceptable, except that simultaneous use should be factored into the sizing process. When effluent is in gpm from a known discharge, base the size on gpm and the equivalent gpm from the fixtures. The pipe shall be sized using the required pitch and a ¾ full pipe.

The laboratory drainage and vent system shall be separate from all other systems until adequately treated and possibly combined on the site with other waste lines. If a manhole is required in the acid waste line, it should be acid resistant.

ACID WASTE TREATMENT

All acid waste requires neutralization to a pH of between 7.5 and 4.0 before it would be permitted to discharge into any public sewer for disposal. Commonly accepted practice permits local authorities to allow primary treated effluent to discharge directly into the public sanitary sewer system after only pH treatment. The most often used primary procedures are direct, continuous contact with limestone chips in an acid neutralizing basin or by means of continuous or batch treatment in an automated neutralization system utilizing chemical feces neutralizing.

An acid neutralizing basin operates on the principle of a chemical reaction between the acid and the limestone chips. Each basin shall be designed by the manufacturer to allow sufficient contact time for the chemical reaction to accomplish complete neutralization based on the maximum flow rate anticipated. Average figures have determined that 100 lbs of limestone chips treat 97 lbs of sulfuric acid and 75 lbs of hydrochloric acid. Effluent consisting mostly of sulfuric acid should be treated with dolomite limestone chips.

For general laboratory waste, several methods for treatment using limestone chips are available. For single isolated sinks, an acid neutralizing trap should be considered. For a small number of sinks in a cluster, a shelf mounted, small diameter basin could be used. It should be confined to treating the discharge of acids from a small number of fixtures and in remote locations or for individual sinks where timely maintenance needed to fill the basin may be questionable. A larger basin, illustrated in Fig. 3.10, is available to treat the effluent from a large number of laboratory sinks. If there is expected discharge of oil or grease in the laboratory waste stream, the installation of an interceptor basin is recommended before the acid sump. Some objectionable contaminants can coat individual chips and prevent proper chemical action to neutralize the acid.

For a larger number of fixtures or equipment and where treatment by limestone chips alone is not practical, a system consisting of single or multiple basins, and/or a mixing tank should be installed. If located at a low level, a pump will be required to discharge up to the level of the sewer. A sophisticated arrangement of probes, chemical feed pumps, level indicators and alarms will be required. An agitator or mixer may be installed in the basin to mix the acid with the caustic. The addition of a recorder may be desired. The acid neutralizing system operates on the principle of automatically adding proper amounts caustic to the incoming acid waste, thereby neutralizing the acid. The probe is connected to an automatic caustic feed pump that introduces the proper amount of neutralizing liquid into the basin or mixing tank. The most commonly used neutralizing chemical is caustic soda. Continuous treatment may also require additional downstream sensing probes and chemical additive locations to assure that the discharge is within acceptable limits. Refer to Fig. 3.11 for an illustration of a typical continuous waste treatment system. Various manufacturers have proven methods of acid treatment.

It is good engineering practice to have the discharge from the neutralizer separately routed into the sanitary house drain outside of a building for dilution prior to ultimate discharge into the public sewer. This may also be necessary for monitoring of the waste stream by local authorities without having to enter a building.

For preliminary determination of the number of sinks that will be required for average laboratories, allow one laboratory sink for each 200 sq. ft of laboratory area. Each sink will discharge one gpm. Cup sinks will discharge 0.5 gpm. For a maximum flow rate, assume that 10% of the cup sinks could discharge simultaneously. Add the two types of sinks together for the maximum discharge.

RADIOACTIVE WASTE DRAINAGE AND VENT SYSTEMS

General

Hospitals and laboratories are generally considered "institutional" facilities characterized by low quantities and levels of radioactive waste and are therefore subject to a less stringent set of regulatory requirements. Those types of facilities having higher quantities of radioactive material and levels of radiation fall under a different, much more stringent set of regulatory requirements. The principles of drainage system design are applicable to all kinds of projects, some of which may have significantly higher levels of radiation than most. The design philosophy is the same, but the submission of documentation for the protection of the public and workers in the event of any accident is considerably more complex. Because of the amount of radioactive material present at industrial facilities, larger storage and treatment systems are provided, as well as severe safety requirements that are not necessary at a site where lesser quantities of radioactive material are present.

With the exception of providing radiation shielding if necessary, the requirements for a radioisotope laboratory are essentially no different from requirements of other laboratories handling toxic chemicals or pathogens. The ultimate aim is to keep the exposure of workers, staff, and the general public to nothing. Since this is not realistic, it is required not only to prevent overexposure but to keep any exposure to radiation as low as reasonably achievable. The design shall implement criteria that will eliminate or reduce to allowable levels the radiation exposure of workers and maintenance personnel and prevent exposure of the general public to unacceptable amounts of radiation by waterborne radioactive waste (radwaste).

THE NATURE OF RADIATION

Radioactivity is the spontaneous emission of "harmful" particles from the unstable nucleus of an atom. *Nuclear radiation* is the propagation of these harmful emissions through space. In an effort to become stable, neutrons are spun out of the orbit of the atom, whereupon they collide with the nucleus of another atom, causing a *fission,* or splitting. This splitting forms a new element and at the same time releases heat, particles (including light), and new neutrons to fly out of orbit and split other atoms. There are many intermediate steps in the stabilization cycle that include the formation of other less complex radioactive by-products called *isotopes.* These by-products in turn decay to form other unstable isotopes as the cycle continues. The end result is an element that is highly stable. As an example, the end product of uranium is lead. One of the intermediate by-products of uranium is radon.

One of the most misunderstood concepts of radioactive materials is the potential for explosion. To clarify this, consider gunpowder. If the grains are spaced widely apart, one grain would ignite and possibly cause the ignition of the grain next to it. There could be not enough energy generated to push a bullet. But, when the same amount of gunpowder is placed in a confined case and ignited, it burns so quickly so as to virtually explode. In the same way, unless the atoms of a highly unstable (radioactive) element are closely packed together, or enriched, there is no possibility of an explosion. For example, the fuel in the cores of many nuclear power reactors is enriched only about 3 percent, which will allow only a self-sustaining reaction.

Radiation is a general term that encompasses any or all of the following: alpha rays, beta rays, gamma rays, neutrons, x-rays, and other atomic particles. There are three general classifications of radiation of concern, namely, alpha, beta, and gamma. *Alpha radiation* is actually a helium atom with a high velocity. *Beta radiation* is an electron with a high velocity. *Gamma radiation* is a particle similar to a photon, which is light. Alpha and beta radiation can generally be stopped by the skin or clothing, paper, or other similar light material. Alpha loses energy very quickly in air and is of no practical concern for distances greater than 12 in. High-energy beta radiation is commonly contained by only 1 in of solid, dense plastic. Beta is denser, carries more energy greater distances than alpha, and will burn bare skin and, in particular, damage the eye, but will generally not penetrate into the body to cause any internal damage. The greatest danger with beta radiation is to the eyes, particularly when the eye is directly exposed close to the source.

Gamma radiation is electromagnetic in nature. It carries the most energy and therefore is the most dangerous to humans. Its wavelength is shorter than light waves. When generated, it is similar to x-rays and behaves in a manner similar to light waves. When released from a source, gamma rays have a mass and velocity that has a measurable energy potential.

The best way to visualize the manner in which radiation harms cells in the body is to think of all the particles as bullets. These particles have velocity and energy. When stopped by the skin, these particles actually bounce off the skin, and the friction produces burns in the same way that sandblasting causes irritation. Sensitive tissue can also be killed by this friction. Gamma rays pass through the body and will change or damage any cell they touch on the way through. All three of the above radiation types are considered *ionizing radiation,* which means that the radioactive particles emitted from the source produce ionization. Ionization is the conversion of neutral particles to charged particles. Since cells are made mostly of water, the most common changes produce radicals that affect the bonds of cell molecules, causing breaks in DNA molecules and abnormalities in cells. Many cells are destroyed outright. Most affected cells die and are naturally replaced by the body. Some cells, particularly in the reproductive system, brain, and eyes, can't be reproduced by the body and are gone forever. Some cells become mutant and reproduce so quickly that they multiply out of control. That is the start of a cancer.

RADIATION MEASUREMENT

Radioactivity is a general term used for the total release of radiation of all types from a source. It is measured in disintegrations per second (dps). This measurement is possible for gamma radiation because in most radioactive materials, disintegrations also produce a known amount of gamma radiation. However, the best manner of measuring gamma radiation is from the energy it produces per kilogram of air. Because the instruments needed to measure radiation in this way are very expensive, they are not widely used outside of the laboratory. The so-called Geiger-Müeller counter is the most common. This instrument measures penetration of the particles that enter into a tube where the particles react with a gas in the tube, creating an electric charge that can be measured. If an amplification device is used, the charge can be heard in the form of static. The more modern instruments have a digital readout.

UNITS OF RADIATION DOSE

Particulate radiation is measured by the number of disintegrations per unit of time. A curie (Ci) is equal to 3.7×10^{10} dps. One millicurie is equal to 0.001 curie, or 3.7×10^{7} dps. One *rad* (*r*adiation *a*bsorbed *d*ose) is defined as the dose corresponding to the absorption of 100 ergs per gram (erg/g) of tissue. A roentgen (R) measures ions carrying a total of 2.58×10^{4} coulombs (C) of electrical energy.

Since the term *radiation* is a general one, a more specific term must be used to indicate the effect of radiation on humans. That measurement is called a *dose*. A dose is defined as the total quantity of radiation absorbed by the body or any portion of the body. Much of the time, the dose will be modified by reference to a unit of time. This differs from radioactivity because all radiation is not absorbed by the body. A rad is a measure of the dose to body tissue in terms of energy absorbed per unit mass. Gamma radiation is the most common of this type.

The most important measurement is the *radiation equivalent to man,* or rem. A rem is the measure of ionizing radiation passing through or absorbed by the body in terms of the biological effect relative to a dose of one roentgen of x-rays. The relation of the rem to other dose units depends upon the actual biological effect to the particular part of the body being studied and the actual conditions and amount of time of the irradiation. One rem is the equivalent of one roentgen due to x or gamma radiation, and also one rad due to x, gamma, or beta radiation. One rem of high-flux neutrons is roughly equivalent to the 14 million neutrons per square centimeter incident to the body.

ALLOWABLE RADIATION LEVELS

There is no exact radiation level that is certain to cause permanent harm to the human body. On the other hand, many scientists believe there is no level below which it is certain that harm will *not* occur. A certain constant amount of naturally occurring radiation, called the *background level of radiation,* exists all over the world. The most common source is the sun, and its radiation is called *cosmic radiation.* In addition, there are many other sources of radiation such as fly ash from burning organic fuels (particularly coal), granite, and other natural substances that contain trace isotopes of elements. One of the most common of these trace elements is carbon 14, used by scientists to date many materials.

The Nuclear Regulatory Commission (NRC) is the government body that has the responsibility for establishing criteria for the field of radioactivity. These criteria appear in the federal government's Code of Federal Regulations. As an example, Title 10, Part 50 is commonly referred to as "10 CFR 50." The levels that have been established by the NRC are as follows (measured in rems for one calendar quarter for the exposure):

Individual in a restricted area:

Whole body (head, trunk, gonads, eye lens, marrow)	1¼ rem*
Hands, forearms, feet, ankles	18¼ rem
Skin of whole body	7½ rem

Individual in an unrestricted area:

Whole body dose for 1 h	2 mrem
Whole body dose for 7 days	100 mrem
Whole body dose for 1 year	0.5 rem

All personnel working at any site that has a possibility of exposure to radiation are required to wear some type of exposure detection device that will allow accurate determination of their actual exposure. The photographic badge is the most common and is used where sensitivity is required. A pen-shaped device called a *dosimeter* is commonly used where there is less need for accuracy. It is used where instant determination of dose is necessary.

An *unrestricted area* is any area within a facility that is not specifically controlled for the purpose of protecting any individual from radiation or radioactive materials. A *restricted area* is access controlled. Another term, *environs,* may also be used to describe areas adjacent to a restricted or high-radiation area. A *high-radiation area* is defined as any accessible area within a facility that is capable of allowing the body to receive 100 mrem of radiation in a 1-h period.

SHIELDING

The purpose of shielding is to reduce or eliminate radiation emanating from any source within the facility. The most effective material has the greatest density, and so lead has been universally used for this purpose. Another material that is commonly used is concrete. In terms of shielding thickness, 4 mm of lead is the equivalent of 12 in of concrete. If it is known that a building will need shielding, concrete is the most often chosen building material as it provides ample structural support as well as shielding from radiation. It is up to the radiation safety officer (RSO) to determine the type of shielding and its placement to lower radiation in specific areas. Radiation travels in a straight line; therefore, if a tank or a length of pipe has to be shielded, the proper manner is to form a labyrinth so that the shine from the tank cannot escape in a straight line.

The most common materials used for shielding purposes are concrete and sheet lead. Other materials that have proven effective are (a) lead-lined concrete blocks, (b) lead-lined lath for plaster, and (c) lead-lined panels and gypsum boards. The barriers set up to reduce radiation levels are primary barriers, which are the first

---

* This can be increased to 3 rem to the whole body provided that no prior dose was received in the previous calendar quarter.

line of defense, and secondary barriers, which are used to eliminate leakage radiation and scattered radiation.

RADIOACTIVE MATERIALS

Radioactive materials are used for the following general types of activity:

1. Imaging sciences
2. Electrical power generation
3. Medical treatments and diagnostic tests
4. Manufacturing
5. Research

Almost all of the materials used are isotopes. An *isotope* is a form of an element with a different (or excess) number of neutrons in its nucleus. Because of this difference, the atom is unstable. These isotopes are identified by their atomic weight, which is the weight of the number of neutrons and protons in the nucleus.
There are a great number of isotopes in use today. Some of the more common are:

Iodine 131(8-day half-life)

Phosphorus 32

Technetium 99 (6-h half-life)

Calcium 45

Carbon 14

Strontium 90

Radium 226

Since any given amount of radioactive materials remains active for different periods of time, it is not possible to predict when any material will become completely stable. The most often used figure is the time required for a specific material to lose one-half of its radioactivity, which is called its *half-life*.

SYSTEM DESIGN CRITERIA

The Approval Process and Application Requirements

The use of any radioactive material requires the licensing of the site for a specific purpose and amount of radioactive material. Application for this license is made either to the NRC or a particular state. Those states that have elected to adopt the NRC regulations and provide their own staff for the purpose of issuing and approving licenses are called *agreement states*. In some cases these states make additional regulations of their own. Those states that rely on the NRC to review and issue licenses are *nonagreement states*. The application in those states is made to the appropriate department of the NRC.

For a hospital type of facility, the granting of the license mandates the presence of some form of a committee assembled for the express purpose of assuring compliance with the terms and conditions of the site license as well as reviewing all proposals for the actual use of the radioactive materials present at the site. This team may consist of a physician specialist in internal medicine, a physician specialist in nuclear medicine, and a radiation safety officer. A health physicist is also commonly included. For a research type of facility, the members of the team will include a member of the research team in lieu of a physician.

The administrative duties of the RSO include monitoring personnel exposure limits and controlling any release of radionuclides into the sewer system. In addition, the RSO works with engineers in the design phase of the facility to assure that the piping runs and all other mechanical work will result in a low exposure to people within the facility. For the most part, this work is directed to assuring that neither facility personnel nor nonstaff members are subjected to more than the maximum permissible radiation dose allowed under the applicable codes for any particular type of radioactive material present. The RSQ is also responsible for the following:

1. Teaching facility staff about the potential dangers
2. Keeping the necessary records for the facility
3. Keeping an inventory of material and records disposal
4. Concentrating materials at the facility
5. Assisting the engineer in design of mechanical systems
6. Designating areas within the facility to be restricted

GENERAL DESIGN CRITERIA

The prime consideration in the design of any facility is a concept controlling the exposure of personnel to radiation called ALARA, an acronym for "as low as reasonably achievable." This requires that the design of the facility must consider every reasonable method to limit the possible exposure of personnel inside the facility and keep the presence of radioactivity in any unrestricted area to a figure as low as reasonably achievable. This must take into account the current state of technology, the economics of further improvements in relation to the benefits to public health and safety, and other aspects of socioeconomic considerations in relation to the utilization of radioactive material in the general public interest. The facility must also make a reasonable effort to eliminate residual radiation. One of the overriding concepts is the "worst-case" possibility, by which contingency plans are made for the worst possible combination of circumstances to determine the possible level and amount of time of radiation exposure. This concept should not be overdone, and a general rule is to have only one "accident" at a time. As an example, a serious spill and a fire would not be considered as occurring simultaneously.

PIPE MATERIAL SELECTION

The pipe selected for the radioactive drainage system depends upon the type of radiation and level of radioactivity, which in turn, depends upon the amount and type of radioactive material at the facility. In general, an ideal radwaste piping should have the following properties:

1. It must be nonporous.
2. It must be easy to clean and decontaminate.
3. It should be acid resistant.
4. It should be nonoxidizing.
5. The joints should not form a crud trap.
6. Joint materials must not be affected by radiation exposure.

It is possible in very high radiation areas to have a pipe affected by the radiation present. The oxides of the pipe can become radioactive, or the pipe itself can be weakened. Another factor is the weakening of elastomeric seals or gaskets because of high levels of radiation. For this reason, Teflon is never used where anything more than a very low level of radiation is present. Other materials should be investigated as to their suitability for use for the levels anticipated.

All the commonly used materials (cast iron, ductile iron, copper, steel, and glass) and the joints normally used to put the pipes together fall far short of the ideal. However, all of them are suitable for low-level waste and radioactive source materials found in conventional health institutions. Plastic piping is not acceptable for radwaste systems, due to the possibility that the plastic may be affected by the radiation. It is only when the radiation levels of the waste materials get into the high-radiation-level category that they fail one or more of these conditions. As a result, stainless steel with welded joints has emerged as the material of choice for all of the industrial types of waste products. Type 316L is the most common.

Welded joints are the only type of joint that meets the criteria for not allowing a crud trap.

GENERAL DESIGN CONSIDERATIONS

Human waste, even that contaminated with radioactivity, is exempt from all NRC regulations, requiring only compliance with local codes as far as disposal, sizing, and all other criteria applicable to standard drainage systems. There are also many isotopes that are exempt from regulations regarding disposal into the public sewer.

Another requirement is that the liquid radwaste to be discharged shall be diluted with the ordinary sanitary system effluent from the rest of the facility before being discharged to the public sanitary sewer system. This usually requires that the radwaste piping first be kept separate from the rest of the facility's piping, then joined to the main system before leaving the building for discharge into a public sewer. There are no restrictions regarding the combining of any radwaste together that are permitted to be discharged separately. A method should be provided for the RSO to take a grab sample of the radwaste stream if desired, such as a valved outlet from both the radwaste line and the combined discharge. Keep the pitch of the piping as steep as p05sible, in order to empty the pipe quickly and to provide a scouring action that will keep the radioactive solids in suspension.

It is common practice in laboratories to have high levels of radiation confined to glove boxes or protected fume hoods. The small amount of liquid waste produced from this equipment would be stored in shielded containers below the equipment that is removed periodically. If storage of larger quantities of low-level radwaste is required, it is piped to a holding tank. A common holding time is 10 half-lives of the effluent. Usually, radwaste is stored for disposal on the site, outside of a building where easy transfer of the liquid is possible. The removal must be done by licensed waste disposal contractors that remove the waste from the holding tank into a special truck that transports the liquid waste to a designated site suitable for disposal of low-level waste. The solid wastes, such as gloves and wipes, are stored in special containers that are removed to the disposal area with the liquid radwaste.

Floor drains are normally not desired in laboratories. If there is a spill of radioactive material, it is wiped up by hand using absorbent material, and the solid containing the spill is put in a special radwaste holding container within the lab. If a floor drain is called for in the design, it should be made of stainless steel, which is available from all the major manufacturers. For testing purposes and to close off a drain when it is not expected to be used, each drain should be supplied with a closure plug. If there are areas where there may be a spill, the floor must be pitched to a floor drain. A generally accepted value for pitch of the floor is 1 in 20 ft. The thickness of the slab must be closely coordinated because the slab is thinnest at the drain and made thicker at the ends of the area served to make up the pitch. It is not practical to cast the slab evenly and add a topping because there is a tendency to chip the topping and possibly have a radioactive spill get under the top coating. Since the slab depth is greater the longer the run to the drain, it is necessary to indicate the top of drain elevation at each drain. This also makes it easier for the shop fabricator to make up accurate pipe spools.

Drains also require special treatment in a high-radiation facility. Like the drains in a low-radiation facility, they should be manufactured of stainless steel. There will be different types of drains in different areas, and they may be installed at different elevations. Because of this and the probability that the piping will be made in spools, it is a good idea to number all of the individual drains on the design drawings. A box next to each drain can be used to provide information regarding type, number, and elevation.

Since fittings are a natural crud trap, avoid running piping in, under, over, or adjacent to unrestricted areas in a facility. If this is not possible, place the line where additional shielding can be added either at the time of construction or after the start of actual use, when the RSO may determine by survey that additional shielding is necessary. Much of the time, the ability to take apart the joint and flush out any crud is an advantage. Any of the popular joints for no-hub or grooved pipe are acceptable, as well as glass pipe if used in a laboratory for chemical resistance.

Be generous with cleanouts. They may be needed to flush out the line to reduce spot high radiation rather than to rod it out.

DECONTAMINATION

Decontamination of both personnel and equipment may also be required. Often, valves, small lengths of piping, or instruments must be taken out of service for repair or replacement. Personnel decontamination areas must be provided in the event that some radioactive contamination may accidentally spill on somebody.

There are three methods generally used to decontaminate equipment:

1. Ultrasonic decontamination
2. Electropolishing decontamination
3. Washing with water and/or brushes

The most commonly used decontamination method is washing equipment with a detergent solution and a stiff wire or bristle brush. This method can be used only if the part is lightly contaminated and the lack of close proximity any person must have to the part being cleaned. If the cleaning is on a system that is unrestricted, the crud relieved can simply be washed down any drain. The RSO must determine that the part is not radioactive enough to cause exposure problems. If the contamination is more than the allowed minimum, the part is placed in a container, and then a high-pressure stream of water is first used to remove the loose crud and then a brush is used to finish the job.

If the part is determined to have a significant amount of radiation and the crud cannot be removed by washing, ultrasonic cleaning is the next choice. Except for very large systems that are built in place, most of the ultrasonic cleaning systems are factory packaged. The cleaning system consists of a tank that contains a detergent cleaning solution with a wetting agent, water heater, ultrasonic generators, filters, and a water circulating system to remove the soil from the detergent solution in the tank. The part to be cleaned is put in the tank, whereupon the filters remove the loosened crud suspended in the water. The filters are in a cartridge that is removed periodically and disposed of along with the other low-level radwaste.

If ultrasonic cleaning is not effective, the final step is to use an electropolishing system. This cleaning system is similar to electroplating except that the part to be cleaned is the sacrificial anode, and a layer of nidal is deposited on the cathode of the system. This system also uses a liquid bath, and the removed crud is filtered out and contained in a filter cartridge.

All of these areas contain floor drains, discharging either to a public sewer if the radiation is from exempt material, or if a higher level, to a holding tank for disposal or to a radwaste treatment system.

Personnel decontamination consists of regular showers, with one exception: hot water cannot be used. Hot water opens the pores of the skin and may allow contaminated particles to enter the body. People go from a dirty area where they shed their protective clothing, then they get scanned, and then they go either to the clean area for a regular shower or to the decon area for a cold shower, during which they use a stiff brush with detergent for scrubbing the contamination off the skin. A rescan determines if it is safe for them to take a regular shower. Again, if the material used in the facility is exempt and the worst case determined by the RSO is acceptable, the effluent from the showers can all discharge into the public sewer system. If not, then it shall be routed to a holding tank or radwaste treatment equipment for disposal. It is highly unlikely that a decon facility will be necessary in any facility using only exempt products.

INFECTIOUS AND BIOLOGICAL WASTE DRAINAGE SYSTEMS

Waste generated in pharmaceutical facilities has the same basic characteristics as that of other laboratory and production facilities, but with the addition of bio-hazardous material. Biohazardous material consists of live organisms suspended in the waste stream that have the potential to cause infection, sickness, and other very serious diseases if not contained. This waste will be discharged by gravity and under pressure from many sources, including:

1. Fermentation tanks and equipment
2. Process centrifuges
3. Sinks, both hand washing and process
4. Containment area floor drains
5. Janitor closet drains
6. Necropsy table drains
7. Autoclave drains
8. Contaminated condensate drains

Containment is the method used to isolate and confine biohazardous material. The facility equipment and design shall conform to acceptable and appropriate containment practices based on the hazard potential. A containment category is used to describe an assembly of both primary and secondary preventive measures that provide personnel, environmental, and experimental protection. Primary barriers are specific pieces of equipment such as the biological safety cabinet (which is the biologist's equivalent of the chemist's fume hood) and glove boxes. Secondary containments are features of facility design surrounding and supporting the primary containment. These features are described and classified in many publications such as those of the National Institutes of Health in Bethesda, Maryland.

The classifications for biological containment in laboratories consist of four bio-safety levels, BL1 through BL4. Publications describe the work practices, equipment, and BL selection criteria based on the activity of that particular laboratory. If the laboratory or production facility produces or uses greater than 10 L of any substance containing viable organisms, the facility may become large scale (LS). This is noted as BL2 LS. Manufacturing standards shall conform to good large-scale production (GLSP) standards. The same standards apply to both small- and large-scale facilities.

CODES AND STANDARDS

Mandated guidelines and regulations include the following:

1. OSHA Bloodborne Pathogen regulations
2. NIH guidelines for the use of recombinant microorganisms
3. FDA cGMP regulations
4. CDC/NIH Guidelines for Biosafety in Microbiological and Biomedical Laboratories

BIOLOGICAL SAFETY LEVELS

CDC/NIH Guidelines for Biosafety in Microbiological and Biomedical Laboratories are summarized in an abbreviated and simplified form for the following laboratory containment levels.

Biosafety Level 1 (BL1) Containment

This is the typical biological research facility classification for work with low-hazard agents. Viable microorganisms not known to cause disease in healthy adults are used at this level. Work activity is done on an open bench, and any hazard present can be controlled by using standard laboratory practice. Standard features consist of easily cleaned, impervious bench surfaces and hand washing sinks, and lab work is separated from general offices, animal rooms, and production areas. Contaminated liquid and solid waste shall be treated with a suitable disinfectant to remove biological hazards before disposal. Wastes containing DNA materials or potentially infectious microorganisms shall be decontaminated before disposal. Facilities to wash hands are required in each laboratory.

Biosafety Level 2 (BL2) Containment

This facility designation is similar to BL1 except that the microorganisms may pose some risk and safety cabinets are often present. Equipment and work surfaces shall be wiped down with a suitable disinfectant. Sinks shall be scrubbed daily with a chlorine-containing abrasive and flushed with a suitable disinfectant. All liquid waste generated shall be immediately decontaminated by mixing with a suitable disinfectant. Nearly all laboratories operate under level 1 or 2 containment. At these levels, the facility is engaged in research or diagnostic or production activities thought to pose little or minimal risk to workers.

Biosafety Level 3 (BL3) Containment

Level 3 activity involves organisms that pose a significant risk or represent a potentially serious threat to health and safety. Biosafety cabinets are required, and all penetrations to outside the facility must be sealed to prevent leakage. These seals must be capable of being cleaned. Liquid waste is kept within the laboratory or facility and steam-sterilized prior to discharge or disposal. Vacuum inlets must be protected by appropriate filters and/or disinfectant traps. Laboratory animals require special housing. Personnel must be appropriately protected with full suits and respirators. A handwashing sink shall be located adjacent to the facility exit and is routed to sterilization. Vents from plumbing fixtures must be filtered.

Biosafety Level 4 (BL4) Containment

This is a rarely used classification that applies to those facilities in which the activities require a very high level of containment. The organisms present life-threatening potential and may initiate a serious epidemic disease. All of the BL3 requirements apply. In addition, showers shall be provided for personnel at the airlock where clothes are changed upon entry or exit. A biowaste treatment system shall be provided within the facility to sterilize liquid waste.

LIQUID WASTE DECONTAMINATION SYSTEM

A liquid waste decontamination system (LWDS) collects and sterilizes (decontaminates) liquid waste. Effluent containing potentially hazardous biomatter is collected in a dedicated drainage system, generally discharging by gravity into a sump below the floor level within the facility. From the sump, effluent is pumped into a "kill tank" where the actual sterilization occurs. A kill tank is a vessel into which steam or chemical disinfectant can be injected to kill any organism. The kill tank system should be qualified to the same biosafety level as the facility that it receives its discharge from. The kill tank system must be a batch process since time is needed to complete the sterilization and decontamination is based on the process used.

System Components

In addition to piping, the system consists of the sump or tank to receive contaminated discharge from the drains and equipment of the facility, a pump to remove the contaminated effluent from the sump and up into the kill tank(s), and the kill tanks that will decontaminate and sterilize the effluent to a point permitting disposal into the same system as the sanitary waste from the facility, generally into a public sanitary sewer.

Sump Pit

The sump pit into which the effluent drains shall have a gasketed, waterproof cover. The controls are similar to that provided on a plumbing sump pump and shall be capable of being chemically or steam sterilized. The pit is sized in conjunction with the sizing of the pump so that the pump stays on for a minimum of 1 mm to avoid too frequent starting. Other considerations, such as having the pit contain one batch of product if necessary, may be considered.

Kill Tank Assembly

The kill tank consists of a duplex tank arrangement, allowing one batch to be decontaminated while the other is filling. The size of the tanks will vary based on the individual facility, but common practice is to have each tank capable of containing 1 day's effluent plus the chemicals used for decontamination. Another consideration is to have sufficient size to hold a catastrophic spill. There is usually an agitator to mix the effluent with the deactivation chemicals. In addition to the kill tanks, tanks containing disinfectant chemicals to be injected are required. A fully automatic control system must be provided to ensure the timely addition of the required chemicals in the correct amounts and for the required duration of deactivation of the biomatter. Alarms and status should be displayed in an appropriate panel located in a facility control room or other area.

Drainage System and Components

The drainage system must be closed, which requires sealed floor drains and valved connections to equipment when not in use. Since the HVAC system maintains a negative pressure, it is important that the traps on all floor drains have a seal 2½ in deeper than the negative difference in air pressure. The traps of floor drains are filled with a disinfectant solution when not used to eliminate the possibility of spreading organisms between different areas served by the same connected sections of the piping system very often the floor drains are sealed.

The choice of drainage piping material is based on the suspected composition of effluent chemicals and the sterilization method. If the local authorities determine that the biowaste is hazardous, a double contained piping system with leak detection may be required. Stainless steel or PTFE pipe is usually chosen where higher-temperature effluent may be discharged or steam sterilization may be required. PVC, CPVC, PP, or lined FRP pipe could be used where effluent temperatures are lower and also where chemicals will provide the method of sterilization.

If waste from pressurized equipment is discharged into a gravity system, the system must be adequately sized to carry away the proposed flow rate with pipe flowing half full and adequate vents provided to equalize the internal pressure and assure that the pipe is always at atmospheric pressure.

Valves shall be of the diaphragm type, capable of being sterilized using the same method as the pipe. After appropriate decontamination, the kill tank effluent shall be discharged to drain. This effluent must be treated prior to discharge into a public sewer system for disposal.

Vents

Vents from pipe, fixtures, sealed sump pits, and kill tanks must be filter-sterilized prior to leaving the system using an HEPA or a 0.2-μm filter. In the event of an accident, OSHA has rules to aid personnel responding to emergencies involving any hazardous material.

SYSTEM DESIGN CONSIDERATIONS

The treated discharge from any containment treatment shall be separately routed to the sanitary system outside the building to allow for monitoring and sampling.

INDUSTRIAL WASTE SYSTEMS

GENERAL

Industrial waste drainage systems could contain an extremely wide variety of waterborne waste. Among them are chemicals, solvents, suspended solids, flammable liquids and wastewater, many of which are considered hazardous. The purpose of the industrial waste drainage system is to collect and transport these wastes from inside a facility to a point on-site where disposal or treatment is accomplished.

CODES AND STANDARDS

A great body of regulations affect the design of any industrial drainage system. Among them are the federal Clean Water Act (CWA) and Resource Conservation and Recovery Act (RCRA), which are administered by the federal Environmental Protection Agency (EPA) as well as state and other local agencies. The local authorities are also empowered to create regulations that are more strict than the federal regulations. Where production and manufacturing facilities discharge waste, it is a general practice to engage the services of professionals experienced in wastewater treatment and environmental issues to assure compliance with all of the latest applicable regulations and maintenance of an acceptable treatment system. The major regulatory factor to be considered is the determination of whether any particular waste stream is hazardous. If so, protective measures, such as double-contained piping systems and leak detection, may be required.

PIPE MATERIAL AND JOINT SELECTION

Because of the vast diversity of manufacturing processes, it is impossible to make any general characterization of industrial wastewater. It is common to have various areas within a plant or industrial complex discharging different types of effluent with greatly varying characteristics.

The largest quantity of effluent in an industrial facility originates from drains. Drains receive discharge from production equipment, floor washdown, process and production machines, and other equipment such as compressors and boilers. The floor drain and discharge pipe from the drain must be capable of resisting chemicals discharged from the production equipment. The selection of the most appropriate piping material can be accomplished only if the nature of the effluent, both present and future, is known and can be allowed for.

SYSTEM DESIGN CONSIDERATIONS

The design of the drainage system is dependent on the location, composition, and quantity of discharged effluent all sources. The layout and engineering of a piping network requires ingenuity and attention to detail.

The selection of the type and location of floor drains is a major aspect of drainage system design. The following are general guidelines for locating and selecting the drains:

1. Wet floors are to be avoided. Drains should be located next to equipment and be large enough to allow easy multiple discharges to spill over them without requiring a run of pipe over the floor or having to spill on the floor and run to the drain. If large flow rates are expected, select a large drain.

2. The use of long trench drains in areas where a number of pieces of equipment are placed will create an easy access to all of the various drains from the equipment. This arrangement is usually less costly than multiple drains.

3. In many cases the discharge from equipment may be under pressure because of the head of water in the piece of equipment such as a tank, being emptied. The drain should be large enough in physical size to accept the largest expected flow. The size of the discharge pipe must be large enough to accept the maximum quantity flowing full by gravity without overflowing. An air gap shall be provided to prevent pressurizing the gravity drainage system.

4. To accept the largest number of multiple, small-sized drainage lines from equipment, a funnel type of drain should be provided. The top of the funnel should be as close to the floor as reasonable in order for an air gap to be provided between the top of the floor drain and the end of the equipment drain. This air gap shall be twice the diameter of the drainage line.

5. Provide adequate cleanouts in drain lines. In lines that are at the ceiling of high floors, extend the cleanouts to the floor above to avoid the need for maintenance personnel to climb ladders in order to clean stoppages.

6. The minimum-size drain line under the slab or underground should be 2 in (50 mm). Floor drains should be a minimum size of 4 in (100 mm).

7. Adequate venting of the drainage line must be provided to allow for smooth flow. These vents shall be connected to the top of the drain line in order to allow air at the top of the pipe to be either vented out (when there is a slug of liquid) or admit air required by the flow of water or due to a partial vacuum created by the liquid flowing full. Vents shall be a minimum size of 2 in.

8. Local regulations may require the use of double-contained piping to prevent potential leakage from discharging into the environment. A leak detection system should be provided that will allow leakage to be annunciated.

FIRE SUPPRESSION WATER DRAINAGE

For industrial facilities, the water used to suppress a fire could become contaminated with the products and raw materials it comes in contact with. It is required that any water, such as sprinkler and fire hose discharge, that has the possibility of being contaminated in this manner, must be routed to holding basins for analysis and possible treatment before being discharged into the environment. If there is no material capable of causing contamination, no special consideration is necessary except to protect other areas of the facility from possible flooding.

SYSTEM DESCRIPTION

The drainage system consists of the drains located to intercept the flow of fire water, the drainage piping, a holding basin on-site to contain and treat the total volume of water, and the necessary treatment system that will neutralize the water prior to discharge into the environment.

The amount of water discharged from the fire suppression system is far greater than wastewater discharged from the facility under normal operating conditions. Overflow floor drains large enough to take the design flow rate shall be installed at points that will intercept the water before it flows out of doorways or drive bays and route it to holding basins. The placement of these overflow drains shall be selected to intercept all of the water discharged and prevent it from damaging other parts of the facility, escaping away from the property or into the ground.

The drainage piping is sized on flow rate and pitch from the facility to the detention basin. The effluent will be essentially clear water with few solids. The flow rate of water required to be disposed of is determined by first calculating the sprinkler water density over the area used for hydraulic calculations. Add to this the flow rate from the number of fire standpipe hose streams possible. Velocity in the drainage pipe is not a major consideration because the system is rarely used. A shallow pitch will give a low velocity that may result in deposit of some material that could be flushed out after the fire event. A high velocity will not affect the life of the piping system because of the short amount of time the system is in operation. Pipe size is selected based on the actual pitch of the pipe and the capacity flowing full. Refer to Table 6.26.

Venting of the system is required in order to allow free flow of the effluent. Each individual drain need not be vented, but each branch should have a loop vent of at least 2 in (50 mm) in size. The vent could be connected to the sanitary vent system or carried through the roof independently. The pipe material selected shall be compatible with the potential chemicals it might carry.

REFERENCES

Georgehegan, R. F., and H. W. Meslar. "Containment Control in Biotechnology Environments," *Pharmaceutical Engineering,* 1993.

Mermel, H., "pH Control of Chemical Waste," *Heating/Piping/Air Conditioning Magazine,* 1988.

FLAMMABLE AND VOLATILE LIQUIDS

Federal, state and local regulations have established standards for the discharge of volatile liquids, particularly oil, into stormwater and sanitary sewers. These standards vary, and the responsible enforcement and code authorities must be consulted to determine the level of treatment required.

The most common flammable liquid is oil. The hazard created is either one of safety, since the vapors could create an explosive condition, the oil will float on water and could be set on fire, or medical, where the breathing of the vapors is dangerous to health and toxic if ingested by humans. The common characteristic of all of these substances is that they are lighter than water. Their removal is similar to that of oil.

OIL IN WATER

Oil in water exists in several forms, and is considered immiscible since it cannot be mixed with water.

1. Free oil.

2. Mechanically dispersed oil is fine droplets ranging in size from microns to fractions of a millimeter. They are stable due to electrical charges and other forces but not due to the presence of surface active agents.

3. Chemically stabilized emulsions are fine droplets that are stable due to surface active agents.

4. Dissolved and dispersed oil is suspended in such a small size (typically 5 microns or smaller) that ordinary filtration is not possible.

5. Oil-wet solids, which are particulates to which oil adheres to their surface.

METHODS OF SEPARATION AND TREATMENT

Oil spills and leaks are best treated in their most concentrated state, which is at their source or as close as reasonable. The primary methods used to separate and remove free oil and oil wet solids are floatation and centrifugation. Secondary treatment, such as chemical treatment/coalescence and filtration, is then used to break up oil water emulsions and remove dispersed oil. Finally, tertiary treatment, such as ultrafiltration, biological treatment and carbon absorption will remove the oil to required levels prior to discharge. This chapter will discuss only the general principles of the primary and secondary separation methods and devices.

The American Petroleum Institute (API) has established criteria for the large scale removal of globules larger than 150 microns. In abbreviated form, they are as follows.

1. The horizontal velocity through the separator may be up to 15 times the rise velocity of the slowest rising globule, up to a maximum of 3 ft per second.

2. The depth of flow in the separator shall be within 3 to 8 in.

3. The width of the separator shall be between 6 and 20 in.

4. The depth to width ratio shall be between 0.3 and 0.5.

5. An oil retention baffle should be located no less than 12 in downstream from a skimming device.

GRAVITY SEPARATORS

Gravity separation is the primary and most often used separation method. It is based on the specific gravity difference between immiscible oil globules and water. Since all of these liquids are lighter than an equal volume of water, gravity separators operate on the principle of floatation. As the water and oil flow through the unit, the oil floats to the top and is trapped inside by a series of internal baffles. Since the oil remains liquid, it is easily drawn off.

FLOTATION DEVICES

For larger scale service, the flotation of oil and oil wet solids to the top of the floatation chamber can be increased by the attachment of small bubbles of air to the surface of the slow rising oil globules. This is done by adding compressed air to the bottom of the floatation chamber in a special manner that will create small bubbles that will mix with, and attach themselves to, the oil globules.

CENTRIFUGAL SEPARATORS

For larger scale service, the centrifugal separator is used. This device operates on the principle of inducing the combined oil and water mixture to flow around a circular separation chamber. The lighter oil globules will collect around the central vortex, which contains the oil removal mechanism, and the clear water will collect at the outer radial portion of the separation chamber. Methods have evolved that will produce effluent water with only 50–70 ppm of oil, and proprietary devices exist that will lower oil content to 10 ppm.

FILTRATION

Chemical methods used to break oil/water emulsions followed by depth-type filters to remove the destabilized mixture have proven effective in removing oil globules in a range of sizes between 1 and 50 microns. The velocity and the flow rate of the mixture must be carefully controlled to allow optimum effectiveness of the system.

SMALLER SYSTEMS

Oil separators for small flows usually take the form of a single unit consisting of a drain grating into which the effluent flows and inside which the oil remains to be drawn off manually. Another type of unit uses an overflow arrangement that sends the trapped oil to a remote oil storage tank.

Because there is the possibility that the vapor given off by the flammable liquid could also ignite, it is important to provide a separator vent that terminates in the open air at an approved location above the highest part of the structure. Some codes require that a flame arrestor be installed on the vent.

The most common material used for an oil interceptor is cast iron, although steel can be used for less severe service. Gratings must have the strength for the type of vehicle expected.

Refer to Fig. 3.6 for an illustration of a typical small oil interceptor. Fig. 3.7 illustrates the installation of a typical oil interceptor with gravity oil drawoff for multiple floor drain inlets.

ADDITIONAL REFERENCES

Grossel, S. F. "Safe Handing of Acids," Chemical Engineering Magazine, July 1998.

Kaminsky, G, "Failsafe Neutralization of Wastewater Effluent," Plant Services Magazine, May 1998.

CHAPTER 11
FACILITY STEAM AND CONDENSATE SYSTEMS

This chapter describes systems concerned with the generation, distribution, installation, and sizing criteria for steam and steam condensate systems to be used for purposes other than comfort or space heating. The systems described here include those providing steam for kitchen and laundry facilities, water purification, humidification, and steam tracing, as well as steam systems for laboratory and sterilizing purposes. Steam for space and comfort heating, although briefly discussed, are outside the scope of this handbook. Heat tracing systems utilizing steam are described in Chap. 5.

SYSTEMS DESCRIPTION

The complete system is actually composed of two codependent, connected subsystems. The steam subsystem includes a source of steam, distribution piping to carry steam to the point of use, and controls to regulate steam flow at the required pressure or temperature. At some point after the steam has been used at the terminal equipment, the steam condenses into a liquid, called *condensate*. The condensate subsystem may be designed in such a way that the condensate is returned to the steam source, absorbed into the process and lost from the system, or simply wasted to a drain. If the condensate is to be recovered for reuse or is to be wasted, then additional components such as steam traps, flash tanks, condensate piping, heat rejection terminal units, and condensate pumps may be required.

CODES AND STANDARDS

The following standards are generally used to design the steam and condensate systems that are the subject of this handbook:

1. ASME Boiler and Pressure Vessel Code
2. ASTM B 31-9, Power Piping Code, Steam Properties

Local building codes and regulations should be carefully checked to ensure that the installation meets their requirements. Local conditions, such as atmospheric pressure, should be checked to ensure the usability of the details and design data.

FUNDAMENTALS

STEAM

Water exists in three states: solid, liquid, and gas. Steam vapor is a combination of liquid and gas. When liquid water is heated to a point where it can no longer exist as a liquid, it has reached the *saturation point.* From the saturation point, any further addition of heat (called the *heat of vaporization*) causes water to *boil,* that is, to rapidly and violently evaporate into vapor, or steam. When the term *steam* is used, it implies *dry steam,* without any suspended water droplets. Steam has all the characteristics of a gas and is normally invisible. It only appears as a mist when it cools and is partially condensed, causing droplets to form in the vapor stream.

Compared to water at any given temperature and pressure, steam occupies about 1600 times greater volume, which is its most useful property. This makes steam an economical heat transfer medium since relatively few pounds per hour of steam flow can transfer a great deal of heat as the steam condenses.

BOILER FEEDWATER

The term *boiler feedwater* is used to describe all water introduced into a boiler for the purpose of being converted into steam.

STEAM TEMPERATURE

Temperature of steam is the measurement of heat of the vapor at any given pressure.

STEAM QUALITY

Steam quality defines the amount of water present in the fluid stream that has not been evaporated into steam. It is calculated by the use of calorimeter tests. It is often defined by its dryness percentage, which is the proportion of completely dry steam present compared to the total amount of steam in the sample being measured. Steam quality, in percent, is measured to quantify flowmeter readings for billing purposes, check boiler practices, and to troubleshoot process operations.

Steam is classified as either wet or dry. Saturated steam has a steam quality of 100 percent.

Saturated Steam

The term *saturated steam* describes steam the moment it is generated in a common boiler where steam and liquid water are in direct contact.

Saturated steam is described as *dry* because all the moisture has been condensed into steam. A pound of wet steam may have the same temperature and pressure as a pound of dry steam, but it will contain less heat since some of the water is in liquid form and therefore contains no latent heat, only sensible heat. In practice, heat loss from the piping may condense a portion of the steam, so that a small amount of liquid may be carried along with the steam, producing a wet steam flow. Superheating the steam will minimize condensation of steam flowing in pipes.

Saturated steam pressure and saturation temperature are directly related. As the pressure of the steam rises, its temperature also rises. Some properties of saturated steam are listed in Table 11.1.

Dry Steam

Dry steam is steam that has been fully evaporated to a point where it contains no droplets of liquid water in suspension.

Wet Steam

Wet steam contains droplets of water in suspension. The droplets can be removed by means of separators and by adding heat.

HEAT

Heat is a form of energy and as such forms a part of the enthalpy of a liquid or gas.

Enthalpy (Sensible Heat)

Sensible heat is the heat energy added to boiler feedwater that raises the feedwater temperature. *Enthalpy*, or heat content, is the total internal and external energy of a fluid or vapor due to both its pressure and temperature. It is expressed in British thermal units (Btu). It's called "sensible heat" because the effect of the heat transfer can be sensed by observing a rise in the water temperature. Sensible heat is the number of British thermal units required to raise the temperature of a pound of melted ice from 32°F to the saturation temperature at a given pressure. The higher the initial temperature of the boiler feedwater, the less sensible heat is required to bring the water to the saturation point.

Latent Heat of Vaporization

Latent heat of vaporization is heat transferred to water that will change its state, at the saturation temperature for a given pressure, from liquid to steam. Also called *latent heat of evaporation* and *enthalpy of evaporation,* it is the number of British thermal units required to change a pound of boiling water into a pound of steam at a given pressure. The additional heat causes water molecules to break the bonds

TABLE 11.1　Properties of Saturated Steam

Pressure, psig	Temperature, °F	Heat, Btu/lb			Specific volume, ft³/lb
		Sensible	Latent	Total	
25*	134	102	1017	1119	142
20	162	129	1001	1130	73.9
15	179	147	990	1137	51.3
10	192	160	982	1142	39.4
5	203	171	976	1147	31.8
0	212	180	970	1150	26.8
1	215	183	968	1151	25.2
2	219	187	966	1153	23.5
3	222	190	964	1154	22.3
4	224	192	962	1154	21.4
5	227	195	960	1155	20.1
6	230	198	959	1157	19.4
7	232	200	957	1157	18.7
8	233	201	956	1157	18.4
9	237	205	954	1159	17.1
10	239	207	953	1160	16.5
12	244	212	949	1161	15.3
14	248	216	947	1163	14.3
16	252	220	944	1164	13.4
18	256	224	941	1165	12.6
20	259	227	939	1166	11.9
22	262	230	937	1167	11.3
24	265	233	934	1167	10.8
26	268	236	933	1169	10.3
28	271	239	930	1169	9.85
30	274	243	929	1172	9.46
32	277	246	927	1173	9.10
34	279	248	925	1173	8.75
36	282	251	923	1174	8.42
38	284	253	922	1175	8.08
40	286	256	920	1176	7.82
42	289	258	918	1176	7.57
44	291	260	917	1177	7.31
46	293	262	915	1177	7.14
48	295	264	914	1178	6.94
50	298	267	912	1179	6.68
55	300	271	909	1180	6.27
60	307	277	906	1183	5.84
65	312	282	901	1183	5.49
70	316	286	898	1184	5.18
75	320	290	895	1185	4.91
80	324	294	891	1185	4.67
85	328	298	889	1187	4.44
90	331	302	886	1188	4.24
95	335	305	883	1188	4.05

TABLE 11.1 Properties of Saturated Steam *(Continued)*

Pressure, psig	Temperature, °F	Heat, Btu/lb Sensible	Heat, Btu/lb Latent	Heat, Btu/lb Total	Specific volume, ft³/lb
100	**338**	**309**	**880**	**1189**	**3.89**
105	341	312	878	1190	3.74
110	344	316	875	1191	3.59
115	347	319	873	1192	3.46
120	350	322	871	1193	3.34
125	**353**	**325**	**868**	**1193**	**3.23**
130	356	328	866	1194	3.12
140	361	333	861	1194	2.92
145	363	336	859	1195	2.84
150	**366**	**339**	**857**	**1196**	**2.74**
155	368	341	855	1196	2.68
160	371	344	853	1197	2.60
165	373	346	851	1197	2.54
170	375	348	849	1197	2.47
175	**377**	**351**	**847**	**1198**	**2.41**
180	380	353	845	1198	2.34
185	382	355	843	1198	2.29
190	384	358	841	1199	2.24
195	386	360	839	1199	2.19
200	**388**	**362**	**837**	**1199**	**2.14**
205	390	364	836	1200	2.09
210	392	366	834	1200	2.05
215	394	368	832	1200	2.00
220	396	370	830	1200	1.96
225	**397**	**372**	**828**	**1200**	**1.92**
230	399	374	827	1201	1.89
235	401	376	825	1201	1.85
240	403	378	823	1201	1.81
245	404	380	822	1202	1.78
250	**406**	**382**	**820**	**1202**	**1.75**
255	408	383	819	1202	1.72
260	409	385	817	1202	1.69
265	411	387	815	1202	1.66
270	413	389	814	1203	1.63
275	**414**	**391**	**812**	**1203**	**1.60**
280	416	392	811	1203	1.57
285	417	394	809	1203	1.55
290	418	395	808	1203	1.53
295	420	397	806	1203	1.49
300	**421**	**398**	**805**	**1203**	**1.47**
305	423	400	803	1203	1.45
310	425	402	802	1204	1.43
315	426	404	800	1204	1.41
320	427	405	799	1204	1.38

(Continued)

TABLE 11.1 Properties of Saturated Steam (*Continued*)

Pressure, psig	Temperature, °F	Heat, Btu/lb			Specific volume, ft³/lb
		Sensible	Latent	Total	
325	**429**	**407**	**797**	**1204**	**1.36**
330	430	408	796	1204	1.34
335	432	410	794	1204	1.33
340	433	411	793	1204	1.31
345	434	413	791	1204	1.29
350	**435**	**414**	**790**	**1204**	**1.28**
355	437	416	789	1205	1.26
360	438	417	788	1205	1.24
365	440	419	786	1205	1.22
370	441	420	785	1205	1.20
375	**442**	**421**	**784**	**1205**	**1.19**
380	443	422	783	1205	1.18
385	445	424	781	1205	1.16
390	446	425	780	1205	1.14
395	447	427	778	1205	1.13
400	**448**	**428**	**777**	**1205**	**1.12**
450	460	439	766	1205	1.00
500	470	453	751	1204	0.89
550	479	464	740	1204	0.82
600	489	475	728	1203	0.74

*Inches vacuum.

holding it to one another and allow each molecule to move freely into the gas phase. No change in temperature can be sensed because the steam forms at constant temperature, and the latent heat simply causes a change of state from liquid to gas. Latent heat from the steam is transferred to the end use device as steam condenses. The heat necessary to evaporate water into steam requires approximately 1000 Btu/lb of water.

Equivalent Direct Radiation (EDR)

The equivalent direct radiation (EDR) is an outmoded rating unit used for selecting boilers and other steam heating equipment like radiators and convectors. Many older steam pipe sizing tables used for designing steam distribution systems show pipe capacity in EDR. For a close, approximate conversion from EDR to pounds per hour flow, use the following formula:

$$\text{Pounds per hour steam flow} = \frac{\text{EDR}}{4} \qquad (11.1)$$

A commonly used EDR value for low-pressure steam systems is 240 Btu/lb steam.

Superheated Steam

If saturated steam is further heated, it is called *superheated*. Superheated steam is dry steam in the sense that it does not carry any liquid water droplets. Superheated steam is generated in boilers that are specially equipped to expose saturated steam to additional heating, which increases its temperature above saturation temperature for that pressure. Superheating can also occur when steam is throttled from a higher pressure to a lower pressure, for example, in passing through a pressure regulator. Superheated steam requires 1 Btu of thermal energy to raise the temperature ½°F.

Boiler Output

The output of a steam generator is often expressed in pounds of steam per hour. Since the output will vary in temperature and pressure, the capacity is more completely indicated as the heat transferred, expressed in British thermal units per hour. An older method of expressing boiler capacity is *boiler horsepower*, which is the equivalent of 33,475 Btu/h at atmospheric conditions.

SPECIFIC VOLUME

The *specific volume* is a volume of unit mass and gives the volumetric space that 1 lb of steam will occupy. It is the reciprocal of density.

SYSTEM CLASSIFICATIONS

Steam Systems

Steam systems are classified in two ways:

1. Pressure rating
2. Piping arrangement

Condensate Systems

Condensate systems are classified by the method used to return condensate back to the boiler:

1. Gravity flow directly into a condensate tank and then into the boiler, with no condensate or vacuum pump. This system is rarely used.
2. Gravity flow to a condensate tank, with a condensate pump used to pump condensate to the boiler.
3. Gravity flow to a condensate tank, with a vacuum pump assisting condensate flow to the tank. A condensate pump is used to pump condensate to the boiler.
4. Condensate lift traps, used to raise condensate from one elevation to another. This system is rarely, if ever, used.

Steam Pressure

Steam pressure is an important factor when steam is to be used to heat another substance. The higher the steam pressure, the higher the temperature of the steam. The steam temperature must be high enough to provide a temperature differential between the steam and the substance being heated. A large differential will reduce the size, and usually the cost, of the heat transfer equipment. On the other hand, unnecessarily high pressure steam can be difficult to control in a temperature-sensitive process, and can actually result in higher operating costs through the waste of flash steam or add to the initial cost of the system by requiring the use of more expensive condensate handling equipment. Selection of the steam pressure to be used is fundamental to the system design. A good practice is to evaluate several alternative pressures and choose the lowest one that will minimize initial and operating costs.

Steam systems are classified by ASHRAE according to operating pressure range as follows:

High pressure: 16 psig (110.1 kPa) and above

Low pressure: 0 to 15 psig (0 to 103.4 kPa)

There is no general agreement as to the assigned limits of medium-pressure steam. Another often-used set of operating pressure ranges is:

Low pressure: 0 to 15 psig (0 to 103.4 kPa)

Medium pressure: 16 to 60 psig (110 to 413 kPa)

High pressure: All pressures above 60 psig

The pressure-heat relationship for steam is as follows:

1. When steam pressure increases:
   a. The total heat increases slightly.
   b. The sensible heat increases.
   c. The latent heat decreases.
2. When steam pressure is reduced:
   a. The total heat decreases slightly.
   b. The sensible heat decreases.
   c. The latent heat increases.

Low-pressure steam is usually associated with space heating, laboratory equipment service, direct laboratory use, and culinary uses, as well as other applications that do not require high-temperature heat transfer. Industrial applications that transfer heat at high temperatures must use higher-pressure steam. In many cases, the codes and standards that govern the design of pressure vessels and steam systems will categorize the system pressure.

STEAM SYSTEM VENTING

Proper venting of the steam piping is important to prevent air binding and to allow steam to fill the system rapidly. *Air binding* is the development of a mass of air inside a piping system at a point from which it cannot escape. This forms a restriction or complete blockage depending on the quantity of air present. An air venting device installed at appropriate locations will minimize or eliminate this situation. The venting devices in steam systems are called *steam traps* or *vents*.

STEAM SUPPLY SYSTEMS

Water Hammer

Water hammer in an undrained or improperly drained steam supply pipe can be caused by the impact of a rapidly moving slug of water. Unless the condensate is removed from the steam mains, it gradually accumulates until the high-velocity steam forms ripples. As condensate builds up, it decreases the area available for steam flow, leading to even higher steam velocities. The slug of water accelerated by the steam can reach velocities in excess of 100 mi/h before it hits some obstruction such as an elbow or other fitting. The rapid change in speed can cause a loud noise or even severe damage to the system. The formation of this slug is illustrated in Fig. 11.1.

Water hammer in the condensate piping system could be caused by turbulent flow creating pockets of lower pressure that cause some of the condensate to flash

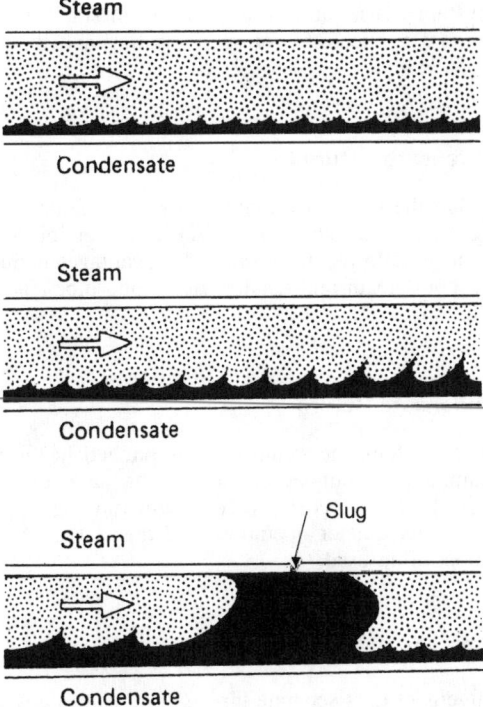

FIGURE 11.1 Formation of condensate water slug in steam piping.

into steam. Improperly operating steam traps may also leak steam into the condensate piping, mg, causing additional condensate to flash into steam. Flashing in this manner can cause a loud noise and increase system pressure to a point where damage is possible.

One-Pipe Steam Distribution

A one-pipe system uses a single pipe through which the steam is distributed and the condensate is returned to the boiler from the terminal unit. Designed to operate at low steam pressures, these systems usually have a gravity condensate return. A condensate pump may be required where gravity return directly to the boiler is not possible. Condensate could be returned to the boiler by either the counterflow or parallel flow method.

The one-pipe system does not have steam traps at the terminal units. Air vents are provided to vent the air, enabling the system to fill with steam. Self-contained thermostatic valves are available to help achieve some degree of terminal unit control. Proper venting is very important to prevent air binding and to allow steam to fill the pipes rapidly. Temperature is controlled by cycling the steam on and off.

An obsolete *vapor system* is a one-pipe steam distribution arrangement that depends on the slow condensation of steam to lower the pressure within the condensate piping into the vacuum range during the time of no or partial load. This system is no longer used because modern boilers do not allow the slow development of system vacuum.

The one-pipe system is used almost exclusively in small warehouses and residential and commercial applications for space heating and finds very limited use in any other type of facility.

Counterflow Condensate Return

This type of return has the steam main pitched back to the boiler opposite the flow of steam, requiring the condensate to also drain in a direction opposite to the flow of steam. The advantage is lower initial cost. Disadvantages include large pipe size (necessary to carry both steam and condensate in one pipe), noise in the system, and difficult installation.

Parallel Flow Condensate Return

In this type of return system, the steam main is pitched in the same direction as the flow of the steam, and the condensate flows in the same direction as the steam. The disadvantage is the higher cost involved with more piping. The advantages include smaller pipe size, quieter operation, and the less pitch required. This arrangement is the most often used.

Two-Pipe Steam Distribution

The two-pipe arrangement uses separate piping systems to deliver the steam to the terminal device and to return the condensate to the boiler, either by gravity or mechanical return. Gravity return is restricted to use in small systems.

A steam trap removes condensate from the steam piping network by allowing condensate to flow from the pipe or terminal unit into the condensate piping system. It also prevents steam from flowing back into the condensate system. It is installed at intervals along the supply pipe distribution system and at the discharge side of all terminal devices. The steam trap is the boundary between the steam supply and the condensate return at each terminal device. A schematic diagram of a two-pipe system is illustrated in Fig. 11.2.

CONDENSATE RETURN SYSTEMS

There are two general categories of condensate return systems: gravity and vacuum. For larger systems with long condensate return distribution piping runs, intermediate condensate or vacuum tank and pump sets may be required to return condensate to the boiler from various points in the system.

Gravity Flow

The gravity return system may be of a nonmechanical type of condensate return that depends on the pitch of the condensate return pipe back to the boiler for the

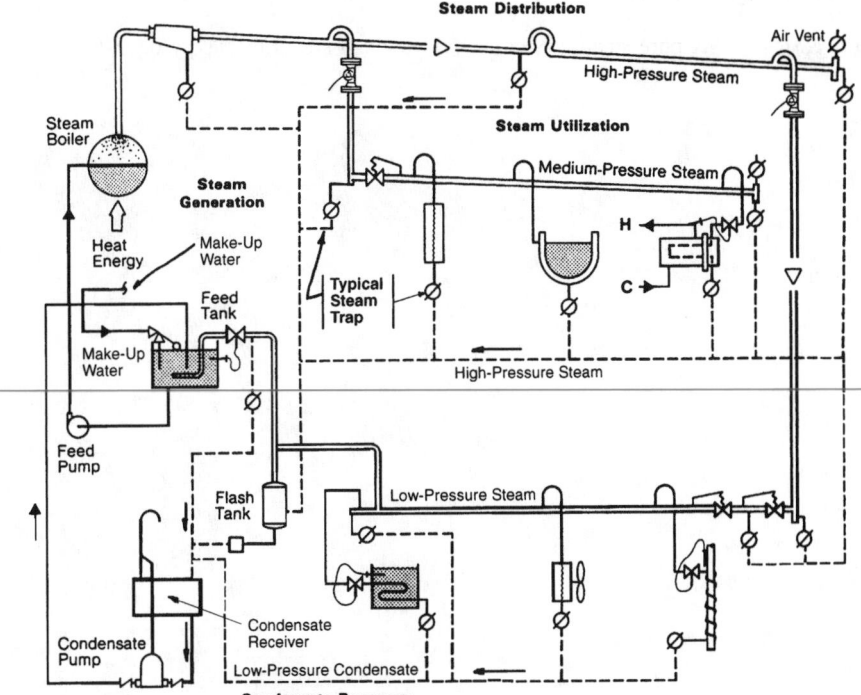

FIGURE 11.2   Schematic diagram of a typical two-pipe steam distribution system.

return of condensate. This system is generally suitable only for small, uncomplicated facilities.

In most gravity return systems, the condensate flows into a condensate storage tank, which is vented to the atmosphere. A condensate pump then transfers the stored condensate to the boiler as required.

Vacuum System

The vacuum condensate return system is similar to the gravity flow system except that a vacuum pump is placed at the condensate storage tank in lieu of the atmospheric vent. The vacuum pump removes air from the condensate return piping system. This allows the condensate to flow more rapidly, thereby reducing corrosion, and allows the use of smaller condensate return pipe sizes.

The vacuum system produces more effective air venting and provides a faster and more even steam distribution. It is often used on larger systems, permitting the steam pressure to be lower than the gravity system and enables quicker filling of the steam pipes with steam.

Special Systems

Clean Steam. A clean steam system provides particulate- and gas-free steam. It is usually produced from high-quality feedwater that is additive free or uncondensed water-for-injection (WFI) steam.

Pure Steam. A pure steam system provides clean steam that is pyrogen free.

STEAM

GENERAL

The steam distribution system consists of components and the piping network from the source of the steam up to and including the inlet of the steam trap at terminal equipment. The condensate return system starts beyond the inlet of the steam trap.

COMPONENT DESCRIPTION

Steam Sources

Steam can be obtained by purchasing it from an outside source such as a utility (if available), from a central internal steam system generator within a facility, or from a dedicated boiler, or generator, within a facility.

Boilers. A boiler is a closed vessel with associated equipment, piping, and controls that uses heat energy from an outside source to generate steam. Boilers that provide facility steam are available using coal, fuel gas, oil, or electricity as the source of heat. In addition, some are designed to use locally available fuel, such as wood shavings or other combustible products.

In general, boilers are designed to provide enough steam for the following:

1. To allow all the steam-condensing devices in the system to operate at their design heat transfer rate
2. To provide for piping and system thermal losses such as raising the mass of metal in the piping system up to operating temperature from some assumed initial ambient temperature
3. To provide for boiler start-stop cycling losses
4. To make up for losses of heat to the ambient air from the piping after it has reached the operating temperature

A small system boiler may be oversized by as much as 133 percent of the connected radiation steam requirement in order to account for these extra "pickup" and heat loss loads because the boiler will probably cycle on and off in response to an interior or outside temperature controller. If the boiler in a facility application is intended to be operated continuously, then oversizing capacity may be reduced. All piping should be well insulated to minimize the heat loss from the piping, especially if the piping is routed through nonheated areas and the condensate is to be returned to the boiler for conversion back to steam.

Boilers must be properly installed with headers and risers sized to allow adequate steam flow from the boiler. Failure to follow the manufacturer's instructions in this "near boiler" piping can lead to reduced output characteristics and *carryover,* that is, liquid droplets and solids from the boiler being suspended in the flow of steam. In any system, carryover should be avoided since it reduces the heat-carrying capacity per pound of fluid, leads to noisy and possibly damaging water hammer in the steam piping, and can cause traps and regulators to foul and clog.

All steam generators require controls to regulate pressure and water level and to prevent unsafe operation.

Electric Steam Generators.   Electric boilers have unique advantages for small-facility steam applications:

1. They do not require fuel piping.
2. They have no burner systems or draft blowers.
3. They don't require flues to carry off the exhaust gas.
4. They are smaller and easier to install.

These boilers must be properly operated and maintained to prevent hot-spots and buildup of solids, which may cause early failure of the boiler.

Clean Steam Generators.   In many facility applications, carryover of boiler treatment chemicals and boiler water is especially prohibited, for example, if steam is to be injected directly into the process or if boiler solids would contaminate the airstream in humidification. For these applications, a "clean steam" generator is appropriate.

Boiler steam or hot water is introduced on one side of a suitable heat exchanger. Clean water on the other side of the heat exchanger is evaporated and sent off to the application as "clean steam." A schematic diagram of a steam-fired clean steam generator is illustrated in Fig. 11.3.

Steam Obtained from a Facility or Outside Source.   In a plant served by a large steam boiler, facility steam could be provided from this source. Because the facility central steam boiler pressure is usually high, a reduced pressure for utility steam is usually required. Reduced steam pressure is obtained by the use of a *pressure-regulating* (*reducing*) *valve* (PRV). A PRV is part of a larger assembly that is usually called a *steam pressure-reducing station*. A typical PRV station is illustrated in Fig. 11.4.

Flash Steam from a Higher-Pressure Process.   Condensate that is formed at high pressure contains a significant amount of sensible heat. If pressure on that condensate is suddenly reduced in a flash tank, part of the condensate will reevaporate to form "flash steam," which could be used in a facility application. It is a common practice to use flash tanks in a high-pressure system to allow high-pressure condensate to flash down to some lower pressure and temperature so that it can be handled by lower temperature-rated equipment. Plant-operating costs are increased if the flash steam is simply allowed to escape from the system since it will carry away with it heat and water. Using flash steam as a facility supply will contribute to energy recovery efforts. If a flash tank is to be the source of facility steam, several issues must be considered, including the maximum pressure available in the flash tank, the amount of flash steam available, and the time that the flash steam will be available. A typical flash tank assembly is illustrated in Fig. 11.5.

Many flash tanks are designed to operate at or slightly above atmospheric pressure. If this is not satisfactory, the flash tank pressure might be raised, assuming that the flash tank is constructed to operate as a pressure vessel. The tank will also require back pressure controls to limit operating pressure and a safety valve, in case the operating controls fail.

It is common practice to provide a PRV to augment or replace the flash steam if the source of high-pressure condensate fails to generate enough flash steam to

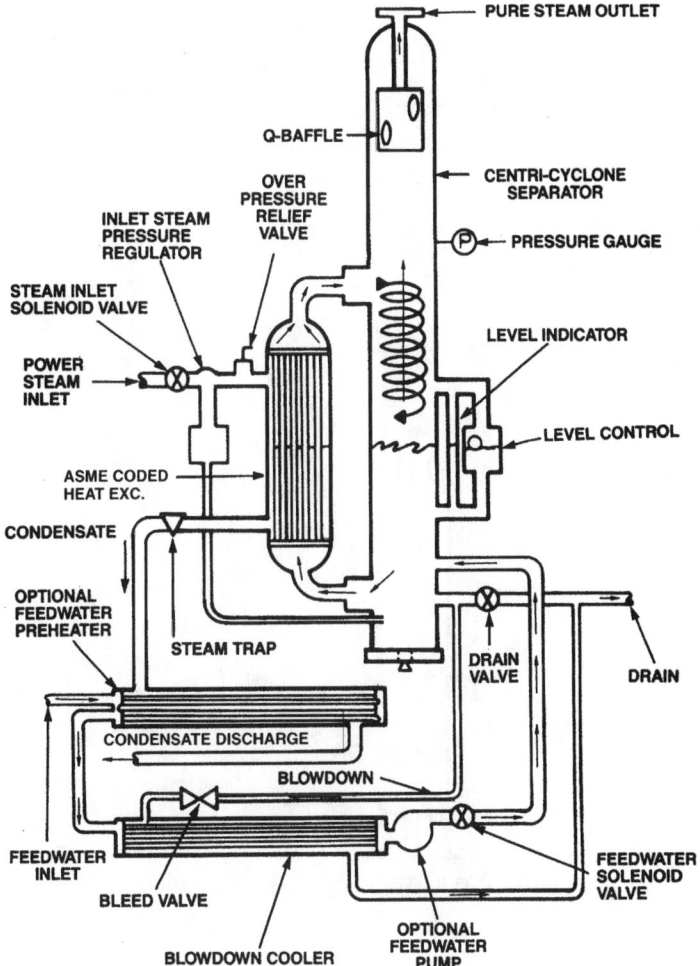

FIGURE 11.3 Schematic diagram of a steam-fired, clean steam generator. (*Courtesy of Mueller/Barnstead.*)

meet the facility demand for low-pressure steam or if it cannot be guaranteed that flash steam will be available. Additional details on flash tank design can be found in "Flash Tanks for Steam and Boiler Systems," published by ASHRAE (1991). The amount of flash steam, in pounds, released from 100 lb of condensate can be found from Table 11.2.

Main Pressure-Regulating Valves

The purpose of the main pressure-regulating valve is to reduce a high inlet steam pressure to a lower pressure under variable flow conditions. Pressure regulators come in a variety of designs. All of them have a valve body to actually contain

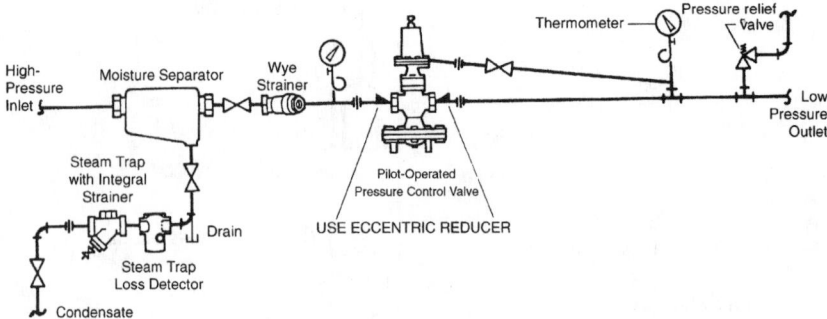

FIGURE 11.4 Pressure-reducing valve station.

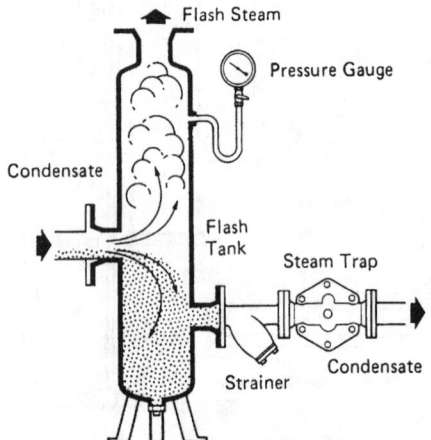

FIGURE 11.5 Typical flash tank assembly.

and control the flow of steam, an actuator that positions the valve stem to allow the proper flow, and a feedback system to determine if the desired downstream pressure is being maintained. The type of actuator and the location of the feedback pressure signal are often used in describing the pressure regulator. There are two types: direct acting and pilot operated.

Direct-Acting Valves. The direct-acting valve, illustrated in Fig. 11.6, is the simplest and least costly design. A reduction in pressure at the inlet acts on the underside of the diaphragm, which in turn, opposes the pressure exerted by the main spring, allowing the closure member to open, thereby increasing pressure. This type of valve is suitable where accurate control is not essential and where the steam flow rate is small and reasonably constant. It has some fluctuation of steam pressure and a relatively low capacity for its size.

Pilot-Operated Valves. The pilot-operated valve is illustrated in Fig. 11.7. A reduction in pressure acts on the underside of the pilot diaphragm either through a pressure control pipe or the drilling. The pilot diaphragm in turn operates the main

TABLE 11.2 Pounds of Flash Steam Released from 100 lb of Condensate

Inlet steam pressure, psig	Discharge, psig 0	Flash tank pressure, psig									
		2	5	10	15	20	30	40	60	80	100
5	1.7	1.0	0								
10	2.9	2.2	1.4	0							
15	4.0	3.2	2.4	1.1	0						
20	4.9	4.2	3.4	2.1	1.1	0					
30	6.5	5.8	5.0	3.8	2.6	1.7	0				
40	7.8	7.1	6.4	5.1	4.0	3.1	1.3	0			
60	10.0	9.3	8.6	7.3	6.3	5.4	3.6	2.2	0		
80	11.7	11.1	10.3	9.0	8.1	7.1	5.5	4.0	1.9	0	
100	13.3	12.6	11.8	10.6	9.7	8.8	7.0	5.7	3.5	1.7	0
125	14.8	14.2	13.4	12.2	11.3	10.3	8.6	7.4	5.2	3.4	1.8
160	16.8	16.2	15.4	14.1	13.2	12.4	10.6	9.5	7.4	5.6	4.0
200	18.6	18.0	17.3	16.1	15.2	14.3	12.8	11.5	9.3	7.5	5.9
250	20.6	20.0	19.3	18.1	17.2	16.3	14.7	13.6	11.2	9.8	8.2
300	22.7	21.8	21.1	19.9	19.0	18.2	16.7	15.4	13.4	11.8	10.1
350	24.0	23.3	22.6	21.6	20.5	19.8	18.3	17.2	15.1	13.5	11.9
400	25.3	24.7	24.0	22.9	22.0	21.1	19.7	18.5	16.5	15.0	13.4

Source: Courtesy of SARCO.

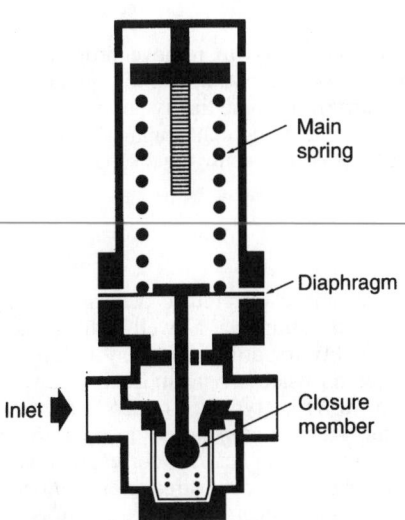

FIGURE 11.6 Direct-acting pressure-reducing valve.

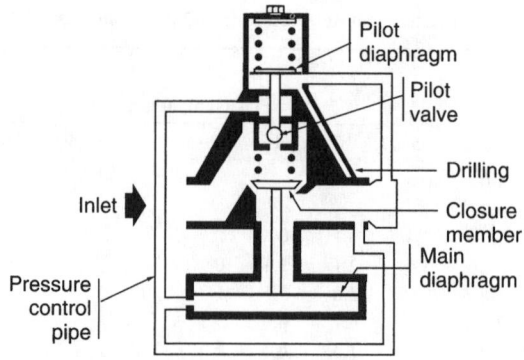

FIGURE 11.7 Pilot-operated pressure-reducing valve.

diaphragm, allowing the closure member to open, thereby increasing pressure. This type of valve is capable of accurate control downstream of the valve and has a relatively high capacity for its size.

Boiler Feed Pump System

A boiler feed pump assembly maintains the water level in the boiler within acceptable limits. The assembly consists of a feedwater receiver, feedwater pump(s), a method to add water to the system, and a level control in the boiler to turn the pump on and off.

Steam Separators

A steam separator is a device used to remove droplets of water from steam in piping and thereby provide equipment with good-quality dry steam. A secondary function is to remove carryover immediately after a boiler outlet. Two types of separators are used: impingement and centrifugal. A typical impingement separator is illustrated in Fig. 11.8a. A typical centrifugal separator is illustrated in Fig. 11.8b.

Steam Meters

It is often required that steam be metered to determine energy efficiency, process control, usage, and equipment efficiency as well as for measurement of the amount of steam used within a facility for utility company billing purposes.

A steam meter must compensate for steam quality as well as pressure and temperature. Performance of different types of meters will vary. Most meters depend on volume measurement, which is dependent on pressure. Measurement must be taken at the appropriate pressure; otherwise corrections have to be applied to the reading. Metering should be done downstream of a properly designed pressure-reducing valve station. A typical steam meter assembly is illustrated in Fig. 11.9.

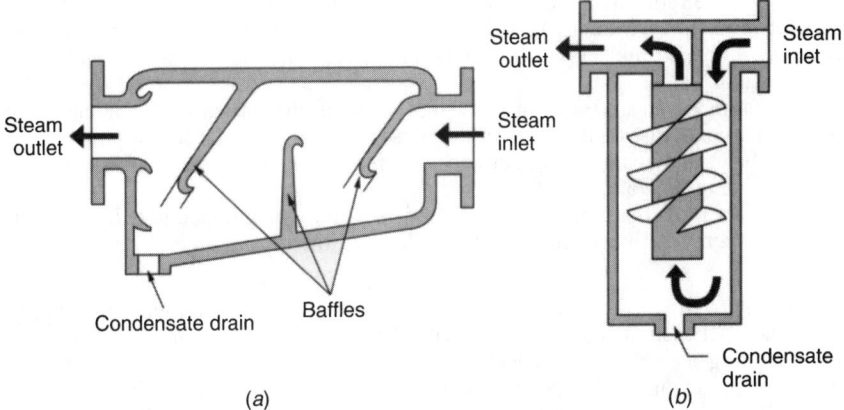

FIGURE 11.8   Steam separators. (*a*) Impingement separator and (*b*) centrifugal separator.

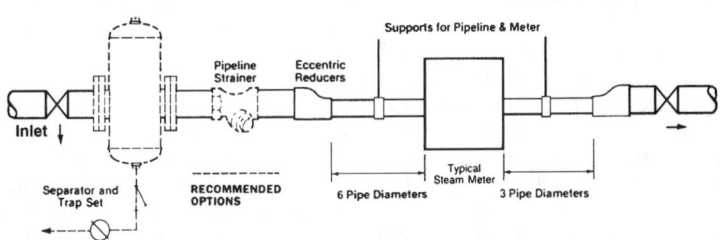

FIGURE 11.9   Steam meter assembly.

Steam Traps

A steam trap removes condensate, noncondensable gases and air from a steam piping network. No single type of trap is suitable for all applications, and most systems require more than one type.

Noncondensable gases, such as carbon dioxide and air, must be removed because they dilute the steam and lower the condensation temperature, as well as reducing efficient heat transfer. Removal is accomplished by either installing vents on top of terminal equipment or through a thermostatic sensor in a steam trap. Thermostatic traps are the most commonly used. The element operates by measuring the steam temperature and determining if it is below the saturation temperature for a given pressure. If it is, then noncondensable gases are present, which must be vented. The trap opens, venting the gas.

Proper condensate removal is essential for efficient plant and process operation. All steam traps shall be able to:

1. Vent air and other gases from piping and equipment
2. Prevent the flow of steam into the condensate piping
3. Allow only condensate into the condensate piping network

Condensate removal can be accomplished either manually or automatically. The manual removal of condensate will require that a valve placed in the network be constantly opened and closed by operating personnel. This method is inefficient and wasteful.

An automatic steam trap is a valve that opens in the presence of condensate or noncondensable gases and closes in the presence of steam in order to remove condensate from the steam piping network and prevent steam from entering the condensate piping network.

There are three main groups of automatic steam traps, grouped together by their mode of operation: mechanical, thermostatic, and thermodynamic.

Mechanical Traps.   Mechanical traps are buoyancy operated and sense the density difference between steam and condensate through the use of a float or bucket as the operating member. Mechanical traps are usually preferred in applications that do not require much air venting, such as high-pressure steam mains and other noncycling systems. Condensate subcooling is not required, and the trap cycles either wide open or dead shut. Mechanical traps are preferred where immediate removal of the condensate is required or where a cooling leg upstream of the trap could interfere with operation of the equipment. This is the typical situation in industrial traps draining condensate from high-pressure steam mains or heat exchangers that can be damaged by flooding.

One disadvantage of all mechanical traps is that the size of the discharge orifice is controlled by the buoyancy power of the float, which is constant. As the pressure of the steam system increases, the size of the orifice should decrease. This requires that the size of the valve seat be different for various pressures.

Loose Float Trap.   The loose float trap, illustrated in Fig. 11.10, is the simplest of traps. When condensate enters the chamber, the float B is lifted off its seat A and allows condensate to discharge. It has no moving parts except for the float and is very inexpensive. Disadvantages are that it cannot vent air, and it is difficult to obtain good seating. A hand cock C has been added to vent air. This type of trap is no longer used but may be found in older facilities.

Float and Lever Trap.   The float and lever trap, illustrated in Fig. 11.11, overcomes some of the problems of the loose float by providing a float arm C connected to an outlet valve D for more effective seating. This type of trap is no longer used in modern systems but may be found in older facilities.

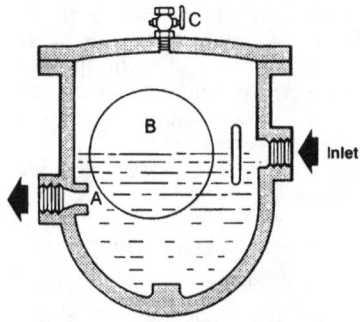

FIGURE 11.10   Loose float steam trap.

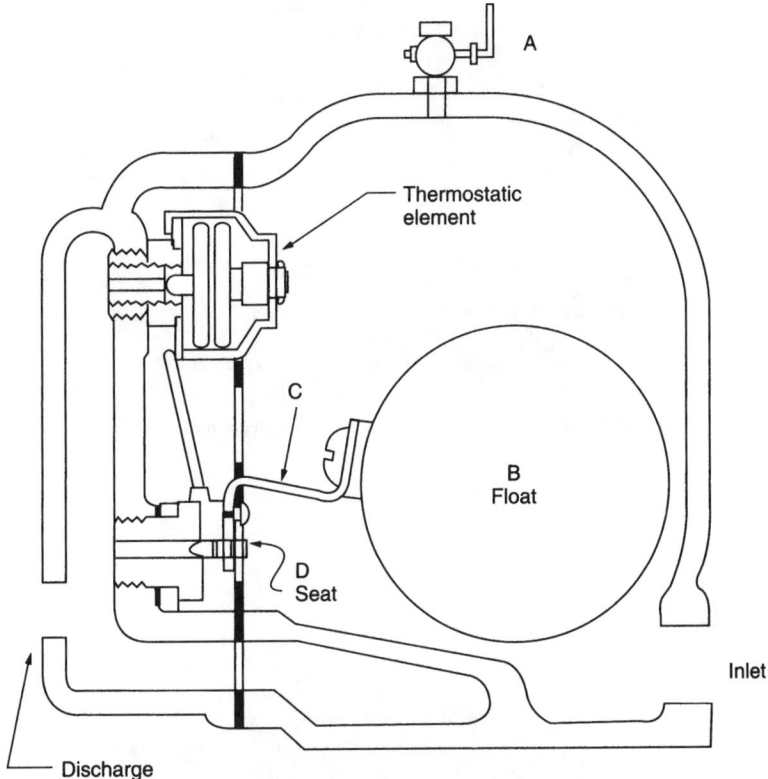

FIGURE 11.11   Float and lever steam trap.

Float and Thermostatic Trap.   The float and thermostatic element trap is illustrated in Fig. 11.12. When condensate enters the trap, the float rises and allows condensate to discharge. A thermostatic element is provided to automatically vent air out of the chamber. This element is open only when air is present to cool the element, thereby allowing air to vent. As the temperature of the chamber rises when no air is present, the element is closed.

Open Bucket Trap.   The open bucket trap, illustrated in Fig. 11.13, uses a bucket instead of a float. The bucket floats when condensate is present and falls when empty. When condensate enters the trap, it first fills the body outside the bucket. The floating bucket holds the seat closed. After the body is full, condensate spills over to fill the bucket causing it to sink drawing the seat open and allowing condensate to discharge.

These traps are mechanically simple and strong, and they are capable of withstanding shock and corrosive condensate. Disadvantages are reduced air-venting capability, heavy weight in relation to their discharge capacity, and susceptibility to damage by freezing.

Inverted Bucket Trap.   The inverted, bucket trap, illustrated in Fig. 11.14, uses an inverted bucket instead of a float. The bucket is normally at the bottom, causing the seat to be open. When condensate enters and fills the chamber, it is able to

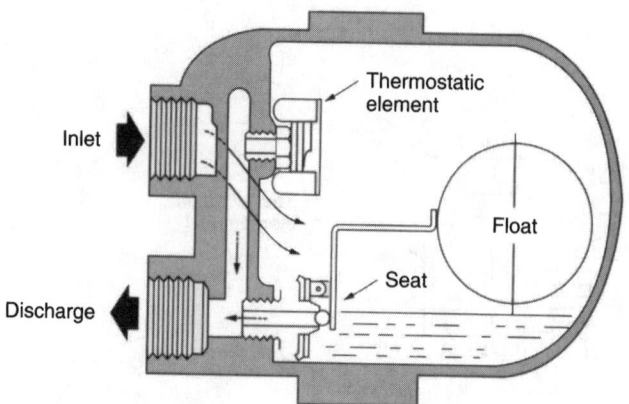

FIGURE 11.12   Float and thermostatic element steam trap.

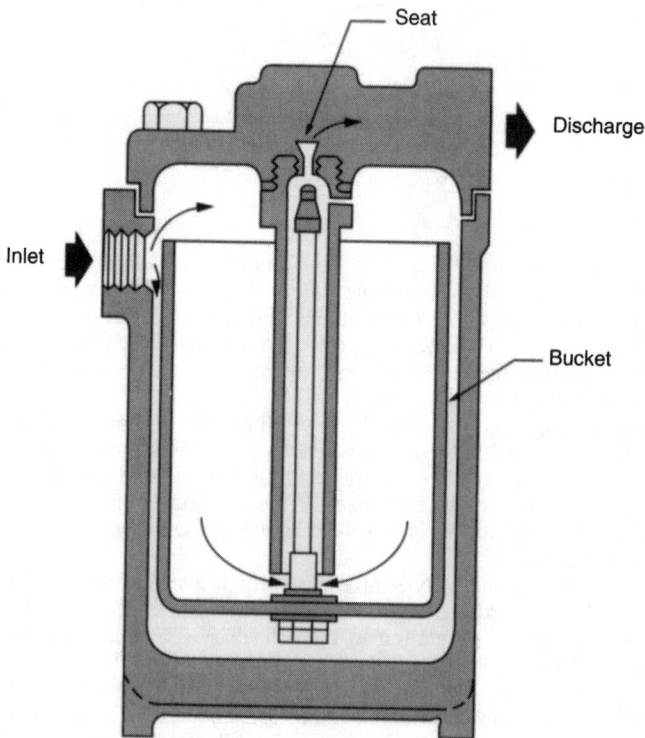

FIGURE 11.13   Bucket steam trap.

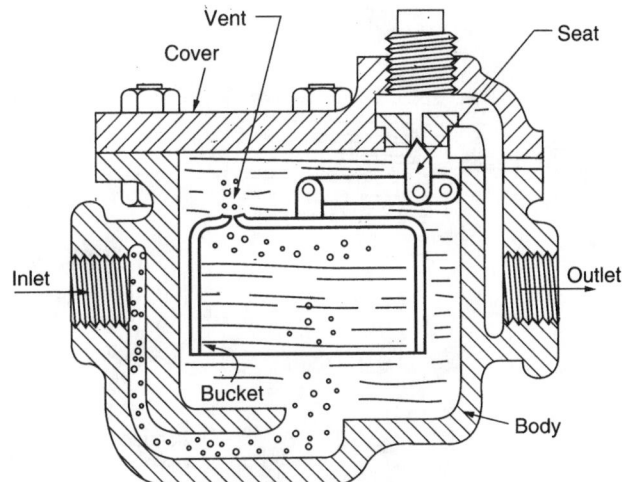

FIGURE 11.14   Inverted bucket steam trap.

discharge through the open seat. When a small amount of steam enters the trap, it escapes through the vent. More steam entering causes the bucket to rise, closing the seat. As the steam escapes through the vent, the bucket will again drop to the bottom, opening the seat.

These traps are mechanically simple and strong, capable of withstanding shock, corrosive condensate, and superheated steam. Disadvantages are the low air-venting capacity and the need to have a steady pressure. Air vent capacity can be increased by oversizing the condensate discharge piping.

Thermostatic Traps.   In this type of trap, a bimetallic or fluid-filled element operates a valve that opens in the presence of condensate and closes in the presence of steam. There are a large number of variations for this type of valve.

The working pressure of the steam does not affect the operation of the trap. Instead, it is the difference in temperature between the steam and the condensate which sets up the difference between the pressure inside and outside the element, that opens and closes the seat.

Thermostatic traps are intended for relatively low condensate removal capacities and typically used in low-pressure steam heating equipment, in equipment that can tolerate condensate collection upstream of the trap in a "cooling leg" where it can reach the subcooled temperature that opens the trap. While the condensate is subcooling, some sensible heat is made available. Sometimes this sensible heat can be useful, for example, in tracing temperature-sensitive components.

Bellows Trap.   Often referred to as a *balanced-pressure trap,* the bellows type is illustrated in Fig. 11.15. The principle of operation is the expansion and contraction of a bellows filled with a liquid that has a lower boiling point than water. The valve seat is open when cold, allowing air to vent and condensate to discharge. As condensate enters and before it reaches the boiling point, the element closes. As the condensate cools, the element opens, discharging the condensate.

Bimetallic Trap.   The bimetallic trap operates by the action of a composite strip of metal that bends when the temperature changes. The principle of operation is illustrated in Fig. 11.16. The seat is open when cold, allowing the free passage of

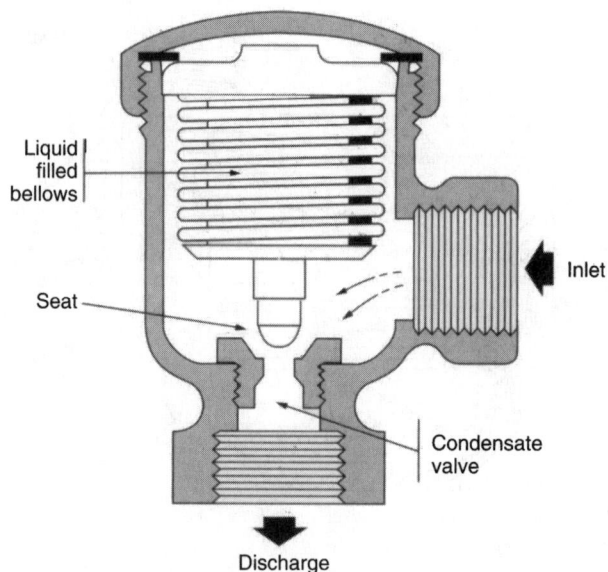

FIGURE 11.15   Bellows steam trap.

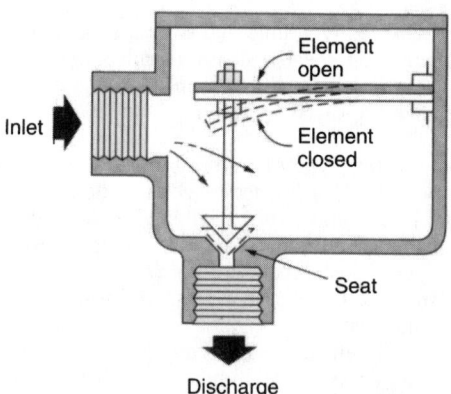

FIGURE 11.16   Bimetallic thermostatic steam trap.

air and condensate. When steam temperature is approached, the element bends and closes the seat.

Liquid Expansion Trap. Another type of thermostatic trap is the liquid expansion type, illustrated in Fig. 11.17.

Kinetic Traps. These types of traps operate on the difference between the flow characteristics of steam and condensate and on the fact that condensate discharging to a lower pressure contains more heat than required to keep it in the liquid phase. The excess heat will cause the condensate to flash into steam at the lower pressure.

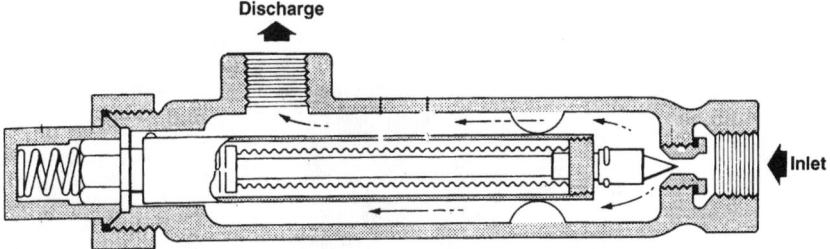

FIGURE 11.17   Liquid expansion steam trap.

Thermodynamic Trap.   The thermodynamic trap, often referred to as a disk trap, is illustrated in Fig. 11.18. This device has only one moving part—the disk. When air or cold condensate enters the trap, it lifts the disk off its seat and is discharged. When steam or hot condensate enters the trap, some of it flashes into steam upon exposure to a lower pressure. The increased velocity of this vapor flow decreases the pressure on the underside of the disk causing it to close.

   Piston Trap.   A piston trap, also called an *impulse trap,* is illustrated in Fig. 11.19. As cooler condensate enters the body, pressure raises the piston to open the seat, allowing air and condensate to discharge. As the condensate nears steam temperature, some flashes into steam, which passes through the gap. The flash steam with the greater pressure forces the cylinder closed, stopping flow.

   Labyrinth Trap.   Another type of kinetic trap is the labyrinth, illustrated in Fig. 11.20.

Terminal Equipment Temperature Control

Temperature controls are used to modulate either the amount of heat or temperature at the terminal equipment. This control will vary either the steam pressure to control

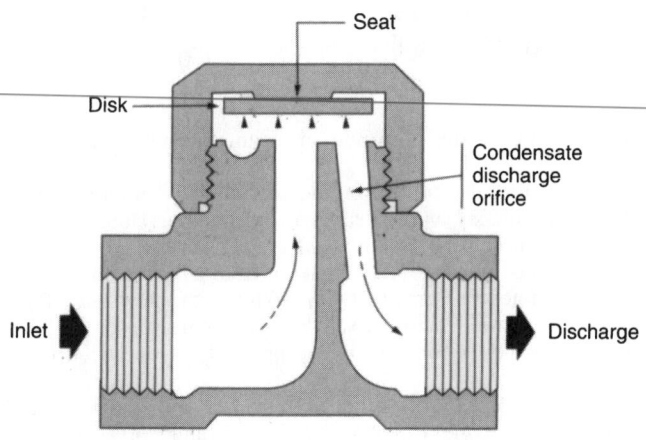

FIGURE 11.18   Thermodynamic steam trap.

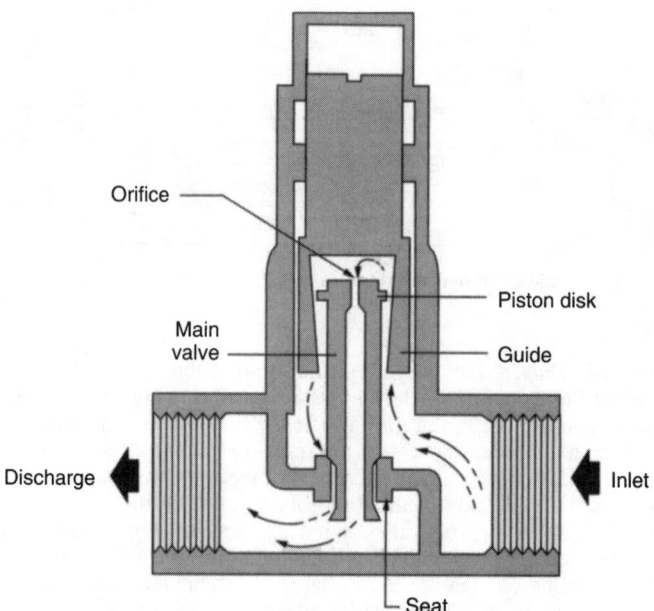

FIGURE 11.19 Piston steam trap.

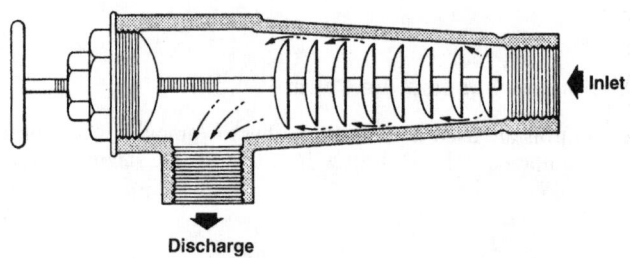

FIGURE 11.20 Labyrinth steam trap.

the steam temperature or the flow rate of steam to control the amount of heat provided.

To vary the pressure, the controlling device will be a pilot or thermostatically operated pressure-regulating valve activated by a temperature sensor in the discharge steam line. This type of valve is different from that of a master type PRV used to provide reduced pressure facility steam from a central source. The principle of operation is illustrated in Fig. 11.21. To modulate the flow rate of steam, a self-acting control valve is needed to activate a temperature sensor (bulb) in the fluid whose temperature is being controlled. The principle of operation is illustrated in Fig. 11.22.

An important consideration in the selection of a temperature control device is the proportional band of the valve, which is the range of accurate temperature regulation. This is obtained from the individual manufacturer.

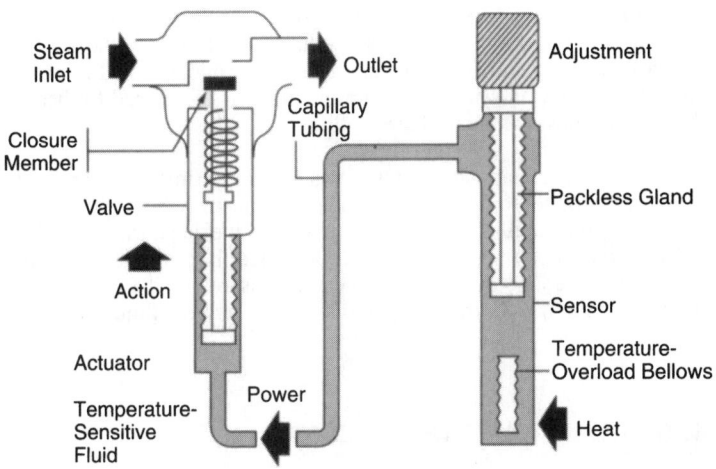

FIGURE 11.21   Self-acting temperature control valve.

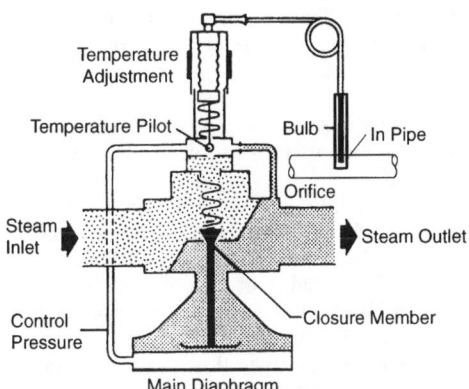

FIGURE 11.22   Pilot-operated temperature control valve.

The type of temperature control may have an effect on the selection of the steam trap necessary to drain condensate from the steam space of the terminal equipment depending on whether the steam flow is constant, variable, or batch.

Other Devices That Use Steam

Nozzles.   Nozzles reduce the high pressure available in the steam to accelerate the steam flow for a number of related applications. The accelerated flow creates a low pressure region in the throat of the nozzle that can be used to pick up a detergent solution, which then mixes with the steam to form a high-velocity jet useful in equipment cleaning. This same principle can be applied as a "jet pump" or air ejector to establish a vacuum.

Direct Steam Injection. Mixing tees and sparging tubes can be used to provide instant heating without the use of a heat exchanger. Of course, the use of mixing tees and sparging tubes implies that the steam will condense in the process fluid and be lost from the steam system. This arrangement is often used for heating water for wash-down purposes in small hose stations.

Steam Tracers. Tracers are small-diameter steam pipes installed parallel to a process pipeline or around some other component that must be kept warm. Special compounds are often applied in order to enhance the heat transfer between the tracer and the item to be traced, and insulation generally covers both the tracer and the item. As steam condenses in the tracer, heat is released to keep fluids from freezing or to maintain a desired temperature in the process fluid. Steam tracing is discussed in Chap. 5.

STEAM SYSTEM COMPONENT SIZING

Steam Pipe Sizing

Steam system piping differs from other piping because it can carry two different fluids: steam and liquid water. The proportions of steam and liquid expected in a given steam system pipe depends upon the type of system and the method of operation.

Pipe Sizing Criteria. The size of steam piping is based on the flow rate and either maximum velocity of the steam or pressure drop, whichever is the greater consideration.
   Flow Rate. The flow rate required depends upon the specific piece of equipment and application. If steam is being used as a heat transfer medium, then the flow rate is determined by the ratio of the heat transfer requirement in British thermal units per hour to the enthalpy of evaporation from the steam tables at the desired steam pressure in British thermal units per pound. For humidification applications, the required steam flow is determined from a psychometric chart for the given and desired conditions. For a given flow rate of dry saturated steam, the pipe size required depends upon the initial pressure of the steam and the pressure drop or velocity of flow.
   Velocity. The velocity of steam is limited to prevent flow noise and pipe erosion. Industrial and facility steam systems generally use velocity as the limiting criteria in the selection of pipe sizes. Most systems are sized using a velocity of between 6000 and 8000 fpm (1800 and 2400 mpm), with a maximum of 10,000 fpm (3000 mpm) in industrial applications where noise is not a consideration.
   Steam Pressure. A pound of high-pressure steam occupies a smaller volume than low-pressure steam. This has a significant effect on steam pipe sizing since higher-pressure systems can use smaller pipes than low-pressure systems for the same steam flow.
   Pressure Loss. Often, low-pressure steam systems will be sized on pressure drop criteria. A range of between ⅛ and ¼ psi drop for 100 ft is recommended. The minimum operating pressure for any equipment shall be obtained from the manufacturer, which relates to the temperature required for the process to proceed efficiently. Under these circumstances the lowest possible pressure of the steam system is desirable. Considerations are:

1. The longest equivalent run of pipe
2. Pressure of the steam required at the terminal equipment and that available at the source
3. Flow rate
4. Velocity of the steam
5. Selection of economical pipe size based on an acceptable range of pressure loss while keeping below the maximum selected velocity

Pressure drop figures in common use are given in Table 11.3.

Pipe-Sizing Method

1. Calculate or obtain the steam flow rate, in pounds per hour.
2. Establish the minimum system steam pressure required to service all connected equipment.
3. Establish the maximum steam velocity for the system.
4. Calculate the allowable friction loss permissible in each branch or main under design.

With all of the above information available, use Fig. 11.23, a basic sizing chart, to size the pipe as follows:

1. The chart is divided into two halves, top and bottom. Using the bottom half, enter with the flow rate and rise vertically until the required steam pressure figure is intersected. From the point where these values intersect on the chart, draw a line parallel to the skewed lines up to the 0 line separating the top half from the bottom half.

TABLE 11.3   Steam Pressure Drops in Common Use for Sizing Steam Pipe*

For corresponding steam pressures

Initial steam pressure, psig	Pressure drop per 100 ft	Total pressure drop in steam supply piping
Vacuum return	2–4 oz/in^2	1–2 psi
0	0.5 oz/in^2	1 oz/in^2
1	2 oz/in^2	1–4 oz/in^2
2	2 oz/in^2	8 oz/in^2
5	4 oz/in^2	1.5 psi
10	8 oz/in^2	3 psi
15	1 psi	4 psi
30	2 psi	5–10 psi
50	2–5 psi	10–15 psi
100	2–5 psi	15–25 psi
150	2–10 psi	25–30 psi

*Equipment, control valves, and so forth must be selected based on delivered pressures.

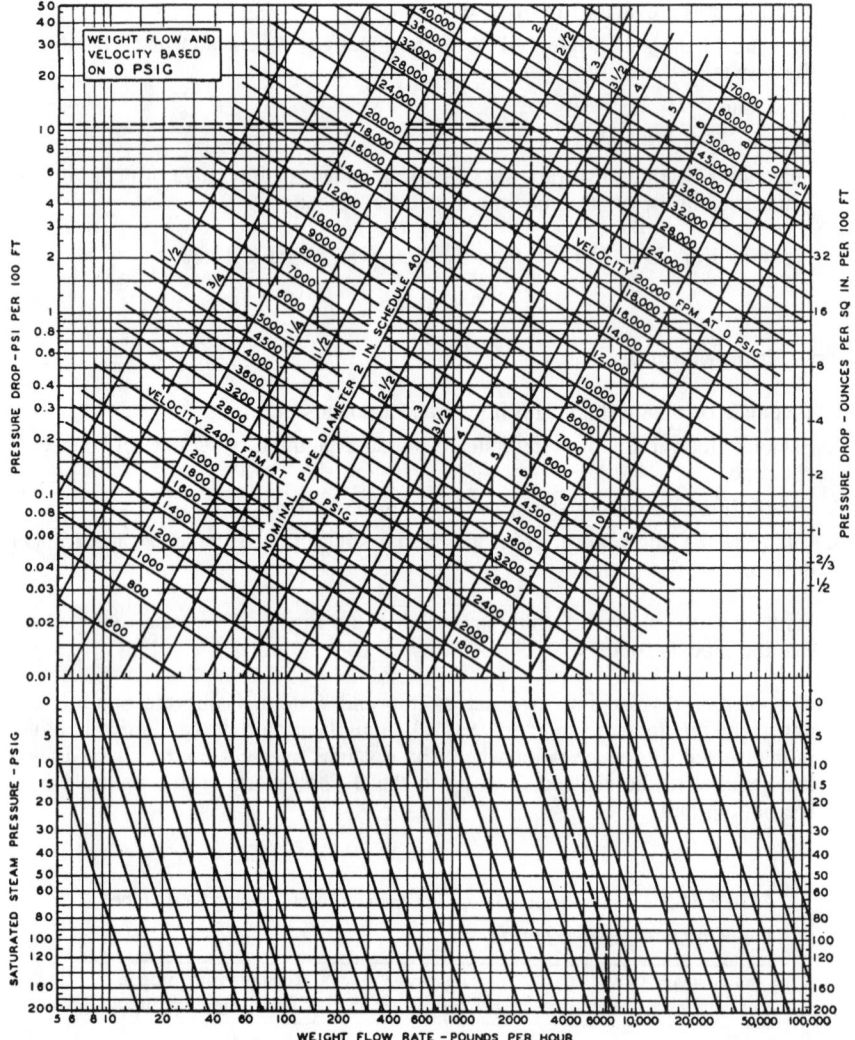

FIGURE 11.23   Basic pipe sizing chart. Chart is based on weight-flow rate and velocity of steam in Schedule 40 pipe, saturation pressure of 0 psig. Based on Moody friction factor where flow of condensate does not inhibit the flow of steam.

2. Where the line intersects the 0 line, proceed straight up the top half of the chart until the line intersects the desired maximum velocity or friction loss established for the pipe under design.

3. At the intersection of either the velocity or friction loss figure, read the pipe size on a skewed line rising up to the right side. If the point falls between two pipe sizes, use the larger size.

For older systems using EDR as the reference, refer to Table 11.4 using the pipe pitch and EDR at the point of design to determine the pipe size.

BOILER FEED PUMPS

A boiler feed unit maintains boiler water level within acceptable limits. Condensate should enter the boiler from only one source in order to maintain the pressure and flow integrity of the system.

If condensate cannot flow directly into a boiler feedwater tank, condensate receivers may be required to transfer condensate to the boiler feedwater tank. Larger commercial or industrial steam systems are likely to have a boiler feed pump in addition to a condensate transfer pump.

A boiler feed pump unit differs from a condensate receiver set in three major respects:

1. The boiler feed pump is controlled by a level-sensing control mounted on the boiler.
2. The boiler feed receiver must be large enough to act as a reservoir of feedwater immediately available in order to maintain the desired level in the boiler. The generally used figure is up to a 1-h supply.
3. The boiler feed receiver has a makeup water supply to add water to the system due to system losses.

Receivers should be manufactured of a corrosion-resistant material since condensate is often highly corrosive. Thick cast iron receivers are common in the industry, though thinner-wall steel receivers are also available. These receivers are usually not rated to handle internal pressure, so they are installed with a vent to the atmosphere.

The most common control used in condensate receiver sets and boiler feed pumps is the float switch, but any level-sensing control compatible with condensate and capable of starting the pump would be satisfactory. Small single-phase pumps may be started directly through the level control, but three-phase pump motors and single-phase motors that draw a large starting current must be started through a motor starter. It is accepted practice to use prepiped and engineered units ready for operation after installation.

The most common pump found in these applications is a centrifugal volute pump. It shall be selected for a discharge head adequate to deliver condensate to the boiler.

This equipment is often supplied with two pumps as a duplex unit so that a second pump is available if one pump fails. A lead-lag control of some type is usually provided to alternate pump operation, providing equal usage, thereby increasing life expectancy. In the event that an unusually heavy condensate load is encountered, the alternator is electrically wired so that both pumps can be operated simultaneously under peak load conditions. Piping and vale arrangements for multiple-pump installations shall be in accordance with recommendations of the specific manufacturer of the installed equipment.

TABLE 11.4 Pipe Sizes Using EDR As Criteria

Pipe size, in	Pitch of pipe, in/10 ft; velocity, ft/s; capacity, ft² , EDR															
	¼ in		½ in		1 in		1½ in		2 in		3 in		4 in		5 in	
	Capacity	Max. vel.	Capacity	Max. vel.	Capacity	Max. vel.	Capacity	Max. vel.	Capacity	Max. vel.	Capacity	Max. vel.	Capacity	Max. vel.	Capacity	Max. vel.
¾	12.8	8	16.4	11	22.8	13	25.6	14	28.4	16	33.2	17	38.6	22	42.0	22
1	27.2	9	36.0	12	46.8	15	51.2	17	59.2	19	69.2	22	76.8	24	82.0	25
1¼	47.2	11	63.6	14	79.6	17	98.4	20	108.0	22	125.2	25	133.6	26	152.4	31
1½	79.2	12	103.6	16	132.0	19	149.6	22	168.0	24	187.2	26	203.2	28	236.8	33
2	171.6	15	216.0	18	275.2	24	333.2	27	371.6	30	398.4	32	409.6	32	460.0	33

Source: From research sponsored by ASHRAE.

STEAM CONDENSATE

At some point after the steam has entered the terminal equipment, the steam condenses into a liquid, called *condensate*. This condensate must be drained away from the point where it is created and removed from the terminal equipment to assure proper equipment operation. There are three elimination alternatives available:

1. The condensate may be absorbed into the process and lost from the steam system.
2. The condensate may be simply wasted to a drain in an approved manner.
3. The condensate may be recovered resulting from the action of the steam and sent back to the boiler as feedwater.

Many systems do not recover condensate at all. Low condensate flow rates, the length of condensate piping required, or the inherent nature of the operation may make it impossible or economically unwise to install condensate return piping. If condensate becomes contaminated with heavy metals, phosphates, oil, dirt, or other chemicals, or if the cost of collecting and returning the condensate is considered to be too high, then it may be wasted to a drain. There are limitations imposed by environmental regulations that should be considered before this is done. Local codes must be consulted to ensure that proper treatment of this waste is accomplished. If clean condensate is to be wasted, additional components, such as condensate coolers, will be necessary to reduce the temperature of the condensate below a level acceptable to code requirements, enabling it to be discharged into a sanitary drainage system. This temperature is generally 140°F (60°C).

If the condensate is to be recovered, then additional components, such as steam traps, flash tanks, contaminant piping, condensate piping, and condensate pumps may be required to return the condensate back to the steam source. The steam trap, which is part of the condensate drainage system, is generally considered the boundary between the steam supply subsystem and the condensate recovery subsystem.

CONDENSATE RECOVERY SYSTEM DESCRIPTION

Condensate shall be drained and recovered from the following directly interconnected portions of the steam supply and condensate drainage system (a diagram of typical condensate drainage system components is illustrated in Fig. 11.24):

1. Collection legs receiving condensate produced in steam mains resulting from cooling of the steam
2. Drain pipes leading to steam traps, often called *drain lines*
3. Steam traps
4. Steam trap discharge pipes
5. The main condensate return pipe

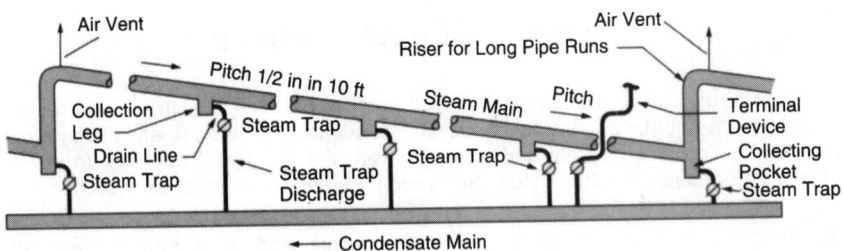

FIGURE 11.24 Components of a condensate drainage system.

COMPONENT DESCRIPTION

Collection Leg

A reservoir, or *collecting leg,* is a fitting or portion of a pipe that collects all the condensate produced from inside a main. One must be provided if the steam trap is to be effective since a steam trap can discharge only the condensate that is brought into it. Collecting legs should be provided at all low points in the system and wherever the condensate can collect such as at the ends of the mains, the bottom of risers, and ahead of expansion joints, separators, pressure-reducing valves, and temperature regulators.

Drain Pipes

The drain pipe conveys condensate from the collection leg to the steam trap. It carries pressurized, high-temperature water.

Steam Trap

The steam trap separates steam from the condensate and discharges condensate only. Steam traps have been previously described.

Steam Trap Discharge Pipe

The steam trap discharge line conveys condensate from the steam trap to the main condensate collecting line. This line must be capable of carrying condensate, flash steam, and noncondensed gases.

Main Condensate Collecting Line

This is the main condensate pipe that collects and returns condensate back to the condensate storage tank.

Condensate Cooler

A condensate cooler is a heat exchanger that lowers the temperature of waste condensate before it can be discharged to a sewer. This can be accomplished by the use of a tempering tank or heat exchanger. The heat exchanger often is used to preheat domestic water prior to heating or to heat water for other purposes. If the water is too hot, a tempering tank, illustrated in Fig. 11.25, mixes cold water with the condensate based on the temperature of the discharge. To determine the quantity of cold water needed, refer to Fig. 9.30 or 9.31.

Condensate Transfer Pump Assembly

The condensate transfer pump assembly collects the condensate and pumps it back to the boiler as quickly as possible in order to minimize heat loss. Higher condensate temperatures at the boiler relate directly to higher system efficiency since the sensible heat of the condensate represents a reduction in fuel required. The assembly usually consists of a vented receiver, a condensate pump, and a level switch to turn the pump on and off. In small systems, condensate receiver sets may be used in place of more complex boiler feed pump systems.

The condensate receiver is relatively small to minimize heat loss as the receiver fills between pumping cycles. The pump is controlled by a float switch mounted in the condensate receiver.

The volume of a condensate transfer receiver should be small compared to the rate of condensate return so that it will fill quickly and start the pump so that the condensate will not cool off excessively. A generally used figure is approximately a 1- to 2-mm supply.

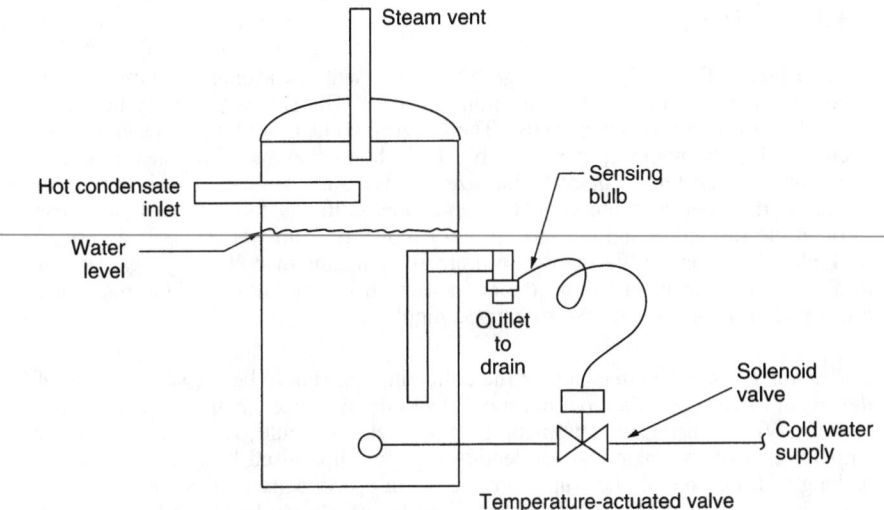

FIGURE 11.25 Detail of condensate cooler.

Flash Tanks

A flash tank is a pressure vessel, often vented to atmosphere, that contains steam produced from hot condensate under pressure that enters the tank. Flash tanks play an important role in condensate return systems. Its name is derived from the sudden evaporation, or "flashing," that occurs when condensate at higher pressure is suddenly released to a lower pressure. When hot condensate under pressure is discharged to a lower pressure, the condensate contains more heat than necessary to maintain itself as a liquid. This excess heat causes some of the liquid to vaporize instantly, or flash, into steam at the lower pressure.

A flash tank may be required to cool condensate before it flows into the condensate receiver tank if large quantities of high-pressure condensate could flow into a relatively small tank.

COMPONENT SELECTION AND SIZING CRITERIA

Steam Trap Ancillary Component Sizing

Collecting Legs Draining Steam Mains. Properly drained mains and care taken in starting up a cold system not only prevent water hammer damage but also improve the quality of the steam and reduce the maintenance required on pressure-reducing valves, temperature controls, and other components.

On long horizontal runs, drain points should be located at intervals of 150 to 400 ft (45 to 120 m). For reverse draining of condensate, intervals of 50 ft (15 m) are recommended. Drain points should also be located at any natural collection point. In general, automatic start-up systems and high-pressure systems will have collection legs located closer together. The diameter of the collecting leg pipe should be 1.5 times the size of the steam main and extend a minimum of 6 to 12 in (18 to 30 cm).

Drain Lines. Drain pipes discharge by gravity, with an average minimum pitch of approximately 1 in per 10 ft (25 mm per 3 m). The size is generally based on twice the full running flow expected. The heaviest condensate load in a steam main occurs during the warm-up period. Table 11.5 shows the amount of steam used to warm up a given length of different size mains. Once the steam main has been raised to the operating pressure, further condensation is the result of heat loss through the insulation and the separation of moisture from wet steam as described in Table 11.6. Condensate mains are drained by means of collecting legs, similar to a tee facing down, that drain the condensate from the bottom of the main into a 6- to 12-in-long (18- to 30-cm) capped nipple.

Collection Legs. The diameter of the collecting leg should be the same as that of the steam main if possible, but not more than one pipe size smaller. The collecting legs for systems using the automatic heat-up method should have a pipe diameter equal to that of the main. The collecting legs for supervised heat-up need not be as long as for automatic heat-up since the warm-up condensate is being eliminated through manually operated drain valves. The length should be about 1.5 times the leg diameter and not less than 8 in.

TABLE 11.5 Condensate Load in Main during Warm-Up

Warm-up load, in pounds of steam per 100 ft of steam main, ambient temperature 70°F*

Steam pressure, psig	Main size, in														0°F correction factor†
	2	2½	3	4	5	6	8	10	12	14	16	18	20	24	
0	6.2	9.7	12.8	18.2	24.6	31.9	48	68	90	107	140	176	207	208	1.50
5	6.9	11.0	14.4	20.4	27.7	35.9	48	77	101	120	157	198	233	324	1.44
10	7.5	11.8	15.5	22.0	29.9	38.8	58	83	109	130	169	213	251	350	1.41
20	8.4	13.4	17.5	24.9	33.8	43.9	66	93	124	146	191	241	284	396	1.37
40	9.9	15.8	20.6	29.3	39.7	51.6	78	110	145	172	225	284	334	465	1.32
60	11.0	17.5	22.9	32.6	44.2	57.3	86	122	162	192	250	316	372	518	1.29
80	12.0	19.0	24.9	35.3	47.9	62.1	93	132	175	208	271	342	403	561	1.27
100	12.8	20.3	26.6	37.8	51.2	66.5	100	142	188	222	290	366	431	600	1.26
125	13.7	21.7	28.4	40.4	54.8	71.1	107	152	200	238	310	391	461	642	1.25
150	14.5	23.0	30.0	42.8	58.0	75.2	113	160	212	251	328	414	487	679	1.24
175	15.3	24.2	31.7	45.1	61.2	79.4	119	169	224	265	347	437	514	716	1.23
200	16.0	25.3	33.1	47.1	63.8	82.8	125	177	234	277	362	456	537	748	1.22
250	17.2	27.3	35.8	50.8	68.9	89.4	134	191	252	299	390	492	579	807	1.21
300	25.0	38.3	51.3	74.8	104.0	142.7	217	322	443	531	682	854	1045	1182	1.20
400	27.8	42.6	57.1	83.2	115.7	158.7	241	358	492	590	759	971	1163	1650	1.18
500	30.2	46.3	62.1	90.5	125.7	172.6	262	389	535	642	825	1033	1263	1793	1.17
600	32.7	50.1	67.1	97.9	136.0	186.6	284	421	579	694	893	1118	1367	1939	1.16

*Loads based on Schedule 40 pipe for pressures up to and including 250 psig and on Schedule 80 pipe for pressures above 250 psig.

†For outdoor temperature of 0°F, multiply the load value in the table for each main size by the correction factor corresponding to steam pressure.

11.37

TABLE 11.6 Condensate Load in Main during Running

Condensation load, in pounds per hour per 100 ft of insulated steam main,* ambient temperature 70°F, insulation 80 percent efficient

Steam pressure, psig	Main size, in														0°F correction factor†
	2	2½	3	4	5	6	8	10	12	14	16	18	20	24	
10	6	7	9	11	13	16	20	24	29	32	36	39	44	53	1.58
30	8	9	11	14	17	20	26	32	38	42	48	51	57	68	1.50
60	10	12	14	18	24	27	33	41	49	54	62	67	74	89	1.45
100	12	15	18	22	28	33	41	51	61	67	77	83	93	111	1.41
125	13	16	20	24	30	36	45	56	66	73	84	90	101	121	1.39
175	16	19	23	26	33	38	53	66	78	86	98	107	119	142	1.38
250	18	22	27	34	42	50	62	77	92	101	116	126	140	168	1.36
300	20	25	30	37	46	54	68	85	101	111	126	138	154	184	1.35
400	23	28	34	43	53	63	80	99	118	130	148	162	180	216	1.33
500	27	33	39	49	61	73	91	114	135	148	170	185	206	246	1.32
600	30	37	44	55	68	82	103	128	152	167	191	208	232	277	1.31

*Chart loads represent losses due to radiation and convection for saturated steam.

†For outdoor temperature of °F, multiply the load value in the table for each main size by the correction factor corresponding to the steam pressure.

STEAM TRAP SELECTION AND SIZING

Selecting and Sizing the Steam Trap

General. Each trap design has specific characteristics. Because there is such a variety of trap designs, one of the key decisions in condensate drainage design is the choice of the right type of trap. Often, this choice is easy; for example, low-pressure heating system radiators are almost always equipped with a thermostatic trap because the characteristics of that type of trap match the condensate drainage and air venting requirements of low-pressure heating equipment. Sometimes, either of two different types of traps could equally well be applied to a given condensate drainage situation because either set of trap characteristics meets the drainage and venting requirements. There are some advantages and disadvantages to both, but either one could do the job. For example, a high-pressure steam main could be equipped with either a thermodynamic or an inverted bucket trap. The choice between them then becomes a nontechnical matter, for example, cost, personal preference. In some cases, a given kind of trap simply would not be able to do the job; for example, a thermodynamic trap requires a significant pressure drop between the steam-condensing device and the condensate pipe. Such a trap installed in a low-pressure heating system would not have a great enough pressure differential to operate. General selection criteria are given in Table 11.7. General characteristics are given in Table 11.8.

Liberal and oversized steam traps do not always provide an efficient and safe steam main drain installation. The following points should also be considered by the design professional:

1. Method of heat-up to be employed
2. Providing suitable reservoirs, or collecting legs, for condensate
3. Ensuring adequate pressure differential across the steam trap
4. Steam trap load safety factor
5. Flow of condensate from the selected trap

TABLE 11.7 General Steam Trap Selection Criteria

Application or condition	Recommended traps
Steam distribution lines	Thermodynamic, balanced pressure
Heat exchangers that cannot tolerate back-up	Mechanical float or inverted bucket
Water hammer	Bimetallic, inverted bucket, or thermodynamic
Batch processes with frequent start-ups	Balanced-pressure or bimetallic thermostatic
Heat exchangers with varying loads and pressures	Mechanical float or inverted bucket
Systems that shut off abruptly	All except inverted bucket traps, which can lose their seals
Freezing	Thermostatic, balanced pressure, bimetallic, or thermodynamic

TABLE 11.8   General Steam Trap Characteristics*

Type of steam trap	Key advantages	Significant disadvantages	Frequently recommended services
Float and thermostatic (F&T)	Continuous condensate discharge Handles rapid pressure changes High noncondensable capacity	Float can be damaged by water hammer. Level of condensate in chamber can freeze, damaging float and body. Some thermostatic air vent designs are susceptible to corrosion.	Heat exchangers with high and variable heat-transfer rates When a condensate pump is required Batch processes that require frequent start-up of an air-filled system
Inverted bucket (IB) Wax or liquid expansion thermostatic (TS)	Rugged Tolerates water hammer without damage	Discharges noncondensables slowly (additional air vent often required). Level of condensate can freeze, damaging the trap body (some models can handle some freezing). Must have water seal to operate. Subject to losing prime. Pressure fluctuations and superheated steam can cause loss of water seal (can be prevented with a check valve).	Continuous operation where noncondensable venting is not critical and rugged construction is important.
	Utilizes sensible heat of condensate Allows discharge of noncondensables at start-up to the set point temperature Not affected by superheated steam, water hammer, or vibration Resists freezing	Element subject to corrosion damage. Condensate backs up into the drain line and/or process.	Ideal for tracing used for freeze protection Freeze-protection, water and condensate lines and traps Noncritical temperature control of heated tanks

TABLE 11.8   General Steam Trap Characteristics* (*Continued*)

Type of steam trap	Key advantages	Significant disadvantages	Frequently recommended services
Balanced-pressure thermostatic (BP)	Small and light-weight Maximum discharge of noncondensables at start-up Unlikely to freeze	Some types damaged by water hammer, corrosion, and super-heated steam. Condensate backs up into the drain line and/or process.	Batch processes requiring rapid discharge of noncondensables at start-up (when used for air vent) Drip-legs on steam mains and tracing Installations subject to ambient conditions below freezing
Bimetal thermostatic (BM)	Small and light-weight Maximum discharge of condensables at start-up Unlikely to freeze; unlikely to be damaged if it does freeze Rugged. Withstands corrosion, water hammer, high pressure, and super-heated steam	Responds slowly to load and pressure changes. More condensate back-up than BP trap. Back-pressure changes operating characteristics.	Drip legs on constant-pressure steam mains Installations subject to ambient conditions below freezing
Thermody-namic (TD)	Rugged. Withstands corrosion, water hammer, high pressure, and superheated steam Handles wide pressure range Compact and simple Audible operation warns when repair is needed	Poor operation with very low pressure steam or high back-pressure. Requires slow pressure build-up to remove air at start-up to prevent air binding. Noisy operation.	Steam main drips, tracers Constant-pressure, constant-load applications Installations subject to ambient conditions below freezing

*This table is intended to serve as a general guide only. Trap design options, materials of construction, and techniques for using the trap in conjunction with other devices (for example, an air vent) can modify some of a trap's inherent characteristics.

Heat-Up Method. The type, size, and installation of the steam trap used to drain steam mains depends upon the method used in bringing the system up to normal operating pressure and temperature. The two methods of system heat-up commonly used are the supervised heat-up and the automatic heat-up.

Supervised Heat-Up. In the supervised heat-up method, manual drain valves are installed at all drainage points in the steam main system. The valves are fully opened to the condensate return before steam is admitted to the system. After most of the heat-up condensate has been discharged, the drain valves are closed, allowing the steam traps to drain the normal operating load. Therefore, the steam traps are sized to handle only the condensate formed due to radiation losses at the system's operating pressure. This heat-up method is generally used for large installations, having steam mains of appreciable size and length, and where the heat-up generally occurs only once a year, such as in large systems where the system pressure is maintained at a constant level after the start-up and is not shut down except in emergencies. A typical installation detail is illustrated in Fig. 11.26.

Automatic Heat-Up. In the automatic heat-up method, the steam boiler brings the system up to full steam pressure and temperature without supervision or manual drainage. This method relies on the traps to automatically drain the warm-up load of condensate as soon as it forms. This heat-up method is generally used in small- and medium-sized installations that are shut down and started up at regular intervals, as in heating systems or in dry cleaning plants, where the boiler is usually shut down at night and started up again the following morning. A typical installation detail is illustrated in Fig. 11.27.

Adequate Pressure Differential across the Trap. The trap cannot discharge condensate unless a pressure differential exists across it, that is, a higher pressure at the inlet compared to a lower pressure in the condensate line. The collecting leg should be of sufficient length to provide a hydrostatic head at the trap inlet so that the condensate can be discharged during warm-up, before a positive steam pressure develops in the steam main. For mechanical traps, not only the minimum differential, but also the maximum allowable differential—that is, the trap "seat pressure rating"—must be considered. In draining devices like heat exchangers controlled by temperature-regulating valves that could possibly operate in a vacuum at part

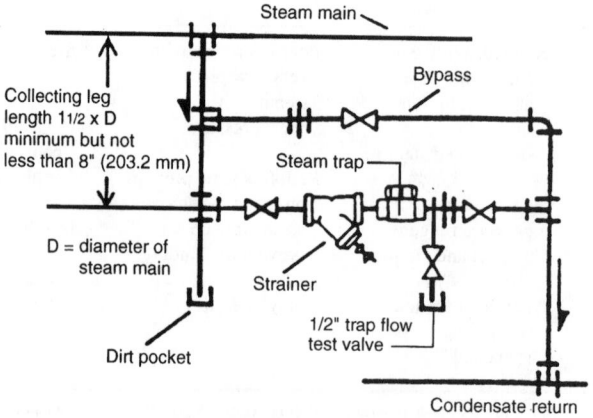

FIGURE 11.26   Steam trap installation for supervised heat-up.

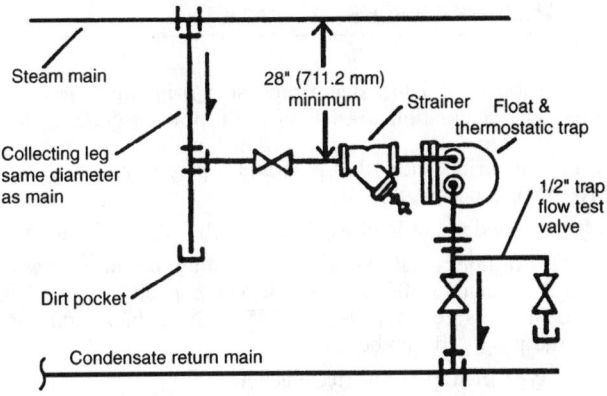

FIGURE 11.27   Steam trap installation for automatic heat-up.

load, install a vacuum breaker to ensure that pressure upstream of the trap cannot fall below atmospheric, and ensure that adequate hydrostatic head is available.

Steam Trap Load Safety Factor.   After determining the actual amount of condensate expected to enter a trap, it is accepted practice to assign a safety factor to increase the amount of condensate. This safety factor is obtained from the manufacturer of the trap. As a guide, Table 11.9 provides recommended safety factors.
Sizing Condensate Pipes to Carry Flashing Condensate

1. Determine the pounds of flash steam from Table 11.2.
2. Multiply total high-pressure condensate flow by the pounds of flash steam to determine the flash steam flow rate.
3. Using Fig. 11.23, size the condensate pipe as if it were a steam pipe carrying nothing but the flash steam flow.

This procedure will oversize the condensate pipe to accommodate the flash steam without generating excess return line pressures. For example, determine the trap requirements for 1000 ft of 10-in horizontal main with a maximum operating pressure of 250 psig. Assume that the supervised heat-up method will be used.

TABLE 11.9   Steam Trap Safety Factors

Type of trap	Safety factor*
Thermostatic	2–4
Float and thermostatic	1.5–2.5
Inverted bucket	2–4
Thermodynamic	1.0–1.2

Note: The safety factor to be applied depends upon the accuracy of the estimates of condensate load and differential pressure, as well as the possibility of any unusual conditions.

PROJECT DESIGN EXAMPLES

EXAMPLE 1 For 1000 ft of 10-in steam pipe at 5 psig, find the condensate load due to heat loss. The calculation consists of the following general steps:

1. Use Table 11.5, entering with 10-in pipe and S psig steam, find the figure of 77 lb/h per 100 ft of pipe length.
2. For 1000 ft, the condensate load is: (77)(1000)I(100) = 770 lb/h.
3. To drain 1000 ft of horizontal main, assume that a minimum of two traps should be used at approximately 500-ft intervals. The capacity required for each trap at 5 psig steam pressure is: 770/2 = 385 lb/h each. If additional traps are desired, use the proposed number of traps.
4. The proposed type of trap is inverted bucket.

   a. From Table 11.3, the recommended safety factor is 2 to 4.

   b. Select a trap with a minimum capacity of (385)(3) = 770 lb/h at S psig.
5. The length of the collecting leg would be:

   a. Diameter: 10 in.

   b. Length: (1.5)(10) = 15 in.

EXAMPLE 2 Determine the size of the trap and the collecting leg size using the automatic heat-up method for 75 ft of 4-in main, maximum steam pressure 15 psig, ambient temperature of 75°F, and heat-up time of 15 mm.

1. Using Table 11.4, entering with 4-in pipe and 15 psig, interpolate to find 23.5 lb/h for 100 ft of pipe.
2. Calculate the actual condensation load for 75 ft of pipe:

$$75 \times \frac{23.5}{100} = 17.6 \text{ lb/h}$$

3. Calculate the average condensate formation rate during the heat-up period:

$$17.6 \times \frac{60}{15} = 70.4 \text{ lb/h}$$

4. The collecting leg size would be:

   a. Diameter: 4 in.

   b. Length: Minimum 28 in to provide a minimum 1 psig hydrostatic head at the inlet of the trap during warm-up.
5. The type and size of the trap might be:

   a. Float and thermostatic.

   b. Capacity: 70.4 lb/h at 1 psig pressure differential.

   c. Maximum operating pressure: The trap must be able to open against differential pressures of at least 15 psig.

   d. Maximum safety factor recommended from Table 11.3 is 1.5.

   e. Trap capacity is 70.4 × 1.5 = 105.6 lb/h.

In determining the heat-up time, safety is the main consideration. Liberal time for heat-up should be allowed to limit the stresses in the piping system caused by thermal expansion, allow drainage in order to minimize the possibility of water hammer, and reduce the size of the trap required. Small bypass valves installed around the large main valves [generally 6 in (15 mm) and larger] are often used to warm up the main slowly.

Maximum Condensate Flow Rate

Maximum Flow Rate from Any Individual Trap. When the trap is installed at a piece of terminal equipment, the PPH steam use of the equipment will discharge the same PPH of condensate. Therefore, 100 PPH of steam will produce 100 PPH of condensate discharged from the trap. The maximum flow rate actually discharged will be determined from published literature for the specific trap selected. For traps placed in steam mains to drain system condensate, the maximum flow rate depends on several variables. Refer to sizing the steam trap as previously discussed.

Diversity Factor. When several traps discharge into a common line, it is not good practice to design a system in the belief that they will all discharge at the same time. Therefore, some diversity factor must be assigned. If the traps are operating on a light loading, a maximum factor of 20 percent should be used. For a medium loading, a 35 to 40 percent factor should be used, and for heavy loads, a 70 percent factor is recommended. Only when using a modulating trap, such as a float and thermostatic trap, can the drainage flow rate be the same as the calculated condensate load.

Condensate Main Pipe Sizing

The following information is required to size the condensate drainage system:

1. All condensate must drain by gravity or be pumped to an elevation from which it can flow by gravity to the final storage point. Determine if the system is to be atmospheric or vacuum assisted.
2. To determine the optimal pitch of the condensate pipe, refer to Table 11.10 for equivalency between sloped and horizontal flows.

TABLE 11.10 Equivalency between Sloped and Horizontal Flows

Pipe slope in 10 ft	Pressure loss, psi/100 ft
$\frac{1}{2}$ in	0.180
1 in	0.361
$1\frac{1}{2}$ in	0.540
2 in	0.722
3 in	1.084
4 in	1.440
5 in	1.805

3. The actual flow rate at the point of design, including the diversity factor, must be known.

The system design, whether atmospheric or vacuum assisted, is determined at the start of the project. The pitch of the pipe is determined by space and project conditions.

To size a condensate main for low-pressure steam, use Table 11.11. To size a condensate main for various steam pressures in a dry, closed system, use Table 11.12.

Condensate Cooler

The condensate cooler size is based on the total maximum expected flow rate in gallons per minute. This information is obtained from the condensate load plus the secondary side fluid requirements to which heat will be transferred. The outlet should be sized to flow half full using Table 9.5. The vessel volume receiving both the condensate and cooling water shall be sized 1.5 times the maximum gallons per minute rate capable of being drained.

Condensate Transfer Pumps

The condensate pump assembly collects the condensate and pumps it back to the boiler as quickly as possible in order to minimize heat loss. Condensate transfer pump assembly should be designed with input from the proposed manufacturer of the equipment. The role of remote condensate transfer pumps is to collect and return the condensate to the boiler as soon as possible.

In higher-pressure steam systems, special pumping equipment might be used to handle higher-temperature condensate (such as from absorption refrigeration machines) without cavitation or other damage. Pressure powered pumps that use compressed air or steam to pump the condensate are also available.

In a two-pipe system the traps vent air and other noncondensable gases into the condensate piping. This piping must be sized to consider the movement of air to the receiver vent in order to actually remove it from the system.

Flash Tanks

A flash tank may be required to cool the condensate before it flows into the receiver. Materials in the condensate pump may not be rated to handle temperatures of the high-temperature condensate, or the pump may cavitate in pumping high-temperature condensate. A flash tank vented to the atmosphere will immediately reduce condensate temperatures to 212°F. A liquid-to-liquid heat exchanger may also be required to further reduce condensate temperatures below 200°F so that the pump will be able to operate without cavitation or other damage.

Cooling condensate so that it can be pumped by low-temperature-rated equipment can add to plant operating costs as much as one-way system operation does, by increasing the use of water and sewage, by adding to energy costs through waste of heat, and by adding to the costs of chemical treatment required at the higher make-up feedwater rates. For these reasons, special low NPSH pumps have been developed to minimize or eliminate the need for condensate cooling when the heat is not needed. A typical low NPSH pump is illustrated in Fig. 11.28.

TABLE 11.11 Condensate Main Size, Low-Pressure Systems

Pipe size, in	$\frac{1}{32}$ psi or $\frac{1}{2}$ oz drop per 100 ft			$\frac{1}{24}$ psi or $\frac{2}{3}$ oz drop per 100 ft			$\frac{1}{16}$ psi or 1 oz drop per 100 ft			$\frac{1}{8}$ psi or 2 oz drop per 100 ft			$\frac{1}{4}$ psi or 4 oz drop per 100 ft			$\frac{1}{2}$ psi or 8 oz drop per 100 ft		
	Wet	Dry	Vac.	Wet	Dry	Vac.	Wet	Dry	Vac.	Wet	Dry	Vac.	Wet	Dry	Vac.	Wet	Dry	Vac.
G	H	I	J	K	L	M	N	O	P	Q	R	S	T	U	V	W	X	Y
Return Main																		
$\frac{3}{4}$	—	—	—	—	—	42	—	—	100	—	—	142	—	—	200	—	—	283
1	125	62	—	145	71	143	175	80	175	250	103	249	350	115	350	—	—	494
$1\frac{1}{4}$	213	130	—	248	149	244	300	168	300	425	217	426	600	241	600	—	—	848
$1\frac{1}{2}$	338	206	—	393	236	388	475	265	475	675	340	674	950	378	950	—	—	1,340
2	700	470	—	810	535	815	1,000	575	1,000	1,400	740	1,420	2,000	825	2,000	—	—	2,830
$2\frac{1}{2}$	1,180	760	—	1,580	868	1,360	1,680	950	1,680	2,350	1,230	2,380	3,350	1,360	3,350	—	—	4,730
3	1,880	1,460	—	2,130	1,560	2,180	2,680	1,750	2,680	3,750	2,250	3,800	5,350	2,500	5,350	—	—	7,560
$3\frac{1}{2}$	2,750	1,970	—	3,300	2,200	3,250	4,000	2,500	4,000	5,500	3,230	5,680	8,000	3,580	8,000	—	—	11,300
4	3,880	2,930	—	4,580	3,350	4,500	5,500	3,750	5,500	7,750	4,830	7,810	11,000	5,380	11,000	—	—	15,500
5	—	—	—	—	—	7,880	—	—	9,680	—	—	13,700	—	—	19,400	—	—	27,300
6	—	—	—	—	—	12,600	—	—	15,500	—	—	22,000	—	—	31,000	—	—	43,800
Riser																		
$\frac{3}{4}$	—	48	—	—	48	143	—	48	175	—	48	249	—	48	350	—	—	494
1	—	113	—	—	113	244	—	113	300	—	113	426	—	113	600	—	—	848
$1\frac{1}{4}$	—	248	—	—	248	388	—	248	475	—	248	674	—	248	950	—	—	1,340
$1\frac{1}{2}$	—	375	—	—	375	815	—	375	1,000	—	375	1420	—	375	2,000	—	—	2,830
2	—	750	—	—	750	1,360	—	750	1,680	—	750	2,380	—	750	3,350	—	—	4,730
$2\frac{1}{2}$	—	—	—	—	—	2,180	—	—	2,680	—	—	3,800	—	—	5,350	—	—	7,560
3	—	—	—	—	—	3,250	—	—	4,000	—	—	5,680	—	—	8,000	—	—	11,300
$3\frac{1}{2}$	—	—	—	—	—	4,480	—	—	5,500	—	—	7,810	—	—	11,000	—	—	15,500
4	—	—	—	—	—	7,880	—	—	9,680	—	—	13,700	—	—	19,400	—	—	27,300
5	—	—	—	—	—	12,600	—	—	15,500	—	—	22,000	—	—	31,000	—	—	43,800

TABLE 11.12 Dry, Closed System Condensate, Main Size

Flow rate (lb/h) for dry-closed returns

D, in	Supply pressure = 5 psig Return pressure = 0 psig			Supply pressure = 15 psig Return pressure = 0 psig			Supply pressure = 30 psig Return pressure = 0 psig			Supply pressure = 50 psig Return pressure = 0 psig		
	$\frac{1}{16}$	$\frac{1}{4}$	1	$\frac{1}{16}$	$\frac{1}{4}$	1	$\frac{1}{16}$	$\frac{1}{4}$	1	$\frac{1}{16}$	$\frac{1}{4}$	1
						$\Delta p/L$, psi/100 ft						
$\frac{1}{2}$	240	520	1,100	95	210	450	60	130	274	42	92	200
$\frac{3}{4}$	510	1,120	2,400	210	450	950	130	280	590	91	200	420
1	1,000	2,150	4,540	400	860	1,820	250	530	1,120	180	380	800
$1\frac{1}{4}$	2,100	4,500	9,500	840	1,800	3,800	520	1,110	2,340	370	800	1,680
$1\frac{1}{2}$	3,170	6,780	14,200	1,270	2,720	5,700	780	1,670	3,510	560	1,200	2,520
2	6,240	13,300	*	2,500	5,320	*	1,540	3,270	*	1,110	2,350	*
$2\frac{1}{2}$	10,000	21,300	*	4,030	8,520	*	2,480	5,250	*	1,780	3,780	*
3	18,000	38,000	*	7,200	15,200	*	4,440	9,360	*	3,190	6,730	*
4	37,200	78,000	*	14,900	31,300	*	9,180	19,200	*	6,660	13,800	*
6	110,500	*	*	44,300	*	*	27,300	*	*	19,600	*	*
8	228,600	*	*	91,700	*	*	56,400	*	*	40,500	*	*

D, in	Supply pressure = 100 psig Return pressure = 0 psig			Supply pressure = 150 psig Return pressure = 0 psig			Supply pressure = 100 psig Return pressure = 15 psig			Supply pressure = 150 psig Return pressure = 15 psig		
	$\frac{1}{16}$	$\frac{1}{4}$	1	$\frac{1}{16}$	$\frac{1}{4}$	1	$\frac{1}{16}$	$\frac{1}{4}$	1	$\frac{1}{16}$	$\frac{1}{4}$	1
						$\Delta p/L$, psi/100 ft						
$\frac{1}{2}$	28	62	133	23	51	109	56	120	260	43	93	200
$\frac{3}{4}$	62	134	290	50	110	230	120	260	560	93	200	420
1	120	260	544	100	210	450	240	500	1,060	180	390	800
$1\frac{1}{4}$	250	540	1,130	200	440	930	500	1,060	2,200	380	800	1,680
$1\frac{1}{2}$	380	810	1,700	310	660	1,400	750	1,600	3,320	570	1,210	2,500
2	750	1,590	*	610	1,300	*	1,470	3,100	6,450	1,120	2,350	4,900
$2\frac{1}{2}$	1,200	2,550	*	980	2,100	*	2,370	5,000	10,300	1,800	3,780	7,800
3	2,160	4,550	*	1,760	3,710	*	4,230	8,860	*	3,200	6,710	*
4	4,460	9,340	*	3,640	7,630	*	8,730	18,200	*	6,620	13,800	*
6	13,200	*	*	10,800	*	*	25,900	53,600	*	19,600	40,600	*
8	27,400	*	*	22,400	*	*	53,400	110,300	*	40,500	83,600	*

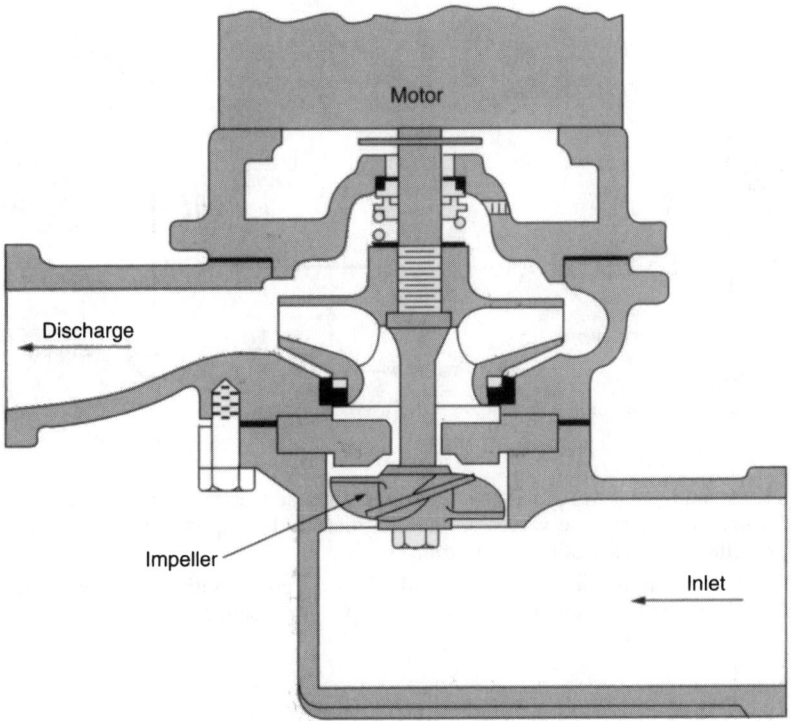

FIGURE 11.28 Typical low NPSH transfer pump.

SYSTEM DESIGN CONSIDERATIONS

Steam Trap Installation

The following recommendations should be observed:

1. The steam trap should be installed as close as possible to the collecting leg.

2. Lifting the condensate or piping condensate directly to a return line under pressure should be avoided. If the condensate must be lifted or discharged into a return system under pressure, the condensate should be collected in a vented flash tank and pumped to the elevated return or return system under pressure. If the trap must discharge to a return line that may be under pressure, then the differential pressure across the trap and the trap capacity will be reduced. A check valve on the discharge side of the trap is always required if the trap discharges into an elevated or pressurized return. A typical detail of discharging to an elevated return line is illustrated in Fig. 11.29.

3. Pipe connections to and from the steam trap should be at least equal to the pipe size of the trap connection, and full-size isolation valves should be installed on each side of the trap to allow service.

4. A strainer equipped with a blowdown valve should be installed before the steam trap.

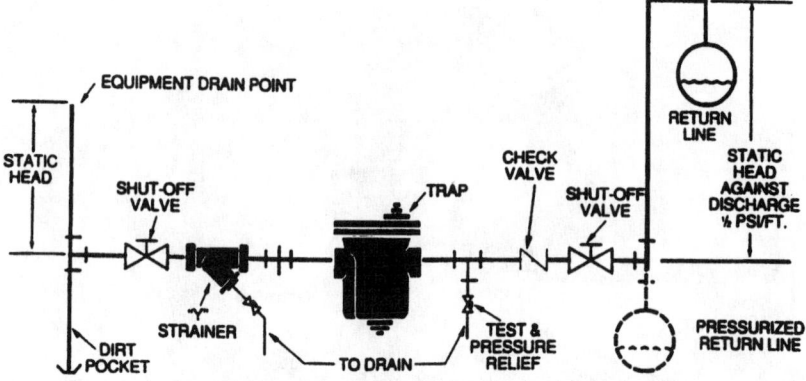

FIGURE 11.29 Steam trap discharging to an overhead return.

5. A test and pressure relief fitting is recommended downstream of the trap to ensure that the service valves are holding before the trap is serviced. It can also provide for quick testing of the trap.

6. All low points of the steam main and wherever condensate can collect, such as ahead of pressure-reducing valves and temperature regulators, should be drained.

High-Pressure System Condensate

Special procedures may be required to size the condensate pipes in high-pressure systems since they can carry a mixture of flash steam and condensate. The production of flash steam is influenced by other components in the system too. For example, thermostatic steam traps open only after the condensate has subcooled below saturation temperature for the given pressure. This subcooled condensate contains less energy than condensate at saturation temperature, so the amount of flash steam for a given drop in pressure will be reduced. Table 11.2 shows the amount of reevaporation, or the percentage of flash steam that will occur for different conditions of initial pressure and pressure drop. In sizing condensate returns for high-pressure systems, the pipe sizing tables shall consider this flash steam since it will provide a great deal more friction loss than would be the case if the flashing' had not occurred and the pipes were carrying only liquid.

Pitching of Steam Piping

In the past some steam systems had "counterflow" steam mains, meaning that the pipe was pitched toward the boiler to return the condensate formed in the system. Counterflow piping designs have some shortcomings. Pipes must be generously sized since steam and water are moving opposite to one another at the same time. Small bubbles of condensate are likely to be surrounded by the steam, rapidly condensing the steam and causing noise in the pipe that sounds like gravel rattling. In addition to the noise, the highly pitched counterflow piping is difficult to install and must be large enough to:

1. Allow for the increase in friction between the fluids traveling in opposite directions

2. Limit the velocity of steam flow to prevent water hammer

For these reasons, it is preferable to pitch the steam mains in the direction of steam flow so that the steam and condensate will flow in the same direction. Short runouts of 3 to 4 ft (0.9 to 1.2 m) may be pitched for counterflow to eliminate the need for traps. These "parallel flow" mains are likely to be quieter and require less pitch and smaller pipe sizes for a given steam flow rate. Modern steam pipe sizing charts usually assume that parallel flow will be maintained to minimize interference between the steam and condensate.

All of these advantages are at least partly offset by the fact that the parallel flow system needs more pipes because *returns* must be installed at the end of the system to bring the condensate back to the boiler. A *dry return* is located above the water line of the boiler, and therefore can carry steam and air in addition to condensate. A *wet return* is located below the boiler water line; it is therefore flooded and doesn't need to be pitched. It does require some hydrostatic head in the form of a column of water to return condensate to the boiler.

Steam-condensing units have separate connections for the steam supply and the condensate return. A steam trap is installed at the condensate connection from each unit and at intervals along the steam main to eliminate any interference between the steam and liquid flow. This trap is the boundary between the steam and the condensate piping.

Costs of Wasting Condensate

There are costs associated with nonrecovery of condensate. Water lost from the steam system must eventually be made up, as well as any water used to lower the condensate temperature before discharge into a sewer. Even in areas that have abundant supplies of fresh water, these costs are rising. In areas that do not have abundant supplies, there may be limits on the amount of water that can be used in such a process.

The loss of condensate will always increase the rate of makeup feedwater introduced to the system. Since it's likely that the makeup feed is colder than the condensate, each pound of makeup water requires significantly more heat to bring it to the boiling temperature. The energy loss of wasted condensate increases with increasing system pressure. In tempering condensate and discharging it, the sensible heat is lost, adding to the energy costs of the process. That's why heat from steam condensate should be used whenever possible to preheat domestic hot water or ventilation air before discharging it.

Problems Resulting from Poor Condensate Drainage

A loss of heat transfer performance will result unless condensate is drained from the steam heating unit. Since heat transfer in a steam system is based upon the large amount of latent heat that becomes available as the steam condenses, it follows that if the heating unit is flooded, only the sensible heat from the condensate will become available as it cools. Table 11.1 shows that the sensible heat is only a small fraction of the latent heat available in a pound of steam at any given pressure. The latent heat is transferred at constant steam temperature as the steam condenses, while the condensate would have to cool down to 32°F (0°C) in order to make all the sensible heat available.

Corrosion

Corrosion generally occurs in steam heating units and condensate piping that are not properly drained. Several corrosion processes have been defined including:

1. Generalized corrosion
2. Oxygen pitting
3. Condensate grooving, which etches away the metal along the path followed by condensate that has become acidic due to dissolved carbon dioxide
4. Fouling or scaling on heat transfer surfaces, which is increased by inadequate condensate drainage and venting

Corrosion is accelerated by the presence of carbon dioxide which can form in the boiler as chemical components like carbonates and bicarbonates decompose. Oxygen introduced by vacuum breakers or from makeup feedwater also increases the corrosion rate. Properly installed steam traps, condensate return piping, and pumping equipment can minimize these problems by draining the condensate and venting noncondensable gases from steam piping and heat exchangers.

Parallel Trapping

For the automatic heat-up method, the largest load that the steam trap must deal with is the warming-up load; consequently, the steam trap must be sized on the warm-up load. Depending on the size of the installation, this condition can result

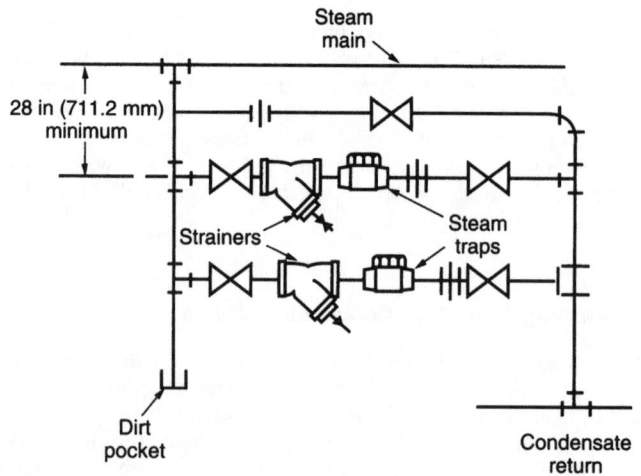

FIGURE 11.30   Parallel steam trapping.

in a considerably oversized steam trap, after the system is up to normal operating pressure. Oversized steam traps can create problems, particularly in high-pressure installations. For example, inverted bucket traps can lose their condensate level and pass live steam. All steam traps wear more rapidly on light loads than on heavy loads. Frequent release of small quantities of condensate at full discharge rate of the steam trap can create undesirable pressures in the condensate return piping system.

These problems can be reduced by installing smaller steam traps in parallel with the same aggregate capacity. Figure 11.30 shows a typical installation. It is recommended that parallel trapping be considered in any situation in which a single large steam trap would otherwise be required. In addition to the economies inherent in a parallel piping hookup, it provides emergency protection in the event that one steam trap becomes clogged or fails closed.

ACKNOWLEDGMENTS

Roy C. E. Ahlgren, ITT Bell & Gossett

Al Greenberg, PE, CEM, Alfred Greenberg & Associates

Jeffery Grant, J. E. Grant & Associates

REFERENCES

American Society of Heating, Refrigerating and Air Conditioning Engineers (ASHRAE): "Flash Tanks for Steam and Boiler Systems," *ASHRAE Journal,* Atlanta, Ga., 1991.

American Society of Plumbing Engineers: "Steam and Condensate Systems," *Data Book,* Westlake, Calif., Chap. 38.

ASHRAE: *Fundamentals Handbook,* New York.

ITT Fluid Handling Division: *Training Manual TES 181, Hoffman Steam Heating Systems Design Manual and Engineering Data,* Morton Grove, Ill.

_____: *Training Manual TES 582, High Temperature Condensate Return,* Morton Grove, Ill.

National Board of Boiler and Pressure Vessel Inspectors: *Recommendations for the Design and Construction of Boiler Blowoff Systems,* Columbus, Ohio, 1991.

Spirax-Sarco: *Design of Fluid Systems-Hook-ups,* Allentown, Pa., 1992.

Spirax-Sarco: *Design of Fluid Systems-Steam Utilization,* Allentown, Pa., 1991.

CHAPTER 12
LIQUID FUEL STORAGE AND DISPENSING SYSTEMS

This chapter will describe the design, selection, and installation requirements for atmospheric storage, distribution, and dispensing of new or replacement systems for liquid-petroleum-based fuels used, for example, in automobiles and other internal combustion motors such as emergency generators and fire pumps. Typical systems have a volume of product storage limited to approximately 6000 gal (22,500 L). Gasoline tanks are installed for commercial purposes and at facility sites primarily intended to serve the public or company-owned vehicles. Similar installations include liquids such as kerosene, motor lubrication oil, and waste oil.

Fuel oil systems for boilers and other heat-producing apparatus and chemical storage facilities are outside the scope of this handbook. Other system aspects not discussed concern existing tanks such as investigation, closure, removal, and repairs of leaking systems.

DEFINITIONS AND LIQUID FUEL CLASSIFICATIONS

A *storage tank for liquid fuel* is any stationary receptacle designed to contain an accumulation of regulated substances. Tanks can be constructed of materials such as steel, concrete, plastic, or various combinations that provide structural support.

A storage tank is considered *underground* if the sum of the total tank volume of single or multiple tanks, including all of the associated and interconnecting piping, is 10 percent or more below grade or covered with earth.

A *regulated substance* is any designated chemical that includes hydrocarbons derived from crude oil such as motor fuels, distillate fuel oils, residual fuels, lubricants, used oils, and petroleum solvents. Kerosene is also a regulated substance. Hydrocarbons are measured in parts per million (ppm) of total petroleum hydrocarbons (TPH). In future discussions, all of these will be called *product*.

OSHA (29 CFR 1926) further defines storage tanks according to the following operating pressure ratings:

1. *Atmospheric tanks.* Atmospheric pressure to 0.5 psig
2. *Low-pressure tanks.* Greater than 0.5 to 15 psig
3. *Pressure tanks.* Greater than 15 psig

Liquid fuels are governed by the requirements of NFPA-30. In this standard, liquids are classified as either flammable or combustible based on their flash points. The *flash point* of a liquid is the temperature at which the liquid could give off vapor in sufficient concentration to form an ignitable mixture with air at or near the surface. In short, the flash point is the minimum temperature at which a fire or explosion could occur. The following definitions are applicable only for the purpose of fire protection.

Liquids are classified by the NFPA as either flammable or combustible and are divided into class I, II, and III. All *flammable liquids* are considered class I. They have a flash point below 100°F (37.8°C) and a vapor pressure no higher than 40 psia (2086 mmHg) at 100°F. Class I liquids are further subdivided into class IA, IB, and IC. Class IA (which includes gasoline and gasoline blends) liquids have a flash point below 73°F (22.8°C) and a boiling point below 100°F (37.8°C).

All combustible liquids are considered class II and III liquids. They have a flash point at or above 100°F (37.8°C). Diesel fuel, light heating oil, and kerosene are class II flammable liquids with a flash point at or above 100°F (37.8°C) but below 140°F (60°C). Class III liquids are further subdivided into class IIIA and IIIB. Class III liquids include motor lubrication and waste oil.

Liquid petroleum or *petroleum products* are defined as any hydrocarbon that is a liquid at atmospheric pressure and at temperatures between 20 and 120°F (−29 and 49°C), or discharged as a liquid at temperatures in excess of 120°F (49°C).

The products discussed in this handbook will be gasoline, gasoline blends, or diesel oil used as fuel for motor vehicles or internal combustion engines. These fuels are classified as hydrocarbons. They are also considered flammable liquids.

Specific gravity is the direct ratio of a liquid's weight to the weight of water at 62°F.

Viscosity is a measure of the internal friction between particles in a liquid that resists any force tending to produce flow. The higher the viscosity, the slower the liquid will flow under gravity conditions. Viscosity is obtained by measuring the amount of time a given quantity of liquid at a specified temperature takes to flow through an orifice. Viscosity is expressed in Seconds Saybolt Universal (SSU) for pump work and also kinematic viscosity centistrokes (centipoises, cP), Seconds Saybolt Furol, and Seconds Redwood. Conversion between these various units and others are given in Table 12.1. A table listing the viscosity and specific gravity of various liquid fuels is given in Table 12.2.

The vapor produced by the evaporation of hydrocarbons are in a category known as *volatile organic hydrocarbons* (*VOC*). Vapor produced by gasoline and gasoline blends are required by code to be recovered. The vapor density of common fuels are given in Table 12.3. *Stage I systems* refer only to storage tanks where vapor is displaced when the tank is filled with product. The recovered vapor is returned to the delivery truck. *Stage II systems* refer only to vapor recovery from automobiles when the tank is filled with product. The recovered vapor is returned to the storage tank. Kerosene and diesel oil storage and dispensing systems do not require vapor recovery.

CODES AND STANDARDS

The U.S. Environmental Protection Agency (EPA) has written basic, minimum regulations to protect the environment and people's health from the leakage of hydro-

TABLE 12.1 Approximate Viscosity Conversions

Seconds Saybolt Universal, SSU	Kinematic viscosity cP	ft²/s	Seconds Saybolt Furol, SSF	Seconds Redwood 1 Standard	Seconds Redwood 2 Admiralty	Degrees Engler	Degrees Barbey
31	1.0	0.00001076		29		1.00	6200
31.5	1.13	0.00001216		29.4		1.01	5486
32	1.81	0.00001948		29.8		1.08	3425
32.6	2.00	0.00002153		30.2		1.10	3100
33	2.11	0.00002271		30.6		1.11	2938
34	2.40	0.00002583		31.3		1.14	2583
35	2.71	0.00002917		32.1		1.17	2287
36	3.00	0.00003229		32.9		1.20	2066
38	3.64	0.00003918		33.7		1.26	1703
39.2	4.00	0.00004306		35.5		1.30	1550
40	4.25	0.00004575		36.2	5.10	1.32	1459
42	4.88	0.00005253		38.2	5.25	1.36	1270
42.4	5.00	0.00005382		38.6	5.28	1.37	1240
44	5.50	0.00005920		40.6	5.39	1.40	1127
45.6	6.00	0.00006458		41.8	5.51	1.43	1033
46	6.13	0.00006598		42.3	5.54	1.44	1011
46.8	7.00	0.00007535		43.1	5.60	1.48	885
50	7.36	0.00007922		44.3	5.83	1.58	842
52.1	8.00	0.00008611		46.0	6.03	1.64	775
55	8.88	0.00009558		48.3	6.30	1.73	698
55.4	9.00	0.00009688		48.6	6.34	1.74	689
58.8	10.00	0.0001076		51.3	6.66	1.83	620
60	10.32	0.0001111		52.3	6.77	1.87	601
65	11.72	0.0001262		56.7	7.19	2.01	529
70	13.08	0.0001408		60.9	7.60	2.16	474
75	14.38	0.0001548		65.1	8.02	2.37	431
80	15.66	0.0001686		69.2	8.44	2.45	396
85	16.90	0.0001819		73.4	8.87	2.59	367
90	18.12	0.0001950		77.6	9.30	2.73	342
95	19.32	0.0002080		81.6	9.71	2.88	321
100	20.52	0.0002209		85.6	10.12	3.02	302
120	25.15	0.0002707		102	11.88	3.57	246
140	29.65	0.0003191		119	13.63	4.11	209
160	34.10	0.0003670		136	15.39	4.64	182
180	38.52	0.0004146		153	17.14	5.12	161
200	42.95	0.0004623		170	18.90	5.92	144
300	64.6	0.0006953	32.7	253	28.0	8.79	96
400	86.2	0.0009278	42.4	338	37.1	11.70	71.9
500	108.0	0.001163	52.3	423	46.2	14.60	57.4
600	129.4	0.001393	62.0	507	55.3	17.50	47.9
700	151.0	0.001625	72.0	592	64.6	20.44	41.0
800	172.6	0.001858	82.0	677	73.8	23.36	35.9
900	194.2	0.002099	92.1	762	83.0	26.28	31.9
1000	215.8	0.002323	102.1	846	92.3	29.20	28.7

(*Continued*)

TABLE 12.1 Approximate Viscosity Conversions (*Continued*)

Seconds Saybolt Universal, SSU	Kinematic viscosity cP	Kinematic viscosity ft²/s	Seconds Saybolt Furol, SSF	Seconds Red-wood 1 Stan-dard	Seconds Red-wood 2 Ad-miralty	Degrees Engler	Degrees Barbey
1200	259.0	0.002786	122	1016	111	35.1	23.9
1400	302.3	0.003254	143	1185	129	40.9	20.5
1600	345.3	0.003717	163	1354	148	46.7	18.0
1800	388.5	0.004182	183	1524	166	52.6	15.6
2000	431.7	0.004647	204	1693	185	58.4	14.4
2500	539.4	0.005806	254	2115	231	73.0	11.5
3000	647.3	0.006967	305	2538	277	87.6	9.6
3500	755.2	0.008129	356	2961	323	102	8.21
4000	863.1	0.009290	408	3385	369	117	7.18
4500	970.9	0.01045	458	3807	415	131	6.39
5000	1078.8	0.01161	509	4230	461	146	5.75
6000	1294.6	0.01393	610	5077	553	175	4.78
7000	1510.3	0.01626	712	5922	646	204	4.11
8000	1726.1	0.01858	814	6769	738	234	3.59
9000	1941.9	0.02092	916	7615	830	263	3.19
10000	2157.6	0.02322	1018	8461	922	292	2.87
15000	3236.5	0.03483	1526	12692		438	1.92
20000	4315.3	0.04645	2035	16923		584	1.44

Source: Cameron hydraulic data.

TABLE 12.2 Viscosity and Specific Gravity of Various Hydrocarbons

Liquid	Specific gravity	Viscosity, SSU 40°F	60°F	80°F	100°F	120°F	140°F	160°F
		Miscellaneous liquids						
Water	1.0	31.5	31.5	31.5	31.5	31.5	31.5	31.5
Gasoline	0.68–0.74	30	30	30	30	30	30	30
Jet fuel	0.74–0.85	35	35	35	35	35	35	35
Kerosene	0.78–0.82	42	38	34	33	31	30	30
Turpentine	0.86–0.87	34	33	32.8	32.6	32.4	32	32
Gasahol 15% gas, 85% alcohol	0.70–0.72	30	30	30	30	30	30	30
		Fuel oil and diesel oil						
No. 1 fuel oil	0.82–0.95	40	38	35	33	31	30	30
No. 2 fuel oil	0.82–0.95	70	50	45	40	—	—	—
No. 3 fuel oil	0.82–0.95	90	68	53	45	40	—	—
No. 5A fuel oil	0.82–0.95	1,000	400	200	100	75	60	40
No. 5B fuel oil	0.82–0.95	1,300	600	490	400	330	290	240
No. 6 fuel oil	0.82–0.95	—	70,000	20,000	90,000	1,900	900	500
No. 2D diesel fuel oil	0.82–0.95	100	68	53	45	40	36	35
No. 3D diesel fuel oil	0.82–0.95	200	120	80	60	50	44	40
No. 4D diesel fuel oil	0.82–0.95	1,600	600	280	140	90	68	54
No. 5D diesel fuel oil	0.82–0.95	15,000	5,000	2,000	900	400	260	160

TABLE 12.3   Specific Gravity of Common Vapors

Air	Diesel fuel	Gasoline	Methanol, 100%	85% methanol, 15% gasoline	Natural gas
1.0	5.5	3.4	1.1	2.0	0.6

carbons and VOCs from USTs, ASTs, and associated piping. The basic purpose of these regulations is to assure proper installation of the various system components, prevent leaks or spills from occurring, and, if a leak or spill does occur, to assure that the leak is quickly found and annunciated. Other organizations regulate component testing and general provisions for system components and installation with regard to fire prevention.

In almost all jurisdictions where these systems are installed, there exist specific requirements mandated by the local and state agencies. These requirements govern permits, registration, fees, and recordkeeping, as well as specific technical rules and regulations concerning system installation, materials, and leak detection. Very often, they may be more stringent than the federal EPA regulations cited here. A thorough code search will be necessary to assure complete compliance with all applicable federal, state, and local regulations.

The following is a list of commonly used codes, regulations, and guidelines. This list is not complete and must be verified in the locality where the project is constructed.

1. NFPA 30 and 30A
2. 40 CFR 112 Spill Containment Control and Countermeasures
3. Regional fire codes such as BOCA and UFC
4. Underwriters Laboratories UL-142 Standard for Safety and UL-2085 Fire Resistance
5. Resource Conservation and Recovery Act (RCRA), Subtitle 1 & C
6. Steel Tank Institute (STI)
7. Public Law (PL) 98 and 616
8. Clear Air Act Amendments (CAAA)
9. SARA, Title III, 1986

SYSTEM DESCRIPTION

The liquid fuel storage and dispensing system is intended to store and dispense liquid fuel used to power internal combustion engines. Other required ancillary functions include systems and devices to recover vapor evaporated from product, measure the level of liquid inside the storage tank, and detect any leakage from the storage tank and piping distribution system.

SYSTEM COMPONENTS

Liquid fuel storage and dispensing, whether in an above-ground storage tank (AST) or underground storage tank (UST), require many interrelated subsystems and components for proper operation and compliance with applicable codes and standards. They are:

1. A storage tank, including:
   a. Tank filling and accidental spill containment
   b. Atmospheric tank venting
   c. Overfill protection
2. Leak detection and system monitoring
3. Vapor recovery systems to prevent the vapor produced by gasoline and gasoline blends from storage tanks and fuel tanks of motor vehicles from escaping into the atmosphere
4. A pump and piping system for dispensing and distributing product from the storage tank into motor vehicles or directly to engines

Because there are significant differences in the materials and the installation and operation of a UST and an AST, they will be discussed in separate sections. In addition, fire pump and emergency generator fuel systems are included in a separate section.

UNDERGROUND STORAGE AND DISTRIBUTION SYSTEMS

STORAGE TANKS

General

Storage tanks are designed and fabricated to prevent product releases due to structural failure and corrosion of the tank from the time of installation to the end of the expected useful life of the system. This requires that the tank manufacturer fabricate the tank in conformance with applicable codes and nationally recognized standards for structural strength and corrosion resistance. Since the tank must be installed in a manner that will prevent distortion and stress, the installation of the specific tank must be done by contractors trained and approved by the manufacturer. The tank foundation, bedding, and backfill shall be done only with material and methods approved by the manufacturer.

The structural integrity of any tank requires manufacturing methods capable of passing an internal pressure and vacuum test. This assures that the tank is capable of resisting a higher internal pressure than would be expected during operation. USTs conforming to codes from ASTM, UL, and STI produce a tank capable of resisting external pressure. The corrosion resistance for USTs is determined by industry consensus codes that provide detailed performance requirements. The corrosion resistance for ASTs is determined by the quality of the protection of the tank exterior. These requirements must be accepted by the implementing agency responsible for the approval of the installation.

Tank Materials

The materials used to manufacture primary and secondary tanks are:

1. Steel, with thin coating or thick cladding based on corrosion protection method selected by the manufacturer. Cathodic protection may be required.
2. Fiberglass reinforced plastic (FRP). No corrosion protection required.
3. Steel and fiberglass reinforced plastic composite.
4. Preengineered, cathodically protected steel. This option is rarely used.

Clad steel is manufactured by applying a layer of plastic, usually FRP, over the exterior surface of the steel tank. This offers the strength of steel with the corrosion protection of plastic. Great care must be taken to prevent damage to the cladding during shipping and installation. Some authorities require the installation of sacrificial anodes.

FRP tanks are manufactured by several proprietary processes from thermoset plastic reinforced by fiberglass. Reinforcing ribs are built into the tank for increased structural strength. Generally, there is a resin-rich layer that contacts the product. The specific plastic materials are listed by the implementing agency as being suitable for the intended product. These tanks are completely resistant to corrosion, but they are more susceptible to damage by mishandling and distortion during installation.

The composite tank is manufactured by having the steel tank "wrapped" in a jacket of high-density polyethylene that is not bonded to the tank itself. This provides a very thin interstitial space that can be monitored. Experience has shown that the jacket is the portion of the assembly that fails most often.

Preengineered steel tanks are constructed by having an insulating coating on a steel tank with sacrificial anodes welded to the tank side. The coating is usually of coal tar epoxy, although FRP and polyurethane are also used. If steel piping is used, it must be isolated from the tank by special bushings. This is the least costly of all materials, but it requires the cathodic protection to be constantly monitored.

Tank Construction

Typical tanks are cylindrical with a round cross section. They are available in either single-wall or double-wall construction. For double-wall construction, the inner tank containing the product is called the *primary tank*. The outer tank is called the *secondary containment tank*. This system of double wall tanks is often referred to as *double contained*. The outer tank could be manufactured from the same material as the primary tank or a different material as approved by the implementing agency.

A space between the primary and secondary tank is called an *interstitial space*. The width of the space varies between different manufacturers and types of construction. It is this space that is monitored for leakage of the primary tank.

Steel primary tanks with steel secondary tanks are available in different configurations. The steel secondary tank could extend partly around the primary tank (300°) with a small interstitial space, extend fully around the primary tank (360°) with a small interstitial space, or extend fully around the primary tank with a larger interstitial space. This last method usually consists of two full tanks, one inside the other.

There are several proprietary methods used to construct the secondary tank around the primary tank. The choice between single- and secondary-contained tank systems as well as the tank material depends on the proposed method of leak detection and corrosion protection. The double-wall tank is the most often used. Typical fiberglass double-contained tank dimensions are listed in Table 12.4. Steel double-contained tanks are comparable in size.

Corrosion Protection

All tanks not manufactured throughout from noncorrosive materials, such as steel, must be protected against corrosion. This also includes distribution pipe of a material subject to corrosion. A bare steel tank can be dialectically coated with a thin material but must also be cathodically protected by means of impressed current or sacrificial anodes. It could also be coated with a thick, noncorrodible cladding that prevents corrosion, and no cathodic protection is necessary. Another type of protection is an exterior wrapping with a noncorrosive material, usually polyethylene. The most often used cladding material is FRP. Both methods are factory applied. There are industry standards detailing performance requirements such as STI-P3 and UL Standard 1746.

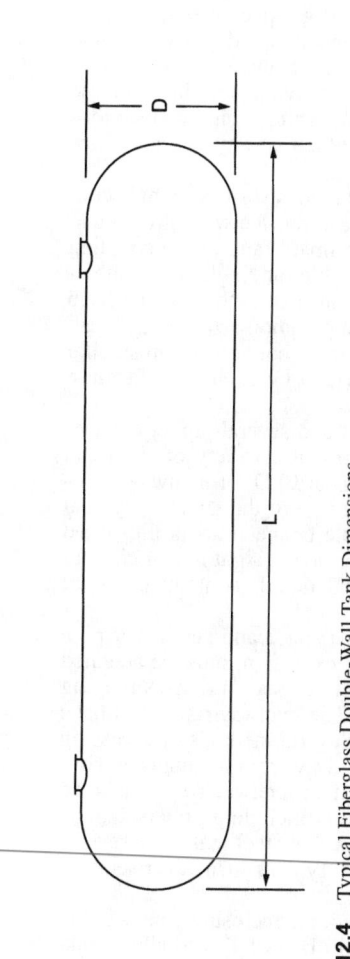

TABLE 12.4 Typical Fiberglass Double-Wall Tank Dimensions

Nominal capacity, gal	Actual capacity, gal	D, nominal diameter, ft	L, overall length	Number of 22-in manways	Number of 4-in NPT service fittings	Number of 4-in NPT monitor fittings	Interstitial volume, gal	Shipping weight, lb
550	553	4	6'-9 1/4"	1	4	1	85	1,100
750	750	4	8'-11"	1	4	1	127	1,250
1,000	1,009	4	11'-7 1/8"	1	4	1	181	1,400
2,500	2,319	6	13'-5 3/4"	1	4	1	48	1,570
3,000	2,900	6	16'-4 1/4"	2	6	2	59	1,950
4,000	3,782	6	20'-8"	2	6	2	76	2,250
6,000	5,562	6	29'-5"	2	6	2	110	3,010
6,000	5,998	8	20'-6 1/2"	2	6	2	114	2,745
8,000	7,841	8	26'-0 1/2"	2	6	2	152	3,460
10,000	9,684	8	31'-6 1/2"	2	6	2	189	4,085
12,000	11,527	8	37'-0 1/2"	2	6	2	227	4,650
10,000	10,369	10	21'-5 1/4"	2	6	2	144	4,170
12,000	11,849	10	24'-0 1/4"	2	6	2	167	4,950
15,000	14,976	10	29'-5 3/4"	2	6	2	213	6,110
20,000	19,703	10	37'-8 3/4"	2	6	2	280	8,210
25,000	25,587	10	48'-0"	—	—	2	1,540	16,000
30,000	30,352	10	56'-3"	—	—	2	1,802	18,500

Source: Courtesy Xerxes.

Tank Connections and Access

There must be a convenient and leakproof method of providing connections directly into the primary tank to allow for filling, venting, product dispensing, gauging, and leak detection. For larger tanks, access directly into the tank for personnel may also be desired. This is done by means of a *manway,* also called a *manhole.* It is formed at the factory during manufacture and provides a large circular opening into the primary tank. The manway is provided with a bolted and gasketed cover.

Two methods are used to connect piping: single connections directly installed in the tank wall and multiple connections in the cover of a manway. Typical manways of 22, 30, and 36 in (53, 72, and 86 cm) diameter are available. This manhole may be a full-size opening for tank access or may have several pipe connections in the manhole cover. A *manway riser* is used to extend the manway to grade. This is also called a *manway extension.*

Standard threaded single connections into steel tanks consist of NPT half couplings welded onto the tank. For FRP tanks, they are cast into the wall of the tank. For double-wall tanks there is a connection from the primary tank to the coupling on the wall of the secondary tank. Connections are available for single- and double-contained piping. Connections for flanged pipe are also available. Sizes range from 2 to 8 in (50 to 200 mm) for most tanks, with the most common sizes being 3 and 4 in. It is generally not practical to have the larger-size pipe connections in smaller tanks. There are standard locations on each tank provided by each manufacturer, but they can be located at custom intervals at extra cost.

If the manway will be used for installing equipment and several piping connections, a manway cover with multiple piping connections in a variety of sizes can be provided. Standard fittings are 3 and 4 in NPT. Standard I.D. of manway openings are generally 22, 30, and 36 in, depending on the size of the tanks. A typical manway into an FRP tank and arrangement of multiple connections is illustrated in Fig. 12.1. These connections could be arranged either in a straight line or circular configuration. Typical manways into a single-wall and double-containment steel tank are illustrated in Fig. 12.2.

In order to extend the manway to grade and provide space and access for pipe fittings and equipment connections, a manway riser, or extension, must be attached to the manway. Manway risers are available in a variety of standard sizes ranging from the manway diameter to larger sizes up to 48 in. The manway riser must have an additional waterproof top that is flush with grade. Several methods of accessing extensions from grade are used. Connection to the manway from the manway riser should be waterproof to guard against the entrance of groundwater or rainwater. Various methods are utilized by different manufacturers including collars around the manway and direct bolted and gasketed connections. For steel tanks, the reverse flange type of manway connection is recommended. Typical manway risers are illustrated in Fig. 12.3.

In cases where only piping connections are provided, an enclosure connected to all connections and including an extension to grade will be installed to allow leakage monitoring and access to the connections themselves for maintenance. This arrangement is commonly called a *containment sump* because there is no direct connection of the sump to the wall of the tank. A containment sump is illustrated in Fig. 12.4.

Tank Filling and Spill Prevention

Tank Filling.   USTs are filled by gravity from delivery trucks using a hose connected to the truck and to a fill connection into the tank called a *port.* This operation

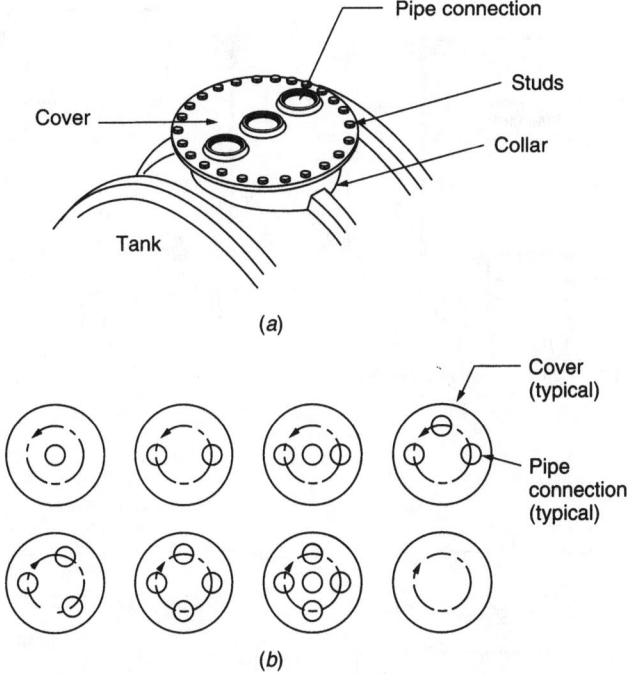

FIGURE 12.1   (*a*) Typical manway into FRP tank with multiple piping connections. (*b*) Typical manway cover fitting configurations. (*Courtesy Xerxes.*)

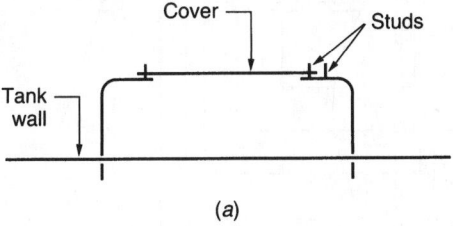

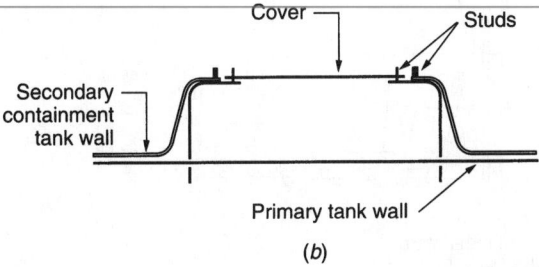

FIGURE 12.2   Typical manway in steel tank. (*a*) Double ring for single-wall tanks. (*b*) Jacketed double ring for total containment tanks. (*Courtesy Highland Tank and Manufacturing Company.*)

22" Dia. (Standard)
32" Dia. (Option)
Dia.
30"
22" Dia.
Dia.

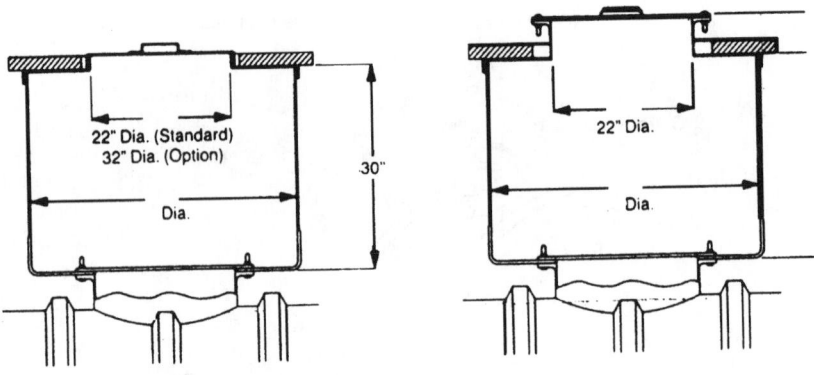

FIGURE 12.3 Typical manway extension.

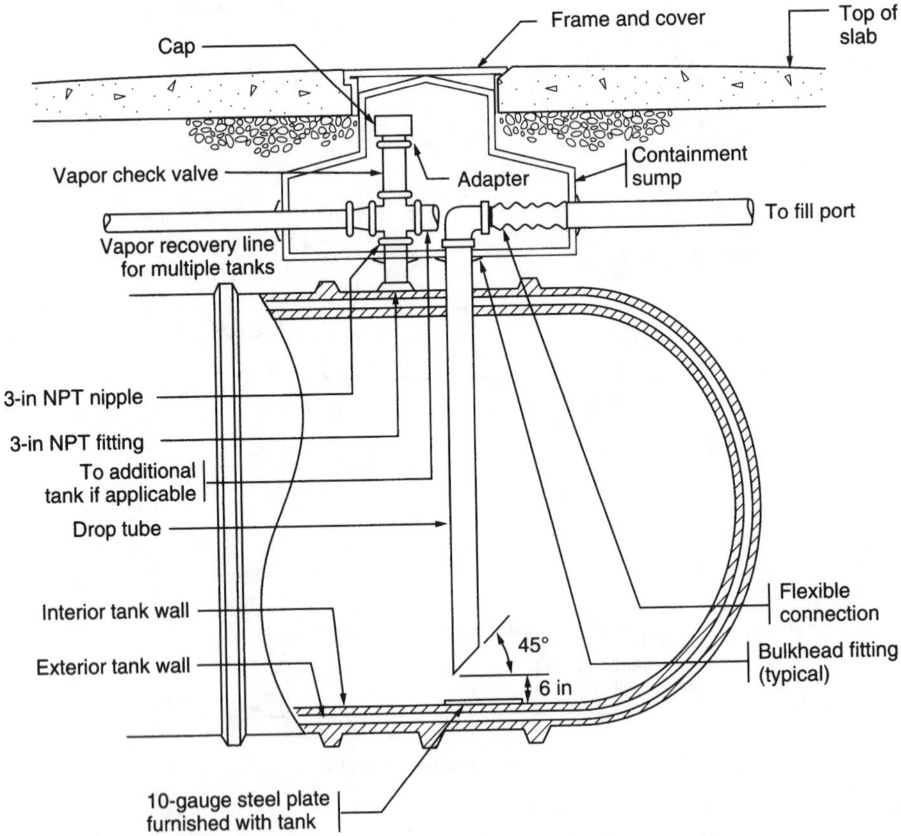

Frame and cover
Top of slab
Cap
Vapor check valve
Adapter
Containment sump
To fill port
Vapor recovery line for multiple tanks
3-in NPT nipple
3-in NPT fitting
To additional tank if applicable
Drop tube
Interior tank wall
Exterior tank wall
45°
6 in
Flexible connection
Bulkhead fitting (typical)
10-gauge steel plate furnished with tank

FIGURE 12.4 Typical containment sump, drop tube, and vapor recovery connection.

is commonly called a *gravity drop* or simply a *drop*. When gravity filling is not possible, truck-mounted pumps are used. If the fill port is located directly over the UST, the assembly is called a *direct fill port*. If there is any horizontal piping from the fill port to the tank, it is considered a *remote fill port*. The fill port covers shall be watertight. Where multiple tanks containing different products are installed, the fill port covers shall be painted different colors to distinguish between the various product ports and also the vapor recovery port if used. Installing the cover plates 1 in (25 mm) above slab level will minimize the possibility of storm water entering the port. A typical fill port assembly is illustrated in Fig. 12.5.

Another integral part of the filling system is the drop tube inside the tank being filled directly connected to the fill line. The drop tube, illustrated in Fig. 12.4, provides a submerged inlet for product during filling. This produces a minimum of turbulence inside the tank, compared to having the product spill directly onto the surface of liquid.

It is common practice for manufacturers to install a wear plate inside the tank under the fill tube outlet to limit internal erosion of the tank bottom by the product stream over the years. Another type of device (not shown) used to divert the product stream for retrofit of existing tanks is a "bottom shield" attached to the bottom of the fill pipe and consisting of four thin rods attached on one end to a horizontal circular plate and the other end to the fill pipe.

Spill Prevention. The purpose of spill prevention is to provide a safe filling method capable of catching spills from delivery hose disconnections. A typical 20 ft, 0 in length of 4-in delivery hose holds 15 gal. Product spilled must be prevented from entering the soil adjacent to the fill port by providing safeguards that are code mandated and recommended by good practice.

This is accomplished by installing a below-grade catchment basin with a capacity of from 3.5 to 15 gal (13.5 to 57 L) to catch spillage of product from truck delivery hose. An optional device that could be included as part of the sump is a drain valve that when opened, will empty product in the sump into the fill line. Water in the sump will be removed manually. Installing a drain valve inside the catchment sump could allow a smaller sump. The containment sump is shown in Fig. 12.5.

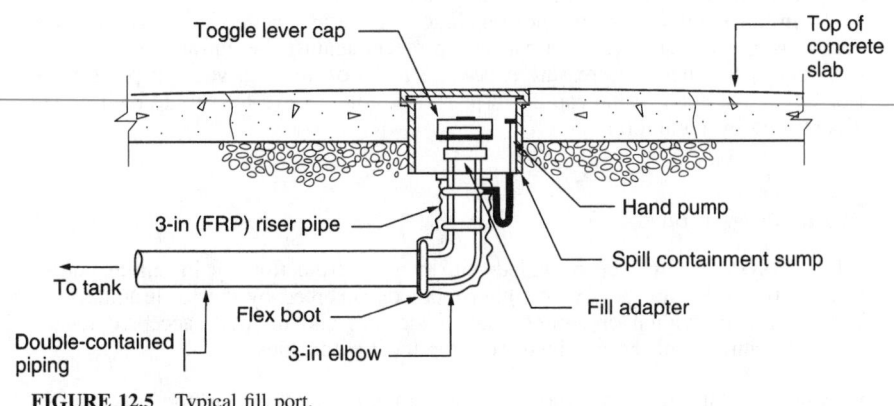

FIGURE 12.5 Typical fill port.

The fill port assembly is designed to accomplish the following:

1. Provide a watertight grade cover, allowing access to the fill hose connection.
2. Provide a fill hose connection for the tanker truck delivery hose. The truck hose end has a standard end connection. This requires that an adapter be installed on the fill pipe leading to the tank. The gravity drop delivers about 200 gpm (750 Lpm). The flow rate of a truck-mounted pump is generally 50 gpm (190 Lpm).
3. Allow any fuel spillage from the fill hose to be contained and returned to the tank and any water to be removed. This is usually accomplished by installing a spill container around the fill port capable of containing from 5 to 15 gal of product. A hand pump removes spilled fuel from the port bottom and returns the fuel into the fill line.

A dry disconnect coupling on the delivery truck hose could also be used to prevent spills.

Another, less commonly used fill method uses a coaxial hose to both fill the tank and recover vapor. This requires only one connection from the truck to the tank fill port and requires a different type of adapter than that used for a fill connection alone.

Atmospheric Tank Venting

The USTs are at atmospheric pressure and require continuous tank venting to assure that no pressure or vacuum is built up inside the tank. This is not to be confused with vapor recovery vents, which serve a different purpose.

Since all of the vapors produced from products are heavier than air, the vapors will not normally escape. Release of vapor can occur as the tank is being filled or as product is removed, or from a buildup of vapor pressure resulting from evaporation of product at times of high temperature. Each tank is vented by means of a dedicated 2-in (50-mm) vent pipe (typical size).

The vent pipe is directly connected to the top of each primary tank and shall be extended to a safe location above the highest level of an adjacent building or a minimum of 12 ft above grade. The vent discharge shall be directed either vertically or horizontally away from other buildings or tanks.

When not in conflict with other regulations, general practice is to have the vent terminate in a pressure-vacuum cap that protects against the entrance of rain and will only open when the pressure exceeds 2 to 15 oz/in^2 or a vacuum pressure of 1 oz/in^2 is exceeded. If the cap is not provided, a flame arrestor should be installed if permitted by regulations.

Overfill Prevention

All UST systems must be provided with overfill protection by installing one or more of the following devices or other methods accepted by the local authorities. The specific individual or combination of devices and methods accepted in any specific location shall be obtained from the local authorities.

1. A device that will alert the operator when the product level in the tank reaches 90 percent full by restricting the flow of product into the tank or alerting the operator by means of an audible alarm actuated by a high-level alarm probe

2. A device to shut down flow into the tank when the product level reaches 95 percent full

3. Equivalent devices or methods accepted and approved by the local implementing authority

The best method is the installation of an automatic device on the fill tube that will reduce flow into the tank when it becomes 90 percent full and stop it entirely when the tank becomes about 95 percent full. There are a number of approved mechanical devices from different manufacturers that perform this function. One such device is illustrated in Fig. 12.6.

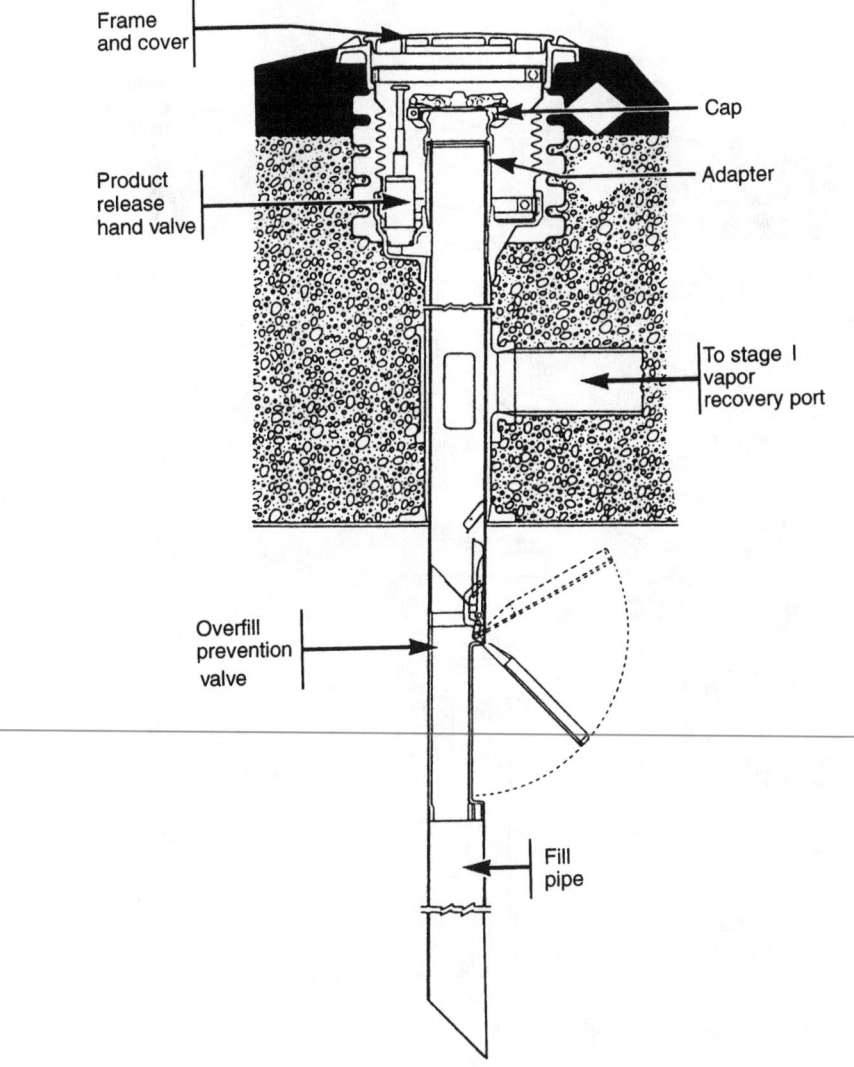

FIGURE 12.6   Automatic overfill protection valve.

Another method is to install an audible alarm actuated by the level gauge that automatically sounds when the liquid level in the tank reaches 90 percent full. Often a visible alarm is included. This alarm must be located in clear sight of the fill port.

In addition, tanks that do not require a vapor recovery system could have a floating ball device installed on the atmospheric vent line that will close the vent when product reaches a predetermined point (usually at the 90 percent full level) when additional filling may cause a spill. By closing, the air pressure will increase inside the tank and restrict the inflow, alerting the operator that the tank is approaching full. An extractor fitting is required that allows access into the line for removal of the float assembly. The float assembly is illustrated in Fig. 12.7.

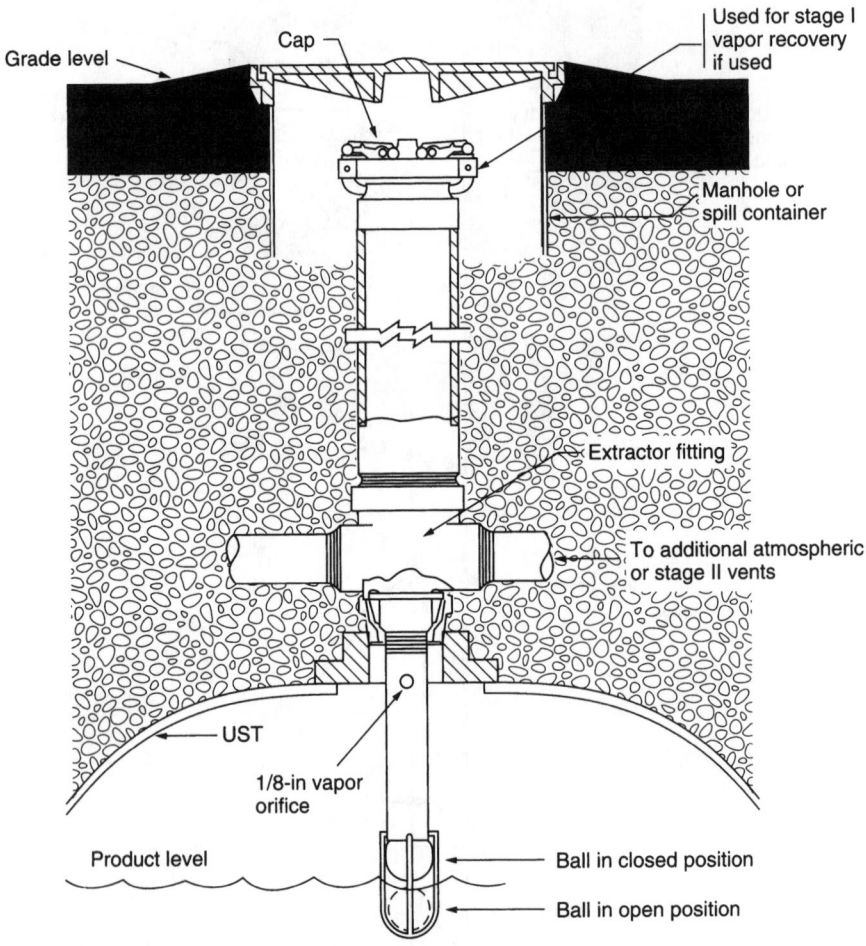

FIGURE 12.7   Ball float overfill protection valve. Stage I vapor recovery vent on UST.

LEAK DETECTION AND SYSTEM MONITORING

General

Leak detection is required by code. The three basic requirements for leak detection are:

1. Leakage must be capable of being detected from any portion of the tank or piping that routinely contains petroleum.
2. The leak detection equipment must meet performance requirements described in Federal Regulations, Sections 280.43 and 280.44.
3. Leak detection equipment must be installed, calibrated, operated, and maintained in accordance with the manufacturer's instructions.

The EPA has established various options, combinations of which must be used depending on project conditions, initial or long-term costs, and product. In addition, state and local requirements may differ in the number and application of these options. A schematic diagram indicating the leak detection methods from tanks and piping is illustrated in Fig. 12.8. These options are:

1. Leakage from tanks
   a. Manual tank gauging
   b. Automatic tank gauging

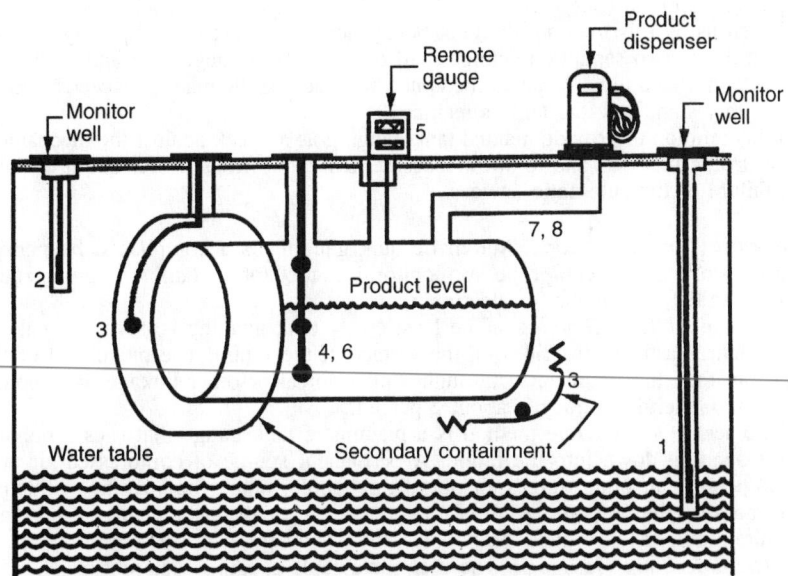

FIGURE 12.8   Leak detection methods for tanks and piping. (1) Groundwater monitoring. (2) Vapor monitoring. (3) Secondary containment with interstitial monitoring. (4) Automatic tank gauging. (5) Tank tightness testing with inventory control. (6) Manual tank gauging. (7) Leak detection for suction piping. (8) Leak detection for pressurized piping.

  c. Tank secondary containment interstitial monitoring
  (1) Vapor monitoring
  (2) Liquid monitoring
  (3) Hydrostatic monitoring
  (4) Pressure monitoring
2. Leakage outside of tanks
  a. Groundwater monitoring
  b. Piping leak detection
3. Tank tightness testing with inventory control

Leakage from Tanks

Measuring leakage from tanks is accomplished by gauging the level of product in a tank and metering the amount of product dispensed and the amount of product delivered. If the dispensed and remaining product figures agree with the amount of product delivered, there is no leakage.

Manual Tank Gauging. Referred to as *sticking,* manual gauging uses a long gauge stick calibrated to ⅛ in (3 mm) lowered directly into the tank until it rests on the bottom. This requires that a straight, direct access into the tank be provided. The wetted level leaves a mark on the stick that is read by the operator after removal from the tank. This method is generally limited to tanks of 2000 gal (7560 L) and smaller, and only in conjunction with tightness testing for tanks with a capacity larger than 550 gal (2080 L).

When using this method, the tank being measured must be completely idle for at least 36 h. Two separate readings shall be taken at the beginning and at the end of that period and must be performed once per week. If the reading exceeds weekly or monthly standards, the tank is leaking.

It is common to provide manual tank gauging as a check against the mechanical or electronic methods and to allow measurement if electrical power is out or there is a failure of the automatic devices.

Automatic Tank Gauging. Automatic tank gauging is accomplished by permanently installing a special probe, or monitor, into the storage tank. The probe could be mechanical, pneumatic, or electronic.

Mechanical Tank Gauging. The least costly tank gauging is a mechanical device, such as a float, that rides on the surface of the liquid. The movement of the float transmits the liquid level through a mechanical or other linkage to a remote indicator. General accuracy is about 2 percent.

Another, more accurate method is a pneumatic tank gauge that uses a bubbler pipe extending down into the liquid. A permanent source of compressed air or a hand-operated air pump forces air out of the bubbler pipe. The operating principle is to measure the pressure required to force air out of the bubbler pipe. The more pressure required to produce the bubbles, the deeper the depth of liquid.

Electronic Tank Gauging. The electronic tank gauging system consists of a probe mounted in a tank opening extending to the bottom of a tank and a remote panel that is microprocessor controlled and could be programmable. The probe is capable of monitoring many parameters and extends from the primary tank bottom to a termination above the tank that both anchors the probe and acts as a junction box for the wiring. For USTs this point is below grade or the slab above the tank.

Access to the box is through a small manhole in the slab. For ASTs, it terminates in a junction box on top of the tank.

The advantage to the electronic gauging system is that it is capable of being programmed to automatically record many items. The probe will monitor such parameters as product and water level inside the tank and product temperature. In addition, probes extending from various locations can be electronically linked together to monitor vapor and liquid leakage from many sources such as monitoring wells, piping, containment sumps, and tank interstitial spaces. Probes from multiple tanks can be linked to a single panel. Overfill and low product levels are also capable of being monitored. The installation of one type of electronic probe is illustrated in Fig. 12.9.

The commonly available types of electronic probes are the magnetostrictive and ultrasonic. The magnetostrictive type uses a change in the magnetic field produced by movable product sensor floats, one for water and one for product. Each float has an integral magnet and is free to ride on the probe shaft as the product (and water) level changes. Signals are sent from a wire in the probe shaft and returned to a device on the top of the probe to gauge levels. The ultrasonic devices use

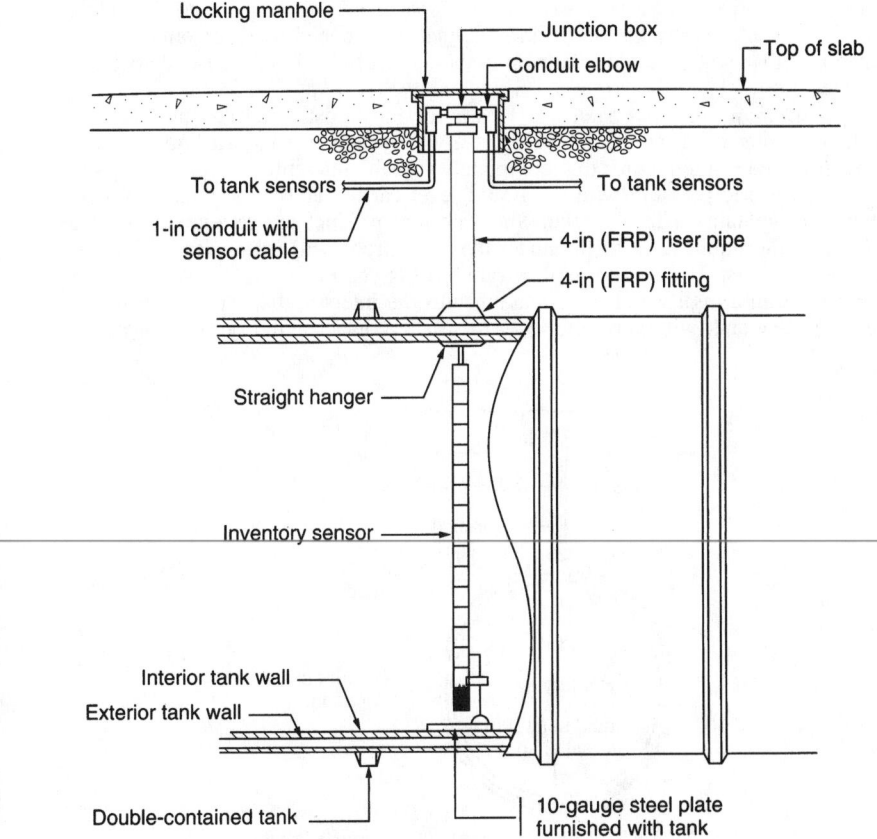

FIGURE 12.9 Electronic fuel gauging detail.

ultrasound waves from the probe on the shaft to a receiver on the top of the probe to signal a change in product level. Capacitance probes are no longer recommended, having been replaced by less expensive and more accurate electronic probes.

Interstitial Monitoring.   The interstitial space between the tanks is used to contain and detect any leakage from the primary tank. Monitoring the interstitial space is divided into two general categories: wet and dry. With dry methods, vapor is monitored for the presence of hydrocarbon vapors; with wet methods, liquid is monitored for the presence of liquid product. Wet methods are also used in pressurized systems using either air pressure or vacuum. Typical locations for dry monitoring devices are shown in Fig. 12.10. The wet system uses hydrostatic monitoring, as illustrated in Fig. 12.11.

Vapor Monitoring.   Vapor monitoring is achieved by placing a probe sensitive to product vapor into the dry interstitial space from a special tank opening. The probe is generally placed one-half the distance into the tank. If leakage occurs, the vapor produced by the product is detected by the probe, and a signal is given that leakage is present. This method has the advantage of not being affected by condensed water.

Liquid Monitoring.   Liquid monitoring is achieved by placing a probe at the bottom of the dry interstitial space that is capable of detecting liquid. If leakage occurs, it is detected by the probe and a signal is produced. A shortcoming of this method is that water condensation into the interstitial space is also detected as leakage.

Hydrostatic Monitoring System.   This is the most costly and the most accurate method of interstitial monitoring. This is a wet system that uses a liquid, usually brine, installed at the factory that completely fills the interstitial space. The system is at atmospheric pressure with the liquid level carried above the tank into a reservoir that contains a level probe. Since normal product temperature differences will cause the liquid level to fluctuate, the level probe will annunciate only unacceptable changes. This system is illustrated in Fig. 12.11.

Water from a high water table leaking into the interstitial space from a hole in the secondary tank will cause the level to rise. Product leaking out of the primary

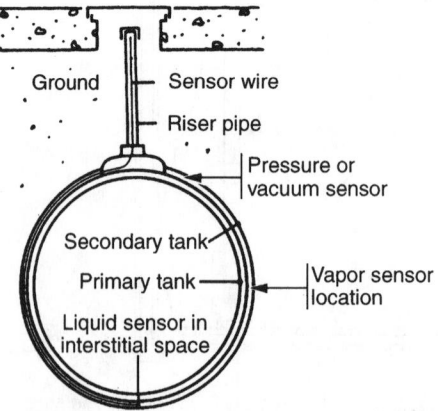

FIGURE 12.10   Dry interstitial monitoring device locations.

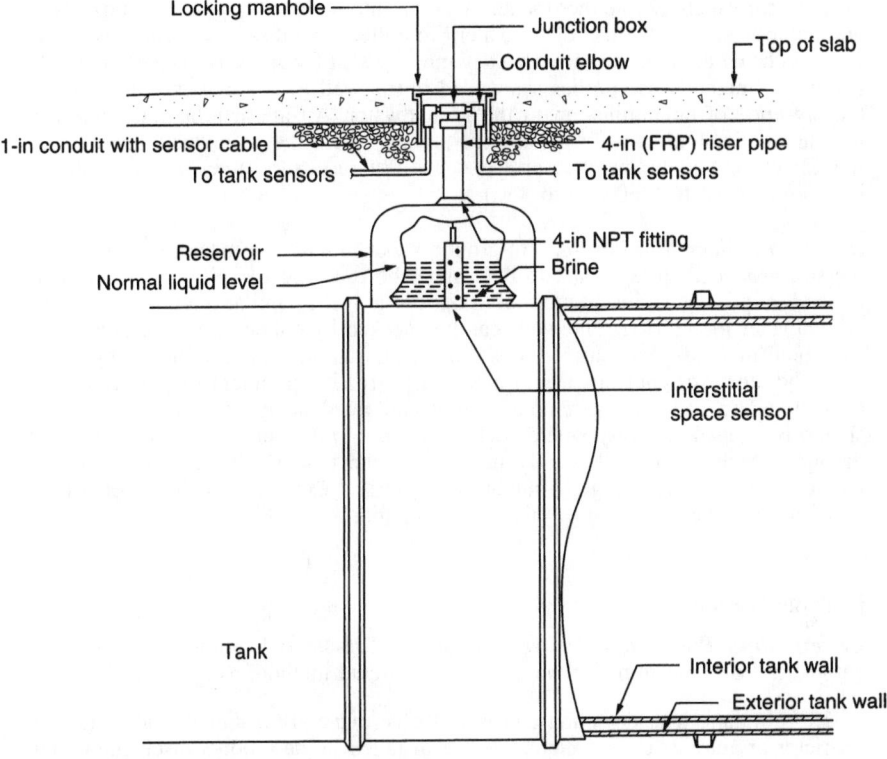

FIGURE 12.11   Detail of hydrostatic interstitial monitoring.

tank into the interstitial space will also cause a rise in level. Conversely, a hole in the secondary tank will allow the brine to leak out of the system.

In order to maintain atmospheric pressure, holes must be made in the standpipe cap. If there is a possibility of a high water table that will allow water to enter the holes, a separate vent line extended to a point safely above grade must be installed.

Pressure Monitoring.   Pressure monitoring of the interstitial is accomplished by injecting light air pressure (1.5 psig) or creating a vacuum in the space. Any increase or reduction of pressure or vacuum indicates a leak that is sent to a remote panel for annunciation. Pressure systems are rarely used due to the increased maintenance required.

Leakage Remote from Tanks

Groundwater Monitoring.   Groundwater monitoring senses the presence of product floating on the surface of groundwater. It can be used only where the products to be measured are immiscible. This method requires the installation of monitoring wells at strategic locations into the groundwater at several locations near the tank. The wells must be properly designed and sealed to eliminate contamination.

They can be checked either manually or by automatic, electronic methods. Manual methods require the use of a "bailer" to collect liquid samples from inside the well and bring them to the surface. They must be sent for analysis once each month. Electronic methods use probes suspended in the well to continuously monitor for the presence of contamination in the groundwater. An alarm is annunciated at a remote panel. Monitor wells are limited to sites where the groundwater level is 20 ft (6.2 m) or less below the surface, with best results obtained from depths of between 2 to 10 ft (0.60 mm to 3 m).

In-Ground Vapor Monitoring.   In-ground vapor monitoring measures "fumes" in the soil around a storage tank to determine the presence of spilled product. This method requires the installation of monitoring wells similar to those used for groundwater monitoring. The wells can be checked by either manual or automatic, electronic methods. Manual methods require air sample gathering and submission to a laboratory for analysis once a month. Electronic equipment uses a probe suspended in the well and a remote, highly calibrated analyzer. Vapor monitoring should be considered only where backfill is sand, gravel, or other similar material through which vapor will readily move from the tank to the monitor. It is not recommended where a high groundwater level and excessive precipitation would interfere with the operation of this system for more than 30 days.

Leakage from Piping

Underground Pressurized Delivery Systems.   Pressurized piping shall have one automatic leak detection method and one additional method, as follows.

1. Automatic leak detection consists of a permanently installed automatic flow restrictor or automatic flow shutoff device installed in the product discharge piping or a continuous alarm system. The continuous alarm could be a probe installed in the secondary pipe continuously monitoring liquid or vapor release of the secondary containment piping system. If the secondary containment terminates in a manway or containment sump, the lower end of the secondary containment pipe shall be open to the sump and allow any leaking liquid to spill into the sump to be detected.

2. Additional methods include one of the following monthly requirements: groundwater monitoring, secondary pipe vapor monitoring, secondary pipe liquid monitoring, or annual pipe tightness testing. A continuous alarm to detect spillage as described above complies with this requirement as an additional method if installed in addition to an automatic detection method.

The following accuracy must be provided for automatic detection: A pressurized shutdown monitoring device that prevents flow of produce when a line tightness test indicates a minimum leakage of 0.10 gph (0.016 Lps) is detected. In addition, the device shall be capable of shutting down the system when a leak of 3 gpm (19 Lps) at 10 psi (68 kPa) is detected. One type of device commonly used is installed on the discharge of the submersible pump in the manway. It must act independently of any other pressurized shutdown monitoring device.

Underground Suction Delivery Systems.   Underground suction piping leak detection is not required if the piping is both sloped back to the tank from the dispenser and if a check valve is installed in the dispenser as close as possible to the vacuum pump. Because of the anticipated increased use of alcohol additives to gasoline, it is general practice to provide double-contained suction piping for gasoline systems. Double-contained piping is not required for diesel fuel, but it is highly recommended.

When the tank is higher than the dispenser, it is necessary to use double-contained piping. A sump is installed at the low point of the piping, and a probe is installed to detect leakage from the secondary pipe.

Tank Tightness with Inventory Control

This is a combination method using periodic tank tightness testing and a monthly inventory control. It is allowed only for the first 10 years of operation. Tank tightness requires that the tank be taken out of service and temporary equipment be installed. One method uses a volumetric test to exactly measure the change of level of product over a period of several hours. Another method uses ultrasound or tracer gas detection techniques to discover a leak in the tank wall. The monthly inventory requires that exact measurement be taken each month of the amount of product delivered compared to the amount stored in the tank and the amount dispensed. If the two figures do not balance, there is a leak.

Leakage into Sumps, Pans, and Other Secondary Containments

Any product leaking from the primary pipe will spill into the secondary containment pipe and flow by gravity to the low point. The piping is pitched downward to a bulkhead, sump, or manway. Double-contained piping penetrates the side of the bulkhead using special fittings called *bulkhead fittings*. These fittings are used to terminate double-contained piping so that the secondary containment is open and free to have liquid spill into the sump to be detected.

Leakage within containment sumps and manways is monitored by means of probes sensitive to liquid or vapor. They are suspended in or attached to the sides of the containment. The liquid probe level is adjusted to signal the presence of liquid. Probes are available to discriminate between water and petroleum products. They are connected by wires to a remote panel for annunciation. A typical containment probe is illustrated in Fig. 12.13.

Inductive sensors are also available that do not require penetration of the secondary containment pipe. This type of sensor is attached directly to the outside of the secondary pipe to detect leakage by interrupting the inductive path of the sensor.

VAPOR RECOVERY SYSTEMS

It is a code requirement that VOC vapors resulting from the displacement of gasoline and gasoline-blended products be prevented from entering the atmosphere. This occurs when storage tanks and motor vehicles are being filled. Diesel fuel, kerosene, waste and motor oil, and heating oil do not require vapor recovery.

Vapor recovery is divided into two phases. Phase I is recovery from gasoline storage tanks, and phase II is recovery from gasoline dispensers.

Phase I Vapor Recovery

Phase I vapor recovery is a separate and independent closed system installed at the storage tank for use only when filling the storage tank. The purpose of phase I vapor recovery is to prevent the escape of VOCs into the atmosphere from the UST. It is used only when filling storage tanks. This can be achieved by means of either a two-point, or coaxial, system.

Two-Point System. The two-point vapor recovery system is a separate and independent closed system. It consists of a separate vent from the UST piped directly to the tanker truck. A separate hose from the truck is connected to an outlet accessed through a vapor recovery fill port adjacent to the product fill port. A typical vapor recovery port is illustrated in Fig. 12.12.

If the vapor recovery connection to the UST is remote, the piping shall pitch back to the tank so that any condensed product will drain back into the tank. The pitch is generally ¼ in by 10 ft (3 m). This piping need not be double contained. If there are multiple USTs, the vapor recovery piping could be manifolded. Connection to the phase II vapor recovery system may be allowed by the responsible code officials.

It is common practice to have the vapor recovery pipe size the same as the fill line. Typically, this size is 3 in (75 mm) and does not require double-contained pipe.

Coaxial Vapor Recovery. This system is a combination system with a single connection point. It is similar in principle to phase II vapor recovery used in dispensers. It consists of a drop tube having a pipe within a pipe and a delivery hose of the same construction. Product is delivered through the center pipe of the delivery hose, and the vapor passes through the outer pipe of the drop tube and continues to the delivery truck through the coaxial delivery hose as the product fills the tank.

Phase II Vapor Recovery

Phase II vapor recovery is a separate and independent closed system installed at the dispenser for use only when dispensing gasoline. Its purpose is to prevent the

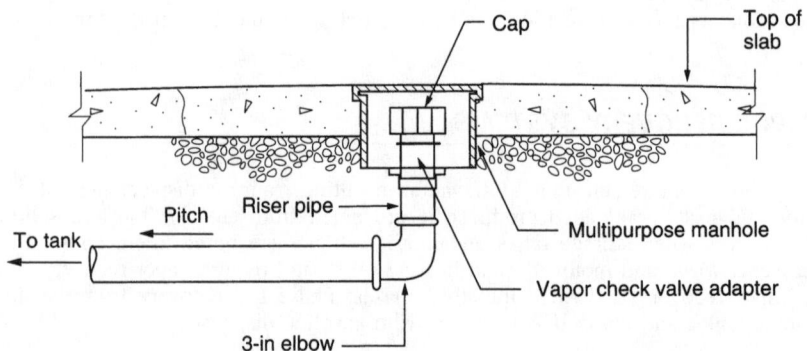

FIGURE 12.12 Typical vapor recovery port.

escape of VOCs from the motor vehicle tank into the atmosphere during tank filling. Phase II vapor recovery can be achieved by either a balanced or vacuum system.

Balanced Vapor Recovery System. A balanced system, which has no moving parts, is used when the tank is lower than the dispenser. The system consists of a coaxial hose from the nozzle to the dispenser, a separate fitting in the dispenser that separates the outer vapor line from the inner product hose and a separate vent pipe from the dispenser routed directly to the UST. The pressure developed by the displacement of air in the vehicle fuel tank is used to force the vapor into the hose. Special coaxial dispensing hoses and a nozzle shroud are ancillary requirements. The dispenser selected for gasoline service should be "vapor recovery ready."

If there are multiple dispensers, the vapor recovery pipe from each dispenser to the UST could be manifolded together. In addition, connection to the phase I vapor recovery system may be allowed by the responsible code officials.

Vacuum Vapor Recovery System. A vacuum system is required when the tank is higher than the dispenser. The system uses a vacuum pump in the dispenser enclosure to draw the vapor from the coaxial dispenser hose and nozzle and provides the pressure necessary to force the vapor and condensed liquid back into the UST.

PRODUCT DISPENSING SYSTEMS

Product dispensing refers only to the transferring of product from the storage tank into the fuel tank of motor vehicles. Other motors, such as those for fire pumps and emergency generators, have integral fuel pumps that supply the engine and often recirculate product from a storage tank.

There are two types of systems used to transfer product from the UST to the motor vehicle: pressure and vacuum. The pressure system uses a pump submerged in the UST to create the pressure needed to transfer the product from the UST to the vehicle. The vacuum system uses a pump installed in the dispenser enclosure to create a suction pressure needed to draw the product from the UST into the vehicle.

For USTs, the pressure system is preferred for the following reasons:

1. Maintenance is much lower compared to vacuum systems.
2. The initial cost is lower.
3. With a vacuum system, there is a practical limit of 10 ft (3 m) suction lift possible from the UST bottom to the highest point of the dispensing system.
4. A submersible pump will supply multiple dispensers. Vacuum pumps must be installed in each individual dispenser.
5. Submersible pumps can deliver much higher flow rates than vacuum pumps.
6. Vapor lock is not possible with a submersible pressure pump system.

Pressure Dispensing System

The pressure system consists of a small-diameter submersible pump installed in the product storage tank that provides the required flow rate and pressure. The submersible pump is sized to fit inside a 4-in (100-mm) tank connection.

Piping extends from the top of the pump directly to the dispenser. Since the pump is screwed into the tank connection, the orientation of the pump outlet is not always in line with the piping connection to the dispenser. A length of flexible hose, usually metallic, is required for the final connection. Larger pumps will fit inside a 6-in (150-mm) tank connection. The assembly shall be enclosed within a manway that will allow access to the electrical work, leak detection options, and the piping connections for future testing and repair. A simplified detail of a typical installation is illustrated in Fig. 12.13.

The system starts operation when the dispenser calls for product. The pump is started, and the product flows to the dispenser nozzle. When the nozzle shuts off, the pump stops. A check valve on the discharge of the pump at the tank prevents product in the piping system from draining back into the tank.

Vacuum Dispensing System

The vacuum dispensing system consists of a drop tube inside the storage tank, a vacuum pump for product installed inside the dispenser housing, piping from the storage tank, and check valves to keep product from draining back into the tank and the dispenser. The primary use of the vacuum dispensing system is for situations in which there is no pitch of the product line from the dispenser to the storage tank.

The system starts operation when the dispenser calls for product. The pump in the dispenser is started, and the product is drawn to the dispenser nozzle. When the nozzle shuts off, the pump stops. A practical limit of 10 ft (3 m) suction lift must not be exceeded from the bottom of the tank to the highest point of the dispenser hose.

A check valve in the product line prevents product in the piping system from draining back into the tank. There are three locations for the check valve: at the base of the product suction line inside the tank (called a *foot valve*), an angle check valve in the suction line mounted above the tank, and a check valve in the dispenser at the vacuum pump. The angle check valve is often used because of its reliability. It is accessible through a manhole installed for accessibility. The foot valve is prone to clogging, and its removal must be made through an extractor filling. It is being used to a lesser extent because clogging is common and removal for servicing is difficult. The check valve in the dispenser enclosure is acceptable by the EPA and is cited as one method that if used, will not require additional leak detection methods if slope back to the tank is possible. It is the most accessible. When used for gasoline, the dispenser-mounted check valve allows product vaporization and resulting vapor lock.

Product Dispenser

Dispensers are commonly available for two flow rates, standard speed for cars, which delivers a flow rate in the range between 7 and 15 gpm, and high speed for trucks which delivers as much as 45 gpm. For passenger cars, the average dispenser discharges approximately 10 gpm (65 Lps) and requires approximately 30 psi (200 kPa). The size hose from the dispenser to the nozzle could be either 5/8 or 3/4 in, with 3/4 in most often used. High flow rates for large fuel tanks require a 1-in (25-mm) size hose.

A wide variety of features can be selected for dispensers. The following are the major components generally found in a dispenser:

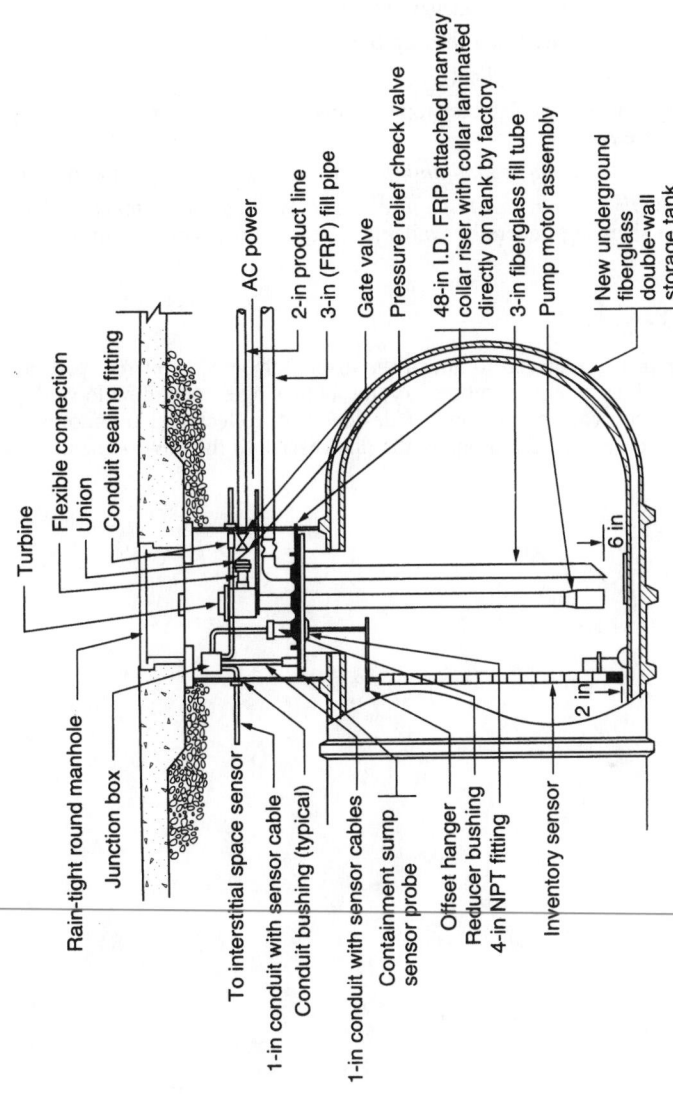

Turbine

Flexible connection
Union
Conduit sealing fitting

Rain-tight round manhole

Junction box

To interstitial space sensor
1-in conduit with sensor cable
Conduit bushing (typical)

1-in conduit with sensor cables

Containment sump
sensor probe

Offset hanger
Reducer bushing
4-in NPT fitting

Inventory sensor

AC power

2-in product line
3-in (FRP) fill pipe

Gate valve
Pressure relief check valve

48-in I.D. FRP attached manway
collar riser with collar laminated
directly on tank by factory

3-in fiberglass fill tube
Pump motor assembly

New underground
fiberglass
double-wall
storage tank

6 in

2 in

FIGURE 12.13 Containment sump with pressure pump and other connections to UST.

12.27

1. *The register.* Used to display the amount of product.
2. *A meter.* Used to register the total gallons.
3. *Location of the hose outlet.* Available in side or front locations.
4. *Hose.* Coaxial type for stage II vapor recovery. A length of 12 ft is average, but 15 ft is available, with breakaway fitting.
5. *A high hose retriever.* Used to keep the hose off the floor.
6. *A high-capacity product filter.*
7. *An emergency shutoff.* Used to stop the flow in the event of a supply or dispenser hose line break.
8. *A dispenser containment mounting pan.* Used when installing the dispenser.
9. *A nozzle with vapor recovery feature where required and automatic shutoff feature.* Two sizes are generally available: ⅝ in and the more common ¾-in.

Dispenser Pan

A dispenser pan is required to attach the dispenser onto a concrete pad or island. The pan provides a liquid-tight entry for both single- and double-wall product piping and electrical conduit and is designed to collect and monitor spills and prevent product leaking from below the dispenser into the environment.

ABOVE-GROUND STORAGE AND DISTRIBUTION SYSTEMS

STORAGE TANKS

Tank Materials

The most often used primary tank is factory constructed of steel and intended for atmospheric pressure conditions. They shall conform to UL-142, Standard for Safety. Other materials, such as FRP, reinforced concrete, and FRP clad steel are seldom used for smaller tanks. Stairs are generally provided to allow inspection and delivery truck operators to reach connections located on top of the tank.

Tank Construction

ASTs are factory fabricated in both round and rectangular configurations. Primary tanks are often manufactured with compartments capable of storing different products. NFPA-30 limits motor fuel AST capacity to 6000 gal (22,680 L). All ASTs must be provided with some form of product leakage and overfill containment conforming to Spill Prevention Containment and Countermeasures (SPCC) regulations. This containment shall be either a dike or impoundment capable of holding 110 percent of tank contents or with integral secondary containment of the primary tank.

Most smaller tanks are provided with an integral secondary containment and an interstitial space. The width of the space varies between different manufacturers. There are several proprietary methods used to construct the secondary containment vault around the primary tank. An often used material is lined concrete, which provides a 2-h fire rating required by UL-2085 if sufficiently thick. Plastic-lined concrete with a wall thickness capable of providing a 2-h fire rating, as required by UL-2085, is generally used for an integral secondary containment. It is a code requirement that the outside of the concrete vault be protected against corrosion, weather, and sunlight. An external secondary containment vault of steel is also used. Insulation between the primary and secondary tank is provided by some manufacturers to protect the primary tank from temperature extremes. This may also be used to meet fire safety requirements for a 2-h rating.

The dimensions of a typical tank with integral secondary containment with 2-h fire rating are given in Fig. 12.14. The dimensions of a typical steel tank with external secondary containment are given in Fig. 12.15. Weight and dimensions are similar to those found accompanying Fig. 12.14.

Corrosion Protection

Since the tank is above ground, the only corrosion protection required for the tank is weather resistance of the tank exterior. This is a code requirement, and each manufacturer has a proprietary method of protecting the outside of the AST. The exposed piping shall be either corrosion-resistant material such as FRP (which must

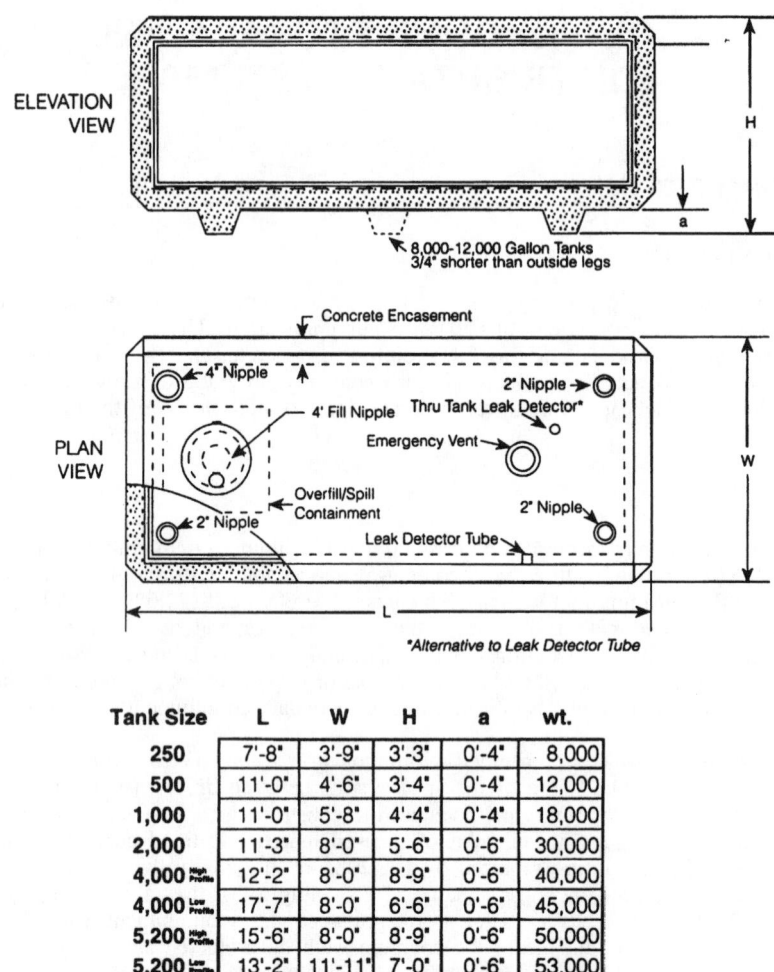

FIGURE 12.14   AST tank weights and dimensions with integral secondary containment. (*Courtesy Convault.*)

Tank Size	L	W	H	a	wt.
250	7'-8"	3'-9"	3'-3"	0'-4"	8,000
500	11'-0"	4'-6"	3'-4"	0'-4"	12,000
1,000	11'-0"	5'-8"	4'-4"	0'-4"	18,000
2,000	11'-3"	8'-0"	5'-6"	0'-6"	30,000
4,000 High Profile	12'-2"	8'-0"	8'-9"	0'-6"	40,000
4,000 Low Profile	17'-7"	8'-0"	6'-6"	0'-6"	45,000
5,200 High Profile	15'-6"	8'-0"	8'-9"	0'-6"	50,000
5,200 Low Profile	13'-2"	11'-11"	7'-0"	0'-6"	53,000
6,000	17'-7"	8'-0"	8'-10"	0'-6"	57,000
8,000	23'-1"	8'-0"	8'-10"	0'-6"	72,000
10,000	28'-7"	8'-0"	8'-10"	0'-6"	87,500
12,000	34'-1"	8'-0"	8'-10"	0'-6"	101,000

also be impervious to UV light), stainless steel, or protected (painted or coated) black steel pipe.

Tank Connection and Access

Connections are located only on top of the tank, and they extend through the vault or secondary containment into the primary tank. There is usually no direct access

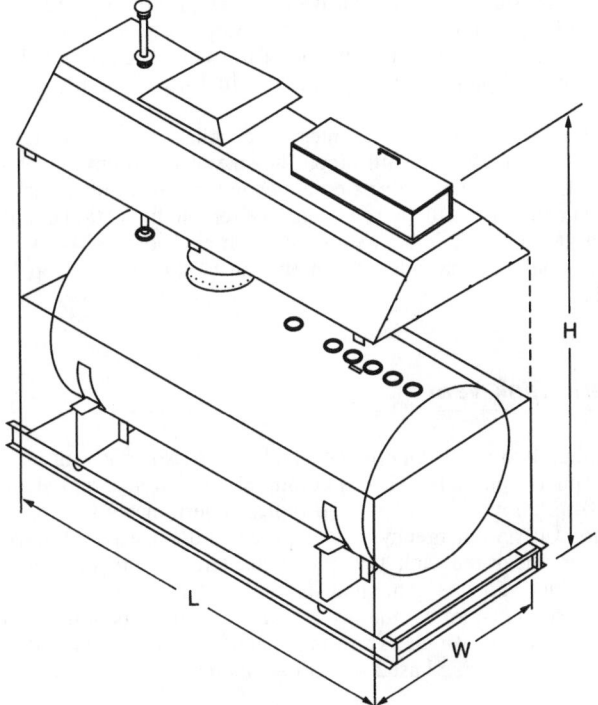

FIGURE 12.15   AST tank with external secondary containment.

for personnel entry into the primary tank except for larger tanks. Standard connections include:

1. Tank vent
2. Emergency vent
3. Product dispenser outlet
4. Product fill (either coaxial or single as required)
5. Phase I vapor recovery (if necessary)
6. Tank gauging
7. Leak detection

Tank Filling and Spill Prevention

Tank Fill.   The tank is filled using a fill port assembly built into the tank encasement or containment at the top of the tank. The delivery truck must have a pump, and the operator is required to climb stairs to make the connection. The fill port cover could be locked if desired. In order to reach the top-mounted connection with the hose, stairs are provided as part of the installed tank. If a coaxial vapor recovery system is used, only one connection to the tank is required. If a two-point system is used, two connections are required. An optional, ground-level remote tank filling station is usually available as a separate piece of equipment to allow convenient filling without having to connect to the top of the tank. The remote fill station could have a self-contained pump or use the built-in fuel truck pump.

Spill Containment. For tanks with integral secondary containment, there is a containment sump surrounding the fill pipe. The size of the sump ranges from 5 to 15 gal (19 to 57 L). For tanks with external secondary containment, spills will enter the containment and be manually removed. The remote fill station could be provided with an integral spill containment sump to catch any hose spills. A small hand or electric pump could be provided to empty the containment sump back into the primary tank.

Atmospheric Tank Venting

An AST requires two vents. One is the standard atmospheric vent used to keep the tank at atmospheric pressure. This is commonly a 2-in size and shall extend to a point 12 ft, 0 in, above grade. The end typically terminates in a pressure-vacuum cap. The second is an emergency vent required to depressurize a tank if there is a fire close to or under the tank that raises the temperature to a point where the amount of product vapor is generated faster than the atmospheric vent is capable of passing. These vents are commonly a 6 or 8 in size, depending on the tank size and number of gallons stored. The tank manufacturer will provide the required emergency vent with a size based on API standards.

Overfill Prevention

Overfilling is prevented by automatic or manual means installed directly on the tank. Automatic overfill prevention uses an overfill preventing valve similar to that previously described in the UST subsection. Manual methods include a direct reading level gauge installed in sight of the operator or an audible high-level alarm activated by a separate probe installed inside the tank. Alarms shall operate when the product level reaches 90 percent of capacity and stopped when the level reaches 95 percent.

LEAK DETECTION AND SYSTEM MONITORING

Leakage from Tanks

Above-ground tanks require a method of containing any possible product release and preventing contamination of the adjacent environment. Releases can result from small leakage from the tank or catastrophic failure of the tank. Required containment methods include providing either a dike completely around the tank, remote secondary containment, or integral secondary containment.

For ASTs without secondary containment, a dike shall be provided. Dikes are required to contain 110 percent of tank capacity and to be constructed of materials such as concrete, steel, or impermeable soil designed to resist the full head of water.

They must be constructed in conformance with NFPA-30. Discharge from the dikes must have a separator along with the necessary control valves that may have to be self-actuating to conform with local codes. It is recommended that an additional impoundment basin be constructed at least 50 ft (16 m) from the AST and at a safe distance from other buildings or tanks. The purpose is to capture and isolate any flammable liquids released during a fire or tank failure and remove it to a safe distance away from the AST. For ASTs, dikes are seldom used because remote and integral secondary containment have a far less initial cost.

Remote secondary enclosures are usually made of steel. They are of the totally enclosed type and are sealed in a manner so as to prevent the entrance of rainwater. They are required by code to provide 110 percent nominal capacity of the primary tank.

Tank integral secondary containment is achieved by enclosing the primary tank with an integral containment, usually steel or reinforced concrete. This type of tank has an interstitial space that is monitored for leakage in the same manner as the USTs discussed previously.

System Monitoring

System monitoring consists of product-level gauging and leakage annunciating. The AST systems can be monitored either manually or electronically.

Product-level gauging in the tank could be achieved by the use of a visual level gauge or an electrical gauge either mounted on or immediately adjacent to the tank or at a remote location. Level gauges similar to those installed in USTs can be used. Remotely mounted electronic gauges capable of recording and placing in memory many functions using probes similar to those installed in USTs are commonly used.

Leak detection for ASTs is much easier than for USTs because leakage from the tank can easily be observed manually. Automatic means of system monitoring include a stand-alone alarm panel or an alarm integral to an electronic panel used for product-level indication.

VAPOR RECOVERY

Stage I and II vapor recovery for gasoline and gasoline blends is required. For stage I recovery, either coaxial or two connections from a delivery truck are necessary during the filling operation. Stage II vapor recovery for tank-mounted dispensers are usually integral. For remote dispensers, a separate vapor recovery line is required, connecting from the dispenser to the tank.

PRODUCT DISPENSING SYSTEMS

For ASTs, the dispenser is usually directly connected to the tank or is located a very short distance away. The vacuum system is usually preferred for these installations because it is lower in initial cost, and due to the short piping runs and single dispenser, most of the objections discussed in the UST subsection do not apply. It is important to include an antisiphon valve to all AST dispensing systems.

For ASTs, the dispenser could be mounted either on the tank or as a separate, remote dispenser (similar to those used for a UST), and available with vacuum or pressure systems (Fig. 12.16a).

The tank-mounted dispensing system consists of a submersible pump, the complete dispenser (nozzle, hose, integral vapor II recovery system, base mounting method, and safety features), product pump, and interconnecting piping. A detail of a typical tank-mounted installation is illustrated in Fig. 12.16b.

TANK PROTECTION

All ASTs located adjacent to a road or subject to a possible collision shall be adequately protected. Accepted means are concrete barriers or bollards. Bollards are the most often used and are similar to those used to protect fire hydrants. Bollards are illustrated in Fig. 6.47.

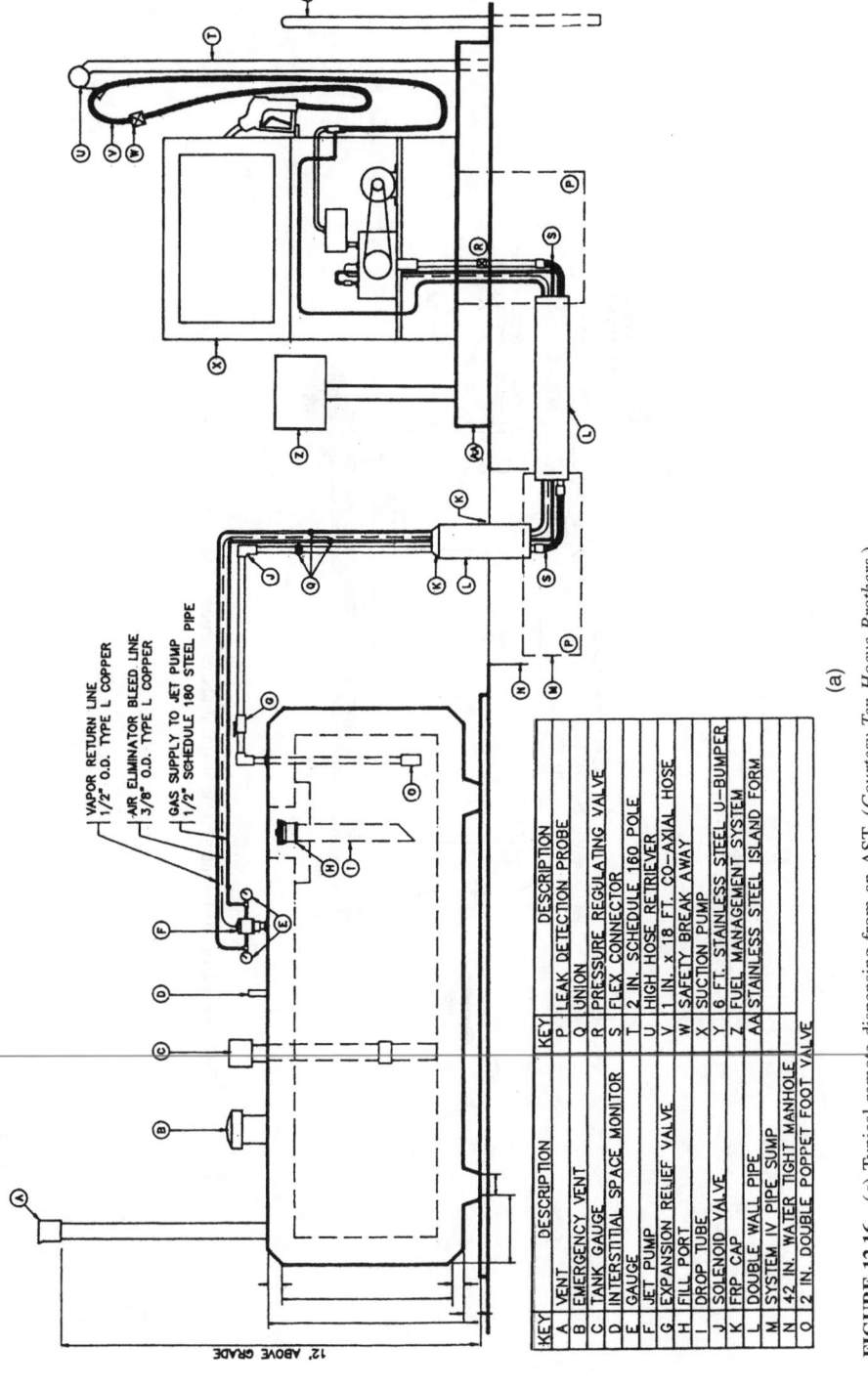

FIGURE 12.16 (*a*) Typical remote dispensing from an AST. (*Courtesy Ten Hoeve Brothers.*)

(a)

VAPOR RETURN LINE
1/2″ O.D. TYPE L COPPER

AIR ELIMINATOR BLEED LINE
3/8″ O.D. TYPE L COPPER

GAS SUPPLY TO JET PUMP
1/2″ SCHEDULE 180 STEEL PIPE

12′ ABOVE GRADE

KEY	DESCRIPTION
A	VENT
B	EMERGENCY VENT
C	TANK GAUGE
D	INTERSTITIAL SPACE MONITOR
E	GAUGE
F	JET PUMP
G	EXPANSION RELIEF VALVE
H	FILL PORT
I	DROP TUBE
J	SOLENOID VALVE
K	FRP CAP
L	DOUBLE WALL PIPE
M	SYSTEM IV PIPE SUMP
N	42 IN. WATER TIGHT MANHOLE
O	2 IN. DOUBLE POPPET FOOT VALVE

KEY	DESCRIPTION
P	LEAK DETECTION PROBE
Q	UNION
R	PRESSURE REGULATING VALVE
S	FLEX CONNECTOR
T	2 IN. SCHEDULE 160 POLE
U	HIGH HOSE RETRIEVER
V	1 IN. x 18 FT. CO−AXIAL HOSE
W	SAFETY BREAK AWAY
X	6 FT. STAINLESS STEEL U−BUMPER
Y	SUCTION PUMP
Z	FUEL MANAGEMENT SYSTEM
AA	STAINLESS STEEL ISLAND FORM

12.35

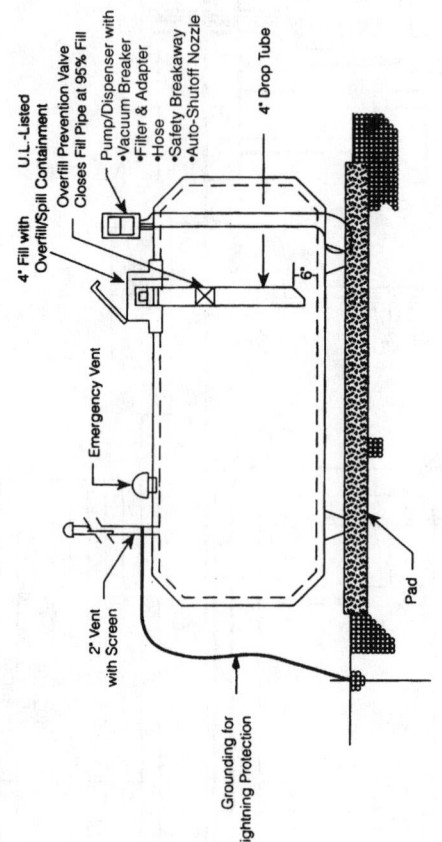

4' Fill with
Overfill/Spill Containment

U.L.–Listed

Overfill Prevention Valve
Closes Fill Pipe at 95% Fill

Pump/Dispenser with
•Vacuum Breaker
•Filter & Adapter
•Hose
•Safety Breakaway
•Auto-Shutoff Nozzle

4' Drop Tube

Emergency Vent

6"

2' Vent
with Screen

Pad

Grounding for
Lightning Protection

(b)

FIGURE 12.16 (b) Typical tank mounted above-ground dispenser. (*Courtesy Convault.*)

MISCELLANEOUS MOTOR FUEL STORAGE

FIRE PUMP FUEL STORAGE

General

Diesel fuel is often used to supply fire pumps and emergency generator motors. This section will discuss storage facilities for these purposes. The driver most often used to power fire pumps other than an electric motor is a diesel fueled engine. This section will discuss the storage of diesel fuel for that purpose.

Codes and Standards

The principal code is NFPA-20, Fire Pumps. In future discussions, this will be referred to as "code." Other codes may be local fire and building codes, as well as other local regulations governing the storage of diesel fuel.

General Design Considerations

The diesel fuel is stored in a tank whose bottom is elevated about 3 ft, 0 in, above the floor. The fuel is fed by gravity to the diesel engine that has an integral fuel pump as part of the engine assembly.

The code recommends that diesel fuel be stored indoors adjacent to the fire pump if allowed by local regulations. The following discussions concern only an indoor tank. If the storage containment is a remote underground or above-ground tank, the requirements are the same as previously discussed for AST or UST installations. The above-ground tank shall be UL approved.

The minimum amount of actual storage mandated is 1 gallon per pump horsepower. Approximately 10 percent additional tank capacity should be added to the calculated storage to allow for freeboard.

The tank should be placed on legs to elevate it as high as reasonable to provide static height for gravity feed to the engine. A generally accepted distance is about 3 ft, 0 in, above the floor. The entire tank shall be protected by a dike high enough to contain the entire contents with an additional allowance of several inches freeboard. It is recommended that a fire extinguisher be placed nearby for safety.

A fill line connection is factory installed at the top of the tank. Since the fill line shall extend outside the building to allow the pumper delivery truck to connect to it, the connection point will be placed lower than the top of the tank. Because of this, an antisiphon valve shall be placed on the high point of the pipe run to the fill line to prevent accidental emptying of the tank. A special "quick disconnect" connection shall be provided. If the fill line is in-ground, a check valve connection on the fill line shall be installed.

Other connections on top of the tank are for the vent and gauge. The vent, usually with 2 in size, must extend from the tank to a point 12 ft, 0 in, above grade and in a safe location. A fuel-level gauge connection is provided that will permit a visual gauge to be mounted directly onto this connection. In this location the gauge must be read directly in front of the tank. If a more convenient gauge location is desired, a remote gauge can be placed on an adjacent wall near the fill line if conditions allow. Interconnection tubing must be installed from the tank connection to the gauge. Although having a higher cost, the remote gauge, located at the fill connection if possible, will allow the delivery person to accurately observe the level of fuel in the tank to avoid overfilling.

Interconnecting fuel supply and return piping from the tank to the engine is generally made of copper tube.

Storage tank dimensions for one manufacturer are given in Fig. 12.17. A diesel fuel storage system for fire pumps from one manufacturer is shown in Fig. 12.18.

EMERGENCY GENERATOR FUEL STORAGE

Emergency generators could have fuel stored either in AST or UST systems. When stored inside buildings, the capacity of the tank is limited to a maximum of 660 gal. This amount of fuel shall be stored in a double-contained tank either remotely or as an integral part of the base of the generator. It is common practice for integral tanks to have an interstitial space capable of containing 110 to 140 percent of the tank capacity. When a remote storage tank is used, it is a good practice to provide

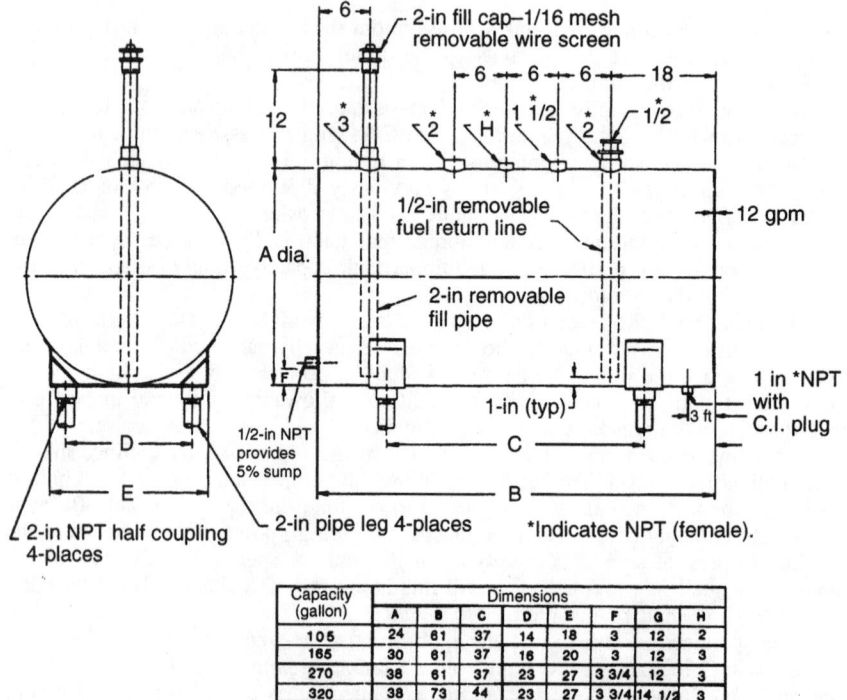

Capacity (gallon)	Dimensions							
	A	B	C	D	E	F	G	H
105	24	61	37	14	18	3	12	2
165	30	61	37	16	20	3	12	3
270	38	61	37	23	27	3 3/4	12	3
320	38	73	44	23	27	3 3/4	14 1/2	3
515	48	73	44	30	34	4 3/4	14 1/2	4

FIGURE 12.17   Typical fire pump fuel storage tank. (*Courtesy Fairbanks Morse.*)

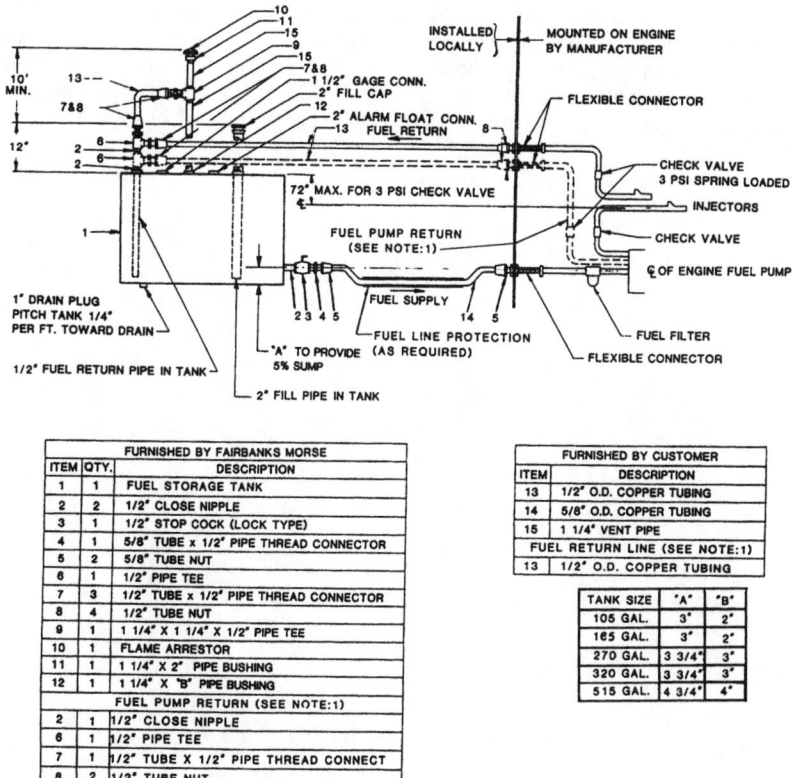

FIGURE 12.18   Typical diesel fire pump fuel system.

a day tank containing 2 h fuel supply immediately adjacent to the generator. The installation must conform to all local rules and regulations previously discussed.

The fill port could be mounted at the generator or remotely. For a remote fill, the inlet should be in a locked box so that the public or a casual passer-by could not have access to the inlet. On a private site, this will avoid tampering by employees. Another method is to place the fill port about 9 ft (3 m) above grade in an adjacent wall. Although this makes it difficult for delivery personnel to gain access, it also makes it much more difficult for the public to try to gain access to the inlet.

Spillage onto grade due to overfilling is prevented by installing visual and audible alarms adjacent to the fill port that signal at the 90 and 95 percent fill levels. The alarms are usually separate to have redundancy, but a single, dual alarm is available. Secondary protection is obtained by providing the nozzle on the delivery hose with an automatic shutoff similar to that on a gasoline dispensing nozzle. Leakage into the interstitial space of integral storage tanks is usually signaled by a float arrangement or probes similar to the previously discussed alarms. The tank vent is routed away from the storage tank, 12 ft (4 m) above grade, either above the generator set or on an adjacent wall.

The fill pipe and other fuel-containing interconnecting pipe should be double contained to eliminate potential leaks. The line should be pitched to a low point. A tee at the low point, with a probe connected to an alarm, will detect any primary pipe leakage. Vent piping is not required to be double contained. The end of the vent should be covered with a mesh screen to prevent entry of bugs and debris.

SYSTEMS DESIGN

PIPING MATERIALS

Piping above ground from an AST is for the vent, product delivery, and for tank fill from a remote location. The most common material is A-53 steel with threaded joints and factory-applied corrosion protection. The pipe shall be coated at the factory with an accepted and proven corrosion paint or coating. A common practice is to use a baked-on powder. Another material used is stainless steel where corrosion protection and strength is required. FRP with ultraviolet protection added to the pipe is another often-used material. Galvanized steel pipe is not considered acceptable. Adapters are used to connect steel pipe to FRP if an underground run to a remote dispenser is necessary.

Because of the requirement to cathodically protect underground steel pipe, interconnecting piping installed for new and replacement USTs is almost exclusively plastic or FRP piping with secondary containment. Requirements of the implementing authority must be verified regarding approval of the specific piping material selected.

Plastic piping could be divided into two general types, flexible and rigid. Flexible pipe is generally manufactured from proprietary materials. If it is approved by UL and/or FM, it is generally acceptable. Flexible piping materials have found limited use and acceptance. In addition, the joints and connection methods should also be closely examined for strength, ease of installation, and corrosion resistance.

Rigid FRP piping with an epoxy interior lining has been widely used and accepted and is considered the piping material of choice. The primary pipe is assembled with socket type fittings and epoxy cement. The outer (secondary containment) pipe is the same material as the primary pipe manufactured in two half sections with a longitudinal flange. It is assembled after the primary pipe is tested using cement placed on the adjacent flanges with nuts and bolts installed to hold the two half sections together until the cement dries. The secondary containment pipe is shown in Fig. 12.19.

Flexible pipe connectors are used to connect piping runs to sumps and manways to allow for settlement. In addition, because submersible pumps are screwed into a tank connection, the product discharge will not always face the direction of the

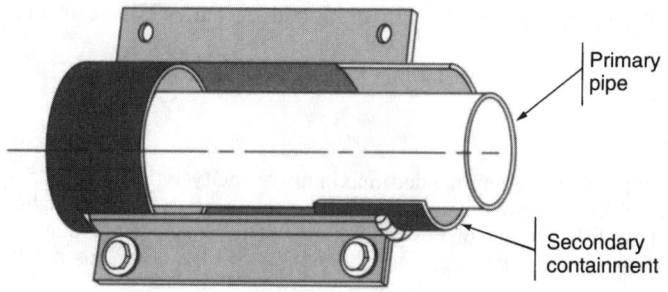

FIGURE 12.19   Secondary containment for FRP piping.

piping run to the dispenser when tightened. Flexible connectors are necessary inside manways to connect submersible pump discharge to the dispenser supply piping.

Because leakage is visible, double-contained piping is not required for ASTs if they are located within a dike or inside a remote containment. If the pipe run is underground, it shall be double contained with leak detection capability.

PIPE SIZING

Pipe sizing is based on the flow rate of the product, the allowable friction loss of the fluid through the system, and fluid velocity. This is an iterative procedure done in conjunction with selecting the size of the product pump.

Information Required

The procedure for sizing the dispensing system involves determining the following before the actual pipe is sized:

1. The dispenser location and type and ancillary devices
2. The pipe material
3. The layout of the piping system
4. The storage tank size and location
5. The type of product pump system, either suction or pressure

Flow Rate

For ordinary applications, the typical discharge flow rate to motor vehicle from average dispensers is 8 to 10 gpm (30 to 38 Lpm). High-rate dispensers for buses and trucks are available with a discharge of up to 45 gpm (285 Lpm).

Simultaneous-Use Factor

The number of dispensers likely to be used at once is usually provided by experience. When no experience exists, multiple dispensers of up to four is normally considered a 100 percent use factor. For more than four, a 75 percent simultaneous-use factor is used.

Velocity

For FRP piping, the recommended maximum velocity is less than 7.5 fps. Maintaining a velocity below this rate keeps the pressure rise from water hammer to a safe level of 150 percent of design pressure. This is necessary due to the quick closing of the dispenser valve. For steel pipe, a velocity of 8 fps has been found acceptable.

Piping Friction Loss

Friction loss of product through piping with 1¼ to 2½ in (125 to 250 mm) size is found from Table 12.5. Using the established flow rate, select the allowable friction loss based on pipe size and the selected product.

For preliminary sizing purposes only and if specific tables are not available, most products are close enough in viscosity to that of water to use standard water charts to obtain a sufficiently accurate friction loss figure. For FRP pipe, decrease the friction loss by 10 percent to obtain a more accurate figure.

TABLE 12.5   Friction Loss of Product through 100 ft Steel Pipe

Pipe size, in	Velocity, ft/s	gpm	Friction loss* Gasoline	Friction loss* Diesel	Pipe size, in	Velocity, ft/s	gpm	Friction loss* Gasoline	Friction loss* Diesel
1½	0.84	6	0.3	0.5	2½	3.92	60	3.1	5.0
	1.26	8	0.7	1.1		4.56	70	4.1	6.6
	1.57	10	0.9	1.5		5.23	80	5.4	8.4
	1.80	12	1.4	2.3		5.80	90	6.5	10.3
	2.36	15	2.1	4.0		6.54	100	8.3	12.4
	3.15	20	3.3	5.7		7.10	110	9.9	14.5
	3.80	25	5.2	8.5		7.84	120	11.7	17.1
	4.72	30	7.2	14.7		8.48	130	13.6	19.7
	6.30	40	12.4	19.2		9.15	140	15.6	22.5
	7.67	50	18.9	28.4		9.51	150	17.8	25.4
	8.44	60	24.7	39.4		11.76	180	25.2	35.3
	11.02	70	36.9	52.5		13.07	200	30.9	42.6
	12.50	80	44.3	63.0		15.69	240	43.8	53.1
	14.71	90	56.5	81.3		16.95	260	59.0	61.1
	15.74	100	71.5	91.0		19.61	300	67.4	86.5
2	2.5	25	1.5	2.6	3	4.5	100	2.6	4.46
	3.0	30	2.1	3.6		5.4	120	3.9	6.18
	3.6	35	2.8	5.2		6.3	140	5.3	8.04
	4.0	40	3.6	5.8		7.2	160	6.7	10.06
	5.1	50	5.4	8.6		8.1	180	8.4	12.44
	6.1	60	7.6	11.9		9.0	200	10.2	15.0
	7.1	70	10.2	15.6		10.2	225	12.9	18.4
	8.2	80	13.1	19.8		11.3	250	15.5	22.3
	9.2	90	16.3	24.4		12.5	275	18.9	26.8
	10.2	100	20.0	29.2		13.6	300	22.2	31.0
	11.2	110	24.3	34.7		14.8	325	25.9	35.6
	12.2	120	28.6	40.7		17.0	375	34.3	41.0
	13.2	130	33.4	46.9		18.2	400	39.0	52.0
	14.3	140	39.5	53.5		19.3	425	43.5	58.0
	15.3	150	43.9	60.6		21.6	475	54.4	71.0
	20.4	200	76.6	101.3		22.7	500	60.3	78.0

*For FRP pipe, reduce friction loss figure by 10 percent.

SUBMERSIBLE PUMP SIZING

The submersible pump is suspended in the product storage tank with the impeller near the bottom of the tank and the motor, piping connections, and electrical work exposed in a manway at the tank top. The pump is sized according to the flow rate of the product and the total head required.

To find the flow rate: Calculate the total gallons per minute from each section of the product line. This is done using the flow rate from the selected dispensers and the simultaneous-use factor for the number of dispensers that may be used at the same time. This will give the gallons per minute (liters per minute) requirement for the pump.

To find the total pump head required:

1. Calculate the height from the bottom of the storage tank to the high point of the dispenser hose, including the elevation of the high hose dispenser.
2. Find the friction loss of the product flow through the distribution piping up to the dispenser, based on the flow rate calculated. This figure must include the equivalent length of run and other losses through fittings and all other connected devices. Most figures are obtained from manufacturers. To find the pressure loss through a submersible pump leak detector, refer to Table 12.6. To find the pressure loss through the dispenser assembly, refer to Table 12.7. To find the pressure loss through the dispenser hose, refer to Table 12.8.
3. Obtain the recommended pressure required for proper operation of the selected dispenser. A typical figure used is 25 to 30 psi (170 to 205 kPa). This figure includes losses through the nozzle, hose, strainer, and so on.
4. Add all of the above figures together to calculate the total head required.
5. For a system head loss calculation sheet containing a checklist of all fittings and devices, refer to Fig. 12.20. This checklist contains items that may not be used for all installations.

From the manufacturer of the selected pump, obtain the pump curves, and select the pump based on the calculated head and flow rate.

Vacuum Pump Sizing. The vacuum pump size is an integral part of the dispenser, installed inside the dispenser above ground level. Since this system serves only one dispenser, the gallons per minute (liters per minute) flow rate and the highest al-

TABLE 12.6 Pressure Loss through Submersible Pump Leak Detector

Flow rate, gpm	Head loss, ft
10	2
20	5
30	6
40	13
50	18
60	22
70	28

Source: Courtesy Marley Pump Co.

TABLE 12.7 Pressure Loss through Typical Dispenser Assembly

Flow rate, gpm	Head loss, ft
3	15
5	23
7	32
9	42
11	53
13	64

Source: Courtesy Marley Pump Co.

TABLE 12.8   Pressure Loss through Dispenser Hose (Friction Loss in Feet Head per 100 ft of Smooth-Bore Rubber Hose)

gpm	\frac{3}{4}	1	1\frac{1}{4}	1\frac{1}{2}	2	2\frac{1}{2}	3	4
				Actual inside diameter, in				
15	70	23	5.5	2.5	0.9	0.2		
20	122	32		4.2	1.6	0.5		
25	182	51	15	6.7	2.3	0.7		
30	259	72	21.2	9.3	3.2	0.9	0.2	
40		132	23	15.5	5.9	1.4	0.7	
50		185	55	23	8.3	2.3	1.2	
60		233	81	32	11.8	3.2	1.4	
70			104	44	15.2	4.2	1.8	
80			134	55	19.6	5.3	2.5	
90			164	70	23.0	7	3.5	0.7
100			203	83	29	9.1	4	0.9
125			305	127	46	12.2	5.5	1.4
150			422	180	62	17.3	8.1	1.6
175				230	85	23.1	10.6	2.5
200				304	106	30	13.6	3.2
250					162	44	21	4.9
300					219	62	28	6.7
350					252	83	39	9.3
400						106	48	11.8
500						163	74	17.1

Project _____ Specific gravity _____ Temp. _____
Liquid _____ gpm Range _____ Viscosity _____
A. Static Losses (ft)
   Tank Diameter   _____
   Bury Depth   _____
   Height High Hose Reel   _____
     Total Static Height   _____ ft
B. Run of Pipe. Pipe Size   _____
   Measured Run   _____ ft
   Fittings   _____
   Valves   _____
   Reducers and Enlargements   _____
     Total Equivalent Run   _____ ft
C. Pipe Friction Loss = Equiv. Run   _____ × friction loss/100 ft
D. Equipment Loss (in feet head)
   Meter   _____
   Dispenser   _____
   Filter and Leak Detector   _____
   Hose and Nozzle   _____
   Strainer   _____
     Total Equipment Losses   _____ ft
E. Total Head Loss (A + C + D)   _____

FIGURE 12.20   System head loss checklist.

lowable suction lift figure must be obtained from the manufacturer of the dispenser selected. Based on the flow rate and pipe size, the suction lift calculated must be below the maximum allowed.

GENERAL DESIGN CONSIDERATIONS

For pressurized product pipe, it is common practice to use a minimum of a 2-in (100-mm) size, increasing it only if the system under design requires a larger size based on a higher flow rate or if the difference in friction loss will allow the selection of a lower horsepower submersible pump. It is generally accepted practice to not use pipe sizes smaller than 1¼ in (30 mm) in size. A larger-size product pipe is generally used in order to lower the head requirements of the pump selected.

For vacuum pump systems, the vacuum pump manufacturer will recommend a pipe size for a specific installation. This is required in order to limit the total static lift plus friction loss figure below that required by the specific vacuum dispenser pump selected.

TESTS

Testing of all piping and the UST at the time of installation is critical to assure that no leakage of product could occur and to check the integrity of the pressure-bearing components. In addition, tests of the UST for deformation after installation and corrosive coating damage are also necessary.

Testing of the Storage Tank

The UST shall be pressure tested prior to placing the tank into the excavation and again after backfilling is complete. If a factory-installed hydrostatic interstitial monitoring system is used, a test before installation is not required since a visual check of the leak detection system will disclose any problem.

The pressure test prior to installation consists of maintaining 5 psig of air pressure for 2 h with no lowering of pressure permitted. For this test all tank openings shall be sealed and a soapy water solution applied to all connections to observe for bubbles, which would indicate leakage. If the tank has a coating for corrosion protection, this shall be checked with an electronic device that will disclose imperfections in the coating, called *holidays*, that must be repaired.

After installation, a hydrostatic test of the tank shall be performed with water at a pressure of 5 psig for a period of 30 min, with all piping isolated so that only the tank is under pressure. It is important to remove all traces of water prior to filling with product.

Testing of the Piping Network

All piping containing product shall be tested hydrostatically at a pressure of 100 psig for a period of 30 min with no leakage allowed. Containment piping shall be tested with air at 10 psig for 30 min with no leakage permitted.

Vent and vapor recovery piping shall be tested hydrostatically at a pressure of 30 psig for 30 min with no leakage permitted.

Tightness Testing

Tightness testing is a general term used for testing and evaluating existing tanks and piping systems that contain product. Tightness testing is performed periodically, as required by local authorities. Generally, periodic tightness testing is not required for the following:

1. Tanks and piping containing No. 5 and No. 6 fuel oil
2. Tanks and piping with a capacity of 1100 gal or less unless the authorities have a reason to believe leakage is occurring
3. Tanks and piping that are corrosion resistant and have an approved leak detection system
4. Tanks and piping installed in conformance with requirements for new construction
5. Tanks larger than 50,000 gal and/or for tanks for which it is technically impossible to perform a meaningful series of tests. An alternative test or inspection approved by authorities shall then be conducted

All tests must conform to EPA and local requirements, and the technicians performing such tests must be trained and qualified by the test equipment manufacturer. The tightness test shall detect a leak of 0.1 gph from the system with a detection probability of 95 percent, and a false alarm probability of 5 percent. Acceptable leakage amounts vary, depending on those values established by the local implementing authorities.

There are many types of tests capable of achieving this precision, and various manufacturers make testing equipment. The type of test chosen must take into consideration the following:

1. Vapor pockets
2. Thermal expansion of product
3. Temperature and temperature stratification
4. Groundwater level
5. Evaporation
6. Tank end deflection
7. Pressure

EVALUATION OF AST OR UST INSTALLATION

For smaller tanks, the decision on whether to install an above-ground or underground tank is based on many other factors in addition to the initial cost. Since the general useful life of any tank system is 20 to 30 years, all of the factors should be included in the calculations to find the actual cost for any specific installation. The following considerations are important to determining which type of installation is better for any specific facility.

AST

Advantages:

1. Easier to visually detect leakage, contain spills, and repair tank
2. Less initial installation cost
3. Easily removed and relocated
4. Fewer regulatory compliance requirements
5. Usually less piping cost if dispenser is integral part of tank

Disadvantages:

1. Occupies space on property, which may be an aesthetic problem
2. Must be protected against physical damage and vandalism
3. More vulnerable to fire damage which would mean higher insurance costs
4. Higher level of maintenance
5. Higher tank initial cost
6. Requires separation from other structures, roads, and property lines
7. May not be accepted by local authorities
8. May require reliable source of water to fight potential fire

UST

Advantages:

1. Superior protection against fire
2. Lower level of maintenance
3. Narrow range of temperature fluctuations, which limits vapor production
4. Buried tank, which allows the use of the ground surface above the tank
5. Not prone to vandalism, which means that physical protection is not required for the tank

Disadvantages:

1. Close supervision and training of installer required for proper installation
2. Cost of repair of tank higher than for an AST
3. Higher leak detection and instrumentation costs
4. Requires separation from well water supply sources
5. Must rely on instruments to detect leakage

Evaluation of New UST Tank Containment System

A primary consideration in selecting a tank containment system is whether to use a single tank or one with secondary containment. Federal rules and most states allow either. The basic advantage to the double-contained option is that it enhances the capability of the leak detection system. For typical installations, the use of a double-contained tank has proven cost effective over the useful life of the system.

SYSTEM DESIGN CONSIDERATIONS

Tank Installation

The installation of the tank is critical to the proper and long-lasting functioning of the system and the prevention of leakage in the later years of its operational life. It is a requirement that the contractor or installer be trained and authorized by the specific tank manufacturer as qualified to install any tank manufactured by that manufacturer only.

FRP underground tanks rely on the quality of the backfill for long-term support to resist elongation and resulting failure. Experience has shown that washed pea gravel, free from any organic matter and having no sharp edges, is the best backfill material. The gravel must be very carefully compacted. FRP tanks should be installed in conformance with API 1615. If there is a remote possibility of the tank's floating, such as in areas subject to flooding or an abnormally high water table, the tanks must be installed over a reinforced concrete ballast pad heavier than the buoyant force and anchored to that pad by means of hold-down straps. The weight of backfill over the pad, which also resists floating, should not be added, and it is used as a safety factor. It is recommended that the calculated load on each hold-down strap be increased by a safety factor of 5. The exact buoyant force shall be obtained for the selected storage tanks from the manufacturer. For preliminary planning purposes, Table 12.9 gives buoyant force values for typical size tanks.

Steel UST tanks are structurally stronger and do not depend as much on backfill for support. The backfill must prevent all but very minor shifting or settling of the tank in the future. The USTs shall be installed in conformance with NFPA-30 and NFPA-31. The requirement for ballast pad and hold down is the same as for FRP tanks.

When steel tanks are used, they are protected from corrosion with a special coating. This coating must be tested for faults (called *holidays*) both on the truck and after initial placement in the excavation. All defects found must be repaired in strict conformity to the manufacturer's recommendations. FRP tanks do not require corrosion protection.

Tanks shall not be stored on the site prior to installation due to the possibility of damage. Arrangements shall be made to deliver the tank on the day installation is to be made. If delays require that the tank not be installed, the tank shall be shipped back to the manufacturer and another, new tank delivered to the site. A detail of a typical UST installation is shown in Fig. 12.21.

The AST shall be placed on a concrete pad of sufficient thickness to adequately support the tank and product. Bollards shall be placed around the tank to protect it from vehicle damage.

TABLE 12.9   Approximate Buoyant Force of Empty Storage Tanks

Tank capacity, gal	Tank diameter	Tank length	Tank weight, lb[1]	Wall thickness	Net upward force (net buoyancy) in pounds totally submerged in water*
560	4'0"	6'0"	480	12 gauge	4,190
560	4'0"	6'0"	828	7 gauge	3,842
1,000	5'4"	6'0"	875	10 gauge	7,475
1,000	4'0"	11'0"	968	10 gauge	7,382
1,000	5'4"	6'0"	1,180	7 gauge	7,459
1,500	5'4"	9'0"	1,575	7 gauge	11,200
2,000	5'4"	12'0"	1,980	7 gauge	14,720
2,500	5'4"	15'0"	2,375	7 gauge	18,479
3,000	5'4"	18'0"	2,765	7 gauge	22,235
3,000	6'0"	14'0"	2,565	7 gauge	22,150
4,000	5'4"	24'0"	3,560	7 gauge	29,790
4,000	6'0"	19'0"	3,310	7 gauge	30,040
4,000	7'0"	14'0"	3,150	7 gauge	30,200
5,000	6'0"	24'0"	5,400	¼ in	36,300
5,000	8'0"	13'6"	4,775	¼ in	36,925
6,000	6'0"	29'0"	6,380	¼ in	43,720
6,000	8'0"	16'0"	5,370	¼ in	44,630
7,500	8'0"	20'0"	6,500	¼ in	56,238
8,000	8'0"	21'6"	7,000	¼ in	59,750
8,000	10'6"	12'5"	6,470	¼ in	60,980
10,000	8'0"	27'0"	8,360	¼ in	74,356
10,000	10'6"	15'6"	7,600	¼ in	75,019
12,000	10'6"	18'6"	8,400	¼ in	91,440
15,000	10'6"	23'3"	12,600	⁵⁄₁₆ in	112,833
20,000	10'6"	13'0"	15,900	⁵⁄₁₆ in	151,345
25,000	12'0"	30'0"	20,875	³⁄₈ in	190,509
30,000	10'6"	46'3"	27,000	³⁄₈ in	222,518
30,000	12'0"	36'0"	24,294	³⁄₈ in	229,392

*Lb × 2.2 = Kg
Source: Courtesy ASPE.

Piping Installation

Pipe should be installed in a trench, far enough underground to prevent damage. Initial backfill should be pea gravel, clean sand, or other acceptable material. A generally accepted minimum depth from grade surface to the top of the pipe where there is no slab is 18 in (45 cm), 12 in (30 cm) below asphalt slabs and 6 in (15 cm) under concrete slabs. It is also accepted practice where electrical conduit is installed in the same trench as piping to protect the conduit with a concrete encasement and to provide 4 in separation.

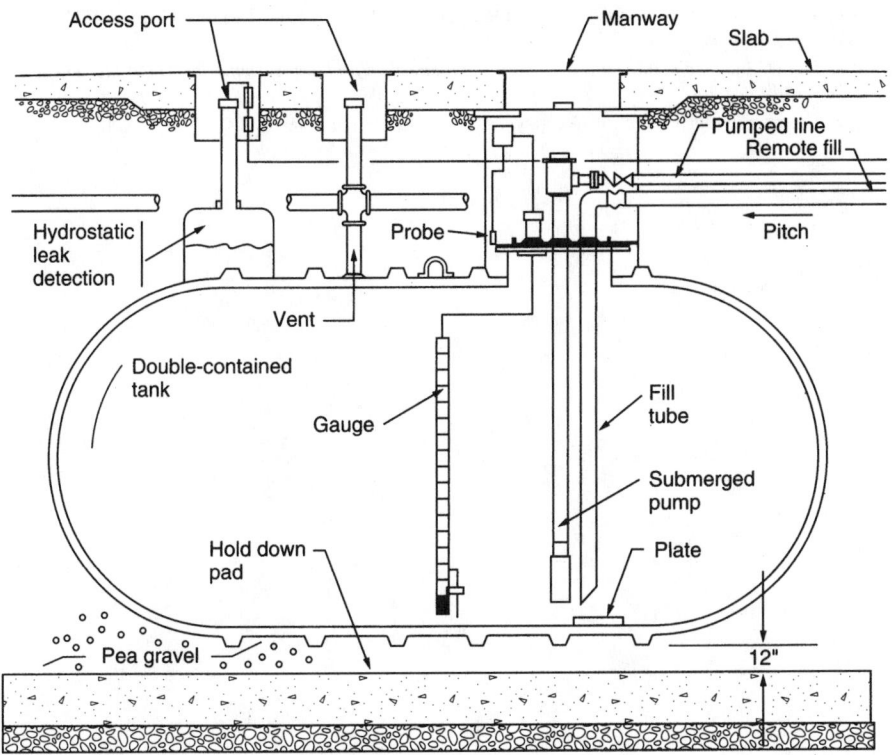

FIGURE 12.21   A typical UST installation.

REFERENCES

American Society of Plumbing Engineers, *Gasoline and Diesel Oil Systems,* Chap. 33, 1986.

Bertin, C. J., "Underground Tank Regulations," *Maintenance Technology Magazine,* August 1992, pp. 42–45.

Cruver, P. C., "An Alternative to Underground Fuel Storage," *Public Works Magazine,* November 1992, pp. 55–57.

Curran, S. D., "Environmental Considerations for Aboveground Storage Tanks," *Public Works Magazine,* November 1993, pp. 42–44.

Deaver, E., "AST vs. UST: A Facility Manager's Dilemma," *The National Environmental Journal,* May/June 1993, pp. 18–22.

Floyd, C. M., "UST Leak Detection: A Must for Owners," *Environmental Protection Magazine,* April 1993, pp. 32–38.

Gryer, W., "Bringing Storage Tanks to the Surface," *Chemical Engineering Magazine,* July 1992, pp. 95–102.

Leiter, J. L. (ed.), *Underground Storage Tank Guide,* Thompson Publishing.

National Fire Protection Association, *Fire Protection Handbook,* 16th ed., NFPA, Quincy, Mass.

National Fire Protection Association, Flammable & Combustible Liquids Code, NFPA-30, Quincy, Mass.

Red Jacket, Inc., *Selection and Sizing of Submersible Pumping Systems for Gasoline, Diesel Fuel and Gasohol.*

U.S. Environmental Protection Agency, *Musts for USTs,* July 1990.

CHAPTER 13
FUEL GAS SYSTEMS

The fuel gas system delivers gas to provide light or heat energy in sufficient volume and pressure as required for the satisfactory operation of all connected devices and for all purposes. This chapter will discuss natural gas and propane only.

FUEL GAS DESCRIPTION

There are many gases used as a fuel gas. Where easily and cheaply available, two major fuel gases, natural gas (NG) and liquefied petroleum gas (LPG), are preferred. Other gases are used because of availability. Refer to Table 13.1 for the physical and combustion properties of the fuel gases most commonly available throughout the world.

NG and LPG are hydrocarbon compounds obtained from the separation of gas mixtures occurring naturally at the wellhead of crude oil or gas producing wells or as a by-product of the oil refining process. NG is primarily composed of methane with minor percentages of other gases that result in variations in heating content. *LPG* is a term applied to a group of hydrocarbons such as propane, butane, iso-butane, and pentane or to various mixtures of each. Propane and butane are the principal constituents of LPG, with propane the most common. Table 13.2 lists the physical properties of NG and propane. Since NG and LPG are odorless, an odorant is added to make detection of the gas possible.

There are variations in the hydrocarbon mixtures that produce variations in specific gravities and British thermal unit contents of NG and LPG obtained from various suppliers. The values presented here are average values and sufficiently accurate for the design of these systems. It is recommended that an analysis of the products actually furnished by the supplier for each specific project be obtained if extreme precision is required.

CODES AND STANDARDS

There are a number of codes and standards governing the manufacture, design, installation, and testing of LPG and NG systems, piping, and components. The principal codes, standards, and regulations are as follows:

TABLE 13.1 Physical and Combustion Properties of Commonly Available Fuel Gases

Name of gas	Heating value Btu/cu ft[1] Gross	Net	Btu/lb[2] Gross	Net	Heat release, Btu[3] Per cu ft air	Per lb air	Specific gravity	Density, lb per cu ft	Specific volume cu ft/lb
1. Acetylene	1498	1447	21,569	20,837	125.8	1677	0.91	.07	14.4
2. Blast Furnace Gas	92	92	1178	1178	135.3	1804	1.02	.078	12.8
3. Butane	3225	2977	21,640	19,976	105.8	1411	1.95	.149	6.71
4. Butylene (Butene)	3077	2876	20,780	19,420	107.6	1435	1.94	.148	6.74
5. Carbon Monoxide	323	323	4368	4368	135.7	1809	0.97	.074	13.5
6. Carburetted Water Gas	550	508	11,440	10,566	119.6	1595	0.63	.048	20.8
7. Coke Oven Gas	574	514	17,048	15,266	115.0	1533	0.44	.034	29.7
8. Digester (Sewage) Gas	690	621	11,316	10,184	107.6	1407	0.80	.062	16.3
9. Ethane	1783	1630	22,198	20,295	106.9	1425	1.06	.060	12.5
10. Hydrogen	325	275	61,084	51,628	136.6	1821	0.07	.0054	186.9
11. Methane	1011	910	23,811	21,433	106.1	1415	0.55	.042	23.8
12. Natural (Birmingham, AL)	1002	904	21,844	19,707	106.5	1420	0.60	.046	21.8
13. Natural (Pittsburgh, PA)	1129	1021	24,161	21,849	106.7	1423	0.61	.047	21.4
14. Natural (Los Angeles, CA)	1073	971	20,065	18,158	106.8	1424	0.70	.054	18.7
15. Natural (Kansas City, MO)	974	879	20,259	18,283	106.7	1423	0.63	.048	20.8
16. Natural (Groningen, Netherlands)	941	849	19,599	17,678	111.9	1492	0.64	.048	20.7
17. Natural (Midlands Grid, U.K.)	1035	902	22,500	19,609	105.6	1408	0.61	.046	21.8
18. Producer (Wellman-Galusha)	167	156	2650	2476	128.5	1713	0.84	.065	15.4
19. Propane	2572	2365	21,500	19,770	108	1440	1.52	.116	8.61
20. Propylene (Propene)	2332	2181	20,990	19,630	108.8	1451	1.45	.111	9.02
21. Sasol (South Africa)	500	443	14,550	13,016	116.3	1551	0.42	.032	31.3
22. Water Gas (bituminous)	261	239	4881	4469	129.9	1732	0.71	.054	18.7

[1] 1 cu ft = 0.0283 m^3
[2] 1 lb = 0.453 kg
[3] 1 BTU = 0.252 kCal

TABLE 13.2   Average Physical Properties of Natural Gas and Propane

	Propane	Natural gas (methane)
Formula	C_3H_8	CH_4
Molecular weight	44.097	16.402
Melting (or freezing) point, °F	−305.84	−300.54
Boiling point, °F	−44	−258.70
Specific gravity of gas (air = 1.00)	1.52	0.60
Specific gravity of liquid 60°F/60°F (water = 1.00)	0.588	0.30
Latent heat of vaporization at normal boiling point, Btu/lb	183	245
Vapor pressure, lb/in^2, gauge at 60°F	92	
Pounds per gallon of liquid at 60°F	4.24	2.51
Gallons per pound of liquid at 60°F	0.237	
Btu per pound of gas (gross)	21591	23000
Btu per ft^3 gas at 60°F and 30 in mercury	2516	1050 ±
Btu per gallon of gas at 60°F	91547	
Cubic feet of gas (60°F, 30 in Hg)/gal of liquid	36.39	59.0
Cubic feet of gas (60°F, 30 in Hg)/lb of liquid	8.58	23.6
Cubic feet of air required to burn 1 ft^3 gas	23.87	9.53
Flame temperature, °F	3595	3416
Octane number (isooctane = 100)	125	
Flammability limit in air, upper	9.50	15.0
Flammability limit in air, lower	2.87	5.0

1 in. Hg = 3.37 kPa
60°F = 15.6°C
1 lb = 2.2 kg
1 L^3 = 4.5 L
1 cu ft = .03 m^3

ANSI/NFPA 30, Flammable and Combustible Liquids Code

ANSI/NFPA 54, National Gas Code

ANSI/NFPA 58, Standard for the Storage and Handling of Liquefied Petroleum Gases

ANSI Z83.3, Standard for Gas Utilization Equipment in Large Boilers

ANSI/UL 144, Pressure Regulating Valves for LP Gas

American Gas Association (AGA) codes and standards

Local utility company rules and regulations

Some insurance carriers, such as Factory Mutual and Industrial Risk Insurers, have standards that in many aspects may be more stringent than those listed above.

SYSTEM OPERATING PRESSURES

The maximum allowable system operating pressure of fuel gas when installed inside a building is governed by NFPA 54, unless local codes or insurance carriers have more stringent requirements. NG systems are not permitted to exceed 5 psig unless all of the following conditions are met:

1. Local authorities permit a higher pressure.
2. All piping is welded.
3. The pipe runs are enclosed for protection or located in a well-ventilated space that will not permit gas to accumulate.
4. The pipe is run inside buildings or areas used only for industrial processes or research or for warehouses or boiler and/or mechanical equipment rooms.

LPG pressures of up to 20 psig are permitted only if all of the following conditions are met:

1. The building is used exclusively for industrial or research purposes.
2. The building is constructed in accordance with NFPA 58, Chap. 7.

NATURAL GAS

GENERAL

Natural gas is usually obtained from a franchised public utility obligated to provide gas to every customer that requests this service. As part of this service, the utility company usually supplies and installs the service line free of charge from the utility main, in addition to providing a regulator-meter assembly in or adjacent to the building.

If the utility company has an existing main in the vicinity of the project—which is most often the case—then the initial start-up cost is fairly low. Sometimes, however, an installation fee is charged if the utility company regulations concerning the time of payback from expected revenue does not justify the cost of installing the service. The same might be true if the use of gas for a commercial or industrial facility is considered too little. It would then be the responsibility of the owner to pay for the design and installation of the complete site service and meter assembly in conformity with utility company regulations.

There are several different types of service that a utility company may provide, each with a different cost (or rate). Specific types of service may be unavailable or known by different names in various localities. They are:

1. *Firm service.* This type of service provides a constant supply of gas under all conditions without exception. This service has the highest rate.

2. *Interruptible service.* This type of service allows the utility company to stop the gas supply to the facility under predetermined conditions, and then to start it again when these conditions no longer exist. The most common condition usually occurs when the outside temperature falls below a certain point. The rate is lower than for firm gas, and users with this service will require a backup source of fuel, such as fuel oil or LPG.

3. *Light or heavy process service.* This type of service is provided for industrial or process use and is reserved for quantities of gas that the utility companies define for this class of service.

4. *Commercial or industrial service.* This type of service is provided for heating and cooling system loads for this class of building usage.

5. *Transportation gas service.* This service is purchased directly from a company other than the public utility, with the gas actually carried to the site by the utility company mains. In addition to the cost of the gas, the utility company will also charge an additional fee for this service. This type of service is always available.

The following criteria and information should be obtained in writing from the public utility company:

1. The British thermal unit content of the gas provided
2. The minimum pressure of the gas at the outlet of the meter
3. The extent of the installation work done by the utility company and the point of connection by the facility construction contractor
4. The location of the utility supply main and the proposed run of pipe on the site by the utility company

5. The acceptable location of the meter and/or regulator assembly and any work required by the owner to allow the assembly to be installed (such as having a meter pit dug or a slab installed on grade)

6. The types of service available and the cost of each

In order for the utility company to provide this data, they require that the following information be provided to them:

1. The total connected load. The utility will use their own diversity factor to calculate the size of the service line. If the design engineer is responsible for the installation, this information is not required since the diversity factor for the facility shall be used.

2. The minimum and maximum pressure requirements for the most demanding device.

3. The site plan indicating the location of the proposed building on the site and the specific area of the building where the proposed NG service will enter the building.

4. Preferred location of the meter-regulator assembly.

5. Expected date for the start of construction.

MAJOR NATURAL GAS SYSTEM COMPONENTS

Drip Pots

Drip pots are necessary to remove water that is in the mains from reaching the building. Modern technology has reduced, but not eliminated, the presence of moisture in the utility company mains. Each utility company has standard methods of installing drip pots depending upon pressure in the main.

Gas Line Filters

Filters for natural gas are installed on the site service to protect the regulator and meter from injury due to particulate clogging, which may damage the equipment inside the building. They should be considered when:

1. Line scale, dirt or rust is known to be present

2. Dirty gas is obtained from a transmission company

3. "Wet" gas is known to be present, such as after large PRVs

The filter consists of a housing and a cartage filter element. Selection of the housing is based on the highest flow rate and pressure expected and the size of the proposed building service. The housing should be capable of having interchangeable elements to allow replacing the original filter if desired. Having a filter oversized is considered good practice since this allows a longer service life if the cost of the larger size is not excessive.

The filter element is usually cellulose or synthetic fiber. Generally, the cellulose type is used for sizes 3 in and less and fiber 4 in and larger. Fiber is stronger and should be considered where the pressure differential may be excessive when recommended by the manufacturer. A filter rating of 10 microns is suggested as a starting point, with actual operating experience being the final criteria.

The size of the filter is based on ACFM and velocity across the filter. The pressure drop across the filter when dirty shall be considered in the actual pressure loss allowed for the service assembly. Since the filter should be changed when the pressure difference approaches 10 percent of the operating pressure, it is good practice to provide pressure gauges on both sides of the filter. If no filter is installed on the service, a spool piece should be provided if there is a possibility that one may be required. Most filters are mounted horizontally.

Gas Meters

Gas meters are part of a service assembly that may consist of filters, valves, regulators and relief valves. The complete assembly is usually supplied and installed by the utility company. For the rare instances where this is not the case, gas meters are selected using the local utility company standards with the size and arrangement of the entire meter assembly based on flow rate and pressure. Because steady pressure is necessary for accurate metering, a regulator is installed upstream of the meter. The utility company will provide sketches and details, with space requirements, of typical meter assemblies for various capacities and pressures. This will permit accurate determination of the dimensional and space requirements for the area where the complete assembly will be installed. Obtain from the utility company all requirements for the specific installation under design.

A contractor other than the utility company will be responsible for providing a pit for the assembly or a concrete slab under the complete meter assembly when installed outside the building. This will include meter slab size for outdoor meter installations and pit sizes and access openings for meters installed in pits. Additional requirements will usually be a weatherproof telephone outlet and a 120 V electrical outlet at the meter site. Facilities with large demands for boilers and other equipment, a meter assembly with dimensions of 6 ft (2 m) wide by 25 ft (8 m) long is not uncommon.

Generally, if the contractor is installing the site service, the utility company will either supply the meter or install the entire assembly at no charge. This must be confirmed.

Requirements of various utility companies differ regarding the placement of the meter assembly. It could be installed either in an underground, exterior meter pit, at an above ground exterior location exposed to the weather, or inside the building in a well ventilated area or mechanical equipment room.

Pressure Regulators

Gas pressure regulators are pressure reducing devices used to reduce a variable high inlet pressure to a constant lower outlet pressure. Two types of regulators are available: direct acting and pilot operated. The direct operated type uses the difference of pressure between the high and regulated side of the regulator to directly move a closure (adjusting) member inside the regulator to adjust the pressure. The pilot operated type uses a primary regulator to sense and magnify differences in pressure of the high and low pressure sides and a second, main valve with the closure member to achieve the desired pressure.

There are several categories of regulators, with the end use determining the nomenclature. The first is the line regulator, which is used to reduce high pressure, often in a range of between 25–50 psig (170–345 kPa), from the gas service provided by the utility company to a lower pressure used for the building service. An intermediate regulator is used to reduce the lower pressure, often in the range of 3–5 psig (21–35 kPa) to a pressure required to supply terminal equipment such as a boiler gas train. The third type is an appliance regulator used at the individual piece of equipment for final pressure.

The line regulator is usually pilot operated and provided with an internal or external relief valve. The regulator is pressure rated to withstand the highest pressure expected. The relief valve is installed downstream from the regulator and is set to trip at a pressure of about 10% higher than the highest set pressure. The line regulator is placed upstream from the meter in order to provide the meter with a constant pressure, allowing accurate measurement. This line pressure regulator is most often selected and installed by the utility company as part of the gas meter assembly.

An intermediate regulator is used within a facility where high pressure used for distribution purposes must be reduced to a lower pressure required by the terminal appliances. There are two types of intermediate regulators used, and the choice is determined by the accuracy desired and the ability to install a relief valve and associated gas vent discharge. One type of regulator has an internal relief valve set to discharge when the pressure rises above the set point. This has the least initial cost and least accuracy. Another type of intermediate regulation, called a monitor type, consists of two pilot operated regulators and does not require a relief valve. They can be installed in two configurations, both of which use regulators in series. The first uses the upstream regulator to initially reduce the inlet pressure to some intermediate value and the downstream regulator to further reduce the pressure to the final set point. This arrangement puts less stress on each of the regulators. The second configuration uses the upstream regulator wide open and the downstream regulator to do all of the pressure reducing. If the gas pressure goes above the set point, the upstream regulator closes to partially lower the pressure. This installation has a high initial cost but must be used where a relief valve cannot be installed. The intermediate type regulator arrangement is the most often used.

The appliance regulator is used to control the pressure of gas directly connected to the terminal appliance or equipment. The appliance regulator is most often provided by the manufacturer as part of the equipment gas train. The gas train is an assembly of piping, valves, regulators, relief vents, etc. used to directly connect the gas supply to the terminal equipment. For larger pieces of equipment such as boilers, etc., the gas train arrangement is dictated by insurance carrier requirements. An additional requirement of most gas trains are small relief vents from various devices. These vents must be piped outside the building to a point where they can be diluted by the outside air and will pose no threat to the public or create a fire hazard. This is usually above a roof or the highest point of the structure.

Another type of appliance regulator is called an atmospheric regulator or a zero governor. This type of regulator is a very sensitive type that works with a very low gas pressure and extremely small differentials.

A differential regulator is a multiple port type that is used to produce a single, uniform outlet pressure when supplied with multiple inlets of different pressures.

A backpressure regulator is a regulator arranged to provide accurate inlet pressure control. It is used as a relief valve where the application requires a higher degree of regulation and sensitivity than possible with a standard poppet type relief valve. When operating as a relief valve it limits inlet pressure to a set point. At pressures below this point, it remains closed. If the pressure rises above the set point, it begins to open and will bleed off only enough pressure to maintain the system set point.

A piped gas vent must be provided from regulators to a point several feet above the roof of any adjacent structure, and it must be properly sized to carry the amount of gas that will be discharged. This is to protect the system from overpressure in the event of a malfunction or failure to fully lockup. Each individual regulator vent must be separately carried to a non-hazardous location away from any potential source of ignition. Common vent lines are not permitted.

Regulator Selection Considerations

It is common practice to oversize the capacity of a regulator by about 15% to provide a margin for accurate regulation.

For large loads, regulators in parallel are often used to keep the pressure drop to a minimum.

The adjustment range of a regulator should be approximately 50% over the desired regulated pressure to 50% under that pressure.

The utility company may require that regulators be of the "lockout" type. This feature will stop regulator operation when the pressure falls below a predetermined set point.

SITE DISTRIBUTION

The site distribution portion of the system starts at the property line of the customer and usually ends above grade (or the concrete pad) at the valve after the meter. The connection to the utility company mains are done by the utility company.

The most often used piping material is High Density Polyethylene (HEPE). The rating for pipe 4 in and smaller should be SDR 11, and for 5 in and larger SDR 13.5. The pipe shall be buried a minimum of 3 ft below ground and a 14 AWG corrosion resistant tracer wire placed in the pipe trench in over the pipe. Another detection method is to put a warning tape containing metallic material with the words "Natural Gas" on it. This is to allow location by a metal detector and to warn of the gas line immediately below the tape if digging has taken place without trying to locate the pipe beforehand.

The pipe above ground shall be metal; therefore, a transition fitting is required if HDPE is used for the supply service. The transition fitting shall comply with ASTM D-2513. Refer to Fig. 13.4.

Testing and Purging

After installation of the pipe is complete, the site system shall be tested, purged of air and then filled with natural gas. The testing phase shall be done with compressed air at a gauge pressure 50% higher than the highest pressure expected in the main. The larger the total volume of the pipe, the longer the test shall last. No loss of pressure shall be allowed. The test period is based on the size of the line and the length of run. Table 13.3 gives the recommended test periods.

TABLE 13.3 Maximum Length Of Pipe Being Tested (ft) Mains To Operate At Less Than 125 PSIG Test Duration (Hours)

Main size	2 hrs.	3 hrs.	4 hrs.	6 hrs.	8 hrs.	10 hrs.	12 hrs.	14 hrs.	16 hrs.	18 hrs.	20 hrs.	22 hrs.	24 hrs.
1" or less	6590	11000	15400	24200									
2"	1700	2830	3960	6220	8490								
3"	743	1240	1730	2730	3720	4710	5700						
4"	439	732	1030	1610	2200	2780	3370	3960	4540	5130			
6"	189	315	442	694	947	1200	1450	1710	1960	2210	2460	2730	2970
8"	108	180	252	396	541	685	829	973	1120	1260	1410	1550	1700
12"	48	80	112	177	241	306	370	435	499	563	628	692	757
14"	41	68	96	151	206	261	316	371	426	482	537	592	647
16"	30	50	70	110	150	191	231	271	311	352	392	432	472
20"	19	31	44	69	95	120	146	171	197	222	247	273	298
24"	13	21	30	48	65	83	100	118	135	153	170	188	205

After successful testing, the line shall be purged of air with dry nitrogen. The reason for this is to prevent a flammable mixture of gas and air when the pipe is filled with natural gas for the first time. Calculate the volume of the piping and introduce an equal volume of nitrogen into the pipe. After the nitrogen purge, natural gas is then introduced until the nitrogen is displaced and left under pressure.

SITE SERVICE SIZING PROCEDURE

Pressure Required

The pressure in the main provided by the utility company and the flow rate is the basis for design. The line is sized using proprietary charts for the specific pressure in the gas main. If these charts are not available, compressed air friction loss tables, using a table having the actual pressure in the gas service main, can be use with adjustment to the flow rate calculated with eq. 1. This will provide sizing within a range acceptable for this system. The pressure drop selected is at the discretion of the design professional, but is generally kept to approximately 10% of the available pressure in psig. To use the pressure loss tables, calculate the pressure drop per 100 ft of pipe.

If the design engineer is responsible for sizing the meter and regulator, the following information must be calculated or established.

1. The adjusted maximum flow rate, in CFM, for all connected equipment in the facility. Typical BTU values for equipment are given in Table 13.8. CFH demand for multiple dwellings are given in Fig. 13.1 (for 50 pts and less) and Fig. 13.2 for more than 50 apts.

2. The highest pressure required for the most demanding equipment inside the building.

3. The pressure in the building service main upstream immediately downstream of the regulator. If a pressure loss through a filter is expected, this will reduce the pressure.

4. The allowable pressure loss for the piping system inside the facility.

5. The pressure differential between the service main pressure and the pressure required inside the building. This is the sum of items 2 and 4 subtracted from item 3.

6. The regulator and meter can now be sized using literature and catalog information from the manufacturers.

Maximum Probable Demand

For some types of buildings, such as multiple dwellings and laboratories, the total connected load is not used to size the piping system since not all of the connected devices will be used at the same time. For design purposes, it will be necessary to apply a diversity factor to reduce the total connected load when calculating the maximum probable demand.

This calculation first requires the listing of every device using gas in the building and the demand in BTU/h for each. The manufacturer of each device should be

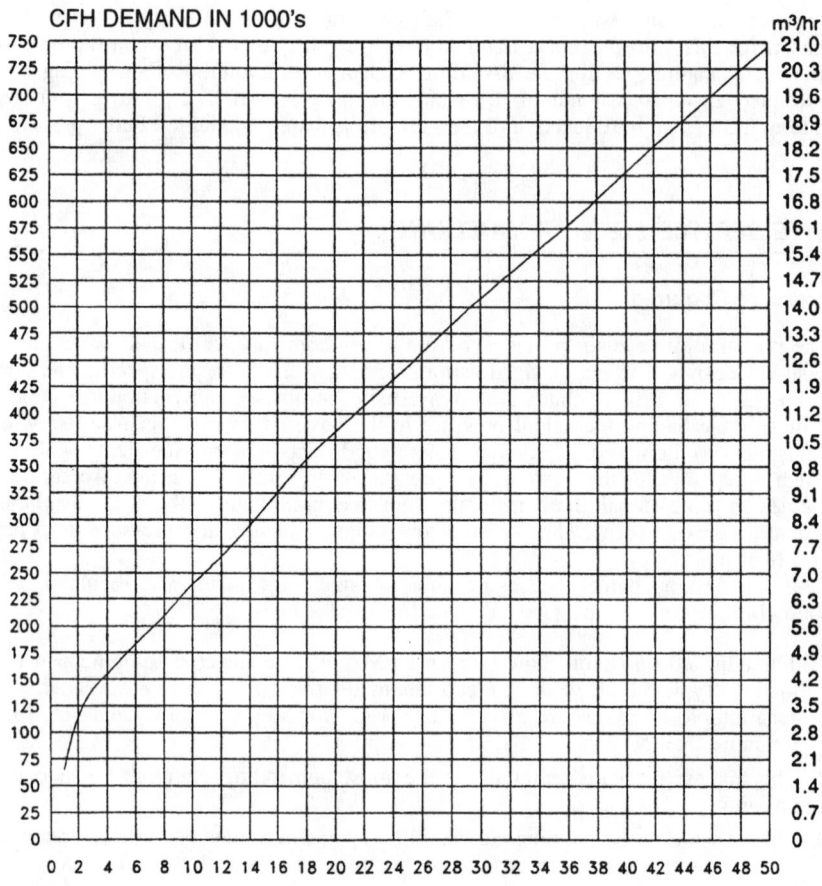

Number of Apartments

FIGURE 13.1   Gas demand for multiple-unit dwellings with less than 50 apartments.

consulted to find its actual input gas consumption. Average gas demand values for typical devices are listed in Table 13.8.

For multiple dwellings, a direct reading of the quantity of gas based on the number of apartments is presented in Fig. 13.1 for buildings up to 50 apartments and Fig. 13.2 for buildings with more than 50 apartments. A diversity factor has been included in creation of the chart. For laboratories, use a figure of 5 cfh based on the use of Bunsen burners. Refer to Table 13.9 for diversity factors. Where laboratories are part of a school, use no diversity for entire rooms, and consult with the school authorities to find the total number of rooms that might be in use at once. If there is no conclusive answer, use no diversity. Demand for trailer parks is given in Table 13.10.

For industrial or process installations, and for major gas using equipment such as boilers and water heaters in all building types, no diversity factor is used because

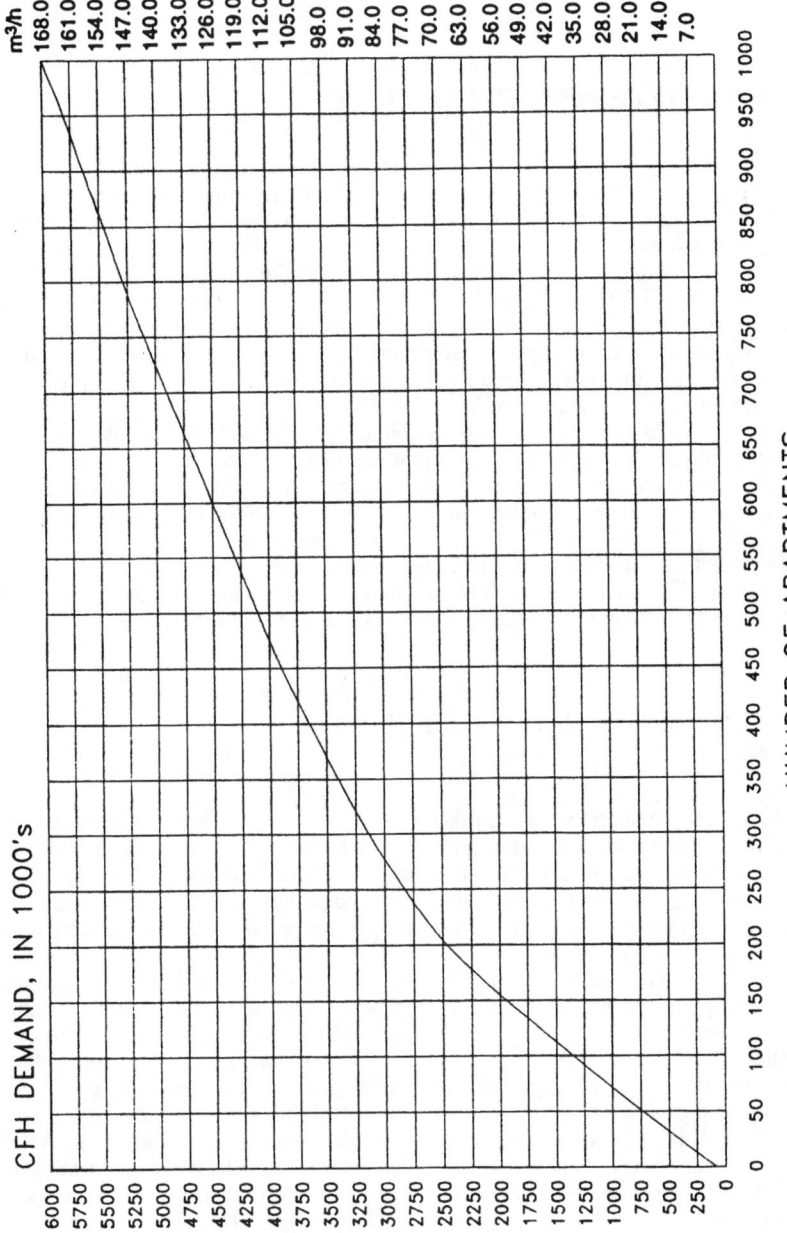

FIGURE 13.2 Gas demand for multiple-unit dwellings with more than 50 apartments.

13.13

it is possible for all connected equipment of this nature to be in use at the same time.

NG SITE SERVICE PIPE SIZING METHODS

The most conservative method for sizing NG piping systems is by the use of tables, such as those prepared by the American Society of Plumbing Engineers (ASPE). Calculators are also available. The calculators are considered more accurate than the prepared tables.

Tables are provided that will allow sizing for a majority of systems. Table 13.4 is for a pressure of 10 psig, Table 13.5 is for a pressure of 20 psig, and Table 13.6 is for a pressure of 50 psi. If higher pressure in encountered, the use of compressed air tables can be used for sizing, reducing the flow rate based on a factor found in Table 13.12 using the specific gravity of air, which is 1.00. Table 13.13 gives the conversion of various low gas pressure designations.

Using the correct chart with the system inlet pressure and the total system pressure drop, find the lowest SCFH figure that equals or exceeds the calculated CFH. Read the size in the correct column. Calculate the equivalent run of pipe from the outlet of the meter or regulator (or point of connection) to the furthest point of use. THIS IS THE ONLY DISTANCE USED. Find the pipe size at the intersection of the above distance column with a CFH figure that equals or exceeds the calculated CFH figure at each design point. Branches are sized using the same distance column but with the CFH figure from the branch to connection with the main.

PIPE AND SYSTEM MATERIALS FOR SITE INSTALLATION

All of the piping materials must be listed in NFPA 54 and other applicable codes. In addition to the codes, the recommendation of the utility company, which has experience in the area of the project, should be considered. The most often used material for underground lines is high density Polyethylene (PE). Larger lines and piping for high pressure (over 100 psig) are steel pipe protected against corrosion by wrapping with a plastic coating.

Codes do not permit plastic pipe to be run above ground. For above ground lines, the piping and jointing methods depend on pipe size and system pressure. For piping up to 3 psig (21 kPa), black steel, ASTM A-53 pipe is used with cast or malleable iron screwed fittings in sizes 3 in (75 mm) and smaller and butt welded joints 4 in (100 mm) and larger. Where natural gas is considered "dry gas," type "L" copper pipe is finding widespread use. All piping for pressures over 3 psig (21 kPa) shall be steel with welded joints. Welded fittings shall conform to ANSI 16.9. Screwed fittings shall be black malleable iron, 150 lb class conforming to ANSI A-197. Check valves shall be cast iron, 316 SS trim, disk type, with a soft seat. In sizes 3 in and smaller use screwed joints. Larger sizes shall be flanged.

Joints for HDPE should be butt type, heat fused joints. Socket type joints have been found to introduce a stiffness in the joint area that is undesirable. Joints for steel pipe shall be screwed for pipe sizes 4 in (100 mm) and smaller. Sizes 5 in (125 mm) and larger should be welded. Where flanged connections are necessary to connect some devices, flanges are heat welded to the HDPE pipe end.

TABLE 13.4 Gas Pipe Sizing, Initial Gas Pressure 10 psig, 1 psig Pressure Loss Capacity of Horizontal Gas Piping (cfh)

Length (ft)	Nominal 0.50 / actual 0.622	0.75 / 0.824	1 / 1.049	1.25 / 1.380	1.5 / 1.610	2 / 2.067	2.5 / 2.469	3 / 3.068	3.5 / 3.548	4 / 4.026	5 / 5.047	6 / 6.065
10	1023	2164	4120	8561	12,914	25,256	40,388	72,079	106,206	148,773	271,820	443,686
20	723	1531	2913	6047	9131	17,780	28,559	50,967	75,164	105,198	191,580	313,733
25	647	1369	2606	5415	8168	15,902	25,543	45,586	67,171	94,093	171,915	280,605
30	590	1293	2379	4943	7456	14,517	23,318	41,615	61,318	85,894	156,936	256,162
35	546	1157	2202	4576	6903	13,440	21,588	38,528	56,770	79,523	145,294	237,411
40	511	1082	2060	4281	6457	12,572	20,194	36,040	53,103	74,386	135,910	221,843
45	482	1020	1942	4036	6088	11,853	19,039	33,978	50,066	70,132	128,137	209,156
50	457	968	1843	3829	5776	11,245	18,062	32,260	47,497	66,533	121,562	198,422
60	417	883	1682	3475	5272	10,265	16,488	29,426	43,358	60,736	110,970	181,133
70	387	818	1558	3236	4881	9503	15,265	27,243	40,142	56,231	102,739	167,698
80	362	765	1457	3027	4566	8889	14,279	25,483	37,550	52,599	96,103	156,868
90	341	721	1373	2854	4304	8381	13,462	24,026	35,401	49,590	90,607	147,895
100	323	684	1303	2707	4084	7952	12,772	22,793	33,585	47,542	85,957	140,306
125	289	612	1165	2422	3653	7112	11,423	20,387	30,040	42,080	76,882	125,493
150	264	559	1063	2210	3334	6492	10,428	18,611	27,422	38,413	70,184	114,559
200	228	484	921	1914	2888	5622	9031	16,117	23,746	33,266	60,780	99,211
300	187	395	752	1563	2358	4591	7374	13,160	19,390	27,162	49,628	81,006
400	162	342	652	1353	2042	3975	6348	11,397	16,793	23,523	42,979	70,153
500	145	306	583	1210	1826	3556	5712	10,193	15,020	21,039	38,441	62,747
1,000	102	216	412	856	1291	2514	4039	7208	10,620	14,877	27,182	44,368
1,500	82	177	336	699	1054	2053	3298	5885	8671	12,148	22,194	36,227
2,000	72	153	291	605	913	1778	2856	5160	7510	10,519	19,221	31,373

TABLE 13.5 Gas Pipe Sizing, Initial Gas Pressure 20 psig, 2 psig Pressure Loss Capacity of Horizontal Gas Piping (cfh)

Length (ft)	Nominal: 0.50 / actual: 0.622	0.75 / 0.824	1 / 1.049	1.25 / 1.380	1.5 / 1.610	2 / 2.067	2.5 / 2.469	3 / 3.068	3.5 / 3.548	4 / 4.026	5 / 5.047	6 / 6.065
10	2040	4329	8240	17,121	25,820	50,285	80,771	144,079	212,400	247,753	543,611	887,323
20	1446	3060	5826	12,332	18,262	35,556	57,113	101,929	150,190	210,385	384,391	629,684
25	1293	2738	5211	10,828	16,334	31,803	51,084	91,168	134,334	188,174	343,810	561,192
30	1180	2499	4757	9885	14,911	29,032	46,633	83,225	122,630	171,779	313,854	512,296
35	1093	2314	4404	9150	13,791	26,878	43,143	77,051	113,532	159,035	290,572	474,294
40	1022	2165	4120	8561	12,913	25,143	40,385	72,674	106,200	148,764	271,806	443,662
45	964	2040	3885	8071	12,175	23,705	38,075	67,953	100,127	140,257	256,260	418,288
50	915	1936	3685	7657	11,550	22,488	36,122	64,466	94,988	133,059	243,110	396,823
60	835	1767	3340	6990	10,544	20,529	32,974	58,848	86,712	121,466	221,928	362,248
70	773	1636	3114	6471	9761	19,006	30,528	54,483	80,276	112,455	205,466	335,377
80	723	1530	2913	6053	9131	17,779	28,556	50,964	75,095	105,193	192,196	313,716
90	682	1443	2746	5707	8609	16,761	26,924	48,050	70,794	99,176	182,750	294,774
100	646	1369	2606	5414	8167	15,902	25,670	45,584	67,167	94,087	171,905	280,596
125	578	1224	2331	4843	7305	14,223	22,845	40,770	60,076	84,153	153,756	250,973
150	528	1118	2128	4421	6668	12,110	20,855	37,219	54,842	76,822	140,360	229,106
200	457	968	1843	3828	5775	11,244	18,061	32,232	47,494	66,529	121,555	198,411
300	374	790	1504	3126	4715	9181	14,747	26,318	38,779	54,322	99,299	162,002
400	324	685	1303	2707	4083	7950	12,771	22,792	33,583	47,043	85,952	140,299
500	289	612	1141	2421	3653	7111	11,423	20,386	30,038	42,077	76,878	125,487
1,000	204	433	824	1712	2583	5028	8076	14,415	21,240	29,753	54,361	88,732
1,500	166	354	673	1398	2109	4106	6595	11,770	17,342	24,293	44,385	72,449
2,000	145	306	583	1210	1826	3557	5712	10,193	15,018	21,039	38,439	62,743

TABLE 13.6 Sizing, NG Pipe with Initial Pressure of 50 psig and 5 psig Pressure Drop

Capacity of horizontal gas piping (CFH)

Length (ft)	Pipe diameter (in.)											
	Nominal	0.5	0.75	1	1.25	1.5	2	2.5	3	3.5	4	5	6
	actual	0.622	0.824	1.049	1.380	1.610	2.067	2.469	3.068	3.548	4.026	5.047	6.065
10		5,850	12,384	23,575	48,984	73,889	143,864	231,083	412,407	607,670	851,220	1,555,251	2,538,598
20		4,137	8,757	16,670	34,637	52,248	101,727	163,400	291,616	429,688	601,903	1,099,729	1,795,060
25		3,700	7,832	14,910	30,980	46,732	90,988	146,150	260,829	384,324	538,359	983,627	1,465,660
30		3,377	7,150	13,611	28,281	42,660	83,060	133,416	238,103	350,839	491,452	897,925	1,356,938
35		3,127	6,619	12,601	26,183	39,495	76,899	123,519	220,441	324,813	454,996	831,317	1,356,938
40		2,925	6,192	11,787	24,492	36,945	71,932	115,541	206,203	303,835	425,610	777,626	1,269,299
45		2,758	5,838	11,113	23,091	34,832	67,818	108,934	194,411	286,459	401,269	733,152	1,196,706
50		2,616	5,538	10,543	21,906	33,044	64,338	103,343	184,434	271,758	380,677	695,529	1,135,295
60		2,388	5,056	9,624	19,998	30,165	58,732	94,339	168,364	248,080	347,509	634,929	1,036,378
70		2,211	4,681	8,910	18,514	27,927	54,376	87,341	155,875	229,678	321,731	587,830	959,500
80		2,068	4,378	8,335	17,319	26,124	50,864	81,700	145,808	214,844	300,952	549,864	897,530
90		1,950	4,128	7,858	16,328	24,630	47,955	77,028	137,469	202,557	283,740	518,417	846,199
100		1,850	3,916	7,456	15,490	23,366	45,494	73,075	130,415	192,162	269,179	491,814	802,775
125		1,655	3,503	6,668	13,855	20,899	40,691	65,360	116,646	171,875	240,761	439,891	718,024
150		1,510	3,197	6,087	12,648	19,078	37,146	59,665	106,483	156,900	219,784	401,564	655,463
200		1,308	2,769	5,271	10,953	16,522	32,169	51,672	92,217	135,879	190,339	347,765	567,648
300		1,068	2,261	4,304	8,943	13,490	26,266	42,190	75,295	110,945	155,411	283,949	463,482
400		925	1,958	3,727	7,745	11,683	22,747	36,537	65,207	96,081	134,590	245,907	401,388
500		827	1,751	3,334	6,927	10,450	20,345	32,680	58,323	85,938	120,381	219,946	359,012
1,000		585	1,238	2,357	4,898	7,389	14,386	23,108	41,241	60,767	85,122	155,525	253,860
1,500		478	1,011	1,925	4,000	6,033	11,746	18,868	33,673	49,616	69,502	126,986	207,276
2,000		414	876	1,667	3,464	5,225	10,173	16,340	29,162	42,969	60,190	109,973	179,506

INTERIOR PIPE SIZING PROCEDURE

In order to size the piping system inside of any facility, the following information must be calculated or established:

1. The minimum gas pressure available from the utility company
2. The allowable friction loss for gas flowing through the piping system
3. The equivalent length of the piping system
4. The maximum probable demand
5. The pipe sizing method acceptable to local codes

Discussion:

Pressure Available from the Utility Company

The minimum pressure that the utility company will guarantee after the meter assembly must be provided upon request, and it is based on the pressure available in the utility supply mains adjacent to the facility under design. In cases where the utility is not providing the meter, the minimum pressure will be given for the main at the tie-in point. If there is a specific requirement for equipment that needs a higher pressure, such as that for a boiler, the utility company must be advised of this requirement and guarantee this pressure in writing. If the higher pressure is not possible, a gas pressure booster will be necessary.

Allowable Friction Loss in the Piping System

The minimum guaranteed pressure supplied by the utility company after the meter-regulator assembly could be as low as 4 to 7 in of water column (WC). This amount of pressure is quite low and requires that the friction loss of NG through the piping system be kept low in order to have sufficient pressure to properly operate the terminal appliance or equipment. A range of values between 0.2 and 0.5 in WC are generally acceptable, depending on the actual range of pressure available. When the available pressure is higher than 7 in WC, a higher friction loss is allowed for economy of pipe sizing depending on the end use pressure requirements.

A pressure loss figure of up to approximately 10 percent of the pressure available is a generally accepted value for distribution main piping with high pressure, depending on the actual pressure in the system. A regulator will then be required at each terminal appliance requiring a lower gas pressure. High-pressure gas distribution may not be practical due to the added cost of regulators and relief vents.

Equivalent Length of Piping

The equivalent length of piping is required to calculate the friction loss in all portions of a piping network. It is common practice not to use the vertical length of piping when calculating the total run for NG systems. Since NG is lighter than air, it expands at the rate of 0.1 in WC for every 15 ft of elevation as the gas rises.

The additional pressure created as the gas rises approximates the pressure lost to friction inside the pipe. LNG is heavier than air; therefore, the entire length of run is used in calculating the pressure loss. Refer to Table 13.7 for equivalent lengths of fittings and valves for the flow of gas in a pipeline.

Maximum Probable Demand

For some types of buildings, such as multiple dwellings and laboratories, the total connected load is not used to size the piping system since not all of the connected devices will be used at the same time. For design purposes, it is necessary to apply a diversity factor to reduce the total connected load when calculating the maximum probable demand.

This calculation first requires the listing of every device using gas in the building and the demand in British thermal units per hour for each. The manufacturer of each device should be consulted to find the actual input gas consumption of each device. Average gas demand values for typical devices are listed in Table 13.12.

For multiple-unit dwellings, a direct reading of the quantity of gas based on the number of apartments is presented in Fig. 13.1 for buildings of up to 50 apartments and in Fig. 13.2 for buildings with more than 50 apartments. A diversity factor has been used to create the charts. For individual risers, refer to Fig. 13.3 for a direct reading of the pipe size by floor. For laboratories, refer to Table 13.9 for diversity factors. Where laboratories are part of a school, use no diversity for entire rooms, and consult with the school authorities to find the total number of rooms that might be in use at once. If there is no conclusive answer, use no diversity.

The maximum demand for trailer parks is given in Table 13.10. For industrial or process installations and for major gas-using equipment such as boilers and water heaters in all building types, a diversity factor is generally not used because it is possible that all connected equipment will be in use at the same time.

NG Pipe Sizing Methods

The most conservative method for sizing NG piping systems is by the use of tables, such as those prepared by the American Gas Association, which are included in App. C of the National Fuel Gas Code, NFPA 54. Other methods using proprietary tables and calculators are available. The calculators are considered more accurate than the prepared tables. Table 13.11 is provided to allow sizing for low pressure systems of up to ½ psi (3.5 kPa).

To use the low-pressure chart, first establish the acceptable friction loss in order to select the proper table. Then calculate the equivalent run of pipe from the outlet of the meter or regulator (or point of connection) to the farthest point of use. *This is the only distance used.* Find the pipe size at the intersection of the above distance column with a cfh figure that equals or exceeds the calculated cfh figure at each design point. Branches are sized using the same distance column but with the cfh figure from the branch to connection with the main.

If a gas fuel with a different specific gravity is being used, the multipliers found in Table 13.12 can be used to adjust the capacity found in Table 13.11.

Refer to Table 13.13 for conversion of gas pressure designations.

TABLE 13.7   Resistance of Valves and Fittings to the Flow of Gas

(Expressed in equivalent feet of pipe)

	Screwed fittings[b]				90° welding elbows and smooth bends[c]					
	45° ell	90° ell	180° close return bends	Tee	R/d = 1	R/d = $1\frac{1}{3}$	R/d = 2	R/d = 4	R/d = 6	R/d = 8
k factor =	0.42	0.90	2.00	1.80	0.48	0.36	0.27	0.21	0.27	0.36
L/d' ratio[e] n =	14	30	67	60	16	12	9	7	9	12

Nominal pipe size, in	Inside diam. d, in, Sched. 40[g]	L = equivalent length in feet of Schedule 40 (standard weight) straight pipe[g]									
$\frac{1}{2}$	0.622	0.73	1.55	3.47	3.10	0.83	0.62	0.47	0.36	0.47	0.62
$\frac{3}{4}$	0.824	0.96	2.06	4.60	4.12	1.10	0.82	0.62	0.48	0.62	0.82
1	1.049	1.22	2.62	5.82	5.24	1.40	1.05	0.79	0.61	0.79	1.05
$1\frac{1}{4}$	1.380	1.61	3.45	7.66	6.90	1.84	1.38	1.03	0.81	1.03	1.38
$1\frac{1}{2}$	1.610	1.88	4.02	8.95	8.04	2.14	1.61	1.21	0.94	1.21	1.61
2	2.067	2.41	5.17	11.5	10.3	2.76	2.07	1.55	1.21	1.55	2.07
$2\frac{1}{2}$	2.469	2.88	6.16	13.7	12.3	3.29	2.47	1.85	1.44	1.85	2.47
3	3.068	3.58	7.67	17.1	15.3	4.09	3.07	2.30	1.79	2.30	3.07
4	4.026	4.70	10.1	22.4	20.2	5.37	4.03	3.02	2.35	3.02	4.03
5	5.047	5.88	12.6	28.0	25.2	6.72	5.05	3.78	2.94	3.78	5.05
6	6.065	7.07	15.2	33.8	30.4	8.09	6.07	4.55	3.54	4.55	6.07
8	7.981	9.31	20.0	44.6	40.0	10.6	7.98	5.98	4.65	5.98	7.98
10	10.02	11.7	25.0	55.7	50.0	13.3	10.0	7.51	5.85	7.51	10.0
12	11.94	13.9	29.8	66.3	59.6	15.9	11.9	8.95	6.96	8.95	11.9
14	13.13	15.3	32.8	73.0	65.6	17.5	13.1	9.85	7.65	9.85	13.1
16	15.00	17.5	37.5	83.5	75.0	20.0	15.0	11.2	8.75	11.2	15.0
18	16.88	19.7	42.1	93.8	84.2	22.5	16.9	12.7	9.85	12.7	16.9
20	18.81	22.0	47.0	105	94.0	25.1	18.8	14.1	11.0	14.1	18.8
24	22.63	26.4	56.6	126	113	30.2	22.6	17.0	13.2	17.0	22.6

Note: For SI units, 1 ft = 0.305 m.

[a] Values for welded fittings are for conditions where bore is not obstructed by weld spatter or backing rings. If appreciably obstructed, use values for "Screwed Fittings."

[b] Flanged fittings have three-fourths the resistance of screwed elbows and tees.

[c] Tabular figures give the extra resistance due to curvature alone to which should be added the full length of travel.

Miter elbows^d (no. of miters)					Welding tees		Valves (screwed, flanged, or welded)			
1–45 in	1–60 in	1–90 in	2–90 in	3–90 in	Forged	Miter^d	Gate	Globe	Angle	Swing check
0.45	0.90	1.80	0.60	0.45	1.35	1.80	0.21	10	5.0	2.5
15	30	60	20	15	45	60	7	333	167	83

L = equivalent length in feet of Schedule 40
(standard weight) straight pipe^g

0.78	1.55	3.10	1.04	0.78	2.33	3.10	0.36	17.3	8.65	4.32
1.03	2.06	4.12	1.37	1.03	3.09	4.12	0.48	22.9	11.4	5.72
1.31	2.62	5.24	1.75	1.31	3.93	5.24	0.61	29.1	14.6	7.27
1.72	3.45	6.90	2.30	1.72	5.17	6.90	0.81	38.3	19.1	9.58
2.01	4.02	8.04	2.68	2.01	6.04	8.04	0.94	44.7	22.4	11.2
2.58	5.17	10.3	3.45	2.58	7.75	10.3	1.21	57.4	28.7	14.4
3.08	6.16	12.3	4.11	3.08	9.25	12.3	1.44	68.5	34.3	17.1
3.84	7.67	15.3	5.11	3.84	11.5	15.3	1.79	85.2	42.6	21.3
5.04	10.1	20.2	6.71	5.04	15.1	20.2	2.35	112	56.0	28.0
6.30	12.6	25.2	8.40	6.30	18.9	25.2	2.94	140	70.0	35.0
7.58	15.2	30.4	10.1	7.58	22.8	30.4	3.54	168	84.1	42.1
9.97	20.0	40.0	13.3	9.97	29.9	40.0	4.65	222	111	55.5
12.5	25.0	50.0	16.7	12.5	37.6	50.0	5.85	278	139	69.5
14.9	29.8	59.6	19.9	14.9	44.8	59.6	6.96	332	166	83.0
16.4	32.8	65.6	21.9	16.4	49.2	65.6	7.65	364	182	91.0
18.8	37.5	75.0	25.0	18.8	56.2	75.0	8.75	417	208	104
21.1	42.1	84.2	28.1	21.1	63.2	84.2	9.85	469	234	117
23.5	47.0	94.0	31.4	23.5	70.6	94.0	11.0	522	261	131
28.3	56.6	113	37.8	28.3	85.0	113	13.2	629	314	157

^{d} Small-size socket-welding fittings are equivalent to miter elbows and miter tees.

^{e} Equivalent resistance in number of diameters of straight pipe computed for a value of f—0.0075 from the relation n—$k/4f$.

^{f} For condition of minimum resistance where the centerline length of each miter is between d and $2\frac{1}{2}d$.

^{g} For pipe having other inside diameters, the equivalent resistance may be computed from the above n values.

TABLE 13.8 Average Gas Demand Values for Typical Equipment (*In British thermal units per hour*)

Appliance	Input-BTU/h (MJ/h)
Commercial kitchen equipment:	
Small broiler	30,000 (31.7)
Large broiler	60,000 (63.3)
Combination broiler and roaster	66,000 (69.6)
Coffee maker, 3 burner	18,000 (19)
Coffee maker, 4 burner	24,000 (25.3)
Deep fat fryer, 45 pounds (20.4 kg) of fat	50,000 (52.8)
Deep fat fryer, 75 pounds (34.1 kg) of fat	75,000 (79.1)
Doughnut fryer, 200 pounds (90.8 kg) of fat	72,000 (76)
2 deck baking and roasting oven	100,000 (105.5)
3 deck baking oven	96,000 (101.3)
Revolving oven, 4 or 5 trays	210,000 (221.6)
Range with hot top and oven	90,000 (95)
Range with hot top	45,000 (47.5)
Range with fry top and oven	100,000 (105.5)
Range with fry top	50,000 (52.8)
Coffee urn, single, 5 gal. (18.9 L)	28,000 (29.5)
Coffee urn, twin, 10 gal. (37.9 L)	56,000 (59.1)
Coffee urn twin, 15 gal. (56.8 L)	84,000 (88.6)
Stackable convection oven, per section of oven	60,000 (63.3)
Residential equipment:	
Clothes dryer	35,000 (36.9)
Range	65,000 (68.6)
Stove top burners	40,000 (42.2)
Oven	25,000 (26.4)
30 gal. (113.6 L) water heater	30,000 (31.7)
40 to 50 gal. (151.4 to 189.3 L) water heater	50,000 (52.8)
Log lighter	25,000 (26.4)
Barbecue	50,000 (52.8)
Miscellaneous equipment	50,000 (52.8)
Commercial log lighter	50,000 (52.8)
Bunsen burner	5,000 (5.3)
Gas engine, per horsepower (745.7 W)	10,000 (10.6)
Steam boiler, per horsepower (745.7 W)	50,000 (52.8)

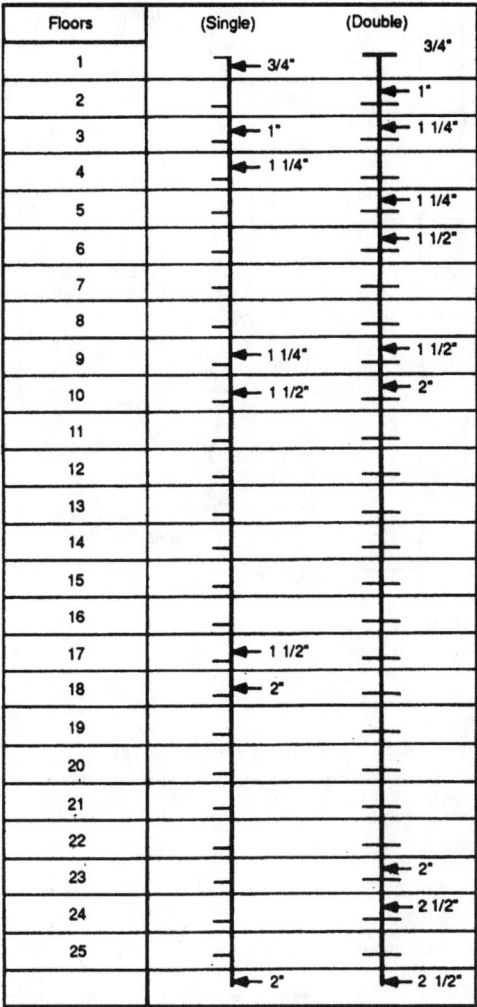

FIGURE 13.3 Gas riser pipe sizing for multiple dwellings.

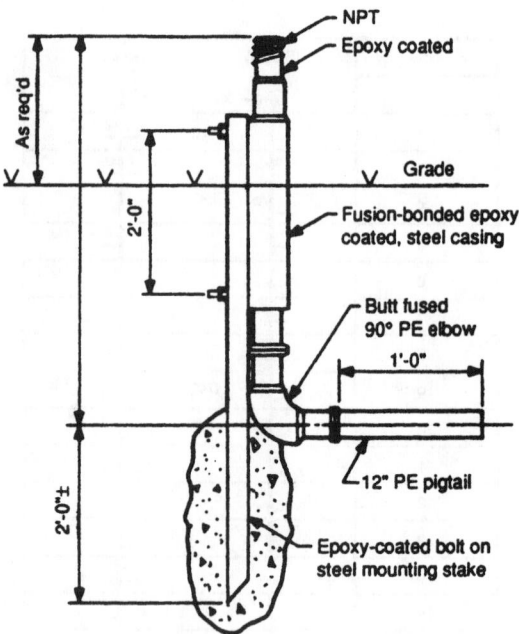

FIGURE 13.4 Transition fitting from underground to above-ground pipe.

TABLE 13.5 Laboratory Diversity Factors

Number of outlets	Average % use	Maximum
1–5	100	
6–10	75	90
11–20	60	75
21–70	40	60
71–150	30	50
Over	20	40

TABLE 13.10 Maximum Demands for Trailer Parks

Number of trailer sites	British thermal units per hour per trailer site	MJ/h
1	125,000	(132)
2	117,000	(123)
3	104,000	(109)
4	96,000	(101)
5	92,000	(97)
6	87,000	(92)
7	83,000	(88)
8	81,000	(86)
9	79,000	(85)
10	77,000	(81)
11–20	66,000	(69)
21–30	62,000	(65)
31–40	58,000	(61)
41–60	55,000	(58)
Over 60	50,000	(53)

Pipe and System Materials

All of the piping materials must be listed in NFPA 54 and other applicable codes. The most often used material for underground lines are high-density polyethylene (PE). For larger lines and high-pressure piping, the most often used material is steel pipe protected against corrosion by wrapping with a plastic coating.

Codes do not permit plastic pipe to be run above ground. For above-ground lines the piping and jointing methods depend on pipe size and system pressure. For piping up to 3 psig (21 kPa), black steel, ASTM A-53, pipe is used with cast or malleable iron screwed fittings in sizes of 3 in (75 mm) and smaller and butt-welded joints of 4 in (100 mm) and larger. Where natural gas is considered "dry gas," type G (ASTM B-837) copper pipe is finding widespread use. It is available in sizes up to 1⅛ in. All piping for pressures over 3 psig (21 kPa) shall be made of steel with welded joints. Welded fittings shall conform to ANSI 16.9. Screwed fittings shall be made of black malleable iron, 150-lb class conforming to ANSI A-197. Check valves shall be cast iron, 316 SS trim, disk type, with a soft seat. In sizes of 3 in and smaller, use screwed joints. In larger sizes, use flanged joints.

Joints for PE should be butt, heat-fused joints. Socket joints have been found to introduce a stiffness in the joint area that is undesirable. Joints for steel pipe shall be screwed for pipe sizes of 4 in (100 mm) and smaller. Sizes of 5 in (125 mm) and larger should be welded.

TABLE 13.11 Low-Pressure Gas Pipe Sizing Schedule

(*Maximum capacity of pipe in cubic feet of gas per hour for gas pressures of 0.5 psig or less and a pressure drop of 0.3 in water column, based on a 0.60 specific gravity gas*)

Nominal iron pipe size, in	Internal diameter, in	Length of pipe, ft													
		10	20	30	40	50	60	70	80	90	100	125	150	175	200
1/4	0.364	32	22	18	15	14	12	11	11	10	9	8	8	7	6
3/8	0.493	72	49	40	34	30	27	25	23	22	21	18	17	15	14
1/2	0.622	132	92	73	63	56	50	46	43	40	38	34	31	28	26
3/4	0.824	278	190	152	130	115	105	96	90	84	79	72	64	59	55
1	1.049	520	350	285	245	215	195	180	170	160	150	130	120	110	100
1 1/4	1.380	1,050	730	590	500	440	400	370	350	320	305	275	250	225	210
1 1/2	1.610	1,600	1,100	890	760	670	610	560	530	490	460	410	380	350	320
2	2.067	3,050	2,100	1,650	1,450	1,270	1,150	1,050	990	930	870	780	710	650	610
2 1/2	2.469	4,800	3,300	2,700	2,300	2,000	1,850	1,700	1,600	1,500	1,400	1,250	1,130	1,050	980
3	3.068	8,500	5,900	4,700	4,100	3,600	3,250	3,000	2,800	2,600	2,500	2,200	2,000	1,850	1,700
4	4.026	17,500	12,000	9,700	8,300	7,400	6,800	6,200	5,800	5,400	5,100	4,500	4,100	3,800	3,500

TABLE 13.12 Multipliers to Be Used for Specific Gravities Different from 0.60

Specific gravity	Multiplier	Specific gravity	Multiplier
0.35	1.31	1.00	0.78
0.40	1.23	1.10	0.74
0.45	1.16	1.20	0.71
0.50	1.10	1.30	0.68
0.55	1.04	1.40	0.66
0.60	1.00	1.50	0.63
0.65	0.96	1.60	0.61
0.70	0.93	1.70	0.59
0.75	0.90	1.80	0.58
0.80	0.87	1.90	0.56
0.85	0.84	2.00	0.55
0.90	0.82	2.10	0.54

TABLE 13.13 Conversion of Gas Pressure to Various Designations

kP	Equivalent inches		Pressure per square inch		Equivalent inches		Pressure per square inch		kPa
	Water	Mercury	Pounds	Ounces	Water	Mercury	Pounds	Ounces	
0.002	0.01	0.007	0.0036	0.0577	8.0	0.588	0.289	4.62	2.0
0.05	0.20	0.015	0.0072	0.115	9.0	0.662	0.325	5.20	2.2
0.07	0.30	0.022	0.0108	0.173					
0.10	0.40	0.029	0.0145	0.231	10.0	0.74	0.361	5.77	2.5
					11.0	0.81	0.397	6.34	2.7
0.12	0.50	0.037	0.0181	0.239	12.0	0.88	0.433		3.0
0.15	0.60	0.044	0.0217	0.346	13.0	0.96	0.469	7.50	3.2
0.17	0.70	0.051	0.0253	0.404					
0.19	0.80	0.059	0.0289	0.462	13.6	1.00	0.491	7.86	3.37
0.22	0.90	0.066	0.0325	0.520	13.9	1.02	0.500	8.00	3.4
					14.0	1.06	0.505	8.08	3.5
0.25	1.00	0.074	0.036	0.577					
0.3	1.36	0.100	0.049	0.785	15.0	1.10	0.542	8.7	3.7
0.4	1.74	0.128	0.067	1.00	16.0	1.18	0.578	9.2	4.0
0.5	2.00	0.147	0.072	1.15	17.0	1.25	0.614	9.8	4.2
0.72	2.77	0.203	0.100	1.60	18.0	1.33	0.650	10.4	4.5
0.76	3.00	0.221	0.109	1.73	19.0	1.40	0.686	10.9	4.7
1.0	4.00	0.294	0.144	2.31					
					20.0	1.47	0.722	11.5	5.0
1.2	5.0	0.368	0.181	2.89			0.903	14.4	6.2
1.5	6.0	0.442	0.217	3.46	25.0	1.84	0.975	15.7	6.7
1.7	7.0	0.515	0.253	4.04	27.2	2.00	1.00	16.0	6.9
					27.7	2.03			

LIQUEFIED PETROLEUM GAS

LPG is supplied to facilities from a tank truck (or railroad tank car if the storage tank is large enough) and is stored on-site as a liquid. Where usage is large, permanent, large tanks are installed on the facility property and refilled directly from a tanker truck. If the usage requirements are small, small propane tanks containing liquid may be installed and the tanks replaced after they are emptied.

The liquid must be vaporized to produce a gas. Depending on the actual installation and flow rate, the liquid can be vaporized in the storage tank (using heat gained through the storage tank wall) or in an auxiliary vaporizer that uses an outside source of heat. The gas may be used as vaporized, or, if it is a substitute supply for natural gas, it may have to be mixed with air to provide a lower British thermal unit content suitable for the specific application and equipment.

STORAGE TANKS

Tanks used for the storage of propane are made of steel. They can be installed either above or below grade. Although underground tanks have an advantage of a more constant environment, most tanks are placed above ground because of the lower initial cost and EPA requirements. Because of the relatively high pressure developed by the propane vapor, all tanks must be designed in conformance with the ASME code for unfired pressure vessels. The vapor pressure developed in above-ground tanks is based on the ambient outside air and can be found from Fig. 13.5. For underground tanks, a figure of 50°F (10°C) is used. Typical capacities of large, standard tanks are shown in Fig. 13.6. The dimensions and capacity of smaller, standard tanks are summarized in Table 13.14. The dimensions vary slightly from manufacturer to manufacturer. For nonstandard tanks, Fig. 13.7 contains a simplified method for calculating the areas of tanks with various configurations.

Although it is common practice to classify containers as either portable or stationary, there are too many exceptions to make these classifications practical. Containers and storage tanks are referred to as either *Department of Transportation* (DOT) *cylinders* (generally portable) or *ASME tanks* (generally stationary). In the following discussions, all capacities referring to gallons are gallons of water, and all weights are of liquid propane, unless indicated otherwise.

DOT cylinders range in size from 1- to 420-lb capacity. Cylinders built in 1966 or earlier may bear the Interstate Commerce Commission (ICC) designation. ASME tanks range in size from 500- to 60,000-gal capacity.

The advantages of an underground tank are:

1. The tank is not visible if aesthetics are a consideration.
2. There is greater vaporization of liquid in the winter due to the constant temperature, which is about 50°F.

Vapor Pressures of Butane-Propane Mixture

Temperature in degrees Celsius

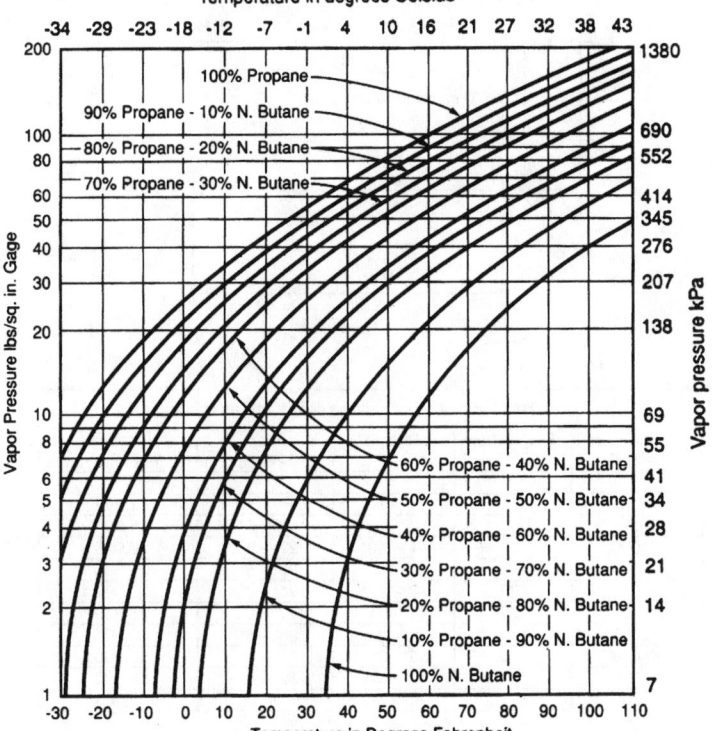

FIGURE 13.5   Vapor pressure of butane-propane mixtures. (*Courtesy Fisher Regulator.*)

The disadvantages are:

1. The tank may have to be anchored to prevent floating in areas of high water tables.
2. The initial cost is higher.
3. Inspection and maintenance are more difficult to perform.

If the underground tank is subject to traffic or potential damage, the top should be a minimum of 2 ft below grade. In remote locations, 6 in of cover is considered adequate. A manhole giving access to the valves, gauges, connections, and so on must be provided for maintenance and inspection.

New tanks must be purged of both air and water prior to being placed in service. A concentration of 6 percent air is the maximum limit acceptable. Water should be removed from the tank by the manufacturer, and the tank should be shipped to the site sealed.

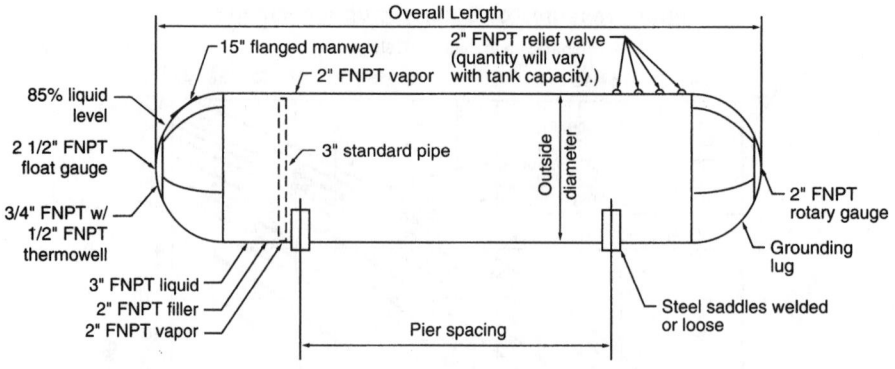

FIGURE 13.6   Dimensions and capacities of larger propane tanks. (*Courtesy Fisher Regulator.*)

TABLE 13.14   Dimensions and Capacities of Small Propane Tanks

Tank size, gal	Net propane capacity, gal, 60°F	Outside diameter, in	Overall length	Total surface	15% full	25% full	33% full	50% full
120	99	24	5 ft 7 in	35.1	7.0	8.75	14.0	17.5
150	124	24	6 ft 11 in	43.5	8.7	10.8	17.4	21.7
250	207	30	7 ft 5 in	58.3	11.6	14.8	23.3	29.1
325	269	30	9 ft 6 in	74.6	14.8	17.6	29.8	37.3
500	414	37	10 ft 0 in	96.9	19.4	24.2	38.1	42.4
1000	827	41	16 ft 1 in	172.6	34.8	43.1	69.1	86.3

Note: The "Dimensions" header spans "Outside diameter, in" and "Overall length"; the "Surface area, ft²" header spans "Total surface" through "50% full"; "Wetted surface" spans "15% full" through "50% full".

1 Gal = 4.5 L
1 Ft = 0.3 m
1 In = 25.4 mm
1 Sq Ft. = 0.092 m^2

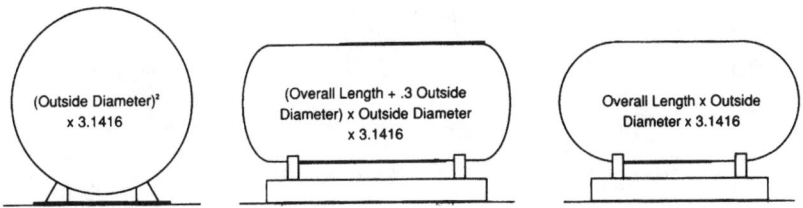

$(Outside Diameter)^2$ x 3.1416

(Overall Length + .3 Outside Diameter) x Outside Diameter x 3.1416

Overall Length x Outside Diameter x 3.1416

FIGURE 13.7 Areas of nonstandard tanks. (*Courtesy Trinity Industries.*)

MAXIMUM CONTENT OF LIQUID IN TANKS

The maximum allowable quantity of LPG permitted in a container is limited by NFPA 58 to a filling density based on tables and formulas provided in that code. In future discussions of tank sizing, this maximum allowable quantity must be compared to that calculated by other methods to make sure that the maximum is not exceeded.

LOCATION OF EQUIPMENT ON THE SITE

The factors to be considered in locating the equipment are:

1. Code required clearance from other buildings, property lines, and any other equipment. These clearances also depend on insurance carrier requirements and client preferences. Factory Mutual requirements, indicated in Fig. 13.8, are considered conservative. Distances specified in NFPA 54 and other insurance carrier codes vary but are not longer.
2. Accessibility for fuel delivery.
3. Location of underground utilities.
4. Site elevations. Avoid placing the equipment at low points.
5. Client preferences.

TANK FOUNDATIONS AND SUPPORT

Tank foundations must provide a stable means of support. Uneven settlement will lead to errors in fuel gauging and stresses leading to broken lines. The weight of the tank should be calculated using water since there may be a requirement for hydrostatic testing during the life of the system. An approximation of the gross weight of a steel tank can be obtained by multiplying the capacity in gallons by 11 for water and 6 for propane. Some larger tanks may require a temporary, intermediate center support during the time they are filled with water for hydrostatic testing.

Minimum Recommended Distance

Dimension	Point to point		Distance, ft	Distance, m
A	1 to	3a	75	23
		4a	150	46
		5b	200	60
		5c	350	105
B		6	20	6
C		12	200	60
D		13	50	15
E	2 to	6	20	6
F		7	50	15
		12	200	60
G		13	75	23
H	3,4,5 to	6	5	1.5
I		7	15	4.5
		8	100	30
J		9	50	15
K		10	20	6
		11	75	23
L		12	75	23
M		13	50	15
N		14	75	23
O	6 to	15	50	15
	7 to 12,13,15		75	23
	13 to 14		75	23

Notes:
a. For single tanks only. Treat multiple tanks as No. 5.
b. For buildings with hydrant protection.
c. For buildings without hydrant protection.
d. 5 ft (1.5 m) for tanks within a group.
e. For tanks smaller than 2000 gal (7.6 m³), 25 ft (7.6 m).

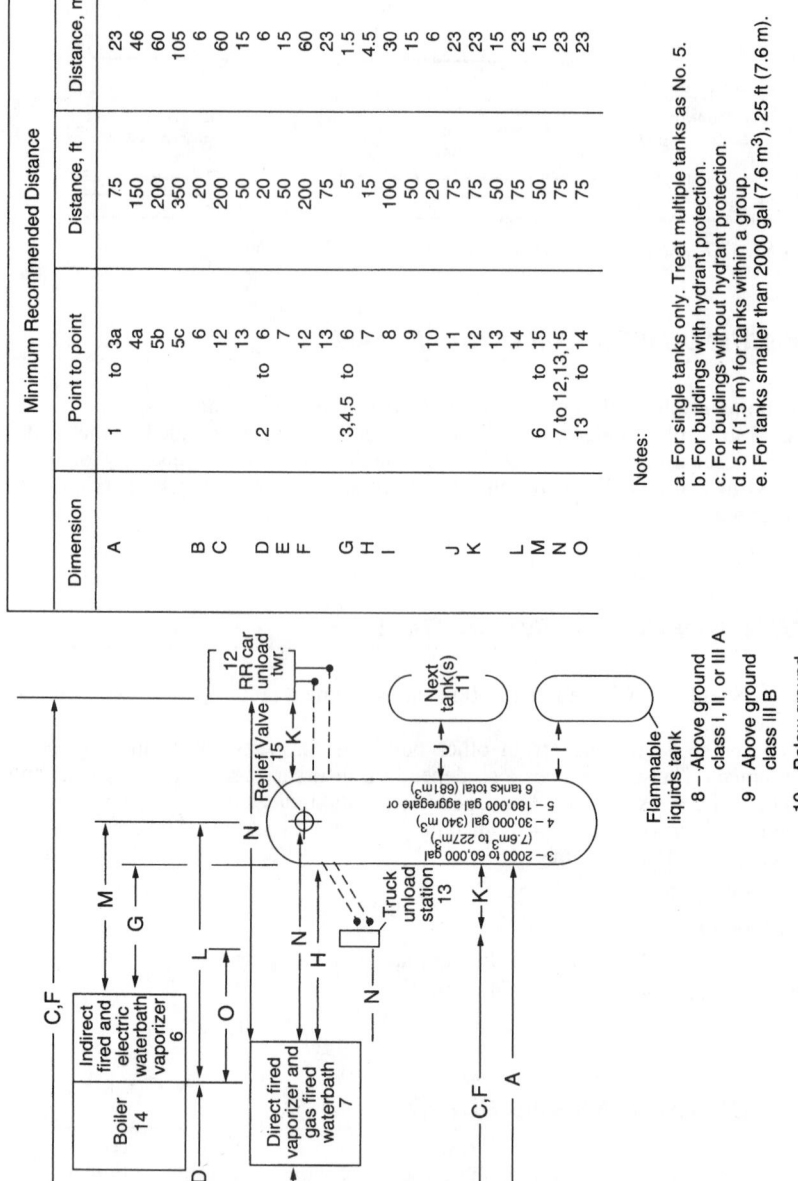

8 – Above ground class I, II, or III A
9 – Above ground class III B
10 – Below ground

FIGURE 13.8 Clearances from propane tank to buildings or other structures.

REGULATORS

The regulator for LPG is different from that used for natural gas service. The purpose of an LPG regulator is to reduce a variable inlet pressure to a constant outlet pressure under variable flow conditions. The following information is required to size a regulator:

1. Minimum significant and maximum flow rate possible
2. Maximum and minimum container pressure
3. Required outlet pressure desired
4. Manufacturer's rating curves for the regulator

The pressure developed inside the storage tank is produced by the vaporization of product inside the storage tank. This pressure is called the *vapor pressure*. This figure varies, depending on the ambient temperature. Figure 13.6 is a direct reading chart to determine the vapor pressure for various mixtures of propane and butane that might be provided by suppliers. Outside temperatures can be found by requesting this information from the National Oceanographic and Atmospheric Administration (NOAA).

Regulators are manufactured in single- and double-stage models and are selected using the capacity and rating curves supplied by the manufacturer. If a relief valve is provided as part of the regulator assembly, refer to Table 13.15 for pressure settings.

Experience has shown that when regulators are kept in service longer than 15 years, the probability of malfunction increases. This figure is only a generalization, with actual climate and service conditions playing an important factor in the service life of a regulator.

PRESSURE RELIEF DEVICES

The purpose of a pressure relief device is to automatically vent propane vapor to the atmosphere upon reaching a predetermined high pressure. It can be an integral part of a pressure regulator, separately installed on the storage tank, part of vaporizers and mixers, or in the piping system itself.

The relief valve flow capacity is calculated on the basis of a full tank and the resultant vaporization of that volume of propane. For an underground tank, the

TABLE 13.15   Final-Stage Relief Valve Settings

Regular delivery pressure	Relief valve start-to-discharge pressure setting (percent of regulator delivery pressure)	
	Minimum, %	Maximum, %
1 psig or less	200	300
Above 1 psig but not over 3 psig	140	200
Above 3 psig	125	200

capacity will be 30 percent of the above-ground value. Refer to Table 13.16 to determine the required flow of propane based on the square foot area of the storage tank or use Fig. 13.7. All larger tanks must have a separate relief valve installed directly on the tank. If a single relief valve does not have the required flow capacity, it will be necessary to have multiple relief valves mounted on the top of the tank to provide the necessary flow rate. A typical tank relief valve is illustrated in Fig. 13.9. The final pressure setting of the relief valve is nominally 88 to 100 percent of the design pressure of the tank, with a permitted error of +10 percent.

TABLE 13.16   Relieving Capacity for Relief Valves on Tanks and Vaporizers

(*Minimum required rate of discharge in cubic feet per minute of gas at 120 percent of the maximum permitted start-to-discharge pressure for relief valves to be used on containers other than those constructed in accordance with Department of Transportation specifications*)

Surface area, ft^2	Flow rate, cfm air	Surface area, ft^2	Flow rate, cfm air	Surface area, ft^2	Flow rate, cfm air
20	626	170	3,620	600	10,170
25	751	175	3,700	650	10,860
30	872	180	3,790	700	11,550
35	990	185	3,880	750	12,220
40	1,100	190	3,960	800	12,880
45	1,220	195	4,050	850	13,540
50	1,330	200	4,130	900	14,190
55	1,430	210	4,300	950	14,830
60	1,540	220	4,470	1,000	15,470
65	1,640	230	4,630	1,050	16,100
70	1,750	240	4,800	1,100	16,720
75	1,850	250	4,960	1,150	17,350
80	1,950	260	5,130	1,200	17,960
85	2,050	270	5,290	1,250	18,570
90	2,150	280	5,450	1,300	19,180
95	2,240	290	5,610	1,350	19,780
100	2,340	300	5,760	1,400	20,380
105	2,440	310	5,920	1,450	20,980
110	2,530	320	6,080	1,500	21,570
115	2,630	330	6,230	1,550	22,160
120	2,720	340	6,390	1,600	22,740
125	2,810	350	6,540	1,650	23,320
130	2,900	360	6,690	1,700	23,900
135	2,990	370	6,840	1,750	24,470
140	3,080	380	7,000	1,800	25,050
145	3,170	390	7,150	1,850	25,620
150	3,260	400	7,300	1,900	26,180
155	3,350	450	8,040	1,950	26,750
160	3,440	500	8,760	2,000	27,310
165	3,530	550	9,470		

Note: Buried tanks require only 30 percent of the vent capacity shown.
Source: Factory Mutual.

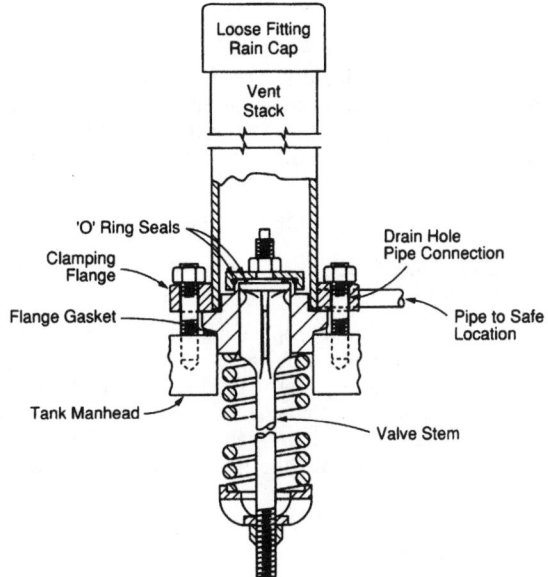

FIGURE 13.9　Tank relief valve. (*Courtesy Factory Mutual.*)

Another pressure relief device within the piping system is the external bypass valve, which relieves excessive pressure in an LPG pumping system upon reaching a high-pressure set point and returns the discharge back to the storage tank instead of to the atmosphere as is done by a relief valve. All LPG pumps require an external bypass valve. Different types of bypass valves are manufactured for different applications.

EXCESS FLOW VALVE

An excess flow valve permits the flow of vapor or liquid in both directions but shuts off the flow of liquid or vapor in only one direction when the flow exceeds a predetermined limit. It is recommended that an excess flow device be placed on all connections to a larger tank except for the safety relief valve. The filler connection should have an integral valve. The capacity is calculated from the largest expected flow and its mounting position (horizontal or vertical). Valves are selected to open at between 150 and 200 percent of the expected maximum flow.

SERVICE LINE VALVES

A service line valve controls the flow of propane from a DOT cylinder. This is a multipurpose valve that could contain a shutoff valve, relief valve, gauge, and filler valve all in one body.

FILLER VALVES

A filler valve permits the flow of liquid in one direction only and is used for filling larger tanks with liquid. This valve contains an integral back pressure check valve that will prevent the loss of vapor or liquid from the tank if the fill hose or a fitting ruptures. A preferred location is at the filling connection used by the delivery truck hose.

VAPOR EQUALIZING VALVES

When large tanks are filled, the liquid propane added to the storage tank compresses the vapor in the tank. A vapor equalizing line is connected between the vapor area of the receiving tank and the vapor area of the delivery truck during propane deliveries to equalize pressure in both containers and to lessen the head requirements of the supply pump. The vapor equalizing valve is mounted in the equalizing line tank truck hose connection and permits the flow of vapor both ways.

LIQUID LEVEL GAUGES

When a larger tank is filled on the basis of volume, it is important that a liquid level gauge be installed on the tank to indicate the exact liquid level inside. There are several types available.

The slip tube gauge is mounted on top of a tank and consists of a vertical rod extending from the liquid level inside to a point outside the tank for measurement. There is a float on one end of the tube that rides on the liquid, automatically rising and falling with the level.

A magnetic float gauge consists of a gauge faceplate mounted on the side of a tank and a connecting rod with a float at the end that rides on the liquid level. As the float rises and falls with the liquid, it automatically moves a magnetic device on the faceplate indicating the level inside. Because there is no direct opening to the tank outside, this is the most widely used gauge. Since it is a mechanical device, it is recommended that this type be used with the slip tube as a backup.

A rotary liquid level gauge is a manually operated gauge that uses an angled dip tube inside the tank connected to a bleeder valve on a calibrated faceplate on the outside of the tank. The bleeder valve is opened, and the dip tube is manually turned until the vapor escaping changes to a liquid. The level is then read from calibrated markings on the faceplate.

The fixed level gauge is an external needle valve connected to a short length of tubing extending inside the tank. There may be a number of these gauges installed at various calibrated levels on the tank. As the filling proceeds, the valve is cracked open until the vapor changes to liquid. When the desired level is reached, the filling operation stops.

MISCELLANEOUS EQUIPMENT

In addition to the devices indicated, it is often desirable to install a liquid temperature gauge and a pressure gauge on a tank to aid in diagnosing any potential problem.

VAPORIZER

A vaporizer is a device that converts liquid propane into a vapor in larger quantities than would be possible from ambient heat alone. It is required when sufficient quantities of liquid propane inside the storage tank cannot be vaporized quickly enough to satisfy the maximum demand.

There are two basic types of vaporizers: direct fired and indirect fired. The direct-fired vaporizer uses the propane itself as fuel for the direct flame. It has the lowest initial cost and no ancillary power requirements. Disadvantages include the problem of pilot light extinguishment in high winds, production of sludge in the propane being vaporized, and a useful life expectancy of only 5 years. Optional accessories to be considered would be an automatic ignition, manual drain, liquid strainer, and propane-air mixer. An indirect type of vaporizer is available that uses steam, glycol-water bath, or the lesser-used electric resistance heaters as a heat source.

PROPANE MIXERS

A *mixer*, also called a *blender* or *proportioner*, is a device used to combine pure propane and air together into a mixture of both gases. It is required when propane will be used as a direct substitute for natural gas because the much higher British thermal unit content of 100 percent propane will not allow the use of the same burner orifices and gas train as natural gas. Experience has shown that a mixed gas content of 1450 Btu/cf with a specific gravity of 1.30 provides the best burning characteristics. Refer to Table 13.17 for properties of various mixtures of propane and air.

Several methods can be used to accomplish mixing. The simplest is the use of a venturi. Intended for exterior installation, it consists of a nozzle inside of a larger body. Propane passing through the nozzle creates a partial vacuum that pulls air from around the opening, thereby creating a mixed gas. Because the venturi principle is accurate only within a very limited flow range, this type of unit is either on or off. If the demand pressure is higher than 5 psi (35 kPa), the use of an air compressor will be required to raise the mixed gas pressure. Generally, this system has the least initial cost. Four conditions have a strong bearing on the accuracy and composition of the mixture:

1. Venturi tube dimensions and throat diameter
2. Nozzle or orifice size and location of venturi in throat
3. Pressure of both gases at the venturi
4. Desired discharge pressure

Another and more costly device is the modulating proportional blender. Suitable only for interior installation, these units are generally used for larger quantities and higher pressures than available from a venturi unit. They use regulators to stabilize both the air and propane in separate supply lines. A proportioning valve is then used to meter both supplies into a common discharge line. An air compressor is often used to supply air at the higher pressure required. The vaporizer and mixer are also available as a factory-assembled package on a single skid, ready for installation.

TABLE 13.17   Properties of Various Propane/Air Mixtures

British thermal unit per cubic foot of mixture	Propane/air mixtures			
	Percentage of propane by volume	Percentage of air by volume	Percentage of oxygen by volume	Specific gravity of the mixture
2550	100.00	0.00	0.000	1.523
2500	98.04	1.96	0.409	1.513
2450	96.08	3.92	0.819	1.502
2400	94.12	5.88	1.288	1.492
2350	92.16	7.84	1.639	1.482
2300	90.19	9.81	2.050	1.472
2250	88.24	11.76	2.458	1.461
2200	86.27	13.73	2.869	1.451
2150	84.31	15.69	3.279	1.441
2100	82.35	17.65	3.688	1.431
2050	80.39	19.61	4.098	1.420
2000	78.43	21.56	4.506	1.410
1950	76.47	23.53	4.918	1.400
1900	74.51	25.49	5.317	1.390
1850	72.55	27.45	5.737	1.379
1800	70.58	29.42	6.149	1.369
1750	68.62	31.38	6.558	1.359
1700	66.67	33.33	6.964	1.349
1650	64.70	35.30	7.378	1.338
1600	62.74	37.26	7.787	1.328
1550	60.78	39.22	8.197	1.318
1500	58.82	41.18	8.606	1.308
1450	56.86	43.14	9.016	1.297
1400	54.90	45.10	9.246	1.287
1350	52.94	47.06	9.835	1.277
1300	50.98	49.02	10.245	1.267
1250	49.02	50.98	10.654	1.256
1200	47.06	52.94	11.064	1.246
1150	45.09	54.91	11.476	1.236
1100	43.13	56.87	11.886	1.226
1050	41.17	58.83	12.295	1.215
1000	39.21	60.79	12.705	1.205
950	37.25	62.75	13.115	1.195
900	35.29	64.71	13.524	1.185
850	33.33	66.67	13.934	1.174
800	31.37	68.63	14.344	1.164
750	29.41	70.59	14.753	1.154
700	27.45	72.55	15.163	1.144
650	25.49	74.51	15.573	1.133
600	23.53	76.47	15.982	1.123
550	21.56	78.44	16.394	1.113
500	19.61	80.39	16.892	1.103

(Continued)

TABLE 13.17   Properties of Various Propane/Air Mixtures (*Continued*)

British thermal unit per cubic foot of mixture	Propane/air mixtures			
	Percentage of propane by volume	Percentage of air by volume	Percentage of oxygen by volume	Specific gravity of the mixture
450	17.65	82.35	17.211	1.092
400	15.69	84.31	17.621	1.082
350	13.73	86.27	18.031	1.072
300	11.76	88.24	18.442	1.062
250	9.80	90.20	18.852	1.051
200	7.84	92.16	19.261	1.041
150	5.88	94.12	19.670	1.031
100	3.92	96.08	20.081	1.021

LPG SYSTEM DESIGN

Prior to sizing components, initially decide what equipment (such as pumps or vaporizers) may be required and if the tank is to be above or below ground, and select the proposed location of the tank and other required equipment on the site. Determine if the propane system will be in constant use or used only periodically, such as for emergency operation or as an occasional substitute for interrupted natural gas. Calculate the maximum hourly and daily fuel gas demand, and determine if this demand is continuous, such as that for a process used all day, or intermittent, such as that used for heating purposes. If only a small part of the demand is continuous, the entire load should be considered continuous.

Storage Tank Sizing and Selection

The storage tank volume is based on one of two factors: a reasonable return schedule for the local supplier or the amount of liquid propane that has to be vaporized by the ambient air in order to satisfy the maximum demand.

Vaporization directly from a tank, if economically feasible, is used to avoid having to add another mechanical device (a vaporizer) that requires constant maintenance. If a single tank sized to optimum criteria will not provide the required vaporization rate, a single oversized tank or two smaller, separate tanks are often installed if space conditions and initial cost are acceptable.

Tank Size Based on Return Schedule.   Using the return schedule, a preliminary starting point for determining actual capacity is a 10-day usable supply for continuous demand and between 3 to 5 days usable supply for intermittent or standby purposes. With the maximum propane level based on the maximum quantity for the allowed filling density and the minimum level between 10 and 15 percent of the capacity, the usable tank capacity is about 75 percent of the total tank capacity.

Tank Size Based on Vaporization Rate of Propane.   When using the vaporization rate as the critical factor in selecting the size of the storage tank, the tank must be large enough to vaporize the maximum flow rate of propane required by the facility when the outside air temperature and liquid level are at their lowest. For aboveground tanks, the rate of vaporization is calculated using the wetted area of liquid propane in the tank. For underground tanks, the entire area of the tank is used even if the tank is partially full.

There are two methods that will be discussed. The first, and most accurate, will use various formulas to calculate the wetted area of a tank that will vaporize the required amount of propane. The second, although slightly less accurate, is the direct reading of the vaporization rate from prepared charts. This is a more conservative method and by far the easier.

Formula Method. Using this method, the wetted area of the proposed aboveground tank necessary to vaporize the required amount of propane will be calculated. If an underground tank is proposed, the area calculated shall be that of the entire tank. The basic formulas are as follows.

TOTAL HEAT OF VAPORIZATION FROM AN ABOVE-GROUND TANK

$$Q = U \times A \times TD \tag{13.1}$$

TOTAL HEAT OF VAPORIZATION FROM A BELOW-GROUND TANK

$$Q = U \times A \times 15 \tag{13.2}$$

Equation (13.1) transposed to find the wetted area is:

$$A = \frac{Q}{U \times (TD \text{ or } 15)} \tag{13.3}$$

ACTUAL AMOUNT OF PROPANE VAPORIZED

$$V = \frac{AQ}{L} \tag{13.4}$$

where Q = the total amount of heat required to vaporize a quantity of liquid propane, Btu.

U = rate of heat transfer through the walls of a steel tank, per square foot of wetted area/temperature difference in °F. Generally accepted figures are 2 Btu/sf/°F for above-ground tanks based on severe conditions and 0.5 Btu/sf/°F for underground tanks.

A = the wetted surface area of the above-ground tank containing liquid propane, or the entire surface area of an underground tank, ft².

TD = the intended difference in temperature between ambient outside air reached during the coldest part of the day and the temperature of propane in the tank reached during the warmest part of the day, in °F. One factor that must be taken into consideration is the formation of frost on the outside of the storage tank. Since frost acts as an insulation, its formation must be avoided. Table 13.18 gives the minimum temperature difference to avoid frost formation. When predicted conditions are outside the figures required to enter the table, use the true temperature difference.

15 = 15°F, which is the generally accepted temperature difference for underground tanks.

V = propane vaporized under design conditions, gal.

TABLE 13.18　Difference between Air Temperature of Frost Formation

°C	Lowest air temp. °F	Relative humidity							
		20	30	40	50	60	70	80	90
−34.4	−30	—	—	—	—	8.0*	5.0*	2.5*	1.0*
−28.9	−20	—	20.0*	15.0*	11.5*	8.5	5.0	3.0	1.5
−23.3	−10	27.5*	20.5	16.0	12.0	9.0	6.0	3.0	1.5
−17.8	0	29.0	21.5	16.5	12.5	9.0	6.0	4.0	2.0
−12.2	10	30.0	22.5	17.0	13.0	9.5	6.5	4.0	2.0
−6.7	20	31.5	24.0	18.0	14.0	10.0	7.0	4.0	2.0
−1.1	30	33.0	25.0	19.5	15.0	11.0	8.0	5.0	3.0
4.1	40	35.0	27.0	21.0	16.5	12.0	9.0	8.0	8.0

*If the full temperature difference is used in these cases, the minimum tank pressure may be too low for satisfactory performance.
1°F = 0.55°C.

L = latent heat of vaporization of propane, Btu/gal. This is the amount of heat required to change liquid propane to a vapor (Table 13.19). Interpolate to find intermediate values when necessary.

The following methodology is used to select a tank with propane vaporization rate as criteria, solving for Eq. (13.3):

Step 1. Establish the lowest predicted ambient air temperature, highest predicted relative humidity, and the required propane gas demand in Btu/h/gal.

Step 2. Convert the British thermal unit per hour demand into gallons per hour. Refer to Table 13.2.

Step 3. Transpose Eq. (13.4) to find Q. Refer to Table 13.19 to find L.

Step 4. Substitute all figures into Eq. (13.3), and solve for the wetted area. After calculating the required minimum area, consult manufacturers' catalogs for a

TABLE 13.19　Latent Heat of Vaporization for Propane

°C	Ambient air temp, °F	Propane	
		Btu/lb	Btu/gal
−40	−40	180.8	765
−34.4	−30	178.7	755
−28.9	−20	176.2	745
−23.3	−10	173.9	735
−17.8	0	171.5	725
−12.2	10	169.0	715
−6.7	20	166.3	704
−1.1	30	163.4	691
4.4	40	160.3	678
10.0	50	156.5	662
15.6	60	152.6	645

size tank providing that figure with the minimum level of propane in the storage tank.

As an example, calculate the selection of a tank using the minimum area of a tank as the criteria. The following project conditions exist:

1. *Lowest predicted temperature:* 30°F
2. *Highest predicted humidity:* 60 percent
3. *Required propane demand:* 132,000 Btu/h
4. *Type of use:* Continuous

First, convert Btu demand per hour into gallons per hour using the figure obtained from Table 13.2:

$$\frac{132,000}{91,547} = 1.44 \text{ gal/h}$$

Second, find L using Table 13.19. Entering the table with 30°F, find 163.4 Btu. Third, calculate Q using transposed Eq. (13.4) to read $Q = L \times V$: Fourth, find TD using Table 13.18. Fifth, substitute the above figures in Eq. (13.3):

$$A = \frac{235.2}{2 \times 11} = \frac{235.2}{22}$$

$$= 10.69 \text{ ft}^2 \text{ of wetted area of tank}$$

The result of the previous calculation is a tank area that will contain the absolute lowest volume of propane that will adequately supply facility needs. A higher volume should be used as an actual "low level." The additional capacity will be required to allow enough time for a delivery before the volume in the tank falls below the level needed to supply the facility. This low level will be a percent of the total capacity of the tank that will be used to select the actual, total capacity of the storage tank in conjunction with the proposed schedule of the supplier. Select a "standard" tank with a slightly greater volume than that calculated. If a single tank selected for adequate vaporization results in a size that is impractical, two tanks can be used. If the tanks are still too large or have too high an initial cost, the only other solution is to provide a vaporizer.

Direct-Reading Method. Table 13.20 provides a nomogram for determining a direct reading of propane vaporization from small tanks. Figure 13.10 provides a table for a direct reading of propane vaporization from larger tanks. Both Table 13.20 and Fig. 13.10 are based on a tank 25 percent full. Reduce the figures by 10 percent to approximate tanks 15 percent full.

LPG Gas Pipe Sizing Methods

A widely used pipe sizing method for low-pressure, 100 percent propane gas systems relies upon the tables used for NG in NFPA 54, such as that appearing in Table 13.7. Proprietary charts are available that are less conservative and more accurate. For use with propane, a conversion factor of 0.63 to reduce the indicated flow rate must be used.

TABLE 13.20 Propane Vaporization Chart for Small Tanks (*Cubic feet per hour of propane*)

Atmos. temp		Cylinder capacity, lb LP gas				
°C	°F	20	100	150	300	420
15.6	60	16	31	50	60	95
−1.1	30	10	20	33	46	63
−12.2	10	7	13	21	30	41
−17.0	0	5	10	15	22	30
−23.3	−10	3	6	10	15	19

Note: The above capacities are based on the assumption that the cylinder is 25 percent filled. It does not include effect of sensible heat. These figures are on the conservative side and should be used as a guide only. They are based on the cylinder being in the shade in still air and on continuous withdrawal. Due to beneficial effect of drafts, sunlight, radiation, and intermittent operations, most actual installations will have a greater capacity than shown in the table.
    1 CFH = 0.03 m³/h.

Propane is often mixed with air to provide a direct substitute for natural gas. Field experience has found that if mixed gas with the same British thermal unit value (1050 Btu/cf) as natural gas is used, the mixture will not burn properly. As a result, it is common practice to provide a mixed gas with a value of 1450 Btu/cf. This mixture has a specific gravity of 1.30. Since gas flow varies inversely as the square root of the specific gravity, a conversion factor can be used with readily available natural gas charts to size the mixed gas–natural gas piping system. The conversion factor for 1450 Btu/cf mixed gas is 0.69, which is used to multiply the capacity found in NG tables in order to convert the chart figure to the actual mixed gas capacity value. The method of using the charts is discussed previously for NG sizing.

LPG Pumps

There are two commonly used pumps in LPG service: sliding vane positive displacement and turbine regenerative centrifugal. In general, the sliding vane positive displacement pump is generally used for flow of from 40 to 350 gpm. The turbine regenerative pump is generally used for constant flows and pressure when the flow is less than 40 gpm, such as those used to fill propane tanks for home use. Either type of pump could be used for any pressure for which it is rated.

To calculate the pump head, add the following:

1. The vapor pressure of the storage tank into which the liquid propane will be pumped. If there is a vapor return line, this will not be a factor.
2. The friction loss of the liquid through the piping network. Refer to Table 13.21 for the friction loss of liquid propane through steel piping. Table 13.22 provides the losses through fittings in equivalent feet of steel pipe.
3. The friction losses of liquid propane through delivery hose in pounds per square inch. Refer to Table 13.23.
4. Static head difference between the supply and receiving points.
5. The resistance to flow from meters, filler valves, and so on. Obtain this information from manufacturers.

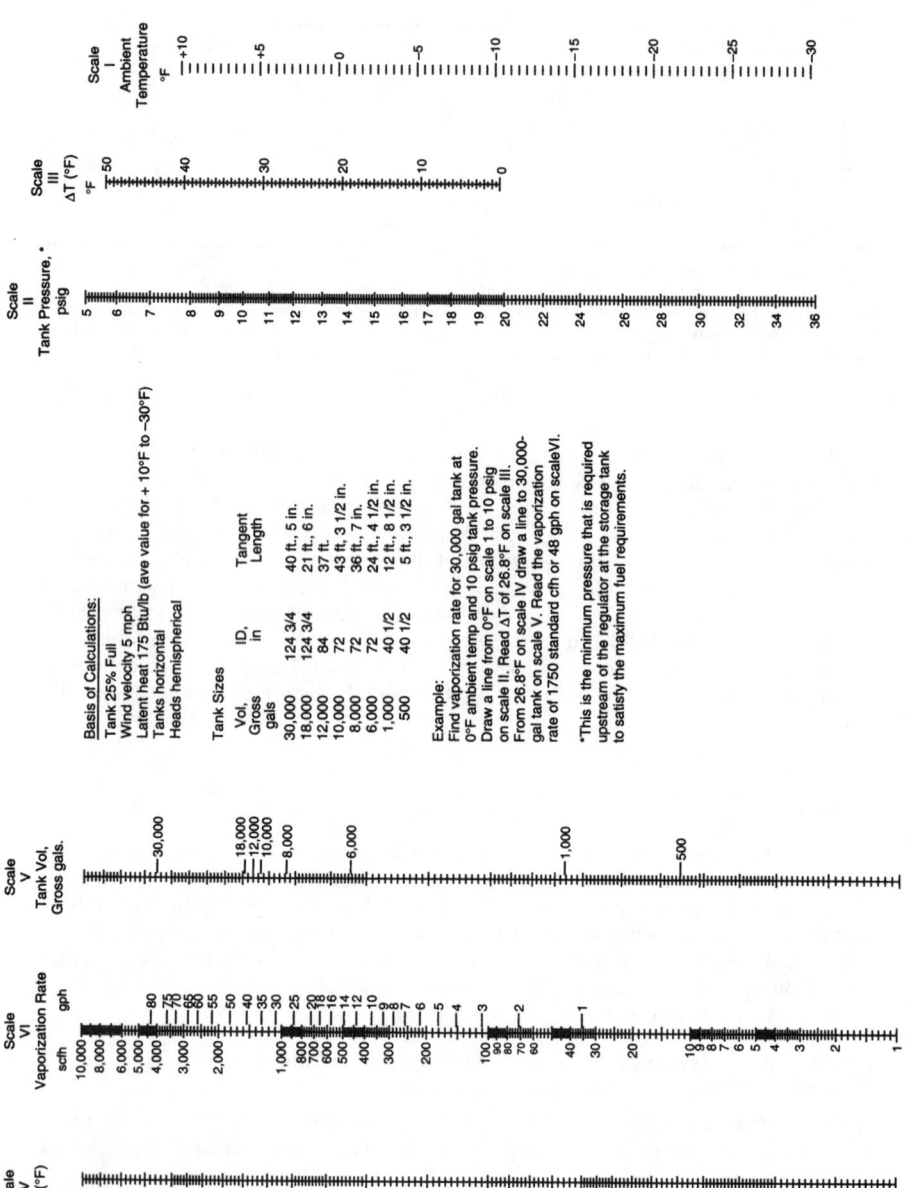

FIGURE 13.10 Propane vaporization of large propane tanks. (*Courtesy Phillips Petroleum Co.*)

Scale I
Ambient Temperature
°F

Scale III
ΔT (°F)
°F

Scale II
Tank Pressure, *
psig

Basis of Calculations:
Tank 25% Full
Wind velocity 5 mph
Latent heat 175 Btu/lb (ave value for +10°F to –30°F)
Tanks horizontal
Heads hemispherical

Tank Sizes

Vol, Gross gals	ID, in	Tangent Length
30,000	124 3/4	40 ft., 5 in.
18,000	124 3/4	21 ft., 6 in.
12,000	84	37 ft.
10,000	72	43 ft., 3 1/2 in.
8,000	72	36 ft., 7 in.
6,000	72	24 ft., 4 1/2 in.
1,000	40 1/2	12 ft., 8 1/2 in.
500	40 1/2	5 ft., 3 1/2 in.

Example:
Find vaporization rate for 30,000 gal tank at 0°F ambient temp and 10 psig tank pressure. Draw a line from 0°F on scale 1 to 10 psig on scale II. Read ΔT of 26.8°F on scale III. From 26.8°F on scale IV draw a line to 30,000-gal tank on scale V. Read the vaporization rate of 1750 standard cfh or 48 gph on scaleVI.

*This is the minimum pressure that is required upstream of the regulator at the storage tank to satisfy the maximum fuel requirements.

Scale V
Tank Vol,
Gross gals.

Scale VI
Vaporization Rate
scfh   gph

Scale IV
ΔT (°F)
°F

13.44

TABLE 13.21 Resistance of Steel Pipe to the Flow of Liquid Propane
(*Expressed in feet of head of liquid LP gas*)

Flow rate, U.S. gpm	Pipe size, schedule 40 steel						
	1 in	2 in	2½ in	3 in	4 in	5 in	6 in
10	0.007						
20	0.028	0.008					
30	0.060	0.017	0.007				
40		0.031	0.012				
50		0.048	0.020	0.004			
60		0.070	0.028	0.009			
70		0.095	0.038	0.013			
80			0.050	0.017			
90			0.063	0.021			
100			0.078	0.025	0.006	0.002	0.001
125				0.036	0.008	0.003	0.001
150				0.051	0.012	0.004	0.002
175				0.088	0.016	0.005	0.002
200				0.088	0.020	0.007	0.003
225					0.025	0.009	0.003
250					0.031	0.010	0.004
275					0.037	0.012	0.005
300					0.043	0.014	0.006

TABLE 13.22   Resistance of Valves and Fittings to Flow of Liquid Propane
(Expressed in equivalent feet of pipe)

Type valves and fittings	Resistance in equivalent feet of pipe of these sizes						
	$1\frac{1}{2}$ in	2 in	$2\frac{1}{2}$ in	3 in	4 in	5 in	6 in
Straight-through ball, and plug valves, same size as pipe	5	6	8	10	14	17	20
Globe valves, wide open, same size as pipe	40	50	60	80	110	130	160
Angle valves, wide open, same size as pipe	20	25	30	40	55	70	80
Swing check valves, same size as pipe	10	13	16	19	25	31	38
90° elbow, same size as pipe	4	5	6	8	11	13	16
45° elbow, same size as pipe	2	$2\frac{1}{2}$	3	$3\frac{1}{2}$	5	6	7
Tee, flow-through side outlet, same size as pipe	8	10	13	16	21	27	33
Tee, flow straight through, same size as pipe	$2\frac{1}{2}$	3	4	5	7	$8\frac{1}{2}$	11
Strainer, same size as pipe	25	60	42	42	50	50	50
Strainer, next larger size than pipe	16	17	14	20	30	30	
Bushing or reducer, to one size larger or smaller	2	$2\frac{1}{2}$	3	4	5	6	7

Note: Figures shown represent average resistance-to-flow values.

TABLE 13.23 Resistance of Delivery Host to the Flow of Liquid Propane, 60°F

Delivery rate		Pressure drop, psi					
L/S	GPM	½ in size	¾ in size	1 in size	1¼ in size	1½ in size	2 in size
0.375	5	8.1	1.1	0.2	0.1	0	0
0.75	10	30.0	4.0	0.9	0.3	0.1	0
1.12	15	64.6	8.5	2.0	0.7	0.3	0
1.5	20	Too high	14.4	3.4	1.0	0.5	0.1
1.9	25	Too high	22.1	5.2	1.7	0.7	0.2
2.25	30	Too high	31.0	7.4	2.4	0.9	0.3
3.0	40	Too high	54.0	12.6	4.2	1.6	0.4
3.75	50	Too high	Too high	19.0	6.4	2.5	0.6
4.5	60	Too high	Too high	26.4	.90	3.5	0.8
5.25	70	Too high	Too high	35.4	11.9	4.6	1.1
6.0	80	Too high	Too high	46.2	15.5	5.9	1.4
6.75	90	Too high	Too high	46.6	19.0	7.4	1.7
7.5	100	Too high	Too high	69.7	23.4	8.9	2.1

1 psi = 7 kPa.

6. Add all pressure obtained from steps 1 through 5 and add 10 psi to have some additional pressure available. This figure is the total head requirement for the pump.

7. The flow rate is determined from the specific application.

PIPING MATERIALS

The piping materials are the same as for the NG system.

FUEL GAS PRESSURE BOOSTER SYSTEM

Occasionally the pressure available in a utility company or site main may not be high enough to supply the required pressure for equipment intended to be installed in a facility. For these conditions, a gas pressure boosting system is required.

SYSTEM DESCRIPTION

The purpose of a gas booster pump system is to raise the pressure of fuel gas supplying equipment or devices that require a higher than available pressure. The gas booster is a low-pressure compressor that has been adapted for flammable gases, and it shall not exceed the 5-psi requirement established in NFPA 54.

CODES AND STANDARDS

In addition to the previously listed codes and standards associated with the fuel gas piping system, there are usually standards that originate with the local utility company in whose jurisdiction the project is located. These standards relate to protecting both the utility company's and the facility's low-pressure piping system from back pressure or suction resulting from the booster pump operation. The utility company will review proposed plans and require the installation of any protective devices they believe are necessary. When starting, the booster may develop a suction pressure of −15 in WC or more on a gas meter. If this situation is possible, protection is required.

SYSTEM COMPONENTS

The components of the gas booster system are the booster pump, heat exchanger (if required), protective components, and the piping, valves, and so on in the distribution network.

Booster Pump

Booster pumps are available in positive displacement, piston, and centrifugal fan types. For general use in facilities, the centrifugal fan is the type most often selected.

The booster pump shall be UL approved and hermetically sealed. The fan is directly coupled to the motor and completely enclosed in an airtight housing that does not require external shaft seals. Access to the fan is made from a gasketed cover plate assembly. The fan shall be manufactured of spark-resistant materials, with the most often used material being aluminum. The design of the pump shall provide different mounting positions that allow easy connection of the distribution piping to the booster outlet.

The booster pump shall be installed in an unconfined area that provides sufficient air to circulate about the unit to provide cooling. The unit shall be mounted so that vibration is minimized.

Heat Exchangers

The purpose of a heat exchanger is to lower the temperature of the fuel gas. This may be necessary for several reasons.

The flow of gas through the booster pump cools the motor. If the flow is too low, the small volume is not capable of removing sufficient heat for proper operation. To relieve this condition, a bypass arrangement that provides partial recirculation of the gas discharge through a heat exchanger to cool it will provide the necessary flow of cool gas.

The passage of gas through the booster pump heats the gas as it passes through the pump. As a general rule, each stage of a booster pump raises the gas temperature 20°F (6°C). If the ambient temperature of the room where the booster is located is hot, the gas temperature downstream of a multistage booster could exceed 140°F. Since most valves, diaphragms, controls, and other equipment are usually rated at 150°F (60°C), it may be necessary to install a heat exchanger to cool the gas.

Heat exchangers can be either air cooled, water cooled, or refrigerated. For general use in facilities, single-pass, modular, air-cooled units are the most common. The most often used material of construction is corrosion-resistant aluminum. This type of unit uses a self-contained, temperature-controlled auxiliary fan to increase the flow of ambient air across the exchange fins. It is considered good practice to have all the gas discharged flow through the exchanger and to use a temperature switch in the discharge gas line to turn on the fan only as required by the temperature of the gas. When the heat exchanger is used to cool the main supply to a facility, it is good practice to install a soft seat check valve in the discharge downstream of the heat exchanger.

The criteria needed to size the heat exchanger are the highest anticipated ambient air temperature and the maximum flow of gas in cubic feet per minute (cubic meters per minute). Each manufacturer has selection criteria based on this information.

Because of the hazardous nature of this system, it is a requirement that the control panels be in an NEMA-7 enclosure.

System Protection

When using a booster pump to increase gas pressure, safety devices and techniques are necessary to protect both the utility company's piping and the facility system from a number of possible accidents and equipment malfunctions. The following is a general discussion of potential failures and methods used to prevent damage to the system. Since not all of these failures are possible for every installation, judgment is required in their application for any specific project.

1. When using pressurized air, oxygen, or another gas for burner nozzles or other similar equipment in conjunction with NG, check valves must be installed on all of the lower-pressure lines to prevent high-pressure gas from entering the lower-pressure system.

2. Burner systems using pressurized air, oxygen, or another gas may cause a suction on the NG line if the NG supply is interrupted. A low-pressure electrically

operated switch shall be installed to stop the operation of the higher-pressure gas system if this situation occurs.

3. Excessive suction may be produced when starting the booster. If starting results in a suction pressure of more than −15 in WC at the utility company's piping, a pressure switch shall be installed to stop the booster.

4. There are usually two independent low-pressure protection devices installed on a burner system to prevent operation if the supply pressure falls too low. The first is a low-pressure switch, usually supplied by burner manufacturers on their gas train, that shuts off the burner under low gas pressure conditions. If this switch proves insensitive, another switch should be installed upstream from the train regulator. The second pressure switch is installed on the utility company's line to prevent booster operation if the pressure in the utility company supply line falls below 3 in WC.

5. Unit heater systems using booster pumps must be rated for the highest pressure capable of being generated by the pump plus the "lock-up pressure." If this arrangement is not possible, a pressure relief must be installed between the pump and the first unit heater. In addition, a low gas pressure switch shall be installed to prevent heater operation in the event that the gas pressure falls too low. The relief valve outlet must be piped outside the building to a safe location.

6. Some utility companies require 20 ft (6.3 m), of piping to be provided between a burner and a check valve when using an alternative fuel with a higher pressure than the booster pump is capable of providing.

7. All of the above may have to be approved by the local utility and building department authorities.

SYSTEM DESIGN CONSIDERATIONS

Gas booster starters should be wired directly to the equipment they are serving, so that if the equipment is not operating, the booster will not run. If multiple pieces of equipment are wired to a booster, isolating relays are required.

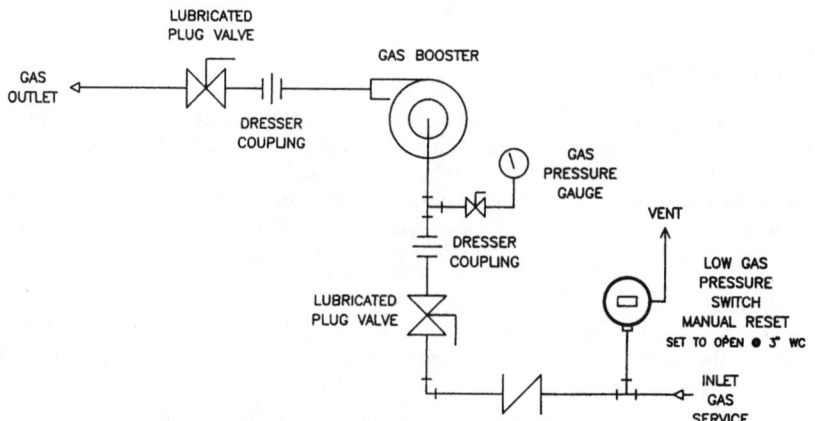

FIGURE 13.11 Schematic piping diagram of a typical booster pump.

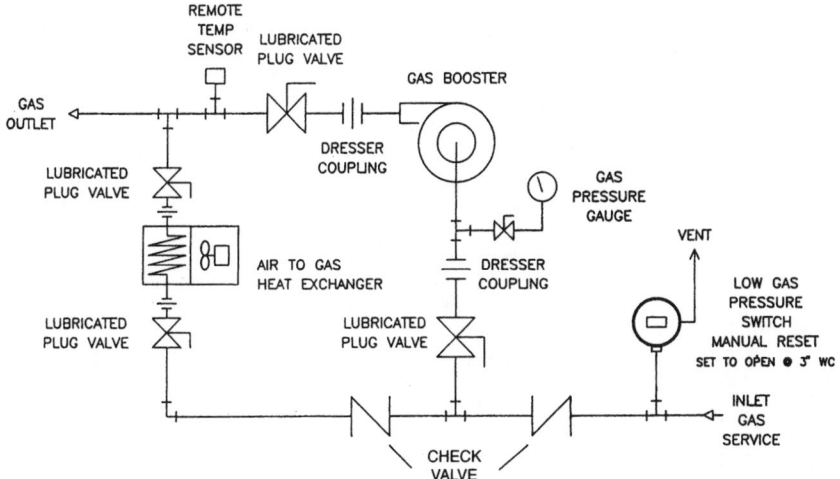

FIGURE 13.12   Schematic piping diagram of a typical booster pump with heat exchanger.

If other gases, such as LPG, are used as an alternate fuel, a three-way valve shall be installed to limit the gas supply to only one source.

If higher pressures are available from the alternate fuel sources, a regulator shall be installed on the higher-pressure sources so that the actual delivery pressure is the same as that supplied by the utility company.

Some utility companies do not permit a large-size booster to be connected to a low-pressure NG service. Contact the utility company to determine any restrictions.

In general, electronic ignition rather than pilot lights should be used. If pilot lights are the only alternative, they are not permitted to be supplied with a pressure higher than approximately 7 in WC.

A schematic detail of a typical simplex booster pump installation without a heat exchanger is illustrated in Fig. 13.11, and a schematic detail of a typical booster pump installation with a heat exchanger is illustrated in Fig. 13.12.

REFERENCES

Clifford, E. A., "A Practical Guide to LP-Gas Utilization," *LP-Gas Magazine,* Duluth, Minn., 1973.

Denny, L. C. et al. (eds.), *Handbook of Butane-Propane Gases,"* 4th ed., Chilton Company, Los Angeles, 1962.

Frankel, M., "Designing Propane Systems," *Plumbing Engineer,* Nov./Dec. 1990, June 1991.

Guide to Corken Liquefied Gas Transfer Equipment, Corken International, 1988.

Chapter 6, Gas Systems American Society of Plumbing Engineers Data Book.

CHAPTER 14
COMPRESSED GAS SYSTEMS

This chapter describes design criteria, production, storage, and central piping distribution methods for various compressed gas systems. Because of the diverse uses and different design criteria for each, this chapter is divided into the following separate sections: utility compressed air for light industrial use, compressed air for instruments and control, specialty gases for laboratories, compressed gases for health care facilities, dental compressed air, and large-scale specialty gas systems for industrial purposes. For purposes of this handbook, a compressed gas is any gas at a pressure higher than atmospheric pressure.

Some of the gases to be discussed could be stored as cryogenic liquids and converted to a gas at the storage location. The storage and vaporization of these gases are discussed in Chap. 19, "Cryogenic Systems." Compressed air used to supply breathing apparatus is discussed in Chap. 18. Ultrapure gases, such as those used by the electronics industry for the manufacture of computer chips and other similar products, are regarded as process gases and are therefore outside the scope of this book.

FUNDAMENTALS

GENERAL

Air is a fluid as compared to a solid. Two kinds of fluids are liquids and gases. In a gas, the molecular structure does not have a lattice type of arrangement, and the cohesive forces that bind the molecules together are not as strong as those for a solid. This means that the molecules are quite mobile and will take the shape of their container. This mobility also allows a gas to expand through space and mix with other gases present.

The actual solid volume that the gas atomic structure occupies in relation to the total volume of a gas molecule is quite small, and so, gases are mostly empty space. This is why gases can be compressed.

Pressure is produced when molecules of a gas in an enclosed space rapidly strike the enclosing surfaces. If this gas is confined into a smaller and smaller volume, molecules strike the container walls more frequently, producing a greater pressure.

For most purposes, air is compressed by the adiabatic process, whereby the heat of compression helps raise the pressure. Because of this, more horsepower is required to obtain the same outlet pressure than that produced under ideal isothermal conditions. Therefore, manufacturers use different methods to reduce power consumption. The use of intercoolers to reduce the temperature during the compression process is the most common.

DEFINITIONS AND PRESSURE MEASUREMENTS

Definition of Compressed Gases

A compressed gas is defined as any gas either stored or distributed at a pressure greater than atmospheric.

Definition of Basic Compressed Air Processes

Isobaric Process. This process takes place under constant pressure.

Isochoric Process. This process takes place under constant volume.

Isothermal Process. This process takes place under constant temperature.

Polytropic Process. A generalized expression for all of the three above processes when variations in pressure, temperature, or volume are allowed to occur during the compression cycle.

Adiabatic Process. This process of compression allows a gas to gain temperature. This process is the most commonly used in facility compressed air production.

Units of Measurement

Pressure measurements are made using force acting upon an area. The most common method of measuring pressure in IP units is in pounds per square inch (psi). In SI units it is in kilograms per square centimeter (kg/cm^2) and kilopascals (kPa). Another common unit of measurement for low-pressure systems is in inches of water column (in wc). To convert in wc to psig refer to Table 13.9.

Standard Reference Points and Measurement

The two basic reference points for measuring pressure are standard atmospheric pressure and a perfect vacuum. When the point of reference is taken from standard atmospheric pressure to a specified higher pressure, this is called gauge pressure, expressed as psig. If the reference pressure level is measured from a perfect vacuum, the term used is absolute pressure, expressed as psia. Local barometric pressure, which is the prevailing pressure at any specific location, is variable and should not be confused with standard atmosphere, which is mean theoretical barometric pressure at sea level. Theoretical standard atmospheric pressure at sea level is equal to 14.696 psia, 101.4 kPa, 0 psig, 760 mmHg and 29.92 inHg. A perfect vacuum has a value of 0 psia, 0 kPa, and 0 inHg. Refer to Fig. 14.1 for the relationship between various methods of measuring air pressure. Pressure expressed only as psi is incomplete.

Theoretical standard atmospheric temperature is 60°F (15.6°C).

For ease of calculations, 14.7 psig is often adjusted to 15 psig and 29.92 inHg is often adjusted to 30 inHg. These minor deviations yield results well within the accuracy required for most engineering calculations.

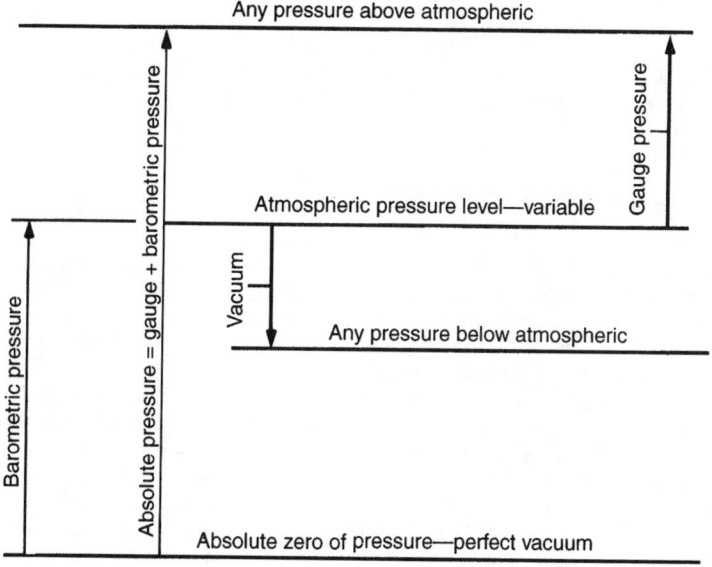

FIGURE 14.1   Relationship between gauge and absolute pressure.

Standard Air. Standard air is dry air with a relative humidity of 0.0 percent, a temperature of 60°F (15.6°C), and a pressure of 14.7 psig (101.4 kPa).

Free Air. Ambient air at a specific location. Temperature, barometric pressure, and moisture content may be different from standard air. The description of free air at a location is not complete unless the ambient temperature, humidity, and barometric pressure conditions at the compressor location are stated.

Flow Rate

The most common measurement of flow rate in IP units is cubic feet per minute (cfm). If the flow rate is low, it is commonly expressed in cubic feet per hour (cfh). For SI units liters per minute (Lpm) and liters per second (Lps) are used. Flow rate must reference standard or actual air.

Standard Cubic Feet (liters) per Minute (scfm), (sLpm). This is a volume measurement of air at standard conditions. One cubic foot of standard air weighs 0.764 lb.

Actual Cubic Feet (liters) per Minute (acfm) (aLpm). This is a volume measurement of standard air after being compressed. To find the acfm equivalent of scfm at pressure, refer to Fig. 14.18. The acfm measurement is not complete unless the pressure is stated.

Inlet Cubic Feet per Minute (icfm). The actual flow rate of free air entering the inlet flange of the compressor, not considering losses through any installed inlet devices or piping. An icfm measurement is not complete unless the ambient temperature, humidity, and barometric pressure conditions at the compressor location are stated.

Free Air Delivered (FAD). The actual volume rate of free air produced at the outlet flange of the compressor when referenced to icfm.

PHYSICAL PROPERTIES OF AIR

Air is the atmosphere surrounding the earth. It is a mixture of many elements and compounds. The composition of dry air is listed in Table 14.1. Pure air is odorless and tasteless unless some foreign matter is suspended in the mixture. The air pressure exerted at the earth's surface is due to the weight of the column of air above that point and is measured barometrically.

Because free air is less dense at higher elevations, a correction factor must be used to determine the equivalent volume of standard air at the higher elevation. The elevation correction factors are given in Table 14.2. By multiplying the volume of air by the correction factor, the actual quantity of standard air will be found.

Temperature is also a consideration. Because an equal volume of free air at a higher temperature will exert a higher pressure than the same volume of standard air at a lower temperature, a correction factor must be used to determine the equiv-

TABLE 14.1 General Composition of Dry Air

Component	Percent by volume	Percent by mass
Nitrogen	78.09	75.51
Oxygen	20.95	23.15
Argon	0.93	1.28
Carbon dioxide	0.03	0.046
Neon	0.0018	0.00125
Helium	0.00052	0.000072
Methane	0.00015	0.000094
Krypton	0.0001	0.00029
Carbon monoxide	0.00001	0.00002
Nitrous oxide	0.00005	0.00008
Hydrogen	0.00005	0.0000035
Ozone	0.00004	0.000007
Xenon	0.000008	0.000036
Nitrogen dioxide	0.0000001	0.0000002
Iodine	2×10^{-11}	1×10^{-10}
Radon	6×10^{-18}	5×10^{-17}

TABLE 14.2 Elevation Correction Factor

Altitude, ft	Meters	Correction factor
0	0	1.00
1600	480	1.05
3300	990	1.11
5000	1500	1.17
6600	1980	1.24
8200	2460	1.31
9900	2970	1.39

alent volume of air at different temperatures. The temperature correction factors are given in Table 14.3. By multiplying the volume of air by the correction factor, the actual quantity of standard air will be found.

WATER VAPOR IN AIR

Both temperature and pressure can affect the ability of air to hold moisture. When a given volume of air is compressed, an increase in temperature occurs. Increased temperature results in an increased ability of air to retain moisture. Conversely, an increase in pressure results in a decreased ability to hold water. With each 20°F increase in temperature, the ability of air to accept water vapor doubles. When air is compressed, the rise in temperature is more critical than the pressure rise. Be-

TABLE 14.3 Temperature Correction Factor

°C	Temperature of intake, °F	Correction factor	°C	Temperature of intake, °F	Correction factor
−46	−50	0.773	4	40	0.943
−40	−40	0.792	10	50	0.962
−34	−30	0.811	18	60	0.981
−28	−20	0.830	22	70	1.000
−23	−10	0.849	27	80	1.019
−18	0	0.867	32	90	1.038
−9	10	0.886	38	100	1.057
−5	20	0.905	43	110	1.076
−1	30	0.925	49	120	1.095

cause of the high temperature rise during the compression cycle, no water will precipitate inside the compressor; but water may, however, precipitate after the cycle has been completed.

Air contains varying amounts of water vapor depending on its temperature and pressure. There are various methods of expressing the amount present.

Saturated Air and Dry Air

Saturated air contains the maximum amount of water vapor possible based on its temperature and pressure. Dry air contains no water vapor. To determine the moisture content of saturated air based on its temperature, refer to Fig. 14.2.

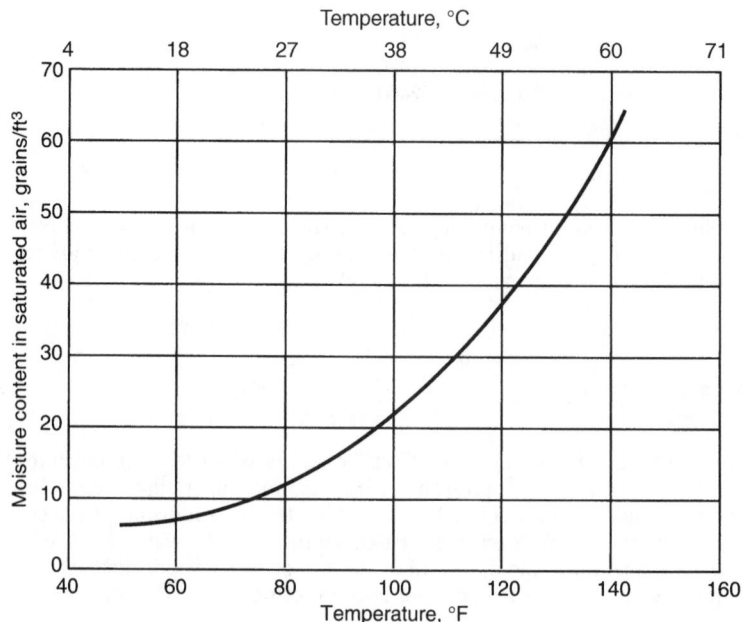

FIGURE 14.2 Moisture content of saturated air.

Relative Humidity

Relative humidity is the amount of water vapor actually present in air expressed as a percent of the total amount capable of being present when the air is saturated. Relative humidity is dependent on pressure and temperature.

Dew Point

The dew point is that temperature at which water in the air will start to condense on a surface, and it is used to express the dryness of the compressed air. The lower the dew point, the dryer the air. Since the dew point of air varies with the air pressure, it must be referred to as the *pressure dew point*. There is a different dew point for air at different pressures. To find the dew point of air at various pressures

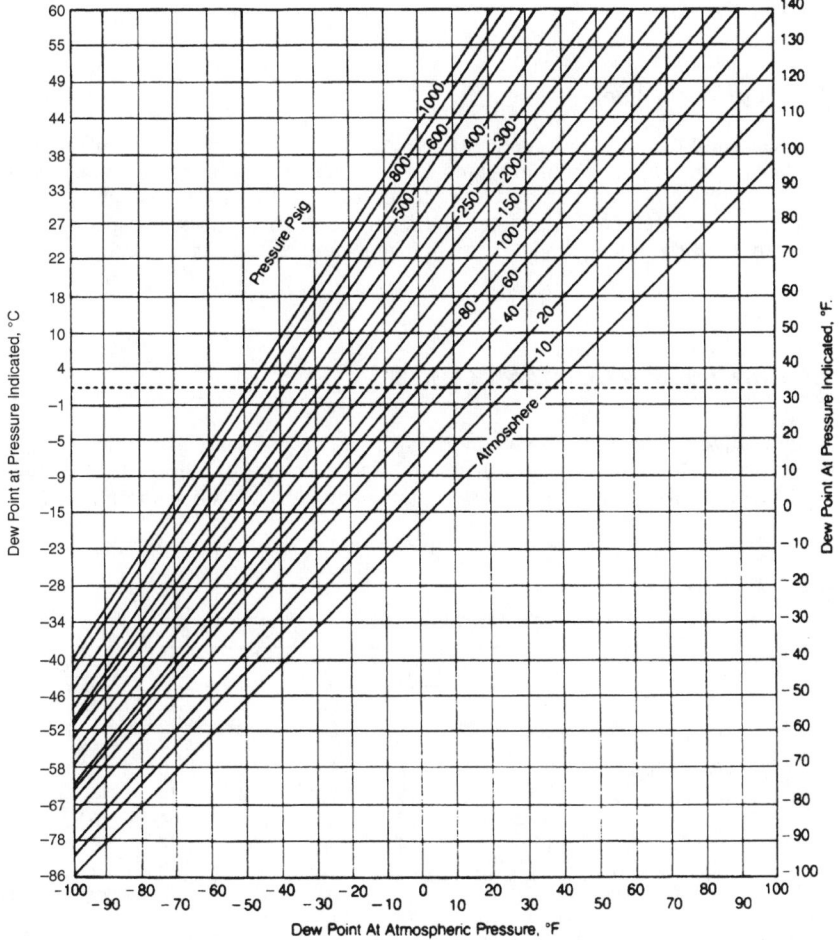

FIGURE 14.3 Dew point conversion chart. (*Courtesy Hankison Corp.*)

and temperatures, refer to the dew point conversion chart in Fig. 14.3. Dew point is the preferred method of expressing the moisture content of compressed air.

To obtain the dew point temperature expected if the gas were expanded to a lower pressure proceed as follows:

1. Using "dew point at pressure," locate this temperature on scale at right hand side of Fig. 14.3.
2. Read horizontally to intersection of curve corresponding to the operating pressure at which the gas was dried.
3. From that point read vertically downward to curve corresponding to the expanded lower pressure.
4. From that point read horizontally to scale on right hand side of chart to obtain dew point temperature at the expanded lower pressure.
5. If dew point temperatures at atmospheric pressure are desired, after step 2 above read vertically downward to scale at bottom of chart which gives dew point at atmospheric pressure.

Weight of Water Vapor in Air

The relationship of the dew point to weight of water per cubic foot of air at a constant temperature is about the same for all different pressures in the range common to facility compressed air systems. Refer to Table 14.4 for the weight of water vapor in air at different temperatures and relative humidity values. For a conversion table giving different methods expressing moisture content of air, refer to Table 14.5.

IMPURITIES AND CONTAMINATION

A knowledge of the various pollutants in the air is necessary when deciding on what equipment is required to effectively reduce or remove them. The required level of protection from the various contaminants depends upon the purpose for

TABLE 14.4 Weight of Water Vapor in Air

(*Grains of moisture per pound of air at standard barometric pressure*)

Temp		RH								
°C	°F	10	20	30	40	50	60	70	80	90
−1	30	3	5	7	9	12	14	17	19	21
4	40	4	7	10	14	16	18	20	22	24
10	50	6	10	14	20	26	32	38	42	48
8	60	8	16	22	30	39	48	54	62	70
22	70	11	21	34	44	55	66	78	88	100
27	80	16	30	46	62	78	92	108	125	140
32	90	21	42	65	85	108	128	158	173	195
38	100	29	58	87	116	147	176	208		

TABLE 14.5 Moisture Content of Air at One Atmosphere*

°C	Dew point, °F	Grains moisture per lb air	Pounds moisture per lb air	Grains moisture per ft³ air	ppm	Vol. percent
−15	0					
			0.0006	0.4	600	0.1
		4				
−23	−10					
			0.0004		400	0.06
		2		0.2		
−28	−20					0.04
			0.0002		200	
−34	−30	1		0.1		
		0.8		0.08		0.02
		0.6	0.0001	0.06	100	
−40			0.00008	0.06	100	
			0.00006		80	
				0.04	60	0.01
						0.008
−46	−50		0.00004		40	
		0.2		0.02		0.006
						0.004
−52	−60					
			0.00002	0.01	20	
		0.1		0.008		
		0.08				0.002
−58	−70		0.00001		10	
		0.06	0.000008	0.006	8	
		0.04	0.000006	0.004	6	0.001
						0.0008
−67	−80					0.0008
			0.000004		4	0.0006
		0.02		0.002		0.0004
−78	−90					
			0.000002		2	
		0.01		0.001		0.0002
−86	−100	0.008				
			0.000001	0.0008	1	

*There are many ways of expressing moisture content of air. The accompanying chart provides a quick comparison of the more frequently used methods. Read straight across to find equivalent moisture contents at one atmospheric pressure.

The relationship of dewpoint to grains per cubic foot does not change much with pressure in the range of 0 to 300 psig. Consequently, grains per cubic foot for elevated pressures can also be read directly from the chart, remembering that actual cubic feet are used.

TABLE 14.5 Moisture Content of Air at One Atmosphere* (*Continued*)

°C	Dew point, °F	Grains moisture per lb air	Pounds moisture per lb air	Grains moisture per ft³ air	ppm	Vol. percent
44	100		0.0600		60,000	
		400		25		9
			0.0500		50,000	8
						7
38	100	300		20		
			0.00400		40,000	
						6
				15		
33	90					5
			0.0300			
		200				
27	80		0.0200			
		150		10	20,000	3
				9		
22	70		0.0150	8		
		100			15,000	2
		90		7		
		80		6		
18	60				10,000	
			0.0100		9,000	
		70	0.0090		8,000	1.5
		60	0.0080			
10	50					
		50	0.0070		7,000	1
		40	0.0060	3	6,000	0.9
4	40				5,000	
			0.0050		4,000	0.8
		30	0.0040	2		0.7
						0.6
−1	30					
		20	0.0030	1.5	3,000	0.5
						0.4
−5	20					
			0.0020	1	2,000	0.3
		10		0.8		
−9	10					
		8	0.0010	0.6	1,000	0.2
		6	0.0008		800	

A pressure correction is necessary for all other measurements listed. For a convenient means of converting dewpoints measured at atmospheric pressure to those measured at an elevated pressure, refer to Fig. 14.3. Use the latter dewpoint on the moisture content chart to read grains per pound, pound per pound, ppm, and volume percent. On the other hand, if moisture content is expressed in these units, read the expanded dewpoint from the moisture content chart and refer to Fig. 14.3 to convert the dewpoint reading to elevated pressure.

Source: Courtesy of Hankison Corp.

which the air will be used. Performance criteria for each individual system must be determined prior to selection of any equipment, along with identification and quantifying of pollutants.

There are four general classes of contaminants:

1. Liquids (oil and water)
2. Vapor (oil, water, and hydrocarbons)
3. Gases
4. Particulates

Liquids

Water enters a system with the intake air, passes through the compressor as a vapor, and condenses afterward into liquid droplets. Most liquid oil contamination originates at the intake location or in an oil lubricated compressor. As the droplets are swept through the system at velocities approaching 4000 fpm (feet per minute) (1200 M/M [meters per minute]) they gradually erode obstructions in their path by repeated collisions. When water settles on pipes, corrosion begins, ultimately ruining machinery and tools, causing product rejection and product contamination. At high temperatures, oils break down to form acids. With particulates, oil will form sludge. Oil can also act like water droplets and cause erosion. Liquid chemicals react with water and corrode surfaces. Water also allows microorganisms to grow.

Vapor

Oil, water, and chemical vapors enter the system in the same manner as liquids and contribute to corrosion of surfaces in contact with the air. Oil vapor reacts with oxygen to form varnish buildup on surfaces. Various chemicals cause corrosion and are often toxic.

Gas

Gases such as carbon dioxide, sulfur dioxide, and nitrogen compounds react with heat and water to form acids.

Particulates

Particulates enter the system from the air intake, originate in the compressor due to mechanical action, or are released from some air drying systems. These particles erode piping and valves or cause product contamination. However, the most harmful effect is that they clog orifices or passages of, for example, tools at the end use points. These particulates include metal fines, carbon and Teflon particles, pollen, dust, rust, and scale. Bacteria enter through the inlet and reproduce in a moist warm environment.

1. There is no safe level of liquids in the airstream. They should be removed as completely as practical.

2. The level of acceptable water vapor varies with end use requirements. A dew point of $-30°F$ ($-34°C$) is required to minimize corrosion in pipelines. For critical applications a dew point of $-100°F$ ($-86°C$) may be required. Oil vapor remaining in the air should be as close to zero as practical. Chemical concentration should be reduced to zero, where practical.

3. Gases in any quantity that are potentially harmful to the system or process requirements should be reduced to zero or to a point that will cause no harm, depending on practical considerations. Condensable hydrocarbons should be removed as completely as practical.

4. Particulate contamination must be reduced to a level low enough to minimize end use machine or tool clogging, cause product rejection, or contaminate a process. These values must be established by the engineer and client and will vary widely. The general range of particles in a typical system are between 10 and 0.01 μm in diameter.

When selecting appropriate and specific air purification components, remember there is no single type of equipment or device that can accomplish the complete job of removing them all. Objective performance criteria must be used to accomplish the desired reduction level. Such criteria must include pressure drop, efficiency, dependability, service life, energy efficiency, and ease of maintenance. There will be further discussion of the contaminant removal processes under individual components.

SYSTEM COMPONENTS

Air Compressors

The selection of an air compressor for any specific application depends primarily upon a knowledge of its performance characteristics as applied to the particular system being designed. Cost, space requirements, and efficiency are other considerations. Increasing attention to energy costs also requires evaluation of total operating costs and ease of maintenance for an extended period of time. A carefully selected air compressor will satisfy systems design and performance criteria while operating in the most cost-efficient manner.

Air Compressor Types. Air compressors are divided into two general categories: displacement and dynamic. Refer to Fig. 14.4 for a listing of various types of compressors according to general categories.

Displacement compressors can be further separated into reciprocating and rotary machines. Typical reciprocating compressors include piston and diaphragm types. Rotary includes such types as sliding vane, liquid ring (or liquid piston), and screw. The most widely used types of dynamic compressors include centrifugal and axial flow.

Piston compressors use a piston within a cylinder to compress the air. When air is compressed in only one direction, it is called *single acting.* Units that compress air in both directions are called *double acting.* See Fig. 14.5. Although these machines are fixed capacity units, the output can be varied. The speed can be lowered, the intake volume can be reduced (by providing adjustable internal clearance that can be selectively cut in or out), cylinder valves can be rendered inoperative to adjust capacity, or a portion of the discarded air can be blown off.

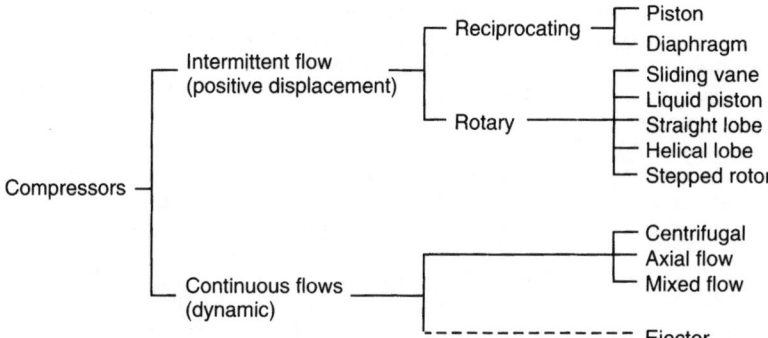

FIGURE 14.4 Air compressor types. Air compressors can be classified by air delivery characteristics as intermittent or continuous flow. Intermittent-flow machines can be distinguished by the type of motion used in the mechanism—reciprocating or rotary. Continuous-flow machines are divided into dynamic types and ejectors.

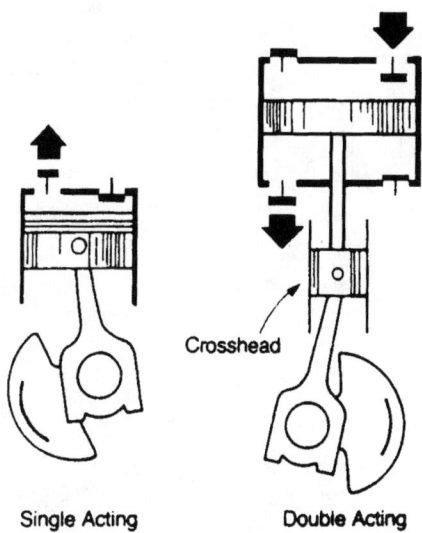

Single Acting   Double Acting

FIGURE 14.5 Piston air compressors.

In general, single-stage compressors are more desirable for pressures of 60 psig (410 kPa) or less, and multistage units are better for pressures of 100 psig or greater. In the 60 to 100 psig (410–690 kPa) range, a single stage is generally recommended for capacities of less than 300 scfm, and multistage for 300 scfm or more.

Piston compressors are also available for oil-free operation by using carbon or Teflon wearing parts that come in contact with the airstream. These parts require no oil due to the low friction.

A water-cooled unit is generally more efficient than an air-cooled one. It has a lower power consumption, but its initial cost is higher. A two-stage piston compressor uses less power than a single-stage unit for equivalent output. Piston compressors are available in an extremely wide selection of capacities and pressures.

Diaphragm compressors use a flexible diaphragm to compress air. These types of units are restricted to light-duty, low scfm, and low-pressure uses where economy is a factor, generally in the range of 50 psig and 25 scfm. A diaphragm compressor is illustrated in Fig. 14.6.

In a sliding vane compressor, the vanes are mounted eccentrically in a cylindrical rotor and are free to slide in and out of slots. As the rotor turns, the space between the compressor casing and the vanes decreases, and the air is compressed. See Fig. 14.7.

These are compact units, well suited for direct connection to a relatively high-speed motor. Their efficiency is usually less than an equivalent piston unit. They are best used where small, low-capacity compressors are required, generally in the range of 100 scfm and up to 75 psig (3 m³/min & 500 kPa).

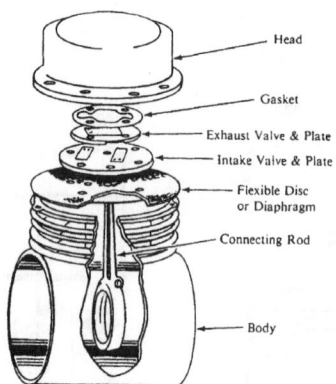

FIGURE 14.6 Detail of diaphragm air compressor.

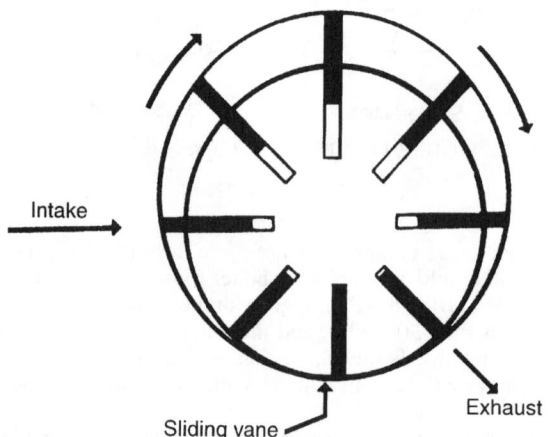

FIGURE 14.7 Detail of sliding vane air compressor.

Liquid ring compressors, sometimes referred to as *liquid piston compressors,* are rotary positive displacement units that use a fixed blade rotor in an elliptical casing. The casing is partially filled with liquid. As the rotor turns, the blades set the liquid in motion. As they rotate, the blades extend deeper into the liquid ring, compressing the trapped air. See Fig. 14.8.

The resulting air is completely oil free. This type of compressor will also handle wet, corrosive, or explosive gases. Different liquids can be used that are compatible with any specific gas to be compressed. This unit is also very well suited for hospital and laboratory use. A practical limitation of 100 psig exists, and there is a higher power consumption than there is for piston units of a similar rating.

Straight lobe compressors function in a manner similar to a gear pump. A pair of identical rotors, each with lobes shaped like the figure 8 in cross section, are mounted inside a casing. As they rotate, air is trapped between the impeller lobes and pump casing, carrying it around without compression. This air is then discharged, using the existing pressure in the system to additionally increase pressure. See Fig. 14.9. In general, this type of compressor has very low operating pressures (around 15 psi or 100 kPa).

Helical lobe, rotary screw, or spiral lobe compressors use a pair of close clearance helical lobe rotors turning in unison. As air enters the inlet, the rotation of the rotors causes the cavity in which air is trapped to become smaller and smaller, which increases the pressure. Designs are available to produce oil-free air. A rotary screw compressor is illustrated in Fig. 14.10.

Varying capacity is obtained by adjusting the speed of the driving motor, reducing the amount of inlet air, or returning a portion of the compressed air discharged back into the inlet. A check valve must be provided on the outlet pipe to prevent air from escaping through the compressor, after the compressor has stopped. Because of its rotary operation, the discharge is almost continuous. This type of compressor is best suited for higher-pressure applications, where, in general, capacities are available from 30 to 26,000 scfm (1 to 780 m³/min) at pressures ranging from 125 to 250 psig (875 to 1750 kPa).

Stepped rotor compressors use two pairs of intermeshing rotors to trap air and reduce its volume. This compressor does not require liquid lubricants in the compressor chamber, thereby producing oil-free air. Clearance between rotors is not

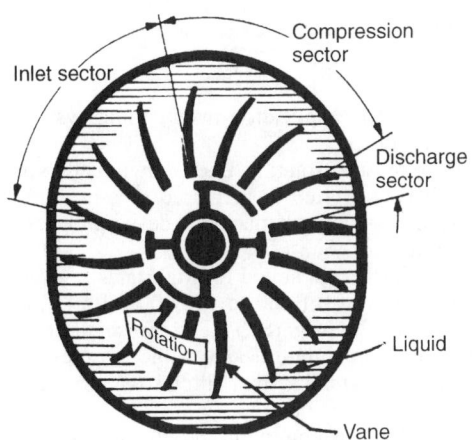

FIGURE 14.8 Detail of liquid ring air compressor.

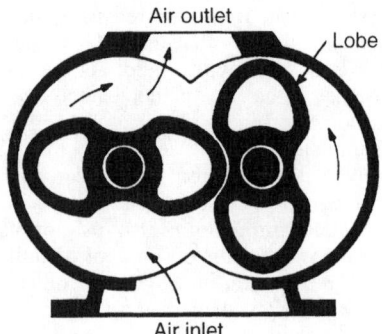

FIGURE 14.9 Straight lobe air compressor.

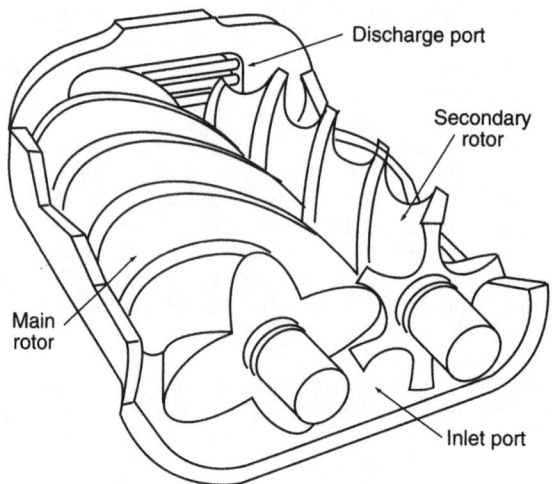

FIGURE 14.10 Rotary screw air compressor.

critical. These units find little application in plumbing system design due to their very limited volume capacities.

Centrifugal compressors use rotating blades or impellers to give velocity (energy) to the air, which is converted to increased pressure inside the pump casing. This classification does not apply to machines developing less than 1 psig, which are usually designated as fans. See Fig. 14.11.

This type of compressor is best suited for high rates at a relatively low pressure. Units are available in capacities of 400 to 170,000 scfm (12 to 5100 m³/min) and pressures of up to about 125 psig (875 kPa). Since these units cannot develop pressures higher than their maximum design value, no pressure relief is necessary. Centrifugal compressors are smaller than similar reciprocating units, but they use more power. The air delivered is oil free.

Axial flow compressors are continuous flow machines that use two rows of blades, one rotating and one stationary. As the air is given velocity by the moving

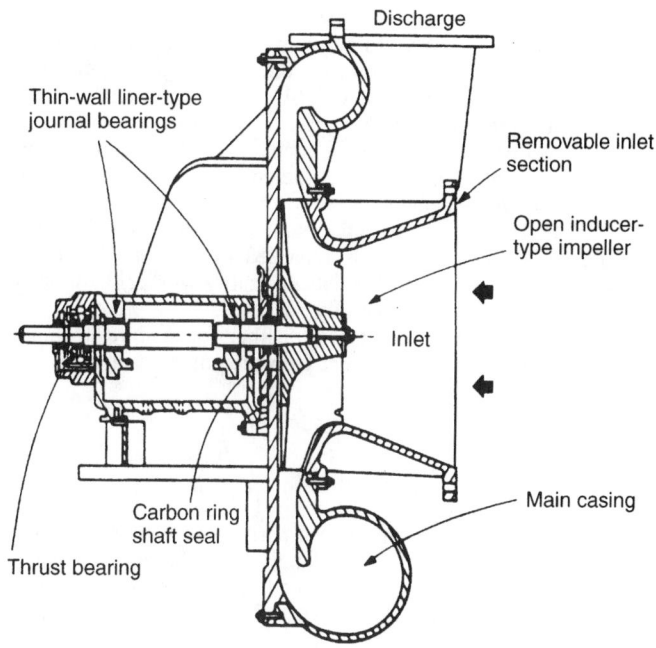

FIGURE 14.11 Centrifugal air compressor.

blades, it passes over the stationary blades, increasing pressure. Many stages are required, with 20 not uncommon. These compressors are well suited for large volumes of air, and they operate best at or near full capacity. Units are available from 10,000 to 800,000 (300 to 24,000 m³/min) scfm at pressures to 125 psig. Selecting between centrifugal or axial compressors depends upon efficiency, size, weight, and initial cost, since their ranges overlap.

Mixed flow compressors use rotating impeller design, combining certain characteristics of both centrifugal and axial units to achieve capacity and pressure requirements. These units find little application in typical plumbing systems.

Ejectors are continuous flow units that use a moving liquid stream across a venturi to draw in and entrap air for later conversion to higher pressures in the ejector housing.

Silencers

When sound becomes too loud, it turns into noise. This noise, depending upon location and circumstances, can be very objectionable. With today's emphasis on noise control, the installation of an intake silencer will probably be necessary for most projects.

There are two types of silencers: reactive and absorptive. The reactive type is used to attenuate (reduce) low-frequency sound in the order of 500 cps (cycles per second), which is most often found on reciprocating compressors. The absorptive silencer is often used on centrifugal and screw compressors where frequencies are above 500 cps. There is no practical limit in scfm for either type.

The selection of a silencer should be made in conjunction with the manufacturer. It will be necessary for the engineer to determine two things: the sound power level of the compressor (which must be obtained from the compressor manufacturer) and the highest level of sound permitted by OSHA, local authorities, or facility personnel. With the establishment of this design criteria, selection of a silencer can be made if the final level of sound desired is included in the specifications. This will allow the various manufacturers to suggest the correct silencer for that purpose, for final acceptance by the engineer.

In general, OSHA has established maximum acceptable sound levels to prevent hearing loss. These levels are generally regarded as excessive. In fact, that noise level, if accepted, will usually disturb adjacent facility workers and the surrounding areas. A level of 85 db has been generally accepted.

Silencers may be combined with the inlet filter for a more economical installation. They could also be mounted directly on the compressor or at the roof level as separate units.

Aftercoolers

An aftercooler is a device, often an integral part of the compressor, used to lower the temperature of compressed air immediately after the compression process. In doing so, large amounts of water are liberated. The primary function of an aftercooler is to remove water vapor rather than to lower the temperature of the compressed airstream.

Air leaving the compressor is very hot. It is desirable to reduce the temperature of air discharged to a range of between 70 and 110°F (21 and 43°C). Refer to Table 14.6 for the average temperature of compressed air leaving the compressor after the compression cycle is completed. A primary reason the temperature is lowered is to remove moisture that would otherwise condense elsewhere in the system as the air cools to ambient conditions. Therefore, it is considered good practice to install a cooling unit as close to the compressor discharge as practical for that reason. Such a unit is called an *aftercooler*. An aftercooler is also useful to first precondition air where additional conditioning is necessary. There are three general types of aftercoolers:

1. Water cooled
2. Air cooled
3. Refrigerant

TABLE 14.6   Temperature of Air Compressor Discharge

Type of compressor	Temperature rise °F over ambient	°C
Piston 1 stage	200	110
Piston 2 stage	250	137.5
Liquid piston	50	27.5
Rotary screw	150	82.5
Centrifugal	150	82.5

If a facility has a plentiful supply of reusable and/or recirculated cooling water, the first choice would be a water-cooled aftercooler. These units are selected on the basis of maximum inlet compressed air temperature, highest temperature and quantity of cooling water available, desired outlet compressed air temperature, and maximum flow in scfm of compressed air. Typical cooling capacity will bring the compressed air to within 10 to 15°F (6 to 9°C) of the water temperature used for cooling.

Air-cooled units are less efficient than water-cooled units. They are selected on the basis of maximum inlet compressed air temperature, highest ambient air temperature, approach temperature desired, and maximum flow in scfm of compressed air. Typical cooling capacity will bring the compressed air to within 20 to 30°F (12 to 18°C) of the air used for cooling.

A refrigerant type of aftercooler is rarely used. If, due to job conditions, one is required, consult the manufacturer's literature for applications. Used for this purpose, a refrigerated aftercooler will follow the same principles for air dryers, which will be discussed later in this chapter.

Since large amounts of water are usually removed from the air in an aftercooler, a moisture separator is usually provided. The separator could be either an integral part of the aftercooler or a separate unit. A typical aftercooler and separator is illustrated in Fig. 14.12.

Additional factors to be considered when selecting an aftercooler are pressure drop through units, space requirements, operation costs, and maintenance.

Filters

The purpose of any filter is to reduce or remove impurities or contaminants in the airstream to an acceptable, predetermined level.

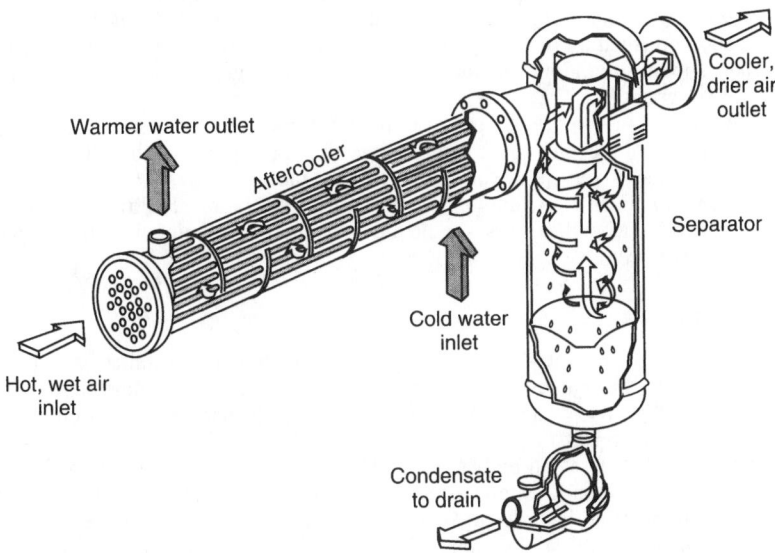

FIGURE 14.12 Typical aftercooler and moisture separator.

Filter Nomenclature.   Filter nomenclature has been developed based on the individual type of filter medium and generally where it is placed in a compressed air system. Inlet filters, prefilters, afterfilters, and point-of-use filters are some examples. Generally speaking, nothing prevents any kind of filter from being used for any application, provided that the required reduction of contaminants is achieved and it is suitable for the purpose intended. The following is a brief discussion in general terms of various types of filters:

Inlet filters remove large amounts of contaminants at the inlet to the air compressor.

Prefilters are generally used before air enters a dryer to remove various contaminants that might foul the unit. These filters are usually of the coalescing type so as to remove particulates and vapors, such as oil, hydrocarbons, and water. When combined with separators at this point, these filters may be called *separator-filters.*

Afterfilters are generally used after the drying process to remove smaller particulates than a prefilter removes. Some dryers produce a very small diameter dust (fines) that must be removed from the airstream. These filters remove particulates only.

Point-of-use filters are generally used immediately prior to any tool or individual piece of equipment that requires additional removal of particulates to a greater extent than an afterfilter.

Oil removal filters are special filters used only to remove an unwanted amount of oil aerosols too small to be removed by coalescing.

Activated carbon filters are used to remove gaseous oil and other hydrocarbons as well as small particulates too small to be removed by coalescing.

Contamination Removal.   Contaminants are removed by mechanical separation (interception), coalescence, adsorption, or a combination of these.

Mechanical Separation.   Mechanical separation is used to remove solid particles from the airstream. The filter element is a thin sheet or membrane whose passages are smaller than the particles to be retained. They intercept and hold dirt, scale, dust, and other solid particles in the matrix of the filter element.

Coalescing.   Coalescing filters are used to remove aerosols from the airstream. This is accomplished by impingement of the small-diameter aerosols onto a medium that causes them to randomly collide and merge into larger droplets that drain from the filter by gravity.

Adsorption.   Adsorption is used to remove vapors from the airstream by causing the contaminant molecules to be trapped into the small pores of the filter medium. This medium has a very high surface-to-volume ratio.

Combination Filters.   Combination filters use two or more filtration principals in a single unit. Manufacturers should be consulted to obtain the most effective combinations to remove specific contaminants.

Specialized Filters

Intake Air Filters.   Intake air filters are required for every installation to protect the compressor from damage. A properly selected air filter will return dividends in the form of reduced wear and maintenance by assuring that sufficiently clean air is supplied to the compressor.

Intake air must be clean and free of foreign matter (such as leaves or insects), solid and gaseous impurities, and abrasive dust particles. In addition, the air should be as cold as possible in order to increase compressor efficiency. In certain urban locations and industrial applications, air is often contaminated with corrosive and acid gases, which might damage the compressor. Unusual volumes of any objectionable contaminant gas in vapor or aerosol form must be considered. Filters should have the following characteristics:

1. High efficiency. The filter should remove large amounts of particulates from intake air and allow them to accumulate on the filter elements while keeping a low-pressure drop.
2. Large storage capacity of particulates before requiring replacement due to reduced flow of intake air.
3. Low resistance to flow of intake air.
4. Mechanical and structural strength. Filters must be capable of withstanding any possible air pressure surge as well as resisting physical damage.
   Filters for intake air fall into the following general categories:

1. *Paper filters.* These are dry and disposable, consisting of corrugated paper, usually impregnated by some material to improve performance. Filter efficiency is high with low-pressure drop when new. Paper is not recommended for inlet air temperatures greater than 150°F or where strong air pulsations may occur as with some piston compressors. They are recommended for air compressors of any capacity.

2. *Felt filters.* These are dry, reusable, pleated felt elements, often reinforced with wire screens. These filters have a large particulate capacity. They are cleaned by using either compressed air or washed according to the manufacturer's instructions. Recommended for oil-free air and other compressors of any capacity.

3. *Oil-wetted labyrinth filters.* These are reusable and are of metal construction. They work based on the principle of separating particulates by rapid changes of direction, causing particles to adhere on surfaces wetted by a film of oil. These filters require careful maintenance to ensure that the oil surface has not dried out or become saturated. These filters are recommended for small-capacity units (up to about 100 scfm) and where large amounts of particulates are present at inlet.

4. *Oil bath filters.* These filters are reusable, with an improved type of wetted labyrinth, using a surface of liquid oil to trap particulates. This filter has a large capacity for particulates, usually equal to the weight of oil in the filter. Careful maintenance is required to regularly change the oil. If an unloader is used on the air compressor, there is a potential for the oil to be blown out of the filter. These are recommended where large amounts of particulates are present at the inlet.

For most installations, the design engineer would select the inlet location and investigate known and potential pollutants. This is ideally obtained from many tests of air at the proposed intake location taken over several months spanning the different seasons and at different times of the day. This is rarely possible. In some urban areas, tests have been taken by some authorities such as state or federal EPA or a health department. In actual practice, during the design phase of a project, tests could be taken and analyzed. Then the purity of air for final use should be determined. With this criteria, the inlet filter, based on the type of compressor used, can be selected. Manufacturers, with a knowledge of their own product line, should be consulted to recommend types of filters capable of meeting the established criteria. The filter best suited for the system under design would then be selected.

For outdoor installations, provide a weatherproof rain cap and an insect screen around and over the actual inlet. Do not locate the filter close to any exhausts or vent pipes. The inlet should be mounted no less than 3 ft, 0 in (1 m) above roof or ground level or above possible snow level.

Separators.   Separators are a type of filter used to remove large quantities of liquid water or oil, individually or in combination with each other, from the airstream. Often, oil and water form an emulsion inside the compressor and are discharged together. Since suspended liquids are present in the airstream leaving the aftercooler or compressor, the most common location is at that point. General design of these units should allow for removal of between 90 and 99 percent by weight, of liquids.

There are two general types of separators: passive and active. The passive separator uses no moving parts and depends on the impaction of the liquid on internal surfaces, along with coalescence, for their effectiveness. Active units use moving internal parts (often centrifugal action) to remove liquid drops.

The purpose of an oil separator is to remove oil present in the airstream, regardless of the quantity or form (drops, aerosol, or vapor). An oil separator can be selected to obtain any degree of removal. Separators can be combined with integral air filters to increase the efficiency of the combined unit. They can also be provided with integral drain traps. If not, a separate drain trap must be provided.

Oil and moisture separators should never be considered similar types of units, as their functions are quite different.

Filter Selection.   Just as the degree of contamination in compressed air varies, so do the requirements for purity of the system at various points of use. These requirements must be established prior to the selection of filters in the system.

Filters should be selected by their ability to meet established design criteria. The manufacturer, who is the most knowledgeable about specific conditions, should be consulted as part of the selection process. The following items should be considered:

1. Maximum flow rate expected
2. Desired pressure drop across filter
3. Temperature of airstream
4. Contaminants to be removed (requirement for filter type and housing material)
5. Pressure rating (ASME stamp requirement)
6. Drain trap requirement (automatic type preferable)
7. Sampling port requirement
8. Filter efficiency

Filter Selection Procedure

1. Calculate the expected maximum flow rate of air. This figure should be expressed in the manufacturer's rating units, usually in scfm.
2. Determine the highest and lowest pressures that the filter can be expected to operate with during operation. This pressure must be used in conjunction with the pressure drop in other system equipment.
3. Based on the filter's position in the system, determine what the contaminant or contaminants to be removed will be.
4. Determine the maximum pressure drop across the filter. Manufacturers often use such values as "wetted pressure drop" and "dry pressure drop" that do not take into consideration dirty elements. Average conditions use a range of between 6 and 10 psig.

5. Will monitoring of the filter element for replacement be required, such as a color change or pressure drop?

6. Select the appropriate filter from a manufacturer's catalog.

Drain Traps

Separators and aftercoolers are not capable of directly discharging any water or oil removed from compressed air to drains. The purpose of a drain trap is to allow for the collection and removal of liquids that have separated from the airstream, with little or no loss of line pressure or compressed air.

Drain traps fall into two general categories: manual and automatic. Automatic traps are by far the more common. Manual traps are simply a drip leg on piping, with a valve that is opened by hand to drain the liquid that has accumulated in the length of pipe making up the drip leg.

Automatic drains fall into three categories:

1. Float
2. Bucket
3. Electronic

Float traps operate on the principle of a sealed float connected to a valve that opens when the float rises and reaches a predetermined level. Bucket traps use an unsealed bucket that moves due to the displacement of water either in or around the bucket. This opens a valve to allow the accumulated liquid to discharge. The electronic trap is a solenoid valve set to open at predetermined programmable intervals for a programmable period of open time. Float and bucket traps are illustrated in Chap. 11.

Selection of an automatic drain valve is based on line pressure, quantity of liquid stored, consistency of stored liquid, and rate of liquid discharge required. We will discuss how to determine the amount of liquid separated later in this section.

Float traps generally allow for higher discharge rates and greater contamination of stored liquid with oil sludge and solid particles than do bucket traps. Bucket traps are usually tighter sealing and can be used for high-pressure applications. They also cost less than float traps. Some facilities have a preference for using a particular type of trap. If traps must be placed outdoors with a potential for freezing, integral heaters are available to keep the stored liquid above freezing. Some traps require a small equalization line from the trap to the compressed-air piping to allow for reliable float operation.

Compressed Air Dryers

Air dryers are devices used to remove water vapor from the airstream. Large volumes of water consisting of droplets are removed by a moisture separator. If additional reduction of water vapor content is desired, it must be accomplished by the use of an air dryer. There are four methods generally used:

1. High pressurization of the compressed air
2. Condensation
3. Absorption

4. Adsorption

5. Heat of compression

High Pressurization.   High pressurization reduces water vapor by compressing air to far greater pressures than required for actual use. When pressure is increased, the ability of air to hold moisture is decreased. Since pressurization requires great amounts of energy, this process is rarely used.

Condensation.   Condensation utilizes the principle of lowering the temperature of the airstream through a heat exchanger, producing a lower dew point. The lower dew point reduces the capacity of air to retain moisture. Moisture then condenses out of the air onto the coils of the dryer. A moisture separator removes the condensate. The cooling medium in the coil could be chilled water, brine, or a refrigerant. The most common type uses a refrigerant and is called a *refrigerated dryer.* A schematic diagram of a refrigerated air dryer is shown in Fig. 14.13.

The greatest limitation is that they cannot practically produce a pressure dew point lower than 35°F. Otherwise, the condensed moisture could freeze on the coils. The advantages are that they have the lowest operating cost, and they do not introduce impurities into the airstream. The initial cost is midrange of the different dryer types. General pressure loss through a refrigerated dryer is approximately 5 psig. To achieve the rated moisture removal, a minimum of 20 percent of rated flow is required.

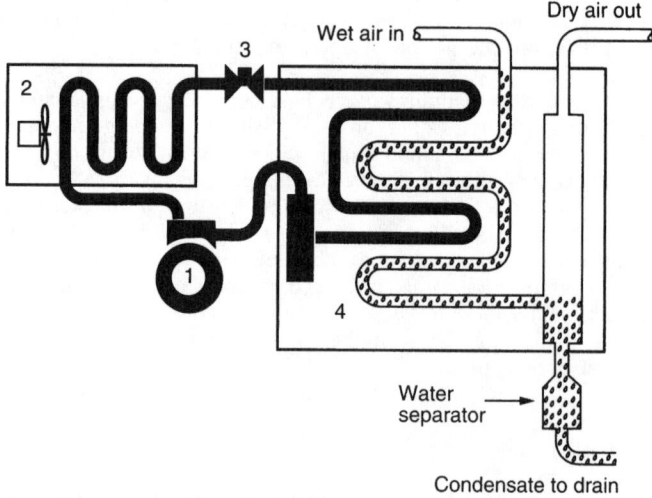

FIGURE 14.13   Refrigerated air dryer. There are four major components. (1) *Refrigeration compressor.* A hermetically sealed motor-driven compressor operates continuously. It generates a high-pressure refrigerant gas. (2) *Hot gas condenser.* The high-pressure refrigerant gas enters an air-cooled condenser where it is partially cooled by a continuously running fan. (3) *Automatic expansion valve.* The high-pressure liquid enters an automatic expansion valve where it thermodynamically changes to a subcooled low pressure liquid. (4) *Heat exchanger.* A system of coils where dry air is produced. (*Courtesy Arrow Pneumatics.*)

Absorption. Absorption dryers use either a solid or liquid medium and operate on the principle of having the airstream pass through or over a deliquescent material. This medium changes state in the presence of water. The solvent is then drained away, removing the water and reducing the amount of material available for absorption. Solid absorbers are much more common than liquid. Refer to Fig. 14.14 for a system schematic.

The liquid absorber is usually a glycol compound. The airstream passes over the liquid, and the water combines with the glycol. The glycol-water solution must be regenerated by having the water distilled off. Refer to Fig. 14.15 for a system schematic.

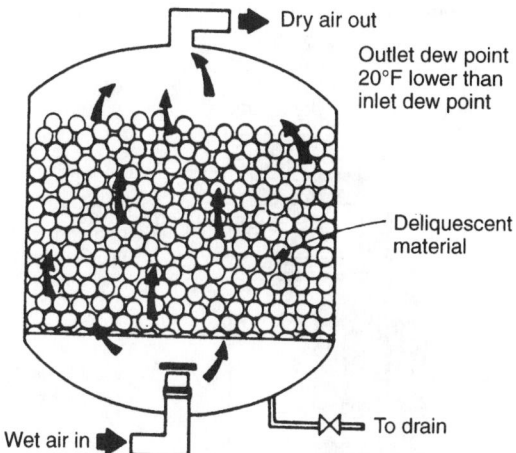

FIGURE 14.14   Absorption (deliquescent) dryer, solid medium.

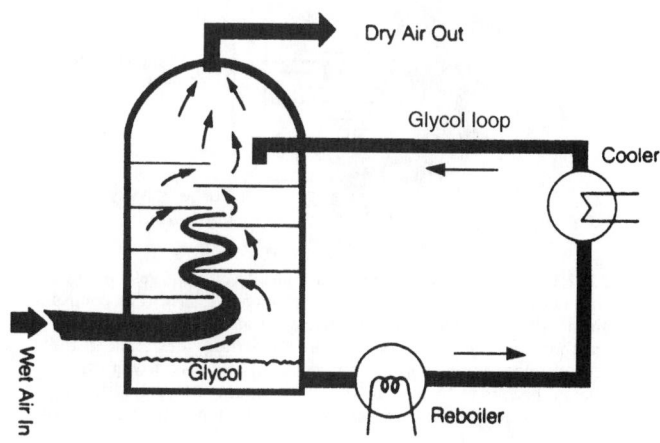

FIGURE 14.15   Absorption (deliquescent) dryer, liquid medium.

This system will produce a 20°F dew point reduction below the inlet air temperature as a general rule. The greatest advantage of this type of dryer is that no external connections of any kind are required for this system to operate except a drain. It has the least initial cost. It is generally used for intermittent flows where a high degree of drying is not required. Disadvantages are that some material used as a medium may be corrosive to the vessel, special treatment of the disposed liquid may be necessary, there is a limited dew point reduction, and since a small amount of glycol is lost in this process, some replacement is periodically required.

Adsorption. Adsorption dryers use a porous, nonconsumable material that causes water vapor to condense as a very thin film on the surface of the material. This material is called a *desiccant.* There is no chemical interaction, and the adsorption process is reversible. For a system schematic, see Fig. 14.16. Desiccant dryers are

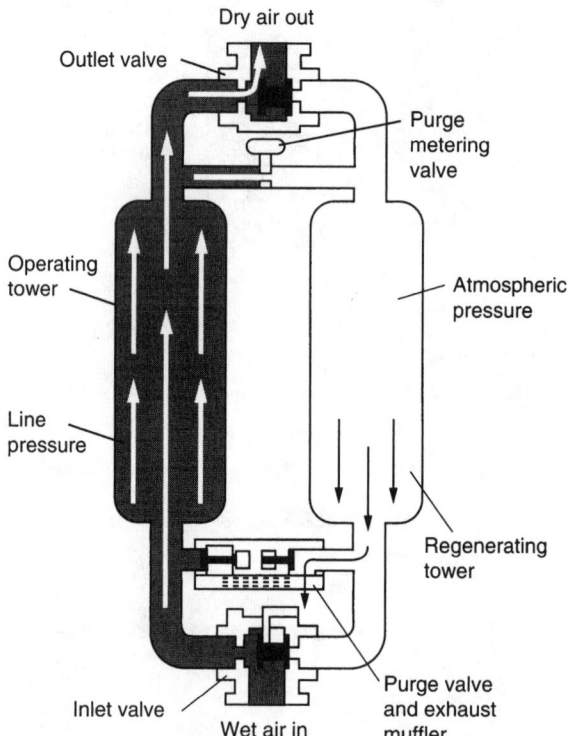

FIGURE 14.16 Absorption (desiccant) dryer. Two vessels filled with desiccant are interconnected by piping and valves that control airflow direction. One tower receives wet air where desiccant collects moisture and dries the air. Simultaneously, desiccant in the second tower is regenerated. Collected moisture is driven off by internal heaters or by dry air from the operating tower. Redirecting the airflow regenerates the first tower while the second dries the main airstream. (*Courtesy Van Air Systems.*)

capable of producing a pressure dew points as low as $-100°F$. The method of regeneration is the primary way of distinguishing between types of desiccant dryers.

Desiccant materials include silica gel, activated alumina, and aluminosilicate (molecular sieve). Each material can also be applied in the removal of specific impurities other than water. Desiccant materials will age in use over a period of years, which may affect capacity. In addition, care must be taken to avoid contamination of the material, particularly by oils. If adequate protection is not provided, the material may have to be replaced, although, if the contamination is not extensive, it might be brought back by removal of the impurities.

The three general types of regeneration methods for desiccant dryers are: pressure swing (heatless), heat activated (internal or external heaters), and heat of compression.

Pressure swing systems regenerate the material by using a portion of dry air from the system to flow in a reverse direction, thus purging the desiccant bed of water. This operation is completed in a short period of time, usually no longer then 10 min. This purge air must be discharged to the atmosphere and thus is lost. This requires that the compressor be oversized in order to accommodate the additional air required for drying.

Thermal swing regeneration using internal heaters is commonly used for units up to 1000 scfm and 100 psig. Purge air from the system is directed to flow in a reverse direction through the unit, with the purge air heated by internal heating elements. Average purge air requirements may be as much as 15 percent of air compressor capacity.

External heat regeneration uses a purge blower, atmospheric air, and an external heater to fulfill regeneration process requirements. The use of system air is not necessary. Because of the external heater, larger units can be selected for faster regeneration time.

In general, pressure swing dryers are lowest in initial cost, among the lowest in operating cost, and produce the most consistent dew point. In addition, they are more efficient in systems where high contamination may be a problem. Both adsorption and absorption dryers require an afterfilter to trap particles (fines) that escape from the dryer into the piping network.

Heat of compression dryers use hot air piped directly from the compressor to dry the desiccant material on a continuous basis. The air is returned to the discharge of the compressor.

Dryer Selection Procedure.   The single most important requirement in the selection process is to determine the lowest required pressure dew point for the intended application. This may eliminate some types of dryers. An excessive regeneration flow rate may eliminate other dryer types. The economics of initial and operating costs between various units will be another determining factor.

The following information should be obtained or calculated in order to select a dryer:

1. Determine the lowest required pressure dew point. This information is usually supplied by the facility based on the equipment that will be used.
2. Obtain from the compressor or aftercooler manufacturer the temperature of the air at the inlet of the dryer. Refer to Table 14.6 when no aftercooler is used.
3. Determine the minimum and maximum system pressure.
4. Determine the minimum and maximum flow rate for the system.

5. Determine whether the required electrical power and drainage are available where the dryer is to be installed.

6. Refer to a manufacturer's catalog to select the required dryer capable of meeting these requirements. Consideration should be given to initial and operating costs.

Lubricators

Lubricators are used to lubricate individual pieces of equipment when operating. There are three basic methods of lubrication, each for unique applications: fixed feed, demand feed, and positive displacement.

Fixed-feed lubricators are used for individual tools only and provide oil at a fixed rate (drops per minute). The amount is controlled by air velocity. This type of lubricator must be adjusted while the tool is running. Any filter and/or regulator must be compatible with the individual installation. This system is best used where there are long operating periods with constant airflow.

Demand feed systems use a wick saturated with oil. As the airstream passes the wick, lubricant is carried into the air. This type of lubricator should not be used where either low air velocity or volume exist.

Positive displacement lubricators use a pneumatic piston to force lubricant to the point of application whenever there is flow. This type is regarded as the most accurate and dependable, but it is the most costly. This system is best used for equipment with short operating cycles.

Air Compressor Drives

Electric motors are the predominant motive force for air compressors. Other methods of powering compressors are diesel, gasoline, or steam engines. Since electric motors are used predominantly, the discussion is limited only to that type. The two most common types of motors are induction and synchronous. Smaller compressors are generally driven by squirrel cage induction motors, with larger units having synchronous motors.

Motors are coupled to compressors either by belt drive, direct coupling, adjustable speed couplings, flexible coupling, or flange coupling. Belt drives are the most common for typical systems. Belt drive couplings lose approximately 4 percent of their power through the drive connection.

Starters used for electric drives are based on the type and size of motor. Magnetic, across-the-line starters are the most commonly used. When an across-the-line starter is used with an induction motor, a starting inrush of current is $5\frac{1}{2}$ times the running current. If this is too much current, and you wish to reduce the initial load, a step starter (reduced voltage) can be used. This type has a series of taps or steps to reduce voltage (and therefore current) to the motor. Care must be taken to select a reduced voltage starter that provides the necessary torque to start the compressor, which is usually 110 percent of running torque. Consult an electrical engineer and manufacturer to check correct selection. The synchronous motor has a starting inrush of $3\frac{1}{2}$ times the running current. Make sure all necessary overload protection is provided to prevent damage to the motor in case the compressor fails to start.

The power factor, which is a charge a utility company makes for peak energy demand, may be a deciding criterion in the selection of an appropriate type of drive motor for larger sizes. The synchronous motor has a lower power factor than an induction motor.

Compressor Regulation

If the total system demand for both air pressure and volume exactly matches the compressor output for as long as the compressor operates, no regulation would be required. Since this does not usually happen, whenever system demand varies, it will be necessary to regulate the compressor. Some manner must be found to adjust output to match the variable demands of the system. The ideal method would be to have an infinitely variable compressor to provide the exact volume and pressure required to satisfy all demands. Again, this is not realistic. We will now discuss the commonly used methods to achieve varying volume of air while maintaining adequate system pressure.

Compressor capacity can be regulated either by continuous or discontinuous methods. Continuous means would require control of the compressor by using either an adjustable speed coupling or by controlling the drive motor speed. Another method would be to bleed compressed air from the discharge either to atmosphere or back into the inlet. This is called either *unloading* or *blowoff* and is wasteful of energy. Finally, the internals of a compressor can be altered to be less efficient by adjusting valves, clearances, and so on. The last method is the least desirable of all because the correct speed and the internal adjustment of the compressor can be determined and accomplished only by the manufacturer for specific projects. This makes it almost impossible for maintenance personnel to repair in the field. Blow-off, however, has several alternatives.

In general, unloading is best used when the compressor operates more than 50 percent of the time or where continuous operation of a motor is desirable. For applications requiring constant compressor speeds, a pilot unloader, pressure-sensing device, or trigger valves would be used. The pilot unloader can operate in one of three ways—it may adjust compressor cylinder suction valves, close valves on the compressor inlet line, or open a bypass in the main discharge line. To summarize, a constant compressor speed unloader operates when upper pressure is reached but the motor continues to run. At the lower pressure limit, the pilot stops working and allows air to be delivered. Of course, the motor is still running.

Discontinuous regulation is the most common method of controlling compressor capacity for relatively small systems. This is accomplished by using a pressure-regulating device (either mechanical or electromechanical) arranged to stop the compressor at a preset high pressure and start it again at a preset low pressure. A receiver (tank) is used to store air. This gives a reserve capacity to keep the compressor from starting too often. Receivers will be discussed later.

There are circumstances when both kinds of regulation may prove beneficial. It is possible to have both types acting together. This is referred to as *dual regulation* or *control*.

Speaking in general terms about continuous regulation, the reciprocating compressor uses the inlet suction valve most often. It is possible to obtain this type of unloader in from one to three gradually increasing stages of operation. Sliding vane and screw compressors commonly use the modulating suction valve, with dual control and a receiver if demand permits. Centrifugal compressors should not be run outside their range, and so a blowoff to atmosphere is used.

Starting Unloader

The starting unloader is used only when starting a compressor. After the first time pressure has been established in the system and the compressor has stopped, the system remains pressurized. When the compressor must start again, it has to overcome the force exerted by the air still under pressure in the casing. There is not enough power in the drive motor to overcome this pressure. Therefore, a means must be provided to vent only the air under pressure in the compressor casing to atmosphere and allow the compressor to start under no load. This is done with a starting unloader, of which there are two types: centrifugal and pressure switch. The pressure switch operates by using a separate switch to open a valve-type mechanism installed in the cylinder head. The centrifugal unloader is an integral part of the compressor and is activated through a connection to the camshaft. The centrifugal unloader is generally preferred because it is more reliable.

Compressed Air Receivers

The primary purpose of a receiver is to store air. The determination as to the need for a receiver is always based on the type of regulation the system will use. If the compressor runs 100 percent of the time and has constant blowoff, an air receiver will not be required.

For most applications, an air compressor is regulated by starting and stopping, with a receiver used to store air and prevent the compressor from cycling too often. Generally accepted practice for reciprocating compressors is to limit starts to about 10 per hour, and a maximum running time of 70 percent. Centrifugal, screw, and sliding vane compressors are best run 100 percent of the time.

An air receiver serves the following purposes:

1. Stores air

2. Equalizes pressure variations (pulsations)

3. Collects residual condensate

Piping connections should be made in such a way that the incoming air is forced to circulate and mix with the air inside the tank before being discharged.

Receivers should be ASME stamped for unfired pressure vessels. Refer to Table 14.7 for standard receiver sizes. Receivers should be sized on the basis of system

TABLE 14.7 Standard ASME Receiver Dimensions

Diameter, in	Length, ft	Volume, ft^3
14	4	4½
13	6	11
24	6	19
30	7	34
36	8	57
42	10	96
48	12	151
54	14	223
60	16	314
66	18	428

$$\frac{1 \text{ in} = 25.4 \text{ mm}}{1 \text{ ft}^3 = 0.03 \text{ m}^3} \quad 1 \text{ ft} = .3 \text{ m.}$$

demand and compressor size, using the starts per hour and running time best suited for the project. The design engineer must keep in mind that the compressor will operate to satisfy the pressure switch rather than the use of air and that the receiver is an integral part of the system that must function in respect to load conditions, amount of storage, and pressure differential.

Cooling Water

Water is used to cool air compressor jackets and also to cool the air passing through intercoolers and aftercoolers. When used to cool compressors, it will result in a lower horsepower motor than similar capacity air-cooled units. Water may also be used to cool the oil used for compressor lubrication.

It is no longer acceptable to waste the water used for cooling. One reason is that in most cases, this water may be fairly expensive if it has been treated. Another is that the wastewater must also be treated. If river water is used with only filtering, it can be discharged back into the source if doing so is acceptable to the local authorities. The use of cooled or chilled water is preferred.

Water temperature is an important consideration. The coldest water should go to the intercooler and aftercooler first because the lower the cooling water temperature, the more efficient that stage will be. If the cooling water supplied to a reciprocating compressor jacket is too cold, it may cause water vapor to condense inside the cylinders and thus wash away some lubrication. This will accelerate wear of pistons, rings, and cylinder walls. A good rule to remember is that water should be the same temperature as the desired discharged air or slightly higher. In general, water over 110°F should not be considered for cooling purposes. A 10°F (6°C) rise inside the compressor is a common average temperature rise. The following should be considered when providing cooling water:

1. Generally, a 5 psig pressure drop can be expected when cooling compressor jackets.

2. A strainer on the water supply to compressor jackets should be provided.

3. A minimum supply pressure of 10 psig should be available.

4. A solenoid valve to start water flow only when compressor starts should be provided.

5. Thermometers should be provided on the inlet and outlet to facilitate troubleshooting.

6. A sight glass or drain funnel should be provided to monitor actual flow.

7. A figure of ½ gpm/hp is the average requirement for supplying cooling water to an air compressor.

The amount of cooling water required for all purposes may be determined from the following formula.

$$\text{gpm} = \frac{\text{BHP} \times H}{T \times 8.33} \tag{14.1}$$

where BHP = brake horsepower of compressor
H = heat dissipation, Btu/h
T = selected temperature rise of cooling water (usually 10°F)

TABLE 14.8 Selection of Supply Hose Size

Air inlet port NPT, in	⅛	¼	⅜	½
Supply hose size I.D., in	¼	⅜	½	¾

1 in = 25.4 mm.

Hose and Fittings

Most tools use flexible hose for connection to the piping system. The hose used is usually larger than the air inlet port on the tool it serves. Table 14.8 indicates generally accepted practice for the selection of supply hoses based on the size of the inlet ports. Where the length of hose may extend beyond 20 ft, hose one size larger than normal should be used to compensate for additional line friction loss. It is good practice to limit friction loss of hose to 5 psig. Refer to Table 14.21 for pressure loss through various sizes and lengths of hose.

UTILITY (LIGHT INDUSTRIAL) COMPRESSED AIR

SYSTEM DESIGN

General

The compressed air system must be controlled, regulated, and sized to ensure that an adequate volume of air, at a pressure and purity necessary to satisfy user requirements, is delivered at any outlet during the period of heaviest use. The design process is an iterative one because the performance of one or several components may have an effect on the performance of other equipment. Therefore, various adjustments will be necessary as the design progresses.

The entire system will be separated into three individual systems:

1. Air compressor, including regulation, intake, and receiver
2. Compressed air conditioning equipment
3. Piping distribution network

Design Sequence

1. Locate and identify each process, workstation, or piece of equipment using compressed air. These components should be located on a plan, and a complete list should be made to simplify record keeping.
2. Determine the volume of air to be used at each location.
3. Determine the pressure range required at each location.
4. Determine the conditioning requirements for each item, such as the allowable moisture content, particulate size, and oil content.
5. Establish how much time the individual tool or process will be in actual use for a 1-min period of time. This is referred to as the *duty cycle.*
6. Establish the maximum number of locations that may be used simultaneously on each branch and main and for the project as a whole. This is known as the *use factor.*
7. Establish the extent of allowable leakage.
8. Establish any allowance for future expansion.
9. Make a preliminary piping layout, and assign preliminary pressure drop.
10. Select the air compressor type, conditioning equipment, equipment location, and air inlet, making sure that either scfm or acfm is consistently used for both the system and compressor capacity rating.
11. Produce a final piping layout, and size the piping network.

Project Air Consuming Device Location. This speaks for itself. In order to accomplish this task, it is recommended that the location of all air consuming devices and their requirements be marked on a plan in order to facilitate the branch piping layout. Prepare a list for future reference of all devices noted on the plans, as well as their location and actual flow rate.

Pressure and Volume Requirement. The information relative to pressure and volume parameters for individual equipment and tools is usually obtained from the tool manufacturer, end user, facility planner, or owner. It is quite common for these facts to be incomplete, with additional investigation required to find the specific values needed. Very often, it is useful to assign preliminary pressure and flow rate requirements of the system, in order to arrange equipment space and give preliminary mechanical data to other disciplines. Table 14.9 lists general air requirements for various tools.

TABLE 14.9 General Air Requirements for Tools

Tools or equipment	Size or type	Air pressure, psig	Air consumed, scfm
Hoists	1 ton	70–100	1
Blow guns		70–90	3
Bus or truck lifts	14,000-lb cap	70–90	10
Car lifts	8,000-lb cap	70–90	6
Car rockers		70–90	6
Drills, rotary	¼-in cap	70–90	20–90
Engine, cleaning		70–90	5
Grease guns	6	70–90	4
Grinders	2-in wheel	70–90	50
Grinders	4-in wheel	70–90	20
Paint sprayers	Production gun	40–70	20
Spring oilers		40–70	4
Paint sprayers	Small hand	70–90	2–10
Riveters	Small to large	70–90	10–35
Drills, piston	½-in cap, 3-in cap	70–90	50–110
Spark plug cleaners	Reach 36–45	70–90	5
Carving tools		70–90	10–15
Rotary sanders		70–90	50
Rotary sanders		70–90	30
Tire changers		70–90	1
Tire inflaters		70–90	1½
Tire spreaders		70–90	1
Valve grinders		70–90	2
Air hammers	Light to heavy	70–90	30–40
Sand hammers		70–90	25–40
Nut setters and runners	¼-in cap to ¾-in cap	70–90	20–30
Impact wrenches / screwdrivers	Small to large	70–90	4–10
Air bushings	Small to large	80–90	4–10
Pneumatic doors		40–90	2
File and burr tools		70–90	20
Wood borers	1–2 in	70–90	40–80
Rim strippers		100–120	6
Body polishers		70–90	2
Carbon removers		70–100	3
Sand blasters	Wide variation	90	6–400

1 psi = 6.9 kPa.
1 cfm = 0.03 m³/min.

All tools use air either through an orifice to do work or to drive a piston. Table 14.10 gives the amount of air in scfm passing through an orifice at various pressures. Table 14.11 provides the actual volume of air required to drive a single-acting piston. The figures should be doubled for double-acting pistons. The best method is to obtain actual equipment cuts from the proposed equipment manufacturer, due to the wide variation in requirements of similar air consuming devices from different manufacturers.

Compressed Air Conditioning. The selection of conditioning equipment depends upon end use requirements, usually obtained when items 2, 3, and 4 are received. Conditioning equipment includes dryers, filters, lubricators, and pressure regulators.

Dryer selection is based on the most demanding user requirement except where special, dedicated equipment may be required. Table 14.12 gives general performance specifications of types of air dryers. If a very low dew point is required, the only selection possible is a desiccant dryer. If, however, a high dew point is acceptable, several different types of dryers can be considered.

Deliquescent Dryer. The deliquescent dryer is the least efficient, but it requires no power to operate, and the initial cost is the lowest of any type of dryer. It has a moderate operating cost since only the drying medium must be replenished at regular intervals. This type of dryer loses efficiency if the inlet air temperature is over 100°F (38°C), and so an efficient aftercooler is mandatory. The type of deliquescent material used will affect the quality of air. For example, a salt material normally reduces dew points about 12 to 20°F (7 to 12°C), while potassium carbonate will lower the dew point about 30°F (18°C). A filter is necessary after the dryer to remove any chemical carryover (fines) into the system.

Refrigerated Dryers. The most often used type of dryer is the refrigerated type. Refrigerated dryers will produce pressure dew points as low as 33°F (1°C), but for practical purposes, a general figure of about 38°F (3°C) has been used. General operating cost is moderate. External requirements are floor drains and electric power. These dryers are sensitive to changes in flow rate and pressure of the airstream. The initial cost is moderate.

TABLE 14.10 Air Volume Passing through an Orifice, CFM

Gauge pressure, psi	Orifice size, inches diameter							
	1/64	1/32	3/64	1/16	3/32	1/8	3/18	1/4
50	0.225	0.914	2.05	3.64	8.2	14.5	32.8	58.2
60	0.26	1.05	2.35	4.2	9.4	16.8	37.5	67
70	0.295	1.19	2.68	4.76	10.7	19.0	43.0	76
80	0.33	1.33	2.97	5.32	11.9	21.2	47.5	85
90	0.364	1.47	3.28	5.87	13.1	23.5	52.5	94
100	0.40	1.61	3.66	6.45	14.5	25.8	58.3	103
110	0.43	1.76	3.95	7.00	15.7	28.0	63	112
120	0.47	1.90	4.27	7.58	17.0	30.2	68	121
130	0.50	2.04	4.57	8.13	18.2	32.4	73	130
140	0.54	2.17	4.87	8.68	19.5	34.5	78	138
150	0.57	2.33	5.20	9.20	20.7	36.7	83	147
175	0.66	2.65	5.94	10.6	23.8	42.1	95	169
200	0.76	3.07	6.90	12.2	27.5	48.7	110	195

1 psi × 6.9 = kPa.
1 cfm = 0.03 m³/min.

TABLE 14.11 Air Volume Requirements for Single-Acting Piston in Cubic Feet

Piston dia., in	Length of stroke, in*											
	1	2	3	4	5	6	7	8	9	10	11	12
1¼	0.00139	0.00278	0.00416	0.00555	0.00694	0.00832	0.00972	0.0111	0.0125	0.0139	0.0153	0.01665
1⅞	0.00158	0.00316	0.00474	0.00632	0.0079	0.00948	0.01105	0.01262	0.0142	0.0158	0.0174	0.01895
2	0.00182	0.00364	0.00545	0.00727	0.0091	0.0109	0.127	0.0145	0.01636	0.0182	0.020	0.0218
2⅛	0.00205	0.0041	0.00615	0.0082	0.0103	0.0123	0.0144	0.0164	0.0185	0.0205	0.0226	0.0244
2¼	0.0023	0.0046	0.0069	0.0092	0.0115	0.0138	0.0161	0.0184	0.0207	0.0230	0.0253	0.0276
2⅜	0.00256	0.00512	0.00768	0.01025	0.0128	0.01535	0.01792	0.02044	0.0230	0.0256	0.0282	0.0308
2½	0.00284	0.00568	0.00852	0.01137	0.0142	0.0171	0.0199	0.0228	0.0256	0.0284	0.0312	0.0343
2⅝	0.00313	0.00626	0.0094	0.01254	0.01568	0.0188	0.0219	0.0251	0.0282	0.0313	0.0345	0.0376
2¾	0.00343	0.00686	0.0106	0.0137	0.0171	0.0206	0.0240	0.0272	0.0308	0.0343	0.0378	0.0412
2⅞	0.00376	0.00752	0.0113	0.01503	0.01877	0.0226	0.0263	0.0301	0.0338	0.0376	0.0413	0.045
3	0.00409	0.00818	0.0123	0.0164	0.0204	0.0246	0.0286	0.0327	0.0368	0.0409	0.0450	0.049
3⅛	0.00443	0.00886	0.0133	0.0177	0.0222	0.0266	0.0310	0.0354	0.0399	0.0443	0.0488	0.0532
3¼	0.0048	0.0096	0.0144	0.0192	0.024	0.0288	0.0336	0.0384	0.0432	0.0480	0.0529	0.0575
3⅜	0.00518	0.01036	0.0155	0.0207	0.0259	0.031	0.0362	0.0415	0.0465	0.0518	0.037	0.062
3½	0.00555	0.0112	0.0167	0.0222	0.0278	0.0333	0.0389	0.0445	0.050	0.0556	0.061	0.0644
3⅝	0.00595	0.0119	0.0179	0.0238	0.0298	0.0357	0.0416	0.0477	0.0536	0.0595	0.0655	0.0715
3¾	0.0064	0.0128	0.0192	0.0256	0.032	0.0384	0.0447	0.0512	0.0575	0.064	0.0702	0.0766
3⅞	0.0068	0.01362	0.0205	0.0273	0.0341	0.041	0.0477	0.0545	0.0614	0.068	0.075	0.082
4	0.00725	0.0145	0.0218	0.029	0.0363	0.435	0.0508	0.058	0.0653	0.0725	0.0798	0.087
4⅛	0.00773	0.01547	0.0232	0.0309	0.0386	0.0464	0.0541	0.0618	0.0695	0.0773	0.0851	0.092
4¼	0.0082	0.0164	0.0246	0.0328	0.041	0.0492	0.0574	0.0655	0.0738	0.082	0.0903	0.0985
4⅜	0.0087	0.0174	0.0261	0.0348	0.0435	0.0522	0.0608	0.0694	0.0782	0.087	0.0958	0.1042
4½	0.0092	0.0184	0.0276	0.0368	0.046	0.0552	0.0643	0.0735	0.0828	0.092	0.101	0.1105

4⅝	0.0097	0.0194	0.0291	0.0388	0.0485	0.0582	0.0679	0.0775	0.0873	0.097	0.1068	0.1163
4¾	0.01025	0.0205	0.0308	0.041	0.0512	0.0615	0.0717	0.0818	0.0922	0.1025	0.1125	0.123
4⅞	0.0108	0.0216	0.0324	0.0431	0.054	0.0647	0.0755	0.0862	0.097	0.108	0.1185	0.1295
5	0.0114	0.0228	0.0341	0.0455	0.0568	0.0681	0.0795	0.091	0.1023	0.114	0.125	0.136
5⅛	0.01193	0.0239	0.0358	0.0479	0.0598	0.0716	0.0837	0.0955	0.1073	0.1193	0.1315	0.1435
5¼	0.0125	0.0251	0.0376	0.0502	0.0627	0.0753	0.0878	0.100	0.1128	0.125	0.138	0.151
5⅜	0.0131	0.0263	0.0394	0.0525	0.0656	0.0788	0.092	0.105	0.118	0.131	0.144	0.158
5½	0.01375	0.0275	0.0412	0.055	0.0687	0.0825	0.0962	0.110	0.1235	0.1375	0.151	0.165
5⅝	0.0144	0.0288	0.0432	0.0575	0.072	0.0865	0.101	0.115	0.1295	0.144	0.1585	0.173
5¾	0.015	0.030	0.045	0.060	0.075	0.090	0.105	0.120	0.135	0.150	0.165	0.180
5⅞	0.0157	0.0314	0.047	0.0628	0.0785	0.094	0.110	0.1254	0.142	0.157	0.1725	0.188
6	0.0164	0.032	0.492	0.0655	0.082	0.963	0.1145	0.131	0.147	0.164	0.180	0.197

Note: in = 25.4 mm, 1 cf = 0.0283 m³.

*These volumes are for single-acting cylinders. For double-acting cylinders, multiply by 2 and subtract the volume of the piston rod.

TABLE 14.12 General Performance of Different Types of Air Dryers

Dryer type	Inlet air capacity at 100°F				Outlet air at 100°F ambient				Prefilter	Afterfilter	Installation
	Flow, scfm	Press., psig	Max. temp., °F	Moisture, % RH	Pressure, psig	Moisture, °F pdp	Cooling	Power required			
Deliquescent	5–30,000	100	100	Saturated	95	80	None	None (requires replenishment of drying medium)	Recommended	Required*	Indoor and outdoor
Refrigerated	5–25,000†	100	130	Saturated	95	35 to 50	Air at 100°F, or water at 85°F	Electrical‡	Recommended	Not required	Indoor
Desiccant, regenerative	1–20,000	100	120	Saturated	95	−40 to −100	None	Electrical + 7% purge air, steam + 7% purge air, or dry air (15 to 35% of system capacity)	Required§	Recommended	Indoor and outdoor

*Some deliquescent dryers have built-in afterfilters. Do not add an additional filter (energy consumer) to the system if unnecessary.

†Higher flow rates will not damage, but air quality is reduced and pressure drop increased. Not sensitive to oil and particulate.

‡The thermal refrigeration type of refrigeration dryer is the only one that does not run continuously. A thermostatically controlled switch turns the refrigeration unit on as needed.

All cycling refrigerated air dryer ratings and capacity are based on National Fluid Power Association recommended Standard T/3.27.2 with saturated entering inlet air at 100°F (37.8°C) and 100 psig (690 kPa), with a 100°F (37.8°C) ambient air temperature. At these standard conditions, the dryer must be capable of producing outlet air with a dewpoint in a range of between 33 and 39°F and a pressure drop of 5 psig (35 kPa) or less. To select a dryer based on pressure drop and rated flow, refer to Fig. 14.17, entering with the actual pressure at the dryer inlet. For correction factors based on temperatures other than 100°F (37.8°C), inlet air temperatures other than 100°F (37.8°C), and ambient air temperatures other than 100°F (37.8°C), refer to Table 14.13. Dew point is also effected by low flow rates.

Desiccant Dryer. The desiccant dryer will produce the lowest dew points. They are the highest in initial cost and the highest in operating cost. Of the three purging methods used, the vacuum type is the most energy efficient. The unheated purge is the fastest, but it uses about 15 percent of system air for purging. Too high an incoming air temperature is detrimental to the desiccant material. An aftercooler is usually recommended for most dryer installations because it is an economical way to reduce the moisture content of air, and it should be selected in conjunction with the dryer. The aftercooler adds cost to the project and is not often used.

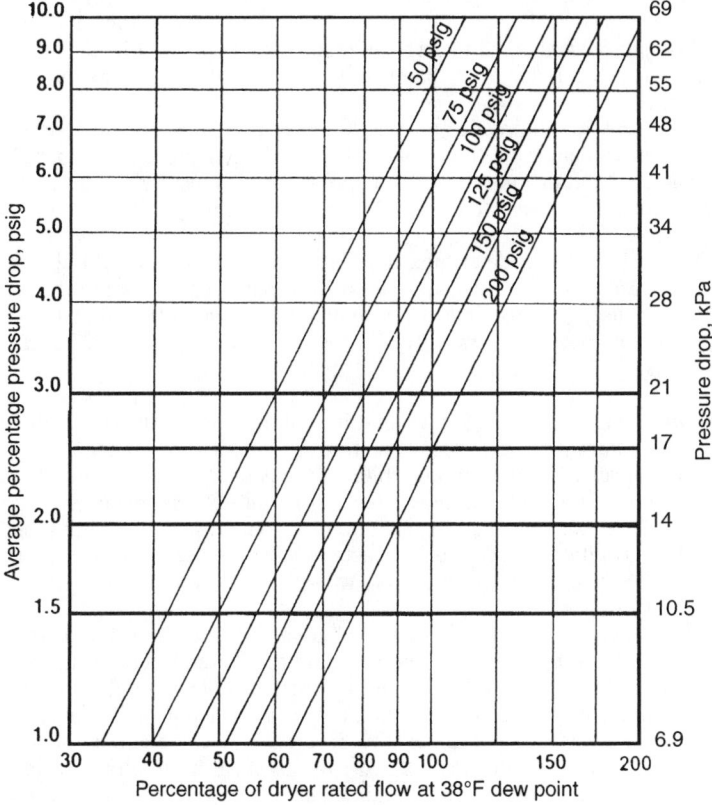

FIGURE 14.17 Pressure loss through refrigerated air dryers.
psi × 6.9 = kPa.

TABLE 14.13 Correction Factors for Refrigerated Air Dryers

Inlet air pressure			Inlet air temperature		Ambient air temperature	
kPa	Pressure, psig	Correction factor	Temperature, °F/°C	Correction factor	Temperature, °F/°C	Correction factor
345	50	1.19	80/27	0.66	80/27	0.92
518	75	1.06	90/32	0.82	90/32	0.95
690	100	1.00	100/38	1.00	100/38	1.00
1035	150	0.95	110/43	1.21	110/43	1.07
1207	175	0.94				
1552	255	0.92	120/49	1.42	120/49	1.16

There are two factors that will be used to select a filter: effectiveness and pressure drop. Do not specify a filter that will produce cleaner air than is actually necessary. If one station or process requires a purity much higher than all other points, use a point-of-use filter for only that area and a less restricted filter for the main supply. It is possible for a filter to have the largest pressure drop of any equipment in the system. In general, a filter will produce a 3 to 10 psig (21 to 70 kPa) pressure drop when dirty. If the actual figure proves to be too much, it would be a good idea to oversize the filter to cut down the pressure drop. In most cases, it would be more economical to pay the added initial cost of a larger filter than to increase energy requirements to compress air to a higher pressure for the life of the system.

It is considered good practice to provide pressure gauges on either side of main filters in order to determine filter condition. Lubricators are selected based on the manufacturer's requirements for tool operation. They must be consulted for type. Pressure regulators are selected using flow rate and pressure drop ratings.

Duty Cycle. In order to determine the duty cycle, the user should be consulted, for in most cases, they are the only authority capable of discussing the length of time air is in use. In most industrial applications, tasks of a similar nature are usually grouped together. This will allow sections or branches to be calculated independently.

Use Factor. Experience has shown that it is almost impossible to accurately determine a simultaneous use factor. Therefore, sufficient receiver capacity or larger compressor capacity to allow for possible variances must be made. For laboratories, air is used mostly for chemical reactions, and is not used as much as in industrial applications. The exceptions are for classrooms, some research facilities, and some areas within hospitals, where the use factor may be quite high. See Table 14.14 and Fig. 14.29 for use factors applicable to typical laboratory projects.

Allowable Leakage. There is no method to accurately determine a reasonable figure. Leakage is a function of the number and type of connections, the age of the system, and quality of pipe assembly. Many smaller tools and operations will generally have a greater leakage of air than a few larger use points. A well-maintained system will have a leakage of about 2 to 5 percent. Average conditions will incur a 10 percent leakage. Poorly maintained systems have been known to have a 25 percent leakage factor. The facility maintenance department should be consulted when selecting a value.

TABLE 14.14 Laboratory Outlet Use Factors

No. of outlets	Percent use factor
1–2	100
3–5	80
6–10	66
11–20	40
21–50	30
Over 50	20

Future Expansion. The facility owner must give guidance as to the possibility and extent of any future expansion. Consideration should be given to oversizing some components in anticipation of expansion, such as filters, dryers, and main pipe sizes, to avoid costly replacement in the future and to save downtime while expansion is under way.

Soot Blowers. Air may be used in large boilers for removing interior soot and dirt. Flow may be either pulsating or steady. Design pressure would range between 100 and 120 psig (700 and 840 kPa) at about 50 scfm (1.5 m²/mh). Check with manufacturer for exact requirements.

Ship Service. When a ship is berthed, air is used to supply pressure for pneumatic controls and some air tools. Expected flow would generally be 100 scfm at 100 psig (3 m³ at 700 kPa). Larger ships may require additional scfm.

Aircraft Starting. Air is used to start jet engines. Pressures range from 45 to 75 psig (315 to 520 kPa) with a maximum flow of 1200 scfm (3.5 m³/min). Air is stored in a receiver at a high pressure. A compressor should fill the receiver in 3 min.

Piping System Design

Piping layout on the plans will now be reasonably complete, with checking for space, clearances, interferences, and securely anchored drops to equipment. Also, the following information must be available:

1. A list of all air consuming devices
2. Minimum and maximum pressure requirements for each device
3. Actual volume of air used by each device
4. Duty cycle and use factor
5. Special individual air-conditioning equipment requirements

It will now be possible to start sizing the piping using the following sequence:

1. In order to use pressure drop tables, it is necessary to find the equivalent length to run from the compressor to the farthest point in the piping system. The reason is that the various pipe sizing tables are based on a pressure drop developed using friction loss for a given length of pipe. Measuring the actual length is the first step. In addition to the actual measured pipe length, the effect of fittings must be considered. This is because fittings and valves create an obstruction to the flow of air. This degree of obstruction has been converted to an equivalent length of pipe in order to make calculations easy. Table 14.15 has been developed to indicate the equivalent pipe length for fittings and valves that should be added to the actual measured run, to establish a total equivalent run. For preliminary calculation purposes, the addition of 50 percent of the actual measured run will give a conservative approximation of the total equivalent run and therefore the means to select a preliminary pipe size, if necessary, before final calculations are made.

   2. Determine the actual pressure drop that will occur only in the piping system. Generally accepted practice is to allow 10 percent of the proposed system pressure for pipe friction loss. So, for a 125 psig (860 kPa) system, a figure of 10 to 12 psig (85 to 90 kPa) will be allowed. Since the air compressor has not been selected yet, this figure is variable. A smaller pipe size may lead to higher compressor horsepower. It is considered good practice to oversize distribution mains to allow for future growth and also to allow for the future addition of conditioning equipment that may add a pressure drop not anticipated at the time of the original design.

   3. Size the piping using the appropriate charts for system pressure. Having calculated the scfm and the allowable friction loss in each section of the piping being sized. Since all pipe sizing charts are formulated on the loss of pressure per some length of piping (usually 100 ft) (30 m), it will be necessary to arrive at the required value for the chart you are using. Tables 14.16 through 14.19 present friction loss of air through Schedule 40 steel pipe in psig for a 100-ft (30 m) length of pipe, at from 50 to 125 psig (345 to 860 kPa) line pressure. Use the highest system working pressure to determine pipe size. The temperature used to calculate the friction loss is 60°F (15.6°C). For 100°F (37.8°C), increase pressure drop figures in the tables by 7.7 percent for greater accuracy. For copper pipe, reduce pressure drop figures by 5 percent. Table 14.20 is a suggested form for sizing stacks and mains. A maximum velocity or 4,000 fpm (1200 m/min) is recommended.

The charts were calculated using the following formula using IP units:

$$P = Q\,\frac{FV^2}{2GD} \tag{14.2}$$

where P = pressure loss due to friction, psi per 100 ft of pipe
   F = friction loss factor (Use 4000 as an average figure.)
   V = velocity, ft/s
   G = acceleration due to gravity, 32.2 ft/s
   D = pipe diameter, ft
   Q = specific weight of air, lb/ft^3

   The following general design parameters can be used for miscellaneous devices as a guide when calculating the total pressure drop of the piping system:

1. Equipment drop leg: 2 psig loss (1 psig if possible)
2. Hose allowance: 2 to 5 psig loss
3. Quick disconnect coupling: 4 psig loss
4. Lubricator: 1 to 4 psig loss
5. Point-of-use filter: ½ to 2 psig loss

   Table 14.21 gives the friction loss of air through hose that would be used to connect tools to the main piping system.

TABLE 14.15 Equivalent Pressure Loss through Valves and Fittings in Feet of Pipe

Nominal pipe size, in	Actual inside diameter, in	Gate valve	Long radius, all or on run of standard tee	Standard ell or on run of tee reduced in size 50 percent	Angle valve	Close return bend	Tee through side outlet	Globe valve
½	0.622	0.36	0.62	1.55	8.65	3.47	3.10	17.3
¾	0.824	0.48	0.82	2.06	11.4	4.60	4.12	22.9
1	1.049	0.61	1.05	2.62	14.6	5.82	5.24	29.1
1¼	1.380	0.81	1.38	3.45	19.1	7.66	6.90	38.3
1½	1.610	0.94	1.61	4.02	22.4	8.95	8.04	44.7
2	2.067	1.21	2.07	5.17	28.7	11.5	10.3	57.4
2½	2.469	1.44	2.47	6.16	34.3	13.7	12.3	68.5
3	3.068	1.79	3.07	6.16	42.6	17.1	15.3	85.2
4	4.026	2.35	4.03	7.67	56.0	22.4	20.2	112.0
5	5.047	2.94	5.05	10.1	70.0	28.0	25.2	140.0
6	6.065	3.54	6.07	15.2	84.1	33.8	30.4	168.0
8	7.981	4.65	7.96	20.0	111.0	44.6	40.0	222.0
10	10.020	5.85	10.00	25.0	139.0	55.7	50.0	278.0
12	11.940	6.96	11.00	29.8	166.00	66.3	59.6	332.0

1 ft = 0.3 m.
1 in = 25.4 mm.

TABLE 14.16 Pressure Drop of Air (in psi through 100 ft) through Steel Pipe, 50 psig and 60°F Temperature (350 kPa and 18°C)

scfm	½	¾	1	1¼	1½	2	2½	3	4	5	6
2	0.024	0.006									
3	0.055	0.012									
4	0.098	0.022	0.006								
5	0.153	0.034	0.009								
6	0.220	0.050	0.013								
8	0.391	0.088	0.023	0.006							
10	0.611	0.138	0.036	0.009							
15	1.374	0.310	0.082	0.020	0.009						
20	2.443	0.551	0.146	0.035	0.016						
25	3.617	0.861	0.227	0.055	0.024	0.007					
30	5.497	1.240	0.328	0.079	0.035	0.010					
35	—	1.688	0.446	0.108	0.047	0.013	0.005				
40	—	2.205	0.582	0.141	0.062	0.017	0.007				
45	—	2.791	0.737	0.178	0.078	0.021	0.009				
50	—	3.445	0.910	0.220	0.097	0.026	0.011				
60	—	4.961	1.310	0.317	0.140	0.038	0.016	0.005			
70	—	—	1.783	0.432	0.190	0.052	0.021	0.007			
80	—	—	2.329	0.564	0.248	0.068	0.028	0.009			
90	—	—	2.948	0.713	0.314	0.086	0.035	0.011			
100	—	—	3.639	0.881	0.388	0.106	0.044	0.014			
125	—	—	5.686	1.376	0.606	0.165	0.068	0.022			
150	—	—	—	1.982	0.872	0.238	0.098	0.031	0.007		
175	—	—	—	2.697	1.187	0.324	0.133	0.043	0.010		
200	—	—	—	3.523	1.550	0.423	0.174	0.056	0.013		
225	—	—	—	4.459	1.962	0.536	0.220	0.070	0.016		
250	—	—	—	5.505	2.423	0.662	0.272	0.087	0.020	0.006	
275	—	—	—	—	2.931	0.801	0.329	0.105	0.024	0.007	

Pipe diam., in

1 in = 25.4 mm.

300	3.489	0.953	0.392	0.125	0.029	0.009
325	4.094	1.118	0.460	0.147	0.034	0.010
350	4.748	1.297	0.533	0.170	0.039	0.012
375	5.451	1.489	0.612	0.195	0.045	0.014
400	6.202	1.694	0.696	0.222	0.051	0.015
425	—	1.912	0.786	0.251	0.057	0.017
450	—	2.144	0.881	0.281	0.064	0.019
475	—	2.388	0.982	0.313	0.072	0.022
500	—	2.646	1.068	0.347	0.079	0.024
550	—	3.202	1.317	0.420	0.096	0.029
600	—	3.811	1.567	0.500	0.114	0.035
650	—	4.473	1.839	0.587	0.134	0.041

Additional column values (rows 375–650):

375	0.005
400	0.006
425	0.007
450	0.006
475	0.009
500	0.010
550	0.012
600	0.014
650	0.016

TABLE 14.17 Pressure Drop of Air (in psi per 100 ft) through Schedule 40 Steel Pipe, 75 psig

(Cubic feet of free air at 60°F, 14.7 psia 18°C and 100 kPa)

scfm	\(1/2\)	\(3/4\)	1	\(1\tfrac{1}{4}\)	\(1\tfrac{1}{2}\)	2	\(2\tfrac{1}{2}\)	3	4	5	6
2	0.018										
3	0.040	0.009									
4	0.070	0.016									
5	0.110	0.025									
6	0.159	0.036	0.009								
8	0.282	0.064	0.017								
10	0.441	0.099	0.026								
15	0.991	0.224	0.059	0.014							
20	1.762	0.398	0.105	0.025	0.011						
25	2.753	0.621	0.164	0.040	0.017						
30	3.965	0.895	0.236	0.057	0.025						
35	5.396	1.218	0.322	0.078	0.034	0.009					
40	7.048	1.590	0.420	0.102	0.045	0.012					
45	8.921	2.013	0.532	0.129	0.057	0.015					
50	—	2.485	0.656	0.159	0.070	0.019	0.008				
60	—	3.579	0.945	0.229	0.101	0.027	0.011				
70	—	4.871	1.286	0.311	0.137	0.037	0.015				
80	—	6.362	1.680	0.407	0.179	0.049	0.020				
90	—	8.052	2.126	0.515	0.226	0.062	0.025	0.008			
100	—	—	2.625	0.635	0.280	0.076	0.031	0.010			
125	—	—	4.101	0.993	0.437	0.119	0.049	0.016			
150	—	—	5.906	1.429	0.629	0.172	0.071	0.023			
175	—	—	8.039	1.946	0.856	0.234	0.096	0.031			
200	—	—	—	2.541	1.118	0.305	0.126	0.040	0.009		
225	—	—	—	3.216	1.415	0.387	0.159	0.051	0.012		
250	—	—	—	3.971	1.747	0.477	0.196	0.063	0.014		
275	—	—	—	4.804	2.114	0.577	0.237	0.076	0.017		

Pipe diam., in

1 in = 25.4 mm.

300	—	—	—	5.718	2.516	0.687	0.283	0.090	0.021	
325	—	—	—	6.710	2.953	0.807	0.332	0.106	0.024	0.008
350	—	—	—	7.782	3.425	0.935	0.385	0.123	0.028	0.010
375	—	—	—	8.934	3.932	1.074	0.441	0.141	0.032	0.011
400	—	—	—	—	4.473	1.222	0.502	0.160	0.037	0.013
425	—	—	—	—	5.050	1.379	0.567	0.181	0.041	0.014
450	—	—	—	—	5.662	1.546	0.636	0.203	0.046	0.016
475	—	—	—	—	6.308	1.723	0.706	0.226	0.052	0.017
500	—	—	—	—	6.990	1.909	0.785	0.251	0.057	0.021
550	—	—	—	—	8.458	2.310	0.950	0.303	0.069	0.025	0.008
600	—	—	—	—	—	2.749	1.130	0.361	0.082	0.029	0.010
650	—	—	—	—	—	3.226	1.326	0.423	0.097		0.012

TABLE 14.18 Pressure Drop of Air (in psi per 100 ft) through Schedule 40 Steel Pipe, 100 psig (700 kPa)

(Cubic feet of free air at 60°F, 14.7 psia)

scfm	Pipe diam., in										
	1/2	3/4	1	1 1/4	1 1/2	2	2 1/2	3	4	5	6
2	0.014										
3	0.031										
4	0.055	0.012									
5	0.086	0.019									
6	0.124	0.028									
8	0.220	0.050	0.013								
10	0.345	0.078	0.021								
15	0.775	0.175	0.046	0.011							
20	1.378	0.311	0.082	0.020							
25	2.153	0.486	0.128	0.031	0.014						
30	3.101	0.700	0.185	0.045	0.020						
35	4.220	0.952	0.251	0.061	0.027						
40	5.512	1.244	0.328	0.079	0.035						
45	6.976	1.574	0.416	0.101	0.044	0.012					
50	8.613	1.943	0.513	0.124	0.055	0.015					
60	12.402	2.799	0.739	0.179	0.079	0.021					
70	—	3.809	1.006	0.243	0.107	0.029	0.012				
80	—	4.975	1.314	0.318	0.140	0.038	0.016				
90	—	6.297	1.663	0.402	0.177	0.048	0.020				
100	—	7.774	2.053	0.497	0.219	0.060	0.025				
125	—	12.147	3.207	0.776	0.342	0.093	0.038	0.012			
150	—	—	4.619	1.118	0.492	0.134	0.055	0.018			
175	—	—	6.287	1.522	0.670	0.183	0.075	0.024			
200	—	—	8.211	1.987	0.875	0.239	0.098	0.031			
225	—	—	10.392	2.515	1.107	0.302	0.124	0.040			
250	—	—	12.830	3.105	1.367	0.373	0.153	0.049	0.011		
275	—	—	—	3.757	1.654	0.452	0.186	0.059	0.014		

300	—	—	—	4.471	1.968	0.537	0.221	0.071	0.016	
325	—	—	—	5.248	2.309	0.631	0.259	0.083	0.019	
350	—	—	—	6.086	2.678	0.731	0.301	0.096	0.022	
375	—	—	—	6.987	3.075	0.840	0.345	0.110	0.025	
400	—	—	—	7.949	3.498	0.955	0.393	0.125	0.029	
425	—	—	—	8.974	3.949	1.079	0.443	0.142	0.032	
450	—	—	—	10.061	4.428	1.209	0.497	0.159	0.036	0.011
475	—	—	—	11.210	4.933	1.347	0.554	0.177	0.040	0.012
500	—	—	—	12.421	5.466	1.493	0.614	0.196	0.045	0.014
550	—	—	—	—	6.614	1.806	0.743	0.237	0.054	0.016
600	—	—	—	—	7.871	2.150	0.884	0.282	0.064	0.020
650	—	—	—	—	9.238	2.523	1.037	0.331	0.076	0.023

TABLE 14.19 Pressure Drop of Air (in psi per 100 ft) through Schedule 40 Steel Pipe, 125 psig

(Cubic feet of free air at 60°F, 14.7 psia)

scfm	½	¾	1	1¼	1½	2	2½	3	4	5
					Pipe diam., in					
3	0.025									
4	0.045									
5	0.071	0.016								
6	0.102	0.023								
8	0.181	0.041	0.017							
10	0.283	0.064	0.038							
15	0.636	0.144	0.067							
20	1.131	0.255	0.105	0.016						
25	1.768	0.399	0.152	0.025						
30	2.546	0.574	0.206	0.037	0.016					
35	3.465	0.782	0.270	0.050	0.022					
40	4.526	1.021	0.341	0.065	0.029					
45	5.728	1.292	0.421	0.083	0.036					
50	7.071	1.596	0.607	0.102	0.045					
60	10.183	2.298	0.826	0.147	0.065	0.018				
70	13.860	3.128	1.079	0.200	0.088	0.024				
80	—	4.085	1.365	0.261	0.115	0.031	0.013			
90	—	5.170	1.685	0.330	0.145	0.040	0.016			
100	—	6.383		0.408	0.180	0.049	0.020			
125	—	9.973	2.633	0.637	0.281	0.077	0.031			
150	—	14.361	3.792	0.918	0.404	0.110	0.045	0.014		
175		—	5.162	1.249	0.550	0.150	0.062	0.020		
200		—	6.742	1.632	0.718	0.196	0.081	0.026		
225		—	8.533	2.065	0.909	0.248	0.102	0.033		
250		—	10.534	2.550	1.122	0.306	0.126	0.040		
275		—	12.746	3.085	1.358	0.371	0.152	0.049		

300	15.169	—	—	3.671	1.616	0.441	0.181	0.058	0.013	
325	—	—	—	4.309	1.896	0.518	0.213	0.068	0.016	
350	—	—	—	4.997	2.199	0.601	0.247	0.079	0.018	
375	—	—	—	5.736	2.525	0.689	0.283	0.090	0.021	
400	—	—	—	6.527	2.872	0.784	0.323	0.103	0.024	
425	—	—	—	7.368	3.243	0.886	0.364	0.115	0.027	
450	—	—	—	8.260	3.635	0.993	0.408	0.130	0.030	
475	—	—	—	9.204	4.050	1.106	0.455	0.145	0.033	
500	—	—	—	10.198	4.488	1.226	0.504	0.161	0.037	
550	—	—	—	12.340	5.430	1.483	0.610	0.195	0.044	0.013
600	—	—	—	14.685	6.463	1.765	0.726	0.232	0.053	0.016
650	—	—	—	—	7.585	2.071	0.852	0.272	0.062	0.019

TABLE 14.20 Suggested System Sizing Forms

			Compressed air sizing main							
Stack	Size stack	Conn. scfm stack no.	Total conn. scfm main	Use factor main	Total scfm main	Size main	Conn. scfm BR	Use factor BR	BR total scfm	Size BR

			Compressed air sizing stack					
Floor	Conn. scfm	Use factor branch	scfm branch	Size branch	Total conn. scfm stack	Stack use factor	Total scfm	Size stack

Air-Conditioning Equipment Selection

Specific performance characteristics of various dryer types have previously been given. All other equipment in this network will now be discussed.

The selection of a cooling medium for aftercoolers has been discussed previously. There are, however, additional points to be considered. Some aftercoolers have a high pressure drop at rated flow. Consider oversizing the unit. Some dryers require inlet air at a low maximum temperature. Selection must be made with this in mind. Provide a bypass around the aftercooler for ease of servicing.

Moisture separators are designed for specific flow conditions and so should be selected based on the actual design of the system. If marginal conditions are encountered, go to the next larger sized unit, but be sure the specified unit is compatible with actual volume. The pressure drop through a properly sized unit is about 3 psig.

The filter selection guidelines have been discussed earlier in this section under design. The sizing parameters must include maximum oil content of the air, maximum particulate size, and total scfm the filters must handle. Pressure loads are also a prime consideration. The magnitude of contaminants depends on the following:

1. Choice of air compressor
2. Presence or absence of an aftercooler
3. Type of dryer used
4. Quality of inlet air

Valves are an often overlooked component of a compressed air system. The selection of valve type and material is important to efficiency and operating life. The following should be considered when selecting valves:

1. The most important valve feature is minimum flow restriction (pressure drop) when the valve is open full. Ball, gate, and plug valves have the lowest pressure drop. It is extremely rare to use a valve for flow restriction; therefore, this is not a consideration.
2. The pressure rating should be suitable for the maximum pressure of compressor.
3. The valve body and seat materials must be compatible with the expected trace gases and contaminants.

TABLE 14.21 Friction Loss for Hose, in psi

Free air flow, scfm	6 ft, 1/8 in	8 ft, 5/32 in	8 ft, 1/4 in	8 ft, 5/16 in	8 ft, 3/8 in	12.5 ft, 1/2 in	25 ft, 1/2 in	50 ft, 1/2 in	25 ft, 3/4 in	50 ft, 3/4 in	8 ft, 5/32 in / 25 ft, 1/2 in	8 ft, 1/4 in / 50 ft, 1/2 in	12.5 ft, 1/2 in / 25 ft, 3/4 in	12.5 ft, 1/2 in / 50 ft, 3/4 in
2	3.5	1.2	—	—	—	—	—	—	—	—	—	—	—	—
3	7.3	2.7	—	—	—	—	—	—	—	—	—	—	—	—
4	12.5	4.4	—	—	—	—	—	—	—	—	1.3	—	—	—
5	—	6.7	—	—	—	—	—	—	—	—	2.8	—	—	—
6	9.3	9.3	—	—	—	—	—	—	—	—	4.6	1.2	—	—
7	—	12.4	1.3	—	—	—	—	—	—	—	6.9	1.6	—	—
8	—	—	1.6	—	—	—	—	—	—	—	9.7	2.1	—	—
10	—	—	2.5	—	—	—	—	—	—	—	12.9	3.2	—	—
12	—	—	3.5	1.3	—	—	—	—	—	—	—	4.5	—	—
15	—	—	5.3	2.0	—	—	—	1.1	—	—	—	6.9	—	—
20	—	—	9.0	3.4	1.4	—	1.0	1.9	—	—	—	11.8	—	—
25	—	—	13.8	5.1	2.2	—	1.5	3.0	—	—	—	—	1.3	1.5
30	—	—	—	7.3	3.1	1.1	2.1	4.2	—	—	—	—	1.8	2.1
35	—	—	—	9.8	4.1	1.5	2.9	5.6	—	—	—	—	2.5	2.8
40	—	—	—	12.5	5.3	2.0	3.7	7.1	—	1.0	—	—	3.2	3.7
45	—	—	—	—	6.6	2.5	4.6	8.9	—	1.2	—	—	4.0	4.6
50	—	—	—	—	8.1	3.0	5.6	10.9	—	1.5	—	—	4.9	5.6
55	—	—	—	—	9.7	3.6	6.7	13.0	1.1	1.8	—	—	5.9	6.8
60	—	—	—	—	11.5	4.3	7.9	—	1.4	2.1	—	—	7.0	8.0
70	—	—	—	—	—	5.7	10.6	—	—	2.8	—	—	9.4	10.7
80	—	—	—	—	—	7.3	13.6	—	1.9	3.6	—	—	12.1	13.9
90	—	—	—	—	—	9.2	—	—	2.3	4.5	—	—	—	—
100	—	—	—	—	—	11.2	—	—	2.8	5.5	—	—	—	—
120	—	—	—	—	—	—	—	—	4.0	7.7	—	—	—	—
140	—	—	—	—	—	—	—	—	5.4	10.3	—	—	—	—
160	—	—	—	—	—	—	—	—	6.9	13.3	—	—	—	—
180	—	—	—	—	—	—	—	—	8.7	—	—	—	—	—
200	—	—	—	—	—	—	—	—	10.6	—	—	—	—	—
220	—	—	—	—	—	—	—	—	12.7	—	—	—	—	—

Note: Based on 95 psig air pressure at hose inlet, includes normal couplings (quick connect couplings will increase pressure losses materially). Hose is assumed to be smooth. Air is clean and dry. If an airline lubricator is upstream from the hose, pressure loss will be considerably higher. Pressure loss varies inversely as the absolute pressure (approximately). Probable accuracy is believed to be ±10 percent. Use one half of indicated value for air at 50 psi.

4. There must be positive shutoff.

5. There should be minimum leakage through the valve stem.

6. The valves used should have been designed for compressed air service. Be careful to examine valve specifications for airway ports or openings smaller than the nominal size indicated or expected.

Selecting the Air Compressor Assembly

There is now enough information to size the compressor assembly. The assembly will include the intake system, compressor and compressor installation, and receiver. To start, the following information must be available:

1. Total connected scfm of all air using devices including flow to the air dryer system if applicable

2. Maximum pressure the assembly devices require

3. Duty and use factors giving maximum expected use of air by devices

4. scfm leakage and future expansion allowance

5. Allowable pressure drops for the entire system including piping and conditioning equipment

6. Altitude, temperature, and contaminant removal corrections

7. Location of air compressor and all ancillary equipment

Having completed the above work, first design the inlet piping system. Since air compressor performance depends on inlet conditions, this system deserves special care. The air intake should provide a supply of air to the compressor that is as clean, cool, and dry as possible. The proposed location should be studied for the presence of any type of airborne contamination and positioned to avoid the probability of contaminated intake. Whenever possible, use outside air.

For an external installation, the inlet should have a rain cap and a screen. An inlet filter should always be provided inside the building. If the manufacturer of the selected compressor indicates that noise may be a problem, a silencer shall be installed. Each compressor (if a duplex) should have an independent air intake. See Table 14.22 for characteristics of air inlet filters.

Uncontrolled piping pulsations can harm inlet piping, damage the building structure, and affect compressor performance. Airflow into a reciprocating compressor pulsates because of the cyclic intake of air into the compressor cylinder. The variable pressure causes the air column in the pipe to vibrate, which creates a traveling wave in the pipe moving at the speed of sound. The inlet pipe itself vibrates at some natural frequency depending on its length. If the air column vibrates at or near the same frequency as the length of pipe, the system is said to be "resonant." High pressures could result when this occurs. Resonant pipe lengths can be calculated by the compressor manufacturers, and the critical length given to the engineer. As an example, with a 600 rpm compressor, avoid a length of pipe 3.2 to 12.5 ft, 16.8 to 26.2 ft, and 32.3 to 41.5 ft. A surge chamber can also be used to eliminate this problem.

The pressure loss of air through the intake piping should be held to a minimum. Suggested inlet pipe size is given in Table 14.23. Velocity of intake air should be limited to about 1000 fpm (300 m/min) to avoid noise problems, and friction loss limited to about 4 in of water. Inlet louver velocity should also be low enough to avoid drawing in rainwater. Standard round duct charts can also be used for sizing.

TABLE 14.22 Inlet Air Filter Characteristics

Filter type	Filtration efficiency, %	Particle size, μm	Maximum drop when clean, in WC	Comments (see key)
Dry	100	10	3–8	(1)
	99	5		
	98	3		
Viscous	100		$\frac{1}{4}$–2	(2) (3)
impingement	95	20		
(oil wetted)	85	10		
Oil bath	98	10	6–10 = nominal	(2) (3) (4)
	90	3	2 = low drop	
Dry with	100	10	5 (5)	
silencer	99	5	7 (6)	
	98	3		

Key to comments:

(1) Recommended for nonlubricated compressors and for rotary vane compressors in a high dust environment.

(2) Not recommended for dusty areas or for nonlubricated compressors.

(3) Performance requires that oil is suitable for both warm and cold weather operation.

(4) Recommended for rotary vane compressors in normal service.

(5) Full flow capacity up to 1600 scfm.

(6) Full flow capacity from 1600 to 6500 scfm.

TABLE 14.23 Recommended Air Inlet Pipe Size

Maximum scfm free air capacity	Minimum size, in
50	$2\frac{1}{2}$
110	3
210	4
400	5
800	6

1 cfm = 0.03 m³/min.
Source: Courtesy James Church.

In general, if air requirements are less than 500 scfm (15 m³/min), the intake can be indoors. Provide an automatic drain on the line leading to the compressor, and pitch the intake piping to the drain point. If indoor air temperature is usually higher than 100°F (37.8°C), the intake should be outdoors.

Many different factors are involved in selection of compressor type:

1. Space limitations

2. Noise limitations

3. Compressor pressure capability

4. Capacity

5. Availability, cost, and quality of cooling water
6. Need for oil-free air
7. Electrical power limitations
8. Cost, both initial and long term

In the following circumstances, a duplex unit should be considered instead of a simplex unit:

1. When the cost of downtime is high. The owner may request two 100 percent capacity machines to eliminate the possibility of a shutdown.
2. Where a facility has a steady flow rate (called a *base load*) and in addition, where there are substantial additional requirements due to periodic or intermittent use.
3. When electrical starting requirements would overload a simplex system. Two units starting at different times would eliminate the problem.
4. Where floor space is not available for one large compressor and ancillary equipment.
5. Where widely separated concentrations of heavy use exist.

Experience has shown that a properly sized, constantly working compressor usually requires less maintenance than one running intermittently.

Most of the power input to a compressor is rejected through the various cooling systems into the space where the compressor is located. This information must be relayed to the HVAC systems engineer for space conditioning if necessary. Good ventilation is mandatory in the area of the compressor.

The selection of the proper type of pump foundation and mounting depends upon the lowest frequency and magnitude of pump vibration and the load bearing requirement of the slab upon which the compressor rests. Metal, rubber, coils, and spring materials are available for use as isolators. The manufacturers of isolators should be consulted to confirm the proper type for the purpose and conditions expected.

Vibration isolation is achieved by the proper selection of resilient devices between the pump base and the building structure. This isolation is accomplished by placing isolators between the pump and the floor, flexible connections on all piping from the compressor, and spring-type hangers on the piping around the compressor for a distance of about 20 ft (6 m). To illustrate the information presented, a typical design problem is presented below.

PROJECT DESIGN EXAMPLE

The project is located in Denver, Colorado, with the following values established:

1. Actual elevation: 5000 ft (1500 m) above sea level
2. Highest inlet air temperature: 90°F (32.2°C)
3. Highest average relative humidity expected: 90 percent
4. Established maximum system volume requirement: 800 scfm free air

5. Maximum required system pressure: 125 psi (860 kPa)
6. The selected air compressor rating units: at scfm 60°F (18°C), dry air at sea level

Adjustment of Required Intake Volume

After calculating the required volume of air that the compressor will have to deliver at pressure, it is time to find the actual compressor requirement for air intake in order to deliver the scfm required for use. As we will see, the required compressor intake volume must be adjusted in order to deliver the necessary volume. The reason adjustment is necessary is that the compressors are rated in scfm, and we have free air at the intake. Since standard air is at 60°F (18°C), dry (0 percent RH) and pressure at sea level. Since all three values from the Denver location do not match standard conditions, all three must be adjusted.

1. Altitude correction. From Table 14.2,

Opposite 5000 ft, read 1.17.

800 scfm × 1.17 factor = 136 scfm
2. Temperature adjustment. From Table 14.3,

Opposite 90°F, read	1.038
Opposite 60°F, read	0.981
Total correction difference	0.057

800 scfm × 0.057 factor = 46 scfm
3. Find the actual volume that the moisture in the inlet air occupies. Since this water will be removed, additional air must replace this water vapor.

a. From Table 14.4, 90 percent RH and 90°F, read 195 gr/lb of air
b. From Table 14.5, 195 gr/lb, read 4 percent volume:

800 scfm × 4 percent = 32 scfm

Adding all adjustments together,

136 scfm—altitude

46 scfm—temperature

32 scfm—removal of water vapor

214 scfm   say, 220 scfm

Therefore, the actual inlet volume would be 1020 scfm free air at the inlet (800 + 220) for a compressor to deliver 800 scfm free air. If some requirements are given in air actually compressed (acfm), see Fig. 14.18 to determine the free air equivalent of compressed air at pressure.

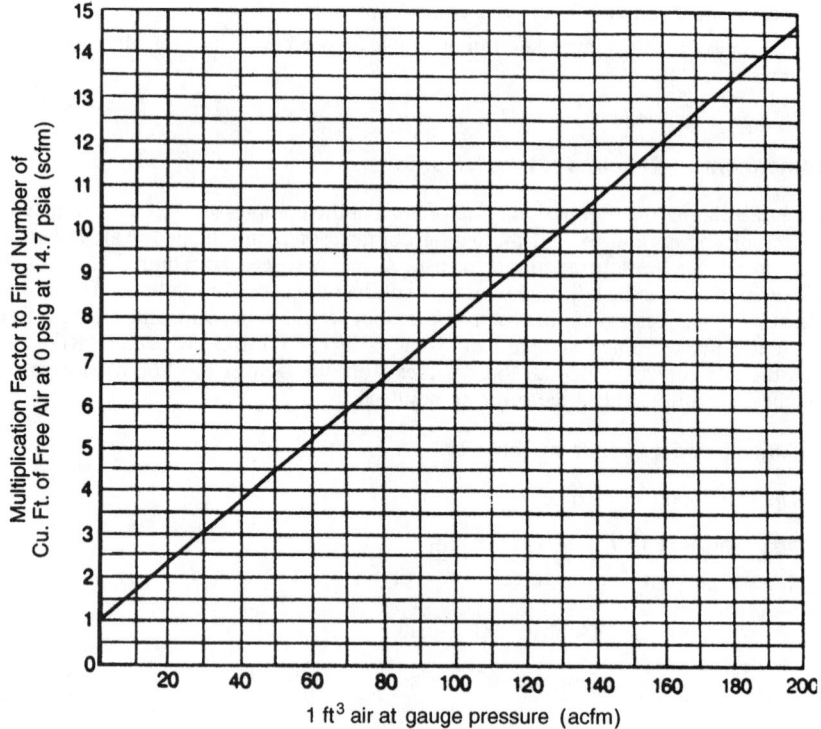

FIGURE 14.18   Ratio of free air to compressed air. To find the free air equivalent for 1 acfm
at 130 psig pressure, find 130 at the bottom, go up to the diagonal, then horizontal to the left,
to find the multiplier of 9.8. Then 1 acfm will equal 9.8 scfm.
1 psig × 7 = kPa.
1 cfm × 0.03 = m³/min.

Sizing Moisture Separators and Traps

Water vapor enters the system through the air intake. The amount depends upon
the temperature and relative humidity of the intake air. After compression and
cooling, the air will not hold the same amount of water.

In order to determine the separator size and drain trap required to discharge the
water condensed out of the air (either by intercooler or aftercooler) immediately
after the compression process, the following is required:

1. Find the amount of water at intake based on the maximum values of temperature
   and relative humidity at the intake location.
2. Find the amount of water compressed air will hold at pressure and discharge
   temperature. The aftercooler must be selected and the discharge temperature
   (approach) known in order to find the correct values.

First, to find the amount of water in the air before compression at the intake,
refer to Table 14.4, and at 90°F and 90 percent RH, read 195 gr/lb air. If the air
is saturated, use Fig. 14.19. Next, determine the amount of water vapor remaining

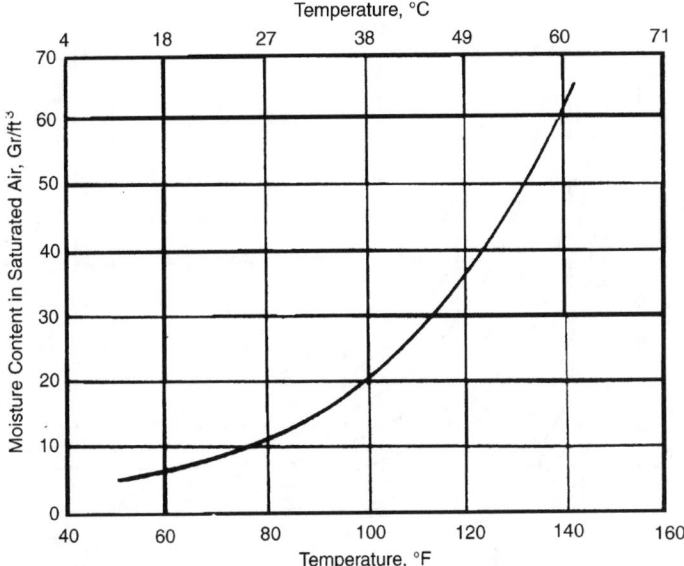

FIGURE 14.19   Moisture content of saturated air at atmospheric pressure.

in the air after compression, and after the aftercooler has cooled the air. The air leaving an aftercooler is saturated with water vapor. The design assumes that the temperature of air at this point is 100°F. Using Fig. 14.20, at the junction of the 125 psig curve and 100°F, read 0.29 lb/1000 cf.

To subtract one from the other, convert 195 gr/lb into gr/cf. Using Table 14.5, 195 gr/lb is 13 gr/cf. To convert 0.29 lb/1000 cf, use the conversion factor of 7000 gr in 1 lb. Therefore, 0.29 × 7000 = 2100 grains per 1000 cubic feet, or 2.1 grains per cubic foot. Therefore, the discharge rate is 13 = 2.1 − 10.9/WFF × 1020 scfm = 11,118 gr, or 1.7 gpm. Size the drain trap accordingly.

The air leaving the aftercooler has been established as having 2.1 gr/cf moisture; using Fig. 14.3, this figure translates to a dew point of +32°F. If a lower dew point is needed, a dryer will be required.

To size the drain trap at the dryer, use Table 14.5 and Fig. 14.3 to determine the amount of moisture lost between +32°F dew point and the actual dew point selected.

Sizing Receivers

Air receivers are used to keep compressors from working continuously. They store air at a higher pressure, allowing the compressor to shut down until the volume used causes the pressure to drop. Then the compressor starts the cycle again.

Air receivers should be placed as close as practical to the compressor. A flexible connection should be used to isolate the vibration of the compressor from the receiver. The size of the receiver is a function of time and pressure. One formula commonly used is:

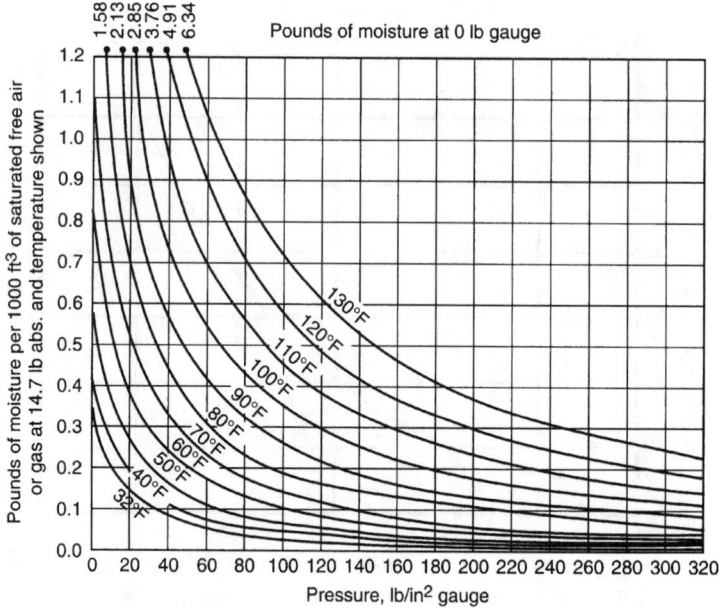

FIGURE 14.20   Moisture remaining in saturated air after compression.
1 psi × 4.9 = kg/m².

$$T = \frac{V(P_1 - P_2)}{CP}$$ (14.3)

where T = Time receiver will supply air from upper pressure to lower pressure,
    min
V = volume of receiver under design, ft³
P_1 = upper pressure of air in receiver, psia
P_2 = lower pressure of air in receiver, psia
C = system air requirements, scfm
P = atmospheric pressure at receiver location, psia

Use the average value of T, which should be about 10 min. Then solve for the
volume, selecting a standard tank as listed in Table 14.7. If the calculated size is
too large, use a smaller T and consult the manufacturer. The receiver should be
provided with an automatic drain trap and a pressure relief valve.

Instrumentation

Pressure and temperature gauges located in the system can help identify problems
and signal the need for maintenance. Put temperature gauges on the discharge of
both the aftercooler and fryer and on the cooling water inlet and outlet. Pressure
gauges on each side of the filters and dryers and on the compressor discharge are
useful for determining buildup and deposits.

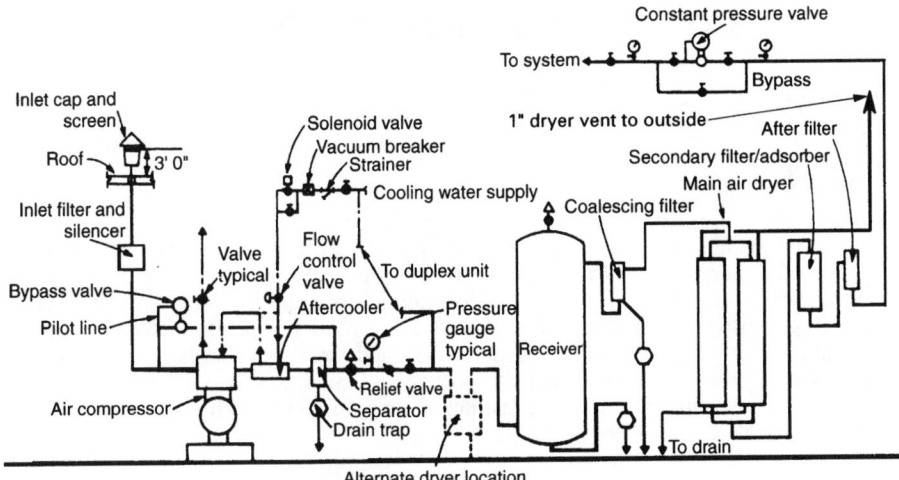

FIGURE 14.21   Typical detail of industrial air compressor assembly.

System Design Considerations

1. Provide for thermal expansion of pipe due to possibility of air reaching a temperature of 350°F.

2. Take all branch connections from top of mains.

3. Pitch pipe in direction of flow. When not possible, increase pipe one size to allow for water obstruction. With the flow, pitch pipe 3 in/100 ft (75 mm/33 m). Opposite flow, use 6 in per 100-ft (150 mm per 33 m) run.

4. Where pressure is over 100 psig, support piping on spring-loaded hangers around the compressor.

5. When quick disconnect fittings are used, provide extra stiff and rigid support (particularly at a ceiling drop). This type of fitting is often subjected to repeated force used in the connecting and disconnecting of hoses.

A detail of a typical industrial compressed air system is given in Fig. 14.21. This illustration shows the location of all equipment; however, not all the equipment shown is used for all installations.

INSTRUMENT AND CONTROL AIR

GENERAL

This subsection will discuss compressed air requirements for pneumatic instruments and control purposes. Instrument air is compressed air used to operate pneumatic controllers, pressure transmitters, pneumatic information transmission systems, pressure transducers, and other similar devices. Control air operates automatic temperature control (ATC) systems and associated heating, ventilating, and air-conditioning (HVAC) devices used to control and condition air for comfort in facilities. This section will describe instrument air quality, production methods of air for use in pneumatic instruments.

CODES AND STANDARDS

The Instrument Society of America (ISA) is the major organization writing and issuing standards. The principal standards concerning the creation and use of air are:

1. ANSI/ISA S7.3, Quality Standard for Instrument Air
2. RP S7.7, Recommended Practice for Producing Quality Instrument Air

AIR QUALITY STANDARDS

There are four elements of quality that must be considered: moisture content, oil content, particulate size, and other toxic contamination.

If any instrument air lines are located outside a building, then the maximum allowable dew point at which the instruments will continuously function satisfactorily is 18°F (10°C) below the lowest temperature that any part of the instrument air system will be exposed to. When the instrument air system is completely indoors, the dew point shall not exceed 35°F (2°C).

The oil content shall be as close to zero as possible, with the maximum allowable content of 1 ppm under normal operating conditions. This requires the use of an oil-free compressor to generate the compressed air rather than the type of compressor that relies on filters to reduce the oil content.

The maximum allowable particulate size shall be 3 μm. The intake shall be free of all corrosive, flammable, and toxic contaminants. If the intake cannot be located in an area free from this kind of contamination, these impurities shall be removed before entering the compressor.

The Instrument Society of America has established the following general requirements:

1. Pressure dew point: Maximum 135°F (2°C)
2. Particulate size: Maximum 3 μm
3. Oil content: Maximum 1 ppm but as close to zero as possible
4. Miscellaneous contaminants: No corrosive or hazardous gases
5. Pressures required: Generally in the 15 to 50 psig (100 to 345 kPa) range

AIR PRESSURE REQUIREMENTS

There are two ranges of pressures used in pneumatic pressure transmission. The preferred nominal pressure is 12 psig (80 kPa), with an operating range (span) of between 3 psig (20 kPa) and 15 psig (100 kPa). The air supplied to this system shall have a range of between 19 psig (130 kPa) and 22 psig (150 kPa).

Another commonly used nominal working pressure is 24 psig (160 kPa), with an operating range of between 6 psig (40 kPa) and 30 psig (200 kPa). The air pressure supplied to this system is in the range of between 38 psig (260 kPa) and 44 psig (300 kPa).

GENERATION OF INSTRUMENT AIR

Instrument air can be generated by means of a dedicated air compressor assembly or obtained from an air compressor serving other purposes. When obtained from other than dedicated sources, the supply air pressure must be adjusted to system requirements and the airstream further purified if necessary to achieve the required purity requirements. It is highly recommended that instrument air be produced by dedicated compressors.

PIPE AND FITTINGS

The most often used material for branch instrument air and pressure sensing lines is PE tubing and fittings. PE tubing is available in long lengths, minimizing joints. Sizes are $\frac{1}{4}$ and $\frac{3}{8}$-in diameter. Also used are copper, aluminum, and PE clad copper. When multiple lines of PE are used, they are often bundled together. When they must be protected, it is common practice to have them run in metallic conduit. Pneumatic lines serving smoke dampers shall be of rigid copper or aluminum tubing.

PE should not be used where the ambient temperatures are greater than 175°F (80°C). PE is also subject to deterioration by solvents and by ultraviolet light.

SPECIALTY GASES FOR LABORATORIES

GENERAL

This subsection will describe various specialty compressed air and gas systems typically used for organic and inorganic chemistry, physics, and biological laboratories, and those used for research and development purposes. The gases used in these types of facilities are characterized by low delivery pressure, low and intermittent volume, and high purity requirements of the gas and of the delivery system. It is extremely rare that the quantity of pure gases used for laboratory and research purposes would justify large bulk storage. For this reason, this section will concentrate on cylinder supply, smaller cryogenic bulk storage tanks, and the generation of such gases. Larger bulk supply and storage systems will be discussed in the specialty industrial gas section of this chapter where production and other uses require large storage volumes.

CODES AND STANDARDS

The building codes and standards impacting the design and installation of the various laboratory gas systems have been put in place to protect the safety and health of operating personnel and building occupants, as have the building code requirements concerning fire and structural consequences of accidents. There are no mandated code requirements concerned with sizing or purity.

Minimum purity requirements are listed in the Compressed Gas Association (CGA) standards for various gases called "Commodity Standards." Often, the actual, onsite purity requirements are higher than those listed in the standards and will be determined by the proposed use of the gas and the standards of the user. The CGA also has material and dimensional standards for pipe connections to terminals. For the gases not covered by the National Fire Protection Association (NFPA) and the CGA, good engineering practice is used to adequately locate the tanks and piping systems.

The NFPA has standards for the storage of flammable gases both inside and outside a building. NFPA-50 covers bulk oxygen at consumer sites, and NFPA-50A and 50B cover the storage of hydrogen. There are also standards for acetylene. NFPA-99 lists the requirements for the storage of flammable and nonflammable gases in cylinders.

Compressed gas systems within any type of facility are often required to conform to requirements of NFPA-99, health care facilities. The decision to adhere to provisions of this standard is dependent on the client, requirements of the client's insurance carrier, and authorities having jurisdiction.

There are EPA health hazard classifications, fire hazard classifications, and sudden release of pressure hazard classifications. All of these ratings are available from a Material Safety and Data Sheet (MSDS). There are gases that fall under a classification of "Reactive Hazard," and these must be kept separate from each other. This is usually done with walls, nonpermanent solid separators available from the supplier of the gas, or gas cabinets. There are also EPA threshold limit values for the degree of concentration of any particular gas in air for breathing purposes.

CLASSIFICATION OF SPECIALTY GASES

Compressed gases are classified into the following general categories:

1. *Oxidizers.* These gases are nonflammable but support combustion. No oil or grease is permitted to be used with any device associated with the use of this gas, and combustibles shall not be stored near these gases. Oxygen is an example.

2. *Inert gases.* These are gases that do not react with other materials. If released into a confined space, they will reduce the oxygen level to a point that asphyxiation could occur. Storerooms should be provided with oxygen monitors and should be well ventilated.

3. *Flammable gases.* These are gases that, when combined with air or oxidizers, will form a mixture that will burn or possibly explode if ignited. Flammable mixtures have a range of concentration below which they are too lean to be ignited and above they are too rich to burn. The most often used figure is the lower explosive level (LEL), which is the minimum percent, by volume, that will form a flammable mixture at normal temperatures and pressures. The high level for alarms is generally one-half of the LEL, with warnings issued at one-tenth of the LEL. The area where flammable gases are stored must be well ventilated, use approved electrical devices suitable for explosive atmospheres, and restrict all ignition sources. Flammability limits for common gases are given in Table 14.24.

4. *Corrosive gases.* These are gases that will attack the surface of rubber, metals, and other substances and also damage human tissue upon contact.

5. *Toxic and poisonous.* These are gases that will harm human tissue by contact or ingestion. Protective clothing and equipment must be used.

6. *Pyrophoric.* These are gases that spontaneously ignite upon contact with air under normal conditions.

7. *Cryogenic.* These gases are stored as extremely cold liquids under moderate pressure and vaporized when used. If the liquid is spilled, bare skin will suffer severe burns, and splashing into the eyes will cause blindness.

The categories and significant values of various gases are given in Table 14.24.

GRADES OF SPECIALTY GASES

There are many grades of pure and mixed gases available. Since there is no industry-recognized standard grade designation for purity, each supplier has its own individual designations. It is possible for the same gas used for different purposes to have a different designation for the same purity. The instrument manufacturer and the end user must be consulted for the maximum acceptable level of the various impurities based on the type of instrument used and the analytical work to be

TABLE 14.24 Specialty Gas Categories

Gas	Compressed gas	Liquefied gas	Oxidant	Inert	Corrosive	Toxic
Acetylene	(1)					
Air	●		●			
Allene		●				
Ammonia		●			●	●
Argon	●			●		
Arsine		●				(3)
Boron trichloride		●			●	●
Boron trifluoride	●		●		●	(3)
1,3-butadiene		●				
Butane		●				
Butenes		●				
Carbon dioxide		●			●	(2)
Carbon monoxide	●					●
Carbonyl sulfide		●			(2)	●
Chlorine		●	●		(2)	(3)
Cyanogen		●				(3)
Cyclopropane		●				
Deuterium	●					
Diborane	●					(3)
Dimethylamine		●			●	(3)
Dimethyl ether		●				
Ethane		●				
Ethyl acetylene		●				
Ethyl chloride		●				
Ethylene	●					
Ethylene oxide		●				(4)
Fluorine	●		●			(3)
Germane	●					(3)
Helium	●				●	
Hydrogen	●					
Hydrogen bromide		●			(2)	(3)
Hydrogen chloride		●			(2)	(3)
Hydrogen fluoride		●			●	(3)
Hydrogen sulfide		●				(3)
Isobutane		●				
Isobutylene		●				
Krypton	●				●	
Methane	●					
Methyl chloride		●				●
Methyl mercaptan		●				(3)
Monoethylamine		●				(3)
Monomethylamine		●			●	(3)
Neon	●				●	
Nitric oxide	●		●		(2)	(3)

(Continued)

TABLE 14.24 Specialty Gas Categories (*Continued*)

Gas	Compresses gas	Liquefied gas	Oxidant	Inert	Corrosive	Toxic
Nitrogen	●			●		
Nitrogen dioxide		●	●		(2)	(3)
Nitrogen trioxide		●	●		(2)	(3)
Nitrosyl chloride		●	●		(2)	(3)
Nitrous oxide		●		●		
Oxygen	●			●		
Phosgene		●				(3)
Phosphine		●				(3)
Propane		●				
Propylene		●				(3)
Halocarbon-12 (dichlorodifluoromethane)		●		●		(3)
Halocarbon-13 (chlorotrifluoromethane)		●			●	
Halocarbon-14 (tetrafluoromethane)	●				●	
Halocarbon-22 (chlorodifluoromethane)		●			●	
Silane	●					(3)
Sulfur dioxide		●			(2)	(3)
Sulfur hexafluoride		●		●		
Sulfur tetrafluoride		●			●	(3)
Trimethylamine		●			●	●
Vinyl bromide		●				●
Vinyl chloride		●				(6)
Xenon	●			●		

Key:
(1) Dissolved in solvent under pressure. Gas may be unstable and explosive above 15 psig.
(2) Corrosive in presence of moisture.
(3) Toxic. It is recommended that the user be thoroughly familiar with the toxicity and other properties of this gas.
(4) Cancer suspect agent.
(5) Pyrophoric; spontaneously flammable in air.
(6) Recognized human carcinogen.
(7) Flammable. However, limits are not known.

performed. The supplier must then be informed of these requirements in order to determine the grade of gas it will supply that meets or exceeds the allowable level of the various impurities.

The following list, although not complete, covers some manufacturers' designations for different grades of gases available. There are additional grades for specific instruments, such as "Hall" grades of gases.

1. Research grade
2. Carrier grade

3. Zero gas

4. Ultra zero

5. Ultrahigh purity plus

6. Ultrahigh purity

7. Purified

8. USP

STORAGE AND GENERATION OF GASES

Cylinder Storage

Where the anticipated gas usage does not require the installation of a cryogenic bulk supply, it is more convenient and less expensive to have gases compressed and stored in cylinders. Cylinders are available in various pressure ratings, with nomenclature differing between the several manufacturers. The high-pressure cylinder has gas stored at pressures ranging to 6000 psig, with the most common pressures between 2000 and 2500 psig. The low-pressure cylinder has gas pressures up to about 480 psig.

Cylinders do not have a standard designation from one supplier to another. If the actual capacity of any gas must be determined, it can be found using the following formula:

$$VG = \frac{CP}{14.7} \times CV \tag{14.4}$$

where VG = volume of gas at pressure, ft^3

CP = actual cylinder pressure (obtained from the supplier), psi

CV = cylinder volume (obtained from Fig. 14.22), ft^3

As an example, find the number of cubic feet of nitrogen stored in a cylinder with a volume of 1.76 ft^3 and a pressure of 2600 psig.

$$VG = \frac{2600}{14.7} \times 1.76$$

$$= 176.87 \times 1.76$$

$$= 311 \ ft^3$$

Cylinders are available in many sizes and pressure ratings. Figure 14.22 illustrates typical sizes. The cylinders themselves are available in four general categories. The first is the *plain carbon steel tank.* The second is called the *ultraclean tank,* which is made of a slightly different alloy steel and in addition, has been completely cleaned, prepared, and dried to reduce contaminants in the cylinder. The third classification is *aluminum tanks.* The tank interior has been specially prepared and the walls treated to maintain stability and reduce particulates. Aluminum is used for cleanliness and for gases that will react with steel. In many cases, the exterior is also treated in order to be more easily kept clean, such as required for clean room installations. The fourth type of cylinder is made of *stainless steel,* which is often used for ultrapure gases.

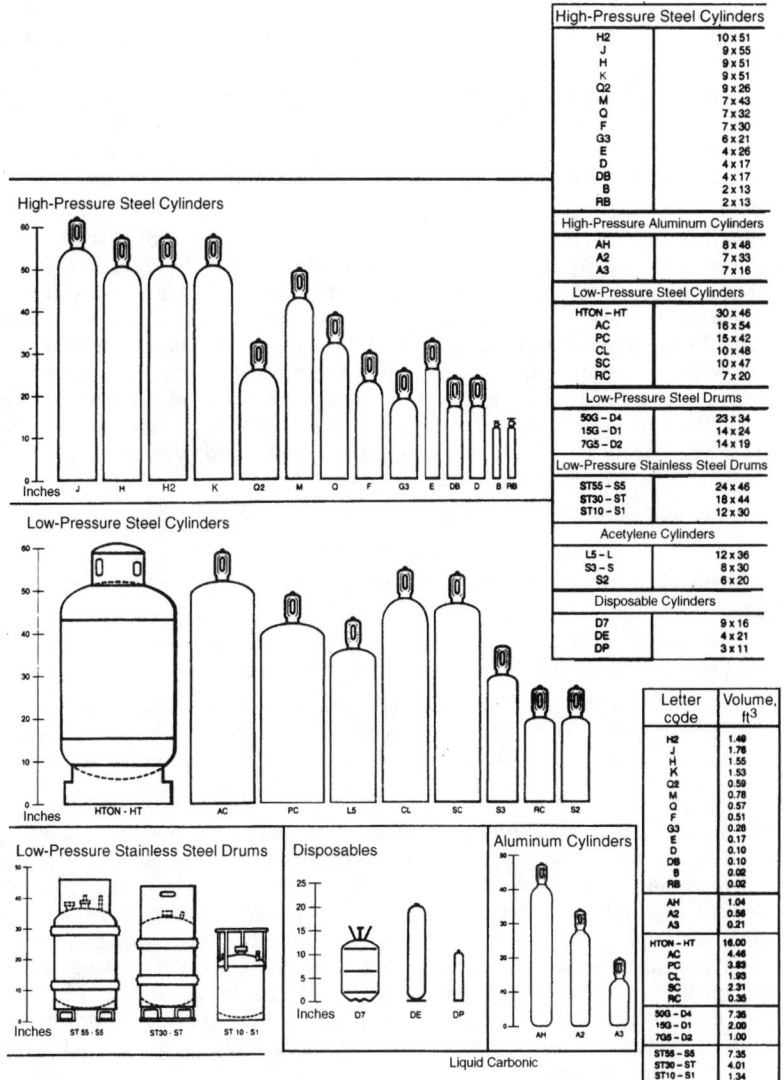

High-Pressure Steel Cylinders	
H2	10 x 51
J	9 x 55
H	9 x 51
K	9 x 51
Q2	9 x 26
M	7 x 43
Q	7 x 32
F	7 x 30
G3	6 x 21
E	4 x 26
D	4 x 17
DB	4 x 17
B	2 x 13
RB	2 x 13

High-Pressure Aluminum Cylinders	
AH	8 x 48
A2	7 x 33
A3	7 x 16

Low-Pressure Steel Cylinders	
HTON – HT	30 x 46
AC	16 x 54
PC	15 x 42
CL	10 x 48
SC	10 x 47
RC	7 x 20

Low-Pressure Steel Drums	
50G – D4	23 x 34
15G – D1	14 x 24
7G5 – D2	14 x 19

Low-Pressure Stainless Steel Drums	
ST55 – S5	24 x 46
ST30 – ST	18 x 44
ST10 – S1	12 x 30

Acetylene Cylinders	
L5 – L	12 x 36
S3 – S	8 x 30
S2	6 x 20

Disposable Cylinders	
D7	9 x 16
DE	4 x 21
DP	3 x 11

Letter code	Volume, ft^3
H2	1.40
J	1.76
H	1.55
K	1.53
Q2	0.59
M	0.78
Q	0.57
F	0.51
G3	0.28
E	0.17
D	0.10
DB	0.10
B	0.02
RB	0.02
AH	1.04
A2	0.56
A3	0.21
HTON – HT	16.00
AC	4.46
PC	3.83
CL	1.93
SC	2.31
RC	0.36
50G – D4	7.38
15G – D1	2.00
7G5 – D2	1.00
ST55 – S5	7.35
ST30 – ST	4.01
ST10 – S1	1.34

FIGURE 14.22 Typical cylinder dimensions. (*Courtesy Liquid Carbonic.*) 1 ft^3 = 0.03 m^3.

The following are general recommendations for the installation and storage of cylinders:

1. The room or area in which cylinders are placed shall have adequate ventilation and be free from combustible material and separated from sources of ignition.
2. Consideration should be given for the storage of additional full and empty cylinders in the same room for convenience.

3. Enough room should be allowed for the easy changing of cylinders. They are brought in on a hand truck or cart, and room should be allowed for their maneuvering.

4. Gas cylinders in active use shall be secured against falling by means of floor stands, wall brackets, or bench brackets. These brackets use straps to attach the cylinder to the bracket. Also available are floor racks and stands that can be provided for the installation and support of cylinders that cannot be located near walls.

5. When toxic or reactive gases are used, the cylinders should be placed in a gas cabinet. The basic purpose of the cabinet is to isolate the cylinder(s) and to contain the gases in the event of a leak and direct those gases away from the immediate vicinity of the cylinder and cylinder storage area to a point outside the building where they are diluted with the outside air. The cabinet could also contain panel-mounted manifolds, purging equipment, and other devices to allow some degree of control of operating parameters. Typical cabinet construction is 11 gauge painted steel or thicker to give a one-half hour fire rating. They could also be provided with vertical and horizontal adjustable cylinder brackets. The following options are available along with the cylinder cabinet:

a. Automatic shutoff of gas in the event of a catastrophic failure (flow limit).

b. Purging of gas lines after cylinder changes.

c. Mechanical cabinet exhaust. A typical system is designed for 13 air changes per minute with the access window open.

d. A sprinkler head for flammable gases. A typical head should be rated at 135°F, with a minimum water pressure of 25 psig.

e. For toxic and reactive gases, a small openable access window could be provided to operate valves without having to open the main door and compromising the exhaust system. A fixed access window is acceptable for inert gases.

When more than one cylinder is used to supply a system, the multiple arrangement is referred to as a *bank of cylinders.* Cylinder banks are classified as primary, secondary, and reserve. They are connected together by a header and controlled by a manifold assembly. The arrangement of the cylinders is chosen by the space available for the installation and the relative ease desired for the changing of cylinders. They can be placed either in a single row, double row, or staggered. The space typically required for various arrangements is shown in Fig. 14.23. Any additional space between banks of cylinders required for specialized devices such as manifold controls, purging devices, filters, and purifiers should be added to the cylinder bank dimensions.

Specialty Gas Generators

In some cases, it is more desirable for a small facility to generate their own high-purity specialty gases rather than having them supplied in cylinders. There are a limited number of gases for which anticipated volume allows this choice in laboratory or research facilities. Among them are nitrogen, hydrogen and helium, and compressed air. The generating units have their own filters and purifiers that can create gases of ultra-high purity. In particular, the use of these units for generation of hydrogen eliminates flammable cylinders in the laboratory or separate storage

ARRANGEMENTS

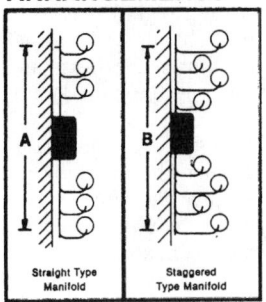

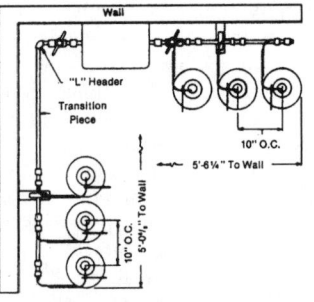

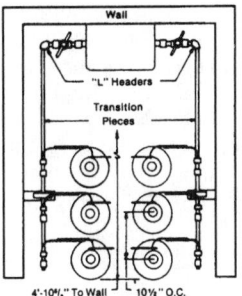

DIMENSIONAL DATA

CYLINDERS PER BANK	DIMENSION "A"	DIMENSION "B"
2 Banks Of 2 Each	5'-0"	4'-6"
2 Banks Of 3 Each	6'-8"	5'-8"
2 Banks Of 4 Each	8'-4"	6'-10"
2 Banks Of 5 Each	10'-0"	8'-0"
2 Banks Of 6 Each	11'-8"	9'-2"
2 Banks Of 7 Each	13'-4"	10'-4"
2 Banks Of 8 Each	15'-0"	11'-6"
2 Banks Of 9 Each	16'-8"	12'-8"
2 Banks Of 10 Each	18'-4"	13'-10"
2 Banks Of 12 Each	21'-8"	16'-2"

FIGURE 14.23   Typical arrangements and dimensions of cylinder installations (9-in diameter).

areas and keeps the actual amount of gas stored below that needed for explosion to take place.

Depending upon the type of generator and the type of gas generated (except compressed air), pressures are available to about 60 psig (415 kPa), and flow rates to 300 cc/min are common. Compressed air generators are available that will deliver up to 20 scfm and 100 psig. This type of unit is ideally suited for analytical purposes in widely separated areas of use, where the installation of cylinders is inconvenient and the changing of cylinders may cause disruption of continuing work. The operating cost is low, but the initial cost is high. However, there is a short payback period compared to cylinder supply. Depending on the type of gas generated, many of these units take their air supply from the room they are installed in, and others require a connection to a separate compressed air supply of known purity.

DISTRIBUTION SYSTEM COMPONENTS

Manifolds

A manifold is an assembly used to connect multiple cylinders together. This assembly could also contain regulators, shutoff valves, gauges, and so on. Manifolds can be specified with manual or automatic changeover, and they can be constructed of high-purity and other special materials compatible with any specific gas being used. A header manifold with individual shutoff valves and connecting pigtail is used to physically connect several cylinders to a changeover manifold. The most often used materials for the header manifold, interconnecting pipe, and fittings are brass and stainless steel, with stainless steel flexible connections connecting the cylinders to the header.

When use is intermittent and the demand is low, a manual, single-cylinder (station) supply is appropriate. The cylinder must be changed when the pressure becomes marginally low. This will require an interruption in supply. The same system could also be used for greater demand where a bank of cylinders is used. When an uninterrupted supply is required, some method of automatic changeover must be used.

The simplest and least costly of the automatic types is the semiautomatic or differential type of changeover manifold. For this type of installation, the regulators for each bank of cylinders are manually set at different pressures. Usually, the secondary bank is set 5 psig lower than the primary bank. When the pressure of the primary bank falls below the lower setting of the reserve bank, the secondary bank automatically becomes the primary supply by default, since it has a higher pressure than the primary bank. A low-pressure alarm or low-pressure gauge reading will indicate that the changeover has taken place. In order to change the cylinders, the empty bank must first be manually isolated. Then, the pressures on the respective primary and secondary regulators must be reset to new settings to reflect the 5 psig difference between the former reserve supply, which is now the primary supply and vice versa. In other types of semiautomatic manifolds, the changeover is fully automatic, but a switch must be manually turned from the reserve position to the primary position when changing cylinders.

The fully automatic changeover manifold uses pressure switches or transducers to sense changes in line and supply pressures. This in turn sends an electric signal to a relay that turns off or on appropriate valves that accomplish the changeover with no variation in system delivery pressure. It also changes the secondary operating bank indicator to primary. For critical applications, connection of the power supply to emergency power should be considered.

A typical manifold assembly is illustrated in Fig. 14.24. Exact manifold dimensions vary and should be obtained from the manufacturer.

Regulators

A regulator is a device used to reduce a variable high inlet pressure to a constant lower outlet pressure. There are two broad categories of regulators: line and cylinder. Line regulators are in-line devices used to reduce a higher to a lower pressure and also used on cryogenic tanks to reduce the pressure, generally in the range of 150 to 250 psig, of the vapor above the vaporized liquid. Cylinder pressure regulators are used on high-pressure cylinders to reduce high-pressure gases, generally in the range of 2000 to 6000 psig, to a lower pressure. The regulator is the first device installed in the distribution system. Depending on the purity of the gas, an integral inlet filter should be considered to keep particulates from entering the regulator.

Regulators are available in two types, single and double stage. The single stage is less costly and less accurate. This type should be chosen if fluctuating pressure is not a major factor in system operation. The double stage is more costly and more

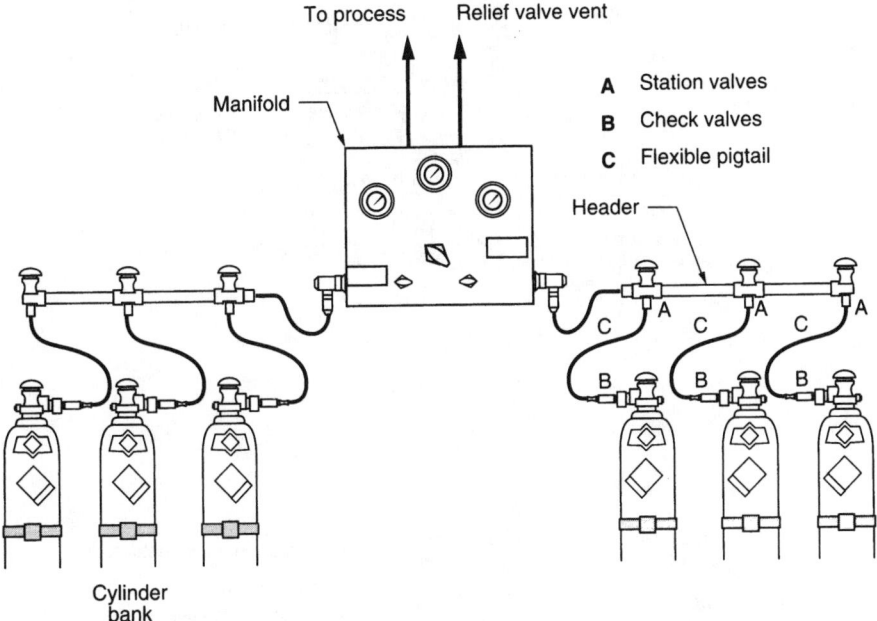

FIGURE 14.24   Typical manifold assembly. (*Courtesy Scott.*)

accurate, and able to achieve a constant outlet pressure within a narrow operating range. The accuracy of the regulator is proportional to the inlet pressure and the flow rate. When selecting a regulator for specific accuracy requirements, obtain the accuracy envelope diagrams from the manufacturer to check the device parameters using actual anticipated system design pressures and flow rates. Typical single- and double-stage regulators are illustrated in Fig. 14.25.

The single-stage regulator reduces pressure in one step. Typical differences in outlet pressure can vary as much as 15 psig when the inlet pressure changes from 2000 to 500 psig. Typical differences in outlet pressure could vary as much as 7 psig from low to high flow rates.

The double-stage regulator reduces the pressure in two steps. Typical differences in outlet pressure can vary as much as 5 psig when the inlet pressure changes from 2000 to 500 psig. Typical differences in outlet pressure could vary as much as 3 psig from low to high flow rates.

Another parameter that may be important in some installations is regulator creep. This is the rise in delivery pressure due to differences in motion of the internal mechanical components caused by aging. Creep is also caused by foreign material interfering with the mechanical operation of the unit. This is the most common cause of unit failure.

The following are other considerations used in the selection of a regulator:

1. The regulator should have a positive gas vent.

2. The regulator must be rated for the highest possible working pressure.

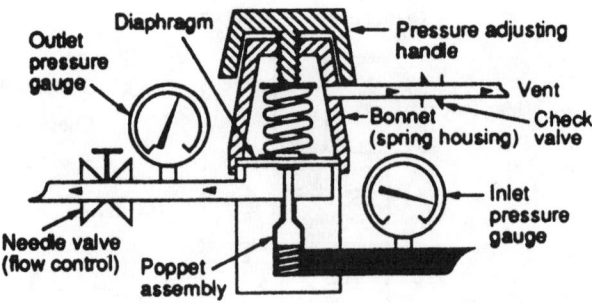

Single-stage regulator

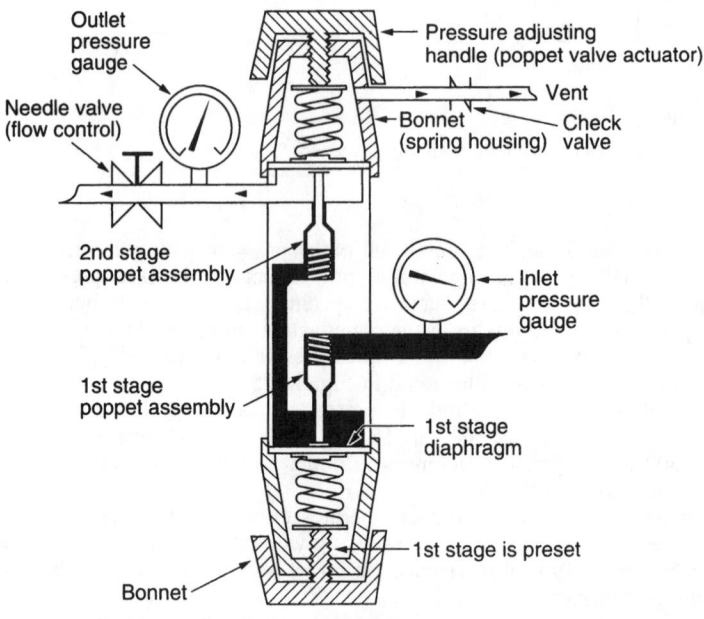

Two-stage regulator

FIGURE 14.25 Typical single- and double-stage regulators. (*Courtesy Scott.*)

3. The delivery pressure range must be adequate.

4. The operating temperature must be compatible for the environment in which the valve is located.

5. The valve body and internal materials should be selected for the specific purity of the desired gas, such as being machine welded, having diffusion-resistant materials and packing, low particulate metals, and flexible diaphragms. High-purity regulators shall have little dead space internally and diaphragm seals consistent with the required purity.

6. The pressure range of the gauges must be compatible with the pressures expected. As an ideal, the working pressure should be half the maximum outlet gauge reading.

One feature that should be considered when only gas is to be used from a bulk liquid supply is an internal tank piping arrangement called an *economizer*. Provided as an integral part of the tank, this allows use of the gas available in the vapor space above the liquid in the tank before the liquid itself has to be vaporized. A special type of pressure regulator shall be provided that will switch from the economizer to the liquid line when the pressure in the vapor space falls below a preset level.

Filters and Purifiers

Filters and purifiers are necessary to reduce or eliminate unwanted contaminants and particulates in the gas stream. The most common purifiers are those used to remove oxygen, water vapor, hydrocarbons, and particulates. They are also used to eliminate other unwanted trace elements. There are a number of materials used for filters:

1. The most often used filter removes particulates 0.2 μm and larger.

2. To remove hydrogen, palladium filters are used.

3. Ceramic, fiberglass, sintered metal, and other adsorbent material are used to remove oil, moisture, and other trace contaminants in order to make the main gas as pure as possible. For some filter mediums, colored materials can be added to change color to indicate when it is time to replace the filter medium.

4. Another type of filter material is the molecular sieve. This is a synthetically produced crystalline metal powder that has been activated for adsorption by removing the water of hydration. This material is manufactured with precise and uniform size and dimensions. The size determines what can be filtered out. Sieves are available as powder, pellets, beads, and mesh. Mesh is not used in laboratories.

5. The 0.2-μm filter for removing particulates is the most commonly used in laboratory service. However, the actual requirements of the end user will dictate the filter medium and type. A filter shall be placed before any flow meter.

6. The housing has to be compatible with the gas being filtered and the pressure involved. None of the filters should be subject to pressures much over the 50 psig (345 kPa) normally used in most laboratories unless specified for a higher pressure.

7. Pressure drop through the filter medium is a critical factor in the selection of the material used. For larger installations, pressure gauges on each side of the filters are used to monitor their effectiveness. Usually, a 5 psig (35 kPa) drop means that replacement is required.

8. It is not possible to improve the purity of a gas with the use of purifiers. If a gas of a certain purity is required, a gas of that grade must be used from the outset.

Refer to Fig. 14.26 for a typical system purifier arrangement.

Gauges

Gauges for pressures of up to 10 psig (70 kPa) are usually the diaphragm sensing element type. For pressures over 10 psig (70 kPa), use the bourdon type. They should be cleaned for oxygen service and the materials must be compatible with the intended gas. Provide a small gas cock between the pipe line and the gauge to shut off the flow and allow the gauge to be replaced without having to shut down the system.

Relief Valves

Relief valves are used to protect a system from overpressure. A relief valve must be provided between the regulator and the first shutoff valve in the system, with the discharge independently piped outdoors, The discharges from a single gas service manifold or regulator may be connected together but shall not be connected to any relief discharge from any other system. The discharge pipe should be a minimum size of ¾ in. The relief valve shall be located at the first point in the system that could be subject to full cylinder pressure if the regulator should fail. There shall be no valve between the relief valve and the regulator. The relief valve release point should be set to 50 percent over working pressure. This is a safe figure because the system test pressure is 150 percent over working pressure. Typical relief venting is illustrated in Fig. 14.27.

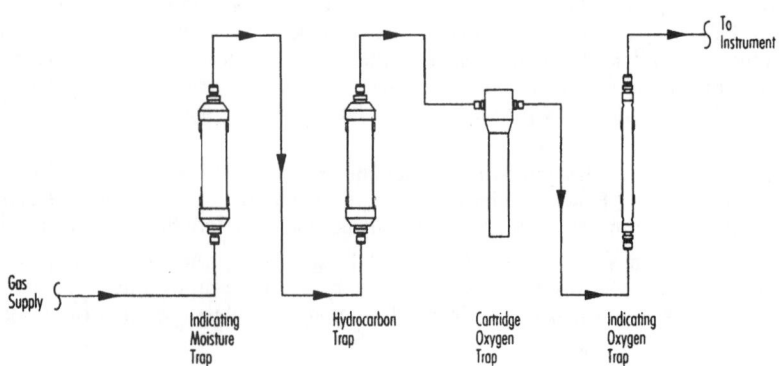

FIGURE 14.26 Typical purifier arrangement.

To process   To vent

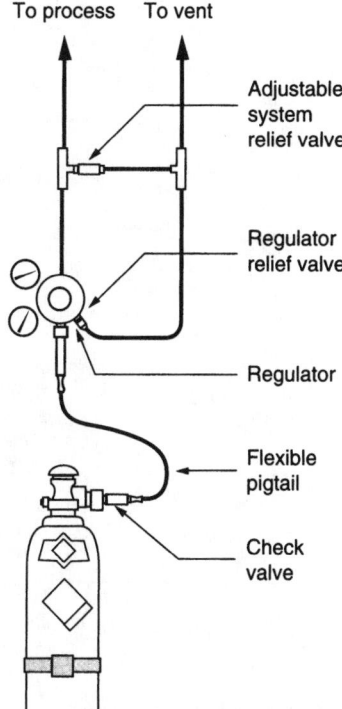

Adjustable
system
relief valve

Regulator
relief valve

Regulator

Flexible
pigtail

Check
valve

FIGURE 14.27 Typical relief venting. When
two-stage regulators are used, a preset first-stage
(or interstage) relief valve is sometimes required
to protect the second stage from overpressure.
Additionally, it is good practice to install an adjustable relief valve on the second stage to protect the system and instruments from damage
from excessive pressure. For outdoor installations involving inert gases, the relief valves can
exhaust directly to atmosphere. For indoor installations, or any installations involving toxic or
flammable gases, the relief valve exhaust should
be captured and vented to a safe location. (*Courtesy Scott.*)

Flow Limit Shutoff Valve

A flow limit shutoff valve automatically shuts off the flow from a cylinder if that
flow rate exceeds a predetermined limit. That limit is usually about 10 times the
highest expected flow rate. This valve must be manually reset after operation. A
typical installation detail is shown in Fig. 14.28.

Check Valves

Check valves are used to prevent the reverse flow of gas in the delivery piping
system. If there is a possibility that one gas at a higher pressure may force its way

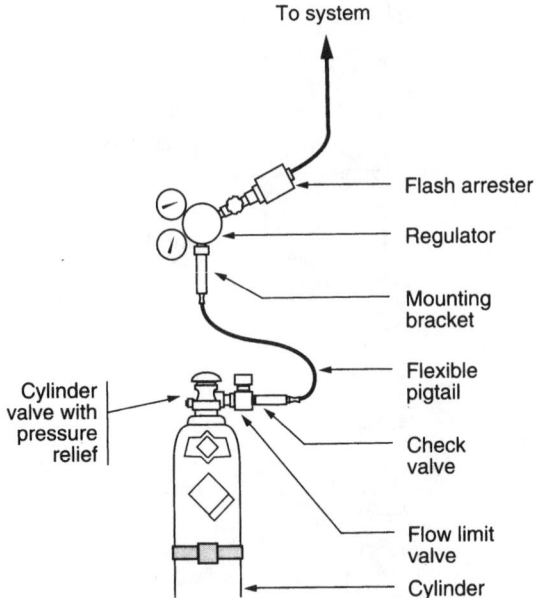

To system

Flash arrester

Regulator

Mounting
bracket

Flexible
pigtail

Cylinder
valve with
pressure
relief

Check
valve

Flow limit
valve

Cylinder

FIGURE 14.28   Typical single-cylinder installation detail.

into another piping system, a check valve shall be installed. A typical single-cylinder installation detail is shown in Fig. 14.28.

Flash Arresters

Flash arresters are required when the gas being used is flammable, particularly hydrogen and acetylene. They are mounted in-line to prevent any flame from going back into the tank in the event that gas in the delivery piping system has ignited. It is standard procedure that a check valve be made an integral part of a flash arrester, although this is not true in all cases. A typical installation detail is shown in Fig. 14.28.

Valves

The most often used shutoff valves are ball valves. Three piece are the most desired because the body can be separated from the end connections when being installed and serviced. For control and modulating purposes, needle valves are used because of the precise level of control permitted. The materials of the valve and seals must be compatible with the gas used.

For specialty applications, there are the diffusion-resistant valves that reduce or eliminate unwanted gases from entering the system through the packing. Where purity is a major consideration, packless and bellows sealed diaphragm valves are available.

Manifold and Regulator Purge Devices

The replacement of cylinders introduces unwanted room air into the piping manifold assembly and the connecting cylinder pigtails. When maintaining a high purity level of the gas is necessary, purge valves are installed to run system gas through the con-taminated parts of the system to replace all such air. The purge valve outlet should be vented outside the building. If the gas is suitable and low enough in volume and the storage room is large enough and well ventilated, it could discharge into the room since the purge volume used is generally quite small. The regulator often requires special purging techniques recommended by the manufacturer.

Flow Measurement

Flow meters can be either of two types: electric or mechanical. The mechanical kind is called a *variable area type* and uses a small ball as an indicator in a variable area vertical tube. The type of mechanical meter most often used has an accuracy of 10 percent full scale. This means that if the flow range is from 1 to 10 scfm, the accuracy is ± 1 acfm. There are also more accurate variable area flow meters available.

The mass flow meters are electronically operated, using the difference in temperature that gas creates when flowing over a heated element. The mass flow meter is quite accurate and quite expensive.

Gas Warmers

On occasion, the gas in cylinders is withdrawn so fast that the regulator could ice up because of the change in temperature. If this occurs, an electrically heated gas warmer is available to be installed in-line, and this warmer would heat the gas out of the cylinder before it reached the regulator. The rule of thumb is to consider a warmer if the use of gas exceeds 35 acfm. The actual figure should be based on experience with the specific type of gas being used. Ask the supplier what his or her experience has been. Carbon dioxide is a particular problem.

Low-Temperature Cutoff

On occasion, the temperature of the delivered gas is a critical factor. If low temperature could harm instruments or interfere with procedures being conducted, a low-temperature cutoff should be installed with a solenoid valve to stop the flow of gas. If this happens often, a gas warmer might be required.

Alarms

Alarms are necessary for the user to be made aware of immediate or potential trouble. They could be visible and/or audible. Usual alarms are high system pressure, low system pressure, and reserve in use. In some installations, a normal light is also requested. Other alarms could be provided that will indicate high pressure loss at filters, low gas temperature, purifiers at limit of capacity, and flow limit valve operation. These alarms are usually installed in an alarm panel. The panel could be mounted in the room where the gases are stored, in a constantly occupied location such as a maintenance shop or receptionist area, or in the laboratory itself depending on the availability and level of maintenance. Often multiple locations are desirable if continued supply of gas is critical. Various devices must be placed in the system for these alarms to function, such as pressure switches, transducers, and auxiliary contacts in a manifold assembly to transmit the alarm signal to the alarm panel.

Toxic and Flammable Gas Monitors

If there is a possibility for a toxic flammable gas to accumulate in an enclosed area or room, it is required that a gas monitor be installed to alarm if the gas percentage rises above a predetermined limit that is considered harmful or dangerous. This should be 50 percent of either the lower flammability limit (LFL) or the concentration that may cause ill effects or breathing problems. The oxygen concentration of ambient air should never be allowed to fall below 19.5 percent. In addition, much lower levels should also be alarmed to indicate that a problem exists well before the evacuation of an area is required because of the leak. Refer to Table 14.25 for the flammability limits of some of the more common gases and to the MSDS for gases not listed.

Gas Mixers

For certain applications gas mixers are available to accurately mix different gases together to produce various proportions. The accuracy of the mixture, flow rates of the various gases, and the compatibility of the piping materials and the gases are considerations in the selection of the mixer.

DISTRIBUTION NETWORK

System Pressure

It is generally accepted practice to use a pressure of 50 to 55 psig (345 to 380 kPa) in the piping distribution system. Accepted practice limits the allowable friction loss in the piping system to approximately 10 percent of initial pressure. These figures should be adjusted for specific conditions or special systems when necessary.

Pipe Material Selection

The piping material must be compatible with the specific gas, capable of delivering the desired gas purity for anticipated usage, and capable of being cleaned and/or sterilized often, if required. Table 14.26 gives the compatibility of various piping materials for the most commonly used gases. For materials or gases not listed, refer to the manufacturer or the supplier of the gas for additional information. The allowable pressure ratings for various piping materials are given in Table 14.27.

The pipe most often used to maintain the highest purity is grade 304L or 316L stainless steel tubing conforming to ASTM A-270. The interior should be electropolished, and the exterior could be mill finished in concealed spaces. In exposed

TABLE 14.25 Flammability Limits and Specific Gravity for Common Gases

Gas	Specific gravity	Flammability in air, percent	
		Low	High
Acetylene	0.906	25	100
Air	1.00	—	—
Ammonia	0.560	15	28
Argon	1.38	—	—
Arsine	2.69	5.1	78
Butane	0.600	1.8	8.4
Carbon dioxide	1.52	—	—
Carbon monoxide	0.967	12.5	74
Chlorine	2.49	—	—
Cyclopropane	0.720	2.4	10.4
Ethane	1.05	3.0	12.4
Ethylene	0.570	2.7	36
Ethyl chloride	2.22	3.8	15.4
Fluorine	1.31	—	—
Helium	0.138	—	—
Hydrogen	0.069	4.0	75
Hydrogen sulfide	1.18	4	44
Isobutane	2.01	1.8	9.6
Isopentane	2.48		
Krypton	2.89	—	—
Methane	0.415	5.0	15
Methyl chloride	1.74	10.7	17.4
Natural gas	0.600		
Neon	0.674	—	—
Nitrogen	0.966	—	—
Nitrous oxide	1.53	—	—
Oxygen	1.10	—	—
Phosgene	1.39	—	—
Propane	1.580	2.1	9.5
Silane	1.11	1.5	98
Sulphur dioxide	2.26	—	—
Xenon	4.53	—	—

locations and where pipe exterior will be sterilized or cleaned, the pipe exterior should have a No. 4 finish. The pipe is joined by orbital welding. This tube should have a minimum wall thickness of 0.65 in in order to be welded. Stainless steel pipe is capable of withstanding repeated sterilization by steam and a variety of chemicals. When welding is not required, a tube wall thickness of 0.28 is commonly used. The installed cost often is less than that of copper tube.

In many laboratory applications, maintaining ultrahigh purity of a gas from storage tank to the outlet is not a requirement. For this type of service, copper tube and fittings that have been cleaned for oxygen service and joined by brazing often has the least initial cost and is the material of choice. The following grades of copper pipe have been used:

1. ASTM B-88
2. ASTM B-819

TABLE 14.26 Compatibility of Pipe to Common Specialty Gases

Gas (common name)	Chemical formula	Metals						Buna-N	Kel-F	Synthetics			
		Aluminum	Brass	Copper	Monel	Carbon steel	Stainless steel			Neoprene	PVC	Teflon	Viton
Acetylene	C_2H_2	Y	Y	N	Y	Y	Y	Y	Y	Y	I	Y	Y
Air	—	Y	Y	Y	Y	Y	Y	Y	Y	Y	Y	Y	Y
Ammonia	NH_3	Y	N	N	Y	Y	Y	Y	Y	Y	Y	Y	N
Argon	Ar	Y	Y	Y	Y	Y	Y	Y	Y	Y	Y	Y	Y
Arsine	AsH_3	I	Y	Y	Y	Y	Y	Y	Y	Y	Y	Y	Y
Boron trichloride	BCl_3	I	Y	Y	Y	Y	Y	I	Y	I	Y	Y	I
Boron trifluoride	BF_3	Y	I	Y	Y	Y	Y	Y	I	Y	Y	Y	Y
1,3-butadiene	C_4H_6	Y	Y	Y	Y	Y	Y	N	Y	Y	Y	Y	Y
n-butane	C_4H_{10}	Y	Y	Y	Y	Y	Y	Y	Y	Y	Y	Y	Y
1-butene	C_4H_8	Y	Y	Y	Y	Y	Y	Y	Y	Y	Y	Y	Y
cis-2-butene	C_4H_8	Y	Y	Y	Y	Y	Y	Y	Y	Y	Y	Y	Y
trans-2-butene	C_4H_8	Y	Y	Y	Y	Y	Y	Y	Y	Y	Y	Y	Y
Carbon dioxide	CO_2	Y	Y	Y	Y	Y	Y	Y	Y	Y	Y	Y	Y
Carbon monoxide	CO	Y	Y	Y	Y	Y	Y	Y	Y	N	Y	Y	Y
Chlorine	Cl_2	N	N	N	Y	Y	Y	N	Y	N	Y	Y	Y
Deuterium	D_2	Y	Y	Y	Y	Y	Y	Y	Y	Y	Y	Y	Y
Dichlorosilane	SiH_2Cl_2	I	I	I	Y	Y	Y	Y	Y	I	I	Y	I
Dimethyl ether	$(CH_3)_2O$	Y	Y	Y	I	Y	Y	N	Y	Y	Y	Y	Y
Dimethylamine	$(CH_3)_2NH$	N	N	N	Y	Y	Y	N	Y	I	I	Y	Y
Disilane	Si_2H_6	I	I	Y	Y	Y	Y	I	Y	I	I	Y	Y
Ethane	C_2H_6	Y	Y	Y	Y	Y	Y	Y	Y	Y	Y	Y	Y
Ethyl acetylene	C_4H_6	Y	I	N	Y	Y	Y	I	Y	Y	I	Y	Y
Ethyl chloride	C_2H_5Cl	Y	Y	Y	Y	Y	Y	Y	Y	Y	N	Y	Y
Ethylene	C_2H_4	Y	Y	Y	I	Y	Y	N	Y	Y	I	Y	Y
Ethylene oxide	C_2H_4O	I	I	N	I	Y	Y	N	Y	N	N	Y	N

Refrigerant	Formula											
Halocarbon 11	CCl_3F	Y	Y	N	Y	Y	Y	Y	Y	Y	Y	Y
Halocarbon 12	CCl_2F_2	Y	Y	N	Y	Y	Y	Y	Y	Y	Y	Y
Halocarbon 13	$CClF_3$	Y	Y	N	Y	Y	Y	Y	Y	Y	Y	Y
Halocarbon 13B-1	$CBrF_3$	Y	Y	N	Y	Y	Y	Y	Y	Y	Y	Y
Halocarbon 14	CF_4	Y	Y	N	Y	Y	Y	Y	Y	Y	Y	Y
Halocarbon 21	$CHCl_2F$	N	Y	N	Y	Y	N	Y	Y	Y	Y	N
Halocarbon 22	$CHClF_2$	N	Y	N	Y	Y	N	Y	Y	Y	Y	N
Halocarbon 23	CHF_3	I	Y	N	Y	Y	I	Y	Y	Y	Y	I
Halocarbon 113	$C_2Cl_3F_3$	Y	Y	N	Y	Y	Y	Y	?	Y	Y	Y
Halocarbon 114	$C_2Cl_2F_4$	Y	Y	N	Y	Y	Y	Y	Y	Y	Y	Y
Halocarbon 115	C_2ClF_5	Y	Y	N	Y	Y	Y	Y	Y	Y	Y	Y
Halocarbon 116	C_2F_6	I	Y	N	Y	Y	I	Y	Y	Y	Y	I
Halocarbon 500	$C_2H_4F_2/CCl_2F_2$	Y	Y	N	Y	Y	Y	Y	Y	I	I	Y
Halocarbon 502	$CHClF_2/C_2ClF_5$	Y	Y	N	Y	Y	Y	Y	Y	I	I	Y
Helium	He	Y	Y	Y	Y	Y	Y	Y	Y	Y	Y	Y
Hydrogen	H_2	Y	Y	Y	Y	Y	Y	Y	Y	Y	Y	Y
Hydrogen bromide	HBr	Y	Y	Y	N	Y	N	Y	Y	Y	N	Y
Hydrogen chloride	HCl	Y	Y	Y	N	Y	N	Y	Y	Y	N	Y
Hydrogen fluoride	HF	N	Y	Y	N	Y	N	Y	Y	I	I	N
Hydrogen sulfide	H_2S	N	Y	Y	Y	Y	Y	Y	Y	I	N	N
Isobutane	C_4H_{10}	Y	Y	Y	Y	Y	Y	Y	Y	Y	Y	Y
Isobutylene	C_4H_8	Y	Y	Y	Y	Y	Y	Y	Y	Y	Y	Y
Methane	CH_4	Y	Y	Y	Y	Y	Y	Y	Y	Y	Y	Y
Methyl bromide	CH_3Br	Y	Y	I	N	Y	Y	Y	Y	I	I	Y
Methyl chloride	CH_3Cl	Y	Y	I	N	Y	Y	Y	Y	Y	Y	Y
Neon	Ne	Y	Y	Y	Y	Y	Y	Y	Y	Y	Y	Y
Nitric oxide	NO	I	Y	Y	Y	Y	I	Y	Y	Y	Y	I
Nitrogen	N_2	Y	Y	Y	Y	Y	Y	Y	Y	Y	Y	Y
Nitrogen dioxide	NO_2	N	Y	N	Y	Y	N	Y	Y	I	I	N
Nitrous oxide	N_2O	Y	Y	Y	Y	Y	Y	Y	Y	Y	Y	Y

(Continued)

TABLE 14.26 Compatibility of Pipe to Common Specialty Gases (*Continued*)

Gases		Metals						Synthetics					
Gas (common name)	Chemical formula	Aluminum	Brass	Copper	Monel	Carbon steel	Stainless steel	Buna-N	Kel-F	Neo-prene	PVC	Teflon	Viton
Oxygen	O_2	Y	Y	Y	Y	Y	Y	Y	Y	Y	Y	Y	Y
Phosgene	$COCl_2$	Y	I	N	Y	Y	Y	I	Y	I	I	Y	I
Phosphine	PH_3	Y	I	I	Y	Y	Y	I	Y	I	I	Y	I
Phosphorous pentafluoride	PF_5	I	I	I	Y	Y	Y	I	Y	I	I	Y	I
Propane	C_3H_8	Y	Y	Y	Y	Y	Y	N	Y	N	Y	Y	Y
Propylene	C_3H_6	Y	Y	Y	Y	Y	Y	Y	Y	Y	Y	Y	Y
Silane	SiH_4	Y	Y	Y	Y	Y	Y	Y	Y	Y	Y	Y	Y
Silicon tetrachloride	$SiCl_4$	Y	Y	Y	Y	Y	Y	Y	Y	Y	Y	Y	Y
Silicon tetrafluoride	SiF_4	Y	Y	Y	Y	Y	Y	Y	Y	Y	Y	Y	Y
Sulfur dioxide	SO_2	Y	N	N	Y	Y	Y	N	Y	N	Y	Y	N
Sulfur hexafluoride	SF_6	Y	Y	Y	Y	Y	Y	Y	Y	Y	Y	Y	Y
Trichlorosilane	$SiHCl_3$	Y	Y	Y	Y	Y	Y	Y	Y	Y	Y	Y	Y
Trimethylamine	$(CH_3)_3N$	N	N	N	Y	Y	Y	N	Y	Y	N	Y	N
Xenon	Xe	Y	Y	Y	Y	Y	Y	Y	Y	Y	Y	Y	Y

Note: Prior to using a gas mixture or a gas that is not listed in the gas compatibility guide, it is strongly recommended that you contact a specialty gas laboratory for information.

Y = Yes, suitable for use with intended gas.

N = No, not suitable for use with intended gas.

I = Insufficient data available to determine compatibility with intended gas.

3. ASTM B-75

4. ASTM B-280

Another pipe material often used for noncritical applications is aluminum tubing ASTM B-210, alloy 6061, T4 or T6 tempers. This pipe is most commonly joined using the patented flare joint.

Maintaining Cleanliness and Purity during Construction

Copper fittings and various valve types can be purchased from the manufacturer specifically cleaned for oxygen service and delivered to the job site capped and bagged to maintain cleanliness. If a fitting becomes dirty prior to being installed, it should be cleaned in accordance with NFPA-99 requirements before being made part of the system. During construction, the greatest threat to cleanliness is dirt and dust entering the pipe because it has not been capped to keep it out.

When brazing joints, the cleanliness of the copper piping system shall be maintained by not using flux and having the joint continuously purged with oil-free, dry

TABLE 14.27   Allowable Pressure Ratings of Pipe and Tube

A. All pressures are calculated from equations in ANSI Code for Pressure Piping ASME/ANSI B31.3.

B. All calculations are based on maximum O.D. and minimum wall thickness.

Example: *¹⁄₂-in O.D. × 0.035-in wall stainless steel tubing purchased to ASTM A269:*

O.D. tolerance ±0.005 in/wall thickness tolerance ±15 percent

Calculations are based on a 0.505 in O.D. × 0.0298-in wall tubing.

C. No allowance is made for corrosion or erosion.

TABLE 14.27a   Aluminum tubing, psig

[Based on ultimate tensile strength 42,000 psig (289,400 kPa). For metal temperatures −20° to 100°F (−29° to 37°C). Allowable working pressure loads calculated from S values (14,000 psi—96,500 kPa) as specified by ANSI B31.3 code.]

Tube O.D., in	Tube wall thickness, in				
	0.035	0.049	0.065	0.083	0.095
¹⁄₈	8600				
³⁄₁₆	5600	8000			
¹⁄₄	4000	5900			
⁵⁄₁₆	3100	4600			
³⁄₈	2600	3700			
¹⁄₂	1900	2700	3700		
⁵⁄₈	1500	2100	2900		
³⁄₄		1700	2400	3100	
⁷⁄₈		1500	2000		
1		1300	1700	2300	2700

Suggested ordering information: High-quality aluminum-alloy seamless tubing ASTM B-210 or equivalent. (Values shown are for alloy 6061-T6.)

TABLE 14.27b   Copper Tubing, psig

[Based on ultimate strength 30,000 psig (206,700 kPa). For metal temperatures −20° to 100°F (−29° to 37°C). Allowable working pressure loads calculated from S values (6000 psi—41,300 kPa) as specified by ANSI B31.3 code.]

Tube O.D., in	Tube wall thickness, in							
	0.028	0.035	0.049	0.065	0.083	0.095	0.109	0.120
1/8	2700	3600						
3/16	1800	2300	3400					
1/4	1300	1600	2500	3500				
5/16		1300	1900	2700				
3/8		1000	1600	2200				
1/2		800	1100	1600	2100			
5/8			900	1200	1600	1900		
3/4			700	1000	1300	1500	1800	
7/8			600	800	1100	1300	1500	
1			500	700	900	1100	1300	1500

Suggested ordering information: High-quality soft annealed seamless copper tubing ASTM B-75 or equivalent. Also soft annealed (Temper 0) copper water tube type K or type L to ASTM B-88.

TABLE 14.27c   Carbon Steel Tubing, psig

[Soft annealed carbon steel hydraulic tubing ASTM A179 or equivalent. Based on ultimate tensile strength 47,000 psi (323,800 kPa). For metal temperatures −20° to 100°F (−29 to 37°C). Allowable working pressure loads calculated from S values (15,700 psi—108,200 kPa) as specified by ANSI B31.3 code.]

Tube O.D., in	Tube wall thickness, in												
	0.028	0.035	0.049	0.065	0.083	0.095	0.109	0.128	0.134	0.148	0.165	0.180	0.228
1/8	8000	10200											
3/16	5100	6600	9600										
1/4	3700	4800	7000	9600									
5/16		3700	5500	7500									
3/8		3100	4500	6200									
1/2		2300	3200	4500	5900								
5/8		1800	2600	3500	4600	5300							
3/4			2100	2900	3700	4300	5100						
7/8			1800	2400	3200	3700	4300						
1			1500	2100	2700	3200	3700	4100					
1 1/4				1600	2100	2500	2900	3200	3600	4000	4600	5000	
1 1/2					1800	2000	2400	2600	2900	3300	3700	4100	5100
2						1500	1700	1900	2100	2400	2700	3000	3700

Suggested ordering information: High-quality soft annealed seamless carbon steel hydraulic tubing ASTM A-179 or equivalent. Hardness Rb72 (HV(VPN)130) or less. Tubing to be free of scratches. Suitable for bending and flaring.

TABLE 14.27d Pressure Rating of Stainless Steel Tubing, psig

[Annealed 304 or 316 stainless steel tubing ASTM A269 or equivalent. Based on ultimate tensile strength 75,000 psig (516,700 kPa). For metal temperature from −20° to 100°F (−29° to 37°C). Allowable working pressure loads calculated from S values (20,000 psig—37,800 kPa) as specified by ANSI B31.3 code.]

For Seamless Tubing

Note: For welded and drawn tubing, a derating factor must be applied for weld integrity: For double-welded tubing multiply pressure rating by 0.85—for single-welded tubing, multiply pressure rating by 0.80.

Tube O.D., in	Tube wall thickness, in															
	0.010	0.012	0.014	0.016	0.020	0.028	0.035	0.049	0.065	0.083	0.095	0.109	0.120	0.134	0.156	0.188
1/16	5600	6800	8100	9400	12000											
1/8						8500	10900									
3/16						5400	7000	10200								
1/4						4000	5100	7500	10200							
5/16							4000	5800	8000							
3/8							3300	4800	6500							
1/2							2400	3500	4700	6200						
5/8								2900	4000	5200	6000					
3/4								2400	3300	4200	4900	5800				
7/8								2000	2800	3600	4200	4800				
1									2400	3100	3600	4200	4700			
1 1/4										2400	2800	3300	3600	4100	4900	4900
1 1/2											2300	2700	3000	3400	4000	4000
2												2000	2200	2500	2900	3600

Suggested ordering information: Fully annealed high quality (Type 304, 316, etc.) (seamless or welded and drawn) stainless steel hydraulic tubing ASTM A200 or A213 or equivalent. Hardness Rb80[HV(VPN)180] or less. Tubing to be free of scratches. Suitable for bending and flaring.

1 psi = 6.9 kPa.

nitrogen, thereby eliminating the formation of copper oxide on the inside of the joint generated by the heat of the brazing process. The flow of purge gas shall be continued until the joint is cool to the touch.

Another consideration in maintaining high purity of the gas is outgasing. This is a phenomenon in which a gas under pressure is absorbed into any porous material. This occurs primarily in elastomers used as gaskets or seals, and to some lesser extent into metallic and plastic pipe and tubing materials. When the pressure is reduced or eliminated, such as when changing cylinder banks or during maintenance, the absorbed gases are spontaneously given off, adding impurities into the gas piping system.

Joints

The most often used joints for copper are brazed. No flux is permitted, and only cast copper fittings should be used, which require no flux. The interior of the joint shall be purged with an inert gas, such as nitrogen type "NF" or argon. For stainless steel pipe, orbital welding leaves the smoothest interior but should be used generally on tubing 0.65 in or thicker. Another type of joint that can be used is the patented flared joint, which is preferable to solder or brazed joints that often leave a residue that contributes particulates into the gas stream. In addition, the flared joint is popular because it can be made up using only a saw and some wrenches. When copper tubing is used with flared joints, the pipe shall not have embossed identification stamped into the pipe because doing so causes leaks at the joint. There is no ASTM designation for the patented flare joints, but they are acceptable for all applications as long as the allowable pressure ratings are not exceeded.

Pipe Sizing

The following is a recommended system sizing procedure:

1. Locate the gas storage area and lay out the cylinders, manifold, and so on.
2. Establish a general layout of the system from the storage area to the farthest outlet or use point. Measure the actual distance along the run of pipe to the most remote terminal, and then add 50 percent of the distance for a fitting allowance. This is the total equivalent run of pipe.
3. Choose all of the filters, purifiers, and so on necessary for system purity in order to establish a combined allowable pressure drop through each of them and the assembly as a whole.
4. Establish the gas pressure required at the farthest outlet, add the pressure required to overcome the drop through the filter-purifier-manifold assembly, and add 10 percent of the supply pressure to allow for friction loss of the gas through the total run of pipe. It is commonly accepted practice for general laboratory use to have a minimum system pressure of 45 to 50 psig (310 to 345 kPa) and to allow 5 psig (35 kPa) as a pressure loss in the pipe. For higher-pressure systems serving specific equipment or tools, start with the actual pressure required.
5. Divide the total run of pipe (in 100s of feet) by the allowable friction loss to calculate the allowable friction loss in psig per 100 ft of pipe, This is to allow the use of the sizing chart provided in this handbook. If other methods are used

to indicate friction loss in the piping system, calculate the loss in that specific method.

6. Determine the total connected flow rate of gas for all parts of the system. For general laboratory use, a figure of 1 scfm (30 Lpm) for each outlet is used. Calculate the scfm (Lpm) of gas through each branch, from the farthest outlet back to the source (or main). For specific equipment, use the flow rate recommended by the manufacturer.

7. Determine the appropriate diversity factor when sizing compressed air pipe in order to allow for the fact that not all outlets will be used at once. This will result in the actual flow rate. A diversity factor for general laboratory use is given in Table 14.14, and a direct reading chart is illustrated in Fig. 14.29. For specific equipment, the diversity factor must be ddetermined from the end user.

8. With all the above information available, the pipe can now be sized. Starting from the most remote point on the branch and then proceeding to the main, calculate the actual flow rate using the diversity factor. Enter Table 14.28 with the actual flow rate and the allowable friction loss. Find the flow rate, and then read across to find a friction loss figure that is equal to or is less than the allowable friction loss. Read up the column to find the size. In some cases, the diversity factor for the next highest range of outlets may result in a smaller-size pipe than the range previously calculated. If this occurs, do not reduce the size of the pipe—keep the larger size previously determined. For equipment using capilary piping, refer to Fig. 14.30 for ⅛ in size, Fig. 14.31 for ¼ in size and Fig. 14.32 for ⅜ in size.

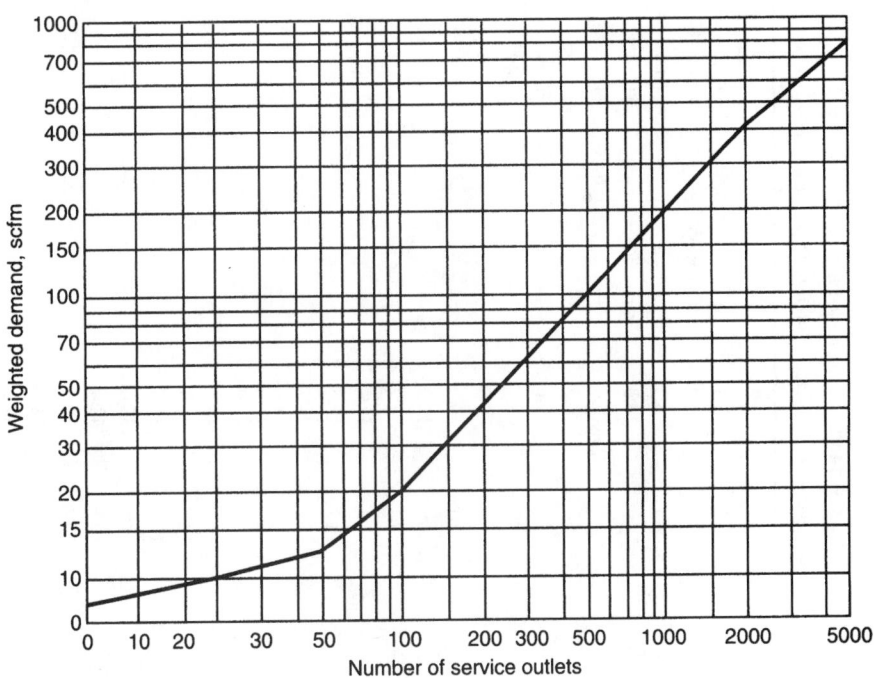

FIGURE 14.29 General laboratory demand for compressed air.

TABLE 14.28 Compressed Air Pipe Sizing Chart

Pressure drop (psi) per 100 ft in 55 psig compressed air system using Darcy's equation with copper tubing

scfm	acfm	½	¾	1	1¼	1½	2	2½	3	4
5	1.1	0.15	0.04	0.01						
10	2.2	0.51	0.13	0.04						
15	3.3	1.04	0.27	0.09	0.01					
20	4.3	—	0.45	0.14	0.02	0.01				
25	5.4	—	0.67	0.21	0.04	0.02	0.01			
30	6.5	—	0.93	0.29	0.06	0.03	0.01			
35	7.6		—	0.39	0.08	0.04	0.02	0.01		
40	8.7		—	0.49	0.10	0.05	0.02	0.01		
45	9.8		—	0.60	0.13	0.06	0.02	0.01		
50	10.9		—	0.73	0.16	0.08	0.03	0.01		
60	13.0		—	1.01	0.20	0.09	0.04	0.02	0.01	
70	15.2			—	0.27	0.13	0.05	0.02	0.01	
80	17.4			—	0.36	0.17	0.07	0.03	0.01	
90	19.5			—	0.45	0.22	0.08	0.03	0.01	0.00
100	21.7			—	0.56	0.27	0.10	0.04	0.02	0.01
110	23.9			—	0.68	0.32	0.12	0.05	0.02	0.01
120	26.0			—	0.81	0.38	0.14	0.06	0.02	0.01
130	28.2			—	0.94	0.45	0.16	0.07	0.02	0.01
140	30.4				1.09	0.52	0.18	0.08	0.03	0.01
150	32.6				—	0.59	0.20	0.09	0.03	0.01
175	38.0				—	0.67	0.27	0.11	0.04	0.01
200	43.4				—	0.89	0.34	0.14	0.05	0.01
225	48.8				—	1.13	0.42	0.18	0.06	0.02
250	54.3				—	—	0.51	0.22	0.08	0.02
275	59.7				—	—	0.60	0.26	0.09	0.02

300	65.1	—	—	—	—	—	0.71	0.30	0.11	0.03
325	70.5	—	—	—	—	—	0.82	0.35	0.12	0.03
350	76.0	—	—	—	—	—	0.94	0.40	0.14	0.04
375	81.4	—	—	—	—	—	1.06	0.45	0.16	0.04
400	86.8	—	—	—	—	—	—	0.51	0.18	0.05
450	97.7	—	—	—	—	—	—	0.63	0.22	0.06
500	108.5	—	—	—	—	—	—	0.76	0.27	0.07
550	119.4	—	—	—	—	—	—	0.90	0.32	0.09
600	130.2	—	—	—	—	—	—	1.06	0.37	0.10
650	141.1	—	—	—	—	—	—	—	0.43	0.12
700	151.9	—	—	—	—	—	—	—	0.49	0.13
750	162.8	—	—	—	—	—	—	—	0.56	0.15
800	173.6	—	—	—	—	—	—	—	0.63	0.17
850	184.5	—	—	—	—	—	—	—	0.70	0.19
900	195.3	—	—	—	—	—	—	—	0.78	0.21
950	206.2	—	—	—	—	—	—	—	—	0.23
1000	217.0	—	—	—	—	—	—	—	—	0.25
1100	238.7	—	—	—	—	—	—	—	—	0.30
1200	260.4	—	—	—	—	—	—	—	—	0.35
1300	282.1	—	—	—	—	—	—	—	—	0.41
1400	303.8	—	—	—	—	—	—	—	—	0.47
1500	325.5	—	—	—	—	—	—	—	—	0.53

Note: Values in the table are for flows not exceeding 4000 fpm vel.

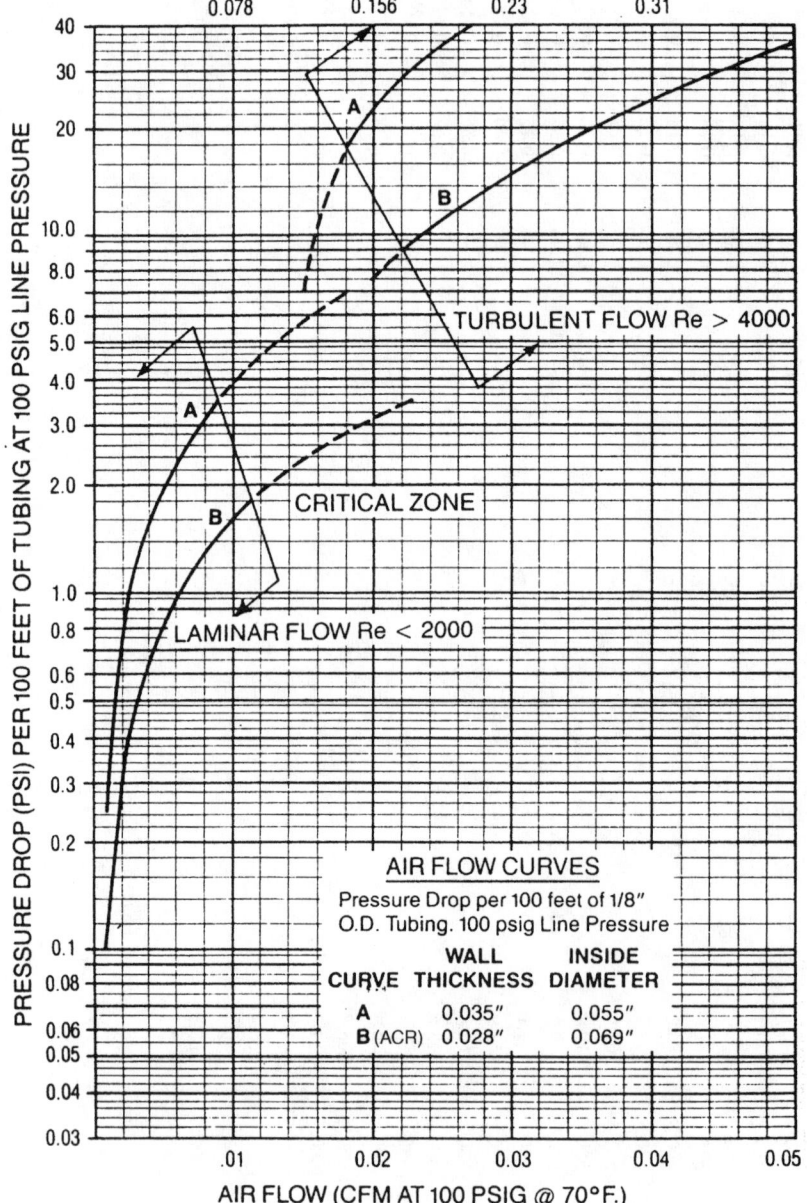

FIGURE 14.30 Compressed air sizing chart ⅛ in (6 mm) copper tubing, 100 psig. (*Courtesy Swagflok.*)

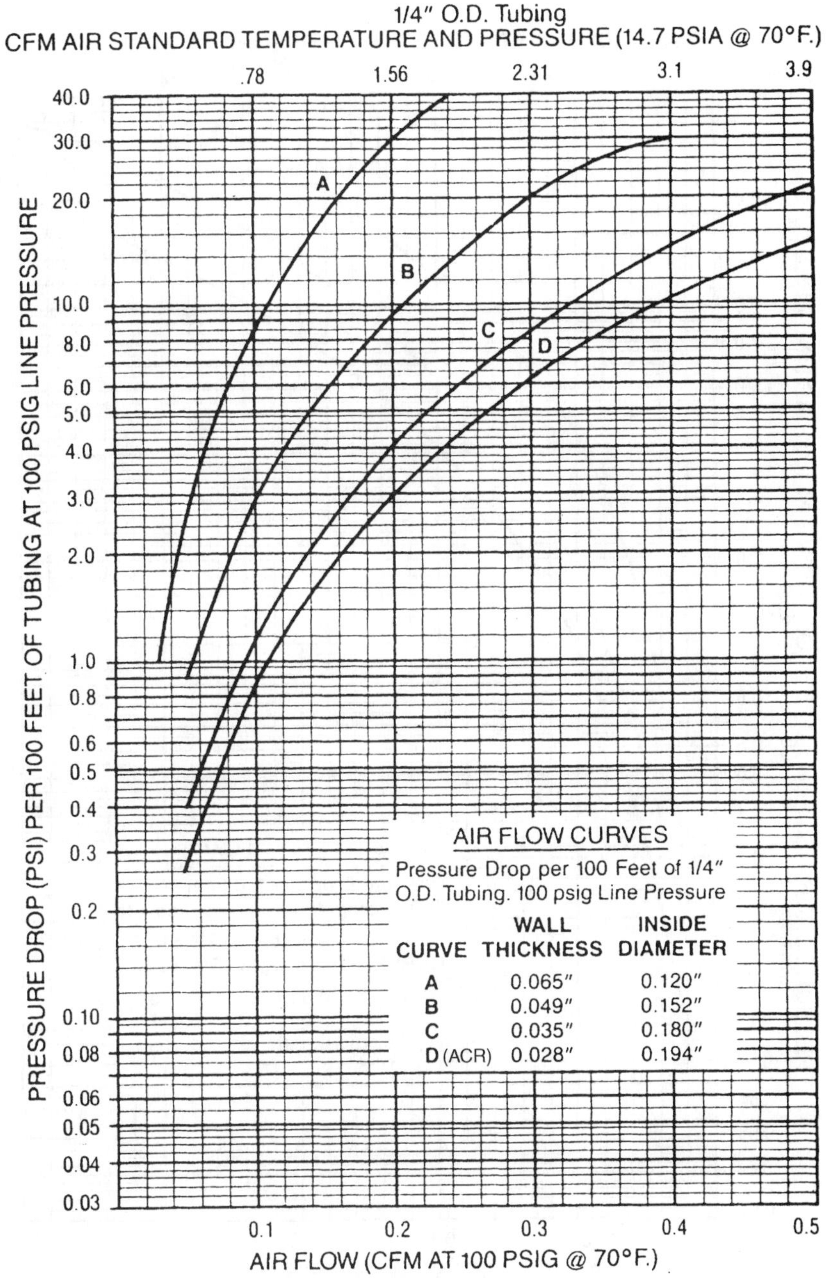

FIGURE 14.31 Compressed air sizing chart ¼ in (8 mm) copper tube, 100 psig.

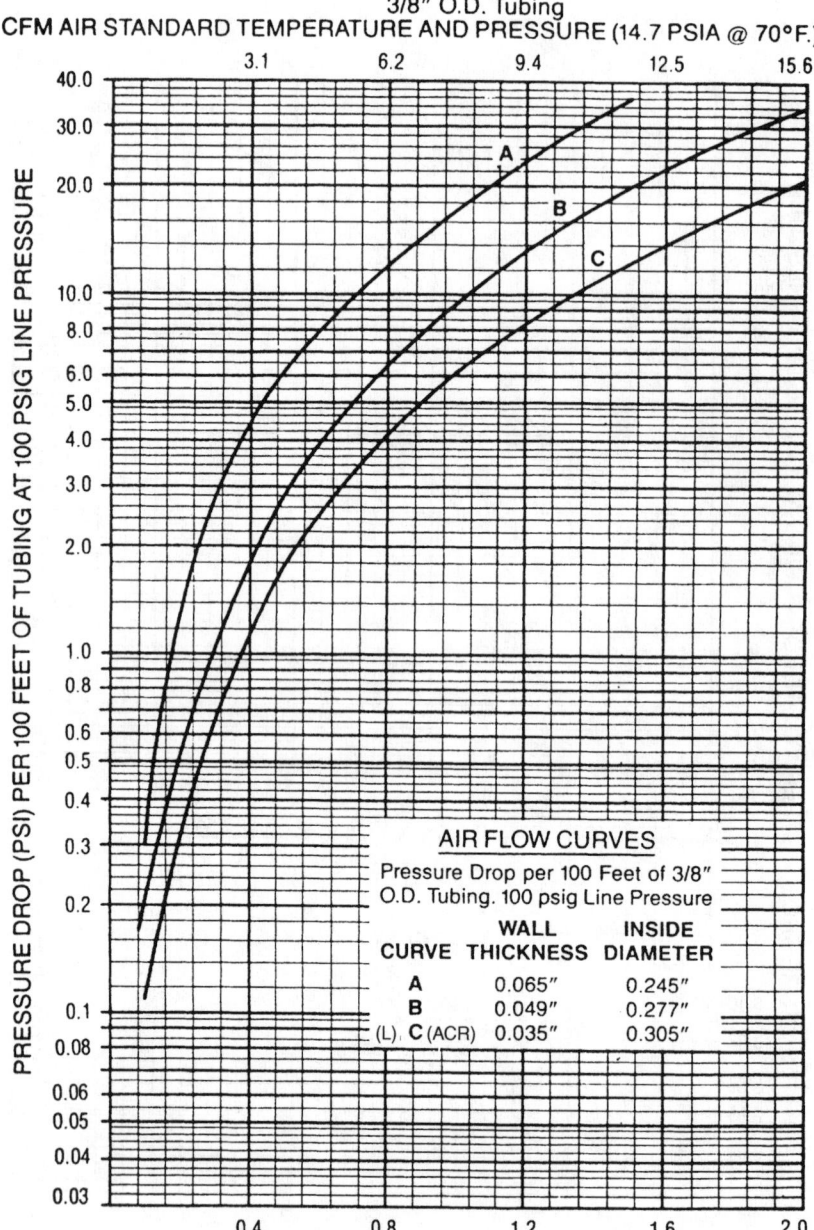

FIGURE 14.32 Compressed air sizing chart ⅜ in (10 mm) copper tubing, 100 psig line pressure.

Discussion:

1. The diversity (or simultaneous use) factor, which determines the maximum number of outlets in use at any one time, has a major influence in the sizing of the piping system. It has no exact method of being determined and is arrived at purely by judgment. Table 14.14 has been developed for general laboratory use and is based on past experience. A direct reading chart for compressed air systems is provided in Fig. 14.29.

2. The sizing chart, Table 14.28, has been calculated specifically for use with compressed air, type L copper pipe, and a pressure of 55 psig. In order to use this chart for other gases, pipe materials, and pressures, the following conversion calculations will be necessary:

   a. When any gas with a specific gravity other than air (1.00) is used, an adjustment to the scfm flow rate will be required. Equation (14.5) shall be used to calculate a factor that will convert scfm in Table 14.28 from compressed air to the equivalent of any other gas or combination of gases. Multiply the calculated factor f by the compressed air flow rate to obtain the new flow rate for the gas in question:

$$f = \sqrt{\frac{1}{g}} \qquad (14.5)$$

   To calculate the specific gravity of any gas, divide the molecular weight of that gas by 29, which is the composite molecular weight of air.

   b. For flow of any compressed gas in steel pipe, use Table 14.16 and Eq. (14.5) to adjust for the type of gas.

   c. For pressures other than 55 psig, use the following formula:

$$PD_a = \frac{P1 + 14.7}{P2 + 14.7} \times PD_r \qquad (14.6)$$

   d. For flow of any compressed gas at temperatures other than 60°F, use the following formula to calculate a factor that, when multiplied by the flow rate, will give the flow rate at the new temperature:

$$f = \frac{460 + t}{520} \qquad (14.7)$$

where $P1$ = 55 (referenced table pressure, psig)
$P2$ = actual service pressure, psig
PD_r = reference pressure drop found in chart for flow rate in table, psi/100 ft
PD_a = adjusted pressure drop for actual pressure, psi/100 ft
g = specific gravity of gas
t = temperature under consideration, °F
f = factor

Another method, applicable only to branch lines with smaller numbers of laboratory outlets used for average purposes, is to use a prepared chart based on the number of outlets with the actual flow of gas not considered. The flow rate and diversity of use is taken into consideration in the sizing chart and assumes that sufficient system pressure is available. When there is a small number of outlets on a branch, this method provides a sufficient degree of accuracy and speed of calculation. Table 14.29 is such a chart for various systems found in a laboratory.

Tests

The bulk storage tanks and dewers are required to be ASME rated and therefore are tested at the factory before shipment. They are not tested after installation. Cylinders are not tested for the same reason. This requires that only the distribution system, from the cylinder valve to the outlets, be subject to pressure tests.

Testing is done by pressurizing the system to the test pressure with an inert, oil-free and dry gas. Nitrogen is often used because of its low cost and availability. For systems with a working pressure up to 200 psig, the entire piping system, including the cylinder manifold, is tested to 300 psig for 1 h with no leakage permitted. If a working pressure higher than 200 psig is required, the system is tested at 150 percent of the system pressure. This pressure testing should be done in increments of 100 psig, starting with 100 psig. This is done to avoid damage due to a catastrophic failure. Leaks are repaired after each increment. After final testing, it is recommended that the piping be left pressurized at system working pressure with system gas if practical.

Flushing, Testing, and Purging the Distribution System

After the system is completely installed and before it is placed in service, the piping system must first be flushed to remove all loose debris, then tested, and finally purged with the intended system gas to assure purity.

TABLE 14.29   Typical Laboratory Branch Sizing Chart

No. of conn.	Cold water, hot water	Air	Gas	Vac.	Oxygen	D.W.	Nitrogen
				Pipe diameter, in			
1	$1/2$	$1/2$	$1/2$	$1/2$	$1/2$	$1/2$	$1/2$
2	$3/4$	$1/2$	$1/2$	$1/2$	$1/2$	$1/2$	$1/2$
3	$3/4$	$1/2$	$1/2$	$3/4$	$1/2$	$1/2$	$1/2$
4	$3/4$	$1/2$	$1/2$	$3/4$	$1/2$	$1/2$	$1/2$
5	$3/4$	$1/2$	$3/4$	$3/4$	$1/2$	$3/4$	$1/2$
6	$3/4$	$1/2$	$3/4$	1	$1/2$	$3/4$	$1/2$
7	1	$1/2$	$3/4$	1	$1/2$	$3/4$	$1/2$
8	1	$1/2$	$3/4$	1	$1/2$	1	$1/2$
9	1	$1/2$	$3/4$	1	$1/2$	1	$1/2$
10	1	$1/2$	$3/4$	1	$1/2$	1	$1/2$
11–20	$1\frac{1}{4}$	$3/4$	1	$1\frac{1}{4}$	$3/4$	1	$3/4$
21 and over	$1\frac{1}{2}$	1	$1\frac{1}{4}$	$1\frac{1}{2}$	1 (21–30) $1\frac{1}{4}$ (31–50) $1\frac{1}{2}$ (over 50)	1	1

1 in = 25.4 mm.

An accepted flushing method is to flow two to five times the volume of the branch or main through each respective part of the system. This is done by connecting the flushing gas under pressure to the piping system and then opening and closing all outlets and valves starting from the closest and working to the most remote.

Tests of the gas at the farthest outlet shall be taken to assure that the gas is the desired purity. This test could be done either by the end user if he or she has the necessary instruments or by an acceptable testing service with the results given to the client to verify that the gas is acceptably pure. A test for particulates is to have the gas flow at a minimum rate of 100 Lpm into a clean white cloth and observe for contamination.

Finally, the system must be capable of providing the desired purity when actually placed in operation. Since flushing and testing leaves the piping system filled with those inert or other gases, they must be removed, or purged. This is accomplished by allowing the system gas to flow through all parts of the piping system, opening all of the valves, and testing the gas purity at various points of the system until the desired purity level is reached.

For high-purity gases, a laboratory specializing in testing for the purity level required shall be used unless the facility is capable of performing the test.

GASES FOR HEALTH CARE FACILITIES

GENERAL

This section will describe centrally distributed oxygen, nitrous oxide, nitrogen, and compressed air systems used specifically in health care facilities. They are used variously for direct patient care, life support, as an anesthetic, or to power medical instruments. These compressed gas systems are characterized by gas purity, pipeline cleanliness, and total system reliability.

The definition of a health care facility includes hospitals, nursing homes, medical and dental offices, clinics, and ambulatory care centers. These facilities fall into two general categories: short-term acute care and long-term care. Short term is considered the typical acute care surgical-medical hospital. Long-term care includes specialty care and nursing home facilities that do not have direct surgical and other specialized capabilities normally associated with hospitals.

The overriding system concept is reliability. The central supply source and piping system shall be designed so that it shall not fail to provide the minimum amount of gas required by the facility no matter how high the demand may be during any condition, even those requiring a much higher than expected usage or experiencing equipment component failure.

The systems consist of a central gas storage or source, the distribution system, station outlets, and alarms.

CODES AND STANDARDS

Building codes that govern building construction and the design and installation of mechanical systems do not have direct provisions for regulating the design of medical gas systems. Such regulation is done by reference to standards that have been so widely accepted throughout the industry and by the various authorities that they have the force of law. These standards are required to be observed for the design, specification, storage, delivery, installation, and testing of the various medical gases. Among them are:

1. NFPA

   a. NFPA-99. Health Care Facilities. This has now become the standard for pressurized medical gases in the United States and some countries throughout the world.

   b. NFPA-50. Bulk Oxygen Systems at Consumer Sites.

2. CGA

   a. C-9.0 Standards for Color Marking of Compressed Gas Cylinders Intended for Medical Use.

   b. G-7.0 Compressed Air for Human Respiration.

   c. DISS (Diameter Index Safety System) for patient and service outlet connections.

3. *JCAHO Accreditation Manual for Hospitals,* 1994. Conformance with the requirements of this manual is mandatory for accreditation by the JCAHO. Accreditation has become necessary for Medicare and Medicaid reimbursement and other licensing requirements. Accreditation is not desired or obtained by all hospitals. The JCAHO manual also refers to other standards.

4. AIA (American Institute of Architects). *Guidelines for Construction and Equipment of Hospital and Medical Facilities,* 1992–1993.

5. CSA (Canadian Standards Association). Since many medical devices and pieces of equipment are sold in Canada, manufacturers commonly conform to the CSA standards when more stringent than U.S. standards.

DESCRIPTION AND GENERAL USES FOR THE COMMON GASES

Oxygen

Oxygen is a colorless and tasteless gas and is one of the most widely used of the medical gases. Its primary use, either undiluted or as part of a mixture, is for respiratory, inhalation, and anesthesia purposes. Although nonflammable, oxygen supports combustion and is therefore considered an oxidant. Oxygen is normally distributed at a pressure of between 50 and 55 psig (340 and 375 kPa).

NFPA distinguishes between small- and large-size systems. Storage of less than 3000 ft³ of oxygen is called a *class I system.* Storage of 3000 ft³ or more is called a *class II system.* Other standards (not NFPA) reference oxygen stored as a gas as *type I,* and when stored as a liquid as *type II.*

Nitrous Oxide

Nitrous oxide is a colorless and tasteless gas that produces a loss of sensitivity to pain when inhaled. Its primary use is in surgical and dental suites as an anesthetic and in far lesser amounts for other specialized applications. Nitrous oxide is nontoxic and nonflammable. It supports combustion to a lesser extent than oxygen but is considered an oxidant. Nitrous oxide is generally stored as a cryogenic liquid in cylinders and normally distributed at a pressure of between 50 and 55 psig (340 and 375 kPa).

Nitrogen

Nitrogen is an inert, colorless, and tasteless gas used primarily to power pneumatic tools in surgical suites. It is also used for inhalation therapy where mixed gases closely matching natural air proportions are desired and rarely for decontamination purposes. When used to power pneumatic tools, nitrogen is distributed at pressures up to 250 psig (1725 kPa) depending on the specific tools used. It is not unusual to have two separate pressure systems in facilities. Some pneumatic power tools require a minimum pressure of 200 psig (1360 kPa) at the tool for proper operation, for which a pressure of 250 psig (1725 kPa) be available at the regulator outlet. Often, a dedicated nitrogen cylinder supply is installed at each location with the required controls and regulator. These tools will operate properly at lower pressures but with reduced efficiency.

Although not directly concerned with health care, nitrogen is also used to test and blow out medical gas piping systems and as a purge to provide an inert atmosphere when brazing copper pipe joints.

Compressed Air

Compressed air is a colorless and tasteless gas. It is separated into three categories depending on specific use with a health care facility: low-pressure surgical-medical compressed air, high-pressure compressed air used to drive dental and other pneumatic tools, and low-pressure compressed air used for other purposes such as laboratories. It is a requirement of NFPA-99 that the compressor used as a source for the surgical-medical air system be a dedicated one and not used for any other purpose.

Medical compressed air is distributed at a pressure of between 50 and 55 psig (345 and 375 kPa) and is used primarily for patient inhalation, anesthesia purposes and to power respirators and ventilators. High-pressure compressed air is distributed at a pressure generally in the range of 160 to 250 psig (1190 to 1725 kPa) and is used to power pneumatic tools. Compressed air for laboratory use is distributed at a pressure of between 50 and 55 psig (340 and 375 kPa) and is used for a variety of general purposes where purified air is not required. For purified laboratory air service, refer to discussions of laboratory compressed air earlier in this chapter.

STORAGE AND GENERATION OF GASES

Oxygen

Bulk oxygen is typically stored as a cryogenic liquid in tanks and as a compressed gas in high-pressure cylinders. Bulk supplies are defined as an assembly of containers, pressure regulators, safety devices, vaporizers, manifolds, and piping that has a storage capacity of 20,000 ft^3 (566,000 L) of gas including unconnected reserves on hand at the same location. The purity of the oxygen must conform to Oxygen USP requirements.

The capacity of commonly used high-pressure cylinders is given in Table 14.30. Typical bulk containers have capacities listed in Table 14.31.

For all but the smallest uses of oxygen, it is recommended that a bulk supply be installed. A generally accepted range of 400 to 600 ft^3 of gas/month/bed (depending on the type of hospital and other specialized treatment activity) is used to find a typical month's usage. If the hospital is in a large city with a large percentage of trauma and surgical patients, or if the hospital is a specialized type that has a large percentage of patients that may require inhalation assistance, use the larger figure. For "average" conditions, use the smaller figure. Consultation with the facility staff regarding past usage and with a potential supplier, along with some judgment, will be necessary to confirm a final decision. The size of the bulk supply tank should allow a minimum of 7 to 10 days between fill cycles, with a longer period if practical.

Because oxygen supports combustion, outdoor storage tanks and cylinders shall be separated from other gases and structures in the event of a fire. Refer to Fig. 14.33 for cylinder and bulk oxygen separation criteria. Indoor storage requires an

TABLE 14.30   Storage Capacity of Typical Medical Gas Cylinders

Type of gas	Standard cubic feet per cylinder H size	G size	Storage mode
Oxygen	244	187	Compressed gas at 2200 psig (15,170 kPa)
Nitrous oxide	558	488	Liquid in cylinder
Air	232	178	Compressed gas at 2200 psig (15,170 kPa)
Carbon dioxide	558	436	Liquid in cylinder
Helium	213	141	Compressed gas at 2200 psig (15,170 kPa)
Argon	244	187	
Xenon	88.8	—	
Nitrogen	226	—	Compressed gas at 2200 psig (15,170 kPa)

TABLE 14.31   Storage Capacity of Typical Bulk Liquid Oxygen Tanks

Gross capacity, gal (L)	Net liquid capacity, gal (L)	Capacity oxygen, scf (10^6 L)	Approximate weight empty vessel, lb (kg)	Approximate weight—vessel loaded with oxygen, lb (kg)
330 (1,249.1)	314 (1,188.5)	36,200 (1.03)	4,000 (1,816)	7,000 (3,178)
575 (2,176.4)	535 (2,025)	61,500 (1.74)	5,800 (2,633.2)	10,900 (4,948.6)
975 (3,690.4)	920 (3,482.2)	105,700 (2.99)	9,300 (4,222.2)	18,100 (8,217.4)
1,625 (6,150.6)	1,533 (5,802.4)	176,100 (4.99)	10,400 (4,721.6)	25,000 (11,350)
3,400 (1,286.9)	3,250 (12,301.3)	374,000 (10.59)	18,500 (8,399)	49,400 (22,427.6)
6.075 (22,993.9)	5,935 (2,246.4)	684,000 (19.37)	27,000 (12,258)	83,500 (37,909)
9,200 (34,822)	8,766 (33,179.3)	1,009,000 (28.57)	34,000 (15,436)	117,500 (53,345)
11,000 (41,635)	10,500 (39,742.5)	1,215,000 (34.41)	40,000 (18,160)	139,750 (63,446.5)

enclosure with a minimum fire rating of 1 h around the cylinders, although a 2-h separation is required by many authorities. Storage of inert gases only are permitted inside this enclosure. A typical schematic diagram of a bulk oxygen supply is illustrated in Fig. 14.33. A typical schematic diagram of a cylinder oxygen supply is illustrated in Fig. 14.34. Liquid oxygen stored in tanks generally requires a vaporizer to convert the liquid to a gas.

Nitrous Oxide

Nitrous oxide is typically stored as a cryogenic liquid in cylinders, with multiple cylinders arranged in banks. The capacity of commonly used storage cylinders is

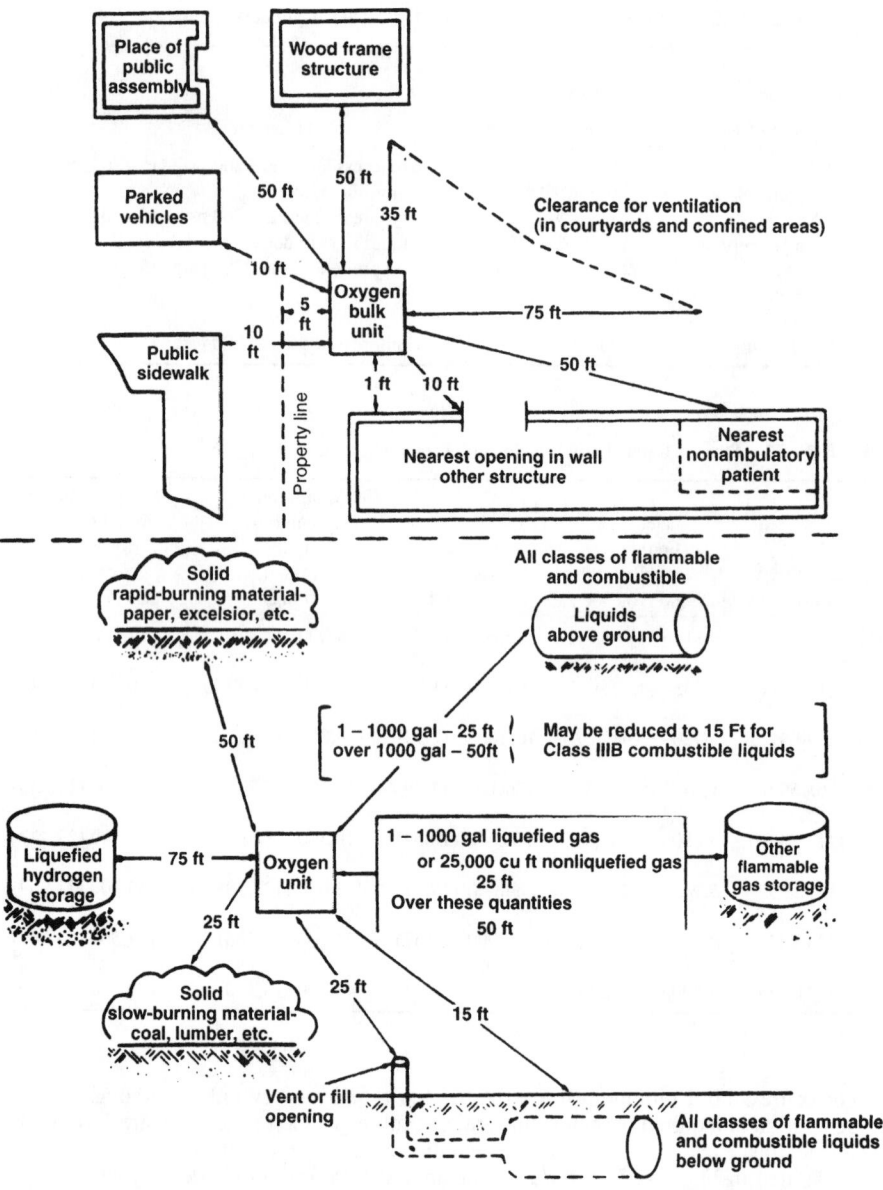

FIGURE 14.33 Separation of bulk oxygen supply.

given in Table 14.32. Vapor pressures for cryogenic storage are given in Fig. 14.35. For reference, the vapor pressure of carbon dioxide, which is also stored as a cryogenic liquid, is included.

A bulk supply of nitrous oxide is defined as an assembly of containers, pressure regulators, safety devices, vaporizers, manifolds, and piping that has a storage ca-

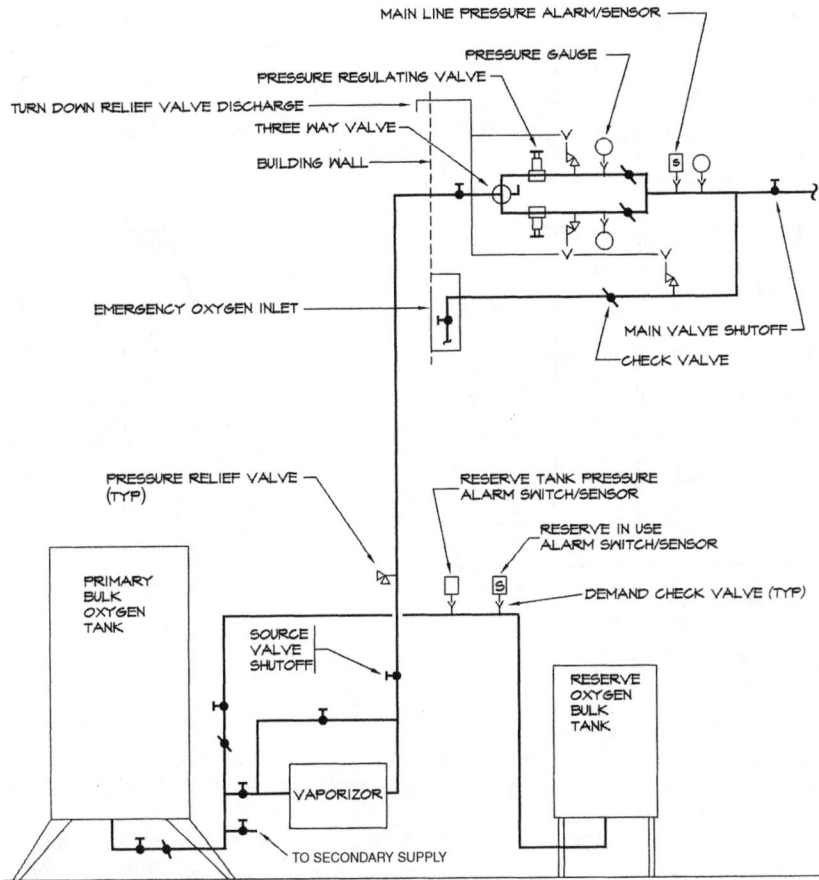

FIGURE 14.34 Schematic detail of bulk oxygen supply on site

TABLE 14.32 Sizing Chart for Nitrous Oxide Cylinder Manifold
(Using 489 ft³; 13.85×10³ L cylinders)

	Duplex manifold size			
	Indoor		Outdoor	
Number of anesthetizing locations	Total cylinders	Cylinders per side	Total cylinders	Cylinders per side
4	4	2	4	2
8	6	3	10	5
10	8	4	12	6
12	10	5	14	7
16	12	6	20	10

Nitrous oxide

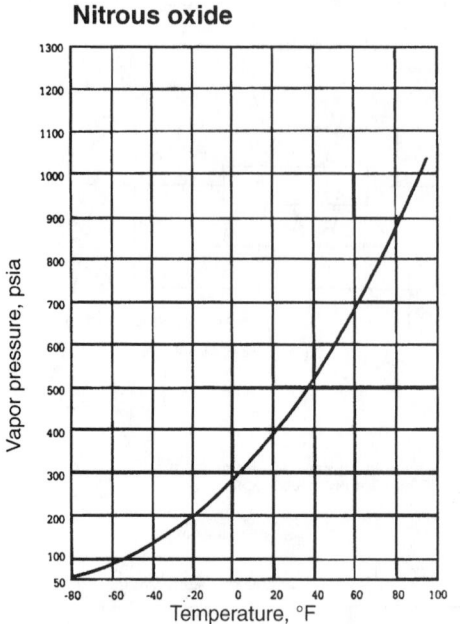

Carbon dioxide

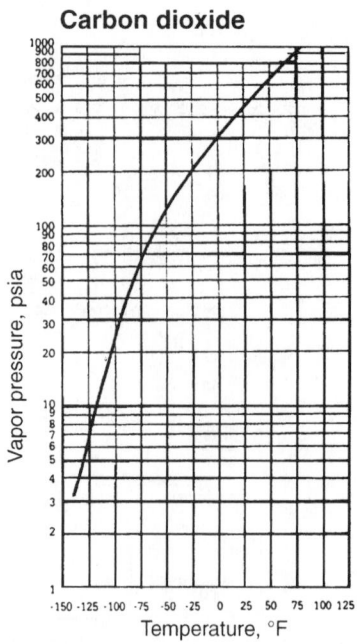

FIGURE 14.35   Vapor pressure versus temperature for nitrous oxide and carbon dioxide.

pacity of 28,000 ft³ (792,000 L) of gas including unconnected reserves on hand at the same location. A bulk supply is usually considered when the use of gas exceeds about 175,000 ft³ per year.

For a suggested starting point to determine the number of primary cylinders for a facility, use Table 14.34 consulting with facility staff to establish past or antici-pated usage. Another generally accepted criterion for the primary supply for smaller facilities is to use one cylinder for each major operating and delivery room or similar location. The same number of cylinders in the primary supply shall be provided as a reserve bank. A reasonable cycle for changing banks should be about 1 week.

It is recommended that volumes of nitrous oxide over 28,000 ft³ (800,000 L) should be located outdoors and above ground. If installed inside a building, ade-quate ventilation and monitors for low oxygen levels should be provided. Precau-tions are necessary if the nitrous oxide storage area is located outdoors. Since nitrous oxide is stored as a liquid, it depends on the heat received from ambient air to vaporize the liquid to a gas. If very high usage is coupled with low outdoor air temperatures, there is a possibility that the liquid in the storage cylinder will not be vaporized fast enough and the supply pressure will fall to a point where adequate pressure would not be maintained in the distribution system. In this case, a vaporizer or some other method of heating the storage location will be required. The potential supplier of the gas should be consulted regarding past experience in the geographical area where the facility is located.

Cylinder and bulk storage location and separation criteria, as well as the typical storage schematic diagram of the gas supply, are the same as those for oxygen.

Dental offices use nitrous oxide usually at the rate of 1 cylinder for each 50 patients. Ten percent of patients use this gas. Usually the flow rate is 3 L/M, with additional oxygen used at a rate of 6 L/min simultaneously.

Nitrogen

Nitrogen is typically stored as a compressed gas in high-pressure cylinders, with an equal number of cylinders (primary and secondary supply) arranged in banks. The capacity of commonly used high-pressure cylinders is given in Table 14.30. A typical nitrogen supply is schematically illustrated in Fig. 14.31.

There is no general consensus of opinion as to the quantity of nitrogen that might be used over an extended period of time in a typical facility because of the constantly changing requirements of tools using nitrogen, the desire of medical staff to use specific instruments, and the degree of use for inhalation therapy, if any. The largest flow rate (generally between 6 and 15 scfm) will be used in facilities that do orthopedic, thoracic, and neurosurgery procedures. Recommended starting point for these facilities would be 1 cylinder as a primary supply for each room where nitrogen is used. If the expected use is not great, three-fourths cylinder per room is acceptable. An equal number of cylinders shall be provided as a secondary bank. The actual quantity should be decided upon after consultation with facility staff to establish past and anticipated usage. Another consideration is to allow a minimum of 1 week, and up to a 10-day cycle for changing cylinder banks based on the anticipated volume of usage.

The grade of nitrogen used for human respiration must comply with National Formulary (NF) specifications. When used for testing, purging, and for pneumatic tool operation, the required grade is referred to as *oil-free, dry nitrogen.*

Compressed Air

Compressed air can be centrally supplied from a high-pressure cylinder manifold system when little use is expected or from air compressors manufactured and installed expressly for health care facilities. The quality of the air used for medical purposes is referenced to United States Pharmacopeia specifications (USP) and shall conform to Quality Verification level (Grade) N, as described in ANSI/CGA G-7, "Commodity Specification for Compressed Air," Table 1, Directory of Limiting Characteristics. Because of the necessity that compressed air for inhalation meet USP requirements, all cylinders must be properly labeled, and the air itself is considered a drug. Air that has been reconstituted from oxygen USP and nitrogen NF with a major portion of other elements eliminated is also available and is called *synthetic air.*

The cylinder manifold system arrangement is similar to installations of other compressed gases. The capacity of commonly used high-pressure cylinders is given in Table 14.30. For central supply installations, the most common and economical method of producing compressed air is by means of an air compressor assembly.

Flammable Gas Storage

Flammable medical and laboratory gases shall be stored in interior rooms or enclosures separate from other gases, or outside of the building. Rooms shall be properly vented and have a fire-resistive rating of 1 h. If local authorities or facility laboratory safety personnel agree that a hazard does not exist, this requirement can be waived.

THE SURGICAL-MEDICAL AIR COMPRESSOR ASSEMBLY

Intakes

The compressor intake must be located in an area that will not contaminate the incoming air with the exhaust from any internal combustion engines, HVAC exhausts, room air, any strong localized odors, or any other undesirable contaminants. Accepted practice is to have the air that is taken into the compressor as clean as the general atmo-sphere in the area of the facility (not the room the compressor is in). The intake may also use a different source of air if it is considered better than the "normal" outside air obtained at the building site, such as that from the HVAC filtered supply used for operating rooms. The only stipulation is that it must be available at all times. The intake line or header can be combined for multiple compressors, but the piping and valve arrangement must be capable of completely isolating one of the compressors while the other(s) continues with uninterrupted operation. Generally accepted practice is to design the size of the intake to have a maximum pressure loss of 4 in WC for the entire line, including filters and inlet louvers. The velocity of the intake air must be low enough to prevent the entrance of rain into the louver.

Air Compressors

It has been found that lubricating oil, when decomposed by a malfunctioning compressor, will introduce odors and carbon monoxide into the air supply. The medical air compressor shall be designed to prevent the introduction of contaminants or liquid into the pipeline by either: (a) the elimination of oil anywhere in the compressor or (b) by separation of the oil-containing section from the compression chamber by an area open to the atmosphere, which allows continuous visual inspection of the interconnecting shaft. Examples of (a) are liquid ring, rotary vane, and permanently sealed bearing compressors and of (b) extended head compressors.

With few exceptions, the liquid ring, rotary vane, and sealed bearing reciprocal compressors are the compressors of choice. The little used diaphragm compressor is considered oil free because the compression chamber is separated from the lubricated portion of the compressor by the diaphragm.

Pumps in the (a) category shall have separate sensors that will shut down the pump and activate an alarm if the temperature of the liquid used for sealing becomes excessive. If permanently sealed bearing compressors are used, a high air temperature monitor, with a setting recommended by the manufacturer, shall be provided at the immediate outlet of each cylinder that will shut down the compressor and activate the local and master alarm. If this compressor has water-cooled heads, a high water level in the receiver shall shut down the compressor and activate the local and master alarm. For pumps in the (b) category, the same sensors are required. In addition, a coalescing filter and a charcoal filter, each provided with a color change indicator for hydrocarbons, shall be provided.

With these restrictions, the engineer is free to select any compressor suitable for the flow rate and pressure required for the facility under design. The most often used compressors are rotary vane and oil-free reciprocating compressors.

Each compressor in a duplex arrangement shall have 100 percent capacity, and two of three compressors in a triplex installation shall have 100 percent system capacity. For duplex systems, a lag compressor-in-use alarm shall be installed in the compressor control cabinet. It shall be both visible and audible. This alarm does not have to be repeated in the master alarm panel.

The interconnecting piping between the compressor(s) and devices must be non-corrosive, such as brass or copper. Black and galvanized steel pipe is expressly prohibited.

Automatic alternation of units is required to evenly wear all compressors. However, if larger units are provided with an automatic means to activate additional units to maintain system pressure, manual lead/lag alternation is permitted. Experience has shown that in a significant number of installations, an equal-wear syndrome may exist that will have both units of a duplex installation break down at the same time because of equal wear provided by automatic alternation. Manual alternators have been shown to be preferable in many cases. Operating personnel of the facility should be consulted prior to selection of the alternation method.

It is a requirement that a main shutoff valve be placed after the entire compressor assembly in the immediate vicinity to isolate the building piping system from the source. It must be properly labeled.

Air Dryers

Water vapor is normally removed by a refrigerated air dryer to produce a system dew point of approximately 36°F (2°C). If a lower dew point is required, a desiccant dryer is recommended. Air dryers shall be duplexed, and if an aftercooler is necessary, it too shall be duplexed and provided with a separate condensate trap. A dew point alarm must be provided to signal a dew point reading of higher than 39°F (4°C).

Receivers

A receiver is required for a medical-surgical air compressor assembly. A receiver evens out any momentary variations in air pressure, provides an additional method of eliminating large drops of water in the airstream and provides a reserve capacity that allows compressors to shut down for a time rather than run continuously.

An automatic receiver drain and a sight glass are required on a receiver. The sight glass will provide a method of visually observing the liquid level in the receiver and checking on the operation of the automatic receiver drain. A receiver is not permitted to become a water separator. The receiver shall also be equipped with a safety relief valve and pressure gauge and shall comply with the ASME code for unfired pressure vessels.

Filters and Other Devices

In order to achieve the low contaminant level required by CGA G-7, it is required that the proper in-line filters and other devices be provided to remove them. These filters and devices are selected and sized on the basis of the degree of contaminants in the intake air supply and the desired level of removal. Charcoal filters remove odor. Coalescing filters remove both oil and water vapor. Carbon monoxide is removed if necessary by a catalytic converter that changes it to carbon dioxide, which is acceptable in larger amounts. A particulate filter is necessary to remove small particles released from the charcoal filter and desiccant dryer.

Each of these filters and devices shall be duplexed, with the piping and valve arrangement capable of completely isolating one of the devices while the other continues uninterrupted operation. If this is not economical or practical, each device must be capable of being bypassed and removed from the system without shutting down the entire system. All final line regulators and filters shall be duplexed.

Installation

All of the above components, along with the control panel and alarms, are commonly prepiped and prewired on a skid at the factory and tested. The skid is shipped to the facility and installed with a minimum number of connections and assembly at the site. A typical surgical-medical air compressor assembly is illustrated in Fig. 14.36.

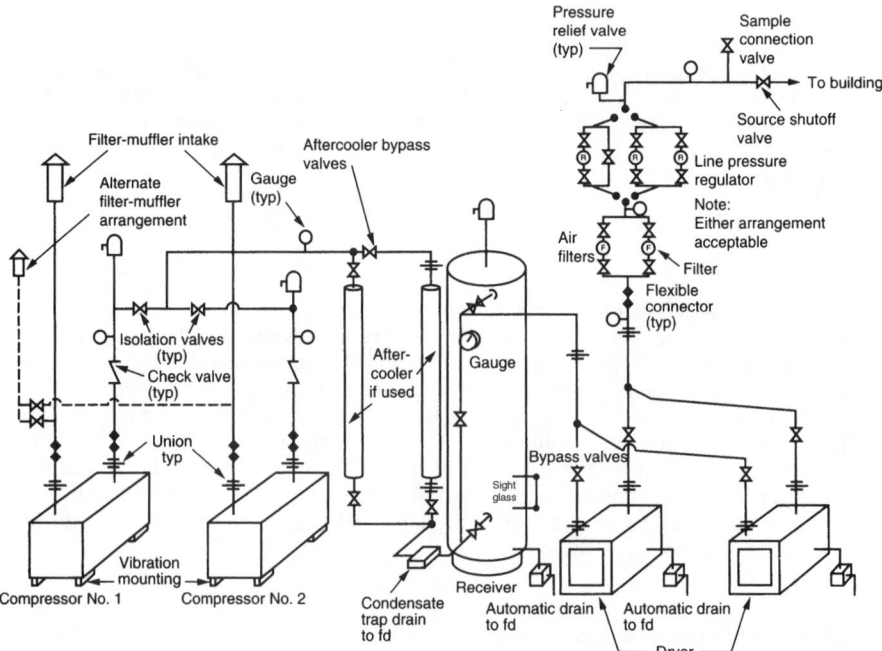

FIGURE 14.36 Typical surgical-medical air compressor assembly. *Note:* Flow schematics that differ may be acceptable as long as they meet the intent of NFPA-99. (*Adapted from NFPA-99.*)

DENTAL AIR COMPRESSOR ASSEMBLY

Any number of compressors can be used. The compressors may be used to power evacuation devices if the system exhaust is closed and is routed directly to the outside. Automatic activation of additional compressors for increased air requirements should be provided.

The intake location is similar, although less restrictive, than the medical compressor. Air intake shall be outside the building when practical, but it could be located in any room where no chemical-based material is stored or used. The system is required to have dry-type intake muffler and/or filters, a receiver, shutoff valves, air dryer(s), and final filter. An appropriate moisture indicator shall be provided downstream of the receiver and upstream of any system pressure regulators. If this compressor is to be used to supply respirable air or otherwise provide life support (such as for a respirator or anesthesia equipment), the compressor shall comply in all respects to the medical-surgical compressor assembly requirements.

The receiver requirements are similar to the medical system. The receiver shall have the necessary controls to assure practical on-off operation of the compressors.

A typical dental air compressor assembly for large facilities is illustrated in Fig. 14.37.

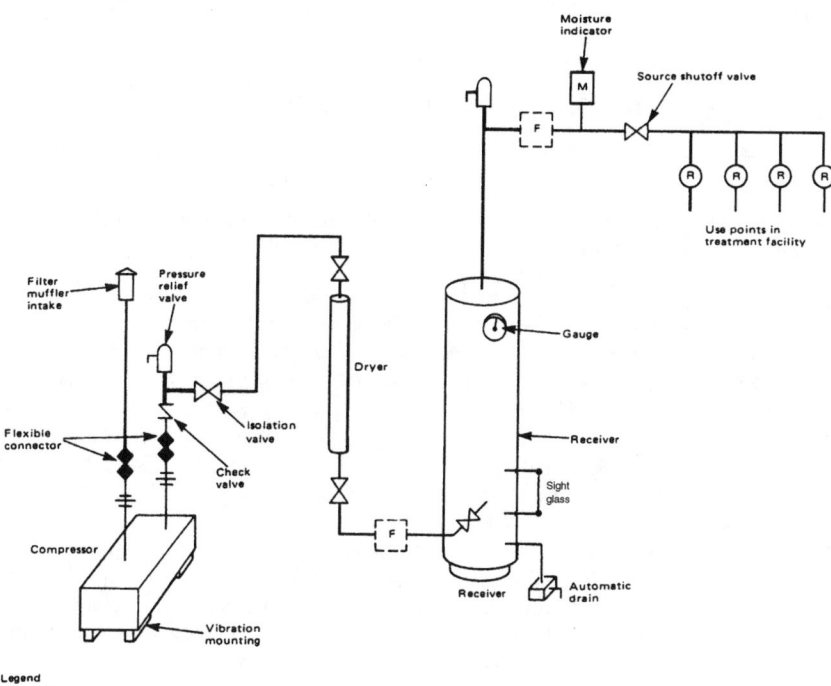

FIGURE 14.37   Typical dental air compressor. (*Adapted from NFPA-99.*)

Alarms

For duplex systems, a lag compressor-in-use alarm shall be installed in the compressor control cabinet. It shall be both visible and audible. This alarm does not have to be repeated in the master alarm panel.

THE DISTRIBUTION NETWORK

System Types

In NFPA-99, there is a reference to level 1, 2 or 3 systems for patient supply. The distinction between them is the amount of gas stored.

Level 1 and 2 systems store an amount larger than 3000 ft^3 (56 m^3) of nonflammable gas, except nitrogen, connected and in storage. Level 3 is a smaller system that does not have more than 3000 ft^3 (56 m^3) of nonflammable gas, except nitrogen, connected and in storage. This can be increased to 5000 ft^3 (143 m^3) if oxygen is used as a liquid in DOT cylinders. Some other restrictions for level 3 systems require that a listed pressure regulator be connected to each cylinder, that it will supply only a single facility or two single treatment facilities. In facilities serving nonhuman use such as veterinary medicine, some provisions need not be followed. In general, most provisions regarding Level 1 installations shall be adhered to.

PIPE MATERIALS, JOINTS, AND INSTALLATION

General

The following general requirements of NFPA-99 regarding the nonflammable piping systems shall be strictly complied with:

1. All piping must be cleaned for oxygen service prior to installation and plugged or capped. Pipe should be purchased in this condition. If piping is contaminated, it must be discarded. Valves, fittings, and other components shall be received cleaned and bagged from the manufacturer or cleaned in accordance with CGA G-4.1 before being installed.
2. Pipe used for pressures of up to 199 psig must be seamless, hard drawn, type K or L, ASTM B-819, and specially marked. When pressures are 200 to 300 psig, only type K shall be used. Other acceptable materials can be used in the manufacture of some medical gas equipment.
3. Joints shall be brazed, and only wrought copper and copper alloy fittings (ASTM B-16.22) are permitted. No flux is allowed. Cast fittings are specifically prohibited. Other jointing methods are permitted, such as memory-metal couplings or other approved joints that have the same properties as a brazed joint. Valves and fittings having flared or compression joints are specifically prohibited.
4. Flexible hoses and connectors shall not be concealed or penetrate the floor, walls, ceilings, or partitions and must have a minimum burst pressure of 1000 psig (7000 kPa).
5. Requirements for hangers and supports are very specific, with distances between supports delineated in tables based on types of service.

6. Extensive detail has been written in NFPA-99 regarding brazing methods, procedures, and qualification of mechanics.

Valves and Valve Locations

Valves provide the capability of shutting down various parts of a system for maintenance or emergencies with the least interruption of other areas or parts of the system as a whole. Valves shall be of the one-quarter-turn type (ball or butterfly) with an indicating handle. All valves shall be properly marked and located so that they are in full view at all times, not behind doors or other movable objects that may hide them. Valves that are placed in locations accessible to other than authorized personnel shall be installed in valve boxes with removable or frangible windows.

Valves intended to isolate existing systems for maintenance or from new systems are permitted. These control valves shall be properly labeled and either locked open or located in a secure area.

Anesthetizing locations and other critical life support areas such as critical and cardiac care and postanesthesia recovery areas shall be supplied directly from a riser or supply with no intervening valves except for zone valves. A zone valve for each individual gas shall be located immediately outside all of the above areas. These zone valves shall not control any other room or area.

VALVES

General

1. All valves shall be 3-piece, full size port bail valves, $\frac{1}{4}$ turn with extensions for brazing, indicating handle.
2. The valve shall be constructed with bronze or brass body.
3. Vacuum valves can be butterfly valves.
4. All valves shall be capable of being locked or latched open or closed as required.
5. Valves for nonflammable medical gas shall not be located in the same zone valve box as flammable gas.

Source Valve

A source valve shall be installed on the immediate outlet of source assembly to permit the entire source of supply, including all accessory devices, to be isolated from the piping system. The source valved shall be located in the immediate vicinity of the source equipment and shall be labeled "source valve for the source name."

Main Valve

The main supply line shall be provided with a main line shutoff valve to permit shut-down of the entire system without having to go to the source. It shall be located downstream of the source valve and outside of the source room, enclosure or where the main line first enters the building. The valve shall be located to permit access by authorized personnel only (similar to in a ceiling or behind a locked access door. The valve shall be labeled "main valve for the source name serving (name of the building)".

Please note that the main valve is not required where the source valve is accessible from within the building.

Riser Valve

Each riser supplied from the main line shall have a shutoff valve adjacent to the connection. They shall be accessible and shall not be obstructed. These valves shall be labeled "RISER FOR NAME OF MEDICAL GAS. SERVING THE FLOORS __ AND THE NAME OF THE AREA SERVED BY THIS PARTICULAR RISER."

Service Valves

A service shutoff valve shall be installed where the lateral branches off the riser prior to any zone valve assembly on that branch. Only one valve is required on each branch no matter how many zone valves are installed on that branch. The main purpose of these valves is to allow facility service personnel to make changes in the piping system on that floor without having to shut down the entire facility. These valves shall be installed in a locked chase or located in a secure area, locked open. They shall be labeled with a tag that says: "CAUTION: NAME OF MEDICAL GAS. DO NOT CLOSE EXCEPT IN CASE OF EMERGENCY. THIS VALVE CONTROLS SUPPLY TO __."

Zone Valves

Station inlets and outlets shall not be supplied directly from a riser unless a manual shutoff valve is located in the same story between the riser and the inlet/outlet. A wall must intervene between the riser valve and the inlet/outlet zone valve. This valve shall be readily operated from a standing position in the corridor on the same floor that it serves. Each lateral branch line serving patient rooms shall be provided with a zone valve that controls the flow of medical gas to the patient rooms, operating room or other anesthesia location. Zone valves shall be so arranged that shutting off the supply to medical gas will not affect the supply to the rest of the system. A pressure/vacuum gauge shall be provided on the patient side of each zone valve. These boxes shall not be located in closed rooms or closets or behind normally open doors, or otherwise hidden from view.

Zone valves shall be located immediately outside each vital life support or critical care area and each anesthesia location. They shall be provided for each medical gas/vacuum location so as to be readily accessible in cases of emergency. Valves shall be protected and marked "CAUTION: NAME OF MEDICAL GAS. DO NOT CLOSE EXCEPT IN CASE OF EMERGENCY. THIS VALVE CONTROLS SUPPLY TO __."

All medical columns, hose reels, ceiling tracks, control panels and other special installations shall be located on the patient side of the valve.

In-Line Valves

In-line valves are provided to isolate piping for maintenance or modification. They shall be located in a secure area, locked open and be identified "CAUTION: NAME OF MEDICAL GAS. DO NOT CLOSE EXCEPT IN CASE OF EMERGENCY. THIS VALVE CONTROLS SUPPLY TO __."

Shutoff Valves

Shutoff valves are provided for the connection of future piping. Pipe ends shall be closed with a brazed cap and tubing allowance that will allow future connections and re-brazing. Valves shall be locked closed and they shall be located in areas accessible only to authorized personnel.

Pressure Gauges

Pressure gauges shall be installed in the following locations mandated by NFPA-99:

1. Gauges shall be provided downstream of each lateral branch line shutoff valve.
2. Gauges shall be provided in the main line adjacent to the sensor unit in cryogenic storage reserve line.
3. Gauges shall be provided downstream of each room shutoff valve for each individual service.
4. It is recommended that pressure gauges be installed on either side of filter assemblies to visually determine pressure drop.
5. It is recommended that pressure gauges be installed on receivers used for compressed air.
6. For nitrogen systems only, it is common practice to provide an adjustable pressure-regulating valve with gauges in the valve box for this service to allow for the use of various instruments in the OR if this is a facility requirement.

Alarms

Because of the importance to life support provided by the various gas systems, alarms are critical in order to warn operating personnel of impending problems and to immediately alert them to problems or emergencies that do arise. Visual and audible warnings strategically located will provide this ability. All of the alarms are both audible and *noncancellable visual* unless indicated otherwise. This means that the visual portion of the alarm cannot be turned off at the alarm location but must be corrected at the source of trouble, and then reset. The audible alarm may be turned off because of the prolonged distraction that it could cause at a switchboard or nurses' station. The master alarm power source shall be connected to the life safety branch of the emergency electrical power system.

Alarms are divided into two categories, master alarms and area alarms. Master alarms monitor the operation and condition of the system supply or sources and respond to trouble that will affect the facility as a whole. They are required to be located in two places to guarantee continuous, 24-hour-a-day, responsible observation and also to provide the ability to immediately notify appropriate personnel that the alarm has tripped. Such locations are usually at the main telephone switchboard, security office, or the principal working area of the individual responsible for facility maintenance. If the alarms are connected to a computer for the building management system, the computer is not considered an acceptable panel. Area alarms provide warnings for anesthetizing and other critical care areas and are intended to be located at or near nurses' stations or in areas that are frequently used by hospital personnel. They must be installed in a location that is in continuous and unobstructed view.

Master alarm functions shall monitor the operation and condition of the source of supply, the reserve, if any, and pressure in the main line of each system. Central liquid or gaseous supply and/or reserve shall have the following alarm functions:

1. Central supply line pressure high: The pressure sensor shall be located on the pipe distribution side of the main line shutoff valve.

2. Central supply line pressure low: The pressure sensor shall be located on the pipe distribution side of the main line shutoff valve.

3. Liquid level of primary supply low, reorder: This is a normal condition and does not constitute an emergency.

4. Emergency reserve in use.

5. Emergency reserve line pressure low: This alarm is considered an extreme emergency condition. For a cryogenic source, it is initiated when the liquid level in the reserve is reduced to one-half of its full volume. If the source is a compressed gas, it is initiated when the pressure lowers to 500 psig.

6. For cryogenic storage used as a reserve, a reserve failure alarm shall be provided. In addition, an alarm shall be installed to indicate when the reserve capacity is reduced to 1 day's supply and if the gas pressure in the reserve unit falls below the set point established by the facility as a minimum necessary for the system to function properly.

7. When check valves are not installed in each cylinder supply of the reserve bank, an alarm should sound when the reserve supply is reduced to 1 day.

8. All alarms for the air compressor source, including the dew point monitor, shall be repeated in the master alarm panel.

9. For air compressors with water-cooled heads, a high water level in the receiver shall shut down the compressor and activate the master alarm and be repeated in the local compressor alarm panel.

Examples of area alarm functions are:

1. An alarm, activated by a pressure switch, shall be installed in each line supplying specific anesthetizing and critical care area branch lines to annunciate 5 percent increases or decreases from that considered as the normal operating pressure range.

2. Alarms for air compressors are required to be included as part of the local compressor panel installed on or near the compressor. For duplex systems, a lag compressor-in-use alarm shall be installed in the local compressor alarm panel. It need not be repeated in the master alarm panel. For air compressors with water-cooled heads, a high water level in the receiver shall shut down the compressor and the alarm repeated in the local compressor alarm panel.

TABLE 14.33 Summary of Medical Gas System Alarm Requirements

Source equipment	Change-over	Reserve in use	Reserve failure	Reserve low	Dew point high
Manifolds	Yes	NA	No	Note	No
Manifolds with reserve	Yes	Yes	No	Note	No
Cryogenic bulk gas units (VIE) with cryogenic reserve	Yes	Yes	Yes	Yes	No
Cryogenic bulk gas units (VIE) with cylinder reserve	Yes	Yes	No	Note	No
Air compressors	No	No	No	No	Yes
Vacuum pumps	No	No	No	No	No

Pipeline	High pressure or vacuum	Low pressure or vacuum
All pressure gas systems	Yes	Yes
Vacuum systems	No	Yes

This table has been added for the convenience of the user of the document.

Note: This signal is required only where cylinder reserves have no check valves for each cylinder lead.

Table 14.33 is a general summary of the master alarm signal requirements for all medical gases. Refer to the text of NFPA-99 for complete provisions.

Testing by Installer

All tests and flushing of compressed medical gas systems shall be conducted with oil-free, dry nitrogen. All of the following tests are mandated by NFPA-99. For a more detailed description of these tests, refer to that standard.

Pressure tests shall be conducted in two stages. The first is done at a pressure of 150 percent of system pressure, but not less than 150 psig (350 kPa) before closing in the walls and installation of alarms, manifold gauges, and pressure relief devices. The source shutoff valve shall be closed, and the entire system is visually observed and each joint manually tested with soapsuds, or other methods compatible with service gas, for leaks. Any leaks are repaired and the system is retested.

After the first-stage test is complete, the piping system is blown out to eliminate particulates. The final test is done to the entire system with all components installed. This is accomplished by disconnecting the test source and allowing the system to stand for 24 h with a pressure of 20 percent above system pressure. No leakage is allowed except to accommodate variations in ambient pressure. Any leaks are repaired, and the system is retested.

After retesting is required, the system is again flushed to remove all traces of debris. Have the full flow of the purge gas pass through a white cloth at the terminals until no trace of particulates is observed. Accepted practice is to use a volume of five times the volume of the pipe being flushed.

After pressure testing, a cross-connection test, to assure that no cross connections with any other gas or vacuum system has occurred, must be done. This requires that all other systems be reduced to atmospheric pressure and the system under test be pressurized to 50 psig (350 kPa). Each station outlet for all medical gas systems shall be operated in turn to assure that the test system gas is not being discharged from another outlet.

System Testing and Purging by Third Parties

After the tests conducted by the installer, a third party experienced in the field of medical gas pipeline testing shall verify system integrity with the following tests.

A cross-connection test, similar to the installer test, shall be conducted. In addition, with all medical-surgical systems in operation, each system installed shall be pressurized to different pressures and the outlets checked to assure that the proper pressure from each individual outlet is read on a gauge.

A valve test shall be performed to verify that the proper rooms or areas are controlled by the valve being tested.

A flow test of all station outlets shall be conducted to assure that each system outlet delivers required pressure and stays within required pressure drop during the test.

All alarm and warning systems shall be tested to assure the proper functioning of all of the separate alarms. This includes observation of pressure gauges in the system to verify that the alarm function is initiated at the desired reduction of pressure where applicable.

In order to ensure that all particulates are removed from inside the piping system, a purge flow of at least 225 sLpm (8 scfm) shall be delivered from each outlet, and the cleanliness shall be verified by filtering a minimum of 35 ft^3 of test gas through a clean, white filter. After this test, the system shall be purged of the test gas and the system gas introduced into the piping system.

The purity of each system gas shall be verified by testing for dew point, total hydrocarbons, and halogenated hydrocarbons. The maximum allowable variation shall be as described in NFPA-99.

An operational pressure test shall be performed for each system at all station outlets, maintaining the system pressure while the flow rate described in the reference standard is flowing.

A concentration test of all medical gas systems shall be conducted to verify that the minimum concentration of each gas is in accordance with the table appearing in the reference standard.

The air compressor discharge shall be tested for purity and to verify the minimum concentration of contaminants specified in the reference standard.

After all of the above have been completed, the systems are now ready for use. It is recommended that the systems remain pressurized after testing.

SYSTEM AND EQUIPMENT SIZING

Following are the procedures necessary to calculate the size of the various system supply sources and the individual piping systems. A discussion of each procedure follows.

1. Locate and count the number of station outlets for all of the individual systems, and separate them into various areas such as operating rooms, critical and cardiac care areas, and postoperating and postanesthesia recovery. This separation will allow a common flow rate and diversity factor for ease of calculations.

2. Determine or select the location of the supply source for the various systems.

3. Determine the operating pressure requirements for the distributing piping network and at the farthest use point for each of the various systems. With this

information known, determine the total allowable pressure loss desired for the entire system under design. The pressure loss is usually 5 psi.

4. Obtain the allowable flow rate for each of the various station outlets, and assign a diversity factor for each area of the facility to determine how many outlets would be used at the same time.

5. Using the above information, select the type, capacity, and physical size of the source equipment assembly. This will include the oxygen tank or manifold size, air compressor assembly, and nitrous oxide and nitrogen manifold assemblies.

6. Route the various piping systems from the source or storage area to all station outlets or use points. Measure the actual distance along the run of pipe to the farthest point of the system. To the actual run add an additional 50 percent of the measured distance as an allowance for valves, fittings, and so on. This is the developed length. Knowing the total allowable pressure drop for the system, calculate the allowable pressure loss through the length of piping required by the specific pipe sizing chart selected for the project. System design considerations are discussed later in this section.

7. Locate the valves and alarm panels. This shall include section, area, and room valves. This is necessary to give the facility planner enough information to provide space for their installation.

Number of Outlets

The first consideration is to locate and count all of the outlets, often called *station outlets,* for each respective medical gas system. This is usually done by consulting a "program" prepared by the facility planner. This program is a list of all rooms and areas in the facility and the services that are required in each room or area. If a program has not been prepared, the floor plans for the proposed facility shall be used.

There is no code that specifically mandates the exact number of station outlets that must be provided in various areas or rooms for all health care facilities. If fact, there is no clear consensus of opinion among medical authorities or design professionals as to how many station outlets are actually required in all the facility areas. Guidelines are published by the AIA, NFPA, and ASPE that recommend the minimum number of station outlets for various services in specific areas.

The recommendations most often used in determining the number of station outlets for hospitals is the desire to be accredited by the JCAHO. Accreditation is required for Medicare and Medicaid compensation. The JCAHO publishes a manual that refers to the AIA guidelines for the minimum number of station outlets for oxygen, compressed air, and vacuum that must be installed in order to obtain accreditation. If this is a factor for the facility, these requirements are mandatory. Other jurisdictions, such as state or local authorities, may require plans to be approved. These approvals may require adhering to the state or local requirements.

If accreditation or approval of authorities is not a factor, the number and area locations of station outlets are optional. The actual count will then depend upon requirements determined by each individual facility or another member of the design team using both past experience and anticipated future use, often using the guideline recommendations as a starting point. Table 14.34 provides the AIA recommendations. Table 14.35 provides generally recommended guidelines for the minimum number of facility station outlets that are a compilation from other

TABLE 14.34a AIA Guidelines for Vacuum and Air Systems for Hospital Facilities Station Outlets for Oxygen, Vacuum (Suction), and Medical Air Systems in Hospitals[1]

Location	Oxygen	Vacuum	Medical Air
Patient rooms (medical and surgical)	1/bed	1/bed	—
Examination/treatment (medical, surgical, and postpartum care)	1/room	1/room	—
Isolation—infectious and protective (medical and surgical)	1/bed	1/bed	—
Security room (medical, surgical, and postpartum)	1/bed	1/bed	—
Critical care (general)	3/bed	3/bed	1/bed
Isolation (critical)	3/bed	3/bed	1/bed
Coronary critical care	3/bed	2/bed	1/bed
Pediatric critical care	3/bed	3/bed	1/bed
Newborn intensive care	3/bassinet	3/bassinet	3/bassinet
Newborn nursery (full-term)	1/4 bassinets[2]	1/4 bassinets[2]	1/4 bassinets[2]
Pediatric and adolescent	1/bed	1/bed	1/bed
Pediatric nursery	1/bassinet	1/bassinet	1/bassinet
Psychiatric patient rooms	—	—	—
Seclusion treatment room	—	—	—
General operating room	2/room	3/room	—
Cardio, ortho, neurological	2/room	3/room	—
Orthopedic surgery	2/room	3/room	—
Surgical cysto and endo	1/room	3/room	—
Post-anesthesia care unit	1/bed	3/bed	1/bed
Anesthesia workroom	1 per workstation	—	1 per workstation
Phase II recovery[3]	1/bed	3/bed	—
Postpartum bedroom	1/bed	1/bed	—
Cesarean/delivery room	2/room	3/room	1/room
Infant resuscitation station[4]	1/bassinet	1/bassinet	1/bassinet
Labor room	1/room	1/room	1/room
OB recovery room	1/bed	3/bed	1/room
Labor/delivery/recovery (LDR)[5]	2/bed	2/bed	—
Labor/delivery/recovery/ postpartum (LDRP)[5]	2/bed	2/bed	—
Initial emergency management	1/bed	1/bed	—
Triage area (definitive emergency care)	1/station	1/station	—
Definitive emergency care exam/ treatment rooms	1/bed	1/bed	1/bed
Definitive emergency care holding area	1/bed	1/bed	—
Trauma/cardiac room(s)	2/bed	3/bed	1/bed
Orthopedic and cast room	1/room	1/room	—
Cardiac catheterization lab	2/bed	2/bed	2/bed
Autopsy room	—	1 per workstation	1 per workstation

[1] For any area or room not described above, the facility clinical staff shall determine outlet requirements after consultation with the authority having jurisdiction.

[2] Four bassinets may share one outlet that is accessible to each bassinet.

[3] If Phase II recovery ara is a separate area from the PACU, only one vacuum per bed or station shall be required.

[4] When infant resuscitation takes place in a room such as cesarean section/delivery or LDRP, then the infant resuscitation services must be provided in that room in addition to the minimum service required for the mother.

[5] Two outlets for mother and two for one bassinet.

TABLE 14.34b Station Outlets for Oxygen, Vacuum, and Medical Air in Outpatient Facilities

Location	Oxygen	Vacuum	Medical air
Examination	0	0	—
Treatment	0	0	—
Isolation	0[1]	0[1]	—
Pre-procedure examination	0[1]	0[1]	—
Operating room		1	
Class A—minor surgical procedure room	1		—
		2	
Class B—intermediate surgical procedure room	2		—
		3	
Class C—major surgical procedure room	2		—
		1	
Post-anesthesia recovery	1	0[1]	—
Step-down recovery area	0[1]	3	—
Cysto procedure	1		—
Emergency		1	
Trauma/cardiac room	1	0[1]	1
Cast room	0[1]	2	—
Catherization room	1	2	2
Birthing room			—
Endoscopy	2	3	
Procedure room		—	—
Decontamination room	2	0[1]	—
Holding/prep/recovery area	—		—
	0[1]		

[1] Portable or hard-piped source should be available for the space.

sources. Information in this table is now considered obsolete but is provided for design engineers who may have to refer to previous guidelines for renovations of existing facilities that used this table.

Locating the Source of Supply

Oxygen supply shall be located in accordance with NFPA-50. The actual location shall be selected by the facility planner in conjunction with the owner and the design team, allowing adequate access for cylinder replacement, cylinder storage, and bulk tank filling by truck.

Nitrous oxide cylinder storage should not be located outdoors in colder climates due to past bad experience with cold weather operating problems. The room where the source is located shall be well ventilated, with adequate space allowed around the installed manifold to easily change cylinders.

The location of the air compressor has been previously discussed.

TABLE 14.35 Generally Recommended Number of Station Inlets and Outlets

Room	O_2	Vac	N_2O	Air	N_2 Evac	Typical uses
Anesthesia workroom	1	1	*	1		Equipment repair testing
Animal oper. (research surgery)	1	1	*			Animal anesthesia and surgery
Animal research lab	1	1		1		Routine animal care
Autopsy	1	1				Suction waste materials from body
Bed, holding	1	1				Cardiac arrest, O_2 therapy
Biochemistry	X	1		1		Standard lab use*
Biochem. lab	X	1		1		Standard lab use*
Biophysics/biochemical	X	1		1		Standard lab use*
Blood processing		1		1		Standard lab use*
Blood receiving (blood donors)	1	1		1		Emergency use
Cardiac catheterization room	1	2				Cardiac arrest and other emerg.
Chem. analysis lab (sm. lab in hosp.)		1		1		Standard lab use*
Chemical lab		1		1		Standard lab use*
Constant temp room (microbiology lab)	1		1	X		Standard lab use*
Cystoscopy	1	3			1	Emergency use
Deep therapy	1	2				Cardiac arrest and other emerg.
Decontamination room (attached to inhalation therapy dept.)	1	1		1		Equipment testing
Demonstration room (inservice training)	1	1				Demo. equip. to new empl. and students
Dental repair	1	1		1	X	Power drills (dental)
Dispensary (minor surgery, first aid, student hlth and exams)	*		*			Emergency use
EEG (electroencephalograms)	1	1				Cardiac arrest and other emerg.
ECG (electrocardiogram)	1	1				Cardiac arrest and other emerg.
EMG (electromyogram)	1	1				Cardiac arrest and other emerg.
Ear-nose-throat exam		1		1		Aspiration, topical spray
Electron microscopy		1		1		Standard lab use*
Examination room	1	1		1		Drive air tools and vacuum cleaning
Emergency room	1	2		1		Cardiac arrest and other emergencies
Exam room and proctoscopic	1	1		1		Cardiac arrest and other emergencies
Experimental lab	X	1		1		Standard lab use*
Eye examination	1	1				Stock and cardiac arrest
Fluoroscopy (x-ray)	1	2				Cardiac arrest and other emergencies
Full-term nursery	1	2		1		Incubators, respirators
General physiology lab	1	1		1		Standard lab use* plus teaching
Heart catheterization lab	1	1		1		Cardiac arrest and other emerg. respir.
Hematology		1		1		Standard lab use*

(*Continued*)

TABLE 14.35 Generally Recommended Number of Station Inlets and Outlets (*Continued*)

Room	O_2	Vac	N_2O	Air	N_2	Evac	Typical uses
High-level radioisotope (x-ray dept.)	1	2		1			Cardiac arrest and other emergencies
Holding pre-OR or	1	1					
Holding nursery	1			1			
Intensive care areas	2	3		1			For critically ill
LDRP (labor, delivery, recovery, postpartum)	1	2		1			
Isolation (infectious and contagious diseases)	1	1		1			Patient care
Isolation room (patient room for contagious)	1	2		1			Oral, gastric, or thoracic
Lab annex		1		1			Pull waste evac. tubing drying apparatus
Lab cleanup area		1		1			Drying glassware
Lab, workroom		1		1			Standard lab use*
Labor rooms, O.B.	1	1	*				Analgesia, patient care
Linear accelerator vault	1	1		1			
Low-level radiation (x-ray dept.)	1	2		1			Cardiac arrest and other emergencies
Microbiology		1		1			Standard lab use*
Multiservice room	1	1					Cardiac arrest and other emergencies
Neonatal	3	4		3			
Neurological pharmacy teaching lab	1		1				Standard lab use*
Neurological physiology teaching lab	1	1		1			Standard lab use*
Nursing floor	1	2		1			Therapy, oral, gastric; IPPB, aerosols
Observation	1	1					Cardiac arrest and other emergencies
Obstetrics (delivery room)	1	3	*				Analgesia, anesthesia, patient care
Operating room (surgery, major and minor)	2	3	1	1	*	1	Patient care
Oral lab (dental)	*	1	*	1	*		Standard lab use*
Orthopedic exam room	1	1					Cardiac arrest and other emergencies
Pathology (doctor's office special lab tests)		1		1			Standard lab use*
Patient room	1	1		1			Patient care
Pharma. room (drug prep.)		1		1			Standard lab use*
Premature nursery and obs.	2	1		1			Incubators, respirators
Private and semiprivate	1	2		1			Patient care
Private recovery room (same as regular recovery)	1	3		1			*Note:* Need 1 more Vac for thoracic
Radiochemical lab		1		1			Standard lab use*
Radioisotope room (research room for animal lab)		1		1			Standard lab use*

TABLE 14.35   Generally Recommended Number of Station Inlets and Outlets   *(Continued)*

Room	O$_2$	Vac	N$_2$O	Air	N$_2$	Evac	Typical uses
Recovery beds (postanesthesia)	2	3		1			2 thoracic, 1 oral, 1 gastric or wound
Respiratory therapy	1	1		1			For outpatient treatments IPPB
Scanning room (part x-ray)	1	2					Cardiac arrest and other emergencies
Security nursing (psychiatric violent patients use lock box)	1	1		1			Patient care
Serology		1		1			Standard lab use*
Standard x-ray rooms	1	2		1			Cardiac arrest and other emergencies
Sterilization (CS or OR)	1	1		1			Equipment testing
Surgical preparation room	1	1		1			Premedication for anesthesia
Teaching lab	1	1		1			Standard lab use*
Treatment room	1	1		1			Special therapy
Urinalysis		1		1			Standard lab use*
Workroom for labs		1		1			Standard lab use*

*One outlet per area.
X Consult owner for number and location.
Source: Courtesy of ASPE.

Flow Rate and Diversity Factor

Each individual station outlet must provide a minimum flow rate for proper functioning of connected equipment under design conditions. The flow rates and diversity factors vary for individual station outlets in each system.

The flow rate from the total number of outlets without regard for any diversity is called the *total connected load.* If the total connected load were used for sizing purposes, the result would be a vastly oversized system since not all of the outlets in the facility will be used at the same time. A diversity, or simultaneous use factor, is used to allow for the fact that not all of the outlets will be used at once. It is used to reduce system flow rate in conjunction with the total connected load for sizing mains and main branch piping to all parts of the distribution system. This factor varies for different areas throughout any facility.

The estimated flow rate and diversity factors for various systems, area station outlets, and pieces of equipment are as follows:

1. For oxygen, refer to Table 14.36.
2. For nitrous oxide, refer to Table 14.37.
3. For low-pressure compressed air, refer to Table 14.38.
4. For high-pressure nitrogen and compressed air, refer to Table 14.39.

System Pressure and Design Pressure Losses

Low-Pressure Oxygen, Nitrous Oxide, Carbon Dioxide, and Compressed Air.   The minimum design pressure required at the most remote outlet with maximum flow rate for low-pressure systems is 50 psig (340 kPa). This figure has been

TABLE 14.36 Flow Rate and Diversity Factor for Oxygen Outlets

Location	Simultaneous use factor, %	Volume, Lpm
First OR (far end of a section of piping and all individual branches to ORs)	100	50 per OR
Each additional OR (on a section of piping)	100	10 per OR
Emergency rooms	100	Same as OR
Trauma rooms	100	Same as OR
LDRP rooms	100	20 per room
Delivery rooms	100	Same as OR
Cystoscopy and special procedures rooms	100	Same as OR
Recovery rooms (postanesthesia recovery)		10 per outlet
1–8 outlets	100	
9–12 outlets	60	
13–16 outlets	50	
additional outlets	45	
Intensive care (ICU) rooms	100	30 per outlet
Neonatal		
Pediatric		
Medical-surgical		
Coronary care (CCU) rooms	100	30 per outlet

Simultaneous use factors for other spaces*

The first outlet on the end section of piping is 20 Lpm. For additional outlets on the section of piping, add 10 Lpm with the fllollowing use factors.

No. of outlets	Simultaneous use factor, %	Volume, minimum Lpm
1–3	100	—
4–12	75	45
13–20	50	115
21–40	33	125
40 and over	25	155

*"Other spaces" include the following: patient rooms (medical and surgical) (bedside outlets), labor rooms, nurseries, examination and treatment rooms, OR bed holding areas, surgical preparation rooms, blood donor rooms, anesthesia workrooms, plaster (fracture) rooms, cardiac and heart catheterization rooms, deep therapy rooms, inhalation therapy rooms, electroencephalogram (EEG) rooms, electrocardiogram (ECG) rooms, electromyogram (EMG) rooms, fluoroscopy rooms, high-level radioisotope rooms, low-level radiation rooms, x-ray rooms, and endoscopy rooms.

TABLE 14.37   Flow rate and Diversity Factor for Nitrous Oxide Outlets

Location	Volume, Lpm
First OR (far end of piping and all individual branches to ORs)	30 per OR
Second OR (on a section of piping)	20 per OR
Each additional OR (on a section of piping)	15 per room
Delivery rooms	20 per room
Emergency rooms	20 per room
Trauma rooms	20 per room
Anesthesia workrooms	15 per room
Plaster (fracture) rooms	20 per room
Endoscopy rooms	15 per room
Dental surgery	15 per room

universally accepted by health care equipment manufacturers and design professionals for system design purposes.

There is a generally accepted "allowable" maximum friction loss for the piping distribution system (after the source) of 5 psig (34 kPa). This figure is considered a reasonable one for design purposes, but there is no code or other mandate that prevents a small deviation from this figure. For short runs of branch piping, another generally used figure is 10 percent of the available pressure. All of the above deviations shall provide a minimum of 50 psig (340 kPa) at the outlet.

High-Pressure Nitrogen and Compressed Air.   The minimum pressure required at the most remote outlet with maximum flow rate is dependent on the specific tools utilized by health care personnel. Currently used surgical room equipment requires a pressure range of from 160 to 200 psig (1000 to 1360 kPa) at the tool. There are also some tools used for precision work that use a much lower pressure of 25 to 50 psig (75 to 345 kPa). Dental pneumatic tools require 30 to 50 psig pressure.

There is a generally accepted average friction loss for the entire piping distribution system (after the source) of 10 percent of the design source pressure psig (136 kPa). Since the source of high-pressure gases is often a cylinder supply, the regulator installed should be capable of being adjusted over a wide range of pressure in the event that higher or lower operating pressures are required in the future. For high-pressure air compressors, the system regulator shall have the same capability.

There are no mandated purity requirements for gases used to operate surgical pneumatic tools. Oil-free air should be used and tool manufacturers consulted regarding maximum particulate size and dew points that are acceptable. NFPA-99 requires copper type "K" for pressures over 200 psig.

Pipe Sizing

The piping network is sized using the flow rate along each pipe run that has been adjusted by the diversity factor and the allowable friction loss for the system. Following is a system sizing procedure:

1. Locate all of the outlets and uses for each system on plans.
2. For each system, start with the most remote point, and work toward the mains. Count the number of outlets, determine the flow rate required from each outlet,

TABLE 14.38 Flow rate and Diversity Factor for Low-Pressure Medical Air Outlets

Air outlet/equipment	Design flow in scfm (free air)				Simultaneous-use factor, %
	Per unit	Per room	Per bed	Per outlet	
Anesthetizing Locations:					
Special surgery and cardiovascular		0.5			100
Major surgery and orthopedic		0.5			100
Minor surgery		0.5			75
Emergency surgery		0.5			25
Radiology		0.5			10
Cardiac catheterization		0.5			10
Ventilators	3.5				100
Delivery rooms		0.5			100
Acute Care Locations:					
Recovery room/surgical (postanesthesia)			2		25
ICU/CCU			2		50
Emergency rooms			2		10
Neonatal ICU			1.5		75
Dialysis units			0.5		10
Recovery rooms/OB		2			25
Ventilators	6				100
Subacute Care Locations:					
Nursery			0.5		25
Patient rooms (where shown)			0.5		10
Exam and treatment	1				10
Preop holding				1.5	20
Respiratory care	1				50
Pulmonary function lab				1	50
EEG and EKG				1	50
Birthing and LDRP	1				50
Patient isolation room			0.5		25
Other:					
Anesthesia workroom		1.5			10
Respirator care workroom	1.5			10	
Nursery workroom		1.5			10
Equipment repair		1.5		1.5	10
Med. laboratory				1.5	25
Autopsy		1			100
Sterile supply		1			10
Plaster room		1			50
Pharmacy		1			10
Dental, high pressure (50 psig)		2 per chair		2	100
Dental, low pressure (30 psig)		1 per chair		3	100

1 scfm = 0.03 m³/min.

TABLE 14.39 Flow Rate and Diversity Factor for High-Pressure Outlets

No. of outlets	Simultaneous use factor, %
1–7	100
8–20	75
21–over	50

Note: Surgical pneumatic tools use a flow rate of up to 15 scfm (425 sLpm) for modern instrument design at 225 psig.

and establish the diversity factor (Tables 14.36 to 14.39) This information should be recorded in a convenient form for later use when using the pipe sizing tables.

3. Calculate the adjusted flow rates from all outlets of each system by multiplying the flow rates by the diversity factors. Add all of the adjusted flow rates for the entire system together to find the required system capacity. It is now possible to select the source equipment capacity and the physical equipment sizes for each of the systems.

4. Locate the gas storage area and physically lay out the cylinders, manifold, tanks, and so on. All necessary separation, ventilation, and venting requirements shall be followed.

5. Establish a general layout of the system from the storage area to the farthest outlet or use point. Measure the actual distance along the run of pipe. To the actual run add an additional 50 percent of the measured distance to allow for fitting allowance. If a more accurate calculation is desired, use Table 14.15 to calculate the loss through each fitting. This is the total equivalent run of pipe.

6. Determine all of the necessary regulators, filters, purifiers, and so on necessary for each system in order to establish a combined allowable pressure drop through each of them and the assembly as a whole.

7. Establish the gas pressure required at the farthest outlet for each system and the allowable pressure loss for the entire system.

8. Dividing the total run of pipe (in 100s of feet) by the allowable system friction loss will establish the allowable friction loss per 100 ft of pipe.

9. Calculate the adjusted flow rate of gas through each branch, starting from the farthest outlet back to the source (or main) using the allowable flow rate from each station outlet and the appropriate diversity factor. For specific equipment, obtain the probable diversity of use from the end user.

10. Depending on the type of compressed gas, enter the appropriate pipe sizing table. Table 14.40 provides pipe sizing data for low-pressure oxygen, nitrous oxide, and compressed air. Table 14.41 provides pipe sizing data for 160 psig (1100 kPa) nitrogen and compressed air. Table 14.42 provides pipe sizing data for 225 psig (1500 kPa) nitrogen and compressed air. Using the adjusted flow rate, read across to find a figure that equals but does not exceed the allowable friction loss. Then, read up to find the size. In some cases, the diversity factor for the next highest range of outlets may result in a smaller size pipe than the range previously calculated. If this occurs, do not reduce the size of the pipe— keep the larger size previously determined.

TABLE 14.40 Pipe Sizing Table for Low-Pressure Oxygen, Nitrous Oxide, and Compressed Air

(In psig loss per 100 ft of pipe at 55 psig)

Gas flow, scfm (Lpm)	Nominal pipe sizes, in								
	1/2	3/4	1	1 1/4	1 1/2	2	2 1/2	3	4
1.8 (50)	0.04 (0.3)								
3.5 (100)	0.16 (1.1)								
4.4 (125)	0.25 (1.7)								
5.3 (150)	0.33 (2.3)	0.04 (0.3)							
6.2 (175)	0.48 (3.3)	0.06 (0.4)							
7.1 (200)	0.63 (4.3)	0.07 (0.5)							
8.8 (250)	0.99 (6.8)	0.11 (0.8)							
10.6 (300)	1.41 (9.7)	0.16 (1.1)	0.04 (0.3)						
14.1 (400)	2.51 (17.3)	0.29 (2.0)	0.07 (0.5)						
17.7 (500)	3.92 (27.0)	0.45 (3.1)	0.11 (0.8)						
26.5 (750)		1.02 (7.0)	0.24 (1.7)						
35.3 (1,000)		1.80 (12.4)	0.42 (2.9)	0.13 (0.9)	0.05 (0.3)				
44.1 (1,250)		2.81 (19.4)	0.66 (4.6)	0.21 (1.5)	0.09 (0.6)				
53.0 (1,500)			0.95 (6.6)	0.30 (2.1)	0.12 (0.8)				
70.6 (2,000)			1.05 (7.2)	0.67 (4.6)	0.22 (1.5)	0.05 (0.3)			
88.3 (2,500)				0.83 (5.7)	0.34 (2.3)	0.08 (0.6)			
105.9 (3,000)				1.19 (8.2)	0.49 (3.4)	0.11 (0.8)			
141.2 (4,000)				2.11 (14.5)	0.88 (6.1)	0.20 (1.4)			
176.6 (5,000)				3.30 (22.8)	1.36 (9.4)	0.32 (2.2)	0.06 (0.4)		
264.8 (7,500)					3.10 (1.4)	0.71 (4.9)	0.10 (0.7)	0.09 (0.6)	
353.1 (10,000)						1.27 (8.8)	0.22 (1.5)	0.16 (1.1)	
529.7 (15,000)						2.82 (19.4)	0.40 (2.8)	0.35 (2.4)	0.08 (0.6)
706.2 (20,000)						5.00 (34.5)	0.89 (6.1)	0.63 (4.3)	0.15 (1.0)
882.8 (25,000)							1.58 (10.9)	0.98 (6.8)	0.23 (1.6)
1059.3 (30,000)							2.47 (17.0)	1.40 (9.7)	0.31 (2.1)
1412.4 (40,000)							3.55 (24.5)	2.48 (17.1)	0.59 (4.1)
1765.5 (50,000)								3.90 (26.9)	0.92 (6.3)

1 psi = 7 kPa.

TABLE 14.41   Pipe Sizing Table for 125 psig Pressure Nitrogen and Compressed Air, scfm
(In psig loss per 100 ft of pipe at 125 psi; Schedule 40 steel pipe)

Gas flow, scfm	Size, in						
	$\frac{1}{2}$	$\frac{3}{4}$	1	$1\frac{1}{4}$	$1\frac{1}{2}$	2	$2\frac{1}{2}$
6	0.102	0.023					
8	0.161	0.041					
10	0.283	0.064	0.017				
15	0.636	0.144	0.038				
20	1.131	0.255	0.067	0.016			
25	1.768	0.399	0.105	0.025			
30	2.546	0.574	0.152	0.037	0.016		
35	3.465	0.782	0.206	0.050	0.022		
40	4.526	1.021	0.270	0.065	0.029		
45	5.728	1.292	0.341	0.083	0.036		
50	7.071	1.596	0.421	0.102	0.045		
60	10.183	2.298	0.607	0.147	0.065	0.018	
70	13.860	3.128	0.826	0.200	0.088	0.024	
80		4.065	1.079	0.261	0.115	0.031	0.013
90		5.170	1.365	0.330	0.145	0.040	0.016
100		6.383	1.685	0.408	0.180	0.049	0.020
125		9.973	2.633	0.637	0.281	0.077	0.031
150		14.361	3.792	0.918	0.404	0.110	0.045
175			5.162	1.249	0.550	0.150	0.062
200			6.742	1.632	0.716	0.196	0.061
225			6.533	2.065	0.909	0.248	0.102
250			10.534	2.550	1.122	0.306	0.126
275			12.746	3.085	1.358	0.371	0.152
300			15.169	3.671	1.616	0.441	0.181
325				4.309	1.896	0.518	0.213

1 psi = 7.0 kPa.

The piping network for a high-pressure gas system is sized using the following system design procedure. First, the following information must be obtained or calculated:

1. Determine the maximum pressure for the system based on the tools expected to be used at the facility. The range of acceptable pressure the tools must have at the tools themselves to operate at maximum effectiveness shall be obtained from the manufacturer. For preliminary estimates, a range within ±5 percent of the recommended pressure will be acceptable. The higher pressure would be used for design purposes to provide a safety factor. Additional pressure, usually 20 psig (140 kPa), shall be available at the terminal connection to provide for final adjustment and losses through the connecting hose, lubricator, and any other accessory equipment at the tool recommended by the manufacturer.

2. Establish the source and type of compressed gas for the system. The most often used gases are either nitrogen or compressed air. Economics should be the deciding factor in the selection of source gas since there is no medical or technical reason to choose one over the other. For existing facilities, if a separate high-

TABLE 14.42   Pipe Sizing Table for 175 psig Pressure Nitrogen and Compressed Air, scfm
(In psig loss per 100 ft of pipe at 175 psi; Schedule 40 steel pipe)

Gas flow, scfm	Size, in						
	$\frac{1}{2}$	$\frac{3}{4}$	1	$1\frac{1}{4}$	$1\frac{1}{2}$	2	$2\frac{1}{2}$
6	0.075						
8	0.173	0.030					
10	0.208	0.047					
15	0.469	0.106	0.028				
20	0.833	0.188	0.050				
25	1.302	0.294	0.078	0.019			
30	1.875	0.423	0.112	0.027			
35	2.552	0.576	0.152	0.037			
40	3.333	0.752	0.199	0.048	0.021		
45	4.218	0.952	0.251	0.061	0.027		
50	5.208	1.175	0.310	0.075	0.033		
60	7.499	1.692	0.447	0.108	0.048		
70	10.207	2.303	0.608	0.147	0.065	0.018	
80	13.331	3.008	0.794	0.192	0.085	0.023	
90	16.872	3.807	1.005	0.243	0.107	0.029	
100	20.830	4.700	1.241	0.300	0.132	0.036	
125		7.344	1.939	0.469	0.207	0.056	0.023
150		10.576	2.793	0.676	0.297	0.081	0.033
175		14.395	3.801	0.920	0.405	0.111	0.045
200		18.801	4.965	1.202	0.529	0.144	0.059
225			6.284	1.521	0.669	0.183	0.075
250			7.757	1.878	0.826	0.226	0.093
275			9.387	2.272	1.000	0.273	0.112
300			11.171	2.704	1.190	0.325	0.134
325			13.110	3.173	1.396	0.381	0.157

1 psi = 7.0 kPa.

pressure compressed air system is not practical or economical, a separate cylinder nitrogen supply is generally provided although compressed air cylinders at a lower cost could also be used.

3. Ascertain the allowable friction loss of gas in the piping system (after all source valves, dryers, and pressure regulators if supplied) to achieve the pressure required at the tool. There is a generally accepted, average friction loss for the entire piping distribution system of 10 percent of the piping system pressure. The terminal regulator installed in each OR should be capable of being accurately adjusted over a wide range of pressure, often as low as 30 psig (210 KpA), in the event that lower operating pressures are required for different tools.

4. Determine the total equivalent run of piping. A conservative estimate is found by adding 50 percent to the measured run of the piping.

5. Determine the simultaneous use factor. This factor is the most problematic of the criteria since there is no method of determining an accurate figure universally applicable to every project. Personnel at each individual facility must be consulted as to the number of ORs in use at any one time, the type of operations performed, and the length of those operations. In this manner some idea of how many tools would be in simultaneous use could be determined. An extreme case is hip replacement, where 90 min of actual running time of a pneumatic tool has been documented.

A project design procedure will now be presented, using as an example a facility with 10 operating rooms, proposed tools requiring 15 scfm at each tool, and a pressure requirement of 200 psig at the tool itself.

1. Locate all of the proposed outlets on plans in conjunction with the owner, architect, or developed program. If there are multiple outlets in a single OR, only one tool will be in use at once unless the facility or program instructs otherwise.

2. Locate the gas source. If it is a storage area, physically lay out the cylinders, manifold, tanks, and so on. All necessary separation, ventilation, and venting requirements shall be provided. Refer to Fig. 14.23 for dimensions and layout of cylinder banks and Fig. 14.22 for cylinder sizes if not 9½-in diameter.

3. Calculate the storage requirements of cylinders or air compressor capacity. For a central compressor, the maximum scfm and pressure has to be calculated. For scfm calculations, add all of the outlets together, assign a maximum scfm value to each, determine the use factor, and calculate the maximum scfm for the system. To this add all other requirements such as dryers, and so on. The pressure is based on the maximum pressure required by the tools, the allowable friction loss, and the additional pressure at the outlet. Select a compressor based on these values with an allowance for additional scfm to accommodate future expansion if desired.

As an example, select a compressor for 10 ORs. First, 10 ORs × 15 scfm each equals 150 scfm. Referring to Table 14.38, a use factor of 75 percent is established, giving an adjusted scfm requirement of 113 scfm for tools. The pressure is based on a requirement for 200 psig at the tool, 20 psig friction loss allowance (10 percent of 200 psi), and an allowance of 20 psig for losses at the tool. This gives the selection criterion for the compressor of 113 scfm at 240 psig only for tools.

The selection of the number of cylinders in a central supply is based on the space available, the OR gas use requirements of the facility, and the supplier's ability or desire to constantly deliver cylinders to the facility, often every day. If the facility uses a large number of cylinders, a central compressed air system is considered very cost effective.

4. Establish the piping layout of the system on plans from the source to the farthest outlet or use point. Measure the actual distance along the run of pipe. To the actual run add an additional 50 percent of the measured distance as a fitting allowance. This figure is the equivalent run of pipe.

5. It is now possible to size the piping network. Pipe is sized using the flow rate and the allowable friction loss. This loss must be in the same units as the pipe sizing chart selected. The charts provided here are in psig loss per 100 ft of pipe run.

Starting with the most remote point, work toward the mains and source, and count the number of outlets on each branch and main. Using the maximum flow rate from each outlet in conjunction with the diversity factor, establish an adjusted flow rate at each connection with any outlet or pipe. These connections are each called a *design point*. This information should be recorded in a convenient form for later use when using the pipe sizing tables. The adjusted scfm for each design point is calculated separately.

Next, calculate the friction loss in psi/100 ft of pipe. This is done by calculating the equivalent run of pipe and dividing the run by 100. This figure in turn is divided into the psig loss allowable for the entire system. As an example, suppose the total equivalent run is 650 ft. This becomes a figure of 6.5. Design criteria has established an allowable loss of 20 psig for the system. Thus 6.5 divided into 20 gives a figure of 3.07 (say 3.0) psi/100 ft of pipe as the allowable friction loss.

Table 14.41 provides pipe sizing data for a 125 psig system, Table 14.41 provides pipe sizing data for a 175 psig system, and Table 14.43 provides pipe sizing data for a 250 psig system. For each design point, find the adjusted flow rate on the appropriate pipe sizing chart and read straight across to find a figure that equals the allowable friction loss. Then, read up to find the pipe size. In some cases, the

TABLE 14.43 Pipe Sizing Table for 250 psig Pressure Nitrogen and Compressed Air, scfm

(In psig loss per 100 ft of pipe at 250 psi; Schedule 40 steel pipe)

Gas flow, scfm	Size, in							
	$\frac{1}{2}$	$\frac{3}{4}$	1	$1\frac{1}{4}$	$1\frac{1}{2}$	2	$2\frac{1}{2}$	3
6	0.054							
8	0.096							
10	0.149	0.034						
15	0.336	0.076						
20	0.597	0.135	0.036					
25	0.933	0.211	0.056					
30	1.344	0.303	0.080					
35	1.829	0.413	0.109	0.026				
40	2.388	0.539	0.142	0.034				
45	3.023	0.682	0.180	0.044				
50	3.732	0.842	0.222	0.054				
60	5.374	1.213	0.320	0.078	0.034			
70	7.315	1.651	0.436	0.105	0.046			
80	9.554	2.156	0.569	0.138	0.061			
90	12.092	2.729	0.721	0.174	0.077			
100	14.928	3.369	0.690	0.215	0.095	0.026		
125	23.325	5.263	1.390	0.336	0.146	0.040		
150		7.579	2.001	0.484	0.213	0.058		
175		10.316	2.724	0.659	0.290	0.079	0.033	
200		13.474	3.558	0.861	0.379	0.103	0.043	
225		17.053	4.503	1.090	0.450	0.131	0.054	
250		21.054	5.559	1.346	0.592	0.162	0.066	
275		25.475	6.727	1.628	0.717	0.196	0.080	0.026
300		30.317	8.006	1.938	0.853	0.233	0.096	0.031
325			9.396	2.274	1.001	0.273	0.112	0.036
350			10.897	2.637	1.161	0.317	0.130	0.042
375			12.509	3.027	1.332	0.364	0.150	0.048
400			14.232	3.445	1.516	0.414	0.170	0.054
425			16.067	3.889	1.711	0.467	0.192	0.061
450			18.013	4.360	1.919	0.524	0.215	0.069

1 psi = 7.0 kPa.

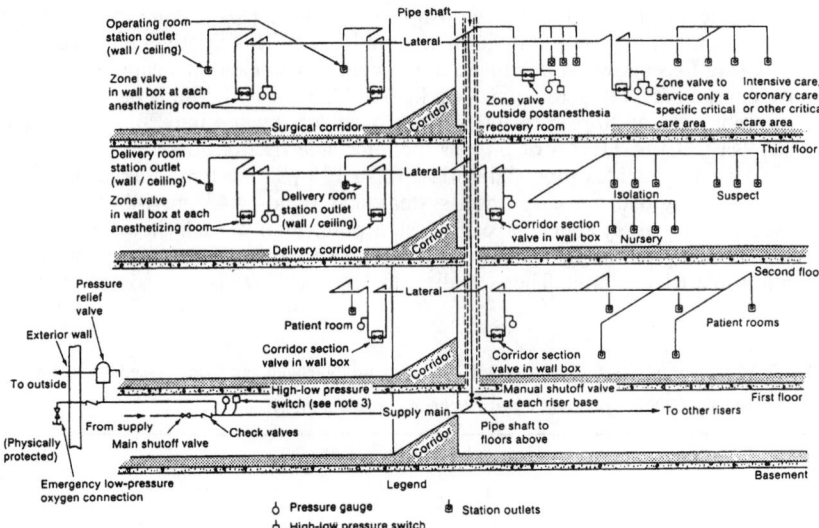

FIGURE 14.38   Schematic detail of typical medical gas systems. (*Adapted from NFPA.*)

diversity factor for the next highest range of outlets may result in a smaller size pipe than the range previously calculated, If this occurs, do not reduce the size of the pipe—keep the larger size previously determined.

Tables 14.41 to 14.43 have been calculated using steel pipe. Reduce pressure by 5 percent for use with copper pipe.

System Design Considerations

The size actually assigned to a pipe is a compromise between "normal" usage and possible "emergency" conditions. This philosophy will certainly result in oversized pipe if "emergency" conditions are used as a basis for design. The main argument for not oversizing the piping system is that it costs more. If this argument is taken in perspective, the small cost of slightly oversizing the medical gas piping system as a percent of the total construction cost of the facility is almost nothing. Another reason for oversizing is to allow for future changes or expansion, where the cost of adding another pipe or replacing a smaller pipe with a larger one will be many times the cost of larger sizing during the initial construction. Good practice is to have the smallest size branch ½ in, submains should be a minimum of ¾ in size, and main size no less than 1 in. This is to allow for future expansion and renovations without the need to replace piping.

A schematic diagram of typical medical gas systems adapted from NFPA-99 is given in Fig. 14.38. Refer to the text of this code for exact requirements.

DENTAL COMPRESSED AIR

GENERAL

Compressed air for dental purposes is used to power pneumatic tools such as drills and for general dentistry purposes such as quickly drying areas to be bonded. Compressed air is usually supplied by a small, dedicated compressor.

CODES AND STANDARDS

NFPA-99, Health Care Facilities, is the code governing the installation of the dental compressed air piping system.

SYSTEM COMPONENTS

Components of the system are illustrated in Fig. 14.34 and depend on the size of the system. Small units serving one or two dental chairs consist of only a compressor that runs continuously with no other ancillary equipment. Larger units serving multiple installations shall conform to the surgical-medical compressed air system.

REQUIRED PRESSURE AND FLOW RATE

Dental tools use both high- and low-pressure air. High-pressure tools such as drills use a flow rate of 2 scfm and a pressure of 50 psig (345 kPa). Low-pressure handpieces used for cleaning and by hygienists use a flow rate of 3 scfm and a pressure of 30 psig (210 kPa). Refer to Table 14.44 for use factors. Surgical tools may require 100 psig (700 kPa).

TABLE 14.44   Flow Rate and Diversity Factor for Dental Air

No. of outlets	Simultaneous use factor, %
4	100
5–10	75
11–over	50

SYSTEM DESIGN CONSIDERATIONS

Dental chairs have an integral air pressure regulating arrangement as part of the chair, and factory-installed distribution lines to the outlets found on the chair. It is generally accepted practice to provide a pressure in a range of 80 to 100 psig (550 to 700 kPa) to the regulator.

INDUSTRIAL SPECIALTY GASES

This subsection will discuss the generation and use of carbon dioxide and nitrogen on a larger scale.

GENERAL

Carbon Dioxide

The vast majority of CO_2 used for industrial purposes is obtained as a by-product of another process. Carbon dioxide is used to freeze foods and for pH control in various industries. It is also used for fumigation, as a replacement for mechanical refrigeration, cleaning, and solvent extraction. The most attractive characteristics of CO_2 are that it is environmentally benign, creates no long-term health hazard, is easy to handle, and requires few safety precautions.

The standard most commonly used is CGA G-6.1, Standard for Low Pressure Carbon Dioxide Systems at Consumer Sites.

There are three grades of purity available. Food, or standard grade, which is 99 percent pure, is the most often used grade. The most pure is that employed for laboratory work and other specialized uses, and it is discussed earlier in this chapter. The least pure, with no purity requirements, is that used for general industrial purposes such as in fracturing rock in oil recovery operations. CO_2 is usually delivered and stored at 0°F (-18°C).

Following are common uses:

1. At 90°F and 1100 psig (32°C and 7600 kPa), CO_2 reaches its supercritical state where it becomes a dense gas with the versatile solvent properties of a liquid.

2. CO_2 is widely used for pH control, replacing sulfuric acid. CO_2 acts like sulfuric acid on a mole-for-mole basis. It is estimated that 44 lb of CO_2 is equivalent to 98 lb of sulfuric acid in neutralizing any solution with a pH above 8.3. The ratio is less as the pH decreases.

3. Fumigation of grain in silos is another common use. CO_2 is stored as a liquid and is introduced into the silo as a gas from the top, from where it migrates down. General use is at the rate of 0.15 to 0.20 lb/bushel of grain.

4. To keep food frozen in trucks and railcars, liquid CO_2 is injected into an overhead bunker or storage area where it expands into a mixture of gas and "dry ice." This melts over the length of the shipment.

5. CO_2 is used for carbonation in the soft drink industry.

Carbon dioxide is stored on-site in bulk tanks that will be discussed later in this section. Other equipment may include high-pressure transfer pumps and recovery systems which compress, purify, and liquefy CO_2 that is to be recycled.

The piping material most commonly used on tanks is Schedule 80 carbon steel, with forged steel fittings and ball valves at the inlet and outlet. Pipe distribution systems for cryogenic service are carbon steel, stainless steel, brass and copper that is insulated with 3 to 4 in of rigid, closed cell urethane and has a jacket. Piping for CO_2 gas is carbon steel in plants where high purity is not required and where physical strength is necessary. For other facilities such as food processing and pharmaceuticals, copper type L or stainless steel is preferred. If sterilization is required, stainless steel is used almost exclusively.

The K factor on the closed cell insulation should not exceed 0.15 for 2 lb/ft^3 density at 75°F. Commonly used jacket materials are PVC, polypropylene, and stainless steel flex. The most often used pipe combinations are type K copper tubing with brazed joints, 4-in polyurethane insulation, and PVC jackets. Allowance must be made for expansion of the pipe due to the large difference in temperature between ambient air temperature and the cold of liquid CO_2 that is about 0°F (-17°C). Valves shall be bubble tight, with operating pressures of 1500 psig for sizes up to ¾ in, 1000 psig for sizes to 1½ in, and 750 psig for sizes to 4 in. Valve materials are recommended to be stainless steel or a combination of steel and brass.

Safety valves must be provided on the tank to vent gas to atmosphere at a pressure of 10 percent over the highest working pressure. Another precaution is to prevent the withdrawal of gas at too high a flow rate. This will lower the pressure in the tank enough to turn the liquid into a solid in the tank or piping system. To prevent this, many installations use a separate pressure-building vaporizer that takes liquid from the tank, vaporizes it, and returns it back to the gas zone in the tank so that the vapor will maintain enough pressure in the tank to prevent solidification.

Precautions must be taken to sound an alarm when the concentration of CO_2 reaches a level of 3 percent in air. OSHA permits a 1 percent level for 8 h and 3 percent for only 15 min. Where large quantities are stored or where pipelines may leak indoors, a continuous CO_2 monitor and an oxygen depletion monitor should be provided to give both an audible and visual alarm. All areas must be properly exhausted.

Nitrogen

Nitrogen is one of the most often used gases for industrial purposes. It is inert, tasteless, and colorless as well as abundant and relatively inexpensive. Common uses for nitrogen gas are inerting, testing, sparging foods, and in the electronics industry for manufacturing various components. In cryogenic form it is used in blood and tissue preservation as well as for biological preservative functions and in various industrial processes.

When the demand for nitrogen exceeds about 15,000 ft^3 per month, on-site generation of nitrogen should be considered. If the demand exceeds 25,000 ft^3 per month, on-site generation is clearly desirable. On-site production will provide a more consistent gas with higher purity than the gas available from bulk liquid supply.

Several on-site production methods are commonly used. Pressure swing adsorption units produce only gas. In this method, air is first compressed, and it then passes through initial filters to remove moisture and other contaminants. It is then fed through a porous, solid molecular sieve adsorption bed that removes oxygen and other trace gases. Typically, the adsorption beds are composed of carbon material, with two separate beds provided. When the pressure is reduced, the adsorption beds give up their impurities to atmosphere. Duplex beds allow continuous operation. After passing through a final filter, the nitrogen is ready for use. Reserves to allow for peaking and backup could be stored in a separate reserve tank.

Another type is the single-column, cryogenic air separation unit, which can utilize either an air-expansion or waste-expansion cycle. Comparing the two basic cycles, waste expansion is more power efficient at product pressures of between 90 and 120 psia (600 and 800 kPa), and air expansion is more efficient at pressures of between 35 and 65 psia (250 and 450 kPa). A hybrid cycle, utilizing a molecular sieve to purify the incoming air, may be used to increase efficiency.

ACKNOWLEDGMENTS

Strachan McCoe Inc.

Hankison Corp.

George Koechine

John Quigg

McCarthy and Robinson Company, Toronto, Canada

Dollinger Silenciers, Peter Sulman

Mirza Z. Hassan, Byrne Compressed Air Equipment Co.

Ingersol Rand, Inc.

Ed Murphy, Pall Pneumatic Equipment Corp.

Charles Levitt, Nash Pump Co., Inc.

Leah Clark, RN, North Shore Medical Center, Vice President/Chief Nursing Officer

Derik Smith

Bill Bledsoe

Steve Waldman, Hill-Rom

REFERENCES

AIA: *Guidelines for Construction and Equipment of Hospital and Medical Facilities,* 1992–1993.

ASPE Data Book, vol. 2. 1981–1982.

Casillo, Antonio, "Sizing Air Compressors," *Plant Engineering,* December 1984.

Compressed Air Fundamentals, Ingersoll-Rand Company.

Compressed Gas Institute: *Compressed Air and Gas Handbook,* 4th ed.

Cunningham, E. R.: "Air Compressors," *Plant Engineering,* May 1980.

Ferrara, A. J.: "Design for Compressed Air," *Air Conditioning, Heating and Ventilating,* 1964.

Foss, R. Scott: "Fundamentals of Compressed Air Systems," *Plant Engineering,* May 1981.

Frankel, M.: "Compressed Air Systems," *Plumbing Engineer,* Sept.–Oct., 1986.

Galus, T.: "How Much Air-Drying Equipment Is Necessary?" *Hydraulics and Pneumatics Magazine,* April 1989.

"Guide To Compressor Selection," *Compressed Air Magazine,* 1978.

NAVFAC DM—3.5. *Compressed Air and Vacuum Systems,* March 1983.

Stanton, W. M.: "Industrial Air Compressors," *Actual Specifying Engineer.*

TM 5—810-4. *Compressed Air,* December 1982.

Ulrich, William B.: "Air and Water Can Be a Nasty Mix," *Machine Design Magazine,* March 1993, pp. 71–75.

Varigas Research, Inc.: *Compressed Air Systems,* 1984.

CHAPTER 15
VACUUM AIR SYSTEMS

This chapter describes criteria, production, and the piping distribution network for various vacuum air systems. Because of the diverse uses and different design criteria for each, this chapter is divided into the following separate sections; health care facilities, laboratory systems, industrial applications, and central vacuum cleaning systems.

FUNDAMENTALS

The performance of any vacuum air system is based on two factors: the flow volume measured in cfm (Lpm) and the maximum vacuum maintained in the system. For most vacuum systems to function, air becomes the transporting medium for any gas or suspended solids, and the pressure provides the energy for transportation. These two essential factors operate in inverse proportion—as the airflow increases, the vacuum pressure decreases. The various systems must be designed to produce specific vacuum pressure and airflow levels that have been determined, often by experience and experimentation, to be most effective in performing their respective tasks. The exception is where vacuum pressure is intended to produce a force used to lift objects or simply to evacuate an enclosed space. For these uses, airflow is only a function of how long it takes the system to achieve its ultimate vacuum pressure.

DEFINITIONS AND PRESSURE MEASUREMENT

Vacuum Definition

Vacuum is defined as an air pressure less than atmospheric. The vacuum level is the difference in pressure between the evacuated system and the atmosphere. Vacuum pressures generally used in the United States fall into three broad categories:

1. Rough (or coarse) vacuum, up to 28 inHg
2. Medium (or fine) vacuum, up to 1 μm
3. Ultrahigh vacuum, greater than 1 μm

In other parts of the world, the categories are often classified as follows:

1. Rough vacuum, 760 to 1 torr
2. Medium vacuum, 1 to 10^{-3} torr
3. High vacuum, 10^{-3} to 10^{-7} torr
4. Ultrahigh vacuum, greater than 10^{-7} torr

While the definition of *vacuum* is straightforward, measuring a vacuum level (or force) is not. Several methods of measurement are used, and each depends on a different reference point.

Units of Measurement and Reference Points

Units of Measurement. In order to compute work forces and changes in volume, conversion to negative gauge pressure (psig) or absolute pressure (psia) will be required. The units used are inches of mercury (inHg) and the millibar (mbar). These units originate from the use of a barometer. The basic barometer is an evacuated vertical tube, the top end of which is closed and the bottom end of which is open and placed in a container of mercury open to the atmosphere. The pressure, or "weight," exerted by the atmosphere on the open container forces the mercury

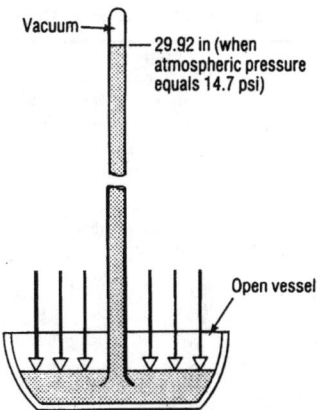

FIGURE 15.1   Basic barometer.

up into the tube. At sea level, this pressure will support a column of mercury 29.92 in high. In pressure units, this becomes 14.69 psi (99.89 kPa). Figure 15.1 illustrates a basic barometer.

The two basic reference points for measuring vacuum are standard atmospheric pressure and a perfect vacuum. When the point of reference is from standard atmospheric pressure to a specified vacuum pressure, this is called *gauge pressure*. If the reference pressure level is measured from a perfect vacuum, the term used is *absolute pressure*. Local barometric pressure, which is the prevailing pressure at any specific location, should not be confused with *standard atmosphere*, which is the mean barometric pressure at sea level.

Standard Reference Points and Conversions.   At standard atmospheric pressure, 0 inHg is equal to 14.7 psig, 101.4 kPa, and 29.92 inHg. For ease of calculations, 14.7 psig is adjusted to 15 psig, and 29.92 inHg is adjusted to 30 inHg. These minor deviations yield results well within the accuracy required for engineering calculations used in this handbook. At the opposite end of the scale, 0 psia (a perfect vacuum) has a value of 0 inHg and 29.92 inHg. Table 15.1 compares vacuum pressure from the two most commonly used reference points. Figure 15.2 gives conversion from various pressure units to another. Table 15.2 gives numerical conversion multipliers for converting torr into various other vacuum pressure units. Table 15.3 gives numerical conversion from inHg to psia and inHg absolute.

On the dials of most pressure gauges, atmospheric pressure is assigned the value of zero. Vacuum measurements must have a value of less than zero. Negative gauge pressure is the difference between the system vacuum pressure and atmospheric. Absolute pressure is the pressure (in psi) above a perfect vacuum and is equal to atmospheric pressure less negative gauge pressure.

Other vacuum units are atmospheres, torr, and micrometers (formerly "micron"). One standard atmosphere equals 14.7 psi, or 29.92 inHg. Any fraction of an atmosphere is a partial vacuum and would equal negative gauge pressure. To calculate atmospheres knowing absolute pressure in psi, divide that figure by 14.7. A torr is $\frac{1}{760}$ of an atmosphere, and a micrometer is 0.001 torr. These units of measurements are very high vacuum pressures and so are generally used for research, industrial, or laboratory use. Conversion factors are given in Table 15.2 and Fig. 15.2.

TABLE 15.1   Basic Vacuum Pressure Measurements

Units			
Negative gauge pressure, P_g, psig	Absolute pressure, P_a, psia	Inches of mercury, P_m	kPa absolute
0	14.7	0	101.4
Atmospheric pressure at sea level			
−1.0	13.7	2.04	94.8
−2.0	12.7	4.07	87.5
−4.0	10.7	8.14	74.9
−6.0	8.7	12.20	59.5
−8.0	6.7	16.30	46.2
Typical working vacuum level			
−10.0	4.7	20.40	32.5
−12.0	2.7	24.40	17.5
−14.0	0.7	28.50	10.0
−14.6	0.1	29.70	1.0
−14.7	0	29.92	0
Perfect vacuum (zero reference pressure)			

Conversion equations:

$$P_a = 0.149 \, P_m$$

$$P_m = 2.04 \, P_a$$

$$P_a = 14.7 - P_g$$

General Vacuum Criteria

Conversion of scfm to acfm.   Vacuum is used by having air at atmospheric pressure enter a piping system that has a lower pressure. Gas at atmospheric pressure will expand to fill the piping system. The air at standard, atmospheric pressure is called *standard cubic feet per hour* (measured as scfm), and the expanded air in the piping system is called *actual cubic feet per minute* (acfm). Another term used to indicate acfm is *inlet cubic feet per minute* or *icfm*. Acfm is greater than scfm.

To convert scfm to acfm with a given pressure of inHg and temperature in °F, use the following formula:

$$\text{acfm} = \text{scfm} \, \frac{29.92}{P} \times \frac{T + 460}{520} \tag{15.1}$$

where P = actual pressure, inHg, for the scfm being converted
  T = actual temperature, in °F, for the scfm being converted
  scfm = standard cubic feet per minute being converted

For practical purposes, a numerical method for solving Eq. (15.1) can be used if the temperature is always 60°F. At that temperature, the second part of the equation becomes unity. Table 15.4 gives numerical values for $29.92/P$. To find acfm, multiply the scfm by the value found in Table 15.4 opposite the vacuum pressure.

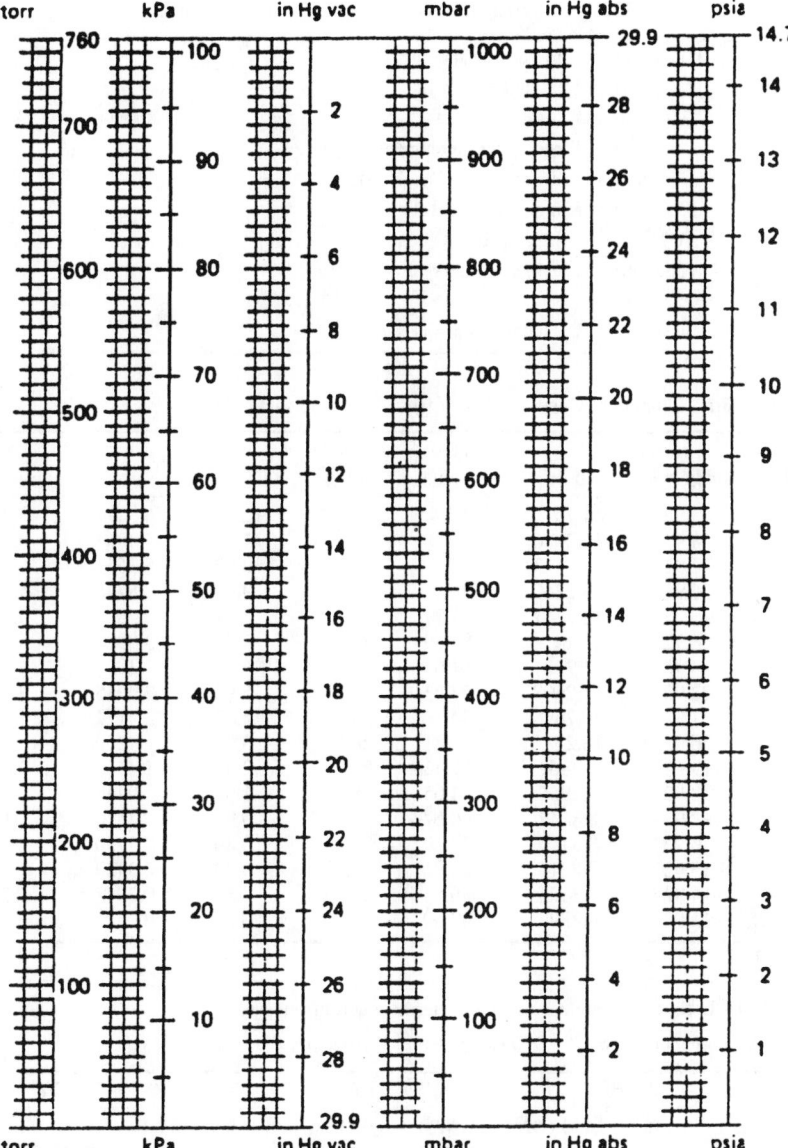

FIGURE 15.2   Conversion of various vacuum pressure units.

TABLE 15.2 Numerical Conversion Multipliers for Vacuum Units

0.0010 torr = 1 micrometer mercury (μmHg)
0.0075 torr = 1 pascal (Pa)
0.7501 torr = 1 millibar (mbar)
1.000 torr = 1 millimeter mercury (mmHg)
1.868 torr = 1 in water at 4°C (in H_2O)
25.40 torr = 1 in mercury (inHg)
51.71 torr = 1 lb/in^2 (psi)
735.6 torr = 1 tech. atmosphere (at)
750.1 torr = 1 bar
760.0 torr = 1 standard atmosphere (atm)

TABLE 15.3 Numerical Conversion of inHg to psia and inHg abs

inHg	inHg abs	psia	kPa absolute	inHg	inHg abs	psia	kPa absolute
0	29.92	14.70	101.4	17	12.92	6.3477	43.71
1	28.92	14.2086	97.9	18	11.92	5.8564	40.33
2	27.92	13.7173	94.5	19	10.92	5.3651	36.95
3	26.92	13.2260	91.5	20	9.92	4.8738	35.57
4	25.92	12.7347	87.77	21	8.92	4.3824	30.20
5	24.92	12.2434	84.39	22	7.92	3.8911	26.82
6	23.92	11.7521	81.01	23	6.92	3.3998	23.37
7	22.92	11.2608	77.63	24	5.92	2.9085	19.99
8	21.92	10.7695	84.19	25	4.92	2.4172	16.61
9	20.92	10.2782	70.81	26	3.92	1.9259	13.23
10	19.92	9.7869	67.43	27	2.92	1.4346	9.85
11	18.92	9.2955	64.05	28	1.92	0.9433	6.48
12	17.92	8.8042	60.67	29	0.92	0.4520	3.10
13	16.92	8.3129	57.29	29.12	0.80	0.3930	
14	15.92	7.8216	53.91	29.22	0.70	0.3439	2.36
15	14.92	7.3303	50.54	29.32	0.60	0.2947	1.52
16	13.92	6.8390	47.09	29.92	0	0	0

TABLE 15.4 Numerical Value to Determine acfm from scfm

[Expanded air ratio (29.92/P) as a function of pressure, P, inHg]

P	29.92/P	P	29.92/P	P	29.92/P	P	29.92/P
29.92	1.00	19.92	1.5020	9.92	3.0161	0.80	37.40
28.92	1.0345	18.92	1.5813	8.92	3.3542	0.70	42.0742
27.92	1.0716	17.92	1.6696	7.92	3.7777	0.60	49.8667
26.92	1.1114	16.92	1.7683	6.92	4.3236	0.50	59.84
25.92	1.1543	15.92	1.8793	5.92	5.0540	0.40	74.80
24.92	1.2006	14.92	2.0053	4.92	6.0813	0.30	99.7334
23.92	1.2508	13.92	2.1494	3.92	7.6326	0.20	149.60
22.92	1.3054	12.92	2.3157	2.92	10.2465	0.10	299.20
21.92	1.3649	11.92	2.5100	1.92	15.5833	—	—
20.92	1.4302	10.92	2.7399	0.92	32.5217	—	—

TABLE 15.5 Direct Ratio for Converting scfm to acfm

inHg	kPa abs	Factor	inHg	kPa abs	Factor
1			16	47.07	2.15
2	94.5	1.1	17	43.71	2.3
3	91.15	1.1	18	40.33	2.5
4	87.77	1.15	19	36.95	2.73
5	84.39	1.2	20	33.57	3
6	81.01	1.25	21	30.20	3.33
7	77.63	1.3	22	26.32	3.75
8	74.19	1.35	23	23.37	4.28
9	70.81	1.4	24	19.99	5
10	67.43	1.5	25	16.61	6
11	67.05	1.55	26	13.23	7.5
12	60.67	1.62	27	9.85	10
13	57.29	1.75	28	6.48	15
14	53.91	1.85	29	3.10	30
15	50.54	2.0	30	0	60

A direct ratio for converting scfm to acfm for various pressures is given in Table 15.5. Multiply the scfm by the factor found opposite the pressure of the system.

Adjusting Vacuum Pump Rating for Altitude. The rating of a pump at altitude is a lower percentage of its rating at sea level. For each 1000 ft increase in altitude, atmospheric pressure drops by approximately 1 inHg. Refer to Table 15.6 for actual barometric pressure at various altitudes. As an example, for the city of Denver (at 5000 ft), the local atmospheric pressure is 24.90 inHg. Dividing 30 into 24.90 gives a percentage of 83.3 percent. If a pump is rated at 25 inHg at sea level, 83.3 percent of 25 equals 20.8 inHg at 5000 ft. This is the required vacuum pressure that would equal 25 inHg at sea level.

At altitudes above sea level, there is a reduction in the scfm delivered because of the difference in local pressure compared to standard pressure. To compensate for this difference, scfm must be increased. Table 15.7 presents a multiplication factor to accomplish this. To find the adjusted scfm, multiply the actual scfm by the factor found opposite the altitude where the project is located.

Time for Pump to Reach Rated Vacuum. The time a given pump will take to reach its rated vacuum pressure depends on the volume of the system in cubic feet and the capacity of the pump in scfm at the vacuum rated pressure. But simply dividing the system volume by the capacity of the pump will not produce an accurate answer. This is because the vacuum pump does not pump the same quantity of air at different pressures. There is actually a logarithmic relationship that can be approximated by the following formula:

$$T = \frac{V}{Q} N \qquad (15.2)$$

where T = time, min
V = volume of system, cf
Q = flow capacity of pump, scfm
N = natural log constant

TABLE 15.6   Actual Barometric Pressure at Various Altitudes

Meters	Altitude (sea level equals zero)		Barometric pressure, inHg	kPa
−3040	−10,000		31.00	104.5
− 152	− 500		30.50	102.8
0	0	Sea level	29.92	100.8
152	+ 500		29.39	99.0
304	1,000		28.87	97.3
456	1,500		28.33	95.5
608	2,000		27.82	93.7
760	2,500		27.31	92.0
912	3,000		26.81	90.3
1064	3,500		26.32	88.7
1216	4,000		25.85	87.1
1368	4,500		25.36	85.5
1520	5,000		24.90	83.9
1672	5,500		24.43	81.9
1824	6,000		23.98	80.8
1967	6,500		23.53	79.3
2128	7,000		23.10	77.8
2280	7,500		22.65	76.3
2432	8,000		22.22	74.9
2584	8,500		21.80	73.4
2736	9,000		21.39	72.1
2888	9,500		20.98	70.7
3040	10,000		20.58	69.3

TABLE 15.7   Multiplication Factor for Adjusting scfm at Altitude

Altitude, ft	Meters	Factor used for required scfm
0	0	1.0
500	152	1.02
1,000	304	1.04
1,500	456	1.06
2,000	608	1.08
2,500	760	1.10
3,000	912	1.12
3,500	1064	1.14
4,000	1216	1.16
5,000	1520	**1.20**
6,000	1824	1.25
7,000	2128	1.30
8,000	2432	1.35
9,000	2736	1.40
10,000	3040	1.45
11,000	3344	1.51

For vacuum up to 15 inHg, $N = 1$.

For vacuum up to 22.5 inHg, $N = 2$.

For vacuum up to 26 inHg, $N = 3$.

For vacuum up to 28 inHg, $N = 4$.

In order to obtain the most accurate answer, obtain pump curves from the manufacturer, and substitute the scfm capacity for the pump at each 5 inHg increment. Add them together to find the total time. The selection of the value for N depends on the highest level of system vacuum pressure and is constant throughout the several calculations.

Adjusting Pressure Drop for Different Vacuum Pressures. The chart for friction loss in a vacuum pipe presented later in this section is based on 15 inHg. For a given scfm and pipe size, the pressure loss at any vacuum pressure other than the 15 inHg the medical-surgical vacuum sizing chart was developed for can be found by dividing the pressure drop in the chart by the ratio found from the following formula:

$$\frac{30 - \text{new vacuum pressure}}{15} \qquad (15.3)$$

Simplified Method of Calculating Velocity. Use the following formula to find the velocity of a gas stream under a vacuum:

$$V = C \times Q \qquad (15.4)$$

where V = velocity, fpm
C = constant based on pipe size (refer to Table 15.8)
Q = flow rate in acfm, based on an absolute vacuum pressure

As an example, calculate the velocity of 100 scfm through a 2 in pipe with a pressure of 20 inHg.

1. First, find the equivalent absolute pressure of 20 inHg. Using Table 15.3, read 9.92 inHg abs.
2. Convert 100 scfm to acfm at a pressure of 9.92 inHg abs by using Table 15.5. Opposite 10 inHg read 1.5:

$$100 \times 1.5 = 150 \text{ acfm}$$

3. Refer to Table 15.8 to obtain C. This table has been developed from flow characteristics of air in Schedule 40 pipe. Opposite 3-in pipe, read 19.53.
4.
$$V = 150 \times 19.53$$
$$V = 2930 \text{ fps}$$
5. When scfm and pressure loss are known, use Fig. 15.22.

Vacuum Work Forces. The total force of the vacuum system acting on a load is based on the vacuum pressure and the surface area on which the vacuum is acting. This is expressed in the following formula:

TABLE 15.8   Factor for Determining Velocity Based on Pipe Size

Sched. 40 pipe NPS, in	DN	C factor	Sched. 40 pipe NPS, in	DN	C factor
³⁄₈	12	740.9	2½	65	30.12
½	15	481.9	3	80	19.53
¾	20	270.0	3½	90	14.7
1	25	168.0	4	100	11.32
1¼	32	96.15	5	125	7.27
1½	40	71.43	6	150	5.0
2	50	42.92	8	200	2.95

Note. Increase velocity by 5 percent for copper tube.

$$F = P + A \tag{15.5}$$

where F = force, psi
P = vacuum pressure, psig
A = area, in^2

Since the above formula is theoretical, it is common practice to use a safety factor in the range of 3 to 5 times the calculated force to compensate for the quality of the air seal and other factors such as configuration of the load and outside forces such as acceleration.

System Leakage

There is a difference between allowable and acceptable leakage in a vacuum system. Ideally, no leakage is desirable. It is common practice to test laboratory vacuum piping systems, section by section, at rated maximum working pressure for 24 hours with no loss of pressure permitted. For large systems, it is almost impossible to install an entire system that does not have small leaks. If such is the case, what is an acceptable amount?

There is no generally accepted value for allowable leakage in a vacuum system. That figure should be related to the volume of the piping network in order to be meaningful. The Heat Exchange Institute has developed a standard based on system volume. This formula transposed to solve for leakage is:

$$L = \frac{\left(\dfrac{0.15 \times V}{T} \right)}{4.5} \tag{15.6}$$

where L = leakage in SCFM
V = total Piping System Volume in cu.ft.
T = time for vacuum pressure to drop 1″ Hg., in minutes.

After calculating the system volume and the leakage from the system, enter Fig. 15.3 to determine if the intersection of the two values falls within the acceptable portion of the chart.

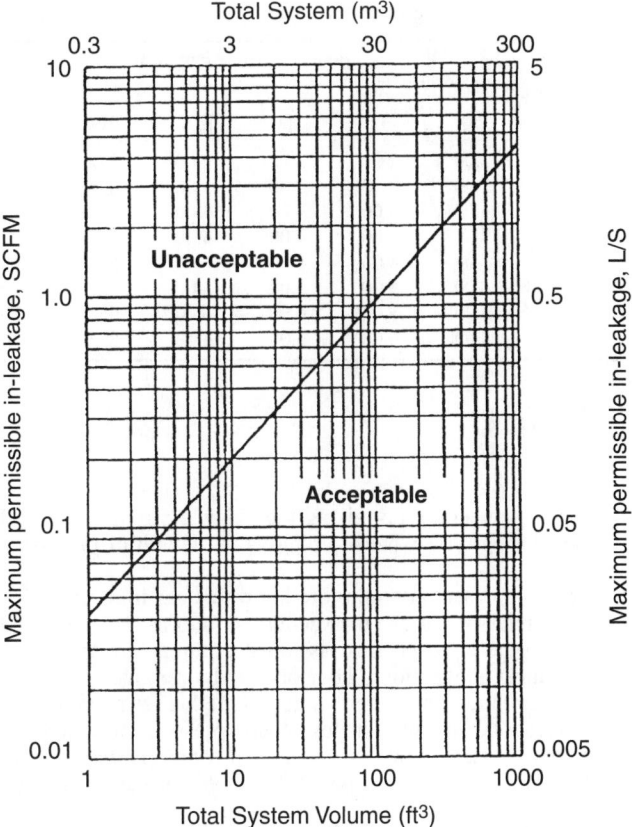

FIGURE 15.3 *(Courtesy of Heat Exchange Institute).*

SYSTEM COMPONENTS

General

Vacuum is produced by a single or multiple vacuum pumps drawing air from remote vacuum inlets or equipment. Except for some industrial applications, vacuum pumps withdraw air from a receiver to produce the vacuum. The piping distribution system is connected to the receiver. The pump(s) are also connected to the receiver and maintain the desired range of vacuum as the demand rises or falls depending on the number of inlets that open or close. When the system vacuum pressure drops to a predetermined level, additional pumps are started. When the desired high level of vacuum is reached, the pumps could be shut off. Larger units may be constantly operated, loading, unloading, or bypassing on demand. Often, there is a timer on the system, allowing the pumps to run for a longer time than required by system pressure to prevent rapid cycling.

Air exhausted from the system must be discharged to the atmosphere by means of an exhaust piping system. The pipe size shall be large enough so as not to restrict

TABLE 15.9   Vacuum Pump Exhaust Pipe Sizing, in.

Total vacuum plant capacity (scfm),* all pumps	Equivalent pipe length, ft						
	50	100	150	200	300	400	500
10	2.00	2.00	2.00	2.00	2.00	2.00	2.00
50	2.00	2.50	3.00	3.00	3.00	3.00	3.00
100	3.00	3.00	3.00	4.00	4.00	5.00	5.00
150	3.00	4.00	4.00	4.00	5.00	5.00	5.00
200	4.00	4.00	4.00	5.00	5.00	5.00	5.00
300	4.00	5.00	5.00	5.00	6.00	6.00	6.00
400	5.00	5.00	6.00	6.00	6.00	8.00	8.00
500	5.00	6.00	6.00	6.00	8.00	8.00	8.00

*SCFM $\times$ 0.03 = m^3/min.
1 in $\times$ 25.4 = mm.
1 ft $\times$ 0.305 = m.

operation of the vacuum pump. For sizing the exhaust piping, refer to Table 15.9, using the equivalent length of exhaust piping as the length of piping.

Gas Transfer Pumps.   Vacuum pumps are known as gas transfer pumps. They are essentially air compressors that use the vacuum system as their inlet and discharge "compressed" air to the atmosphere. Gas transfer pumps are the greater majority of pumps used for most applications. They operate by removing gas from the lower pressure in the system and conveying this gas to the higher pressure of the free-air environment through one or more stages of compression provided by a vacuum pump. These pumps are also known as *mechanical rotary pumps* and are most often used for industrial and laboratory purposes. Examples of gas transfer pumps are:

1. Rotary vane [once-through oil (OTO) type or oilless]
2. Reciprocating (rotary) piston pumps
3. Rotary lobe (roots), ordinary lobe or claw
4. Screw
5. Liquid ring
6. Diaphragm
7. Centrifugal (turbo)

The operation of the above pumps are described in Chap. 14, "Compressed Air."

8. Vacuum ejector pump. Technically not a pump, it operates on the venturi principle. Ejector operation is described and illustrated in Fig. 15.4.

Gas transfer pumps are classified as either positive displacement or kinetic. The various types are illustrated in Fig. 15.5.

Capture Pumps.   Capture pumps operate on the principal of having the molecules of the gas retained in the pump itself by sorption or condensation on internal surfaces. Examples are the diffusion pump, sorption pump, sublimation pump, sputter-ion pump, and the cryopump. Capture pumps typically are low-volume, ultrahigh

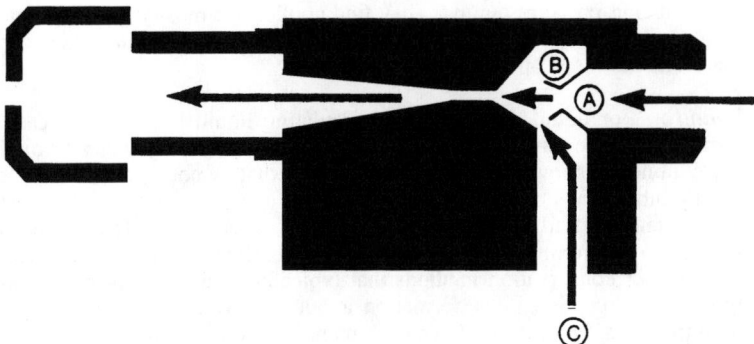

FIGURE 15.4 Vacuum ejector pump (gas jet). Ejectors operate on the venturi principle. Compressed air enters at *A*, and orifice *B* causes the airstream to increase in velocity, which creates a vacuum at *C*.

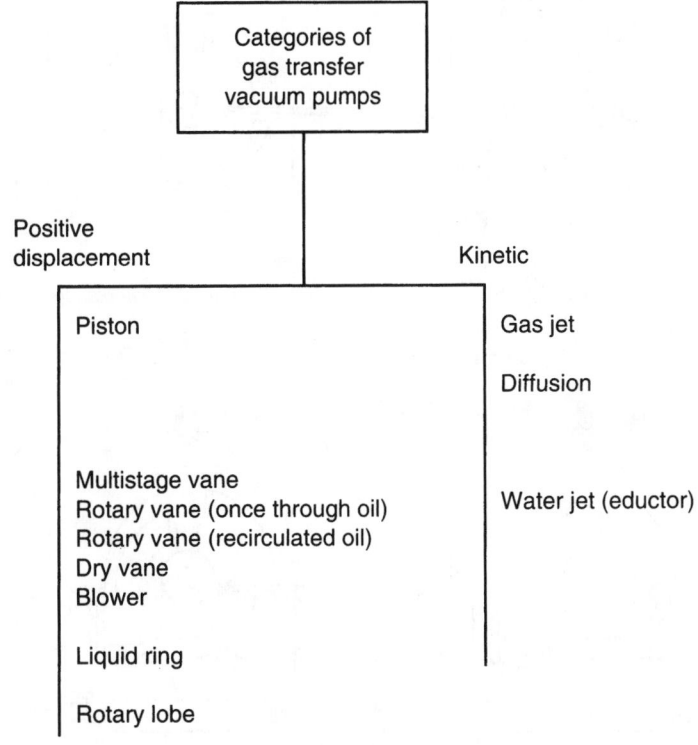

FIGURE 15.5 Categories of gas transfer vacuum pumps.

vacuum-producing pressure pumps. They find application mostly in semiconductor and unique research facilities. Because of their extremely limited use, capture pumps are outside the scope of this handbook.

Seal Liquids.   For liquid ring pumps, a circulating liquid in the pump casing is an integral part of the pump operation. This liquid is commonly water or oil. This liquid is commonly known as *seal liquid,* a term that is not intended to refer to shaft or any other kinds of sealing.

Water commonly used for sealing purposes must be continuously replaced. With no conservation, approximately 0.5 gpm/hp is used. Manufacturers have developed proprietary water conservation methods that typically reduce the usage to approximately 0.1 gpm/hp. Specific information about any water usage and additional space required must be obtained from each manufacturer. Various seal water piping methods are shown in Fig. 15.6.

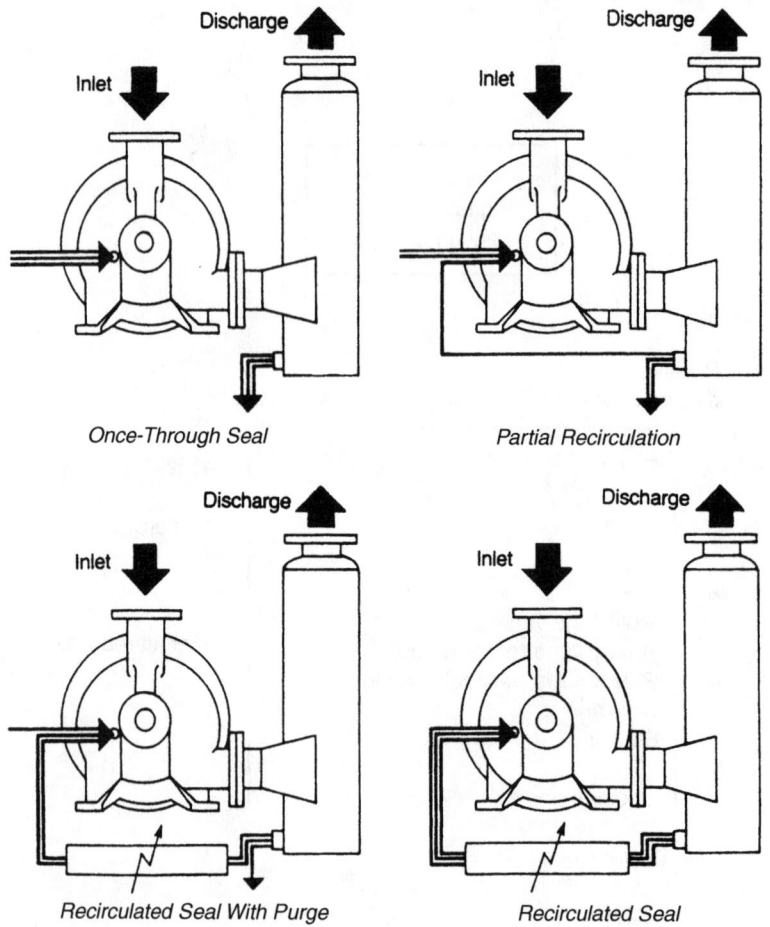

FIGURE 15.6   Seal water piping arrangements.

Oil used for sealing purposes is recirculated and may have to be cooled. The pump does not require any water to operate. The oil eventually becomes contaminated and must be replaced on a regular basis. Typically, a running time of 1500 to 2000 h is the useful life of the seal oil. Specific information about additional space required must be obtained from each manufacturer. It may be necessary to install a running time meter on these pumps to aid in maintenance. Pumps using oil often require more installation space than other types.

Operating Ranges of Various Pumps. Refer to Fig. 15.7 for typical operating ranges of the various types of pumps.

Vacuum Pressure Gauges

Manometer. A manometer is used to measure relative pressure between the system and local barometric pressure. It consists of a cylindrical U tube partially filled with liquid. One end is connected to the system being measured, and the other end

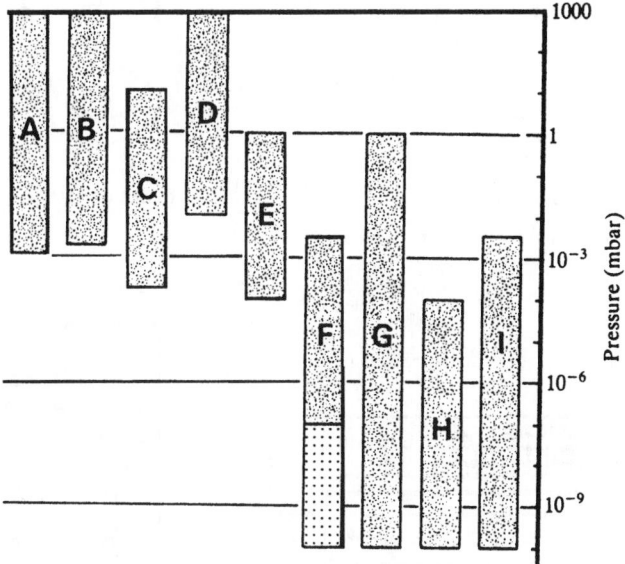

KEY PUMP
A   Sorption
B   Mechanical rotary
C   Mechanical booster (normally used with a rotary pump)
D   Dry pump (oil-free rotary)
E   Vapor booster
F   Diffusion (low pressures obtained with accessories)
G   Turbomolecular
H   Ion
I   Cryo

FIGURE 15.7 Operating ranges for various pump types.

could be open or closed. The difference between liquid levels in each tube is used to calculate the pressure. A manometer is illustrated in Fig. 15.8. A McLeod gauge is a variation of the manometer and is considered more accurate than a simple manometer. The manometer is used in laboratory work and is rarely found in industrial applications. A McLeod gauge is illustrated in Fig 15.9.

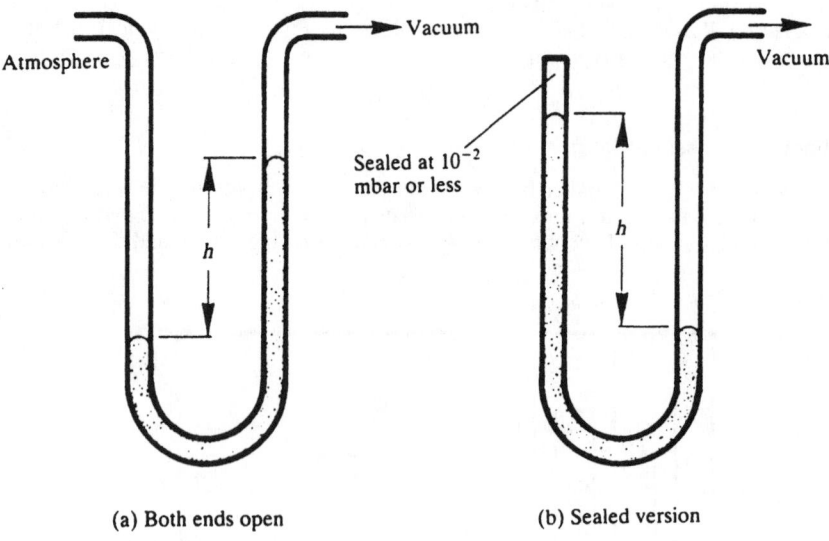

(a) Both ends open   (b) Sealed version

FIGURE 15.8   Manometer vacuum gauge.

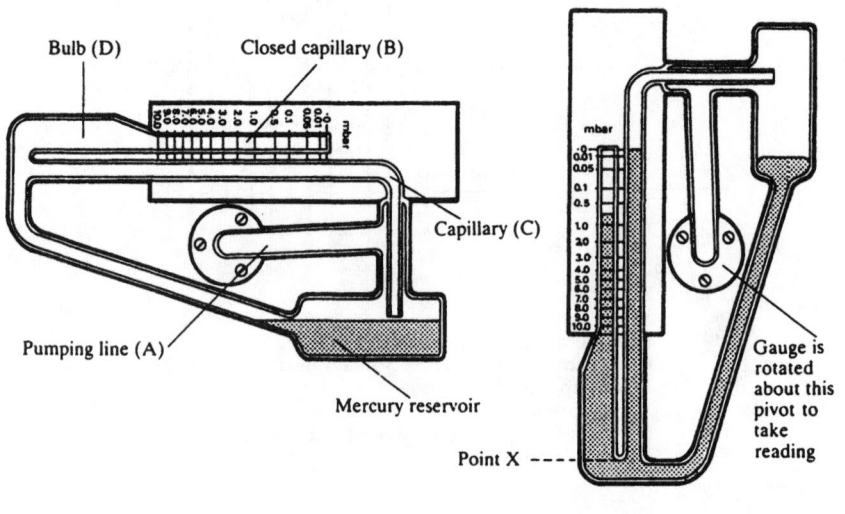

(a) In 'ready' position   (b) Pressure reading position

FIGURE 15.9   McLeod vacuum gauge.

Bourdon Gauge.   An often used type of mechanical gauge is the Bourdon gauge. This type of gauge is used to measure the difference between relative pressure between the system and local barometric pressure. Mechanical gauges are simple, inexpensive, and rugged and are the most widely used type of gauge. The heart of the gauge is the Bourdon tube that is closed at one end and open to the vacuum at the other. As the vacuum pressure varies, the tube changes shape. A pointer attached to the tube moves, indicating the pressure on a dial. A Bourdon gauge is illustrated in Fig. 15.10.

Diaphragm Gauges.   The diaphragm gauge measures the pressure difference by sensing the deflection of a thin metal diaphragm or capsular element. Similar to the bourden gauge, their operation relies on the deformation of an elastic metal under pressure.

A capacitance meter is, in essence, an electronic diaphragm gauge. Instead of a mechanical linkage, it uses a change in a variable capacitance sensor to detect changes in pressure, which are transmitted electronically. The response time is fast, and the signal can be remotely transmitted.

Strain Gauges.   Strain gauges also use the deflection of a diaphragm to produce a change in electrical resistance of the attached strain gauge. The response time is fast, and the signal can be remotely transmitted. A strain gauge is illustrated in Fig. 15.11.

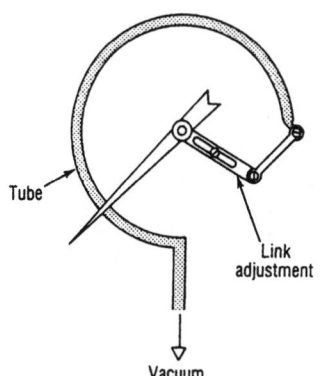

FIGURE 15.10   Bourdon tube vacuum gauge.

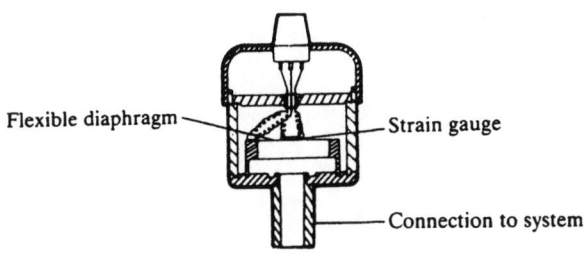

FIGURE 15.11   Strain gauge.

Ancillary Equipment. A coalescing, or oil mist, filter should be used on the exhaust of any pump that uses oil, to prevent the discharge of that oil into the atmosphere. It can also be used to recover solvents from the discharged airstream.

A knock-out pot is a device that removes entrained liquid or slugs of liquid from entering the inlet of mechanical pumps used in industrial applications. It can also be combined with an inlet filter in one housing.

Inlet filters are used to remove solids or liquids that may be present in the inlet airstream prior to the air entering the pump. Various filter elements are available to remove particulates approximately 0.3 mm in size.

In some cases, it is desired to lower the vacuum pressure to a branch where the system as a whole has a high vacuum pressure. This is done with an air bleed valve on the branch where the lower vacuum pressure is desired. The valve is opened, and air is allowed to enter the system. For precise control, a needle type of valve is used.

HEALTH CARE FACILITIES (SURGICAL/MEDICAL)

GENERAL

This section will describe vacuum air systems used in health care facilities. These systems include medical-surgical, dental, and waste anesthesia gas disposal. Vacuum air is the nomenclature used to separate this system from the plumbing of sanitary vent system.

CODES AND STANDARDS

The building codes that govern the construction and installation of plumbing, mechanical, and fire protection systems generally do not have specific standards directly regulating vacuum systems. The authorities having jurisdiction do regulate these systems by referring to standards originated by other organizations. These organizations have developed standards that are so widely accepted throughout the industry and by the various authorities that by reference, they have the force of law. These standards are required to be observed for the design, specification, storage, delivery, and testing of the various systems. Principal among them are:

1. National Fire Protection Association (NFPA)

   a. NFPA 99, Health Care Facilities

2. Compressed Gas Association (CGA)

   a. P 2.1, Recommendations for Medical/Surgical Vacuum Systems in Health Care Facilities

3. Heat Exchange Institute (HEI)

   a. Performance Standard for Liquid Ring Pumps

4. JCAHO, Accreditation Manual for Hospitals. Conformance with the requirements of this manual is mandatory for accreditation by the JCAHO. Accreditation has become necessary for Medicare and Medicaid reimbursement and other licensing requirements. Accreditation is not desired or obtained by all hospitals. This manual also refers to other standards.

5. AIA, Guidelines for Construction and Equipment of Hospital and Medical Facilities.

6. CSA (Canadian Standards Association). Since many medical devices and equipment are sold in Canada, manufacturers commonly conform to the CSA standards when they are more stringent than U.S. standards.

Sometimes local codes refer to specific standards that include a dated issue. In many cases, the date referred to in the code is not the latest issue. It is essential to become acquainted with the differences in the standards between the latest issue and the one referenced and discuss them with the local authorities. It is probable that the reference in the code will be used.

Some anesthesia equipment and devices manufactured in the United States for sale in Canada are not manufactured to U.S. standards but rather to Canadian standards because they are more stringent. The larger equipment manufacturers may also supply other foreign markets where requirements are stricter than those for the U.S. market.

MEDICAL-SURGICAL VACUUM AIR SYSTEMS DESCRIPTION

Medical-surgical vacuum systems, sometimes referred to as *patient vacuum systems,* are used in operating rooms, intensive care areas, medical and surgical suites, and patient rooms to assist in the removal of fluids. They are also used to remove waste anesthesia gases and for laboratory purposes. The operating pressure of this system is in a range of 13 to 20 inHg vacuum. It is commonly reduced when used.

SYSTEM COMPONENTS

Vacuum systems for health care facilities may consist of a vacuum source (pump), a receiver, separator (optional), exhaust to atmosphere system, inlets, gauges, the piping network and alarm systems. Vacuum supply for small dental offices often consists of liquid ring vacuum pumps draining directly to the sanitary sewer. Each component will be discussed as it relates to specific systems.

VACUUM SOURCE

The vacuum source consists of two or more pumps that are designed to operate as system pressure demand requires, a receiver, a separator (optional) used to remove liquids from the vacuum airstream, the interconnecting piping around the pumps and the receiver, and alarms. There is a requirement that the pumps selected must have the capacity selected when the largest pump is out of service. There is no requirement that the surgical-medical vacuum pump be used exclusively for this purpose. Waste anesthesia gas removal and laboratory vacuum use are also permitted to be generated by this source. If there is only one source of vacuum available for both surgical and laboratory services, the laboratory system shall connect separately to the receiver, but it must have its own separator and valving arrangement before such connection.

A schematic detail of a typical surgical-medical source is illustrated in Fig. 15.12. Since there is such a wide diversity of seal or cooling fluid piping, this is not shown. The fluid piping arrangement should be obtained from the pump manufacturer.

Pumps

There are two types of pumps most often used for vacuum service in health care facilities: liquid ring and sliding vane. A reciprocating pump is also available, but

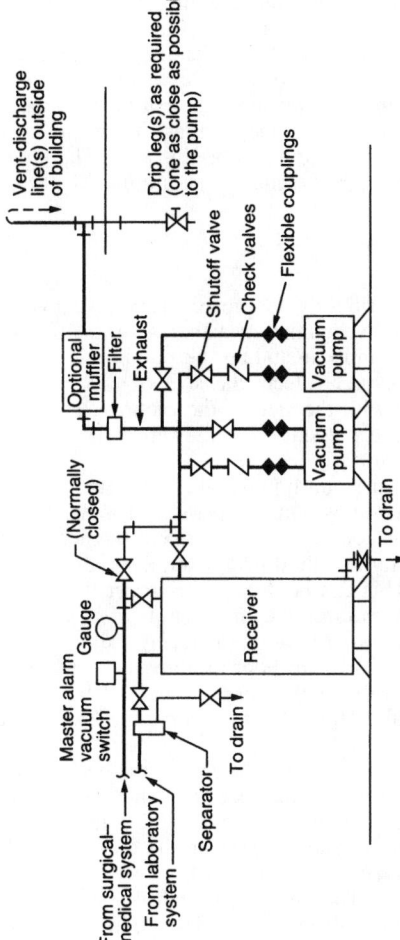

FIGURE 15.12 Schematic detail of surgical-medical vacuum source.

Vent-discharge line(s) outside of building

Drip leg(s) as required (one as close as possible to the pump)

Shutoff valve

Check valves

Flexible couplings

Optional muffler

Filter

Exhaust

Vacuum pump

Vacuum pump

(Normally closed)

Master alarm vacuum switch

Gauge

Receiver

To drain

From surgical-medical system

From laboratory system

Separator

To drain

15.21

it is rarely used because of its noise and higher operating and maintenance expense. Liquid ring pumps are relatively vibration free and have the advantage of a lower operating temperature that allows for the safe handling of any rarely used flammable anesthetic gas mixtures. It has a reputation for high reliability and low maintenance requirements. Lastly, it is oil free and will not have to be checked for oil consumption. The disadvantage of this pump is that it requires a continuous supply of water to operate, which is wasted if a recirculated cooling water supply is not available. Options are available to reduce the quantity of water used by recirculating the seal water and replacing only the amount necessary to maintain proper pump temperature. Other methods use a different fluid for the liquid ring, such as oil, that is recirculated and which may require cooling. When used to produce a higher vacuum pressure, the temperature of the seal water has an effect on pressure. The colder the water, the higher the vacuum that can be produced.

Rotary vane pumps are usually the smallest in physical size, have low starting and running torque, and are relatively vibration free. The typical rotary vane pump is capable of achieving a higher vacuum pressure than the typical liquid ring pump.

Receivers

Receivers are installed primarily to prevent excessive cycling of the pumps by providing a "reserve" of vacuum. Since the majority of vacuum pumps are furnished as a prepiped assembly including the pump, receiver, interconnecting pipe, and valves, the manufacturer selects the size of the receiver based on previous acceptable performance with the size of the pump selected. The receiver shall be rated as an ASME unfired pressure vessel. There is no requirement that a receiver be provided.

Fluid filter traps that pass no liquids when operated properly are used at every vacuum inlet in patient rooms and in operating rooms. However, experience has shown that some liquid does enter the surgical-medical vacuum system piping. Since there is a possibility that fluid may enter, a suitable method shall be provided to drain the receivers. This can be done either manually or automatically.

The manual method requires the installation of a valved drain line from the bottom of the receiver. In order to drain the receiver, it must be isolated and brought to atmospheric pressure. This procedure removes the receiver from service for a limited time. The drain valve is then opened and the liquid discharged into a floor drain. A sight glass shall be provided to observe the level of accumulated liquid in the receiver.

Automatic methods are rarely used because of the higher initial cost, but they will permit uninterrupted use of the receiver. Automatic drainage requires the installation of a separate, smaller drain tank with a level switch inside. The drain tank is installed adjacent to, and lower than, the receiver. A drain line, installed at the bottom of the receiver and controlled by a solenoid valve, shall connect the receiver to the drain tank. The drain tank shall be capable of being isolated from the receiver by means of a solenoid valve. The drain line shall also have a solenoid valve. All valves shall be set to operate in the correct sequence when the level of liquid in the drain tank reaches its set point and sends a signal to each solenoid valve. The effluent is then discharged to a floor drain.

Separators

When vacuum pumps use a constant flow of seal water, a separator is often necessary to remove water from the exhaust airstream. The effluent is commonly discharged to a floor drain. If the facility will be constantly introducing water into the vacuum piping system, a separator will be necessary before the piping system can be connected to the receiver. The receiver is not intended to be a separator, although a small amount of liquid is expected. In order to provide for draining and service of the receiver without interrupting the vacuum system, a valved bypass shall be provided.

Interconnecting Piping

The interconnecting piping between the vacuum pumps and devices of the central supply system shall be corrosion-resistant pipe such as copper, brass, stainless steel, or galvanized steel pipe.

Vacuum Pump Exhaust

Multiple vacuum pump exhausts may be manifolded together if there is a method of isolating individual exhausts so that one pump can be removed from service without affecting the other. Exhaust piping shall be sloped back to the pump. The minimum size of the exhaust pipe should be at least the same size as the vacuum pump exhaust port. The exhaust piping assembly shall be sized to limit the pressure loss to 1 psi or less. The exhaust should have an in-line muffler to lessen the noise and the exhaust line routed outside the facility. The exhaust line end shall have a louver and screen to prevent rain and insects from entering. Systems that serve research laboratories and patient treatment areas shall have a duplex filter on the exhaust. The pressure drop through this filter shall be added to the friction loss through the exhaust pipe in order to calculate total friction loss through the line. For sizing of the exhaust pipe, refer to Table 15.9.

Alarms at Vacuum Source

For duplex systems, when the second pump (called a *lag pump*) must be started because of low vacuum pressure produced only from the lead pump, a lag-pump-in-use alarm shall be installed in the pump control cabinet. It shall be both visible and audible. This alarm does not have to be repeated in the master alarm panel.

Inlets

Inlets are any terminals that receive any vacuum device or equipment. Connections to terminals in patient care areas shall be either threaded with a DISS adapter or a quick coupling adapter which is noninterchangeable with any other quick coupling for any other system. A secondary check valve may be required for some equipment but is not required on any patient care terminal. Laboratories generally use plastic hose connected to a serrated end of a laboratory cock.

Gauges

A main line gauge shall be located upstream (on the inlet side) of the main line shutoff valve. Gauges shall also be located at all anesthesia area locations at the room control valve. The gauges shall read from 0 to 30 inHg. These gauges are usually located in the valve box serving that area.

Valves

For discussion of valves, see "Valves" in compressed air chapter.

DISTRIBUTION NETWORK SIZING AND ARRANGEMENT

Pipe Material

Piping at the source could be seamless copper tube type L, M, or ACR or other corrosion-resistant material such as stainless steel or galvanized steel pipe (usually Schedule 40 ASTM A-53). Piping for the distribution system should be copper tubing. Copper tube shall be hard temper when installed in exposed locations and soft temper when installed in concealed locations and underground. Whenever piping passes under areas subject to surface loads such as roadways and parking lots, it shall be protected by ducts or casing enclosing the pipe.

Permanent fittings for copper tube shall be copper, brass, or bronze. Nonpermanent fittings, such as unions and flare connections, shall be installed so as to be readily accessible. Other joints, such as threaded and welded, are permitted. If proprietary joints or fabricating processes other than those commonly used are considered for use, they are permitted if they are listed and approved as equal to those joints made by brazing or soldering.

The most often used piping is copper tube type L with brazed joints for piping 4 in and smaller, and galvanized steel pipe with threaded joints for pipe 5 in and larger. Although solder joints are permitted for copper tube, brazed joints are used because all other compressed medical gas systems require this type of joint. The use of brazed joints will eliminate the possibility that other piping systems will be installed using solder. The fittings used shall be of the long turn type to cause as little restriction of the flow as possible.

Cleanliness of Piping System

The need to have the interior of the pipe clean is not a service requirement for this piping system. If the vacuum system is installed along with other medical gases, as is usually the case, there is a possibility of a cross connection or the inadvertent switching of pipes. This has the potential for contaminating other clean systems or creating a fire hazard. When vacuum lines are installed, it is required that the vacuum piping be either well identified and labeled or cleaned as other compressed medical gas systems. It is far less costly to have the pipe properly labeled or marked than to have it cleaned.

The exterior of the pipe shall be cleaned with soap and water.

System Sizing Procedures

Number of Inlets. The first consideration is to locate and count all of the inlets for each respective vacuum system that uses the same source. This is usually done by consulting a "program" prepared by the facility planner. This program is a list of all rooms and areas in the facility and the services that are required in each room or area. If a program has not been prepared, the floor plans for the proposed facility showing inlet locations shall be used.

There are guidelines published by the AIA, and ASPE that recommend the minimum number of station outlets for various services in specific areas. The most often used recommendations to determine the number of inlets for hospitals are those specified by the JCAHO since their accreditation is required for Medicare and Medicaid compensation. The JCAHO publishes a manual that refers to the AIA guidelines for the minimum number of station outlets for oxygen, compressed air, and vacuum that must be installed in order to obtain accreditation. If this is a factor for the facility, these requirements are mandatory. Other regulators, such as state or local authorities, may require that their approval of plans be obtained. These approvals may require adhering to the state or local requirements.

If accreditation or approval of authorities is not a factor, the number and area locations of station outlets are optional. The actual count will then depend upon requirements determined by each individual facility or another member of the design team using both past experience and anticipated future use, often using the guideline recommendations as a starting point. Vacuum inlets are included in Table 14.32 of the AIA recommendations. Table 14.34 provides recommended guidelines for the minimum number of facility station inlets and outlets that are a compilation from the AIA. Table 15.10 is the recommended number of inlets based on AIA recommendations.

Flow Rate. Each individual station inlet must provide a minimum flow rate for proper functioning of connected equipment under design conditions. The estimated flow rates for various system inlets and equipment are given in Table 15.10.

Many of the design parameters used to size the medical-surgical vacuum system have been adapted from information arrived at after extensive field surveys of hospital systems.

Diversity Factor. The flow rate from the total number of inlets connected to the system, without regard for any diversity, is called the *total connected load.* If the total connected load were used for sizing purposes, the result would be a vastly oversized system since not all of the outlets in the facility will be used at the same time. To allow for this fact, a diversity, or simultaneous use, factor has been developed. It is used to reduce system flow rate in conjunction with the total connected load for sizing the piping distribution system. This factor varies for different areas throughout any facility.

All areas (except WAGD and laboratory) of the facility are divided into separate usage groups: high usage called group A and lower usage called group B. Refer to Table 15.10 for the usage group assigned to each location. The actual diversity factor, or percent of usage, is a function of the total connected load and is found using Fig. 15.13. Enter the figure knowing the use group and the number of inlets. Read the percent use factor on the left. Table 15.11 is a direct numerical reading of Fig. 15.13 for convenience when sizing a large system. For discussions of the WAGD and laboratory diversity factors, refer to the following subsections.

TABLE 15.10   Recommended Number of Vacuum Inlets for Health Care Facilities

	Minimum number of station inlets	Usage group	Demand in scfm
Anesthetizing Locations			
Operating room	3/rm	A	
Cystoscopy/Endoscopy	3/rm	A	1.5
Delivery	3/rm	A	per
Special procedures	3/rm	A	room
Other anesthetizing locations	3/rm	A	
Acute Care Locations (Nonanesthetizing Locations)			
Neonatal	4/bed	A	
Recovery room (postanesthesia)	3/bed	A	
Critical care	3/bed	A	
Special procedures	2/rm	A	
Emergency rooms	1/bed	A	
Emergency rooms—cardiac	2/bed	A	
Cardiac ICU (CCU)	2/bed	A	
Catheterization lab	2/rm	B	
Surgical excision rooms	1/rm	B	
Dialysis unit	(½)/bed	B	0.25
Birthing rooms (LDRP or LDR)	2/rm	A	per
Postpartum bedroom	1/rm	B	outlet
Subacute Care Areas (Nonanesthetizing Locations)			
Nurseries	1/bed	B	
Infant resuscitation station	1/bassinet	A	
Exam and treatment rooms	1/bed	B	
Respiratory care	Convenience		
Other			
Autopsy	1/table	B	
Central supply	Convenience	B	
Equipment repair, calibration, and teaching	Convenience	B	
Laboratory*	—		1.0

* Author's recommendation—not in original table.

SYSTEM SIZING

Vacuum Pump (Source) Sizing Procedure

The vacuum pump size and equipment arrangement is found as follows:

1. Determine the total number of inlets throughout the facility, and categorize them based on groups A or B. If the number of inlets is unknown, an estimate can be made by using Table 15.10. Use the scfm listed for each inlet. For each WAGD and laboratory inlet, allow 1 scfm.

2. Calculate the flow rate from all of the connected outlets in each use group, using values obtained from Table 15.10.

3. For each separate group A, B, and anesthetizing locations, enter Fig. 15.13 or Table 15.11 with the total inlets in each group, and read the use factor. WAGD

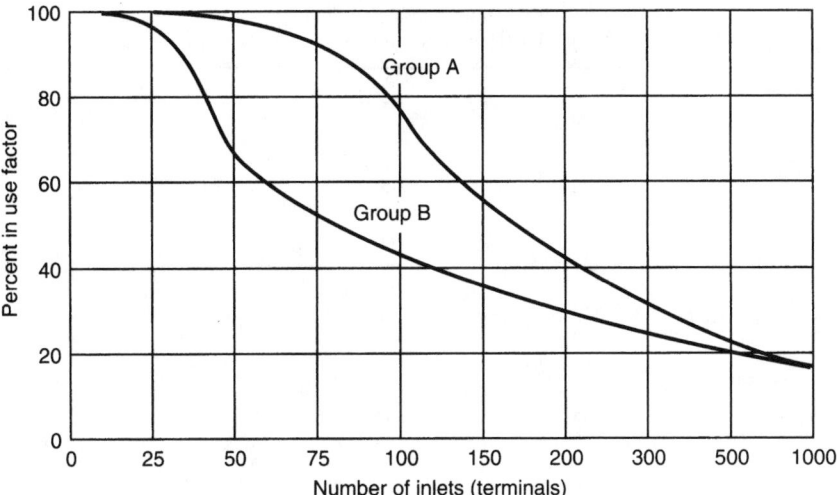

FIGURE 15.13   Simultaneous use curves for health care facilities.

outlets use the total connected load and have no diversity factor. For laboratory outlets, use the diversity figure found in Table 15.12.

4. To calculate the actual total system scfm, multiply the scfm from each of the individual groups of inlets by the appropriate use factors. Add the A use group, the B use group, the anesthetizing locations, and scfm together to find the required medical requirement. To this, add the laboratory and WAGD figures to find the total system scfm. Adjust the scfm figure for altitude if necessary.

5. The pump scfm ratings are selected based on a multiple-pump arrangement. If duplex pumps are selected, each pump shall be sized for 100 percent of the load. For a three-pump arrangement, 100 percent of the load shall be available with one pump out of service. In some cases, greater operating and horsepower efficiencies can be obtained with a three- or possibly four-pump arrangement. This must be analyzed for each project. The pumps should be sized for an operation based on the lead pump start setting pressure and system scfm. Since most manufacturers actually rate their pumps on acfm, scfm must be converted to acfm using Eq. (15.4).

6. The vacuum pressure at the pump should be based on 20 inHg. A minimum vacuum pressure at the farthest outlet should be 14 inHg. An often used figure of 5 inHg friction loss in the piping system would allow for a margin of safety to allow for greater-than-expected use. For a duplex pump installation, the following start and stop conditions for the pump are recommended for 14 inHg terminal pressure:

	Start, inHg	Stop, inHg
Lead pump	18	21
Lag pump	17	20

For a triplex installation, the following start and stop conditions for the pump are recommended for 14 inHg terminal pressure:

TABLE 15.11 Direct Reading of Simultaneous
Use Factors for Health Care Facilities

No. of inlets	Diversity	
	A	B
15	100	100
20	100	99
25	100	96
30	100	92
35	99	86
40	99	78
45	99	70
50	98	66
57	97	62
60	96	59
65	95	56
70	94	54
75	92	52
80	90	50
85	87	48
90	84	46
95	80	44
100	75	42
110	70	40
120	66	39
130	62	38
140	58	36
150	55	35
160	53	34
170	50	33
180	47	32
190	44	30
200	42	29
220	38	28
240	36	27
260	34	27
280	32	26
300	31	25
340	30	24
380	28	23
420	26	22
460	24	21
500	22	21
600	21	20
700	20	20
800	19	19
900	19	19
1000	18	18

Note: Diversities are based on an average hospital.
Specialty hospitals may require higher diversity.

TABLE 15.12 Vacuum Pipe Sizing Chart

(Maximum velocity: 5000 fpm; pipe material: L tubing; pressure drop: inHg per 100 ft)

		Pipe diameter, in							
sLpm	scfm	¾	1	1¼	1½	2	2½	3	4
28.3	1	0.08							
56.6	2	0.27	0.08						
85.0	3	0.53	0.15						
113.3	4	0.88	0.25	0.09					
141.6	5	1.3	0.36	0.14					
169.9	6	1.8	0.50	0.19					
198.2	7		0.65	0.24	0.11				
226.6	8		0.82	0.30	0.13				
254.9	9		1.01	0.37	0.16				
283.2	10		1.22	0.45	0.20				
424.8	15			0.91	0.40	0.11			
566.4	20				0.66	0.18			
708.0	25					0.26	0.09		
850.0	30					0.36	0.13		
990.0	35					0.47	0.17		
1130.0	40						0.21	0.09	
1275.0	45						0.26	0.11	
1415.0	50						0.32	0.14	
1700	60						0.44	0.19	
2000	70							0.25	0.06
2250	80							0.31	0.08
2550	90								0.10
2830	100								0.12
3540	125								0.18
4250	150								0.25
4950	175								0.35
5665	200								0.44

	Start, inHg	Stop, inHg
Lead pump	19	21
Second pump	18	20
Third pump	17	19

7. The pump selected will have a capacity based on the total acfm calculated in step 4 and capable of producing the highest vacuum pressure established.

Distribution Piping Sizing Procedure

Establishing System Design Criteria.   There are two figures needed to size the piping network. The first is the allowable pressure drop (which includes friction loss in the piping system, exhaust pipe losses, and losses through in-line devices such as filters or traps) for the entire network, in inHg per 100 ft of pipe (33 m). This is found by dividing the allowable pressure loss into the total equivalent length of pipe, in 100s of feet. The second is the adjusted scfm (connected scfm multiplied by a diversity factor) at the design point of the piping network.

To find the allowable friction loss for the network, first find the equivalent length of pipe. This is the actual, measured run of pipe, in feet, added to a distance, also in feet, that allows for the additional friction loss through fittings, filters, valves, and so on. The usual method of finding the equivalent run is to add 50 percent to the actual run. If the measured run is 300 ft (90 m), add 150 ft (45 m) to obtain the equivalent run, which is 450 ft.

The allowable friction loss is found by subtracting the design vacuum pressure desired at the most remote inlet from the highest pressure at the source, which is the lead pump stopping pressure. Generally accepted low pressure is 14 inHg at the most remote outlet, with 15 inHg being a more conservative figure. The previous discussion on pumps will determine the lead pump stop pressure.

There is no single accepted system to determine friction loss (pressure drop) criteria. For most systems, and for ease of calculations, it is generally accepted practice to allow a total friction loss of 3 to 5 inHg for the piping system, using any remaining pressure loss available as an allowance to compensate for in-line devices and vacuum pump source interconnecting piping.

Using the previous equivalent run of 450 ft (135 m) and a total friction loss of 3 inHg, the friction loss for the piping system is calculated by dividing 4.5 (450 ft divided by 100 because the chart uses loss per 100 ft) by 3 (inHg). This gives a figure of 0.66 inHg per 100 ft of pipe as the allowable friction loss for the piping network.

Pipe Sizing Procedure.   The piping distribution network is sized starting from the most remote point and working forward to the source. The scfm criteria, the use groups for each individual inlet, and the determination of the use factor is the same as that used to size the pump.

Starting at the most remote inlet on each branch, add together all the individual outlets by use group, starting with the first inlet and continuing toward the source. At each design point, find the scfm of each inlet. Add inlets with different use groups and specific uses separately. Multiply the inlet scfm by the appropriate diversity factor. Add this to other separate specific inlet scfm. This will result in calculating the design scfm in that particular pipe at the point being sized. This is continued separately at each design point to connection with the submain and finally the main. Determine the size pipe by using the vacuum sizing chart, Fig. 15.12. Entering the chart with the design scfm and the allowable system friction loss, find the intersection of the two values and select the larger pipe size. Entering Table 15.12 with the adjusted scfm, find a friction loss figure that is equal to or less than the allowable figure calculated; then read up to find the pipe size at the top of the column where the value is found. This chart is based on a vacuum pressure of 15 inHg, which is a commonly used and generally accepted value. To find pressure losses for values other than 15 inHg, use Eq. (15.5).

General Design Considerations.   There are several basic recommendations regarding the sizing of a system that shall be followed when sizing a complete network:

1. Because of the use of the diversity factor, it may be possible for a branch line to have a greater size than the main it is fed from. Always use the largest-size pipe calculated at any junction.
2. Do not use any diversity factor for OR's.

3. The smallest-size pipe shall be ½ in to any individual inlet. Use ¾ in as a minimum size for any branch and 1-in size minimum for any main or riser.

Valves and Valve Locations. The required valves and their locations are similar to those provided for the compressed gas systems in health care facilities. In short, valves must be provided at the base of all risers serving more than one floor and to isolate sections of the piping network for repair, maintenance, and expansion without interference to the remainder of the system. The valves shall be metallic and of a type that will not create more friction loss than the pipe itself. Refer to the compressed air, health care facilities section, of Chap. 14 for additional discussion.

System Alarms. In addition to the lag-pump-in-use alarm, the only other code-mandated alarm is low vacuum pressure for both the entire system and the anesthetizing systems. The low-pressure alarm is activated when the system vacuum pressure decreases to 12 inHg or to a point considered below the "normal" operating pressure range established by the facility. The pressure sensor for the system shall be located on the facility (inlet) side of the main line shutoff valve.

The pressure sensor for specific facility areas outside the pump room shall be set to activate below a pressure of 12 inHg. Anesthetizing areas shall have the sensor and a gauge located on the inlet side of any room control valve.

A high vacuum pressure alarm is not required. Refer to compressed air, Chap. 14, health care section, for further discussion of the location of annunciation and location of pressure sensor alarms.

PURGE AND TESTS

After installation, the entire system shall be blown clear of all debris with oil-free nitrogen. After this purge, conduct the following tests:

1. Visually inspect all joints at atmospheric pressure to assure penetration of solder or brazing alloy into the joint.

2. Pressure test the piping system at 15 psig (105 kPa) with oil-free nitrogen, and again observe joints for leakage.

3. Test only the piping system prior to the pumb being installed with nitrogen at 60 psig (35 kPa) and allow it to stand for 24 h. A maximum loss of 5 psig is permitted.

4. Test for cross connection as discussed in Chap. 14.

5. Test the entire system, including the pumps, gauges, and alarms, with vacuum at 12 inHg for 1 h. A maximum loss of 1.5 inHg is allowed.

6. Test the alarm system to assure activation when pressure falls below 12 inHg at the farthest outlet.

7. Test the piping to assure that the outlets, when evacuating 3 scfm (0.9 m^3/m), shall not reduce an adjacent outlet to a pressure below 12 inHg.

After all of the tests have been successfully completed, the system is then ready for operation.

WASTE ANESTHETIC GAS DISPOSAL

This section will describe vacuum systems used to remove waste anesthetic gas expelled from patients within operating rooms and other anesthetizing locations.

GENERAL

A brief description of anesthetic gas delivery will aid in the understanding of the waste anesthetic gas disposal (WAGD) system. Anesthetic gas is a mixture composed of air, oxygen, nitrous oxide, and anesthetic in various proportions. It is mixed in an anesthesia machine to which all of the compressed gases are connected and mixed. Often a liquid anesthetic is used, and the compressed gas is bubbled through the liquid and then delivered to the patient. The tube transporting the anesthetic gas is connected to a "circle breathing system" that is worn by the patient and allows inhalation. The mixture exhaled by the patient is directed into an adsorber, purified, and rebreathed by the patient. This line contains an expired-gas valve that acts as a check valve. A gas scavenger interface is installed on the exhaust of the adsorber that contains a popoff valve (a safety valve), which has the connection to the WAGD system and is provided with both a pressure and vacuum relief valve. It is this valve that protects the patient against full vacuum, and it has proven to be extremely reliable. Approximately one-half of the volume of anesthesia mixture will be expelled to the WAGD system, which is approximately 1 to 2 Lpm. A schematic illustration of a typical anesthesia machine is shown in Fig. 15.14.

Anesthetic gases, if allowed to accumulate within any room, can produce conditions that are capable of causing numerous problems. If constantly inhaled, any of the gases may cause health problems or reduce the effectiveness of the surgeon and other members of the operating team. Regardless of the problem, it is required that the gas be removed from the area as quickly as they are released and disposed of. There are both active and passive methods used for removal.

There are differences of opinion regarding acceptable concentrations. The most often used guideline is the AIA's *Guidelines for Construction and Equipment of*

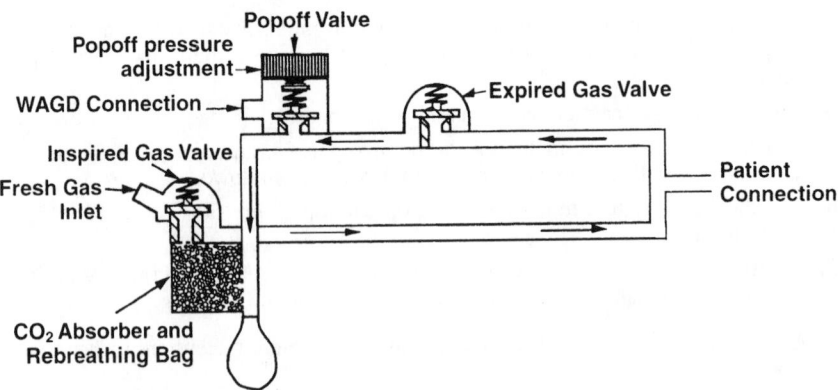

FIGURE 15.14   Schematic illustration of a typical anesthesia machine.

Hospitals and Medical Facilities. It states: "Acceptable concentrations of anesthetizing agents are unknown at this time. The absence of specific data makes it difficult to set up specific standards. However, any scavenging system must remove as much of the gas as possible from the room environment."

The most severe problem mentioned in some standards is flammable anesthesia gases, although this problem has mostly disappeared. Modern practice has phased out the use of these gases. The WAGD method most commonly used is to have a dedicated vacuum inlet to collect the vapors and transport them away for disposal.

METHODS OF ANESTHESIA GAS REMOVAL

There are four basic systems used for WAGD:

1. Room air changes
2. High vacuum active
3. Low vacuum active
4. Passive

Room Air Changes

One method is to increase the number of air changes inside the room to prevent gases from accumulating. This is not considered a very reliable or effective solution because of the problems controlling turbulence and avoiding dead spots in the individual rooms. Other problems are that since the air is contaminated, it cannot be recirculated. The wasted cost of once-through heating and cooling of the air makes operating costs very high.

Passive

A second method is to use only the pressure generated by the anesthetic machine itself to force the waste gas into a small-sized hose or tube for disposal outside the building or into an HVAC exhaust duct. This type of system is mostly used when no provision has been made for a dedicated WAGD system and has the advantage of being relatively free of danger to the patient. Past history of installed systems have been plagued by collapsed hoses, inadequate airflow, and improperly sized tubing. Maintenance is also very high. These systems work best when there is a good airflow to carry away the gases. The cost of this system is low, and the gas removal capabilities are considered poor.

Low Vacuum Active

This active method uses a low-pressure fan and small-diameter ductwork as a dedicated system to draw off the gases and exhaust them to outside the building or into the HVAC exhaust system. Because of the low vacuum pressure involved, usually inthe range of 6 to 12 in WC, the ducts must be relatively large. If the facility is large, the system becomes very complex. This system also requires accurate balancing in order to operate properly. The cost of this system is moderately high, and the gas removal capabilities are good. Because of the low pressure of the vacuum, this system is increasingly being used. In some codes it is mandated. The reason for this is that if an accident occurs and the full force of the vacuum system is exposed to the patient, it is not a life-threatening condition as may happen when using the medical-surgical vacuum system with a pressure of 15 to 20 inHg. It must be pointed out that the modern equipment in use has been manufactured to superior standards and has not caused this problem.

High Vacuum Active

The most often used system in the United States is either a dedicated WAGD system with its own dedicated pump or connection of the WAGD from the anesthesia machine directly into a dedicated outlet connected to the medical-surgical vacuum system. This type is called a *shared system.* Current good practice recommends that a dedicated WAGD system be used, although it is not mandated by code at this time. One problem associated with a shared system is that some of the anesthesia gases are not compatible with many gasket and seal materials and some types of piping used in standard medical-surgical vacuum systems. The most potentially serious problem is the possibility that the patient could be exposed to the full vacuum of the system upon failure of interface device safeguards on the anesthesia machine. Although a very remote possibility, this is a life-threatening condition. A schematic diagram of the interface from the WAGD inlet with the anesthesia machine is illustrated in Fig. 15.15.

SYSTEM SIZING

For a dedicated, piped system, a minimum vacuum pressure of 14 inHg is recommended. For a shared system, the inlets are connected to the surgical-medical sys-

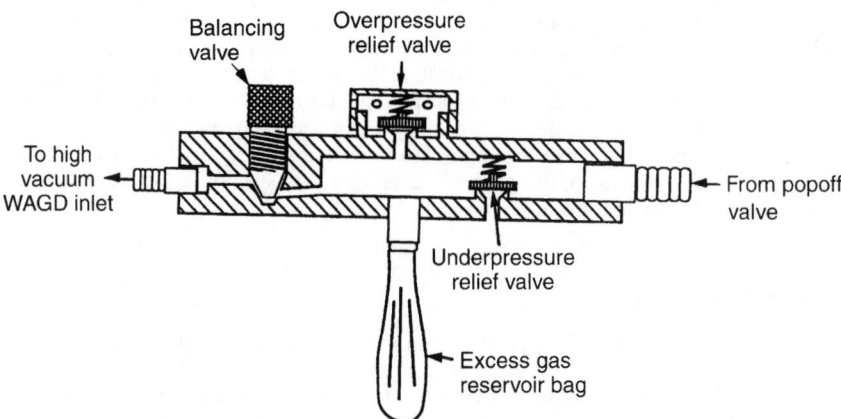

FIGURE 15.15   Interface of high-pressure WAGD inlet with the anesthesia machine.

TABLE 15.13 Advantages and Disadvantages of WAGD Systems

System type	Patient safety	Cost of installation	Ease of use	Ability to prevent gas accumulation
Air changes	Good	Low	Good	Poor
High vacuum	Poor	Mod. to high	Moderate	Good
Low vacuum	Good	Moderate	Good	Good
Passive	Good	Low	Moderate	Poor

tem, which has a minimum pressure of approximately 15 inHg. Vacuum pressure developed at the pump should be 19 or 20 inHg. Calculate the flow rate, determine the allowable friction loss per 100 ft of pipe, and use Table 15.12 to find the size pipe.

The flow rate shall be 1 scfm per WAGD inlet. No diversity factor is used. The total connected scfm shall be added separately to the surgical-medical system calculations if it is used as a source.

For the dedicated piped system, a larger pressure drop in the system can be allowed because there is only gas to be evacuated. A recommended loss of 6 inHg/100 ft will allow for smaller-sized pipes while keeping the velocity below 4000 scfm.

The ducted system uses 1 scfm per inlet with a pressure of about 12 in WC. Connection from the duct outlet is made with either a flexible hose or with copper pipe using appropriate adapters. Use standard HVAC criteria for sizing the main ducts allowing 1 scfm per inlet.

SYSTEM OPERATING CHARACTERISTICS

Advantages and disadvantages of the various systems are listed in Table 15.13.

DENTAL VACUUM SYSTEMS

GENERAL

The dental vacuum system provides suction for the removal of fluids and suspended residue from oral cavities during operative and other dental procedures.

CODES AND STANDARDS

NFPA-99 is the standard for design and installation for this system. There is very little specific material on dental vacuum systems, and the reference is made to the surgical-medical vacuum system for a majority of the requirements.

SYSTEM COMPONENTS

The components of this system are the vacuum pumps, separators (if required), and the exhaust system. As an ancillary part of the liquid ring vacuum pump assembly, water conservation methods are used by many manufacturers to reduce the amount of water used for this type of pump.

Vacuum Pumps

The two most often used types of pumps is the liquid ring and the centrifugal blower. Of these types, the liquid ring is commonly used for installations up to 5 hp motors. It also has the smallest installation space requirement since there is no need for a separator. The single disadvantage is that the pump must use water to seal the pump and is an integral part of its operation. The amount of water used is approximately 0.5 gpm/hp. Water conservation methods developed by manufacturers can often reduce the water flow rate to 0.1 gpm/hp, but the filtering arrangement requires more space. It is strongly recommended.

The blower is a centrifugal type of fan that will provide a larger flow rate and is generally used for motors 5 hp and larger. Since it cannot tolerate any moisture, a separator is mandatory. This source does not use any water, but it requires a larger amount of space for its installation because of the separator. The disadvantage is that it is not generally capable of producing as high a vacuum as the liquid ring pump.

Receivers are generally not used except for large installations because the vacuum system runs continuously.

Separators

A separator removes liquids and suspended solids from the vacuum airstream. It could be placed upstream of the vacuum pump to remove system liquids before they could enter the pump, or placed downstream on the exhaust line to remove liquids and oil before being discharged to the environment.

Because typical systems are shut down at night, the separators can be drained at this time. Air is prevented from entering the separator by a check valve held closed by the negative air pressure. When the system is returned to atmospheric pressure, the check valve opens and allows the effluent to drain to the sanitary system.

Separators could be provided with an overflow. Often, an automatic switch is provided to shut down the system if the liquid level rises too high. In this case, it should be drained manually.

For vacuum systems used in dental laboratories only, the debris in the system may be dry. It is common practice to use a cyclonic dry separator using a dry filter bag for this type of system.

DESIGN CRITERIA

General

Dental work is divided into three general categories: dental surgery, general dentistry, and laboratory uses.

It is recommended that the vacuum source for surgical uses should be separate from the other systems because the surgical vacuum could be considered a life support system. The reason is that very often the patient may be under anesthesia and the vacuum prevents accumulation of fluids that may choke the patient. For general dentistry, the patient is conscious and is capable of communicating to the dentist that fluids are accumulating.

Vacuum Pressure

The recommended vacuum pressure for the various systems are as follows:

1. *Dental surgery:* 12 to 17 inHg
2. *General dentistry:* 8 to 12 inHg
3. *Laboratory:* 5 to 9 inHg
4. For small practices, experience has shown that a vacuum pressure of 10 to 12 inHg will serve both surgery and general dentistry requirements.

The connection to the dental surgical vacuum system is often made using a typical filter bottle on the inlet similar to that used for the medical-surgical vacuum system.

Flow Rate

The commonly used dental equipment has the following average flow rates. There is no difference in the instruments between dental surgery and general dentistry.

1. *Saliva ejector:* 2 to 3 scfm (depending on tip opening)
2. *High-volume ejector:* 5 to 10 scfm (used to remove drill water)
3. *Hygienist:* 5 scfm
4. *One dentist and one hygienist:* 15 scfm
5. *Dental laboratory inlet:* 20 to 30 scfm. This is necessary to capture a wide variety of dust from grinding operations by the use of a "fishmouth" installed at each station.

Other often used flow rates are set according to the following selection criteria:

1. Two chairs: 15 scfm
2. Four chairs: 22 scfm
3. Five chairs: 30 scfm
4. Eight chairs: 44 scfm

When establishing the flow rate for dental offices, experience has shown that a figure of 5 scfm/chair is the lowest marginally acceptable flow rate, 10 scfm/chair is a good, average flow rate, and 15 scfm/chair is the flow rate used for offices that will have additional special-purpose equipment.

Diversity Factor

There is no generally accepted method or criteria for determining the diversity factor for dental vacuum. Since each facility is different, the largest anticipated simultaneous use shall be obtained from the end user. Information obtained from successful working systems in facilities and from designers-installers of dental equipment have established the following general criteria for both dentists and laboratory inlets:

1. 1 to 2 chairs: 100 percent
2. 3 to 4 chairs: 75 percent
3. 5 to 10 chairs: 60 percent

VACUUM GENERATION

Although any type of pump could be used to produce dental vacuum, the liquid ring and blower types are the most often used. A major consideration is the small amount of space available in which to put all the mechanical systems, such as air compressors and vacuum pumps, in a small facility. The space available will often dictate the type of vacuum pump selected.

Information obtained from various manufacturers indicates that the liquid ring pumps are preferred for pumps up to 3 to 5 hp units. For one dental unit a simple and less costly liquid ring type of pump, referred to as a *water flood machine,* could be used. The liquid ring pump uses a constant water supply that mixes with the system fluids that are discharged directly to the sanitary system drain. Since all liquids are run through the pump, no separator is required.

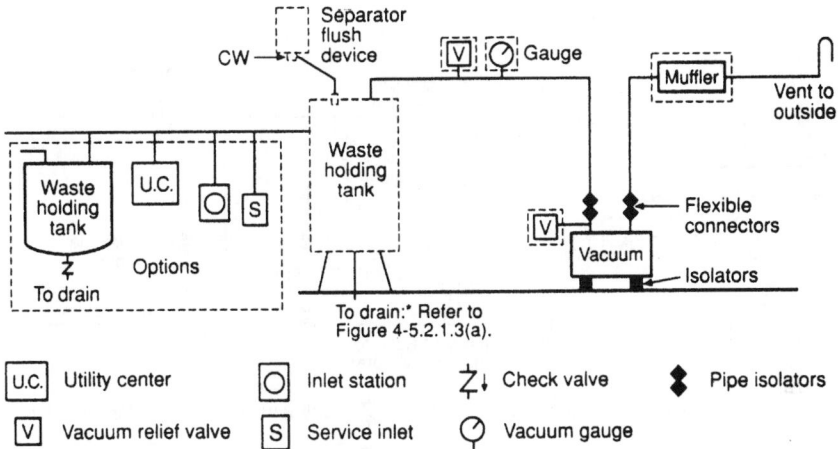

Note 1: Does not have to be below floor.
Note 2: Dotted lines indicate optional items.

FIGURE 15.16 Typical Level 3 wet or dry piping system with single vacuum source. (*Courtesy of NFPA.*)

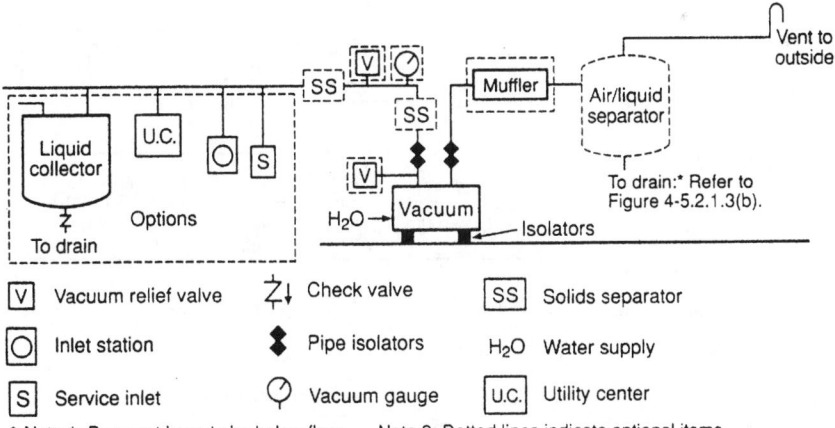

* Note 1: Does not have to be below floor.   Note 2: Dotted lines indicate optional items.

FIGURE 15.17 Typical Level 3 wet or dry piping system with single vacuum pump source.

A centrifugal pump is often used for motors larger than 3 to 5 hp (depending on the manufacturer) and is capable of supplying 10 to 12 inHg vacuum pressure for general dentistry requirements. A typical schematic based on the various pump arrangements is illustrated in Figs. 15.16, 15.17, 15.18 and 15.19. Typical receiver drainage is illustrated in Figs. 15.20 and 15.21.

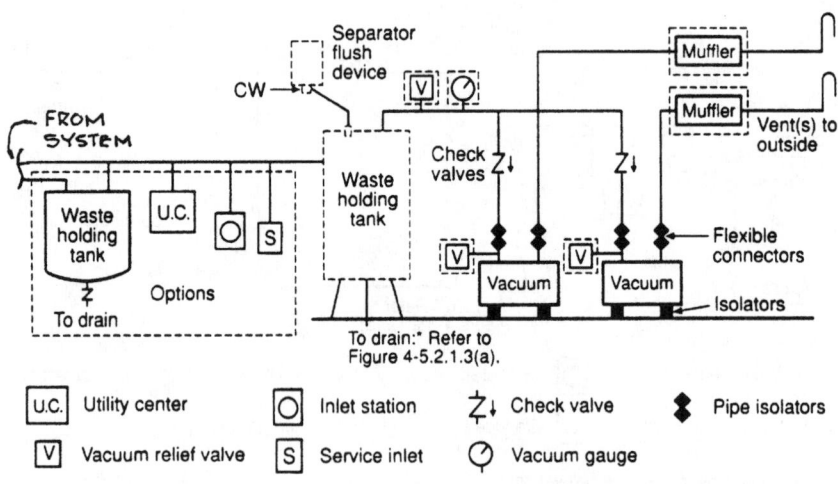

* Note 1: Does not have to be below floor.   Note 2: Dotted lines indicate optional items.

FIGURE 15.18 Typical Level 3 wet or dry piping system with duplex vacuum source.

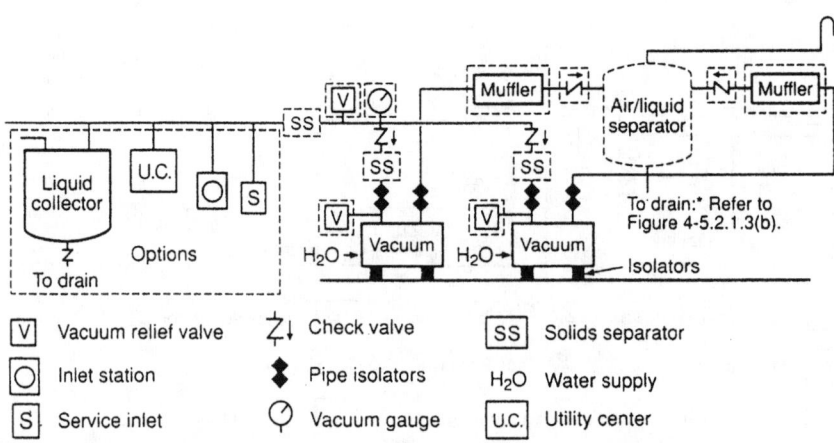

* Note 1: Does not have to be below floor.   Note 2: Dotted lines indicate optional items.

FIGURE 15.19 Typical Level 3 wet or dry piping system with duplex vacuum source.

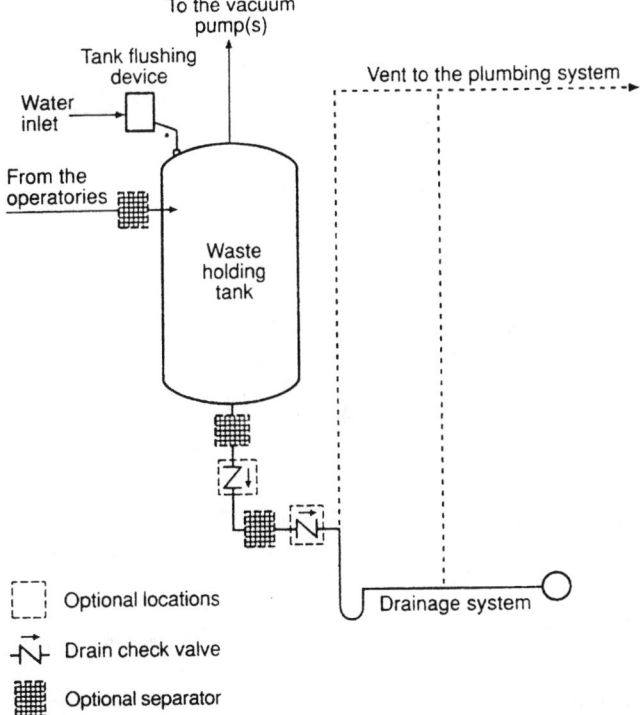

FIGURE 15.20   Drainage from a gravity-drained, liquid collector tank.

Another feature of many systems is an automatic wash. This consists of a potable water supply provided at one point of the piping system or only at the separator. Controlled by a timer and a solenoid valve, the solenoid is opened at a set time during the night for a 5- to 10-min period. Often, the wash is started at night when the system is shut down.

A large majority of installations are set to run continuously during the day and are shut down at night. For this type of system, a common method used to control vacuum pressure is to have an automatic line pressure relief valve controlled by a pressure switch set to the desired pressure. This valve allows air to enter the piping system if the vacuum pressure gets too high and closes when the design pressure is reached.

Exhaust

The pump exhaust shall be independently piped outside the facility, as similarly required for medical-surgical vacuum pumps. The exhaust pipe size is found using Table 15.9.

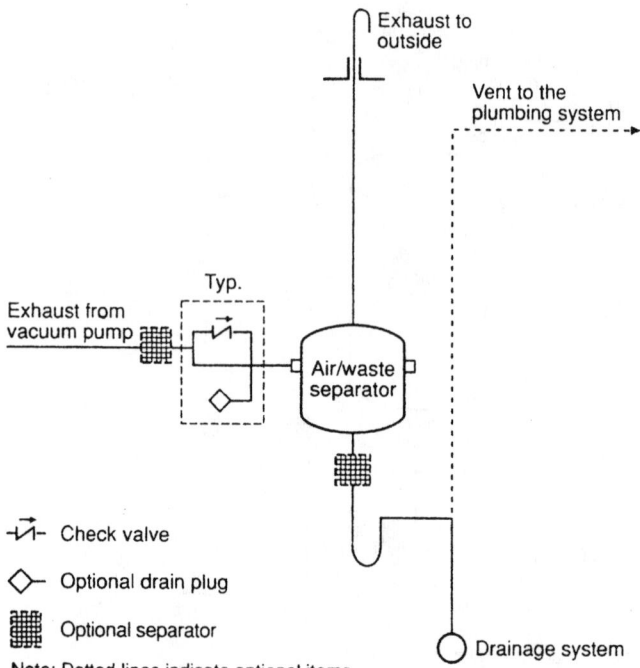

−⊢⤚ Check valve

◇⤚ Optional drain plug

▦ Optional separator

Note: Dotted lines indicate optional items.

FIGURE 15.21   Drainage from a positive discharge vacuum pump through an air/liquid separator.

VACUUM SOURCE DESIGN CONSIDERATIONS

All pumps should be duplexed, each capable of 100 percent capacity. If a separator is required, the size is a function of the pump capacity, with manufacturers recommending the size based on the horsepower of the pumps. Sizes of the separator range from approximately 10 to 60 gal, with larger sizes used with the larger pumps. Space conditions also play an important part in sizing the separator, since mechanical space is often very limited in small dental offices. This may require that the size separator be reduced to fit the space or eliminated by using a liquid ring pump.

DISTRIBUTION NETWORK

Pipe Material, Joints, and Installation

The pipe, materials, joints, and installation are the same as for medical-surgical systems. The most often used material is copper tube type L. Plastic is not recommended as a piping material. Joints are usually soldered when using copper pipe.

System Sizing

1. Obtain the layout of the facility, count all the inlets and chairs, and determine the length of run.

2. Determine the maximum number of simultaneous users.

3. Calculate the total connected scfm, and select the maximum vacuum pressure required.

4. Establish the range of vacuum pressure required. Using the allowable vacuum pressure drop for the system and the equivalent length of run, calculate the allowable pressure drop per 100-ft run of pipe using the lowest acceptable pressure.

5. Select the vacuum source and arrangement. This is based on adjusted scfm (using the connected load multiplied by the diversity factor), highest vacuum pressure desired, and available space.

6. Starting at the most remote point in the system, at each design point, calculate the adjusted scfm by multiplying the total connected load by the diversity factor. Entering Fig. 15.28 with the adjusted scfm and the allowable pressure drop, determine the size pipe by finding the highest allowable friction loss figure without exceeding the value, and finding the pipe size at the top of the table. Proceed to the source, using the same methods.

GENERAL DESIGN CONSIDERATIONS

It is very important to have the piping pitch back to the source in order to allow any liquid drain back to the floor drain provided at the source equipment. Pitch should not be less than ⅛ in per foot, with a ¼ in pitch preferred. Another important consideration is to have the piping properly supported to avoid liquids accumulating at low points of the pipe system. In general, piping shall be oversized. Branches shall not be less than ¾ in with mains and rises not less than 1 in in size. Provide cleanouts to allow for easy cleaning and removal of stoppages.

LABORATORY VACUUM SYSTEMS

The laboratory vacuum system serves general chemical, biological, and physics laboratory purposes, principally drying, filtering, transferring fluid, and evacuating air from apparatus. The usual working pressure of standard vacuum systems is in the range of 15 to 20 inHg. In some cases, there is a need for "high" vacuum in the range of 24 to 29 inHg, which is usually produced with a separate vacuum pump. The major difference between laboratory and surgical-medical systems is that the laboratory vacuum system does not normally carry liquids, but some invariably are introduced into the system. It is used primarily for pumping down and maintaining a vacuum rather than the transport of liquids or solids back to the source.

CODES AND STANDARDS

There are no codes and standards that are required to be used directly in the design of vacuum systems for laboratories. The most important requirements are those of the end user and good engineering practice. Often, NFPA-99 is used as a guide.

Laboratories conducting biological work where airborne pathogens could be released are required to follow the appropriate biological level criteria established by the NIH. For most biological installations it is recommended that check valves be installed in each branch line to every room or area to prevent any cross discharge. In addition, the vacuum pump exhaust shall be provided with duplex 0.2-μm filters on the exhaust to eliminate all pathogenic particulates.

VACUUM SOURCE

The vacuum source usually consists of two or more pumps that are designed to operate as system demand requires, a receiver used to provide a vacuum reservoir and to separate liquids from the vacuum airstream, the interconnecting piping around the pumps and receiver, and alarms. A duplex pump arrangement is usually selected if the system is critical to the operation of the laboratory. In some smaller installations where the vacuum system is not critical, it may be acceptable to have a single vacuum pump. The pumps selected should be oil free.

The principal types of vacuum pumps are divided into two general groups: gas transfer and capture. Capture pumps are outside the scope of this chapter. The types of pumps fall into the category of gas transfer and are as follows:

1. Rotary vane
2. Liquid ring
3. Rotary screw
4. Reciprocal

Rotary vane pumps include oil lubricated dry vane and oil flooded types. The oil lubricated is the oldest type requiring the oil to be dripped on the vanes for easy sliding. The oil flooded is a later improvement over the drip type. The dry vane type is the most often used.

If the pump manufacturer expresses the flow rate in ACFM, SCFM is calculated from the total connected inlets, the diversity factor and eq. 5. The stopping point of the pump is the required vacuum level, with levels similar to that discussed in the pump selection. Refer to the discussion in the medical-surgical section for advantages and disadvantages of each type of pump. The receiver, interconnecting piping, and exhaust arrangement are also similar. The basic difference is the absence of mandated alarms. None are required except for those necessary for maintenance by the end user. A detail of a typical laboratory vacuum pump assembly is illustrated in Fig. 15.22.

REDUNDANCY

The redundancy of laboratory systems is based on the system capacity required with one pump being worked on or otherwise out of service. Only one fault is postulated. If there is a fire along with a power outage, additional provisions must be made. Mixed sizes are also not included in this chapter.

There are two scenarios to be considered. The first is that there is no excess capacity in the system. The second is that the system will provide full capacity if one pump is down or out of service. The only way to find out is to question the client (or end user) as to their needs and comply with them. The question to ask is how important is it to have 100% redundancy.

It is normal to have multiple pumps unless the client is determined to have the least first cost. If the client chooses no excess, write a letter giving your reasons for disagreeing with the lack of capacity. This will allow your firm to have something in the file if things go wrong. If the number of pumps is to be more than one, the answer is not simple. The decision has to be made as to how many, with the understanding that the various pump combinations would be almost infinite. With the decision not to have excess capacity, the number of pumps must be considered. Is it to be 50% and 50%, three at 33% or four at 25%. For no excess capacity, it is normal to have two 50% pumps. This will have at least the capability of some usefulness during an outage.

It is most normal to have excess capacity. This will allow the pumps to keep up with the full load with one of the pumps out of action. This means that the pump assembly will consist of two pumps or more, with the combination of pumps having the full load capacity. The range of pump combinations is almost infinite. It would be up to the manufacturer to explain how the pump assembly would be set up in order to make a decision. The most common would be two 60% or two 75% pumps. This will allow almost full operation except during "design" operating conditions. If design parameters are exceeded, the second pump will start giving additional vacuum production. It is common to have three 50% pumps mounted on a receiver. This way it will be able to handle full design load with one pump out of service.

A similar consideration should be given to the compressed air source pumps.

DISTRIBUTION NETWORK

Pipe Material and Joints

Piping for the distribution system shall be a corrosion-resistant material such as copper tube type K, L, or M, stainless steel, or galvanized steel pipe (usually

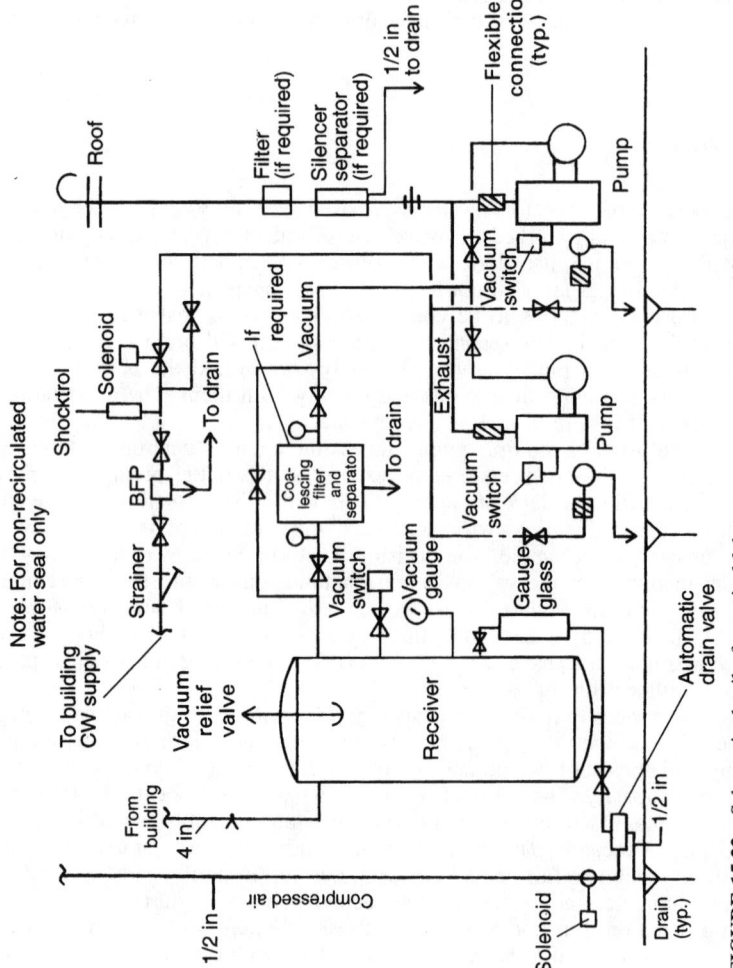

FIGURE 15.22 Schematic detail of a typical laboratory vacuum pump assembly.

Schedule 40 ASTM A-53). Copper tube shall be hard temper except when installed underground, when soft temper should be used. Whenever piping passes under areas subject to surface loads such as roadways and parking lots, it shall be protected by ducts or secondary containment. Although cost has a major influence on the selection of the piping material, the most commonly used is copper tube type L, ASTM B-88 up to 4 in in size, with soldered joints. Pipe 5 in and larger is usually Schedule 40 galvanized steel pipe with threaded joints. Fittings shall be long turn drainage pattern so as not to impede the flow of fluids in the pipe.

SIZING CRITERIA

Number of Inlets

There is no code or other mandated requirements for specific locations of laboratory vacuum inlets. The number of inlets is determined by the user, based on a program of requirements for all rooms and areas and equipment used in the facility. Inlets for laboratory stations, fume hoods, and so on shall be appropriate for the intended use, based on requirements of the end user.

Flow Rate. The basic flow rate from each laboratory inlet shall be 1.0 scfm. This is an arbitrary number and is based on experience. This flow rate is used in conjunction with the diversity factor.

Diversity Factor. The diversity factor established for general laboratories is based on experience. It has been found to be slightly more than that used for compressed air because the vacuum is often left on for longer periods of time. Refer to Fig. 15.23 for a direct reading chart to determine the adjusted general laboratory vacuum

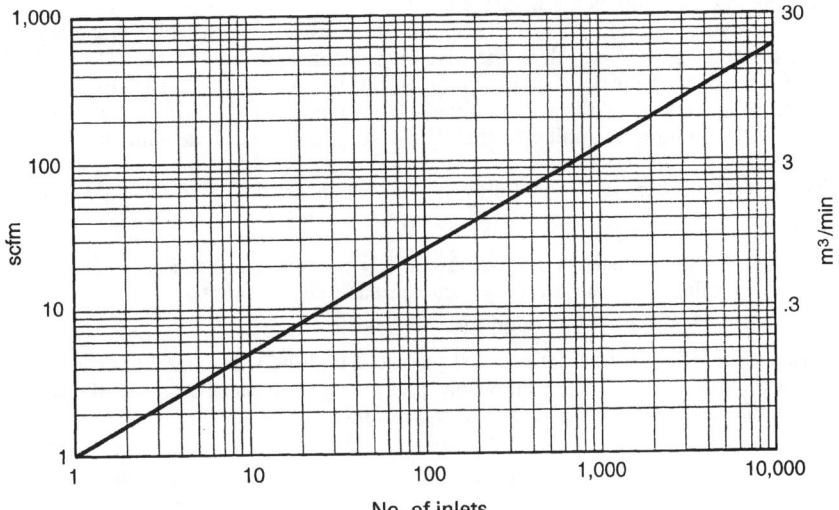

FIGURE 15.23 Laboratory vacuum demand, direct reading.

TABLE 15.14 Diversity Factor for Laboratory
Vacuum Air Systems

Number of inlets	Percent use factor
1–2	100
3–5	80
6–10	66
11–20	35
21–100	25

flow rate using the number of connected inlets. Refer to Table 15.14 for the diversity factor for up to 100 inlets.

For the design of classrooms, the diversity factor for one and two classrooms on one branch is 100 percent. For more than two classrooms, use a diversity factor double for that of compressed air in Table 14.13 but never less than the largest scfm calculated for the first two rooms. Since the above flow rates and diversity factors are arbitrary, they must be used with judgment and modified if necessary to adjust for special conditions and owner requirements. Always consult the user for definitive information regarding the maximum probable simultaneous usage.

Allowable Friction Loss.   A generally accepted figure used to size a piping system is to allow a friction loss of 3 inHg for the entire system (after the source assembly) and a maximum velocity of 5000 fpm. If noise may be a problem, use 4000 fpm. For smaller systems, use a figure of 1 inHg for each 100 ft of pipe.

Vacuum Pump Sizing.   The source pump for laboratories is selected using the flow rate of gas calculated using all inlets, the diversity factor for the whole facility, and the required vacuum pressure. In general, it has been found that in most facilities, the vacuum pumps are oversized.

If the pump manufacturer expresses the flow rate in acfm, scfm is calculated from the total connected inlets, the diversity factor, and Eq. (15.1) or Table 15.4 or 15.5. The stopping point of the pump is similar to that previously discussed in the pump selection for the surgical-medical vacuum system. The exhaust and interconnecting piping is similar to that of the surgical-medical vacuum system.

Piping Network Sizing.   The following method is used to size the pipe at each design point:

1. Calculate the adjusted scfm (sLpm) at each point using the connected scfm (sLpm) reduced by the diversity factor at each design point.
2. Calculate the allowable friction loss per 100 ft of pipe.
3. Enter Table 15.12 with the scfm (sLpm), and find the value equal to or less than the previously determined allowable pressure loss. Read the size at the top of the column where the selected value is found.

VACUUM CLEANING SYSTEMS

This section will discuss vacuum systems used for removing unwanted solid dirt, dust, and liquids from floors, walls, and ceilings. This can be accomplished by the use of either a permanent, centrally located system or a portable, self-contained, electric powered unit. The central system will transport the dirt to a central location where it can be easily disposed of or recovered. Portable units can be easily moved throughout all areas of a facility. They are outside the scope of this handbook.

TYPES OF SYSTEMS AND EQUIPMENT

There are three types of permanent systems: dry, wet, and a combination system. The dry system is intended exclusively for free-flowing, dry material. It is the most commonly used, with cleaning capabilities ranging from cleaning carpets to removing potentially toxic and explosive product spills from floors in an industrial facility. Equipment consists of a vacuum producer, one or more separators that remove collected material from the airstream, tubing to convey the air and material to the separator, and inlets located throughout the facility. A wide variety of separators are available to allow disposal and recovery of the collected material.

The wet system is intended exclusively for liquid handling and pickup. It is commonly found in health care, industrial, and laboratory facilities where sanitation is important and frequent washings are required. Equipment consists of a vacuum producer, a wet separator constructed to resist the chemical action of the liquids involved, piping or tubing of a material resistant to the chemical action of the liquid, and inlets located throughout the facility. A typical wet vacuum cleaning pump assembly is illustrated in Fig. 15.24.

A combination system is capable of both wet and dry pickup. Equipment consists of a vacuum producer, a wet separator constructed to resist the chemical action of the liquid mixtures involved, pipe or tubing of a material resistant to the chemical action of the combined solid-liquid, and inlets located throughout the facility. Another method uses a portable wet separator attached to the dry piping system at the inlet location.

CODES AND STANDARDS

There are no codes or standards governing the design and installation of vacuum cleaning systems.

SYSTEM COMPONENTS

Vacuum Producer (Exhauster)

Vacuum producers for typical vacuum cleaning systems consist of a single or multistage centrifugal unit powered by an electric motor. The housing can be con-

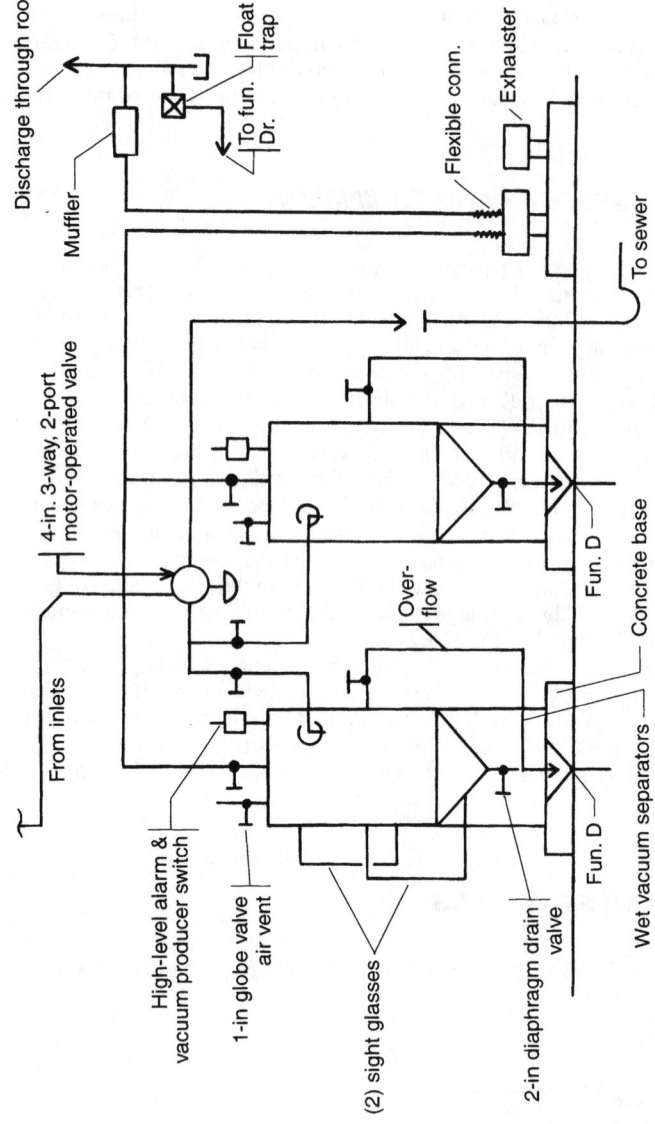

FIGURE 15.24 Schematic of a typical wet vacuum cleaning pump assembly.

structed of various materials to handle special chemicals such as nonsparking aluminum for potentially explosive dust. The discharge of the unit can be positioned at various points to accommodate requirements of the exhaust piping system.

Separators

Separators are used to remove the solid particulates in the airstream generated by the vacuum producers. For dry systems, tubular bag and centrifugal separators can be used. If only dust and other fine materials are expected, a tubular bag type is adequate. The bag(s) are permanently installed and removed only when replacement is necessary. They function as an air filter for fine particles and collect a majority of the dirt. This dirt eventually falls into a hopper or dirt can at the bottom of the unit. To empty the entire unit, the system must be shut down. The bag(s) must be shaken to remove as much of the collected material as practical and emptied into the dirt can. The dirt can is removed (or the hopper is emptied into a separate container) in order to clean out the unit. The dirt can should be sized to contain at least one full day's storage. Units are available with multiple bags to increase filter bag area. Shaking can be done either manually or automatically, with a motor-operated shaker. The motor-operated shaker has adjustable time periods to start operation after a variable length of time from shutdown of the system and for a variable length of time for the bags to be shaken. If continuous operation is required, compressed air can be used to blow through the bags and remove the dirt without requiring a shutdown.

The centrifugal separator is designed to remove coarser dry particles from the airstream. It is also used when additional dirt storage is needed and also when more than six simultaneous operators are anticipated. The air enters the separator tangentially to the unit, forcing the air containing particulates into a circular motion within the unit. Centrifugal force accomplishes separation.

The wet separator system collects the liquid, separates the water from the airstream, and discharges the waste to a drain. This type of separator can be equipped with an automatic overflow shutoff that stops the system if the water level reaches a predetermined high water level and can be equipped with automatic emptying features.

Immersion separators are used to collect explosive or flammable material in a water compartment. If there is a potential for explosion, such as in a grain- or flour-handling facility, the separator shall be provided with an integral explosion relief-rupture device that is vented to the outside of the building.

Filters

Vacuum producers are normally exhausted to the outside air, and usually do not require any filtration. However, when substances removed from the facility are considered harmful to the environment, an HEPA filter must be installed in the discharge line to eliminate the possibility of contamination of the outside air. The recommended location is between the separator and vacuum producer.

Silencers

When the exhaust from the vacuum producer is considered too noisy, a silencer shall be installed in the exhaust to reduce the noise to an acceptable level. Pulsating airflow will require special design considerations. Connection to silencers shall be made with flexible connections. Separate supports for silencers are recommended.

Inlets

Inlets are female inlet valves and are equipped with a self-closing top-hinged cover. They provide a quick connection for any male hose or equipment. The cover can be locked closed as an option. Many different inlet types are available in various materials and in sizes ranging from 1½ to 4 in.

Control and Check Valves

Valves for the vacuum cleaning system are different than standard valves. They are used to control the flow or stop the reverse flow of air in the vacuum cleaning system. When a valve is used as a regulating valve, it is called a *wafer butterfly valve* or an air gate valve. An air gate is illustrated in Fig. 15.25. A less costly substitute for an air gate is called a blast gate, and it operates using a sliding plate in a channel. An air gate is illustrated in Fig. 15.26. The plate has a hole that matches the size of the opening in the channel, with room to close off the opening completely. Air gates can be used only in low-pressure systems and are generally available in sizes from 2 to 6 in.

Check valves are typically spring-loaded, swing-type and are hinged in the center. A typical check valve is illustrated in Fig. 15.27.

Air Bleed Control

If the exhauster is constantly operated with low or no inlet air, there is a possibility that the exhauster motor will become hot enough to require shutdown due to overheating. To avoid this, an air bleed device can be installed on the inlet to the

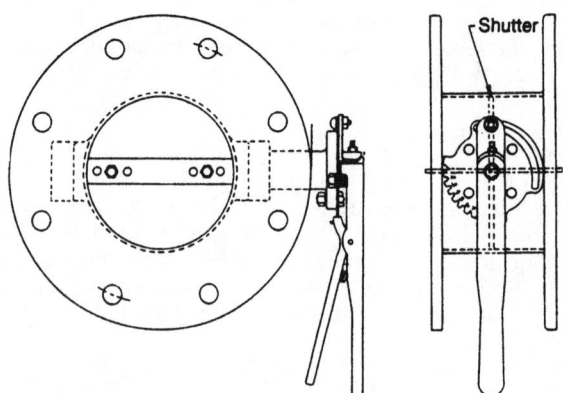

FIGURE 15.25 Air gate valve.

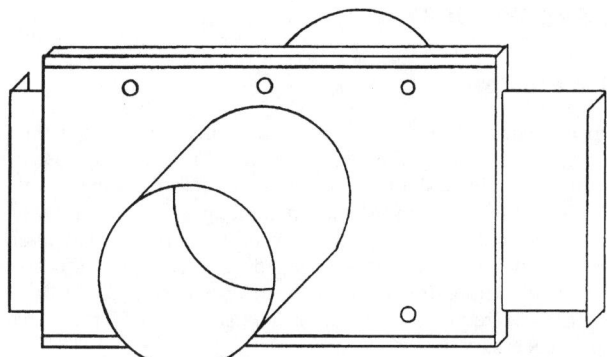

FIGURE 15.26   Typical blast gate. (*Courtesy Spencer Turbine.*)

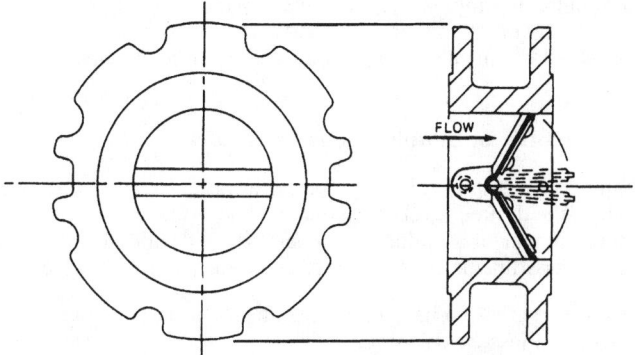

FIGURE 15.27   Check valve. (*Courtesy Spencer Turbine.*)

exhauster that will automatically allow air to enter the piping system. If the facility indicates that overheating may be a possibility, the manufacturer of the unit should be consulted to determine the need for this device for the system selected.

Pipe and Fittings

The most often used material is thin-wall tubing, generally in a range of 12 to 16 gauge. This tubing is available in plain carbon steel, zinc coated steel, aluminum, and stainless steel. Fittings are special, designed for the vacuum cleaning system. Tubing is normally joined using shrink sleeves over the joints. Compression fittings and flexible rubber sleeves and clamps are also used. Tubing shall be supported every 8 to 10 ft, depending on size, under normal conditions. Standard steel pipe is often used in areas where the additional strength is required. In special areas where leakage and strength are mandatory, the tubing joints can be welded if required.

DETAILED SYSTEM DESIGN

Inlet Location and Spacing

The first step in system design is to locate the inlets throughout the facility. The spacing of inlets depends on the length of hose selected for use. After this is decided, the inlet locations shall be planned in such a manner that all areas are capable of being reached by the selected hose length. This must take into account furniture, doorways, columns, and all obstructions. Some small overlap must be provided to allow for hoses not being able to be stretched to the absolute end of their length. Consideration should be given to providing a 25 ft, 0 in, spacing for areas where spills are frequent, heavy floor deposits may occur, and where frequent spot cleaning is necessary.

Generally, there are several alternate locations possible for any given valve. Inlets should be placed near room entrances. Wherever possible, try to locate inlets in a constant pattern on every floor. This allows for the location of common vertical risers since the distance between floors is less than the distance between inlets. In any system, minimizing piping system losses by a careful layout will be reflected in reduced power requirements of the exhauster.

The inlets should be located between 24 and 36 in above the floor.

Determining Number of Simultaneous Operators

This is a major consideration for design purposes because an underdesigned system will not produce the desired level of vacuum and an oversized system will be costly. The maximum number of simultaneous operators is decided by the housekeeping or maintenance departments of the facility and depends on a number of factors:

1. Is the preferred method to have gang cleaning? Is it possible to alter this practice in order to use a less costly system?
2. What is the maximum number of operators expected to use the system at the same time?
3. Is the work done daily?

For commercial facilities where there may be no available information, the following guidelines are based on experience and can be used to estimate simultaneous use based on productivity. These figures consider the greater efficiency of using a central system compared to portable units, often in the order of 25 percent. They must be verified and based on actual methods anticipated:

1. For carpets, one operator will be expected to cover 20,000 ft^2 of area for regular carpeting in an 8-h shift. For long or shag carpets, the figure is about 15,000 ft^2. Another generally accepted figure is 3000 ft^2/h for standard floors and 2500 ft^2/h for shag and long carpets.
2. For hotels, an average figure of 100 rooms, including adjacent corridors, per 8-h shift would be expected. For long or shag carpets, the figure is about 75 rooms.
3. For theaters, use the number of seats divided by 1000 to establish the number of simultaneous operators.
4. For schools, an average figure is 12 classrooms per day for a custodian to clean in addition to other duties normally accomplished.

Inlet Valve and Hose Sizing

Experience has shown that the use of $1\frac{1}{2}$-in size hose and tools for cleaning floors, walls, and ceilings is the most practical size. Smaller, 1-in size tools are used for cleaning production tools, equipment, and benches. Larger hoses and tools are used for picking up expected large spills and used to clean large tanks, boxcars, and the holds of ships. Refer to Table 15.15 for general recommendations for tool and hose sizes.

Standard hoses are available in 25-, 37.5-, and 50-ft lengths. For general cleaning, the location of inlet valves should allow for convenient cleaning with 50 ft as the maximum hose length. This represents a labor saving by halving the times an operator has to change outlets. This length should not be exceeded except for occasional cleaning because of excessive pressure drop.

Locating the Vacuum Producer Assembly

The vacuum producer assembly consists of the vacuum producer, commonly called an *exhauster*, and separators. The following shall be considered in locating the vacuum equipment:

1. Provide enough headroom for the piping above the equipment and to allow the various pieces to be easily brought into the room or area where it is to be installed.
2. Ideally, the vacuum producer assembly should be installed on the floor below the lowest inlet of the building or facility and in a central location to minimize the differences at remote inlet locations.
3. A convenient means to dispose of the dirt should be close by. If a separator is used, an adequately sized floor drain will be required.
4. Enough room around the separators shall be provided to allow for easy inspection, and where the dirt bins must be emptied, room must be provided for the carts needed to move them. Dry separators could also be located outside the building for direct truck disposal of the dirt if sufficiently protected.

TABLE 15.15   Recommended Sizes of Hand Tools and Hose

Nominal size, in	Average floor cleaning and moderate spills	Close hand work	For removing heavy spills or large quantities of materials	Overhead vacuum cleaning	Standard hose length, ft
1	Not used	Yes	Inadequate	Not used	8
$1\frac{1}{2}$	Excellent	Yes	Fair	Preferred	25 and 50
2	Good	No	Good	Poor	25 and 50
$2\frac{1}{2}$	Not used	No	Excellent	Not used	25 and 50

Source: Courtesy Hoffman.

Sizing the Piping Network

General.   After the inlets and vacuum equipment have been located, the layout of the piping system accomplished, and the number of simultaneous operators decided upon, the process of system sizing can begin. Cleaning systems using hose and tools shall have sufficient capacity so that only one pass over the area being cleaned is all that is necessary. With adequate vacuum, light to medium dirt deposits shall be removed as fast as the operator moves the floor tool across the surface. The actual cleaning agent is the velocity of the air sweeping across the floor.

Inlet Tool Size.   The recommended inlet size for hand tools and hose is given in Table 15.15.

Vacuum Pressure Requirements and Hose scfm.   In order to achieve the necessary air velocity, the minimum recommended vacuum pressure for ordinary use is 2 inHg. For hard-to-clean and industrial materials, 3 inHg vacuum pressure is required. The flow rate must be enough to bring the dirt into the tool nozzle. Refer to Table 15.16 to determine the minimum and maximum recommended flow rate of air and the friction losses of each hose size for the flow rate selected. For ordinary carpeting and floor cleaning purposes, a generally accepted flow rate of 70 scfm is recommended.

Recommended Velocity.   The recommended velocity in the vacuum cleaning piping system depends on the orientation of the pipe (horizontal or vertical) and the size. Since the velocity of the air in the pipe conveys the suspended particles, it should be kept in a recommended range. Refer to Table 15.17, which indicates recommended velocity based on pipe size and the horizontal or vertical orientation of the pipe. The higher velocity is recommended for dense material or for material considered difficult to move. It is the air velocity that moves the dirt in the system. Oversizing the pipe will lead to low velocity and poor system performance.

Pipe Sizing

Selecting the Number of Outlets in Simultaneous Use.   Facilities may have many inlet valves, but only a few will be used at once. Under normal operating conditions, these inlets will be chosen at random by the operators. In order to aid in the determination of simultaneous usage, the following conditions should be expected:

1. Adjacent inlet valves will not be used simultaneously.
2. For calculation of simultaneous use, the most remote inlet on the main and the inlet closest to the separator will be assumed to be in use along with other inlet valves between these two.
3. Where mains and outlets are located on several floors, the use of inlets will be evenly distributed along a main on one floor or on different floors.
4. For long horizontal runs on one floor, allow for two operators on that branch.

Sizing the Piping Network.   Refer to Table 15.18 for selecting the initial pipe size based on the number of simultaneous operators. This table has been calculated to achieve the minimum velocity of air required for adequate cleaning. In this table, "line" refers to permanently installed pipe from inlet to separator, and "hose" is the hose connecting the tool to the inlet. A 1½-in-diameter hose is recommended except where the size of the material to be cleaned will not pass through the hose or a large volume of material is expected.

TABLE 15.16 Flow Rate and Friction Loss for Vacuum Cleaning Tools and Hoses

Use	Nominal size of tools and hose, in	Minimum volume and pressure drop		Maximum volume and pressure drop	
		Volume, scfm	Pressure drop, in Hg	Volume, scfm	Pressure, in Hg
Bench use	1-in diam., 8-ft 1-in flexible hose	40	1.20	50	1.90
White rooms or areas with very low dust content	1½-in diam., 50-ft 1½-in flexible hose	60	2.25	90	4.10
Usual industrial	1½-in diam., 50-ft 1½-in flexible hose	70	2.80	100*	4.80
Fissionable materials or other heavy metallic dusts and minute particles of copper, iron, etc.	1½-in diam., 50-ft 1½-in flexible hose	100	2.50	120	4.20
Heavy spills cleaning railroad cars and ship holds	2-in diam., 50-ft 2-in flexible hose	120	2.60	150	3.80

Note: The pressure drop in flexible hose is $2\frac{1}{2}$ times the pressure drop for the same length and size of Schedule 40 pipe.
*Can be exceeded by 10 percent if necessary.
Source: Courtesy Hoffman.

TABLE 15.17 Recommended Velocities for Vacuum Cleaning System

Nominal tubing size, in	Horizontal runs of branches and mains and vertical down-flow risers		Vertical up-flow risers	
	Minimum velocity, ft/min	Recommended max. velocity, ft/min	Minimum velocity, ft/min	Recommended max. velocity, ft/min
1½	1800	3000	2600	3800
2	2000	3500	3000	4200
2½	2200	3900	3200	4700
3	2400	4200	3800	5100
4	2800	4900	4200	6000
5	3000	5400	4800	6500
6	3400	6000	5000	7200

1 ft/min = .3 m/min.
Source: Courtesy Hoffman.

TABLE 15.18 Pipe Diameter Based on Simultaneous Usage

Line dia., in	No. of operators 70 scfm, 1.5-in hose	No. of operators 140 scfm, 2-in hose
2	1	—
2½	2	1
3	3	2
3½	4	2
4	5	3
5	8	4
6	12	6
8	20	10

Source: Courtesy Spencer Turbine.

After the initial selection of the pipe sizes, the actual velocity and friction loss based on anticipated flow rates in each section of the piping system should be checked by using Fig. 15.28. This chart is a more accurate method of determining the pipe size, friction loss, and velocity of the system. Enter the chart with the adjusted scfm and allowable pressure loss. Read the pipe size at the point where these two values intersect. If this point is between lines, use the larger pipe size. If any parameter is found to be outside any of the calculated ranges, the pipe size should be revised.

Pipe sizing is an iterative procedure, and the sizes may have to be adjusted to reduce or increase friction loss and velocity as design progresses.

Piping System Friction Losses. With the piping network sized, the next step is to precisely calculate the worst-case total system friction losses, in inches of mercury, in order to size the exhauster. This is calculated by adding together all of the following values, starting from the most remote inlet from the exhauster and continuing to the source.

1. *Initial level of vacuum required.* For average conditions the generally accepted figure is 2 inHg. For hard-to-clean material, industrial applications, and long shag carpet, the initial vacuum should be increased to 3 inHg.

2. *Pressure drop through the hose and tool.* Refer to Table 15.16 for the friction loss through individual tools and hose based on the intended size and length of hose and the flow rate selected for the project.

3. *Loss of vacuum pressure due to friction of the air in the pipe.* Losses in the straight runs of the piping system are based on the flow rate of air in the pipe at the point of design. Refer to Fig. 15.18. Fittings are figured separately, using an equivalent length of pipe to be added to the straight run. Refer to Table 15.19 to determine the equivalent length of run for each type and size of fitting. Starting from the farthest inlet, use the scfm, the pipe size, fitting allowance, and the pipe length along the entire run of pipe to find the total friction loss.

4. *Loss through the separator.* A generally accepted figure is 1 inHg loss through all types of separators. The exact figure must be obtained from the manufacturer.

5. *Exhaust line loss.* This can usually be ignored except for long runs. Allow 0.1 inHg as an average figure for a run of 100 ft.

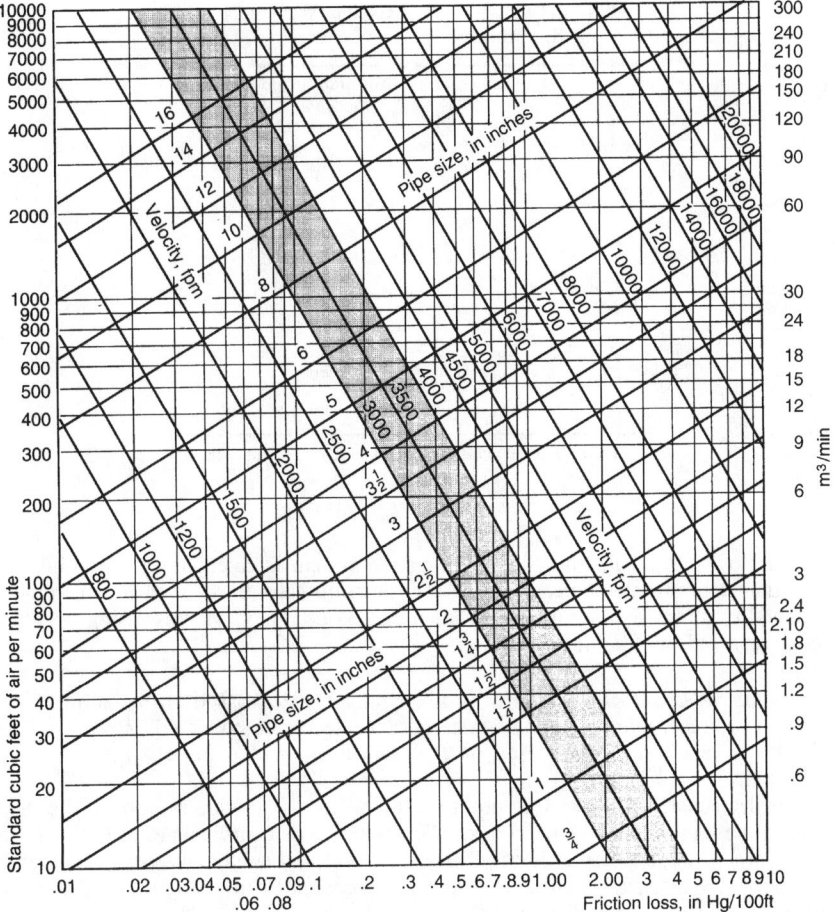

FIGURE 15.28 Vacuum cleaning piping friction loss chart. Friction loss, in inches of mercury per 100 ft of line with inlet air at 70°F and 14.7 psia. (*Courtesy Spencer Turbine.*) 1 fpm = .3 M/ min.

Vacuum Producer (Exhauster) Sizing

Exhauster Inlet Rating Determination. It is now possible to size the exhauster. There are two exhauster ratings that must be known in order to select the size and horsepower. They are: (1) worst-case piping system vacuum pressure losses and (2) the flow rate, in scfm, of air required by the system.

The vacuum pressure required from the exhauster is the total necessary to overcome all piping system losses. This consists of the total pressure drop from all components in the piping network from the inlet farthest from the exhauster. These are the initial inlet vacuum level required, the pressure lost through the tool and hose selected, the friction loss of air flowing through the piping system, the pressure lost through separators, filters, and silencers, and finally the exhaust pressure to be overcome, if required, These values are added together to establish the vacuum rating of the exhauster.

TABLE 15.19   Equivalent Length of Vacuum
Cleaning Pipe Fittings

Nominal size of pipe, in	Using cast iron drainage fittings	
	90° change in direction	45° change in direction
$1\frac{1}{4}$	3	$1\frac{1}{2}$
$1\frac{1}{2}$	4	2
2	5	$2\frac{1}{2}$
$2\frac{1}{2}$	6	3
3	7	4
4	10	5
5	12	6
6	15	$7\frac{1}{2}$
8	20	10

Note: For smooth-flow fittings, use 90 percent of these values.

The flow rate of air, in scfm, entering the system is calculated by multiplying the number of simultaneous operators by the scfm selected as appropriate for the intended clean-up requirements. For smaller, less complex systems, using only the actual selected inlet scfm is sufficient.

Separator Selection and Sizing.   The separator is sized based on the scfm of the vacuum producer and the type of material expected to be collected. Refer to Table 15.20 to select a separator based on the classification of material expected to be collected.

For dry separators, a starting point for sizing would provide a ratio of filter bag area to the bag volume of 6:1 for smaller volumes of coarse material and 3:1 for fine dust and larger quantities of all material. Wet and centrifugal separator sizing is proprietary to each manufacturer and is dependent on the quantity and type of material expected to be removed.

Some automatic separator cleaning systems use compressed air to aid in dislodging the dust. The air pressure recommended is generally in the range of 100 to 125 psig (700 to 875 kPa).

Exhauster Discharge.   The discharge from the exhauster is usually routed through steel pipe to be vented outside the building. It is also possible to route the exhauster discharge into an HVAC exhaust duct that is routed directly to outside the building.

For a piped exhaust, if the end is elbowed down, it shall be a minimum of 8 ft, 0 in, above grade. If the end is vertical, an end cap shall be installed to prevent rain from entering the pipe. A screen will prevent insects from entering. The size shall be equal to or one size larger than the size pipe into the exhauster. Use HVAC ductwork sizing methods to find the size of the exhaust piping while keeping the air pressure loss to a minimum.

The pressure loss through the exhaust pipe shall be added to the exhauster inlet pressure drop, the total of which will be calculated into the pressure that the exhauster must overcome. For short runs of about 20 ft, 0 in, this can be ignored.

TABLE 15.20 Classification of Material for Separator Selection

	Very fine		Fines		Granular		Lumpy		Irregular
Volume of material	Recommended sep. (*S*)	Ratio volume to bag area	Recommended sep. (*S*)	Ratio volume to bag area	Recommended sep. (*S*)	Ratio volume to bag area	Recommended sep. (*S*)	Ratio volume to bag area	Separator selection & bag area dependent on material
Small	CENT.	Not appl.	CENT.	Not appl.	CENT.	Not appl.	CENT.	Not appl.	
Medium	TB	6:1	CENT. and TB	6:1	CENT.	Not appl.	CENT.	Not appl.	
Large	CENT. and TB	3:1	CENT. and TB	6:1	CENT. and TB	6:1	CENT.	Not appl.	

Definition of terms:

Small: Light accumulations such as found in clean rooms, white rooms, laboratories, and so on

Medium: Average accumulations such as found in classrooms, motels, assembly areas, and so on

Large: Heavy accumulations such as found in foundries, spillage from conveyor belts, waste from processing machines, and so on

Fines: 100 mesh to $\frac{1}{8}$ in

Very fine: Less than 100 mesh

Granular: $\frac{1}{8}$ to $\frac{1}{2}$ in

Lumpy: Lumps $\frac{1}{2}$ in and over

Irregular: Fibrous, stringy, and so on

Note: Centrifugal separators do not utilize bags.
Abbreviations: CENT = centrifugal; TB = tubular bag.
Source: Courtesy Spencer Turbine Co.

Exhauster Rating Adjustments

scfm Adjustment for Long Runs. For systems with long runs or complex systems with both long and short runs of piping, some adjustment in the selected inlet scfm shall be made. This is necessary because the actual scfm at the inlets closest to the exhauster will be greater than the scfm at the end of the longest run due to the lesser friction loss. The adjustment will establish an average inlet scfm flow rate for all inlets that will be used for ease of system sizing instead of the actual inlet scfm.

In order to establish the adjusted scfm, it will be necessary to separately calculate the total system friction loss for each branch line containing inlets nearest and farthest from the exhauster. Following the procedures previously explained will result in minimum and maximum system friction loss figure. The following formula will calculate the adjusted inlet scfm used for design purposes:

$$\text{Adjusted scfm} = \frac{\text{farthest inlet friction loss, inHg}}{\text{closest inlet friction loss, inHg}} \times \text{selected scfm} \qquad (15.6)$$

To size the exhauster, the adjusted scfm figure will be used instead of the selected scfm, and it will be multiplied by the number of simultaneous operators.

Adjustment Due to Elevation.   All of the above calculations are based on scfm. If the location of the project is at a higher elevation than sea level, the scfm should be adjusted to allow for the difference in barometric pressure. Refer to Table 15.7 for the factor. This factor shall be multiplied by the scfm figure to calculate the adjusted scfm to be used in sizing the exhauster.

Adjustment for Different scfm Standards.   Another adjustment to the scfm figure used to size the exhauster that may be required is if the equipment manufacturer uses the inlet cfm (icfm) instead of scfm. *Inlet cfm* is the actual volume of air at the inlet of the exhauster using local temperature and barometric conditions. Previously discussed temperature and barometric conversions shall be used.

To convert scfm to acfm, refer to Eq. (15.1).

Preliminary Sizing Method

At the start of a project, it may be necessary to establish the preliminary horsepower rating of the exhauster for the electrical discipline and physical size of the vacuum producer assembly to establish preliminary space conditions. The following method will allow this selection.

First, find the longest equivalent piping run, in feet. Then estimate the number of simultaneous operators. Using Fig. 15.29, draw a vertical line up from the bottom where the length is indicated until it intersects the length of hose that will probably be selected. From that point of intersection, draw a horizontal line to the left side of the figure. At the intersection of the line with the left side, read the vacuum

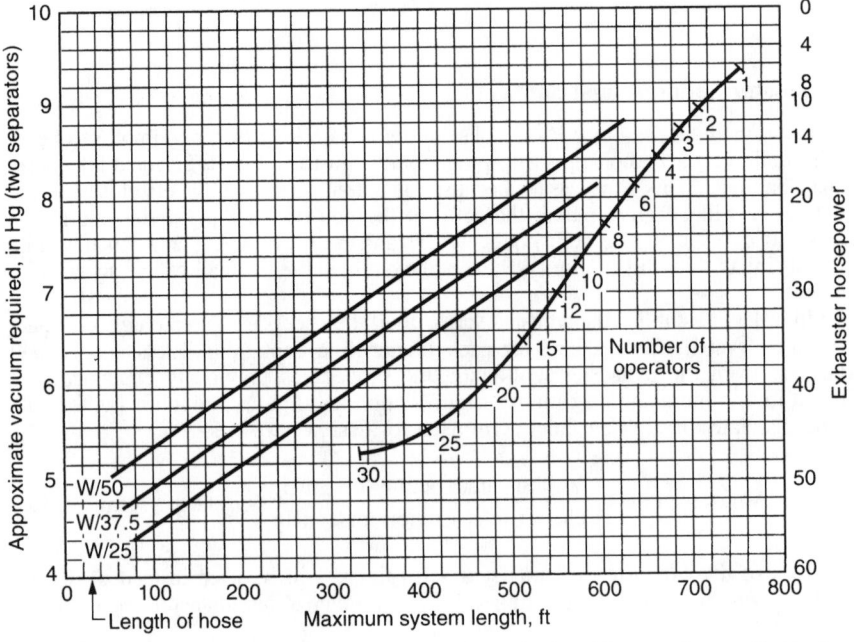

FIGURE 15.29   Preliminary exhauster horsepower determination. (*Courtesy Spencer Turbine.*)

requirement. From the intersection of the horizontal line with the left edge, draw another line from the left side intersection point to where it intersects the line indicating the number of simultaneous operators. Continuing this line to the right edge of the nomograph will establish the horsepower.

With the horsepower and vacuum parameters now established, calculate the preliminary flow rate from the number of simultaneous operators and the scfm used for each inlet. With this information, select the equipment from a manufacturer's catalog.

General Design Considerations

Abrasion is the wearing away of the interior of the pipe wall by large, hard particles at the point where these particles strike the pipe. The effects are greatest at changes of direction of the pipe, such as elbows and tees, and under bag plates of separators. When abrasive particles are expected, it is recommended that either cast iron drainage fittings or Schedule 40 steel pipe fittings using sanitary pattern sweeps and tees be substituted for the normally used tubing materials.

It is good practice to provide a safety factor scfm to assure that additional capacity is available from the exhauster without affecting the available vacuum. This should not exceed 5 percent of the total scfm, and it is used only when selecting the exhauster, not for sizing the piping system. Select the exhauster size, and then add the safety factor. The unit selected should have that extra flow available.

The piping shall be pitched toward the separator. Install plugged cleanouts at the base of all risers and at 90° changes in direction to allow any blockages to be easily cleared.

Piping geometry in the design of wet system piping could become critical. Every effort shall be made to keep the piping below the inlet valves to prevent any liquid from running out of the inlet after completion of the cleaning routines and to ease the flow of the liquid into the pipe. The wet system pipe should pitch back to the separator at about ⅛ in/ft. All drops should be no larger than 2 in in size, and only one inlet shall be placed on a single drop. Each drop should terminate in a plugged tee facing down. This will allow any liquid still clinging to the sides of the pipe to collect at the bottom of the riser and be carried away the next time the system is used.

ACKNOWLEDGMENTS

D. P. McSweeney, Spencer Turbine Company
Nash Engineering Company
Steve Waldman, Ohmeda Inc.
Ronald A. Creed, Wintek Corp.

REFERENCES

Albern, W. F., PE: "Vacuum Piping Systems," *Building Systems Design.* 1972.
Bellia, S. J.: "Medical Vacuum System Design," *Building Systems Design.* 1974.

Farruggia, S., and K. Diamond: *Vacuum Systems,* Seminar notes, 1986.

Grashorn, Tom: "Selecting the Right Vacuum Pump," *Plant Engineering Magazine,* 1988.

Harris, Nigel: *Modern Vacuum Practice,* McGraw-Hill, New York/United Kingdom, 1989.

Hesser, Henry H.: "Vacuum Sources," *Pumps and Systems Magazine,* August 1993.

Hoffman Industries: *Design of Hoffman Industrial Vacuum Cleaning Systems.*

McSweeney, D. P., and R. Glidden: "Vacuum Cleaning Systems," Unpublished article, 1993.

Moffat, R.: "Putting Industrial Vacuum to Work," *Hydraulics and Pneumatics Magazine,* 1987.

National Fire Protection Association (NFPA): *Health Care Facilities.*

Ohmeda, Inc.: *Medical Gas Design Guide.*

Roper, D. L., and J. L. Ryans: "Select the Right Vacuum Gauge," *Chemical Engineering Magazine.* 1989.

Spencer Turbine Co., The: *How to Design Spencer Central Vacuum Systems.*

U.S. Air Force: *Central Dental Evacuation Systems,* Review 3-82, 1982.

Stein, Ivan DDS, Northfield Dental Group, NV.

CHAPTER 16
ANIMAL FACILITY PIPING SYSTEMS

This chapter will discuss various piping systems uniquely associated with the physical care, health, and well-being of laboratory animals. Included will be various utility systems for animal watering, water treatment, room and floor cleaning, equipment, washing, cage flushing and drainage, and other specialized piping required for laboratory and experimental work within the facility. Other systems involved with general laboratory and facility work, such as compressed gases and plumbing, are discussed in their respective chapters.

GENERAL

It is expected that a facility involved with long-term studies will have different operating and animal drinking water quality requirements than one used for medical research. For critical studies, the various utility systems shall incorporate design features necessary to ensure reliability and provide a consistent environment. This will eliminate as many variables as practical (or desirable) to ensure accuracy of the ongoing experiments being conducted. Regardless of the type of facility, different users and owners have individual priorities based on experiences, operating philosophies and corporate cultures that must be established prior to the start of the final design phase of a project.

CODES AND STANDARDS

1. The local codes applicable to plumbing systems must be observed in the design and installation of ordinary plumbing fixtures, potable water, and drainage lines for the facility.
2. lO-CFR-58 is the code for Good Laboratory Practice for Nonclinical Laboratory Studies.
3. 2l-CFR-2 11, cGMP, requires compliance with FDA protocols for pharmaceutical applications.
4. NIH Publication 86-23, *Guide for the Care and Use of Laboratory Animals*.
5. American Association for Accreditation of Laboratory Animal Care (AAALAC). Inspection and accreditation by the AAALAC is accepted by the National In-

stitutes of Health (NIH) as assurance that the facility is in compliance with Public Health Service (PHS) policy.

ANIMAL DRINKING WATER SYSTEM

The purpose of the animal drinking water system is to produce, distribute, and maintain an uninterruptible supply of drinking water within a specific and consistent range of purity for all animals in a facility. There are two general types of systems: an automated central distribution system and individual water bottles.

System Types

The far greater majority of animals used by laboratories for medical and product research are mice, rats, guinea pigs, rabbits, cats, dogs, and primates. Smaller animals and primates are kept in stacked cages, often on racks. Medium-sized animals, such as dogs, goats, and pigs are kept in kennels or pens. Larger floor areas are required for barnyard animals such as cows. Watering can be done either by an automatic, reduced pressure central system piped from the source directly to each cage, kennel, or pen or separate drinking bottles or watering devices manually placed in individual cages or pens.

System Description

Automated Central Supply and Distribution System.   The purpose of an automated central drinking water supply system is to automatically treat and distribute drinking water. Ancillary devices are used to flush the system and maintain a uniform and acceptable level of purity.

The system consists of a raw or otherwise treated water source, purification system, medicinal and disinfection injection equipment if necessary, pressure-reducing stations, and a distribution piping network consisting of a low-pressure room distribution piping system and a rack manifold pipe terminating in a drinking valve for each cage or pen for the animals. Also necessary is an automated flushing system for the room distribution piping activated by a flush sequence panel and a monitoring system to automatically provide monitoring of such items as drinking water pressure, flow, and possible leakage, among other parameters.

Animals in cages are kept in animal rooms. Cages are usually placed in multi-tiered portable or permanent cage racks that contain a number of cages. The cage rack has an integral piping system installed, called a *rack manifold,* that distributes the water to all cages. The rack manifold could be installed by the manufacturer or in the facility by operating personnel. The rack manifold receives its water from the room distribution piping. The connection between the room distribution piping and the rack manifold is made by means of a detachable recoil hose generally manufactured from PP, nylon, or EPDM. This hose is flexible, generally $\frac{3}{8}$ in (12 mm) in size and coiled to conserve space. It will stretch to a length of about 6 ft, 0 in (2 m). Each end is provided with a quick disconnect fitting used to attach the hose to both the room distribution piping and the rack manifold.

To maintain drinking water quality, a method of flushing the room distribution piping and the rack manifold shall be provided. Ancillary equipment includes flushing and sanitizing systems to wash the recoil hose and the cage rack piping interior.

Water Bottles.   Drinking water bottles are individual units with an integral drinking tube that are placed by hand on a bracket in each cage. These bottles could be filled either by hand or automatically in a bottle filler. Automatic bottle fillers should be considered to reduce the time necessary to fill bottles and minimize water spillage. Bottle fillers are available with manifolds to fit any size bottles. They can be supplied with purified water from a central water supply, and separate programmable proportioners could acidify, chlorinate, and medicate the water as required. The bottle filler automates the filling procedure so that the bottles are correctly positioned during filling and so that the flow is stopped when the water reaches a predetermined level.

Flushing System.   In order to maintain drinking water quality, the drinking water distribution system should be flushed periodically. This is accomplished by having the same drinking water that is normally distributed to the animals flow through the piping system at an elevated flow rate, pressure, and velocity. The water is sent to drain and not recovered. This is initiated automatically at the drinking water pressure-reducing station by the addition of separate regulating valves and pressure-regulating arrangements.

Different flushing arrangements are possible, depending on the cost, facility protocol, and purity desired. One method flushes only the main runs by adding a solenoid valve at the end of the main run and providing a return line to drain from this point. Another method would be to flush the mains and the room distribution piping by adding a solenoid valve at the end of each room distribution branch with the return line to drain from each room. A third method flushes the entire system, including the rack manifold, by adding a solenoid valve on each cage connection to the room distribution pipe, which flushes the recoil hose and the rack manifold.

It is accepted practice to replace all the drinking water in the room distribution piping system at regular intervals, at a minimum of twice daily. An approximation of the amount of water in the pipe is to allow 1 gal (4 L) for each 33 ft, 0 in (10 m) of pipe. General practice is to flush the system with water at about 15 psi (90 kPa) at a rate of 15 gpm (60 Lpm). If the drinking water is not purified, it is recommended that the piping be flushed at least twice daily for about 2 mm. For purified water, flush once daily for about 1 mm. Flushing can be done manually by means of a valve in the pressure-reducing station enclosure or automatically by adding a bypass and solenoid valve to the pressure-reducing station around the low-pressure assembly. The sequence and duration of the automatic flush cycle is controlled from a flush sequencer panel.

DRINKING WATER TREATMENT

The purpose of the drinking water treatment system is to remove impurities from the raw water supply to achieve the water quality required by the animals in the facility. In addition, disinfectant and medication will be added to the water at this time if required.

System Description

There are no generally recognized and accepted standards for animal drinking water quality. Purity and consistency requirements depend on the incoming water quality, established protocol of the end user, the importance of either initial or operating cost of the proposed system, the species of animals housed in the facility, and the animal housing methods. The overall objective is to eliminate as many variables as possible for the entire period of time the study or experiments are conducted.

The most often used treatment for drinking water is reverse osmosis (RO). Other possible treatment methods are distillation and deionization (DI). A discussion of these purification methods appears in Chap. 4. Refer to Table 16.1 for commonly used water treatment options and generalized contaminant removal properties.

Reverse Osmosis (RO).   When a higher-quality water is required and other types of purified water are not available in a facility, the RO treatment method is normally selected. Since the amount of water is usually small, a package type of unit mounted on a skid is provided and connected directly to the water supply. The RO system is flexible and when used in combination with DI water supply, will provide water that is virtually contamination free.

Disinfection and Medication of Drinking Water

Disinfection chemical mixtures are added to the animal drinking water supply to eliminate and control bacterial contamination in the central and room distribution piping system. Medication is added to conform with experimental protocols if necessary. These mixtures are usually introduced into the piping system by a self-contained central proportioning (injector) unit using facility water pressure. All equipment is available in a wide range of sizes and materials. A schematic detail of a typical central proportioner is illustrated in Fig. 16.1.

Chlorination.   Chlorination is a recognized biocidal treatment that leaves a residual of chlorine in the entire central distribution system. Hyperchlorinated water is not as corrosive as acidified water and could be used with brass-copper distribution system components. Accepted practice is to provide a pH higher than 4, with a residual range of free chlorine between 5 to 12 ppm. Free chlorine in water dissipates in time with light, heat, and reaction with organic contaminants, making it ineffective when water bottles are used. Chlorine creates toxic compounds in reaction with some water contaminants and medication.

Acidification.   Acidification has an advantage over chlorination in that it is more stable and lasts longer in the system. The disadvantage is that corrosion-resistant materials must be used. The pH range should be between 2.5 and 3 in order to be effective. A pH lower than 2.5 will cause the water to become "sour," and the animals will not drink it. Above 3 the mixture is not considered an effective germicide.

TABLE 16.1   Water Treatment Methods

	Reverse osmosis	Distilled	Deionized	Chlorination	Acidification	Ultraviolet
Dissolved inorganics	X	X	X			
Dissolved organics	X	X	X			
Suspended particles	X	X				
Microorganisms	X	X		X	X	X

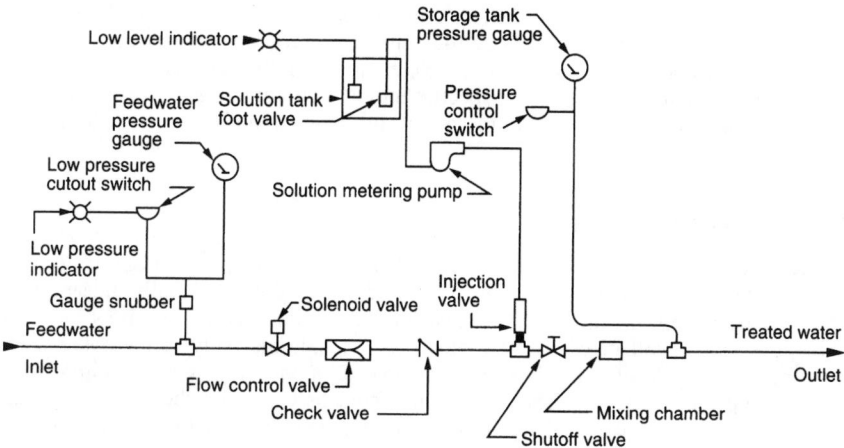

FIGURE 16.1   Typical central proportioner unit.

Adding Medication

Medication is added to the drinking water using the same proportioning equipment that adds disinfectant.

DRINKING WATER SYSTEM COMPONENTS AND SELECTION

Pressure-Reducing Station

The pressure-reducing station reduces the normal pressure of the raw water supply to a lower level required for the room drinking water distribution system serving the animals. As an option, another secondary system can be added to provide a higher pressure in the room distribution system for flushing purposes.

The pressure and flow rate depend on the type and number of animals to be supplied. Also usually included is a 5 μm water filter, pressure gauge, and backflow preventer. Timing devices to automatically control flushing duration is provided by a remote flush sequencer panel that controls all flushing sequencing operations. The recommended pressures for animal room piping distribution to various animals are:

1. Small animals such as rats and mice: 3 to 5 psi (20.4 to 34 kPa)
2. Primates: 3 to 5 psi
3. Dogs and cats: 3 to 5 psi
4. Swine and piglets: 6 to 12 psi (41 to 81.6 kPa)

The secondary pressure-reducing assembly used to automatically provide a room distribution pipe flushing water operates at a pressure of 15 psi (102 kPa). This assembly is installed as a bypass around the low-pressure assembly. Manual operation at a lesser cost could also be provided. This additional pressure for a short period of time will not cause the animals any difficulty if they decide to drink during the flushing cycle. One pressure-reducing station can be connected to as many as 35 individual small-animal rack manifolds, often referred to as *drops*. This will allow one station to control more than one animal room. The pressure-reducing station is a preassembled unit that is complete with all of the various valves, fittings, and reducing valves required for a specific project. All components are installed in a cabinet that requires only mounting and utility connections.

Drinking Valves

Drinking valves are used by the animals to obtain water from the distribution system piping. There is an internal mechanism that keeps the valve normally closed, and the animal drinking from the valve must open it by some action, such as moving the entire valve itself or operating a small lever inside the body of the valve with its tongue. Many kinds of valves are available to supply any type of animal that may be kept in the facility. They can be mounted on cages, on the rack manifold, or on the walls of pens and kennels at varying heights with the use of special brackets.

ANIMAL RACK MANIFOLD CONFIGURATIONS

The configuration of the piping on the animal rack plays an important part in the effectiveness and efficiency of drinking water system filling and flushing. The two most often used configurations are the reverse S and the H.

The reverse S, illustrated in Fig. 16.2, is the most often used configuration. It has two basic styles based on the valve location in the flush drain line. One style has a control valve at the top, and the other has a drain valve at the bottom. Either location is optional, with the deciding factor being the ease of operating the valve where the rack is installed. This configuration has the advantage of eliminating dead legs and offers more convenience to the facility personnel when filling the piping after washing. The vent is a manually operated air bleed that is used when the cage rack is reconnected to the room distribution pipe. It is opened until water is discharged, thereby eliminating any air pockets in the manifold. This manifold style provides a positive exchange of water during flushing with a minimum of time and water usage. This configuration is used far more than any other manifold style. It is easily converted to automatic flushing by installing solenoid devices on the valve. It is recommended when microisolator cage systems are installed. The complete online rack manifold flushing system is illustrated in Fig. 16.3. This cage system has the advantage of complete isolation of individual cages, with the accompanying capability for additional flushing and disinfection of the piping system.

One variation of the reverse S is the standard S, illustrated in Fig. 16.4. This configuration has the advantage of complete online flushing and lesser initial cost of the manifold. Disadvantages are the need for extra supports on the cage rack and the need for venting to be done manually or by the animals after being placed in service. This configuration is no longer recommended.

The H style, illustrated in Fig. 16.5, although rigidly installed and with positive venting, is not suitable for online flushing. Because of this, it is rarely used except for larger animals that will consume all the water in the rack piping manifold.

The most common piping materials are C PVC and 304L stainless steel. CPVC conforms to ASTM D 2846 and has an 0.875 O.D. with a 0.188 minimal wall

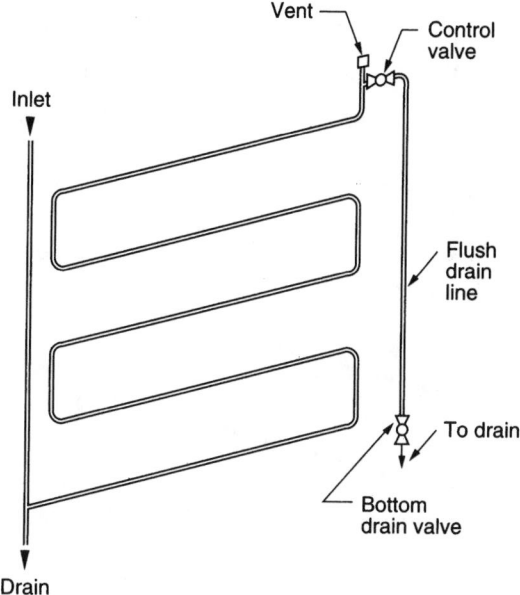

FIGURE 16.2   Reverse-S-rack watering manifold.

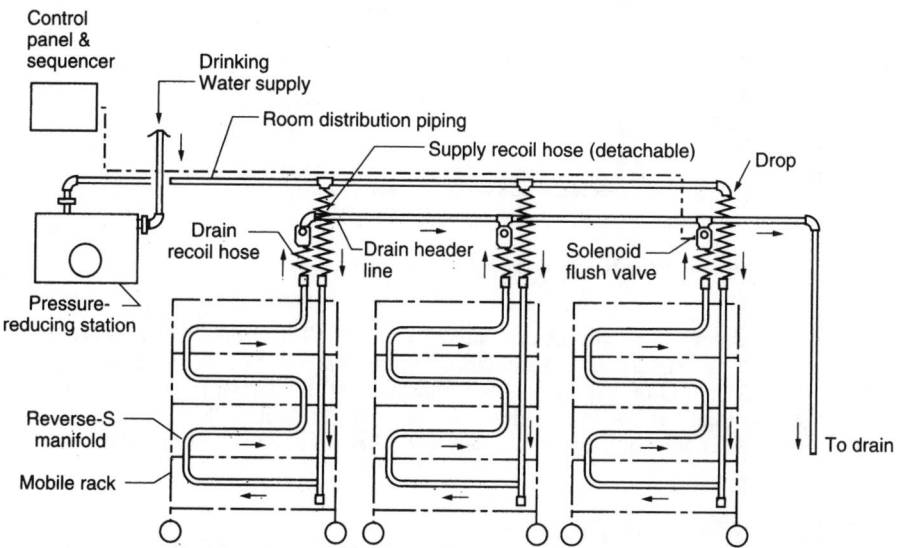

FIGURE 16.3   Typical room distribution on-line rack manifold flushing system. (*Courtesy Edstrom Industries.*)

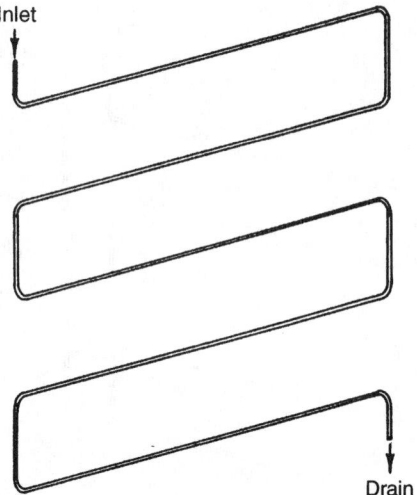

FIGURE 16.4   Standard S-rack watering manifold.

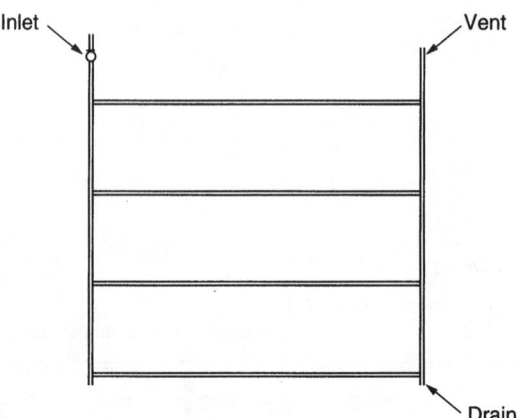

FIGURE 16.5   Standard H-rack watering manifold.

thickness. Jointing uses solvent cement socket joints. The drinking valves are installed with a proprietary drilled and tapped fitting. Stainless steel (SS) tubing, type 304L, has an 0.50 O.D. with a 0.036-in minimal wall thickness. Fittings are made with O ring joints and socket fittings or compression fittings. Mounting of both pipe materials is accomplished with the use of 304 SS clamps and fasteners.

SYSTEM SIZING METHODS

The water consumption of small animals in cages is usually very low. Table 16.2 provides animal laboratory information on the amount of water consumed daily

TABLE 16.2 Laboratory Animal Information

Species	Age or weight	Room temp., °F	Relative humidity, %	Ventilation changes per hour	Comfort zone temperature, °F	Amount of food required daily	Amount of water required daily	Amount of urine excreted daily	Amount of feces excreted daily	Average Btu heat produced per animal per hour	Lighting requirements
Baboon	Young adult weighing more than 5 kg	70–78	Not less than 50	12–16	65–83	2–3 lb	300–500 mL	25–45 mL/kg B.W.	150–300 mg/kg B.W.	60–140	Light cycle 12–14 h
Cat	2–4 kg	70–72	45–60	10–20	60–75	4–8 oz	100–200 mL	22–30 mL/kg B.W.	2–8 oz	25–30	Light cycle 12–14 h
Cattle	Adult	35 minimum	40–80	10–20	50–80	16–28 lb	10–14 gal	3–5 gal	60–90 lb	800	Light cycle 6–12 h
	Calves	60 minimum	No greater than 70	10–20	60–70	4–15 lb	2–4 gal	°1–3 gal	3–14 lb	350	Light cycle 6–12 h
Chimpanzee	Adult	70–78	40–60	12–15	68–84	2–4 lb	600–1500 mL	½–1 L	140–410 mg/kg B.W.	40–220	Light cycle 10–14 h
	Juveniles										
Sheep	Adult	35 minimum	No greater than 70	10–20	69–77	2–4.5 lb	½–1½ qt	1–2 qt	3–6 lb	800	
Swine	Adult	60–75	55–75	10–20	42–70	4–8 lb	1–1½ gal	½–1 gal	6–7 lb		
	Sow with litter					6–15 lb	5 gal	1–2 gal	8–11 lb		
	Miniature swine	65–72	60–75	10–15	50–72	0.5–no more than 2 lb	¼–½ gal	¼–½ gal	1–2 qt	2–4 lb	250–450

(Continued)

TABLE 16.2 Laboratory Animal Information *(Continued)*

Species	Age or weight	Room temp., °F	Relative humidity, %	Ventilation changes per hour	Comfort zone temperature, °F	Environmental considerations Amount of food required daily	Amount of water required daily	Amount of urine excreted daily	Amount of feces excreted daily	Average Btu heat produced per animal per hour	Lighting requirements
Rat	50 g	70–80	50–55	10–20	81–86	$\frac{1}{3}$–$\frac{2}{3}$ oz	20–45 mL	10–15 mL	$\frac{1}{4}$–$\frac{1}{2}$ oz	4.0	Equal light or dark periods or longer light than dark
Opossum	Adult	70–76	45–65	10–12		3–5 oz	100–150 mL	50–75 mL			
Pigeon	Adult	40–60	45–70	10–15	60–80	1–3 oz	Ad lib		Droppings, urine, and feces 2–6 oz/day	1–2	Breeding birds 12–14 h light daily
Rabbit	3–5 lb	60–70	40–60	10–20	82–84	1–3 oz	60–140 mL/kg B.W.	40–100 mL/kg B.W.	$\frac{1}{2}$–2 oz/day	34	
Monkey (Macaca muletta)	Adult	70–78	40–60	8–15	68–84	0.25–2 lb	200–950 mL	110–550 mL	110–320 mg/kg B.W.	65–200	Minimum of 8 h light per day
Mouse	Adult	70–80	50 as minimum up to 80	6–12	82–87	0.10–0.25 oz	4–7 mL	1–3 mL	0.05–0.10 oz	0.6	14 h light–10 h darkness
Guinea pig	Adult / Young	64–69	45–60	2–8	84–88	0.5–1.0 oz; require source of vitamin C	85–150 mL	15–75 mL	0.75–3 oz	5.6	Up to 15 h light per day

16.10

Hamster	Adult	70–76	45–65	6–10	82–88	0.1–0.8 oz	8–12 mL	6–12 mL	0.2–0.8 oz	2.5	12-h cycle light-darkness
Horse	450–550 kg, 4–7 years old	35 minimum	40–70	6–12	50–70	17–36 lb	5–12 gal	½–3 gal	25–50 lb	550–750	
Mink	Adult	40–65	40–65	8–15		5–7 oz	75–100 mL	25–50 mL	½–4 oz	30–60	
Gerbil	Adult	70–75	55–65	8–10	81–86	0.3–0.5 oz	None if greens fed	2–3 drops	Minimal 0.1 g or less	4.0	Even tight darkness ratio; protect from direct sunlight
Goat	Adult	45 minimum	55–65	6–12	55–70	1.5–10 lb	1000–4000 mL	700–2000 mL	3–6 lb	350–550	
	Kids	60 minimum	60–70								
Chinchilla	Adult	60–70	40–60	10–15	81–86	1–2 oz/kg	65–130 mL	25–50 mL	½–1 oz	34	Light 12 h
Chicken	Adult	60–70	45–70	10–15	62–82	3.4 oz/day	Ad lib		4–8 oz/day urine and feces (droppings 4–12 oz)	30	13 h daylight or artificial light
Dog	Adult	65–75	45–55	8–12	68–70	½ lb–10-lb dog 2½ lb–100-lb dog	25–35 mL/kg B.W.	65–400 mL		80–150	

along with other useful criteria for the design of animal facilities. It is also probable that the animal room may not be used to full capacity. Because of this low flow, the flushing water flow rate of the system is the critical factor in sizing the piping. Typically, the animal room piping distribution network is a header uniformly sized at ½ in (50 mm) throughout the animal room.

The pipe sizes in other areas of the animal facility are determined from requirements of maximum flow rate at the necessary pressure to supply the flushing velocity. Maximum flow rate depends on the flush sequencing, and the pressure drop depends on overcoming pressure loss through the equipment connected to the branch being sized according to such factors as the number of pressure-reducing stations, solenoid valves, and recoil hoses, as well as friction loss through the piping network. Allowance must be made for a flow rate and water velocity sufficiently high to efficiently provide the flushing action desired.

CLEANING AND DRAINAGE PRACTICES

General

Keeping the animal rooms and cages clean is an extremely important facet of facility practice. Cleaning of the animal room is accomplished either by sponging the walls, floors, and ceiling or hosing down the room. Cage racks can be cleaned by means of washing with a hose or in a large washing machine. Cages are cleaned in a cage washer. Pen and kennels are hosed down. Floors in pens are cleaned with hoses, and the feces with bedding are pushed into trenches with floor drains.

In specialized areas such as holding or isolation rooms where only small animals are kept, it is common practice to have permanent cage racks or have the portable racks remain in the animal room. The litter is put in bags and brought to other areas for disposal. The cage racks are manually wiped down, and no rack washer is required. A sink is usually provided in the animal room for the convenience of the cleaning personnel. Individual water bottles, if provided, could be washed in the sink. The cages are removed and washed separately in a cage washer. This type of animal room usually does not require a floor drain if the entire room will be sponged down. If hosing is practiced, a floor drain is required.

Rabbits and guinea pigs have a tendency to spray urine and feces. This requires that the racks be hosed down in the room. A wash station containing a hose reel and detergent injection capability to hose down the cage racks and the room itself is usually placed in individual rooms. Citric acid is often used as a cleaning agent for rabbits.

Hose Stations. Hose stations usually consist of a mixing valve with cold water and steam to make hot water or hot water alone, a length of flexible hose, and an adjustable spray nozzle. It can be exposed or provided with an enclosure when an easily cleaned surface is required.

Cleaning Agent Systems. Cleaning agents are used to clean and/or disinfect the walls, ceiling, and floor of a room and to add agent to the cage washwater. When used to clean rooms, it is commonly called a *facility detergent system.* When used to add agent to the cage washing water, it is often called a *cage washing detergent system.* These are separate systems that are not capable of providing agent to each other.

A single-station, detergent-dispensing system is used when rooms are cleaned with mops or squeegees. It consists of a wall-mounted unit having a holder for detergent concentrate and an injector unit. A container filled with detergent concentrate is placed in the holder and used to supply agent to the injector that dispenses a metered amount of agent when a hose bibb is opened to fill the pail or container. These rooms usually have sinks and mop racks inside for use only for these rooms. A typical schematic detail of a single-station detergent system is illustrated in Fig. 16.6.

When used to supply a single- or multiple-spray hose for cleaning floors and walls, a central system could be installed to supply several rooms within a facility by means of a detergent pump to dispense agent. A 55-gal drum of agent should be used to reduce the number of times the supply has to be changed. A typical central supply detergent-dispensing system is illustrated in Fig. 16.7.

The cage washing detergent system is usually located in the wet area of the cage washing facility, and with the use of a detergent pump, it could be used as a central system to supply cage and bottle washers. A typical schematic detail of a cage washing system is illustrated in Fig. 16.8.

It is common practice to have a central system or a wall-mounted cleaning agent dispenser unit along with the hose station. Separate, portable units could be used when cross contamination between animal rooms is a consideration. A typical wall-

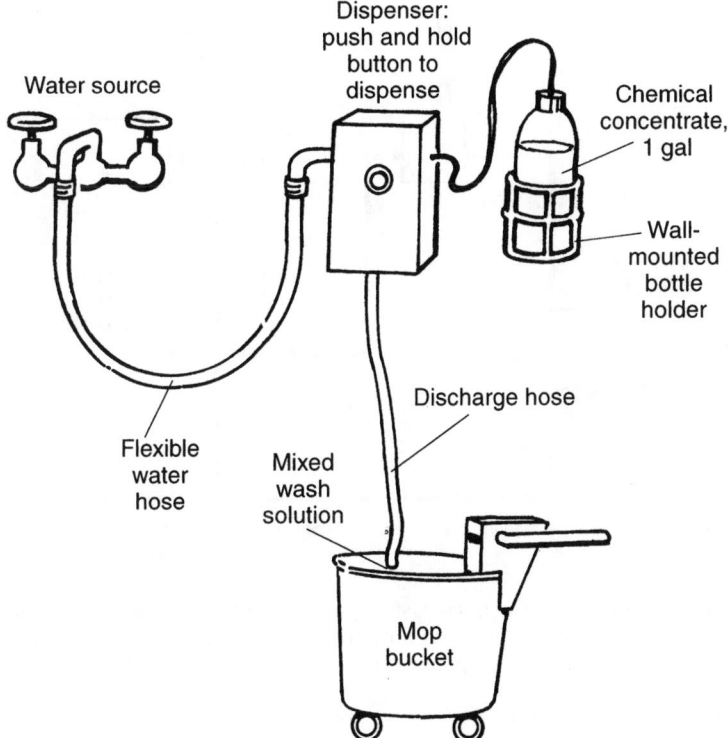

FIGURE 16.6 Typical single-station detergent-dispensing system detail.

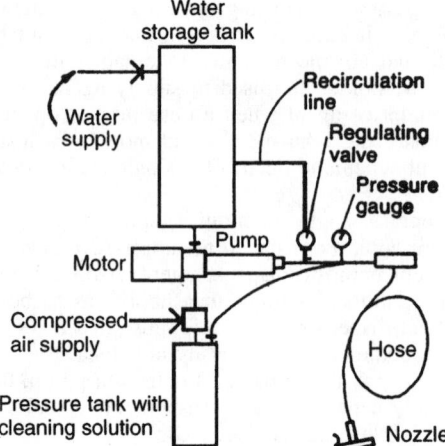

FIGURE 16.7 Typical central supply detergent-dispensing system detail.

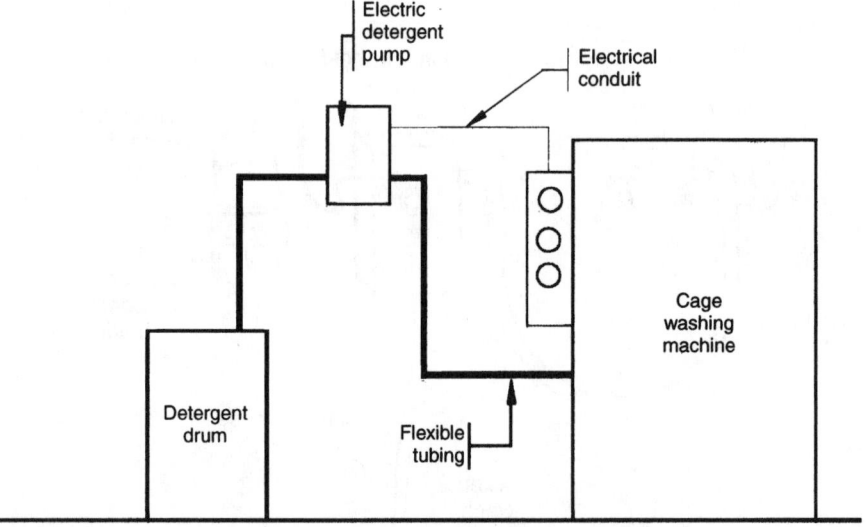

FIGURE 16.8 Typical cage washing detergent-dispensing system detail.

mounted cleaning agent system consists of separate water and cleaning agent tanks, a water pump and a special, coaxial hose that sprays a proportioned mixture of the water and cleaning agent. Compressed air is often used to provide pressure.

Cage Flushing Water System. The removal of animal waste from cages can be done by several methods. One method removes the waste along with bedding at the time cages are removed from the animal room to be washed. Another method uses an independent rack flush system to automatically remove animal waste from cages on racks while the animals and cages remain in the animal room.

The independent rack flush is a separate system that uses chlorinated water automatically distributed to each animal room. The cages and racks are constructed so that the animal droppings fall through the cage floor onto a sloping pan below each tier of cages. Each tier is cascaded at the end onto the sloping pan below. Eventually, the lowest pan spills into a drain trough in the animal room. The flushing schedule is decided by facility.

The water supply could be a reservoir placed on the rack that is filled with water, which automatically discharges onto the pans at preset intervals determined by experience. Generally water is discharged from one to three times daily. Another method uses a solenoid valve to automatically discharge water onto the pans sequenced by a timer set to alternate fill and dump cycles. The timer could be either centrally located or installed separately in each animal room. Larger cages, such as those for primates, are usually one or two cages high. Current practice is to have these gages manually cleaned by personnel who hose down the pans directly into floor or wall troughs.

Water is supplied to each cage rack by means of a recoil hose that has a different quick disconnect end than that of the drinking water recoil hose to avoid cross connection. Refer to Fig. 16.9 for a detail of a typical cage rack utility connection arrangement.

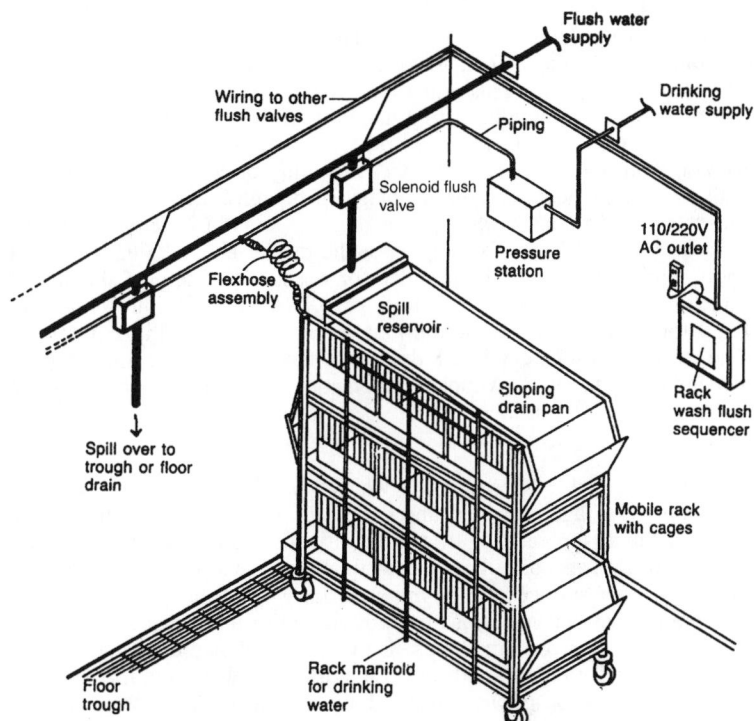

FIGURE 16.9   Typical cage rack utility connection arrangement.

Solid Waste Disposal. Solid waste consists of bedding, feces, animal carcasses, and other miscellaneous waste including straw and sawdust used for larger farm animals. Bedding makes up the largest quantity of this solid waste. It is necessary to determine the quantity of bedding before a decision is made as to the most cost-effective method used to dispose of it.

Bedding can be disposed of into an incinerator, a regular garbage bin, or into the sewer system. Incinerators are costly and require compliance with many regulatory agencies and multiple permits, and often they must be built over the objections of adjoining property owners. Incineration is the preferred method of disposing of carcasses and large quantities of contaminated waste. Carcasses could also be autoclaved and disposed of as regular garbage. Regular garbage disposal is the most common and involves the collecting, moving, and storing of the waste into large containers until regular garbage collection is made. This is very labor intensive.

Discharge into the drainage system must first be accepted by the local authorities and responsible code officials. The bedding shall be water soluble and shall not float, and it shall be made to thoroughly mix with water. This mixture is called a *slurry.* Experience has shown that if done properly, discharge into an adequately sized drain line with a minimum size of 6 in (150 mm) has caused no problems, since the slurry has the same general characteristics of water.

A self-contained waste disposal system is available that is capable of disposing of animal bedding and waste. The system consists of a pulping unit that grinds the waste into a slurry and sanitizes it, a water extractor that removes most of the water from the slurry, and the interconnecting piping system that transports the slurry from the pulper to the extractor and recirculates the water removed from the extractor back to the pulping unit for reuse. The solid waste is removed as garbage. Manufacturers are available for assistance in the design and equipment selection for this specialized system. This system has the advantages of reducing water use, reducing operating cost by eliminating handling of the waste by operating personnel, compacting the waste to about 20 percent of the space required for standard garbage not compacted, and reducing the possibility of contamination by isolation of the disposal equipment. The disadvantage is its high initial cost.

This system could consist of single or multiple units of different capacities and requires water intermittently for pulping at the rate of about 10 to 30 gpm (63 to 190 Lpm). Hose bibbs should be installed for washdown. The pipe should be sized for a maximum velocity of 8 fps (1.75 mps), with typical slurry lines ranging between 2 (50 mm) and 4 in (200 mm) and return lines generally 2 in (50 mm) in size. The extractor discharges into a drain that should be 4 (100 mm) or 6 in (150 mm) depending on the flow. A typical schematic diagram of a multiple installation is illustrated in Fig. 16.10.

Room Waste Disposal. The rooms in which animals are kept must be designed to allow proper drainage practices, and the design must be in accordance with the anticipated cleaning procedures of the facility. Providing floor drains, drainage trenches (or troughs) at room sides, and adequate and consistent floor pitch to drains or troughs and floor surfaces are all important considerations.

There are several considerations to be reviewed in locating floor drains. Experience has shown that placing drains in the center of a room is not acceptable because it is difficult to hose solids down the drain in this location. Another reason is that the floor must be pitched to the drain, and if a cage rack is defective, it should roll to the side of the room. The best location is in a corner or at the side. Floor drains, without troughs, are considered if the floors will only be squeegeed rather than hosed down. This is also a consideration in contagious areas where

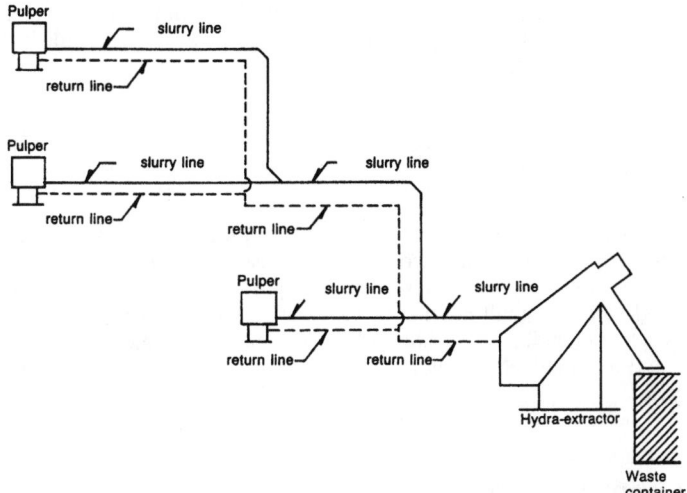

FIGURE 16.10   Typical schematic of a waste disposal system. Valves, fittings, and cleanouts not shown. (*Courtesy Somot.*)

contamination between rooms must be avoided. Gratings must have openings smaller than the wheels of racks or cages.

In rooms where washdown and cage rack flushing are expected, the provision of a floor trough should be considered. Troughs are often provided at opposite ends of the room to minimize the amount of floor drop due to pitch. Accepted practice uses a minimum floor pitch of ⅛ in/ft of floor run. The floor is pitched to the troughs to facilitate cleaning and also to provide an easy method to dispose of waste generated from the rack flush system. It is common practice to provide an automatic or manual trough flushing system with nozzles or jets to wash down the trough sides and eliminate as much remaining contamination in the trough as possible. Wall troughs, similar to roof gutters, are located at a higher elevation. This type of trough arrangement is sometimes provided in addition to or in lieu of floor troughs if the arrangement of elevated cages and racks make this an effective drainage method. Experience has shown that prefabricated drain troughs in floors are preferred over those built on the wall as part of the architectural construction.

The floor troughs are drained by means of a floor drain placed in a low point at one end. The troughs are usually pitched at ¼ in/ft of run to the drain. The drain should be constructed of acid-resistant materials and have a grate that can be easily removed. For small-animal rooms where bedding is not disposed of in the room, a 4-in (100-mm) size drain is considered adequate. In most other locations, it is recommended that a 6-in (150-mm) size drain be provided. A flushing rim type of drain should be considered to flush all types of waste into the drainage system.

Floor drains should have the capability of being sealed by replacing the grates with solid covers during periods in which the room may not be in service.

Drainage System Sizing

For individual animal rooms where bedding is not disposed of in the drainage system, a 4-in size drain is acceptable. In general, a 6-in size drain is considered good practice. The drainage system piping should be a minimum 6-in size, with a ¼-in pitch when possible, and the piping sized to flow ½ to ⅔ full in order to accommodate unexpected inflow.

EQUIPMENT WASHING

Most facilities contain washing and sanitizing machines to wash cages, cage racks, and bottles, if used. There are two commonly used types of cage washers used: batch and conveyer (tunnel). Batch washers require manual loading and unloading and are used where a small number of cages and racks are washed. A conveyer is similar to a commercial dishwasher where the cages and racks are loaded on a conveyer and automatically moved through the machine for the washing and sanitizing cycles.

EQUIPMENT SANITIZING

Maintaining drinking water quality requires that the recoil hoses and rack manifolds be internally sanitized, not merely washed. This is most often done at the same time the cages are washed. Separate rack manifold and recoil hose flush stations are available for this purpose, and they are usually installed in the cage wash area. Washing could be done manually or automatically. The hoses are flushed for 1 to 2 min with 4 gpm (16 Lpm) of water. Chlorine is injected into the water by a chlorine injection station (proportioner) set to deliver 10 to 20 ppm into the flush water. Hoses are dried by blowing 10 scfm of oil-free compressed air at 60 psig. If chlorine is used as a disinfectant, a contact time of 30 min is recommended before evacuation and drying.

Periodic sanitizing of the room distribution piping system is required for maintaining good water quality. Sanitizing is done prior to system flushing. To accomplish this, a portable sanitizer is used to manually inject a sanitizing solution directly into the piping system. In order to do this, an injection port is required at the inlet to the pressure-reducing station. The portable sanitizer usually consists of a 20-gal (90-L) polyethylene tank with a submersible pump inside, and a flexible hose is used to connect the tank to the injection port. The disinfecting solution is a mixture of chlorine and water to make 20 ppm of chlorine. The mixture should maintain a contact time in the piping for 30 to 45 mm. Refer to Fig. 16.11 for a detail of the portable sanitizer and injection port.

MONITORING SYSTEMS

The monitoring of various animal utility systems is critical to keeping within a range of values consistent with the protocol of the experiments being conducted at the facility from time to time. This is accomplished by a central monitoring system that includes many measurements from HVAC and electrical systems. For the animal drinking water system, parameters such as water pressure, flow rates, leakage, pH, and temperature from various areas of the facility will be helpful for maintenance, monitoring, and alarms.

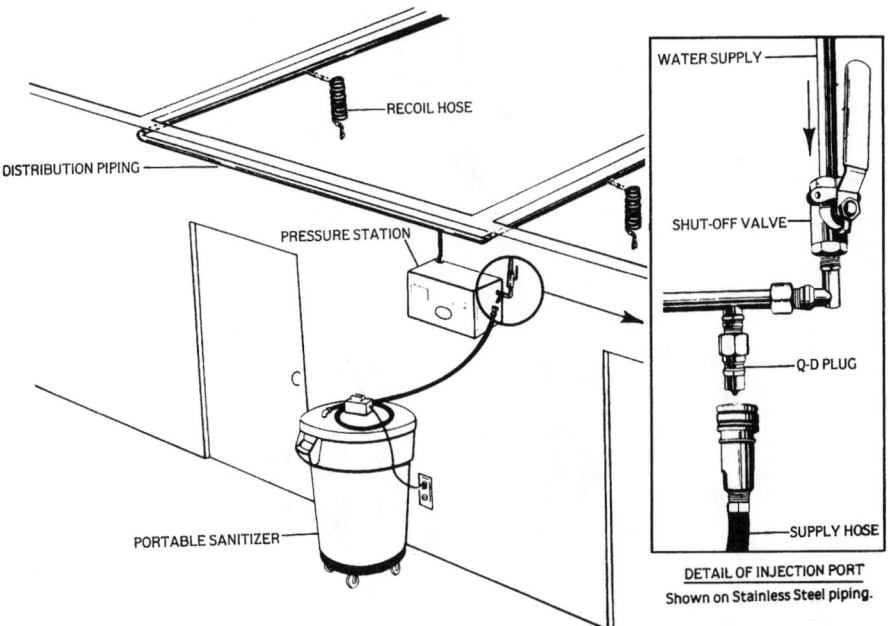

WATER SUPPLY

RECOIL HOSE

DISTRIBUTION PIPING

SHUT-OFF VALVE

PRESSURE STATION

Q-D PLUG

PORTABLE SANITIZER

SUPPLY HOSE

DETAIL OF INJECTION PORT
Shown on Stainless Steel piping.

FIGURE 16.11   Detail of portable sanitizer and injection port. (*Courtesy Edstrom Industries.*)

GENERAL SYSTEMS DESIGN CONSIDERATIONS

The amount of exposed piping inside any animal room should be minimized. The exception is the animal drinking water system, which is usually exposed on the walls of the room. This piping should be installed using standoffs to permit proper cleaning of the wall and around the pipe. The piping material used for all systems should be selected with consideration given to the facility cleaning methods and type of disinfectant. Where sterilization is required and cleaning very frequent, stainless steel pipe should be considered.

If insulation is used on piping, it should be protected with a stainless steel jacket to permit adequate cleaning. Pipe penetrations should be sealed with a high-grade, impervious, and fire-resistant sealant. Escutcheons should not be used because they will allow the accumulation of dirt and bacteria behind them.

SWINE COOLING SYSTEMS

Swine do not sweat and become stressed when the temperature reaches 75°F and higher. The effect could be death. It is therefore necessary to cool down the swine when these conditions are possible.

Two methods are used. The first is a drip hose used generally for sows when farrowing and for animals kept in individual pens. A small amount of water is dripped out of a hose mounted above the animals onto their neck and shoulders.

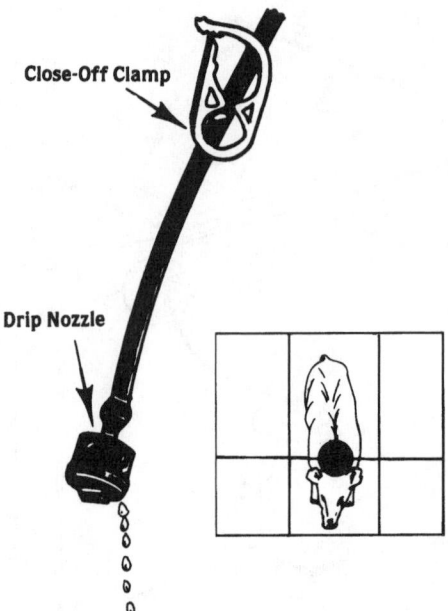

FIGURE 16.12   Detail of drip cooling outlet.

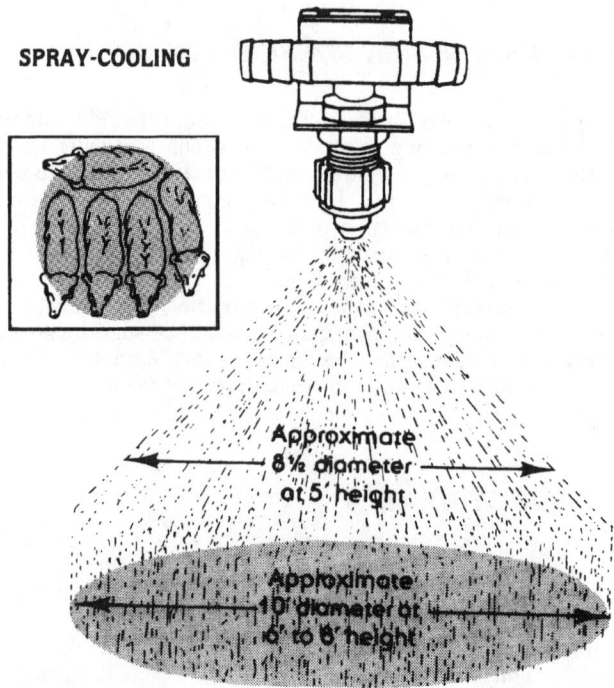

FIGURE 16.13   Detail of spray cooling nozzle.

They learn to move into position if this system is used. Little or no runoff is produced. Refer to Fig. 16.12 for a detail of the drip hose outlet.

The second method uses a spray nozzle that intermittently produces a coarse spray of water onto animals housed in group pens. Refer to Fig. 16.13 for a detail of the spray nozzle installation.

Both systems are operated from a controller. The controller is capable of operating multiple nozzles and drip hoses. It has adjustable settings for temperature at which water releases, intervals between release, and the duration of wetting.

ACKNOWLEDGMENT

Mr. H. David Dinkins, Edstrom Industries, Inc.

REFERENCES

National Institutes of Health (NIH): *Guide for the Care and Use of Laboratory Animals,* NIH Publication No. 86-23.

Ruys, T.: *Handbook of Facilities Planning,* Vol. 2, Van Nostrand Reinhold, Florence, Ky., 1991.

CHAPTER 17
LIFE SAFETY SYSTEMS

GENERAL

An often present threat to personnel safety in facilities is accidental exposure and possible contact with toxic gases, liquids, and solids. This chapter describes water-based emergency drench equipment and systems commonly used as a first aid measure to mitigate the effects of such an accident, and breathing air systems that supply air to personnel for escape and protection when exposed to either a toxic environment resulting from an accident or normal working conditions that make breathing the ambient air hazardous.

EMERGENCY DRENCH EQUIPMENT

GENERAL

When toxic or corrosive chemicals come in contact with the eyes, face, and body, flushing with water for 15 min with clothing removed is the most recommended first aid action that can be taken by nonmedical personnel prior to medical treatment. Emergency drench equipment is intended to provide a volume of water sufficient to effectively reach any area of the body that has been exposed to or has come into direct contact with any injurious material. Within facilities, this is accomplished by means of specially designed emergency drench equipment, such as showers, drench hoses, and eye and face washes located adjacent to all such hazards. Although the need to protect personnel is the same for any facility, specific requirements will differ widely because of architectural, aesthetic, location, and space constraints necessary for various industrial and laboratory installations.

SYSTEM COMPONENTS

Emergency drench equipment consists of showers, eyewash units, facewash units, and drench hoses along with interconnecting piping and alarms if required. Each of these units is available either singly or in combination with each other.

SYSTEM CLASSIFICATIONS

Drench equipment is classified into two general kinds of systems based on the source of water. They are *plumbed systems,* which are connected to a permanent water supply, and *self-contained,* or *portable, equipment,* which contains its own water supply. Self-contained systems can be either *gravity fed* or *air pressurized.*

Another type of self-contained eyewash unit is available that does not meet code requirements for storage or delivery flow rate. These units are called *personnel eyewash stations* and are selected only to supplement, not replace, standard eyewash units. They consist of solution-filled bottle(s) in a small cabinet. This cabinet is small enough to be installed immediately adjacent to a high hazard. If an accident occurs, the bottle containing the solution is removed and used without delay to flush the eyes while waiting for the arrival of trained personnel and during travel to a code-approved eyewash or first aid station.

CODES AND STANDARDS

1. ANSI Z-358.1: Emergency shower and eyewash equipment
2. OSHA: Various regulations for specific industries pertaining to location and other criteria for emergency eyewashes and showers

3. SEI: Safety Equipment Institute certified equipment that meets ANSI standards
4. Applicable plumbing codes.
5. Code of Federal Regulations (CFR), Chapter 29.

In future discussions in this section, the reference to "code" on drench equipment refers to ANSI Z-358.1.

TYPES OF DRENCH EQUIPMENT

Each piece of equipment is designed to perform a specific function. Each piece is not intended to be a substitute for another but rather, to complement the other pieces to provide additional availability of water to specific areas of the body where required.

Emergency Showers

Plumbed Showers. Plumbed emergency showers are permanently connected to the potable water piping and designed to supply enough water continuously to drench the entire body. The unit consists of a large-diameter showerhead intended to distribute water over a large area. The most commonly used type has a control valve with a handle extending down from the valve on a double chain or single rod that is used to turn water manually off to on in 1 second. Code requires that the shower be capable of delivering a minimum of 20 gpm (75.7 Lpm) of evenly dispersed water at a velocity low enough so as not to be injurious to the user. The minimum spray pattern shall have a diameter of 20 in (58.8 cm) diameter, measured at 60 in (152.4 cm) above the surface upon which the user stands. This requires a minimum pressure of 30 psi (4.47 kPa). Emergency showers can be ceiling mounted, wall mounted, or floor mounted on a pipe stand, with the center of the spray at least 16 in (40.6 cm) from any obstruction. Showers should be chosen for the following reasons:

1. When large volumes of potentially dangerous materials are present
2. Where a small volume of material could result in large affected areas, such as in laboratories and schools
3. Where an accident involving corrosive material may result in full body exposure
4. It shall be located within 10 seconds or 100 ft (33 m) of the hazard.

A typical emergency showerhead is mounted in a hung ceiling is illustrated in Fig. 17.1. Since free-standing emergency showers only are rarely installed, a freestanding combination showerhead and eye-face wash is illustrated in Fig. 17.2.

Self-Contained Showers. A self-contained emergency shower has a storage tank for water. Often this water is heated. The shower shall be capable of delivering a minimum of 20 gpm (75.5 Lpm) for 15 min. The mounting height and spray pattern requirements are the same as for plumbed showers.

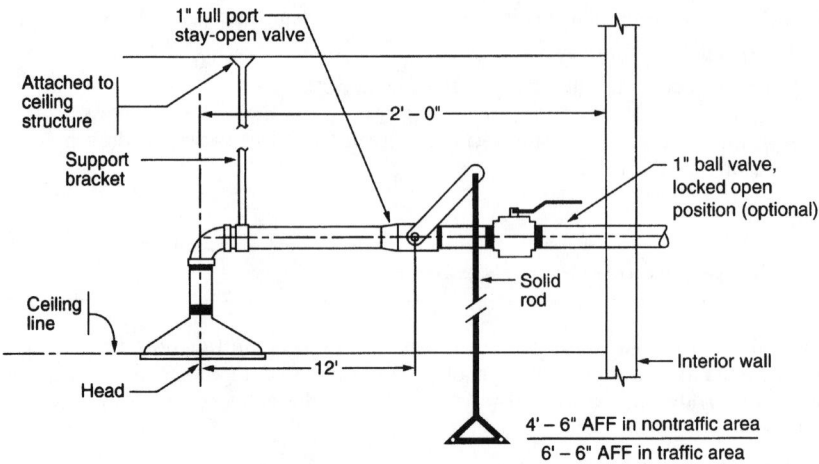

FIGURE 17.1 Detail of typical emergency shower.

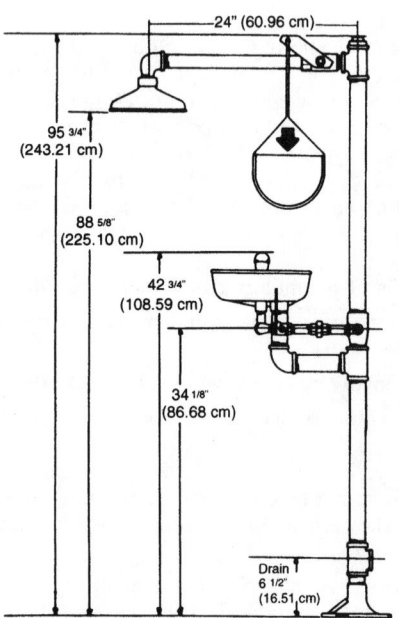

FIGURE 17.2 Detail of typical combination shower, eyewash, and drench shower.

Emergency Eyewash

Plumbed Eyewash. Emergency eyewashes are specifically designed to irrigate and flush both eyes simultaneously with a dual stream of water. The unit consists of dual heads in the shape of a U specifically designed to deliver a narrow stream of water, and a stay-open valve usually controlled by a large push plate. The stream configuration is illustrated in Fig. 17.3. Code requires that the eyewash be capable of delivering a minimum of 0.4 gpm (1.5 Lpm). Many eyewashes of recent manufacture deliver approximately 3 gpm (11.4 Lpm). Once started, the flow must be continuous and designed to operate without the use of the hands, which shall be free to hold open the eyelids. The flow of water must be soft to avoid additional injury to sensitive tissue. To protect against airborne contaminants, each dual stream head must be protected with a cover that is automatically discarded when the unit is activated. The head covers shall be attached to the heads by a chain. The eyewash can be mounted on a counter, on a wall, or as a free-standing unit secured to the floor. The eyewash could be provided with a bowl. The bowl does not increase efficiency or usefulness of the unit but aids in identification by personnel and provides a drain path to the floor.

The code recommends (but does not require) the use of a buffered saline solution to wash the eyes. This could be accomplished with a separate dispenser filled with concentrate that will introduce the proper solution into the water supply prior to reaching the device head. A commonly used device is a wall-mounted, 5- to 6-gal (20- to 24-L) capacity solution tank connected to the water inlet that dispenses a measured amount of solution when flow to the eyewash is activated. A backflow device shall be installed on the water supply. A typical free-standing eyewash is illustrated in Fig. 17.4.

Self-Contained Eyewash. A typical self-contained eyewash has a storage tank with a minimum 15-min water supply and is illustrated in Fig. 17.5. The mounting height and spray pattern requirements are the same as for plumbed eyewashes.

FIGURE 17.3 Stream from typical eyewash.

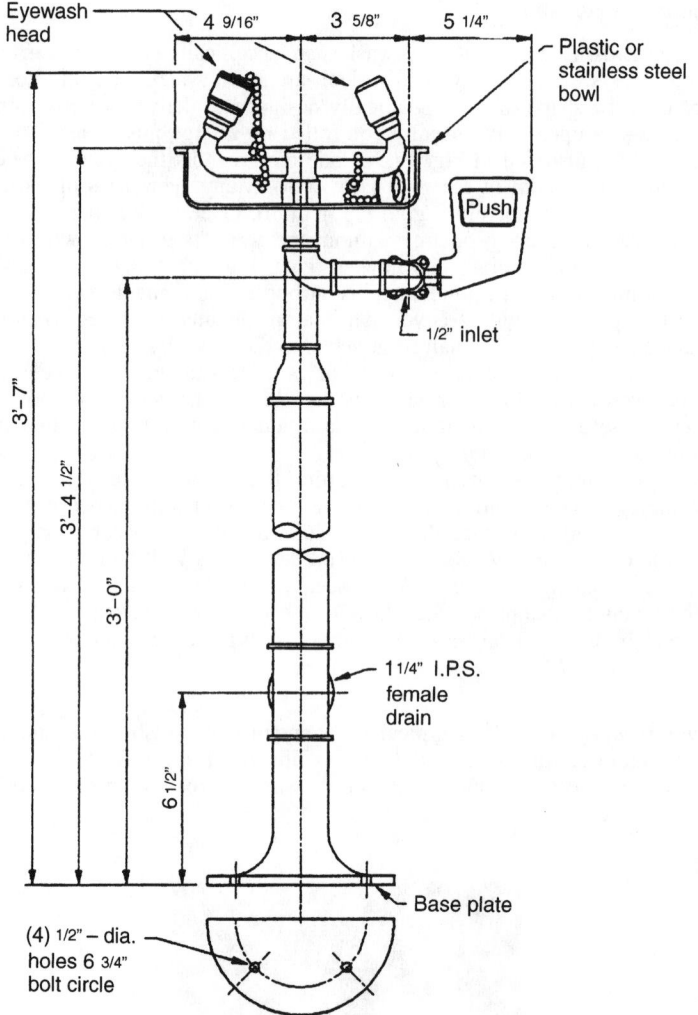

FIGURE 17.4   Detail of free-standing eye or face wash.

Emergency Facewash

The facewash is an enhanced version of the eyewash. It has the same design requirements and configuration except that the spray heads are specifically designed to deliver a larger water pattern and volume that will flush the whole face and not just the eyes. The facewash should deliver approximately 5–8 gpm (36–55 Lpm). The stream configuration is illustrated in Fig. 17.6. Very often, the facewash is chosen for combination units. In general, the facewash is more desirable than the eyewash because it is very likely that an accident will affect more than just the eyes. All dimensions and requirements of the free-standing facewash are similar to the eyewash, illustrated in Fig. 17.4.

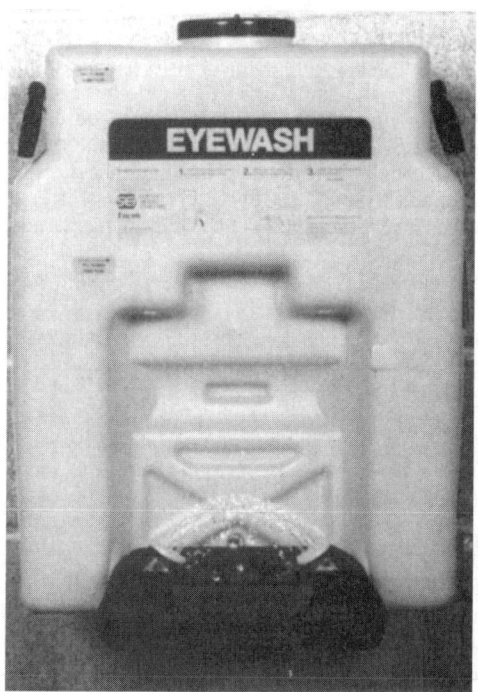

FIGURE 17.5 Portable eyewash.

FIGURE 17.6 Stream from typical face-wash.

Drench Hoses

A drench hose is a single-head unit connected to a water supply with a flexible hose. The head is generally the same size as a single head similar to that found on a eye-face wash. Code requires that the drench hose be capable of delivering a minimum of 0.4 gpm (1.5 Lpm). It is controlled either by a squeeze handle near the head or a push plate ball valve located at the connection to the water source. It is used as a supplement to a shower and eye-face wash to irrigate specific areas of the body. Drench hoses are selected for the following purposes:

1. To spot drench a specific area of the body when the large volume of water delivered by a shower is not called for
2. To allow irrigation of an unconscious person or a victim who is unable to stand
3. To irrigate under clothing prior to having the clothing removed while simultaneously using the emergency shower

A drench hose is not considered as a substitute for any emergency equipment.

Combination Equipment

Combination equipment consists of multiple-use units with a common water supply and supporting frame. Combinations are available that consist of shower, eye-face wash, or drench hose in any configuration. The reason for the use of combination equipment is usually economy, but the selection should consider the type of irrigation potential for injuries that might occur at a specific location. For combination units, the water supply must be larger and capable of delivering the flow rate of water required to satisfy two devices concurrently rather than only a single device.

The most often used combination is the drench shower and facewash. Figure 17.7 illustrates a combination shower, eye-face wash, and drench hose.

INSTALLATION REQUIREMENTS FOR DRENCH EQUIPMENT

The need to provide emergency drench equipment is determined by an analysis of the hazard by design professionals, health or safety personnel, and the use of common sense in conformance to OSHA, CFR, and other regulations for specific occupations. Judgment is necessary in the selection of equipment and its location. Very often, facility owners have specific regulations for their need and location.

Dimensional Requirements

The standard range of mounting heights as illustrated in ANSI Z-358.1 are shown in Fig. 17.8. If the showerhead is free-standing, the generally accepted dimension for the mounting height is 7 ft, 0 in (2.17 m) above the floor. Figure 17.13 illustrates a wheelchair-accessible, free-standing combination unit. Generally accepted clearance around showers and eye-face washes is illustrated in Fig. 17.9.

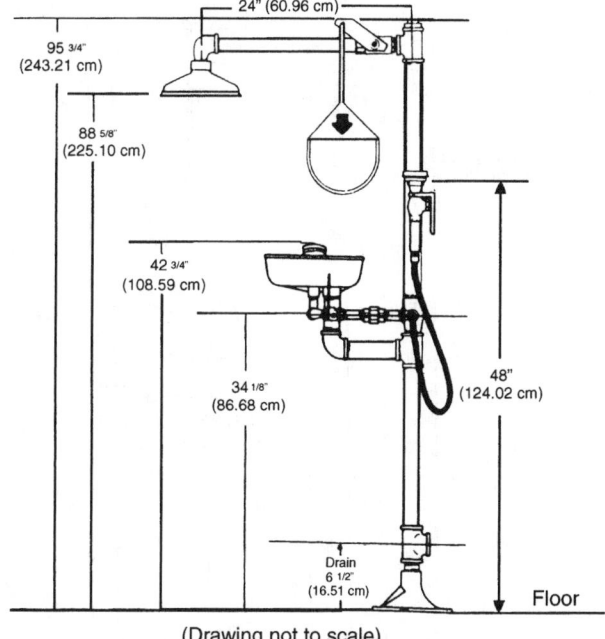

(Drawing not to scale)

FIGURE 17.7   Detail of combination emergency shower, eye-face wash, and drench hose.

Equipment Location

The location of the emergency drench equipment is crucial to the immediate and successful first aid treatment of an accident victim. It should be located as close to the hazard as practical without being affected by the hazard itself or by potential accidental conditions such as a large release or spray of chemicals resulting from are explosion or a pipe and tank rupture. In addition, drench equipment must not be placed adjacent to electrical equipment. Location along normal access and egress paths in the work area will reinforce the location to personnel as they see it each time they pass it.

There are no requirements in any code pertaining to the location of any drench equipment in terms of specific, definitive dimensions. ANSI code Z-358.1 requires that emergency showers be located a maximum distance of either 10 seconds' travel time by an individual or no more than 100 ft (30.5 m) from the protected hazard, whichever is shorter. It is recommended that a distance of 75 ft (21 m) be used. If strong acid or caustic is used, the equipment should be located within 10 ft (3 m) from the source of the hazard. The path to the unit from the hazard shall be clear and unobstructed, so that impaired sight or panic will not prevent clear identification and access. There is no regulation as to what distance could be covered by an individual in 10 s. There is also no specific provisions for the handicapped.

Since there are no specific code requirements for locating drench equipment, good judgment is required. Accepted practice is to have the equipment accessible

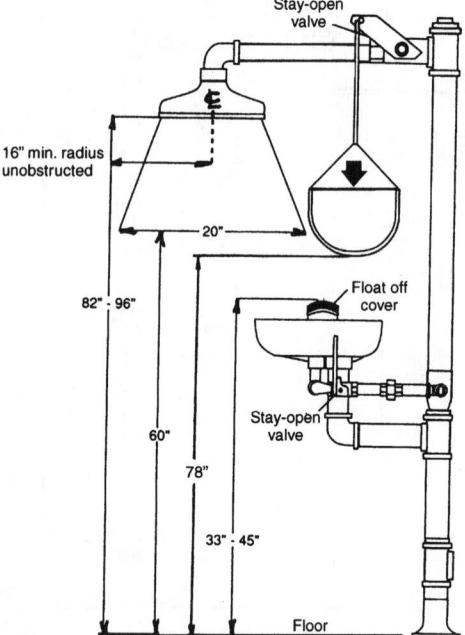

FIGURE 17.8 Code dimensions for wheelchair-accessible shower and eyewash mounting height.

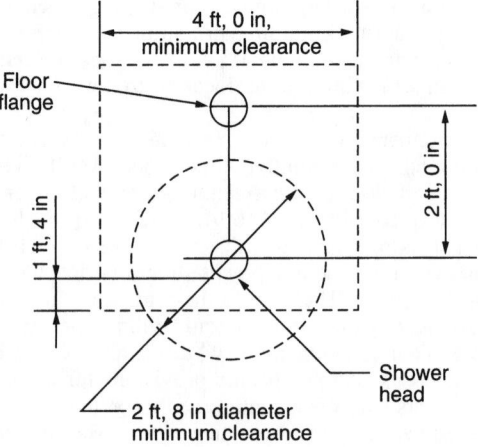

FIGURE 17.9 Clearance dimensions in plan for shower and eye-face wash.

from three sides. Anything less generally creates a "tunnel" effect that makes it more difficult for the victim to reach the equipment. It should be located on the same level as the hazard when possible. Traveling through rooms that may have locked doors to reach equipment shall be avoided, except that placing emergency showers in a common corridor, such as outside individual laboratory rooms, is accepted practice. Care should be taken not to locate the shower in the path of the swinging door to the protected room to avoid personnel who are coming to the aid of the victim from knocking the victim over.

Emergency eye-face washes should be located close to the source of hazard. In laboratories, accepted practice is to have one sink in a room fitted with an eyewash on the counter adjacent to a sink. The sink cold water supply provides water to the unit. It could be designed to swing out of the way of the sink.

Number of Stations

The number of drench equipment devices provided in a facility is a function of the type of hazard and number of people in rooms and areas exposed to any particular hazard at any one time, using a worst-case scenario. It is rare that more than one combination unit will be installed. It is important to consider that if a group of individuals may be exposed to a specific hazard, more than one drench unit may be required. Consulting with the end user and safety personnel will provide a good basis for the selection of the type and number of equipment.

Generally, one shower can be provided between an adjacent pair of laboratories, with emergency eye-face washes located inside each individual laboratory. In open areas, it is common practice to locate emergency equipment adjacent to columns for support.

Water Temperature

Code mandates the temperature of water supplied to equipment shall be "tempered." A range of 60 to 95°F (15 to 35°C) is suggested. Medical authorities recommend irrigation of chemical burns with 78 to 92°F (26 to 33°C) water. For a dedicated indoor system, this temperature range is achieved because the interior of a facility may be heated in the winter and cooled in the summer to approximately 70°F (20°C). Since the water in the emergency drench system is stagnant, it assumes the temperature of the ambient air. A generally accepted temperature of between 80 and 85°F (29°C) has been established as a "comfort zone" and is the most desirable water temperature.

The body will attempt to generate body heat lost if the drenching fluid is below the comfort zone. The common effect is shivering and increased heart rate. In fact, most individuals are uncomfortable taking a shower with water below 60°F (15°C). With the trauma induced by the accident, the effect is escalated.

Another consideration is the potential chemical reaction and/or acceleration of reaction with flushing water or water at a particular temperature. Where the hazard is a solid, such as radioactive particles that can enter the body through the pores, a cold water shower shall be used to prevent pores from opening in spite of its being uncomfortable. It is necessary to obtain the opinions of medical and hygiene personnel where any doubt exists about the correct use of water or water temperature in specific facilities.

Where showers are installed outdoors, or indoors where heating is not provided, the stagnant water supplying the showers may become too cold and must be tempered. Manufacturers offer a variety of tempering methods, including water temperature maintenance cable similar to that used for domestic hot water, fail-safe blending systems, and heat exchangers. In remote locations, complete self-contained units are available with tanks storing and maintaining heated water. OSHA reserves the right to cite in cases where individuals run away from drench water streams due to cold water temperatures.

Protection against Temperature Extremes

In areas where freezing is possible and water drench equipment is connected to an above-ground plumbed water supply, freeze protection is required. This is most often accomplished by electric heating cable and providing insulation around the entire water supply pipe and the unit itself. Recommended water temperature should be maintained at 75°F (24°C). Refer to Chap. 5 for the design of the freeze protection system.

For exterior showers located where freezing is possible, the water supply shall be installed below the frost line and a frost-proof shower installed. This type of shower has a method for draining the water above the frost line when the water to the drench equipment is turned off. A typical frost-proof emergency shower and eyewash is illustrated in Fig. 17.10.

When a number of drench equipment devices are located where low temperature is common, a circulating tempered water supply should be considered (Fig. 17.11). This uses a hot water heater and a circulating pump to supply the drench equipment. The heater shall be capable of generating 30 gpm (or more if more than one shower could operate simultaneously) of water from 40 to 85°F (4 to 29°C) with a low temperature of 78°F (26°C).

In areas where the temperature may get too high, it is accepted practice to insulate the water supply piping to maintain the temperature as long as practical.

DRENCH EQUIPMENT COMPONENTS

Controls

Often referred to as *activation devices,* controls cause water to flow at an individual device. Stay-open valves are required by code in order to leave the hands free to remove clothing or hold eyelids open. The valves most often used are ball valves with handles modified to provide for the attachment of chains, rods, and push plates. In very limited situations, such as in schools, valves that automatically close (quick closing) are permitted if acceptable to the facility and authorities having jurisdiction.

Valves are operated by different means to suit the specific hazard, location, durability, and visibility requirements. Operators for valves are handles attached to pull rods, push plates, and foot-operated treadle plates and triangles. A solid pull rod is often installed on concealed showers in order to push the valve closed after operation. Another method could be to have two handles attached to chains that extend below the hung ceiling, one to turn the valve on and another handle to turn it off. This is illustrated in Fig. 17.12. Chains are used if the handle might be accidentally struck, enabling the handle to move freely and not injure the individual striking the hanging operator.

Operating handles for a handicapped-accessible unit are mounted lower than for a standard unit. In many cases, this will require that operating handles be placed

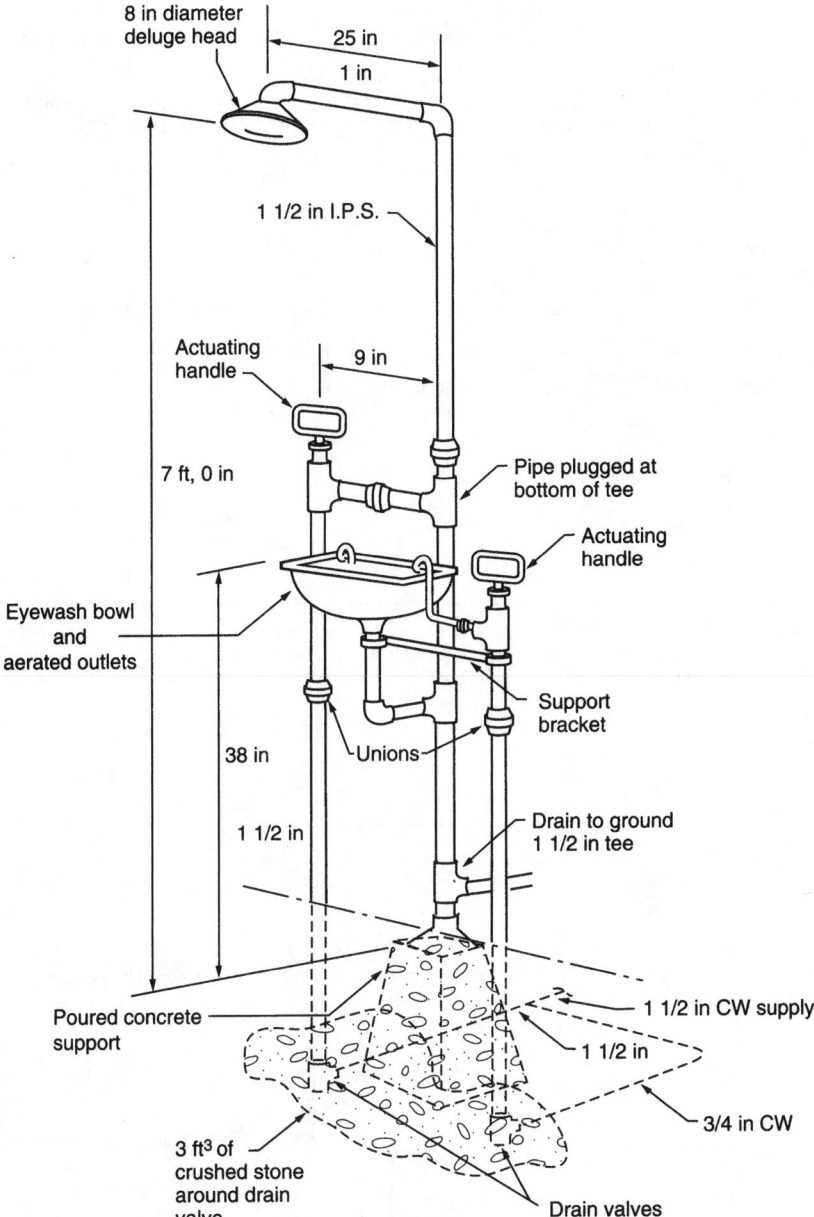

8 in diameter
deluge head

25 in

1 in

1 1/2 in I.P.S.

Actuating
handle

9 in

7 ft, 0 in

Pipe plugged at
bottom of tee

Actuating
handle

Eyewash bowl
and
aerated outlets

Support
bracket

38 in

Unions

1 1/2 in

Drain to ground
1 1/2 in tee

Poured concrete
support

1 1/2 in CW supply

1 1/2 in

3/4 in CW

3 ft³ of
crushed stone
around drain
valve

Drain valves

FIGURE 17.10 Typical frost-proof emergency shower and eyewash. *Note:* Provide shutoff valve in supply line in adjacent building. Where there is no building nearby, provide valve in supply line underground with curb box.

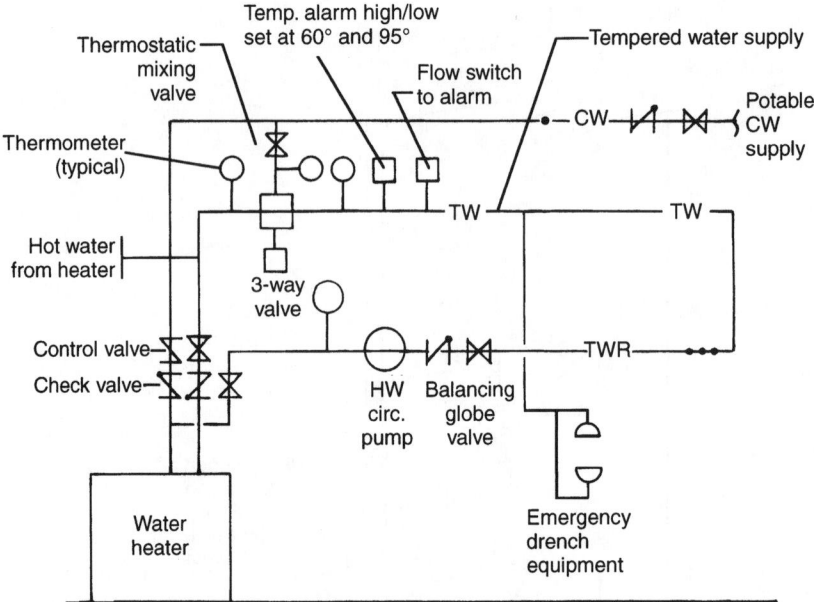

FIGURE 17.11 Schematic detail of central tempered water emergency shower supply. Note that one alarm could be a horn or light in occupied area. Normal temperature is 70°F (18°C). High temperature is 95°F (38°C). Low temperature is 60°F (15°C).

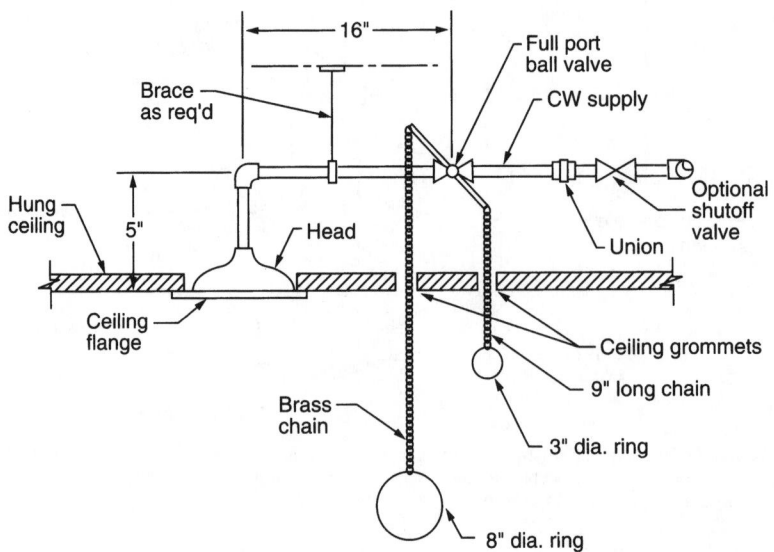

FIGURE 17.12 Typical emergency shower with two control rods in hung ceiling.

near walls to keep them out of traffic patterns where they would be an obstruction to ablebodied people passing under them. A free-standing, handicapped-accessible combination shower and eyewash using handles hung from the ceiling is illustrated in Fig. 17.13. The handle must be located close enough to the center of the shower to be easily reached, which is about 2 ft, 0 in from the center of the shower. Another method using a wall attached chain is illustrated in Fig. 17.14.

Alarms

Alarms are often installed to alert security or other rescue personnel that emergency drench equipment has been operated and to guide them rapidly to the scene of the accident. Commonly used alarms are audible and visual devices, such as flashing or rotating lights on top of, or adjacent to, a shower or eyewash and electronic alarms wired to a remote security panel. Remote areas of a plant are particularly at risk if personnel often work alone. Alarms are most often operated by a flow

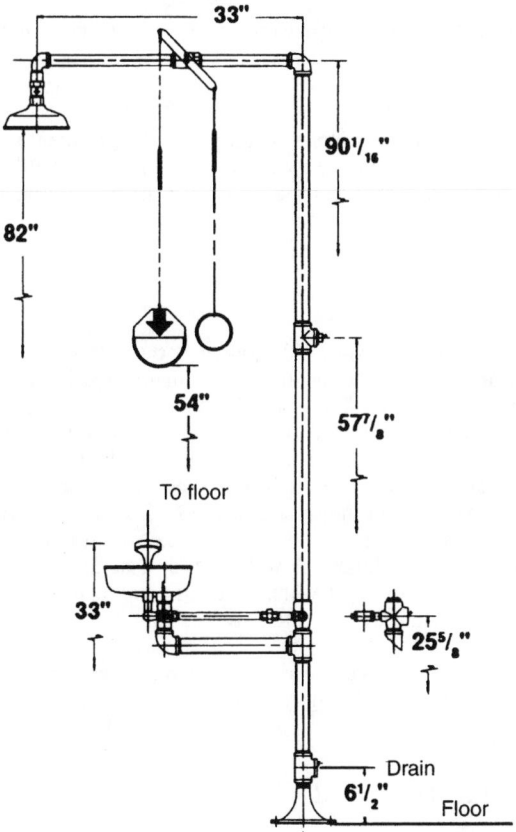

FIGURE 17.13 Handicapped-accessible free-standing combination emergency shower and eyewash.

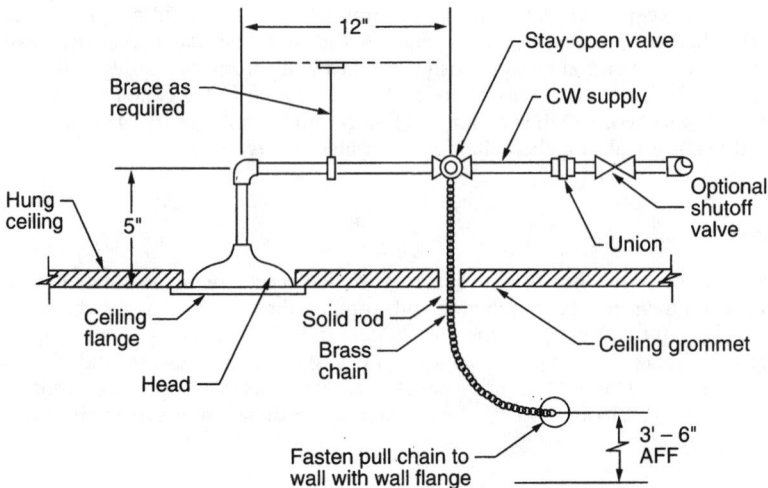

FIGURE 17.14 Detail of combination emergency shower in hung ceiling with wall attached chain operator.

switch activated by the flow of water when a piece of equipment is used, proximity switches, switches mounted on push plates, and on shower pull chains or rods. When tempered water systems are used to supply drench equipment, a low water temperature of 60°F shall cause an alarm annunciation.

Flow Control

If water pressure exceeds 80 psig (550 kPa) or if the difference in water pressure between the first and last shower head is more than 20 psig (140 kPa), it is recommended that a self-adjusting flow control device be installed in the water supply pipe. The purpose of this device is to limit the flow to just above the minimum required by the specific manufacturer to function properly. Flow control devices are considered important because a shower installed at the beginning of a long run will have a much greater flow than the device at the end. During operation, the higher pressure could cause the flow rate to be as much as 50 gpm (189 Lpm). If there is no floor drain provided, the higher flow for 15 mm at the higher pressure could produce a much greater amount of water that must be cleaned up and disposed of afterward. Drench hoses and eye- and face-washes are not affected because of their lower flow rates and flow head design.

Where a pressure-reducing device is required for an entire system, it should be set to provide approximately 50 psig (345 kPa) at the most remote shower.

FLUSHING WATER DISPOSAL

Water from emergency drench equipment is mainly discharged onto the floor. Individual eye-face washes mounted on sinks discharge most of the water into the adjacent sink. Combination units that have an attached eye-face wash will also discharge that water onto the floor. There are different methods of disposing of the water resulting from an emergency device depending on the facility. The basic consideration is whether to provide a floor drain adjacent to a device to route that water from the floor to a drainage system.

It is accepted practice not to provide a floor drain at an emergency shower. Experience has shown that in most cases, particularly in schools and laboratories, it is easier to mop up water from the floor in the rare instances that emergency devices are used. Considerations are:

1. Is the drain in an area where frequent cleaning is done, so that the trap will not dry out, allowing odors to be emitted?
2. Is there an available drainage line in the area of the device?
3. Can the chemical, even in a diluted state, be released into the sanitary sewer system, or must it be routed to a chemical waste system for treatment?
4. Must purification equipment be specially purchased for this purpose?

VISIBILITY OF DEVICES

High visibility must be considered in the selection of any device. Usually selected recognition methods are high-visibility signs mounted at or on the device, surrounding floors and walls painted a contrasting, bright color, specifically colored indicator lights, and a bright, well-lit area on a plant floor to help a victim identify the area and help in first aid activities. Color blindness should also be considered in color selection.

SYSTEM DESIGN

General

It is a requirement that a plumbed system be connected to a potable water supply as the sole source of water. This system is therefore subject to filing with a plumbing or other code official for approval and inspection of the completed facility as required for standard plumbing systems.

The water supply must provide an adequate size pipe and sufficient pressure to overcome system and device operating pressure requirements in order to function satisfactorily. One maintenance requirement is to regularly flush the water in the piping system to avoid bacterial growth.

It is common practice to add antibacterial and saline products into a self-contained eyewash unit, and an antibacterial additive to an emergency shower. Water is also commonly used if it can be changed every week. It is well established that no preservative will inhibit bacterial growth for an extended period of time. Self-contained equipment must be checked regularly to determine if the quality of the stored water has deteriorated to a point where it is not effective or safe to use.

If valves are placed in the piping network for maintenance purposes, they should be locked open to prevent unauthorized shutoff.

Water Supply Pressure and Flow Rates

Emergency showers require a minimum of 20 gpm (76 Lpm), with 30 gpm recommended. The minimum pressure required is 30 psig (4.5 kPa) at the farthest unit, with a generally accepted maximum pressure of 70 psi (485 kPa). Code mentions a high pressure of 90 psig (612 kPa), which is generally considered to be excessive. Most plumbing codes do not permit water pressures as high as 90 psig. Generally accepted practice limits the high water pressure to between 70 and 80 psig (480 and 620 kPa).

Most eyewash units require a minimum operating pressure of 15 psig (105 kPa) with a minimum flow rate of 3 gpm (12 Lpm) at the furthest unit. Maximum pressure is similar to that for showers. Facewash and drench hoses require a minimum operating pressure of 30 psig (210 kPa) with a minimum flow rate of between 5–8 gpm (36–55 Lpm) at the farthest unit.

System Selection

Plumbed System.   The advantages of the plumbed systems are:

1. The systems are permanently connected to a supply of water, and minimum testing of the devices is needed to assure proper operation.
2. The systems provide an unlimited supply of water, often at larger volumes compared to self-contained units.

Disadvantages are:

1. The first cost is higher than for a self-contained system.
2. These systems are maintenance intensive. They require flushing, often into a bucket, to remove stagnant water in the piping system, which is replaced with fresh water.

Self-Contained System.   Advantages of the self-contained system:

1. Lower first cost compared to a plumbed system.
2. Can be filled with a buffered, saline solution, which is recommended for washing eyes.
3. Available with a container to catch wastewater.
4. Units are portable and can be moved to areas of greatest hazard with little difficulty.
5. Gravity eyewash is more reliable. The water supply can be installed where there is room above the unit. If not, a pressurized unit mounted remotely should be selected.

Disadvantages:

1. Only a limited supply of water at a lesser flow rate is available.
2. Stored liquid must be changed on a regular basis to maintain purity.

The plumbed system is the most often selected type of system because of the unlimited water supply.

Pipe Sizing and Material

In order to supply the required flow rate to a shower, a minimum pipe size of 1 in (25 mm) is required by code, with 1¼ in (32 DN) recommended. If the device is a combination unit, 1½-in (40 DN) size should be considered as minimum. Emergency eye-face wash requires a minimum ¾-in (20 DN) pipe size.

Except in rare cases where multiple units are intended to be used at once, the piping system size should be based on only one unit operating. The entire piping system is usually a single-size pipe based on the requirements of the most remote fixture. Appropriate pressure loss calculations should be made to assure that the hydraulically most remote unit is supplied with adequate pressure with the size selected. Adjust sizes accordingly to meet friction loss requirements. PRVs shall be installed if required.

The pipe material should be copper to minimize clogging the heads of the units in time with the inevitable corrosion products released by steel pipe. Plastic pipe (PVC) should be considered where excessive heat and the use of closely located supports will not permit the pipe to creep in time.

Emergency drench equipment shall be sized based on the single highest flow rate, which is 20 gpm (75 Lpm) for an emergency shower. Piping is usually a 1½ or 2 in (40 or 50 DN) header of copper pipe, for the entire length of a plumbed system.

BREATHING AIR

GENERAL

Breathing air systems supply air of specific minimum purity to personnel for purposes of escape and protection when exposed to a toxic environment resulting from an accident or during normal work where conditions make breathing the ambient air dangerous. As defined by 30 CFR 10, a *toxic environment* has air that "may produce physical discomfort immediately, chronic poisoning after repeated exposure, or acute adverse physiological symptoms after prolonged exposure."

This section will discuss the production, purification, and distribution of a low-pressure breathing air and individual breathing devices used to provide personnel protection only when used with supplied air systems.

Low pressure for breathing air refers to compressed air pressures of up to 250 psig (1725 kPa) delivered to the respirator. The most common operating range for systems is between 90 and 110 psig (620 and 760 kPa).

Much of the equipment used in the generation, treatment, and distribution of compressed air for the breathing air system is common to that for medical-surgical air discussed in Chap. 14, "Compressed Gases."

CODES AND STANDARDS

1. OSHA, 29 CFR 1910
2. CGA, Commodity Specification G-7 and G-7.1
3. CSA, Canadian Standards Association
4. NIOSH, National Institute of Occupational Safety and Health
5. MSHA, Mine Safety and Health Act
6. NFiPA, NFPA-99 Medical Compressed Air
7. DOD, Where applicable
8. ANSI, Z-88.2 Standard for Respiratory Protection
9. CFR, Code of Federal Regulations

SYSTEM COMPONENTS

The supplied air category of respirators uses compressed air cylinder(s) or an air compressor, air purifying equipment at the source, and central piping system. Respirators are connected to the cylinder or distribution piping system.

TYPES OF SYSTEMS

There are three basic types of breathing air systems: constant flow, demand flow, and pressure demand.

Constant Flow

Also known as *continuous flow,* constant flow systems provide an uninterrupted flow of purified air through personal respirators to minimize leakage of contaminants into the respirator and to ventilate the respirator with cool or warm air depending on conditions. This system could be used in wide variety of areas ranging from least harmful to most toxic depending on the type of respirator selected.

Demand Flow

The demand flow system provides an interrupted supply of purified air to respirators during inhalation. Upon exhalation, the flow of air is shut off until the next breath. Demand flow systems automatically adjust to an individual's breathing rate. This system requires tight-fitting respirators. Their application is generally limited to less harmful areas because the negative pressure in the respirator during inhalation may permit leakage of external contaminants. These systems are designed for economy of air use during relatively short duration tasks and are usually supplied from cylinders.

Pressure Demand

A pressure demand system delivers purified air continuously through personal respirators with increased airflow during inhalation. By continuously providing a flow of air above atmospheric pressure, leakage of external contaminants is minimized.

This system also uses tight-fitting respirators, but the positive pressure aspect allows them to be used in more toxic applications.

TYPES OF PERSONAL RESPIRATORS

There two general categories of respirators used for individual protection: air purifying and supplied air. The air purifying category of respirator is portable and has self-contained filters that purify the ambient air on a demand basis. The advantages to their use are that they are less restrictive to movements and they are light in weight. Disadvantages are that they must not be used where gas or vapor contamination cannot be detected by odor or taste or where the atmosphere is oxygen deficient. This type of respirator is outside the scope of this book, and is mentioned only because of its availability.

The type of respirator selected depends on the expected breathing hazards. When choosing a respirator, the highest expected degree of hazard, applicable codes and standards, manufacturer recommendations, suitability for the intended task, and comfort of the user are all important considerations.

The EPA Office of Emergency and Remedial Response has identified four levels of hazards at cleanup sites involving hazardous materials and lists guidelines for the selection of protective equipment:

1. Level A calls for maximum available protection requiring a positive pressure, selfcontained suit, generally with a self-contained breathing apparatus worn inside the protective suit.

2. Level B protection is required when the highest level of respiratory protection is needed but a lower level of skin protection is acceptable.

3. Level C uses full facepiece, air-purifying respiratory protection with chemical-resistant, disposable garments. This is required when the contaminant is known and the level is relatively constant. Typical of the uses are asbestos removal projects.

4. Level D protection is used where special respiratory or skin protection is not required but a rapid increase of contaminant level or degradation of ambient oxygen content is possible.

If the hazard cannot be identified, it must be considered an immediate danger to life and health (IDLH). This condition exists when the oxygen content of the air falls below 12.5 percent (95 ppm O_2) or where the air pressure is less than 8.6 psi (450 mm Hg), which is the equivalent of 14,000 ft (4270 m).

There are five general types of respirators available as follows.

Mouthpiece Respirators

Used only with demand-type systems, mouthpiece respirators are designed only to deliver breathable air. They offer no protection to the skin, eyes, or face. Their use is limited to only those areas where there is insufficient oxygen and no other contaminants that could affect the eyes and skin. A mouthpiece respirator is illustrated in Fig. 17.15.

Half Facepiece Respirators

Half facepiece respirators cover the nose and mouth and are designed primarily for demand and pressure systems. They are usually tight-fitting and provide protection for extended periods of time in atmospheres not harmful to eyes and skin. Often worn with goggles, these respirators are limited to areas of relatively low toxicity. A half facepiece respirator is illustrated in Fig. 17.16.

Full Facepiece Respirators

Full facepiece respirators cover the entire face and are designed for use with constant flow and pressure demand systems. They are tight-fitting and suitable for

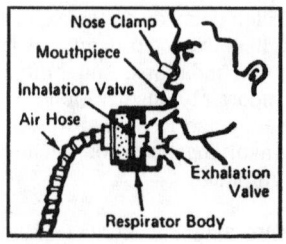

FIGURE 17.15 Illustration of mouthpiece respirator.

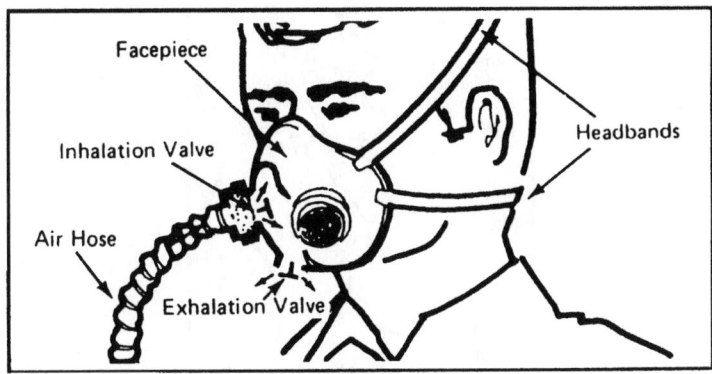

FIGURE 17.16   Illustration of half facepiece respirator.

atmospheres of moderate and high toxicity. They are usually used in conjunction with full protective clothing for such tasks as chemical tank cleaning where corrosive and toxic gas, mist, and liquids may be present. Since the face mask provides protection to the face and eyes, they are also suitable for other tasks such as welding and inspection of tanks and vessels where there is an oxygen-deficient atmosphere. A full facepiece respirator is illustrated in Fig. 17.17.

Hood and Helmet Respirators

Hood and helmet respirators cover the entire head and are normally used with a constant flow system. They are loose-fitting and suitable only for protection against contaminants such as dust, sand, powders, and grit. Constant flow is necessary to ventilate the headpiece and to provide sufficient air pressure to prevent contaminants from entering the headpiece. A hood and helmet respirator is illustrated in Fig. 17.18.

Full Pressure Suits

Full pressure suits range in design from loose-fitting, body protective clothing to completely sealed, astronautlike suits that provide total environmental life support.

FIGURE 17.17   Illustration of full facepiece respirator.

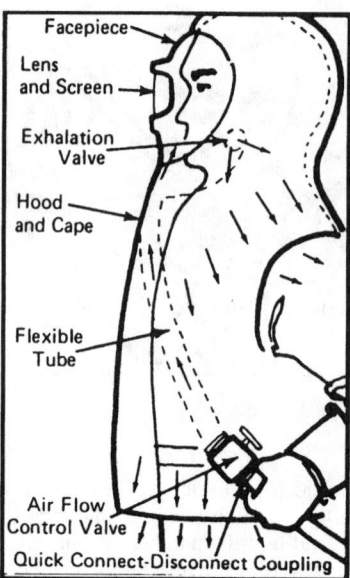

FIGURE 17.18 Hood and helmet respirator.

They are designed to be used only with constant flow systems and are suitable for the most toxic and dangerous environments and atmospheres.

BREATHING AIR PURITY

Air for breathing purposes supplied from compressors or from a pressurized tank must comply as a minimum with quality verification level grade D in CGA G-7. 1 (ANSI Z-86.1). Table 1 from ANSI/CGA G-7.1 is reproduced here as Table 17.1 and lists the maximum contaminant levels for various grades of air. For grade D quality air, individual limits exist for condensed hydrocarbons, carbon monoxide, and carbon dioxide. Particulates and water vapor, whose allowable quantities have not been established, must also be controlled because of the effect they may have on different devices of the purification system, in the piping system, and by the end user of the equipment.

Contaminants

Condensed Hydrocarbons. Oil is a major contaminant in breathing air. It causes breathing discomfort, nausea, and in extreme cases, pneumonia. It can also create an unpleasant taste and odor and interfere with an individual's desire to work. In addition, oxidation of oil in overheated compressors could produce carbon monoxide. A limit of 5 ppm has been established. Some types of reciprocating and rotary screw compressors put oil into the airstream as a result of their operating

TABLE 17.1 Listing of the Maximum Contaminant Level for Various Grades of Air

(Units in ppm (mole/mole) unless shown otherwise)

Limiting characteristics	A	K	L	D	E	G	J	M	N
Percent O_2 balance predominantly N_2 (Note 2)	atm/ 19.5– 23.5	atm/ 19.5– 23.5	atm/ 19.5– 23.5	atm/ 19.5– 23.5	atm/ 20– 22	atm/ 19.5– 23.5	atm/ 19.5– 23.5	atm/ 19.5– 23.5	atm/ 19.5– 23.5
Water, ppm (v/v) (Note 3)		200	50					1	3
Dew point, °F (Note 3)		−33	−54				−104	−92	
Oil (condensed) (mg/m³ at NTP)				5/[4]	5/[4]				None*
Carbon monoxide				10/[5×6]	10	5	1	1	10
Odor									None
Carbon dioxide				1000/[6]	500	500	0.5	1	500
Total hydrocarbon content (as methane)	25				25	15	0.5	1	
Nitrogen dioxide						2.5	0.1	0.5	2.5
Nitric oxide									
Sulfur dioxide						2.5	0.1		5
Halogenated solvents						10	0.1		
Acetylene							0.05		
Nitrous oxide							0.1		
USP									Yes

*Includes water.

1. The last edition of CGA G-7.1-1973 listed nine quality verification levels of gaseous air lettered A to J and two quality verification levels of liquid air lettered A and B. Some of those letter designations have been dropped from this edition (1989) since they no longer represent major volume usage by industry. Four new letter designations, K, L, M, and N have been added to reflect current specifications. To get a listing of quality verification levels dropped, see CGA-7.1-1973 or contact the Compressed Gas Association.

2. The term "atm" (atmospheric) denotes the oxygen content normally present in atmospheric air; the numerical values denote the oxygen limits for synthesized air.

3. The water content of compressed air required for any particular quality verification level may vary with the intended use from saturated to very dry. For breathing air used in conjunction with a self-contained breathing apparatus in extreme cold where moisture can condense and freeze, causing the breathing apparatus to malfunction, a dew point not to exceed −50°F (63 ppm v/v) or 10 degrees lower than the coldest temperature expected in the area is required. If a specific water limit is required, it should be specified as a limiting concentration in ppm (v/v) or dew point. Dew point is expressed in °F at one atmosphere pressure absolute, 101 kPa abs. (760 mm Hg).

4. Not required for synthesized air whose oxygen and nitrogen components are produced by air liquefaction.

5. Not required for synthesized air when nitrogen component was previously analyzed and meets *National Formulary* (NF) specification.

6. Not required for synthesized air when oxygen component was produced by air liquefaction and meets *United States Pharmacopeia* (USP) specification.

Source: ANSI/CGA G-7.1, ANSI 2-86.1, Table 1.

characteristics. Accepted practice is to use only oil-free air compressors in order to eliminate the possibility of introducing oil into the airstream.

Carbon Monoxide. Carbon monoxide is the most toxic of the common contaminants. It enters the breathing air system through the compressor intake or is produced by oxidation of heated oil in the compressor. Carbon monoxide easily combines with the hemoglobin in red blood cells and replaces oxygen. The lack of oxygen causes dizziness, loss of motor control, and loss of consciousness. A limit of 10 ppm in the airstream has been established based on NIOSH standards.

Carbon Dioxide. Carbon dioxide is not considered one of the more dangerous contaminants. Although the lungs have a concentration of approximately 50,000 ppm, a limit of 1000 ppm has been established for the breathing airstream.

Water and Water Vapor. Water vapor enters the piping system through the air compressor intake. Since no upper or lower limits have been established by code, the allowable concentration is governed by specific operating requirements by the most demanding device in the system, which is usually the CO converter or the requirement of having the dew point 10°F lower than the lowest temperature where piping is installed.

After compression, water vapor is detrimental to the media used to remove CO. The dew point of the airstream must be greatly lowered at this point in order to provide the highest efficiency possible for this device. Water vapor is removed to such a low level that breathing air with this level of humidity will prove uncomfortable to users.

After purification, too much humidity will fog the faceplate of a full face mask. It will also cause freeze-up in the pipeline if the moisture content of the airstream in the pipe has a higher dew point than the ambient temperature of the area where the compressed air line is installed.

Solid Particles. Known as *particulates,* solid particles can enter the system through the intake and are released from nonlubricated compressors as a result of friction from carbon and Teflon material used in place of lube oils. No limits have been established by code.

Odor. There is no standard for odor measurement. A generally accepted requirement is that there be no detectable odor in the breathing air delivered to the user. This requirement is subjective and will vary with individual users.

SYSTEM COMPONENTS

The breathing air system consists of a compressed air source, purification devices and filters to remove unwanted contaminants from the source airstream, humidifiers to introduce water vapor into the breathing air, the piping distribution network, respirator outlet manifolds, and respirator hose and the individual respirators used by personnel. Alarms are needed to monitor the quantity of contaminants and other parameters of the system as a whole.

Generation and Storage of Breathing Air

The source of air for the breathing air system is an air compressor and/or high-pressure air stored in cylinders. The air in cylinders uses ambient air which is purified to reduce or eliminate impurities to the required level and compresses it to the desired pressure. A typical schematic detail is shown in Fig. 17.19.

Air Compressor. The required standard for air compressors used to supply breathing air shall comply with the oil-free medical gas discussed in Chap. 14, "Compressed Gases." Medical gas compressor systems are used because, as a whole, they generate far fewer contaminants than other types. When a liquid ring

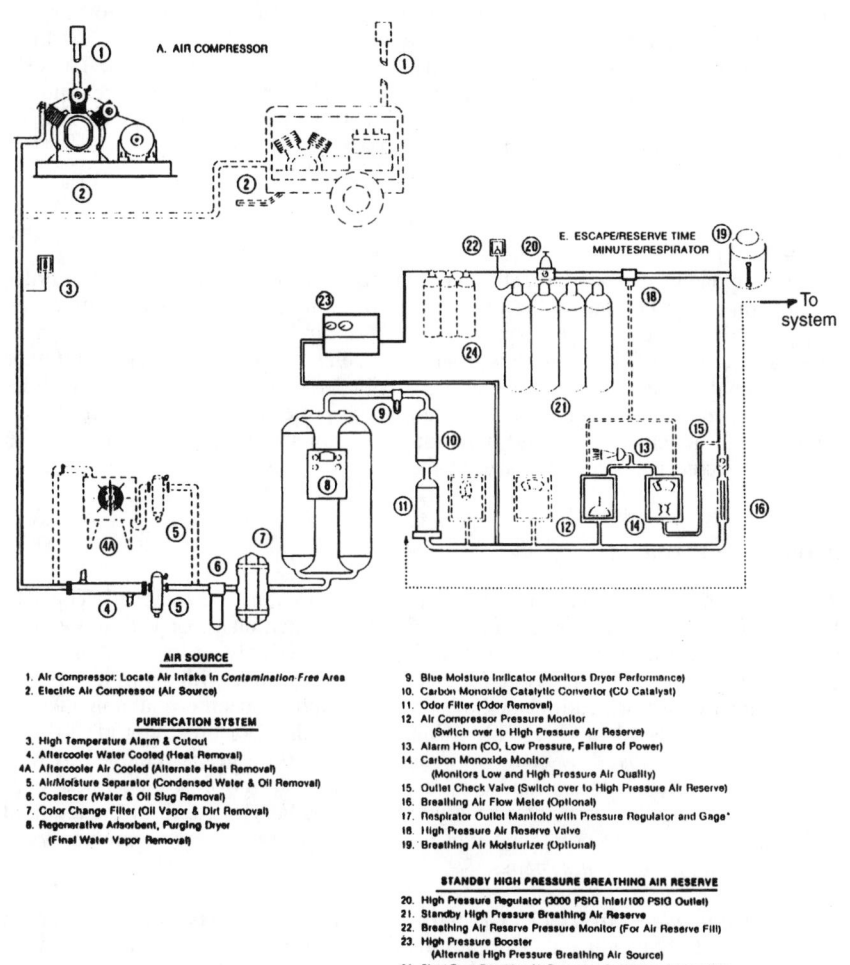

AIR SOURCE

1. Air Compressor: Locate Air Intake in *Contamination-Free* Area
2. Electric Air Compressor (Air Source)

PURIFICATION SYSTEM

3. High Temperature Alarm & Cutout
4. Aftercooler Water Cooled (Heat Removal)
4A. Aftercooler Air Cooled (Alternate Heat Removal)
5. Air/Moisture Separator (Condensed Water & Oil Removal)
6. Coalescer (Water & Oil Slug Removal)
7. Color Change Filter (Oil Vapor & Dirt Removal)
8. Regenerative Adsorbent, Purging Dryer
   (Final Water Vapor Removal)
9. Blue Moisture Indicator (Monitors Dryer Performance)
10. Carbon Monoxide Catalytic Convertor (CO Catalyst)
11. Odor Filter (Odor Removal)
12. Air Compressor Pressure Monitor
   (Switch over to High Pressure Air Reserve)
13. Alarm Horn (CO, Low Pressure, Failure of Power)
14. Carbon Monoxide Monitor
   (Monitors Low and High Pressure Air Quality)
15. Outlet Check Valve (Switch over to High Pressure Air Reserve)
16. Breathing Air Flow Meter (Optional)
17. Respirator Outlet Manifold with Pressure Regulator and Gage*
18. High Pressure Air Reserve Valve
19. Breathing Air Moisturizer (Optional)

STANDBY HIGH PRESSURE BREATHING AIR RESERVE

20. High Pressure Regulator (3000 PSIG Inlet/100 PSIG Outlet)
21. Standby High Pressure Breathing Air Reserve
22. Breathing Air Reserve Pressure Monitor (For Air Reserve Fill)
23. High Pressure Booster
   (Alternate High Pressure Breathing Air Source)
24. Short Term Breathing Air Reserve – Supervisory & Inspection

FIGURE 17.19   System with standby high-pressure breathing air reserve. (*Courtesy Nomonox.*)

compressor is used, it has the advantage of keeping the temperature of the air leaving the unit low. It is also possible to use any type of compressor for this service provided that the purification system is capable of producing air meeting all the requirements of code.

The air compressor assembly consists of the intake assembly (including the inlet filter), the compressor and receiver, aftercooler, and interconnecting water seal supply and the other ancillary piping. All of these components are discussed in Chap. 14, in the subsection called "Medical Gases."

Air compressors have a high first cost and are selected if the use of air for breathing is constant and continuous, making the use of cylinders either too costly or too maintenance intensive in the frequent changing of cylinders.

Cylinder Storage.   When high-pressure cylinders are used either as a source or as an emergency supply of breathing air, they shall be filled with air conforming to breathing air standards. The regulator should be set to about 50 psi (340 kPa) depending on the pressure required to meet system demands and losses. The cylinders have a low initial cost and are not practical to use if there is continuous demand. Cylinders are best suited to intermittent use for short periods of time or as an emergency escape backup for a compressor if it should fail.

Aftercoolers

Some components of the purification system require a specific temperature in order to function properly. Depending on the type of compressor selected and the type of purification necessary, the temperature of the air leaving the compressor may have to be reduced. This is done with an aftercooler installed on an air compressor by the manufacturer. Aftercoolers can be supplied with cooling water or use air as the cooling medium. Water, if recirculated, is the preferred method. The manufacturer of both the compressor and purification system should be consulted as to the criteria used and recommended size of the unit.

Purification Methods

The contaminants of concern to breathing air systems must be removed. This can be done by separate devices to remove individual contaminants or with a prepiped assembly of all necessary purification devices, requiring only an inlet and outlet air connection. For breathing air systems, this is commonly done with a purifier system.

The individual purification methods used to remove specific contaminants are the same as discussed in Chap. 14, "Compressed Air." For breathing air, oil and particulates are removed by coalescing and other filters, water is removed by desiccant or refrigerated dryers, and carbon monoxide is removed by chemical conversion to carbon dioxide using a catalytic converter. Very often all of the required purification devices are combined into a single prepiped assembly commonly referred to as a *purification system.*

Carbon Monoxide Converter.   The purpose of the converter is to oxidize carbon monoxide and convert it into carbon dioxide, which is tolerable in much greater quantities. This is typically accomplished by the use of a catalyst usually consisting of manganese dioxide, copper oxide, cobalt, and silver oxide in various combinations and placed inside a single cartridge. The material is not consumed, but it does become contaminated. The conversion rate greatly decreases if any oil or moisture is present in the airstream. Therefore, moisture must be removed before air enters the converter. Catalyst replacement is recommended generally once a year since it is not possible to completely control all contaminants that contribute to decreased conversion.

Moisture Separator. Water and water vapor are removed by two methods, desiccant and refrigerated dryers. The most common desiccant drying medium is activated alumina. For a discussion of air drying methods, refer to Chap. 14.

Odor. Activated, granular charcoal in cartridges is used for removal of odors.

Particulates. Particulates are removed by means of in-line filters. Generally accepted practice eliminates particulates in the piping system 1 μm and larger. Refer to Chap. 14 for a discussion of filter types.

Humidifiers

When water is removed from the compressed airstream prior to catalytic conversion, the dryer produces very dry air. If the breathing air system is intended to be used for long periods of time, very low humidity will dry the mucous membranes of the eyes and mouth. Therefore, moisture must be added to the airstream to maintain recommended levels. Humidifiers, often called *moisturizers,* are devices that inject the proper level of water vapor into an airstream. Some require a water connection. A recommended level of moisture is 50 percent relative humidity in the compressed airstream. Care must be taken not to route the air distribution piping through areas capable of having temperatures low enough to cause condensation. If this routing is impossible to change, a worker will have shorter periods of time on the respirator.

Respirator Manifold and Pressure Reducer

This is a combination device used to provide a single component that has multiple quick disconnect outlets, providing a convenient place to both reduce the pressure of the distribution network and a connection point for several hoses. A pressure gauge should be installed on the manifold to assure that the outlet pressure is within limits required by the respirator.

Respirator Hose

The respirator hose is flexible and is used to connect the respirator worn by an individual to the central distribution piping system. Code allows a maximum hose length of 300 ft (93 m). A quick disconnect is the most common method of connecting a hose to the supply piping.

COMPONENT SELECTION AND SIZING

Source of Breathing Air

Air Compressor. The air compressor size is based on the highest flow rate in cfm (Lpm) required by the number and type of respirators intended to be used simultaneously and the minimum pressure required by the purification system.

The following general flow rates are provided as a preliminary estimate for various types of respirators. Since there is a wide variation in the pressure and flow rates required for various types of respirators, the actual figures used to size the system must be based on the manufacturer's recommendations for the specific respirators selected.

1. 4 scfm (113 Lpm) for pressure demand respirators
2. 6 scfm (170 Lpm) for constant flow regulators
3. Up to 16 scfm (453 Lpm) for flooded hood respirators
4. Up to 35 scfm (990 Lpm) for flooded suits
5. Add 15 scfm (425 Lpm) of air for suit cooling if used

High-Pressure Cylinder Storage. High-pressure cylinders are used either to supply air for normal operation to a limited number of personnel for short periods of time or as an emergency supply to provide a means of escape from a hazardous area if the air compressor fails. The main advantage to using cylinders is that the air in the cylinders is prepurified, and no further purification of the air is necessary.

The number of cylinders is based on the simultaneous use of respirators, the cfm (Lpm) of each, and the duration, in minutes, that the respirators are expected to be used, plus a 10 percent safety factor. The total amount of compressed air in the cylinders should not be allowed to decrease too low. This requires that a low-pressure alarm be activated when the pressure falls to 500 psig (3450 kPa) in a cylinder normally pressurized to 2400 psig (16,500 kPa) when refilled to capacity.

As an example, establish the number of cylinders required for an emergency supply of air for eight people using constant flow respirators that require 15 min to escape the area:

1. $8 \times 6 \times 15 = 720$ scfm + 72 (10 percent) = 792 scfm total required.
2. Next, find the actual capacity of a single cylinder at the selected high pressure, generally 2400 psi (16,500 kPa), and divide the capacity of each cylinder into the total scfm required to find the number of cylinders required. Refer to Eq. (14.4) for the cylinder capacity.
3. If 1 cylinder has a capacity of 225 cf of air, 792 divided by 225 = 3.5. Use 4 cylinders.

Selection of Purification Components

The air used to fill breathing air cylinders is purified before being compressed. Breathing air produced by air compressors requires purification to meet minimum code standards for breathing air.

Prior to the selection of the purification equipment, several samples of the air where the compressor intake is to be located should be taken so that specific contaminants and their amounts can be identified. Ideally, the tests are taken at different times of the year and at different times of the day. These tests will quantify the type and amount of contaminants present at the intake. Knowing this, the purification systems needed to meet code criteria can be chosen. The other requirement is the highest flow rate that could be expected. With these two criteria, the appropriate size and types of purifiers can be selected.

The most commonly used method of purification is by an assembly of devices called a *purification system* specifically chosen and based on the previously selected criteria. The manufacturer's recommendations are commonly followed in the selection and sizing of the assembly.

Carbon Monoxide Converter. Based on experience, the requirement for installation of a carbon monoxide converter in rural areas is rare. The need for a converter is based on tests of the intake air at the proposed location of the compressor intake. Another source of information is the EPA, which has conducted tests in many urban areas throughout the country. Another indication for installation of a converter is the use of a non-oil-free compressor. Good practice requires the installation of a converter if there is a possibility that the level of carbon monoxide may rise above the 10 ppm limit required by code. The converter is sized on the flow rate of the system.

Coalescing Filter-Separator. The coalescing filter-separator is a single unit that removes large-size oil, water, and other particulates from the airstream before the air enters the rest of the system. It is selected on the basis of maximum system pressure, flow rate, and the expected level of contaminants leaving the air compressor using the manufacturer's recommendations. If an oil-free compressor is used, a simple particulate filter could be substituted for the coalescing filter.

Dryers (Moisture Separators)

Desiccant Dryers. The two types of desiccant medium dryers most commonly used are the single-bed dryer, which is a disposable cartridge, or a continuous duty, two-bed dryer. When two-bed dryers are used, a portion of the air from the compressor is used for drying one bed while the other is in service. The compressor must be capable of producing enough air for both the system and dryer use.

The single bed has a lower first cost but has a higher operating cost. The disposable cartridge often is combined with other purification devices into a single, prepiped unit. An indicator is often added to the media so that the need for replacement is indicated by a color change.

Disposable units are best suited for short durations or occasional use such as replacement for a main unit during periods of routine service. Because of their generally small size, only a limited number of respirators can be supplied from a single unit. Other considerations are that these disposable units have a limited capacity in total cfh that it can process. Manufacturers' recommendations must be used in the selection of the size and number of replacement cartridges required for any application.

The two-bed unit, commonly called a *heatless dryer,* is similar in operating principle to that discussed in Chap. 14. These units are used for continuous duty.

The two factors contributing to the breakdown of the medium are fast-drying cycles and high air velocity. In selecting the desiccant dryer, the velocity of air through the unit shall conform to manufacturers' recommendations. Velocity should be as low as practical to avoid fluidizing the bed. High velocity requires more cycles for drying, which means more air is wasted. If the size of the dryer is a concern, the more drying cycles mean smaller dryer beds. Longer drying cycles reduce component wear.

Refrigerated Dryers.   Refrigerated dryers are used if there is no requirement for a nitrous oxide converter and if the 35 to 39°F dew point produced is 10°F below the lowest ambient air temperature where any pipe will be installed. The refrigerated dryer is less efficient than the desiccant dryer. The advantages of the refrigerated dryer is that all of the air produced by the compressor is available to the system and it has a lower pressure loss.

When refrigerated dryers are preferred, several purification devices are often combined into a single unit, combining the refrigeration unit, filter-separator for oil and water, and a charcoal filter for odor removal. This unit produces air that is lower in temperature than that of the inlet air.

If the breathing air distribution piping is to be routed through an area of lower temperature, the pressure dew point of the air must be reduced to 10°F lower than the lowest temperature that could be expected.

Odor Removal

Odor is not usually a problem, but its removal is provided for as a safeguard. The activated charcoal cartridges that remove odors are selected by using manufacturers' recommendations based on the maximum calculated flow rate of the breathing air system. The cartridges must be replaced periodically.

Humidifiers

Often called a *moisturizer;* a humidifier is required to increase the relative humidity of the breathing air to approximately 50 percent if required. The unit is selected using the increase in moisture required for the airstream and the flow rate of air. Caution must be used so as not to increase the dew point of the compressed air below a temperature 10°F higher than the lowest temperature in any part of the facility the pipe is routed through.

Respirator Hose

The respirator hose size most often used to connect the respirator worn by an individual to the central distribution piping system is ⅜ in (10 mm) in size. Code allows a maximum hose length of 300 ft (93 m). The most common lengths are between 25 ft (7.75 m) and 50 ft (15.5 m).

System Sizing Criteria

System Pressure.   The outlet pressure of the compressor shall be within the range required by the purification system. Typically, the pressure is approximately 100 psi (70.3 kg/cm^2). The precise range of pressure and flow rate shall be obtained from the purification system manufacturer selected for the project.

The pressure in the distribution system should be as high as possible to reduce the size of the distribution piping network. Code requires that the pressure be kept below 125 psi (88 kg/cm^2). The distribution piping pressure range is usually 90 to 110 psig (620 to 760 kPa) available in the system after the purifier.

The pressure required at the respirator ranges from approximately 15 psig for pressure demand respirators to 80 psig to full flooded suits that require cooling. The actual requirements can be obtained only from the manufacturer of the proposed equipment because of the wide variations possible. Pressure regulating valves shall be installed to reduce the pressure to the range acceptable to the respirator used. Often, this is done at the respirator manifold if one is used or, if a single respirator type with a single pressure is used throughout the facility, a single regulator could be installed to centrally reduce the pressure.

Pipe Sizing and Materials. The most commonly used pipe is type L copper tubing, with wrought copper fittings and brazed joints. For pipe sizing, follow the sizing procedure discussed in Chap. 14, "Compressed Gas," in the medical air subsection, and Table 14.15 to 14.18 for the proper size pipe. The number of simultaneous users must be obtained from the facility. Use no diversity factor.

Alarms and Monitors

The following alarms and monitors are often provided.

CO Monitor. Usually included as a built-in component, this monitor will measure the CO content of the airstream and sound an alarm when the level reaches a predetermined high set point.

Oxygen Deficiency Monitor. Used as a precautionary measure in an area where respirators are not normally required, the oxygen monitor measures the oxygen content of the air in a room or other enclosed area and will sound an alarm to alert personnel when the level falls below a predetermined level. Usually, there are several alarm points that are annunciated prior to reaching a level low enough to require the use of respirators.

Low Air Pressure. The low air pressure monitor must sound an alarm when the pressure in the cylinder supply reaches a predetermined low point. This set point will allow the users of the breathing air system to immediately leave the area while still being able to breathe from the system. For cylinder storage, this set point is about 500 psig (3450 Lpm) in the cylinders. For a compressor system, the alarm should sound when the pressure falls to a point 10 psig (70 kPa) below the pressure set to start the compressor. This should also switch over to the emergency backup supply if one is used. If no backup is used, the pressure set point shall be approximately 5 psig (35 Lpm) higher than the minimum required by the respirators being used.

Dew Point Monitor. A dew point monitor is used to measure the dew point and sound an alarm if it falls to a low point as previously set by a health officer as potentially harmful to the users. The alarm is required to be activated if the dew point reaches a point high enough to freeze in some parts of the system.

High-Temperature Air Monitor. Some purifiers or purifier components will not function properly if the inlet air temperature is too high. The set point is commonly set at 120°F (49°C) and will vary between different manufacturers and components.

Failure to Shift. This monitor is placed on desiccant dryers to alarm the users if the unit fails to shift from the saturated dryer bed to the dry bed when regeneration is required.

REFERENCES

Bollas, Chris, and Jim Coffey: "In case of Emergency," *QH&S Magazine,* Canada, 1991.

Deltech Engineering L.P.: *How to Purify Low Pressure Breathing Air,* New Castle, Del., 1992.

Fendall Company: *Emergency Eyewash Handbook,* Arlington Heights, Ill.

Miller/Bowers Company, Inc.

CHAPTER 18
WATER DISPLAY FOUNTAINS AND POOLS

This chapter will describe the equipment and design criteria associated with water display fountains and reflecting pools.

GENERAL

Water fountains and reflecting pools are divided into four general types:

1. A *self-contained fountain* consists of a complete fountain including a pool, pump, fountain hardware, lighting, and pump. This kind of fountain is shipped complete so that only the excavation for the pool, adding water, and connecting the electrical supply are required to place the fountain into full operation. No permanent connection to the plumbing system is required.

2. A *fountain kit* is a preengineered fountain complete with pipes, pumps, nozzles, and all required hardware. This type requires installation into a previously constructed pool and permanent connection to a water source and electrical supply.

3. A *custom fountain* is designed to include many unusual decorative features and displays, including automatic and timed features such as "dancing waters" and other programmed effects. It is intended to be installed in a pool built specifically for the project.

4. A *reflecting pool* is a custom-designed body of water intended to achieve an architectural reflecting treatment without spray nozzles or other functions.

TYPES OF INSTALLATIONS

There are three general types of pool design: flow through, fill and drain, and recirculated.

The *flow-through design* has water passing through the pool continuously. This type should be used only where there is an almost unlimited supply of reasonably clear water and an easy means of supplying the water to the pool and returning the water to its source. The water supply should be filtered and the backwash returned to the source.

The *fill-and-drain* design has the pool filled with clear water, which is kept in service until it becomes turbid enough to be unfit for further use. At this point the water is drained, and the pool is refilled with water. This type of pool is usually limited to a capacity of approximately 500 gal and less.

The *recirculation design* is the most common. This design uses the same water continuously, with only small amounts added to make up for water lost through evaporation and wind drift. Water is pumped through filters to keep it clean with an algicide added if necessary to keep the pool free of algae. Often a vacuum cleaning system is provided to remove debris. It is generally accepted practice in pools of 1000-gal capacity or more to have a separate pumped pool circulation line that is filtered and treated and a second pumped spray nozzle water system that is not treated.

SYSTEM COMPONENTS

The equipment consists of pool spray (display) nozzles, lights, pool circulation system, water display system, pool makeup and fill methods, pool vacuum cleaning system, overflows, water treatment system, heaters, filters, and strainers.

EQUIPMENT

Spray and Jet Nozzles

Jet nozzles can direct a smaller, solid stream of water high into the air with pressure from a pump. Mushroom-type nozzles use a large volume of water to form a wider pattern of water lower to the pool surface. Spray nozzles are designed to produce a thin sheet of water in many possible patterns. Any of the above types of nozzles can be designed to produce an unlimited variety of patterns and spray height. If the design of a decorative fountain is to be made, become familiar with the patterns and limitations of the available nozzles in order to select a nozzle for a specific purpose. Each spray nozzle will produce the desired pattern only when supplied with the volume and pressure of water it requires. The manufacturer will provide the necessary nozzle information for each specific nozzle in its line.

By using a timing motor and automatic valves that are turned on and off in programmed sequence, very special shows that include lighting effects also timed to the nozzle operation can be achieved.

Lights

Lights, when provided, should be installed under the surface of the water, generally 2 in lower than the water level. The lights are used for both general pool illumination and specific direction. Different colored lenses are available for different effects. The lights can be directed either horizontally or vertically up toward a nozzle. Dimmers can be used to control the intensity of the lighting, and timers to vary the sequence of operation of individual lights to obtain special effects. The programmed sequencing units are not waterproof and should be located indoors and near the source of power for ease of wiring.

Pool Recirculation System

Used to keep the pool water clean and warm if desired, the recirculation system functions by drawing water from various pool drains, circulates the water through a strainer, main filter, and filter pump, and returns the water to the pool through separate inlets. In a fountain with special or programmed spray jets, the spray jets are usually separated from the pool circulation system. If the spray jets are uncomplicated, the filtered water can be returned to the pool through the jets.

A strainer is the first item in the recirculation system. It is used to protect the pump from damage and nozzles from becoming clogged by removing large-sized debris. A basket strainer is preferred for larger pools because of its large capacity. Where vacuum cleaning is an integral part of the circulating system and continuous operation of the fountain is desired, duplex strainers may be considered. Generally accepted practice is for the strainer to have openings one-half the size of the spray nozzle orifice.

The selection of a pool circulating pump is based on the pressure drop through all the components of the system and the gpm required to circulate the contents of the pool in 8 h. A centrifugal pump is most often used for this service.

The filter is usually a high-rate pressure sand filter. It consists of a pressure vessel containing sand of different sizes. A distribution header inside the vessel distributes water evenly over the sand bed. When the filter becomes clogged with dirt, water is reversed (backwashed) through the filter and the discharge is routed to the sewer. Experience has shown that this type of filter has the least initial cost and the least maintenance. An average filter size has a capacity of approximately 15 gpm/ft^2 of filter area. The backwash rate is the same as the filter rate. The need for backwashing is determined by reading pressure gauges installed on the inlet and outlet of the filter. The source of backwash water is usually separate from the display pool system, often originating from the potable water supply. If not separate, the pool water must be used for backwash and an equal volume used to fill the pool.

Another type of filter is a diatomaceous earth filter, which could be either pressure or vacuum type. For both, the filter medium (a fine powder) is distributed evenly on a fine mesh screen, and the water passes through the screen with the filter medium on it. This type of filter is more costly, and it requires costly controls, but it will produce clearer water than the sand filter. The vacuum type requires a different piping arrangement. Both types are maintenance intensive. For all of these reasons, this type of filter is rarely used.

In some installations, it may be sufficient to filter a smaller percentage of the capacity of the circulation pump. This would permit a smaller filter and a separate pump for the filter but will result in operating cost reduction. Smaller pools may use a cartridge filter that is discarded if the flow of water and the amount of debris are small.

Water is drawn into the recirculation system from drains and inlets located in various parts of the pool. Each pump suction creates a vortex, which is a funnel-shaped element below the water level having the larger open end at the water surface and extending into the inlet that has the smaller end of the vortex. This vortex allows air to be drawn into the piping, which is detrimental to pumps and causes sputtering in the display nozzles. In order to prevent this, drains shall be provided with an antivortex cover or inlet with an integral antivortex arrangement. In addition, inlet capacities to drains should be limited to 80 percent of the rated capacity, thereby reducing the velocity of the water entering the inlet.

Water Display System

Used to control the operation of the display spray nozzles, this system operates by using a separate drain in the pool that is connected to the suction side of the dedicated display pump. The pump discharge is connected to the display nozzle assembly. This assembly consists of piping, display control valves, and the display nozzles.

The pump is selected by finding the maximum flow requirements of the nozzles and calculating the total dynamic head by adding together the pressure loss from the water spray nozzle and pressure losses through all components of the piping system. The characteristics of each nozzle is obtained from the manufacturer.

Since it is almost impossible to obtain a pump with the exact pressure requirements for a perfectly matched display, part of the piping to the nozzles should include both in-line control valves and bypass valves at the nozzles to divert water to the pool and provide the exact gpm and head required. Valves installed on branch lines to individual nozzles will allow them to be isolated if repair or adjustments are found to be necessary. Plug, ball, and globe valves have proven to be satisfactory for regulating flow. The main drain used to obtain or discharge water from the bottom of the pool may have to be installed in a depression to avoid drawing in air along with the water if the pool is shallow. To avoid excessive noise and erosion of the piping, generally accepted practice is to limit the velocity of water in the system to 6 fps.

In order to avoid wind-carried water spray from wetting passersby, a wind control device could be installed that would shut down or lower the height of some or all spray nozzles that, because of the height of water discharge, would cause the drifting of water beyond that desired.

Pool Water Makeup

Water is lost from a pool by evaporation and wind drift from the water spray nozzles. The level of water in a pool should be kept within a narrow range so that the lights are covered and the nozzles are the same distance below water level. This can be done either manually or automatically.

The manual method has the least initial cost and requires a maintenance person to be present most of the time. When the maintenance person becomes aware that the level is low, the valve on the water supply shall be opened to fill the pool until it reaches the high water level. Manual operation shall be limited to small pools that cannot be damaged by a low water level.

If done automatically, it is possible to maintain two different levels of water. The first is the normal water level. When water falls below this level, water shall be added to the pool. The second level is the emergency low level that when reached will shut down lights or other devices that require that minimum height of water to function properly. The level device is placed in the pool within a stilling tube. This avoids premature starts and stops by wave action. This device sends a signal to an automatically operated valve that opens on low water and closes on high water and cuts off electricity at the emergency low level.

When potable water is used as the initial fill and for makeup, the pool supply should be over the rim to create an air gap. This dimension shall be either 2 in or 2 pipe diameters of the fill line, whichever is the larger dimension. If any below-water-level method is used, a reduced pressure zone backflow preventer must be installed in the inlet to the pool.

Vacuum Cleaning

Vacuum cleaning of the pool is required if the pool is outdoors where the public may litter the water. It is also easier to clean algae from the pool with a vacuum system. This can be done either with a built-in system or portable unit.

A built-in system has special outlets located below the surface of the water with the piping connected to the suction side of the pool circulation pump. The outlets have hose connector ends that make the attachment of cleaning hoses easy. The hoses have a brush or other special end and are of a length so that they can be extended to all parts of the pool. An average spacing of the outlets is 1 for every 40 ft, 0 in, of the pool perimeter.

Overflows

An overflow is required to prevent water from spilling over the top of the pool onto the surrounding area. There are two reasons an unwanted amount of water enters a pool: (1) accidental fresh water makeup from a public water supply and (2) rainfall, if the pool is located outdoors. The size of the overflow should be two sizes larger than the makeup water supply or sized to take the maximum amount of rainfall, whichever is larger. The local code must be followed to determine if the discharge is routed to the storm water or sanitary sewer system. The height of the overflow above the water level must allow for wave action from the pool splashing over the side.

EPA or local regulations may prevent the discharge of chlorinated water into the storm water drainage system, dictating the use of the sanitary system for discharge of the overflow and to empty the pool.

Water Treatment

The growth of algae in pool water causes the water to become cloudy and pool surfaces to be covered with slime and causes additional load on the filter. Algae grows faster in sunlight. To prevent algae growth, the pool water must be treated with an algicide or disinfectant. In small installations, they could be added to the water by hand. In larger pools, it is more easily done automatically with a chemical proportioning system. The two most often used chemicals are sodium or calcium hypochlorate, with sodium hypochlorate being the most common.

The chemical proportioning system consists of a chemical supply in liquid form and a proportioning pump that delivers the correct amount of chemical into the recirculating water stream.

Chemicals can be bought either as a powder that must be mixed by hand or as premixed solutions. The solution must be pumped into the water by means of an *injector pump,* commonly called a *hypochlorinator:* The suction is connected to the chemical supply, and the discharge into the circulated water line as close to the pool inlet as practical. The pump is arranged to be running only when the circulating water pump is running.

The dosage generally recommended for pool water is 5 ppm for indoor pools and 10 ppm for outdoor installations. The capacity of the pump can be found from the following formula:

$$P = 0.0005 \times \text{gpm} \times \text{ppm} \qquad (18.1)$$

where P = pounds of chemical per hour
    ppm = required dosage
    gpm = capacity of the circulating water pump

For example, the following calculation is for a feed rate with a 15 gpm circulating pump for an indoor pool. Substituting in Eq. (18.1):

$$P = 0.0005 \times 15 \times 5$$

$$= 0.0375 \text{ pound of chemical per hour}$$

In order to be effective, the pH of the water should be within the limits established by the supplier of the chemicals used.

Pool Heater

A pool water heater is necessary if there is a possibility that the water in the pool may freeze. Frozen water will damage the nozzle piping and crack the sides of the pool and must therefore be avoided at all costs.

There are two types of heating systems. One type heats the pool circulation water. This requires only the additional piping to and from the heater. A second type, which has a much higher initial cost, is an independent, closed-loop system. It circulates a chemical solution (such as ethylene glycol in water) through a pump and a separate piping system embedded in the pool floor. This system also requires a compression tank. This type of system is selected for large pools that are intended to be shut down, but not emptied, for periods of time.

Although the heater can be oil, gas, steam, or electric fired, the most common is electric because of the low initial cost and the fact that no vent is required. Oil- and gas-fired heaters require vents that may not be possible to install in the area where the pool mechanical room is located.

The heater size is based on maintaining a pool water temperature in a range between 40 and 45°F. For outdoor pools, a wind velocity of 15 mph is assumed. The formula for finding the kW requirement of the heater is:

$$\text{Kilowatts} = \frac{\text{square feet of pool area}}{405} (40 - \text{lowest air temperature}) \qquad (18.2)$$

To convert kilowatts to British thermal units, multiply kilowatts by 3.4.

Another important feature to consider installing is an emergency pool drain that will send an alarm and/or automatically operate a valve to empty the pool when the water temperature reaches 35°F. If there is no place that is occupied 24 h/day to put an alarm, install an automatic drain valve on the pool drain to the sewer.

SYSTEM CONFIGURATIONS

For a schematic detail of a typical large display pool, refer to Fig. 18.1. If multiple display pools are installed at different levels, refer to Fig. 18.2 for a typical arrangement. Some approximate dimensions for aspects of water spray nozzles are illustrated in Fig. 18.3.

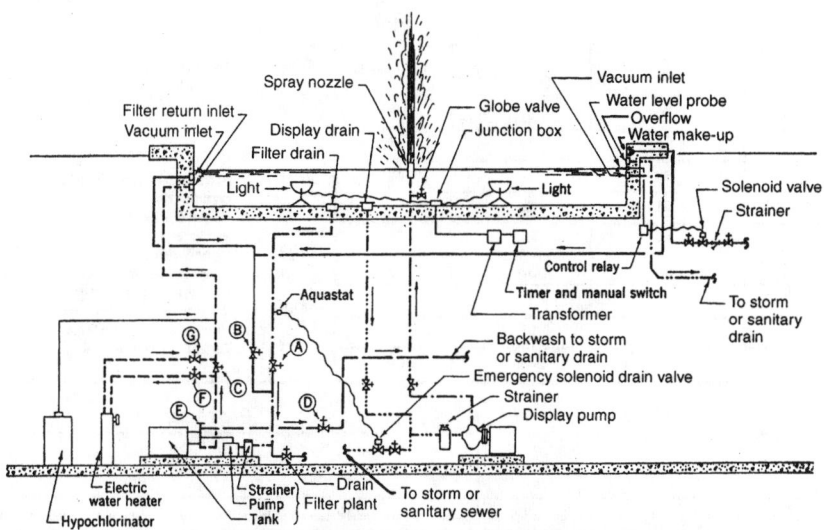

Lettered valve schedule; A, normally open, closed when vacuuming; B, normally closed, open when vacuuming; C, normally open, partially closed in winter when heater is on; D, normally closed, opens in conjunction with E for backwashing; E, two-position valve is set for either filtering or backwashing; F and G, normally closed, open when water heater in on.

FIGURE 18.1   Piping diagram of typical decorative display fountain.

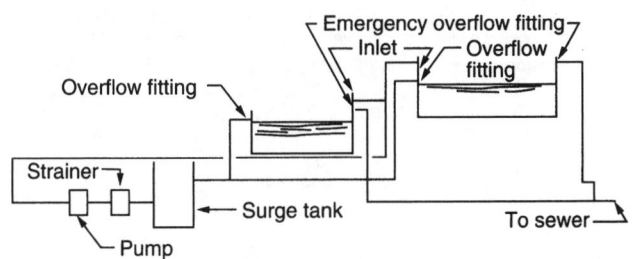

FIGURE 18.2   Piping diagram for display pools at different levels.

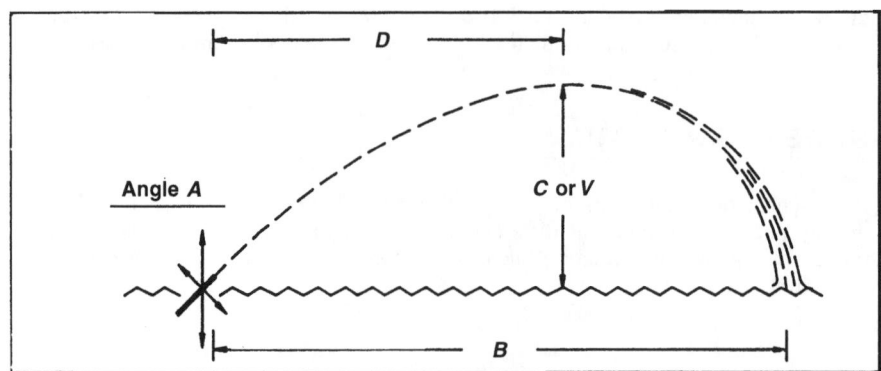

Design factors

A	E	C,% of B	D,% of B
5°	0.90	6	36
15°	1.33	11	46
25°	1.83	17	49
35°	1.94	22	51
45°	2.10	27	52
55°	1.80	36	53
65°	1.50	50	56
75°	0.90	99	59
85°	0.40	245	64

A	Angle of nozzle elevation.
B	Horizontal distance of throw from nozzle
C	Height of trajectory in percentage of B
D	Highest point of trajectory in horizontal distance from nozzle, measured in percentage of B
E	Multiplying or dividing factor for spray design calculations
V	Vertical spray height

	To find:	Procedure:
1.	B	$B = V \times E$
2.	Performance requirement of a spray pattern with known angle of nozzle discharge or the equivalent vertical spray height performance requirements	Establish horizontal distance of throw from nozzle, and divide by factor E on same line as shown discharge angle of nozzle. This will give vertical spray height, which is then used to find performance requirements. $A \div B$ E = vertical spray height
3.	Trajectory of a steam or jet of water V	Establish horizontal distance of throw (factor B) then calculate factors D and C therof, and combine the results with B to lay out the trajectory.
4.	The jet elevation angle A	Establish horizontal distance of throw (factor B), calculate highest point of trajectory (factor C) thereof, and read on the factors table the angle of elevation (factor A) on the same line as the result of the calculated height of trajectory (Factor C).
5.	Nozzle pressure to achieve V	Multiply vertical spray height (factor V) x 1.22 + 10%

FIGURE 18.3   Dimensions for aspects of water spray nozzles. (*Courtesy PIM.*)

CHAPTER 19
NONPOTABLE WATER SYSTEMS

This chapter will discuss nonpotable water systems used in various types of facilities. Systems included are recycled (gray) water and salt water.

RECYCLED AND GRAY WATER

DEFINITIONS

There is no specific, universally accepted definition of gray and recycled water. In many cases, the expressions have been used interchangeably. In general, the term *gray water* is intended to include appropriately treated water that has been recovered from typical fixtures such as lavatories, bathtubs, showers, and clothes washers. Waste potentially containing grease such as that from kitchens and dishwashers are excluded, as well as waste from food disposals in kitchens. *Recycled water* is intended to include "clean water" discharged from sewage treatment or other waste treatment facilities that has been additionally treated to remove bacteria, heavy metals, and organic material. *Black water* is water recovered from plumbing fixtures, such as water closets and urinals, that discharge human excrement. This handbook will limit discussions to gray water only, unless it specifically mentions otherwise.

GENERAL

Gray water systems have been in use for long periods of time in various areas of the world. In localities where the underground aquifers are in danger of depletion and where adequate supplies of water are dwindling, the reuse of water offers a considerable saving of water resources. Wastewater management is also a significant reason for gray water system use.

Onsite reclamation and recycling of relatively clean, nonpotable water is considered for the following situations:

1. In areas where code mandates that gray water be used because potable water is in short supply or its use is restricted
2. For projects where public liquid sewage disposal capacity is either limited or inadequate
3. For economic reasons when obtaining potable water or disposing of liquid waste is very costly
4. For economic reasons alone, where the time of payback will be economically feasible and will result in substantial operating cost savings

COMMON USES

Appropriately treated gray water is commonly used for the following purposes:

1. Flushing water for water closets and urinals
2. Lawn irrigation

3. Cooling tower makeup
4. Decorative pools and fountain fill water
5. Floor and general hard surface washdown
6. Laundry prerinse water

The most often used purpose is to provide water for flushing of urinals and water closets.

CODES AND STANDARDS

There are no nationally or regionally established model codes mandating or requiring the use of gray water. Conversely, the use of gray water is not prohibited. Many specific local areas have established requirements for the use of gray water in facilities and homes. Where gray water use is permitted, local health departments have established minimum treatment standards. These localities must be contacted for applicable regulations in the same manner as for plumbing and building codes. NSF International has established recycled water quality standards in certification standard No. 41.

SYSTEM DESCRIPTION

A gray water system collects the dilute wastewater discharged from lavatories, service sinks, baths, laundry tubs, showers, and other similar types of fixtures. This water is then filtered or treated to a level of quality consistent with its intended reuse. The piping network distributes it to sources not used for human consumption in a safe and distinctive manner.

A gray water system requires modifications to the standard plumbing systems throughout the total facility to accommodate duplicate drainage systems. Instead of all of the liquid discharged from all the plumbing fixtures going to the sanitary sewer, selected fixtures have their effluent routed for recovery by the gray water treatment system and the remainder to the sanitary sewer. There is also a duplicate water supply: potable water to lavatories, sinks, showers, and so on and gray water to water closets, urinals, and other fixtures permitted by the quality of gray water treatment.

SYSTEM COMPONENTS

The following components are generally used for most systems. Their arrangement and type depend on the specific treatment system selected.

1. The gray water collection piping system, which is a separate drainage system
2. The primary waste treatment system consisting of turbidity removal, storage, biological treatment, and filtering

3. Disinfection systems, consisting of ozone, ultraviolet irradiation, chlorine, or iodine

4. Treated water storage and system distribution pressure pumps, and piping

DESIGN FLOW

It is estimated that two-thirds of the wastewater discharged from a typical household in 1 day is gray water. Of this, one-half is from water closets. The discharge from the separate piping system supplying the gray water system should be sized on the applicable plumbing code. The following should be considered in the design of a system:

1. The design flow is based on the number of people in a facility.

2. Lavatory use is estimated at 0.25 gallon per toilet use.

3. Men use the urinal 75 percent of the time and water closets 25 percent of the time.

4. The average person uses a toilet 3 times per day.

TREATMENT SYSTEMS

Treatment systems vary widely. The treatment system conditions the recovered water to a degree consistent with both the intended use of the conditioned water and the design requirements of the design engineer, applicable code, or the responsible code official, whichever is the most stringent. Typical flow sheets used for various types of projects are shown in Fig. 19.1. The size of the treatment systems available vary from those installed for individual private dwellings to those serving multiple-family facilities.

The selection of a treatment system must also depend on the quality and type of the influent water. In order to decide on the most appropriate treatment, a decision must be made as to which kind of fixture discharge is to be used for reclaiming and the requirements of the authorities for treatment.

One example of water quality standards used for a multiple-family dwelling project is listed in Table 19.1. Normal process efficiency obtained from commonly used treatment methods is listed in Table 19.2.

PRECAUTIONS

Since gray water is a potential health hazard, a great deal of care must be exercised once the system is installed. One of the greatest dangers is the possibility that the gray water will be used for drinking purposes or a cross connection inadvertently will be made so that the gray water system is connected to the potable water system.

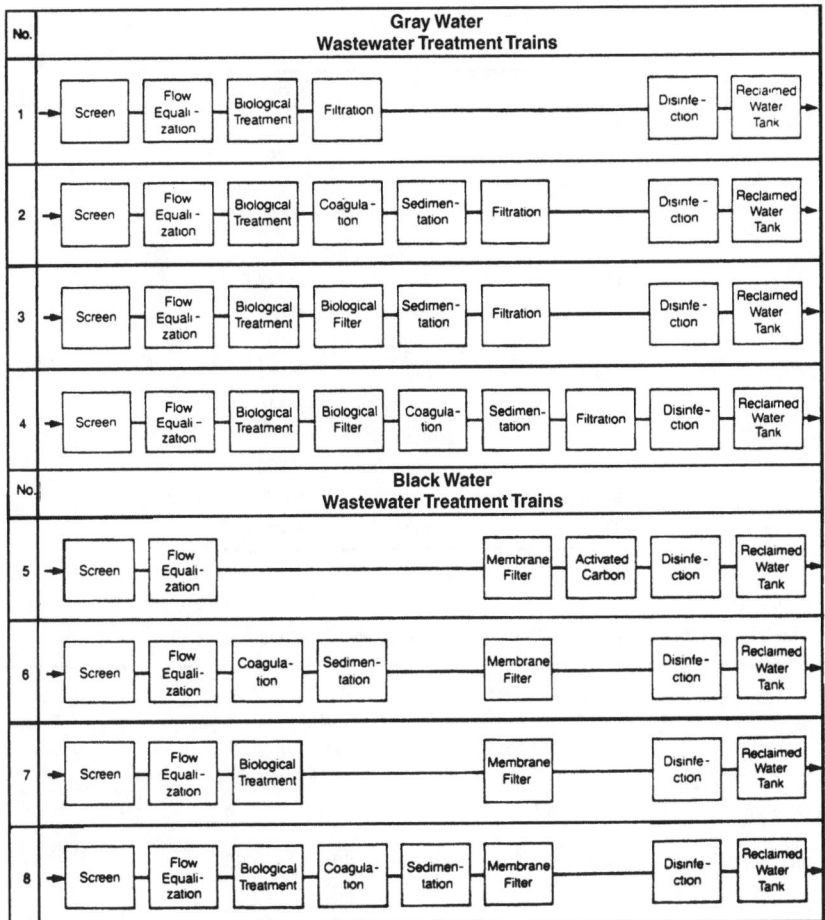

FIGURE 19.1 Typical flow sheets for on-site reuse systems.

To avoid this possibility, the water itself and the piping system must be made easily distinguishable.

Treated water could be colored by food dye that is biodegradable. Fixtures could be bought in the color of the water if the water color is found to be objectionable. The piping system itself must be clearly identified with labels. If possible, the piping material should be different so that the possibility of mistaking the two systems will be unlikely.

The most important consideration is education of individuals and the staff of any facility. Informing the staff of the dangers, as well as the proper operating instructions, will ensure that the system will be operated and maintained in a correct manner.

TABLE 19.1 Water Quality Standard for Multiple-Family Dwelling

Item	Unit	Criterion
Odor	—	
Color	Unit	<10
Turbidity	Unit	<5
TDS	mg/L	<1,000
SS	mg/L	<5
pH	Unit	5.8–8.6
COD	mg/L	<20
BOD_5	mg/L	<10
PO_4^{3-}	mg/L	<1.0
ABS	mg/L	<1.0
Coliform	count/mL	
General bacteria	count/mL	<100
Residual chlorine	mg/L	>0.2
TOC	mg/L	<15

TABLE 19.2 Normal Process Efficiency for Gray Water Treatment Methods

	Percent removal					
Process	Suspended solids	Biological oxygen demand	Chemical oxygen demand	Phosphates, PO_4	Nitrogen	Total dissolved solids
Filtration	80	40	35	0	0	0
Coagulation/filtration	90	50	40	85	0	15
Chlorination	0	20*	20*	0	0	0
Water treatment	95	95	90	15–60	50–70	80
Absorption (carbon filtration)	0	60–80	70	0	10	5

*Nominal, additional removals possible with superchlorination and extended contact time.

PUBLIC ACCEPTANCE

Although the use of gray water is a proven, safe, and cost-effective alternative for using potable water in various systems, there is a reluctance on the part of authorities to allow approval. Some other reasons are:

1. There is no generally accepted standard for the quality of recycled water. Several states in the United States, Japan, and the Caribbean have adopted codes and guidelines, but, for most of the world, there is no universal standard. This has

resulted in disapproval of the systems or long delays of a project during the approval process where the quality of the water is in question.

2. Even when regulatory and plumbing codes do not have any specific restriction against using gray water or have ambiguous language that could be interpreted for its use, some officials still impose special standards due to their lack of experience.

TYPICAL INSTALLATIONS AND DETAILS

1. A typical black water treatment system is illustrated in Fig. 19.2.
2. A typical gray water treatment system is illustrated in Fig. 19.3.
3. A simplified gray water dual-piping distribution system is illustrated in Fig. 19.4.
4. A simplified high-rise water and waste piping distribution system is illustrated in Fig. 19.5.
5. A typical laundry wastewater recovery system is illustrated in Fig. 19.6.

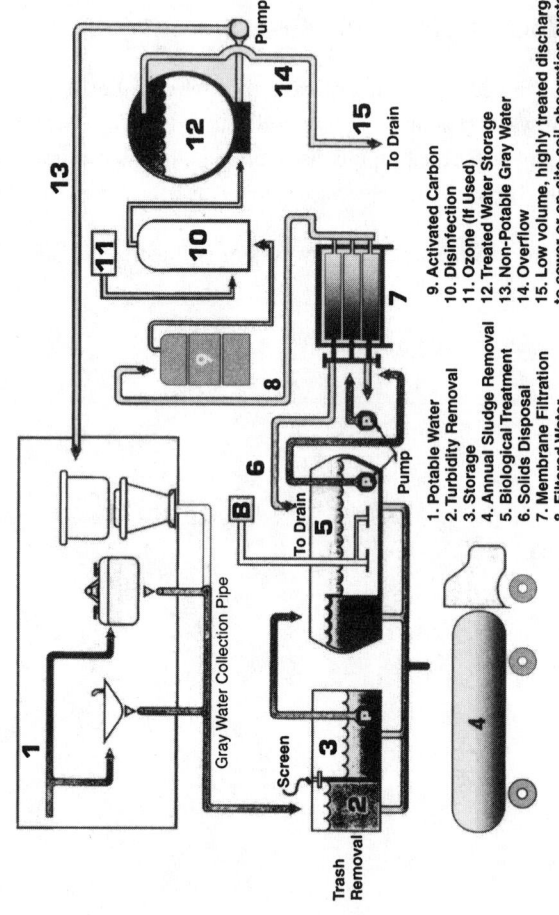

FIGURE 19.2 Typical black water treatment system.

1. Potable Water
2. Turbidity Removal
3. Storage
4. Annual Sludge Removal
5. Biological Treatment
6. Solids Disposal
7. Membrane Filtration
8. Filtered Water

9. Activated Carbon
10. Disinfection
11. Ozone (If Used)
12. Treated Water Storage
13. Non-Potable Gray Water
14. Overflow
15. Low volume, highly treated discharge to sewer or on-site soil absorption system.

Gray Water Collection Pipe

Screen

Trash Removal

Pump

To Drain

To Drain

Pump

Pump

19.8

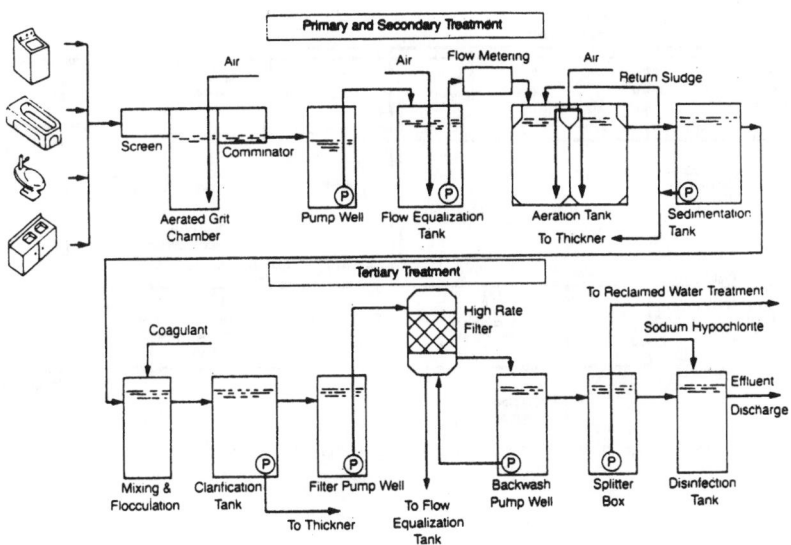

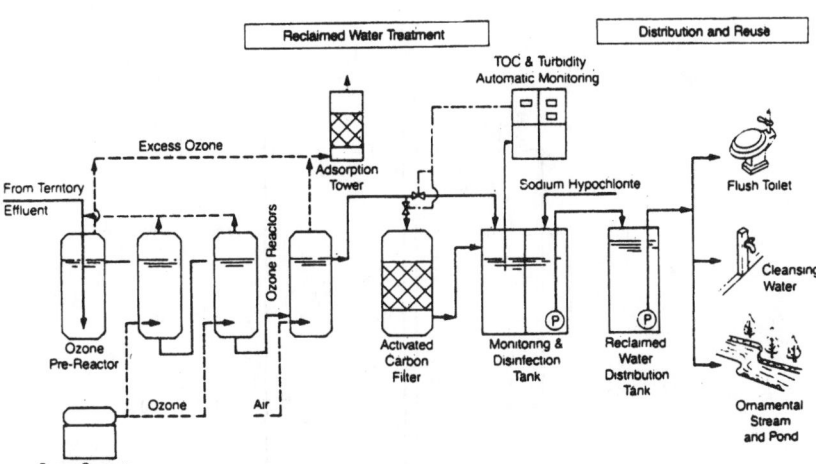

FIGURE 19.3 Typical gray water treatment system.

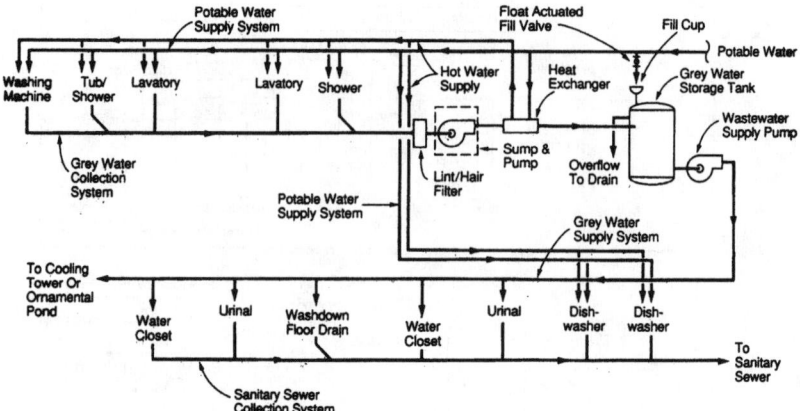

FIGURE 19.4   Typical gray water dual piping distribution system.

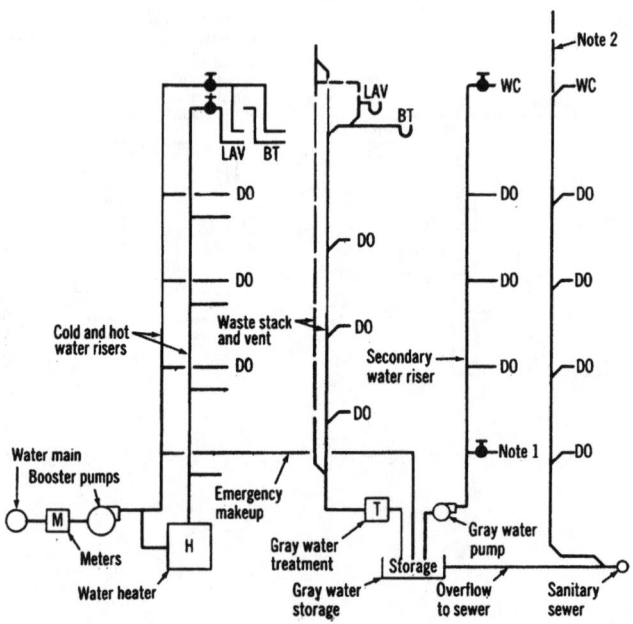

FIGURE 19.5   Typical high-rise dual distribution system. Notes: (1) Gray water can also be utilized for other uses such as irrigation, and cooling tower makeup, providing that treatment is adequate. (2) Common vent for both drainage stacks.

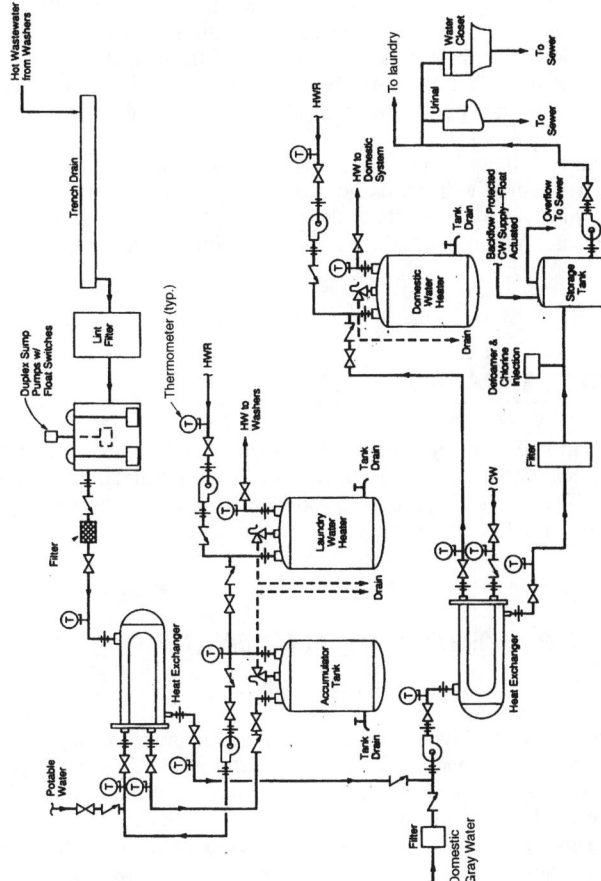

FIGURE 19.6 Laundry waste recovery system.

SEAWATER

GENERAL

Seawater is used in many situations including desalinization systems and saltwater aquariums; it is also used as a heat exchange medium and as flushing water for water closets. The major differences between seawater and other types of nonpotable water are that the water is reasonably clean, there are living organisms present in the water source, and the seawater accelerates corrosion to metallic pipes and valves.

Living organisms include those that swim in the water and those that are permanently attached, some with hard and soft shells, no shells, and some that move. They attach themselves to the walls of the pipe or intake structure and can multiply at an alarming rate. When dead, their remains adhere to the surface they attached themselves to.

CODES AND STANDARDS

There are no nationally or regionally established model codes concerning the use of salt water. If salt water is to be used, say, for flushing toilets, then the gray water regulations established in the plumbing and health codes for the locality where the project is located must be followed.

METHODS OF PREVENTING ATTACHMENT OF ORGANISMS

The greatest problem in saltwater systems is the attachment of organisms onto the piping in the distribution system. There are several methods used to prevent their attachment to the pipe, control their presence, and eliminate them. The method depends on the final use of the water.

Copper-bearing pipe materials discourage attachment. The drawback is that copper is known to damage the ecology and should not be used except in a closed system where the seawater will not be released back to the ocean.

Toxic chemicals will kill all living things when introduced into the water. In aquariums, killing organisms with toxic cleaning agents is not possible since, in the event of an accident, the toxins could kill the exhibited fish.

Having redundant piping systems and allowing one of them to be isolated for a period of time will permit two methods of control. One process would be to valve off the pipe and starve them with a lack of food and oxygen. This method takes several weeks to accomplish and is very costly. Another method would be to flood the piping with warm freshwater. This takes a shorter period of time based on the type of organisms present.

In the case of once-through seawater used for cooling purposes, a high water velocity (10 fps) is enough to keep any organism from attaching itself to the pipe wall. This velocity has to be balanced against erosion of the pipeline.

PIPELINE CLEANING

Once shells are attached, they remain firmly attached after death and have to be removed. Inside a pipe, this can be done only by a mechanical scraping operation. One method popular for smaller lines uses a plumber's snake in the same way that a sanitary sewer line is cleaned. Another method is called *pigging*. A pig is an object that fits inside a pipe, and its circumference is large enough to scrape the inside as it passes. The pig is powered by flowing water in the pipe. There are many configurations of pigs possible, and some are made with stiff wire brushes or cone-shaped scrapers for cleaning the pipe walls.

Because of pigging, unique design considerations are required in the piping system. There should be no control valves, no fittings greater than 45°, and no protrusions such as thermometers or flow meters inserted into the pipe line to be pigged. An adequate number of cleanouts throughout the system is very important. In lines that cannot be pigged or cleaned, pipe sections capable of being removed from the system should be provided in order to gain easy access. Lines smaller than 2 in cannot be pigged, but generally accepted practice is not to pig a line smaller than 4 in. The smaller lines are cleaned with rotating plumbing snakes. A pig launching piping arrangement is detailed in Fig. 19.7, and a recovery arrangement is detailed in Fig. 19.8. Pigging is very often used to clean piping for other systems. Clearing intake structures such as screens usually requires a diver who must clean them by hand.

FILTERS

Filters are used to remove virtually all living things from the source of water and to improve clarity of the water when this is a consideration. Sand filters are the most commonly used, with FRP or rubber-lined tanks. Readers inside the tanks are PVC, and the filter medium support is FRP. Backwash is piped back to the ocean.

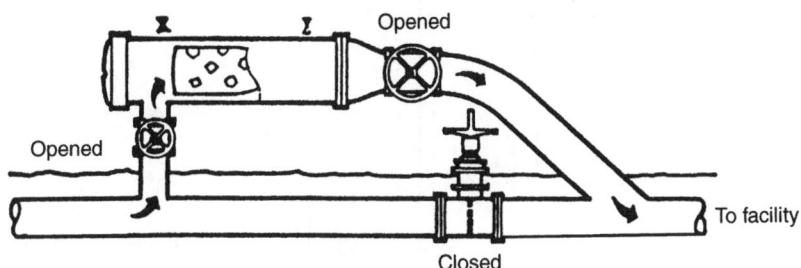

FIGURE 19.7   Typical pig launching detail.

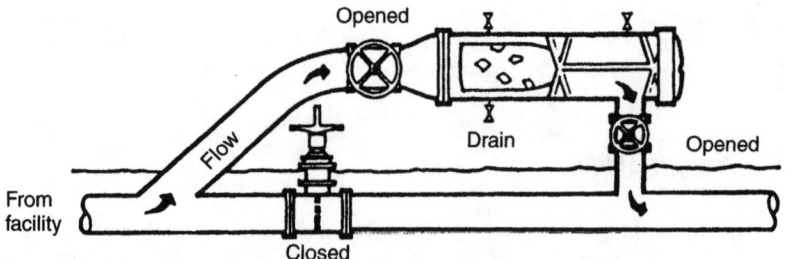

FIGURE 19.8 Typical pig recovery piping detail. (*Courtesy Girard Industries Incorporated.*)

SEAWATER INTAKE

Seawater taken directly from the ocean is susceptible to the intake of debris, sand, fish, and other living things. The intake structure must be designed to act as a primary screen to eliminate them. A typical intake structure is schematically illustrated in Fig. 19.9.

An intake structure provides a reservoir for the facility's main seawater supply pump. The natural level of the ocean at low tide must be high enough to provide a sufficient amount of water and allow for an additional lowering of the water level during pumping.

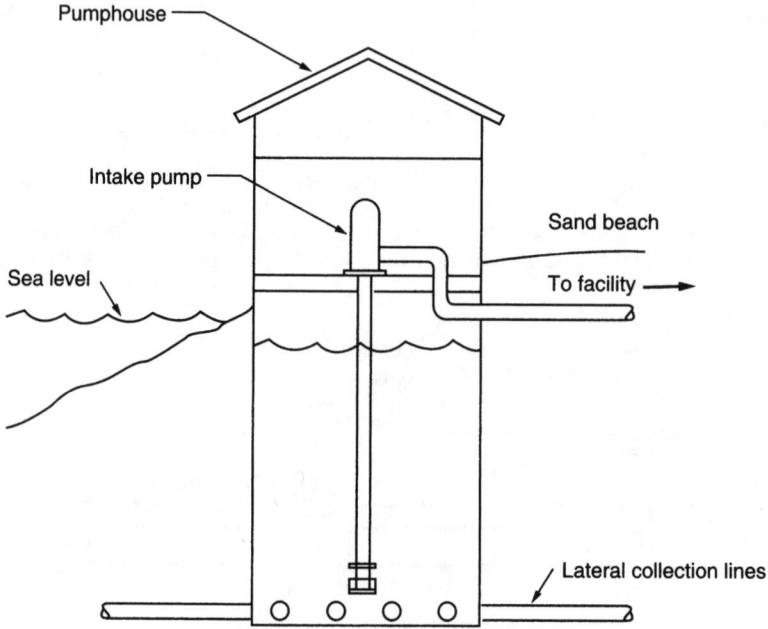

FIGURE 19.9 Typical intake structure.

Where the supply must provide the full range of living things, the inlet piping may have to extend out along the ocean floor several hundred feet to permit obtaining the desired water quality. The actual inlet should be covered with a bar screen or a cylindrical stainless steel mesh screen. A 1½ × ¾-in size mesh has worked well and is light enough to be easily handled by a diver using a flotation device. Experience has shown that bar screens foul faster than mesh screens, and in addition, are much heavier. A typical schematic of an intake pipe installation is shown in Fig. 19.10.

If relatively clear water free from living organisms is acceptable, an intake structure built in the sand close to shore should be considered. The water is obtained from perforated pipes run in the sand, using the sand itself as a filter medium before it reaches the inlet structure. This type of installation requires a large and deep sand beach. A typical intake structure is illustrated in Fig. 19.9.

Another solution to controlling the fouling of salt water intakes and piping is the introduction of chemicals at the intake structure. These chemicals must be acceptable to regulatory agencies and could include sodium hypochlorate, hydrogen peroxide plus iron, potassium permanganate, chlorine dioxide, chloramine, and ammonia. This system could include space for chemical storage, chemical metering pumps, and transfer pumps.

PIPING MATERIALS

The material of choice for most piping installations is plastic. PVC, CPVC, and PVDF are used based on temperature and the chemical nature of any cleaning solution. PVDF is useful at elevated temperatures. Another consideration is the abrasion to the pipe that would be caused by pigging. FRP pipe is used when physical strength is a factor, such as that used for intake piping. Special epoxies can be used in the manufacture of the FRP pipe to make it more abrasion resistant. When environmental considerations are not a factor and physical strength is important such as when salt water is used in heat exchangers, 90-10 copper nickel

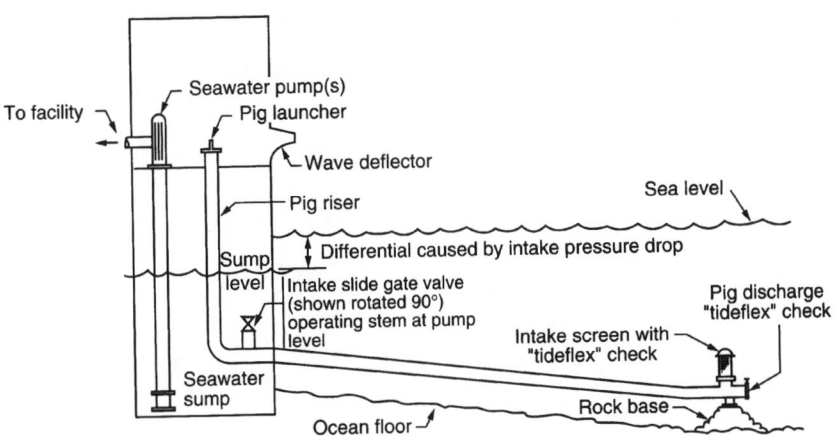

FIGURE 19.10 Intake pipe schematic detail.

alloy pipe is recommended. Experience has shown that stainless steel is not effective as a pipe material. If its use is unavoidable, design that component to be easily replaced. Titanium has often been used for critical applications but does not find general use because of cost.

Valves should be made of plastic, FRP, monel, copper-nickel, or marine bronze with a proven resistance to seawater. Coated surfaces have not proven effective, particularly epoxy-coated steel and cast iron.

Other materials such as rods and bolts should be made of FRP, as well as structural shapes and gratings. Cementing FRP together is effective, providing that the surfaces to be mated are clean and rough. FRP reinforcing rods are available for concrete exposed to seawater.

EQUIPMENT SELECTION AND MECHANICAL ROOM DESIGN CONSIDERATIONS

Pigging requires extra space to launch and recover the pigs. Space and a method to collect debris must be allowed for. Adequate space for the removal of filters is important. Pumps made of plastic materials are available, as well as those constructed with rubber lining and of FRP material. Large pumps can be obtained with removable end plates if desired. Because of the uncertainty of system conditions, duplex equipment is required for any critical operation.

REFERENCES

County of Santa Barbara Graywater Technical Advisory Committee: *How to Use Graywater,* Santa Barbara, Calif., 1990.

Irwin, John, PE: "Wastewater Recycling in Commercial Buildings," *Plumbing Engineer;* September 1989.

Lehr, Valentine A., PE: "Gray Water Systems," *Heating/Piping/Air Conditioning,* January 1987.

Whitworth, Patrick L., CIPE: "Designing Gray Water Recovery Systems," *Plumbing Engineer;* September 1991.

Wistort, Robert A., PE: "Sea Water Piping Systems," *Plumbing Engineer;* October 1992.

CHAPTER 20
DRINKING WATER SYSTEMS

Drinking water systems provide either chilled or ambient temperature water used only for drinking purposes. This chapter describes the drinking water systems, selection of components, design criteria and the piping distribution system used to supply and distribute chilled water to all units of the system.

CODES AND STANDARDS

1. Americans With Disabilities Act. (ADA) Accessibility of all drinking fountain types.
2. ANSI/NSF 61. Testing criteria recommendations for the Clean Drinking Water Act.
3. ASHRE Standard 18. Ratings and test methods.
4. American Refrigeration Institute (ARI) Standard 1010. Standard rating conditions.
5. American Refrigeration Institute (ARI) Standard 1020. Standard Rating Criteria
6. Clean Drinking Water Act. Contaminant levels for potable water.
7. Applicable plumbing code that regulates the number of drinking fountains.
8. Federal Specifications for government agency projects.

SYSTEM COMPONENTS

Drinking Water Fixtures

Bottled Water Unit. Bottled water units are not considered plumbing fixtures because they have no connection to any plumbing system. These units are completely self contained and do not require a piped water supply or a drain. Water is supplied from bottles, generally containing 5 gallons of water, that must be replaced when empty. Water is discharged from faucets and dispensed into cups because there is not enough pressure to use a spout. Spills are contained in a receptacle on the unit under the faucet and must be emptied when full. If not emptied, the water overflowing the spill receptacle will spill onto the floor. If electricity is available, the water could be cooled or heated.

Drinking Fountains. (DF) Drinking fountains are considered plumbing fixtures and are directly connected to plumbing potable water and drainage systems. Water to the unit is supplied either from a remote, central chilled water unit, adjacent chiller unit for a single fixture only or ambient temperature water from the potable cold water system in the facility. Water is dispensed under pressure through a bubbler or spout and does not require a cup to drink from. Excess water not consumed is piped direct to the sanitary drainage system. Pedestal units are used for isolated outdoor use; the non-freeze type is used where freezing is a possibility.

Electric Water Cooler. (EWC) An electric water cooler is a completely self contained plumbing fixture consisting of an integral, directly connected drinking fountain and water chilling unit serving only this unit. These units are supplied with pressurized ambient temperature water, which is chilled by the self contained refrigeration unit. Water is dispensed under pressure through a bubbler or spout and does not require a cup to drink from. Excess water not consumed is discharged directly into the sanitary drainage system generally through a coil that additionally cools the incoming water.

If a refrigeration unit is intended to serve more than one DF, a maximum of 10 ft distance from the unit is recommended.

Drinking Water Fixture Accessories

There are many different accessories and components that can be provided for many models of drinking fountains.

1. Bubblers of many different configurations are available
2. Glass fillers
3. Hot water supply
4. Different filters for removal of various contaminants
5. Foot pedal control

Central Water Chiller Assembly

The central chilled water assembly is a mechanical refrigeration unit specifically designed to chill incoming water and distribute the chilled water to multiple drinking fountains. It consists of a compressor and condenser only used with the refrigeration cycle of the cooling medium, and an evaporator, which is the device that actually removes the heat from the cold water supply.

The *central chilling unit assembly* consists of a *compressor and condenser,* converting the cooling medium, which is usually R-20, from liquid to vapor states. The condenser can be cooled either with building chilled water or air.

The evaporator is a heat exchanger that uses the vapor state of coolant and is actually the device that produces the chilled drinking water by drawing out the heat of the water inlet to convert the cooling medium from vapor back to liquid. The evaporator could either be remotely located or installed in the drinking water storage tank.

A *storage tank* is necessary to even out fluctuations in demand that might cause chiller overloading short cycling of the chiller and to keep the size of the chiller unit to a minimum while providing the capacity to adequately supply the facility with enough chilled drinking water.

A recirculation pump is necessary to keep the chilled water at the lowest possible temperature by continuously cooled the water in the piping as it gains temperature in the piping network.

Filters on the water supply to the chilled water unit may be necessary if there are objectionable particulates in the water supply.

System controls usually are freeze protection cutouts, high and low pressure controls, and a thermostatic control to stop the compressor when there is no flow through the chilling unit.

Chilled water distribution piping is insulated to prevent heat gain and condensation. Refer to Chap. 5 for appropriate selection of insulation.

DRINKING FOUNTAIN INSTALLATION

Individual drinking fountains are capable of being mounted in many ways depending on the requirements of the facility and the availability of a pipe space or recess. These are illustrated in Fig. 20.1. In many cases, provisions for handicapped persons

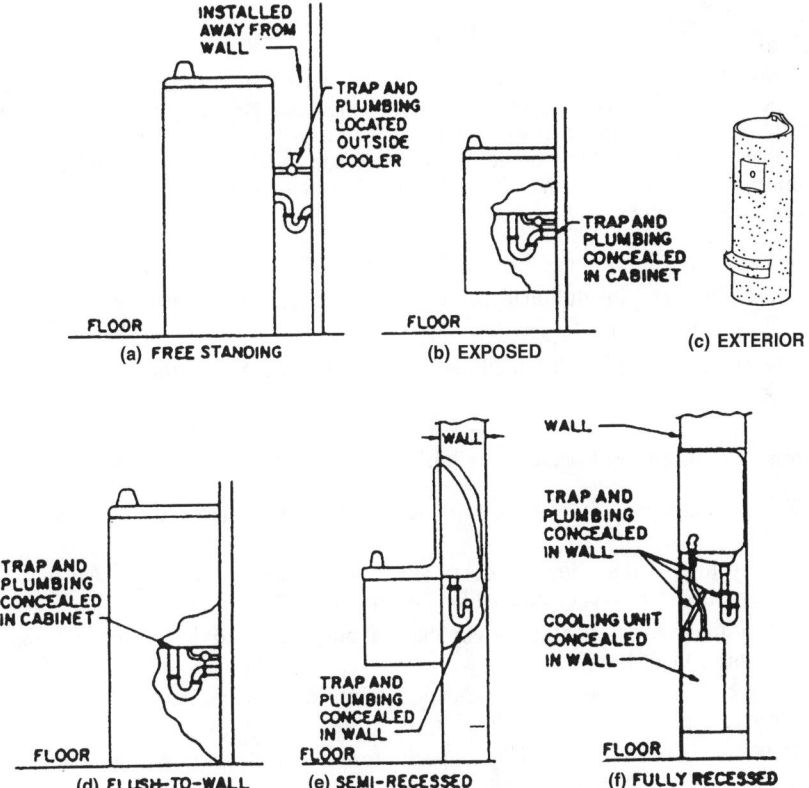

FIGURE 20.1   Types of installation for drinking water fountains.

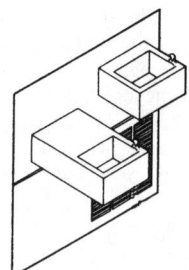

FIGURE 20.2 Dual drinking fountains

or children will require that drinking fountains be installed next to each other with one being mounted lower that the other. Such an installation is shown in Fig. 20.2.

SYSTEM DESIGN

Number of Drinking Fountains Required

The first step is to determine the minimum number of drinking fountains required for the specific facility. The applicable plumbing code generally has requirements for the minimum number of DFs required based on the amount of people served and the facility type. Very often, convenience for building occupants will dictate additional drinking fountains. The following information has been compiled from various code sources.

1. Places of assembly range from 1 DF for each 200 to 1000 people.
2. Educational and institutional facilities require 1 DF per 100 people.
3. Factories require 1 DF per 75 people.
4. Mercantile and residential facilities require 1 DF per 1000 people.

Types of Drinking Fountains Available

1. For normal temperature ranges, an air cooled EWC it the most economical choice.
2. If very high ambient temperatures are expected, and where excessive dust is present, a water cooled model EWC is recommended.
3. Where there is a possibility of explosive atmospheres and dust, select a hazard-ous duty model EWC.
4. For potentially corrosive atmospheres, a corrosion protected model will be necessary.
5. Determine from the client if any special features, such as glass fillers, refrigeration space or hot water capability is requested.
6. Determine requirements for children and ADA compliant fixtures. This is usually done by an architect, but advise if deficient.

Available Location and Space for DF Installation

1. Is there enough space to install any specific type based on available room in corridors, pipe spaces and recesses?
2. Where in the facility will the DFs be located? Will they be one above the other or widely separated?
3. Are domestic water, drainage and electricity available?

System Space Requirements

1. Is space available for installation of a central chiller unit in a conveniently located Mechanical Equipment Room?
2. Free standing units require about 18 in (460 mm) square, with a recess of 30 in (760 mm) wide by 24 in (600 mm) deep.
3. The chiller pack for an EWC requires a recess under the unit of about 24 in (600 mm) square and 9 in (230 mm) deep. The recess for the DF itself varies with the model selected.

Storage Tank Sizing

The storage tank is normally sized for one half of the calculated system hourly demand. For example, if a 100 gph (380 L/h) demand is calculated, a 50 gallon (190 L) tank would be selected. Use a standard size tank from the manufacturer.

PIPE AND INSULATION SIZING AND SELECTION

Central System Pipe Material

The most often used piping is copper water tube, conforming to ASTM B88 and soldered joints.

Central System Insulation

Fiberglass insulation with a vapor barrier and a recommended thickness of 1 in (25 mm) is used to keep the heat gain to a minimum and also to prevent condensation. The minimum thickness to prevent condensation is found in Table 5.1 but the 1 in (25 mm) thick insulation is used for economic reasons. Refer to Table 20.6 for the heat gain from copper pipe with 1 in (25 mm) insulation.

SYSTEM AND COMPONENT SIZING

General Criteria

The standard used by manufacturers for the capacity of chilled water produced from a single EWC is based on ARI standard 1010. This standard uses an inlet water temperature of 80°F (27°C) a discharge water temperature of 50°F (10°C) and the DF installed in an area with an ambient temperature of 90°F (35°C). Table 20.1 lists standard rating conditions for a variety of installations. If any conditions are different, use Table 20.2 to determine a factor that will calculate the flow at those nonstandard conditions.

Tests have shown that water at 50°F (10°C) is considered too cold by most people tested, and the flavor buds of the tongue are numbed by the cold. Based on this study, a higher temperature of discharged water will not be considered as detrimental.

When using unit capacity tables, it is up to the design engineer to determine which condition is applicable before deciding on the criteria to be used for unit capacity sizing.

The distribution pipe for a central project is a looped header originating at the chiller, serving all of the drinking fountains and returning to the chiller.

TABLE 20.1   Standard Rating Conditions

Type of cooler	Temperature, F (°C)				
	Ambient	Inlet water	Cooled water	Heated potable water*	Spill, %
Bottle types	90 (32.2)	90 (32.2)	50 (10)	165 (73.9)	None
Pressure types					
Utilizing precooler (bubbler service)	90 (32.2)	80 (26.7)	50 (10)	165 (73.9)	60
Not utilizing precooler	90 (32.2)	80 (26.7)	50 (10)	165 (73.9)	None
Compartment-type coolers	During the Standard Capacity Test, there shall be no melting of ice in the refrigerated compartment, nor shall the average temperature exceed 46 F (7.8°C).				

Note: For water-cooled condenser *water coolers* the established flow of water through the condenser shall not exceed 2.5 times the Base Rate Capacity and the outlet condenser water temperature shall not exceed 130 F (54.4°C). Base Rate Capacity of a pressure water cooler having a precooler is the quantity of water cooled in one hour, expressed in gallons per hour, at the Standard Rating Condition, with 100% diversion of spill from the precooler.
*This temperature shall be referred to as the Standard Rating Temperature (Heating).
Source: Reprinted by permission from ARI Standard 1010.

TABLE 20.2   Factor for Determining Drinking Fountain Capacity for Non Standard Conditions

	G.P.H. of 50° drinking water at various temperatures and water inlet temperatures.											
Factor	1.58	1.15	0.84	1.53	1.08	0.80	1.40	1.00	0.75	1.25	0.92	0.65
Room Temprature °F	70	70	70	80	80	80	90	90	90	100	100	100
Water Inlet Temperature °F	70	80	90	70	80	90	70	80	90	70	80	90

Example: Water Cooler size—8.9 G.P.H. (at Standard rating 90°–80°). Desired condition—Ambient 80°F, water inlet 70°F. Estimated capacity = factor × G.P.H. (rated).
$$= 1.53 \times 8.9$$
$$= 13.6 \text{ GPH}$$

The maximum recommended velocity for circulating water is approximately 3 ft per minute (FPM) or 1 meter per minute (1 m/min).

The friction loss through the piping should be limited to approximately 10 ft (3 m) of head for each 100 ft (33 m) of pipe.

Dead end piping should be limited to approximately 10 ft (3 m) in length.

The maximum recommended pressure for any system is 125 psig (870 kPa). If more than this, a second zone should be established.

Individual Unit Sizing Criteria

The primary criteria in the sizing process, whether individual EWCs or a central system, is the number of persons served either at a specific location or by the entire system. Table 20.3 is the recommended number of persons served per gallon (liter) of water from a single DF. This table is used by determining the number of people served by the unit and dividing the figure in the table by the number of people served. The answer will give the capacity for the unit.

Another often used table often found in manufacturers' literature is Table 20.4, which is the number of persons served per hour by a unit.

There is a wide discrepancy between information for offices, schools and hospitals found in Tables 20.3 and 20.4. The use of Table 20.3 will generally result in

TABLE 20.3 Drinking Water Requirements (Based on Standard Rating Conditions)

Location	Bubbler service Persons served per gallon (litre) of standard rating capacity	Cup service Persons served per gallon (litre) of base rate capacity	
Offices	12 (3)	30 (8)	
Hospitals	12 (3)	—	
Schools	12 (3)	—	
Light manufacturing	7 (2)	—	
Heavy manufacturing	5 (2)	—	
Hot heavy manufacturing	4 (2)	—	
Restaurants		10 (3)	
Cafeterias		12 (3)	
Hotel	0.08 (0.02) per room		

		Required rate capacity per bubbler gph (L/h)	
		One bubbler	Two or more bubblers
Retail stores, hotel lobbies, office building lobbies	12 (3)	5 (20)	5 (20)
Public assembly halls, amusement parks, fairs, etc.	100 (26)	20 to 25 (80 to 100)	15 (60)
Theaters	19 (5)	10 (40)	7.5 (30)

Source: Reprint by permission from ARI Standard 1010.

TABLE 20.4 Number of Persons Served per Hour

Installation	Standard rating bubbler service*	Installation	Standard rating bubbler service*
Offices	25	Light Mfg.	15
Hospitals	25	Heavy Mfg.	12
Schools	25	Hot Heavy Mfg.	10

* Based on Standard Rating Conditions of ARI Standard 1010. IRA Standard 1020.

one half the size unit that results from the use of Table 20.4. Experience has shown that a value of between 5 and 8 gallons (19 and 30 L) per hour for average conditions is sufficient for DFs found in these areas.

Individual fluid requirements for various work loads given in Table 20.5 should be compared to information found in Tables 20.3 and 20.4 and used if necessary. Sufficient unit capacity should be allowed for the required amount of water replenishment provided at the unit based on the number of people the unit is serving.

Individual Unit Selection

1. Determine the number of units required from where shown on the plans, required from the code or from the architect.
2. Determine the ambient environment in which the units will be installed.
3. Find the specific area where the installation will be made to see if there is water, drain and space necessary for the style of unit to be installed.
4. Determine children or ADA requirements (for mounting height).
5. Size the refrigeration requirements for individual units based on Tables 20.3, 20.4 and 20.5 using judgment as to the various conditions encountered.

Central Water Chiller Sizing

1. Calculate or obtain the total number of DFs.
2. Determine a single fountain usage by using Tables 20.3, 20.4 and 20.5, as applicable, to find the number of gallons per hour required for one unit.

TABLE 20.5 Individual Fluid Replenishment for Heavy Work

Work load	Amount prior to task	Amount during task	Amount at the end of the task
Light	8 oz	As needed	8 oz
Moderate	8 oz	8 oz/60 min	8 oz
Heavy	8 oz	8 oz/30 min	8 oz

1 oz = L.

3. Calculate the total water usage (make up) for all units served by the central chiller by multiplying the figure found in step 1 by the figure calculated in step 2.

4. Calculate a portion of the chiller cooling load, in BTUs, necessary to cool the make up water for drinking purposes found in step 3. First, select the inlet water temperature. Next, use Table 20.6 by entering with the water temperature and read the make up heat gain in BTU/h per gallon (W/h/L) for each 100 ft (30 m) of pipe.

5. Select a preliminary size for the distribution loop header. The size will be selected to reduce the friction loss in order to keep the circulation pump horsepower as low as possible. A general starting point is generally a 1 in (DN 100) size line. This is an iterative procedure to find the final pipe size.

6. Calculate the capacity of the chilled water circulation pump. First, select the gpm of the pump. This is accomplished using Table 20.7. Entering the table with the pipe size found in step 5 and ambient room temperature, find the gpm required to limit the heat loss from the pipe to 5° F (2.8° C) at the intersection of the two figures. The actual pump selection must overcome the friction loss of the flow in the distribution loop using the GPH (LPH) and can be obtained from Fig. 9.28 or other standard engineering text. Selecting the size is an iterative procedure using manufacturers pump curves to select a cost efficient pump.

7. Calculate the heat gain from the piping distribution system by using Table 20.8. Entering with the pipe size and ambient room temperature, read the BTU (W)

TABLE 20.6 Make Up Water Chiller Cooling Requirements

	Btu/h per gallon (W/L) cooled to 45 F (7.2°C)					
Water inlet temp. F (°C)	65 (18.3)	70 (21.1)	75 (23.9)	80 (26.7)	85 (29.4)	90 (32.2)
Btu/gal	167 (13)	208 (17)	250 (20)	291 (23)	333 (27)	374 (30)

TABLE 20.7 Circulating Pump Capacity

Gph per 100 ft (L/h per 30 m) of pipe including all branch lines necessary too circulate to limit temperature rise to 5 deg F (2.8°C) [water at 45 F (7.2°C)]

Pipe size	Room temperature, F (°C)		
in. DN	70 (21.1)	80 (26.7)	90 (32.2)
½ (15)	8.0 (99)	11.1 (138)	14.3 (177)
¾ (20)	8.4 (104)	11.8 (146)	15.2 (188)
1 (25)	9.1 (113)	12.8 (159)	16.5 (205)
1¼ (32)	10.4 (129)	14.6 (181)	18.7 (232)
1½ (40)	11.2 (139)	15.7 (195)	20.2 (250)

Add 20% for safety factor. For pump head figure longest branch only. Install pump on the return line to discharge into the cooling unit. Makeup connection should be between the pump and the cooling unit.

TABLE 20.8 Heat Gain from Copper Pipe with 1 in Fiberglass Insulation

| Pipe size | Btu/h per ft per deg F | Btu/h per 100 ft (watt per 100 m) [45 F (7.2°C) circulating water] | | |
| | | Room temperature, F (°C) | | |
in. DN	(W·°C/m)	70 (21.1)	80 (26.7)	90 (32.2)
½ 15	0.110 (0.190)	280 (269)	390 (374)	500 (480)
¾ 20	0.119 (0.206)	300 (288)	420 (403)	540 (518)
1 25	0.139 (0.240)	350 (336)	490 (470)	630 (605)
1¼ 32	0.155 (0.268)	390 (374)	550 (528)	700 (672)
1½ 40	0.174 (0.301)	440 (422)	610 (586)	790 (758)
2 50	0.200 (0.346)	500 (480)	700 (672)	900 (864)
2½ 65	0.228 (0.394)	570 (547)	800 (768)	1030 (989)
3 80	0.269 (0.465)	680 (653)	940 (902)	1210 (1162)

TABLE 20.9 Circulating Pump Heat Input

Motor Hp (kW)	¼ (0.19)	⅓ (0.25)	½ (0.37)	¾ (0.56)	1 (0.75)
Btu/h (W)	636 (186)	850 (249)	1272 (373)	1908 (559)	2545 (746)

per hour heat gain. To calculate the total heat loss from the pipe distribution system, use the total measured feet (m) run of piping, in increments of 100 ft (30 m), by the heat loss value found in Table 20.8.

8. Calculate the heat generated by the circulating pump that must be made up. Use Table 20.9 entering with the circulating pump horsepower from the size selected in step 6.

9. Heat gain from the chilled water storage tank; 1½ in fiberglass insulation is usually selected. This thickness has a conductivity of 0.13 BTU/ft2 (0.4 W/M2). Since the tanks vary in size, obtain the area of the tank from the manufacturer.

10. The actual central chiller size in BTU (W) is calculated by adding the results of steps 4, 7, 8 and 9. Add an additional 10% as a safety factor.

REFERENCES

ASPE Data Book Chapter 27, Water Coolers, 1994.
American Refrigeration Institute Standards.

ACKNOWLEDGEMENTS

Elkay Manufacturing Company
Oasis Water Coolers
Halsey Taylor Incorporated

CHAPTER 21
HEAT EXCHANGERS

This chapter will describe the basic operation, construction, configuration and design criteria for heat exchangers (HX) commonly used to heat water and will discuss their advantages, disadvantages and application. Included will be all types of heat exchangers intended to recover waste heat, heat water for domestic purposes and other types of heat transfer purposes.

INTRODUCTION

The basic process behind the heating of all water is heat exchange, where the heat from a hot fluid (the heating medium) is given up to a colder fluid (water). This heat exchange takes place between a heating medium at a higher temperature and another medium, in our case water, at a lower temperature, in a piece of equipment called a heat exchanger specifically manufactured and designed to efficiently and cost effectively transfer the heat from one medium to another. A heat exchanger, if properly selected, installed, and maintained, could be the most trouble free piece of equipment in a water heating system.

Heat exchangers have been used to heat water for domestic and other water heating purposes in commercial and industrial applications for many years. In addition, the ever-increasing cost of energy finds heat exchangers used more and more to extract and conserve energy from hot liquid and gaseous by-products of processes that was previously wasted.

CODES AND STANDARDS

Plumbing Codes

There has been a code interpretation by many plumbing inspectors and local authorities of one section of the Uniform Plumbing Code regarding unlawful connections that affect the construction of water heaters. In a condensed explanation, this section is concerned with the contamination of potable water during normal use or as a result of excess pressure by any fluid while flooded in a tank or heat exchanger. This has led to the introduction of double wall heat exchangers used to generate domestic hot water. Some controversy exists regarding the interruption of this requirement, and it may not be necessary for this requirement to be enforced. Refer to the local AHJ regarding this matter.

Tubular Exchanger Manufacturers Association (TEMA)

TEMA, Tarrytown, N.Y., is an association that has established heat exchanger standards and nomenclature for industrial applications. Every shell-and-tube device has a three-letter designation; the letters refer to the specific type of stationary head at the front end, the shell type, and the rear-end head type, respectively (a fully illustrated description can be found in the TEMA standards).

Heat exchangers used specifically for domestic water heating purposes in plumbing systems are not required to be approved by or comply with TEMA standards.

ASTM B-31.9, CODE FOR PRESSURE PIPING, ASHRAE B-90.1, CONSERVATION OF ENERGY

DEFINITIONS

Heating Medium. A heating medium is any fluid used to heat water or fuel used to heat a fluid to a higher temperature than the raw water to be heated. The only exception is electrical energy, which uses a hot solid wire, or element, to directly transfer heat to the water by contact. Examples of heating mediums are:

1. Steam
2. Water
3. Fuel gas
4. Liquid fuel
5. Electricity
6. Solar energy
7. Geothermal

Approach. The term "approach" is used to describe how close the outlet temperature of the water being heated (the colder fluid) comes to (or approaches) the inlet temperature of a fluid heating medium.

Heat Exchanger. A device specifically constructed and designed to efficiently transfer heat energy from a hot fluid to a colder fluid. In generally accepted usage, the heat exchanger is intended to mean any device used to recover or reclaim waste heat, as compared to a water heater, which is a specific piece of equipment intended to generate hot water.

Countercurrent. A term used when the liquid heating medium in any heat exchanger flows in the opposite direction to that of the liquid being heated.

Temperature Cross. A temperature cross occurs when the liquid being heated has an outlet temperature that falls between the inlet and outlet temperature of the heating medium; this is only possible when flows are 100 percent countercurrent.

GENERAL HEAT EXCHANGER TYPES

General

The two most common types of heat exchangers used for utility type service are the shell and tube and the plate type, also known as plate and frame.

Operating conditions, ease of access for inspection and maintenance, and compatibility with heating medium fluids are just some of the variables engineers must consider when assessing heat exchanger options. Other factors include:

1. Maximum design pressure and temperature

2. Heating or cooling applications

3. Material compatibility with various fluids

4. Cleanliness of the heating medium and liquid to be heated

5. Approach temperature

In recent years, the plate-and-frame has emerged as a viable alternative to the shell-and-tube. Air-cooled exchangers have gained popularity because of their ability to reduce water consumption.

Shell and Tube

The shell-and-tube HX can be found in almost every type of application. Mechanically simple in design and relatively unchanged for more than 60 years, the shell-and-tube HX offers a low-cost method of heat exchange.

This type of HX consists of a number of tubes that can be varied in diameter and flow path encased in an outer manifold or shell pipe. The tubes are fabricated together into an assembly called a tube bundle. A typical shell and tube heat exchanger is illustrated in Fig. 21.1.

Heat is transferred by either having the heating medium flow through the shell and the liquid to be heated flow through the tube or vise versa.

There are many types of shell and tube heat exchangers, distinguished by the head and tubesheet configurations.

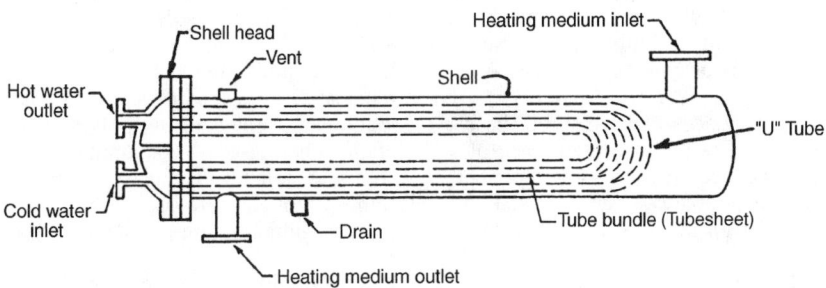

FIGURE 21.1   Typical U-tube heat exchanger.

U-tube, Removable Bundle. The U-tube design consists of straight length tubes bent into a U-shape; hence, its name. The U-bend tubes are then mechanically rolled into a common header or tubesheet. Depending on the fluid outside the tubes, the bundle is fitted with either tube supports or flow baffles along its length. The tubesheet, tubes, and tube supports/flow baffles make up the bundle assembly. The bundle assembly is then placed in a shell (a length of pipe that contains inlet and outlet connections and a pipe size flange at one end for insertion of the tube bundle; the other end of the shell is fitted with a cap) to contain the fluid on the outside of the tube bundle. A head assembly (usually a casting that contains the inlet and outlet connections for directing a fluid into the tube bundle) is then bolted to the shell flange to complete the heat exchanger.

The head assembly contains one or more pass partitions for controlling tube velocity and, hence, the tube-side heat transfer coefficient and pressure drop.

Where the presence of a condensing vapor (such as steam) is a possibility, the tube bundle shall have tube supports designed to support the tubes along their length and provide for proper flow and drainage of condensate out of the shell. When a liquid is circulated outside the tube bundle, flow baffles are used to support the tubes and direct flow across the bundle. In this case, the number and spacing of the flow baffles controls the shell-side heat transfer coefficient and its pressure drop.

The U-tube heat exchanger is well suited for large domestic water heating applications using either boiler water or steam as the heating medium. The nature of the U-tube construction allows for large temperature differences between the tube-side and shell-side fluids with the U-tubes expanding or contracting independently of the shell assembly. In addition, the tube bundle assembly is removable for easy and economically replacement of the heat transfer surface should a failure or leak develop in the bundle.

A variation of the single pass tube and shell heat exchanger is the *hairpin* design, often referred to as a "G" fin, double pipe or multitube heat exchanger. This design is a single pass shell and tube unit that has been folded in half, giving it a hairpin appearance. What distinguishes it from the traditional tube and shell exchangers are the end closures, which allow for expansion without expansion joints and removal of the tubes. This design is used when there are solids in the process stream, high pressure is present in tubes, and high flow rate ratios exist between tube and shell fluids.

The U-tube design does have its limitations. First, because of the U-bends, the tube-side fluid always makes multiple passes down the length of the unit. This reduces the economical use of the design on close temperature approaches and eliminates using a single U-tube unit for temperature cross applications such as those found in energy reclamation projects. Also, because of the U-bend, this unit cannot be totally cleaned by mechanical means when the tube-side fluid is dirty or prone to scaling/fouling. The basic U-tube design can be modified to meet a number of special type applications. A typical U-tube is illustrated in Fig. 21.1.

Double-wall Heat Exchangers. Over the last few years, revisions to many plumbing codes have required the use of double-wall protection on potable water systems. The purpose is to warn of a tube failure before cross-contamination can occur between the tube-side and shell-side fluids. A number of manufacturers now produce double-wall units in a U-tube design. Although there are differences in some design features, the basic design is fairly common among double-wall manufacturers.

The double-wall U-tube unit consists of a tube-within-a-tube design. Fins or grooves are used on one of the tubes to create a leak path between the tubes when they are mechanically bonded to enhance heat transfer characteristics. The outside tube is machined back at each end, bent into the U-tube, and either mechanically rolled or brazed into a double-tubesheet arrangement. Should either of the tubes fail, its respective fluid would be channeled through the leak path between the tubes to the space between the tubesheets. The appearance of the fluid from between the tubesheets is evidence of tube failure.

While the double-wall design is very expensive compared to the single-wall unit it replaces, local plumbing codes for the most part are responsible for its growing use. Double-wall exchangers are being used for applications where a failed tube bundle creates a greater loss to the customer than the initial cost of the double-wall unit.

A severe disadvantage is the loss of efficiency in transferring heat from the heating medium to the water.

Straight-tube Designs

This exchanger design is often used to handle heavy fouling fluids or applications where a temperature cross condition may exist.

Because of the straight tubes, the head assemblies can be removed and the tubes can be mechanically cleaned. (This is especially important if the tube-side fluid is prone to heavy fouling/scaling.) In addition to having the capability for multiple tube-side passes, the fixed tubesheet unit can be designed for a single tube-side pass of the fluid through the unit. This means that 100 percent countercurrent flow can be achieved between the tube-side and shell-side fluids.

The fixed tubesheet construction, however, does limit the design's ability to handle large temperature differences between the fluids. Because the tube bundle and the shell assembly are not independent, any differential expansion or contraction between the two will result in stress being transferred to the tube-tubesheet joint. The forces involved can be sufficient to cause a break in the mechanical bond between the tube and tubesheet and, consequently, a failure of the unit. While an expansion joint can be incorporated into the shell to absorb the stresses, this is an expensive alternative subject to fatigue of its own based on the thermal cyclic rate. As a general rule, differences between the average fluid temperatures greater than 75°F should be checked by the heat exchanger manufacturer for excessive stress in the unit. A note of caution, however, even though the difference between average operating temperatures may not indicate a differential expansion problem, the heating up and cooling down from high operating temperatures may, in itself, cause excess stress at the tubesheet interface.

Finally, the bundle assembly is nonremovable. This increases the cost of replacement since the shell must be replaced along with the tube bundle.

Straight Tube, Fixed Tubesheet. The fixed-tubesheet exchanger is the most common used heat exchanger, and typically has the lowest capital cost per square foot of heat-transfer surface area. Fixed-tubesheet exchangers consist of a series of straight tubes sealed between flat, perforated metal tubesheets. The straight tubes are mechanically rolled into a header at each end of the shell pipe. The headers are integral to the shell and act as both a tubesheet and mounting flange for each of the tubeside head assemblies. As a result of its welded design, the fixed tubesheet unit can be designed to accommodate pressures well over 1000 psig.

Because there are neither flanges nor packed or gasketed joints inside the shell, potential leak points are eliminated, making the design suitable for higher-pressure or potentially lethal service. However, because the tube bundle cannot be removed, the shellside of the exchanger (outside the tubes) can only be cleaned by chemical means. The inside surfaces of the individual tubes can be cleaned mechanically, after the channel covers have been removed. The fixed-tubesheet exchanger is limited to applications where the shellside fluid is non-fouling; fouling fluids must be routed through the tubes. The straight tube, fixed tubesheet HX is illustrated in Fig. 21.2.

Straight Tube, Floating Tubesheet. The floating tubesheet design is a removable bundle with a stationary tubesheet at one end of the unit and a floating (pull through) tubesheet at the opposite end.

The floating head design is similar to the floating tubesheet design in that it incorporates both the stationary tubesheet and floating (pull-through) tubesheet. The difference is that an internal head is bolted, with the use of a gasket, to the floating tubesheet. The elimination of the packing removes the associated temperature and pressure limitations.

While the floating head unit provides all the advantages of a straight tube unit without the disadvantages of either the fixed or floating tubesheet designs, the floating head unit is very expensive and rarely used in commercial/HVAC heat exchanger applications.

The floating tube-sheet is a removable bundle design with a stationary tubesheet at one end of the unit and a floating (pull-through) tubesheet at the opposite end. The floating tubesheet is independent of the shell and fits inside the shell pipe and head assembly diameters. To contain the shell-side and tube-side fluids at the floating tubesheet end, packing and a packing retainer are used.

The ability of the tubesheet to move or float within the shell and head means that the unit can handle large temperature differences without creating excessive stresses in the unit. By the same token, the packing is also responsible for the relatively low design pressure and temperature capabilities of the floating tubesheet unit. Generally, these limits are about 800 psi (5500 kPa) and 875°F (466° C). In addition, the packing, as in other packed devices, is a maintenance item and may require periodic replacement. This is especially true when steam is used as the heating medium.

While the floating tube sheet design is available in single-pass on the tube-side and, therefore, capable of handling temperature cross applications, the design is limited to a maximum of two tube-side passes. The reason is that the head assembly does not float with the tubesheet and, therefore, cannot contain any tube-side pass partitions. This, in turn, may limit the design from achieving optimum heat transfer rates in some applications. The floating tubesheet HX is illustrated in Fig. 21.3.

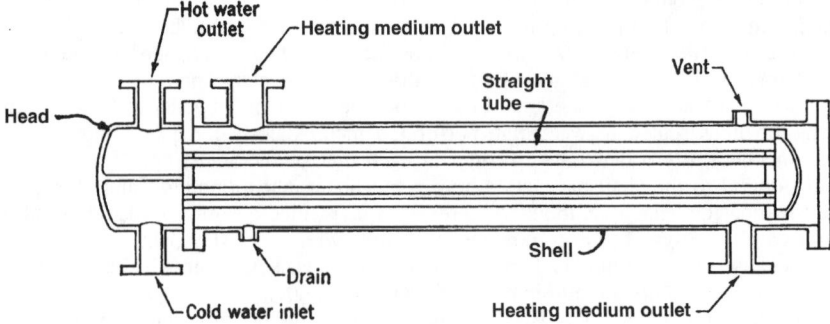

FIGURE 21.2 Typical straight tube, fixed tubesheet heat exchanger.

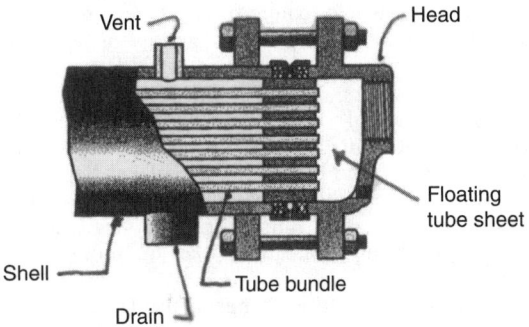

FIGURE 21.3   Straight tube, floating tubesheet.

Removable Bundle, Externally Sealed, Floating Head.   Floating-head exchangers are so named because they have one tube-sheet that is fixed relative to the shell, and another that is attached to the tubes, but not to the shell, so it is allowed to "float" within the shell. Unlike fixed tubesheet designs, the dimensions of which are fixed at a given dimension relative to the shell wall, floating-head exchangers are able to compensate for differential expansion and contraction between the shell and the tubes.

Since the entire tube bundle can be removed, maintenance is easy and inexpensive. The shellside surface can be cleaned by either steam or mechanical means. In addition to accommodating differential expansion between the shell and tubes, the floating tubesheet keeps shellside and tubeside process fluids from intermixing.

This exchanger has some design limitations: both shellside and tubeside fluids must be non-volatile or non-toxic, and tubeside arrangements are limited to one or two passes. In addition, the packing used in this exchanger limits design pressure and temperature to 800 psig (5500 kPa) and 300°F (149°C).

Removable Bundle, Outside Packed, Floating Head.   This design is especially suited for applications where corrosive liquids, gases or vapors are circulated through the tubes, and for air, gases or vapors in the shell. Its design also allows for easy inspection, cleaning and tube replacement, and provides large bundle entrance areas without the need for domes or vapor belts.

Unlike the previous design, only shellside fluids are exposed to packing, allowing high-pressure, volatile or toxic fluids to be used inside the tubes. The packing in the head does, however, limit design pressure and temperatures.

Removable Bundle, Internal Clamp Ring, Floating Head.   This design is useful for applications where high-fouling fluids require frequent inspection and cleaning. And, because the exchanger allows for differential thermal expansion between the shell and tubes, it readily accommodates large temperature differentials between the shellside and tubeside fluids.

This design has added versatility, however, since multi-pass arrangements are possible. However, since the shell cover, clamp ring and floating-head cover must be removed before the tube bundle can be removed, service and maintenance costs are higher than in "pull through" designs.

Removable Bundle, Pull Through, Floating Head.   In the pull-through, floating-head design, the floating-head cover is bolted directly to the floating tubesheet. This allows the bundle to be removed from the shell without removing the shell or floating-head covers, and this eases inspection and maintenance.

   This is ideal for applications that require frequent cleaning. However, it is among the most expensive designs. And, the pull through design accommodates a smaller number of tubes in a given shell diameter, so it offers less surface area than other removable bundle exchangers.
   Removable Bundle, U-tube.   In the U-tube exchanger, a bundle of nested tubes, each bent in a series of concentricity tighter U-shapes, is attached to a single tube-sheet. Each tube is free to move relative to the shell, and relative to one another, so the design is ideal for situations that accommodate large differential temperatures between the shellside and tube-side fluids during service. Such flexibility makes the U-tube exchanger ideal for applications that are prone to thermal shock or intermittent service.
   As with other removable-bundle exchangers, the U-tube bundle can be withdrawn to provide access to the inside of the shell and to the outside of the tubes. However, unlike the straight tube exchanger, whose tube internals can be mechanically cleaned, there is no way to physically access the U-bend region inside each tube, so chemical methods are required for tubeside maintenance. As a rule of thumb, non-fouling fluids should be routed through the tubes, while fouling fluids should be reserved for shellside duty.
   This inexpensive exchanger allows for multi-tube pass arrangements. However, because the U-tube cannot be made single pass on the tubeside, true countercurrent flow is not possible.
   Tank Suction Heater.   A tank suction heater is another variation of the U-tube design. In this application, the portion of the shell opposite the head is removed and a tank mounting flange is welded into the shell near the head of the unit. Tank suction heaters are designed for heating high-viscosity liquids to permit pumping them from their storage tanks. Typical applications include lube oil, heavy fuel oils, tar, road oil, and asphalt. They are also used for preheating oil for oil burners in industrial plants or other large installations.
   The main advantage of a tank suction heater over other types of heaters is that it heats the tank fluid on demand and thereby eliminates the loss of energy associated with a system that maintains the entire tank at a temperature above ambient. Flapper plates or valves can be fitted on the open end of the shell to allow maintenance of the bundle without having to drain the entire tank of its contents. A typical tank suction heater is illustrated in Fig. 21.4.
   Tank Heater.   Replacing the shell assembly with a tank mounting collar allows the U-tube heat exchanger to function as a storage heater. In this application, the tank heater uses hot water for a heating medium that is pumped through the tubes, thus maintaining the tank system water at a set temperature. Steam can also be used as the heating medium when a special head assembly that allows for proper condensate drainage of the unit is installed.

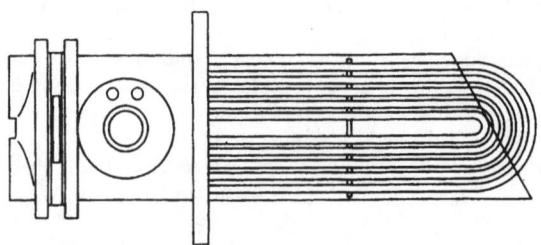

FIGURE 21.4   Typical tank suction heater.

The tank heater uses natural convection for transferring heat to the tank-side system. This is quite different from most other heat exchangers, where pumped or forced convection is used. The significance is that natural convection produces much lower rates of heat transfer. The result is that tank heaters, for a given capacity, require more heat transfer surface area than heat exchangers utilizing forced convection. In addition, it is very important for proper natural convection that the relationship between the size of the tank heater and the size of the tank be within specific limits. The guideline for this relationship is to have the tube bundle extend into the tank from 50 to 75 percent of its length in a horizontal tank or nearly its full diameter if the tank is to be installed vertically.

There are times when a steel tank requires a lining of either cement or epoxy on a domestic water system. When this is done, special consideration must be given by the tank heater manufacturer to insure that the tube bundle fits inside the tank mounting collar. In addition, every tank heater should have adequate support inside the tank to eliminate stress on the tube-tubesheet rolled joint. Inadequate support often leads to leaks of the tube bundle in this area.

A typical tank heater arrangement in a domestic hot water generator is illustrated in Fig. 21.5.

CHOOSING OFF-THE-SHELF SHELL AND TUBE EXCHANGERS

Fixed tubesheet and U-tube shell-and-tube exchangers are the most common types of off-the-shelf heat exchangers available today. Such stock models are typically used as components in vapor condensers, liquid to liquid exchangers, reboilers and gas coolers.

Standard fixed tubesheet units, the most common shell-and tube heat exchangers, range in size from 2 to 8 in dia. Materials of construction include brass or copper, carbon steel, and stainless steel. Even though this exchanger is one of the least expensive available, it is constructed to standards specified by the manufacturer. If

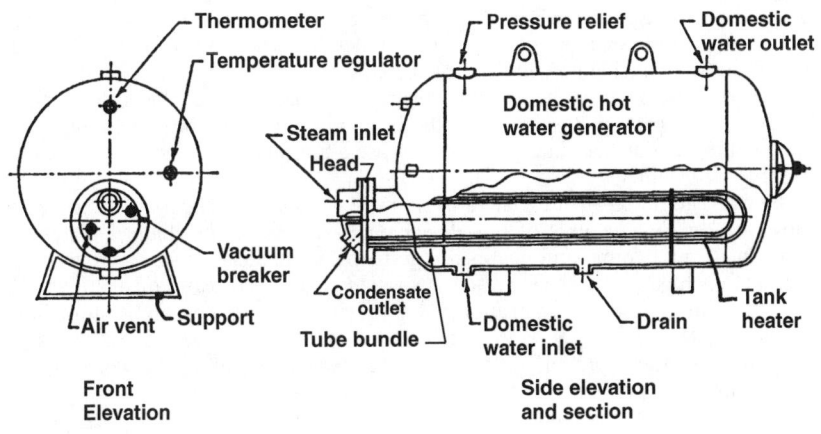

FIGURE 21.5 Tank heater installation.

the user desires, stock exchangers can be constructed to American Society of Mechanical Engineers (ASME) codes.

U-tube heat exchangers are commonly used in steam heating applications, or heating and cooling applications that handle chemical fluids as opposed to water. While the U-tube is generally the lowest-priced heat exchanger available, service and maintenance costs tend to be higher than other exchangers, since the nested, U-bend design makes individual tube replacement difficult.

PLATE TYPE HEAT EXCHANGERS

In recent years, plate type heat exchangers have emerged as an alternative to shell-and-tube heat exchangers. With their ability to optimize thermal performance, they have made possible a number of close approach and temperature cross applications that previously were not economical or practical with a shell-and-tube design. These units are efficient, easy to maintain, less prone to foul and take up little space.

In its most basic form, plate HX consist of corrugated plates compressed in a frame. Plate type heat exchangers are characterized by having heat transfer occur via metal, plastic, glass or ceramic barriers between fluids. One stream heats (or cools) the other by means of conduction (or radiation) through or from the barrier. Inside the heat exchanger the fluids heat by convection. There are two types of plate type units: prime surface and plate and frame. They could be either gasketed or welded. Double-wall heat exchanges could also be provided.

In general, the prime surface units are best suited to small heat loads and batch operations. Because of their very limited use in utility systems, they will only be briefly discussed. The plate and frame units are most efficient when used for larger heat loads and continuous duty.

A typical plate and frame heat exchanger is illustrated in Fig. 21.6.

PLATE AND FRAME HEAT EXCHANGERS

Plate and frame units are fabricated from a series of channel plates that are pressed together to form a plate pack, with the holes at the corners of the plates forming a continuous passage or manifold. This distributes the heat transfer media from the inlet of the heat exchanger into the plate pack for each of the fluids. The media are then distributed into the narrow channels formed by the plates. The gasket arrangement on each plate distributes the hot and cold media into alternating flow channels throughout the plate pack. In all cases, hot and cold media flow countercurrent to each other.

The most common of the plate and frame type heat exchangers is the gasketed plate unit. Heat exchangers of this design include a series of channel plates that are mounted on a frame and clamped together. Each plate is made from pressable materials, such as stainless steel, and is formed with a series of corrugations. The most common pattern of corrugation is the herringbone or chevron. Also included with each plate is an elastomer gasket. This gasket is used for sealing purposes and to distribute the fluids properly in the plate heat exchanger. Spaces between adjacent plates form flow channels for the hot and cold fluids.

A corrugated herringbone or chevron style pattern is pressed into each plate to produce highly turbulent fluid flows. The high degree of turbulence results in high

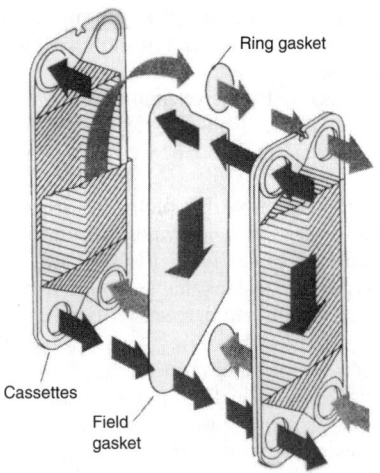

Flow Path of Typical Gasketed Plate Heat Exchanger

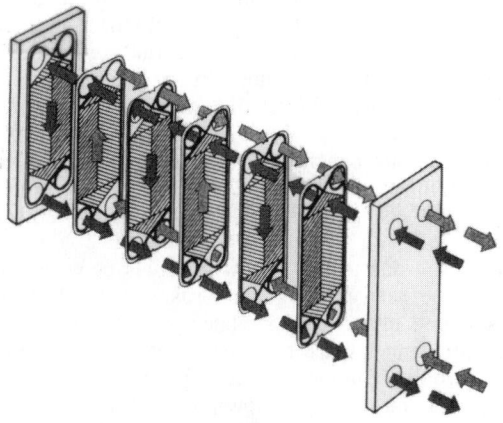

Flow Path of Typical Welded Plate Heat Exchanger

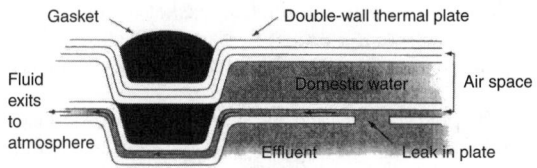

Double Wall Plate Heat Exchanger

FIGURE 21.6 Plate and frame heat exchangers.

heat transfer coefficients and keeps fouling to a minimum. In addition, the corrugations add rigidity to each channel plate. This allows the use of thinner plate material and improves heat transfer.

The basic design of the gasketed plate exchanger allows for the opening of the frame either to add or remove channel plates to optimize heat exchanger performance or to allow for service and maintenance of the channel plates, all with a minimum of downtime.

Taking into account all the benefits of a plate and frame heat exchanger (100 percent countercurrent flow, high turbulence, and thin plate material), we find it to be a highly efficient device that typically yields heat transfer rates three to five times greater than other types of heat exchangers. As a result, a more compact design is possible for a given application relative to other heat exchanger types. Ideal operating conditions are those involving temperature crosses and close approach temperatures between the hot and cold media.

While gasketed plate and frame heat exchangers can be used in almost any application, there are limitations that must be considered. These limitations are primarily focused on the design pressures and temperatures of the unit. Practical design pressures are limited to 800 psig, while design temperatures are a function of the gasket material used in the exchanger. The most popular and widely applied gasket is nitrile rubber (NR). Its temperature limit is 280 F. NR is followed by EPDM (ethylene propylene diene monomer) with a temperature limit of 820 F. EPDM gaskets can be used as a substitute for NR (for higher temperature ratings) on all applications except those involving oil heating or cooling since EPDM will swell in the presence of most oils. Other gasket materials, such as hypalon and viton, are also available. These gasket materials are more prevalent in industrial applications.

Gasketed exchangers are benefiting from improvements in the quality and diversity of elastomer materials and gasket designs. The use of exchangers with welded connections, rather than gaskets, is also reducing the likelihood of process-fluid escape.

Other limitations are due to the narrow channels between adjacent plates. If a fluid entering the plate heat exchanger has suspended solids or is susceptible to depositing large amounts of scale on the plate surfaces, careful consideration should be given to the free channel space between the plates. Also, the narrow channels and resultant high turbulence of the fluid flows produce high-pressure drops, making the plate exchanger incompatible with low-pressure applications.

Until recently, a major limitation to the gasketed plate heat exchanger was the method used to attach the gaskets to the channel plates. In the past, gaskets were glued to the channel plates. Since gaskets are a replaceable part, removal and reinstallation of the new gaskets was a very time consuming and labor intensive procedure. Most manufacturers now use a glueless gasket design. Clip-type and snap-type are the two most common forms off glue-less gasket. Both simplify the regasketing procedure, making it possible for on-site service and thus reducing service downtime.

Recent advances in plate design and technology have produced two variations to gasketed plate and frame heat exchangers: double-wall and welded plate.

Double Wall Plate and Fame Exchangers.   In double-wall plate and frame exchangers, two standard channel plates are welded together at the four corner ports to form one assembly. An air space or leak path is created between the plates for the passage of a fluid should a plate fail. The appearance of this fluid is evidence of plate failure.

The purpose of the double-wall plate and frame exchanger, like that of the double-wall shell-and-tube heat exchanger, is to warn of a plate failure before cross contamination can occur between the heating medium and the colder water.

Welded Plate and Frame Exchangers. In welded plate and frame exchangers, two standard channel plates are welded together at the periphery of the plates. In this design, the two welded plates (usually called a cassette) form a flow channel where the elastomer gasket has been replaced by the welded joint. This configuration may be necessary when there is no elastomeric gasket compatible with the fluid or more positive containment is required. Typical applications include refrigerant evaporators/condensers, ammonia refrigeration, and wherever aggressive or corrosive fluids are present.

Brazed Plate and Frame Exchangers. The latest addition to the plate and frame type exchanger line is the brazed plate exchanger. This type of unit shares the same features, benefits, and method of operation as the gasketed plate and frame exchanger. What it does not share are the gaskets and heavy frame components.

In this design, the elastomer gaskets are replaced with a brazed material (copper or nickel) that greatly increases its pressure and temperature capabilities. The brazed plate design is typically rated at pressures up to 450 psig with temperatures up to 500 F.

Most brazed plate exchangers are very compact in size and are lightweight. They are suitable for most OEM applications where package size is a major consideration or, for that matter, any application where space consideration is a factor. Typical applications include water heating/cooling, refrigerant evaporators/condensers, heat recovery systems, steam applications, district heating systems, oil cooling, and air drying.

The limitations of a brazed plate have to do with the size of plate that can be successfully and reliably brazed. Currently, flow rates top out at about 850 gpm in a single unit. In addition, the brazed plate design is a sealed unit and not serviceable, as such, and it should be considered a throw-away unit in the event the unit becomes fouled or fails altogether.

Wide Gap Plate Exchangers. Compared with traditional plate and frame exchangers, this design relies on a more loosely corrugated chevron pattern, which provides exceptional resistance to clogging. The plates are designed with few, if any, contact points between adjacent plates to trap fibers or solids. Some styles of this exchanger use wide gap plates on the process side and conventional chevron patterns on the coolant side, to enhance heat transfer.

Prime Surface

These units are fabricated from two die-formed sheets welded together. One or both sheets are die or pressure formed (cold formed) to create a series of well defined passages through which the heating (or cooling) medium flows. Most common metals that could be cold worked and resistance welded are used, the most common being carbon steel, stainless steels, monel, titanium and Hasteloy. They are a single circuit design and could be used as shelves, immersed, clamped on or built into tanks and otherwise used where plate and frame exchangers are not suitable even when the media are the same. Maximum operating parameters are generally limited to a temperature of 650°F (343°C) and a pressure of 500 psig (3450 kPa).

IMMERSION HEAT EXCHANGERS

An immersion heat exchanger consists of a tube bundle (coil) containing fluid to be heated and an atmospheric basin or tank in which the tube bundle is immersed that contains the heating medium, which could be any liquid. The heat is transferred directly from the heating medium to the water in the coil. A typical immersion heater installation is shown in Fig. 21.7.

AIR-COOLED EXCHANGERS

In the air-cooled exchanger, a motor and fan assembly forces ambient air over a series of tubes, to cool or condense the fluids carried within. The tubes are typically assembled in a coiled configuration.

Air is cheap and abundant, but it is a relatively poor heat transfer medium. To increase the heat transfer rates of the system, the tubes in air-cooled exchangers are typically given fins, which extend the surface area, increase heat transfer, and give such systems the nickname fin-tube coils.

The diameter and materials specified for the tubes and fins depend on system requirements. The fins are commonly made from aluminum or copper, but may be fabricated of stainless or carbon steel. Tubes are generally copper, but can be made from most any material, and they range in size from 5/8 in to 1/4 in outer diameter. The design of the air-cooled exchanger is such that individual coils can be removed independently for easy cleaning and maintenance.

Aluminum Brazed-fin Exchangers

In this design, corrugated plates and fins are added to a brazed-composite core, to create alternating air and fluid passages. This compact, lightweight design is considered to most cost-effective air-cooled unit available. Turbulence created in the fluid channels boosts efficiency.

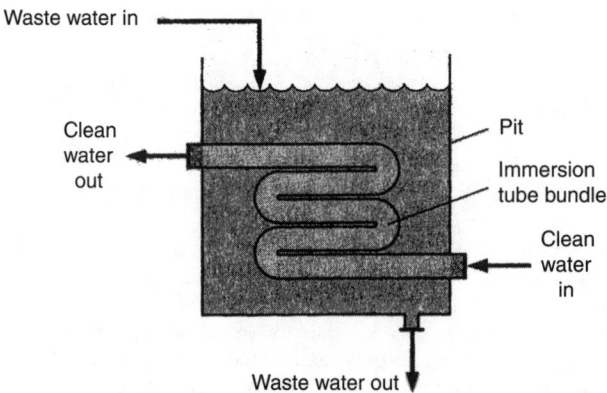

FIGURE 21.7   Immersion heat exchanger.

Aluminum Plate-fin Exchangers

This type of exchanger is constructed with traditional heat exchanger tubing. Stacked, die-formed aluminum plates extend the surface to maximum air-side heat transfer. Like the brazed-fix exchanger, this unit is also used for oil and glycol cooling, but its higher flow-rate expands its capabilities. Built from standard components, aluminum-fin exchangers are designed with a more solid construction than their brazed-fin counterparts.

Fin-tube Exchangers

In this design, one continuous fin is wrapped spirally around a series of individual tubes. Often referred to as a "heavy duty coil," this air exchanger has fin-tube attachments that can be built either to ASME and API standards, or to customer specifications. Often used in air-heating applications, the heavy-duty coil is available with several different fin variations, including the tapered fin, footed "L" fin overlapped-footed fin and the embedded fin, which describe the geometry at the fin-tube interface. The method of attaching the fin to the tube is critical, since the loosening os this bond may hinder heat exchange.

HEAT EXCHANGER SELECTION

One of the most asked questions when it comes to selecting a type of heat exchanger for a particular application is "Which is the most appropriate- shell-and-tube or plate type?" Assuming that the application is within the pressure and temperature limits of both designs, the issue usually centers around initial cost, maintenance cost, and future operating conditions.

The shell-and-tube heat exchanger has the following advantages:

1. Greatest flexibility of design and configuration
2. Large choice of shell and tubes material
3. High temperature and pressure limitations and specific design features
4. Ability to handle high levels of particulate material
5. All-welded construction and the absence of gaskets contribute to a longer exchanger life.

The plate and frame heat exchanger has the following advantages:

1. The highly efficient heat-transfer surface and pure countercurrent flow yield smaller and less costly heat exchangers that require less space.
2. All heat transfer surfaces can be easily opened and mechanically cleaned on a unit with gaskets.
3. All-bolted construction allows for easy maintenance.

Initial cost is usually dictated by the approach temperatures of the application. Close approach temperatures and temperature crosses favor the plate heat exchanger, while wide temperature approaches favor the shell and tube. Materials of construction can influence this relationship, especially if the application requires the use of stainless steel. With the extensive use of computerized selection programs, it requires little effort to obtain prices for each types of unit and be able to compare initial cost quickly.

With respect to maintenance cost, much depends on the properties of the fluids involved. If the fluid has a tendency to foul, the plate heat exchanger offers somewhat easier and direct access to the heat transfer surface for the purpose of cleaning. In addition, because of the high turbulence in plate units, they have less of a tendency to scale or foul compared to the shell-and-tube design.

If a plate and frame type heat exchanger has a weakness compared to the shell-and-tube, it lies in the amount of gasketing in the unit. Compared to the shell-and-tube, the amount of gasketing is magnitudes larger, and therefore the potential for leakage is much higher. In addition, the gaskets are elastomers, which means they do have a service life. On average, the life of a gasket on a plate and frame heat exchanger is approximately 6 to 7 years with operating temperatures having a significant effect on the average. Units operating close to the temperature limit of the gasket will experience shorter gasket life. There is one other aspect of an elastomer gasket that must be considered: the phenomenon of cold leakage. Cold leakage is due to the cooling down of a plate heat exchanger from high operating temperatures when there is differential pressure between the hot and cold media in the unit. The plate and frame unit has a tendency to weep through the gasket interface. The weeping normally stops after the gaskets reset or the unit is brought back up to operating temperatures. Basically, if the application requires a high probability against leakage, the better choice is a prime surface or shell-and-tube design rather than plate and frame.

While gaskets may be a weakness in a plate and frame unit, the ability to expand its thermal capacity by merely adding channel plates to the existing unit is one of its major strengths. If it is known that a particular application needs to be expanded in the future, a plate unit is by far the easiest and most economical design for such an expansion.

Certain exchanger designs operate better at different approach temperatures. Plate-and-frame exchangers, for example, work well at a very close approach, on the order of 2°F (1°C). For shell-and-tube exchangers, however, the lowest possible approach is on the order of 10°F (6°C).

As for cleanliness, shell-and-tube exchangers have tube diameters that can accommodate a certain amount of particular matter without clogging or fouling. Plate-and-frame exchangers, however, have narrow passageways, making them more susceptible to damage from precipitation or particulate fouling.

HEAT EXCHANGER DESIGN

SHELL AND TUBE HEAT EXCHANGERS

Shell and tube HX use tubes arranged inside a shell in such a way that one fluid, known as the tube fluid, flows within the tubes while the other fluid, known as the shell fluid, flows outside the tubes and within the shell. Heat is transferred through the tube wall.

The shell fluid enters one of the shell nozzles, passes around the tubes, follows the path formed by the shell baffles and finally leaves through the other tube nozzle. The tube fluid enters one of the tube nozzles at one end, generally flows in only one direction and leaves through the other tube nozzle. This is called a single pass because the fluid passes through the tube only once. By adding partitions, the tube fluid could have additional passes.

This section will discuss a simplified method of calculating the area of the tube bundle for only single pass HX units. The heat transfer rate depends on the basic flowpath of each stream. HX design depends on the overall heat transfer coefficient, flow rate of the heating medium, the size and length of the tube bundle and the difference in temperature between the tube fluid and the shell fluid.

To calculate the inside area of coils for heating water use the following formula, where the heating medium is inside the HX shell:

$$A = \frac{Q \times 8.33 \times (t2 - t)}{U \times DT} \qquad (21.1)$$

where: A = surface area of coil, sq. ft.
    Q = quantity of water, gph
    T2 = outlet temperature of tube fluid, °F
    T1 = Inlet temperature of tube fluid, °F
    U = Heat transfer coefficient, (overall conductance in btu/hr)
    DT = mean temperature difference between shell fluid and tube fluid
    TS = Temperature of shell fluid

$$DT = TS - \frac{(T2 - T1)}{2} \qquad (21.2)$$

Design Considerations for Tube Bundles

1. Table 21.1 gives the area (A) and other properties for often used copper tubes.
2. "U", the heat transfer coefficient, is found in Table 21.2.
3. The heat transfer coefficient is reduced by a fouling factor that reduces the amount of heat transferred between the tube bundle and the shell fluid. Typical fouling factors vary depending on the fluid can be found in Table 21.3.

The basic water heating rate for conventional storage type domestic water heaters used by many manufacturers is 20 gallons per hour per square ft of bundle outside area when heating water from 40°F to 180°F with saturated steam condensing inside the coils at 0 psig (212°F). The heat transfer rate therefore, is 280 BTU/h/sq.ft coil/°F.

TABLE 21.1 Properties of Copper Tube

Nominal or standard size, inches	Nominal dimensions, inches		Calculated values (based on nominal dimension)	
	Outside diameter	Wall thickness	Cross sectional area of bore, sq inches	External surface, sq ft per linear ft
⅛	.125	.030	.00332	.0327
³⁄₁₆	.187	.030	.0127	.0490
¼	.250	.030	.0284	.0654
⁵⁄₁₆	.312	.032	.0483	.0817
⅜	.375	.032	.076	.0982
	.375	.030	.078	.0982
½	.500	.032	.149	.131
	.500	.035	.145	.131
⅝	.625	.035	.242	.164
	.625	.040	.233	.164
	.750	.035	.363	.196
¾	.750	.042	.348	.196
	.750	.042	.348	.196
⅞	.875	.045	.484	.229
	.875	.045	.484	.229
1⅛	1.125	.050	.825	.295
	1.125	.050	.825	.295
1⅜	1.375	.055	1.26	.360
	1.375	.055	1.26	.360
1⅝	1.625	.060	1.78	.425
	1.625	.060	1.78	.425
2⅛	2.125	.070	3.09	.556
2⅝	2.625	.080	4.77	.687
3⅛	3.125	.090	6.81	.818
3⅝	3.625	.100	9.21	.949
4⅛	4.125	.110	12.0	1.08

TABLE 21.2 Typical Heat Transfer Coefficients

Controlling fluid and apparatus	Type of exchanger	U free convection	U forced convection
Air-flat plates	Gas to gas[a]	0.6–2	2–6
Air-bare pipes	Steam to air[a]	1–2	2–10
Air-fin coil	Air to water[a]	1–3	2–10
Air-HW radiator	Water to air[a]	1–3	2–10
Oil-preheater	Liquid to liquid	5–10	20–50
Oil-preheater	Steam to liquid	10–30	25–60
Brine-flooded chiller	Brine to R12, R22		30–90
Brine-flooded chiller	Brine to NH_3		45–100
Brine-double pipe	Brine to NH_3		50–125
Water-double pipe	Water to NH_3		50–150
Water-Baudelott cooler	Water to R12, R22		60–150
Brine-DX chiller	Brine to R12, R22, NH_3		60–140
Brine-DX chiller	E glycol to R12, R22		100–170
Water-DX Baudelot	Water to R12, R22, R502		100–200
Water-DX shell & tube	Water to R12, R22, NH_3		130–190
Water-shell & int finned tube	Water to R12, R22		160–250
Water-shell & tube	Water to water		150–300
Water-shell & tube	Condensing vapor to water		150–800

Notes: U factor = Btu/h − ft^2 • °F
    Liquid velocities 3 ft/sec or higher
    a At atmospheric pressure
    b At 100 psig
    Values shown are for commercially clean equipment.
Adapted from "Numbers", Bill Holladay and Cy Otterhelm, 1985.

An important consideration in the design of tube fluid when it is water is the effect of buildup of scale on the inside of the coil if it is copper. See Fig. 21.8.

SIZING PRESSURE AND TEMPERATURE RELIEF VALVE

The purpose of the temperature relief valve is to prevent the water in a heater from reaching 210°F (99°C). The temperature rating equals the maximum heat rate from heat input to the heater on which thee valve is installed. It shall be water rated on the basis of 1250 BTU for each gallon per hour discharge at 30 psi (210 kPa). The formula is:

$$\text{BTU capacity of valve} = \frac{\text{Heated GPH} \times 8.33 \times \text{Temp. rise}}{0.8} \quad (21.3)$$

The purpose of the pressure relief valve is prevent the pressure in the heater from rising higher than 10 percent in excess of the set opening pressure of the valve. It shall be set at a pressure not exceeding the working pressure of the tank or heater. Water shall be discharged to a safe location.

TABLE 21.3 Typical Fouling Factors

Recommended minimum fouling allowances for water flowing at 3 ft/sec* or higher:

Distilled water		0.0005
Water, closed system		0.0005
Water, open system		0.0010
Inhibited cooling tower		0.0015
Engine jacket		0.0015
Treated boiler feed (212°F)		0.0015
Hard well water		0.0030
Untreated cooling tower		0.0033

Steam:

Dry, clean and oil free	0.0003
Wet, clean and oil free	0.0005
Exhaust from turbine	0.0010

Brines:	Non-ferrous tubes	Ferrous tubes
Methylene chloride	none	none
Inhibited salts	0.0005	0.0010
Non-inhibited salts	0.0010	0.0020
Inhibited glycois	0.0010	0.0020

Vapors and gases:

Refrigerant vapors	none
Solvent vapors	0.0008
Air, (clean) centrifugal comp	0.0015
Air, reciprocating compressor	0.0030

Other liquids:

Organic solvents (clean)	0.0001
Vegetable oils	0.0040
Quenching oils (filtered)	0.0050
Fuel oils	0.0060
Sea water	0.0005

*Lower velocities require higher f values.
Adapted from "Numbers," Bill Holladay and Cy Otterhelm, 1985.

REFERENCES

Clark, Jack "Domestic Water Heater Fundamentals", Air Conditioning, Heating and Ventilating Magazine, October 1968.

Plate Heat Exchangers, Process Heating Magazine, 1999.

Madejczyk, J and Stephan, M, "Shell and Tube vs. Plate Heat Exchangers, Heating, Piping, Air Conditioning Magazine, 1994.

Trumpfheller, G. "Selecting the Appropriate type Heat Exchanger", Chemical Processing Magazine, January, 1998.

CHAPTER 22
MEASUREMENT INSTRUMENTATION AND METHODS

INTRODUCTION

This chapter will discuss the various methods, components and criteria that might reasonably be expected to be found in typical utility piping systems that are used to provide a variety of commonly used measurements and metering methods for liquids and gases and level measurements for fluids and solids.

For most systems described in this chapter, different forms of measurement are very important for maintenance and control purposes. These systems are not generally as demanding in terms of extreme accuracy except where required by healthcare, laboratory and other specific reasons. Health care facilities depend on many forms of measurement to assure correct flow, continued reliability and performance of life safety equipment and piping distribution networks; and for laboratories, correct pressures and flow are critical for the correct performance of various equipment and also in calibrating and analysis types of instruments and equipment.

GENERAL

The subject of measurement is important to many aspects of facility piping systems. The information necessary to maintain proper system conditions, confirm design parameters and to obtain actual operating conditions when repair is necessary can only be obtained from various instruments installed in the distribution system or at the equipment. Subjects to be discussed in this chapter are:

1. Flow measurement
2. Level measurement
3. Temperature measurement
4. Pressure measurement
5. pH measurement
6. Metering pumps
7. Measuring flow in open channels

8. Miscellaneous
Sight glass
Thermowells
Space for piled materials

CODES AND STANDARDS

1. ISA Specification S-20, Instrument Data Sheet
2. British Standard BS 7405, Selection Standard for Meters
3. AWWA 700 series, standards for water meters
4. Various standards for almost all of the devices discussed too numerous to mention

DEFINITIONS

The following are definitions of terms used in the following discussions.

Reynolds Number

The performance of flowmeters is influenced by a dimensionless unit called the Reynolds number. It is defined as the ratio of the liquid's inertial forces to its drag forces. The equation is:

$$R = \frac{3160 \times Q \times G}{D \times \mu} \qquad (22.1)$$

where R = Reynolds number
Q = measured liquid's flow rate, gpm
G = measured liquid's specific gravity
D = pipe inside diameter, inches
μ = measured liquid's viscosity, centipoise

The flow rate and the specific gravity are inertia forces, and the pipe diameter and viscosity are drag forces. The specific gravity and pipe diameter remain constant for most applications.

R values below 2000 are considered laminar flow and R values above 3000 are considered turbulent flow. Values between 2000 and 3000 is a transition zone where flow could be either laminar or turbulent and is based on piping and installation conditions. Most applications involve turbulent flow.

Viscosity

Absolute or dynamic viscosity for a liquid is defined as the ratio of sheer stress to the sheer rate. If a cubic volume of liquid were isolated, the sheer stress is the relative force between the top and bottom divided by the length between them. The units of sheer stress are given as dines per square centimeter (dyne sec/cm2) or centipoise. More simply stated, it is a measure of the internal friction that exists as the liquid flows. The more it resists the tendency to flow, the higher the viscosity.

Kinematic viscosity is the dynamic viscosity divided by the density of the liquid. The units of kinematic viscosity are given as centistokes. The conversion from dynamic viscosity to kinematic viscosity is calculated by dividing the dynamic viscosity by the fluid density in grams per cm^2.

Care must be taken when comparing manufacturers' specifications for flowmeter capacity.

NEWTONIAN FLUIDS

A Newtonian fluid is one in which the viscosity does not depend on the sheer rate and which exhibits sheer stress proportional to sheer rates, such as water and mineral oil. No matter what sheer rate is applied, the viscosity stays the same. A non-Newtonian fluid will change viscosity as the liquid is sheared at a greater rate.

Another classification of non-Newtonian fluid is called a *plastic fluid.* This type will behave as a solid until a critical sheer rate is achieved (called the yield value), at which time the fluid will start to flow.

Water is a Newtonian fluid. Paint is a non-Newtonian fluid because it will decrease its viscosity with an increasing flow rate. Ketchup is an example of a plastic fluid since it is difficult to pour until an appropriate shear rate is reached. Viscosity changes will cause misregistration in turbine and positive displacement flowmeters meters and other level measuring methods.

SENSORS

A sensor is a device that measures some condition—temperature, pressure or flow—and provides a signal to an instrument package that converts that signal to a useful indication of the condition being measured.

HYDRAULIC RADIUS

The hydraulic radius, a term used for open channel flow, is the ratio of cross section area to the wetted perimeter of a channel. It is not a radius in the geometrical sense. It relates two important parameters that influence the movement of water in a channel. The first is the cross sectional area of the water, which is directly proportional to the volume of water (flow rate) carried in the channel. The second is the length of the solid surface of the channel (wetted perimeter), which resists the movement of the water.

The cross sectional area is defined as the actual area of water flowing in the channel. The wetted perimeter is the linear surface contacted by the flowing water excluding the surface of the water.

FLOW MEASUREMENT

Flow is defined as fluid volume per unit of time.

Flowrate of fluids is one of the most widely measured variables in industry. The measurement is made by means of some type of flowmeter (meter). These discussions will concern the terms, principles of operation, advantages and limitations of the most commonly used meter types for facility type applications.

FLOWMETER CLASSIFICATION

Flowmeters are often divided into two large, major categories: those that measure quantity or those that measure flow rate. The meters will be divided into additional sub-categories that further define their method of operation. A simple classification of the various flowmeters will help visualize the operating principles of the various meter types.

Displacement meters, often referred to as positive displacement meters, measure or count successive and discreet quantities of fluids. They divide the flow stream into discrete units of volume and capture these volume units in a measuring chamber. The meter then measures these captured volume units passing through the meter. Positive displacement meters include piston, oval gear, notating disk and rotary vane types.

Velocity type meters measure the rate of flow and operate linearly. There is no square root relationship as with differential meters, so their rangeability is greater. In these meters, the measuring element is placed in the flowing stream and either converts potential energy into kinetic energy (where energy is lost due to pressure drop) or extracts energy in the form of work done to the object in the flow path, such as a turbine or rotor. Examples of velocity meters that convert energy are the differential pressure, positive displacement, turbine and vortex meters. Velocity meters that add energy require adding some form of energy to the meter in order to obtain a flow measurement. The effects of either the fluid on the added energy or the added energy on the fluid are observed and related to the actual flowrate. Examples of the of additive energy meters are magnetic flowmeters, sonic meters, ultrasonic Doppler type, Coriolis and thermal mass meters.

Differential pressure flowmeters, or head-type flowmeters, are among the oldest and most common meters. They operate on Bernoulli's principle of energy conservation, where the sum of static energy (pressure head), kinetic energy (velocity head) and potential energy (elevation head) is constant for flow across a restriction in a pipe. The derived basic flow formula states that the flowrate is proportional to the square root of the differential pressure developed across the restriction. Examples of differential meters are orifice plates, venturi tubes, pitot tubes, flow nozzles and variable flow meters.

Mass flow meters measure volumetric rate of flow and include Coriolis and thermal types. Because the mass does not change, this type of meter is linear with no adjustment required for variations in liquid properties.

Electronic flowmeters that do not require intrusion into the flow stream.

Open channel flowmeters include weirs and flumes.

PRINCIPLES OF FLOWMETER OPERATION

Variable Area Flowmeters

The variable area flowmeter, sometimes called a rotometer, measures volumetric rate of flow. This type is mainly used for gases and is one of the oldest technologies available. It is constructed of a tapered tube, usually of glass or plastic, and a free moving float. Fluid moving through the meter causes a pressure drop across the float producing an upward force that causes the float to move up and down the tube. As this occurs, the cross sectional area between the tube walls and the float varies, giving the meter its name. The displacement of the float is proportional to the volume of fluid flowing through the device. The accuracy of this type of meter is 2 to 4 percent full scale.

Two types of meter are available: direct reading and correlated. The direct reading meter has a scale numbered in the flowrate and units desired and allows the actual flow to be read directly off the scale. The correlated meter is divided into a unitless scale that requires a separate data sheet to allow conversion of the scale into units of flow.

Another variable is the direction of flow of fluid through the meter; this can be through the tube from top to bottom or straight through the meter only at the bottom. In the top to bottom type, illustrated in Fig. 22.1, flow is directed from the bottom to the top. The float is displaced by the volume of the fluid. In the straight through type, illustrated in Fig. 22.2, the float moves up and down in proportion to the flow rate and the area of the tapered tube.

The variable area flowmeter is well suited for measuring liquid or gas flow for plants, laboratories and health care facilities and for purging of gas lines. The effect of pressure and temperature variations for gases will cause a deviation from the actual indicated flowrate depending on the set point of the meter. If accurate flow rate is an important consideration, this deviation should be obtained from the manufacturer.

Thermal Mass Flowmeters

A mass flowmeter measures volumetric rate of flow and is an often used gas measurement technology. A typical thermal mass flowmeter is illustrated in Fig. 22.3.

The inlet gas stream enters the meter chamber and is immediately split into two separate paths. Most of the gas will pass through the bypass tube but a fraction of the gas will enter the sensor tube, which is isolated from the main fluid flow path. The sensor tube contains two temperature coils that introduce heat into the sensor tube. When gas passes through the sensor tube, it carries heat from the upstream coil to the downstream coil. The difference in temperature creates a proportional resistance change in the sensor windings in the sensor tube. The resistance change created by the temperature difference is amplified and calibrated to give a digital readout of the actual flow. The accuracy of this type of meter is ± 1.5 to 2 percent fullscale.

Another type of mass flowmeter is a single probe device often used for lower flows that occur in facility and laboratory type operations. This single probe meter uses two elements in the one probe: a reference RTD and active RTD. As the gas flows past the sensor, the flow causes a difference in temperature between the two sensor parts that is measured and converted into flow indication.

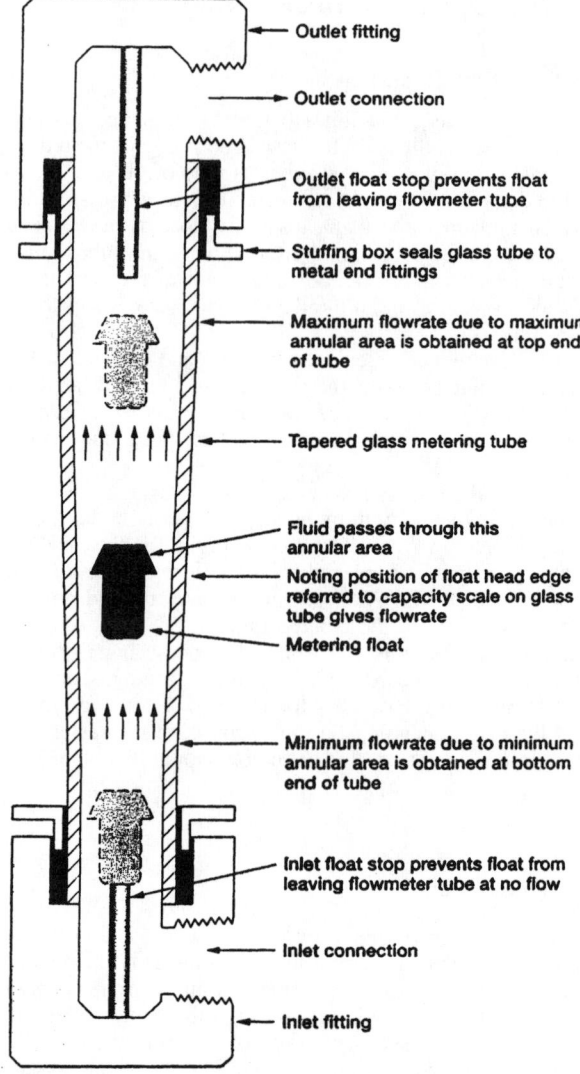

Outlet fitting

Outlet connection

Outlet float stop prevents float from leaving flowmeter tube

Stuffing box seals glass tube to metal end fittings

Maximum flowrate due to maximum annular area is obtained at top end of tube

Tapered glass metering tube

Fluid passes through this annular area

Noting position of float head edge referred to capacity scale on glass tube gives flowrate

Metering float

Minimum flowrate due to minimum annular area is obtained at bottom end of tube

Inlet float stop prevents float from leaving flowmeter tube at no flow

Inlet connection

Inlet fitting

FIGURE 22.1   Typical top to bottom flowmeter.

The greatest advantage of the mass flowmeter is the ability to lower any varia-tions of flow due to temperature and pressure changes to a point where they can be considered almost negligible. Typical differential values of flow for temperature are 0.10 percent full flow per degree Celsius and 0.02 percent per psi. Where the measure of costly gases is required, the mass flowmeter is a more appropriate choice because of its greater accuracy.

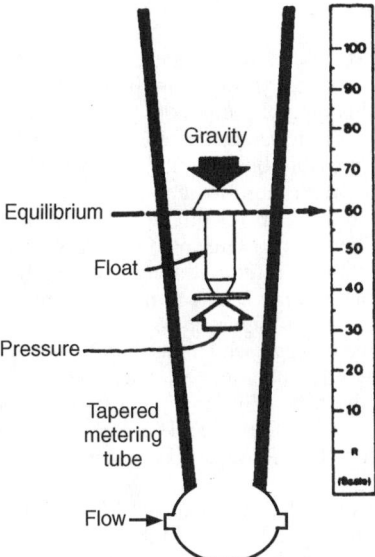

FIGURE 22.2 Typical straight through flowmeter.

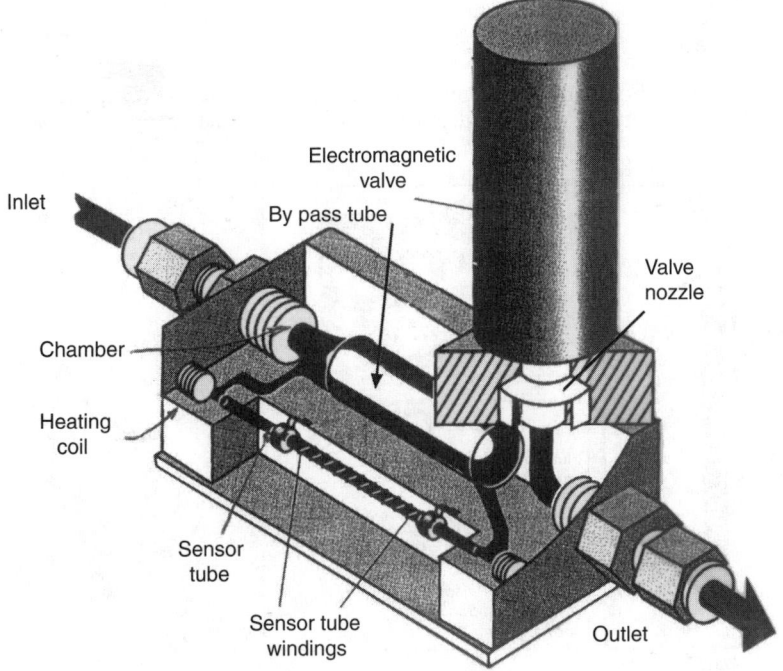

FIGURE 22.3 Typical mass flowmeter.

Coriolis Effect Flowmeters

A Coriolis effect meter is a velocity-type meter that directly measures mass rate of flow. It is unaffected by changes in liquid properties such as pressure, temperature, viscosity and density. This type of meter is used where accuracy, which approaches 0.05 percent of full flow, is required and where the previously mentioned parameters are useful. It is used to measure both liquids & gases where only the total volume of liquid is desired rather than the flowrate.

The meter operates on the Coriolis effect, which is an inertial force discovered by the 19th century mathematician Gustave-Gaspard Coriolis. The basic operation concerns one or more vibrating tubes through which the liquid to be measured flows. Tubes are manufactured in various forms, including straight line, "U", S and Z shaped and helix or coiled configurations.

Simply stated, the tube through which the fluid to be measured flows is vibrated. As the fluid flows through the tube, it accelerates and opposes the vibrating motion and imparts a twist to the tube. The amount of twist is proportional to the fluid mass flowrate flowing through the tube. The amount of twist is measured and a direct reading is produced. This flowmeter is best for clear fluids installed in piping 6″ and smaller.

One often used type of unit consists of a "U" shaped tube through which the liquid flows, illustrated in Fig. 22.4a. The tube is vibrated by an electro-magnetic device located at its bend, similar to that of a tuning fork. The oscillation occurs even when there is no flow. As the liquid flows through the tube, it is forced to take on the vertical movement of the vibrating tube. A straight line tube is illustrated in Fig 22.4b, and a helix arrangement is illustrated in Fig 22.4c.

Orifice Meter

An orifice is a differential-pressure type meter that measures flow rate. It consists of a flat plate with a specific size hole or opening and has no moving parts. In

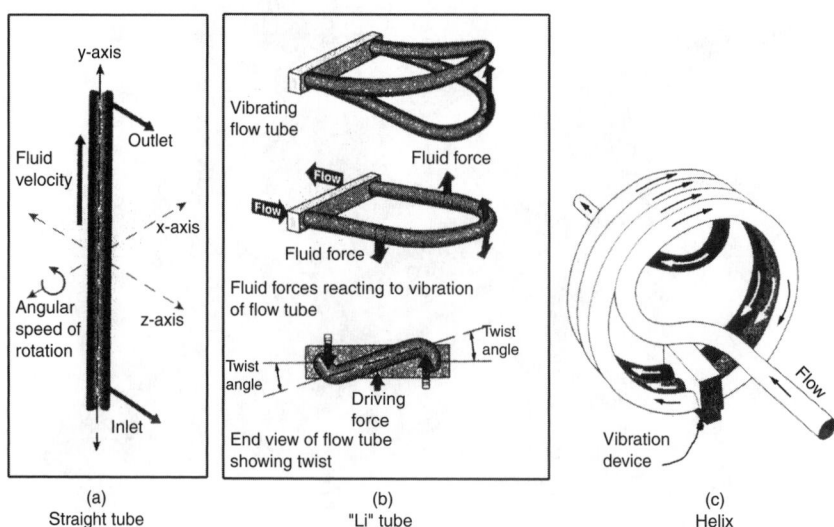

| (a) | (b) | (c) |
| Straight tube | "Li" tube | Helix |

FIGURE 22.4   Typical Coriolis flowmeter.

practice, the orifice plate is often installed between two flanges downstream from a section of straight and uniform run of pipe of approximately 20 pipe diameters. Straightening vanes installed in the downstream piping will shorten the length of straight pipe runs. A typical orifice meter installation is illustrated in Fig. 22.5.

The position of the hole in the plate can be concentric, eccentric, or segmental type. The cross section shape of the hole can be either square edged or conical. These holes are illustrated in Fig 22.6.

The basic operation requires a primary and secondary element. The primary element, or orifice, constricts the flow of liquid. This produces a difference of

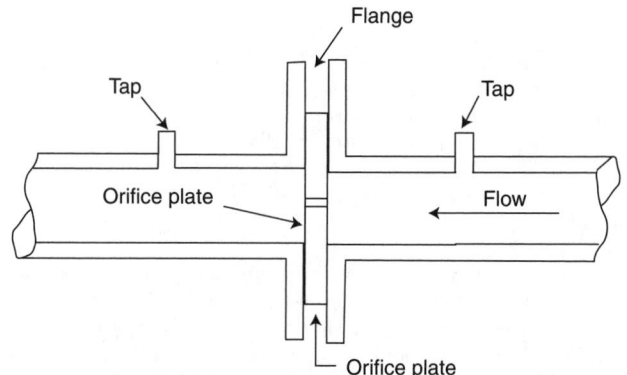

FIGURE 22.5 Typical orifice meter installation.

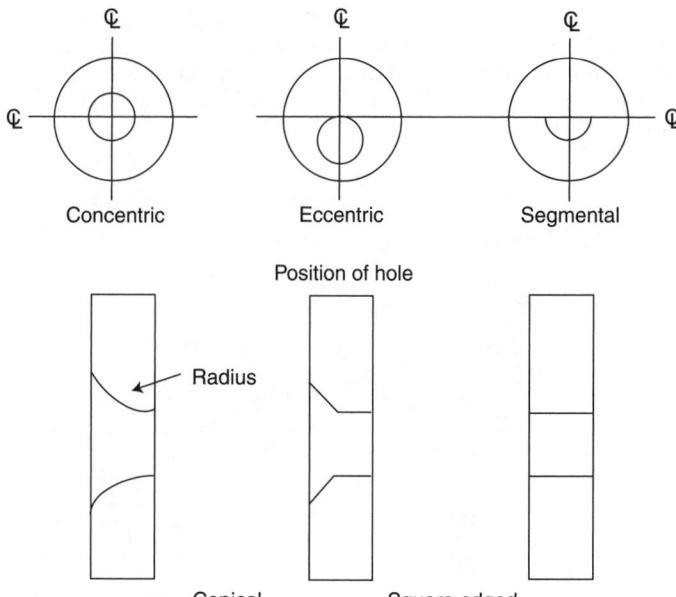

FIGURE 22.6 Hole shape.

pressure on both sides of the orifice plate. The secondary element measures the differential pressure and provides the signal or readout that is converted to the actual flowrate value. The instrumentation package is installed into taps on both sides of the plate that is used to measure the different pressures and provide the readout. The most common position of these taps are 2 pipe diameters upstream and 8 pipe diameters downstream of the flange containing the orifice plate.

The formula for flow rate is derived from the area of the pipe and the velocity of the liquid. A simplified formula for determining flow rate is:

$$W = 275 \times d^2 \times \sqrt{\frac{hp}{1 - b^4}} \tag{22.2}$$

where W = flow rate, lbs per hour
$\quad\ d$ = orifice diameter, inches
$\quad\ h$ = differential pressure, inches water
$\quad\ p$ = density of fluid, lbs per cu. ft
$\quad\ b$ = ratio of orifice diameter to pipe diameter

Venturi Tubes

A venturi tube meter, often referred to as a tube meter, is a differential-pressure-type meter that measures flow rate. The meter is a manufactured unit consisting of a tapered converging section, a throat and a tapered diverging section intended to be installed inline. It has no moving parts. The angle of the entrance section is usually between 20 to 25 degrees. The angle of the diverging section is usually between 5 to 15 degrees. This meter is used to measure large flows with a low pressure loss. A typical venturi meter is illustrated in Fig. 22.7.

The liquid to be measured enters the converging section and increases in velocity as it enters the throat. As this occurs, part of the static head of the liquid is converted into velocity head in the throat. This causes a pressure differential between the converging section and the throat. The difference in head is proportional to the flow. The instrument package converts the difference in head to flow rate. This is given in the following formula:

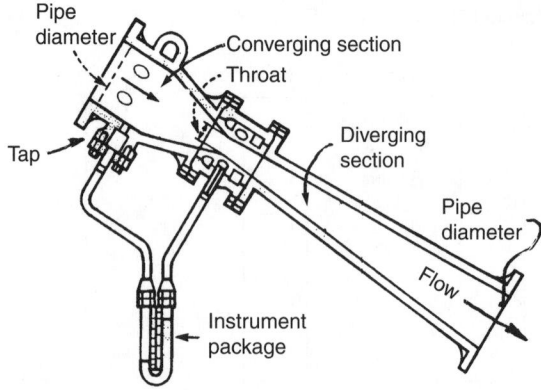

FIGURE 22.7 Typical venturi meter.

$$Q = C \times A \times k \sqrt{\Delta H} \tag{22.3}$$

where Q = volumetric flow rate, cfs
$\quad C$ = discharge coefficient (0.98 for this type meter)
$\quad A$ = area of pipe, square feet
$\quad H$ = differential head, ft
$\quad k$ = is given as:

$$k = \frac{2g}{1 - \dfrac{d2^4}{d1}} \tag{22.4}$$

where $d1$ = upstream pipe diameter, ft
$\quad d2$ = venturi throat diameter, ft
$\quad g$ = gravity constant (32 ft/sec/sec

Flow Tube Meter

A flow tube meter is similar to the venturi tube except for the lack of an entrance throat. The throat is tapered and the exit portion is smooth and elongated.

Pitot Tube Meter

A pitot tube meter is a differential-pressure-type meter that measures flow rate. It has no moving parts. The original multi-port averaging device was invented by Henri Pitot. This meter type is suitable for measurement of liquids, steam and gases.

The pitot tube is an assembly of two isolated tubes, or chambers, within a single probe. Multiple sensing holes are drilled into each chamber to sample both the high velocity at the center of the probe and the lower pressure at the outer chamber. The probe is positioned in the fluid stream so that the ports in one chamber are facing upstream and the others are facing downstream. The inner chamber is called the impact tube and the outer one is called the static tube.

The probe's obstruction to flow creates an impact pressure on the upstream-facing tube that is continuously sampled and averaged, and the downstream facing holes sample and average the static pressure. A typical pitot tube meter assembly is illustrated in Fig. 22.8. A schematic detail of the two often used probe arrangements, straight and curved, is illustrated in Fig. 22.9.

The pitot tube assembly is installed into the flow stream by means of a pipe tap or welded coupling on the pipe. In operation, the liquid to be measured flows past the pitot tube assembly, which senses two pressures simultaneously, impact and static. An instrument package senses the differences between the static and impact pressures and converts it into flow rate.

The basic equation for determination of flow is:

$$Q = A \, K \sqrt{\frac{2g \, DP}{DE}} \tag{22.5}$$

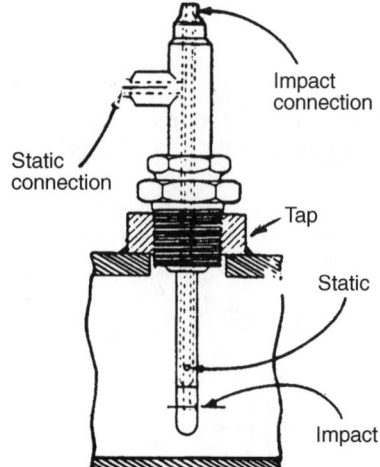

FIGURE 22.8 Typical pitot tube.

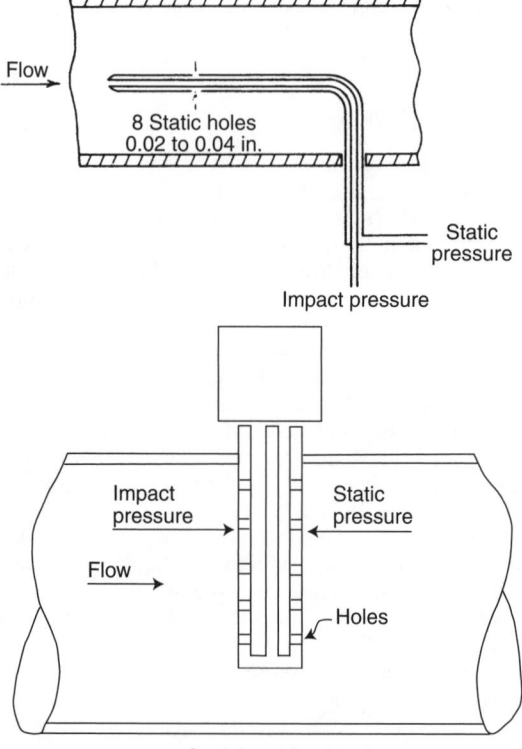

FIGURE 22.9 Pitot tube probe.

where Q = volumetric flow rate, cfs
$\quad K$ = flow coefficient, obtained from probe manufacturer
$\quad A$ = area of pipe, sq ft
$\quad G$ = gravity constant (32 ft/sec/sec)
$\quad DP$ = differential pressure, psi
$\quad D$ = liquid density, lbs per cu. ft

Elbow Taps

An elbow tap is a differential-pressure-measurement-type meter that measures flow rate. It has no moving parts. This method of measuring flow rate uses the difference in pressure exerted by the centrifugal force of the flowing liquid around the outside walls or inside and outside walls of a 90 degree elbow that forms a regular part of the piping distribution system. A typical elbow tap arrangement is illustrated in Fig. 22.10.

Pressure measurements are obtained by placing taps at the center (45°) position on both the inside and outside wall or on the outside wall only of the elbow. The differences in pressure can be measured by a pressure sensing instrument package and converted into flow rate.

Flow is calculated by the following formula:

$$W = 244 \sqrt{r\,h\,D^3\,p} \qquad (22.6)$$

where W = flow rate, lbs per hour
$\quad r$ = elbow radius, inches
$\quad h$ = differential pressure, inches water column
$\quad D$ = elbow diameter, inches
$\quad p$ = density of fluid being measured, lbs per ft

Flow Nozzle

A flow nozzle is a differential-pressure-type meter that measures flow rate. It has no moving parts. It is similar to an orifice meter except it has a flow nozzle instead of an orifice plate with a large tapered throat. This enables the flow nozzle meter to measure liquids with moderate suspended solids and has the same pressure drop with 60 percent higher flow rates than orifice plates. A typical flow-nozzle-type meter is illustrated in Fig. 22.11.

The basic operation requires a primary and secondary element. The primary element, or flow nozzle, constricts the flow of liquid. This produces a difference in

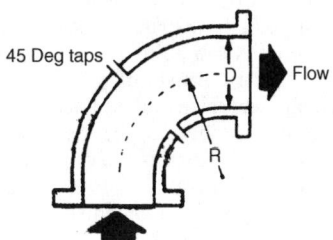

FIGURE 22.10 Typical elbow tap.

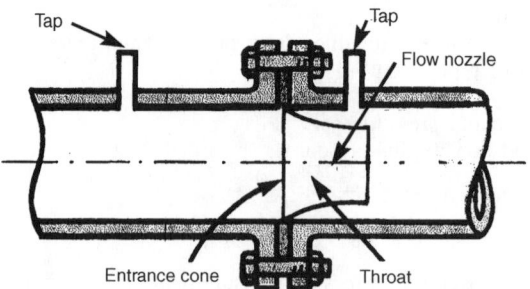

FIGURE 22.11   Typical flow nozzle.

pressure on both sides of the nozzle. The secondary element measures the differential pressure and provides the signal or readout that is converted to the actual flowrate value. The instrumentation package that is used to measure the different pressures and provide the readout is installed into taps on both sides of the nozzle.

The flow nozzle will measure approximately 60 percent greater flow with the same pressure drop through the meter as an orifice meter.

Flow is calculated by using eq. 22.2, the same as that for orifice meters.

Rotary Vane

The rotary vane meter is a positive displacement meter and is available in several designs. The basic unit consists of an equally divided rotating impeller that contains two or more compartments mounted inside the meter's housing. The impeller is in continuous contact with the casing. A fixed volume of liquid is swept into the meter's outlet from each compartment as the impeller rotates. The revolutions of the impeller are counted and registered as volumetric units.

Helix Flowmeter

The helix flowmeter is a positive-displacement-type meter that uses helical rotors geared together to displace liquid axially from one end of the chamber to the other.

Disk Meter

The disk meter, with variations of the basic design known as a notating disk and oscillating piston meter, is a positive-displacement-type meter. Many sizes and capacities are available, and the units can be made from a wide selection of materials.

The notating meter mechanism consists of a movable disk mounted on a concentric sphere located in a spherical, side walled chamber. The pressure of the liquid passing through the measuring chamber causes the disk to rock in a circular path without rotating on its own axis. The disk is the only moving part in the measuring chamber. A typical disk meter is illustrated in Fig. 22.12.

A pin extending perpendicularly from the disk is connected to a mechanical counter that monitors the disk's rocking motion. Each cycle is proportional to a specific quantity of flow.

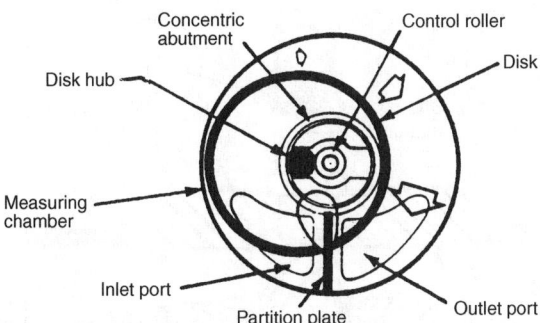

FIGURE 22.12 Typical disk meter.

The oscillating piston meter operates by means of a magnetic drive, so that the liquid does not come in contact with the meter body.

Turbine Meters

The turbine meter (often called a current meter) is a velocity-type meter that uses a multiple bladed rotor (similar to a propeller on a boat) mounted in the flow stream perpendicular to the liquid flow. The rotor has blades (vanes) that rotate as the fluid passes through the vanes. The speed of the rotor is a direct function of the flow rate and can be counted in a number of ways. It can also be used to determine volume. A typical turbine meter is illustrated in Fig. 22.13.

A variation of the turbine meter is called a paddlewheel meter, which uses a multiple flat bladed rotor installed parallel to the liquid flow that rotates in the flow stream. The rotation speed varies with velocity of the fluid.

The device can be used in both open channels and closed piping.

Vortex Flowmeters

A vortex meter, also known as a vortex shedding meter, is a velocity-type meter with no moving parts. It makes use of a natural phenomenon that occurs when a liquid flows around an element, known as a bluff body, that is suspended in the flow stream. Eddies, or vortices, are created (shed) and flow downstream of the bluff body. The frequency of the vortex shedding is directly proportional to the velocity of the liquid flowing through the meter. Its use for slurries or high viscosity liquids is not recommended. A vortex meter is illustrated in Fig. 22.14.

The three components of a vortex meter are the bluff body strut mounted in the flow stream, a sensor to detect the presence of vortices and to generate an electrical impulse, and a signal amplification package, the output of which is proportional to the flow rate. Straightening vanes may be required.

Swirl Meter

A swirl meter is a velocity-type meter that is a variation of the vortex meter. The basic operating principle is similar, except that the fluid entering the meter is forced

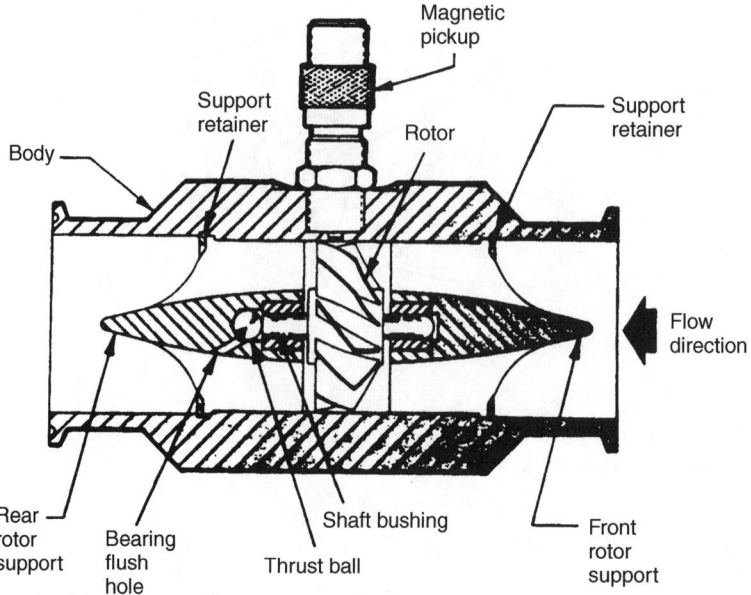

FIGURE 22.13 Typical turbine meter.

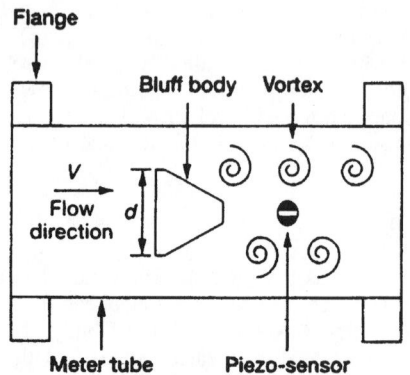

FIGURE 22.14 Typical vortex meter.

into a continuous swirl pattern instead of periodically producing vortices. Due to the meter geometry, straight run pipe requirements upstream of the meter are minimal. A typical swirlmeter is illustrated in Fig. 22.15

The swirl pattern is then conducted into a helical path, turning the swirl pattern on its side, similar to that of a tornado being converted into a screw type thread pattern. This creates a precession-like motion that is linear in proportion to the flow rate. The frequency of the precession is detected by a sensor and instrument package and converted into flow rate.

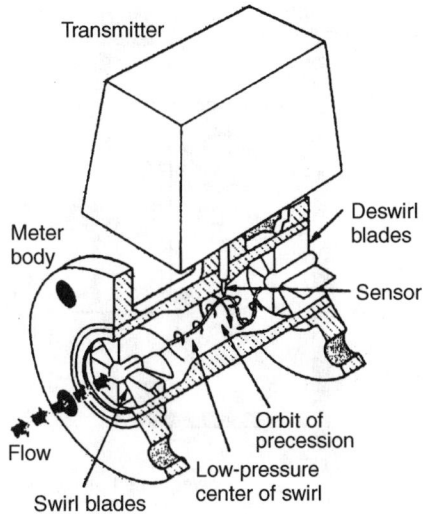

FIGURE 22.15   Typical swirlmeter.

Target Meter

The target meter is a volumetric-type meter. It senses and measures forces caused by a flowing liquid that impact on a drag disk, or target, suspended in the flow stream. A typical target meter is illustrated in Fig. 22.16.

This meter can measure liquids, gases, steam and slurries and is most useful in metering dirty or corrosive liquids. Accuracy is generally between $\frac{1}{2}$ and 1 percent full flow. It has a wide range of flow rates, and the disk can be optimized to suit any range of application. There is a pressure loss due to the suspended target. A use for a simple target meter is to determine if a fluid is moving at all, or if it is moving slowly or with high velocity with no requirement for quantity.

The target is typically a circular disk mounted on a hinged swinging shaft installed concentrically in the pipe perpendicular to the direction of flow. It is supported on a circular shaft that extends from the target through a shaft seal to a force-measuring secondary element, usually a strain gauge, and instrument package. Volumetric flowrate is inferred from fluid velocity and the known open area of the meter.

Oval Gear Flowmeter

The oval gear meter is a displacement type meter. The basic operating principle is the use of oval shaped gear toothed rotors that rotate in a chamber of specially designed geometry. As the rotors turn, they sweep out and trap very precise amounts of liquid with none of the liquid passing through the rotors. As the rotors turn, each rotation is a measured amount of fluid and is counted by an instrument package. The pulsation of flow is minimal. It operates best when there is little backpressure.

This meter is well suited to measure higher viscosity liquids such as lubricating oil, diesel fuel, and syrups, but a higher pressure is required to operate the meter

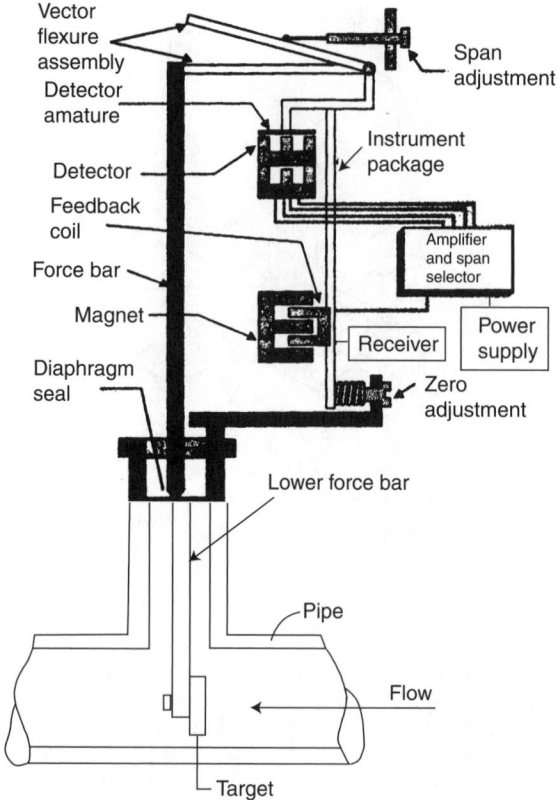

FIGURE 22.16   Typical target meter.

the higher the viscosity. No straight runs are required for accuracy. It is not recommended for water because of slippage between gears and is unsuitable for gases.

ELECTRONIC (ELECTROMAGNETIC) FLOWMETERS

Electronic meters are divided into two general categories: magnetic and Doppler.

Magnetic Flowmeters

A magnetic flowmeter is essentially a single package with wire coils mounted on or outside a suitably insulated piece of pipe (flowtube) and the electronics necessary for flow measurement. It operates on Faraday's Law of electromagnetic induction where a voltage will be induced when a conductor (fluid) moves through a magnetic field. It has no moving parts or obstructions to flow, so the pressure drop through the meter is equivalent to a straight piece of pipe. It is well suited to measure corrosive fluids and slurries and is unaffected by changes in temperature, pressure, fluid density or viscosity. It is not suited for measurement for non-conductive liquids such as hydrocarbons. To function correctly the pipe must be full. The wetted transducer mounting is preferred if the fluid is low density, such as a gas. The clamp-on is preferred for liquids. A hybrid device is also available to combine the advantages of both types. The operation of a typical magnetic flowmeter is illustrated in Fig 22.17.

The magnetic flowmeter consists of three primary components: the flowtube, the flow transmitter, and an instrument package that measures the induced voltage and converts the voltage readings to flowrates.

These meters measure the flowrate of any electrically conductive liquid using Faraday's Law of electromagnetic induction. Faraday's Law is expressed as:

$$E = B \times V \times D \tag{22.7}$$

where E = generated voltage
  B = magnetic flux density
  V = fluid velocity
  D = pipe diameter or distance between electrodes

Since B and D are constant, E is then proportional to V.

Ultrasonic Flowmeters

Ultrasonic meters consist of two main types: Doppler and transit time. The Doppler flowmeter is a velocity type flowmeter. An ultrasound beam is generated at a common frequency of 500 kHz at an oblique angle to the flow. The flowmeters use what is technically called the contrapropagation method of ultrasonic flow measurement, which is sound pulse reflection rates. Ultrasound is generated by means

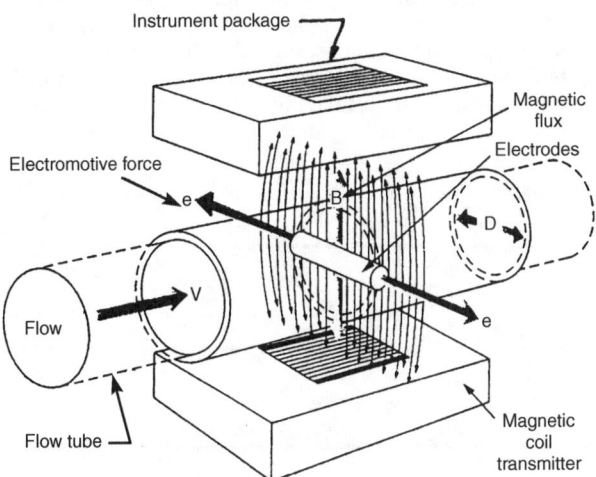

FIGURE 22.17 Typical magnetic flowmeter.

of piezoelectric transducers installed either inside the pipe (wetted) or outside (clamp-on). Transit times are measured in the direction of flow, and later (or sometimes simultaneously) against the direction of flow. From the two measured transit times of the beams, the velocity can be calculated.

The generated beam is reflected from suspended particles or bubbles in the fluid stream being measured. The reflected beam shifts frequency and mixes with the original transmitted frequency. This reflected and transmitted beam is then collected and measured by an instrument package (receiving transducer) that converts the frequency shift into fluid flow rate. A simplified illustration of Doppler operation is given in Fig. 22. 18.

Doppler flowmeters are not generally used for clean fluids. It is most useful for any liquid containing suspended droplets such as slurries, sludges, emulsions, dispersions and pulps.

Identifying suitable fluids for Doppler meters is a complex combination of four basic criteria. How well these criteria are met determines the suitability and accuracy of using a Doppler flowmeter. These criteria are:

1. The scattering material must have a sonic impedance different from that of the fluid that is being metered.
2. There must be some particles large enough to cause longitudinal scattering.
3. For any given pipe size the longitudinal scattering must have sufficient energy to overcome the energy wasted (Rayleigh scattering) that is caused by smaller particles.
4. The scattering velocity must travel at the same velocity as the fluid for good accuracy.

The Transit Time Flowmeter

The transit time flowmeter is a velocity type flowmeter and is a variation of the Doppler flowmeter. The difference is that the fluid must be relatively free of entrained gas or solids to minimize or eliminate signal scattering or absorption.

The meter operates by having ultrasound signal transducers mounted on each side of the pipe. This configuration is such that the ultrasound waves traveling between the transducers are at a 45 degree angle to the direction of liquid flow. A time differential relationship proportional to the flow can be obtained by transmit-

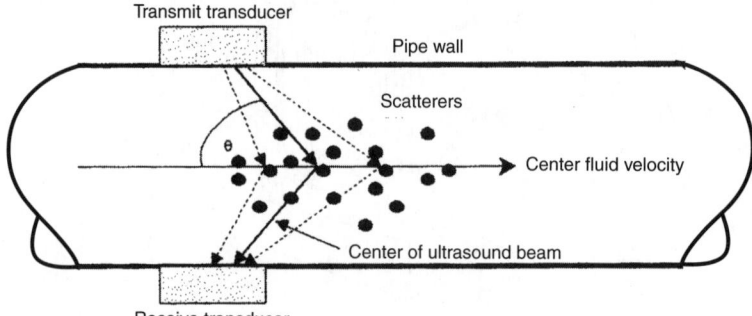

FIGURE 22.18 Simplified Doppler flowmeter operation.

ting the signal alternately in both directions. When flow begins in either direction, the pulse traveling in the same direction as flow reaches its destination faster than the pulse traveling in the opposite direction. The difference in transit time across the pipe is proportional to the flow velocity.

FLOWMETER SELECTION

There are some primary factors that should be considered prior to the selection of any particular method of flow metering. Table 22.1 is an inclusive list, all of which may not apply of any specific installation. It is intended for review in order to provide a checklist.

Table 22.2 is intended to present a general selection guide giving primary characteristics of the flowmeters discussed. Consult the manufacturers for a complete technical discussion of parameters not mentioned and for any unusual conditions.

To convert fluid velocity into flowrate, use the following formula:

$$\text{GPM} = \frac{\text{ID}^2 \text{ (in feet)} \times \text{velocity (fps)}}{0.408} \tag{22.8}$$

TABLE 22.1   Factors Influencing Flowmeter Choice

Performance factors	Viscosity
Accuracy	Lubricity
Repeatability	Chemical properties
Linearity	Surface tension
Rangeability (turndown)	Compressibility
Pressure drop	Abrasiveness
Output signal characteristics	Pressure or other phrases
Response time	Presence of other components
Purpose of measurement	
Suspended particles	**Environmental factors**
	Ambient temperature effects
Installation factors	Humidity effects
Available space	Safety factors
Flow direction	Pressure effects
Upstream and downstream pipe work	Electrical interference
Line size	
Location for servicing	**Economic factors**
Effects of local vibration	Purchase price
Location of valves	Installation costs
Electrical connections	Operation costs
Provision of accessories	Maintenance costs
Hazardous atmosphere	Calibration costs
Effect of pulsations/unsteady flow	Meter life
Ease of installation	Spares cost and availability
	Pumping power and head loss
Fluid property factors	Technical optimization
Liquid or gas	
Temperature and pressure	
Density	
Specific gravity	

TABLE 22.2 Flowmeter Selection Guide

Flowmeter element	Recommended service	Turndown (rangeability)[1]	Pressure loss	Typical accuracy, percent	Required upstream pipe, diameters	Viscosity effect	Relative cost
Orifice	Clean, dirty liquids; some slurries	4 to 1	Medium	±2 to ±4 of full scale[2]	10 to 30	High	Low
Venturi tube	Clean, dirty, and viscous liquids; some slurries	4 to 1	Low	±1 of full scale	5 to 20	High	Medium
Flow nozzle	Clean, and dirty liquids	4 to 1	Medium	±1 to ±2 of full scale	10 to 30	High	Medium
Pitot tube	Clean liquid	3 to 1	Very low	±3 to ±5 of full scale	20 to 30	Low	Low
Elbow meter	Clean, dirty liquids; some slurries	3 to 1	Very low	±5 to ±10 of full scale	30	Low	Low
Target meter	Clean, dirty, viscous liquids; some slurries	10 to 1	Medium	±1 to ±5 of full scale	10 to 30	Medium	Medium
Variable meter	Clean, dirty, viscous liquids	10 to 1	Medium	±1 to ±10 of full scale	None	Medium	Low
Positive displacement	Clean, viscous liquids	10 to 1	High	±5 of rate[3]	None	High	Medium
Turbine	Clean, viscous liquids	20 to 1	High	±0.25 of rate	5 to 10	High	High
Vortex	Clean, dirty liquids	10 to 1	Medium	±1 of rate	10 to 20	Medium	High
Electromagnetic	Clean, dirty viscous conuctive liquids and slurries	40 to 1	None	±0.5 of rate	5	None	High
Doppler	Dirty, viscous liquids and slurries	10 to 1	None	±5 of full scale	5 to 30	None	High
Time-of-travel	Clean, viscous liquids	20 to 1	None	±1 to ±5 of full scale	5 to 30	None	High
Coriolis	Clean, dirty, viscous liquids; some slurries	10 to 1	Low	±0.4 of rate	None	None	High
Thermal	Clean, dirty, viscous liquids; some slurries	10 to 1	Low	±1 of full scale	None	None	High

[1] For given transmitter span setting
[2] Percent of the flowmeter's full range
[3] Percent of liquid flow rate

LEVEL MEASUREMENT

GENERAL

This subsection will discuss the measuring of liquid levels (product) in atmospheric and pressurized tanks to allow an operator to The evolution of level measurement has advanced from the manual methods of using a stick or rope that exposed an operator to potentially hazardous conditions to multiple technologies that are very accurate and repeatable.

There are two general categories of level measurement devices: continuous system and point measurement. The continuous system indicates the level of product in a vessel over a specified range of measurement. It is the most flexible because the sensing element is fixed and any number of control points can be adjusted at will. It can also be provided with a method for recording the levels for inventory management or to establish demand. In general, the point measurement indicates if product is present at those points, or not. The point method is less costly and can be more accurate at control points. Control points can be provided to indicate high level, low level, pump control and spill prevention.

LEVEL TECHNOLOGIES

Codes and Standards

Some of the federal regulations regarding fugitive emissions and spills for level technologies are summarized in Table 22.3.

Float-based Devices

Float-based devices are a continuous system and the most basic means of measuring liquid level. It uses a float that remains on top of the liquid level in an atmospheric vessel. The position of the float or floats are coupled to switches that indicate the level at any point. The float can be coupled to a switch by mechanical, magnetic or a cable suspension link. A multiple point float is illustrated in Fig. 22.19a and a single point float in Fig. 22.19b.

The float switches can also be used to start and stop pumps and to perform alarm functions with the proper electronics and alarm signals.

Radio Frequency Admittance and Capacitance

A capacitance level gauge and the radio frequency admittance gauge are similar technologies. Both are continuous and point measuring systems depending on the instrument package. The two capacitance concepts of measurement both employ the same technology with enhancements for the different types. The basic system

TABLE 22.3 Regulations Covering Fugitive Emissions and Spills

Hazardous Organic National Emisson Standard for Hazardous Air Pollutants (NESHAP) [HON]	Requires regular monitoring of process connections, such as flanges, fittings, valves, etc., for fugitive emissions in processes containing hazardous volatile organic compounds (VOCs)
Spill Prevention Control and Countermeasure (SPCC) Regulation	Requires registering of tanks, reporting of spills, and submission of a downstream notification plan (for which overfill protection is critical)
Oil Pollution Prevention Act of 1990	Requires companies that transfer oil over navigable waters or areas with runoff accessible to navigable waters to have a plan to respond to "worst case scenarios"
Superfund Amendments and Reauthorization Act (SARA) Title III (also known as the Emergency Planning and Community Right-to-Know Act)	Requires companies to file annual reports detailing spills, leaks, emissions, and other releases of hazardous materials, as well as activities designed to reduce releases, such as overfill protection, automatic shutoffvalves, new procedures, etc.
American Petroleum Institute (API) Recommended Practice 2350	Recommends a high level alarm independent of any gauging system for spill prevention; a test on the spill prevention system should simulate an actual high level condition but should not require filling the vessel above its normal fill level
National Fire Protection Association (NFPA) NFPA 30, Section 10, Preventing Overfilling of Tanks	Requires spill prevention for flammable and combustible liquids; requirements include frequent visual observations by plant personnel, high level devices with audible alarms independent of tank gauging equipment, and the ability to test these systems

operates by means of a long cylindrical probe immersed in a conductive product that contains no moving parts.

The radio frequency admittance probe is two plates of a capacitor supplied with a radio frequency (rf) signal, with the dielectric between the two plates being the product. As the impedance between the plates changes with a change in liquid level, an instrument package converts the changing signal into liquid level. An rf admittance gauge is illustrated in Fig. 22.20. This gauge can handle a wide range of temperature and pressure differences.

The plain capacitance level gauge is similar except that there is no rf generator but rather a capacitance measurement instrument. The capacitance at the probe is measured and sent to an instrument package for conversion into level indication.

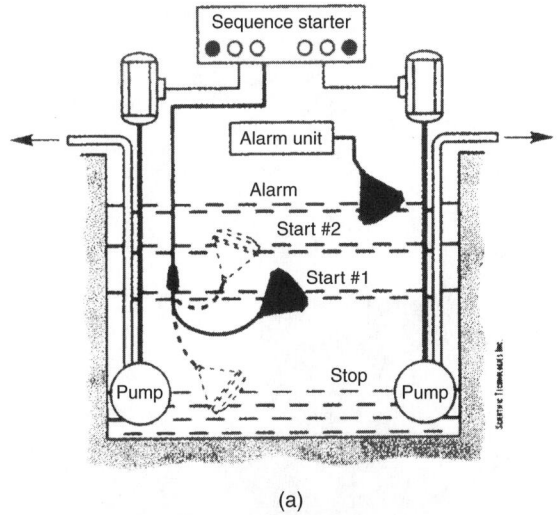

(a)

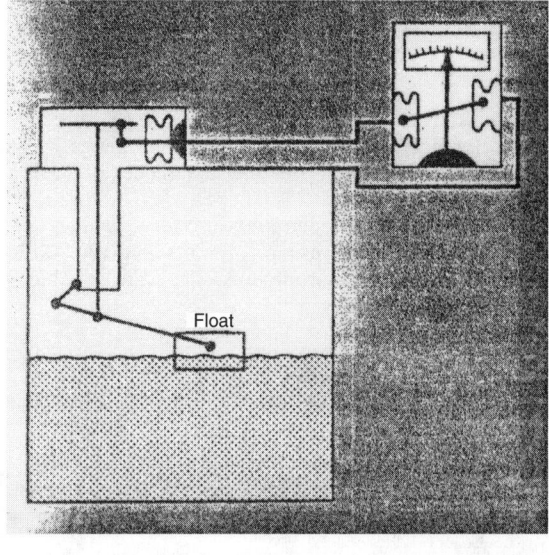

(b)

FIGURE 22.19 (*a*) Multiple point float. (*b*) Single point float.

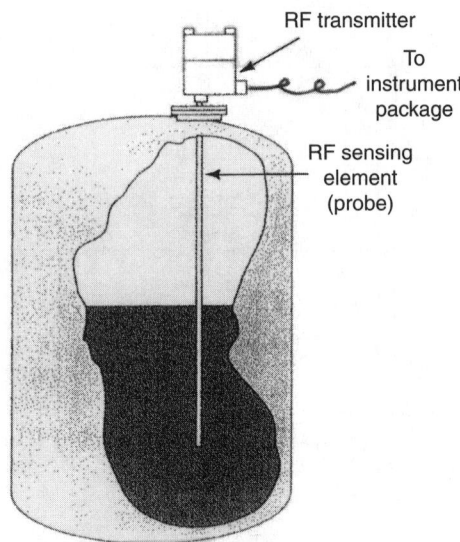

FIGURE 22.20 RF admittance gauge.

For non-conductive product, the vessel wall or an auxiliary reference plate provides the other capacitor reference plate. A capacitance level gauge is illustrated in Fig. 22.21. Wide ranging temperature and pressure differences do have an effect on the installation.

These systems require only one penetration into the vessel and are effective with high temperature liquids and pressures up to 3,000 psig (20,700 kPa). They also have an advantage in hazardous environments. They are suitable for slurries and

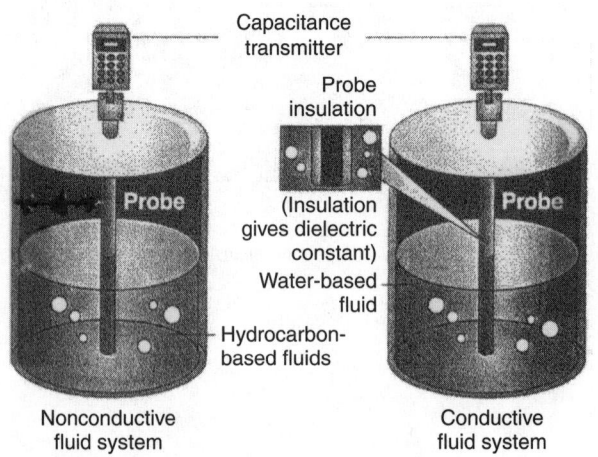

FIGURE 22.21 Capacitance gage.

where there may be above average turbulence. Another advantage is the probe can be set up to measure the level of different interface levels of products with different impedances.

Disadvantages are that the probe is sensitive to different product dielectric properties. Also, changes to vapor space content above the product also effect output so that the instrument package must be readjusted.

Pressure Gauge

Level measurements by pressure gauges can be accomplished by means of either single pressure or differential pressure (DP) measurements. This type of gauge is the most popular method of tank measurement. It works by means of measuring the pressure from product in a tank or vessel applied to a sensing element and transmitter (sensor) installed outside the tank in a housing directly connected at the bottom of the product tank. It does not measure the level directly, but only the head (height) or pressure exerted by the product. The head multiplied by the product density yields the level measurement. To measure levels in an open or vented tank, a single sensor/pressure transmitter is used. When installed in a pressurized tank, a second sensor/pressure transmitter is necessary to provide a differential pressure referenced to the pressure above the liquid level. This is illustrated in Fig. 22.22.

DP works best with clean liquids. Product that may coat the sensor is not recommended because of maintenance problems.

Bubblers

Bubblers operate on the principle of forcing air through a tube immersed in the liquid. The basic system consists of air pipe immersed in the product to the bottom, a regulated air supply, and differential pressure transmitter. As air bubbles escape from the open bottom of the tube, the air pressure in the tube corresponds to the hydraulic head inside the tank and varies with any change in level. Air is introduced

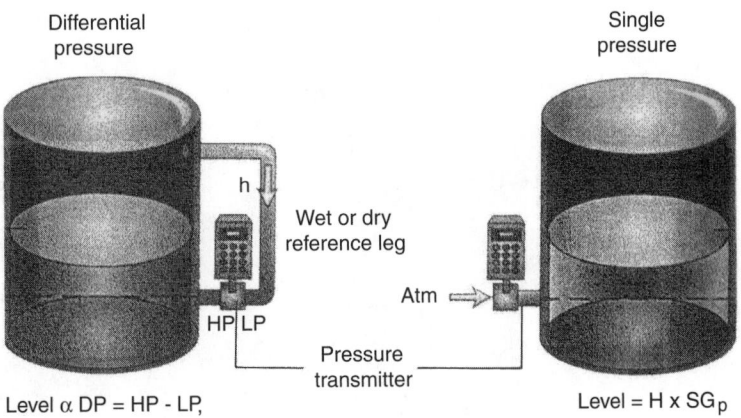

FIGURE 22.22 Pressure gauges.

into the system from a facility air supply that is regulated. An instrument package measures the back pressure of that air, and an instrument package converts the changing signal into liquid level. A typical bubbler system is illustrated in Fig. 22.23.

Accuracy is about 10 percent of full scale and depends on a stable and regulated air supply. This is less precise that many other systems.

Bubbler systems are relatively simple and inexpensive, easy to install and can be placed inside a full tank. In general, they are limited to atmospheric tanks. They are well suited for underground tank installations.

Disadvantages include difficulty in calibration and inaccuracies due to changing product density. Product with suspended solids has a tendency to clog the tube.

Ultrasonic Systems

Ultrasonic devices operate by having a transmitter produce a continuous high frequency, ultrasonic sound pulse. The pulse is directed toward the liquid level surface where it is reflected back as an echo. The time lapse for the echo to return and be received by a transducer is converted by an instrument package into a measured level of liquid. This system is capable of both continuous or point measurements. A typical ultrasound installation is illustrated in Fig. 22.24.

Ultrasound is non-invasive, contains no moving parts and measurement is unaffected by changes in density, pH and dielectric properties of the liquid. It is suitable for liquids and slurries. Acceptable temperature for installation is limited to about 300°F (152°C) and pressure to 50 psig (340 kPa). Only one penetration is necessary, and the penetration is above the level of product. The transmitter and transducer do not come into contact with product. It is not susceptible to fouling where petroleum products or other hazardous material is present.

It should not be considered for high temperature, stratified vapor, steam, excessive foam, as turbulence and dust may cause the signal to be distorted.

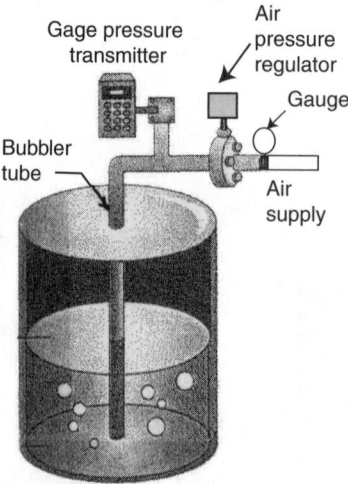

FIGURE 22.23 Bubbler level system.

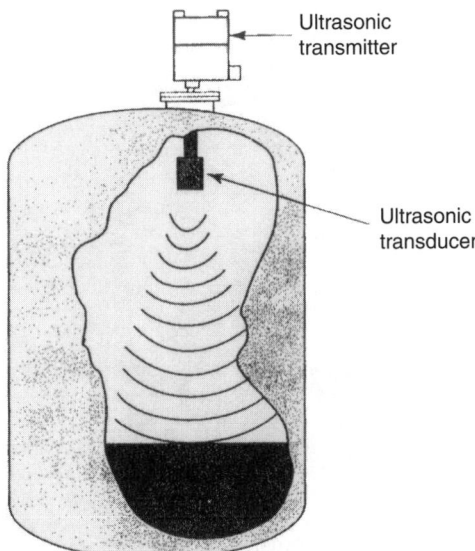

FIGURE 22.24   Ultrasonic measurement.

Radar Systems

Radar is similar to the ultrasonic systems. It produces and transmits high-frequency electromagnetic waves to the surface of the liquid and reflects the waves back to a receiver. The return time is proportional to the level of the liquid.

Radar is non-invasive, contains no moving parts and measurement is unaffected by changes in density, pH and dielectric properties of the liquid. It is suitable for liquids and slurries. It is not affected by any interference caused by any gas in the vapor space. Turbulence presents no problems for the signal.

Radar cannot measure interface levels.

Nuclear Measurement

Nuclear devices use the absorption or attenuation of gamma rays released from radioisotopes as they pass through the tank and product from one side to the other. Gamma radiation is a high energy, short wavelength energy, similar to that of x-rays, that have great penetrating power. Different isotopes are used based on the amount of penetration required to pass through the tank and product. A gamma ray source is mounted externally on one side of the tank and a sensitive detector is placed externally on the other. The strength of the signal indicates the level of the liquid. The percentage of gamma transmission decreases as the level increases. A typical nuclear measurement system is illustrated in Fig. 22.25.

Nuclear is non-invasive and can be used as a point, continuous or interface indicator. It is not affected by hazardous and corrosive material and is considered more reliable than electronic methods. Measurements are totally unaffected by temperature, pressure, or product corrosiveness. Readings may be influenced by

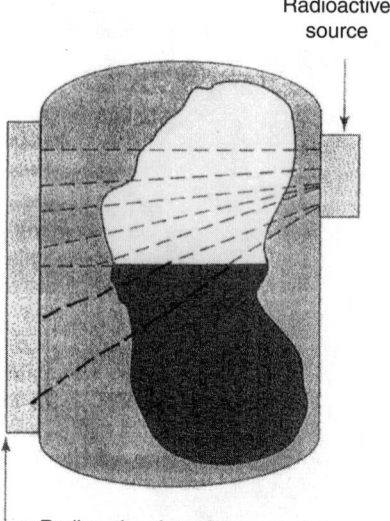

Radioactive
source

Radioactive detectior or receiver

FIGURE 22.25 Nuclear measurement system.

changes in product specific gravity. Nuclear technology is often used when other types of measurement fail.

Nuclear measurement has a high initial cost, often two to four times that of other methods. Spent radiation sources are difficult and expensive to replace and dispose of. Licenses, permits, approvals and inspections are required during the useful life of the system. The system requires continuous monitoring and produces discomfort for the maintenance personnel that is difficult to determine.

Displacement System

A displacement system is most often a continuous measurement system. It uses a sealed, heavy body immersed in the product, as compared to floats that are intended to stay on the liquid's surface. It is based on Archimedes' principle that a submerged body is buoyed up by a force equal to that of the weight of liquid being displaced. Displacement forces are expressed as:

$$F = V \times D \tag{22.9}$$

where F = displacement force
V = volume of submerged body
D = density of fluid

The operating principle of a submerged system is buoyancy, where the buoyant force of the liquid to be measured is exerted on a sealed body (displacer). As the fluid rises in the vessel, the displacer weight is buoyed up in proportion to the rise in liquid level. This weight difference is measured and converted to level measurement by an instrument package. This is illustrated in Fig. 22.26. There are several variations in use, including the use of a displacing chamber installed on the outside of the vessel similar to that of a sight glass.

The displacement system is primarily used in pressurized or closed tanks to measure slurries, liquid-vapor interfaces and liquid-liquid differences where the liquids have different specific gravities. Installed in a stilling well, it is suitable for turbulent situations.

It is not recommended where product could deposit material on the displacer. This can be overcome with coatings.

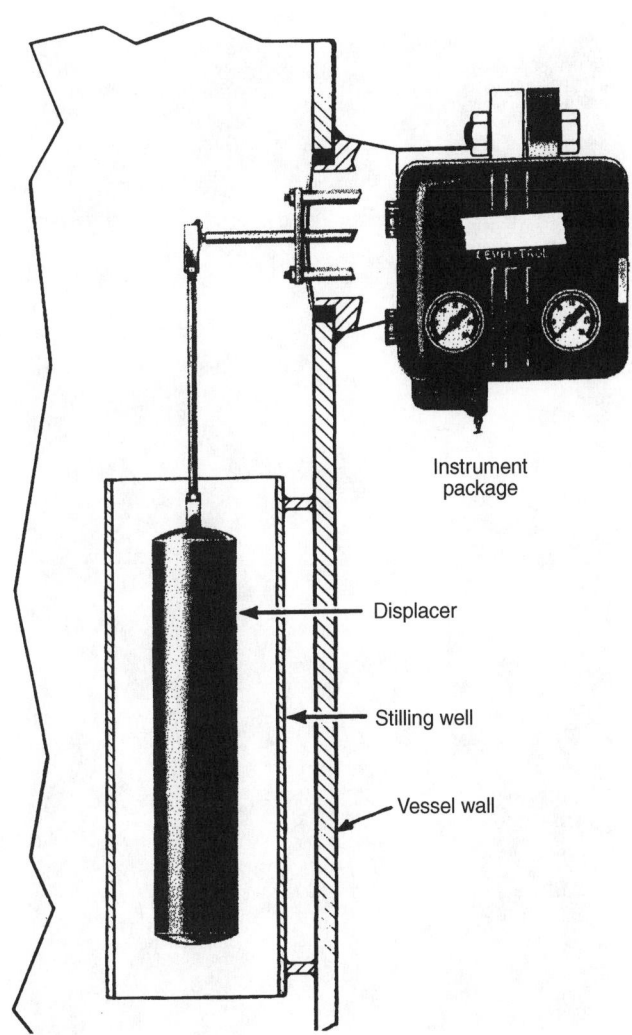

Instrument package

Displacer

Stilling well

Vessel wall

FIGURE 22.26   Displacement system.

Magnetostrictive Liquid Level Measurement

The magnetostrictive level theory is based on the principle of magnetostriction, which is the ability of some metals to expand or contract in the presence of a magnetic field. When a wire, (the magnetostrictive "waveguide") is installed inside a guidetube that is surrounded by an axial magnetic field becomes electromagnetized, it twists. It is this amount of twisting that allows a measure of the level to take place. This is illustrated in Fig. 22.27.

The system consists of a wire inside a guidetube, a float (or floats) and a permanent magnet attached to the float. The float is free to move up and down with the level of product in a tank. When an electrical interrogation pulse is sent down

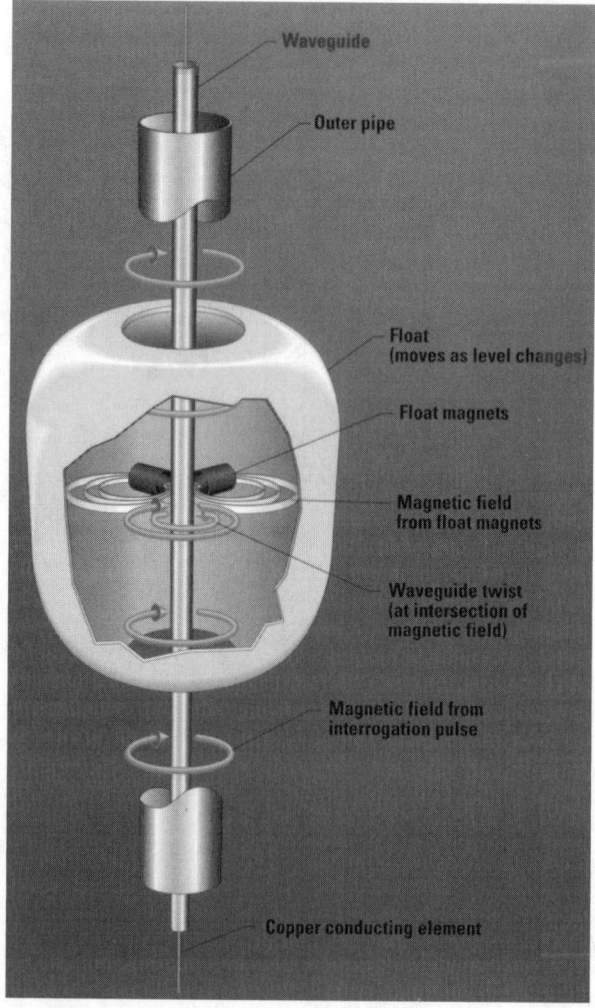

FIGURE 22.27 Magnetostrictive level device.

the wire, this pulse reaches the magnetic field produced by the magnet in the float and produces a twist in the wire. When the pulse travels back to a pickup device, the time of travel is converted by an instrument package into a level measurement.

The main advantage of this system is extreme accuracy. For this reason it is used mostly in liquid fuel tanks to measure product levels. Multiple floats can be installed where there are two liquids of different specific gravity to give separate indications for each, such as water in an oil tank. It has a wide temperature range, a low sensitivity to temperature changes and is not sensitive to pressure, product conductivity, foam, vapor and other atmospheric conditions.

LEVEL MEASUREMENT SELECTION

Table 22.4 is intended to present a general selection guide giving primary characteristics of the level measurement systems discussed. Consult the manufacturers for complete technical details of parameters not mentioned and for any unusual conditions.

TABLE 22.4 Selection Guide for Level-measurement Technologies

Technology	Advantages	Disadvantages	Successful applications	Problem applications
Pressure, D/Ps, bubblers	• Familiarity • Reasonable cost • Simple design • Easy to install	• Affected by density changes • Multiple fugitive-emission points	• Clean liqids with stable densities • Some light slurry use	• Heavy slurry • Coating-buildup applications
Displacer	• Limited motion • Few moving parts	• Affected by density changes • Most require bottom of tank penetration	• Clean liquids with stable densities • Some light slurry use • Interface use	• Small density changes • interfaces • Coating buildup
Float	• Unlimited tank height • Can achieve high accuracy • Low cost if not remote reading	• Moving parts exposed to process • Limited pressure rating • High maintenance	• Most liquids • Light slurries	• Heavy slurries • Granular solids • Interfaces
Ultrasonic	• High accuracy • Non-contacting • Easy calibration	• Position sensitive • Limited temperature and pressure ratings	• Total level of liquids, slurries, some granular solids	• Interfaces measurements • Vapors can cause calibration shifts • Heavy agitation • Vacuum service
RF Capacitance	• Good accuracy • No moving parts • Suitable for wide pressures, temperature range	• Contact measurement • If material properties change, calibration can be affected	• Most liquids • Most slurries • Most liquid-liquid interface measurements • Some granular solids	• Stratified liquids • Granular solids with wide moisture changes
Microwave	• Relatively low cost • Ignores vapors, dusts, and most process variables	• Contact measurement • Not suitable for light powders or some liquefied gases	• Most granular solids • Most slurries • Some stratified liquids	• Low-dielectric-constant materials, • Corrosive materials • High temperature • High pressure
Radar	• Non-contacting • Suitable for vacuum and pressure service • Ignores vapors and dusts	• Complicated setup • FCC license required by some • Can have high cost • Position sensitive	• Most liquids • Most slurries • Some granular solids	• Low-dielectric-contant materials, (granular solids, liquids) • Heavy agitation • Process that tend to leave heavy coating deposits
Nuclear	• Non-invasive • Reliable • Ignores hazardous and corrosive service	• High cost • License, permits & approval required • Disposal costly	• Corrosive & hazardous materials	• Interface measurement

TEMPERATURE MEASUREMENT

Temperature is defined as the energy level of matter, which is evidenced by some change in that matter. There is a wide variety of temperature sensors and they have one thing in common: they all measure temperature by sensing some change in a physical characteristic of the matter being measured.

The seven most often used types of measurement devices will be discussed.

LIQUID EXPANSION THERMOMETERS

Thermometers are well known liquid expansion devices. In general, they come in two basic classifications: mercury and organic. They operate on the principle that the liquid material in the unit takes on the temperature of the fluid being measured and expands proportionally to the amount of heat sensed by the unit. They are independent of a power supply.

The mercury filled devices have limitations due to environmental concerns. For example, breakage can be hazardous. Therefore, shipping and transportation should be carefully controlled.

Another type of liquid expansion device has a bulb filled with liquid that is connected by a capillary tube to a Bourdon tube. The bulb is immersed in the fluid to be measured. When the liquid expands and contracts, the pressure on the Bourdon tube causes the indicator linkage to rotate and indicate the temperature.

BIMETALLIC DEVICES

These devices operate on the principle of expansion of metals when they are heated. When two metals are bonded together and then heated, one metal expands more than the other causing the bonded metals to rotate. When mechanically linked to a pointer, the rotation will indicate temperature.

They are independent of a power supply and easily portable. They are not as accurate as electrical sensors, and the recording of temperatures is not possible.

CHANGE OF STATE SENSORS

The change-of-state sensors indicate that a change of temperature has occurred. The change of state of these devices cause the device to change color or disappear. Some are capable of reversing color. Commercially available devices include labels, pellets, crayons and lacquers. These devices are slow in response and accuracy is not high.

Labels are used on steam traps that need adjustment. A dot on the label changes color indicating a preselected temperature. This dot color is irreversible for less costly indicators. They are also valuable when conformation is needed that the temperature did not exceed a predetermined temperature.

Liquid crystal devices are available that do reverse color. Some liquid crystal models will have the ability to show different colors for different temperature ranges. Crayons used as an indicator of temperature simply disappear after indication. Pellets become visually deformed.

Although not perfectly precise, these sensors have an advantage when a small, rugged, non-electrical indicator is necessary.

SILICON DIODE

This sensor is a linear device most often used in cryogenic service. It operates on the principle that causes the conductivity to change in direct proportion with a change in product temperature. This change is measured to indicate temperature.

INFRARED SENSORS

An infrared sensor is the only non-contact temperature measuring device discussed. Infrared radiation is emitted from an object corresponding to the temperature. The radiation is received by a sensor (detector) that converts the signal into temperature indications by using a visual screen that converts temperature into color. The color corresponds to temperature.

There are two detectors used: thermal and photon type. The thermal detectors convert incoming radiation into heat, raising the temperature of the thermal detector. The change in temperature is converted to an electrical signal, which is displayed and amplified. Photon detectors react to the photons emitted by the object that cause changes to the electrical properties of the detector unit. These changes are monitored as an output signal.

Infrared units in facilities-type operations are not used for point measurements but rather for maintenance purposes to detect heat emissions from objects where heat is lost due to problems.

THERMOCOUPLES

Thermocouples are electrical devices that measure temperature by measuring a change in voltage. As the temperature goes up, the output voltage of the thermocouple rises. The output voltage is not linear. The National Institute of Standards has established criteria for thermocouples to insure that users can count on reliability and repeatability for many classes of thermocouples.

Thermocouples are basically two dissimilar metals (sensing elements) joined together at a sensing junction. The sensing element is specifically designed to have a specific temperature resistance at a specific temperature. As the temperature increases at the sensing junction, the thermocouple generates an electromotive force proportional to the heat applied. An instrument package converts this signal into temperature measurements. For each thermocouple type, the properties of the metals used dictate the upper range of temperature capable of being sensed.

Thermocouples are generally easy to handle and low cost. They can be installed in rugged environments. When selecting the type, temperature range, sensitivity, accuracy and reliability are important factors to consider.

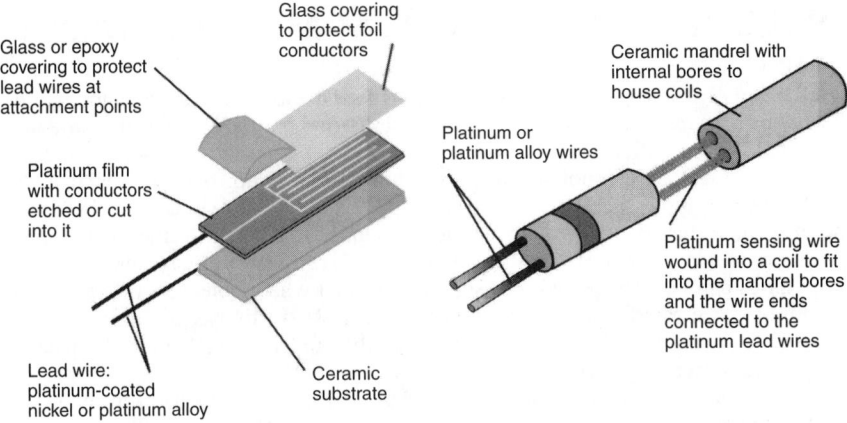

FIGURE 22.28   Typical thermocouples.

TABLE 22.5   Sensing Elements and Temperature Limits

Material	Usable temperature range
Platinum	−450 to 1,200°F
Nickel	−150 to 600°F
Copper	−100 to 300°F
Nickel/Iron	32 to 400°F

RESISTIVE TEMPERATURE DEVICES

Resistive temperature devices (RTD) are electrical devices. Rather than using a change in the voltage as the thermocouple does, this device measures a change in resistance resulting from a change in temperature. This change in voltage is sensed by an instrument package and converted to temperature readings. The two most often used arrangements of RTDs are the wire wound and the film element. These are illustrated in Fig 22.28.

The materials used to construct the RTD have limits that impact their use. The temperature limits of these materials are given in Table 22.5. Another consideration to check when selecting RTDs is the output over the applicable temperature range.

PRESSURE AND VACUUM MEASUREMENT

Pressure and measurements to be discussed will be only for gases and liquids. Vacuum measurements will be only for gases. For a gas, the molecular structure does not have a lattice type of arrangement, and the cohesive forces that bind the molecules together are not as strong as those for a solid. This means that the molecules are quite mobile, and will take the shape of their container. The actual solid volume that the gas atomic structure occupies in relation to the total volume of a gas molecule is quite small, and so, gases are mostly empty space. This is why gases can be compressed. Pressure is produced when molecules of a gas in an enclosed space rapidly strike the enclosing surfaces. If this gas is confined into a smaller and smaller volume, molecules strike the container walls more frequently, producing a greater pressure.

With water, the molecules are quite mobile and also will conform to the shape of the container. However, the molecular structure of water is such that it is practically incompressible. Therefore, added pressure from any source will exert pressure on all parts of a closed system open to the added pressure.

The same gauges discussed can be used for both positive and negative pressures.

MANOMETER

A manometer measures relative pressure between the system and local barometric pressure. It consists of a cylindrical "U" tube partially filled with liquid. One end is connected to the system being measured and the other end could be open or closed. The difference between liquid levels in each tube is used to calculate the pressure.

It finds greatest use in determination of leakage rather than being used to measure either pressure or actual flow.

BOURDON GAUGE

An often used type of mechanical gauge is the Bourdon gauge. This type of gauge measures pressure by the amount of deflection that an oval tube bent in an arc and closed at one end exerts under internal pressure. This mechanical gauge is simple, inexpensive and rugged, and is the most widely used type of pressure gauge.

The heart of the gauge is the Bourdon tube that is closed at one end and open to either pressure or vacuum at the other. As the pressure varies, the tube changes shape. A pointer attached to the tube moves, indicating the pressure on a dial. A Bourdon gauge is illustrated in Fig 22.29. The face of the gauge can be filled with oil to dampen pointer vibration.

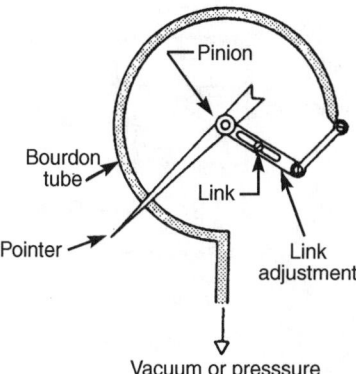

FIGURE 22.29   Bourdon tube gauge.

A Bourdon gauge should not be exposed to temperatures above 150°F (68°C) unless special material is specified.

DIAPHRAGM GAUGES

The diaphragm gauge measures the pressure difference by sensing the deflection of a thin metal diaphragm or capsular element. Similar to the Bourden gauge, its operation relies on the deformation of an elastic metal under pressure that is connected by means of a linkage to a pointer.

Diaphragm gauges are limited to low pressure applications, generally to a limit of 10 psig (70 kPa) full scale.

CAPACITANCE METERS

Capacitance meters are, in essence, electronic diaphragm gauges. Instead of a mechanical linkage, they use a change in a variable capacitance sensor to detect changes in pressure, which is transmitted electronically. The response time is fast and the signal can be remotely transmitted.

STRAIN GAUGES

Strain gauges also use the deflection of a diaphragm to produce a change in electrical resistance of the attached strain gauge. The response time is fast and the signal can be remotely transmitted. A typical strain gauge is illustrated in Fig. 22.30.

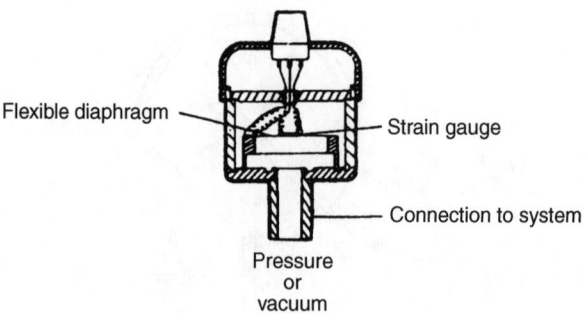

Flexible diaphragm

Strain gauge

Connection to system

Pressure
or
vacuum

FIGURE 22.30 Strain gauge.

PH MEASUREMENT

pH measures the level of acidity or alkalinity of a solution. The method most used for measurement of liquid in a liquid flow stream is an immersed sensor. The selection of the proper sensor for this purpose is based on accuracy and sensor performance.

A pH sensor consists of a galvanic cell with two electrodes—one that is sensitive to hydrogen ion activity (pH sensor) and a second reference sensor. These electrodes can be either separate or combined into a single probe. The single probe is most often used for facility systems in the measurement of acid waste prior to and after the neutralizing process. A single pH probe is illustrated in Fig. 22.31.

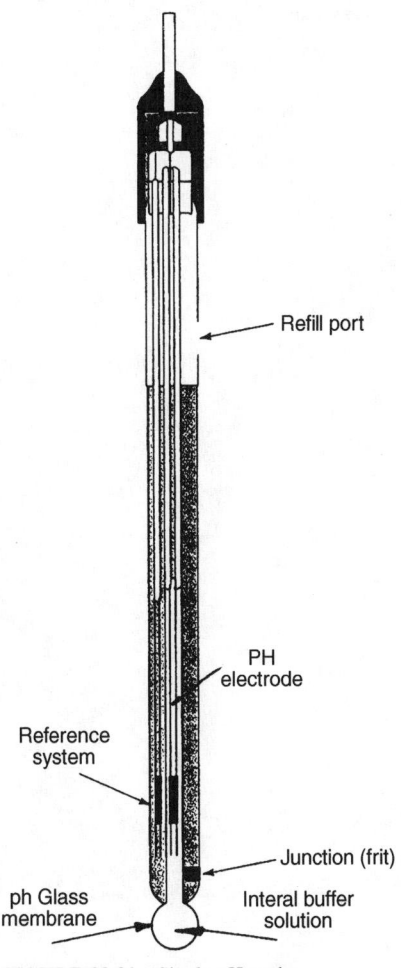

FIGURE 22.31   Single pH probe.

The pH sensitive electrode (or system) is enclosed in a special glass membrane that is sensitive to changes in pH. Such an electrode consists of a metal wire conductor suspended inside a specially constructed glass tube filled with an electrolyte (or buffered) solution. The sensing area in the tip is most often sphere-shaped to provide the largest area for liquid contact and is a membrane especially formulated from pH sensitive material. The fluid being measured permeates the tip and a potential is created between the fluid and the buffered solution.

The reference electrode is constructed of a non-conducting glass envelope containing a reference element. The reference electrode is also filled with an electrolyte. The junction (also called a frit) provides entrance for the liquid being measured. The junctions are made from a variety of non-conductive materials.

The electrodes change electrical potential when immersed in the solution being measured. An instrument package measures this change and converts it into a pH reading or sends a signal to a remote location. The life expectancy of a pH sensor depends on the temperature of the liquid it is immersed in. The higher the temperature, the lower the life. It is recommended that the velocity in the pipe be kept below 9 ft per second to reduce wear.

METERING PUMPS

INTRODUCTION

A metering pump is used to inject a constant, accurate and adjustable flow rate of a liquid chemical additive into a larger process. The primary objective of a metering pump is to effect a predictable and adjustable flow rate that remains stable even when system pressure conditions change.

The most commonly used category of metering pump is a positive displacement pump, which will be the only pump type discussed. What differentiates a metering pump from most positive displacement pumps is its accuracy, which is typically ± 1.0 percent. There are several metering pump types; the most often used are diaphragm, piston, peristaltic and gear pumps. They all have one feature in common, and this is a fixed volume cavity to deliver the same volume with each pumping cycle. The primary challenges of pump design are to control cavity dimensions, minimize leaks and eliminate dead volumes.

METERING PUMP DECRIPTION

Diaphragm Pump

This pump operates by means of a flexible disk, called a diaphragm, that moves back and forth in response to a reciprocating plunger connected to a drive shaft. The discharge is pulsating. A schematic detail showing operation of the diaphragm pump is illustrated in Fig. 22.32.

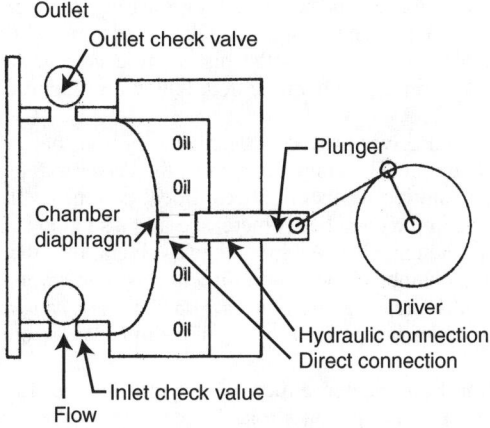

FIGURE 22.32   Typical diaphragm pump.

The operation of the diaphragm pump centers around the pump chamber. One side of the chamber is the pump body and the other side is a flexible diaphragm. A vacuum created by one half of the plunger stroke enlarging the chamber area creates a vacuum that allows the liquid to be pumped to enter the larger chamber. On the other half of the plunger stroke the area of the chamber is compressed, forcing the liquid out of the discharge port. Inlet and outlet check valves are required to prevent short circuiting through the pump. Chemicals to be pumped contact only one side of the diaphragm, which can be lined with a material compatible to the chemical.

The diaphragm pump is available in two configurations of plunger diaphragm connections. One method is to have the plunger directly connected, which creates stress on the diaphragm at the point of attachment. This is less costly but results in a shorter pump life cycle. Since the driver action is fixed, adjustment is made by means of an adjustable stop on the return travel mechanism.

The other method has the space between the plunger and diaphragm filled with hydraulic oil. This is commonly called a hydraulic diaphragm pump. As the plunger moves, it displaces the oil, and this moves the diaphragm. The amount of chemical product pumped is equal to the amount of displaced oil. Flow rate is adjusted by means of a bypass port connected to a reservoir that is adjustable to release oil from the space, thus reducing the amount of oil that is displaced.

The solenoid pump is another method of driving the diaphragm. It operates from a timing mechanism that energizes an electromagnet that slides the plunger into the discharge position. When the magnet is de-energized, it drives the magnet back into the suction position. A system of check valves keeps the fluid flowing in only one direction.

Diaphragm pumps are designed to produce a discharge pressure of approximately 150 psig (1,050 kPa). Pulsation can be reduced by using a pulsation damper installed on the discharge piping of the pump.

Piston Pump

The piston pump operates by means of a reciprocating plunger moving inside a machined cylindrical cavity. The plunger is called a piston and the cavity is called the cylinder. A driver is connected to the piston by eccentric cam. The discharge from the pump is pulsating. A schematic detail showing operation of a gear pump is illustrated in Fig. 22.33.

The power train and driver provides the rotary motion that drives the piston in and out inside the cylinder. The connection from the driver to the piston is eccentric to convert the rotary motion to a straight reciprocal action. As the piston is driven rearward, the volume in the chamber is increased and a suction is created that draws liquid chemical into the chamber. An inlet check valve allows the fluid to flow only one way—inside the chamber. The piston direction is then reversed, decreasing the chamber volume and increasing the pressure on the fluid causing it to discharge. The outlet check valve now opens, allowing the fluid to flow out of the discharge line.

The flow rate can be regulated either by varying the speed of the driver or by making an adjustment to the piston stroke length to vary the displacement of the piston in the cylinder.

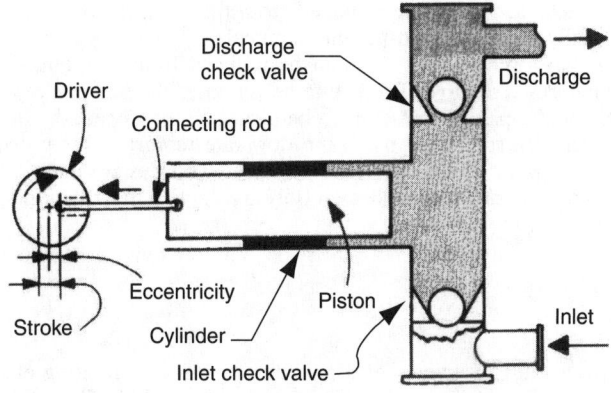

FIGURE 22.33 Typical piston pump.

Peristaltic Pump

This pump operates by means of a revolving roller compressing a smooth wall, flexible tube with fluid inside the tube. As the roller rotates, the fluid is forced through the tube from inlet to outlet. A schematic detail showing operation of a peristaltic pump is illustrated in Fig. 22.34.

The operating principle is to have the tube squeezed and released by a roller along a predetermined length positively displacing the fluid inside. The tube recovers its shape after the squeezing action that produces a vacuum drawing more chemical into the tube. This provides a gentle pumping action that causes little damage to the chemical being pumped. The volume of chemical discharged per cycle is proportional to the tube diameter and the distance between rollers. The amount of fluid between the rollers is called the "pillow volume."

There are several advantages to this pump. With the chemical being pumped totally enclosed in the flexible tubing, no path for leakage or fugitive emissions is

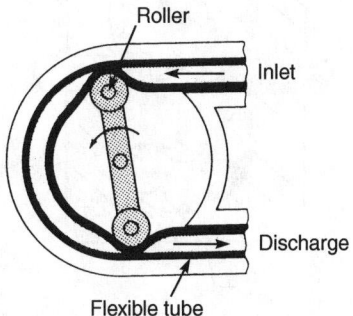

FIGURE   22.34 Typical   peristaltic pump.

provided. The hose can easily be removed from the pump body for cleaning. The tube can be easily changed for different chemicals.

The discharge is pulsed, but this can be reduced by using multiple rollers to produce an almost steady flow. The key to selection of the pump is to optimize the tube material for compatibility with the chemical being pumped, the tube diameter and distance between rollers to match the flow rate required. Pump head is determined by the characteristics of the tube selected, which is typically 50 psig (340 kPa). The pump volume can be adjusted only by varying the rotation speed.

Gear Pump

A gear pump operates by means of a pair of continuously rotating elements with gear teeth. The design of the teeth create a cavity that when filled with liquid and rotated moves liquid through the pump. Flow is determined by the amount of fluid in the teeth multiplied by the number of teeth and the rotation speed. A gear pump is illustrated in Fig 22.35.

The operating principle is to have fluid introduced into the inlet and find its way between the gears of the rotating element. It is then carried around the pump between the teeth of the gears and the centrifugal action gives the fluid its added pressure. No check valves are required because of the very close tolerances of the tips of the gear teeth inside the pump body retaining the liquid. Because of the number of teeth on the rotating element, the flow is virtually continuous. Flow rate is adjusted by varying the rotating speed of the driver.

These pumps are well suited for liquids of wide range of viscosity. Other advantages include non-pulsating flow, simple piping arrangement, low first cost, and less tendency to leak. The major disadvantage is that the gear pump has a decreasing output when pumping against an increasing backpressure.

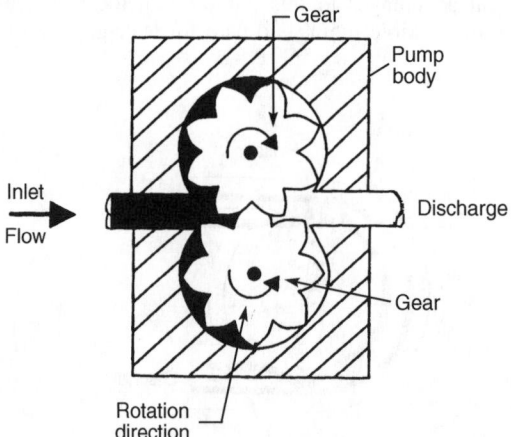

FIGURE 22.35 Typical gear pump.

METERING PUMP CONSIDERATIONS

The following considerations and ancillary devices may be required for systems to avoid problems. Not all of the devices will be necessary for all installations. A typical metering pump system for a larger scale chemical injection is illustrated in Fig 22.36.

Controlling Flow Rate

There are two primary methods used to control the flow rate: an adjustable speed motor and adjustment of the pump's displacement or stroke. If the pump is to be operated in the top 50 percent of its range, either method can be used. Experience has shown that the variable speed method is more accurate if the pump is to be operated in the lower portion of the range.

Backpressure Valve

A backpressure valve is another name for a check valve. It is installed on the outlet of a metering pump to assure that flow occurs in only one direction. Metering pumps may also require that the pressure at the discharge of the pump be greater than the pressure at the inlet of the pump. There are two reasons for this. First, if sufficient discharge pressure is not present, liquid could flow freely through the pump; second, flow rate accuracy is lost.

A similar check valve is required on the inlet side of the pump.

Pulsation Dampeners

A pulsation dampener smooths out the pulsations from a pump, and in doing so the pressure and force exerted on the piping is reduced as well as water hammer. If a flowmeter is used, the indications will be more accurate. The following is considered good engineering practice for installation of a pulsation dampener.

1. Locate the pulsation dampener near the discharge of the pump. The reason is that the pulsation dampener only dampens the pulses downstream of itself. It does not dampen flow between itself and the pump.
2. Install a pressure gauge downstream of the pulsation dampener.
3. Do not reduce the pipe size between the pump and the pulsation dampener.
4. Precharge the dampener with dry nitrogen to a pressure of 80 to 85 percent of the lowest normal operating pressure. The pressure rating of the dampener shall equal system pressure and temperature.
5. Select the size of the pulsation dampener to equal to approximately 15 pump strokes.

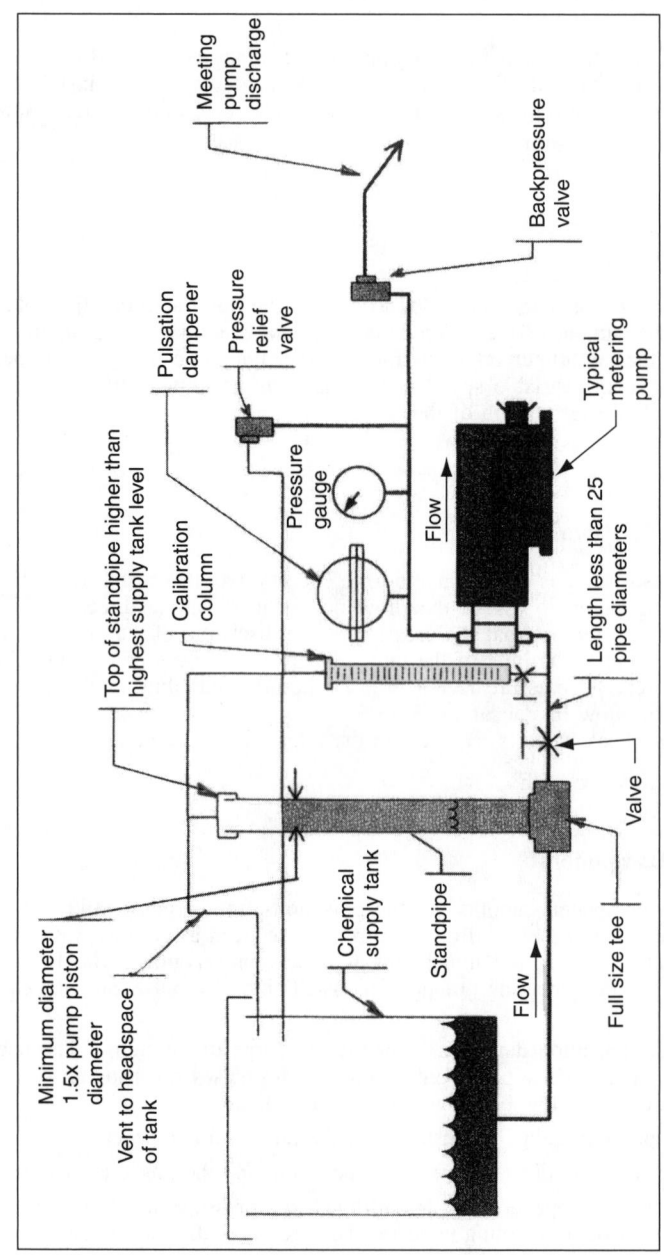

FIGURE 22.36 Typical metering system.

Suction Pipe Size

The suction pipe size probably has the most effect on successful pump operation. Generally accepted practice sizes the suction pipe at one size larger than the pump inlet connection. To actually size the inlet where there may be a problem, an NPSH analysis should be performed, understanding that the peak flow rate from each stroke will be at least three times the average flow rate.

Discharge Pipe Size

If a pulsation dampener is used, it is common practice to size the discharge piping ½ to ¾ the size of the pump outlet size. If no pulsation dampener is used, the discharge shall be sized at least as large, if not one size larger. Acceleration pressure drop is another consideration.

Standpipe

A standpipe, which is nothing more than a straight length of pipe, will eliminate flow pulsations between the pump and chemical supply, will eliminate acceleration pressure drop and reduce the frictional pressure drop by a factor of three. It should be capped and vented back to the supply or a separate tank. It should be installed in a full size tee and extend higher than the highest level of the chemical supply tank.

Calibration Column

A calibration column is used to measure the pump flow rate. It is essentially a small length of glass pipe with calibration marks. It is used by filling the column with chemical and cutting off the supply from the main source. When the mump is started, the exact amount of chemical delivered can be observed.

FLOW IN OPEN CHANNELS

Open channel flow is the flow of liquids in a channel the geometry of which has one liquid surface free of solid boundaries. A channel is often called a conduit. This section will discuss flow measurement techniques for open channels other than in pipes. Gravity, non-pressurized flow in pipes can be found in Chap. 6, Storm Water System.

The flow in open channels, or conduits, is classified into two types: steady and unsteady flow. Steady flow is a constant rate of discharge, and unsteady flow is characterized as variable rate of discharge over time. A flow is uniform if velocity and depth are constant along the conduit. If velocity and depth change in the conduit, the flow is unsteady.

CALCULATING THE HYDRAULIC RADIUS OF AN OPEN CHANNEL

Often the hydraulic radius of the conduit must be found in order to calculate the flow rate. This is accomplished by first calculating the area of the water flowing in the channel in square feet. Next, determine the contact surface of the flowing water contacting the bottom and sides of the channel, in feet. To calculate the hydraulic radius, divide the area by the length of contact surface. The answer is in feet.

CALCULATING THE FLOW OF AN OPEN CHANNEL

A widely used method used to find the flow in open channels is the Manning formula. This is very similar to, and yields the same results as, Kutter's formula.

Fig. 22.37, a graphical solution of the Manning formula, has been developed for use in open channels. In order to enter the figure, most of the following must be determined or calculated.

1. *Hydraulic radius.* Refer to previous discussion.

2. *"n" value* for friction coefficient of the channel. Refer to Table 22.6.

3. *The average velocity* of the flow stream can be found by placing a floating object in the stream and observing the amount of time it takes to move a given distance. Generally accepted practice requires that the surface velocity be reduced by 15 percent to give the average velocity.

4. *Loss of head* can only be determined from other criteria. If the hydraulic radius and average velocity is known, the loss of head per 1,000 feet can be found in Fig. 22.38. This chart is calculated for an "n" value of 0.01, which is an average value.

5. To use the chart, first connect the hydraulic radius point with the head loss point with a straightedge. Where the imaginary line crosses the pivot point, rotate

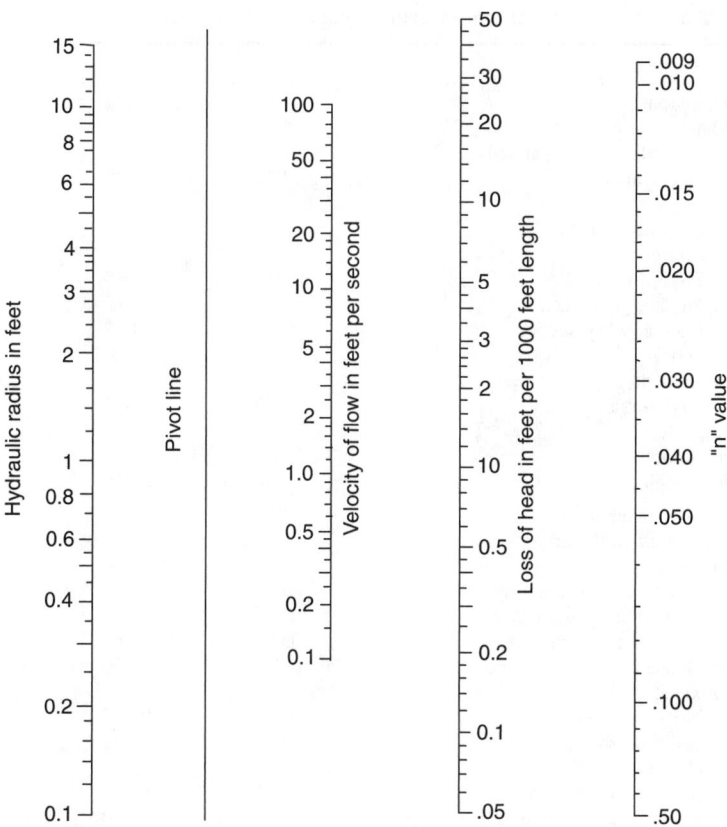

FIGURE 22.37   Graphical solution to Mannings formula.

the straightedge from the pivot point just found to cross the established "n" value on the proper scale. Read the velocity on the velocity scale.

6. Multiply the velocity by the area of the flowing water to find the flow rate in cfs.

METERS FOR MEASURING FLOWRATE IN OPEN CHANNELS

Current Meter

The operating principles of a current meter are similar to those of the previously discussed turbine meter. This type of meter should be considered when the use of a weir is not practical because of insufficient head.

TABLE 22.6 "n" Value for Use in Mannings Formula. For Open Channels.

	Min	Avg	Max
A. Lined channels	0.011	0.012	0.014
1. Metal			
a. Smooth steel (unpainted)			
b. Corrugated	0.021	0.025	0.030
2. Wood			
a. Planed, untreated	0.010	0.012	0.014
3. Concrete			
a. Float, finish	0.013	0.015	0.016
b. Gunite, good section	0.016	0.019	0.023
c. Gunite, wavy section	0.018	0.022	0.025
4. Masonry			
a. Cemented	0.017	0.025	0.030
b. Dry rubble	0.023	0.032	0.035
5. Asphalt			
a. Smooth	0.013	0.013	
b. Rough	0.016	0.016	
B. Unlined channels			
1. Excavated earth, straight and uniform			
a. Clean, after weathering	0.018	0.022	0.025
b. With short grass, few weeds	0.022	0.027	0.033
c. Dense weeds, high as flow depth	0.050	0.080	0.120
d. Dense brush, high stage	0.080	0.100	0.140
2. Dredged earth			
a. No vegetation	0.025	0.028	0.033
b. Light brush on banks	0.035	0.050	0.060
3. Rock cuts			
a. Smooth and uniform	0.025	0.035	0.040
b. Jagged and irregular	0.035	0.040	0.050

Flume

A flume is a specially designed open channel flow section that has engineered restrictions in the area that produces an increase in velocity of the fluid being measured. The most widely used flumes are the *Parchall Flume,* illustrated in Fig. 22.39, and the *Palmer-Bowles* flume.

The flume measures water flow because the design of the flume has the head of water surface in the converging section proportional to the flow through the flume. The increase in height is measured by instrument packages and converted into flow rates.

Flumes generally require relatively smooth flow of approximately 10 diameters upstream for proper operation. They should not be installed near a sharp change of slope or adjacent to obstructions in the flow path. They are selected where low head loss is important and where large amounts of suspended solids may be present.

Weirs

A weir is a dam or obstruction placed in the flow path of a partially filled pipe or channel. Because of the obstruction, water will back up behind the barrier creating

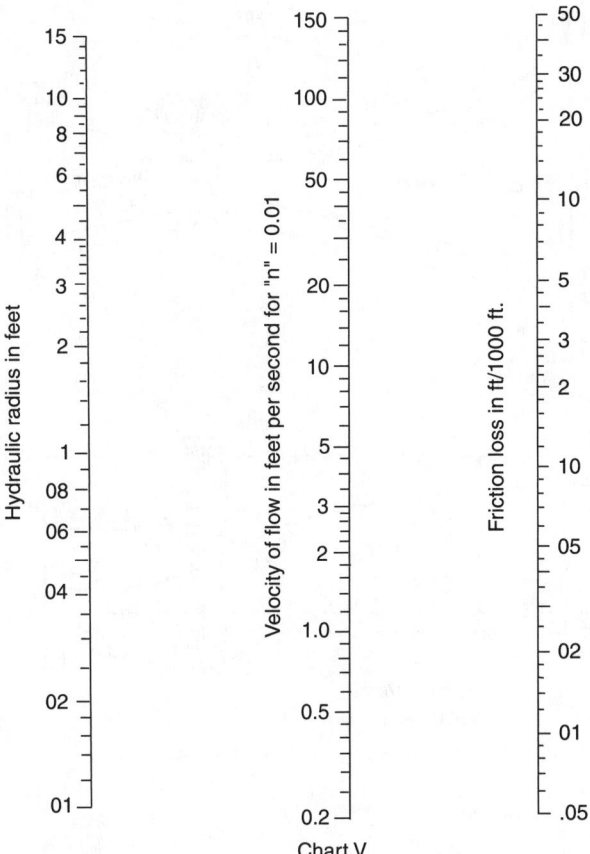

FIGURE 22.38   Chart to find head loss in Mannings formula for open channels.

a head of water. That head is a function of flow velocity, and therefore the flowrate through the weir. There is an opening that allows water to freely flow through the opening. A cross section through a weir is illustrated in Fig. 22.40. This opening has three shapes: rectangular, triangular and Cippolletti. These shapes are illustrated in Fig. 22.41. The difference in geometry between the Cippolletti and rectangular or triangular shape is that the Cippolletti weir has a 4 to 1 slope ratio of the sides. The advantage of the Cippolletti weir is that the discharge occurs as if there is no end contraction and no correction to the flow rate is necessary.

Discharge rates are found by obtaining the measurement of the vertical distance from the weir crest to the surface of the water. In order to assure accuracy, the measurement must be taken some distance upstream of the weir. Generally accepted practice used for measuring the head of water places the staff gauge at distance of at least four times the head of water upstream of the weir crest.

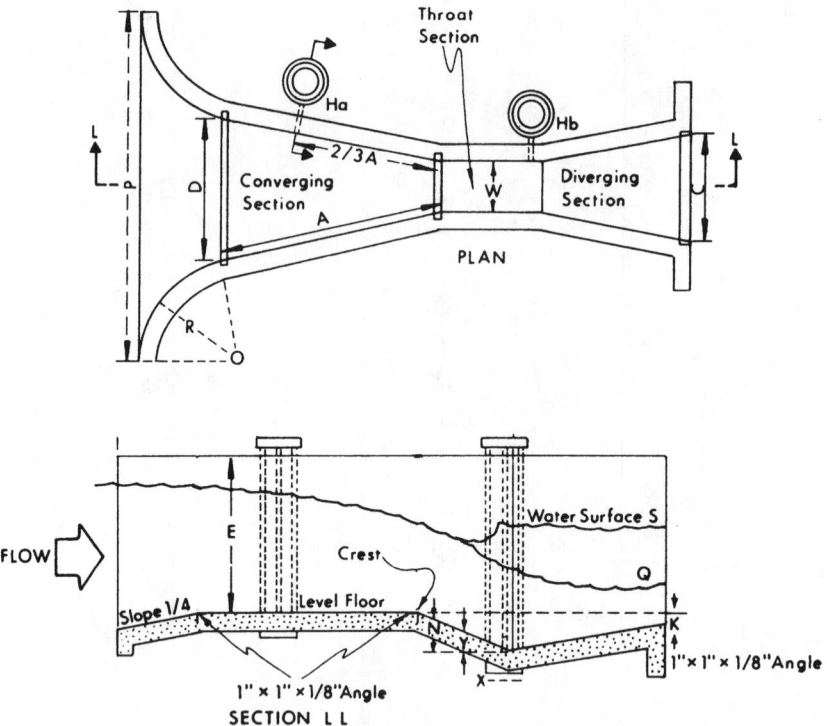

FIGURE 22.39 Plan and sectional views of a Parshall flume.

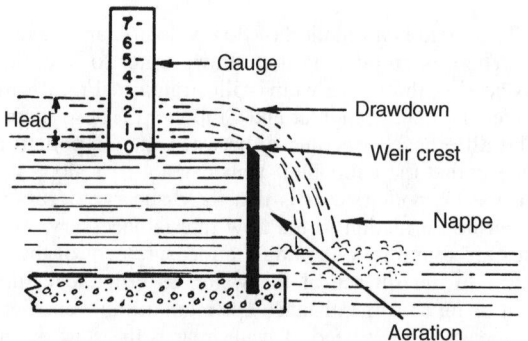

FIGURE 22.40 Cross section through a weir.

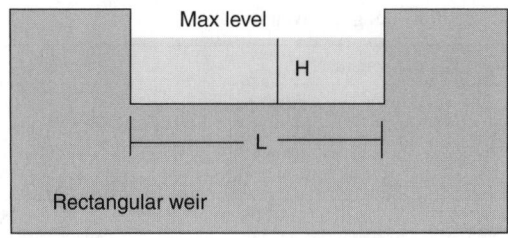

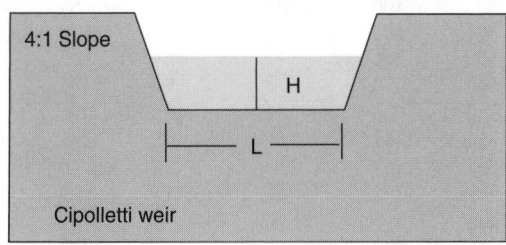

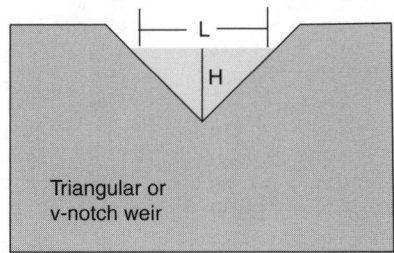

FIGURE 22.41 Types of sharp-crested weirs.

The basic formulas for flow through the various weirs are:

1. For rectangular or square weirs $Q = 3.33 \times L \times H^{3/2}$ (22.10)

2. For Cippolletti weirs $Q = 3.37 \times L \times H^{3/2}$ (22.11)

3. For triangular weirs (90°) $Q = 2.44 \times H^{5/2}$ (22.12)

4. For triangular weirs (60°) $Q = 1.41 \times H^{5/2}$ (22.13)

where Q = flow rate in cfs
     L = for rectangular weir, length of weir, ft
         For Cippolletti weir, length at base, ft
         For triangular weir, length of water at flowing level above notch.
     H = measured head, ft

Table 22.7 is a tabular solution for discharge from a rectangular weir. Cippolletti weirs are calculated using Table 22.7 with a length equal to that at the bottom of

TABLE 22.7 Flow-through Rectangular Weirs—GPM

Head (H) in inches	Length (L) of weir in feet				Head (H) in inches	Length (L) of weir in feet		
	1	3	5	Additional gpm for each ft over 5 ft		3	5	Additional gpm for each ft over 5 ft
1	35.4	107.5	179.8	36.05	8	2338	3956	814
1¼	49.5	150.4	250.4	50.4	8¼	2442	4140	850
1½	64.9	197	329.5	66.2	8½	2540	4312	890
1¾	81	248	415	83.5	8¾	2656	4511	929
1	98.5	302	506	102	9	2765	4699	970
2¼	117	361	605	122	9¼	2876	4899	1011
2½	136.2	422	706	143	9½	2985	5098	1051
2¾	157	485	815	165	9¾	3101	5288	1091
3	177.8	552	926	187	10	3216	5490	1136
3¼	199.8	624	1047	211	10½	3480	5940	1230
3½	222	695	1167	236	11	3716	6355	1320
3¾	245	769	1292	261	11½	3960	6780	1420
4	269	846	1424	288	12	4185	7165	1495
4¼	293.6	925	1559	316	12½	4430	7595	1575
4½	318	1006	1696	345	13	4660	8010	1660
4¾	344	1091	1835	374	13½	4950	8510	1780
5	370	1175	1985	405	14	5215	8980	1885
5¼	395.5	1262	2130	434	14½	5475	9440	1985
5½	421.6	1352	2282	465	15	5740	9920	2090
5¾	449	1442	2440	495	15½	6015	10400	2165
6	476.5	1535	2600	528	16	6290	10900	2300
6¼		1632	2760	560	16½	6565	11380	2410
6½		1742	2920	596	17	6925	11970	2520
6¾		1826	3094	630	17½	7140	12410	2640
7		1928	3260	668	18	7410	12900	2745
7¼		2029	3436	701.5	18½	7695	13410	2855
7½		2130	3609	736	19	7980	13940	2970
7¾		2238	3785	774	19½	8280	14460	3090

the weir. Table 22.8 is the tabular solution for discharge from triangular weirs. Interpolation should be used to calculate discharges for intermediate dimensions.

Sluice Gates

A sluice gate is a device used in controlling the flow in open channels, canals or rivers. It consists of an opening installed in a full obstruction in a channel that is controlled by an operable gate. The gate can be raised or lowered by means of a manual gear-operated drive that opens the gate from the bottom of the channel, allowing water to flow out.

TABLE 22.8 Flow-through Triangular Weirs

Head (H) in inches	Flow in gallons per min 90° notch	60° notch	Head (H) in inches	Flow in gallons per min 90° notch	60° notch	Head (H) in inches	Flow in gallons per min 90° notch	60° notch
1	2.19	1.27	6¾	260	150	15	1912	1104
1¼	3.83	2.21	7	284	164	15½	2073	1197
1½	6.05	3.49	7¼	310	179	16	2246	1297
1¾	8.89	5.13	7½	338	195	16½	2426	1401
2	12.4	7.16	7¾	367	212	17	2614	1509
2¼	16.7	9.62	8	397	229	17½	2810	1623
2½	21.7	12.5	8¼	429	248	18	3016	1741
2¾	27.5	15.9	8½	462	267	18½	3229	1864
3	34.2	19.7	8¾	498	287	19	3452	1993
3¼	41.8	24.1	9	533	308	19½	3684	2127
3½	50.3	29.0	9¼	571	330	20	3924	2266
3¾	59.7	34.5	9½	610	352	20½	4174	2410
4	70.2	40.5	9¾	651	376	21	4433	2560
4¼	81.7	47.2	10	694	401	21½	4702	2715
4½	94.2	54.4	10½	784	452	22	4980	2875
4¾	108	62.3	11	880	508	23½	5268	3041
5	123	70.8	11½	984	568	23	4565	3213
5¼	139	80.0	12	1094	632	23½	5873	3391
5½	156	89.9	12½	1212	700	24	6190	3574
5¾	174	100	13	1337	772	24½	6518	3763
6	193	112	13½	1469	848	25	6855	3958
6¼	214	124	14	1609	929			
6½	236	136	14½	1756	1014			

Free Discharge of Water into the Atmosphere.   Use the illustration shown in Fig 22.42a. To find the flow rate of water discharged through a sluice gate, the area of the opening necessary to discharge a given flow rate or head of water above the as, refer to Fig. 22.43, entering the chart with any two values to find the third.

Submerged Discharge of Water.   Use the illustration shown in Fig. 22.42b. The solution is similar to that of free discharge except that the head of water used in the calculations is the difference of water levels as shown in the illustration.

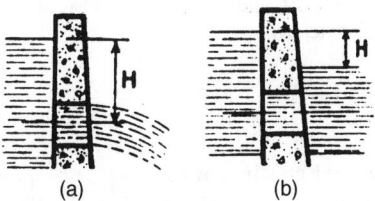

(a)   (b)

FIGURE 22.42 Sluice gates.

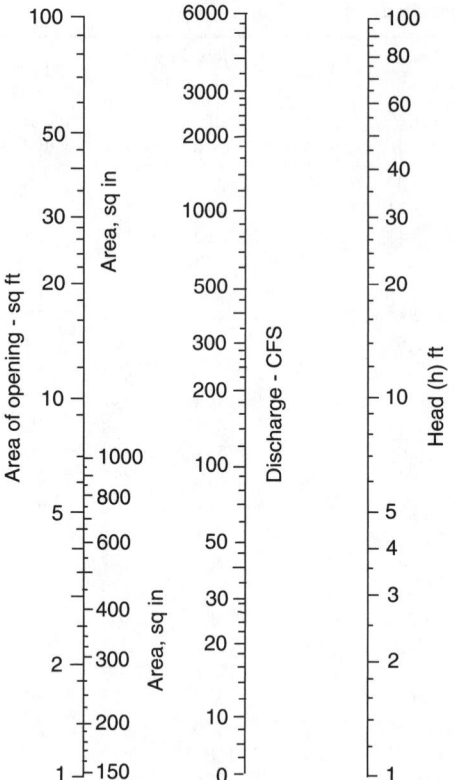

FIGURE 22.43   Sluice gate discharge.

MISCELLANEOUS DEVICES AND MEASUREMENT METHODS

Sight Glass

A sight glass is a "U" shaped transparent device that allows the contents of the vessel to freely move inside it. It is connected to the side of a vessel and measures the level of a liquid inside. A connection is made at the extreme top and bottom of the vessel, and as the level of liquid inside the vessel changes, it changes inside the sight glass. This is the only method that allows direct visual observation of levels and can be the most accurate.

Thermowells

A thermowell is an enclosed attachment to a pipe that permits the insertion of a temperature measuring device. It is important that the thermowell be installed in that portion of the pipe run capable of producing the least measurement error. The most common method of thermowell attachment is by means of a threaded con-

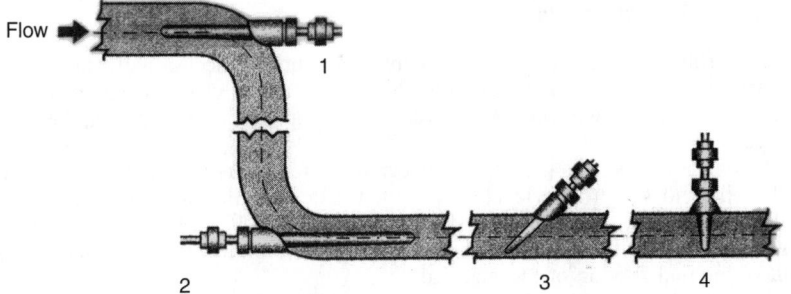

FIGURE 22.44   Preferred thermowell locations.

TABLE 22.9   Angles of Repose

Material	Angle, deg
Anthracite ash	45
Cinders, bit. coal	25–41
Coal, anthracite	27
Coal, bituminous	35
Coke (piled loose)	30–45
Gravel, round	30
Gravel, sharp	40
Clay, soft	10
Clay, compacted	20–25
Sand-clay, compacted	40–50
Sand, dry	25–35
Sand, moist	30–45
Sand, wet	20–40
Silt, compacted	25–40
Silt, loose	20–30
Soda ash, dense	42
Soda ash, light	59
Sulfur, granules	35
Salt	36
Cement	37.5
Iron ore	35–45
Scrap metal	35–45
Plastic resin grains:	
Cellulose acetate	16
Fluorocarbon	6
Nylon	9
Vinyl	10
Viscose	10

nection. The distance the thermometer intrudes into the pipe is called the immersion. For the most accurate measurement, the immersion shall be at least into the center of the pipe.

The location of the thermowell is important. Fig. 22.44 illustrates the four most preferred methods of installing a thermowell in a pipeline.

Space for Piled Materials

When working with volumes of piled materials, one is occasionally faced with calculating the amount of material in a pile or how much a given amount of ground area a piled material will occupy. In order to accomplish this calculation, the angle of repose must be known.

The *angle of repose* is the angle between the naturally occurring sloping surface of the pile and the ground level that a piled material will naturally assume. The angle of repose for common materials is given in Table 22.9. The angle of repose is generally found by observation or laboratory tests and is based on the lowest limit of internal friction of the material.

Knowing the angle of repose and the shape of the pile (conical, wedge, etc.), the area can be calculated using trigonometric functions found in standard engineering texts. With the area and weight of the material known, the actual quantity of the material can be found.

REFERENCES

Cheremosinoff, N. P. and P. N., "Wastewater Flow Measuring Devices," Pollution Engineering Magazine, September, 1980.

Cheremosinoff, N. P., "Fluid Flow," Ann Arbor Science Publishers, Ann Arbor Michigan, 1982.

Constantinescu, S., "Space for Piled Materials," Chemical Engineering Magazine, January, 1998.

Desmarais, R., "Things You Need to Know About RTD's," Process Heating Magazine, March, 1997.

Ginesi, D., "Choices Abound in Flow Measurement," Chemical Engineering Magazine, April, 1991.

Hays, J. W., "Select the Right pH Measurement System," Chemical Engineering Progress Magazine, October, 1995.

Ingersoll-Rand, "Cameron Hydraulic Data," 16th edition.

Madden, J. R, "Pick the Right Thermocouple for Your Application," Process Heating Magazine, January, 1997.

Moss, R. A., "Doppler Dilemma," Flow Control Magazine, October, 2000.

Owen, R. E., "Selecting Flowmeters for Viscous Fluids," Plant Engineering Magazine.

Peace, R. L., "Piston and Diaphragm Pumps," Plant Engineering Magazine, December, 1979.

Poe, C., "In's and Out's of Metering Pumps," Plant Engineering Magazine, June, 2000.

Reif, D., "Matching the Flowmeter to the Job," Flow Control Magazine, May, 1997.

Swearington, C., "High Viscosity Flowmeters," Flow Control Magazine, May, 1998.

Thomas, D. and Angelo, R., "Know Your Pitot," Flow Control Magazine, September, 2000.

Warwick, E., "Metering Pump System Design," Pumps and Systems Magazine, September, 2000.

APPENDIX A
PIPE DISTRIBUTION SYSTEMS

The following pipe materials, jointing methods, and system components have been successfully used for various projects. Where trade names are mentioned, they are used only as an example to indicate general type, style, or construction. In no manner does this represent any endorsement or recommendation for use. There are materials suitable for applications other than those mentioned.

All materials used in a piping network must be carefully studied for pressure, compatibility, code acceptance, and suitability for use in any proposed project. The examples presented must not be used without all considerations outlined above, with the final choice of materials being the complete responsibility of the design professional.

Services:
  High-Pressure Steam (Saturated)
  Atomizing Steam
  Boiler Feedwater
  High-Pressure Condensate
  Blowoff and Continuous Blowdown
  Chemical Feed Direct to Boiler
Pressure/Temperature Limits: 250 psig at 400°F
General Materials: Carbon Steel, Cast Iron

Item	Pipe size	Description	Reference standards
Pipe	2 in and smaller	Sch. 40, electric-resistance welded (ERW), or seamless steel (except for all condensate piping which shall be Sch. 80, ERW, or seamless steel)	ASTM A 53, Grade B
	2½–12 in	Sch. 40, ERW, or seamless steel (except for all condensate piping which shall be Sch. 80, ERW, or seamless steel)	ASTM A 53, Grade B
	14–18 in	Sch. 30, ERW, or seamless steel	ASTM A 53, Grade B
Joints	2 in and smaller	Socket-welded connections with 3000-lb forged steel socket weld fittings	ASTM A 105
	2½ in and larger	Butt welded connections	
Fittings	2 in and smaller	3000-lb forged steel socket weld	ASTM A 105 ANSI B 16.11
	2½ in and larger	Standard seamless butt weld carbon steel, end, long radius ells	ASTM A 234
Flanges	All	300-lb forged steel weld neck, raised face, bore to match pipe I.D.	ASTM A 181, Class 60 ANSI B 16.5
Flange bolting	All	Alloy steel bolt studs, thread full length, ANSI class 2A threads	ASTM A 193, Grade B7
		Heavy hex nuts	ASTM A 194, Class 2H ANSI B 18.2

Item	Pipe size	Description	Reference standards
Unions	2 in and smaller	300-lb malleable iron, threaded brass seat, ground joint OR 3000-lb socket weld, steel to steel seat, ground joint	
Gate valves	2 in and smaller	300-lb, bronze, union bonnet, rising stem, OS&Y, threaded connections	
	2½ in and larger	250-lb, cast iron, bronze trim, solid wedge, OS&Y, flanged ends	
Globe and angle valves	2 in and smaller	300-lb, bronze renewable disc, union bonnet, threaded connections	
	2½ in and larger	250-lb, cast iron, renewable seat and disc, OS&Y, flanged ends	
Check valves	2 in and smaller	300-lb, bronze, swing check, Y pattern, renewable discs and side plugs, screwed ends	
	2½ in and larger	250-lb, cast iron, swing check, trimmed, bronze renewable seat, discs and side plugs, flanged ends	
Gaskets	All	Nonasbestos with synthetic fibers and SBR binder	ASTM F 104

Services:
  Medium-Pressure Steam (Saturated)
  Low-Pressure Steam (Saturated)
  Medium-Pressure Condensate and Pumped Condensate
  Low-Pressure and Humidification Condensate
  Gravity Condensate
  Safety Valve Vents
  Deaerator Chemical Feed and Accessory Piping
Pressure/Temperature Limits: 125 psig at 350°F
General Materials: Carbon Steel, Cast Iron, Bronze

Item	Pipe size	Description	Reference standards
Pipe	2 in and smaller	Sch. 40, ERW, or seamless steel (except for all condensate piping, which shall be Sch. 80, ERW, or seamless steel)	ASTM A 53, Grade B
	2½ in and larger	Sch. 40, ERW, or seamless steel (except for all condensate piping, which shall be Sch. 80 seamless steel)	ASTM A 53, Grade B
Joints	2 in and smaller	Threaded connections	ANSI B 2.1
	2½ in and larger	Welded connections	
Fittings	2 in and smaller	125-lb, cast iron, banded type, screwed	ASTM A 126
		OR	
		3000-lb forged steel socket weld	ASTM A 105 ANSI B 16.11
	2½ in and larger	Standard seamless steel, butt weld, beveled ends, long radius ells	ASTM A 234
Flanges	All	150-lb forged steel weld neck, flat face when matching cast iron flanges, bore to match pipe I.D.	ASTM A 181, Class 60 ANSI B 16.5
Flange bolting	All	Alloy steel bolt studs, thread full length with heavy hex nuts	ASTM A 193, Grade B7

Item	Pipe size	Description	Reference standards
Unions	2 in and smaller	150-lb, malleable iron, brass trim, ground joint, threaded ends	ASTM A 197 ANSI B 16.3
		OR	
		3000-lb socket weld, steel to steel seat, ground joint	ASTM A 105 ANSI B 16.11
Gate valves	2 in and smaller	125-lb, bronze, union bonnet, solid wedge disc, rising stem, threaded connections	ASTM B 62
	2½ in and larger	125-lb, cast iron, OS&Y, bronze mounted, solid wedge, rising stem, flanged ends	ASTM A 126, Class B
Globe and angle valves	2 in and smaller	150-lb, bronze, union bonnet, renewable disc, threaded connections	ASTM B 62
	2½ in and larger	125-lb, cast iron, OS&Y, bolted bonnet, bronze mounted, renewable disc, flanged ends	ASTM A 126, Class B
Check valves	2 in and smaller	125-lb, bronze, horizontal swing check, renewable plugs and discs, threaded connections	ASTM B 62
	2½ in and larger	125-lb, cast iron, swing check, bolted cap, renewable plug and discs, flanged ends	ASTM A 126, Class B
Vertical silent lift checks	1 in and larger	125-lb, brass, flanged wafer type	
Gaskets	All	Nonasbestos with synthetic fibers and SBR binder, full face gaskets on flat face flanges	ASTM F 104

Services:
  Nonpotable Fresh Water above Ground
  Pond Water above Ground
  River Water above Ground
Pressure/Temperature Limits: to 200 psig, 120°F (49°C)
General Materials: Copper, Steel

Item	Pipe size	Description	Reference standards
Pipe	2–4 in	Copper type L, seamless hard drawn	ASTM B-88
Fittings	All	Wrought copper Cast copper	ANSI 16.22 ANSI 16.18
Joints	All	Solder (50-50 or 95-5 alloy)	ASTM B-32
Pipe	5–12 in	Carbon steel, Sch. 40	ASTM A 106 or ASTM A 53 Grade B
Valves	All	Same as potable water	
Fittings	All	Ductile iron grooved fittings Malleable iron grooved fittings	ASTM A 536 ASTM A 47
Joints	All	Cut grooved, 800 psi rated	Victaulic Style 77 EPDM gasket

Services:
  Clean Steam and Condensate

Item	Pipe size	Description	Reference standards
Steam pipe	½–4 in	Stainless steel, Sch. 10S beveled ends	Type 316L, welded, ASTM A-312, TP316L
Fittings	½–4 in	Stainless steel, Sch. 10S butt weld	Type 316L, ASTM A 403, Grade WP 316L
Flanges	½–4 in	Forged stainless steel, 150# slip-on or weld neck	ASTM A-182, Grade F-316L/ASA B16.5
Gaskets bolting	All	Service type solid reinforced Teflon, ⅛-in thick	
		Stainless steel, hex head machine bolts	Type 304, and monel nuts

Item	Pipe size	Description	Reference standards
Condensate pipe	½–4 in	Stainless steel, Sch. 40S threaded ends	Type 316L, ASTM A-403, seamless or welded
Fittings	½–4 in	Stainless steel, 150# threaded	Type 316L, ASTM A 182
		Stainless steel, 3000 psi at 100°F forged	ASTM A 182-F316
Unions flanges	½–4 in	Forged stainless steel, 150# slip-on or weld neck	ASTM A-182, Grade F-316L/ASA B16.5
Gaskets	½–4 in	Service type, solid reinforced Teflon, ⅛-in thick	
Bolting		Stainless steel, hex head machine bolts with heavy hex nuts	Type 304, and monel nuts
Steam and condensate		Pressure: 6000 at 200 psi gauge valve with bleed	
Instrument valves			
Ball valves	½–1½ in	Stainless steel body, Type 316L, ball and stem with reinforced Teflon seals and seats butt weld ends	
Butterfly valves	2½–4 in	Stainless steel, 150# lug	ASTM A351-CF8M
Gate valves	½–2 in	Stainless steel, 150 psi, bolted flanged yoke, bonnet outside screw, rising stem, threaded, 200 psi at 500°F	Type 316, ANSI B2.1, B16.34-1977: ASTM A351/296, Grade CF8M
	2–4 in	Stainless steel, 150# flanged, bolted flanged yoke bonnet outside screw, rising stem	Type 316; ANSI B16.10 and B16.34; ASTM A351, Grade CF8M

Item	Pipe size	Description	Reference standards
Check valves	½–2 in	Stainless steel, 150# raised face flanges	Swing check, Type 316L, integral seat, bolted cover with Teflon gasket, ANSI B16.5
Steam traps	½ or ⅜ in	Thermodynamic steam trap, renewable disc, stainless steel body and inner parts, threaded ends	ANSI/ASME PTC-39. 1 and ANSI B1.20.1
Gauge		Stainless steel, Sch. 10 butt welded, diaphragm, Type 316, and Teflon for all wetted parts, filled with high-temperature silicon	
Pigtail siphons		Stainless steel, ¼-in MPT connections, 500 psi at 400°F	Type 316
Strainers	½–2 in (threaded)	Stainless steel, 600 psi at 842°F, Y type with standard ¹⁄₃₂-in screen	Type 316
	½–2 in (welded)	Stainless steel, 600 psi at 842°F	Type 316, ANSI B16.11

Services:
   Instrument Air (All Sizes)
Pressure/Temperature Limits: 125 psig at 100°F
General Materials: Copper, Brass

Item	Pipe size	Description	Reference standards
Tubing	¼-¾ in	Hard-drawn seamless copper tubing, type L	ASTM H 23.1 ASTM B-88
Joints	All	Soldered, 95/5 (tin/antimony) or Swagelok three-piece compression fittings	ASTM B 32
Fittings	All	Wrought copper solder or brass Swagelok three-piece compression tube fittings	ANSI B 16.22
Valves	⅜ in and smaller	Forged body (brass) regulating and shutoff, KEL-F tip stem with Swagelok end connections	
	½ and ¾ in	150-lb, brass or carbon steel three-piece ball valve	
Unions	⅜ in and smaller	Swagelok union connections	
	½ and ¾ in	Cast brass, ground joint, with solder ends	

Services:
  Hot Water (Heating)
  Glycol-Water Solution
  HVAC Chilled Water
  Process Chilled Water
  Water Chemical Treatment Lines for Closed Systems (Sample and Chemical Injection
    Lines)
  Condenser Water
  Refrigerant Vent
  Water Chemical Treatment Lines for Open Systems (Sample and Chemical Injection
    Lines)
Pressure/Temperature Limits: 150 psig at 20 to 250°F for Steel Pipe
                                150 psig at 100°F for Copper Pipe
General Materials: Carbon Steel, Copper

Item	Pipe size	Description	Reference standards
Pipe	2½ in and smaller	Sch. 40 seamless or welded carbon steel	ASTM A 53, Grade B ANSI B 36.10
		OR	
	4 in and smaller	Seamless copper tubing, type L	ASTM B 88
	3–12 in	Sch. 40, ERW, or seamless carbon steel	ASTM A 53
	14–18 in	Sch. 30, ERW, or seamless steel	ASTM A 53
	20–30 in	Sch. 20, ERW, or seamless steel	ASTM A 53
Joining materials	2½ in and smaller	Screwed joints with Teflon tape or paste	
	3 in and larger	Welded	ANSI B 16.4
		OR	
	Copper	95/5 (tin/antimony) solder	ASTM B 32
Fittings	2½ in and smaller	Malleable iron threaded fittings, class 150	ANSI B 16.3
		OR	
	Copper	Wrought copper solder joint pressure fittings	ANSI B 16.22

Item	Pipe size	Description	Reference standards
Fittings	3 in and larger	Steel butt welding fittings	ASTM A 234 ANSI B 16.9
		OR	
		For chiller and condenser water dedicated loop pipings in central plant (no main headers) and local building secondary chilled water distribution piping (downstream of local building pumps) only, use malleable or ductile iron grooved end fittings (not permitted in inaccessible chases)	ASTM A 47 ASTM A 536
Flanges	All	150-lb, forged steel, slip-on, flat or raised face to match equipment	ASTM A 105, Grade 2
Flange bolting	All	Alloy steel bolt studs; thread full length with heavy hex nuts	ASTM A 193, Grade B7
Dielectric fittings	All	Isolation flanges, unions and couplings	
Unions	2 in and smaller	Brass solder unions	
Ball valves (shutoff service only)	2½ in and smaller	150-lb, bronze, serviceable in-line, quarter-turn lever, soldered connections	ASTM B 62
		OR	
	2½ in and smaller	200-lb, cast steel body, chromed steel ball and stem, double TFE seal, threaded ends	
	3 in and larger	150-lb, ball valve, cast steel body, stainless steel ball and stem, double TFE seal, flanged ends	
Ball valves (balancing and shutoff service)	2½ in and smaller	250-lb, bronze body, ball type, threaded ends, with pressure taps, with memory stop slot	
	3 in and larger	150-lb, ball valve, cast steel body, stainless steel ball and stem, double TFE seal, flanged ends	

Item	Pipe size	Description	Reference standards
Butterfly valves (balancing and shutoff service)	3 in and larger	150-lb, high-performance butterfly valve, cast steel body, stainless steel disc and stem, TFE seal, threaded lug (for balancing, add memory stop slot)	
Swing check valves	2½ in and smaller	125-lb, bronze body, bronze disc and seat, soldered ends	
	3 in and larger	125-lb, cast iron body, bronze trim disc and seat, flanged ends	
Silent check valves	2½ in and smaller	150-lb, cast steel body, springloaded stainless steel disc, renewable stainless steel seat, flanged ends	
	3 in and larger	150-lb, cast steel body, spring-loaded stainless steel disc, renewable stainless steel seat, flanged ends	
Triple-duty valves	2½ in and larger	Straight pattern, cast iron body combining the functions of a silent check valve, a balancing valve, and a positive shutoff valve. Include a calibrated stem and pointer to indicate valve position. Valve stem shall be capable of being repacked under pressure. 175 psig working pressure at 300°F	
Safety relief valves	All	150-lb, bronze body, threaded connections, ASME rated	
Joints*	2½ in and smaller (On or below computer areas, use solder only.)	Threaded OR Soldered	
	3 in and larger	Butt welded OR Mechanical coupling grooved end (use only for chilled water and condenser water where permitted)	

*For piping installed on floor below finished raised-floors of computer and telecommunications spaces, use only soldered or butt-welded joints. Threaded joints are not acceptable.

Service:
Chilled Water "Underground"

Item	Pipe size	Description	Reference standards
Pipe	4–24 in	Mechanical joint, ductile iron cement lining	ANSI A21.51, ANSI A21.4
Fittings	3–12 in	250# mechanical joint, ductile cast iron, with cement lining	ANSI A21.10, ANSI A21.4
	14–24 in	150# mechanical joint, ductile cast iron, with cement lining	ANSI A21.10, ANSI A21.4
Valve box		2-piece screw or slip with water cap	
Valves	4–12 in	125# mechanical joint gate valve	ANSI A21.11
		125# flanged gate valve	ANSI A21.11
	14–24 in	125# mechanical joint gate valve	ANSI A21.11
Flanges	4–24 in	Ductile iron, 125# FF	ANSI A21.10
Boltings (for ANSI mechanical joints) (for ANSI flanged joints)		Carbon steel, tee head bolts with hevy hex nuts Carbon steel, hex head machine bolts with heavy hex nuts	ASTM A307, Grade B
Gaskets (for ANSI mechanical joints) (for ANSI flanged joints)		Plain rubber 150# FF neoprene service gasket	

Note: 125# cast iron fittings may be substituted for ductile iron according to availability.

Services:
  Above Ground Only
  Fuel Oil Supply and Return to Boiler and Day Tank
  Fuel Oil Tank Fill, Vent, and Overflow
  Day Tank Vent, Overflow, Supply, and Return for Diesel Generator
  Kerosene
Pressure/Temperature Limits: 150 psig at 300°F
General Materials: Carbon Steel

Item	Pipe size	Description	Reference standards
Pipe	10 in and smaller	Sch. 40, ERW, or seamless steel	ASTM AS3, Grade B
Joints	All	Welded, except where noted	ASTM A 53
Fittings	2 in and larger	3000-lb, forged steel, socket welded	ASTM A 234 82a ANSI B 16.9 78 B 16.11 80 B 16.28 78
	2½ in and larger	150-lb steel, butt welded	ASTM A 234 82a ANSI B 16.9 78 B 16.11 80 B 16.28 78
Flanges	2½ in and larger	150-lb forged steel, weld neck or socket weld, raised face	ASTMA 181 81 ANSI B 16.5 81
Unions	2 in and smaller	3000-lb forged steel with bronze-steel joints, socket welded	ASTM A 105 Grade
Stop valves	All	Butterfly valve, 150 psi rated carbon steel trim, Teflon seat and seal	
Check valves	2 in and smaller	150-lb class, forged steel, lift check, socket weld, SS trim	

Note: All piping shall be housed in a secondary containment of Schedule 40, transparent rigid PVC pipe and fittings, solvent-welded socket joints, fuel resistant and sloped to detection points.

Services:
   Compressed Medical Gases, Interior, Above Ground
   Gases for Pneumatic Tools
   Vacuum Air (WAGD, Medical Vacuum)
   General Laboratory Use
Pressure/Temperature Limits: Up to 200 psig
General Materials: Copper

Item	Pipe size	Description	Reference standards
		For pressures up to 200 psig	
Tubing	Up to 4 in	Hard-drawn copper tubing, seam-less type L	ASTM B 819
Fittings	All	Wrought copper solder end	ASTM B 16.22
Joints	All	Brazed, BCup filler metal, no flux, purged with nitrogen or argon	ASTM B 32
Valves	All	Three-piece ball valves, distribution piping only, 300 psi rated	
		For pressures from 201 to 300 psig	
Tubing	Up to 4 in	Hard-drawn copper tubing, seam-less type K	ASTM B 819
Valves	All	Three-piece ball valves, distribution piping only, 400 psi rated	
		All other components similar to above	

Service:
  Safety Valve Drains
Pressure/Temperature Limits: S psig at 100°F
General Materials: Carbon Steel, Copper

Item	Pipe size	Description	Reference standards
Pipe	¾–2 in	Sch. 40 welded carbon steel	ASTM A 53, Grade B
		OR	
		Hard-drawn seamless copper tubing, type L	ASTM B 88
Joints	¾–2 in	125-lb cast iron, banded, screwed connections	ASTM A 126, Class B
		OR	
		Soldered	ASTM B 32
Fittings	¾–2 in	125-lb cast iron, banded, screwed connections	ASTM A 126, Class B
		OR	
		Wrought copper solder joint fittings	ANSI B 16.22
Unions	¾–2 in	125-lb malleable iron, brass to iron seat, screwed connections	ASTM A 197
		OR	
		Wrought copper solder unions	
Valves	¾–2 in	125-lb bronze ball valve, screwed connections	

Services:
    Domestic Cold, Hot, and Hot Water Return (Above Ground)
    Gravity Condensate Drains for AHUs, Ductwork, and Dehumidifiers
    Humidification Water
    Emergency Drench Equipment
    Trap Primers
    Safety Showers
    Slurry Return and Drain Water
Pressure/Temperature Limits: 90 psig at 40 to 140°F
General Materials: Copper, Steel

Item	Pipe size	Description	Reference standards
Pipe	4 in and smaller	Hard-drawn seamless copper tubing, type L	ASTM B 88
	6 in and larger	Sch. 40 seamless carbon steel, galvanized	ASTM A 53, Grade B
Joints	4 in and smaller	Soldered	ASTM B 32, lead free
	6 in and larger	Cut grooved split coupling	ASTM A 47 ASTM A 536
Fittings	4 in and smaller	Wrought copper solder joint fittings	ANSI B 16.22
	6 in and larger	Malleable or ductile iron, cut grooved	ASTM A 47 ASTM A 536
Flanges	6 in and larger	150-lb, malleable or ductile iron	ANSI B 16.5 ASTM A 47 ASTM A 536
Unions	1½ in and smaller	Red bronze	ASTM 75
Flange bolting	6 in and larger	Hex machine bolts with heavy hex nuts	ASTM A 307, Grade B
Unions	3 in and smaller	Wrought copper solder unions	ANSI B 16.22
		AND	
		Dielectric, copper to steel	
Ball valves	2 in and smaller	150-lb, bronze, serviceable in-line, quarter-turn lever, soldered connections	
Gate valves	2 in and smaller	125-lb, bronze, nonrising stem, soldered ends	
	2½ in and larger	125-lb, cast iron, nonrising stem, flanged ends	

Item	Pipe size	Description	Reference standards
Check valves	2 in and smaller	125-lb, bronze, swing check, renewable bronze disc and plugs, solder end connections	
	2½ in and larger	125-lb, cast iron swing check, bolted cap, renewable discs and plugs, flanged ends	
Balancing valves	2 in and smaller	250-lb, bronze body, Teflon seats and seals, solder ends	
Gaskets	All	Nonasbestos with synthetic fibers and SBR binder, full face on flat face flanges	

Services:
  Domestic Water, Underground
  Potable Water
  City Water, "Underground"

Item	Pipe size	Description	Reference standards
Pipe	4–24 in	Ductile cast iron, mechanical joint, 250 psi rated, 18-ft lengths, class 52, cement lining	ANSI/AWWA, C151 A21.51
Joints	All	Bolted mechanical joints	ANSI/AWWA, C104/A21.4
Fittings	4–24 in	Ductile cast iron, 250 psi pressure rated, mechanical joint with glands, bolts, and gaskets	ANSI/AWWA, C110/A21.10, and ANSI/AWWA, C111/A21.11
		Cement lining	ANSI/AWWA C104/A21.4
Joint restraint and anchors	All	Ductile iron set-screw retainer glands on joints at all directional changes and at dead ends, also at tees, hydrants, and valves	
		Alternate: Tie rods and nuts or Victaulic style 37	
Thrust blocks		Thrust blocks must be provided in addition to retainer glands and/or tie rods	

Item	Pipe size	Description	Reference standards
Valves	hydrants	5¼-in main valve, 6-in mechanical joint inlet connection, one (1) 4½-in pumper connection, two (2) 2½-in hose connections, automatic drains and bronze pentagon operating nut, measuring 1½-in point to flat	
	6-in hydrant shutoff valves	Mechanical joint, gate pattern, square operating nut, nonrising stem with curb box. Valve to be located within 5 ft of hydrant served.	AWWA C111-53
Valve box (4) (curb box)		Cast iron, 2-piece telescoping, screw adjustable for depth. Cover to be loose type with water cast iron top. Provide one (1) crowfoot wrench for every (6) valves with curb boxes; painted	
Valves	4–12 in	Bronze-mounted gate valve with mechanical joint ends, nonrising stem, parallel seats and operating nut. W.P. 175# nonshock, cold water, UL listed, FM approved	AWWA C111-53
	4–12 in	Bronze-mounted 125# flanged gate valve, nonrising stem, parallel seats and operating nut, W.P. 150# nonshock, cold water, UL listed, FM approved	ANSI A21.11
	14–24 in	Iron, gate valve with mechanical joint ends, nonrising stem, inside screw parallel seats, double disc, W.P. 150# cold water	ANSI 21.11, AWWA C-111
	14–24 in	Iron, gate valve with 125# flanged ends, nonrising stem inside screw, parallel seats, double disc, W.P. 150# cold water	ANSI 21.11 AWWA C500

Item	Pipe size	Description	Reference standards
Flanges	4–24 in	125# flat faced flange	Ductile iron, ANSI A21.10
Boltings (for ANSI mechanical joints)		Tee head bolts with heavy hex nuts	High-strength corrosion-resistant alloy
(for ANSI flanged joints)		Hex head machine bolts with heavy hex nuts	ANSI B16.1
Gaskets (for ANSI mechanical joints)		Plain rubber	Clow #F-1950
(for ANSI flanged joints)		150# FF service gasket to 12 in	Clow #F-1950

Services:
Storm Water Drainage Interior Aboveground
Pressure/Temperature Limits: Gravity drainage/140°F, PVC Only—120°F
General Materials: Cast Iron, PVC

Item	Pipe size	Description	Reference standards
Pipe	2–15 in	Cast iron, soil pipe, service weight hub and spigot	ASTM A 74
		No-hub cast iron soil pipe	CISPI 301
		PVC plastic drainage pipe Sch. 40	ASTM D 2665
Fittings		Cast iron hub and spigot	ASTM A 74
		No-hub cast iron fittings	CISPI 310
		PVC socket-type fittings	ASTM D 2665
Joints		Poured lead and oakum	ASTM B 29
		No-hub compression coupling, stainless steel cover and neoprene gasket	CISPI 310
		PVC solvent cement	

Service:
Storm Drainage "Underground"

Item	Pipe size	Description	Reference standards
Pipe	3–12 in	XH (extra-heavy) cast iron, plain end hub and spigot soil pipe	ASTM A 74
	12–30 in	Reinforced concrete, bell and spigot	ASTM C 76, Class IV, Wall B
Fittings	3–15 in	XH (extra-heavy) cast iron, plain end hub and spigot soil fittings	ASTM A 74
	15–30 in	Reinforced concrete, bell and spigot	ASTM C 76, Class IV, Wall B
Gaskets	3–15 in (for cast iron joints)	One-piece neoprene compression gasket	ASTM C 564
	15–30 in (for concrete joints)	Push-on rubber O ring	ASTM C 443

Services:
  Above Ground Interior
  Natural Gas
  Propane
  Mixed Gas
Pressure Limit: Medium Pressure up to 20 psig
General Materials: Carbon Steel, Iron, Brass

Item	Pipe size	Description	Reference standards
Pipe	All	Sch. 40 welded black steel	ASTM A 53
	Up to 1⅛ in	Copper tubing type G	ASTM B 837
Joints	3 in and smaller	Screwed	
	4 in and larger	Welded	
	Copper pipe	Soldered	
Fittings	3 in and smaller	Screwed malleable iron	ANSI B 16.3
	4 in and larger	Steel butt weld	ANSI B 16.9
	Fittings	Wrought copper	ASTM B 16.22
Joining materials	3 in and smaller	Teflon tape or paste, Du Pont	
	4 in and larger	Welded	
Valves	1 in and smaller	Brass squarehead gas cocks	
	Larger than 1 in	Steel lubricated plug gas cocks	

Service:
  Below Ground Fuel Gases, 1- to 12-in sizes; Natural Gas, Propane, Mixed Gases
  Pressure/Temperature Limits: 100 psig, Ambient
  General Materials: Steel, PE

Item	Pipe size	Description	Reference standards
Steel pipe	All	Black steel, Sch. 40, with shop or field-applied coating	ASTM A 53 Grade B
Pipe coating		Shop coating of extruded PE, min. 36 mils thickness, with adhesive undercoat	Type 1, Fed. Spec. L-C-530
		Shop coating of thermoset epoxy, mm. 10 mils thick	Type 2, Fed. Spec. L-C-530
		Coal tar enamel 0.129 in thick with felt or fiberglass wrap, mm. 120 mils thick	AWWA C 203
		Field-applied plastic tape, min. 10 mils thick	Fed. Spec. L-T 1512, Type 1
Fittings		Welded steel	ANSI B 16.11
		Threaded steel	None
		Dresser	ANSI B 16.5
		Flanged steel	
Joints		Welded	API-1104
		Threaded	ANSI B 2.1
		Flanged	ANSI B 31.8
		Mechanical	AWWA C 111/A 21.11
Plastic pipe and fittings		High-density PE, SDR 13.5 Black, grades 2306, 3306 or 3408	ASTM D 2513
Joints		Butt heat fusion (preferred) or other mechanical couplings approved by manufacturer	
Valves	All	Lubricated plug type, high head extension, dresser or flanged end min. rating 175 psig wog	

Service:
Sanitary Soil and Vent (Above Ground)
Pressure/Temperature Limits: Atmospheric/Ambient
General Materials: Cast Iron

Item	Pipe size	Description	Reference standards
Pipe	All	Service-weight cast iron soil pipe	ASTM A 74
		OR	
		No-hub cast iron soil pipe	CISPI 301
Joints	All	Poured lead and oakum, bell and spigot ends	
		OR	
		No-hub, plain end	
		OR	
		Puch-on, bell and spigot ends	AWWA C 111/ ANSI A 21.11
Fittings	All	Cast iron hub and spigot fittings	
		OR	
		Cast iron no-hub fititngs	CISPI 310
Joining materials	All	Molten lead and oakum fiber packing, plumbing grade	ASTM B 29
		OR	
		No-hub neoprene gasket with CISPI Series 310 stainless steel shield and retaining clamps	
		OR	
		Neoprene compression gasket	

Service:
Sanitary Drainage "Underground"

Item	Pipe size	Description	Reference standards
Pipe	2–15 in	Service-weight cast iron, plain end, hub and spigot soil pipe	ASTM A 74
Fittings	2–15 in	Service-weight, plain end, hub and spigot-type soil fitting	ASTM A 74
Gaskets	3–15 in	Neoprene, one-piece compression gasket	ASTM C 443

Services:
Nonsanitary Waste and Vent (Above Ground) Other Than Laboratories
Pressure/Temperature Limits: Atmospheric/120°F
General Materials: Copper, Carbon Steel, Cast Iron

Item	Pipe size	Description	Reference standards
Pipe	2½ in and smaller	DWV seamless copper drainage tube, hard drawn	ASTM B 306
	3 in and larger	Sch. 40 welded carbon steel, galvanized	ASTM A 120
		No-hub cast iron soil pipe	CISPI 301
Joints	2½ in and smaller	Soldered	
	3 in and larger	Screwed, threaded ends	
		No-hub, plain end	
Fittings	2½ in and smaller	Cast and wrought copper DWV fittings	ANSI B 16.23
	3 in and larger	Galvanized cast iron threaded drainage fittings	ANSI B 16.12
		Cast iron no-hub fittings	CISPI 310
Joining materials	2½ in and smaller	95/5 (tin/antimony) solder	ASTM B 32
	3 in and larger	Teflon tape or paste	
		No-hub neoprene gasket with CISPI Series 301-85 stainless steel shield and retaining clamps	

Services:
  Nonsanitary Waste and Vent (Above Ground) Other Than Laboratories
  Pressure/Temperature Limits: Atmospheric/120°F
  General Materials: Copper, Carbon Steel, Cast Iron

Item	Pipe size	Description	Reference standards
		OR	
Pipe	1½–4 in	PVC Sch. 40	ASTM D 2665
Fittings	1½–4 in	Socket solvent joints	ASTM D 2665
Joining materials	All	Solvent/cement	ASTM D 2855

Service:
  Slurry System
  Pressure/Temperature Limits: 100 psig at 75°F
  General Materials: Copper

Item	Pipe size	Description	Reference standards
Pipe	All	Hard-drawn seamless copper tubing, type K	ASTM B 88
Joints	All	Soldered	
Fittings	All	Wrought copper	ANSI B 16.23

Service:
  Sump Pump and Sewage Ejector Pump Discharge (Above Ground)
  Pressure Limit: 125 psig Ambient
  General Materials: Carbon Steel, Ductile Iron

Item	Pipe size	Description	Reference standards
Pipe	All	Sch. 40 carbon steel, galvanized	ASTM A 53, Grade B
Joints	All	Threaded end	ASTM B 2.1
Fittings	All	Galvanized cast iron threaded drainage fittings	ANSI B 16.12
Joining materials	All	Teflon tape or paste	
Gate valves	2 in and smaller	125-lb, bronze, nonrising stem, soldered ends	
	2½ in and larger	125-lb, cast iron, nonrising stem, flanged ends	
Check valves	2 in and smaller	125-lb, bronze, swing check, renewable bronze disc and plugs, soldered end connections	
	2½ in and larger	125-lb, iron body, bronze disc and seat, flanged ends	

Service:
USP Purified Water Feedwater
Pressure/Temperature Limits: 100 psig at 180°F
General Materials: Polypropylene

Item	Pipe size	Description	Reference standards
Pipe	All	Sch. 40 virgin unpigmented pipe grade polypropylene specifically for deionized water	ASTM D 2146, Type I
Joints	All	Heat-fusion butt welded	
Fittings	All	Sch. 40 virgin unpigmented pipe, grade polypropylene heat-fusion pipe fittings	ASTM D 2146
Valves	All	150-lb, polypropylene body, ball and stem compatible with pipe, socket-welded ends	
		OR	
Pipe and fittings	All	Type 316L sanitary tubing 0.065 wall thickness with square ends. Fittings shall be long turn for machine welding with 240 grit finish, electropolished.	ASTM A 270
Joints	All	Automatic machine butt weld. All equipment shall be joined with sanitary clamp type similar to Tri-Clover, Inc. Clamp shall be 304SS with metal wing nut or SS bolts depending on size.	
Valves	All	Diaphragm, wier type, clamp ends Type 316L, 150 psig rated top entry sanitary diaphragm of Grade R-2 Teflon, finish same as pipe.	

Service:
   Diesel Engine Exhaust
Pressure/Temperature Limits: 5 psig at 1000°F
General Materials: Carbon Steel

Item	Pipe size	Description	Reference standards
Pipe	All	Sch. 20 welded carbon steel	ASTM A 53, Grade A
Joints	All	Welded	
Fittings	All	Butt weld, carbon steel, beveled end, long radius, with thickness to match pipe	
		For drain piping, use 3000-lb forged steel socket weld connections.	ASTM A 105
Flanges	All	125/150 lb. forged steel, slip-on	
Flange bolting	All	Alloy, steel bolt studs, thread full length	ASTM A 193, Grade B7
		Heavy hex nuts	ASTM A 194, Class 2H ANSI B 18.2
Drain valves	All	Forged steel gate valves, solid wedge, bolted bonnet and gland, OS&Y, socket weld connections, Vogt Figure SW12111	ASTM A 105, Grade 1
Gaskets	All	Garlock style 1000	

Service:
  Vacuum Cleaning Systems
  Pressure/Temperature Limits: 10 in Hg Vacuum/Ambient
  General Materials: Zinc-Coated Steel

Item	Pipe size	Description	Reference standards
Tubing	2⅛ in and larger	No.16 gauge minimum zinc-coated steel	
Joints	All	Sealed joints with heat-shrink sleeves	
Fittings	All	Galvanized steel, long-radius type, with expanded ends	
Joining materials	All	Heat-shrink sleeves	
Valves	2 in	Nickel-plated ball valve forged brass type D	Spencer Turbine
	4–12 in	125# FF blast gate, cast iron body, SS shaft C/S shutter	

Service:
  Refrigerant Piping Systems
  Pressure/Temperature Limits: 225 psig at 115°F (Condensing)
  General Materials: Copper

Item	Pipe size	Description	Reference standards
Pipe	2 in and smaller	Seamless copper air-conditioning and refrigeration tubing, type ACR, hard drawn	ASTM B 280
Joints	2 in and smaller	Soldered or brazed	
Fittings	2 in and smaller	Wrought copper solder joint pressure fittings or wrought copper brazing fittings	ANSI B 16.22

Item	Pipe size	Description	Reference standards
Joining materials	2 in and smaller	Solder, 95/5 (tin/antimony) brazed silver solder, AWS Classification BCuP5 alloy (15% silver, 5% phosphorous, 80% copper) Silfos 15 by Handy and Harmon; Flux, Handy Flux by Handy and Harmon	ASTM B 32, Alloy, Grade 95TA or ANSI/AWS A5.3
Shutoff valves	2 in and smaller	Diaphragm type, forged brass body and bonnet, positive backseating when fully open, raised seat with nylon seat disc, stainless steel spring, flared or soldered connections, soldered connections with extended ends, UL listed	
Check valves	2 in and smaller	Forged brass body, Teflon seat, guided piston, stainless steel spring, accessible internal parts, operable in all positions. Rated for 300°F and 500 psig.	
Relief valves	2 in and smaller	Automatic, forged brass body, reseating characteristics, stem guide, factory sealed, ASME constructed and certified.	

Services:
  Process Waste and Vent (Above Ground)
  Solvent Waste and Vent (Above Ground)
  Oxidizable Waste and Vent (Above Ground)
  Isolation Waste and Vent (Above Ground)
  Lab Waste and Vent (Above Ground)
Pressure/Temperature Limits: Atmospheric/Ambient
General Materials: Borosilicate Glass

Item	Pipe size	Description	Reference standards
Pipe	All	Borosilicate glass	ASTM E 438-83
Joints	All	Compression couplings (bead-to-bead and bead-to-plain-end compression couplings)	
Fittings	All	Compression coupling fittings	
		OR	
Pipe and fittings	All	Flame-retardant polypropylene, Sch. 40	
Joints	2 in and smaller	Proprietary mechanical-screwed	
	2½ in and larger	Socket, heat-fused joint	
		OR	
	2–10 in	Bell and spigot	Silicon iron, ASTM A 518 Duriron Co.
Fittings	2–10 in	Bell and spigot floor and drainage fittings	Silicon iron, ASTM A 518 Duriron Co.
Joints		Caulked	
Joint packing		Acid resistant, asbestos free	Oakum No. 312 sealtite red stripe or BMS approved equal

Service:
  Process Drainage "Underground"

Item	Pipe size	Description	Reference standards
Pipe	3–15 in	Plain end, hub and spigot soil pipe, Tyler pipe or equal	XH (extra-heavy) cast iron, ASTM A 74
Fittings	3–15 in	Plain end, hub and spigot, soil fitting	XH (extra-heavy) cast iron, ASTM A 74
Gaskets	3–15 in	Plain end, hub and spigot, soil fitting	XH (extra-heavy) cast iron, ASTM A 74
Gaskets	3–15 in	One-piece compression gasket	Neoprene, ASTM C 443

Service:
  Severe Service Biohazard Waste and Vent System
  Pressure/Temperature Limits: 75 psig at 280°F
  General Materials: Polyvinylidene Fluoride (PVDF)

Item	Pipe size	Description	Reference standards
Piping	1–4 in	Unpigmented, natural tube, polyvinylidene fluoride (PVDF) pipe	ASTM D 1785
Joints	1–4 in	Socket heat-fusion welded	ASTM D 2657
Fittings	1–4 in	Unpigmented, natural PVDF, socket-fusion	ASTM D 2467

Service:
  Ordinary Service Biohazard Waste and Vent System
  Pressure/Temperature Limits: Atmospheric/Ambient
  General Materials: Chlorinated Polyvinyl Chloride (CPVC)

Item	Pipe size	Description	Reference standards
Piping	All	Sch. 80 chlorinated polyvinyl chloride (CPVC) pipe	ASTM F 441
Joints	All	Socket solvent cement	
Fittings	All	Sch. 80 CPVC, socket ends	ASTM F 439
Primer for solvent cement	All	Recommended by manufacturer	
Solvent cement	All	Do	ASTM F 493

Services:
  Compressed Medical, Laboratory, and Process Gases and Vacuum:
  Vacuum, Nitrogen, Compressed Air, High-Pressure Compressed Air, and Cylinder Gases
  Carbon Dioxide
  Pressure/Temperature Limits:
  100 psig/Ambient (Breathing Air)
  29 in Hg/Ambient (Vacuum)
  200 psig/Ambient (Nitrogen and Compressed Air)
  General Materials: Copper, Carbon Steel

Item	Pipe size	Description	Reference standards
Pipe	4 in and smaller	Seamless copper tubing, Type L, hard drawn	ASTM B 88, Grade B
	High pressure gases	Type K copper tubing	
	4 in and larger (process only)	Sch. 40 seamless carbon steel	ASTM A 53
	4 in and smaller	Brazed	AWS 5.2
Joints	4 in and larger (process only)	Welded connections	

Item	Pipe size	Description	Reference standards
Fittings	2 in and smaller	Wrought copper brazed fittings	ANSI B 16.22
	3 in and larger	Standard seamless steel, butt welded	ASTM A 234, Grade WPB
Joining materials	All	15% silver solder, AWS classification BCuP5 alloy (15% silver, 5% phosphorous, 80% copper)	ANSI/AWS A 5.3
Ball valves	½-¾ in	150-lb, bronze, three-piece ball valve	
	1–2 in	150-lb, bronze serviceable in-line, quarter turn, lever operated, 3 piece, full port, threaded connections	
	2½–4 in	400-lb, bronze body ball valve, chrome-plated bronze ball, full flow with Teflon seats and double Teflon stem seal	
Pressure-regulating valves	¼–1 in	400-lb maximum inlet pressure rated, standard relieving diaphragm, pressure adjustment range 5 to 125 psig, zinc body, aluminum bonnet and brass valve assembly with pressure gauge (0–160 psig)	
Gaskets	All	Homogeneous nitrile rubber, full face for flat face flanges	

Service:
Cryogenic Liquid
Pressure/Temperature Limit: 150 psig at $-320°F$
General Materials: Preinsulated Copper

Item	Pipe size	Description	Reference standards
Pipe and fittings	Up to 2 in	Preinsulated, type K copper tubing, hard drawn. 3-in polyurethane foam insulation with minimum 0.06-in PVC jacket	ASTM B 88
Joints	All	Soldered, 45% silver	
Ball valves	All	Cryogenic, extended stem, screwed end, bronze body, stainless steel ball	

Service:
  Cylinder Gases, Pharmaceutical Use, High Purity
  Pressure/Temperature Limits: 100 psig/Ambient
  General Materials: Stainless Steel

Item	Pipe size	Description	Reference standards
Pipe	All No. 4 finish	304L or 316L stainless steel tube 180 grit electropolished I.D., no. 4, mill finish O.D.	ASTM A 270
Joints	All	Machine orbital welded	
Fittings	All	304L or 316L stainless steel tube welded fitting, 180 grit electropolished I.D., mill finish O.D.	
Valves	All	Ball valve, 150-lb, 304L or 316L SS, stainless steel serviceable in-line, quarter turn, lever operated, 3 piece, full port, butt-welded connections, oxygen cleaned	ASTM A 351
		OR	
		Diaphragm valve, wier type	

Service:
   Reverse-Osmosis Water
Pressure/Temperature Limits: 100 psig at 180°F
General Materials: Polypropylene

Item	Pipe size	Description	Reference standards
Pipe	All	Type 1, Sch. 80, unpigmented polypropylene	ASTM D 2146
Joints	All	Heat-fusion welded	
Fittings	All	Type 1, Sch. 80, polypropylene, socket ends, heat-fusion welded	ASTM D 2146
Flanges	All	150-lb, socket ends	ASTM D 2146
Gaskets	All	Viton	
Couplings and adapter couplings	All	Type 1, Sch. 80, polypropylene, socket ends or socket-to-thread ends	ASTM D 2146
Ball valves (balancing and shutoff)	All	150-lb, Type I polypropylene body, ball and stem compatible with pipe, socket-welded ends, true union design, Hayward True Union Series	ASTM D 2146
Check valves	All	Type 1 polypropylene, body, true union ball type, Viton or EPDM O-ring seals and seat, socket connection, Nibco Ball Check Valve	ASTM D 2146

Service:
Distilled, Deionized, USP Purified
USP, up to 2-in Pipe Size Only

Item	Pipe size	Description	Reference standards
Pipe	½–2 in	Stainless steel, Sch. 10S welded	Type 316, ASTM A 312
	¼ in and smaller	Stainless steel, Sch. 40S seamless	Type 316, ASTM A 312
Fittings	½–2 in	Stainless steel, Sch. 105 butt weld	Type 316, ASTM A 403
	¼ in and smaller	Forged stainless steel, screwed end	ASTM A 182 F316
Flanges		Forged stainless steel, 150# slip-on or raised face well neck	ASTM A 182 F-316L
Gaskets	All sizes	Teflon and neoprene, 1 50# envelope with neoprene insert	
Bolting	All sizes	Stainless steel, stud bolts	
	All sizes	Stainless steel, hex nuts	
Valves	2 in and smaller	Stainless steel, butt weld ball reinforced Teflon seats, solid ball	Type 316

Service:
Chlorichlorinated Distilled, Deionized, USP Purified
USP, up to 2-in Pipe Size Only

Item	Pipe size	Description	Reference standards
Sanitary fittings		Stainless steel, Sch. 5 or tube size, clamped fittings	Type 316L
Gauges		Stainless steel, Sch. 10 butt welded, diaphragm and Teflon for all wetted parts, filled with high-temperature silicone	Type 316

Service:
  Water for Injection
  Up to 2-in Pipe Size Only

Item	Pipe size	Description	Reference standards
Tubing	½–2 in	Welded stainless steel, 0.065-in wall thickness, interior polished to 180 grit min., manufactured to 3A standard	Type 316L, ASTM A-270
Fittings	½–¾ in	Stainless steel, ½-in O.D. × 0.065-in wall thickness	Type 316L, ASTM-270
		¾-in × 0.065-in wall thickness, 180 grit polished I.D. min. Manufactured to 3A standard. Fittings must have extended ends for automatic welding	Type 316L
Sanitary fittings	1–2 in	Stainless steel, Sch. 5 or tube size, clamped fittings	Type 316L
Gauges		Stainless steel, Sch. 10 butt welded, diaphragm, filled with high-temperature silicone	Type 316
Gaskets	½ in		Use Velex No. VL40T, 0.757 O.D.
	1–2 in	Standard molded, one-piece gasket Teflon envelope	
Joints	½–¾ in	Stainless steel, 2-in long male ferrule and 2-in long female ferrule. 180 grit polished I.D. min.	Type 316L
	1–2 in	Stainless steel, manufactured to 3A standards, 3-in long ferrule butt weld, Type 180, grit polished I.D. min.	Type 316L

Item	Pipe size	Description	Reference standards
Valves	½–2 in	Ball valve 316L stainless steel body with tube ends of 0.065-in wall thickness, ball and stem 316 stainless steel externals (nuts, bolts, etc.). Reinforced Teflon seats and S-gasket body seals. Valves to be passivated interior of valve and ball, to be polished to 180 grit.	
		OR	
		For use where tubing spec is not practicable or not adaptable to existing plant, butt-welded WFI Systems.	
Pipe	½–2 in	Stainless steel, Sch. 10S welded	Type 316, ASTM A-312
	¼ in	Stainless steel, Sch. 40S seamless	Type 316, ASTM A-312
Fittings	½–2 in	Stainless steel, Sch. 10S butt weld	WP-316, ASTM A-403
	¼ in	Forged stainless steel, screwed	ASTM A-182-F316L
Flanges		Forged stainless steel, 150# slip-on or RF weld neck	ASTM A-182-F316L
Gaskets	All sizes	150# envelope, Teflon and neoprene, with neoprene insert	
Bolting	All sizes	Stainless steel, stud bolts	
	All sizes	Stainless steel, hex nuts	
Valves	½–2 in	Stainless steel, reinforced Teflon seats, S-gasket body seals and solid balls, butt weld ball, with 4-in stainless steel extension	Type 316
Sanitary fittings		Stainless steel, Sch. 5 or tube size clamped fittings	Type 316L
Gauges		Stainless steel, Sch. 10 butt-welded, diaphragm, Type SM	Type 316, and Teflon for all wetted parts, filled with high-temperature silicon

Service:
  Deionized Water
  PVC Piping (Polyvinyl Chloride)

Item	Pipe size	Description	Reference standards
Pipe	¼–3 in	PVC, Sch. 80, Type 1, plain end	Grade 1
Fittings	¼–3 in	PVC, Sch. 80 socket type ends	
Unions	¼–3 in	PVC, unions, socket type ends for solvent cementing and Viton O rings, Sch. 80	ASTM D-2467 and ANSI B2.1
Flanges	½–3 in	PVC, socket type flanges for solvent cementing, Sch. 80, 150#, F.F.	ASTM D-2467
Bolting		Heavy hex head machine bolts with heavy hex nuts and flat C/S washers	ASTM A-307, Grade B
Gaskets		150#, ⅛-in thick Gylon full face, ANSI dimensions	
Valves	½–3 in	PVC, ball valve socket ends, 150#, one-quarter turn	Type 1, Grade 1, ASTM D-1784
	½–3 in	PVC, check valve, socket ends, 150#, with Viton O rings	Type 1, Grade 1, ASTM D-1784
	½–3 in	PVC, diaphragm valve socket ends (Teflon diaphragm) with standard weir-type bonnet assembly	Type 1, Grade 1

Note 1: PVC adapters shall be used at all threaded connections.

TRADE ASSOCIATIONS

Acoustical Society of America (ASA)
Huntington Quandrangle, Suite 1N01
Melville, NY 11707
(516) 576-2360

Adhesive and Sealant Council (ASC)
1500 Wilson Blvd., Suite 515
Arlington, VA 22209
(703) 841-1112

Air and Waste Management
Association (AWMA)
P.O. Box 2861
Pittsburgh, PA 15230
(412) 232-3444

Air Conditioning and Refrigeration
Institute (ARI)
4301 N. Fairfax Drive, Suite 425
Arlington, VA 22203
(703) 524-8800

Air Conditioning Contractors of
America (ACCA)
2800 Shirlington Rd., Suite 300
Arlington, VA 22206
(703) 575-4477

Air Diffusion Council (ADC)
104 S. Michigan Ave., Suite 1500
Chicago, IL, 60603
(312) 201-0101

Air Distribution Institute (ADI)
4415 West Harrison Street, Suite 242-C
Hillside, IL 60162
(708) 449-2933

Air Movement and Control Association
(AMCA)
30 West University Drive
Arlington Heights, IL 60004
(847) 394-0150

Alliance for Engineering in Medicine
and Biology
1101 Connecticut Avenue NW, Suite 700
Washington, DC 20038
(202) 857-0666

Alliance of American Insurers (AAI)
1501 Woodfield Avenue
Schaumburg, IL 60173
(708) 330-8500

Alliance to Save Energy
1725 K Street NW, Suite 914
Washington, DC 20006
(202) 857-0666

Aluminum Association (AA)
900 19th Street NW, Suite 300
Washington, DC 20006
(202) 862-5100

Amateur Athletic Union (AAU)
3400 West 86th Street
Indianapolis, IN 46468
(317) 872-2900

American Architectural Manufacturers
Association (AAMA)
2700 River Road, Suite 118
Des Plaines, IL 60018
(708) 699-7310

American Association for Accreditation
of Laboratory Animal Care (AAALAC)
208A North Cedar Road
New Lenox, IL 60451
(815) 485-7101

American Association for
Laboratory Accreditation
656 Quince Orchard Road
Gaithersburg, MD 20878
(301) 670-1377

American Association of Cost Engineers
(AACE)
P.O. Box 1557
Morgantown, WV 26507
(304) 296-8444

American Association of Engineering
Societies (AAEC)
415 Second Street NE, Suite 200
Washington, DC 20002
(202) 546-2237

American Association of Nurserymen
(AAN)
1250 I Street NW, Suite 500
Washington, DC 20005
(202) 789-2900

American Association of State Highway
and Transportation Officials (AASHTO)
444 North Capitol Street NW, Suite 225
Washington, DC 20001
(202) 624-5811

American Backflow Prevention
Association (ABPA)
P.O. Box 3051
Bryan, TX 77805
(979) 486-7606

American Boiler Manufacturers
Association (ABMA)
950 North Globe Road, Suite 160
Arlington, VA 22203
(703) 522-7350

American Builders and Contractors
(ABC)
729 15th Street, NW
Washington, DC 20005
(202) 637-8800

American Chemical Society (ACS)
1155 16th Street, NW
Washington, DC 20038
(202) 872-4600

American Concrete Institute International
(ACI)
P.O. Box 19151
Detroit, MI 48219
(313) 532-2600

American Concrete Pipe Association
(ACPA)
8300 Boone Boulevard, Suite 400
Vienna, VA 22182
(703) 821-1990

American Concrete Pressure Pipe
Association
8300 Boone Boulevard, Suite 400
Vienna, VA 22182
(703) 821-3054

American Conference of Government
Industrial Hygienists
6500 Glenway Avenue
Bridgetown, OH
(513) 661-7881

American Consulting Engineers Council
(ACEC)
1015 15th Street NW, Suite 802
Washington, DC 20005
(202) 347-7474

American Contractors Association
(ACA)
1004 Duke Road
Alexandria, VA 22314
(703) 684-3450

American Council of Independent
Laboratories (ACIL)
1725 K Street NW, Suite 412
Washington, DC 20006
(202) 887-5872

American Design Drafting Association
(ADDA)
966 Hungerford Drive, Suite 10-B
Rockville, MD 20850
(301) 294-8712

American Engineering Model Society
(AEMS)
P.O. Box 2066
Aiken, SC 29802
(803) 649-6710

American Filtration Society
2360 Highway 59
Kingswood, TX
(713) 540-2116

American Fire Sprinkler Association
(AFSA)
12959 Jupiter Road, Suite 142
Dallas, TX 75328
(214) 349-5965

American Galvanizers Association
6881 S. Holly Cir.
Engelwood, CO 80112
(800) 468-7723

American Gas Association (AGA)
1515 Wilson Blvd.
Arlington, VA 22209
(800) 336-4795
(703) 841-8400

American Hardboard Association (AHA)
520 North Hicks Road
Palatine, IL 60067
(708) 934-8800

American Health Care Association
(AHCA)
1201 L Street NW
Washington, DC 20005
(202) 842-4444

American Hospital Association (AHA)
840 North Lake Shore Drive
Chicago, IL 60611
(800) 242-2626
(312) 180-5223

American Industrial Hygiene Association
475 Wolf Ledges Parkway
Akron, OH 44311
(216) 762-7294

American Institute of Architects (AIA)
1735 New York Avenue NW
Washington, DC 20006
(202) 626-7300

American Institute of Chemical
 Engineers (AICE)
345 East 47th St.
New York, NY 10017
(212) 705-7338

American Institute of Plant Engineers
(AIPE)
3975 Erie Avenue
Cincinnati, OH 45208
(513) 561-6000

American Institute of Timber
 Construction (AITC)
11818 Southeast Mill Plains Blvd.,
 Suite 415
Vancouver, WA 98484
(206) 254-9132

American Institute of Steel Construction
(AISC)
400 North Michigan Avenue, 8th Floor
Chicago, IL 60611
(312) 670-2400

American Insurance Association
(AIA)
1130 Connecticut Avenue NW,
 Suite 1000
Washington, DC 20036
(202) 828-7100

American Iron and Steel Institute
(AISI)
1000 16th Street NW
Washington, DC 20036
(216) 835-3040

American Lumber Standards Committee
(ALSC)
P.O. Box 210
Germantown, MD 20874
(301) 972-1700

American National Metric Council
(ANMC)
1010 Vermont Avenue NW, Suite 320
Washington, DC 20005
(202) 628-5757

American National Standards Institute
(ANSI)
25 West 43rd Street, 4th Floor
New York, NY 10036
(212) 642-4900

American Nuclear Insurers (ANI)
The Exchange, Suite 245
Farmington, CT 06032
(203) 677-7305

American Nuclear Society (ANS)
555 North Kensington Avenue
La Grange Park, IL 60525
(708) 352-6611

American Petroleum Institute (API)
1220 L Street NW
Washington, DC 20005
(202) 682-8000

American Pharmaceutical Association
(APA)
2215 Constitution Avenue NW
Washington, DC 20037
(202) 628-4410

American Pipe Fittings Association
(APFA)
203 Old Keene Mill Court
Springfield, VA 22152
(703) 644-0001

American Plywood Association
(APA)
7011 South 19th Street, P.O. Box 11700
Tacoma, WA 98411
(206) 565-6600

American Production & Inventory
Control Society
500 West Annandale Road
Falls Church, VA 22046
(703) 237-8344

American Public Gas Association
(APGA)
P.O. Box 1426
Vienna, VA 22183
(703) 281-2910

American Public Health Association
1015 15th Street NW
Washington, DC 20005
(202) 789-5600

American Public Power Association
(APPA)
2301 M Street NW
Washington, DC 20037
(202) 775-8300

American Public Works Association
(APWA)
1313 East 60th Street
Chicago, IL 60637
(312) 667-2200

American Railway Engineering
Association
50 F Street NW
Washington, DC 20001
(202) 639-2190

American Rental Association (ARA)
1900 19th Street
Moline, IL 61265
(309) 764-2475

American Road and Transportation
Builders Association (ARTBA)
525 School Street, ARTBA Building
Washington, DC 20024
(202) 488-2722

American Society for Engineering
Societies
1111 19th Street NW
Washington, DC 20036
(202) 296-2237

American Society for Health Care
Engineering
1 North Franklin Street
Chicago, IL 60606
(312) 422-3807

American Society for Non-Destructive
Testing (ASNT)
4153 Arlington Plaza
Columbus, OH 43228
(614) 274-6003

American Society for Quality Control
(ASQC)
P.O. Box 3005
Milwaukee, WI 53201-3005
(414) 272-8575

American Society for Testing and
Materials (ASTM)
1916 Race Street
Philadelphia, PA 19103
(215) 299-5400

American Society for Training and
Development
1630 Duke Street, P.O. Box 1443
Alexandria, VA 22313
(703) 683-8100

American Society of Agricultural
Engineers (ASEA)
2950 Niles Road
St. Joseph, MI 49085
(616) 429-0300

American Society of Anesthesiologists
(ASA)
520 North Northwest Highway
Park Ridge, IL 60068
(708) 825-5586

American Society of Association
Executives (ASAE)
1575 I Street NW
Washington, DC 20005-1168
(202) 626-2723

American Society of Gas Engineers
(ASGE)
P.O. Box 936
Tinley Park, IL 60477
(708) 532-5707

American Society of Heating,
Refrigeration and Air Conditioning
Engineers (ASHRAE)
1791 Tullie Circle NE
Atlanta, GA 30329
(404) 636-8400

American Society of Horticultural
Sciences (ASHS)
113 South West Street
Alexandria, VA 22307
(703) 836-4606

American Society of Hospital Engineers
840 North Lake Shore Drive
Chicago, IL 60611
(312) 280-6144

American Society of Interior Designers
(ASID)
1430 Broadway
New York, NY 10018
(212) 944-9220

American Society of Landscape
Architects (ASLA)
4401 Connecticut Avenue NW, 5th Floor
Washington, DC 20008
(202) 686-2752

American Society of Mechanical
Engineers (ASME)
345 East 47th Street
New York, NY 10017
(212) 705-7722

American Society of Microbiology
(ASM)
1913 I Street NW
Washington, DC 20006
(202) 737-3600

American Society of Plumbing Engineers
(ASPE)
8614 W. Catalpa Ave., Suite 1007
Chicago, IL 60656
(773) 693-2773

American Society of Professional
Estimators (A.S.P.E.)
6911 Richmond Highway, Suite 230
Alexandria, VA 22306
(703) 765-2700

American Society of Safety Engineers
1800 East Oakton Street
Des Plaines, IL 60018
(708) 692-4121

American Society of Sanitary
Engineering (ASSE)
901 Canterbury Road, Suite A
Westlake, OH 44145
(216) 835-3040

American Solar Energy Society (ASES)
2400 Central Avenue
Boulder, CO 80301
(303) 443-3130

American Subcontractors Association
(A.S.A.)
1004 Duke Street
Alexandria, VA 22314
(703) 684-3450

American Supply Association
20 North Wacker Drive, Suite 2260
Chicago, IL 60606
(312) 236-4082

American Water Works Association
(AWWA)
6666 West Quincy Drive
Denver, CO 80235
(303) 794-7711

American Welding Society (AWS)
550 Le June Road NW, P.O. Box 351040
Miami, FL 33135
(305) 443-9353

American Wood Preservers' Association
(AWPA)
P.O. Box 5283
Springfield, VA 21666
(703) 339-6660

American Wood Preservers' Bureau
(AWPB)
P.O. Box 6058
2772 South Randolph Street
Arlington, VA 22206
(703) 931-8180

Anthracite Industry Association
1110 Penn Plaza, Suite 1000
New York, NY 10001
(212) 279-3580

Arbitration Forums
200 White Plains Road
Tarrytown, NY 10591-0066
(914) 332-4770

Architectural Woodwork Institute
  (AWl)
2310 South Walter Reed Drive
Arlington, VA 22206
(703) 671-9100

Asbestos Abatement Council (AAC)
1600 Cameron Street
Alexandria, VA 22314
(703) 684-2924

Asbestos Abatement Equipment
  Distributors Association
5875 Peachtree Industrial Blvd.,
  Suite 370
Norcross, GA 30092
(800) 222-5252

Asbestos Information Association
1745 Jefferson Davis Highway
Arlington, VA 22202
(703) 979-1150

Asbestos Litigation Group
151 Meeting Street, Suite 600
Charleston, SC 29402
(803) 577-6747

Asphalt Institute (AI)
Asphalt Institute Building
College Park, MD 20740
(301) 277-4258

Associated Air Balance Council
  (AABC)
1518 K Street NW, Suite 503
Washington, DC 20005
(202) 737-0202

Associated Equipment Distributors
  (AED)
615 West 22nd Street
Oak Brook, IL 60521
(708) 574-0650

Associated General Contractors of
  America (AGC)
1957 East Street NW
Washington, DC 20006
(202) 393-2040

Associated Laboratories (ALI)
8 Brush Street
Pontiac, MI 48053
(312) 358-7400

Associated Specialty Contractors
  (ASC)
7315 Wisconsin Avenue, 13th Floor
Bethesda, MD 20814
(301) 657-3110

Association for the Advancement of
  Medical Instrumentation (AAMI)
3330 Washington Blvd., Suite 400
Arlington, VA 22201
(703) 525-4890

Association of American Railroads
  (AAR)
50 F Street
Washington, DC 20001
(202) 639-2190

Association of Construction Equipment
  Managers (ACEM)
P.O. Box 43859
Louisville, KY 40243
(502) 244-2574

Association of Diesel Specialists (ADS)
9140 Ward Parkway, Suite 200
Kansas City, MO 64114
(816) 444-3500

Association of Energy Engineers
4025 Pleasantdale Road, Suite 420
Atlanta, GA 30340
(404) 447-5083

Association of Home Appliance
  Manufacturers (AHAM)
20 North Wacker Drive
Chicago, IL 60606
(312) 984-5800

Association of Manufacturing
  Technology
7901 Westpark Drive
McLean, VA 22101-4269
(703) 827-5520

Association of Physical Plant
  Administrators, University and Colleges
1446 Duke Street
Alexandria, VA 22314
(703) 684-1446

Association of School Business Officials,
  International
11401 North Shore Drive
Reston, VA 22090-4232
(703) 478-0405

BCR National Laboratory (BCR)
500 William Pitt Way
Pittsburgh, PA 15238
(412) 826-3030

Better Heating-Cooling Council
P.O. Box 218
35 Russo Court
Berkeley Heights, NJ 07922
(201) 464-8200

Brick Institute of America (BIA)
11490 Commerce Park Drive,
  Suite 300
Reston, VA 22091
(703) 620-0010

Builders Hardware Manufacturers
  Association (BHMA)
60 East 42nd Street, Room 511
New York, NY 10016
(212) 682-8142

Building Officials & Code
  Administrators International (BOCA)
4051 West Flossmoor Road
Country Club Hills, IL 60477-5795
(708) 799-2300

Building Owners and Managers
  Association International
1201 New York Avenue NW, Suite 300
Washington, DC 20005
(202) 289-7000

Building Research Board
2101 Constitution Avenue NW
Washington, DC 20418
(202) 334-3376

Bureau of Land Management
Interior Building
18th and C Streets
Washington, DC 20245
(202) 343-1801

Cast Iron Soil Pipe Institute (CISPI)
5959 Shallowford Road, Suite 419
Chattanooga, TN 37412
(423) 892-0137

Casting Industry Suppliers Association
  (CISA)
6990 Rieber Street
Worthington, OH 43085
(614) 848-8199

Center for Disease Control (CDC)
1600 Cliffon Road
Atlanta, GA
(404) 639-3311

Certified Ballast Manufacturers
  Association (CBMA)
Hanna Building
1422 Euclid Avenue, Suite 772
Cleveland, OH 44115
(216) 241-0711

Chemical Manufacturers Association
  (CMA)
2501 M Street NW
Washington, DC 20037
(202) 887-1100

Chemical Specialties Manufacturers
  Association (CSMA)
1001 Connecticut Avenue NW
Washington, DC 20036
(202) 872-8110

Coastal Engineering Research Board
  (CERB)
P.O. Box 631
Vicksburg, MS 39180
(601) 634-2513

Cold Regions Research & Engineering
  Laboratory (CRREL)
U.S. Department of Defense
72 Lyme Road
Hanover, NH 03755
(603) 646-4200

Color Association of the United States
  (CAUS)
343 Lexington Avenue
New York, NY 10016
(212) 683-9531

Combustion Institute
5001 Baum Blvd.
Pittsburgh, PA 15213
(412) 687-1366

Compressed Air and Gas Institute
(CAGI)
1300 Sommer
Cleveland, OH 44115
(216) 241-7333

Compressed Gas Association (CGA)
1725 Jefferson Davis Highway, Suite 1004
Arlington, VA 22202
(703) 979-0900

Concrete Reinforcing Steel Institute
(CRSI)
933 Plum Grove Road
Schaumburg, IL 60195
(708) 517-1200

Construction Engineering Research
Laboratory
U.S. Department of Defense
P.O. Box 4005
Champaign, IL 61820
(217) 373-7201

Construction Industry Manufacturers
Association (CIMA)
111 East Wisconsin Avenue, Suite 1700
Milwaukee, WI 53202
(414) 272-0943

Construction Specifications Institute
(CSI)
601 Madison Street
Alexandria, VA 22314
(703) 884-0300

Consumer Product Safety Commission
(CSPC)
5401 Westbard Avenue
Bethesda, MD 20814
(301) 492-6800

Consumer Products Safety Council
(CPSC)
1111 18th Street NW
Washington, DC 20207
(202) 634-7700

Controlled Release Society (CRS)
16 Nottingham Drive
Lincolnshire, IL 60069
(708) 940-4277

Cooling Tower Institute (CTI)
P.O. Box 73383
Houston, TX 77273
(713) 583-4087

Copper Development Association (CDA)
260 Madison Avenue
New York, NY 10016

Corps of Engineers (COE)
U.S. Department of Defense
Washington, DC 20314
(202) 272-0660

Cosmetic, Toiletry & Fragrance
Association (CTFA)
1110 Vermont Avenue NW, Suite 800
Washington, DC 20005
(202) 331-1770

Council of American Building Officials
(CABO)
5203 Leesburg Pike, Suite 708
Falls Church, VA 22041
(703) 931-4533

Cryogenic Society of America
c/o Huget Advertising
1033 South Blvd.
Oak Park, IL 60302
(312) 383-8848

Canadian Standards Association (CSA)
178 Rexdale Blvd.
Toronto, ONT, M9W 1R3, Canada
(416) 747-4000

Deep Foundations Institute (DFI)
P.O. Box 281
Sparta, NJ 07871
(201) 729-9679

Delaware River Basin Commission
1100 L Street NW
Washington, DC 20240
(202) 343-5761

Ductile Iron Research Association
(DIRA)
245 Riverchase Parkway East, Suite D
Birmingham, AL 35244
(205) 988-9870

Economic Development Administration
15th & Constitution Avenues
Washington, DC 20203
(202) 377-5081

Edison Electric Institute (EEl)
1111 19th Street NW
Washington, DC 20207
(202) 778-6400

Electric Power Research Institute
 (EPRI)
P.O. Box 10412
Palo Alto, CA 94303
(415) 855-2000

Electrical Apparatus Service Association
 (EASA)
1331 Baur Boulevard
St. Louis, MO 63132
(314) 993-2220

Electronic Industries Association (EIA)
1722 I Street NW
Washington, DC 20006
(202) 457-4900

Engine Manufacturers Association
 (EMA)
111 East Wacker Drive
Chicago, IL 60601
(312) 644-6610

Environmental Industry Council (EIC)
1825 K Street NW, Suite 210
Washington, DC 20006
(202) 331-7706

Environmental Management
 Association (EMA)
1019 Highland Avenue
Largo, FL 34640
(813) 586-5710

Environmental Protection Agency (EPA)
Waterside Mall
401 M Street NW
Washington, DC 20460
(202) 382-2080

Environmental Resource Center (ERC)
3679 Rosehill Road
Fayetteville, NC 28311
(800) 537-2372

Equipment Maintenance Council (EMC)
113 Highland Lake Road
Lewisville, TX 70567
(214) 436-9257

Expansion Joint Manufacturers
 Association (EJMA)
25 North Broadway
Tarrytown, NY 10591
(914) 332-0040

Factory Mutual Engineering and
 Research (FM)
1151 Boston-Providence Tpk.
Norwood, MA 02062
(617) 762-4300

Federal Aviation Administration
 (FAA)
800 Independence Avenue SW
Washington, DC 20591
(202) 267-3111

Federal Communications Commission
 (FCC)
1919 M Street NW
Washington, DC 20554
(202) 632-7000

Federal Highway Administration (FHA)
400 7th Street SW
Washington, DC 20590
(202) 366-0650

Federal Specifications (FS)
Superintendent of Documents
U.S. Government Printing Office
Washington, DC 20402
(202) 783-3238

Fire Apparatus Manufacturers
 Association (FAMA)
c/o Spartan Motors Inc.
P.O. Box 440
Charlotte, MI 48813
(517) 543-6400

Fire Equipment Manufacturers and
 Services Association
1776 Massachusetts Avenue NW
Washington, DC 20036
(202) 659-0600

Fire Information Research and Education
 Center (FIRE)
550 Cedar Street
St. Paul, MN 55101
(612) 296-6516

Fish and Wildlife Service
Interior Building
18th and C Streets
Washington, DC 20245
(202) 343-4717

Fluid Controls Institute (FCI)
P.O. Box 9036
Morristown, NJ 07960
(201) 829-0990

Fluid Power Institute (FPI)
31 South Street, Suite 303
Morristown, NJ 07960
(201) 829-0990

Fluid Sealing Association (FSA)
2017 Walnut Street
Philadelphia, PA 19103
(215) 569-3650

Food and Drug Administration (FDA)
5600 Fishers Lane
Rockville, MD 20853
(301) 443-3170

Forest Service
14th and Independence Avenues SW
Washington, DC 20250
(202) 446-6661

Forging Industry Association (FIA)
1121 Illuminating Building
Cleveland, OH 44115
(216) 781-6260

Foundation for Cross Connection Control
and Hydraulic Research
University of Southern California
KAP-200 University Park, MC-2531
Los Angeles, CA 90089-2531
(213) 743-2032

Gas Appliance Manufacturers
Association (GAMA)
1901 North Moore Street, Suite 1100
Arlington, VA 22209
(703) 525-9565

Gas Processors Association (GPA)
6526 East 60th Street
Tulsa, OK 74145
(918) 493-3872

Gas Research Institute (GRI)
8500 West Bryn Mawr Avenue
Chicago, IL 60631
(312) 399-8100

General Services Administration
(GSA)
18th and F Streets, NW
Washington, DC 20405
(202) 566-0628

Geological Survey Department National
Center
12201 Sunrise Valley Drive
Reston, VA 22091
(703) 860-7411

Geothermal Resources Council
(GRC)
P.O. Box 1350
Davis, CA 95617
(916) 758-2360

Government Printing Office
North Capitol and G Streets NW
Washington, DC 20402
(202) 783-3238

Gypsum Association (GA)
101 South Wacker Drive
Chicago, IL 60606
(312) 606-4000

Hardwood Manufacturers Association
(HMA)
805 Sterick Building
Memphis, TN 38103
(901) 525-8221

Hardwood Plywood Manufacturers
Association (HPMA)
1825 Michael Faraday Drive,
P.O. Box 2789
Reston, VA 22090
(703) 435-2900

Hazardous Waste Research and
Information Center
c/o University of Illinois,
Urbana-Champaign
1 East Hazelwood Drive
Champaign, IL 61820
(217) 333-8940

Health Industry Manufacturers
Association (HIMA)
1030 15th Street NW
Washington, DC 20036
(202) 452-8240

Historical Construction Equipment
Association (HCEA)
6604 Breeds Hill Road
Indianapolis, IN 46237
(317) 782-3612

Hydraulic Institute
14600 Detroit Avenue
Cleveland, OH 44107
(216) 226-7700

Hydraulic Tool Manufacturers
Association (HTMA)
P.O. Box 1337
Milwaukee, WI 53201
(414) 639-6770

Hydronics Institute
35 Russo Place, Box 218
Berkeley Heights, NJ 07922
(201) 464-8200

Illuminating Engineering Society of
North America (IESNA)
235 East 47th Street
New York, NY 10017
(212) 705-7900

Industrial Biotechnology Association
(IBA)
1625 K Street NW
Washington, DC 20006
(202) 875-0244

Industrial Health Foundation (IHF)
34 Penn Circle West
Pittsburgh, PA 15206
(412) 363-6600

Industrial Heating Equipment
Association (IHEA)
1901 North Moore Street
Arlington, VA 22209
(703) 525-2513

Industrial Risk Insurers (IRI)
85 Woodland Street
Hartford, CT 06012
(203) 520-7300

Industrial Safety Equipment Association
(ISEA)
1901 North Moore Street, Suite 501
Arlington, VA 22209-1706
(703) 525-1695

Institute of Business Designers (IBD)
341 Merchandise Mart
Chicago, IL 60654
(708) 467-1950

Institute of Electrical and Electronic
Engineers (IEEE)
345 East 47th Street
New York, NY 10017
(212) 705-7900

Institute of Environmental Sciences
(IES)
940 East Northwest Highway
Mt. Prospect, IL 60056
(312) 255-1561

Institute of Gas Technology
3424 South State Street
Chicago, IL 60616
(708) 768-0500

Institute of Industrial Engineers
25 Technology Park
Atlanta, GA 30092
(404) 449-0460

Instrument Society of America (ISA)
P.O. Box 12277
67 Alexander Drive
Research Triangle Park, NC 27709
(919) 549-8411

Insulated Cable Engineers Association
(ICEA)
P.O. Box P
South Yarmouth, MA 02664
(617) 394-4424

Insulating Glass Certification Council
(IGCC)
Route 11, Industrial Park
Cortland, NY 13045
(607) 753-6711

Insurance Information Institute
10 Williams Street
New York, NY 10038
(212) 669-9200

Insurance Services Office (ISO)
160 Water Street
New York, NY 10038
(212) 487-5000

International Air Transport Association
(IATA)
1001 Pennsylvania Avenue NW,
Suite 285 North
Washington, DC 20004
(202) 624-2977

International Association of Fire Chiefs
(IABPFF)
1329 18th Street NW
Washington, DC 20036
(202) 833-3420

International Association of Heat and
Frost Insulators and Asbestos Workers
1300 Connecticut Avenue NW, Suite 505
Washington, DC 20036
(202) 785-2388

International Association of Mechanical
and Plumbing Officials (IAPMO)
20001 Walnut Drive, South
Walnut, CA 91789-2825
(909) 595-8449

International Compressor
Remanufacturers Association (ICRA)
P.O. Box 33092
Kansas City, MO 64114
(816) 822-8818

International Conference of Building
Officials (ICBO)
5360 South Workman Mill Road
Whittier, CA 90601
(213) 699-0541

International District Heating and
Cooling Association (IDHCA)
1101 Connecticut Avenue NW, Suite 700
Washington, DC 20036
(202) 429-5111

International Electrical Testing
Association
221 Red Rocks Vista
P.O. Box 687
Morrison, CO 80465
(303) 467-0526

International Facility Management
Association
1 East Greenway Plaza, 11th Floor
Houston, TX 77046-0194
(800) 359-4362
(713) 623-4362

International Institute of Ammonia
Refrigeration (IIAR)
111 East Wacker Drive, Suite 600
Chicago, IL 60601
(312) 644-6610

International Management Institute
(IMI)
P.O. Box 266695
Houston, TX 77207
(713) 481-0869

International Mobile Air Conditioning
Association
3003 LBJ Freeway, Suite 219
Dallas, TX 75324
(214) 484-5750

International Municipal Signal
Association (IMSA)
P.O. Box 539
Newark, NY 14513
(315) 331-2182

International Ozone Association (IOA)
83 Oakwood Avenue
Norwalk, CT 06850
(203) 847-8169

International Sanitary Supply Association
(ISSA)
7373 North Lincolnwood Avenue
Lincolnwood, IL 60646-1799
(708) 982-0800

International Society of Fire Service
Instructors (ISFSI)
30 Main Street
Ashland, MA 01721
(508) 881-5801

International Society of Pharmaceutical
Engineers (ISPE)
3816 West Linebaugh Avenue, Suite 412
Tampa, FL 33624
(813) 960-2105

International Thermal Storage Advisory
Council (ITSAC)
3769 Eagle Street
San Diego, CA 92103
(619) 295-6267

Interstate Commerce Commission
 (ICC)
12th Street & Constitution Avenue NW
Washington, DC 20036
(202) 275-7119

Irrigation Association
1911 North Ft. Meyer Drive
Arlington, VA 22209
(703) 524-1200

Joint Commission on Accreditation of
 Healthcare Organizations (JCAHO)
1 Renaissance Boulevard
Oakbrook Terrace, IL 60181
(708) 916-5600

Joint Council of Health Care
 Organizations
1 Renaissance Boulevard
Oakbrook Terrace, IL 60181
(708) 916-5600

Land Improvement Contractors of
 America (LICA)
1300 Maybrook Drive, P.O. Box 9
Maywood, IL 60153
(708) 344-0700

Lead Industries Association (LIA)
292 Madison Avenue
New York, NY 10017
(212) 578-4750

Lightning Protection Institute (LPI)
P.O. Box 458
Harvard, IL 60033
(815) 943-7211

Manufacturers Standardization Society of
 the Valve and Fittings Industry (MSS)
127 Park Street NE
Vienna, VA 22180
(703) 281-6613

Manufacturing Chemical Association
 (MCA)
1825 Connecticut Avenue NW
Washington, DC 20009
(202) 887-1100

Mechanical Contractors Association of
 America (MCAA)
1385 Picard Drive
Rockville, MD 20832
(301) 869-5800

Metal Building Manufacturers
 Association (MBMA)
1230 Keath Building
Cleveland, OH 44115
(216) 241-7333

Metal Lath/Steel Framing Association
 (ML/SFA)
600 South Federal Street, Suite 400
Chicago, IL 60605
(312) 346-1600

Mineral Insulation Manufacturers
 Association (MIMA)
1420 King Street
Alexandria, VA 22314
(703) 684-0084

Mississippi River Commission
P.O. Box 80
Vicksburg, MS 39180
(601) 634-5750

National Academy of Sciences (NAS)
2101 Constitution Avenue NW
Washington, DC 20418
(202) 334-2100

National Aeronautics and Space
 Administration (NASA)
400 Maryland Avenue SW
Washington, DC 20546
(202) 453-1010

National Asbestos Council (NAB)
1777 Northeast Expressway, Suite 150
Atlanta, GA 30329
(404) 633-2622

National Asphalt Pavement Association
 (NAPA)
6811 Kenilworth Avenue, Suite 620
Riverdale, MD 20737
(301) 779-4880

National Association of Architectural
 Metal Manufacturers (NAAMM)
600 South Federal Street, Suite 400
Chicago, IL 60605
(312) 922-6222

National Association of Corrosion
 Engineers (NACE)
P.O. Box 218340
Houston, TX 77218
(713) 492-0535

National Association of County
Engineers
326 Pike Road
Ottum, WA 52501
(515) 684-6928

National Association of Demolition
Contractors (NADC)
4415 West Harrilson Street
Hillside, IL 60162
(708) 449-5959

National Association of Fire Equipment
Distributors
111 East Wacker Drive, 1 Illinois Center
Chicago, IL 60601
(312) 644-6610

National Association of Home Builders
Technical Services
15th and M Streets
Washington, DC 20005
(202) 822-0200

National Association of Oil Heating
Service Managers
P.O. Box 380
Elmwood Park, NJ 07407
(201) 796-8121

National Association of Mutual Insurance
Companies
7931 Castleway Drive
Indianapolis, IN 46250
(317) 875-5250

National Association of Plumbing,
Heating & Cooling Contractors
(NAPHCC)
P.O. Box 6068
180 South Washington Street
Falls Church, VA 22046
(703) 237-8100

National Association of Sewer Service
Companies (NASSC)
101 Wymore Road, Suite 101
Altamonte, FL 32714
(305) 774-0304

National Association of Trade and
Technical Schools (NATTS)
2251 Wisconsin Avenue NW,
Suite 200
Washington, DC 20007
(202) 333-1021

National Association of Women in
Construction (NAWIC)
327 South Adams Street
Fort Worth, TX 76104
(817) 877-5551

National Board of Boiler and Pressure
Vessel Inspectors
1055 Crupper Avenue
Columbus, OH 43229
(614) 888-8320

National Building Material Distributors
Association (NBMA)
1417 Lake Cook Road, Suite 130
Deerfield, IL 60015
(708) 945-7201

National Bureau of Standards (NBS)
Gaithersburg, MD 20899
(301) 975-2000

National Cargo Bureau (NCB)
1 World Trade Center, Suite 2757
New York, NY 10048
(212) 571-5000

National Certified Pipe Welding Bureau
5410 Grosvenor Lane, Suite 120
Bethesda, MD 20814
(301) 897-0770

National Clay Pipe Institute (NCPI)
P.O. Box 759
Lake Geneva, WI 53147
(414) 248-9094

National Coal Association (NCA)
1130 17th Street NW
Washington, DC 20036
(202) 463-2625

National Computer Graphics Association
(NCGA)
2722 Merrilee Drive, Suite 200
Fairfax, VA 22031
(703) 698-9600

National Concrete Masonry Association
(NCMA)
P.O. Box 781
Herndon, VA 22070
(703) 435-4900

National Construction Software
Association (NCSA)
7430 East Caley Avenue, Suite 350
Engelwood, CO 80111
(303) 740-8647

National Council of Acoustical
Consultants (NCAC)
66 Morris Avenue, Box 359
Springfield, NJ 07081
(201) 379-1100

National Council on Radiation Protection
and Measurement (NCRPM)
7910 Woodmont Avenue, Suite 1016
Bethesda, MD 20814
(301) 657-2652

National Corrugated Steel Pipe
Association
2011 I Street NW
Washington, DC 20006
(202) 223-2217

National Electrical Contractors
Association (NECA)
7315 Wisconsin Avenue
Bethesda, MD 20814
(301) 657-3110

National Electrical Manufacturers
Association (NEMA)
2101 L Street NW, Suite 300
Washington, DC 20037
(202) 457-8400

National Elevator Industry (NEII)
630 Third Avenue
New York, NY 10016
(212) 986-1545

National Energy Information Center
(NEIC)
1000 Independence Avenue SW
Washington, DC 20585
(202) 586-8800

National Energy Specialist Association
(NESA)
518 NW Gordon
Topeka, KS 66608
(913) 232-1702

National Environmental Health
Association (NEHA)
South Tower, Suite 970
720 South Colorado Blvd.
Denver, CO 80222
(303) 756-9090

National Fire Protection Association
(NFPA)
P.O. Box 9101
One Batterymarch Park
Quincy, MA 02269-9101
(617) 770-3000

National Fire Sprinkler Association
(NFSA)
P.O. Box 1000
Robin Hill Corporate Park,
Route 22
Patterson, NY 12563
(914) 878-4200

National Fluid Power Association
3333 North Mayfair Road
Milwaukee, WI 53222
(414) 778-3344

National Food Processors Association
1401 New York Avenue NW
Washington, DC 20005
(202) 639-5900

National Forest Products Association
(N.F.P.A.)
1250 Connecticut Avenue NW
Washington, DC 20036
(202) 463-2700

National Geothermal Association (NGA)
P.O. Box 1350
Davis, CA 95617
(916) 758-2360

National Hardwood Lumber Association
(NHLA)
P.O. Box 34518
Memphis, TN 38184
(901) 377-1818

National Institute for Certification in
Engineering Technologies (NICET)
1420 King Street
Alexandria, VA 22314-2715
(703) 684-2835

National Institute for Occupational
Safety and Health (NIOSH)
NIOSH Building
Morgantown, WV 26505-2888
(304) 291-4126

National Institute of Building Sciences
(NIBS)
1015 15th Street NW, Suite 700
Washington, DC 20005
(202) 347-5710

National Institute of Environmental
Health Sciences
Research Triangle Park, NC 27709
(919) 541-3345

National Institute of Standards and
Technology (NIST)
Gaithersburg, MD 20899
(301) 975-2000

National Institutes of Health (NIH)
9000 Rockville Pike
Bethesda, MD 20816
(301) 496-4000

National Insulation Contractors
Association (NICA)
1025 Vermont Avenue NW, Suite 410
Washington, DC 20005
(202) 783-6277

National Insulation and Abatement
Contractors Association (NIAC)
99 Canal Center Plaza, Suite 222
Alexandria, VA 22314-1538
(703) 683-6422

National Liquified Petroleum Gas
Association (NLPGA)
1600 Eisenhower Lane
Lisle, IL 60532
(708) 515-0600

National Oceanic and Atmospheric
Administration (NOAA)
3300 Whitehaven Street NW
Washington, DC 20235
(202) 842-7460

National Paint and Coatings Association
(NPCA)
1500 Rhode Island Avenue NW
Washington, DC 20005
(202) 462-6272

National Park Service
Interior Building
18th and C Streets
Washington, DC 20245
(202) 343-4621

National Petroleum Refiners Association
1899 L Street, NW
Washington, DC 20036
(202) 457-0480

National Pool and Spa Institute (NPSI)
2111 Eisenhower Avenue
Alexandria, VA 22314
(703) 838-0083

National Research Council (NRC)
2101 Constitution Avenue NW
Washington, DC 20418
(202) 334-2000

National Restaurant Association
(NRA)
1200 17th Street NW
Washington, DC 20001
(202) 331-5900

National Roofing Contractors
Association (NRCA)
6250 River Road
Rosemont, IL 60018
(708) 318-6722

National Safety Council (NSC)
444 North Michigan Avenue
Chicago, IL 60611
(312) 527-4800

National Sanitation Foundation,
International (NSF)
789 Dixboro Road
Ann Arbor, MI 48105
(734) 769-8010

National Science Foundation (NSF)
800 G Street NW
Washington, DC 20550
(202) 357-94889

National Society of Professional
Engineers (NSPE)
1420 King Street
Alexandria, VA 22314
(703) 684-2810

National Solid Waste Management
Association (NSWMA)
1730 Rhode Island Avenue NW,
Suite 1000
Washington, DC 20036
(202) 659-4813

National Swimming Pool Foundation
(NSPF)
10803 Gulfdale, Suite 300
San Antonio, TX 78216
(512) 525-1227

National Technical Information Service
(NTIS)
5285 Fort Royal Road
Springfield, VA 22161
(703) 487-4650

National Transportation Safety Board
(NTSB)
800 Independence Avenue SW
Washington, DC 20594
(202) 382-6500

National Truck Equipment Association
(NTEA)
38705 Seven Mile Road, Suite 345
Livonia, MI 48152
(313) 462-2190

National Utility Contractors Association
(NUCA)
1235 Jefferson Davis Highway, Suite 606
Arlington, VA 22202
(703) 486-5555

National Water Well Association
(NWWA)
6375 Riverdale Drive
Dublin, OH 43017
(614) 761-1711

Naval Facilities Engineering Command
200 Stoval Street
Alexandria, VA 22332-2300
(703) 325-0589

Naval Publications and Forms Center
5801 Tabor Avenue
Philadelphia, PA 19120
(215) 697-2000

Nuclear Regulatory Commission (NRC)
1717 H Street NW
Washington, DC 20555
(301) 492-7000

Occupational Safety and Health
Administration (OSHA)
Regional Offices:

OSHA Region 1 (MI, MA, NH, RI, VT)
U.S. Department of Labor, OSHA
133 Portland Street
Boston, MA 02114
(617) 565-7164

OSHA Region 2 (NY, NJ, PR)
U.S. Department of Labor, OSHA
201 Varick Street, Room 670
New York, NY 10014
(212) 337-2325

OSHA Region 3 (PA, DE, DC, MD,
VA, WV)
U.S. Department of Labor, OSHA
Gateway Building, Suite 2100
3535 Market Street
Philadelphia, PA 19104
(215) 596-1201

OSHA Region 4 (AL, FL, GA, KY,
MS, NC, SC, TN)
U.S. Department of Labor, OSHA
1375 Peachtree Street NE, Suite 587
Atlanta, GA 30367
(404) 347-3573

OSHA Region S (IN, IL, MN, MI,
OH, WI)
U.S. Department of Labor, OSHA
230 South Dearborne Street, 32nd Floor,
Suite 3244
Chicago, IL 60604
(312) 353-2220

OSHA Region 6 (AR, LA, NM, OK, TX)
U.S. Department of Labor, OSHA
525 Griffin Street, Room 602
Dallas, TX 75202
(214) 767-4731

OSHA Region 7 (KS, IA, MO, NE)
U.S. Department of Labor, OSHA
911 Walnut Street, Room 406
Kansas City, MO 64106
(816) 426-5861

OSHA Region 8 (CO, MO, ND, SD, UT, WY)
U.S. Department of Labor, OSHA
Federal Building, Room 1576
1961 Stout Street
Denver, CO 80204
(303) 844-3016

OSHA Region 9 (AZ, CA, NV)
U.S. Department of Labor, OSHA
71 Stevenson Street, 4th Floor
San Francisco, CA 94105
(415) 995-5896

OSHA Region 10 (AK, ID, OR, WA)
U.S. Department of Labor, OSHA
Federal Office Building, Room 6003
909 1st Avenue
Seattle, WA 89174
(206) 553-5930

Parenteral Drug Association (PDA)
1 Penn Plaza, Suite 640
Philadelphia, PA 19103
(215) 564-6466

Petroleum Equipment Institute (PEI)
P.O. Box 2380
Tulsa, OK 74101
(918) 494-9696

Petroleum Marketing Education
Foundation (PMEF)
5600 Rosewell Road, Prado North
#318
Atlanta, GA 30342
(404) 255-7600

Pharmaceutical Manufacturers
Association (PMA)
1100 15th Street NW
Washington, DC 20036
(202) 835-3400

Pipe Fabricators Institute (PFI)
P.O. Box 173
Springdale, PA 15144
(412) 274-4722

Pipe Line Contractors Association
(PLCA)
4100 First City Center
Dallas, TX 75201-4618
(214) 969-2700

Plastic Pipe and Fittings Association
(PPFA)
800 Roosevelt Road
Building C, Suite 20
Glen Ellyn, IL 60137
(630) 858-6540

Plastics Pipe Institute (PPI)
(202) 371-5306

Plumbing and Drainage Institute
International (PDI)
45 Bristol Drive
South Easton, MA 02375
(800) 589-8956

Plumbing, Heating and Cooling
Contractors (PHCC)
180 S. Washington Street
P.O. Box 6808
Falls Church, VA 22040
(703) 237-8100

Plumbing Manufacturers Institute (PMI)
1340 Remington Road, Suite A
Schaumburg, IL 60173
(847) 884-9764

Portland Cement Association (PCA)
5420 Old Orchard Road
Skokie, IL 60077
(708) 966-6200

Prestressed Concrete Institute (PCI)
201 North Wells Street
Chicago, IL 60606
(312) 346-4071

Project Managers Institute (PMI)
P.O. Box 43
Drexel Hill, PA 19026
(215) 622-1796

Property Loss Research Bureau (PLRB)
1501 East Woodfield Avenue
Schaumburg, IL 60194
(708) 330-8650

Public Health Service (PHS)
200 Independence Avenue SW
Washington, DC 20201
(202) 245-7000

Refrigerating Engineers and
Technicians Association (RETA)
111 East Wacker Drive, Suite 600
Chicago, IL 60601
(312) 644-6610

Refrigeration Research Foundation
(RRF)
7315 Wisconsin Avenue, Suite 1200
North Bethesda, MD 20814
(301) 652-5674

Refrigeration Service Engineers
Society (RSES)
1666 Rand Road
Des Plaines, IL 60016
(708) 297-6464

Resilient Floor Covering Institute
(RFCI)
966 Hungerford Drive, Suite 12-B
Rockville, MD 20805
(301) 340-8580

Rivers and Harbors Board of Engineers
Kingman Building
Fort Belvoir, VA 22060
(202) 355-2453

Robotic Industries Association
P.O. Box 3624
900 Vickers Way
Ann Arbor, MI 48106
(313) 994-6088

Rubber Manufacturers Association
(RMA)
1400 K Street NW
Washington, DC 20005
(202) 682-4800

Safe Building Alliance (SBA)
655 15th Street NW, Suite 1200
Washington, DC 20005
(202) 879-5120

Safety Equipment Distributors
Association (SEDA)
111 East Wacker Drive, Suite 600
Chicago, IL 60610
(312) 644-6610

Safety Glazing Certification Council
(SGCC)
Route 11, Industrial Park
Cortland, NY 13045
(607) 753-6711

Scaffolding, Shoring and Forming
Institute (SSFI)
1230 Kieth Building
Cleveland, OH 44152
(216) 241-7333

Scientific Apparatus Makers
Association (SAMA)
1101 16th Street NW
Washington, DC 20036
(202) 223-1360

Sealed Insulating Glass Manufacturers
Association (SIGMA)
111 East Wacker Drive
Chicago, IL 60601
(312) 644-6610

Sheet Metal and Air Conditioning
Contractors National Association
(SMACNA)
P.O. Box 70
Merrifield, VA 22116
(703) 790-9890

Single Ply Roofing Institute (SPRI)
104 Wilmot Road, Suite 201
Deerfield, IL 60016
(708) 940-8800

Society of American Value Engineers
(SAVE)
60 Revere Drive, Suite 500
Northbrook, IL 60062
(708) 480-1730

Society of Automotive Engineers
(SAE)
400 Commonwealth Drive
Warrendale, PA 15096
(412) 776-4841

Society of Fire Protection Engineers
(SFPE)
60 Batterymarch Street
Boston, MA 02110
(617) 482-0686

Society of Industrial Microbiology
(SIM)
P.O. Box 12534
Arlington, VA 22209
(703) 941-5373

Society of Insurance Research (SIR)
P.O. Box 933
Appleton, WI 54912
(414) 730-8858

Society of Manufacturing Engineers
(SME)
One SME Drive
Dearborn, MI 48121
(313) 271-1500

Society of Petroleum Engineers (SPE)
P.O Box 833836
Richardson, TX 75083
(214) 669-3377

Society of Plastics Engineers (SPE)
14 Fairfield Drive
Fairfield, CT 06804
(203) 775-0471

Society of Plastics Industries
Composites Institute
355 Lexington Avenue
New York, NY 10017
(212)351-5410

Society of the Plastics Industry (SPI)
1275 K Street NW, Suite 400
Washington, DC 20005
(202) 371-5200

Society of Tribologists and Lubrication
Engineers
840 Busse Highway
Park Ridge, IL 60068-2367
(708) 825-5536

Society of Women Engineers (SWE)
345 East 47th Street, Room 305
New York, NY 10017
(212) 705-7855

Soil Conservation Service (SCS)
14th and Independence Avenues SW
Washington, DC 20250
(202) 447-4525

Solar Energy Industries Association
(SEIA)
1730 North Lynn Street
Arlington, VA 22209
(703) 524-6100

Southern Building Code Congress
International (SBCCI)
900 Montclair Road
Birmingham, AL 35213-1206
(205) 591-1853

Southwest Research Institute (SWRI)
6220 Culebra Road
San Antonio, TX 78284
(512) 684-5111

Standards Engineering Society (SES)
11 West Monument Drive, Suite 510
Dayton, OH 45410
(513) 223-2410

Steel Deck Institute (SDI)
P.O. Box 9506
Canton, OH 44711
(216) 493-7886

Steel Door Institute (S.D.I.)
c/o A. P. Wherry & Associates
712 Lakewood Center North
Cleveland, OH 44107
(216) 226-7700

Steel Joist Institute (SJI)
1205 48th Street, North, Suite A
Myrtle Beach, SC 29577
(803) 449-0487

Steel Structures Painting Council (SSPC)
4400 Fifth Avenue
Pittsburgh, PA 15213
(412) 578-3327

Steel Tank Institute (STI)
728 Anthony Trail
Northbrook, IL 60062
(708) 498-1980

Submersible Wastewater Pump
Association
1866 Sheridan Road, Suite 210
Highland Park, IL 60035
(847) 681-1868

Sump and Sewage Pump
Manufacturers Association
P.O. Box 298
Winnetka, IL 60093
(312) 446-4434

Superintendent of Documents
U.S. Government Printing Office
North Capitol and G Streets NW
Washington, DC 20235
(202) 512-1800

Technical Association of the Pulp and
  Paper Industry (TAPPI)
15 Technology Parkway South NW
Atlanta, GA 30092
(404) 446-1400

Tennessee Valley Authority (TVA)
400 West Summit Hill Drive
Knoxville, TN 37902
(615) 623-3554

Thermal Insulation Manufacturer's
  Association (TIMA)
7 Kerby Plaza
Mt. Kisco, NY 10549
(914) 241-2284

Tile Council of America (TCA)
P.O. Box 326
Princeton, NJ 08542
(609) 921-7050

Tissue Culture Association (TCA)
19110 Montgomery Village Avenue,
  Suite 300
Gaithersburg, MD 20879
(301) 869-2900

Topographic Laboratory
U.S. Department of Defense
Cude Building
Fort Belvoir, VA 22060
(202) 355-2600

Truss Plate Institute (TPI)
583 D'Onofrio Drive, Suite 200
Madison, WI 53719
(608) 833-5900

Tubular Exchange Manufacturers
  Association (TEMA)
25 North Broadway
Tarrytown, NY 10591
(914) 332-0040

Underground Contractors Association
  (UCA)
2720 River Road, Suite 222
Des Plaines, IL 60018
(708) 299-6930

Underwriters Laboratories (UL)
333 Pfingsten Road
Northbrook, IL 60062
(847) 272-8800

Uni-Bell PVC Pipe Association (UNI)
2655 Villa Creek Drive, Suite 155
Dallas, TX 75234
(214) 243-3902

U.S. Army Chief of Engineers (COE)
20 Massachusetts Avenue NW
Washington, DC 20314
(202) 272-0001

U.S. Bureau of Mines
2401 E Street NW
Washington, DC 20245
(202) 634-1004

U.S. Department of Agriculture
14th and Independence Avenue SW
Washington, DC 20250
(202) 447-3631

U.S. Department of Commerce (DOC)
Commerce Building
15th and Constitution Avenue NW
Washington, DC 20230
(202) 377-2112

U.S. Department of Defense (DOD)
The Pentagon
Washington, DC 20301
(202) 695-5261

U.S. Department of Energy (DOE)
1000 Independence Avenue SW
Washington, DC 20585
(202) 586-6120

U.S. Department of Health and Human
  Services (DHHS)
200 Independence Avenue SW
Washington, DC 20201
(202) 245-7000

U.S. Department of Housing and
  Urban Development (HUD)
HUD Building
541 7th Street SW
Washington, DC 20410
(202) 755-6417

U.S. Department of the Interior (DOI)
Interior Building
18th and C Streets NW
Washington, DC 20245
(202) 343-7351

U.S. Department of Justice
10th and Constitution Avenue NW
Washington, DC 20530
(202) 633-2001

U.S. Department of State
2201 C Street NW
Washington, DC 20520
(202) 647-4910

U.S. Department of Transportation
(DOT)
400 7th Street SW
Washington, DC 20590
(202) 366-4000

U.S. Fire Administration (USFA)
16825 South Seton Avenue
Emmitsburg, MD 21727
(301) 447-1080

U.S. Metric Association (USMA)
10245 Andesol Avenue
Northridge, CA 91325
(818) 363-5606

United States Pharmacopoeia
(301) 881-0666

Urban Mass Transportation
  Administration
400 7th Street SW
Washington, DC 20590
(202) 366-4040

Utility Location and
  Coordination Council
1313 East 60th Street
Chicago, IL 60637
(312) 667-2200

Valve Manufacturers Association of
  America (VMAA)
1050 17th Street NW, Suite 701
Washington, DC 20036
(202) 331-8105

Vibration Institute
6262 South Kingery Highway
Willowbrook, IL 60514
(630) 654-2254

Vinyl Institute
155 Route 46 West
Wayne, NJ 07470
(201) 890-9299

Wall Covering Manufacturers
  Association
66 Morris Avenue
Springfield, NJ 07081
(201) 379-1100

Water Pollution Control Federation
  (WPCF)
601 Wythe Street
Alexandria, VA 22314
(703) 684-2400

Water Quality Association (WQA)
4151 Naperville Road
Lisle, IL 60532
(708) 505-0160

Water Systems Council (WSC)
600 South Federal Street, Suite 400
Chicago, IL 60605
(312) 922-6222

Western Society of Engineers (WSE)
176 West Adams Street, Suite 1734
Chicago, IL 60603
(312) 372-3760

Wire Reinforcement Institute (WRI)
8361-A Greensboro Drive
McLean, VA 22102
(703) 790-9790

Wood Heating Alliance (WHA)
1101 Connecticut Avenue NW,
  Suite 700
Washington, DC 20036
(202) 857-1181

Woven Wire Products Association
  (WWPA)
2515 North Nordica Avenue
Chicago, IL 60635
(312) 637-1359

Zinc Institute
292 Madison Avenue
New York, NY 10017
(212) 578-4750

APPENDIX C
GLOSSARY AND ABBREVIATIONS

AA (arithmetic average) A numerical measure of metal surface roughness.

ABS Acrylonitrile butadiene styrene.

absorption The soaking up of a gas or liquid into a solid substance.

ac Alternating current.

acfm Actual cubic feet per minute for compressed or vacuum air.

actuator A movable component of a valve that when operated, causes the closure element to open or close.

adsorption The condensation of gas or liquid onto the surface of a solid.

AEC Architect, engineer, and constructor.

aerobic Bacteria requiring free oxygen for their growth.

aerosol Particles, solid or liquid, suspended in air.

aftercooler In compressed-air service, the aftercooler is a device used to lower the temperature of the compressed air immediately after the compression process.

AI Aggressiveness index (a guideline parameter to find the corrosive tendency of potable water).

air gap An unobstructed separation between a source of potable water and any source of contamination.

AISI American Iron and Steel Institute.

anaerobic Bacteria living in the absence of free oxygen, deriving it instead from breaking down complex substances. Bacteria requiring no free oxygen for their growth.

anchor A pipe support that restrains a pipe against all movement.

anion A negatively charged atom attracted to an anode electrode.

ANSI American National Standards Institute.

API American Petroleum Institute.

approach A term, expressed in degrees Fahrenheit (°F), used for fluid system dryers to indicate how close the outlet temperature of the water being heated comes to, or approaches the fluid heating medium to the cooling medium.

approved Accepted for the intended purpose, as an appropriate design or for installation into a piping system, by a responsible code official or other agency having jurisdiction for a specific project.

APTE American pipe thread external.

APTI American pipe thread internal.

aquifer A water-bearing formation or stratum capable of storing or transmitting water in sufficient quantities to permit development.

areaway An enclosed excavated area below grade level open to the weather.

ASME American Society of Mechanical Engineers.

AST Above ground storage tank.

ASTM American Society for Testing and Materials.

ATC Automatic temperature control.

autoignition temperature The lowest temperature at which a material will ignite and sustain combustion in the absence of a spark or flame. This value is influenced by such factors as the size, shape, and material of the heated surface.

backfill Material placed from the pipe haunch up to grade.

backflow Any reversal of the flow of water from its intended direction.

back pressure Backflow caused by an increase of normal pressure.

backseating A part of a valve, the backseat is a second seat in the bonnet used in the fully open position to seal the valve stem against leakage into the packing. A bushing on the stem provides the mating surface.

back siphonage Backflow caused by a lowering of normal pressure.

backwater valve A commonly used term for a type of check valve used in a drainage system.

bearings Machine parts placed at various locations on the shaft of a pump to reduce friction and carry radial and thrust loads.

bedding That material in contact with the pipe that is beneath and up the haunches of a pipe.

BFP Backflow preventer.

bituminous Of or containing bitumen; as asphalt or tar.

block valve A commonly used term for a shut-off valve.

bonnet A valve component that provides a leakproof closure for the body through which the stem passes and is sealed.

booster hot water system A secondary water heating system used to heat water to a temperature higher than that of the primary water heating system.

branch A horizontal run of pipe not considered a house drain or stack.

branch interval The distance measured along the stack, within which horizontal drainage branches are connected to a drain stack. This distance is usually one story high, but never less than 8 ft.

branch vent A vent that connects one or more individual or common vents to a vent stack or a stack vent.

brazing A means of joining pipe where the filler metal holding the pipe together melts at a temperature higher than 940°F.

BT Bathtub.

Btu British thermal unit, which is the quantity of heat required to raise the temperature of one pound of water one degree Fahrenheit.

building drain   The lowest horizontal part of the drainage piping system, considered the principal pipe conveying sanitary effluent by gravity to a point outside the building.

building sewer   The continuation of the house drain from a point outside the building wall to the actual connection to an adequate and approved point of disposal, such as a public sewer or private sewage disposal system.

building trap   A trap installed on the house sewer to prevent the circulation of sewer gas between the building sewer and the building drain.

CAB   Cellulose acetate butyrate (Celcon).

CAD   Computer-aided design and/or drafting.

CAE   Computer-aided engineering.

CAM   Computer-aided manufacturing.

canopy   A small roof protecting a window or entrance.

CAP   College of American Pathologists.

carryover   Water droplets in steam immediately downstream of a boiler.

casing   (1) The stationary covering around the impeller of a pump that gives direction to the discharge and converts velocity energy into pressure energy. (2) When used in water wells, the casing is a thin-walled cylinder placed in the borehole of the well.

catch basin   A receptacle designed to collect wastewater from the floor surface of an open structure unit.

cation   A positively charged atom attracted to a cathode electrode.

cavitation   A phenomenon of flowing water caused by the rapid formation and collapse of air cavities, which results in the pitting of surfaces on which they occur.

CB   Catch basin (a site structure that admits storm water into the piping system with an integral storage space for catching and holding debris).

CDI   Continuous deionization.

cf   Cubic foot.

CFC   Chlorofluorocarbon.

cfh   Cubic feet per hour.

cfm   Cubic feet per minute.

CFR   Code of Federal Regulations.

cfs   Cubic feet per second.

CGA   Compressed Gas Association.

chemical waste   Any substance that may cause harm to the sanitary piping system, treatment facility, or environment without being treated or neutralized prior to discharge into the sanitary drainage system.

cGMP   Current good manufacturing practice.

CI   Cast iron.

CII   Chlorinated isobutene isoprene.

CIP   Clean in place.

circuit vent   A branch vent that serves two or more traps and extends from a connection to a drainage line in front of the last fixture to a connection with a vent stack.

city water   A commonly used term for potable water.

CLA   Centerline average.

class   Designation given to pipe, flanges, and fittings to replace psi rating, e.g., class 150 instead of 150 psi.

cleanout   A gas-tight, water-tight pipe fitting with a removable plug that is used to obtain access to the inside of a drainage pipe for cleaning or maintenance.

clean steam   Steam that has been generated using additive-free feedwater or uncondensed WFI steam.

closure element   A valve component that when moved, opens or closes to allow the passage of fluid through the valve.

CO   Clean out.

CODP   Clean out deck plate.

colloid   A very small, electrically charged particle suspended in water.

combined drainage system   A drainage system that combines sanitary effluent and storm water runoff in a single piped system.

common vent   A single vent line serving two fixtures.

conductor   Storm water piping inside of a building.

connected load   The sum of the rated input of every device connected to the entire fuel gas system, expressed in either British thermal units or cubic feet per hour.

contaminant   Any impurity or toxic substance that, when introduced into a potable water supply, will create a health hazard or threaten the well-being of a consumer.

continuous vent   A vertical vent that is a continuation of the waste line from a fixture to which it is connected.

counter current   A term used in heat exchanger design to indicate that the heating medium flows in the opposite direction to the fluid being heated.

coupling   When referring to pumps, any device used to connect the driver to the shaft of a pump. When referring to a pipe, a coupling is a fitting that joins two pieces of pipe when continuing in a straight line.

CPE   Chlorinated polyethylene.

CPI   Chemical and petroleum industry.

CPVC   Chlorinated polyvinyl chloride.

CR   Chloroprene rubber (neoprene).

cross connection   Any physical connection between a potable water system and any potential source of contamination not protected by an approved device specifically designed to prevent flow between the two.

cryogenic liquid   A refrigerated liquid gas having a boiling point below $-130°F$ ($-90°C$) at atmospheric pressure.

CS   Carbon steel.

CSA   Canadian Standards Institute.

CSP   Chlorine sulphonyl polyethylene (Hypalon).

cu ft   Cubic foot.

cycles of concentration   In water treatment, this term indicates the number of times the dissolved solids concentration has increased as a result of evaporation comparing makeup water to condenser water.

dc   Direct current.

DCV Double check valve (a means of backflow prevention).

deliquescent A material that changes state in the presence of water.

demand Estimated flow rate expected under specific operating conditions.

demand respirator An atmosphere-supplying respirator that admits respirable gas to the face piece only when a negative pressure is created by inhalation.

density The ratio of the weight of a substance to its volume.

desiccant material A material that easily adsorbs water vapor.

design point The specific point in the piping network where pipe size is calculated.

developed length The total length of a vent pipe measured along the centerline of that pipe, from point to point.

device Any appliance or piece of equipment utilizing fuel gas to produce light, heat, or heat energy.

dewpoint The temperature at which water in the air will start to condense on a surface.

DF Drinking fountain.

DFU Drainage fixture unit (used in plumbing systems to size sanitary drainage and vent lines).

DI Deionized, ductile iron.

D.I. Drainage inlet (a site structure that allows the entrance of storm water).

DIN Deutsches Institute für Normung (German Institute for Standardization).

direct fired A water heater whose primary heat source is an integral part of the water heater assembly.

disk The closure element of some types of valves.

dissociation The separation of compounds dissolved in water into ions.

dissolved gases Oxygen, carbon dioxide, and hydrogen sulfide that are released upon heatmg or pressure reduction within a water supply system.

dissolved mineral salts Commonly used as a measure of hardness, dissolved mineral salts are bicarbonates, sulfates, chlorides, and nitrates that form ion components when in solution. Positively charged ions are called cations; negatively charged ions are called anions.

dissolved organic materials Nonionic solids that form covalent bonds with water molecules.

diversity factor An estimate of the maximum probable simultaneous use of the connected devices, outlets, or equipment; expressed either as a decimal or percentage.

DN Nominal dimension used for conversion of inch/pound system to SI units.

domestic water Potable water primarily intended for direct human use, such as that supplied to plumbing fixtures.

DOT Department of Transportation.

downspout A vertical pipe attached to gutters installed on the outside of a building.

DR Dimensional ratio.

drain A receptacle for the collection and removal of storm water that accumulates on surfaces exposed to the weather and flows into the storm water drainage piping network.

drawdown The distance between the static level and the dynamic level in a water well.

dry gas When used in reference to fuel gas, it is a gas having a moisture and hydrocarbon dewpoint below any normal temperature to which the gas piping will be exposed.

dry return   A condensate return that has the piping above the waterline of the boiler.

duration   A commonly used term for time of concentration in storm water drainage.

dust   An aerosol consisting of mechanically produced solid particles derived from the breaking up of larger particles. Dusts generally have a larger particle size when compared to fumes.

duty cycle   The actual amount of time that a device is in use during a measured period of time; generally expressed as a percent.

dynamic water level   The elevation to which water in a well falls during pumping at a given flow rate.

earth load   The weight of all earth backfill over the pipe.

ECTFE   Ethylenechlorotrifluoroethylene.

EDR   Equivalent direct radiation; expressed in square feet [the heat output of 240 Btu/h (70.3 W) from a device when filled with steam at 215°F (102°C) (1 psig) and surrounded with air at 70°F (21°C)].

effluent   A general term describing any substance entering, or carried in, a drainage system.

elastomer   An elastic rubberlike substance that stretches at low stress to at least twice its length at ambient temperature and returns to its approximate original shape upon release.

electrolyte   A dissolved impurity in water.

electropolishing   An electrochemical process that removes surface atoms from a metal surface for the purpose of producing a smooth finish.

enthalpy   The total heat content above some base temperature.

EP   Epoxide.

EPA   Environmental Protection Agency.

EPDM   Ethylene propylene-diene monomer.

EPM   (1) Ethylene propylene terpolymer. (2) Equivalents per million.

equivalent run   The actual measured length of a pipe including an additional allowance for resistance to fluid flow resulting from valves, fittings, devices, etc.

equivalent weight   The weight, in pounds, of any element that could combine with one pound of hydrogen.

erosion   The gradual destruction of material by abrasive action of liquids and/or solids.

ERW   Electric resistance welded for manufacture of pipe.

ET   Evapotranspiration.

evacuation type plumbing fixtures   Plumbing fixtures, such as water closets and urinals, used to receive and discharge waterborne human bodily waste.

exchange capacity   Expressed as kilograms of ions removed per cubic foot of resin before breakthrough.

exfiltration   Liquid leaking out of a sewer.

exhaustion   Depletion of ion exchange capacity to the point that an acceptable purity of product water can no longer be obtained, and regeneration becomes necessary.

exposure limit   The maximum allowable concentration of a contaminant in the air to which an individual may be exposed. These allowable concentrations may be time-weighted averages, short-term limits, or ceiling limits.

FAD   Free compressed air delivered.

FD   Floor drain.

FDA   Food and Drug Administration.

feedwater   Water received directly from the supply source. A generic term used to describe the water intake into any device, system, or treatment process.

FH   Fire hydrant.

filling density   The percent ratio of the weight of gas in a container to the weight of water that the container will hold at 60°F (15.6°C).

filter   A component used in respirators to remove solid or liquid aerosols from the inspired air.

filtration   The use of a porous medium to retain solids while allowing a fluid to pass through the medium.

fitting   A device used to connect one or more pipes together and/or to change the direction of a straight run of pipe.

fixture battery   Any group of two or more fixtures that discharge into a common horizontal waste or soil branch.

flammable gas   Any gas that will ignite easily and burn rapidly in the presence of air or an oxidizer.

flammable limits   The minimum concentration of vapor in air or oxygen below which propagation of a flame does not occur on contact with a source of ignition, and the maximum proportion of vapor or gas in air above which propagation of a flame does not occur; usually expressed in terms of percentage by volume of gas or vapor in air. A change in temperature or pressure may vary the flammable limits of a gas.

flash arrestor   A device that prevents any flame from going back into a storage tank or supply source.

flashback   A phenomenon characterized by vapor ignition and flame travel back to the vapor source.

flash point   The lowest temperature at which a liquid will give off enough flammable vapor at or near its surface to form an ignitable mixture with air.

floor drain   A plumbing fixture that removes liquid effluent from the surface of floors and other areas.

flow rate   The measurement of a volume of water over time, such as cubic feet per second.

fluid   A substance, such as air or water, that takes the shape of its container.

FM   Factory mutual.

FMA   Free mineral acidity (the total amount of strong acid in effluent from a hydrogen exchanger).

FNPT   Female national pipe thread.

force main   A pumped sanitary line under pressure.

fpm   Feet per minute.

fps   Feet per second.

FPT   Female pipe thread.

frame   Any external part of a pump that assists in the mounting and support of the pump to the structure.

frequency   The estimated number of years that elapse between the reoccurrence of storms with a specific intensity.

ft   Foot.

FTU   Formazin turbidity unit.

FU   Fixture units (used for plumbing systems to size sanitary and potable water systems).

FV   Flush valve.

fume   Solid aerosols formed by condensation of a gas or vapor. Fumes generally have a smaller particle size when compared to dusts.

GAC   Granulated activated charcoal.

gas   A fluid that has neither independent shape nor volume and tends to expand indefinitely.

GC   Gas chromatograph.

GLP   Good laboratory practices.

GMAW   Gas metal arc welding.

GMP   Good manufacturing practices.

gpd   Gallons per day.

gpg   Grains per gallon.

gpm   Gallons per minute.

grade   The surface elevation of the ground.

granular material   Coarse-grained noncohesive soil usually well graded. Consists mainly of sands and gravels. Compacts best by vibration.

gravel   Coarse-grained soil of sizes ⅜ to 3 in.

groundwater   Water found below the water table. Groundwater is obtained from wells or other aquifers originating underground. This term also applies to streams that intercept an aquifer and are found on the surface of the ground.

grout   A fluid mixture of cement, sand, and water that can be easily placed or pumped.

GTAW   Gas tungsten arc welding.

guide   A pipe support attachment that allows axial pipe movement only.

gutter   An open horizontal channel used to collect storm water; usually made of sheet metal or wood and attached to the lowest point of a pitched roof.

hanger   A pipe support consisting of an attachment to a structure, connection rod, or support and pipe; a device to secure the pipe to the connection.

haunch   The portion of a sewer pipe below the spring line.

hazardous atmosphere   An atmosphere that contains a contaminant(s) in excess of the exposure limit or is oxygen deficient.

HDPE   High-density polyethylene.

header   A pipe that does not diminish in size.

heat exchanger   A device specifically designed and constructed to efficiently transfer heat energy from a hot fluid to a cooler fluid.

hood   A respiratory inlet covering that completely covers the head and neck and that may also cover portions of the shoulders.

hot water   Water at a temperature higher than ambient; established by generally accepted practice or code as being suitable for a specific application.

house drain   A commonly used term for a building drain.

house sewer   A commonly used term for a building sewer.

house trap   A commonly used term for a building trap.

HVAC   Heating, ventilating, and air-conditioning.

hydraulically remote   Farthest from the source of supply in terms of total pressure lost through the entire water supply piping system.

hydrologic soil group   Groups of soils that have the same runoff potential under similar storm conditions.

icfm   Inlet cubic feet per minute for compressed air.

I.D.   Inside diameter.

IE   Invert elevation (an elevation taken at the inside bottom of a pipe).

IEEE   Institute of Electrical and Electronic Engineers.

IIR   Isobutene isoprene (butyl) rubber (an elastomer).

immiscible   A liquid incapable of being dissolved in water, such as oil.

impeller   A rotating part of a pump that imparts velocity to the liquid being pumped by means of centrifugal force.

imperviousness factor   A number indicating the percent of rainfall available as runoff and not absorbed into the ground, absorbed by plants, left as puddles, or lost to evaporation during the rainstorm; expressed as a decimal.

impurity   Any physical, chemical, or biological substance found in water making it undesirable for a specific use or degrading it as a source of potable water.

indirect fired   A water heater whose primary heat source is generated remotely from the water heater.

indirect waste   Any waste pipe not connected directly into the drainage system, that discharges through an air gap into a fixture, interceptor, trap, or drain.

individual vent   A vent that connects directly to only one fixture and extends to either a branch vent or vent stack.

inert   Materials that do not react with other materials at normal pressure and temperature.

infiltration   When used in reference to gravity piping systems, it is groundwater leaking into a sewer. It is also a term used to describe the rate at which water travels deeper into soil.

inflow   Surface water flowing into a manhole or collecting device.

influent   Sewage flowing into a pipe, basin, or waste treatment plant.

initial backfill   That material from the top bedding to 12 in above the pipe.

inlet filter   For compressed-air service, an inlet filter is any filter installed on the inlet, or intake, to the air compressor.

inlet time   A frequently used term for overland flow time.

inorganic   Chemical substances of mineral origin.

intensity   The rate at which rain falls as considered for design purposes; measured in inches per hour.

interceptor   A device that separates, retains, and allows removal of specific harmful material suspended in the waste stream, while permitting the remaining acceptable liquid effluent to be discharged into the drainage system.

input   The total amount of fuel gas required for proper operation at the inlet to a device.

invert   The elevation of the inside bottom of a drainage pipe.

ion   An atom or group of atoms that has an electrical charge.

ion exchange capacity   A measure of the mass of ionic impurities that can be removed by a demineralizer before exhaustion occurs and regeneration becomes necessary. Ion exchange capacity is typically given as grains of calcium carbonate or grains of sodium chloride.

IP   Inch pound, a reference to units of measurement where inches and pounds are used.

IPS   Iron pipe size.

IQ (Installation qualification)   Documented verification that all key aspects of equipment installation adhere to appropriate codes and approved design intentions, and that recommendations of the manufacturer have been considered.

ISA   (1) Instrument Society of America. (2) Industry Standard Architecture.

ISO   International Organization for Standardization.

JTU   Jackson turbidity unit.

latent heat of vaporization   The amount of heat required to change state from liquid to vapor or vice versa.

lav   Lavatory.

lb   Pound.

LDR   A single labor, delivery, and recovery room in a hospital.

LDRP   A single labor, delivery, recovery, and postpartum room in a hospital.

leader   A vertical pipe carrying storm water either inside or outside the building.

leakage   When used in reference to water treatment systems, *leakage* is the presence of undesired ions in the final treated water.

LEL   Lower explosive limit. The lowest percent of a gas mixture in air that will allow an explosion to occur under normal temperature and pressure conditions.

LFL   Lower flammability limit. The lowest percent of a gas mixture in air that will support combustion under normal temperature and pressure conditions.

liquefied compressed gas   A gas that, under the charged pressure, is partially liquid at a temperature of 70°F (21.1°C).

loop vent   A branch vent that serves two or more traps and extends from a point in front of the last fixture connection to a stack vent.

LPG   Liquefied petroleum gas.

Lpm   Liters per minute.

Lps   Liters per second.

LSI   Langelier saturation index (a measure of the tendency of water to form deposits of mineral scale and to corrode substances).

main vent   A main vent is the principal vent of a building, remaining undiminished in size from the connection with the drainage system to its terminal.

manifold   An assembly used to connect multiple supplies together.

maximum acceptable pressure   The highest pressure that will not cause a nuisance or produce premature and accelerated damage to any component.

maximum building demand   The estimated flow of water from the maximum fixture demand plus the highest water demand from various equipment throughout a building.

maximum fixture demand   The greatest estimated flow of water resulting from the probable maximum simultaneous use of intermittently operated plumbing fixtures.

maximum LPG liquid tank capacity   To allow space for propane vaporization, 85 percent is the maximum permitted filling level.

maximum probable demand   The estimated maximum amount of fuel gas per unit of time that is expected to be in simultaneous use; expressed as either Btu or cubic feet per hour. This is the connected load multiplied by the diversity factor.

mechanical seal   Used in place of a pump's stuffing box, this provides a mechanical assembly capable of preventing leakage around the shaft by means of very close tolerances of mating parts.

mg/L   Milligrams per liter.

mgd   Millions of gallons per day.

MH   Manhole.

microorganisms   Bacteria, algae, and other similar living microscopic organisms.

mill coated pipe   Factory applied plastic coating for underground steel piping.

minimum acceptable pressure   The lowest pressure permitting safe, efficient, and satisfactory operation of the most remote fixture, device, or component.

minimum LPG liquid tank capacity   To allow time for resupply, 10 to 15 percent is recommended. Absolute low level is 5 percent.

mist   An aerosol composed of liquid particles.

monitor   A permanently mounted fire protection nozzle assembly, connected to a water main or FH and capable of being rotated and elevated.

monomer   A chemical compound capable of reacting to form a polymer.

mpm   Meters per minute.

mps   Meters per second.

MPT   Male pipe thread.

MSDS   Material safety and data sheet.

MTBF   Mean time between failures.

MW   Molecular weight.

NACE   National Association of Corrosion Engineers.

NBS   National Bureau of Standards.

NC   Normally closed.

NCCLS   National Committee for Clinical Laboratory Standards.

NEMA   National Electrical Manufacturers Association.

NF   National Formulary.

NFPA   National Fire Protection Association.

NG   Natural gas.

NO   Normally open.

normal pressure   The design or expected force per unit area at any point in a water system; usually expressed as pounds per square inch.

NTP   Normal temperature and pressure [68°F (20°C) and 14.7 psia (760 torr)].

NTU   Nephelometric turbidity unit (a measure of turbidity in water).

NPS   Nominal pipe size.

NPT   National pipe thread.

O.C.   On center.

O.D.   Outside diameter.

OEM   Original equipment manufacturer.

offset   Any change in direction of a stack from vertical, or any change in direction of a horizontal drainage line.

operational qualification (OQ)   Documented verification that systems and equipment perform as intended throughout the design or anticipated operating range.

OSHA   Occupational Safety and Health Administration.

O.S. & Y.   Outside screw and yoke.

output   When used in reference to a gas appliance, the actual number of Btu's available to perform the intended function of the device; usually expressed as a percent of the input and taking into consideration the efficiency.

overflow   A positive and fail-safe outlet for removal of liquids that have reached a predetermined height above a normally expected level.

overland flow time   The time rainwater takes to travel on the ground from the farthest point of an outside area to a drain; measured in minutes.

oxidant   A substance that can remove electrons from another substance (oxidize it) and is itself reduced (gains electrons).

oxidizers   A nonflammable gas that supports combustion.

P & ID   Piping and instrumentation diagram.

PA   Polyamide.

packing   Material inserted into a pump's stuffing box that surrounds the shaft and prevents liquid from forcing its way past the shaft. For a valve, it is the material that surrounds the stem and prevents liquid from forcing its way past the stem to the outside of the valve.

PAEK   Polyaryl etherketone.

PB   Polybutylene.

PC   Polycarbonate.

PCTFE   Polychlorotrifluoroethylene (Halar).

PCU   Platinum cobalt unit.

PE   Polyethylene.

PEEK   Polyether etherketone.

PET   Potential evapotranspiration.

Percolation   The rate at which water travels deeper into soil. Also known as infiltration.

perm   An abbreviation for permeance, the transmission of water vapor through insulation.

PF   Phenol-formaldehyde.

PFA   Perfluoroalkoxy.

Ph   A measurement of the hydrogen ion concentration of a solution.

pig   A flexible device propelled through pipelines to clean the interior.

piping network   The entire piping system, including all pipe, valves, and appurtenances from the source of supply or connection to the farthest fixture, device, or point of disposal.

PIR   Polyisopropane (an elastomer).

pitch   The distance that one end of a pipe is lower than the other end; expressed as a percent of the total length of run or as a dimension, in inches or feet per foot of run.

PIV   Poat indicator valve.

plastic   A material whose essential ingredient is an organic substance of large molecular weight, which at some stage in its manufacture can be shaped by flow and becomes solid in its finished state.

plug   The closure element for some types of valves.

plumbing fixture   Any approved receptacle or device specifically designed to receive human or other waterborne waste and discharge that waste directly into the sanitary drainage system, often with the addition of water.

POE   Point of entry.

point-of-use heater   Locations immediately adjacent to the fixtures and/or equipment requiring hot water, as compared to a remote, centralized location serving an entire building, area, or project.

polishing   The process by which the purity of water pretreated by reverse osmosis, deionization, or distillation, is increased by the addition of posttreatment equipment either immediately after the central system or at the points of use. When referring to metal finishing, it is a process that produces a smooth surface.

pollutant   Any nontoxic impurity that may create a moderate or minor hazard to the water supply.

polymer   A material consisting of molecules with a high molecular weight.

polymerization   A chemical reaction by which a large number of monomer molecules are linked together to form a chainlike molecule or polymer. When two or more monomers are used, the process is called *copolymerization.*

pore   The space between individual particles of a soil.

positive pressure respirator   A respirator in which the pressure inside the respiratory inlet covering is normally positive with respect to the ambient air pressure.

potable water   Water of sufficient purity to meet standards established as being fit for human consumption.

powered air purifying respirator   An air purifying respirator that uses a blower to force the ambient atmosphere through air purifying elements to the inlet covering.

PP   Polypropylene.

ppb   Parts per billion.

PPH   Pounds per hour.

ppm   Parts per million.

PPS   Polyphenylene sulfide.

ppt   Parts per thousand.

PQ (Performance qualification)   Replacing the term validation, PQ is documented testing confirming that a system or equipment will achieve the desired and intended results.

P.R.   Pressure rated.

pressure demand respirator   A positive pressure, atmosphere supplying respirator that admits respirable gas when the positive pressure is reduced inside the face piece by inhalation.

pressure zone   A water distribution system within any area of a building having a common source of water supply or pressure origin.

product water   Purified water obtained with any one of several treatment technologies or obtained from a system employing a combination of treatment techniques.

PRV   Pressure reducing valve, pressure regulating valve, pressure relief valve.

psi   Pounds per square inch.

psia   Pounds per square inch, absolute.

psig   Pounds per square inch, gauge.

PSM   An arbitrary designation for plastic piping products having certain dimensional characteristics unique to a very specific product.

PTFE   Polytetrafluoroethylene (Teflon).

pure steam   Pyrogen-free steam that is generated from additive-free feedwater.

PVC   Polyvinyl chloride.

PVDC   Polyvinylidene chloride.

PVDF   Polyvinylidene fluoride.

pyrophoric   A substance that will spontaneously ignite upon contact with air under normal pressure and temperature.

QC   Quality control.

R & D   Research and development.

RA   Roughness average (a numerical measure of pipe wall roughness).

rainfall intensity   The rate of rainfall measured in inches per hour.

rate of rainfall   A commonly used term for rainfall intensity.

raw water   Water used as the intake to any device, equipment, or treatment process. This term is generally used to describe water obtained from a natural source such as a river, lake, or well; it is also used to describe water received directly from the supply source.

recovery rate   The amount of water capable of being heated to the design temperature per unit of time in a water heater.

RD   Roof drain.

regenerable deionizer   A water-purification system including ion-exchange resin, piping, valving, and controls to permit chemical flush of the resin bed for the purpose of reactivating the resin upon exhaustion.

regeneration   The process by which the cation resin is reactivated with acid and the anion resin reactivated with a caustic substance, thereby permitting reuse of ion-exchange resin for water purification.

regulator   A device used to reduce a variable inlet pressure to a constant outlet pressure under variable flow conditions.

relative humidity   The amount of water vapor actually present in air; expressed as a percent of the amount of water capable of being present when the air is saturated.

relief vent   An auxiliary vent that connects the vent stack to the soil or waste stack in multistory buildings and that is used to equalize pressure between them. This connection will occur at offsets and at set vertical intervals determined by code.

required pressure   The minimum pressure necessary for satisfactory operation of any device.

residual water pressure   The pressure of water available in a piping system when water is flowing at a referenced flow rate.

resin, anion   A beaded, insoluble polymer, normally of styrene, chemically activated to exchange negatively charged ions.

resin, cation   A beaded, insoluble polymer, normally of styrene, chemically activated to exchange positively charged ions.

resin, scavenger   A highly porous (macroreticular) resin that is used for removing organic or colloidal material. An anion resin, which can be chemically regenerated, is most often used for organic removal in a pretreatment application or for post-DI colloid removal in ultrapure system applications. In addition to its unique adsorptive properties, the resin retains ion exchange capability.

respirator   A personal device designed to protect the wearer from the inhalation of hazardous atmospheres.

return period   A commonly used term for frequency. The statistical period of years that must elapse to produce the most severe design storm once in that period of time.

revent   Another name for an individual vent.

RH   Relative humidity.

RI (Ryzner stability index)   An empirical description of the scale-forming tendencies of water.

riprap   Rough stone of various sizes placed irregularly to prevent scouring or erosion by water or debris.

RMS   Root mean square (a numerical measure of pipe wall roughness).

RO   Reverse osmosis.

RPZ BFP   Reduced pressure zone backflow preventer.

RQ   Roughness quotient (a numerical measure of pipe wall roughness).

runout   A commonly used term for the horizontal portion of a line directly connected to a vertical pipe at its lowest level.

safe yield   In a water well, safe yield is the quantity of water that can be withdrawn annually without the ultimate depletion of the aquifer.

sanitary   When used for plumbing, denotes a system or component intended to convey any effluent containing bodily waste. When used for pharmaceutical work, denotes a clean or sterile system or component.

sanitization   The removal of contaminants and the inhibiting of the agents that cause infection or disease.

scfm   Standard cubic feet per minute for compressed and vacuum air.

scupper   A penetration through a parapet above the roof level serving as an overflow.

SDI   Silt density index (a measure of the fouling potential of a feedwater source).

SDR   Standard dimensional ratio (used to find wall thickness for plastic pipe).

seat   A valve component that provides a surface capable of sealing against the flow of fluids in a valve when contacted by a mating surface on the disk. The seat is attached to the valve body.

secondary containment tank   A tank having an inner and an outer wall with an interstitial space (annulus) between the walls and having means for monitoring the interstitial space for a leak. Underground secondary containment tanks are of either Type I or Type II construction:

Type I. A primary tank wrapped by an exterior shell that is in direct contact with it. The exterior shell might or might not wrap the 360° circumference of the primary tank.

Type II. A primary tank wrapped by an exterior shell that is physically separated from it by stand-offs and that wraps the full 360° circumference of the primary tank.

self-extinguishing   The ability of a material to resist burning when the source of heat or flame that ignited it is removed.

separator   Any filter intended to remove large volumes of a contaminant. Separates individual components of a soil such as sand, silt, and clay.

separates   A term used to describe individual components of a soil.

service hot water   Hot water intended for commercial, industrial, or domestic use within a facility.

SH   Shower.

shaft   A rotating pump member connecting the driver to the impeller that transmits power from one to the other.

shaft sleeve   A cylindrical protection around the shaft of a pump where it passes through the stuffing box. It is not required when mechanical seals are used.

SI (International System of Units)   System of metric units recommended for universal use.

sidewall area   Vertical surfaces that contribute runoff to the storm water drainage system.

slope   A commonly used term for pitch.

SMAW   Shielded metal arc welding.

soil line   Any drainage pipe that conveys human waste.

solder   A filler metal for jointing pipe that melts at a temperature of less than 940°F.

solvent cement   An adhesive that contains a chemical that dissolves or softens the surface being bonded so that the assembly, upon drying, becomes essentially one piece of the same plastic.

sorbent   A material that is contained in a cartridge or canister and that removes specific gases and vapors from the inhaled air.

source water   Water used as the intake to any device, equipment, or treatment process. Water received directly from the supply source.

specific conductance   A measurement of the ability of a solution to allow the free flow of an electric current.

specific gravity   As applied to fuel gas, it is the ratio of the weight of a given volume of gas to the same volume of air under the same conditions.

specific resistance   A measure of the amount of electrolytes in water.

springline   The horizontal centerline of a sewer pipe.

SS   Stainless steel.

stack   A vertical drainage line, usually more than three floors in height.

stack vent   The extension of a soil or waste stack above the highest horizontal drainage connection to that stack. It is also the name of a method of venting using the stack as a branch vent connection.

standard proctor test   A test for measuring the degree of compaction (density) of soil.

static water level   The level to which water exists in a well under atmospheric conditions when the well is not being pumped.

static water pressure   The pressure of water in a piping system during the time that no water is flowing.

stel   Short term exposure limit, generally 15 minutes or less.

stem   A movable component of a valve that connects the actuator to the closure element.

storm water   Liquid effluent resulting from any form of precipitation, such as rain, snow, hail, or sleet.

stp   Standard temperature and pressure.

street pressure system   Used in water systems, a piping system supplied from a public water main, using only the pressure available in that main.

stuffing box   The interior area of the valve between the stem and the bonnet that contains the packing.

sublimination   The direct passage of a substance from solid to vapor without having an intermediate state, such as dry ice vaporizing into gaseous carbon dioxide.

suds pressure zone   An area of a waste stack where the formation of soap suds could create a pressure higher than atmospheric pressure.

suds vent   A method of venting where there is a suds pressure zone.

suspended solids   Commonly called *turbidity*, it consists of insoluble particulate material, soluble material exceeding its solubility limits, and immiscible liquids such as oil and grease.

sustained yield   In a water well, sustained yield is the maximum rate at which water can be withdrawn from an aquifer on a continuing basis for beneficial use without developing undesired results.

SWP   Steam working pressure.

SWRO   Spiral wound reverse osmosis membranes.

TDS   Total dissolved solids.

TE   Top elevation.

tee   A fitting with the branch at a 90° angle to the run.

temperature cross   When a liquid being heated has an outlet temperature that falls between the inlet and outlet temperature of the heating medium.

thermoplastic   A plastic that is capable of repeated physical change and softening by heat and hardened by cooling and that while in the softened state, can be extruded or shaped by flow.

thermoset   A plastic material that is cured by application of heat or chemical means into a substantially infusible and insoluble product that is not repeatable.

time in pipe   The length of time storm water will take to reach one design point from another design point while inside the piping network.

time of concentration   The length of time a rainstorm will persist for design purposes, calculated by adding the site storm water overland flow time to the time in pipe.

TLV   Threshold limit value (the airborne concentration of substances that should never be exceeded, not even instantaneously).

TOC   Total organic carbon.

total exchangeable ions   Those ions capable of being removed from water by ion exchange.

toxic substance   A commonly used term for a substance that will harm human tissue by contact or ingestion.

trap   A device that maintains a water seal, preventing the passage of sewer gas, vermin, air, and odors originating from inside the drainage system while permitting the unrestricted passage of liquid waste into the drainage system.

trap arm   That portion of the drain pipe between the trap and the vent.

TWA   Time weighted average, usually an 8 hour day or 40 hour week.

two-bed   A deionizer consisting of two vessels, one containing cation resin, the other containing anion resin.

UL   Underwriters Laboratories.

ULF   Ultra low flush [water closets requiring a low volume of water (1.6 gpm) for discharge into the sanitary drainage system in lieu of the 3.5 gpm currently the standard].

underground piping   Piping in contact with the earth below grade.

UNS   Unified Numbering System.

UR   Urinal.

use factor   The number of devices that may be used at the same time.

UST Underground storage tank.

UV Ultraviolet.

valence A measure of the chemical combining power of an atom or compound compared to that of a hydrogen atom.

validation *See* PQ.

valve body The housing for all of the internal working components of a valve; contains the mechanism for joining the valve to the piping system.

vapor The gaseous phase of matter that normally exists in a liquid or solid state at room temperature.

vapor pressure The pressure exerted by the vapor above a pure liquid when the two phases are in equilibrium. The value depends on the temperature of the system, but at any temperature it is independent of the amount of liquid present.

vent extension The height of the vent above the roof at its terminal.

vent header A single pipe at the highest level of a building connecting the top of vent stacks in order to penetrate the roof only once.

vent stack A vertical pipe, extending one or more stories and terminating in the outside air.

vent terminal The open air location where the end of the vent stack is placed, generally above the roof.

VOC Volatile organic compound.

V/V Volume per volume.

WAGD Waste anesthetic gas disposal.

WC Water closet.

wearing rings An easily replaceable component found on the interior of a pump casing, opposite the impeller.

well development The process that removes finer material from the natural formation around the well intake, enlarging it and having only larger gravel and stones around the well screen.

well points Devices used to remove groundwater from an excavation or trench.

well pump A pump used to bring well water in an aquifer to the surface.

wet return A condensate return that has the piping below the waterline of the boiler.

wet vent A vent line that may also serve as a drain pipe.

WFI Water for injection.

WFU Water fixture units (used in plumbing systems to size the potable water system).

WOG Water, oil, and gas working pressure.

working pressure The maximum allowable pressure for which a pipe or system is designed. Also referred to as maximum operating pressure.

W/W Weight per weight.

WWP Water working pressure.

wye Wye branch (a sewer fitting with the branch at a 30°, 45°, or 60° angle to the run).

yield In water wells, the yield is the minimum desired output of the well and the conditions under which the flow is desired.

zeolites Processed natural green-sand minerals with ion-exchange properties. In common usage, an outmoded word used to describe all ion-exchange materials used for water-softening purposes.

APPENDIX D

METRIC UNIT CONVENTIONS AND CONVERSIONS

The modern metric system, called the "Système International d'Unités" (or the International System of Units in English), was finalized in its current form by international agreement in 1960 and is the standard international language of measurement. It has been mandated as the preferred system of measurement in the United States since the "Metric Conversion Act" was passed by congress in 1975. This was further enhanced by an Executive Order in 1991 that required all government publications to be revised and that a transition plan for conversion be adapted. All federal government construction, and most states, now require metric units to be used for their projects, but not much progress has been made in the United States for total implementation of metric units for material and design used only in this country.

The United States is the last industrialized country that does not completely use the Metric system, but rather a system called inch-pound units. This appendix will provide the most currently accepted means of expressing the various metric measurement units and their abbreviations. They have been obtained from experience with various code bodies and from vendors who supply equipment to both the United States and foreign markets. In addition, a limited table of conversion factors is provided for the most often referred to units used by engineers in the construction industry. For a complete list of conversion factors, refer to standard engineering references. The abbreviations used will refer to either IP for the inch-pound units currently in use in the United States or SI for the International System of Units.

BASIC METRIC FUNDAMENTALS

There are three classes of SI units: base units, derived units, and supplementary units.

A. Base units consist of seven well-defined units which, by convention, are regarded as dimensionally independent. They are:
1. Length............................... meter (m)
2. Mass .... kilogram (kg)
3. Time .... second (s)
4. Electrical current..................... Ampere (A)
5. Thermodynamic temperature .... Kelvin (K)

  6. Amount of a substance...............mole (m)
  7. Luminous intensity...................candela (cd)
B. Derived units are formed by combining base units and other derived units.
C. Supplementary units consist of only two units which are not used in general engineering practice.

Pronunciation

The following are the correct pronunciation for often mispronounced words:

  Joule rhymes with jewel
  Pascal rhymes with rascal
  pico rhymes with peek-oh (not pie-ko)
  kilo rhymes with kill-oh (not keel-oh)

Basic Conventions

A. All symbols shall be lowercase except for the following:
  1. The liter is always capitalized when abbreviated. When written out no capitalization is used. This is to avoid any confusion with the number one or the lowercase "l" in other unit designations.
  2. All abbreviated units that refer to, or are derived from, a person's name are always capitalized (such as Pa for Pascals or °C for degrees Celsius).
B. No periods are used after any symbol or abbreviation except at the end of a sentence.
C. For technical writing, always use abbreviations in conjunction with numerals (such as 10 m). For nontechnical writing, the use of numerals could be combined with the written out unit names (such as 10 meters), if desired.
D. No plurals are used for any abbreviation. Plurals may be used for written out units.
E. The use of square and cubic designations in SI units is always a superscript, such as m^3 or m^2. In IP units, abbreviations are generally used, such as sq ft or cu ft although it is not incorrect to use superscripts.
F. When written out, no capitalization is used on any measurement unit except where standard punctuation or proper grammar requires a capital letter.
G. Do not mix written names and abbreviations.
H. A soft conversion uses an exact equivalent. A hard conversion uses a rationalized, rounded number that is created to be easy to work with and remember.
I. Insert a space between the numerical value and the abbreviation (23 m). Do not insert a space between a prefix and an abbreviation (km).
J. When writing formulas, use a dot ($\cdot$) instead of a second slash for multiplication; for example, $L/min \cdot m^2$, not $L/min/m^2$.

ACCURACY

When converting from IP to SI units or vice versa, it is important that the precision of the most accurate figure be retained. Use the least significant figure (or the same number of digits) as a basis for accuracy. As an example, if a conversion from

50 kPa to psi is desired, a conversion factor of 6.894 is found. If the soft conversion factor of 7.0 is used, the degree of precision falls well within the accuracy required of most calculations and does not exceed the least significant digit of one. When extreme accuracy is required, hard conversions should be used.

The identification of the significant digit is possible only through a knowledge of the circumstances for which the conversion is used. Numbers that are intended to be exact are treated as though there are an infinite number of significant digits.

ABBREVIATIONS

The following are abbreviations for the most often used units:

second	s	liter	L	milli	m
minute	min	meter	m	centi	c
hour	h	gram	g	kilo	k
day	d	Pascal	Pa	pound	lb
week	wk	Newton	N	Watt	W
month	mo	inch	in	Joule	J
year	yr	foot	ft		
Celsius	°C	yard	yd		

PIPE SIZES

The pipe sizes currently used in the United States are nominal sizes roughly based on inches and fractions of an inch referred to as "Iron Pipe Size" or IPS. The current standard most often used for SI units are "Nominal Dimension" or DN roughly based on the pipe size in millimeters. The conversions from one to the other are as follows:

IPS	DN	IPS	DN
¼	10	6	150
½	15	8	200
¾	20	10	250
1	25	12	300
1¼	32	14	350
1½	40	15	375
2	50	18	400
2½	65	20	500
3	80	24	600
4	100	30	750
5	125	36	900

Pitch

in/ft	1/16	1/8	¼	½
cm/m	0.5	1.0	2.0	4.0

PREFERRED USAGE

A. Although the base unit of temperature is the degree kelvin (°K), the degree Celsius (°C) is used much more extensively and is the preferred unit of measurement. The °C is equal to °K; only the starting point is different.

B. The preferred unit of area is the square meter (m^2) when converting square feet. Large areas should be measured in square kilometers (km^2) and for smaller areas use the square millimeter (mm^2).

C. The term "weight" is commonly used as a synonym for mass. These terms can be used interchangeably.

D. IP customary units express reference to gauge or absolute pressure as either psig or psia. No such conventions are officially permitted in SI units. For expressing a pressure differential, the use of kilopascals (kPa) is preferred. Common usage of the term kPa is generally understood and intended by common usage to indicate the equivalent of psig. If kPa is intended to indicate absolute pressure, it should be written either as "an absolute pressure of 456 kPa" or "456 kPa abs."

E. The standard designation for pressure is the Pascal (Pa) which is the equivalent of Newtons per square meter (Nm^2). Other designations, such as kilograms per square centimeter (kg/cm^2) is no longer preferred. The bar and torr are still commonly used in vacuum and pressure calculations.

F. The preferred designation for velocity in SI units is meters per second (m/s). In the United States the foot per minute is the preferred unit of measure in IP units.

G. The commonly accepted practice of expressing standard conditions for gas flow rate in both IP and SI units is to use the prefix "s" and is written as scfm to indicate standard cubic feet per minute and sLpm to indicate standard liters per minute. Some foreign manufacturers are also using the prefix "n" (for normal) to indicate standard conditions.

H. The use of centimeters is discouraged except for measurements of the human body. The use of millimeters is preferred for linear measurements where practical, generally up to 9'-0". As an example, a 4'-0" × 8'-0" sheet of plywood is written as 1200 × 2400 mm. The use of meters and a decimal add-on is also common practice for larger dimensions, such as 4.6 m.

I. For water and fire protection flow-rate calculations within facilities, the use of liters per minute (L/min) is preferred when converting SI units in the United States, rather than per second. This is to conform with the U.S. standard of gallons per minute. In other countries, liters per second may be the preferred unit of measure.

J. When more exact calculations are desired, it is common practice to round off to three places to the right of the decimal point.

CONVERSION FACTORS

Where applicable, all IP references are in U.S. customary units. In Table D.1, numbers in parentheses have been rounded (hard conversions) to conform to current usage.

To calculate the reverse of the units, simply divide the conversion factor by 1. For example, feet per second multiplied by 0.305 equals meters per second. To convert meters per second to feet per second, divide 0.305 into 1 and get 3.28, which is the conversion factor.

TABLE D.1 Conversion Factors

To convert from	To	Multiply by
Pressure		
psi	kPa	6.894 (7)
psi	bar	0.069 (0.07)
ft lb	Nm	1.36
in Hg	kPa	3.4
ft head	kPa	3.03 (3)
in water	Pa	250
psi	kg/cm^2	0.07
psf	kg/m^2	4.9
psf	Pa	47.9
Length		
in	mm	25.4
yd	m	0.91
km	yd	1093
km	ft	3333.3
ft	m	0.3048 (0.3)
mi	km	1.609 (1.6)
Area and spatial volume		
cu ft	m^3	0.028 (0.03)
cu ft	L	28.32
cu yd	m^3	0.764
cu in	mm^3	16.39
sq ft	m^2	0.093 (0.1)
sq yd	m^2	0.836
sq mi	km^2	2.59 (2.6)
acre	m^2	4047
sq in	mm^2	6.5
cu in	cm^3	16.4
Liquid volume		
gal	L	3.79
quart	L	0.95
Fluid flow rate		
gpm	L/min	3.785 (3.8)
gpm	L/s	0.063
gpm	L/h	227
gpm	m^3/h	0.227
gph	L/s	0.0013
gph	L/min	0.08
cfm	m^3/s	0.0005
cfm	m^3/h	0.03
cfm	L/s	0.47
cfm	L/min	28.316 (28.3)
cfm	m^3/s	2119
cfm	m^3/h	1.7
cfs	L/min	1697.0 (1700)
cfs	m^3/s	0.03
cfh	L/s	0.008
cfh	m^3/h	0.03

TABLE D-1   Conversion Factors (*Continued*)

To convert from	To	Multiply by
Velocity		
fps	m/s	0.305
mph	m/s	0.45
mph	km/h	1.6
fpm	km/h	0.018
fps	km/h	1.1
Weight (mass)		
lb	kg	0.45
lb/cu ft	kg/m³	16
psf	kg/m²	4.9 (5.0)
Heat and temperature		
°F (measured temperature)	°C	°F − 32 × 0.56
°F (temperature difference)	°C	°F × 0.56
Btu/h	mJ/h	0.0010
Btu/cu ft	mJ/m³	0.037
Btu/h	W/h	0.3
Btu/sq ft/h	W/m²/h	3.15

INDEX

ABOUT THE AUTHOR

Michael Frankel, CIPE, CPD, is president of Utility Systems Consultants, a mechanical and electrical consulting engineering firm. A graduate of the City University of New York (CUNY) with more than 44 years of experience, he is an often-quoted and recognized authority in the field of plumbing and piping engineering.

A frequent lecturer and author, his articles have appeared primarily in *Plumbing Engineer,* the journal of the American Society of Plumbing Engineers (ASPE), and he serves as a member of the editorial advisory board. He has contributed to several handbooks, including McGraw-Hill's *Piping Handbook,* now in its Sixth Edition.

Mr. Frankel is a faculty member of CUNY, teaching extension division courses in plumbing design and specification writing. He is a member of ASPE and former president of the New Jersey chapter. He is Certified in Plumbing Engineering (CIPE) and Certified in Plumbing Design (CPD) by ASPE. In addition, he is ASPE code liaison to the National Fire Protection Association (NFPA), and is a member of the Technical Committee on Piping Systems for NFPA-99 (Health Care Facilities).